Multidimensional Residue Theory and Applications

Mathematical
Surveys
and
Monographs

Volume 275

Multidimensional Residue Theory and Applications

Alekos Vidras

Alain Yger

AMS AMERICAN MATHEMATICAL SOCIETY

Providence, Rhode Island

2020 *Mathematics Subject Classification.* Primary 13Pxx, 14Q25, 32-02, 32A05, 32A10, 32A26, 32A27, 32Cxx, 32C05, 42Bxx.

For additional information and updates on this book, visit
www.ams.org/bookpages/surv-275

Library of Congress Cataloging-in-Publication Data

Cataloging-in-Publication Data has been applied for by the AMS.
See http://www.loc.gov/publish/cip/.
DOI: https://doi.org/10.1090/surv/275

∞ The paper used in this book is acid-free and falls within the guidelines
established to ensure permanence and durability.
Visit the AMS home page at https://www.ams.org/

10 9 8 7 6 5 4 3 2 1 28 27 26 25 24 23

To the memory of our teacher and friend, Carlos A. Berenstein, who shared with us enthusiastically his ideas about the project of the present monograph from the day we started it nearly fourteen years ago. Such ideas inspired and motivated us since then by influencing the content and organization of the text.

Contents

Preface xiii

Chapter 1. Residue calculus in one variable 1
1.1. An analytic approach 1
1.1.1. Laurent expansions and meromorphic functions 1
1.1.2. Residue of a $(1,0)$ meromorphic form 4
1.1.3. Jordan's lemma and Mellin transform 5
1.2. The geometric point of a view 11
1.2.1. Factorization of integration currents through residue currents 12
1.2.2. Residue currents and Lagrange's interpolation formula 21
1.2.3. Residue currents and Bergman–Weil developments 25
1.2.4. Residue currents and multiplicative calculus 27
1.2.5. Residues of sections of hermitian bundles 32
1.3. The algebraic point of view 38
1.3.1. Cauchy's formula in $A^1_{\mathbb{C}} = \mathbb{C}$ or $\mathbb{P}^1_{\mathbb{C}}$ and division in $\mathbb{K}[X]$ 38
1.3.2. Currential Cauchy's formula on compact Riemann surfaces 44
1.3.3. Green currents and heights of arithmetic cycles 49
1.3.4. Residues, residue currents, and Gorenstein $\mathbb{K}$-algebras 58

Chapter 2. Residue currents: a multiplicative approach 63
2.1. De Rham complex and iterated residues 63
2.1.1. Division of forms and de Rham's lemma 63
2.1.2. Leray coboundary morphism, cohomological residue formulae 69
2.1.3. Meromorphic multilogarithmic forms 72
2.2. Coleff–Herrera sheaves of currents 75
2.2.1. Integration currents, pseudo-meromorphicity, and holonomy 76
2.2.2. The sheaf $\mathrm{CH}_{\mathcal{X},V}$ 82
2.2.3. Coleff–Herrera currents with prescribed polar set 90
2.2.4. Extension-restriction of pseudo-meromorphic currents 93
2.3. Coleff–Herrera's original construction revisited 95
2.3.1. Essential intersection and Coleff–Herrera's original construction 95
2.3.2. Coleff–Herrera currents attached to sections of line bundles 101
2.3.3. Coleff-Herrera theory in the complete intersection setting 110

Chapter 3. Residue currents: a bundle approach 117
3.1. A toric complex geometry digest 117
3.1.1. Rational fans 117
3.1.2. Complete simplicial fans and homogeneous coordinate rings 119
3.1.3. Kähler cone and moment maps 122
3.1.4. Projective toric manifolds 123

3.2. Bochner–Martinelli currents attached to bundle sections 124
3.2.1. The currents $R^s_{\| \|}$ and $P^s_{\| \|}$ for $s \in \mathcal{O}(\mathcal{X}, \overline{E})$, $\overline{E} = (E, \| \|)$ 125
3.2.2. The complete intersection setting revisited 128
3.2.3. Transformation laws from the geometric point of view 131
3.3. Bochner–Martinelli currents and generically exact complexes 136
3.3.1. Generically exact Koszul and Eagon–Northcott complexes 136
3.3.2. The currents $R^{\overline{\mathbf{F}_\bullet}}$ and $P^{\overline{\mathbf{F}_\bullet}}$ for a metrized complex $\overline{\mathbf{F}_\bullet}$ 139
3.3.3. Residual obstruction for exactness and duality theorems 145
3.4. Bochner–Martinelli residue currents: structural results 150
3.4.1. Decomposition of $R^s_{\| \|}([V])$ along distinguished varieties 150
3.4.2. $R^{\overline{\mathbf{F}_\bullet}}$ and the Buchsbaum–Eisenbud–Fitting sequence 153
3.4.3. Residue currents and structure forms 158
3.4.4. The Cohen–Macaulay case 161

Chapter 4. Bochner–Martinelli kernels and weights 167
4.1. Diagonal submanifold and Koszul complex $\mathscr{K}_\bullet^{\mathscr{D}}$ 167
4.1.1. Koszul complex over the duplication of a good manifold 167
4.1.2. Nabla operators and $\mathcal{L}^\nu$ sheaves of currents for good manifolds 169
4.2. Affine and projective Bochner–Martinelli kernels 171
4.2.1. The affine case $\mathcal{X} = \mathbb{C}^N$ 171
4.2.2. The projective case $\mathcal{X} = \mathbb{P}^N_{\mathbb{C}}$ 173
4.3. Chern connection and Koppelman's formulae on good manifolds 175
4.4. Bochner–Martinelli weights, definition, and constructions 179
4.4.1. Global weights 179
4.4.2. Local weights 182
4.4.3. Global weights attached to metrized complexes 185
4.5. Bochner–Martinelli weighted integral representation formulae 188
4.5.1. The affine case $\mathcal{X} = \Omega \subset \mathbb{C}^N$ 188
4.5.2. The projective case $\mathcal{X} = \mathbb{P}^N_{\mathbb{C}}$ 193
4.5.3. Weighted Koppelman's formulae in the general setting 194
4.6. Solving the $\mathbf{F}_\bullet$-Hefer problem in specific situations 196
4.6.1. Complexes of trivial bundles over Stein manifolds 196
4.6.2. Projective Hefer forms and Koszul complex over $\mathbb{P}^N_{\mathbb{C}}$ 201
4.6.3. Hefer problem, syzygies for homogeneous polynomial ideals 205

Chapter 5. Integral closure, Briançon–Skoda type theorems 211
5.1. Integral closure of ideals and valuative criterion 211
5.1.1. Integral closure of ideals, the analytic reduced case 211
5.1.2. The nonreduced case, valuative criterion, nœtherian operators 212
5.2. Briançon–Skoda theorem in $\mathbb{A} = \mathcal{O}_0$ 215
5.3. Briançon–Skoda theorem in $\mathbb{A} = (\mathcal{O}/\mathcal{I}_V)_0$ 221
5.4. Briançon–Skoda theorem for purely dimensional $\mathbb{A} = (\mathcal{O}/\mathcal{I})_0$ 230
5.5. Primary ideals in $\mathcal{O}_0$ 236

Chapter 6. Residue calculus and trace formulae 243
6.1. Trace in analytic polyhedra 243
6.1.1. Analytic Weil polyhedron, nondegeneracy, skeleton 243
6.1.2. The case when $m = n$ 245
6.1.3. The case when $m > n$ 250

6.1.4. Hardy spaces of strongly nondegenerate analytic Weil polyhedra 253
6.2. Algebraic approach to residue symbols 254
6.2.1. Quasi-regular sequences in a commutative $\mathbb{K}$-algebra $\mathbb{A}$ 254
6.2.2. Algebraic residue symbols as traces 257
6.2.3. Wiebe's theorem and the algebraic transformation law 260
6.3. Residue symbols in $\mathbb{K}$-polynomial algebras 263
6.3.1. Residue symbols $\mathrm{Res}_{\mathbb{K}[X]/\mathbb{K}}$ 264
6.3.2. Residue symbols $\mathrm{Res}_{\mathbb{K}_{\mathbb{J}}[X,Y]/\mathbb{K}}$ 268
6.3.3. Residues symbols in $\mathbb{K}[X_1^{\pm 1}, \ldots, X_n^{\pm 1}]$ 269
6.4. Trace in $\mathbb{K}[X]$ or $\mathbb{K}_{\mathbb{J}}[X,Y]$ 270
6.4.1. The algebraic trace function in the $\mathbb{K}[X]$-setting 270
6.4.2. The algebraic trace function in the $\mathbb{K}_{\mathbb{J}}[X,Y]$ setting 277
6.5. The characteristic zero case 283
6.5.1. Algebraic residue symbols realized as currents 284
6.5.2. Properness and algebraic residue symbols in characteristic 0 289
6.5.3. Properness and toric residue symbols (characteristic 0) 297
6.6. Residue symbols and arithmetics 300
6.6.1. The univariate case 301
6.6.2. Geometric and arithmetic markers for complexity 302
6.6.3. Size estimates of rational algebraic residue symbols 310
6.6.4. Rational algebraic residue symbols: a dynamic approach 316

Chapter 7. Miscellaneous applications: intersection, division 323
7.1. Hilbert's nullstellensatz in $\mathbb{K}[X_1, \ldots, X_n]$ and residue calculus 323
7.1.1. Rabinowitsch's trick and Grete Hermann's algorithmic procedure 323
7.1.2. Lipman–Teissier theorem and N-solvability of Bézout identity 325
7.1.3. N-solvability of Bézout identity in $\mathbb{C}[X]$ and Bochner–Martinelli currents 328
7.1.4. N-solvability in $\mathbb{C}[X]$ of Bézout identity via a diagram chase 331
7.1.5. Sharp geometric nullstellensatz 333
7.1.6. An arithmetic Perron theorem in its parametric form 337
7.1.7. Hilbert's nullstellensatz and $P = NP$ 340
7.2. Jacobi–Lagrange–Kronecker (JLK) parametric identities 346
7.2.1. Parametric Bézout identity of the JLK form 347
7.2.2. Parametric membership identities of the JLK form 353
7.3. Effective geometric Briançon–Skoda–Huneke type theorems 361
7.3.1. Global algebraic Briançon–Skoda–Huneke exponents $\mu(\mathbb{X}:\mathbb{P}^N_{\mathbb{C}})$ 361
7.3.2. An effective Briançon–Skoda–Huneke type theorem on $\mathbb{X} \subset \mathbb{C}^N$ 365
7.3.3. Global Briançon–Skoda or Lipman–Teissier theorems in $\mathbb{K}^n$ 369
7.4. Algebraic residues and tropical considerations 372
7.4.1. Rational polyhedral complexes and tropical cycles 372
7.4.2. The tropical current $\mathrm{trop}^*\overline{\mathscr{X}}$ 376
7.4.3. Tropicalization and (p,q)-supercurrents on $\mathbb{R}^n$ 378
7.5. Jacobian and socle, radical and top radical 385
7.6. Multivariate residue calculus and the exponential function 389
7.6.1. Weighted algebras of entire functions, tempered currents 390
7.6.2. Trace formulae in weighted algebras of entire functions 392
7.6.3. Ehrenpreis–Palamodov's fundamental principle revisited 401
7.7. Residue calculus outside the commutative setting 410

7.7.1. Univariate quaternionic setting and relevant regularity concepts 410
7.7.2. Principal value and residue currents in the $\mathbb{H}$-setting 417

Appendix A. Complex manifolds and analytic spaces 421
A.1. Complex manifold, structural sheaf $\mathcal{O}_{\mathcal{X}}$, sheaves of $\mathcal{O}_{\mathcal{X}}$-modules 421
A.1.1. Definitions 421
A.1.2. Examples ($\mathbb{C}^N$, $\mathbb{P}^N_{\mathbb{C}}$, $\mathcal{X}(\Sigma)$) 423
A.2. Coherence, Stein manifolds, free resolutions 425
A.2.1. Coherent sheaves of $\mathcal{O}_{\mathcal{X}}$-modules 425
A.2.2. Stein manifolds and Cartan Theorems A and B 425
A.2.3. Free resolutions of coherent sheaves 427
A.3. Closed analytic subsets of a complex manifold 429
A.3.1. Local dimension 429
A.3.2. Splitting $V = V_{\text{sing}} \cup V_{\text{reg}}$, global dimension 430
A.3.3. Free resolutions of sheaves and codimension of supports 430
A.3.4. Local presentation in the purely dimensional case 431
A.3.5. Weakly holomorphic functions and Oka universal denominator 432
A.3.6. Example: closed analytic subsets of $\mathbb{P}^N_{\mathbb{C}}$, Chow's theorem 434
A.4. Complex analytic spaces, normalization, and log resolutions 435
A.4.1. Complex analytic spaces 435
A.4.2. Normalization and blowup of complex analytic spaces 436
A.4.3. Log resolution and the Hironaka theorem 441

Appendix B. Holomorphic bundles over complex analytic spaces 443
B.1. Analytic cocycles, holomorphic bundles, isomorphism classes 443
B.1.1. Cocycles on a complex analytic space, examples 443
B.1.2. Linear algebra and construction of cocycles 444
B.1.3. Holomorphic bundles over a complex analytic space 444
B.1.4. Isomorphism classes of holomorphic bundles 446
B.1.5. Local holomorphic frames and sheaves of holomorphic sections 447
B.1.6. The concepts of local versus global complete intersection 448
B.1.7. The bundles $T^{*(p,q)}_{\mathcal{X}}$ over a complex manifold $\mathcal{X}$ 449
B.2. Sheaves of bundle-valued differential forms or currents 451
B.2.1. Differential forms and currents on a complex manifold 451
B.2.2. Differential forms and currents on a complex analytic space 456
B.2.3. Dolbeault complex, Serre duality on a complex manifold 458
B.2.4. Strongly holomorphic $(p,0)$ forms on $V \hookrightarrow \mathcal{X}$ 459
B.2.5. The Barlet sheaves on $V \hookrightarrow \mathcal{X}$ 460
B.2.6. Dolbeault complex, Serre duality on a complex analytic space 462
B.3. Hermitian bundles 463
B.3.1. Hermitian complex bundles over a complex manifold 463
B.3.2. Hermitian holomorphic bundles over a complex analytic space 464
B.3.3. Chern and Segre forms and classes 465

Appendix C. Positivity on complex analytic spaces 469
C.1. Positive forms, positive currents 469
C.1.1. Positivity of differential forms on a $\mathbb{C}$-vector space 469
C.1.2. Positive currents on a complex manifold 470
C.1.3. Trivial extension of positive closed (p,p)-currents 472

C.2. Lelong numbers and integration currents 473
C.2.1. Lelong numbers of positive closed currents on Stein manifolds 473
C.2.2. Lelong numbers of positive closed currents on analytic spaces 474
C.2.3. Analytic subvarieties and corresponding integration currents 475
C.2.4. Integration currents attached to sections of holomorphic bundles 477
C.2.5. Holonomy of integration currents and consequences 478
C.3. Positivity of holomorphic hermitian bundles 479

Appendix D. Various concepts in algebraic or analytic geometry 483
D.1. Cycles on a complex analytic space 483
D.1.1. Cycles, effectivity, splitting along a dimension and cutting out 483
D.1.2. Lelong–Poincaré equation and classes of generalized cycles 484
D.1.3. Local multiplicities of classes of generalized cycles 488
D.2. Cartier divisors and Čech cohomology 489
D.2.1. Cartier divisors and Picard group on a complex analytic space 489
D.2.2. Toric Cartier divisors on a toric variety 490
D.2.3. Picard group in terms of Čech cohomology with values in $\mathcal{O}_{\mathcal{X}}^*$ 491
D.2.4. The first Chern characteristic class of a complex line bundle 492
D.3. Weil divisors and Chow groups 492
D.3.1. Weil divisors on complex analytic spaces or algebraic varieties 492
D.3.2. Chow groups on complex analytic spaces or algebraic varieties 493
D.4. Ampleness versus positivity for holomorphic line bundles 494
D.4.1. Globally generated holomorphic bundles 494
D.4.2. Semi-ampleness of holomorphic bundles 495
D.4.3. Very ampleness/ampleness of line bundles (compact setting) 495
D.4.4. Very ampleness and ampleness of higher rank bundles 496
D.4.5. Examples 497
D.5. Local versus global intersection theory 498
D.5.1. Local extended multi-index of intersection, Crofton formula 498
D.5.2. Crofton's averaging and metrized toric divisors 500
D.5.3. The join product in the projective setting 502
D.5.4. Global intersection in $\mathbb{P}_{\mathbb{C}}^N$ and classes of generalized cycles 503

Bibliography 507

Index 527

Preface

The first glimpses of multi-dimensional residue theory within the framework of complex analytic geometry and its applications were already apparent in the work of Henri Poincaré [**P1885, P1887, P1906**]. The concluding sentence of his 1887 memoir [**P1887**],

> We may summarize the preceding by saying that the integral taken along a closed surface S only depends on the singular curves contained in the interior of this surface...,

announces the birth not so much of the concept of residue but of a new way of thinking that nearly half a century later would lead Georges de Rham [**DR32, DR36**] and Jean Leray [**Le59**] to consider in terms of homology and cohomology groups the problem of integrating a singular differential form.

Let us describe Leray's approach in the simplest possible case of a product manifold $\mathcal{X} = \mathbb{C} \times \mathcal{Y}$, where $\mathcal{Y}$ is a complex $(N-1)$-dimensional manifold ($N \geq 2$), in which one considers the smooth complex hypersurface $S = \{0\} \times \mathcal{Y} \subset \mathcal{X}$. Let

$$U = \mathbb{D}_{\mathbb{C}}(0,1) \times \mathcal{Y} = \bigcup_{y \in \mathcal{Y}} \mathbb{D}_y$$

be the simplest open neighborhood of S considered here with its trivial fibration over S. For every $y \in \mathcal{Y}$ the fiber $\mathbb{D}_y = \{(\zeta, y) : |\zeta| < 1\} \subset \mathbb{C}$ is a complex disc, whose boundary is the circle $\Gamma_y = \{(\zeta, y) : |\zeta| = 1\}$. Furthermore, the closures $\overline{\mathbb{D}_y}$ of all fibers are pairwise disjoint. Consider the projection map defined by $\mu : (\zeta, y) \in \overline{U} \longmapsto (0, y) \in S$ and the map

$$\nu : (\zeta, y) \in \mathcal{X} \setminus S \longmapsto \begin{cases} (\zeta, y) \text{ if } (\zeta, y) \notin \overline{U} \\ (\zeta/|\zeta|, y) \text{ if } (\zeta, y) \in \overline{\mathbb{D}_y} \setminus \{(0, y)\} \end{cases}$$

with values in $\in \mathcal{X} \setminus U$. Both maps are deformation retracts. For every $k \in \mathbb{N}$ the deformation retract ν induces isomorphisms between homology groups

$$\nu_* \ : H_k(\mathcal{X} \setminus S) \longrightarrow H_k(\mathcal{X} \setminus U).$$

The same is true for the deformation retract μ inducing isomorphisms between homology groups

$$\mu_* \ : H_k(S) \longrightarrow H_{k+2}(\mathcal{X}, \mathcal{X} \setminus U)$$

for every $k \in \mathbb{N}$. Here the construction is more subtle and is described as follows. Let $c = \sum_\iota \nu_\iota \{\sigma_\iota\} \in C_k(S)$ be a k-chain with coefficients ν_ι in $\mathbb{Z}$ representing $\bar{c} \in H_k(S)$. Then $\mu^{-1}(\operatorname{Supp} \sigma_\iota) = \overline{\mathbb{D}_{\mathbb{C}}(0,1)} \times \operatorname{Supp} \sigma_\iota$ for each ι. The set $\overline{\mathbb{D}_{\mathbb{C}}(0,1)} \times \sigma_\iota$ inherits the orientations of $\mathbb{D}_{\mathbb{C}}(0,1)$ and σ_ι and thus defines a $(k+2)$-chain in $\mathcal{X}$ with support in $\overline{U}$. Therefore, the class of the $(k+2)$-chain $\sum_\iota \nu_\iota \{\overline{\mathbb{D}_{\mathbb{C}}(0,1)} \times \sigma_\iota\}$ in $C_{k+2}(\mathcal{X})$ modulo $C_{k+2}(\mathcal{X} \setminus U)$ defines an element $\check{\mu}_*(c)$ in

$C_{k+2}(\mathcal{X}, \mathcal{X}\setminus U) := C_{k+2}(\mathcal{X})/C_{k+2}(\mathcal{X}\setminus U)$. It follows that $\partial(\breve{\mu}_*(c)) = \breve{\mu}_*(\partial c)$, since $\partial\mathbb{D}_{\mathbb{C}}(0,1)\times\sigma_\iota \in C_{k+1}(\mathcal{X}\setminus U)$ for any ι. Thus $\breve{\mu}_*$ induces a homomorphism μ_* from $H_k(S)$ to $H_{k+2}(\mathcal{X}, \mathcal{X}\setminus U)$. The support of each $(k+2)$-simplex in $\mathcal{X}$ intersects S transversally, ensuring that the intersection of a chain $C \in C_{k+2}(\mathcal{X})$ with S leads to a k-chain $S\cap C \in C_k(S)$ satisfying $\partial(S\cap C) = S\cap\partial C$. The transversal intersection with S induces, at the level of homology classes, the construction of the inverse homomorphism μ_*^{-1}. Thus, μ_* is an isomorphism. Now, consider the Mayer–Vietoris long exact sequence

$$\cdots \xrightarrow{\iota_*} H_{k+2}(\mathcal{X}) \xrightarrow{\pi_*} H_{k+2}(\mathcal{X}, \mathcal{X}\setminus U) \xrightarrow{\partial_*} H_{k+1}(\mathcal{X}\setminus U) \xrightarrow{\iota_*} H_{k+1}(\mathcal{X}) \longrightarrow \cdots .$$

Most interesting is the connecting homomorphism

$$\delta = \nu_*^{-1}\circ\delta_*\circ\mu_* : H_k(S)\to H_{k+1}(\mathcal{X}\setminus S)$$

and its dual homomorphism

$$\mathrm{Res} = \delta_*^\star \ : H^{k+1}(\mathcal{X}\setminus S)\longrightarrow H^k(S).$$

The dual homomorphism $\delta_*^\star$ on the de Rham cohomology groups corresponds precisely to *taking* the residue on S of closed differential forms ϕ on $\mathcal{X}\setminus S$ representing classes in $H^{k+1}(\mathcal{X}\setminus S)$. If $\phi\in Z^{k+1}(\mathcal{X}\setminus S)$ is a meromorphic form with poles of order 1 along S and $\gamma\in Z^k(S)$, then the residue formula

$$\frac{1}{2i\pi}\int_{\delta\gamma}\phi = \int_\gamma \mathrm{Res}\,\phi$$

holds. This procedure iterates when an ordered sequence $(S_1,\ldots,S_m)$ of smooth hypersurfaces intersecting transversally on $\mathcal{X}$ replaces S. The important monograph by L. A. Aizenberg and A. P. Yuzhakov [**AY**] elaborates on applications of Leray iterated residues to the cohomological reduction of semimeromorphic and rational forms. It is of crucial importance when solving, for example, systems of nonlinear algebraic equations, and it applies to elimination theory.

Thinking in terms of geometry, one realizes that the theory of distributions, hence of currents, brought a different, promising approach towards the concept of residue in complex analytic geometry since the early 1970s. Its origin lies in the efforts directed at division by holomorphic or real analytic functions as described in the work of Laurent Schwartz [**Sc55**], Stanisłas Łojasiewicz [**Lo59**], and, more recently of Michael Atiyah [**At70**]. This point of view also profits from the fundamental result by Heisuke Hironaka [**Hir64**] about the resolution of singularities over a field with characteristic 0. We should mention here the pioneering works in this direction by Miguel Herrera and David Liebermann [**HL71**], Pierre Dolbeault and Jean-Baptiste Poly [**Pol72, Pol74**], and François Norguet [**Nor74, Nor96**]. The ground-breaking monograph by Nicolas Coleff and Miguel Herrera [**CoH78**] describes what is meant by taking a residue of a semimeromorphic form $\psi/(f_1\cdots f_m)$, where the smooth form ψ is compactly supported, along an analytic set

$$Z(f) = \{f_1 = \cdots = f_m = 0\}$$

defined by a holomorphic map $f = (f_1,\ldots,f_m) : \mathbb{C}^N \longrightarrow \mathbb{C}^m$. That is, in this monograph, instead of residue, iterated residue, or Leray morphism, the central object is the *residue current*, together with its *principal value current companion.* The work of J. Solomin [**Sol77**], completing the relevant circle of ideas, becomes a junction point for their continuation in a highly stimulating thesis by Mikael Passare [**Pas88A**] and his subsequent work [**Pas88B**] more than ten years later.

Let s be a section of a holomorphic line bundle L over a complex analytic space $\mathcal{X}$. Assuming that the line bundle L is smoothly metrized as $\overline{L} = (L, |\ |)$, the two fundamental objects in the Coleff–Herrera multiplicative calculus are currents P^s and R^s (principal value and residue currents, respectively) defined by

$$(0.1)\quad P^s : \varphi \in \mathcal{D}^{(N,N)}(\mathcal{X}, L) \longmapsto \lim_{\varepsilon\to 0^+} \int_{|s|\geq\varepsilon} \langle \frac{s^*}{2i\pi|s|^2}, \varphi\rangle = \Big[\int_{\mathcal{X}} |s|^{2\lambda} \langle \frac{s^*}{2i\pi|s|^2}, \varphi\rangle\Big]_{\lambda=0} = \big[\langle P^{s,\lambda}, \varphi\rangle\big]_{\lambda=0}$$

and

$$(0.2)\quad R^s = \overline{\partial} P^s : \varphi \in \mathcal{D}^{(N,N-1)}(\mathcal{X}, L) \longmapsto \lim_{\varepsilon\to 0^+} \int_{|s|=\varepsilon} \langle \frac{s^*}{2i\pi|s|^2}, \varphi\rangle = \Big[\int_{\mathcal{X}} \langle \overline{\partial}|s|^{2\lambda} \frac{s^*}{2i\pi|s|^2}, \varphi\rangle\Big]_{\lambda=0} = \big[\langle R^{s,\lambda}, \varphi\rangle\big]_{\lambda=0},$$

where s^* denotes the conjugate section of s with respect to the smooth hermitian metric on L. Mellin transform relates here both of the proposed representations. A multiplicative formalism involving such objects is relatively easy to describe thanks to the second representation (0.2) involving Atiyah's approach to division by holomorphic functions through the analytic continuation principle. Keeping in mind the problem of division by holomorphic or real analytic functions, we point out the key role that Michael Atiyah's view plays in our presentation from the beginning of our monograph, unifying his novel approach with the point of view of Poincaré, de Rham, and Leray. This is a deliberate choice. Let us explain briefly why. Firstly, this approach makes it relatively easier to use multiplicative formalism, as explained briefly above. Secondly, our choice is essentially motivated by the following important result. Let $\mathcal{O}_{\mathbb{C}^N}\langle \partial/\partial\zeta\rangle$ be the sheaf of differential operators with analytic coefficients on $\mathbb{C}^N$. Given $f_1, \dots, f_m \in \mathcal{O}_{\mathbb{C}^N,0}$, a crucial fact is that there exists a system of formal identitities [**Bj74, Sab87, Gy93**]

$$(0.3)\qquad b(\lambda_1, \dots, \lambda_m)\, f^\lambda = \mathcal{Q}_j\Big(\lambda, \zeta, \frac{\partial}{\partial\zeta}\Big)(f_j\, f^\lambda), \quad j = 1, \dots, m,$$

where $\lambda_1, \dots, \lambda_m$ are independent parameters, $\mathcal{Q}_j \in \mathcal{O}_{\mathbb{C}^N,0}\langle\partial/\partial\zeta\rangle \otimes_{\mathbb{C}} \mathbb{C}[\lambda]$ for any $j = 1, \dots, m$, $f^\lambda = \prod_{j=1}^m f_j^{\lambda_j}$, and $b \in \mathbb{Q}[\lambda]$, realizing a transcendental basis of $\mathbb{C}[\lambda]$. The polynomial $b \in \mathbb{Q}[\lambda]$ is of the form

$$(0.4)\quad b(\lambda) = \prod_{\iota} \big(n_{\iota,0} + \textstyle\sum_{j=1}^m n_{\iota,j}\lambda_j\big) \quad (n_{\iota,0} \in \mathbb{N}^*,\ (n_{\iota,1}, \dots, n_{\iota,m}) \in \mathbb{N}^m \setminus \{0\}).$$

The archetypal example for such a system of formal identities is realized by

$$\Big(\prod_{\kappa=1}^N (\lambda_\kappa + 1)\Big) \prod_{\kappa=1}^N \zeta_\kappa^{\lambda_\kappa} = \Big(\Big(\prod_{\substack{\kappa=1,\dots,N\\ \kappa\neq j}} (\lambda_\kappa + 1)\Big) \frac{\partial}{\partial\zeta_j}\Big)\Big(\zeta_j \prod_{\kappa=1}^N \zeta_\kappa^{\lambda_\kappa}\Big),\ j = 1, \dots, N,$$

after substituting $f_j = \zeta_j$ for $j = 1, \ldots, N$ in (0.3), thus taking $m = N$. Formal identities (0.3) lead to true identities

(0.5)

$$
\begin{aligned}
b(\lambda_1, \ldots, \lambda_j - r_j, \ldots, \lambda_m)\, \frac{|f|^{2\lambda}}{f_j^{r_j}} &= \mathcal{Q}_j\Big(\lambda, \zeta, \frac{\partial}{\partial \zeta}\Big)\Big(\frac{|f|^{2\lambda}}{f_j^{r_j - 1}}\Big), \\
b(\lambda_1, \ldots, \lambda_j - r_j, \ldots, \lambda_m)\, \frac{|f|^{2\lambda}}{\overline{f_j}^{r_j}} &= \overline{\mathcal{Q}_j}\Big(\lambda, \overline{\zeta}, \frac{\partial}{\partial \overline{\zeta}}\Big)\Big(\frac{|f|^{2\lambda}}{\overline{f_j}^{r_j - 1}}\Big), \quad j = 1, \ldots, m,\ r_j \in \mathbb{N},
\end{aligned}
$$

involving germs of distributions at the origin of $\mathbb{C}^N$, complex parameters $\lambda_1, \ldots, \lambda_m$, and using the notation $|f|^{2\lambda} = \prod_{j=1}^m |f_j|^{2\lambda_j}$. Then, division by the f_j's, by the $\overline{f_j}$'s, or even by $\|f\|^2 = \sum_{j=1}^m |f_j|^2$ (or their powers) is solved using integration by parts, combined with the analytic continuation principle, thanks to true identities (0.5) in the sense of germs of distributions. In the global algebraic setting, where $\mathcal{O}_{\mathbb{C}^N,0}$ is replaced by $\mathbb{K}[X_1, \ldots, X_N]$, when $\mathbb{K}$ is a field with characteristic 0, analogues of (0.3) persist with polynomial $b \in \mathbb{K}[\lambda]$ and the operator $\mathcal{Q}$ belonging the Weyl algebra $\mathbb{K}[\lambda, X]\langle d/dX \rangle$; see [**Be72, Lic88**]. Using algebraic substitutes to formal equations (0.3), we shall be able to express the action of principal value or residue currents arising from elements in $\mathbb{Q}[X_1, \ldots, X_N]$ or $\mathbb{Q}[X_0, \ldots, X_N]$, respectively, on Kähler differential forms (similarly on $\mathbb{A}^N_{\mathbb{Q}}$ or $\mathbb{P}^N_{\mathbb{Q}}$), as values at the origin of arithmetic ζ-functions, thus making them quite special complex numbers; see for example [**BY98, CaM00**]. Finally, we mention another reason, of more prospective flavor, for why we keep using Atiyah's approach. Although the pairing of holomorphic coordinates with ghost antiholomorphic ones (which is essential for us in the present monograph) may be lost, some of the ideas behind such an approach may persist in settings totally different from that of complex analytic geometry. For example, this is the case with Igusa calculus [**Igu**] on $\overline{\mathbb{K}}$-prehomogeneous vector spaces, where $\mathbb{K}$ is a p-adic field.

The above considerations lead us to pursue consistently Atiyah's approach to representing principal values and residues from the current point of view. We will focus on principal value and residue currents attached to smoothly metrized, generically exact complexes of holomorphic bundles. Among those of particular interest to us are the algebraically tractable complexes of Koszul and of Eagon and Northcott. Alternatively, we apply our approach to (unfortunately) untractable syzygies for $\mathbb{K}[X_1, \ldots, X_N]/\mathbb{J}$, where $\mathbb{J}$ is an ideal in $\mathbb{K}[X_1, \ldots, X_N]$, or to the untractable syzygies $\mathbb{K}[X_0, \ldots, X_N]/\mathbb{J}$, where $\mathbb{J}$ is a homogeneous ideal in $\mathbb{K}[X_0, \ldots, X_N]$. In both cases, $\mathbb{K}$ is a subfield of $\mathbb{C}$. Note that the field $\mathbb{K}$ could be $\mathbb{C}$, $\mathbb{Q}$, or a number field, which justifies the need to keep track of Atiyah's approach to profit from the closed-form expressions of formal identities that are global analogues of (0.3). Furthermore, bundle valued currents introduced using this point of view proved to be quite flexible, since one profits from $2N$ parameters $\zeta_1, \ldots, \zeta_N, \overline{\zeta_1}, \ldots, \overline{\zeta_N}$ instead of just the holomorphic coordinates $\zeta_1, \ldots, \zeta_N$ to metrize holomorphic bundles or complexes. We intend in this monograph to illustrate this flexibility by adapting it to study questions ranging from multi-dimensional complex analysis and commutative algebra to intersection theory. In some sense, we update here the content of our text *Residue Currents and Bézout Identities* ([**BGVY**]) with Carlos Alberto Berenstein and Roger Gay, published almost 30 years ago (see also [**BGY89**]), by incorporating research developments that took place in multi-dimensional residue theory over these years.

Chapter 1 of the monograph contains the presentation in the univariate setting of most of the tools and methods extended in this monograph to the multi-variate frame. Section 1.1 presents in one complex variable the machinery of analytic continuation used in a more complex form later in the book. We elaborate there on Mellin transformation leading us consistently to Atiyah's approach. Formal identities (0.3), followed by true identities (0.5) in the sense of distributions, are of course trivial in the univariate context. In §1.2.1, §1.2.4, and §1.2.5, we present in the univariate setting the machinery that will be extensively carried to the multi-variate situation throughout Chapters 2 and 3. The contents of §1.2.2, §1.2.3, §1.3.1, §1.3.3, and §1.3.4 in the univariate case foreshadow the development of research themes of a more algebraic nature which will be carried out later on starting from Chapter 5. Even though the triviality of some results in the univariate context could make them a bit disappointing, we highly advise the reader to start with this preliminary chapter before entering the inherently highly technical world of those that follow.

Chapter 2 establishes the bridge between the Poincaré–de Rham–Leray geometric approach towards the concept of residue and its more analytic, by nature, realization through division by holomorphic or antiholomorphic functions leading to the construction of principal value and residue currents as proposed by N. Coleff, M. Herrera and D. Liebermann [**HL71**, **CoH78**]; see §2.1 and §2.3.1. As always, this point of view is realized through Atiyah's approach. Although we keep to division by holomorphic functions, we will nevertheless comment on the alternative construction of such employing integration on semianalyic chains. We will point out then that the Mellin transformation, as presented in Chapter 1 in the univariate setting (see §1.1.3 and §1.2.1), connects the two presentations (see the end of §2.3.2). The class of so-called Coleff–Herrera currents is identified and studied in §2.2. The contribution of Jan-Erik Björk is towering here. Since all the currents involved in the present monograph are pseudo-meromorphic and they obey the crucial dimension principle, we describe them, together with their properties, in §2.2.1. The case of complete intersections, important in the sequel, will be detailed in 2.3.3.

The aim of Chapter 3 is to present a fundamental pairing of smoothly metrized generically exact complexes $\overline{\mathbf{F}_\bullet}$ of holomorphic bundles over a complex manifold or over a complex analytic space $\mathcal{X}$ (think for example about Koszul or Eagon–Northcott complexes, or complexes related to syzygies) with a couple of objects $\{$residue current, principal value$\}$ $=(R^{\overline{\mathbf{F}_\bullet}}, P^{\overline{\mathbf{F}_\bullet}})$. When $\mathcal{X}$ is a Stein manifold and the complex of sheaves of sections of holomorphic bundles corresponding to $\mathbf{F}_\bullet^*$ happens to be exact, each current $R^{\overline{\mathbf{F}_\bullet}}$ will play the role of a duality marker. This construction and its consequences culminate in §3.3. The complementary role of $P^{\overline{\mathbf{F}_\bullet}}$ as a marker for the quotient (instead of the remainder) in the realization through closed-form division formulae of such duality will be exemplified later in Chapter 4. It is in Chapter 3, namely in §3.2.3 and §3.4.4, that a key tool in multi-variate residue calculus appears. This tool is the geometric transformation law and its variants, extensively used in the part of this monograph devoted to applications, namely in Chapters 6 and 7. The extrinsic and intrinsic structures of $R^{\overline{\mathbf{F}_\bullet}}$ are described in §3.4.2 and §3.4.3 when the complex $\mathbf{F}_\bullet$ is induced by a minimal resolution of $(\mathcal{O}_\mathcal{X}/\mathcal{I}_V)_0$ for a purely dimensional closed analytic subset V singular at 0. Such an intrinsic description proves to be essential in §5.3 and §5.4, as well as in §7.3.2 in the global algebraic affine setting.

Chapter 4 has a pivotal role. Suppose $\mathcal{X}$ is a complex manifold (which embeds as the diagonal submanifold in $\mathscr{X} = \mathcal{X} \times \mathcal{X}$) and that the diagonal submanifold is defined (at the order exactly one) as the zero set of a section Δ of a holomorphic bundle, whose rank equals $\dim \mathcal{X}$. It implies that the Koszul complex $\mathscr{K}_\bullet^\Delta$ is exact (except at order 0). This class of manifolds includes two main examples, namely $\mathbb{C}^N$ and $\mathbb{P}^N_\mathbb{C}$, which are of major importance for algebraic questions. Bochner–Martinelli kernels on such manifolds, whose construction was initiated by Rudolf Fueter [**Fu39**, **Mar43**], enter the stage in §4.2. Weights are then introduced, as was already done in the univariate case for the Cauchy–Pompeiu formula (see §1.2.2 and §1.2.3), and they lead to the weighted integral representation formula. The contributions of Mats Andersson, Bo Berndtsson, and Mikael Passare [**AndB82**, **AndP88**, **And03**, **And06A**] are essential here. Solving the so-called Hefer problem attached to a generically exact complex of holomorphic bundles over such complex manifold $\mathcal{X}$ will be the almost constant objective from §4.4.3 and on in this chapter. Once this is done (if possible), explicit weighted division formulae, where the principal value current $P^{\overline{\mathbf{F}_\bullet}}$ is a marker for the quotient and the residue current $R^{\overline{\mathbf{F}_\bullet}}$ is a marker for the remainder, will follow either on a Stein manifold, in particular on $\mathbb{C}^N$, or on the compact manifold $\mathbb{P}^N_\mathbb{C}$; see Theorems 4.44, 4.51, and 4.54. These closed-form division formulae realize explicitly division, whenever $R^{\overline{\mathbf{F}_\bullet}}$ is a marker for duality. They will play a dominant role in §7.6.1 and §7.6.3, which nearly concludes the core of this monograph.

The point of view inherited from the current approach towards the concept of residue opened new inroads for its use. It allows us to answer questions related to ideals in polynomial rings and to solve problems related to under-determined systems of equations. It applies to many other problems formulated in $\mathbb{C}^N$ or in complex manifolds as described in the highly engaging monograph by August Tsikh published in 1988 (Russian edition) [**Ts**], having almost the same title as the monograph we present here. Starting with Chapter 5, we focus on different, but still relevant, aspects residue. This time, we view it through an *algebraic* instead of a *geometric* (or *analytic*) point of view. The fundamental result established by Joël Briançon and Henri Skoda in [**BriS74**]—which states that given any ideal I in the local ring $\mathcal{O}_{\mathbb{C}^N,0}$, one has that $\overline{I^{N+k-1}} \subset I^k$ for any $k \in \mathbb{N}^*$ where $\overline{J}$ denotes the integral closure of an ideal in a commutative ring—came as a surprise. It took nearly seven years before an alternative proof of a somewhat more algebraic nature, that did not rely on a $\overline{\partial}$-resolution, appeared in [**LipS81**, **LipT81**]. The residue, this time with its algebraic interpretation, plays a decisive role in the results of the Briançon–Skoda type. The algebraic point of view on residue theory was developed and elaborated in the work by E. Kunz, R. Waldi, M. Kreutzer, and R. Hübl [**Ku92**, **KuW88**, **KK87**, **HK90**] and of J. Lipman [**Lip87**], and we will consider it as well in the sequel of this monograph. Our goal in Chapter 5 is to present effective realizations of the Briançon–Skoda theorem in a regular local ring and of its companion theorem by Briançon, Skoda, and Huneke in a nœtherian nonregular local ring $(\mathcal{O}/\mathcal{I}_V)_0$, where V is a germ of the purely dimensional singular analytic subset at the origin. We also present some effective realizations in the nonreduced setting, once the formulation of the statement is correctly settled. The effectiveness of the proofs is provided here by the tools introduced in Chapter 4. The results presented in this chapter are due to Mats Andersson and Jakob Sznajdman.

The notion of trace appears as a common denominator between the algebraic, analytic, and geometric interpretations of the concept of residue. The aim of Chapter 6 is to make this evident. Throughout this chapter, we will manipulate the three visions, either as algebraists or analysts and geometers. Cauchy's formula, established by André Weil in 1935 [**We35**], is our starting point. We will recall the interpretation of residues as traces by Joseph Lipman [**Lip87**] and then illustrate this interpretation in polynomial algebras $\mathbb{K}[X_1, \dots, X_N]$ or in their quotient by a purely dimensional radical ideal $\mathbb{J}$. We chose to as a guideline in this chapter the notion of properness, especially through §6.3–§6.5. It seems to be a good setting for showing how to profit from the smoothness of tools inspired by the current approach. The Euler–Jacobi formula, relative to the vanishing of total sums of residues in the affine (nonprojective) setting, illustrates how powerful such a flexible approach can be when compared to a more rigid algebraic one. Chapter 6 ends with arithmetic considerations about the arithmetic complexity of generalized Taylor-type developments involved in expanded versions of the algebraic Cauchy–Weil formula. We consider here the residue calculus in $\mathbb{Q}[X_1, \dots, X_N]$ or in $\mathbb{Q}[X_1, \dots, X_N]/\mathbb{J}$ and introduce markers for both the geometric and arithmetic complexities. We continue analyzing how multi-variate residue calculus behaves with respect to both geometric and arithmetic intersection theory in §6.6.

In Chapter 7, we pursue questions in interpolation and elimination theory, as in the previous chapter, and focus again on the role of multi-variate residue calculus there. It means that we return to a circle of ideas, whose origin lies in the contributions of Carl Jacobi, Leopold Kronecker, Arthur Cayley, Francis Macaulay, and David Hilbert. Hilbert's nullstellensatz, as well as the Briançon–Skoda global theorem in $\mathbb{C}^N$, are presented under various facets: Lagrange–Kronecker's parametric closed form formulae, diagram chase, explicit division formulae that can be interpreted as averaged ones (through Crofton's averaging formula) respecting the field over which polynomial entries are defined. We provide in §7.6 trace or division formulae in the affine setting, where the transcendental function exp is involved. For example, we present trace or division formulae in the weighted Paley–Wiener algebra of entire functions. In the same section, we also reformulate the Euler–Ehrenpreis–Palamodov fundamental principle asserting that any solution of a system of partial differential equations (PDEs) in a convex open subset of $\mathbb{R}^N$ is a "sum" of elementary solutions. The same strategy applies to representations of mean-periodic functions according to the same principle under some specified hypothesis. Summarizing, one could say that Chapter 7 contains applications of multi-dimensional residue theory to a broad spectrum of mathematical problems. Moreover, it also contains a short insight (§7.7) into the noncommutative setting of the univariate quaternionic algebra $\mathbb{H}$. This could be interpreted both as a modest attempt to transpose some pieces of residue machinery to the noncommutative world and as a tribute to Rudolf Fueter, who introduced Bochner–Martinelli kernels

[**Fu39**] that are present throughout this monograph. In some sense, the material contained in this chapter is intended to illustrate our point of view that residue currents are a powerful, flexible tool to be applied in many, seemingly unrelated, areas of mathematical research.

The monograph concludes with four appendices containing the material needed to make the presentation of the main subjects as self-contained as possible. It required that we rework already known notions and adapt them into our setting. We hope that the material presented in the monograph illustrates the strength and the applicability of the methods developed in this particular area of multi-variate complex analysis.

The authors are indebted to the referees for essential comments on and corrections to earlier versions of the manuscript. Their critical reading of the original text contributed much to its clarity and presentation. Many thanks are also due to Ina Mette for her patience and support.

CHAPTER 1

Residue calculus in one variable

In this introductory chapter, we recall briefly the classical residue in complex analysis, analytic geometry, and commutative algebra. We focus in particular on some analytic aspects of this notion that will be crucial when we introduce and study its multi-dimensional analogue in several complex variables. Even though some of the points about residue theory that are discussed below look irrelevant in the univariate setting, they are extremely important in the multi-variable framework. We conclude the chapter with some nontypical applications preparing the reader for the problems to be studied in the following chapters.

1.1. An analytic approach

1.1.1. Laurent expansions and meromorphic functions. Let U be an open connected subset of $\mathbb{C}$, α a point in U, and h a holomorphic function in $U \setminus \{\alpha\}$. Let also $r > 0$ be small enough so that the disk $D = D(\alpha, r)$ is relatively compact in U. For all $\zeta \in D(\alpha, r)$, $\zeta \neq \alpha$, one can then write the Laurent series

$$h(\zeta) = \sum_{k \in \mathbb{Z}} a_k (\zeta - \alpha)^k, \tag{1.1}$$

where the Laurent coefficients a_k, $k \in \mathbb{Z}$, are given by

$$a_k = \frac{1}{2\pi r^k} \int_0^{2\pi} h(\alpha + re^{i\theta}) e^{-ik\theta} d\theta. \tag{1.2}$$

Note that the coefficients a_k do not depend on r.

Let $A = \{k \in \mathbb{Z} : a_k \neq 0\}$. Then the *order* of h at α is defined to be the element $\theta_\alpha(h) \in \mathbb{Z} \cup \{-\infty, +\infty\}$:

$$\theta_\alpha(h) = \begin{cases} +\infty & \text{if } A = \emptyset, \\ \inf A & \text{if } A \neq \emptyset \text{ and is bounded from below,} \\ -\infty & \text{if } A \neq \emptyset \text{ and is not bounded from below.} \end{cases} \tag{1.3}$$

The singularity of h is *removable* if $\theta_\alpha(h) \geq 0$. In this case, there is a unique holomorphic extension g of h on U. In particular, if $\theta_\alpha(h) = +\infty$, the analytic continuation principle ensures that $g \equiv 0$ on U. Note that the singularity α is removable if and only if $|h|$ is bounded in $\overline{D(0,r)} \setminus \{\alpha\}$ for $r > 0$ small enough. If $\theta_\alpha(h)$ is a strictly positive (resp., strictly negative) integer, then we say that α is a zero (resp., a *pole*) of order $\theta_\alpha(h)$ (resp., $-\theta_\alpha(h)$) for h. The singularity α is *essential* if $\theta_\alpha(h) = -\infty$. Otherwise, if $\theta_\alpha(h) = \nu \in \mathbb{Z}$, then one can factorize h in the open subset $U \setminus \{\alpha\}$ as $h(\zeta) = (\zeta - \alpha)^\nu u(\zeta)$, u being holomorphic on U with $u(\alpha) \neq 0$.

Assume that U is an open connected subset of $\mathbb{C}$, W is a closed discrete subset of U, and h is a holomorphic function in $U \setminus W$. We say that h is *meromorphic* in U if, for any α in W, $\theta_\alpha(h) > -\infty$. This amounts to saying that h is the restriction to $U \setminus W$ of a (necessarily) unique holomorphic function from U to the complex projective line $\mathbb{P}^1_{\mathbb{C}}$. We adopt here an alternative point of view: a holomorphic function $h : U \setminus W \to \mathbb{C}$ is meromorphic in U if and only if it extends as a distribution in U. If this is the case, then the extension of h is not unique. One way to construct a particular one (denoted as $h_{|U\setminus W}$ and called a *standard extension*) in a neighborhood of $\alpha \in W$ is to set

$$\langle h_{|U\setminus W}, \varphi\rangle := \Big[\int_U |(\zeta-\alpha)|^{2\lambda}\, h(\zeta)\varphi(\zeta,\overline{\zeta})\frac{d\overline{\zeta}\wedge d\zeta}{2i}\Big]_{\lambda=0}, \tag{1.4}$$

where the support of the smooth test function $\varphi(\zeta,\overline{\zeta})$ is contained in a neighborhood D_α of α satisfying $D_\alpha \cap W = \{\alpha\}$. If $\operatorname{Re}\lambda \gg 1$, the interior part of the brackets in (1.4) is perfectly well defined. Then one applies to it the analytic continuation as a holomorphic function of λ to some half-plane $\operatorname{Re}\lambda > -\eta_0$, $\eta_0 > 0$. Its value at $\lambda = 0$ is described by (1.4). Here the integration is understood with respect to the complex variable ζ. The ability to obtain such a suitable analytic continuation is based on Stokes's theorem, as we shall explain next. If $W = \{z \in U : f(z) = 0\}$, where f is holomorphic in U, then the standard extension $h_{|U\setminus W}$ is defined as the *principal value* distribution, denoted by $\mathrm{PV}[h]$,

$$\mathrm{PV}[h] : \varphi \in \mathcal{D}(U) \mapsto \langle h_{|U\setminus W}, \varphi\rangle := \Big[\int_U |f(\zeta)|^{2\lambda}\, h(\zeta)\varphi(\zeta,\overline{\zeta})\frac{d\overline{\zeta}\wedge d\zeta}{2i}\Big]_{\lambda=0}. \tag{1.5}$$

This definition does not depend on the choice of f. Any other holomorphic function g in U that satisfies $W = \{g = 0\}$, without taking into account possible multiplicities, will generate the same principal value distribution. The right-hand side in (1.5) amounts in fact to taking the limit

$$\lim_{\epsilon\to 0}\int_{|f|\geq\epsilon} h(\zeta)\,\varphi(\zeta,\overline{\zeta})\,\frac{d\overline{\zeta}\wedge d\zeta}{2i}$$

when $\epsilon > 0$. The approach in (1.5) through the analytic continuation process has an advantage that will be more evident later on in the multi-variate context. It also appears to be more coherent with the classical definition of meromorphic functions as holomorphic functions from U to $\mathbb{P}^1_{\mathbb{C}}$ given the role played by analytic continuation. To compute the value at $\lambda = 0$ in (1.4), or more generally in (1.5), one can profit (locally, about each point $z \in U$, in particular about each $z = \alpha \in W$) from the formal *local Bernstein–Sato equation* for f, identified here with its germ in $\mathcal{O}_{\mathbb{C},z}$,

$$\mathcal{Q}_z(\lambda,\zeta,\partial)[f^{\lambda+1}] = \mathcal{Q}_z(\lambda,\zeta,\partial)[f\otimes f^\lambda] = b_z(\lambda)\, f^\lambda\,, \tag{1.6}$$

where $b_z \in \mathbb{C}[\lambda]$ is a polynomial with strictly negative rational roots, and $\mathcal{Q}_z$ is a differential operator in ∂ whose coefficients are polynomial in λ and analytic in ζ in a neighborhood of z. Here the differential operator $\partial = \partial/\partial\zeta$ acts on $f\otimes f^\lambda$ according to Leibniz's rule in $\mathcal{O}_{\mathbb{C},z}$. Furthermore, $\partial(f^\lambda) = \lambda(f'/f)\otimes f^\lambda$ belongs to $\mathcal{M}_{U,z}[\lambda]\otimes f^\lambda$, where $\mathcal{M}_U$ denotes the sheaf of meromorphic functions on U. Any element in $\mathcal{O}_{U,z}\langle\partial\rangle$ acts then from $\mathcal{O}_{U,z}\otimes f^\lambda$ to $\mathcal{M}_{U,z}[\lambda]\otimes f^\lambda$. In particular $\partial(f^{\lambda+1}) = \partial(f\otimes f^\lambda) = (\lambda+1)\, f'\otimes f^\lambda$. The monic polynomial b_z, which is minimal among all $b \in \mathbb{C}[\lambda]$ involved in such formal identity, is called the *local Bernstein–Sato polynomial* of f at z (or of the germ $f_z \in \mathcal{O}_{\mathbb{C},z}$). Observe that the existence of formal equation (1.6) about $z = \alpha \in W$ follows trivially from the fact that f

can be considered there as $f(\zeta) = (\zeta - \alpha)^{\mu_\alpha(f)}$ (up to some local biholomorphism about α) with $\mu_\alpha(f) \in \mathbb{N}^*$ and that we have for any $\alpha \in \mathbb{C}$, $q_1, q_2 \in \mathbb{Z}$, the formal relations

(1.7)

$$|\zeta - \alpha|^{2\lambda}(\zeta - \alpha)^{q_1}(\overline{\zeta} - \overline{\alpha})^{q_2}\, d\zeta = \frac{1}{\lambda + q_1 + 1}\,\partial\Big[|\zeta - \alpha|^{2\lambda}(\zeta - \alpha)^{q_1+1}(\overline{\zeta} - \overline{\alpha})^{q_2}\Big],$$
$$|\zeta - \alpha|^{2\lambda}(\zeta - \alpha)^{q_1}(\overline{\zeta} - \overline{\alpha})^{q_2}\, d\overline{\zeta} = \frac{1}{\lambda + q_2 + 1}\,\overline{\partial}\Big[|\zeta - \alpha|^{2\lambda}(\zeta - \alpha)^{q_1}(\overline{\zeta} - \overline{\alpha})^{q_2+1}\Big].$$

If $\mathcal{Q}_z^*(\lambda, \zeta, \partial/\partial\zeta) = \mathcal{Q}_z(\bar{\lambda}, \zeta, -\partial/\partial\zeta)$, then for any test function φ, whose is support sufficiently close to z, one has that

$$(1.8)\quad \langle h_{|U\setminus W}, \varphi\rangle = \Big[\frac{1}{b(\lambda)}\int_U f^\lambda \overline{f}^{\lambda+1} h\, \overline{\mathcal{Q}_z^*(\lambda, \zeta, \partial/\partial\zeta)(\varphi)}\, \frac{d\overline{\zeta}\wedge d\zeta}{2i}\Big]_{\lambda=0} = \cdots$$
$$= \left[\frac{1}{\prod_{j=0}^{M-1} b(\lambda + j)}\int_U |f|^{2\lambda}\overline{f}^M\, h\,\Big(\prod_{j=0}^{M-1}\overline{\mathcal{Q}_z^*(\lambda + j, \zeta, \partial/\partial\zeta)}\Big)(\varphi)\,\frac{d\overline{\zeta}\wedge d\zeta}{2i}\right]_{\lambda=0}$$
$$= \frac{1}{\prod_{j=0}^{M-1} b(j)}\int_U \overline{f}^M\, h\,\Big(\prod_{j=0}^{M-1}\overline{\mathcal{Q}_z^*(j, \zeta, \partial/\partial\zeta)}\Big)(\varphi)\,\frac{d\overline{\zeta}\wedge d\zeta}{2i},$$

where $M \in \mathbb{N}^*$ and satisfies $M \geq \max_\alpha(\max(0, -\theta_\alpha(h)/\mu_\alpha(f)))$. Here $\max_\alpha$ denotes the maximum with respect to $\alpha \in W \cap \operatorname{supp}\varphi$. Let us point out what will be the main advantage of the representation (1.8) for the standard extension $h_{|U\setminus W}$ if one compares it to (1.5). If $\mathbb{K}$ is a field of characteristic 0 and $p \in \mathbb{K}[X]$, then there is an algebraic formal *global Bernstein–Sato* equation

$$(1.9)\qquad \mathcal{Q}(\lambda, X, \partial/\partial X)[p^{\lambda+1}] = \mathcal{Q}(\lambda, X, \partial/\partial X)[p\otimes p^\lambda] = b(\lambda)\, p^\lambda.$$

In (1.9) the Weyl algebra $\mathbb{K}\langle X, \partial/\partial X\rangle$ acts from $\mathbb{K}[X]\otimes p^\lambda$ to $\mathbb{K}(X)\otimes p^\lambda$ according to Leibniz's rule and $(\partial/\partial X)(p^\lambda) = \lambda\,(p'(X)/p(X))\otimes p^\lambda$, while $b \in \mathbb{K}[\lambda]$. The polynomials $b \in \mathbb{K}[\lambda]$ involved in the formal identity (1.9) define a principal ideal, whose monic generator is called the *global Bernstein–Sato polynomial* of $p \in \mathbb{K}[X]$ [**Be72**], see also [**Bor87**, V.1.16.4]. Whenever $\mathbb{K}$ embeds to $\mathbb{C}$, such minimal polynomial b belongs to $\mathbb{Q}[\lambda]$ and factorizes into a product of first order factors $X + \gamma$, with $\gamma \in \mathbb{Q}^+$ since b is the least common multiple of all local Bernstein–Sato polynomials b_α, $\alpha \in p^{-1}(\{0\})$ [**BiM98**]. Using the global Bernstein–Sato equation (1.9) to represent $h_{|U\setminus W}$ then allows us to keep track of some algebraic, or even arithmetic, information. Consider, for example, the case when $U = \mathbb{C}$, $h = H = g/p \in \mathbb{Q}(X)$, and $f = p \in \mathbb{Z}[X]$ with coprime coefficients. It means that $\mathcal{Q}$ and b in (1.9) can be taken with integer coefficients. In this case, using (1.9), one expresses the action of the duality bracket as a polynomial in z with explicit convergent integrals over $\mathbb{C}$ as coefficients

$$\Big\langle \frac{\mathrm{PV}[H]}{2\pi}, \frac{(1 + \overline{\zeta}z)^N}{(1 + |\zeta|^2)^{N+1}}\Big\rangle = \lim_{\epsilon\to 0^+}\frac{1}{2\pi}\int_{|p|\geq\epsilon}\frac{g(\zeta)}{p(\zeta)}\Big(\frac{(1 + \overline{\zeta}z)^N}{(1 + |\zeta|^2)^{N+1}}\Big)\frac{d\overline{\zeta}\wedge d\zeta}{2i}$$
$$= \frac{1}{2\pi}\Big[\int_{\mathbb{C}} |p(\zeta)|^{2\lambda}\frac{g(\zeta)}{p(\zeta)}\Big(\frac{(1 + \overline{\zeta}z)^N}{(1 + |\zeta|^2)^{N+1}}\Big)\frac{d\overline{\zeta}\wedge d\zeta}{2i}\Big]_{\lambda=0}$$

for $N \in \mathbb{N}$ large enough and $z \in \mathbb{C}$. If $\mathcal{Q}^*(\lambda, X, \partial/\partial X) = \mathcal{Q}(\bar{\lambda}, X, -\partial/\partial X)$, then such an expression is the coefficient a_0 in the Laurent expansion with respect to λ

about the origin of

$$
\begin{aligned}
(1.10)\quad \lambda \longmapsto & \frac{1}{2\pi}\int_{\mathbb{C}} \overline{p}^{\lambda} p^{\lambda-1}\Big(g(\zeta)\frac{(1+\overline{\zeta}z)^N}{(1+|\zeta|^2)^{N+1}}\Big)\frac{d\overline{\zeta}\wedge d\zeta}{2i} \\
&= \frac{1}{2\pi}\frac{1}{b(\lambda-1)}\int_{\mathbb{C}} \overline{p}^{\lambda} p^{\lambda}\,\overline{\mathcal{Q}^*(\lambda-1,\zeta,\partial/\partial\zeta)}\Big(g(\zeta)\frac{(1+\overline{\zeta}z)^N}{(1+|\zeta|^2)^{N+1}}\Big)\frac{d\overline{\zeta}\wedge d\zeta}{2i} \\
&= \frac{1}{b(\lambda-1)}\int_{\mathbb{C}} \frac{e^{\lambda \log|p(\zeta)|^2}}{2\pi}\,\overline{\mathcal{Q}^*(\lambda-1,\zeta,-\partial/\partial\zeta)}\Big(g(\zeta)\frac{(1+\overline{\zeta}z)^N}{(1+|\zeta|^2)^{N+1}}\Big)\frac{d\overline{\zeta}\wedge d\zeta}{2i}.
\end{aligned}
$$

Convergent integrals of the form

$$
\begin{aligned}
(1.11)\quad & \frac{1}{2\pi}\int_{\mathbb{C}} (\log|p(\zeta)|)^{\nu}\frac{\overline{\zeta}^{\nu'}}{(1+|\zeta|^2)^{\nu''}}\frac{d\overline{\zeta}\wedge d\zeta}{2i} \\
&= \int_0^{\infty}\Big(\frac{1}{2\pi}\int_0^{2\pi}(\log|p(\rho e^{i\theta}|)^{\nu}e^{-i\nu'\theta}\,d\theta\Big)\frac{\rho^{\nu'+1}}{(1+\rho^2)^{\nu''}}\,d\rho
\end{aligned}
$$

(with $\nu,\nu',\nu''\in\mathbb{N}$, $\nu'-2\nu''<-2$), which appear (once multiplied by rational numbers) as coefficients of the polynomial expression in z defined by (1.10), are indeed of analytic nature. Nevertheless, one could recognize in the particular case where $\nu=1$, $\nu'=0$, and $\nu''>2$ that the integral expression (1.11) involves the Ronkin function

$$\rho\in\mathbb{R}^+\mapsto \frac{1}{2\pi}\int_0^{2\pi}\log|p(\rho e^{i\theta})|\,d\theta,$$

whose value at $\rho=1$ is the so-called *Mahler measure* of the polynomial p; see also §D.5.2. Mahler measure is a marker for the arithmetic complexity of the algebraic hypersurface $p^{-1}(\{0\})$ considered here as an algebraic subvariety of the affine algebraic variety $\mathbb{A}_{\mathbb{Q}}$.

1.1.2. Residue of a $(1,0)$ meromorphic form. Let h be a meromorphic function on a simply connected domain U, let α be a pole of h, and let $\epsilon_\alpha>0$ be small enough so that $\overline{D(\alpha,\epsilon_\alpha)}\subset U$. Then the *local residue* of $h(\zeta)d\zeta$ at the pole α is, by definition, the quantity

$$(1.12)\qquad \operatorname{Res}[h(\zeta)d\zeta,\alpha] := \frac{1}{2\pi i}\int_{|\zeta-\alpha|=\epsilon_\alpha} h(\zeta)d\zeta.$$

Denoting by γ a continuous loop in U which does not contain any pole of h, and writing $i_\alpha(\gamma)=(1/2\pi i)\int_\gamma \frac{d\zeta}{\zeta-\alpha}$ for the index of γ about α, we have the classical residue formula

$$(1.13)\qquad \frac{1}{2\pi i}\int_\gamma h(\zeta)d\zeta = \sum_{\substack{\alpha\in U\\ \alpha \text{ pole of } h}} i_\alpha(\gamma)\operatorname{Res}[h(\zeta)d\zeta,\alpha].$$

Indeed, from the homological equivalence

$$\gamma \cong \sum_{\substack{\alpha\in U\\ \alpha \text{ pole of } h}} i_\alpha(\gamma)\partial D(\alpha,\epsilon_\alpha)$$

(where $\partial D(\alpha,\epsilon_\alpha)$ has the usual orientation), it follows that

$$\int_\gamma h(\zeta)d\zeta = \sum_{\substack{\alpha\in U\\ \alpha \text{ pole of } h}} i_\alpha(\gamma)\int_{|\zeta-\alpha|=\epsilon_\alpha} h(\zeta)d\zeta.$$

Since

$$\int_{|\zeta-\alpha|=\epsilon_\alpha} h(\zeta)d\zeta = 2\pi i\text{Res}\,[h(\zeta)d\zeta,\alpha]\,,$$

we recover (1.13).

It is also interesting to point out that the expression Res $[h(\zeta)d\zeta,\alpha]$ does not seem to have any true connection with the integration. Besides the fact that ϵ_α has to be taken small enough, the choice of ϵ_α is irrelevant in (1.12). We just use the integration symbol to materialize the duality of homology-cohomology. More precisely, if c is a cycle (i.e., a smooth chain without boundary) and ω is a cocycle (i.e., a C^1 closed differential form), then the duality bracket $\langle c,\omega\rangle$ is expressed by

$$\langle c,\omega\rangle := \int_c \omega\,.$$

Before going on in this brief expository survey of the theory, let us recall how one can classically perform the computation of residues. Let D be a disk centered at α, let h be a holomorphic function in $D\setminus\{\alpha\}$ which has a simple pole at α, i.e.,

$$h(\zeta) = \frac{1}{\zeta-\alpha}f(\zeta)$$

with f holomorphic in a neighborhood of α, and let $f(\alpha)\neq 0$. The Taylor expansion of $f(\zeta)$, $f(\zeta)=\sum_{k\geq 0}a_k(\zeta-\alpha)^k$ gives us $a_0=f(\alpha)$ for the coefficient of $1/(\zeta-\alpha)$ in the Laurent expansion of h. If α is a pole of h of order ν, then we have the formula,

$$\text{Res}\,[h(\zeta)d\zeta,\alpha] = \frac{1}{(\nu-1)!}\times\left[\left(\frac{\partial}{\partial\zeta}\right)^{\nu-1}\Big((\zeta-\alpha)^\nu h(\zeta)\Big)\right]_{\zeta=\alpha}. \tag{1.14}$$

This formula can be somehow misleading since it creates the impression that division by the factor $(\nu-1)!$ happens to be integrated within the computation. This could be a stumbling block when one works in fields of positive characteristic (as we will do at some point). It is much better to forget about (1.14) and just interpret the residue of $hd\zeta$ at α as the Laurent coefficient a_{-1} in the Laurent development of h about α. Note anyway that computing directly the Laurent development of h about α is usually more efficient from the computational point of view than using (1.14).

1.1.3. Jordan's lemma and Mellin transform. We describe in this section some very classical applications of residue calculus in a global context that we will need later on, and we refer the reader to **[BG]** for complex analysis prerequisites.

Let h be a nonconstant meromorphic function in an open set U, and let ∂K be the oriented boundary of a compact set $K\subset U$, where $|\partial K|$ denotes its support. If h has no poles on ∂K and a is a complex number outside $h(|\partial K|)$, then for any holomorphic function f in U, the integral

$$\frac{1}{2\pi i}\int_{\partial K} f(\zeta)\frac{h'(\zeta)}{h(\zeta)-a}d\zeta$$

equals the difference between the sum of the values of f at all zeroes of $h-a$ in K (counted with their multiplicities) and the sum of the values of f at all poles of $h-a$, that is, all poles of h in K (counted also with their orders). In particular, if

$f(\zeta) = 1$, then we have

$$\frac{1}{2\pi i} \int_{\partial K} \frac{h'(\zeta)}{h(\zeta) - a} d\zeta = N_h^+(a) - N_h^-(a),$$

where $N_h^+(a)$ is the sum of the multiplicities of the distinct roots of the equation $h(z) - a = 0$ contained in K, and $N_h^-(a) = N_h^-(0)$ is the sum of the orders of the poles of $h - a$ contained in K.

Let f be a continuous function on $\tau + i\mathbb{R}$, with $\tau \in \mathbb{R}$. We set

$$\mathrm{PV} \int_{\tau + i\mathbb{R}} f(\zeta) t^{-\zeta} d\zeta := \lim_{\rho \to \infty} \int_{\tau - i\rho}^{\tau + i\rho} f(\zeta) t^{-\zeta} d\zeta$$

for any $t > 0$ for which the limit exists. Let us recall the classical Jordan lemma.

LEMMA 1.1 (Jordan's lemma in a half-plane). *Let f be a continuous function on $\tau + i\mathbb{R}$, where $\tau \in \mathbb{R}$. Suppose that f can be extended to a meromorphic function, denoted also by f, on the half-plane* $\mathrm{Re}\ \zeta < \tau$. *that is also continuous up to the boundary* $\mathrm{Re}\ \zeta = \tau$. *Furthermore, we assume that this extension f satisfies the following condition:*

$$\lim_{n \to \infty} \max_{\{\zeta\,:\, |\zeta - \tau| = \rho_n, \mathrm{Re}\ \zeta < \tau\}} |f(\zeta)| = 0\,,$$

for some increasing sequence of strictly positive numbers $\{\rho_n\}$, so that $\rho_n \longrightarrow +\infty$. If $0 < t < 1$ and

$$\mathrm{PV} \int_{\tau + i\mathbb{R}} f(\zeta) t^{-\zeta} d\zeta$$

exists, then we have

$$\mathrm{PV} \int_{\tau + i\mathbb{R}} f(\zeta) t^{-\zeta} d\zeta = 2\pi i \lim_{n \mapsto +\infty} \sum_{\substack{\alpha \text{ pole of } f \\ \mathrm{Re}\ \alpha < \tau\,,\, |\alpha - \tau| < \rho_n}} \mathrm{Res}[f(\zeta) t^{-\zeta} d\zeta, \alpha]\,.$$

PROOF. For $\rho > 0$, we have

$$\int_{\tau - i\rho}^{\tau + i\rho} f(\zeta) t^{-\zeta} d\zeta = t^{-\tau} \int_{\tau - i\rho}^{\tau + i\rho} t^{-(\zeta - \tau)} f(\zeta) d\zeta\,.$$

For any $n \in \mathbb{N}$, we consider the loop realized as the concatenation of the vertical segment $[\tau - i\rho_n, \tau + i\rho_n]$ with the half-circle

$$\gamma_{\tau, \rho_n} := \left\{ \zeta = \tau + \rho_n e^{i\theta} \,:\, \frac{\pi}{2} \leq \theta \leq \frac{3}{2}\pi \right\}.$$

Cauchy's theorem gives

$$\int_{\tau - i\rho_n}^{\tau + i\rho_n} f(\zeta) t^{-(\zeta - \tau)} d\zeta + \int_{\gamma_{\tau, \rho_n}} f(\zeta) t^{-(\zeta - \tau)} d\zeta$$
$$= 2i\pi \sum_{\substack{\alpha \text{ pole of } f \\ \mathrm{Re}\ \alpha < \tau, |\alpha - \tau| < \rho_n}} \mathrm{Res}[f(\zeta) t^{\tau - \zeta} d\zeta, \alpha].$$

It remains to show that $\int_{\gamma_{\tau,\rho_n}} f(\zeta)t^{-(\zeta-\tau)}d\zeta$ vanishes as n tends to $+\infty$. Indeed

$$\begin{aligned}\int_{\gamma_{\tau,\rho_n}} f(\zeta)t^{\tau-\zeta}d\zeta &= \int_{\pi/2}^{3\pi/2} f(\tau+\rho_n e^{i\theta})t^{-\rho_n e^{i\theta}} i\rho_n e^{i\theta}d\theta \\ &= \int_{\pi/2}^{3\pi/2} f(\tau+\rho_n e^{i\theta})e^{-(\log t)\rho_n(\cos\theta+i\sin\theta)} i\rho_n e^{i\theta}d\theta \\ &= \int_{\pi/2}^{3\pi/2} f(\tau+\rho_n e^{i\theta})e^{|\log t|\rho_n(\cos\theta+i\sin\theta)} i\rho_n e^{i\theta}d\theta.\end{aligned}$$

So

$$\Big|\int_{\gamma_{\tau,\rho_n}} f(\zeta)t^{\tau-\zeta}d\zeta\Big| \le \rho_n \max_{\{\zeta\,:\,|\zeta-\tau|=\rho_n,\mathrm{Re}\,\zeta\le\tau\}} |f(\zeta)| \int_{-\pi/2}^{\pi/2} e^{-|\log t|\rho_n\sin\theta}d\theta.$$

From the standard estimate

$$\frac{2\theta}{\pi} \le \sin\theta \quad \text{for } \theta\in[0,\pi/2],$$

it follows that

$$\rho_n\int_{-\pi/2}^{\pi/2} e^{-|\log t|\rho_n\sin\theta}d\theta \le \rho_n\int_{-\pi/2}^{\pi/2} e^{-2\frac{|\log t|\rho_n\theta}{\pi}}d\theta$$

remains bounded when n tends to ∞. The conclusion of the lemma follows. □

Now, we will present a result that will be very important for us later on. It describes the behavior of the so-called inverse Mellin transform.

Let $\theta :]0,\infty[\to\mathbb{C}$ be any locally integrable function with compact support. We assume that there are two constants c and $\tau_0>0$ such that, near the origin, we have

$$|\theta(t)|\le ct^{-\tau_0} \text{ for } t>0\,.$$

Then we define the *Mellin transform* M_θ of the function θ as follows: for any complex number λ with $\mathrm{Re}\,\lambda>\tau_0$, put

$$M_\theta(\lambda)=\lambda\int_0^{+\infty} t^{\lambda-1}\theta(t)dt\,.$$

The function M_θ is holomorphic in the half-plane $\mathrm{Re}(\lambda)>\tau_0$. We focus here on some properties of the function θ (essentially related to its behavior near the origin) that are deduced from the study of its Mellin transform M_θ.

The Fourier inversion formula provides one way to recover the function θ from the knowledge of its Mellin transform M_θ. Let τ be a real number, $\tau>\tau_0$, such that the function

$$\omega\mapsto f(\omega)=M_\theta(\tau+i\omega)$$

belongs to the space $L^2(\mathbb{R})$. Changing the variable $t=e^s$, we can write

$$f(\omega)=(\tau+i\omega)\int_{-\infty}^{+\infty} e^{i\omega s}e^{\tau s}\theta(e^s)ds\,.$$

Then the function $\omega\mapsto f(\omega)/(\tau+i\omega)$ appears as the Fourier transform of the function $s\mapsto e^{\tau s}\theta(e^s)$, belonging to the space $L^1(\mathbb{R})$. The Fourier transform inversion formula applied to the function $\omega \mapsto f(\omega)/(\tau + i\omega)$, belonging to the space $L^1(\mathbb{R})$, implies that the function $s \mapsto e^{\tau s}\theta(e^s)$ is continuous in $\mathbb{R}$ and that one has

$$e^{\tau s}\theta(e^s) = \frac{1}{2\pi}\int_{-\infty}^{\infty} \frac{M_\theta(\tau + i\omega)}{\tau + i\omega} e^{-is\omega}\, d\omega,$$

for any $s \in \mathbb{R}$ and for any $\tau > \tau_0$. For any real number $\tau > \tau_0$, the last identity is equivalent to

$$\theta(t) = \frac{1}{2i\pi}\int_{\tau + i\mathbb{R}} \frac{M_\theta(\zeta)}{\lambda} t^{-\zeta}\, d\zeta\,. \tag{1.15}$$

The formula (1.15) is known as the *Mellin transform inversion formula.*

PROPOSITION 1.2. *Assume that there exist two constants $C > 0$ and $\eta > 0$ such that the Mellin transform M_θ extends to the holomorphic function f in the half-plane* Re $\lambda > -\eta$ *and*

$$-\eta/2 \le \operatorname{Re} \lambda \le \tau_0 + \eta \Longrightarrow |f(\lambda)| \le C,$$
$$-\frac{\eta}{2} \le \tau \le \tau_0 + \eta \Longrightarrow \int_{\mathbb{R}} \frac{|f(\tau + i\omega)|}{1 + |\omega|} d\omega < +\infty.$$

Then the function θ extends to $[0, \infty[$ continuously as $\theta(0) = f(0)$.

PROOF. Let t be a real strictly positive number. The function $\lambda \mapsto f(\lambda)t^{-\lambda}/\lambda$ is meromorphic in the half-plane Re $\lambda > -\eta$ and has a unique simple pole at the origin. The residue of $\lambda \mapsto f(\lambda)t^{-\lambda}/\lambda$ at the origin is equal to $f(0)$. Mellin's inversion formula implies for $\tau = \tau_0 + \eta$ that

$$\frac{1}{2i\pi}\int_{\tau + i\mathbb{R}} \frac{f(\zeta)}{\zeta} t^{-\zeta} d\zeta = \theta(t).$$

Note that the second hypothesis implies the convergence of the integral and the validity of the Mellin inversion formula. One can use the residue theorem as follows. Let C_ρ be the loop consisting of the concatenation of the paths $C_{\rho,1} = [\tau - i\rho, \tau + i\rho]$, $C_{\rho,2} = [\tau + i\rho, -\eta/2 + i\rho]$, $C_{\rho,3} = [-\eta/2 + i\rho, -\eta/2 - i\rho]$, $C_{\rho,4} = [-\eta/2 - i\rho, \tau - i\rho]$ (for $\rho > 0$). For $t > 0$ we have that

$$\frac{1}{2i\pi}\int_{C_\rho} \frac{f(\zeta)t^{-\zeta}}{\zeta} d\zeta = f(0)\,.$$

One can see easily that the integrals on $C_{\rho,2}$ and $C_{\rho,4}$ of the differential form $f(\lambda)t^{-\lambda}d\lambda/\lambda$ tend to zero when ρ tends to infinity. The integral along $C_{\rho,1}$ tends to $\theta(t)$ when ρ tends to $+\infty$. Therefore, one has

$$\theta(t) = \frac{1}{2i\pi}\int_{-\frac{\eta}{2} + i\mathbb{R}} \frac{f(\zeta)}{\zeta} t^{-\zeta} d\zeta + f(0)\,.$$

Since the integral in the formula above is bounded by (constant) $t^{\eta/2}$, we have the result we wanted to prove. □

In the same vein, we can obtain an asymptotic expansion for θ when t tends to 0^+. The precision of the expansion increases as τ moves to the left.

PROPOSITION 1.3. *Assume that there exist three constants $C > 0$, $R > 0$, and $\eta > 0$ such that the Mellin transform M_θ extends to a holomorphic function f in the half-plane* $\mathrm{Re}\,\lambda > -\eta$ *and*

$$-\eta/2 \leq \mathrm{Re}\,\lambda\,,\ |\mathrm{Im}\,\lambda| \geq T \Longrightarrow |f(\lambda)| \leq C$$
$$-\frac{\eta}{2} \leq \tau \leq \tau_0 + \eta \Longrightarrow \int_{|\omega|\geq R} \frac{|f(\tau + i\omega)|}{1+|\omega|} d\omega < +\infty.$$

Then, for all $\tau \in \,]-\eta/2, \tau_0]$ such that f has no poles on $\tau + i\mathbb{R}$,

$$\theta(t) = \sum_{\substack{\alpha \text{ pole of } f \\ \mathrm{Re}\,\alpha > \tau}} \mathrm{Res}[t^{-\zeta} f(\zeta) d\zeta/\zeta, \alpha] + \frac{1}{2i\pi} \int_{\tau + i\mathbb{R}} f(\zeta) t^{-\zeta} \frac{d\zeta}{\zeta} \tag{1.16}$$

with

$$\lim_{t\to 0} \frac{\displaystyle\int_{\tau+i\mathbb{R}} f(\zeta)t^{-\zeta} d\zeta/\zeta}{\displaystyle\sum_{\substack{\alpha \text{ pole of } f \\ \mathrm{Re}\,\alpha > \tau}} \mathrm{Res}[t^{-\zeta} f(\zeta) d\zeta/\zeta, \alpha]} = 0. \tag{1.17}$$

The following two propositions involving the Gamma function naturally complete this subsection devoted to the Mellin transform and its inversion.

PROPOSITION 1.4. *Let $t > 0$, let β be a complex number such that* $\mathrm{Re}\,\beta > 0$, *and let* $0 < \tau < \mathrm{Re}\,\beta$. *Then, we have the formula*

$$(1+t)^{-\beta} = \frac{1}{2i\pi\Gamma(\beta)} \int_{\tau+i\mathbb{R}} \Gamma(\zeta)\Gamma(\beta - \zeta) t^{-\zeta} d\zeta. \tag{1.18}$$

PROOF. To prove (1.18), we notice that the function

$$\zeta \mapsto \Gamma_\beta(\zeta) = \Gamma(\zeta)\Gamma(\beta - \zeta)$$

is a meromorphic function in $\mathbb{C}$, with poles are $0, -1, -2, \ldots$ and $\beta, \beta + 1, \ldots$. On the other hand, Stirling's formula

$$\Gamma(\zeta) \sim \sqrt{2\pi} e^{-\zeta} \zeta^{\zeta - 1/2}$$

ensures the rapid decrease of Γ_β at infinity along any vertical line in the complex plane. Let n be a strictly positive integer and let $\eta \in \,]0, 1/2[$ such that Γ_β has no poles on the line $-(n + \eta) + i\mathbb{R}$. Using Cauchy's formula as in the proof of Proposition 1.2 implies that

$$\begin{aligned}
\frac{1}{2i\pi\Gamma(\beta)} \int_{\tau+i\mathbb{R}} \Gamma(\zeta)\Gamma(\beta-\zeta) t^{-\zeta} d\zeta &= \frac{1}{2i\pi\Gamma(\beta)} \int_{-n-\eta+i\mathbb{R}} \Gamma(\zeta)\Gamma(\beta-\zeta)t^{-\zeta} d\zeta \\
&\quad + \sum_{k=0}^{n} \mathrm{Res}[\Gamma(\zeta)\Gamma(\beta-\zeta)t^{-\zeta}d\zeta, -k] \\
= \frac{1}{2i\pi\Gamma(\beta)} \int_{-n-\eta+i\mathbb{R}} &\Gamma(\zeta)\Gamma(\beta-\zeta)t^{-\zeta} dt \\
&+ 1 + \sum_{k=1}^{n} \frac{(-\beta)(-\beta-1)\cdots(-\beta-k+1)t^k}{k!}.
\end{aligned}$$

As we notice immediately

$$R_n(t) = 1 + \sum_{k=1}^{n} \frac{(-\beta)(-\beta-1)\cdots(-\beta-k+1)t^k}{k!}$$

represents the principal part of the Taylor development at order n for the function $t \mapsto (1+t)^{-\beta}$ about $t=0$. If for $t>0$ we denote by

$$\Psi(t) = \frac{1}{2i\pi\Gamma(\beta)} \int_{\tau+i\mathbb{R}} \Gamma(\zeta)\Gamma(\beta-\zeta)t^{-\zeta}d\zeta,$$

then it follows from

$$\left|\frac{1}{2i\pi\Gamma(\beta)} \int_{-n-\eta+i\mathbb{R}} \Gamma(\zeta)\Gamma(\beta-\zeta)t^{-\zeta}d\zeta\right| \leq C_n t^{n+\eta}$$

that

$$\Psi(t) - R_n(t) = o(t^n),$$

when t tends to 0^+ for any $n \in \mathbb{N}^*$. Using the functional equation of Γ, one has, for any $n \in \mathbb{N}^*$, that

$$\Gamma(-n-\eta+it) = \frac{\Gamma(1-\eta+it)}{\prod\limits_{k=0}^{n}(-k-\eta+it)} \tag{1.19}$$

and

$$\Gamma(\beta+n+\eta-it) = \Gamma(\beta+\eta-it)\prod_{k=0}^{n-1}(\beta+k+\eta-it). \tag{1.20}$$

For any $\epsilon>0$, there exists $T(\epsilon)$ in $\mathbb{N}^*$ such that, for any $(t,k) \in \mathbb{R}\times\mathbb{N}$ satisfying $|t|+k \geq T(\epsilon)$, one has

$$\left|\frac{\eta-it+\beta+k}{\eta-it+k}\right| \leq (1+\epsilon).$$

Using (1.19) and (1.20), one obtains, as soon as $n > T(\epsilon)$, the estimate

$$\begin{aligned}\int_{|t|\geq T(\epsilon)} &|\Gamma_\beta(-n-\eta+it)|dt \\ &\leq (1+\epsilon)^{n-1}\int_{|t|\geq T(\epsilon)} \frac{|\Gamma(\beta+n-it)\Gamma(1-\eta+it)|}{|\eta-it+n|}dt \leq C(\epsilon)(1+\epsilon)^{n-1}.\end{aligned}$$

Therefore, one has, for $n>T(\epsilon)$, a more precise estimate

$$\left|\frac{1}{2i\pi\Gamma(\beta)} \int_{-n-\eta+i\mathbb{R}} \Gamma(\zeta)\Gamma(\beta-\zeta)t^{-\zeta}d\zeta\right| \leq C(\epsilon)t^{n+\eta}(1+\epsilon)^n,$$

which allows us to conclude that

$$\lim_{n\mapsto\infty}\left|\frac{1}{2i\pi\Gamma(\beta)} \int_{-n-\eta+i\mathbb{R}} \Gamma(\zeta)\Gamma(\beta-\zeta)t^{-\zeta}d\zeta\right| = 0$$

whenever $t(1+\epsilon)<1$. Therefore, for any such t, one has

$$\lim_{n\to\infty} R_n(t) = \Psi(t),$$

and the proposition is proved when $0<t<1$. To prove the result when $t>1$, we just need to move the contour of integration to the right instead of moving it to the left. In order to get the result for $t=1$, it is enough to apply Lebesgue's theorem and verify that the function Ψ is continuous at $t=1$. □

PROPOSITION 1.5. *Let $m \in \mathbb{N}$ so that $m \geq 2$, let $t_1, \ldots, t_m$ be m strictly positive numbers, let β be a complex number with positive real part* $\operatorname{Re} \beta > 0$, *and let $\tau_1, \ldots, \tau_{m-1}$ be $m-1$ strictly positive numbers satisfying $\tau_1 + \cdots + \tau_{m-1} < \operatorname{Re} \beta$. Then one has*

$$(1.21)\quad (t_1 + \cdots + t_m)^{-\beta} \\ = \frac{1}{(2i\pi)^{m-1}\Gamma(\beta)} \int_{\tau_1+i\mathbb{R}} \cdots \int_{\tau_{m-1}+i\mathbb{R}} \Gamma_m^*(\zeta) t_1^{-\zeta_1} \cdots t_{m-1}^{-\zeta_{m-1}} t_m^{\zeta^*} d\zeta_1 \cdots d\zeta_{m-1},$$

where

$$\Gamma_m^*(\zeta) = \Gamma(\zeta_1) \cdots \Gamma(\zeta_{m-1}) \Gamma(\beta - \zeta_1 - \cdots - \zeta_{m-1}), \; \zeta^* = \sum_{k=1}^{m-1} \zeta_k - \beta.$$

PROOF. If t_1, t_2 are two strictly positive numbers, then for any complex number β such that $\operatorname{Re} \beta > 0$ and any τ satisfying $0 < \tau < \operatorname{Re} \beta$, Proposition 1.4 implies that

$$(t_1 + t_2)^{-\beta} = t_2^{-\beta} \Big(1 + \frac{t_1}{t_2}\Big)^{-\beta} = \frac{1}{2i\pi\Gamma(\beta)} \int_{\tau+i\mathbb{R}} \Gamma(s)\Gamma(\beta - s) t_1^{-s} t_2^{s-\beta} ds.$$

Proposition 1.5 is then proved when $m = 2$. This procedure can be iterated, which leads to (1.21). □

REMARK 1.6. Formula (1.21) is a continuous version of the binomial formula

$$(t_1 + \cdots + t_m)^k = \sum_{\kappa_1 + \cdots + \kappa_{m-1} = m}^{k} \frac{k!}{\kappa_1! \cdots \kappa_m!} t_1^{\kappa_1} \cdots t_m^{\kappa_m}, \quad k \in \mathbb{N}^*, t_1, \ldots, t_m > 0,$$

when the positive integer exponent k is replaced by the complex exponent with strictly negative real part $-\beta$. It transforms the additive operation between the t_j in $(t_1 + \cdots + t_m)^{-\beta}$ into a multiplicative one in $t_1^{-\zeta_1} \cdots t_{m-1}^{-\zeta_{m-1}} t_m^{\zeta^*}$ in the integrand. Multi-dimensional residue theory is intrinsically based on this multiplicative scheme (think about the formal expression $f_1^{\lambda_1} \otimes \cdots \otimes f_m^{\lambda_m}$). On the other hand, the geometric point of view naturally appeals to an additive scheme. For example, if $f = (f_1, \ldots, f_m)$ is a section of the trivial bundle $\mathbb{C}^m$ equipped with its canonical hermitian metric over some complex manifold $\mathcal{X}$, then the additive expression $\|f\|^2 = \sum_1^m |f_j|^2$ plays the central role, instead of the multiplicative one $f_1^{\lambda_1} \otimes \cdots \otimes f_m^{\lambda_m}$. This observation explains why Proposition 1.5 plays an interesting algebraic role and therefore will have consequences for us later, despite its analytic nature.

1.2. The geometric point of a view

In this section, we start from an analytic point of view. We analyze how the *integration current* on the analytic cycle $[V(F)]$, where $F = (f_1, \ldots, f_m)$ is a vector of holomorphic functions in some domain U in $\mathbb{C}$, can be factorized through *residue currents* and the differential forms df_j. Such factorization will be a crucial operational tool in multi-variate residue calculus. Then we slightly change our point of view and adopt a more geometric one. We think of U as a connected open subset in a Riemann surface $\mathcal{X}$. In this case, we consider the collection $F = (f_1, \ldots, f_m)$ to be a holomorphic section of the trivial hermitian m-bundle $(\mathbb{C}^m, |\;|) \to \mathcal{X}$ over $\mathcal{X}$.

This vector bundle is assumed to be equipped with its canonical metric, inherited from the canonical scalar product on $\mathbb{C}^m$

$$(Z,W) \in \mathbb{C}^m \times \mathbb{C}^m \longmapsto Z \cdot W = \sum_{j=1}^{m} Z_j \overline{W_j}.$$

We need also to introduce the $\mathbb{C}$-bilinear form

$$(Z,W) \in \mathbb{C}^m \times \mathbb{C}^m \longmapsto \langle\!\langle Z, W \rangle\!\rangle = \sum_{j=1}^{m} Z_j W_j \tag{1.22}$$

corresponding to the duality bracket when one of the two copies of $\mathbb{C}^m$ is identified with its dual. The necessity of the introduction of $\|F\|^2 = \sum_{j=1}^m |f_j|^2$ within this more invariant frame becomes more evident in the present section.

1.2.1. Factorization of integration currents through residue currents. It is well known that the field of meromorphic functions defined in a domain $U \subset \mathbb{C}$, equipped with the natural addition and multiplication of functions is the quotient field of the ring $H(U)$ of holomorphic functions in U. Consider a meromorphic function $h = g/f$, where $f \not\equiv 0$ and g are holomorphic functions in some domain $U \subset \mathbb{C}$, and let α be a zero of f. We denote by χ^{-1} any bi-holomorphic transform between a neighborhood D_α of α in the ζ-plane and a disk Δ centered at $\chi^{-1}(\alpha) = 0$ in the w-plane such that its pullback $\chi^*(f)$ equals

$$\chi^*(f)(w) = f(\chi(w)) = w^\nu,$$

where $\nu = \mu_\alpha(f)$ is the multiplicity of α as a zero of f. Its inverse is denoted by χ. It is clear that for any $\epsilon > 0$ sufficiently small,

$$\{\zeta \in \mathbb{C} :\ |f|^2 = \epsilon\} \cap D_\alpha = \chi(\{|w|^2 = \epsilon\})$$

is the support of a cycle homologous to some circular loop $\partial D(\alpha, \epsilon_\alpha)$ about α. Therefore we have

$$\begin{aligned}\operatorname{Res}[h(\zeta)d\zeta, \alpha] &= \frac{1}{2i\pi}\int_{\{|f|^2=\epsilon\}} \frac{g(\zeta)}{f(\zeta)} d\zeta = \frac{1}{2i\pi}\frac{1}{\epsilon}\int_{\{|f|^2=\epsilon\}} g(\zeta)\overline{f(\zeta)} d\zeta \\ &= \lim_{\epsilon\to 0^+}\Big(\frac{1}{2i\pi}\frac{1}{\epsilon}\int_{\{|f|^2=\epsilon\}} g(\zeta)\overline{f(\zeta)} d\zeta\Big).\end{aligned} \tag{1.23}$$

Now let $F \in (H(U)\backslash\{0\})^m$, where U is a domain in $\mathbb{C}$. Let α be a zero of F, that is a common zero of the entries of F, and let $\nu = \mu_\alpha(F) = \nu_\alpha$ be the multiplicity of α as a zero of F. That is, ν_α is the minimum of the multiplicities $\nu_{\alpha,1}, \dots, \nu_{\alpha,m}$ of α respectively as a zero of the entries of F. Since $V(F) = F^{-1}(\{(0,\dots,0)\})$ is a discrete subset of U, Weierstrass's theorem implies that there exists $f \in H(U)$ such that $f^{-1}(\{0\}) = V(F)$ and $\mu_\alpha(f) = \nu_\alpha$ for any $\alpha \in V(F) = V(f)$. The analytic cycle associated with the proper ideal generated by the entries of F in $H(U)$ is

$$\operatorname{div}(f) = \sum_{\alpha\in V(F)} \nu_\alpha \{\alpha\}.$$

It is also known as the intersection cycle $\operatorname{div}(f_1) \bullet \cdots \bullet \operatorname{div}(f_m)$ if one refers to the entries f_j of F in $H(U) \setminus \{0\}$, but one needs to observe that our situation is a situation of improper intersection as soon as $m > 1$. Here $\{\alpha\}$ denotes the complex number α considered as a 0-cycle (i.e., a *Weil divisor*[1] in the one variable context) in U. The *integration current* on this cycle is the $(1,1)$-*current* in U which associates to any test form $\varphi \in \mathcal{D}(U) = \mathcal{D}^{(0,0)}(U)$ the value

$$\langle [F=0], \varphi\rangle = \langle [\mathrm{div}(f)], \varphi\rangle = \sum_\alpha \nu_\alpha \varphi(\alpha),$$

where $\nu_\alpha = 0$ when $\alpha \notin V(F)$. More information regarding currents (that is, differential forms with distributions as coefficients) is to be found in §B.2.1. In particular, §C.2.3 is devoted to the notion of *integration current*. For any $\alpha \in V(F)$ and any j in $\{1,\dots,m\}$, we may write the jth entry of F as

$$f_j(\zeta) = (\zeta-\alpha)^{\nu_\alpha} q_{F,\alpha,j}(\zeta). \tag{1.24}$$

where $q_{F,\alpha} = (q_{F,\alpha,1},\dots,q_{F,\alpha,m})$ is a vector of holomorphic functions in U such that at least one of its coordinates does not vanish at α. Alternatively, we may also write

$$f_j(\zeta) = f(\zeta)\, u_j(\zeta), \tag{1.25}$$

where $u_1,\dots,u_m \in H(U)$ and the set of zeroes of $u = (u_1,\dots,u_m)$ is empty, i.e., $V(u) = \emptyset$. We fix $\alpha \in V(F)$ for the moment and set $\nu = \nu_\alpha$, $q = q_{F,\alpha}$ for the sake of simplicity. We write

$$\|F\|^2 = \sum_{j=1}^m \overline{f_j}\, f_j\ ,\ \|q\|^2 = \sum_{j=1}^m \overline{q_j}\, q_j,$$

so that $\|F\|^2 = |(\zeta-\alpha)|^{2\nu}\|q\|^2$. The function $\|F\|^2$ has an isolated zero at the point α, while the function $\|q\|^2$ is strictly positive in a neighborhood of the point α. We define formally two differential forms of bidegree (1,0) in U by setting

$$\partial \log(\|F\|^2) = \|F\|^{-2}\partial\|F\|^2\ ,\ \partial\log(\|q\|^2) = \|q\|^{-2}\partial\|q\|^2.$$

Observe that the first one is smooth in $U\setminus V(F)$, while the second one is smooth in $U\setminus V(q)$. In particular, it means that it is smooth in a neighborhood of the point α in U. Note also that

$$\partial\log(\|F\|^2) = \frac{\nu d\zeta}{\zeta - a} + \partial\log\|q\|^2 .$$

Define now a real change of variables by setting

$$w = (\zeta-\alpha)\|q\|^{1/\nu}.$$

This induces a real diffeomorphism $\zeta \to w$ between some open neighborhood D_α of α and some open disk Δ about the origin in the w-plane. Recall that we denote by χ its inverse diffeomorphism. That is,

$$\zeta = \chi(w) = \alpha + wA(w)\,,\ w\in\Delta, \tag{1.26}$$

where $A(w) = \|q(\chi(w))\|^{-1/\nu}$. We have

$$d\zeta = A(w)\,dw + w\,dA(w)\ ,\ d\overline{\zeta} = A(w)\,d\overline{w} + \overline{w}\,dA(w),$$

[1]That is, a formal sum $\sum \nu_\alpha\{\alpha\}$ of points α in U, where each of them is equipped with a multiplicity $\nu_\alpha \in \mathbb{Z}$. This formal sum is locally finite in the following sense: for any compact K in U, $\mathrm{card}\{\alpha\in K : \nu_\alpha \neq 0\} < \infty$. The divisor is called *effective* when all ν_α are positive. The *support* of a Weil divisor is the set of points α where $\nu_\alpha \neq 0$. For more details, see §D.3.1.

and therefore the pullback of the form $\partial \log \|F\|^2$ equals

$$\chi^*(\partial \log \|F\|^2) = \frac{\nu(A(w)\,dw + w\,dA(w))}{w\,A(w)} + \chi^*(\partial \log \|q\|^2) = \frac{\nu dw}{w} + \omega,$$

where ω is a smooth form in Δ. For any test function in the $\mathbb{C}$ vector space $\mathcal{D}(D_\alpha) = \mathcal{D}^{(0,0)}(D_\alpha)$ of smooth complex-valued functions with compact support in D_α and for any $\epsilon > 0$ small enough, let

$$\begin{aligned}\theta(\varphi;\epsilon) &= \frac{1}{2i\pi}\int_{\|F\|^2=\epsilon} \varphi(\zeta,\overline{\zeta})\partial \log \|F(\zeta)\|^2 \\ &= \frac{1}{2i\pi}\int_{|w|^{2\nu}=\epsilon} (\varphi\circ\chi)(w,\overline{w})\chi^*(\partial \log \|F\|^2).\end{aligned}$$

Following previous remarks, we have

$$\lim_{\epsilon\to 0^+} \theta(\varphi;\epsilon) = \nu\,\varphi(\alpha)\,.$$

It means that for any test function $\varphi \in \mathcal{D}(D_\alpha)$,

$$\begin{aligned}\nu\,\varphi(\alpha) &= \lim_{\epsilon\to 0^+} \frac{1}{2i\pi}\int_{\|F\|^2=\epsilon} \frac{\varphi(\zeta,\overline{\zeta})\partial\|F\|^2}{\|F\|^2} \\ &= \lim_{\epsilon\to 0^+} \frac{1}{2i\pi\epsilon}\int_{\|F\|^2=\epsilon} \varphi(\zeta,\overline{\zeta})\partial\|F\|^2.\end{aligned} \tag{1.27}$$

We state the following lemma.

LEMMA 1.7. *Let $F \in (H(U)\setminus\{0\})^m$, where U is a domain in $\mathbb{C}$. For any test function $\varphi \in \mathcal{D}^{(0,0)}(U)$, we have*

$$\begin{aligned}\langle [F=0]\,,\varphi\rangle &= \lim_{\epsilon\to 0^+} \frac{1}{2i\pi}\int_{\|F\|^2=\epsilon} \varphi(\zeta,\overline{\zeta})\,\partial \log \|F\|^2 \\ &= \lim_{\tau\to 0^+} \frac{\tau}{2i\pi}\int_U \varphi(\zeta,\overline{\zeta})\frac{\overline{\partial}\|F\|^2\wedge\partial\|F\|^2}{\|F\|^2(\|F\|^2+\tau)^2}.\end{aligned} \tag{1.28}$$

PROOF. The proof is rather simple. If φ is a test function, we can consider the function

$$\epsilon \to \theta(\varphi;\epsilon) = \frac{1}{2i\pi}\int_{\|F\|^2=\epsilon} \varphi(\zeta,\overline{\zeta})\partial \log \|F\|^2\,. \tag{1.29}$$

It is clear that

$$\lim_{\epsilon\to 0^+} \theta(\varphi;\epsilon) = 0$$

if the support of φ does not intersect $V(F)$. On the other hand, if the support of φ lies in D_α for some $\alpha \in V(F)$, then one has from (1.27) that

$$\lim_{\epsilon\to 0^+} \theta(\varphi;\epsilon) = \nu_\alpha\varphi(\alpha) = \langle [F=0]\,,\varphi\rangle.$$

To recover the second formula in (1.28), we just notice that, for any integer $\kappa \geq 1$, one has, for $\tau > 0$

$$\tau\int_0^\infty \frac{t^{\kappa-1}}{(t+\tau)^{p+1}}dt = \frac{1}{\kappa}\,. \tag{1.30}$$

Moreover, if σ is a locally integrable function on $[0,\infty[$, which is continuous at the origin and compactly supported, then, it follows from (1.30), that

$$\sigma(0) = \lim_{\tau\to 0^+} \tau \int_0^\infty \frac{t^{\kappa-1}\sigma(t)}{(t+\tau)^{\kappa+1}} dt\,.$$

Apply this for $\kappa = 1$ and $\sigma(\cdot) = \theta(\varphi;\cdot)$ defined in (1.29), where φ is a test function in U whose support lies in an arbitrary small neighborhood of $\alpha \in V(F)$. We get

$$\langle [F=0]\,,\varphi\rangle = \theta(\varphi;0) = \lim_{\tau\to 0^+} \tau \int_0^\infty \frac{\theta(\varphi;t)}{(t+\tau)^2} dt.$$

Fubini's theorem implies that

$$\tau \int_0^\infty \frac{\theta(\varphi;t)}{(t+\tau)^2}\, dt = \frac{\tau}{2i\pi} \int_{D_\alpha} \varphi(\zeta,\overline{\zeta}) \frac{\overline{\partial}\|F\|^2 \wedge \partial\|F\|^2}{\|F\|^2(\tau+\|F\|^2)^2}. \tag{1.31}$$

Let us give a detailed proof of the last claim. Using the change of variables (1.26), that is

$$\zeta = \alpha + w\|q(\zeta)\|^{-1/\nu},\ w\in\Delta,\ \zeta\in D_\alpha,$$

where $\nu = \nu_\alpha$, we have

$$\overline{\partial}\|F\|^2 = \nu\,|w|^{2\nu-2}\, w\, d\overline{w}\ ,\ \partial\|F\|^2 = \nu\,|w|^{2\nu-2}\, \overline{w}\, dw.$$

Thus, the last integral in (1.28) becomes

$$\begin{aligned}\frac{\tau}{2i\pi}\int_U \varphi(\zeta,\overline{\zeta})\frac{\overline{\partial}\|F\|^2\wedge\partial\|F\|^2}{\|F\|^2(\|F\|^2+\tau)^2} &= \frac{\tau\nu^2}{2i\pi}\int_\Delta \frac{(\varphi\circ\chi)(w,\overline{w})|w|^{2(2\nu-2)+1} d\overline{w}\wedge dw}{|w|^{2\nu}(|w|^{2\nu}+\tau)^2}\\ &= \frac{\tau\nu^2}{2i\pi}\int_\Delta (\varphi\circ\chi)(w,\overline{w})|w|^{2\nu-3}\frac{d\overline{w}\wedge dw}{(|w|^{2\nu}+\tau)^2}.\end{aligned}$$

Passing to the polar coordinates $w = r^{1/\nu}e^{i\xi}$, $r>0, \xi\in[0,2\pi]$, so that $|w|^{2\nu} = r$, and remarking that

$$d\overline{w}\wedge dw = \frac{i}{\nu}\big(r^{\frac{1}{2\nu}}\big)^{2(1-\nu)} e^{i\xi} dr\wedge d\xi,$$

we imply that

$$\begin{aligned}&\frac{\tau\nu^2}{2i\pi}\int_D (\varphi\circ\chi)(w,\overline{w})|w|^{2\nu-3}\frac{d\overline{w}\wedge dw}{(|w|^{2\nu}+\tau)^2}\\ &= \frac{\tau\nu}{2i\pi}\int_0^{+\infty}\frac{\big(r^{\frac{1}{2\nu}}\big)^{2\nu-3+2-2\nu}}{(r+\tau)^2}\Big(\int_0^{2\pi}(\varphi\circ\chi)\big(r^{\frac{1}{2\nu}}e^{i\xi}\big)ie^{i\xi}d\xi\Big)dr\\ &= \tau\int_0^{+\infty}\frac{\theta(\varphi;t)}{(t+\tau)^2}dt.\end{aligned}$$

This proves (1.31), and so we have proved the second formula in (1.28) when the support of φ lies in an arbitrary small neighborhood of some $\alpha \in V(F)$. If the support of φ does not intersect $V(F)$, then it is clear that

$$\lim_{\tau\to 0^+}\frac{\tau}{2i\pi}\int_U \varphi\frac{\overline{\partial}\|F\|^2\wedge\partial\|F\|^2}{\|F\|^2(\|F\|^2+\tau)^2} = 0 = \langle [F=0]\,,\varphi\rangle.$$

The proof of the lemma is complete. □

Inspired by the result stated in Lemma 1.7, one can associate to a vector $F \in (H(U) \setminus \{0\})^m$, where U is a domain in $\mathbb{C}$, a vector $R^F = (R_1^F, \ldots, R_m^F)$ of $(0,1)$-currents in U called *residue currents*, whose action can be expressed in terms of F. This is a vector of continuous linear functionals, namely $(0,1)$-currents, on the space of smooth $(1,0)$ differential forms with compact support in U, which is denoted by $\mathcal{D}^{(1,0)}(U)$, while its dual is denoted by $'\mathcal{D}^{(0,1)}(U)$. The next proposition provides such construction.

PROPOSITION 1.8. *Let U be a domain in $\mathbb{C}$, let $F \in (H(U)\setminus\{0\})^m$ with entries $f_1, \ldots, f_m$, and let $\varphi \in \mathcal{D}^{(1,0)}(U)$. For each $j \in \{1, \ldots, m\}$, the limit*

$$\lim_{\epsilon \to 0^+} \frac{1}{2i\pi} \frac{1}{\epsilon} \int_{\|F\|^2 = \epsilon} \overline{f_j(\zeta)} \varphi(\zeta, \bar{\zeta}) \tag{1.32}$$

exists, and the map

$$R_j^F : \varphi \in \mathcal{D}^{(1,0)}(U) \longmapsto \lim_{\epsilon \to 0^+} \frac{1}{2i\pi} \frac{1}{\epsilon} \int_{\|F\|^2 = \epsilon} \overline{f_j(\zeta)} \varphi(\zeta, \bar{\zeta}) \tag{1.33}$$

defines a $(0,1)$-current on $\mathcal{D}^{(1,0)}(U)$. Moreover, for any $\varpi \in \mathcal{D}^{(0,0)}(U)$, the function

$$\lambda \longmapsto \frac{\lambda}{2i\pi} \int_U \|F\|^{2(\lambda-1)} \frac{\overline{f}_j \, d\|F\|^2}{\|F\|^2} \wedge \varpi(\zeta, \bar{\zeta}) \, d\zeta$$

extends from $\{\lambda \in \mathbb{C} : \operatorname{Re} \lambda \gg 1\}$ as a meromorphic function in $\mathbb{C}$ with (possible) poles in $-\mathbb{Q}^+$, and one has

(1.34)

$$\begin{aligned}\langle R_j^F, \varpi d\zeta \rangle &= \Big[\frac{\lambda}{2i\pi} \int_U \|F\|^{2(\lambda-1)} \frac{\overline{f}_j \, d\|F\|^2}{\|F\|^2} \wedge \varpi(\zeta, \bar{\zeta}) \, d\zeta \Big]_{\lambda=0} = [\langle R_j^{F,\lambda}, \varpi d\zeta \rangle]_{\lambda=0} \\ &= \sum_{\alpha \in V(F)} \frac{1}{(\nu_{F,\alpha} - 1)!} \Big(\frac{\partial}{\partial \zeta} \Big)^{\nu_{F,\alpha}-1} \Big(\varpi \frac{\overline{q}_{F,\alpha,j}}{\|q_{F,\alpha}\|^2} \Big)(\alpha).\end{aligned}$$

PROOF. Let $\varphi \in \mathcal{D}^{(1,0)}(U)$ and take $j = 1$. We compute first the Mellin transform M_{θ_1} of the function

$$\theta_1(\varphi; \epsilon) : \epsilon \mapsto \frac{1}{2i\pi} \frac{1}{\epsilon} \int_{\|F\|^2 = \epsilon} \overline{f_1(\zeta)} \, \varphi(\zeta, \bar{\zeta}).$$

When the support of φ lies in some arbitrary small neighborhood of a zero α of $\|F\|^2$, the computation of this Mellin transform can be performed using Fubini's theorem and a local change of coordinates as before. It is always possible to reduce our problem to this situation since

$$\lim_{\epsilon \to 0^+} \frac{1}{2i\pi} \frac{1}{\epsilon} \int_{\|F\|^2 = \epsilon} \overline{f_1(\zeta)} \varphi(\zeta, \bar{\zeta}) = 0$$

when the support of φ does not intersect $V(F)$. For $\operatorname{Re} \lambda \gg 1$ large enough, one has

$$M_{\theta_1}(\lambda) = \frac{\lambda}{2i\pi} \int_U \|F\|^{2(\lambda-1)} \frac{\overline{f}_1 \, d\|F\|^2}{\|F\|^2} \wedge \varphi(\zeta, \bar{\zeta}).$$

It is clear from the definition that the function M_{θ_1} is holomorphic in $\{\operatorname{Re} \lambda > 1/2\}$. We assume from now on that the support of φ lies in D_α for some $\alpha \in V(F)$. Suppose also that $\nu = \nu_\alpha$, D_α and Δ have been defined correspondingly in (1.24) or according to (1.26). We want to show that the function M_{θ_1} can be analytically continued as a meromorphic function of λ to the whole complex plane $\mathbb{C}$ with (possible) poles in $-\mathbb{Q}^+$ and hence is holomorphic in the half-plane Re $\lambda > -\tau$ for some $\tau \in \mathbb{Q}^+$. This is easily done after one notices that $\|F\|^2 = |\zeta - \alpha|^{2\nu} \times \|q\|^2$, where $\|q\|^2 = \|q_\alpha\|^2$ is a smooth nonvanishing function in U. One then applies the Bernstein–Sato equation (1.6) for $f = (\zeta - \alpha)^\nu$. This will be the procedure one will systematically employ when dealing with several variables, because it opens a way to bypass the difficulties inherent in division thanks to repeated use of Stokes's theorem (integration by parts really). Let us give here an elementary and explicit approach leading in particular to (1.34). We use again the local real change of variables $w = (\zeta - \alpha)\|q\|^{1/\nu} = \chi^{-1}(\zeta)$ and write $\chi^*((\overline{q}_1/\|q\|^2)\varphi) = \xi\, dw + \eta\, d\overline{w}$. Since $d|w|^{2\nu} = \nu|w|^{2\nu-2}(w\, d\overline{w} + \overline{w}\, dw)$, we have

$$\begin{aligned}
M_{\theta_1}(\lambda) &= \frac{\lambda}{2i\pi}\int_{D_\alpha} \|F\|^{2(\lambda-1)}\overline{f}_1 \frac{d\|F\|^2}{\|F\|^2} \wedge \varphi(\zeta,\overline{\zeta}) \\
&= \frac{\lambda}{2i\pi}\int_{\Delta} |w|^{2\nu(\lambda-1)}\overline{w}^\nu \frac{d|w|^{2\nu}}{|w|^{2\nu}} \wedge \chi^*\Big(\frac{\overline{q}_1}{\|q\|^2}\varphi\Big)(w,\overline{w}) \\
&= \frac{\nu\lambda}{2i\pi}\int_{\Delta} |w|^{2(\nu\lambda-\nu-1)}\overline{w}^\nu (w\, d\overline{w} + \overline{w}\, dw) \wedge (\xi\, dw + \eta\, d\overline{w}) \\
&= \frac{\nu\lambda}{2i\pi}\int_{\Delta} |w|^{2(\nu\lambda-\nu-1)}\overline{w}^\nu (w\,\xi - \overline{w}\,\eta)\, d\overline{w} \wedge dw \\
&= \frac{\nu\lambda}{2i\pi}\Big(\int_{\Delta} w^{\nu(\lambda-1)}\overline{w}^{\nu\lambda-1}\xi\, d\overline{w} \wedge dw - \int_{\Delta} w^{\nu(\lambda-1)-1}\overline{w}^{\nu\lambda}\,\eta\, d\overline{w} \wedge dw\Big) \\
&= \lambda(I(\lambda) - J(\lambda)).
\end{aligned}$$

The possible poles for the function

$$I : \lambda \mapsto \frac{\nu}{2i\pi}\int_{\Delta} w^{\nu(\lambda-1)}\overline{w}^{\nu\lambda-1}\,\xi\, d\overline{w} \wedge dw$$

are the points λ satisfying $2\nu\lambda - \nu - 1 = -|1-\nu| - 2\ell$, $\ell \in \mathbb{N}^*$, or equivalently, the points $\lambda = (1-\ell)/\nu$, $\ell \in \mathbb{N}^*$. In particular, there exists a simple pole at $\lambda = 0$. Similarly, the possible poles for the function

$$J : \lambda \mapsto \frac{\nu}{2i\pi}\int_{\Delta} w^{\nu(\lambda-1)-1}\overline{w}^{\nu\lambda}\,\eta\, d\overline{w} \wedge dw$$

are the points λ such that $2\nu\lambda - \nu - 1 = -\nu - 1 - 2\ell$, $\ell \in \mathbb{N}^*$ or, which is the same, the points $\lambda = -\ell/\nu$, $\ell \in \mathbb{N}^*$. Therefore, the function M_{θ_1} can be continued as a meromorphic function having poles in the half-plane Re $\zeta \leq -1/\nu$, and we have $M_{\theta_1}(0) = \text{Res}\,[I(\lambda)d\lambda, \alpha]$. We can even be more precise and get

$$M_{\theta_1}(0) = \text{Res}\,[I(\lambda)d\lambda, \alpha] = \frac{1}{(\nu-1)!}\Big(\frac{\partial}{\partial\zeta}\Big)^{\nu-1}\Big(\varpi\frac{\overline{q}_1}{\|q\|^2}\Big)(\alpha)$$

if $\varphi = \varpi d\zeta$. The proof of the proposition is complete. □

REMARK 1.9. When $m = 1$, the term with index α in the right-hand side of formula (1.34) becomes that of formula (1.14), where the meromorphic function h is replaced by the semimeromorphic[2] one ϖ/f . When $m > 1$, one can rewrite

[2]The terminology used here comes from the fact that the denominator f is holomorphic while the numerator ϖ is smooth (more precisely here in $\mathcal{D}(U)$).

(1.34) for $j = 1, \ldots, m$ as

$$R_j^F = \frac{\overline{u_j}}{\|u\|^2} R^f. \tag{1.35}$$

Here $f \in H(U)$ is such that $\mathrm{div}(f) = \sum_{\alpha \in V(F)} \nu_{F,\alpha} \{\alpha\}$, $R^f = R_1^{(f)}$, while the holomorphic functions u_j in U, satisfying $V(u) = \emptyset$, have been introduced in (1.25). Also, note that $\|u\|^2 = \sum_{j=1}^m |u_j|^2$. The equality (1.35) is understood in the sense of currents, that is, as an equality between distributions which appear as coefficients of differential forms. The key point here is that the action of R^f, and hence the action of all residue currents R_j^F, $j = 1, \ldots, m$, involves only computations of holomorphic derivatives. Expressions depending only on the antiholomorphic coordinate $\overline{\zeta}$, though they are essential to keep track of positivity for $\|F\|^2 \geq 0$ or $\|q\|^2 > 0$, are treated as *constants* in such residue computations. Since *positivity* is a basic concept in multi-variate complex analysis, this remark will play a major role all through this monograph.

Let us conclude this section with the fundamental observation summarizing the results obtained in Lemma 1.7 and Proposition 1.8.

PROPOSITION 1.10. *Let $F = (f_1, \ldots, f_m)$ as in Proposition 1.8. The integration current on the intersection cycle $[F = 0] = \mathrm{div}(f_1) \bullet \cdots \bullet \mathrm{div}(f_m)$ factorizes through the formula*

$$[F = 0] = \sum_{j=1}^m R_j^F \wedge df_j. \tag{1.36}$$

REMARK 1.11. For the sake of simplicity, we use the same notation for an analytic 0-cycle and the corresponding integration $(1,1)$-current. The factorization formula (1.36) is known as *Lelong–Poincaré formula.* It still holds in the more general geometric context where F is a section of a holomorphic bundle as in §1.2.5; see §C.2.4 (in particular Proposition C.32).

The vector of currents $R^F = (R_1^F, \ldots, R_m^F)$ that was introduced in Proposition 1.8 inherits a very important property: for any $h \in H(U)$ vanishing on $V(F)$, one has

$$\forall \varphi \in \mathcal{D}^{(1,0)}(U),\ \langle R^F, \overline{h}\varphi \rangle = 0. \tag{1.37}$$

This can be written in the abridged way $\overline{h}R^F = 0$ as $\mathbb{C}^m$-valued $(0,1)$-current on U. Moreover, let h be a holomorphic function in U such that $|h| \leq C\|F\|$. This inequality is equivalent to saying that h vanishes at any point $\alpha \in V(F)$ with multiplicity at least equal to $\nu_{F,\alpha}$. It then follows from (1.34) that

$$\forall \varphi \in \mathcal{D}^{(1,0)}(U),\ \langle R^F, h\varphi \rangle = 0, \tag{1.38}$$

or, in an abridged way, $hR^F = 0$ as $\mathbb{C}^m$-valued $(0,1)$-current on U.

In the particular case when $m = 1$ and $F = (f)$, the action of the current R^F (also denoted then R^f) on a test form $\varphi \in \mathcal{D}^{(1,0)}(U)$ is given as

$$\lim_{\epsilon \to 0+} \frac{1}{2i\pi} \frac{1}{\epsilon} \int_{|f|^2 = \epsilon} \overline{f}\varphi = \lim_{\epsilon \to 0+} \frac{1}{2i\pi} \int_{|f|^2 = \epsilon} \frac{\varphi}{f} = \frac{1}{2i\pi} \Big[\int_U \overline{\partial}\Big(\frac{|f|^{2\lambda}}{f} \Big) \wedge \varphi \Big]_{\lambda = 0}.$$

Stokes's formula allows rewriting this action as

$$\langle R^f, \varphi\rangle = -\lim_{\epsilon\to 0+}\frac{1}{2i\pi}\frac{1}{\epsilon}\int_{|f|^2\geq\epsilon}\overline{f}\,\overline{\partial}\varphi = -\frac{1}{2i\pi}\Big[\int_U \Big(\frac{|f|^{2\lambda}}{f}\Big)\wedge\overline{\partial}\varphi\Big]_{\lambda=0}.$$

If $\varphi = \varpi\, d\zeta$, where $\varpi \in \mathcal{D}^{(0,0)}(U)$, then one recognizes from (1.5) that

$$\langle R^f, \varpi\, d\zeta\rangle = -\frac{1}{\pi}\Big\langle \mathrm{PV}[1/f], \frac{\partial\varpi}{\partial\overline{\zeta}}\Big\rangle. \tag{1.39}$$

Denote as $P_0^f \in {}'\mathcal{D}^{(0,0)}(U)$ the $(0,0)$-current on U acting on smooth $(1,1)$-forms $\theta\in\mathcal{D}^{1,1}(U)$ as

$$\langle P_0^f, \theta\, d\overline{\zeta}\wedge d\zeta\rangle = \frac{1}{\pi}\langle \mathrm{PV}[1/f], \theta\rangle.$$

It follows from (1.39) that one has

$$\overline{\partial}(P_0^f) = R^f \tag{1.40}$$

in the sense of currents in U. If we denote alternatively P_0^f as $[1/f]$, it makes sense to denote R^f as $\overline{\partial}(1/f)$. Proposition 1.12 below extends from the case where $m=1$ to the case where $m\in\mathbb{N}^*$ the definition of $P_0^f = [1/f]$.

PROPOSITION 1.12. *Let $F = (f_1,\dots,f_m)\in (H(U)\setminus\{0\})^m$, where U is a domain in $\mathbb{C}$ and $\varphi\in\mathcal{D}^{1,1}(U)$. For each $j\in\{1,\dots,m\}$, the limit*

$$\lim_{\epsilon\to 0+}\frac{1}{2i\pi}\int_{\|F\|^2\geq\epsilon}\frac{\overline{f_j(\zeta)}}{\|F(\zeta)\|^2}\varphi(\zeta,\overline{\zeta}) \tag{1.41}$$

exists and the map

$$P_{0,j}^F : \varphi\in\mathcal{D}^{(1,1)}(U)\longmapsto \lim_{\epsilon\to 0+}\frac{1}{2i\pi}\int_{\|F\|^2\geq\epsilon}\frac{\overline{f_j(\zeta)}}{\|F(\zeta)\|^2}\varphi(\zeta,\overline{\zeta}) \tag{1.42}$$

defines a $(0,0)$-current on U. Moreover, for any $\varphi\in\mathcal{D}^{(1,1)}(U)$, the function

$$\lambda\longmapsto \frac{1}{2i\pi}\int_U \|F\|^{2(\lambda-1)}\,\overline{f_j}\,\varphi$$

extends from $\{\lambda\in\mathbb{C}; \mathrm{Re}\,\lambda\gg 1\}$ as a meromorphic function with (possible) poles in $-\mathbb{Q}^+$, and one has

$$\langle P_{0,j}^F, \varphi\rangle = \frac{1}{2i\pi}\Big[\int_U \|F\|^{2(\lambda-1)}\,\overline{f_j}\,\varphi\Big]_{\lambda=0} = [\langle P_{0,j}^{F,\lambda},\varphi\rangle]_{\lambda=0}. \tag{1.43}$$

PROOF. It is enough to take $j=1$. If $\mathrm{Supp}(\varphi)\cap V(F)=\emptyset$, then the function

$$\lambda\in\mathbb{C}\longmapsto \frac{1}{2i\pi}\int_U \|F\|^{2(\lambda-1)}\,\overline{f_1}\,\varphi$$

is entire and Lebesgue's theorem implies that

$$\begin{aligned}\lim_{\epsilon\to 0+}\frac{1}{2i\pi}\int_{\|F\|^2\geq\epsilon}\frac{\overline{f_1(\zeta)}}{\|F(\zeta)\|^2}\varphi(\zeta,\overline{\zeta}) &= \frac{1}{2i\pi}\int_U \frac{\overline{f_1(\zeta)}}{\|F(\zeta)\|^2}\varphi(\zeta,\overline{\zeta})\\ &= \frac{1}{2i\pi}\Big[\int_U \|F\|^{2(\lambda-1)}\,\overline{f_1}\,\varphi\Big]_{\lambda=0}.\end{aligned}$$

One can therefore assume that $\mathrm{Supp}(\varphi) \subset D_\alpha$, where $\alpha \in V(F)$. From now on $\nu = \nu_\alpha$, D_α, χ, Δ, $q = q_\alpha$ are taken as in the proof of Proposition 1.8, according to (1.24) and (1.26). For any $\lambda \in \mathbb{C}$ with $\mathrm{Re}\,\lambda \gg 1$, one has then that

$$\begin{aligned}(1.44)\qquad \frac{1}{2i\pi}\int_U \|F\|^{2(\lambda-1)}\,\overline{f_1}\,\varphi &= \frac{1}{2i\pi}\int_{D_\alpha} \frac{|\chi^{-1}(\zeta)|^{2\nu\lambda}}{(\chi^{-1}(\zeta))^\nu}\,\frac{\overline{q_1(\zeta)}}{\|q(\zeta)\|^2}\,\varphi(\zeta)\\ &= \frac{1}{2i\pi}\int_\Delta \frac{|w|^{2\nu\lambda}}{w^\nu}\,\chi^*\Big(\frac{\overline{q}_1}{\|q\|^2}\,\varphi\Big)(w).\end{aligned}$$

On the other hand, for any $\varepsilon > 0$ sufficiently small one has that

$$\begin{aligned}\frac{1}{2i\pi}\int_{\|F\|^2\geq\epsilon} \frac{\overline{f_1(\zeta)}}{\|F(\zeta)\|^2}\,\varphi(\zeta) &= \frac{1}{2i\pi}\int_{\{\zeta\in\mathbb{D}_\alpha\,:\,|(\chi^{-1}(\zeta))^\nu|^2\geq\varepsilon\}} \frac{1}{(\chi^{-1}(\zeta))^\nu}\,\frac{\overline{q_1(\zeta)}}{\|q(\zeta)\|^2}\,\varphi(\zeta)\\ &= \frac{1}{2i\pi}\int_{\{w\in\Delta\,:\,|w^\nu|^2\geq\varepsilon\}} \frac{1}{w^\nu}\,\chi^*\Big(\frac{\overline{q}_1}{\|q\|^2}\,\varphi\Big)(w).\end{aligned}$$

It follows then from (1.44) and the Bernstein–Sato equation (1.6) with $f = w^\nu$ that the function

$$\lambda \in \{\lambda\in\mathbb{C}\,:\,\mathrm{Re}\,\lambda >> 1\} \longmapsto \frac{1}{2i\pi}\int_U \|F\|^{2(\lambda-1)}\,\overline{f_1}\,\varphi$$

extends as a meromorphic function with rational poles in $]-\infty,-1/\nu]$. On the other hand, one can see using polar coordinates $w = \rho e^{i\theta}$ that

$$\lim_{\varepsilon\to 0_+}\frac{1}{2i\pi}\int_{\{w\in\Delta;|w^\nu|^2\geq\varepsilon\}} \frac{1}{w^\nu}\chi^*\Big(\frac{\overline{q}_1}{\|q\|^2}\varphi\Big)(w) = \Big[\frac{1}{2i\pi}\int_\Delta \frac{|w|^{2\nu\lambda}}{w^\nu}\,\chi^*\Big(\frac{\overline{q}_1}{\|q\|^2}\varphi\Big)(w)\Big]_{\lambda=0}.$$

This concludes the proof of the proposition. □

REMARK 1.13. It follows from (1.35) and from the definition of P_0^f, where $f \in H(U)$ satisfies $\mathrm{div}(f) = \sum_{\alpha\in V(F)} \nu_{F,\alpha}\,\{\alpha\}$, that

$$(1.45)\qquad P_{0,j}^F = \frac{\overline{u_j}}{\|u\|^2}\,P_0^f$$

in the sense of currents for any $j = 1,\dots,m$. Furthermore, the holomorphic functions $u_j \in H(U)$ satisfying $V(u) = \emptyset$ have been introduced in (1.25).

COROLLARY 1.14. *For any $j = 1,\dots,m$ and any $\varphi \in \mathcal{D}^{(1,0)}(U)$, the function*

$$(1.46)\qquad \lambda \longmapsto \frac{1}{2i\pi}\int_U \|F\|^{2\lambda}\,\overline{\partial}\Big(\frac{\overline{f_j}}{\|F\|^2}\Big)\wedge\varphi$$

extends from $\{\lambda;\mathrm{Re}\,\lambda\gg 1\}$ as a meromorphic function in $\mathbb{C}$ with (possible) poles in $-\mathbb{Q}^+$. Its value

$$\varphi\in\mathcal{D}^{(1,0)}(U)\longmapsto\langle S_j^F,\varphi\rangle = [\langle S_j^{F,\lambda},\varphi\rangle]_{\lambda=0}$$

at $\lambda = 0$ defines the action of a current S_j^F such that

$$(1.47)\qquad \bar{\partial}P_{0,j}^F = R_j^F + S_j^F,$$

which extends (1.40) to the case $m > 1$.

PROOF. It follows from Leibniz's formula and Stokes's theorem that for $\lambda \in \mathbb{C}$ such that $\operatorname{Re}\lambda \gg 1$,

$$\frac{1}{2i\pi}\int_U \|F\|^{2\lambda}\,\overline{\partial}\Big(\frac{\overline{f_j}}{\|F\|^2}\Big)\wedge\varphi = \frac{1}{2i\pi}\int_U \Big(\overline{\partial}(\|F\|^{2(\lambda-1)}\overline{f_j}) - \frac{\overline{f_j}}{\|F\|^2}\,\overline{\partial}\|F\|^{2\lambda}\Big)\wedge\varphi$$
$$= -\frac{1}{2i\pi}\Big(\int_U \|F\|^{2(\lambda-1)}\overline{f}_j\,\overline{\partial}\varphi + \int_U \overline{\partial}\|F\|^{2\lambda}\wedge\frac{\overline{f_j}}{\|F\|^2}\varphi\Big).$$

It follows from Proposition 1.12 that the function

$$\lambda \in \{\lambda \in \mathbb{C} : \operatorname{Re}\lambda \gg 1\} \longmapsto -\frac{1}{2i\pi}\Big(\int_U \|F\|^{2(\lambda-1)}\overline{f}_j\,\overline{\partial}\varphi\Big)$$

extends as a meromorphic function with poles in $-\mathbb{Q}^+$, whose value at $\lambda = 0$ equals $\langle\overline{\partial}P_j^F, \varphi\rangle$. On the other hand, one has

$$\frac{1}{2i\pi}\int_U \overline{\partial}\|F\|^{2\lambda}\wedge\frac{\overline{f_j}}{\|F\|^2}\varphi = \langle R_j^{F,\lambda}, \varphi\rangle;$$

see (1.34). Such a function of λ extends also by Proposition 1.8 as a meromorphic function with poles in $-\mathbb{Q}^+$, whose value at $\lambda = 0$ equals $\langle R_j^F, \varphi\rangle$. Corollary 1.14 is proved. □

REMARK 1.15. Let $f \in H(U)$ satisfy $\operatorname{div}(f) = \sum_{\alpha\in V(F)} \nu_{F,\alpha}\{\alpha\}$ and let $u_1, \dots, u_m$ in $H(U)$ be such that $V(u) = \emptyset$, that is, they have been chosen as in (1.25). It follows from (1.35), (1.45), and Corollary 1.14 that for any $j = 1, \dots, m$,

$$S_j^F = \overline{\partial}\Big(\frac{\overline{u}_j}{\|u\|^2}\Big)\,P_0^f \tag{1.48}$$

in the sense of currents.

1.2.2. Residue currents and Lagrange's interpolation formula. Suppose that U is a bounded domain in $\mathbb{C}$ with piecewise C^1-boundary and that f is a function holomorphic in U that extends continuously to $K = \overline{U}$. Assume also that $f \neq 0$ on the boundary of U. That is, we assume that $f \neq 0$ on the support of the piecewise C^1 cycle ∂K equipped with the counterclockwise (positive) orientation. The function

$$(z,\zeta) \in U \times K \longmapsto b_f(z,\zeta) = \frac{f(\zeta) - f(z)}{\zeta - z}$$

belongs to $C^1(U \times K)$ and is holomorphic in $U \times U$ as a function of two complex variables. According to the terminology used in the algebraic context (see Corollary 1.37 in §1.3.1), such a function b_f is called a *bézoutian* attached to f. It is well known that any function h, holomorphic in U and continuous up to K, admits in U the Cauchy integral representation formula. What is less known is that this integral representation formula can be transformed in order to become *Lagrange's interpolation formula*

(1.49)

$$
\begin{aligned}
h(z) &= \frac{1}{2\pi i}\int_{\partial K} \frac{h(\zeta)\,d\zeta}{\zeta - z} \\
&= \frac{1}{2\pi i}\Big(\int_{\partial K} h(\zeta)\,\frac{b_f(z,\zeta)\,d\zeta}{f(\zeta)} + f(z)\int_{\partial K} \frac{h(\zeta)}{f(\zeta)(\zeta - z)}\Big) \\
&= \sum_{\{\alpha\in U\,:\,f(\alpha)=0\}} \operatorname{Res}\Big[\frac{h(\zeta)b_f(z,\zeta)}{f(\zeta)}\,d\zeta, \alpha\Big] + f(z)\Big(\frac{1}{2\pi i}\int_{\partial K}\frac{h(\zeta)}{f(\zeta)(\zeta-z)}\,d\zeta\Big) \\
&= \Big\langle \overline{\partial}\Big(\frac{1}{f(\zeta)}\Big), h(\zeta)\,b_f(z,\zeta)\,d\zeta\Big\rangle + f(z)\Big(\frac{1}{2i\pi}\int_{\partial K}\frac{h(\zeta)}{f(\zeta)(\zeta-z)}\,d\zeta\Big).
\end{aligned}
$$

Formula (1.49) will play an essential role throughout this monograph. As a consequence, one gets the *trace formula*

$$
h(z) = \Big\langle \overline{\partial}\Big(\frac{1}{f(\zeta)}\Big)\,,\, h(\zeta)\,b_f(z,\zeta)\,d\zeta\Big\rangle \ \text{modulo}\, fH(U) \tag{1.50}
$$

for any $h \in H(U)$.

Let us consider in this section, instead of $f \in H(U) \cap C^1(K)$, which does not vanish on the boundary ∂U, a vector $F = (f_1,\dots,f_m)$ of nonidentically zero functions belonging to $H(U) \cap C^1(K)$ with set of common zeroes $V(F) \subset U$. As a consequence $V(F)$ is finite. For any invertible element u of $H(U)\cap C^1(K)$ the divisor $\operatorname{div}(f)$ of the function

$$
f : \zeta \in U \longmapsto u(\zeta) \prod_{\alpha\in V(F)} (\zeta - \alpha)^{\nu_{F,\alpha}} \in H(U) \in C^1(K) \tag{1.51}
$$

equals the intersection cycle $\operatorname{div}(f_1) \bullet \cdots \bullet \operatorname{div}(f_m)$ in U. From now on we fix such a function f and rewrite $F = (f_1,\dots,f_m)$ as $F = (fu_1,\dots,fu_m)$, where $f_j = fu_j$, for $j = 1,\dots,m$ and $u = (u_1,\dots,u_m)$ is a vector of functions in $H(U)\cap C^1(K)$, whose components do not share any common zero in $K = \overline{U}$. The corresponding bézoutian is then equal to $b_F = (b_{f_1},\dots,b_{f_m})$. Let also $\sigma_j = \overline{u}_j/\|u\|^2 \in C^\infty(U)\cap C^1(K)$ for any $j = 1,\dots,m$ with the respective vector map $\sigma = (\sigma_1,\dots,\sigma_m)$. If λ denotes a complex parameter with sufficiently large real part $\operatorname{Re}\lambda \gg 1$ and $N \in \mathbb{N}$, then one defines a function $\Phi_N^{F,\lambda} \in C^1(U\times K)$ as

$$
\begin{aligned}
\Phi_N^{F,\lambda}(z,\zeta) &= (\Phi^{F,\lambda}(z,\zeta))^N = \Big(1 + (z-\zeta)\,\|F(\zeta)\|^{2\lambda}\,\Big(b_F(z,\zeta)\cdot\frac{F(\zeta)}{\|F(\zeta)\|^2}\Big)\Big)^N \\
&= \Big(1 - \|F(\zeta)\|^{2\lambda} + \|F(\zeta)\|^{2\lambda}\,\Big(F(z)\cdot\frac{F(\zeta)}{\|F(\zeta)\|^2}\Big)\Big)^N \\
&= \Big(1 - \|u(\zeta)\|^{2\lambda}\,|f(\zeta)|^{2\lambda} + \|u(\zeta)\|^{2\lambda}\langle\!\langle \sigma(\zeta), F(z)\rangle\!\rangle\,\frac{|f(\zeta)|^{2\lambda}}{f(\zeta)}\Big)^N.
\end{aligned} \tag{1.52}
$$

The function (1.52) is holomorphic in its first variable z. Observe that for any $N \in \mathbb{N}^*$

$$
\begin{aligned}
\frac{\overline{\partial}_\zeta(\Phi_{N+1}^{F,\lambda}(z,\zeta)\,d\zeta)}{2i\pi} &= -(N+1)\,\Phi_N^{F,\lambda}(z,\zeta)\,\overline{\partial}\,\langle\!\langle P_0^{F,\lambda}(\zeta), b_F(z,\zeta)\rangle\!\rangle \\
&= -(N+1)\,\Phi_N^{F,\lambda}(z,\zeta)\langle\!\langle\, R^{F,\lambda}(\zeta) + S^{F,\lambda}(\zeta), b_F(z,\zeta)\rangle\!\rangle,
\end{aligned} \tag{1.53}
$$

where $P_0^{F,\lambda}$, $R^{F,\lambda}$, $S^{F,\lambda}$ denote respectively the vectors of C^1 differential forms $P_{0,j}^{F,\lambda}$, $R_j^{F,\lambda}$, and $S_j^{F,\lambda}$ defined in K by Proposition 1.8, Proposition 1.12, and Corollary 1.14. In the above statements, these forms are described as corresponding currents in U. The starting point in order to transform Cauchy's formula and involve the vector of holomorphic functions $F = (f_1, \dots, f_m)$ instead of a single holomorphic function is the Cauchy–Pompeiu integral formula in U [**Pom1909**]. This was previously already done within establishing (1.49) in the case when $m = 1$. Namely, if ϕ is a C^1 function in K, then for any $z \in U$,

$$\phi(z) = \frac{1}{2i\pi}\int_{\partial K}\frac{\phi(\zeta)\,d\zeta}{\zeta - z} - \frac{1}{2i\pi}\int_K \frac{\overline{\partial}(\phi(\zeta)\,d\zeta)}{\zeta - z}. \tag{1.54}$$

Weighted Koppelman's formulae (in either the analytic or geometric context), which we will study later on (in §4.3 and §4.5.3), can be derived from the following modification of (1.54). Let $\Phi : (z,\zeta) \longmapsto \Phi(z,\zeta)$ in $C^1(U \times K)$ be such that $\bar{\partial}_z \Phi(z,\zeta) = 0$, which means that Φ is holomorphic in z. Then, for any $h \in H(U) \cap C^1(K)$ and for any $z \in U$, one has that

$$\begin{aligned} h(z)\,\Phi(z,z) &= \frac{1}{2i\pi}\Big(\int_{\partial K} h(\zeta)\,\frac{\Phi(z,\zeta)\,d\zeta}{\zeta - z} - \int_K h(\zeta)\,\frac{\overline{\partial}_{z,\zeta}(\Phi(z,\zeta)\,d\zeta)}{\zeta - z}\Big) \\ &= \frac{1}{2i\pi}\Big(\int_{\partial K} h(\zeta)\,\frac{\Phi(z,\zeta)\,d\zeta}{\zeta - z} - \int_K h(\zeta)\,\frac{\partial \Phi}{\partial \overline{\zeta}}(z,\zeta)\frac{d\overline{\zeta}\wedge d\zeta}{\zeta - z}\Big). \end{aligned} \tag{1.55}$$

Thus one can formulate the following Lagrange-type division formula.

PROPOSITION 1.16. *Let U, $K = \overline{U}$, $F = (f_1, \dots, f_m)$ be as before. We assume that f is of the form* (1.51), *where $f_j = f u_j$ and $\|u\| = \|(u_1, \dots, u_m)\| > 0$ in K. If $\sigma = (\sigma_1, \dots, \sigma_m) = \overline{u}/\|u\|^2$, then for any $h \in H(U) \cap C(K)$, the function*

$$\lambda \longmapsto \int_K h(\zeta)\,\Phi^{F,\lambda}(z,\zeta)\,\overline{\partial}\langle\!\langle P^{F,\lambda}(\zeta), b_F(z,\zeta)\rangle\!\rangle$$

extends from the half-plane $\{\lambda \in \mathbb{C} : \ \mathrm{Re}\,\lambda \gg 1\}$ to the complex plane $\mathbb{C}$ as a meromorphic function with (possible) poles in $-\mathbb{Q}^+$. Moreover, for any $z \in U$,

$$\begin{aligned} h(z) = 2\Big[\int_K h(\zeta)\,\Phi^{F,\lambda}(z,\zeta)\,\overline{\partial}\,\langle\!\langle P_0^{F,\lambda}(\zeta), b_F(z,\zeta)\rangle\!\rangle\Big]_{\lambda=0} \\ + \frac{1}{2i\pi}\int_{\partial K} h(\zeta)\,\Big(\frac{F(z)\cdot F(\zeta)}{\|F(\zeta)\|^2}\Big)^2\frac{d\zeta}{\zeta - z}. \end{aligned} \tag{1.56}$$

The first term in the sum on the right-hand side of (1.56) *splits as the sum of the residue term*

$$\Big\langle \bar{\partial}\Big(\frac{1}{f}\Big) + \langle\!\langle \sigma, F(z)\rangle\!\rangle\,\bar{\partial}\Big(\frac{1}{f^2}\Big)\,,h\,\langle\!\langle \sigma, b_F(z,\cdot)\rangle\!\rangle\,d\zeta\Big\rangle \tag{1.57}$$

with the spread integral

$$2\Big\langle \chi_K\,\langle\!\langle \sigma, F(z)\rangle\!\rangle\,P_0^{f^2}\,,\,h\,\overline{\partial}(\langle\!\langle \sigma, b_F(z,\cdot)\rangle\!\rangle\,d\zeta)\Big\rangle. \tag{1.58}$$

PROOF. Observe that one has for any $\zeta \in K$ and any $\lambda \in \mathbb{C}$ that the identity

$$\|u(\zeta)\|^{2\lambda} = e^{\lambda \log \|u(\zeta)\|^2} = 1 + \lambda\sum_{\kappa=0}^{\infty}\frac{\lambda^\kappa}{(\kappa+1)!}(\log\|u(\zeta)\|^2)^{\kappa+1} = 1 + \lambda\,\rho_\lambda(\zeta)$$

holds. Moreover, the function $(\lambda, \zeta) \mapsto \rho_\lambda(\zeta)$ is continuous on $\mathbb{C} \times K$ and is holomorphic in λ. We rewrite the last equality in (1.52) as

$$\Phi^{F,\lambda}(z,\zeta) = 1 - |f(\zeta)|^{2\lambda} + \langle\!\langle \sigma(\zeta), F(z) \rangle\!\rangle \frac{|f(\zeta)|^{2\lambda}}{f(\zeta)} + \lambda\, \rho_\lambda(\zeta)\, |f(\zeta)|^{2\lambda} \Big(\frac{\langle\!\langle \sigma(\zeta), F(z) \rangle\!\rangle}{f(\zeta)} - 1 \Big) \tag{1.59}$$

and split each $R_j^{F,\lambda}$ and $S_j^{F,\lambda}$ as

$$R_j^{F,\lambda} = \frac{\sigma_j}{2i\pi} \Big(\overline{\partial}\Big(\frac{|f|^{2\lambda}}{f} \Big) + \lambda\, \overline{\partial} \rho_\lambda \frac{|f|^{2\lambda}}{f} \Big), \quad S_j^{F,\lambda} = \big(\bar{\partial} \sigma_j + \lambda\, \sigma_j\, \overline{\partial} \rho_\lambda \big) \frac{|f|^{2\lambda}}{2i\pi\, f}. \tag{1.60}$$

Let $\theta \in \mathcal{D}(U)$ be such that $\theta \equiv 1$ about $V(F)$. For each continuous $(1,0)$-form φ on K, and each $j = 1, \dots, m$, the functions

$$\lambda \longmapsto \int_K (1 - \theta(\zeta))\, h(\zeta)\, \Phi^{F,\lambda}(z,\zeta)\, \bar{\partial} P_{0,j}^{F,\lambda}(\zeta) \wedge \varphi(\zeta)$$

and

$$\lambda \longmapsto \int_K (1-\theta)\, \frac{\sigma_j}{2i\pi} \Big(1 - |f|^{2\lambda} + \langle\!\langle \sigma, F(z) \rangle\!\rangle \frac{|f|^{2\lambda}}{f} \Big) \overline{\partial}\Big(\frac{|f|^{2\lambda}}{f} \Big) \wedge \varphi + \int_K (1-\theta) \langle\!\langle \sigma, F(z) \rangle\!\rangle\, \overline{\partial} \sigma_j\, \frac{|f|^{4\lambda}}{2i\pi\, f^2} \wedge \varphi.$$

are entire, since F, and hence f, does not vanish on the support of $1-\theta$. Moreover, relations (1.59), (1.60), and the fact that $\lambda \longmapsto \rho_\lambda(\zeta)$ is entire for any $\zeta \in K$ show that they share the same value at $\lambda = 0$. Repeating the proofs of Proposition 1.8 and Proposition 1.12, one shows that whenever the $(1,0)$-form φ is in addition smooth in U, the two functions

$$\lambda \longmapsto \int_K \theta(\zeta)\, \Phi^{F,\lambda}(z,\zeta)\, \overline{\partial} P_{0,j}^{F,\lambda}(\zeta) \wedge \varphi(\zeta)$$

and

$$\lambda \longmapsto \int_K \theta\, \frac{\sigma_j}{2i\pi} \Big(1 - |f|^{2\lambda} + \langle\!\langle \sigma, F(z) \rangle\!\rangle \frac{|f|^{2\lambda}}{f} \Big) \overline{\partial}\Big(\frac{|f|^{2\lambda}}{f} \Big) \wedge \varphi + \int_K \theta \langle\!\langle \sigma, F(z) \rangle\!\rangle\, \overline{\partial} \sigma_j\, \frac{|f|^{4\lambda}}{2i\pi\, f^2} \wedge \varphi.$$

extend from the half-plane $\{\lambda \in \mathbb{C} : \operatorname{Re} \lambda \gg 1\}$ to meromorphic functions in $\mathbb{C}$ with (possible) poles in $-\mathbb{Q}^+$ and still share the same value at $\lambda = 0$. This proves the first assertion in the proposition. To prove the second assertion, one observes that for any $z \in U$ and any complex number λ satisfying $\operatorname{Re} \lambda \gg 1$ it holds that $\Phi^{F,\lambda}(z,z) = 1$. Then it follows from the Cauchy–Pompeiu formula (1.55) that for any such λ and any $z \in U$,

$$h(z) = \frac{1}{2i\pi} \Big(\int_{\partial K} h(\zeta)\, \Phi_2^{F,\lambda}(z,\zeta)\, \frac{d\zeta}{\zeta - z} - \int_K h(\zeta) \frac{\overline{\partial}_{z,\zeta}(\Phi_2^{F,\lambda}(z,\zeta)\, d\zeta)}{\zeta - z} \Big).$$

To obtain the integral representation formula (1.56), one needs to use (1.53) for $N = 2$ first and then follow the analytic continuation of both sides of this identity as meromorphic functions of λ. Finally, one identifies the evaluations of both sides at $\lambda = 0$. The final assertion follows from the following equalities between $(0,1)$-currents:

$$\begin{aligned}(1.61)\qquad \Big[(1-|f|^{2\lambda})\,\overline{\partial}\Big(\frac{|f|^{2\lambda}}{2i\pi\, f}\Big)\Big]_{\lambda=0} &= \frac{1}{2}\,\overline{\partial}\Big(\frac{1}{f}\Big)\\ \Big[\frac{|f|^{2\lambda}}{f}\,\overline{\partial}\Big(\frac{|f|^{2\lambda}}{2i\pi\, f}\Big)\Big]_{\lambda=0} &= \frac{1}{2}\,\overline{\partial}\Big(\frac{|f^2|^{2\lambda}}{2i\pi\, f^2}\Big) = \overline{\partial}\Big(\frac{1}{f^2}\Big). \qquad \square\end{aligned}$$

REMARK 1.17. The right-hand side of the integral representation formula (1.56) is congruent, as a function of z, to

$$z \longmapsto \Big\langle \overline{\partial}\Big(\frac{1}{f}\Big)\,,\, h(\zeta)\,\langle\!\langle \sigma, b_F(z,\cdot)\rangle\!\rangle\, d\zeta \Big\rangle = \langle\!\langle R^F(\zeta)\,,\, h(\zeta)\, b_F(z,\zeta)\, d\zeta \rangle\!\rangle$$

modulo the ideal generated by $f_1,\dots,f_m$ in $H(U)$. The major difference between the Lagrange division-interpolation formula (1.49) (when $m=1$) and the formula (1.56) is contained in the description of the defect term

$$h(z) - \langle\!\langle R^F(\zeta)\,,\, h(\zeta)\, b_F(z,\zeta)\, d\zeta \rangle\!\rangle = \langle\!\langle F(z), q(z) \rangle\!\rangle.$$

When $m=1$, q is expressed in terms of the restriction of h to the boundary of K, while in the case when $m>1$, the expression of q involves all values of h in K given the contribution (1.58).

1.2.3. Residue currents and Bergman–Weil developments. Let U, K, F, f, σ be as in the previous subsection. Let $\rho_F > 0$ be the minimum of $\|F\|$ on the boundary of U and let $U' = \{\zeta \in U : \|F(\zeta)\| < \rho_F\}$. Cauchy's integral representation formula for $h \in H(U)\cap C(K)$ (equation (1.49), line 1) can be transformed, when $z \in U'$, to

$$\begin{aligned} h(z) &= \frac{1}{2i\pi}\int_{\partial K}\Big(\sum_{j=1}^{m} \frac{\overline{f_j(\zeta)}}{(F(\zeta)-F(z))\cdot F(\zeta)}\Big(h(\zeta)\,\frac{f_j(\zeta)-f_j(z)}{\zeta - z}\, d\zeta\Big)\Big)\\ &= \frac{1}{2i\pi}\sum_{k=0}^{\infty}\int_{\partial K}\Big(\frac{F(z)\cdot F(\zeta)}{\|F(\zeta)\|^2}\Big)^k\Big(\sum_{j=1}^{m}\frac{\overline{f_j(\zeta)}}{\|F(\zeta)\|^2}\Big(h(\zeta)\,\frac{f_j(\zeta)-f_j(z)}{\zeta-z}\, d\zeta\Big)\Big)\\ (1.62)\qquad &= \Big[\frac{1}{2i\pi}\sum_{k=0}^{\infty}\int_{\partial K}|f^{k+1}|^{2\lambda}\,\frac{\langle\!\langle \sigma, F(z)\rangle\!\rangle^k}{f^{k+1}}\, h\,\langle\!\langle \sigma, b_F(z,\cdot)\rangle\!\rangle\, d\zeta\Big]_{\lambda=0}. \end{aligned}$$

It follows from the definition of the principal value and residue currents for $k \in \mathbb{N}$ (see (1.39) and (1.40)),

$$P_0^{f^{k+1}} = \Big[\frac{|f^{k+1}|^{2\lambda}}{2i\pi f^{k+1}}\Big]_{\lambda=0},\quad R^{f^{k+1}} = \overline{\partial}\Big(\frac{1}{f^{k+1}}\Big) = \Big[\overline{\partial}\Big(\frac{|f^{k+1}|^{2\lambda}}{2i\pi f^{k+1}}\Big)\Big]_{\lambda=0},$$

and the use of Stokes's formula, that the right-hand side of the integral representation formula (1.61) truncates as follows: for any $N \in \mathbb{N}^*$ and $z \in U$,

$$\begin{aligned}(1.63)\quad h(z) = \sum_{k=0}^{N-1}&\Big(\Big\langle \langle\!\langle \sigma, F(z)\rangle\!\rangle^k\,\overline{\partial}\Big(\frac{1}{f^{k+1}}\Big)\,,\, h\,\langle\!\langle \sigma, b_F(z,\cdot)\rangle\!\rangle\, d\zeta\Big\rangle\\ &+\Big\langle \chi_K\, P_0^{f^{k+1}}\,,\, h\,\overline{\partial}(\langle\!\langle \sigma, F(z)\rangle\!\rangle^k\,\langle\!\langle \sigma, b_F(z,\cdot)\rangle\!\rangle\, d\zeta)\Big\rangle\Big)\\ &\qquad + \frac{1}{2i\pi}\int_{\partial K} h(\zeta)\Big(\frac{F(z)\cdot F(\zeta)}{\|F(\zeta)\|^2}\Big)^N\frac{d\zeta}{\zeta - z}.\end{aligned}$$

When $m = 1$ and $F = f$, formula (1.63) simplifies as

$$\begin{aligned} h(z) &= \sum_{k=0}^{N-1} \Big\langle \overline{\partial}\Big(\frac{1}{f^{k+1}}\Big)\,,\, h\, b_f(z,\cdot)\, d\zeta \Big\rangle\, f^k(z) + \Big(\frac{1}{2i\pi}\int_{\partial K} \frac{1}{f^N(\zeta)}\,\frac{d\zeta}{\zeta - z}\Big)\, f^N(z) \\ &= L_N[h;f](z) + \Big(\frac{1}{2i\pi}\int_{\partial K} \frac{1}{f^N(\zeta)}\,\frac{d\zeta}{\zeta - z}\Big)\, f^N(z). \qquad (1.64) \end{aligned}$$

When $m > 1$, the following proposition provides a generalization of (1.64) that respects the fact that the remainder term $L_N[h;F]$ modulo the Nth power of the ideal $\sum_1^m f_j H(U)$ involves only the action of $(0,1)$-currents supported by $V(F)$. This was not the case in the N-truncated version (1.63) of (1.62).

Proposition 1.18. *Let U, K, F, f, σ be as in Proposition 1.16 and let N in $\mathbb{N}^*$. For any $h \in H(U) \cap C^1(K)$, the function*

$$\lambda \longmapsto \int_K h(\zeta)\, \Phi_{N+1}^{F,\lambda}(z,\zeta)\, \overline{\partial}\, \langle\!\langle P_0^{F,\lambda}(\zeta), b_F(z,\zeta)\rangle\!\rangle$$

extends from the right half-plane $\{\lambda \in \mathbb{C} : \operatorname{Re}\lambda \gg 1\}$ to $\mathbb{C}$ as a meromorphic function with (possible) poles in $-\mathbb{Q}^+$. Moreover, for any $z \in U$,

$$\begin{aligned} h(z) = (N+1)\Big[\int_K h(\zeta)\, \Phi_N^{F,\lambda}(z,\zeta)\, \overline{\partial}\, \langle\!\langle P_0^{F,\lambda}(\zeta), b_F(z,\zeta)\rangle\!\rangle\Big]_{\lambda=0} \\ + \frac{1}{2i\pi}\int_{\partial K} h(\zeta)\, \Big(\frac{F(z)\cdot F(\zeta)}{\|F(\zeta)\|^2}\Big)^{N+1} \frac{d\zeta}{\zeta - z}. \end{aligned} \qquad (1.65)$$

The first term in the sum on the right-hand side of (1.65) splits as the sum of the residue term

$$\sum_{k=0}^{N} \Big\langle \langle\!\langle \sigma, F(z)\rangle\!\rangle^k\, \overline{\partial}\Big(\frac{1}{f^{k+1}}\Big)\,,\, h\, \langle\!\langle \sigma, b_F(z,\cdot)\rangle\!\rangle\, d\zeta \Big\rangle \qquad (1.66)$$

and of the spread integral

$$(N+1)\, \Big\langle \chi_K\, \langle\!\langle \sigma, F(z)\rangle\!\rangle^N\, P_0^{f^{N+1}}\,,\, h\, \overline{\partial}\big(\langle\!\langle \sigma, b_F(z,\cdot)\rangle\!\rangle\, d\zeta\big) \Big\rangle. \qquad (1.67)$$

Proof. The first assertion is proved in the same way as the first assertion in Proposition 1.16. To prove the integral representation formula (1.65), one applies again the Cauchy–Pompeiu formula (1.55) with $\Phi(z,\zeta) = \Phi_N^{F,\lambda}(z,\zeta)$, where the complex parameter λ has sufficiently large real part $\operatorname{Re}\lambda \gg 1$. Formula (1.65) follows by applying the analytic continuation principle up to $\lambda = 0$. For the last assertion, one first observes that the equalities between $(0,1)$-currents (1.61) can be easily generalized as follows: for any $N \in \mathbb{N}^*$ and $0 \le k \le N$, one has

$$\begin{aligned} \binom{N}{k}\Big[(1 - |f|^{2\lambda})^{N-k} \frac{|f|^{2k\lambda}}{f^k}\, \overline{\partial}\Big(\frac{|f|^{2\lambda}}{2i\pi\, f}\Big)\Big]_{\lambda=0} &= \binom{N}{k}\Big(\int_0^1 t^k (1-t)^{N-k}\, dt\Big)\, \overline{\partial}\Big(\frac{1}{f^{k+1}}\Big) \\ &= \frac{1}{N+1}\, \overline{\partial}\Big(\frac{1}{f^{k+1}}\Big). \end{aligned} \qquad (1.68)$$

This implies that

$$(N+1)\Big[\int_K \Phi_N^{F,\lambda}(z,\cdot)\,\langle\!\langle R^{F,\lambda}, b_F(z,\cdot)\rangle\!\rangle\,d\zeta\Big]_{\lambda=0}$$
$$= \sum_{k=0}^{N} \Big\langle \langle\!\langle \sigma, F(z)\rangle\!\rangle^k\, \overline{\partial}\Big(\frac{1}{f^{k+1}}\Big)\,,\, h\,\langle\!\langle \sigma, b_F(z,\cdot)\rangle\!\rangle\,d\zeta\Big\rangle.$$

Since one has also

$$(N+1)\Big[\int_K \Phi_N^{F,\lambda}(z,\cdot)\,\langle\!\langle S^{F,\lambda}, b_F(z,\cdot)\rangle\!\rangle\,d\zeta\Big]_{\lambda=0}$$
$$= (N+1)\Big\langle \chi_K\,\langle\!\langle \sigma, F(z)\rangle\!\rangle^N\, P_0^{f^{N+1}}\,,\, h\,\overline{\partial}(\langle\!\langle \sigma, b_F(z,\cdot)\rangle\!\rangle\,d\zeta)\Big\rangle$$

because

$$\Big[(1-|f|^{2\lambda})^{N-k}\Big(\frac{|f|^{2(k+1)\lambda}}{f^{k+1}}\Big)\Big]_{\lambda=0} = 0 \quad \text{for } k=0,\dots,N-1$$
$$\Big[\frac{|f|^{2(N+1)\lambda}}{2\pi i f^{N+1}}\Big]_{\lambda=0} = P_0^{f^{N+1}} \qquad \text{(with singular support } V(F)\subset U\text{)},$$

the last assertion of the proposition follows. □

REMARK 1.19. The right-hand side of the integral representation formula (1.65) is congruent to the reminder contribution, as a function of z,

$$z \longmapsto L_N[h;F](z) := \sum_{k=0}^{N-1} \Big\langle \langle\!\langle \sigma, F(z)\rangle\!\rangle^k\, \overline{\partial}\Big(\frac{1}{f^{k+1}}\Big)\,,\, h\,\langle\!\langle \sigma, b_F(z,\cdot)\rangle\!\rangle\,d\zeta\Big\rangle, \tag{1.69}$$

modulo the Nth power of the ideal generated by the F in $H(U)$. When $m>1$ one observes in the expression for the quotient contribution of $h - L_N[h;F]$ as an N homogeneous form

$$h - L_N[h;F] = \langle\!\langle F, q_N[h]\rangle\!\rangle^{[N]} = \sum_{|\kappa|=N} \Big(\prod_1^m f_j^{\kappa_j}\Big)\, q_{N,\kappa}[h], \tag{1.70}$$

deduced from (1.65), that the quotient vector $q_N[h] = (q_{N,\kappa}[h])_{|\kappa|=N}$ fails to be described in terms of boundary values of h. This occurs because for $m>1$ the spread integral (1.67) contributes to $\langle\!\langle F, q_N[h]\rangle\!\rangle^{[N]}$.

DEFINITION 1.20. Let U, K, F, f, σ be as in Proposition 1.16. The sequence $(L_N[h;F], q_N[h])_{N\geq 1}$, defined by (1.69) and (1.70), is a marker for iterated euclidean division in $H(U)$ with respect to the ideal generated by F. Such a sequence is called a *Bergman–Weil expansion* for h in $H(U)$.

1.2.4. Residue currents and multiplicative calculus. Let U be a domain in $\mathbb{C}$. In this section, we formalize a multiplicative commutative tensorial calculus involving as fundamental bricks (or objects) the vector-valued residue currents R^F and principal value currents $2i\pi P_0^F$, where $F \in (H(U)\setminus\{0\})^m$. These currents are defined in Propositions 1.8 and 1.12, respectively. This calculus has been already used in the proofs of Proposition 1.16 and Proposition 1.18 in the previous subsections. Although multiplication of distributions, hence of currents, remains a stumbling block, one profits here from the fact that our fundamental objects R^F or $2i\pi P_0^F$ are realized by employing analytic continuation as $R^F = [R^{F,\lambda}]_{\lambda=0}$ or

$2i\pi P_0^F = [2i\pi P_0^{F,\lambda}]_{\lambda=0}$. Let us start with a result that ensures the robustness of the multiplication of principal value currents $2i\pi P_0^F$.

PROPOSITION 1.21. *Let $N \in \mathbb{N}^*$, $(m_1,\dots,m_N) \in (\mathbb{N}^*)^N$, $F_\ell \in (H(U)\setminus\{0\})^{m_\ell}$ for $1 \le \ell \le N$, and $M = m_1\cdots m_N$. The function*

$$\boldsymbol{\lambda} \in \{\boldsymbol{\lambda} \in \mathbb{C}^N \,:\, \operatorname{Re}\lambda_\ell \gg 0 \text{ for } \ell = 1,\dots,N\} \longmapsto 2i\pi P_0^{F_1,\lambda_1} \otimes \cdots \otimes 2i\pi P_0^{F_N,\lambda_N} \in (C(U))^M$$

induces a ${}'\mathcal{D}^{(0,0)}(U,\mathbb{C}^M)$-valued map which extends to $\mathbb{C}^N$ as a meromorphic function in N complex variables. Its polar set is the union of affine hyperplanes defined over $\mathbb{Q}$ that are disjoint from $\{\boldsymbol{\lambda} \in \mathbb{C}^N \,:\, \operatorname{Re}\lambda_\ell \ge 0 \text{ for } \ell = 1,\dots,N\}$. Moreover

$$\big[2i\pi P_0^{F_1,\lambda_1} \otimes \cdots \otimes 2i\pi P_0^{F_N,\lambda_N}\big]_{\boldsymbol{\lambda}=0} = 2i\pi\, P_0^{F_1\otimes\cdots\otimes F_N}. \tag{1.71}$$

PROOF. Let $\mathfrak{f}_\ell \in H(U)$ be such that $[F_\ell = 0] = [\operatorname{div}(\mathfrak{f}_\ell)]$ every $\ell = 1,\dots,N$. That is, $F_\ell = \mathfrak{f}_\ell U_\ell$, where $U_\ell \in H(U)$ and $V(U_\ell) = \emptyset$. Let also $\Sigma_\ell = \overline{U_\ell}/\|U_\ell\|^2$. For any $\boldsymbol{\lambda}$ in $\mathbb{C}^N$ satisfying $\operatorname{Re}\lambda_\ell >> 0$ for $\ell = 1,\dots,N$, one has

$$\bigotimes_{\ell=1}^N 2i\pi P_0^{F_\ell,\lambda_\ell} = \Big(\prod_{\ell=1}^N \frac{|\mathfrak{f}_\ell|^{2\lambda_\ell}}{\mathfrak{f}_\ell}\Big)\,\big(1 + \langle\!\langle r_{\boldsymbol{\lambda}}(\zeta),\boldsymbol{\lambda}\rangle\!\rangle\big) \bigotimes_{\ell=1}^N \Sigma_\ell,$$

where $r_{\boldsymbol{\lambda}}(\zeta) \in \mathbb{C}^N$ depends smoothly on $\zeta \in U$ and holomorphically on $\boldsymbol{\lambda} \in \mathbb{C}^N$. Here, following the convention for notation (1.22), $\langle\!\langle r_{\boldsymbol{\lambda}}(\zeta),\boldsymbol{\lambda}\rangle\!\rangle := \sum_{\ell=1}^N r_{\boldsymbol{\lambda},\ell}(\zeta)\,\lambda_\ell$. Let $\alpha \in \bigcup_{\ell=1}^N V(F_\ell) = \bigcup_{\ell=1}^N V(\mathfrak{f}_\ell)$ and let $\nu_{\ell,\alpha}$ be the order of f_ℓ at α for any $\ell = 1,\dots,N$. Denote also by $\nu = \nu_\alpha := (\nu_{1,\alpha},\dots,\nu_{N,\alpha})$. Then one has in a small disk D_α centered about α that

$$\prod_{\ell=1}^N \frac{|\mathfrak{f}_\ell|^{2\lambda_\ell}}{\mathfrak{f}_\ell} = \frac{|\zeta-\alpha|^{2\langle\!\langle \nu,\boldsymbol{\lambda}\rangle\!\rangle}}{(\zeta-\alpha)^{|\nu|}}\,\frac{1 + \langle\!\langle \boldsymbol{\lambda},\rho_{\boldsymbol{\lambda}}(\zeta)\rangle\!\rangle}{\prod_{\ell=1}^N v_\ell},$$

where $\langle\!\langle \nu,\boldsymbol{\lambda}\rangle\!\rangle = \sum_{\ell=1}^N \nu_\ell\lambda_\ell$, $|\nu| = \sum_{\ell=1}^N \nu_\ell$, the v_ℓ are invertible holomorphic functions in D_α with $\mathfrak{f}_\ell(\zeta) = (\zeta-\alpha)^{\nu_\ell} v_\ell(\zeta)$ for $\ell = 1,\dots,N$, and $\rho_{\boldsymbol{\lambda}}(\zeta) \in \mathbb{C}^N$ depends smoothly on $\zeta \in D_\alpha$ and holomorphically on $\boldsymbol{\lambda} \in \mathbb{C}^N$. Using repeatedly the second identity in (1.7) to perform integration by parts in D_α, one gets the first assertion of the proposition. It is immediate to check that (1.71) holds. □

Let us now incorporate into our finite collection of elementary principal value currents $2i\pi P_0^{F_\ell}$, $\ell = 1,\dots,N$ (as in Proposition 1.21), the residue current R^{F_0}, where $F_0 \in (H(U)\setminus\{0\})^{m_0}$. Given $G \in (H(U)\setminus\{0\})^M$ and $\alpha \in V(F_0)$, we need to define first the *localization* of $R^{F_0\otimes G}$ at α by

$$(R^{F_0\otimes G})_\alpha = \theta\, R^{F_0\otimes G}, \tag{1.72}$$

where $\theta \in \mathcal{D}(U)$ equals identically 1 about α and $\operatorname{Supp}(\theta) \cap V(F_0 \otimes G) = \{\alpha\}$. Observe that the vector-valued current defined by (1.72) is supported by $\{\alpha\}$ and does not depend on the choice of θ. Although the robustness of the multiplicative approach within the context of Proposition 1.21 does not subsist anymore within this new context, the following useful result needs to be mentioned.

PROPOSITION 1.22. *Let $N \in \mathbb{N}^*$, $(m_0,\dots,m_N) \in (\mathbb{N}^*)^{N+1}$, let for $0 \le \ell \le N$, $F_\ell \in (H(U)\setminus\{0\})^{m_\ell}$ and let $M = m_1\cdots m_N$. The function*

$$\lambda \in \{\lambda \in \mathbb{C};\, \operatorname{Re}\lambda \gg 0\} \longmapsto R^{F_0,\lambda} \otimes 2i\pi P_0^{F_1,\lambda} \otimes \cdots \otimes 2i\pi P_0^{F_N,\lambda} \in (C(U))^{m_0 M}$$

induces a $'\mathcal{D}^{(0,1)}(U,\mathbb{C}^{m_0 M})$-valued map which extends to $\mathbb{C}$ as a meromorphic function of λ with poles in $-\mathbb{Q}^+$. Moreover, if $G = F_1 \otimes \cdots \otimes F_N$, then

$$(1.73)\qquad \left[R^{F_0,\lambda} \otimes 2i\pi P_0^{F_1,\lambda} \otimes \cdots \otimes 2i\pi P_0^{F_N,\lambda}\right]_{\lambda=0} = \sum_{\alpha \in V(F_0)} \frac{\nu_{F_0,\alpha}}{\nu_{F_0\otimes G,\alpha}} (R^{F_0 \otimes G})_\alpha .$$

PROOF. Let $\mathfrak{f}_\ell \in H(U)$ be such that $[F_\ell = 0] = [\mathrm{div}(\mathfrak{f}_\ell)]$, for each $\ell = 0,\dots,N$. Then $F_\ell = \mathfrak{f}_\ell U_\ell$, where $U_\ell \in H(U)$ and $V(U_\ell) = \emptyset$. Define $\Sigma_\ell = \overline{U_\ell}/\|U_\ell\|^2$. Let us now introduce $N+1$ complex parameters λ_ℓ, $\ell = 0,\dots,N$, and then we set $\boldsymbol{\lambda} = (\lambda_0,\lambda_1,\dots,\lambda_N) = (\lambda_0,\boldsymbol{\lambda}')$. If $\operatorname{Re}\lambda_\ell \gg 0$ for $\ell = 0,\dots,N$, one has

$$(1.74)\quad R^{F_0,\lambda_0} \otimes \Big(\bigotimes_{\ell=1}^N 2i\pi P_0^{F_\ell,\lambda_\ell}\Big) = \frac{\lambda_0}{2i\pi}\Big(\prod_{\ell=0}^N \frac{|\mathfrak{f}_\ell|^{2\lambda_\ell}}{\mathfrak{f}_\ell}\Big)\frac{\overline{d\mathfrak{f}_0}}{\overline{\mathfrak{f}_0}}\Big((1+\langle\!\langle r_{\boldsymbol{\lambda}}(\zeta),\boldsymbol{\lambda}\rangle\!\rangle)\bigotimes_{\ell=0}^N \Sigma_\ell\Big),$$

where $r_{\boldsymbol{\lambda}}(\zeta) \in \mathbb{C}^{N+1}$ depends smoothly on $\zeta \in U$ and holomorphically on the parameter $\boldsymbol{\lambda} \in \mathbb{C}^{N+1}$. It follows from Proposition 1.21 that the current-valued function

$$\boldsymbol{\lambda} \longmapsto \Big(R^{F_0,\lambda_0} \otimes \Big(\bigotimes_{\ell=1}^N 2i\pi P_0^{F_\ell,\lambda_\ell}\Big)\Big)_{|U\setminus V(F_0)} \in {}'\mathcal{D}^{(0,1)}(U \setminus V(F_0))$$

extends to $\mathbb{C}^{N+1}$ as a meromorphic map, whose polar set is the union of $\mathbb{Q}$-hyperplanes (in the $\boldsymbol{\lambda}'$ coordinates) that are disjoint from

$$\{\boldsymbol{\lambda} \in \mathbb{C}^{N+1} : \operatorname{Re}\lambda_j \geq 0 \text{ for } j = 0,\dots,N\}.$$

Moreover, because of the factor λ_0 in front of the right-hand side of (1.74), one has

$$\Big[\Big(R^{F_0,\lambda_0} \otimes \Big(\bigotimes_{\ell=1}^N (2i\pi P_0^{F_\ell,\lambda_\ell}\Big)\Big)_{|U\setminus V(F_0)}\Big]_{\boldsymbol{\lambda}=0} = 0.$$

It remains to study what happens about a point $\alpha \in V(F_0)$. Let α be such a point and let $\nu_0 = \nu_{0,\alpha}$ be the order of $\mathfrak{f}_0$ at α. Let also $\nu' = \nu'_\alpha = (\nu_{1,\alpha},\dots,\nu_{N,\alpha})$ where $\nu_{\ell,\alpha}$ is the order of $\mathfrak{f}_\ell$ at α, $\nu = (\nu_0,\nu')$. Then one has in a small disk D_α about α that

$$(1.75)\quad \begin{aligned} \frac{\lambda_0}{2i\pi}\Big(\prod_{\ell=0}^N \frac{|\mathfrak{f}_\ell|^{2\lambda_\ell}}{\mathfrak{f}_\ell}\Big)\frac{\overline{d\mathfrak{f}_0}}{\overline{\mathfrak{f}_0}} &= \frac{\nu_0\lambda_0}{\langle\!\langle \nu,\boldsymbol{\lambda}\rangle\!\rangle}\Big(\frac{1+\langle\!\langle \boldsymbol{\lambda},\rho_{\boldsymbol{\lambda}}(\zeta)\rangle\!\rangle}{\prod_{\ell=0}^N v_\ell}\Big)\bar\partial\Big(\frac{1}{2i\pi}\frac{|\zeta-\alpha|^{2\langle\!\langle \nu,\boldsymbol{\lambda}\rangle\!\rangle}}{(\zeta-\alpha)^{|\nu|}}\Big) \\ &\quad + \frac{\lambda_0}{2i\pi}\frac{|\zeta-\alpha|^{2\langle\!\langle \nu,\boldsymbol{\lambda}\rangle\!\rangle}}{(\zeta-\alpha)^{|\nu|}}\Big(\frac{1+\langle\!\langle \boldsymbol{\lambda},\rho_{\boldsymbol{\lambda}}(\zeta)\rangle\!\rangle}{\prod_{\ell=0}^N v_\ell}\Big)\frac{\overline{dv_0}}{\overline{v_0}}, \end{aligned}$$

where the v_ℓ are invertible holomorphic functions in D_α with $\mathfrak{f}_\ell(\zeta) = (\zeta-\alpha)^{\nu_\ell}v_\ell(\zeta)$, and $\rho_{\boldsymbol{\lambda}}(\zeta)$ depends smoothly on $\zeta \in D_\alpha$ and holomorphically on $\boldsymbol{\lambda} \in \mathbb{C}^{N+1}$. If one takes all λ_ℓ to be equal to a complex parameter λ satisfying $\operatorname{Re}\lambda \gg 0$, then the right-hand side of (1.75) becomes

$$\Big(\frac{1+\lambda\rho_\lambda(\zeta)}{\prod_{\ell=0}^N v_\ell}\Big)\Big(\frac{\nu_0}{|\nu|}\bar\partial\Big(\frac{1}{2i\pi}\frac{|(\zeta-\alpha)^{|\nu|}|^{2\lambda}}{(\zeta-\alpha)^{|\nu|}}\Big) + \frac{\lambda}{2i\pi}\frac{|(\zeta-\alpha)^{|\nu|}|^{2\lambda}}{(\zeta-\alpha)^{|\nu|}}\frac{\overline{dv_0}}{\overline{v_0}}\Big).$$

This expression, considered a current-valued function in λ, extends the whole complex plane to a meromorphic function with a polar set in $-\mathbb{Q}^+$. The first assertion of the proposition follows. Moreover one has

$$\Big[\Big(\frac{1+\lambda\rho_\lambda(\zeta)}{\prod_{\ell=0}^N v_\ell}\Big)\Big(\frac{\lambda}{2i\pi}\frac{\big|(\zeta-\alpha)^{|\nu|}\big|^{2\lambda}}{(\zeta-\alpha)^{|\nu|}}\frac{\overline{dv_0}}{\overline{v_0}}\Big)\Big]_{\lambda=0}=0$$

in view of the factor λ. Formula (1.73) follows from (1.74), (1.75), together with the fact that

$$\Big[\frac{1}{\prod_{\ell=0}^N v_\ell}\overline{\partial}\Big(\frac{1}{2i\pi}\frac{\big|(\zeta-\alpha)^{|\nu|}\big|^{2\lambda}}{(\zeta-\alpha)^{|\nu|}}\Big)\wedge\bigotimes_{\ell=0}^{N}\Sigma_\ell\Big]_{\lambda=0}=(R^{F_0\otimes G})_\alpha,$$

where $G=F_1\otimes\cdots\otimes F_N$. □

We present below another multiplicative procedure (which happens to be this time dis-symmetric) to multiply the residue current R^{F_0} with the principle value current $2i\pi P_0^{F_\ell}$ for $\ell=1,\dots,N$. One needs to introduce for that purpose $N+1$ strictly positive integer weights $\gamma_0,\dots,\gamma_N$ such that

$$\gamma_0<\min(\gamma_1,\dots,\gamma_N)\ \text{or}\ \gamma_0>\max(\gamma_1,\dots,\gamma_N). \tag{1.76}$$

Proposition 1.23. *Let $N\in\mathbb{N}^*$, $(m_0,\dots,m_N)\in(\mathbb{N}^*)^{N+1}$, let for $0\le\ell\le N$, $F_\ell\in(H(U)\setminus\{0\})^{m_\ell}$ and let $M=m_1\cdots m_N$ be as in the previous proposition. Let also $\boldsymbol{\gamma}=(\gamma_0,\dots,\gamma_N)\in(\mathbb{N}^*)^{N+1}$ satisfying (1.76) and $\eta<(\|\boldsymbol{\gamma}\|_\infty)^{-1}$. The function*

$$\lambda\in\{re^{i\theta}\in\mathbb{C}\,:\,|\theta|<\eta\frac{\pi}{2},r\gg 1\} \tag{1.77}$$
$$\longmapsto R^{F_0,\lambda^{\gamma_0}}\otimes 2i\pi P_0^{F_1,\lambda^{\gamma_1}}\otimes\cdots\otimes 2i\pi P_0^{F_N,\lambda^{\gamma_N}}\in(C(U))^{m_0M}$$

induces a ${}'\mathcal{D}(U,\mathbb{C}^{m_0M})$-valued map extending to an open neighborhood of the closed sector $\overline{S}$ with aperture $\pi\,\|\boldsymbol{\gamma}\|_\infty^{-1}$, bisected by the positive real axis, as a holomorphic map. In particular, it means that the map (1.23) extends holomorphically up to a neighborhood of the origin in the complex plane. Moreover, one has

$$\Big[R^{F_0,\lambda^{\gamma_0}}\otimes 2i\pi P_0^{F_1,\lambda^{\gamma_1}}\otimes\cdots\otimes 2i\pi P_0^{F_N,\lambda^{\gamma_N}}\Big]_{\lambda=0} \tag{1.78}$$
$$=\begin{cases}\sum_{\alpha\in V(F_0)}(R^{F_0\otimes G})_\alpha & \text{when } \gamma_0<\min(\gamma_1,\dots,\gamma_n)\\ 2i\pi P_0^G\otimes\Big(\sum_{\alpha\in V(F_0)\setminus V(G)}(R^{F_0})_\alpha\Big) & \text{when } \gamma_0>\max(\gamma_1,\dots,\gamma_n),\end{cases}$$

where $G=F_1\otimes\cdots\otimes F_N$.

Proof. One repeats the proof of Proposition 1.22 up to (1.75) about the local extension of the mapping for some $\alpha\in V(F_0)$). From now on we keep the notation used from this stage in the proof of Proposition 1.22. If one takes $\lambda_\ell=\lambda^{\gamma_\ell}$ for $\ell=0,\dots,N$, then the right-hand side of (1.75) splits as the sum of two expressions, both multiplied by a factor

$$\frac{1+\lambda\,\tilde\rho_\lambda(\zeta)}{\prod_{\ell=0}^N v_\ell}.$$

Within the above factor v_ℓ are the invertible holomorphic functions in D_α present in (1.75) and the function $\tilde\rho_\lambda(\zeta)$ still depends smoothly on $\zeta\in D_\alpha$ and holomorphically on $\lambda\in\mathbb{C}$, but now depends also on the choice of $\boldsymbol{\gamma}$. One of these expressions in λ is

$$\frac{\lambda^{\gamma_0}}{2i\pi}\frac{|\zeta-\alpha|^{2\langle\!\langle\nu,\lambda^{\boldsymbol{\gamma}}\rangle\!\rangle}}{(\zeta-\alpha)^{|\nu|}}\frac{\overline{dv_0}}{\overline{v_0}}.$$

Once considered as a $(0,1)$-current valued function of the complex parameter λ, it extends holomorphically to an open neighborhood of $\overline{S}$ as a holomorphic function vanishing at the origin. The second expression in λ is

$$\frac{\nu_0 \lambda^{\gamma_0}}{\langle\!\langle \nu, \lambda^\gamma \rangle\!\rangle} \overline{\partial}\Big(\frac{1}{2i\pi} \frac{|\zeta-\alpha|^{2\langle\!\langle \nu, \lambda^\gamma\rangle\!\rangle}}{(\zeta-\alpha)^{|\nu|}}\Big).$$

If $\gamma_0 < \min_{1\leq \ell \leq N} \gamma_\ell$, then one has

$$\frac{\nu_0 \lambda^{\gamma_0}}{\langle\!\langle \nu, \lambda^\gamma \rangle\!\rangle} = \frac{1}{1+\lambda \sum_{\ell=1}^N (\nu_\ell/\nu_0)\lambda^{\gamma_\ell-\gamma_0-1}}.$$

The above quotient extends holomorphically as a function of λ to an open neighborhood of the closed sector $\overline{S}$ that takes value 1 at the origin. Since

$$\mu \longmapsto \overline{\partial}\Big(\frac{1}{2i\pi} \frac{|\zeta-\alpha|^{2\mu}}{(\zeta-\alpha)^{|\nu|}}\Big)$$

extends from $\{\mu \in \mathbb{C}; \operatorname{Re}\mu \gg 0\}$ as a meromorphic current valued map with poles in $-\mathbb{Q}^+$, the first assertion of the proposition follows in this case. Moreover, one can check easily that the value of the holomorphic continuation of the map (1.77) at $\lambda = 0$ is equal to the current

$$\Big[\frac{1}{\prod_{\ell=0}^N v_\ell} \Big(\frac{1}{2i\pi} \frac{|\zeta-\alpha|^{2\mu}}{(\zeta-\alpha)^{|\nu|}}\Big) \wedge \bigotimes_{\ell=1}^N \Sigma_\ell\Big]_{\mu=0} = (R^{F_0\otimes G})_\alpha$$

in D_α. When $\gamma_0 < \min_{1\leq\ell\leq N}\gamma_\ell$, the formula (1.78) follows. Suppose now that $\gamma_0 > \max_{1\leq\ell\leq N}\gamma_\ell$. If $\alpha \in V(F_0)\cap V(F_\ell)$ for some $\ell \in \{1,\dots,N\}$ with γ_ℓ minimal, then one has

$$\frac{\nu_0 \lambda^{\gamma_0}}{\langle\!\langle \nu, \lambda^\gamma \rangle\!\rangle} = \frac{\nu_0 \lambda^{\gamma_0-\ell}}{\chi + \lambda \sum_{\{\ell' \in \{0,\dots,N\} : \gamma_{\ell'} > \gamma_\ell\}} \nu_{\ell'} \lambda^{\gamma_{\ell'}-\gamma_\ell-1}}$$

for some $\chi \geq \nu_\ell$ in $\mathbb{N}^*$. Such a function of λ is holomorphic in a neighborhood of $\overline{S}$ and vanishes at the origin. As a consequence, the assertions in the proposition follow provided U is replaced by D_α. It remains to consider the case where U is replaced by D_α with $\alpha \in V(F_0)\setminus V(G)$, in which case the conclusion also holds. The proposition is therefore also proved when $\gamma_0 > \max_{1\leq\ell\leq N}\gamma_\ell$. □

Given Propositions 1.22 and 1.23, let us propose the following definition and thus summarize the main results obtained in this subsection.

DEFINITION 1.24. Let F, G be two vectors of nonidentically zero holomorphic functions in a domain $U \subset \mathbb{C}$. The currents $P_0^G \circledast R^F = R^F \circledast P_0^G$, $R_F \otimes P_0^G$, and $P_0^G \otimes R^F$ are vector-valued currents defined and realized respectively through analytic continuation by

(1.79)

$$P_0^G \circledast R^F = R^F \circledast P_0^G = \Big[P_0^{G_0,\lambda} \otimes R^{G,\lambda}\Big]_{\lambda=0} = \sum_{\alpha\in V(F)} \frac{\nu_{F,\alpha}}{\nu_{F\otimes G,\alpha}} (R^{F\otimes G})_\alpha,$$

$$R^F \otimes P_0^G = \Big[R^{F,\lambda} \otimes P_0^{G,\lambda^2}\Big]_{\lambda=0} = \sum_{\alpha\in V(F)} (R^{F\otimes G})_\alpha,$$

$$P_0^G \otimes R^F = \Big[P_0^{G,\lambda} \otimes R^{F,\lambda^2}\Big]_{\lambda=0} = \sum_{\alpha\in V(F)\setminus V(G)} (R^{F\otimes G})_\alpha = P_0^G \otimes (R^F)_{|U\setminus V(G)}.$$

The commutator $[R^F, P_0^G]$ for the $\otimes$ operation equals

$$[R^F, P_0^G] = R^F \otimes P_0^G - P_0^G \otimes R^F = \sum_{\alpha \in V(F) \cap V(G)} (R^{F\otimes G})_\alpha.$$

1.2.5. Residues of sections of hermitian bundles. Let us consider now a hermitian holomorphic bundle $\overline{E} = (E, |\ |)$ with rank m over some Riemann surface $\mathcal{X}$. Recall that $|\ |$ denotes a smooth hermitian metric on E. The reader, who is not familiar with the geometric point of view of hermitian bundles with finite rank on a complex manifold, is urged to refer to §B.3. To any global holomorphic section $s \in \mathcal{O}_{\mathcal{X}}(\mathcal{X}, E)$ one attaches its (smooth) conjugate section s^* with respect to the hermitian metric $|\ |$; see (B.59) in §B.3.2.1. We adopt the notation $\bigwedge^\ell E^* = 0$ if $\ell \in \mathbb{Z}^-$ and $\bigwedge E^* = \bigoplus_{\ell\in\mathbb{Z}} \bigwedge^\ell E^* = \mathbb{C} \oplus \bigoplus_{\ell=1}^m \bigwedge^\ell E^*$ otherwise. Contraction or *interior multiplication* $\rfloor_s : \bigwedge E^* \to \bigwedge E^*$ is the morphism between complex bundles, whose action is described locally as follows:

- $\rfloor_s \bigwedge^\ell E^* = 0$ for any $\ell \leq 0$;
- if $\ell \geq 1$ and $(e_1, \dots, e_m)$ is a local holomorphic frame for E, then, for any ordered multi-index $J : 1 \leq j_1 < \cdots < j_\ell \leq \ell$,

$$\rfloor_s e_J^* = \rfloor_s (e_{j_1}^* \wedge \cdots \wedge e_{j_\ell}^*) = \sum_{\kappa=1}^{\ell} (-1)^{\kappa-1} \langle e_{j_\kappa}^*, s\rangle\, e_{j_1^*} \wedge \cdots \wedge \widehat{e_{j_\kappa}^*} \wedge \cdots \wedge e_{j_\ell}^*, \tag{1.80}$$

where the symbol $\widehat{\ }$ means omission and the brackets $\langle\,,\,\rangle$ stand for the description of the duality (E^*, E).

The contraction morphism $\rfloor_s : \bigwedge E^* \to \bigwedge E^*$, whose action is defined above, presides over the definition of the Koszul complex $K_\bullet^s$.

DEFINITION 1.25. Let $\mathcal{X}$, E, s, and $\rfloor_s$ be as above and let $V(s) = s^{-1}(\{0\})$. The *Koszul complex* $\mathbf{K}_\bullet^s$ over $\mathcal{X}$ is the complex of holomorphic bundles over $\mathcal{X}$ given by

$$0 \longrightarrow \bigwedge^m E^* \xrightarrow{\rfloor_s} \bigwedge^{m-1} E^* \xrightarrow{\rfloor_s} \cdots \xrightarrow{\rfloor_s} \bigwedge^{\ell} E^* \xrightarrow{\rfloor_s} \bigwedge^{\ell-1} E^* \xrightarrow{\rfloor_s} \cdots \xrightarrow{\rfloor_s} E^* \xrightarrow{\rfloor_s} \mathbb{C}, \tag{1.81}$$

which is exact over the open subset $\mathcal{X} \setminus V(s)$.

Let

$$\mathcal{L}_E = \Big({}'\mathcal{D}^{(0,\bullet)}(U, \bigwedge E^*)\Big)_{U \overset{\text{open}}{\subset} \mathcal{X}} = \Big({}'\mathcal{D}^{(0,0)}(U, \bigwedge E^*) \oplus {}'\mathcal{D}^{(0,1)}(U, \bigwedge E^*)\Big)_{U \overset{\text{open}}{\subset} \mathcal{X}}$$

be the sheaf of $\bigwedge E^*$-valued $(0,\bullet)$-currents on $\mathcal{X}$. The action of $\rfloor_s$ as a homomorphism from the sheaf $\mathcal{L}_E$ into $\mathcal{L}_E$ is described locally, in terms of representations with respect to a local holomorphic frame, by

$$\begin{aligned} &\rfloor_s \Big(\sum_{j=1}^m T_j e_j^*\Big) = \varpi \sum_{j=1}^n s_j T_j, \\ &\rfloor_s \Big(\Big(\sum_{j=1}^m T_j e_j^*\Big) \wedge e_J^*\Big) = \varpi \Big(\Big(\sum_{j=1}^n s_j T_j\Big)\, e_J^* - \Big(\sum_{j=1}^m T_j e_j^*\Big) \wedge \rfloor_s e_J^*\Big) \end{aligned} \tag{1.82}$$

for any ordered multi-index J with length $0 \leq \ell \leq m$. If T has pure bidegree $(0,0)$, then $\varpi = 1$. Otherwise, $\varpi = -1$, if T has pure bidegree $(0,1)$. The sheaf $\mathcal{L}_E$ is decomposed into a direct sum as

$$\mathcal{L}_E = \bigoplus_{\nu \in \mathbb{Z}} \mathcal{L}_E^\nu = \bigoplus_{\nu=-m}^{1} \mathcal{L}_E^\nu,$$

where

$$\begin{aligned} \mathcal{L}_E^\nu &= \bigoplus_{\gamma \in \mathbb{Z}} \Big({}'\mathcal{D}^{(0,\gamma+\nu)}(U, \bigwedge^{\gamma} E^*)\Big)_{U \overset{\text{open}}{\subset} \mathcal{X}} \\ &= \Big({}'\mathcal{D}^{(0,0)}(U, \bigwedge^{-\nu} E^*)\Big)_{U \overset{\text{open}}{\subset} \mathcal{X}} \oplus \Big({}'\mathcal{D}^{(0,1)}(U, \bigwedge^{1-\nu} E^*)\Big)_{U \overset{\text{open}}{\subset} \mathcal{X}}, \end{aligned} \tag{1.83}$$

for each $\nu \in \mathbb{Z}$. The decomposition (1.83) leads to the following definition of the homorphism (operator) ∇_s along the section $s \in \mathcal{O}(\mathcal{X}, E)$.

DEFINITION 1.26. The sheaf homomorphism $\nabla_s : \mathcal{L}_E \to \mathcal{L}_E$ is defined to be the difference between the anticommuting homomorphism $2i\pi \rfloor_s$ and Dolbeault homomorphism $\bar\partial$.[3] That is,

$$\nabla_s = 2i\pi \rfloor_s - \bar\partial. \tag{1.84}$$

The homomorphism ∇_s acts from $\mathcal{L}_E^{\nu-1}$ into $\mathcal{L}_E^\nu$ for each $\nu \in \mathbb{Z}$.

The goal of the following proposition is to extend and complete the definitions of the vector-valued currents $R^F = [R^{F,\lambda}]_{\lambda=0}$ and $P_0^F = [P_0^{F,\lambda}]_{\lambda=0}$ in the framework of hermitian holomorphic bundles using the homomorphism (1.84). These currents were previously described in Propositions 1.8 and 1.12, respectively, for the particular case where $s = F$ was a section of the trivial bundle $U \times \mathbb{C}^m$ over a domain $U \subset \mathbb{C}$.

PROPOSITION 1.27. *Let $\overline{E} = (E, |\ |)$ be a hermitian holomorphic bundle with rank m over a Riemann surface $\mathcal{X}$ and let $s \in \mathcal{O}_{\mathcal{X}}(\mathcal{X}, E)$. The holomorphic $\mathcal{L}_E^0(\mathcal{X})$ (respectively $\mathcal{L}_E^{-1}$)-valued maps*

$$\begin{aligned} &\lambda \longmapsto R_{|\ |}^{s,\lambda} = \bar\partial\, |s|^{2\lambda}\, \frac{s^*}{2i\pi |s|^2}, \\ &\lambda \longmapsto P_{|\ |}^{s,\lambda} = P_{|\ |,0}^{s,\lambda} + P_{|\ |,1}^{s,\lambda} = |s|^{2\lambda}\, \frac{s^*}{2i\pi|s|^2} + |s|^{2\lambda} \frac{s^*}{2i\pi|s|^2} \wedge \bar\partial\Big(\frac{s^*}{2i\pi|s|^2}\Big) \end{aligned} \tag{1.85}$$

defined in $\{\lambda \in \mathbb{C} : \operatorname{Re} \lambda \gg 1\}$ extend to $\mathbb{C}$ as meromorphic functions with poles in $-\mathbb{Q}^+$. Moreover, their values $R_{|\ |}^s$ and $P_{|\ |}^s$ at $\lambda = 0$ satisfy

$$\begin{aligned} &\nabla_s R_{|\ |}^s = 0, \\ &\nabla_s P_{|\ |}^s = [1] - R_{|\ |}^s. \end{aligned} \tag{1.86}$$

The current $R_{|\ |}^s$ is called the residue current attached to the section s of $\overline{E}$, while the current $P_{|\ |}^s$ is known as the principal value current attached to the section s of the bundle $\overline{E}$.

---

[3]Since $\bigwedge E^*$ is a holomorphic bundle with finite rank, the Dolbeault sheaf homomorphism $\bar\partial : \mathcal{L}_E \to \mathcal{L}_E$ is well defined.

PROOF. Because of the correspondence between Cartier divisors on $\mathcal{X}$ and holomorphic line bundles over $\mathcal{X}$ (see Example B.2), the section s factorizes as $s = \mathfrak{f} \otimes \mathfrak{u}$. In this factorization, $\mathfrak{f}$ is a holomorphic section of the line bundle L corresponding to the Cartier divisor attached to s and $\mathfrak{u}$ is a nonvanishing section of the holomorphic bundle $L^{-1} \otimes E$ with rank m. The nonvanishing section $\mathfrak{u}$ induces an isomorphism between L and a subbundle of E having rank 1. This isomorphism allows us to carry the metric $|\ |$ on E as a metric $|\ |$ on L such that $|\mathfrak{f}| = |s|$. Intuitively it means that one finds locally a holomorphic function, whose zero set realizes $V(s) = s^{-1}(0)$. Let $s = f\, u_1 e_1 + \cdots + u_m e_m = f\, u$ along a local holomorphic frame $e = (e_1, \dots, e_m)$ for E over a trivialization local chart U. Suppose that the metric $|\ |$ on E is defined over U by

$$\forall x \in U,\ \forall \xi \in E_x,\ \|\xi\|_x^2 = \sum_{j=1}^{m} \sum_{k=1}^{m} h_{j,k}(x) \xi_j \overline{\xi_k},$$

so that $s^* = \overline{f} \sum_{j=1}^m \big(\sum_{k=1}^m h_{j,k} \overline{u_k}\big) e_j^*$ in U. For $\operatorname{Re} \lambda \gg 1$ one has in U that

$$\begin{aligned}
|s|^{2\lambda} \frac{s^*}{2i\pi |s|^2} &= |\mathfrak{f}|^{2\lambda} \frac{s^*}{2i\pi |\mathfrak{f}|^2} = \Big(\frac{|f|^{2\lambda}}{2i\pi f}(1 + \lambda \rho_\lambda)\Big)\, \chi_{\mathfrak{u}}^*, \\
(\overline{\partial} |s|^{2\lambda})\, \frac{s^*}{2i\pi |s|^2} &= (\overline{\partial} |\mathfrak{f}|^{2\lambda})\, \frac{s^*}{2i\pi |\mathfrak{f}|^2} \\
&= \Big((1 + \lambda \rho_\lambda)\, \overline{\partial}\Big(\frac{|f|^{2\lambda}}{2i\pi f}\Big) + \lambda \frac{|f|^{2\lambda}}{2i\pi f}\, \overline{\partial} \rho_\lambda\Big)\, \chi_{\mathfrak{u}}^*, \\
|s|^{2\lambda} \frac{s^*}{2i\pi |s|^2} \wedge \overline{\partial}\Big(\frac{s^*}{2i\pi |s|^2}\Big) &= |\mathfrak{f}|^{2\lambda} \frac{s^*}{2i\pi |\mathfrak{f}|^2} \wedge \overline{\partial}\Big(\frac{s^*}{2i\pi |\mathfrak{f}|^2}\Big) \\
&= \Big(\frac{|f|^{2\lambda}}{2i\pi f^2}(1 + \lambda \rho_\lambda)\Big)\, \chi_{\mathfrak{u}}^* \wedge \overline{\partial}\Big(\frac{\chi_{\mathfrak{u}}^*}{2i\pi}\Big),
\end{aligned}$$

where

$$\forall x \in U,\ \chi_{\mathfrak{u}}^*(x) = \frac{\sum_{j=1}^m \big(\sum_{k=1}^m h_{j,k}(x) \overline{u_k(x)}\big) e_j^*(x)}{\sum_{j=1}^m \sum_{k=1}^m h_{j,k}(x)\, u_j(x) \overline{u_k(x)}}, \tag{1.87}$$

while $\rho_\lambda(x)$ depends smoothly on $x \in U$ and holomorphically on $\lambda \in \mathbb{C}$. Similar arguments to those used in the proofs of Propositions 1.8 and 1.12 show that these three functions of λ, considered respectively as $'\mathcal{D}^{(0,0)}(\mathcal{X}, E^*)$, $'\mathcal{D}^{(0,1)}(\mathcal{X}, E^*)$, $'\mathcal{D}^{(0,1)}(\mathcal{X}, E^* \wedge E^*)$-valued, extend to the whole complex plane as meromorphic functions with poles in $-\mathbb{Q}^+$. The first assertion of the proposition follows. Moreover, the identities

(1.88)

$$R^s_{|\,|} = [R^{s,\lambda}_{|\,|}]_{\lambda=0} = \overline{\partial}\Big(\frac{1}{f}\Big)\, \chi_{\mathfrak{u}}^* \quad \text{and} \quad P^s_{|\,|} = [P^{s,\lambda}_{|\,|}]_{\lambda=0} = P^f\, \chi_{\mathfrak{u}}^* + P^{f^2}\, \chi_{\mathfrak{u}}^* \wedge \overline{\partial}\Big(\frac{\chi_{\mathfrak{u}}^*}{2i\pi}\Big)$$

are valid in the local chart U, where $\chi_{\mathfrak{u}}^*$ is defined in (1.87). Note that analytic continuation commutes with the action of ∇_s. Thus, it follows from (1.82) that

$$\Big[\nabla_s R^{s,\lambda}_{|\,|}\Big]_{\lambda=0} = -\Big[\overline{\partial}\, |s|^{2\lambda} \rfloor_s \Big(\frac{s^*}{|s|^2}\Big)\Big]_{\lambda=0} = \big[\overline{\partial}\, |s|^{2\lambda}\big]_{\lambda=0} = \overline{\partial}[1] = 0.$$

Since

$$\overline{\partial} P^{s,\lambda}_{|\,|} = \overline{\partial} P^{s,\lambda}_{|\,|,0} = R^{s,\lambda}_{|\,|} + |s|^{2\lambda}\,\overline{\partial}\Big(\frac{s^*}{2i\pi|s|^2}\Big),$$

$$2i\pi\rfloor_s(P^{s,\lambda}_{|\,|}) = |s|^{2\lambda}\Big(1+\overline{\partial}\Big(\frac{s^*}{2i\pi|s|^2}\Big)\Big),$$

one has that

$$\nabla_s(P^{s,\lambda}_{|\,|}) = |s|^{2\lambda} - R^{s,\lambda}_{|\,|}.$$

Hence

$$\nabla_s(P^{s}_{|\,|}) = [1] - R^{s}_{|\,|},$$

when evaluating the analytic continuation at $\lambda = 0$. The proposition is proved. □

REMARK 1.28. When s is a section of a line bundle, the L^{-1}-valued current $R^s_{|\,|}$ is independent of the choice of the metric. In this case, it is also denoted as $R^s = \overline{\partial}(1/s)$. Moreover, if $s' \in \mathcal{O}(\mathcal{X}, L')$ and $s'' \in \mathcal{O}(\mathcal{X}, L'')$ are such that $s'' = as'$, where a is a global holomorphic section of $\mathrm{Hom}_{\mathbb{C}}(L', L'')$, then one has $R^{s'} = aR^{s''}$ as $(L')^{-1}$-valued $(0,1)$-currents on $\mathcal{X}$. This is a formulation of the so-called *transformation law* in the univariate setting. Such transformation law will be later a crucial tool for multi-variate residue calculus.

The Dolbeault complex of sheaves bundles

$$0 \longrightarrow \Big(\mathcal{O}_{\mathcal{X}}(U, \bigwedge^{\ell} E^*)\Big)_U \stackrel{\iota}{\longrightarrow} \Big('\mathcal{D}^{(0,0)}(U, \bigwedge^{\ell} E^*)\Big)_U \stackrel{\overline{\partial}}{\longrightarrow} \Big('\mathcal{D}^{(0,1)}(U, \bigwedge^{\ell} E^*)\Big)_U \stackrel{\overline{\partial}}{\longrightarrow} 0$$

is locally exact as a consequence of Dolbeault's lemma, for any $\ell = 0, \dots, m$; see §B.2.3. The next proposition relies on the following chasing diagram of sheaves.

$$\begin{array}{ccccc}
\Big('\mathcal{D}^{(0,0)}(U, \bigwedge^{m} E^*)\Big)_U & \stackrel{\rfloor_s}{\to} \cdots & \Big('\mathcal{D}^{(0,0)}(U, \bigwedge^{\ell} E^*)\Big)_U & \stackrel{\rfloor_s}{\to} \cdots & \Big('\mathcal{D}^{(0,0)}(U)\Big)_U \\
\overline{\partial}\downarrow & & \overline{\partial}\downarrow & & \overline{\partial}\downarrow \\
\Big('\mathcal{D}^{(0,1)}(\mathcal{X}, \bigwedge^{m} E^*)\Big)_U & \stackrel{\rfloor_s}{\to} \cdots & \Big('\mathcal{D}^{(0,1)}(\mathcal{X}, \bigwedge^{\ell} E^*)\Big)_U & \stackrel{\rfloor_s}{\to} \cdots & \Big('\mathcal{D}^{(0,1)}(U)\Big)_U \\
\overline{\partial}\downarrow & & \overline{\partial}\downarrow & & \overline{\partial}\downarrow \\
0 & & 0 & & 0
\end{array}$$

PROPOSITION 1.29. *Suppose that $\mathcal{X}$ is a noncompact Riemann surface and that s is a holomorphic section of a hermitian holomorphic bundle $\overline{E} = (E, |\,|)$ with rank m. Given a holomorphic function $h \in \mathcal{O}_{\mathcal{X}}(\mathbb{C})$, one has*

$$h\, R^s_{|\,|} = 0 \Longleftrightarrow h \in \rfloor_s(\mathcal{O}_{\mathcal{X}}(\mathcal{X}, E^*)). \tag{1.89}$$

PROOF. The implication ($\Longleftarrow$) follows from the local expression of $R^F_{|\,|}$ with respect to a local orthonormal frame $(\epsilon_1^*, \dots, \epsilon_m^*)$ for E^*; see (1.88) and (1.38). Since any noncompact Riemann surface is a Stein manifold (see §A.2.2) it is enough to prove the direct implication ($\Longrightarrow$), when $\mathcal{X}$ is replaced by U. Here, $(U, \zeta : U \to \mathbb{C})$ is a local chart over which E is a trivial bundle. Since the Dolbeault complex of sheaves of currents $('\mathcal{D}^{(0,\bullet)}(U, E^*))_U$ is exact in this case, there exists $q \in {}'\mathcal{D}^{(0,0)}(\mathcal{X}, E^*)$ such that $\overline{\partial} q = P^s_{|\,|,1}$. Let

$$a = h \cdot P^s_{|\,|,0} + h \cdot (2i\pi\rfloor_s q).$$

One has

$$2i\pi \rfloor_s a = h\cdot (2i\pi \rfloor_s (P^s_{|\,|,0})) = [h]. \tag{1.90}$$

On the other hand, one has, according to the computation rules (1.82) and to the first identity in (1.86), that the assumption $h\cdot R^s_{|\,|}=0$ implies

$$\begin{aligned}\overline{\partial} a &= h\cdot \big((2i\pi \rfloor_s - \nabla^s)(P^s_{|\,|,0})\big) - h\cdot \big(2i\pi \rfloor_s(\overline{\partial} q)\big)\\ &= [h] - h\cdot (\nabla_s P^s_{|\,|} - \nabla_s P^s_{|\,|,1}) - h\cdot \big(2i\pi \rfloor_s P^s_{|\,|,1}\big)\\ &= [h] - h\cdot \big([1] - R^s_{|\,|} - 2i\pi \rfloor_s P^s_{|\,|,1}\big) - h\cdot \big(2i\pi \rfloor_s (P^s_{|\,|,1})\big) = 0.\end{aligned}$$

This shows that $2i\pi a$ is represented as a $(0,0)$-current by a holomorphic section of E^*. Hence $h\in \rfloor_s(\mathcal{O}_{\mathcal{X}}(\mathcal{X},E^*))$ in view of (1.90). □

Proposition 1.10 can be reformulated within the geometric context of this subsection. One needs to introduce for such purpose a (smooth) connection D on the holomorphic bundle E; see §B.2.1.2. The connection D is acting on $C^\infty(\mathcal{X},E)$ and has values in the space $C^\infty_1(\mathcal{X},E) = C^\infty_{1,0}(\mathcal{X},E)\oplus C^\infty_{0,1}(\mathcal{X},E)$ of E-valued one-differential forms on the underlying two-dimensional differentiable manifold $\mathcal{X}_{\mathbb{R}}$. Given such connection D on E, one can consider as well the left action $\rfloor_D$ of D as a contraction on E^*-valued currents with degree 1 on $\mathcal{X}$. An important example is the following: the metrized bundle $\overline{E}$ inherits a Chern connection $D_{\overline{E}} = D_{|\,|}$, which is compatible with both the hermitian metric $|\,|$ and the complex structure on E; see Definition B.44.

PROPOSITION 1.30. *Let $\overline{E}=(E,|\,|)$ be a hermitian holomorphic bundle with finite rank m over a Riemann surface $\mathcal{X}$ and let $s\in\mathcal{O}_{\mathcal{X}}(\mathcal{X},E)$. Let also $s=\mathfrak{f}\otimes\mathfrak{u}$ be any factorization of s as the product of a holomorphic section of the line bundle L corresponding to the Cartier divisor attached to s with a nonvanishing section of the holomorphic bundle $L^{-1}\otimes E$ having rank m. The $(1,1)$ integration current $[s=0]$ corresponding to the divisor* $\mathrm{div}(\mathfrak{f})$ *factorizes as*

$$[s=0] = [\mathrm{div}(\mathfrak{f})] = \rfloor_{D(s)} R^s_{|\,|}. \tag{1.91}$$

PROOF. Let (U,ζ) be a local chart about $\alpha\in V(s)$ over which E is trivialized and let $(e_1,\dots,e_m)$ be a local holomorphic frame for E over U. Let us start with the local representation $s=f\,u_1e_1+\cdots+u_me_m$ of the section s in U, where $f\in H(U)$, $u\in (H(U))^m$ with $V(u)=\emptyset$. One has $D(s)=\omega_1e_1+\cdots+\omega_me_m$, where

$$\omega_j = (df)\,u_j + f\Big(du_j + \sum_{k=1}^m \omega_{j,k}\,u_k\Big)$$

for $j=1,\dots,m$. Furthermore, the smooth 1-forms $\omega_{j,k}$ are the entries of the matrix $\mathcal{M}(D;e;e)$ of the connection D expressed in the local frame e; see (B.18) and (B.19) in §B.2.1.2. Since

$$R^s_{|\,|} = \overline{\partial}\Big(\frac{1}{f}\Big)\chi^*_{\mathfrak{u}}$$

(see (1.88)) and $f\,\overline{\partial}(1/f)=0$ in the sense of currents in U, one has that

$$\begin{aligned}\rfloor_{D(s)}(R^s_{|\,|}) &= \rfloor_{D(s)}\Big(\overline{\partial}\Big(\frac{1}{f}\Big)\,\chi^*_{\mathfrak{u}}\Big) = \rfloor_{(df)\,\sum_1^m u_j\,e_j}\Big(\overline{\partial}\Big(\frac{1}{f}\Big)\,\chi^*_{\mathfrak{u}}\Big)\\ &= \overline{\partial}\Big(\frac{1}{f}\Big)\wedge df = [\mathrm{div}(f)] = [\mathrm{div}(\mathfrak{f})]\end{aligned}$$

in U, taking into account Proposition 1.10. □

One may transpose within the above geometric context the multiplicative calculus of currents as introduced in §1.2.4. Thus, Definition 1.24 can be reformulated as follows.

DEFINITION 1.31. Let u, v be respective holomorphic sections of two hermitian holomorphic bundles $(E', |\ |)$, $(E'', \|\ \|)$, satisfying $E = E' \oplus E''$. One defines three global sections of $\mathcal{L}_E^{-1}$ through analytic continuation as follows:

$$(1.92)\qquad \begin{aligned} P^v_{\|\ \|} \circledast R^u_{|\ |} &= R^u_{|\ |} \circledast P^v_{\|\ \|} = \Big[P^{v,\lambda}_{\|\ \|} \otimes R^{u,\lambda}_{|\ |}\Big]_{\lambda=0},\\ R^u_{|\ |} \otimes P^v_{\|\ \|} &= \Big[R^{u,\lambda}_{|\ |} \otimes P^{v,\lambda^2}_{\|\ \|}\Big]_{\lambda=0},\\ P^v_{\|\ \|} \otimes R^u_{|\ |} &= \Big[P^{v,\lambda}_{\|\ \|} \otimes R^{u,\lambda^2}\Big]_{\lambda=0}. \end{aligned}$$

Considerations leading to constructions in Proposition 1.27, when s is a holomorphic section of a hermitian holomorphic bundle $\overline{E} = (E, |\ |)$, motivate the following definition.

DEFINITION 1.32. Let $\mathcal{X}$ be a connected Riemann surface and let E be a holomorphic bundle with a finite rank over $\mathcal{X}$. Let also W be a polar set, that is, a discrete subset of $\mathcal{X}$ so that there exists a holomorphic line bundle L and a meromorphic section $\mathfrak{f}$ of L defined on $\mathcal{X}$ satisfying $V(\mathfrak{f}) = |\mathrm{div}^+(\mathfrak{f})| = W$. A *semimeromorphic* section h of E over $\mathcal{X}$ is a smooth section of E over the maximal open subset $\mathcal{X} \setminus W$, denoted by $\mathrm{def}(h)$, so that $\mathfrak{f} \otimes h$ extends from $\mathcal{X} \setminus W$ to $\mathcal{X}$ as a smooth section of $L \otimes E$. The set W is denoted by $\mathrm{Pol}(h)$ in this case.

Any semimeromorphic section h of E admits a *standard extension* $h_{|\mathcal{X}\setminus\mathrm{Pol}(h)}$ as an element of ${}'\mathcal{D}^{(0,0)}(\mathcal{X}, E)$. Such extension is defined in a similar way to the extension of $h_{|U\setminus W}$ in (1.4).

DEFINITION 1.33. Let h be a semimeromorphic section of E with W and $\mathfrak{f}$ as in Definition 1.31. Let also $\mathrm{div}^-(\mathfrak{f}) = \mathrm{div}(\mathfrak{f}) - \mathrm{div}^+(\mathfrak{f})$, with support $|\mathrm{div}^-(\mathfrak{f})|$ being equal to the polar set of $\mathfrak{f}$. The standard extension $h_{|\mathcal{X}\setminus W} \in {}'\mathcal{D}^{(0,0)}(\mathcal{X}, E)$ is the $(0,0)$-current defined as

$$(1.93)\qquad \begin{aligned} &\forall \varphi \in \mathcal{D}^{(1,1)}(\mathcal{X} \setminus |\mathrm{div}^+(\mathfrak{f})|, E^*),\ \langle h_{|\mathcal{X}\setminus W}\,,\, \varphi\rangle = \int_{\mathcal{X}} \rfloor_\varphi h,\\ &\forall \varphi \in \mathcal{D}^{(1,1)}(\mathcal{X} \setminus |\mathrm{div}^-(\mathfrak{f})|, E^*),\ \langle h_{|\mathcal{X}\setminus W}\,,\, \varphi\rangle = \Big[\int_{\mathcal{X}} \rfloor_\varphi\big(\|\mathfrak{f}\|^{2\lambda} h\big)\Big]_{\lambda=0}, \end{aligned}$$

where $\|\ \|$ is an arbitrary smooth metric on $L_{|\mathcal{X}\setminus|\mathrm{div}^-(\mathfrak{f})|}$ and $\rfloor_\varphi$ denotes the interior contraction by the E^*-valued $(1,1)$-form φ to an E-valued $(0,0)$-current. The definition depends neither on the choice $\mathfrak{f}$ nor on the choice of the metric $\|\ \|$ on $L_{|\mathcal{X}\setminus|\mathrm{div}^-(\mathfrak{f})|}$, and the two definitions (1.93) coincide on $\mathcal{X} \setminus |\mathrm{div}(\mathfrak{f})|)$.

When h is a meromorphic section over $\mathcal{X}$ of a holomorphic line bundle L, one has that

$$(1.94)\qquad \mathrm{Supp}\Big(\overline{\partial}\Big(\frac{h_{|\mathcal{X}\setminus|\mathrm{div}^-(h)|}}{2i\pi}\Big)\Big) = |\mathrm{div}^-(h)|.$$

When h is a global meromorphic section of the canonical bundle, here it is the cotangent holomorphic bundle $T^{*(1,0)}_{\mathcal{X}} = (T^{(1,0)}_{\mathcal{X}})^*$, whose global sections are $(1,0)$-differential forms (see Example B.13), one has moreover that

$$\overline{\partial}\Big(\frac{h_{|\mathcal{X}\setminus|\mathrm{div}^-(h)|}}{2i\pi}\Big) \in {}'\mathcal{D}^{(1,1)}(\mathcal{X}).$$

If $\mathcal{X}$ is compact, then the identity

$$\int_{\mathcal{X}} \overline{\partial}\Big(\frac{h_{|\mathcal{X}\setminus|\mathrm{div}^-(h)|}}{2i\pi}\Big)\Big) = 0, \tag{1.95}$$

derived using Stokes's theorem, is known as *the residue formula*. Assume that α is a pole of h. Let (U,ζ) be a local chart about α, with centered local coordinate $\zeta\,:\,\zeta(\alpha)=0$. If

$$h(\zeta)\,d\zeta = \Big(\sum_{k\in\mathbb{Z}} a_k\zeta^k\Big)\,d\zeta \tag{1.96}$$

is the expression of h in the local centered coordinate ζ, then none of the Laurent coefficients a_k (for $k\neq -1$) is geometrically intrinsic. These coefficients depend on the choice of the local coordinate ζ. This is not the case for the coefficient

$$a_{-1} = \int_{\mathcal{X}} \Big(\overline{\partial}\Big(\frac{h_{|\mathcal{X}\setminus|\mathrm{div}^-(h)|}}{2i\pi}\Big)\Big)_\alpha,$$

where $(\ )_\alpha$ means, as it was already the case in (1.72), the localization at α. That is,

$$\Big(\overline{\partial}\Big(\frac{h_{|\mathcal{X}\setminus|\mathrm{div}^-(h)|}}{2i\pi}\Big)\Big)_\alpha = \theta\,\overline{\partial}\Big(\frac{h_{|\mathcal{X}\setminus|\mathrm{div}^-(h)|}}{2i\pi}\Big), \tag{1.97}$$

where $\theta\in\mathcal{D}(\mathcal{X})$, $\theta\equiv 1$ about α and $\mathrm{Supp}\,\theta\cap|\mathrm{div}^-(h)| = \{\alpha\}$. It is important to observe that all Laurent coefficients a_k for $k\neq -1$ fail to be geometrically intrinsic. However, the geometrically intrinsic, localized current (1.97) encodes the Laurent coefficients a_k, $k<0$, involved in the representation (1.96) of h along the local centered coordinate ζ. The essence of this approach will be exemplified in the algebraic setting; see §1.3.2, in particular, Lemma 1.43 in the case $\mathcal{X}=\mathbb{C}$.

1.3. The algebraic point of view

1.3.1. Cauchy's formula in $A^1_{\mathbb{C}} = \mathbb{C}$ or $\mathbb{P}^1_{\mathbb{C}}$ and division in $\mathbb{K}[X]$. Let $\mathbb{K}$ be a subfield of $\mathbb{C}$. Observe that any field of characteristic 0 and cardinality at most $2^{\aleph_0}$ is isomorphic to such a subfield. Let us also assume throughout this section that the polynomial $p\in\mathbb{K}[X]$ is fixed, has degree $d=\deg p\geq 1$ and that its dominant coefficient equals γ_0. For any $a\,dX$, with $a\in\mathbb{K}[X]$, the Kähler differential $\omega(X) = a(X)\,dX/p(X)\in\Omega_{\mathbb{K}(X)/\mathbb{K}}$ induces a meromorphic section

$$h = \frac{a(\zeta)\,d\zeta}{p(\zeta)}$$

of the canonical bundle of $\mathbb{P}^1_{\mathbb{C}}$. One can express h in homogeneous coordinates as

$$h = \frac{A(X_0,X_1)}{P(X_0,X_1)}\,\frac{X_1dX_0 - X_0dX_1}{X_0^{\deg a-(d-2)}}, \tag{1.98}$$

where $P\in\mathcal{O}_{\mathbb{P}^1_{\mathbb{C}}}(d)$ and $A\in\mathcal{O}_{\mathbb{P}^1_{\mathbb{C}}}(\deg a)$ denote the homogenizations of p and a, respectively. Let $h_{|\mathbb{C}\setminus p^{-1}(\{0\})}$ be the standard extension of h, when considered as a meromorphic form on the open chart $U = \mathbb{C}$ Let also $h_{|\mathbb{P}^1_{\mathbb{C}}\setminus|\mathrm{div}^-(h)|}$ be its standard extension, when considered as a meromorphic form on $\mathbb{P}^1_{\mathbb{C}}$.

PROPOSITION 1.34. *One has in* $'\mathcal{D}^{(1,1)}(\mathbb{C})$ *that*

$$\overline{\partial}\Big(\frac{h_{|\mathbb{C}\setminus p^{-1}(\{0\})}}{2i\pi}\Big) = \overline{\partial}\Big(\frac{1}{p}\Big) \wedge a(\zeta)\, d\zeta. \tag{1.99}$$

If $\deg a \leq d-2$, *then one has in* $'\mathcal{D}^{(1,1)}(\mathbb{P}^1_{\mathbb{C}})$ *that*

$$\overline{\partial}\Big(\frac{h_{|\mathbb{P}^1_{\mathbb{C}}\setminus|\mathrm{div}^-(h)|}}{2i\pi}\Big) = \Big(\overline{\partial}\Big(\frac{h_{|\mathbb{P}^1_{\mathbb{C}}\setminus|\mathrm{div}^-(h)|}}{2i\pi}\Big)\Big)_{|\mathbb{C}} = \sum_{\{\alpha\in\mathbb{C}\,;\,p(\alpha)=0\}} \Big(\overline{\partial}\Big(\frac{1}{p}\Big)\Big)_\alpha \wedge a(\zeta)\, d\zeta,$$

where $(\ )_\alpha$ *denotes, as usual, the localization at* α. *Therefore, using Stokes's formula on the compact projective space* $\mathbb{P}^1_{\mathbb{C}}$ *implies the Euler–Jacobi formula*

$$\sum_{\{\alpha\in\mathbb{C}\,:\,p(\alpha)=0\}} \Big\langle \Big(\overline{\partial}\Big(\frac{1}{p}\Big)\Big)_\alpha, a(\zeta)\, d\zeta\Big\rangle = \int_{\mathbb{C}} \overline{\partial}\Big(\frac{1}{p}\Big)\wedge a(\zeta) d\zeta = \int_{\mathbb{P}^1_{\mathbb{C}}} \overline{\partial}\Big(\frac{h_{|\mathbb{P}^1_{\mathbb{C}}\setminus|\mathrm{div}^-(h)|}}{2i\pi}\Big) = 0. \tag{1.100}$$

PROOF. The first assertion is an immediate consequence of the definition of $R^f = \overline{\partial}(1/f)$. The second assertion follows from the fact that, if $\deg a - (d-2) \leq 0$, then the support of $\mathrm{div}^-(h)$ does not contain $\infty = [0:1]$; see (1.98). □

COROLLARY 1.35. *Let* $\mathbb{K}$ *be a subfield of* $\mathbb{C}$. *Let*

$$a = pq + r = pq + \Big(\sum_{\kappa=0}^{d-1} \rho_\kappa(a;p) X^{d-1-\kappa}\Big)$$

be the euclidean division formula of the polynomial $a \in \mathbb{K}[X]$ *by the polynomial* $p = \sum_{\kappa=0}^d \gamma_\kappa X^{d-\kappa} \in \mathbb{K}[X]$, *when* $d \geq 1$ *and* $\gamma_0 \in \mathbb{K}^*$. *If* h *is the meromorphic* $(1,0)$*-form on* $\mathbb{P}^1(\mathbb{C})$, *whose restriction to* $\mathbb{C}$ *equals* $a(\zeta)\, d\zeta/p(\zeta)$, *then one has*

$$\begin{aligned}\sum_{\{\alpha\in\mathbb{C}\,:\,p(\alpha)=0\}} \Big\langle \Big(\overline{\partial}\Big(\frac{1}{p}\Big)\Big)_\alpha, a(\zeta)\, d\zeta\Big\rangle &= -\int_{\mathbb{P}^1(\mathbb{C})} \Big(\overline{\partial}\Big(\frac{h_{|\mathbb{P}^1(\mathbb{C})\setminus|\mathrm{div}^-(h)|}}{2i\pi}\Big)\Big)_\infty \\ &= \frac{\rho_{d-1}(a;p)}{\gamma_0} \in \mathbb{K}. \end{aligned} \tag{1.101}$$

PROOF. The first equality in (1.101) follows from Stokes's theorem. The fact that $p \cdot \overline{\partial}(1/p) = 0$ in $'\mathcal{D}^{(1,0)}(\mathbb{C})$ (see (1.38) with $F = (f)$), and Proposition 1.34 imply that to prove the second equality, it is enough to prove that

$$\sum_{\{\alpha\in\mathbb{C}\,:\,p(\alpha)=0\}} \Big\langle \Big(\overline{\partial}\Big(\frac{1}{p}\Big)\Big)_\alpha, \zeta^{d-1}\, d\zeta\Big\rangle = \frac{1}{\gamma_0}.$$

Observe that $\zeta^{d-1}\, d\zeta = dp/(\gamma_0 d) - b(\zeta)\, d\zeta$, when $\deg b \leq d-2$. Thus, Propositions 1.34 and 1.10 imply that

$$\begin{aligned}\sum_{\{\alpha\in\mathbb{C}\,:\,p(\alpha)=0\}} \Big\langle \Big(\overline{\partial}\Big(\frac{1}{p}\Big)\Big)_\alpha, a(\zeta)\, d\zeta\Big\rangle &= \frac{1}{\gamma_0} \times \frac{1}{d} \int_{\mathbb{C}} \overline{\partial}\Big(\frac{1}{p}\Big)\wedge dp \\ &= \frac{1}{\gamma_0} \times \frac{\sum_{\{\alpha\in\mathbb{C}\,:\,p(\alpha)=0\}} \nu_{p,\alpha}}{d} = \frac{1}{\gamma_0},\end{aligned}$$

when $\mathrm{div}_{\mathbb{C}}(p) = \sum_{\{\alpha\in\mathbb{C}\,:\,p(\alpha)=0\}} \nu_{p,\alpha}\, \{\alpha\}$, which concludes the proof. □

It follows from (1.14) that for any fraction $a \in \mathbb{K}(X)$ and any $\alpha \in p^{-1}(\{0\})$, which is not a complex pole of a,

$$\Big\langle \Big(\overline{\partial}\Big(\frac{1}{p}\Big)\Big)_\alpha, a(\zeta)\, d\zeta \Big\rangle \in \overline{\mathbb{K}},$$

where $\overline{\mathbb{K}}$ denotes the integral closure of $\mathbb{K}$ in $\mathbb{C}$. In the particular case where a is regular on $p^{-1}(\{0\})$, it is convenient to introduce the following definition and notation.

DEFINITION 1.36. Let $p \in \mathbb{K}[X]$ with $d = \deg p \geq 1$ and let $a \in \mathbb{K}(X)$ be regular on $p^{-1}(\{0\})$. The global residue of the Kähler differential $a(X)\, dX$ with respect to p is the element of $\mathbb{K}$ defined and denoted[4] by

$$\operatorname{Res}_{\mathbb{K}[X]/\mathbb{K}} \begin{bmatrix} a\, dX \\ p \end{bmatrix} := \sum_{\{\alpha \in \mathbb{C} \,:\, p(\alpha) = 0\}} \Big\langle \Big(\overline{\partial}\Big(\frac{1}{p}\Big)\Big)_\alpha, a(\zeta)\, d\zeta \Big\rangle. \tag{1.102}$$

One can reformulate and at the same time extend Corollary 1.35 as follows.

COROLLARY 1.37. *Assume that $p \in \mathbb{K}[X]$ has degree $d = \deg p \geq 1$. Let the fraction $a = a_1/a_0 \in \mathbb{K}(X)$, where $a_0, a_1 \in \mathbb{K}[X]$, be such that a_0 is coprime with p. If $u \in \mathbb{K}[X]$ satisfies $a_0 u \equiv 1$ modulo $p\,\mathbb{K}[X]$ in $\mathbb{K}[X]$, then*

$$\operatorname{Res}_{\mathbb{K}[X]/\mathbb{K}} \begin{bmatrix} a(X)\, dX \\ p \end{bmatrix} = \operatorname{Res}_{\mathbb{K}[X]/\mathbb{K}} \begin{bmatrix} a_1 u\, dX \\ p \end{bmatrix} = \frac{\rho_{d-1}(a_1 u; p)}{\gamma_0} \in \mathbb{K}. \tag{1.103}$$

PROOF. The assertion follows from Corollary 1.35, once one invokes the fact that $p \cdot \overline{\partial}(1/p) = 0$ in $'\mathcal{D}^{(0,1)}(\mathbb{C})$. □

Taking into account the material from §1.2.5, Cauchy's formula in the affine space $\mathbb{A}^1_{\mathbb{C}} = \mathbb{C}$ admits the following current formulation. For any $z \in \mathbb{C}$,

$$\Big(2i\pi \rfloor_{(\zeta - z)\partial/\partial\zeta} - \overline{\partial}\Big)\Big(\Big(\frac{d\zeta}{2i\pi\,(\zeta - z)}\Big)_{|\mathbb{C}\setminus\{z\}}\Big) = [1] - [\{z\}], \tag{1.104}$$

where $(\zeta, z) \longmapsto (\zeta - z)\,\partial/\partial\zeta$ plays the role of a section of a holomorphic line bundle over $\mathbb{C}_\zeta \times \mathbb{C}_z$. The zero set of this section equals precisely the diagonal subset $\zeta = z$ and is of multiplicity 1. As usual, $[\{z\}]$ denotes the integration $(1,1)$-current attached to the divisor $\{z\}$, and $[1]$ describes the $(0,0)$-current induced by the constant function 1. Given $p = \sum_{\kappa=0}^{d} \gamma_\kappa X^{d-\kappa}$ in $\mathbb{K}_d[X]$ with $\gamma_0 \in \mathbb{K}^*$ as before, let

$$b(X,Y) = \frac{p(Y) - p(X)}{Y - X} = \sum_{\kappa=0}^{d-1} \Big(\sum_{\kappa'=0}^{\kappa} \gamma_{\kappa'} Y^{\kappa - \kappa'}\Big) X^{d-1-\kappa} \in \mathbb{K}[X,Y]. \tag{1.105}$$

For a fixed $z \in \mathbb{C}$ one has that

$$\Big(2i\pi \rfloor_{(\zeta - z)\partial/\partial\zeta} - \overline{\partial}_\zeta\Big)\big(b(z,\zeta)\, d\zeta\big) = 2i\pi \rfloor_{(\zeta - z)\partial/\partial\zeta}\big(b(z,\zeta)\, d\zeta\big) = 2i\pi\,[p(\zeta) - p(z)]. \tag{1.106}$$

According to the terminology already used in §1.2.2 within the analytic context, the polynomial $b(X,Y)$ is called a *bézoutian* for p. The tautological identity

$$\Big(\frac{d\zeta}{2i\pi\,(\zeta - z)}\Big)_{|\mathbb{C}\setminus\{z\}} = \Big(\frac{b(z,\zeta)\, d\zeta}{2i\pi(p(\zeta) - p(z))}\Big)_{|\mathbb{C}\setminus p^{-1}(\{p(z)\})},$$

---

[4] The notations used here were introduced by R. Hartshorne and J. Lipman [**Ha66, Lip87**].

allows Cauchy's formula to bridge euclidean division in $\mathbb{K}[X]$ with algebraic residue calculus with $\mathbb{K}$ as field of constants.

PROPOSITION 1.38. *Let $p \in \mathbb{K}_d[X]$, the $\mathbb{K}$ vector space of polynomials with degree $d \in \mathbb{N}^*$, and let $b \in \mathbb{K}[X,Y]$ be as in (1.105). For any $a \in \mathbb{K}[X]$ and $N \in \mathbb{N}^*$, the quotient and the remainder in the euclidean division formula $a = p^N q_N + r_N$ in $\mathbb{K}[X]$ can be expressed by*

$$\begin{aligned} q_N(X) &= \sum_{k=0}^{[\frac{\deg a}{d}]-N} \mathrm{Res}_{\mathbb{K}[Y]/\mathbb{K}} \left[\begin{matrix} a(Y)\, b(X,Y)\, dY \\ p^{k+N+1} \end{matrix} \right] p^k(X) \\ &= \sum_{k=0}^{[\frac{\deg a}{d}]-N} \sum_{\kappa=0}^{d-1} \mathrm{Res}_{\mathbb{K}[Y]/\mathbb{K}} \left[\begin{matrix} a(Y) \Big(\sum_{\kappa'=0}^{\kappa} \gamma_{\kappa'} Y^{\kappa-\kappa'} \Big)\, dY \\ p^{k+N+1} \end{matrix} \right] X^{d-1-\kappa} p^k(X) \end{aligned} \tag{1.107}$$

and

$$\begin{aligned} r_N(X) &= \sum_{k=0}^{N-1} \mathrm{Res}_{\mathbb{K}[Y]/\mathbb{K}} \left[\begin{matrix} a(Y)\, b(X,Y)\, dY \\ p^{k+1} \end{matrix} \right] p^k(X) \\ &= \sum_{k=0}^{N-1} \sum_{\kappa=0}^{d-1} \mathrm{Res}_{\mathbb{K}[Y]/\mathbb{K}} \left[\begin{matrix} a(Y) \Big(\sum_{\kappa'=0}^{\kappa} \gamma_{\kappa'} Y^{\kappa-\kappa'} \Big)\, dY \\ p^{k+1} \end{matrix} \right] X^{d-1-\kappa} p^k(X), \end{aligned} \tag{1.108}$$

respectively.

PROOF. Let $R > 0$ be such that $p^{-1}(\{0\}) \subset D(0,R)$. As seen in §1.2.3 (see (1.62) applied for $F = (p)$ and $h = a$), one has that for any $z \in D(0,R)$ the identity

$$a(z) = \sum_{k=0}^{\infty} \mathrm{Res}_{\mathbb{K}[Y]/\mathbb{K}} \left[\begin{matrix} a(Y)\, b(z,Y)\, dY \\ p^{k+1} \end{matrix} \right] p^k(z) \tag{1.109}$$

holds, provided that $|p(z)| < \min_{|\zeta|=R} |p(\zeta)|$. On the other hand, it follows from Proposition 1.34 that

$$kd > \deg a \Longrightarrow \mathrm{Res}_{\mathbb{K}[Y]/\mathbb{K}} \left[\begin{matrix} a(Y)\, b(X,Y)\, dY \\ p^{k+1} \end{matrix} \right] \equiv 0.$$

Then the analytic identity (1.109), valid for $z \in D(0,R)$ which satisfies the inequality $|p(z)| < \min_{|\zeta|=R} |p|$, truncates as the algebraic identity

$$a(X) = \sum_{k=0}^{[\frac{\deg a}{d}]} \mathrm{Res}_{\mathbb{K}[Y]/\mathbb{K}} \left[\begin{matrix} a(Y)\, b(X,Y)\, dY \\ p^{k+1} \end{matrix} \right] p^k(X).$$

The assertion of the proposition follows from the uniqueness of the quotient and remainder in the euclidean division. □

COROLLARY 1.39. *Let p_0 and p_1 be two coprime polynomials in $\mathbb{K}[X]$, with respective degrees $d_0 \geq 1$, $d_1 \geq 1$. Bézout's identity*[5] $1 = u_0 p_0 + u_1 p_1$, *where*

[5]Such identity will play a central role in the multi-variate setting later on in the present monograph; see Chapters 6 and 7. Although it goes back to Euclid and is not explicitly formulated in such a way by Étienne Bézout (1730–1783) himself in his work (see, for example, the excellent E. Bézout biography by Liliane Alfonsi [**Alf**]), its pivotal role in Bézout's contributions justifies the terminology we will use from now on.

$\deg u_0 \leq d_1 - 1$ *and* $\deg u_1 \leq d_0 - 1$ *is equivalently expressed in* $\mathbb{K}[X]$ *as*

$$1 = \operatorname{Res}_{\mathbb{K}[Y]/\mathbb{K}} \begin{bmatrix} \dfrac{1}{p_1(Y)} \begin{vmatrix} b_0(X,Y) & p_0(X) \\ b_1(X,Y) & p_1(X) \end{vmatrix} dY \\ p_0 \end{bmatrix}$$

$$(1.110) \qquad = -\operatorname{Res}_{\mathbb{K}[Y]/\mathbb{K}} \begin{bmatrix} \dfrac{1}{p_0(Y)} \begin{vmatrix} b_0(X,Y) & p_0(X) \\ b_1(X,Y) & p_1(X) \end{vmatrix} dY \\ p_1 \end{bmatrix}.$$

The bézoutians $b_0, b_1 \in \mathbb{K}[X,Y]$ *for the polynomials* p_0 *and* p_1*, respectively, are determined by* (1.105).

PROOF. One has in $\mathbb{K}(X,Y)$ that

$$\frac{1}{p_1(Y)} \begin{vmatrix} b_0(X,Y) & p_0(X) \\ b_1(X,Y) & p_1(X) \end{vmatrix} = \begin{vmatrix} b_0(X,Y) & p_0(Y)/p_1(Y) \\ b_1(X,Y) & 1 \end{vmatrix},$$

$$\frac{1}{p_0(Y)} \begin{vmatrix} b_0(X,Y) & p_0(X) \\ b_1(X,Y) & p_1(X) \end{vmatrix} = \begin{vmatrix} b_0(X,Y) & 1 \\ b_1(X,Y) & p_1(Y)/p_0(Y) \end{vmatrix}.$$

Since $p_j \cdot \overline{\partial}(1/p_j) = 0$ in $'\mathcal{D}^{(0,1)}(\mathbb{C})$, we have that

$$\operatorname{Res}_{\mathbb{K}[Y]/\mathbb{K}} \begin{bmatrix} b_0(X,Y)\, dY \\ p_0 \end{bmatrix} = \operatorname{Res}_{\mathbb{K}[Y]/\mathbb{K}} \begin{bmatrix} \dfrac{1}{p_1(Y)} \begin{vmatrix} b_0(X,Y) & p_0(X) \\ b_1(X,Y) & p_1(X) \end{vmatrix} dY \\ p_0 \end{bmatrix},$$

$$\operatorname{Res}_{\mathbb{K}[Y]/\mathbb{K}} \begin{bmatrix} b_1(X,Y)\, dY \\ p_1 \end{bmatrix} = -\operatorname{Res}_{\mathbb{K}[Y]/\mathbb{K}} \begin{bmatrix} \dfrac{1}{p_0(Y)} \begin{vmatrix} b_0(X,Y) & p_0(X) \\ b_1(X,Y) & p_1(X) \end{vmatrix} dY \\ p_1 \end{bmatrix}.$$

The assertion of the corollary then follows from Proposition 1.38 with $a = 1$ and $N = 1$. □

Let us realize the above ideas in the setting of the compact Riemann surface $\mathbb{P}^1_{\mathbb{C}}$. This compact surface is the Riemann sphere, whose genus is equal to 0 and whose first homology group is trivial $H^1(\mathbb{P}^1_{\mathbb{C}}) = 0$. Its canonical bundle[6] is $\mathcal{O}(-2)$. The diagonal subspace of the product space $(\mathbb{P}^1_{\mathbb{C}})_z \times (\mathbb{P}^1_{\mathbb{C}})_\zeta$, equipped with the projections π_z and π_ζ on the coordinate spaces, is defined once again as the zero set of a section of a holomorphic line bundle $\mathbf{L}$. Namely, let

$$\begin{aligned} \mathbf{L} &= \operatorname{Hom}_{\mathbb{C}}((\mathcal{O}(1))_\zeta, (\mathcal{O}(1))_z) \otimes (T^{(1,0)}_{\mathbb{P}^1_{\mathbb{C}}})_\zeta \\ &\simeq (\mathcal{O}(1))_z \otimes (\mathcal{O}(-1))_\zeta \otimes (\mathcal{O}(2))_\zeta = (\mathcal{O}(1))_z \otimes (\mathcal{O}(1))_\zeta, \end{aligned}$$

where $L_z = \pi_z^* L$ and $L_\zeta = \pi_\zeta^* L$ for any holomorphic line bundle over $\mathbb{P}^1_{\mathbb{C}}$. The section of $\mathbf{L}$ expressed in homogeneous coordinates as

$$(1.111) \qquad \boldsymbol{s}([z_0 : z_1], [\zeta_0 : \zeta_1]) = -z_0 \frac{\partial}{\partial \zeta_0} - z_1 \frac{\partial}{\partial \zeta_1}$$

is such that $V(\boldsymbol{s}) = \{(z,\zeta) \in \mathbb{P}^1_{\mathbb{C}} \times \mathbb{P}^1_{\mathbb{C}} : z_0\zeta_1 - z_1\zeta_0 = 0\}$. Thus $V(\boldsymbol{s})$ is the diagonal subspace of $\mathbb{P}^1_{\mathbb{C}} \times \mathbb{P}^1_{\mathbb{C}}$ and the vanishing order of the section $\boldsymbol{s}$ is equal to 1. The

[6] We denote $\mathcal{O}_{\mathbb{P}^1_{\mathbb{C}}}(k)$ by $\mathcal{O}(k)$ for any $k \in \mathbb{Z}$.

$\mathbf{L}^* \simeq \mathbf{L}^{-1}$-valued meromorphic form over the open subset

$$U = \{(z,\zeta) \in \mathbb{P}^1_{\mathbb{C}} \times \mathbb{P}^1_{\mathbb{C}}\,;\, z_0\zeta_1 - z_1\zeta_0 \neq 0\},$$

defined by

$$(z,\zeta) \in U \longmapsto \frac{1}{2i\pi}\frac{\zeta_1\, d\zeta_0 - \zeta_0\, d\zeta_1}{\zeta_0 z_1 - \zeta_1 z_0} = \frac{\zeta_0^2}{\zeta_0 z_1 - \zeta_1 z_0}\, d\Big(\frac{\zeta_1}{\zeta_0}\Big), \tag{1.112}$$

is such that for all $z = [z_0 : z_1] \in \mathbb{P}^1_{\mathbb{C}}$ the equality

$$\Big(2i\pi\rfloor_{\boldsymbol{s}} - \overline{\partial}_\zeta\Big)\Big(\Big(\frac{1}{2i\pi}\frac{\zeta_1\, d\zeta_0 - \zeta_0\, d\zeta_1}{\zeta_0 z_1 - \zeta_1 z_0}\Big)_{|\mathbb{P}^1_{\mathbb{C}}\setminus\{z\}}\Big) = [1] - [\{z\}] \otimes \mathrm{Id}\big((\mathcal{O}(-1))_z\big) \tag{1.113}$$

holds. The relation (1.113) is the current formulation of Cauchy's formula on $\mathbb{P}^1_{\mathbb{C}}$ to be compared with its current analogue (1.104) on the affine space $\mathbb{A}^1_{\mathbb{C}}$. For $d \geq 1$, let P be a global section of $\mathcal{O}(d)$ defined over $\mathbb{K}$. It means that P is a homogeneous polynomial with degree d in X_0, X_1. Let $\boldsymbol{B} = (B_0, B_1)$ be a pair of (nonunique) global sections $B_j(X_0,X_1,Y_0,Y_1)$ of $\mathcal{O}_{\mathbb{P}^3_{\mathbb{C}}}(d-1)$ defined over $\mathbb{K}$ such that

$$P(Y_0,Y_1) - P(X_0,X_1) = B_0(X,Y)(Y_0 - X_0) + B_1(X,Y)(Y_1 - X_1). \tag{1.114}$$

This amounts to saying that B_0 and B_1 are homogeneous polynomials with degree $d-1$ in four variables with coefficients in $\mathbb{K}$ satisfying (1.114). At this point, one needs to equip holomorphic bundles, among them the $\mathcal{O}(1)$, with hermitian metrics. One considers the line bundle $\mathcal{O}(1)$ equipped with the Fubini–Study metric[7] having its first Chern form equal to

$$\begin{aligned} c = dd^c \log\|\zeta\|^2 &= \frac{1}{2i\pi}\frac{\overline{(\zeta_1\, d\zeta_0 - \zeta_0\, d\zeta_1)} \wedge (\zeta_1\, d\zeta_0 - \zeta_0\, d\zeta_1)}{\|\zeta\|^4} \\ &= -\frac{1}{2i\pi}\Big(d\zeta_0 \wedge \overline{\partial}\Big(\frac{\overline{\zeta_0}}{\|\zeta\|^2}\Big) + d\zeta_1 \wedge \overline{\partial}\Big(\frac{\overline{\zeta_1}}{\|\zeta\|^2}\Big)\Big) \end{aligned}$$

and Chern connection D satisfying

$$D\zeta_j = d\zeta_j - \zeta_j\, \partial \log\|\zeta\|^2$$

for $j = 0,1$; see §B.3.2.2. For $(z,\zeta) \in \mathbb{P}^1_{\mathbb{C}} \times \mathbb{P}^1_{\mathbb{C}}$, let

$$w(z,\zeta) = \frac{z_0\,\overline{\zeta_0} + z_1\,\overline{\zeta_1}}{\|\zeta\|^2} + c(\zeta).$$

One defines in this way a smooth form with values in $\mathbb{C} \oplus \mathrm{Hom}\big((\mathcal{O}(1))_\zeta, (\mathcal{O}(1))_z\big)$. This form is holomorphic in z, but unfortunately not in ζ. Nevertheless, it satisfies

$$\begin{aligned} (2i\pi\rfloor_{\boldsymbol{s}} - \overline{\partial}_\zeta)(w) &= -z_0\,\overline{\partial}\Big(\frac{\overline{\zeta_0}}{\|\zeta\|^2}\Big) - z_1\,\overline{\partial}\Big(\frac{\overline{\zeta_1}}{\|\zeta\|^2}\Big) \\ &\quad - \rfloor_{\boldsymbol{s}}\Big(d\zeta_0 \wedge \overline{\partial}\Big(\frac{\overline{\zeta_0}}{\|\zeta\|^2}\Big) + d\zeta_1 \wedge \overline{\partial}\Big(\frac{\overline{\zeta_1}}{\|\zeta\|^2}\Big)\Big) = 0. \end{aligned} \tag{1.115}$$

For a fixed z in $\mathbb{P}^1_{\mathbb{C}}$, the algebraic identity in $z, \zeta, \overline{\zeta}, d\zeta, d\overline{\zeta}$

$$\Big(2i\pi\rfloor_{\boldsymbol{s}} - \overline{\partial}_\zeta\Big)\Big(B_0(z, w\,\zeta)\, D\zeta_0 + B_1(z, w\,\zeta)\, D\zeta_1\Big) = 2i\pi\big(w^d P(\zeta) - P(z)\big) \tag{1.116}$$

[7] In some sense, as will be specified in this monograph, such a metric is the most "algebraic" that is available on $\mathcal{O}(1)$ given Crofton's formula; see §D.5.1 and §D.5.2. The canonical metric $\|(\zeta_0,\zeta_1)\|_\infty$ instead of $\|(\zeta_0,\zeta_1)\|_2$ would certainly be a better choice given diophantine geometry. Unfortunately, it is just continuous and fails to be smooth as the Fubini–Study metric is.

holds. The field $\mathbb{K}$ is still considered the field of coefficients since $B_j \in \mathbb{K}[X,Y]$ for $j = 0, 1$. The identity (1.116) resulting from (1.115) and from (1.114) plays in $(\mathbb{P}^1_{\mathbb{C}})_\zeta$ the same role as the identity (1.106) plays in the affine space $(\mathbb{A}^1_{\mathbb{C}})_\zeta$, when $z \in \mathbb{A}^1_{\mathbb{C}}$ is fixed. It allows Cauchy's formula in $\mathbb{P}^1_{\mathbb{C}}$ in its current formulation (1.113) to bridge geometric residue calculus on $\mathbb{P}^1_{\mathbb{C}}$, as introduced in §1.2.5, and division problems. The major difference is the necessity to introduce the Fubini–Study metric on $\mathcal{O}(1)$, its Chern connection, and the related weight w, to keep track of homogenization. Unlike what happens in $\mathbb{A}^1_{\mathbb{C}}$, the introduction of the above machinery in our calculus prevents us to respect in this context the basis field $\mathbb{K}$. One needs to invoke averaging to interpret algebraically closed-form division formulae that are obtained. This approach will be carried through in $\mathbb{P}^N_{\mathbb{C}}$, see §4.4, but we wanted to introduce it briefly in the univariate setting as soon as possible.

1.3.2. Currential Cauchy's formula on compact Riemann surfaces. We consider in this section a connected compact Riemann surface $\mathcal{X}$ with genus $g \geq 1$ and Euler characteristic $2(1-g) \leq 0$. To formulate Cauchy's formula in terms of bundle-valued currents on $\mathcal{X}$, as we did in §1.3.1 for $\mathcal{X} = \mathbb{C}$ or $\mathcal{X} = \mathbb{P}^1_{\mathbb{C}}$ (the case of genus $g = 0$), we need to recall some basic facts about compact Riemann surfaces. For more details, we suggest to the reader the following references [**Fa73, Mu84, FK, Bos92**].

A connected compact Riemann surface $\mathcal{X}$ inherits the structure of an algebraic projective nonsingular curve embedded in $\mathbb{P}^3_{\mathbb{C}}$. If $g = 0$ the Riemann surface $\mathcal{X}$ equals the Riemann sphere, which may be also identified with the projective line $\mathbb{P}^1_{\mathbb{C}}$. When the genus $g = 1$, $\mathcal{X}$ is the complex torus, while $\mathcal{X}$ is a hyperbolic compact Riemann manifold, when $g > 1$. Since we assume here $g \geq 1$, Riemann's uniformization theorem implies the existence of a universal covering

$$\pi : \widehat{\mathcal{X}} \to \mathcal{X}, \quad \text{where } \widehat{\mathcal{X}} = \begin{cases} \mathbb{C} \text{ for } g = 1 \\ \mathbb{D} \text{ for } g > 1. \end{cases} \tag{1.117}$$

The canonical bundle $T^{*(1,0)}_{\mathcal{X}} = \omega_{\mathcal{X}}$ has a degree equal to $2(g-1)$. The first object one needs to introduce is a *canonical* $\mathbb{Z}$*-basis* $\{\mathfrak{a}, \mathfrak{b}\}$ for the $2g$-lattice $H^1(\mathcal{X}, \mathbb{Z})$. The notation means that $\mathfrak{a} = \{\mathfrak{a}_1, \dots, \mathfrak{a}_g\}$ and $\mathfrak{b} = \{\mathfrak{b}_1, \dots, \mathfrak{b}_g\}$ are two families of g elements in the first homology group $H^1(\mathcal{X}, \mathbb{Z})$ such that the intersection numbers[8] of $(\mathfrak{a}_\kappa, \mathfrak{b}_{\kappa'})$, $(\mathfrak{a}_\kappa, \mathfrak{a}_{\kappa'})$, $(\mathfrak{b}_\kappa, \mathfrak{b}_{\kappa'})$ for any $1 \leq \kappa, \kappa' \leq g$ are equal to $\delta_\kappa^{\kappa'}$, 0, 0 respectively. We let also $\{\varpi_1, \dots, \varpi_g\}$ be a basis of the g-dimensional $\mathbb{C}$-vector space $\Omega(X) = \mathcal{O}(\mathcal{X}, \omega_{\mathcal{X}})$ of abelian differentials on $\mathcal{X}$ normalized with respect to $\mathfrak{a}$. That is,

$$\frac{1}{2i\pi} \int_{\alpha_{\kappa'}} \varpi_\kappa = \delta_\kappa^{\kappa'}, \tag{1.118}$$

for each $1 \leq \kappa, \kappa' \leq g$. Let also

$$H = \left[\frac{1}{2i\pi} \int_{\mathfrak{b}_{\kappa'}} \varpi_\kappa \right]_{1 \leq \kappa, \kappa' \leq g}$$

be the symmetric Riemann period matrix with positive definite imaginary part. Let $J(\mathcal{X}) = \mathbb{C}^g / (\mathbb{Z}^g + H\mathbb{Z}^g)$ (with projection quotient morphism π_J) be the jacobian

[8]For intersection theory between elements in $H^1(\mathcal{X}_{\mathbb{R}}, \mathbb{Z})$, we refer the reader to [**FK**, III.1].

of $\mathcal{X}$, o be the marked origin on $\mathcal{X}$ and $\mu : \mathcal{X} \to \Gamma$ be the Abel–Jacobi map defined by

$$\mu(x) = \Big(\int_o^x \varpi_1, \ldots, \int_o^x \varpi_g\Big) \quad \text{modulo} \quad \mathbb{Z}^g + H\,\mathbb{Z}^g. \tag{1.119}$$

The map (1.119), acting on Weil divisors by $\mu(\sum_\iota \nu_\iota\{\alpha_\iota\}) = \sum_\iota \nu_\iota\, \mu(\alpha_\iota)$, realizes a homomorphism from the Picard group $\mathrm{Pic}(\mathcal{X})$ into jacobian $J(\mathcal{X})$ of $\mathcal{X}$. This homomorphism induces an isomorphism between the subgroup $\mathrm{Pic}^0(\mathcal{X})$ of elements with degree[9] 0 and $J(\mathcal{X})$, thanks to Abel's inversion theorem. The Abel–Jacobi map μ is lifted to the universal covering $\widehat{\mathcal{X}}$ as the morphism $\widehat{\mu} : \widehat{\mathcal{X}} \to \mathbb{C}^g$ satisfying

$$\mu \circ \pi = \pi_J \circ \widehat{\mu}. \tag{1.120}$$

The fundamental entire function of g complex variables attached to the positive definite matrix H, hence attached to the canonical $\mathbb{Z}$-basis $\{\mathfrak{a}, \mathfrak{b}\}$ of $H^1(\mathcal{X}, \mathbb{Z})$, is its thêta function $\theta = \theta_H$ defined by

$$\theta(w) = \sum_{\kappa \in \mathbb{Z}^g} \exp\Big(i\pi \sum_{\ell=1}^{g}\sum_{\ell'=1}^{g} h_{\ell,\ell'}\kappa_\ell\kappa_{\ell'} + 2i\pi\,\langle\!\langle \kappa\,,\, w\rangle\!\rangle\Big), \quad \forall w \in \mathbb{C}^g. \tag{1.121}$$

Let $w_0 = v_0 + H u_0 \in \mathbb{C}^g$, where $u_0, v_0 \in \mathbb{R}^g$ be such that $\theta(w_0) \neq 0$. The normalized *thêta function with characteristics* $\theta[w_0]/(\theta[w_0](0))$ is the entire function on $\mathbb{C}^g$ defined by

$$\frac{\theta[w_0]}{\theta[w_0](0)}(w) = \exp\big(2i\pi\,\langle\!\langle u_0, w\rangle\!\rangle\big)\,\frac{\theta(w + w_0)}{\theta(w_0)}, \quad \forall w \in \mathbb{C}^g. \tag{1.122}$$

Normalization, which will be important for us, corresponds to the fact that

$$\frac{\theta[w_0]}{\theta[w_0](0)}(0) = 1.$$

Let us make the first observation.

PROPOSITION 1.40. *Let $\mathfrak{j} \in J(\mathcal{X})$ be such that $\theta(w_0) \neq 0$ for some w_0 in $\pi_J^{-1}(\{\mathfrak{j}\})$. Then $\theta(w) \neq 0$ for any $w \in \pi_J^{-1}(\{\mathfrak{j}\})$. Moreover, θ defines a Cartier divisor $\mathrm{div}_{J(\mathcal{X})}(\theta)$ on the g-dimensional complex manifold J. All entire functions in g variables defined by*

$$w \in \mathbb{C}^g \longmapsto \frac{\theta[w_0]}{\theta[w_0](0)}(w), \ \textit{where } w_0 \in \pi_J^{-1}(\{\mathfrak{j}\})$$

coincide with the entire function denoted by

$$w \in \mathbb{C}^g \longmapsto \frac{\theta[\mathfrak{j}]}{\theta[\mathfrak{j}](0)}(w).$$

For any such $\mathfrak{j}$, the multi-valent function

$$\zeta \in \mathcal{X} \longmapsto \frac{\theta[\mathfrak{j}]}{\theta[\mathfrak{j}](0)}(\zeta) := \frac{\theta[\mathfrak{j}]}{\theta[\mathfrak{j}](0)}(\widehat{\mu}(\xi)), \quad \textit{where } \xi \in \pi^{-1}(\{\zeta\}) \tag{1.123}$$

is interpreted as a section of the unitary flat[10] *line bundle with class $\mu^{-1}(\{\mathfrak{j}\})$ belonging to* $\mathrm{Pic}^0(\mathcal{X})$.

---

[9] The degree of $L \in \mathrm{Pic}(\mathcal{X})$ is defined as the degree of the Weil divisor attached to any Cartier divisor representing L.

[10] An holomorphic line bundle L over $\mathcal{X}$ is said to be unitary flat if and only if it can be equipped with a hermitian metric $|\ |$ with first Chern class $c_1(\overline{L}) = 0$; see (B.3.3).

PROOF. We give here a sketch of the proof. Its details can be found, for example, in [**Fa73**, I]. One has the quasi-periodicity relations
(1.124)

$$\theta(w+k'')=\theta(w),\quad \theta(w+H\cdot k')=\exp\Big(-i\pi\sum_{\ell=1}^{g}\sum_{\ell'=1}^{g}h_{\ell,\ell'}k'_\ell k'_{\ell'}-2i\pi\langle\!\langle k',w\rangle\!\rangle\Big)\theta(w)$$

for any $w\in\mathbb{C}^g$ and $k',k''\in\mathbb{Z}^g$. The first assertion follows. Take now two elements ξ_0 and ξ_1 in $\pi^{-1}(\{\zeta\})$. Then it follows from (1.120) that

$$\widehat{\mu}(y_0)-\widehat{\mu}(y_1)=k''+H\cdot k'\in\mathbb{Z}^g+H\cdot\mathbb{Z}^g.$$

Quasi-periodicity relations (1.124) imply that, if $w_0=v_0+H\cdot u_0$, where $u_0,v_0\in\mathbb{R}^g$, then

$$\frac{\theta[w_0]}{\theta[w_0](0)}(\widehat{\mu}(\xi_1))=\exp\big(2i\pi(\langle\!\langle k',v_0\rangle\!\rangle-\langle\!\langle k'',u_0\rangle\!\rangle)\big)\,\frac{\theta[w_0]}{\theta[w_0](0)}(\widehat{\mu}(\xi_0)).$$

As a consequence, the multi-valent function (1.123) can be understood as a meromorphic section over $\mathcal{X}$ of a unitary flat bundle representing an element of $\mathrm{Pic}^0(\mathcal{X})$. Observe finally that the image of a such element of $\mathrm{Pic}^0(\mathcal{X})$ by the Abel–Jacobi map μ equals the class of $v_0+H\cdot u_0$ modulo $\mathbb{Z}^g+H\cdot\mathbb{Z}^g$, that is, j. □

Riemann's theorem for half-periods, see [**Fa73**, Theorem 1.1], asserts that there exists an element $\sqrt{\omega_{\mathcal{X}}}\in\mathrm{Pic}(\mathcal{X})$, whose degree equals

$$\deg\sqrt{\omega_{\mathcal{X}}}=g-1=\mu(\sqrt{\omega_{\mathcal{X}}}),$$

so that $\sqrt{\omega_{\mathcal{X}}}\otimes\sqrt{\omega_{\mathcal{X}}}=\omega_{\mathcal{X}}$ in $\mathrm{Pic}(\mathcal{X})$. Meromorphic sections of $\sqrt{\omega_{\mathcal{X}}}$ are naturally called *(meromorphic) half-order differentials* on $\mathcal{X}$. Furthermore, there are 2^{2g} elements in $\mathrm{Pic}(\mathcal{X})$, whose meromorphic sections have the properties that allow them to be also called (meromorphic) half-order differentials on $\mathcal{X}$. We are ready now to formulate the following definition.

DEFINITION 1.41. An element of $\mathrm{Pic}(\mathcal{X})$, whose meromorphic section is called (meromorphic) half-order differentials on $\mathcal{X}$, is of the form $\sqrt{\omega_{\mathcal{X}}}\otimes(\mu^{-1}(\{\mathrm{j}\}))$, where $\mu^{-1}(\{\mathrm{j}\})\in\mathrm{Pic}^0(\mathcal{X})$ is attached by Abel's inversion theorem to any $\mathrm{j}\in J(\mathcal{X})$ so that $2\,\mathrm{j}=\dot{0}$. There are 2^{2g} elements of this form in $\mathrm{Pic}(\mathcal{X})$, all of them having their degree equal to $g-1$ and all of them satisfying

$$(1.125)\quad \big(\sqrt{\omega_{\mathcal{X}}}\otimes(\mu^{-1}(\{\mathrm{j}\}))\big)\otimes\big(\sqrt{\omega_{\mathcal{X}}}\otimes(\mu^{-1}(\{\mathrm{j}\}))\big)=\omega_{\mathcal{X}}\otimes(\mu^{-1}(\{2\,\mathrm{j}\}))=\omega_{\mathcal{X}}.$$

A fundamental fact is that the line bundle

$$(1.126)\qquad \mathbf{L}=(\pi_z^*(\sqrt{\omega_{\mathcal{X}}}))^{-1}\otimes(\pi_\zeta^*(\sqrt{\omega_{\mathcal{X}}}))^{-1}\otimes\rho^*(\mathbf{0}_{J(\mathcal{X})}),$$

over $\mathcal{X}_z\times\mathcal{X}_\zeta$, where

$$\rho:(z,\zeta)\in\mathcal{X}\times\mathcal{X}\longmapsto\mu\big(\{z\}-\{\zeta\}\big)\in J(\mathcal{X})$$

and $\mathbf{0}_{J(\mathcal{X})}$ is the null bundle over $J(\mathcal{X})$, admits an holomorphic section E, whose zero set is equal exactly to the diagonal $\{(z,\zeta)\in\mathcal{X}\times\mathcal{X}\,:\,z=\zeta\}$ and it vanishes to order 1 on this diagonal. To be more specific, given a local holomorphic coordinate t about z, one has, for ζ close to z in $\mathcal{X}$, that

$$(1.127)\qquad E(z,\zeta)\simeq\frac{t(\zeta)-t(z)}{\sqrt{dt(z)}\sqrt{dt(\zeta)}}\Big(1+O\big((t(\zeta)-t(z))^2\big)\Big).$$

Moreover, E is antisymmetric. Namely,

$$E(z,\zeta) = -E(\zeta,z), \quad \forall (\zeta,z) \in \mathcal{X} \times \mathcal{X}. \tag{1.128}$$

We refer to [**Mu84**, IIIb-§1] for construction of E, together with its list of properties, among them (1.127); see also [**Fa73**, II]. Note in particular, that this construction can also be carried out in the case where $g = 0$. That is, it can be realized in the case when $\mathcal{X} = \mathbb{P}^1_{\mathbb{C}}$, as explained on page 3.209 in [**Mu84**]. Moreover, it can be considered as an alternative way for the construction proposed in §1.3.1. Such section E of $\mathbf{L}$ is called the *prime form* of $\mathcal{X}$ with respect to the $\mathbb{Z}$-canonical basis $\{\mathfrak{a},\mathfrak{b}\}$, the corresponding matrix H, and thêta function $\theta = \theta_H$.

Then a Cauchy-type formula in terms of distributions holds for smooth sections of any line bundle $\sqrt{\omega_{\mathcal{X}}}\otimes\mu^{-1}(\{\mathfrak{j}\})$, when $\mathfrak{j} \in J(\mathcal{X})$ is such that the θ does not vanish on $\pi_J^{-1}(\{-\mathfrak{j}\})$.[11] This holds in particular for smooth sections of the line bundles introduced in Definition 1.41, provided that $\mathfrak{j} \in J(\mathcal{X})$, satisfies $2\mathfrak{j} = \dot{0}$ and is such that θ does not vanish on $\pi_J^{-1}(\{\mathfrak{j}\})$.

PROPOSITION 1.42. *Let $\mathfrak{j} \in J(\mathcal{X})$ be such that θ does not vanish on $\pi_J^{-1}(\{-\mathfrak{j}\})$ and let φ be a smooth section of $\sqrt{\omega_{\mathcal{X}}}\otimes\mu^{-1}(\{\mathfrak{j}\})$ on $\mathcal{X}$. For any $z \in \mathcal{X}$, the section*

$$\zeta \in \mathcal{X} \longmapsto \Big(\frac{\theta[-\mathfrak{j}]}{\theta[-\mathfrak{j}](0)}(\zeta - z)\Big) \otimes \frac{1}{E(z,\zeta)}$$

is a meromorphic section of $\sqrt{\omega_{\mathcal{X}}} \otimes \mu^{-1}(\{-\mathfrak{j}\})$. Furthermore, the section

$$\zeta \in \mathcal{X} \longmapsto \varphi(\zeta) \otimes \Big(\frac{\theta[-\mathfrak{j}]}{\theta[-\mathfrak{j}](0)}(\zeta - z)\Big) \otimes \frac{1}{E(z,\zeta)}$$

is also a semimeromorphic section of

$$\big(\sqrt{\omega_{\mathcal{X}}} \otimes \mu^{-1}(\{\mathfrak{j}\})\big) \otimes \big(\mu^{-1}(\{-\mathfrak{j}\}) \otimes \sqrt{\omega_{\mathcal{X}}}\big) = \omega_{\mathcal{X}},$$

whose standard extension to $\mathcal{X}$ satisfies

$$\varphi(z) = -\frac{1}{2i\pi}\int_{\mathcal{X}} \overline{\partial}_\zeta\Big(\Big(\varphi(\zeta) \otimes \Big(\frac{\theta[-\mathfrak{j}]}{\theta[-\mathfrak{j}](0)}(\zeta - z)\Big) \otimes \frac{1}{E(z,\zeta)}\Big)_{|\mathcal{X}\setminus\{z\}}\Big). \tag{1.129}$$

PROOF. Proposition 1.40 implies that

$$\zeta \longmapsto \frac{\theta[-\mathfrak{j}]}{\theta[-\mathfrak{j}](0)}(\zeta)$$

is a section of $\mu^{-1}(\{-\mathfrak{j}\})$. The first assertion follows from the definition of the bundle $\mathbf{L}$, see (1.126), whose section $E(z,\zeta)$ is. Formula (1.129) follows from the fact the meromorphic section

$$\zeta \longmapsto \Big(\frac{\theta[-\mathfrak{j}]}{\theta[-\mathfrak{j}](0)}(\zeta - z)\Big) \otimes \frac{1}{E(z,\zeta)}$$

of $\sqrt{\omega_{\mathcal{X}}} \otimes \mu^{-1}(\{-\mathfrak{j}\})$ admits z as a single simple pole, that the prime form is expressed in terms of a local coordinate t about z as (1.127), and that the entire function

$$w \in \mathbb{C}^g \longmapsto \frac{\theta[-\mathfrak{j}]}{\theta[-\mathfrak{j}](0)}(w)$$

[11] Such condition on $\mathfrak{j}$ is equivalent to the fact that $\mathcal{O}_{\mathcal{X}}(\mathcal{X}, \sqrt{\omega_{\mathcal{X}}}\otimes\mu^{-1}(-\{\mathfrak{j}\})) = 0$. Meromorphic sections of some $\sqrt{\omega_{\mathcal{X}}} \otimes \mu^{-1}(\{\mathfrak{j}\})$, with $\mathfrak{j} \in \mathcal{X}$, are sometimes called *multiplicative half-order differentials* on $\mathcal{X}$; see [**AlPV20**, §2.1] for such terminology.

has been specifically normalized so that its value at 0 equals 1. Observe that for ζ fixed in $\mathcal{X}$,

$$z \longmapsto \frac{\theta[-\mathrm{j}]}{\theta[-\mathrm{j}](0)}(\zeta - z) \otimes \frac{1}{E(z,\zeta)} = -\frac{\theta[-\mathrm{j}]}{\theta[-\mathrm{j}](0)}(\zeta - z) \otimes \frac{1}{E(\zeta,z)}$$

is a section of $\sqrt{\omega_{\mathcal{X}}} \otimes \mu^{-1}(\{\mathrm{j}\})$, which is coherent with the current Cauchy representation (1.129) of $\varphi \in C^\infty(\mathcal{X}, \sqrt{\omega_{\mathcal{X}}} \otimes \mu^{-1}(\{\mathrm{j}\})$. □

Let h be a meromorphic function on $\mathcal{X}$, and let U be an open subset of $\mathcal{X}$. If W is the finite polar set of h in U, then $h_{|U\setminus W} \in {}'\mathcal{D}^{(0,0)}(U)$ is the standard extension of h from $U \setminus W$ to U. Let also $f \in \mathcal{O}_{\mathcal{X}}\big(U, \sqrt{\omega_{\mathcal{X}}} \otimes \mu^{-1}(\{\mathrm{j}\})\big)$, where j is such that θ does not vanish on $\pi_J^{-1}(\{-\mathrm{j}\})$. For each $z \in U \setminus W$ we consider the $(1,0)$-meromorphic form in U having single, simple pole at z:

$$\zeta \longmapsto f(\zeta) \otimes \Big(\frac{\theta[-\mathrm{j}]}{\theta[-\mathrm{j}](0)}(\zeta - z)\Big) \otimes \frac{1}{E(\zeta,z)}. \tag{1.130}$$

The residue current $\overline{\partial}(h_{|U\setminus W}/(2i\pi)) \in {}'\mathcal{D}^{(0,1)}(U)$ with support W acts by duality on the $(1,0)$ meromorphic form (1.130), since this form is holomorphic in the neighborhood $U \setminus \{z\}$ of W in U. Moreover

$$z \in U \setminus W \longmapsto \Big\langle \overline{\partial}\Big(\frac{h_{|U\setminus W}}{2i\pi}\Big)(\zeta)\,,\, f(\zeta) \otimes \Big(\frac{\theta[-\mathrm{j}]}{\theta[-\mathrm{j}](0)}(\zeta - z)\Big) \otimes \frac{1}{E(\zeta,z)}\Big\rangle \tag{1.131}$$

defines a meromorphic section of $\sqrt{\omega_{\mathcal{X}}} \otimes \mu^{-1}(\{\mathrm{j}\})$, with polar set contained in W. Observe that the order of such meromorphic section at a point $\alpha \in W$ equals, at most, to the order of h at α. In the particular case where any $\alpha \in W$ is a pole with order 1 of h, it follows from the normalization $\theta[\{-\mathrm{j}\}]/\theta[\{-\mathrm{j}\}](0) = 1$ at the origin in $\mathbb{C}^g$ and from (1.127) that the meromorphic extension of (1.131) to U coincides with the *polar part* of hf in U. In other words, the section

$$z \in U \setminus W \longmapsto h(z)f(z) - \Big\langle \overline{\partial}\Big(\frac{h_{|U\setminus W}}{2i\pi}\Big)(\zeta)\,,\, f(\zeta) \otimes \Big(\frac{\theta[-\mathrm{j}]}{\theta[-\mathrm{j}](0)}(\zeta - z)\Big) \otimes \frac{1}{E(\zeta,z)}\Big\rangle$$

is the restriction to $U \setminus W$ of a holomorphic section of $\sqrt{\omega_{\mathcal{X}}} \otimes \mu^{-1}(\{\mathrm{j}\})$ in U. This idea has been extensively used in operator theory concerning the concept of *resolvent*; see for example [**Vin98**, §3], [**AlV02**] or [**AlPV20**]. The situation is by far more involved in the general case when the poles of h in U are not supposed to be simple anymore. If such is the case, then the Laurent coefficients $a_k(h;\alpha)$ for $k \leq -1$ involved in the expansion of $h = \big(\sum_{k\in\mathbb{Z}} a_k(h;\alpha)t^k\big)\, dt$ in a local coordinate t centered at the pole $\alpha \in W$ with order > 1 depend precisely on the choice of the local coordinate t about α. This prevents them to be defined intrinsically, while the definition of the meromorphic function (1.131) is intrinsic. Let us mention also here the following elementary local lemma about germs of meromorphic functions at the origin in $\mathbb{C}$.

LEMMA 1.43. *Let $\mathfrak{f} \in \mathcal{O}_{\mathbb{C},0}$ be a germ of holomorphic function at the origin $0 \in \mathbb{C}$. Let also $\mathfrak{h}$ be a germ of meromorphic function at the origin such that $o(\mathfrak{h};\alpha) = \nu \in \mathbb{N}^*$, with standard extension $[\mathfrak{h}] \in ({}'\mathcal{D}^{(0,0)}(\mathbb{C}))_0$ as a germ of $(0,0)$-current. Then*

$$z \in \mathbb{C}^* \longmapsto \Big\langle \big(\overline{\partial}[\mathfrak{h}]\big)_0(\zeta)\,,\, \mathfrak{f}(\zeta)\,\frac{d\zeta}{z-\zeta}\Big\rangle$$

extends as a meromorphic function with single pole $\zeta = 0$, whose germ at this pole is exactly the polar part of $\mathfrak{h}\mathfrak{f}$ about the origin.

PROOF. Let $\mathfrak{f} = \mathfrak{u}/\zeta^\nu$, where $\mathfrak{u}$ is an invertible germ in $\mathcal{O}_{\mathbb{C},0}$. Then, for any $z \in \mathbb{C}^*$,

$$\Big\langle \big(\overline{\partial}[\mathfrak{h}]\big)_0(\zeta)\,,\, \mathfrak{f}(\zeta)\,\frac{d\zeta}{z-\zeta}\Big\rangle = \Big\langle \Big(\overline{\partial}\Big(\frac{1}{\zeta^\nu}\Big)\Big)_0\,,\, (\mathfrak{u}\mathfrak{f})(\zeta)\,\frac{d\zeta}{z-\zeta}\Big\rangle$$
$$= \frac{1}{z}\sum_{\kappa=0}^{\nu-1}\Big\langle \Big(\overline{\partial}\Big(\frac{1}{\zeta^{\nu-1-\kappa}}\Big)\Big)_0\,,\,(\mathfrak{u}f)(\zeta)\,d\zeta\Big\rangle\frac{1}{z^\kappa}$$
$$= \frac{1}{z}\sum_{\kappa=0}^{\nu-1}\frac{a_{\nu-\kappa-1}(\mathfrak{u}\,\mathfrak{f};0)}{z^\kappa} = \frac{1}{z^\nu}\sum_{\kappa=0}^{\nu-1} a_\kappa(\mathfrak{u}\,\mathfrak{f};0)\,z^\kappa.$$

□

1.3.3. Green currents and heights of arithmetic cycles. This section is inspired by S. Lang's presentation of Arakelov's theory of arithmetic surfaces [**Ar74**, **L88**]. In what follows, $\mathcal{X}$ is a connected compact nonsingular Riemann surface, which inherits a structure of the complex algebraic projective nonsingular curve. We will focus mainly on the case $\mathcal{X} = \mathbb{P}^1_{\mathbb{C}}$ or on the case of the arithmetic scheme $\mathbb{X} = \mathrm{Spec}\big(\mathrm{Proj}(\mathbb{Z}[X_0, X_1])\big)$, when arithmetic considerations are of importance. A 0-cycle (or Weil divisor) on $\mathcal{X}$ is a formal finite sum $Z = \sum_{\alpha\in\mathcal{X}} \nu_\alpha\{\alpha\}$, where $\alpha \in \mathcal{X}$ and $\nu_\alpha \in \mathbb{Z}^*$. Its support $|Z|$ equals the finite subset

$$|Z| = \{\alpha \in \mathcal{X} \,:\, Z = \sum_\alpha \nu_\alpha\,\{\alpha\} \text{ with } \nu_\alpha \in \mathbb{Z}^*\}. \tag{1.132}$$

This 0-cycle is decomposed to the sum of a positive and negative parts as

$$Z = Z^+ + Z^- = \Big(\sum_{\{\alpha:\,\nu_\alpha\in\mathbb{N}^*\}} \nu_\alpha\,\{\alpha\}\Big) + \Big(\sum_{\{\alpha:\,\nu_\alpha\in-\mathbb{N}^*\}} \nu_\alpha\,\{\alpha\}\Big).$$

It is said to be *effective* if $Z^- = \emptyset$. The *geometric complexity* of the cycle Z is defined to be

$$\deg Z = \sum_\alpha \nu_\alpha.$$

If $L \in \mathrm{Pic}(\mathcal{X})$, then $\deg L$ is the degree of the Weil divisor (or 0-cycle) $Z = Z_D$, corresponding to any representant D of L as a Cartier divisor; see §D.2. The degree $\deg L$ does not depend on the representative D. Let $\overline{D} = (D, |\ |)$ be the line bundle corresponding to D, equipped with an hermitian continuous metric $|\ |$, whose Chern current is the $(1,1)$-current $c_1(\overline{D})$; see Definition B.46.

PROPOSITION 1.44. *Let L, D, and $\overline{D}$ be as above. For any meromorphic section s of $\overline{D}$, one defines the current $G(\overline{D}; s) \in {}'\mathcal{D}^{(0,0)}(\mathcal{X})$ as follows:*[12]

$$\begin{aligned} G(\overline{D}; s) &= [-\log|s|^2] \text{ about } \zeta \notin |Z_D|\\ G(\overline{D}; s) &= \Big[\frac{-|s|^{2\lambda}}{\lambda}\Big]_{\lambda=0} \text{ about } \zeta \in |Z_D^+|\\ G(\overline{D}; s) &= \Big[\frac{|s|^{-2\lambda}}{\lambda}\Big]_{\lambda=0} \text{ about } \zeta \in |Z_D^-|. \end{aligned} \tag{1.133}$$

[12]The almost everywhere (a.e.) defined section $-\log|s|^2$ is locally integrable on $\mathcal{X}$ and the current $G(\overline{D}; s)$ defined in (1.133) equals the current $[-\log|s|^2]$ everywhere. Nevertheless, it will be convenient to exploit in the multivariate situation the analytic continuation approach presented here.

The notation $[\ ]_{\lambda=0}$ *means that one considers the meromorphic continuation in* λ *from* $\{\lambda \in \mathbb{C} : \operatorname{Re}\lambda \gg 1\}$ *to a complex plane* $\mathbb{C}$ *of a* $'\mathcal{D}^{(0,0)}(\mathcal{X})$*-valued map* (1.133) *and then takes the coefficient* a_0 *in the Laurent development of such continuation about* $\lambda = 0$. *The current* $G(\overline{D}, s)$ *satisfies in* $'\mathcal{D}^{(1,1)}(\mathcal{X})$ *the Lelong–Poincaré–Green equation*

$$dd^c\big(G(\overline{D}; s)\big) + [\operatorname{div}(s)] = \frac{1}{2i\pi}\overline{\partial}\partial\big(G(\overline{D}; s)\big) + [\operatorname{div}(s)] = c_1(\overline{D}). \tag{1.134}$$

PROOF. Let us examine the second case in (1.133). In particular, we are interested to see in detail what happens in a local chart about $\alpha \in \mathcal{Z}^+$. One has, for $\operatorname{Re}\lambda \gg 1$, that

$$\frac{-|s|^{2\lambda}}{\lambda} = -\frac{\exp(\lambda \log|s|^2)}{\lambda} = -\frac{1}{\lambda} - [\log|s|^2] - \sum_{\kappa=1}^{\infty}\frac{[(\log|s|^2])^{\kappa}}{\kappa!}$$

in $'\mathcal{D}^{(0,0)}(\mathcal{X})$, because all functions $(\log|s|^2)^{\kappa}$ are locally integrable on $\mathcal{X}$ and hence integrable, since $\mathcal{X}$ is compact, for $\kappa \in \mathbb{N}^*$. Let t be a local coordinate centered at α ($t(\alpha) = 0$) and e be a local frame for $\mathcal{X}$ about α. Then s is factorized as $s(t) = h(t)e(t)$, where h is a meromorphic function. One has

$$\begin{aligned} dd^c\Big(\frac{-|s|^{2\lambda}}{\lambda}\Big) &= dd^c\Big(-\frac{|h|^{2\lambda}}{\lambda}|e|^{2\lambda}\Big) = -\frac{\overline{\partial}}{2i\pi}\Big(\Big(|h|^{2\lambda}\frac{\partial h}{h} + \partial\log|e|^2\Big)|e|^{2\lambda}\Big)\\ &= -|e|^{2\lambda}\Big(\lambda\frac{|h|^{2\lambda}}{2i\pi}\frac{\overline{\partial h}}{\overline{h}} \wedge \frac{\partial h}{h} + dd^c\log|e|^2\\ &\qquad + \lambda\Big(|h|^{2\lambda}\frac{\partial h}{h} + \partial\log|e|^2\Big) \wedge \frac{\overline{\partial}\log|e|^2}{2i\pi}\Big). \end{aligned}$$

The meromorphic continuation of such current-valued section s of λ is holomorphic up to the origin and its value at $\lambda = 0$ equals $-[\operatorname{div}(h)] - dd^c\log|e|^2$, given Proposition 1.10 (with $m = 1$). Since $c_1(\overline{D}) = -dd^c\log|e|^2$ (see Definition B.46), the Lelong–Poincaré formula (1.134) is proved. □

Proposition 1.44 suggests the following definition.

DEFINITION 1.45. Let L be a holomorphic line bundle over $\mathcal{X}$ and let s be a meromorphic section of L. A *Green current* with respect to (L, s) is an element $G \in {'\mathcal{D}^{(0,0)}}(\mathcal{X})$ satisfying

$$dd^cG + [\operatorname{div}(s)] = [\varphi] \quad \text{where} \quad \varphi \in C^{\infty}_{1,1}(\mathcal{X}). \tag{1.135}$$

Given a continuous metric $|\ |$ on L, a Green current G with respect to (L, s) is said to be *normalized with respect to* $(\overline{L}, s)$, where $\overline{L} = (L, |\ |)$, if and only if $[\varphi] = c_1(\overline{D})$, where $c_1(\overline{D})$ is the first Chern current of $\overline{L}$.

EXAMPLE 1.46. Let s be a meromorphic section of a metrized holomorphic bundle $\overline{E} = (E, |\ |)$ with $\operatorname{rank}(E) > 1$. Assume also that s is factorized in the form $s = \mathfrak{f} \otimes \mathfrak{u}$, where $\mathfrak{f}$ is a meromorphic section of the line bundle L corresponding to the Cartier divisor attached to s and $\mathfrak{u}$ is a nonvanishing section of $L^{-1} \otimes E$; see Proposition 1.30. Let $|\ |_s$ be the unique hermitian metric on L such that $|\mathfrak{f}|_s = |s|$. One has

$$dd^c([-\log|s|^2]) = dd^c\Big[-\frac{|s|^{2\lambda}}{\lambda}\Big]_{\lambda=0} = -[\operatorname{div}(\mathfrak{f})] - c_1\big(L, |\ |_s\big). \tag{1.136}$$

Given two holomorphic line bundles (L, s) and (L', s') with respective meromorphic sections s and s', the 0-cycles $\mathrm{div}(s)$ and $\mathrm{div}(s')$ are said to *intersect properly* on $\mathcal{X}$ (or *define a proper intersection* on $\mathcal{X}$) if and only if the intersection of their supports $|\mathrm{div}(s)| \cap |\mathrm{div}(s')| = \emptyset$. We describe the proper intersection in terms of 0-cycles or $(1,1)$-currents respectively as

$$\begin{aligned} &\mathrm{div}(s) \bullet \mathrm{div}(s') = \mathrm{div}(1) \quad (\text{with } |\mathrm{div}(1)| = \emptyset) \\ &[\mathrm{div}(s)] \wedge [\mathrm{div}(s')] = 0 \quad (\text{for dimension reasons}). \end{aligned} \tag{1.137}$$

Let now G and G' be normalized Green currents with corresponding line bundles $(\overline{L}, s)$ and $(\overline{L'}, s')$. Recall that the hermitian metrics on L and L' are smooth. It appears illuminating[13] to express the fact that the $(1,1)$-current

$$G \smile G' := G\,[\mathrm{div}(s')] + c_1(\overline{L})\, G', \tag{1.138}$$

satisfies the trivial identity $dd^c(G \smile G') = 0$ (for dimension reasons), as

$$\begin{aligned} dd^c(G \smile G') &= dd^c G \wedge [\mathrm{div}(s')] + c_1(\overline{L}) \wedge dd^c G' \\ &= \big(c_1(\overline{L}) - [\mathrm{div}(s)]\big) \wedge [\mathrm{div}(s')] + c_1(\overline{L}) \wedge \big(c_1(\overline{L'}) - [\mathrm{div}(s')]\big) \\ &= -[\mathrm{div}(s)] \wedge [\mathrm{div}(s')] + c_1(\overline{L}) \wedge c_1(\overline{L'}) \\ &= -[\mathrm{div}(1)] + c_1(\overline{L}) \wedge c_1(\overline{L'}). \end{aligned}$$

Thus $G \smile G'$ is interpreted as a normalized Green current for the section $s \equiv 1$ of the trivial bundle $\mathcal{X} \times \mathbb{C}$, equipped with the flat hermitian structure induced by the canonical hermitian metric on $\mathbb{C}$. This leads to the following definition.

DEFINITION 1.47. Let $\overline{L}$ and $\overline{L}'$ be two holomorphic line bundles equipped with smooth metrics. Let also G and G' be normalized Green currents corresponding to the line bundles $(\overline{L}, s)$ and $(\overline{L'}, s')$. The *Chow arithmetic*[14] *product* of the pairs $(\mathrm{div}(s), G)$ and $(\mathrm{div}(s'), G')$ is defined by

$$\begin{aligned} \big(\mathrm{div}(s), G\big) \smile \big(\mathrm{div}(s'), G'\big) &= \big(\mathrm{div}(1), G \smile G'\big) \\ &= \Big(\mathrm{div}(1), G\,[\mathrm{div}(s')] + c_1(\overline{L})\, G'\Big). \end{aligned} \tag{1.139}$$

From now on let us consider the case of $\mathcal{X} = \mathbb{P}^1_{\mathbb{C}}$. We equip the line bundle $L^{[0]} = \mathcal{O}(1) = \mathcal{O}_{\mathbb{P}^1_{\mathbb{C}}}(1)$ with the Fubini–Study metric $\|\ \|_1$ defined as follows: for any $(\xi_0, \xi_1) \in \mathbb{C}^2$ and $\zeta \in \mathbb{P}^1_{\mathbb{C}}$,

$$\|\xi_0\zeta_0 + \xi_1\zeta_1\|_1 = \frac{|\xi_0\,\zeta_0 + \xi_1\,\zeta_1|}{\sqrt{|\zeta_0|^2 + |\zeta_1|^2}}. \tag{1.140}$$

---

[13]This formal reformulation of a known fact will be justified in the higher dimensional setting. There, multiplication between currents is realized using analytic continuation method, inspired by those described in §1.2.4. It also makes sense in the univariate setting, but within the arithmetic frame that will be presented later in this subsection.

[14]The use of the word *arithmetic* here is not out of place, although there is no arithmetic involved up to this point, since algebraic projective curves are considered over the basis field $\mathbb{C}$. The justification of the term arithmetic used is based on the fact that such an operation will play its full role within the arithmetic setting, for example, when algebraic projective curves considered could be as well assumed as defined over $\mathbb{Q}$ or $\overline{\mathbb{Q}}$.

Then the first Chern form of $\widehat{L^{[0]}} = (L^{[0]}, \| \, \|_1)$ equals

$$\widehat{c} = dd^c(\log(|\zeta_0|^2 + |\zeta_1|^2)) = \frac{1}{2i\pi} \frac{d(\overline{\zeta_1/\zeta_0}) \wedge d(\zeta_1/\zeta_0)}{\left(1 + |\zeta_1/\zeta_0|^2\right)^2}. \tag{1.141}$$

It coincides with the Kähler form on $\mathbb{P}^1_{\mathbb{C}}$ generating the $\mathbb{C}$-vector space of harmonic $(1,1)$-forms on $\mathbb{P}^1_{\mathbb{C}}$. Proposition 1.44, together with Stokes's theorem, implies that

$$\begin{aligned} \int_{\mathbb{P}^1_{\mathbb{C}}} \widehat{c} &= \int_{\mathbb{P}^1_{\mathbb{C}}} dd^c(-\log \|\zeta_0\|_1^2) + \int_{\mathbb{P}^1(\mathbb{C})} [\mathrm{div}(\zeta_0)] \\ &= \int_{\mathbb{P}^1(\mathbb{C})} [\mathrm{div}(\zeta_0)] = \deg(\mathrm{div}(\zeta_0)) = 1. \end{aligned} \tag{1.142}$$

More generally, for any $\delta \in \mathbb{N}^*$, the (Fubini–Study) metric $\| \, \|_\delta$ on the line bundle $\mathcal{O}(\delta)$ (with degree δ) is defined as follows: for $(\xi_0, \dots, \xi_\delta) \in \mathbb{C}^{\delta+1}$ and $\zeta \in \mathbb{P}^1_{\mathbb{C}}$,

$$\Big\| \sum_{\kappa=0}^{\delta} \xi_\kappa \, \zeta_0^\kappa \zeta_1^{\delta-\kappa} \Big\|_\delta = \frac{\left| \sum_{\kappa=0}^{d} \xi_\kappa \zeta_0^\kappa \zeta_1^{\delta-\kappa} \right|}{(\sqrt{(|\zeta_0|^2 + |\zeta_1|^2})^\delta}. \tag{1.143}$$

For any $\delta \in \mathbb{N}^*$, we denote by $\widehat{L_\delta}$ the metrized line bundle $(\mathcal{O}(\delta), \| \, \|_\delta)$, so that $\widehat{L^{[0]}} = \widehat{L_1}$.

Given any $T \in {}'\mathcal{D}^{(0,0)}(\mathbb{P}^1_{\mathbb{C}})$, the current

$$T^{\perp \widehat{c}} = T - \langle T, \widehat{c} \rangle \tag{1.144}$$

satisfies

$$\langle T^{\perp \widehat{c}}, \widehat{c} \rangle = \langle T, \widehat{c} \rangle - \langle T, \widehat{c} \rangle \int_{\mathbb{P}^1_{\mathbb{C}}} \widehat{c} = 0, \tag{1.145}$$

taking into account (1.142). The current (1.144) is said to be the projection of T *orthogonally to harmonic forms.* The following lemma provides a complement to Definition 1.45, when one works on a kählerian compact one-dimensional manifold such as $(\mathbb{P}^1_{\mathbb{C}}, \widehat{c})$.

Lemma 1.48. *Let $\overline{L} = (L, | \, |)$ be a holomorphic line bundle on $\mathbb{P}^1_{\mathbb{C}}$ equipped with a continuous metric and let s be a meromorphic section of L. The current*

$$[-\log |s|^2]^{\perp \widehat{c}} = [-\log |s|^2] - \Big(\int_{\mathbb{P}^1_{\mathbb{C}}} \big(-\log |s(\zeta)|^2 \big) \widehat{c}(\zeta) \Big) \widehat{c}$$

is the unique normalized Green current with respect to $(\overline{L}, s)$ orthogonal to harmonic $(1,1)$-forms.

Proof. The difference T between two such Green currents is necessarily a dd^c-closed current, hence a harmonic current. Since it satisfies $\langle T, \widehat{c} \rangle = 0$, one has $T = 0$. ☐

Now, fix an integer $d \in \mathbb{N}^*$ and let $s = P$ be a holomorphic section of the line bundle $L = \mathcal{O}_{\mathbb{P}^1_{\mathbb{C}}}(d) = \mathcal{O}(d)$. That is, P is a homogeneous polynomial (for the time being with complex coefficients), with degree d in homogeneous variables $z = (z_0, z_1)$. From now on we shall use the notation $z = [z_0 : z_1]$ for points in $\mathbb{P}^1_{\mathbb{C}}$

expressed in terms of homogeneous coordinates. We will assume that $P(X_0, X_1)$ is the homogenization of a polynomial $p \in \mathbb{C}[x]$ with $\deg p = d \in \mathbb{N}^*$, that is,

$$(1.146)\quad p(x) = \sum_{\kappa=0}^{d} \gamma_\kappa x^{d-\kappa} \text{ with } \gamma_0 \in \mathbb{C}^*, \quad P(X) = P(X_0, X_1) = \sum_{\kappa=0}^{d} \gamma_\kappa X_0^\kappa X_1^{d-\kappa}.$$

One has then

$$\operatorname{div}(P) = (\operatorname{div}(P))^+ = \sum_{\{\xi \in \mathbb{C}\,:\, p(\xi)=0\}} \nu_{p,\xi}\, \{[1:\xi]\},$$

where $\nu_{p,\xi}$ denotes the multiplicity of ξ as a zero of p. Thus

$$(1.147)\quad p(x) = \gamma_0 \prod_{\{\xi \in \mathbb{C}\,:\, p(\xi)=0\}} (x-\xi)^{\nu_{p,\xi}}, \quad P(X) = \gamma_0 \prod_{\{\xi \in \mathbb{C}\,:\, p(\xi)=0\}} (X_1 - \xi X_0)^{\nu_{p,\xi}}.$$

Therefore, the geometric complexity of the 0-cycle $\operatorname{div}(P)$ is

$$(1.148)\qquad \deg(\operatorname{div}(P)) = \sum_{\xi \in V(p)} \nu_{p,\xi} = d.$$

Another important geometric object attached to a 0-cycle on $\mathbb{P}^1_{\mathbb{C}}$ is its Chow cycle on a dual copy of $\mathbb{P}^1_{\mathbb{C}}$. See, for example, Remark D.13 for the related concept of *sheaf of Chow ideals* of a cycle Z on a complex manifold of arbitrary dimension. Naturally, the above observation applies to $\operatorname{div}(P)$.

DEFINITION 1.49. Let $(\mathbb{P}^1_{\mathbb{C}})^\star_u$ be a copy of $(\mathbb{P}^1_{\mathbb{C}})_z$, whose homogeneous variables are denoted by (u_0, u_1). That is, the current point is denoted by $u = [u_0 : u_1]$ instead of $z = [z_0 : z_1]$. If

$$Z = \sum_{\alpha = [\alpha_0:\alpha_1] \in \mathbb{P}^1_{\mathbb{C}}} \nu_\alpha \{[\alpha_0 : \alpha_1]\}$$

is a 0-cycle on $(\mathbb{P}^1_{\mathbb{C}})_z$, then its *Chow cycle* $Z^\star$ is the cycle $\operatorname{div}(\Phi^\star)$ on $(\mathbb{P}^1_{\mathbb{C}})^\star_u$, where $\Phi^\star$ is the holomorphic section of $\mathcal{O}_{(\mathbb{P}^1_{\mathbb{C}})^\star}(\deg Z)$ defined, up to multiplication by some $\chi \in \mathbb{C}^*$, as

$$(1.149)\qquad \Phi^\star([u_0 : u_1]) = \prod_{\alpha=[\alpha_0:\alpha_1] \in |Z|} (u_0\alpha_0 + u_1\alpha_1)^{\nu_\alpha} = \Phi^\star(u_0, u_1).$$

$\Phi^\star$ is taken to be a *Chow form* for $\operatorname{div}(P)$.

EXAMPLE 1.50. If $Z = \operatorname{div}(P)$, where $P(X_0, X_1)$ is the homogenization of $p \in \mathbb{C}[x]$ with degree exactly $d \in \mathbb{N}^*$, then one can take $\Phi^\star(u_0, u_1) = P(u_1, -u_0)$.

REMARK 1.51. Consider the product space $\mathbf{X} = (\mathbb{P}^1_{\mathbb{C}})_\zeta \times (\mathbb{P}^1_{\mathbb{C}})^\star_u$. Its subset consisting of points $(z, u) = \big([z_0 : z_1], [u_0, u_1]\big)$ such that the supports of divisors $|Z|$ and $|\operatorname{div}_{(\mathbb{P}^1_{\mathbb{C}})_\zeta}(u_0\,\zeta_0 + u_1\,\zeta_1)|$ intersect at $[z_0 : z_1]$ on $(\mathbb{P}^1_{\mathbb{C}})_\zeta$ is called an incidence set and the points $[z_0 : z_1] \in |Z|$ and $[u_0 : u_1] \in (\mathbb{P}^1_{\mathbb{C}})^\star_u$ are then said to be *incident.* Taking into account the multiplicity $\nu_{Z,[z_0:z_1]}$ for such a pair of incident points, one gets a 1-cycle Inc_Z on the two-dimensional product manifold $\mathbf{X}$. Then the Chow cycle $Z^\star$ is the direct image of Inc_Z through the proper projection $(\zeta, u) \mapsto u$; see §D.1.1 for the definition of the direct image of a cycle.

The concept of a Chow cycle is of major interest in the arithmetic setting. To illustrate this, we assume for the rest of the current section that the polynomial p in (1.147), whose homogenization equals P, belongs to $\mathbb{Z}[X]$, with $\gcd(\gamma_0, \dots, \gamma_d) = 1$. Moreover, one has that

$$\Phi^\star(U_0, U_1) = P(-U_1, U_0) \in \mathbb{Z}[U_0, U_1]$$

with jointly coprime coefficients. Then $P \in \mathbb{Z}[X_0, X_1]$ defines an arithmetic cycle $\mathrm{div}_{\mathrm{arithm}}(P)$ with pure codimension 1 on the arithmetic scheme

$$\mathbb{X} = \mathrm{Spec}\big(\mathrm{Proj}(\mathbb{Z}[X_0, X_1])\big).$$

Note that arithmetic cycles with pure codimension 2 (or dimension -1) on $\mathbb{X}$ are formal (finite) linear combinations

$$\sum_{\mathfrak{p}\ \text{prime}} \nu_{\mathfrak{p}}\{\mathfrak{p}\} \quad \text{with } \nu_{\mathfrak{p}} \in \mathbb{Z}^*.$$

Let us now explain how to measure the *arithmetic complexity* of $\mathrm{div}_{\mathrm{arithm}}(P)$ for the metrized line bundle $\widehat{L^{[0]}}$. For that purpose, we will need first the following lemma.[15]

LEMMA 1.52. *Let $\delta \in \mathbb{N}^*$. For any holomorphic section*

$$Q(z) = \prod_{\kappa=1}^{\delta} (\alpha_{\kappa_0} z_1 - \alpha_{\kappa_1} z_0), \ \textit{with}\ \alpha_\kappa = (\alpha_{\kappa,0}, \alpha_{\kappa,1}) \in \mathbb{C}^2 \setminus \{(0,0)\},$$

of the line bundle $\mathcal{O}(\delta)$, one has

$$\int_{\mathbb{P}^1_{\mathbb{C}}} \big(-\log \|Q(\zeta)\|^2_\delta\big)\, \widehat{c}(\zeta) = \delta - \sum_{\kappa=1}^{\delta} \log \|\alpha_\kappa\|^2. \tag{1.150}$$

PROOF. It is enough to prove the assertion for $\delta = 1$. The assertion for $\delta \in \mathbb{N}^*$ follows from the fact that log is a homomorphism from $(\mathbb{R}^+, \times)$ to $(\mathbb{R}, +)$. Let then $(u_0, u_1) \in \mathbb{C}^2 \setminus \{(0,0)\}$ and $Q(z) = \langle\!\langle u, z \rangle\!\rangle = u_0 z_0 + u_1 z_1$. It follows from the invariance of the Kähler form $\widehat{c}$ under the action of $U(2)$ on $\mathbb{P}^1_{\mathbb{C}}$ that one can assume that $(u_0, u_1) = \|u\|(0,1)$. Then, (1.142) implies that

$$\begin{aligned}
\int_{\mathbb{P}^1_{\mathbb{C}}} \big(-\log \|\langle\!\langle u, \zeta \rangle\!\rangle\|^2_1\big)\, \widehat{c}(\zeta) &= \frac{1}{2i\pi} \int_{\mathbb{C}} \Big(-\log \frac{|t|^2}{1+|t|^2}\Big) \frac{d\bar{t} \wedge dt}{(1+|t|^2)^2} - \log \|u\|^2 \\
&= -\int_0^\infty \Big(\log \frac{r}{1+r}\Big) \frac{dr}{(1+r)^2} - \log \|u\|^2 \\
&= -\int_0^1 \log \rho \, d\rho - \log \|u\|^2 = 1 - \log \|u\|^2. \qquad \square
\end{aligned}$$

Let us introduce the unit sphere $\mathbb{S} = \mathbb{S}^3$ in $\mathbb{C}^2 \simeq \mathbb{R}^4$ with its canonical normalized measure $\sigma_{\mathbb{S}}$. We now state the following proposition.

[15]This lemma corresponds to one, among many, of the formulations of Crofton's averaging formula in the univariate setting; see §D.5.2.

PROPOSITION 1.53. *Let P be the homogenization of $p \in \mathbb{Z}_d[x]$ as in (1.147). Let G be the unique $(0,0)$ Green current normalized with respect to $(\widehat{L_d}, P)$. Furthermore, assume that G is orthogonal to harmonic forms, that is, $\langle G, \widehat{c}\rangle = 0$. Let $\Phi^\star(u) = P(u_1, -u_0)$ as Chow form for* div(P). *Then*

$$(1.151)\quad \frac{d}{2} + \int_{\mathbb{S}} \log |P(\varpi_0, \varpi_1)|^2 \, d\sigma_{\mathbb{S}}(\varpi) \\ = \log|\Phi^\star(u)| + \frac{1}{2}\int_{\mathbb{P}^1_{\mathbb{C}}} ([-\log \|\langle\!\langle u, \zeta\rangle\!\rangle\|_1^2] \smile G)(\zeta) \geq 0.$$

PROOF. Since $\langle G, \widehat{c}\rangle = 0$, one has

$$(1.152)\quad \int_{\mathbb{P}^1_{\mathbb{C}}} ([-\log \|\langle\!\langle u, \zeta\rangle\!\rangle\|_1^2] \smile G)(\zeta) = \int_{\mathbb{P}^1_{\mathbb{C}}} (-\log \|\langle\!\langle u, \zeta\rangle\!\rangle\|_1^2) \, [\mathrm{div}(P)] \geq -d \log \|u\|^2.$$

The final inequality follows from the positivity of

$$(u, \zeta) \longmapsto -\log \frac{\|\langle\!\langle u, \zeta\rangle\!\rangle\|_1^2}{\|u\|^2}$$

on $(\mathbb{P}^1_{\mathbb{C}})^\star \times \mathbb{P}^1_{\mathbb{C}}$ because of the Cauchy–Schwarz inequality. Moreover,[16] one has

$$(1.153)\quad \log|\Phi^\star(u)|^2 + \int_{\mathbb{P}^1_{\mathbb{C}}} (-\log (\|\langle\!\langle u, \zeta\rangle\!\rangle\|_1^2) \, [\mathrm{div}(P)] = d + \int_{\mathbb{S}} \log |P(\varpi_0, \varpi_1)|^2 \, d\sigma_{\mathbb{S}}(\varpi).$$

The equality in (1.151) follows. Observe now that the left-hand side does not depend on the choice of u. The positivity of the expression in (1.151) follows then from the inequality in (1.152) with a convenient choice of u. We need to use here only the following immediate observation: since $p \in \mathbb{Z}[x]$, any nonzero coefficient γ of P is such that $|\gamma| \in \mathbb{N}^*$, hence $|\gamma| \geq 1$. □

Proposition 1.53 suggests the following definition.

DEFINITION 1.54. Let $p \in \mathbb{Z}[X]$ be as in (1.146) with jointly coprime coefficients γ_κ, $\kappa = 0, \ldots, d$. The *projective logarithmic height* of the codimension 1 0-cycle $\mathrm{div}_{\mathrm{arithm}}(P)$ in $\mathrm{Spec}\big(\mathrm{Proj}(\mathbb{Z}[X_0, X_1])\big)$ is the positive real number defined by

$$(1.154)\quad h_{\mathrm{proj}}\big(\mathrm{div}_{\mathrm{arithm}}(P)\big) = h_{\mathrm{proj}}\big(\mathrm{div}_{\mathrm{arithm}}(p)\big) = d + \int_{\mathbb{S}} \log |P(\varpi_0, \varpi_1)|^2 \, d\sigma_{\mathbb{S}}(\varpi).$$

REMARK 1.55. Let $u = [u_0 : u_1] \in (\mathbb{P}^1_{\mathbb{Q}})^\star$, where $(u_0, u_1) \in \mathbb{Z}^2 \setminus \{(0,0)\}$, be such that $\mathrm{div}\big(z \mapsto \langle\!\langle u, z\rangle\!\rangle\big)$ and div(P) intersect properly on $\mathbb{P}^1_{\mathbb{C}}$. This implies that the corresponding arithmetic cycles intersect properly along a codimension 2 cycle on $\mathrm{Spec}\big(\mathrm{Proj}(\mathbb{Z}[X_0, X_1])\big)$. For such $u = [u_0 : u_1] \in (\mathbb{P}^1_{\mathbb{Q}})^\star$ the first contribution $\log|\Phi^\star(u_0, u_1)|$ on the right-hand side of (1.151) is a marker for the degree of the intersection for codimension 2 (or dimension -1) cycle of $\mathrm{div}_{\mathrm{arithm}}(P)$ and of $\mathrm{div}_{\mathrm{arithm}}\big(z \mapsto \langle\!\langle u, z\rangle\!\rangle\big)$. Both entries involved in the cup product (1.139), where $L = L_d$, $L' = L^{[0]}$, $s = P$, $s' = \langle\!\langle u, z\rangle\!\rangle$ and the metrics are Fubini–Study metrics, contribute to the splitting (1.151) of the projective logarithmic height. This interpretation of (1.151) is valid, provided that the intersection is understood on

[16] Formula (1.151) is the incarnation in the univariate situation of a more general important statement; see [**Ph91**, Theorem 2].

$\mathrm{Spec}(\mathrm{Proj}(\mathbb{Z}[X_0, X_1]))$ instead of on $\mathbb{P}^1_{\mathbb{C}}$. The above construction, as described here, is the univariate version of the approach developed in [**GS90, BGS94**] in the multivariate projective setting.

The use of the metrized line bundle $\widehat{L^{[0]}}$ as a standard measurement tool (a marker, see Definition 1.54), for the arithmetic complexity of the codimensional cycle $\mathrm{div}_{\mathrm{arithm}}(P)$, where $P \in \mathbb{Z}[X_0, X_1]$ is the homogenization of $p \in \mathbb{Z}_d[X]$ as in (1.146) having jointly coprime coefficients, has an alternative. This alternative is based on the natural choice to equip the holomorphic line bundle $L^{[0]} = \mathcal{O}(1)$ with its *canonical metric* defined by

$$|\xi_0\zeta_0 + \xi_1\zeta_1|_{\mathrm{can}} = \frac{|\xi_0\,\zeta_0 + \xi_1\,\zeta_1|}{\max(|\zeta_0|, |\zeta_1|)}. \tag{1.155}$$

The metric $|\ |_{\mathrm{can}}$ is not smooth as the Fubini–Study metric $|\ |_1$ was. It is just continuous and the first Chern current of the metrized line bundle $\overline{L^{[0]}} = (L^{[0]}, |\ |_{\mathrm{can}})$ has its support included in $\mathbb{C}$. Furthermore, it equals, when expressed in the affine complex coordinate $t = \zeta_1/\zeta_0$,

$$c_{\mathrm{can}} = c_1(\overline{L^{[0]}}) = \sigma_{\mathbb{S}^1}\,\frac{d\bar{t} \wedge dt}{2i}, \tag{1.156}$$

where $\sigma_{\mathbb{S}^1}$ is the normalized Haar measure on the unit circle $\mathbb{S}^1$ in $\mathbb{C} \simeq \mathbb{R}^2$. Observe that the positive $(1,1)$-current c_{can} is also realized as the limit of positive $(1,1)$-currents on $\mathbb{P}^1_{\mathbb{C}}$

$$c_{\mathrm{can}} = \lim_{N\to+\infty} \frac{[T_N^*(\widehat{c})]}{N} = \lim_{N\to+\infty} \frac{\left[dd^c\left(\log\left(|\zeta_0|^{2N} + |\zeta_1|^{2N}\right)\right)\right]}{N}, \tag{1.157}$$

where

$$T_N([\zeta_0 : \zeta_1]) = [\zeta_0^N : \zeta_1^N].$$

For any $N \in \mathbb{N}^*$, the direct image of the arithmetic, codimension 1 cycle $\mathrm{div}_{\mathrm{arithm}}(P)$ of $\mathrm{Spec}(\mathrm{Proj}(\mathbb{Z}[X_0, X_1]))$ by T_N is the codimension 1 cycle $\mathrm{div}_{\mathrm{arithm}}(P^{[N]})$. Recall that $P^{[N]}$ is the holomorphic section of $\mathcal{O}(Nd)$ defined by

$$P^{[N]}(\zeta) = \gamma_0^N \prod_{\{\xi\in\mathbb{C}\,:\,p(\xi)=0\}} \prod_{\kappa=0}^{N-1} \left(\zeta_1 - \exp(2i\pi\kappa/N)\,\xi\,\zeta_0\right) \in \mathbb{Z}[X_0, X_1]. \tag{1.158}$$

Formula (1.157) suggests the following notion for *normalized logarithmic projective height*, known also as *canonical logarithmic height.*

DEFINITION 1.56. Let $p \in \mathbb{Z}_d[x]$, $P \in \mathbb{Z}_d[X_0, X_1]$ be as in (1.146), and let their coefficients be jointly coprime. The following limit

$$\lim_{N\to+\infty} \frac{1}{N}\,\frac{h_{\mathrm{proj}}\left((T_N)_*(\mathrm{div}_{\mathrm{arithm}}(P))\right)}{N\,d} = \lim_{N\to+\infty} \frac{1}{N}\,\frac{h_{\mathrm{proj}}\left(\mathrm{div}_{\mathrm{arithm}}(P^{[N]})\right)}{Nd}$$

exists and the positive real number $h_{\text{can}}(\text{div}_{\text{arithm}}(P)) = h_{\text{can}}(\text{div}_{\text{arithm}}(p))$ is defined asymptotically by

$$\begin{aligned}\frac{h_{\text{can}}(\text{div}_{\text{arithm}}(P))}{d} &= \frac{h_{\text{can}}(\text{div}_{\text{arithm}}(p))}{d} \\ &= \lim_{N\to+\infty} \frac{1}{N}\,\frac{h_{\text{proj}}\big((T_N)_*(\text{div}_{\text{arithm}}(P))\big)}{N\,d} \\ &= \lim_{N\to+\infty} \frac{1}{N}\,\frac{h_{\text{proj}}\big(\text{div}_{\text{arithm}}(P^{[N]})\big)}{Nd}.\end{aligned} \tag{1.159}$$

The canonical logarithmic height of $\text{div}_{\text{arithm}}(P)$ (or $\text{div}_{\text{arithm}}(p)$) is a marker for the arithmetic complexity of the codimension 1 arithmetic cycle $\text{div}_{\text{arithm}}(P)$, if one takes as a standard measurement tool the metrized line bundle $\overline{L^{[0]}}$. What we described above is an Arakelov-type approach to complexity through a normalization process operating on the projective logarithmic height (Definitions 1.54 and 1.56) in the univariate setting. Another approach, involving Berkovich analytifications of $\mathbb{P}^1_{\mathbb{Q}}$ at $\mathfrak{p}$-adic (ultrametric) places, as companions of the algebraic Riemann surface $\mathbb{P}^1_{\mathbb{C}}$ (at the archimedean place), will be briefly presented in §6.6.2. The construction of the projective logarithmic height in the multivariate setting (see [**GS90**, **BGS94**]), together with the realization of normalized Green currents through the analytic continuation approach, provides the opportunity to express in some particular cases the projective logarithmic height of a hypersurface of $\mathbb{P}^n_{\mathbb{C}}$ defined over $\mathbb{Q}$ as the value at the origin of an Igusa zeta function; see for example [**CaM00**, **BY98**]. On the other hand, the normalized projective logarithmic height of an arithmetic codimension 1 cycle in $\text{Spec}\big(\text{Proj}(\mathbb{Z}[X_0,\dots,X_n])\big)$ is related to a very important convex (or concave, depending of the point of view one adopts) function on the real affine space $\mathbb{R}^n$; see [**Ro00**, **PasR04**, **Mai00**, **Gua18**, **GuaS23**]. Let us state here this result in the univariate case.

PROPOSITION 1.57. *Let $p \in \mathbb{Z}_d[X]$ be as in* (1.146), *whose coefficients are jointly coprime. The functions*

$$\begin{aligned}R_p &: x \in \mathbb{R} \longmapsto \frac{1}{2\pi}\int_0^{2\pi} \log|p(e^{x+i\theta})|\,d\theta, \\ \rho_p &: x \in \mathbb{R} \longmapsto -\frac{1}{2\pi}\int_0^{2\pi} \log|p(e^{-x-i\theta})|\,d\theta\end{aligned} \tag{1.160}$$

are respectively convex and concave, piecewise affine functions. Furthermore, one has

$$\begin{aligned}h_{\text{can}}(\text{div}_{\text{arithm}}(p)) &= R_p(0) = -\rho_p(0) \\ &= \log|\gamma_0| + \sum_{\{\xi\in\mathbb{C}\,:\,p(\xi)=0\}} \nu_{p,\xi}\, \max(0,\log|\xi|).\end{aligned} \tag{1.161}$$

The real number $\exp\big(h_{\text{can}}(\text{div}_{\text{arithm}}(p))\big) = \exp(R_p(0)) = \exp(-\rho_p(0))$ *is called the Mahler measure of* $p \in \mathbb{Z}_d[x]$, *whose coefficients are coprime.*

PROOF. The function R_p is piecewise affine, increasing, and hence convex. This claim follows from the fact that for any $\alpha \in \mathbb{C}^*$ and any $r > 0$,

$$\frac{1}{2i\pi}\int_0^{2\pi} \log|re^{i\theta} - \alpha|\,d\theta = (\log|\alpha|)\,\chi_{]0,|\alpha|]}(r) + (\log r)\,\chi_{]|\alpha|,+\infty[}(r).$$

The last equality in (1.161) is a direct consequence of the above identity. The first equality in (1.161) follows from the definition of the projective logarithmic height (see Definition 1.54), then of its normalization (see Definition 1.56), combined with (1.157). □

1.3.4. Residues, residue currents, and Gorenstein $\mathbb{K}$-algebras. According to the algebraic point of view we adopt in this section, it appears that the concept of residue appeals to that of Gorenstein $\mathbb{K}$-algebra.

DEFINITION 1.58. Let $\mathbb{K}$ be a field and let $\boldsymbol{P}$ be a commutative $\mathbb{K}$-algebra with finite dimension as a $\mathbb{K}$-vector space. The $\mathbb{K}$-algebra $\boldsymbol{P}$ is said to be a *Gorenstein* $\mathbb{K}$-*algebra* if and only if there exists a $\mathbb{K}$-linear map $\rho \ : \boldsymbol{P} \to \mathbb{K}$ such that the symmetric bilinear form

$$(a,b) \in \boldsymbol{P} \longmapsto \rho(ab) \in \mathbb{K} \tag{1.162}$$

(where $\boldsymbol{P}$ is considered with its structure of $\mathbb{K}$-vector space) is nondegenerate.

We present in this section some examples of Gorenstein $\mathbb{K}$-algebras, where the symmetric bilinear nondegenerate form ρ is specified in terms of residue theory in the univariate case. Such examples illustrate either local, global, or semilocal situations.

PROPOSITION 1.59. *Let f be a nonzero element in the maximal ideal $\mathfrak{M}$ of $\mathcal{O} = \mathcal{O}_{\mathbb{C},0}$. The $\mathbb{C}$-algebra $\mathcal{O}/(f)$, with finite dimension $\nu_{f,0} \in \mathbb{N}^*$, is Gorenstein, with the symmetric bilinear form ρ taken as*

$$\rho \ : (a,b) \in \mathcal{O}/(f) \otimes \mathcal{O}/(f) \longmapsto \rho(ab) = \Big\langle \Big(\overline{\partial}\Big(\frac{1}{f}\Big)\Big)_0, (AB)(\zeta)\, d\zeta \Big\rangle, \tag{1.163}$$

where A, B are arbitrary representatives of the classes a and b modulo f in $\mathcal{O}$.

PROOF. The result is an immediate consequence of Proposition 1.16 (see in particular Remark 1.17), in the particular case $m = 1$. In fact, it is enough to invoke here formula (1.50). □

REMARK 1.60. Since

$$\Big\langle \Big(\overline{\partial}\Big(\frac{1}{f}\Big)\Big)_0, (AB)(\zeta)\, d\zeta \Big\rangle$$

depends only on the germ of 1-cohomology class of $\zeta \mapsto (AB)(\zeta)\, d\zeta$ at the origin, the duality bracket realized by the symmetric nondegenerate bilinear form ρ in (1.162) is also called the $\mathcal{O}/(f)$ *dualizing local cohomological residue.*

When $\mathbb{K}$ is a field, elements in $\mathbb{K}[X, X^{-1}]$ are called *Laurent polynomials* with coefficients in $\mathbb{K}$. Moreover, if $\mathbb{K}$ is a subfield of $\mathbb{C}$, then let $p \in \mathbb{K}[X]$ be a polynomial with degree $d \in \mathbb{N}^*$ and let $F \in \mathbb{K}[X, X^{-1}]$ be a Laurent polynomial with $\text{card}(\text{Supp}(F)) \in \mathbb{N}^*$. Then the finite sets $V(p) = \{\alpha \in \mathbb{C} : p(\alpha) = 0\}$, respectively $V_{\mathbb{T}}(F) = \{re^{i\theta} \in \mathbb{C}^* : F(re^{i\theta}) = 0\}$, are nonempty subsets of the algebraic closure of $\mathbb{K}$ in $\mathbb{C}$, and one has the following proposition.

PROPOSITION 1.61. *The $\mathbb{K}$-algebras $\mathbb{K}[X]/(p)$ and $\mathbb{K}[X, X^{-1}]/(F)$ are finitely dimensional as $\mathbb{K}$-vector spaces and Gorenstein, with symmetric bilinear form ρ*

defined by

$$\begin{aligned}\rho(a,b) &= \sum_{\alpha\in V(p)} \Big\langle \Big(\overline{\partial}\Big(\frac{1}{p}\Big)\Big)_\alpha, (AB)(\zeta)\,d\zeta\Big\rangle \\ &= \mathrm{Res}_{\mathbb{K}[X]/\mathbb{K}}\begin{bmatrix}(AB)(X)\,dX \\ p\end{bmatrix}\end{aligned} \tag{1.164}$$

for any representatives A, B of $\bar{A}, \bar{B}$ in $\mathbb{K}[X]$ in the first case and by

$$\begin{aligned}\rho(a,b) &= \sum_{re^{i\theta}\in V_{\mathbb{T}}(F)} \Big\langle \Big(\overline{\partial}\Big(\frac{1}{F}\Big)\Big)_{re^{i\theta}}, (AB)(\zeta)\,\frac{d\zeta}{\zeta}\Big\rangle \\ &:= \mathrm{Res}_{\mathbb{K}[X,X^{-1}]/\mathbb{K}}\begin{bmatrix}(AB)(X)\,dX \\ F\end{bmatrix}_{\mathbb{T}}\end{aligned} \tag{1.165}$$

for any representatives A, B of $\bar{A}, \bar{B}$ in $\mathbb{K}[X, X^{-1}]$ in the second case.

PROOF. In the first case, when $p \in \mathbb{K}[X]$, the vector space $\boldsymbol{P} = \mathbb{K}[X]/(p)$ is finite dimensional, with dimension $\dim \boldsymbol{P} = \deg p = d$. Moreover, it follows from Corollary 1.35 that

$$\mathrm{Res}_{\mathbb{K}[X]/\mathbb{K}}\begin{bmatrix}(AB)(X)\,dX \\ p\end{bmatrix} := \sum_{\alpha\in V(p)} \Big\langle \Big(\overline{\partial}\Big(\frac{1}{p}\Big)\Big)_\alpha, (AB)(\zeta)\,d\zeta\Big\rangle$$

(see Definition 1.36) belongs to $\mathbb{K}$ and depends only on the classes a and b of A and B in $\mathbb{K}[X]/(p)$. The fact that the symmetric bilinear map defined by (1.164) is nondegenerate follows from Proposition 1.38. Any $F \in \mathbb{K}[X, X^{-1}]$ is factorized as the product $F = p(X)/X^\mu$, with $p \in \mathbb{K}[X]$, $p(0) \in \mathbb{K}^*$, and $\mu \in \mathbb{Z}$. The algebra $\mathbb{K}[X, X^{-1}]/(F)$ is a finitely dimensional $\mathbb{K}$-algebra with dimension $\deg p$, that is, the length of the closed convex hull of $\mathrm{Supp}(F)$ in $\mathbb{R}$. For any Laurent polynomials $A, B \in \mathbb{K}[X, X^{-1}]$, let $(AB)(X) = q(X)X^\nu$ with $q \in \mathbb{K}[X]$, $q(0) \in \mathbb{K}^*$ and $\nu \in \mathbb{Z}$. Then

$$\begin{aligned}&\sum_{re^{i\theta}\in V_{\mathbb{T}}(F)} \Big\langle \Big(\overline{\partial}\Big(\frac{1}{F}\Big)\Big)_{re^{i\theta}}, (AB)(\zeta)\,\frac{d\zeta}{\zeta}\Big\rangle \\ &\qquad = \sum_{re^{i\theta}\in V_{\mathbb{T}}(F)} \Big\langle \Big(\overline{\partial}\Big(\frac{1}{p}\Big)\Big)_{re^{i\theta}}, q(\zeta)\zeta^{\mu+\nu-1}\,d\zeta\Big\rangle.\end{aligned}$$

Let us discuss the following alternatives:

- if $\mu + \nu - 1 \in \mathbb{N}$, then

$$\begin{aligned}&\sum_{re^{i\theta}\in V_{\mathbb{T}}(F)} \Big\langle \Big(\overline{\partial}\Big(\frac{1}{F}\Big)\Big)_{re^{i\theta}}, (AB)(\zeta)\,\frac{d\zeta}{\zeta}\Big\rangle \\ &\qquad = \mathrm{Res}_{\mathbb{K}[X]/\mathbb{K}}\begin{bmatrix}q(X)\,X^{\mu+\nu-1}\,dX \\ p\end{bmatrix} - \Big\langle \Big(\overline{\partial}\Big(\frac{1}{p}\Big)\Big)_0, q(\zeta)\zeta^{\mu+\nu-1}\,d\zeta\Big\rangle\,;\end{aligned}$$

- if $\mu + \nu - 1 = -\chi \in \mathbb{Z}^-$, then

$$\begin{aligned}&\sum_{re^{i\theta}\in V_{\mathbb{T}}(F)} \Big\langle \Big(\overline{\partial}\Big(\frac{1}{F}\Big)\Big)_{re^{i\theta}}, (AB)(\zeta)\,\frac{d\zeta}{\zeta}\Big\rangle \\ &\qquad = \mathrm{Res}_{\mathbb{K}[X]/\mathbb{K}}\begin{bmatrix}q(X)\,dX \\ X^\chi p\end{bmatrix} - \Big\langle \Big(\overline{\partial}\Big(\frac{1}{\zeta^\chi p}\Big)\Big)_0, q(\zeta)\,d\zeta\Big\rangle.\end{aligned}$$

In both cases, the complex number $\rho(a,b)$ defined by (1.165) is independent of the choice of the representatives A, B of a, b modulo (F) respectively and belongs to $\mathbb{K}$ according to Corollary 1.35 and formula (1.14). The nondegeneracy of the bilinear symmetric form ρ follows again from Proposition 1.38, since F divides G in $\mathbb{K}[X, X^{-1}]$ if and only if $G(\zeta)/F(\zeta) \in H(\mathbb{C}^*)$. $\square$

The fact that for $F \in \mathbb{K}[X, X^{-1}]$ satisfying $\mathrm{card}\big(\mathrm{Supp}(F)\big) \geq 1$ the vector space $K[X, X^{-1}]/(F)$ is a Gorenstein algebra with symmetric bilinear form given by (1.165) suggests the following definition.

DEFINITION 1.62. Let $\mathbb{K}$ be a subfield of $\mathbb{C}$. For any pair of Laurent polynomials (F, G) assumed to be coprime in $\mathbb{K}[X, X^{-1}]$, the element of $\mathbb{K}$ defined by

$$\mathrm{Res}_{\mathbb{K}[X,X^{-1}]/\mathbb{K}} \begin{bmatrix} G(X)\, dX \\ F \end{bmatrix}_{\mathbb{T}} := \sum_{re^{i\theta} \in V_{\mathbb{T}}(F)} \Big\langle \Big(\overline{\partial}\Big(\frac{1}{F}\Big)\Big)_{re^{i\theta}}, G(\zeta)\, \frac{d\zeta}{\zeta} \Big\rangle \tag{1.166}$$

is called the (global) *toric residue* with respect to $F \in \mathbb{K}[X, X^{-1}]$ of the Kähler differential $G(X)\, dX \in \Omega_{\mathbb{K}[X,X^{-1}]/\mathbb{K}}$.

Now, let $\overline{E} = (E, |\ |)$ be a hermitian holomorphic bundle with rank m over a noncompact Riemann surface $\mathcal{X}$ and $s \in \mathcal{O}_{\mathcal{X}}(\mathcal{X}, E)$. For h in $\mathcal{O}_{\mathcal{X}}(\mathcal{X}, \mathbb{C})$ and $u \in \mathcal{D}(\mathcal{X}, E)$, let

$$\rho(u, h) = \big\langle R^s_{|\ |}, h\, u \big\rangle,$$

where the $(0,1)$-current

$$R^s_{|\ |} = \Big[\overline{\partial} |s|^{2\lambda}\, \frac{s^*}{2i\pi\, |s|^2}\Big]_{\lambda=0} \tag{1.167}$$

has been introduced in Proposition 1.27. On observes that $\rho(u,h)$ depends only on the class $\mathfrak{u}$ of u modulo $\rfloor_s \mathcal{D}^{(1,0)}(\mathcal{X}, E^* \wedge E^*)$ and on the class $\mathfrak{h}$ of h modulo $\rfloor_s \mathcal{O}_{\mathcal{X}}(\mathcal{X}, E^*)$. Let

$$\rho\ :\ \frac{\mathcal{D}^{(1,0)}(\mathcal{X}, E)}{\rfloor_s \mathcal{D}^{(1,0)}(\mathcal{X}, E^* \wedge E^*)} \times \frac{\mathcal{O}_{\mathcal{X}}(\mathcal{X})}{\rfloor_s \mathcal{O}_{\mathcal{X}}(\mathcal{X}, E^*)} \longrightarrow \mathbb{C}$$

be the bilinear form defined by

$$\rho(\mathfrak{u}, \mathfrak{h}) = \big\langle R^s_{|\ |}, h\, u \big\rangle \in \mathbb{C}, \tag{1.168}$$

where $u \in \mathcal{D}^{(1,0)}(\mathcal{X}, E^*)$ is any representative of $\mathfrak{u}$ modulo $\rfloor_s \mathcal{D}^{(1,0)}(\mathcal{X}, E^* \wedge E^*)$ and h is any representative of $\mathfrak{h}$ modulo $\rfloor_s \mathcal{O}_{\mathcal{X}}(\mathcal{X}, E^*)$. It follows from Proposition 1.29 that

$$\Big(\rho(\mathfrak{u}, \mathfrak{h}) = 0 \quad \text{for any } \mathfrak{u} \in \frac{\mathcal{D}^{(1,0)}(\mathcal{X}, E^*)}{\rfloor_s \mathcal{D}^{(1,0)}(\mathcal{X}, E^* \wedge E^*)}\Big) \Longrightarrow \mathfrak{h} = 0. \tag{1.169}$$

The ring $\mathcal{O}_{\mathcal{X}}(\mathcal{X})/\rfloor_s \mathcal{O}_{\mathcal{X}}(\mathcal{X}, E^*)$ is always Cohen–Macaulay as a zero-dimensional ring; see §A.3.3 for the notion of Cohen–Macaulay ring or sheaf of ideals. Nevertheless, in the case where it is finitely dimensional as a $\mathbb{C}$-vector space, namely when $\mathrm{card}\,(s^{-1}(\{0\})) < +\infty$, the $\mathbb{C}$-algebra $\mathcal{O}_{\mathcal{X}}(\mathcal{X})/\rfloor_s \mathcal{O}_{\mathcal{X}}(\mathcal{X}, E^*)$ fails, in general, to be Gorenstein when $m > 1$.[17] Duality is materialized instead by the implication (1.169), where the bilinear form ρ acts this time asymmetrically according to (1.168). The current $R^s_{|\ |}$ is not $\overline{\partial}$-closed when $m > 1$, but a positive point is that its expression (1.167) in terms of the holomorphic section s does not require any information about $s^{-1}(\{0\})$.

The situation will be drastically different in the higher dimensional setting. Let us consider the case of quotients of $\mathbb{C}[X_1, \dots, X_n]$ as an example. It remains true that if $p_1, \dots, p_m \in \mathbb{C}[X_1, \dots, X_n]$ are such that $\mathbb{C}[X_1, \dots, X_n]/(p_1, \dots, p_m)$ is finitely dimensional as a $\mathbb{C}$-vector space, then $\mathbb{C}[X_1, \dots, X_n]/(p_1, \dots, p_m)$ is Cohen–Macaulay. But such a $\mathbb{C}$-algebra fails in general to be $\mathbb{C}$-Gorenstein, since being $\mathbb{C}$-Gorenstein relies on the quasi-regularity of the sequence $(p_1, \dots, p_m)$, which, geometrically speaking, appeals to the concept of complete intersection; see §6.2.1. The situation is much worse if $\dim(\mathbb{C}[X_1, \dots, X_n]/(p_1, \dots, p_m)) > 0$, since in this case, the existence of embedded associated primes could prevent in general $\mathbb{C}[X_1, \dots, X_n]/(p_1, \dots, p_m)$ from being Cohen–Macaulay; see §A.3.3. For all these reasons, the current materialization of duality, as in (1.169) or at least some intermediate result but with a tractable current $R^s_{|\ |}$, will be of major interest. It is therefore now time to jump from the univariate situation to the multivariate one.

[17] Observe nevertheless that if $\mathcal{X} = U$ is a domain in $\mathbb{C}$ and s is a section of the trivial m bundle $\mathbb{C}^m$, then one has $\rfloor_s \mathcal{O}(U, (\mathbb{C}^m)^*) = p\,\mathcal{O}(U)$ for some polynomial map p. This implies that the $\mathbb{C}$-algebra $\mathcal{O}(U)/\rfloor_s \mathcal{O}(U) = \mathcal{O}(U)/p\,\mathcal{O}(U)$ is Gorenstein. Similarly, $\mathbb{K}[X]/(p_1, \dots, p_m)$ and $\mathbb{K}[X, X^{-1}]/(F_1, \dots, F_m)$ remain Gorenstein $\mathbb{K}$-algebras when the p_j (respectively the F_j) are polynomials (respectively Laurent polynomials) with coefficients in $\mathbb{K}$.

CHAPTER 2

Residue currents: a multiplicative approach

2.1. De Rham complex and iterated residues

We recall briefly in this section the cohomological theory of residue (through the formulation of residue formula (2.23)), which goes back to the pioneering work of H. Poincaré [**P1887**]. This theory was developed later by G. de Rham [**DR32, DR36**], and more recently by J. Leray [**Le59**]. For an updated overview of H. Poincaré's contribution to multivariate residue calculus, accompanied by updated bibliography, the reader may refer to [**Y10**]. Let us quote here once more one of the fundamental remarks of H. Poincaré [**P1887**]:

> the integral over a closed surface depends only on singular curves which lie on this surface.

This remark is our guideline throughout the whole section. We focus here on bridging the cohomological and current approaches to the residue. The current approach to the residue is developed and presented in §2.2 and §2.3.3.

2.1.1. Division of forms and de Rham's lemma. Let $\mathcal{X}$ be a complex N-dimensional manifold.[1] Let also $U \subset \mathcal{X}$ be an open subset in $\mathcal{X}$ and let S be a codimension 1 submanifold in $\mathcal{X}$. That is, for any $z \in S$, one can find a neighborhood $U \subset \mathcal{X}$ of z in $\mathcal{X}$, a holomorphic function s on U such that $ds \neq 0$ in U, and $S \cap U = \{z \in U \,:\, s(z) = 0\}$. The function $s \in \mathcal{O}_{\mathcal{X}}(U)$ is called a *defining function* for the complex submanifold S (with codimension 1) in U.

DEFINITION 2.1. Let $\omega \in C^\infty_k(\mathcal{X} \setminus S)$ be a $\mathbb{C}$-valued, *smooth* differential k-form on the real, $2N$-dimensional, underlying differentiable manifold $(\mathcal{X} \setminus S)_{\mathbb{R}}$, $0 \leq k \leq 2N$. The k-form ω is called *semimeromorphic on $\mathcal{X}$ with polar set contained in S* if and only if, for any $z \in S$ there exist a neighborhood U of z and a defining holomorphic function s for S in U, together with a positive integer $\nu = \nu(z)$ such that the k-form $s^\nu \omega$ extends as an element in $C^\infty_k(U)$, that is, $s^\nu\omega$ is a smooth differential k-form on $U_{\mathbb{R}}$. When $\nu(z) = 1$ for any $z \in S$, ω is called *semimeromorphic on $\mathcal{X}$, with at most simple poles along the hypersurface S.*

Let us recall here the crucial division lemma, due to G. de Rham (see [**DR32, DR36**]) but used as a crucial operational tool in [**Le59**].

[1] From now on in this chapter, $\mathcal{X}$ will denote either an N-dimensional complex manifold or an irreducible reduced complex analytic space with pure dimension N; see respectively §A.1.1 or §A.4.1. We will treat similar semilocal problems (think of $\mathcal{X}$, for example, as a neighborhood of a specified point, or more generally of a closed analytic subset V in some ambient manifold or irreducible reduced analytic space) or global ones. Of course, obstructions for the definition of non-trivial global objects (such as holomorphic functions on $\mathcal{X}$, or more generally global holomorphic sections for some holomorphic vector bundle $E \to \mathcal{X}$) have always to be taken into account.

LEMMA 2.2. *Let $\mathcal{X}$, S be as above. Let also ω be a semimeromorphic k-form ($k \geq 1$) on $\mathcal{X}$ with at most simple poles along the hypersurface S, such that $d\omega = 0$ in $\mathcal{X} \setminus S$. Then, for any $z \in S$ one can find an open neighborhood $U = U_z \subset \mathcal{X}$ of z, a holomorphic function $s = s_z$ on U, such that one has simultaneously $ds \neq 0$ in U and $S \cap U = \{z \in U \, : \, s(z) = 0\}$. Furthermore, one can also find a smooth $(k-1)$-form $r[\omega] = r_z[\omega]$ and a smooth k-form $\sigma[\omega] = \sigma_z[\omega]$ on U, such that*

$$\omega = \frac{ds}{s} \wedge r[\omega] + \sigma[\omega] \,. \tag{2.1}$$

Moreover, $(r[\omega])_{|S\cap U} \in C^\infty_{k-1}(S \cap U)$ and satisfies $d((r[\omega])_{|S\cap U}) = 0$. For each $z \in S$, let

$$(\mathrm{res}_{S\cap U_z}[\omega])_z := (r[\omega])_{|S\cap U_z} \tag{2.2}$$

be a locally defined, smooth, $(k-1)$-form. All locally defined forms $(\mathrm{res}_{S\cap U_z}[\omega])_z$ on S fit together as a d-closed, smooth $(k-1)$-form on S, called the Poincaré–Leray residue of ω on S, and is denoted as $\mathrm{Res}_{[\mathcal{X};S]}(\omega)$.

PROOF. Let s be the local defining function of S in a neighborhood U of a point $z \in S$, with $ds \neq 0$ in U. Since ω is d-closed, one has $d(s\,\omega) = ds \wedge \omega$ in $U \setminus S$, which implies that $d(s\,\omega) \wedge ds = 0$ in $U \setminus S$. This holds also in U, since $s\,\omega$ extends as a smooth k-form in U, as a consequence of the assumption that the differential form $\omega \in C^\infty_k(U \setminus S)$ is semimeromorphic with at most simple poles along the hypersurface S. This implies, since $ds \neq 0$ on U, that there exists $\sigma[\omega] \in C^\infty_k(U)$ such that

$$d(s\,\omega) = ds \wedge \sigma[\omega] \tag{2.3}$$

in U. Thus, one has in U that

$$ds \wedge (s\,\omega - s\,\sigma[\omega]) = s\,(d(s\,\omega) - ds \wedge \sigma[\omega]) = 0.$$

This in turn implies, following the same argument as above and using again the fact that $ds \neq 0$ on U, that there exists $r[\omega] \in C^\infty_{k-1}(U)$ satisfying the identity

$$s\,\omega - s\,\sigma[\omega] = ds \wedge r[\omega] \tag{2.4}$$

in U. This leads to the division formula (2.1). Next, applying the differential operator d to (2.4), we get

$$ds \wedge \omega - ds \wedge \sigma[\omega] - s\,d(\sigma[\omega]) = -ds \wedge d(r[\omega]),$$

that is

$$d(s\,\omega) = ds \wedge \sigma[\omega] + s\,d(\sigma[\omega]) - ds \wedge d(r[\omega]).$$

If one uses (2.3), then it follows that $s\,d(\sigma[\omega]) = ds \wedge d(r[\omega])$ in U holds, implying that the restriction $(\mathrm{res}_{S\cap U}[\omega])_{|S\cap U}$ of $r[\omega]$ to the submanifold $S\cap U$ is d-closed. The same reasoning shows that, if z and w are two distinct points in S with $U_z \cap U_w \neq \emptyset$, then

$$(ds_z)_{|S} \wedge (r_z[\omega] - r_w[\omega])_{|S} = 0 \tag{2.5}$$

in $U_z \cap U_w$. Therefore, one has in $U_z \cap U_w$ that

$$(r_z[\omega])|_{S\cap U_z} = \mathrm{res}_{S\cap U_z}[\omega] = (r_w[\omega])|_{S\cap U_w} = \mathrm{res}_{S\cap U_w}[\omega],$$

which shows that the collection of $\{r_z[\omega],\ z \in S\}$, represents the collection of restrictions to different U_z, $z \in S$, of a smooth d-closed, $(k-1)$-form on the complex submanifold S of $\mathcal{X}$. This completes the proof of the lemma. $\square$

When ω is a semimeromorphic k-form on $\mathcal{X}$ with poles of order at most ν_0, $\nu_0 > 1$ along S, a division formula similar to the first assertion in Lemma 2.2 exists.

LEMMA 2.3. *Let $\mathcal{X}$, S be as above. Let also ω be a semimeromorphic k-form ($k \geq 1$) on $\mathcal{X}$, with poles with order at most ν_0, $\nu_0 > 1$, along the hypersurface S, such that $d\omega = 0$ in $\mathcal{X} \setminus S$. For any point $z \in S$, one can find an open neighborhood $U = U_z \subset \mathcal{X}$ of z, a holomorphic defining function $s = s_z$ for S in U which satisfies $ds \neq 0$ there, and differential forms $r_{\nu_0}[\omega] = r_{z,\nu_0}[\omega] \in C^\infty_{k-1}(U)$, $\sigma_{\nu_0}[\omega] = \sigma_{[z,\nu_0]}[\omega] \in C^\infty_k(U)$ such that one has in U*

$$\omega = \frac{ds}{s^{\nu_0}} \wedge r_{\nu_0}[\omega] + \frac{\sigma_{\nu_0}[\omega]}{s^{\nu_0 - 1}}\,. \tag{2.6}$$

PROOF. Take U and s in a neighborhood of z to be as in the proof of Lemma 2.2. One has in $U \setminus S$ that $d(s^{\nu_0}\,\omega) = \nu_0\, d(s^{\nu_0 - 1}) \wedge \omega$. Hence, $ds \wedge d(s^{\nu_0}\omega) = 0$ in U, since $s^{\nu_0}\omega$ extends to U as a smooth k-form. Since $ds \neq 0$ on U, it follows that $d(s^{\nu_0}\omega) = \nu_0\, ds \wedge \sigma_{\nu_0}[\omega]$ for some $\sigma_{\nu_0}[\omega] \in C^\infty_k(U)$. Then

$$ds \wedge \big(s^{\nu_0}\,\omega - s\,\sigma_{\nu_0}[\omega]\big) = s\Big(\frac{d(s^{\nu_0}\omega)}{\nu_0} - ds \wedge \sigma_{\nu_0}[\omega]\Big) = 0$$

in U. This implies, since again $ds \neq 0$ there, the existence of $r_\nu[\omega] \in C^\infty_{k-1}(U)$ such that $s^{\nu_0} - s\,\sigma_{\nu_0}[\omega] = ds \wedge r_{\nu_0}[\omega]$. □

Lemma 2.2 suggests strengthening the concept of semimeromorphic form with polar set along a smooth complex hypersurface S and with at most simple poles along S (see Definition 2.1) as follows.

DEFINITION 2.4. Let S be a smooth complex hypersurface in $\mathcal{X}$. A semimeromorphic k-form $\omega \in C^\infty_k(\mathcal{X} \setminus S)$ with at most simple poles along S is said to be *S-logarithmic* if and only if $d\omega \in C^\infty_{k+1}(\mathcal{X} \setminus S)$ is a semimeromorphic $(k+1)$-form with at most simple poles on S.

REMARK 2.5. If a semimeromorphic form $\omega \in C^\infty_k(\mathcal{X} \setminus S)$ happens to be S-logarithmic, then one can repeat the argument used to prove Lemma 2.2 and express ω in a neighborhood U of $z \in S$ as in (2.1), using the defining function s for S in the open neighborhood U of z. Moreover, the local forms $(r_z[\omega])_{|S \cap U_z}$, for $z \in S$, still correspond to the local expressions of $\mathrm{Res}_{[\mathcal{X};S]}(\omega) \in C^\infty_{k-1}(S)$, which is also called the *Poincaré–Leray* (in short *Leray*) *residue of ω on S*; see [**Pol72**]. In this case, the Poincaré–Leray residue is not d-closed anymore as in Lemma 2.2, unless $d\omega = 0$ in $U \setminus S$. This remark justifies the terminology used in Definition 2.4.

EXAMPLE 2.6 (Poincaré–Leray residue of $(N,0)$-meromorphic forms). Let Ω be a domain in $\mathbb{C}^N$ and let $f \in \mathcal{O}_{\mathbb{C}^N}(\Omega)$ be such that $df \not\equiv 0$ on any irreducible component of the reduced hypersurface $f^{-1}(\{0\})$. Consider

$$\mathcal{X} := \Omega \setminus (f^{-1}(\{0\}))_{\mathrm{sing}} = \Omega \setminus \{z \in \Omega;\ f(z) = df(z) = 0\}.$$

Then the set $S \subset \mathcal{X}$ defined by

$$S = \{z \in \mathcal{X} \,:\, f(z) = 0\}$$

is a complex submanifold of $\mathcal{X}$. For any holomorphic $(N,0)$-form

$$\omega = h(z)\, dz_1 \wedge \cdots \wedge dz_N$$

in Ω one can realize

$$\mathrm{Res}_{[\mathcal{X}:S]}\Big(\frac{\omega(z)}{f(z)}\Big)$$

in the open set

$$u_j = \{z \in S \, : \, (\partial f/\partial z_j)(z) \neq 0\},$$

for any $j = 1, \dots, N$, as follows:

$$\Big(\mathrm{Res}_S\Big(\frac{\omega(z)}{f(z)}\Big)\Big)_{|u_j} = (-1)^{j-1}\Big(\frac{h\, dz_1 \wedge \cdots \wedge \widehat{dz_j} \wedge \cdots \wedge dz_N}{\partial f/\partial z_j}\Big)_{|u_j}. \tag{2.7}$$

EXAMPLE 2.7 (The *Bochner–Martinelli kernel*). Let $\widehat{c}(z) = dd^c \log \|z\|^2$ be the Kähler form on the projective space[2] $\mathbb{P}^N_{\mathbb{C}}$, where z_j's denote the homogeneous coordinates, namely $z = [z_0 : \cdots : z_N]$. The Kähler form $\widehat{c}(z)$ is also the first Chern form of the metrized line bundle $\widehat{\mathcal{O}(1)} = (\mathcal{O}(1), \|\ \|_1)$, where the metric is the Fubini–Study metric, as defined by (1.140) in the univariate setting. The Fubini–Study normalized volume form $\widehat{c}^{\,N}$ factorizes in $\mathbb{P}^N_{\mathbb{C}}$ as

$$(\widehat{c}(z))^N = g(z,\overline{z})\, E(z) \wedge \overline{E(z)}\,, \tag{2.8}$$

where

$$E(z) = \sum_{j=0}^{N} (-1)^{j-1} z_j dz_0 \wedge \cdots \wedge \widehat{dz_j} \wedge \cdots \wedge dz_N$$

denotes the *Euler form* in $\mathbb{P}^N_{\mathbb{C}}$. The coefficient $(z,\overline{z}) \to g(z,\overline{z})$ is a smooth $(-2N)$-homogeneous function in $\mathbb{C}^{N+1} \setminus \{(0,\dots,0)\}$, which contributes to make the form ω, defined below, invariant under the torus action $(z_0,\dots,z_N) \mapsto (tz_0,\dots,tz_N)$, $t \in \mathbb{C}^*$, on $\mathbb{C}^{N+1} \setminus \{(0,\dots,0)\}$. In the punctured projective space $\mathcal{X} = \mathbb{P}^{N+1}_{\mathbb{C}} \setminus \{(0,\dots,0)\}$ which splits as $(\mathbb{C}^{N+1} \setminus \{(0,\dots,0)\}) \cup \mathbb{P}^N_{\mathbb{C}}$ consider, for each $j = 0,\dots,N$, the open set U_j defined by

$$U_j = \bigcup_{\substack{(z_0,\dots,z_N) \in \mathbb{C}^{N+1} \\ z_j \neq 0}} \{t(z_0,\dots,z_N) \, : \, t \in \mathbb{C}^*\} \cup \{[z_0 : \cdots : z_N]\}.$$

One interprets[3] here the point $[z_0 : \cdots : z_N]$ as the intersection of the projective line $\overline{\{(tz_0,\dots,tz_N) \, : \, t \in \mathbb{C}^*\}}$ with the hyperplane at infinity in $\mathbb{P}^{N+1}_{\mathbb{C}}$, materialized here as $\mathbb{P}^N_{\mathbb{C}}$. One has $\mathcal{X} = \bigcup_{j=0}^N U_j$. In each $U_j \setminus \mathbb{P}^N_{\mathbb{C}}$, consider the $(N+1,N)$ semimeromorphic form with at most simple poles along $\mathbb{P}^N_{\mathbb{C}}$ defined by

$$\omega_j = \frac{dz_j}{z_j} \wedge (\widehat{c}(z))^N.$$

Since

$$\frac{dz_j}{z_j} \wedge E(z) = dz_0 \wedge \cdots \wedge dz_N$$

in $U_j \setminus \mathbb{P}^N_{\mathbb{C}}$, it follows from (2.8) that

$$\omega_j(z) = dz_0 \wedge \cdots \wedge dz_N \wedge g(z,\overline{z})\, \overline{E(z)}$$

---

[2] We refer to Example A.1.2(2), for a first introduction to the (complex) projective space $\mathbb{P}^N_{\mathbb{C}}$, based on the point of view that will be developed in this monograph.

[3] This reflects the point of view on which the concept of *perspective* during the Renaissance was based; see [**ShcTY06**].

there. Thus the forms ω_j, $j = 0, \dots, N$, fit together in $\mathcal{X} \setminus \mathbb{P}^N_{\mathbb{C}}$ as the $(N+1, N)$ closed, smooth differential form

$$\omega(z) = \frac{\overline{z_j} dz_j}{|z_j|^2} \wedge (\widehat{c}(z))^N = \frac{\sum_{j=0}^N \overline{z_j} dz_j}{\|z\|^2} \wedge (\widehat{c}(z))^N = \partial[\log \|z\|^2] \wedge (\widehat{c}(z))^N.$$

This smooth closed form $\omega \in \mathbb{C}^\infty_{2N+1}(\mathcal{X} \setminus \mathbb{P}^N_{\mathbb{C}})$ is semimeromorphic in $\mathcal{X}$, with at most simple poles along $\mathbb{P}^N_{\mathbb{C}}$. Moreover, one has that

$$\operatorname{Res}_{[\mathbb{P}^{N+1}_{\mathbb{C}};\mathbb{P}^N_{\mathbb{C}}]}(\omega) = \widehat{c}^{\,N}. \tag{2.9}$$

The $(N+1, N)$-differential form $\omega/(2i\pi) \in C^\infty_{2N+1}(\mathbb{C}^{N+1} \setminus \{(0, \dots, 0)\})$ extends as a locally integrable form to $\mathbb{C}^{N+1}$. This extension is known as the *Bochner–Martinelli kernel* in $\mathbb{C}^{N+1}$ and it will play an essential role throughout this monograph; see in particular §4.2.1. The nonstandard approach to the derivation of the Bochner–Martinelli kernel presented above was presented in [**ShcTY06**, §3].

It will be important for us to deal later on with the case of the closed hypersurface $S \subset \mathcal{X}$ failing to be smooth. Namely, it will be the case when the closed analytic subset $S_{\text{sing}} \subset S$, with codimension at least 2 (see §A.3.2), fails to be empty. Since we deal with meromorphic $(p, 0)$-forms for the moment, as in our two previous Examples 2.6 and 2.7, we need to specify the notion of *meromorphic logarithmic form on $\mathcal{X}$ with poles along a closed analytic hypersurface* S; see [**AlT01, AlT08**]. We use for that purpose the terminology already used in Definition 2.4.

DEFINITION 2.8. Let $\mathcal{X}$ be an N-dimensional complex manifold, and let S be a closed analytic hypersurface in $\mathcal{X}$. A meromorphic $(p, 0)$-form on $\mathcal{X}$, $1 \le p \le N$, that is, a meromorphic section of the sheaf Ω^p, is said to be *S-logarithmic* if and only if, for any $z \in S$, one can find an open neighborhood $U = U_z$ of z in $\mathcal{X}$ and a *reduced defining holomorphic function*[4] s in U, such that the meromorphic forms $(s\,\omega)_{|U}$ and $(s\,d\omega)_{|U}$ (or, which is equivalent, the meromorphic forms $(s\,\omega)_{|U}$ and $(ds \wedge \omega)_{|U}$) are pole-free in U.

Let $\mathcal{X}^\circ = \mathcal{X} \setminus S_{\text{sing}}$ and $S^\circ = S \setminus S_{\text{sing}} = S_{\text{reg}}$. Then S° is a codimension 1 submanifold of the N-dimensional complex manifold $\mathcal{X}^\circ$. Given a meromorphic S-logarithmic form, one can define (see Remark 2.5) the Poincaré–Leray residue $\operatorname{Res}_{[\mathcal{X}^\circ;S^\circ]}(\omega_{|\mathcal{X}^\circ})$ as a holomorphic $(p-1, 0)$-form on $S^\circ \subset \mathcal{X}^\circ$. Such holomorphic $(p-1, 0)$-form on S° induces the definition of a $(p, 1)$-current on $\mathcal{X}^\circ$, supported by S°, and such an action is defined as

$$\begin{aligned} &(2.10)\quad \big[\operatorname{Res}_{[\mathcal{X}^\circ;S^\circ]}(\omega_{|\mathcal{X}^\circ})\big] : \\ &\varphi \in \mathcal{D}^{(N-p,N-1)}(\mathcal{X}^\circ) \longmapsto \int_{S^\circ} \operatorname{Res}_{[\mathcal{X}^\circ;S^\circ]}(\omega_{|\mathcal{X}^\circ}) \wedge \varphi = \int_S \operatorname{Res}_{[\mathcal{X}^\circ;S^\circ]}(\omega_{|\mathcal{X}^\circ}) \wedge \varphi. \end{aligned}$$

PROPOSITION 2.9. *Let $\mathcal{X}$, S and ω be as in Definition 2.8. Let $\mathcal{X}^\circ = \mathcal{X} \setminus S_{\text{sing}}$ and $S^\circ = S_{\text{reg}}$. The $(p, 1)$-current $T^\circ = \big[\operatorname{Res}_{[\mathcal{X}^\circ;S^\circ]}(\omega_{|\mathcal{X}^\circ})\big] \in {}'\mathcal{D}^{(p,1)}(\mathcal{X}^\circ)$ defined in (2.10) extends as a current $T \in {}'\mathcal{D}^{(p,1)}(\mathcal{X})$ supported by S and $\overline{\partial}$-closed. The extension T is defined locally around S, in terms of any reduced defining function*

[4]By this we mean a defining function s for S in U, such that $ds \not\equiv 0$ on any irreducible component of $S \cap U$.

s for S in a neighborhood U of $z \in S$, by

$$T = R_S \wedge [\omega] : \varphi \in \mathcal{D}^{(N-p,N-1)}(U) \longmapsto \Big[-\int_U |s|^{2\lambda} \frac{\omega}{2i\pi} \wedge \overline{\partial}\varphi \Big]_{\lambda=0}, \tag{2.11}$$

where the meaning of $[\ ]_{\lambda=0}$ was introduced in Chapter 1, through the analytic continuation method from $\{\lambda \in \mathbb{C};\ \mathrm{Re}\,\lambda \gg 1\}$. In other words

$$T_{|U} = \Big[\overline{\partial}\Big(\frac{|s|^{2\lambda}}{2i\pi}\Big) \wedge \omega\Big]_{\lambda=0} = \Big[\frac{\lambda}{2i\pi}\,|s|^{2\lambda}\,\overline{\partial}\log|s|^2 \wedge \omega\Big]_{\lambda=0}. \tag{2.12}$$

PROOF. For $z \in S$ let U be an open neighborhood of z in $\mathcal{X}$, and let s be a reduced holomorphic defining function for S in U. One may also assume that $\zeta : U \to \mathbb{D}^N$ is a centered system of coordinates on $\mathcal{X}$ with $\zeta(z) = 0$ so that there is a formal Bernstein–Sato equation[5]

$$\mathcal{Q}(\lambda, \zeta, \partial/\partial\zeta)(s(\zeta))^{\lambda+1} = b(\lambda)(s(\zeta))^{\lambda}, \tag{2.13}$$

where $b(\lambda)$ is a polynomial, called minimal *local Bernstein–Sato polynomial* for the germ of $s(\zeta)$ at the origin. In this case the roots of $b(z)$ belong $-\mathbb{Q}^+$. The differential operator $\mathcal{Q}$ is a differential operator in $(\partial/\partial\zeta_1, \dots, \partial/\partial\zeta_N)$ (in short $\partial/\partial\zeta$), whose coefficients are polynomial in λ and analytic in local coordinates ζ in a neighborhood of the origin. Namely, for any meromorphic coefficient $w(\zeta)/s(\zeta)$ of $\omega(\zeta)$ along the $d\zeta_J$ with cardinality $|J| = p$, one deduces from (2.13) the following equality in the sense of distributions:

$$\Big(\overline{\mathcal{Q}}(\lambda, \overline{\zeta}, \partial/\partial\overline{\zeta})\Big)\Big(|s(\zeta)|^{2\lambda}\, w(\zeta)\, \frac{\overline{s(\zeta)}}{s(\zeta)}\Big) = b(\lambda) \times \Big(|s(\zeta)|^{2\lambda} \frac{w(\zeta)}{s(\zeta)}\Big). \tag{2.14}$$

Since the roots of b lie in $-\mathbb{Q}^+$, the $'\mathcal{D}^{(p,0)}(U)$-valued map

$$\lambda \in \{\lambda \in \mathbb{C} : \mathrm{Re}\,\lambda \gg 1\} \longmapsto \Big[|s|^{2\lambda}\omega_{|U}\Big]$$

extends meromorphically to $\mathbb{C}$, with (possible) poles in $\mathbb{Q}^-$. Its value at the origin $\lambda = 0$ is well defined. Moreover, such value does not depend on the choice of the reduced holomorphic function s, since any two such defining functions differ by a multiplicative factor consisting of an invertible holomorphic function χ in U, such that $\lambda \mapsto [|\chi(\zeta)|^{2\lambda}]$ is an entire, current-valued function of λ that is equal to $[1]$ at $\lambda = 0$. It remains to prove that

$$\Big(\overline{\partial}\Big(\Big[|s|^{2\lambda}\omega_{|U}\Big]_{\lambda=0}\Big)\Big)_{|U\setminus S_{\mathrm{sing}}} = (T^{\circ})_{|U\setminus S_{\mathrm{sing}}}.$$

Fix a point $z^{\circ} \in S^{\circ} \cap U$ and take $s_{|U^{\circ}}$ as a holomorphic defining equation for S° in a neighborhood of z°. Let us assume that $U^{\circ} \subset U$ with $ds \neq 0$ in U°. Since ω is a meromorphic logarithmic form with poles on S, one can express (see Remark 2.5) ω in U° as

$$\omega = \frac{ds}{s} \wedge r^{\circ}[\omega] + \sigma^{\circ}[\omega],$$

where $r^{\circ}[\omega]$ and $\sigma^{\circ}[\omega]$ are holomorphic forms with respective bidegrees $(p-1, 0)$ and $(p, 0)$ in U°. Therefore, for λ satisfying $\mathrm{Re}\,\lambda \gg 1$, one has in U° that

$$\overline{\partial}(|s|^{2\lambda}\omega) = \overline{\partial}|s|^{2\lambda} \wedge \frac{ds}{s} \wedge r^{\circ}[\omega] + \overline{\partial}|s|^{2\lambda} \wedge \sigma^{\circ}[\omega].$$

[5]For the existence of such a local formal equation, analogous to (1.6), valid in $\zeta(U)$ provided U is small enough around 0, we refer to [**Be72, Bj74, Ka76**].

For any $\varphi^\circ \in \mathcal{D}^{(N-p,N-1)}(U^\circ)$, one can check immediately that

$$\begin{aligned}\big[\langle \overline{\partial}(|s|^{2\lambda}\omega), \varphi^\circ\rangle\big]_{\lambda=0} &= \Big[\Big\langle \overline{\partial}|s|^{2\lambda} \wedge \frac{ds}{s} \wedge r^\circ[\omega]\, , \varphi^\circ \Big\rangle\Big]_{\lambda=0} \\ &= 2i\pi \int_{U^\circ \cap S^\circ} r^\circ[\omega] \wedge \varphi^\circ = \langle T^\circ, \varphi^\circ\rangle,\end{aligned}$$

which concludes the proof of the proposition. □

Proposition 2.9 suggests the following definition.

DEFINITION 2.10. Let ω be a logarithmic $(p,0)$-form in $\mathcal{X}$, with polar set along the closed hypersurface S. The $(p,0)$-current, expressed locally as

$$\Big[\int_U |s|^{2\lambda}\omega\Big]_{\lambda=0}, \tag{2.15}$$

where s is any reduced holomorphic defining function for S in U, is called the *standard extension of* $[\omega]$ *from* $\mathcal{X} \setminus S$ and is denoted by $\mathbf{1}_{\mathcal{X}\setminus S} \cdot [\omega]$. Note that one can take $s = 1$ in case $S \cap U = \emptyset$. One has then

$$\overline{\partial}\Big(\frac{\mathbf{1}_{\mathcal{X}\setminus S} \cdot [\omega]}{2i\pi}\Big) = R_S \wedge [\omega], \tag{2.16}$$

according to Proposition 2.9.

We point out here that the standard extension current $\mathbf{1}_{\mathcal{X}\setminus S} \cdot [\omega]$ can also be realized locally as

$$\big\langle \mathbf{1}_{\mathcal{X}\setminus S} \cdot [\omega]\, , \varphi \big\rangle = \lim_{\varepsilon \to 0^+} \Big(\int_{U \cap \{|s| \geq \varepsilon\}} \omega \wedge \varphi\Big), \quad \forall \varphi \in \mathcal{D}^{(N-p,N)}(U), \tag{2.17}$$

where s is any reduced defining holomorphic function for S in U. The Mellin transform realizes the bridge between the locally explicit formulas (2.15) and (2.17) describing the action of the current $\mathbf{1}_{\mathcal{X}\setminus S} \cdot [\omega]$.

2.1.2. Leray coboundary morphism, cohomological residue formulae. As a preamble let us first introduce some pieces of notation. Given a complex manifold $\mathcal{Y}$, we denote by $\mathscr{C}_k^\infty(\mathcal{Y}, \mathbb{C})$, for any $0 \leq k \leq 2\dim \mathcal{Y}$, the group of k-dimensional smooth chains $c = \sum_\iota \nu_\iota \{\gamma_\iota\}$, $\nu_\iota \in \mathbb{C}$ on the corresponding differentiable manifold $\mathcal{Y}_{\mathbb{R}}$. Similarly, we denote by

$$\begin{aligned}\mathscr{Z}_k^\infty(\mathcal{Y}, \mathbb{C}) &= \mathrm{Ker}\big(\partial : \mathscr{C}_{k+1}^\infty(\mathcal{Y}, \mathbb{C}) \to \mathscr{C}_k^\infty(\mathcal{Y}, \mathbb{C})\big), \\ \mathscr{B}_k^\infty(\mathcal{Y}, \mathbb{C}) &= \partial\big(C_{k+1}^\infty(\mathcal{Y}, \mathbb{C})\big)\big) \subset \mathscr{Z}_k^\infty(\mathcal{Y}, \mathbb{C})\end{aligned}$$

the subgroups of (smooth) k-cycles and k-boundaries respectively, where ∂ denotes the boundary morphism. The support of a k-chain $c = \sum_\iota \nu_\iota\{\gamma_\iota\} \in \mathscr{C}_k^\infty(\mathcal{Y})$ is defined as the union of closed subsets $\gamma_\iota(\Delta_k)$, which are smooth images of the simplexes $\Delta_k = \{t \in \mathbb{R}^k \, : \, t_1 + \cdots + t_k \leq 1\}$. The support of the k-chain c is denoted by $|c|$. The notions of boundary and coboundary of a k-chain lead to the description of the kth homology group

$$H_k(\mathcal{Y}, \mathbb{C}) \simeq \frac{\mathscr{Z}_k^\infty(\mathcal{Y}, \mathbb{C})}{\mathscr{B}_k^\infty(\mathcal{Y}, \mathbb{C})},$$

for any $0 \leq k \leq 2\dim \mathcal{Y}$. Given a complex N-dimensional manifold $\mathcal{X}$, together with a codimension 1 complex submanifold $S \subset \mathcal{X}$, the *Leray coboundary morphism* from $H_\bullet(S, \mathbb{C})$ to $H_\bullet(\mathcal{X} \setminus S, \mathbb{C})$, is defined as follows.

DEFINITION 2.11. Let $\bar{\gamma} \in H_{k-1}(S, \mathbb{C})$ be a $(k-1)$-cycle for $1 \le k \le 2N-1$. A k-cycle $\Gamma \in \mathscr{Z}_k(\mathcal{X} \setminus S, \mathbb{C})$ represents $\delta^{[\mathcal{X};S]}(\bar{\gamma})$, that is, $\delta^{[\mathcal{X};S]}(\bar{\gamma}) = \overline{\Gamma}$ if and only if there exists $c \in \mathscr{C}^{\infty}_{k+1}(\mathcal{X}, \mathbb{C})$ satisfying $|\partial c| \subset \mathcal{X} \setminus S$ and a representative γ of $\bar{\gamma}$ in $\mathscr{Z}^{\infty}_{k-1}(S, \mathbb{C})$, such that

- $|c|$ intersects transversally S along $|\gamma|$,
- $\partial\sigma - \Gamma \in \mathscr{B}_k(\mathcal{X} \setminus S, \mathbb{C})$.

EXAMPLE 2.12.

(1) When $\mathcal{X} = \mathbb{C}$ and $S = \{0\}$, $\delta^{[\mathbb{C};\{0\}]}(\overline{\{0\}})$ is the 1-homology class of the cycle $\{t \in [0,1] \mapsto e^{2i\pi t}\}$.

(2) When $\mathcal{X} = \mathbb{P}^{N+1}_{\mathbb{C}}$ and $S = \mathbb{P}^{N}_{\mathbb{C}}$ is its hyperplane at infinity (see Example 2.7) a representative for the class $\delta^{[\mathbb{P}^{N+1}_{\mathbb{C}};\mathbb{P}^{N}_{\mathbb{C}}]}(\overline{[\mathbb{P}^{N}_{\mathbb{C}}]})$ in $H_{2N+1}(\mathbb{P}^{N+1}_{\mathbb{C}} \setminus \mathbb{P}^{N}_{\mathbb{C}})$ is the $(2N+1)$-cycle $\{\Gamma\}$. Here $\Gamma : \Delta_{2N+1} \to \mathbb{S}^{2N+1}(1)$ is a parametrization of the unit sphere centered at the origin in $\mathbb{C}^{N+1} \simeq \mathbb{R}^{2(N+1)} = \mathbb{P}^{N+1}_{\mathbb{C}} \setminus \mathbb{P}^{N}_{\mathbb{C}}$.

Fix now $m \in \{1, \dots, N\}$. Let $S_1, \dots, S_m$ be a collection of m codimension 1 complex submanifolds (called also hypersurfaces) in $\mathcal{X}$ *intersecting there transversally* . It means that for any $z \in S_1 \cup \cdots \cup S_m$ there exists an open neighborhood U of z in $\mathcal{X}$, together with holomorphic defining functions $s_1, \dots, s_m$ for $S_1, \dots, S_m$ respectively such that, for any subset $J \subset \{1, \dots, m\}$,

$$\Big\{z \in \Big(\bigcap_{j \in J} S_j\Big) \cap U \, : \, \Big(\bigwedge_{j \in J} ds_j\Big)(z) = 0\Big\} = \emptyset. \tag{2.18}$$

An ordering of $\{S_1, \dots, S_m\}$ given by $(S_1, \dots, S_m)$ and setting $S_0 = \mathcal{X}$ lead to the embedding of

$$\Sigma_k = \Big(\bigcap_{\kappa \le k} S_\kappa\Big) \setminus \Big(\bigcup_{\kappa > k} S_\kappa\Big) \hookrightarrow \mathcal{Y}_k = \Big(\bigcap_{\kappa \le k-1} S_\kappa\Big) \setminus \Big(\bigcup_{\kappa > k} S_\kappa\Big) \tag{2.19}$$

as a codimension 1 complex submanifold into a $((N-(k-1))$-dimensional complex manifold for any $k = 1, \dots, m$. One has also that $\mathcal{Y}_k \setminus \Sigma_k = \Sigma_{k-1}$ by construction. Then any $\bar{\gamma} \in H_k(\Sigma_m, \mathbb{C})$,

$$\delta^{[\mathcal{Y}_m;\Sigma_m]}(\bar{\gamma}) \in H_{k+1}(\mathcal{Y}_m \setminus \Sigma_m) = H_k(\Sigma_{k-1}, \mathbb{C}),$$

for any $1 \le k \le 2N - m$. If $\boldsymbol{S} = (S_1, \dots, S_m)$, then one can iterate this construction and define

$$\begin{aligned} \delta^{[\mathcal{X};\boldsymbol{S}]}(\bar{\gamma}) &= \big(\delta^{[\mathcal{Y}_1;\Sigma_1]} \circ \cdots \circ \delta^{[\mathcal{Y}_m;\Sigma_m]}\big)(\bar{\gamma}) \\ &\in H_{k+m}(\Sigma_0, \mathbb{C}) = H_{k+m}\big(\mathcal{X} \setminus (S_1 \cup \cdots \cup S_m), \mathbb{C}\big). \end{aligned} \tag{2.20}$$

As we introduced for semimeromorphic form in $\mathcal{X}$ the concept of being S-*logarithmic* when S was a smooth complex hypersurface of $\mathcal{X}$ (see Definition 2.4), one needs to introduce the concept of being $\boldsymbol{S}$-multilogarithmic, when one has $m \le N$ and $S_1, \dots, S_m$ are m codimension 1 complex submanifolds which intersect transversally on $\mathcal{X}$. Let us for that denote, for $j = 1, \dots, m$,

$$\widehat{|\boldsymbol{S}|}_j = S_1 \cup \cdots \cup \widehat{S_j} \cup \cdots \cup S_m.$$

DEFINITION 2.13. Let $m \le n$, $S_1, \dots, S_m$ be m codimension 1 complex submanifolds intersecting transversally on $\mathcal{X}$, according to condition (2.18) around each point $z \in |\boldsymbol{S}|$. Given $0 \le k \le 2N$, a *semimeromorphic k-form with polar set contained in* $|\boldsymbol{S}|$ is an element in $C^{\infty}_k(\mathcal{X} \setminus |\boldsymbol{S}|)$ such that each $z \in |\boldsymbol{S}|$ admits a neighborhood U, with reduced holomorphic defining functions $s_1, \dots, s_m$ respectively for $S_1, \dots, S_m$ in U, such that $(s_1 \cdots s_m)^\nu \omega$ extends from $U \setminus |\boldsymbol{S}|$ as a smooth form in U for some positive integer $\nu = \nu(z)$. Such a semimeromorphic form with polar set contained in $|\boldsymbol{S}|$ is said to be $\boldsymbol{S}$*-multilogarithmic*, if and only if, for each $z \in |\boldsymbol{S}|$, there exists an open neighborhood of z in $\mathcal{X}$, together with holomorphic defining functions $s_1, \dots, s_m$ respectively for $S_1, \dots, S_m$ in U, such that, for any $j = 1, \dots, m$, the restrictions of $s_j\omega$ and $s_j d\omega$ to U are of the form

$$\sum_{\kappa=1}^{m} \omega_{j,\kappa}, \quad \omega_\kappa \text{ semimeromorphic, with } \operatorname{Pol}(\omega_{j,\kappa}) \subset \widehat{|\boldsymbol{S}|}_\kappa. \tag{2.21}$$

Given $S_1, \dots, S_m$ m codimension 1 complex submanifolds which intersect transversally in $\mathcal{X}$, and a k semimeromorphic form ω, with $m \leq k \leq 2N$, which is $\boldsymbol{S}$-multilogarithmic and such that $d\omega = 0$ in $\mathcal{X} \setminus |\boldsymbol{S}|$, one can define iteratively, starting from $\omega \in C_k^\infty(\mathcal{X} \setminus |\boldsymbol{S}|)$,

(2.22)
$$\begin{aligned}\operatorname{Res}_{[\mathcal{X};\boldsymbol{S}]}(\omega) = \Big(\operatorname{Res}_{[\mathcal{Y}_m;\Sigma_m]} \circ \operatorname{Res}_{[\mathcal{Y}_{m-1};\Sigma_{m-1}]} \circ \cdots \circ \operatorname{Res}_{[\mathcal{Y}_2;\Sigma_2]} \circ \operatorname{Res}_{[\mathcal{Y}_1;\Sigma_1]}\Big)(\omega) \\ \in C_{k-m}^\infty(S_1 \cap \cdots \cap S_m),\end{aligned}$$

where the Σ_k and the $\mathcal{Y}_k$ were defined in (2.19). Such construction depends on the ordering of the collection $\{S_1, \dots, S_m\}$. Within this geometric context, we formulate the cohomological residue formula.

PROPOSITION 2.14 (Cohomological residue formula). *Let $\mathcal{X}$ be an N-dimensional complex manifold, and let $\boldsymbol{S} = (S_1, \dots, S_m)$ with $1 \leq m \leq N$ be an ordered sequence of codimension 1 complex submanifolds which intersect transversally on $\mathcal{X}$. Let ω be a semimeromorphic k-form which is $\boldsymbol{S}$-multilogarithmic and moreover is supposed to be d-closed on $\mathcal{X} \setminus |\boldsymbol{S}|$, with $m \leq k \leq 2N$. Let also $\gamma \in \mathscr{Z}_{k-m}(S_1 \cap \cdots \cap S_m, \mathbb{C})$ and suppose that $\Gamma \in \mathscr{Z}_k(\mathcal{X} \setminus |\boldsymbol{S}|, \mathbb{C})$ represents $\delta^{[\mathcal{X};\boldsymbol{S}]}(\bar{\gamma}) \in H_k(\mathcal{X} \setminus |\boldsymbol{S}|, \mathbb{C})$. Then,*

$$\int_\gamma \operatorname{Res}_{[\mathcal{X};\boldsymbol{S}]}(\omega) = \frac{1}{(2i\pi)^m} \int_\Gamma \omega, \tag{2.23}$$

where the iterated morphisms $\delta^{[\mathcal{X};\boldsymbol{S}]}$ and $\operatorname{Res}_{[\mathcal{X};\boldsymbol{S}]}$ have been defined in (2.20) *and* (2.22)*, respectively.*

PROOF. The fact that the hypersurfaces S_j intersect transversally makes possible the construction of the iterated operator $\operatorname{Res}_{[\mathcal{X};\boldsymbol{S}]}$ through an iteration of Lemma 2.2. The proof of the proposition reduces then to the proof in the case $m = 1$, which is a direct consequence of Lemma 2.2, together with the fact that $\overline{\partial}(1/\zeta) = \delta_0(d\overline{\zeta} \wedge d\zeta)/(2i\pi)$ in the univariate setting. For more details, we refer the reader to [**AY**]. □

Let us conclude this subsection with a few comments relative to the case $m = 1$. It follows from Proposition 2.14 and de Rham's duality theorem that $\operatorname{Res}_{[\mathcal{X};S]}(\omega)$ depends only on the cohomology class of ω in $\mathcal{X} \setminus S$. When $1 \leq k \leq 2N$ and ω is any semimeromorphic k-form with at most poles with order $\nu_0 > 1$ along S and such that $d\omega = 0$ in $\mathcal{X} \setminus S$, one can write, using a partition of unity $(\theta_\iota)_\iota$ in $\mathcal{X}$

subordinate to a covering of $\mathcal{X}$ by open sets U_ι, where a division formula such as (2.6) holds,

$$\omega = \sum_\iota \theta_\iota \Big(\frac{ds_\iota}{s_\iota^{\nu_0}} \wedge r_{\nu_0,\iota}[\omega] + \frac{\sigma_{\nu_0,\iota}[\omega]}{s_\iota^{\nu_0 - 1}} \Big)$$
$$= \frac{1}{1-\nu_0} d\Big(\sum_\iota s_\iota^{1-\nu_0} \theta_\iota r_{\nu_0,\iota}[\omega] \Big) - \frac{1}{1-\nu_0} \Big(\sum_\iota \frac{d(\theta_\iota r_{\nu_0,\iota}[\omega])}{s_\iota^{\nu_0-1}} \Big) + \sum_\iota \theta_\iota \frac{\sigma_{\nu_0,\iota}[\omega]}{s_\iota^{\nu_0-1}}.$$

Iterating this procedure, one gets a representant $\omega_{\log}$ of the cohomology class of ω in $H^k(\mathcal{X} \setminus S, \mathbb{C})$, which is semimeromorphic with simple poles along S. The residue formula (2.23) now becomes

$$\int_\gamma \mathrm{Res}_{[\mathcal{X};S]}(\omega_{\log}) = \frac{1}{2i\pi} \int_\Gamma \mathrm{Res}_{[\mathcal{X};S]}(\omega_{\log}). \tag{2.24}$$

In fact any smooth d-closed k-form ω in $\mathcal{X} \setminus S$ with $1 \leq k \leq 2N$, where S is a smooth complex hypersurface in $\mathcal{X}$, admits in its cohomology class in $H^k(\mathcal{X} \setminus S, \mathbb{C})$ a representative $\omega_{\log}$ with simple poles along S. This can be proved indirectly, using a duality argument based on de Rham's theorem [**DR55**]; see for example [**AY**, Theorem 16.5]. Since such a proof fails to be constructive, we will admit the result here. As a consequence, one can attach a Leray residue $\mathrm{Res}_{[\mathcal{X};S]}(\bar{\omega})$ to each cohomology class $\bar{\omega} \in H^k(\mathcal{X} \setminus S, \mathbb{C})$, namely $\mathrm{Res}_{[\mathcal{X};S]}(\omega_{\log})$. Residue formula (2.24) remains valid whenever Γ represents $\delta^{[\mathcal{X};S]}(\bar{\gamma})$ in $H_k(\mathcal{X} \setminus S, \mathbb{C})$.

2.1.3. Meromorphic multilogarithmic forms. For $m \leq N$, we consider now an ordered sequence $\boldsymbol{S} = (S_1, \dots, S_m)$ of complex hypersurfaces $S_1, \dots, S_m$ in an N-dimensional complex manifold $\mathcal{X}$. Let, as before, $|\boldsymbol{S}| = S_1 \cup \cdots \cup S_m$ and

$$\widehat{|\boldsymbol{S}|}_j = S_1 \cup \cdots \cup \widehat{S_j} \cup \cdots \cup S_m, \text{ for, } j = 1, \dots, m.$$

It is important to realize here that the closed analytic subsets $(S_1)_{\mathrm{sing}}, \dots, (S_m)_{\mathrm{sing}}$, whose codimension is greater than or equal to 2, are not assumed anymore to be empty, as was the case for $m = 1$ in Definition 2.8 and Proposition 2.9.

If $0 \leq p \leq N$, then the $(p,0)$ meromorphic forms ω, whose polar set $\mathrm{Pol}(\omega)$ is contained in $|\boldsymbol{S}|$, are characterized among elements ω in $C^\infty_{p,0}(\mathcal{X} \setminus |\boldsymbol{S}|)$ by the following two facts:

- one has that $\overline{\partial}\omega = 0$ in $\mathcal{X} \setminus |\boldsymbol{S}|$;
- any $z \in |\boldsymbol{S}|$ admits an open neighborhood $U = U_z$, with reduced holomorphic defining functions $s_1, \dots, s_m$ in U for hypersurfaces $S_1, \dots, S_m$ respectively, so that for some positive integer $\nu = \nu_z$ the $(p,0)$ holomorphic differential form $(s_1 \cdots s_m)^\nu \omega_{U \setminus |\boldsymbol{S}|}$ extends as a $(p,0)$ holomorphic form in U.

This characterization follows from Hilbert's nullstellensatz in the local nœtherian ring $\mathcal{O}_{\mathbb{C}^N,0}$. The notion of *$\boldsymbol{S}$-multilogarithmic meromorphic form* was introduced in [**Pol72, AlT01**] as an extension of Definition 2.8 to the case where $m > 1$. This notion does not contradict Definition 2.13 in the case where the hypersurfaces S_j were smooth.

DEFINITION 2.15. A meromorphic $(p,0)$-form with polar set contained in $|\boldsymbol{S}|$ is said to be $\boldsymbol{S}$-*multilogarithmic* if and only if any $z \in |\boldsymbol{S}|$ admits an open neighborhood $U = U_z$, with reduced holomorphic defining functions $s_1, \dots, s_m$ in U for the hypersurfaces $S_1, \dots, S_m$ respectively, such that:

- the meromorphic forms $(s_j\omega)_{|U}$ and $(s_j d\omega)_U$ or, equivalently, the forms $(s_j\omega)_{|U}$ and $(ds_j \wedge \omega)_{|U}$ are of the form

$$\sum_{\kappa=1}^{m} \omega_{j,\kappa} \quad \text{with} \quad \mathrm{Pol}(\omega_{j,\kappa}) \subset \widehat{|\boldsymbol{S}|}_{\kappa} \cap U, \tag{2.25}$$

  for any $j = 1, \dots, m$;
- the differential form $ds_J = \bigwedge_{j\in J} ds_j$ does not vanish identically on any irreducible component of $\boldsymbol{S}_J \cap U$, where $\boldsymbol{S}_J = \bigcap_{j\in J} S_j$ for every ordered subset J of $\{1, \dots, m\}$.

Let $\mathcal{X}^\circ = \mathcal{X} \setminus \bigcup_{J\subset\{1,\dots,m\}} (\boldsymbol{S}_J)_{\mathrm{sing}}$ and let the meromorphic $(p,0)$-form ω be $\boldsymbol{S}$-multilogarithmic as in Definition 2.15. Then $\omega_{|\mathcal{X}^\circ}$ is $\boldsymbol{S}^\circ$-multilogarithmic in $\mathcal{X}^\circ$, where $\boldsymbol{S}^\circ$ is defined as the ordered sequence $\boldsymbol{S}^\circ = (S_1^\circ, \dots, S_m^\circ)$ of hypersurfaces, satisfying $S_j^\circ = S_j \cap \mathcal{X}^\circ$ for every $j = 1, \dots, m$. When $m \le p \le N$ and the meromorphic $(p,0)$-form ω is $\boldsymbol{S}$-multilogarithmic form in $\mathcal{X}$, its restriction to $\mathcal{X}^\circ$, with polar set contained in $|\boldsymbol{S}^\circ|$, is $\boldsymbol{S}^\circ$-multilogarithmic as a meromorphic $(p,0)$-form in $\mathcal{X}^\circ$. Then one can attach to it its Leray residue $\mathrm{Res}_{[\mathcal{X}^\circ;\boldsymbol{S}^\circ]}(\omega_{|\mathcal{X}^\circ})$, which is a holomorphic $(p-m,0)$ form on $\bigcap_{j=1}^m S_j^\circ$; see (2.22). Similarly one can attach to it the (p,m)-current on $\mathcal{X}^\circ$ supported by $\bigcap_{j=1}^m S_j^\circ \subset \boldsymbol{S}_{\{1,\dots,m\}}$, whose action is defined as follows:

$$\begin{aligned}(2.26)\quad \big[\mathrm{Res}_{[\mathcal{X}^\circ;\boldsymbol{S}^\circ]}(\omega_{|\mathcal{X}^\circ})\big] &: \varphi \in \mathcal{D}^{(N-p,N-m)}(\mathcal{X}^\circ) \\ &\longmapsto \int_{\bigcap_{j=1}^m S_j^\circ} \mathrm{Res}_{[\mathcal{X}^\circ;\boldsymbol{S}^\circ]}(\omega_{|\mathcal{X}^\circ}) \wedge \varphi = \int_{\boldsymbol{S}_{\{1,\dots,m\}}} \mathrm{Res}_{[\mathcal{X}^\circ;\boldsymbol{S}^\circ]}(\omega_{|\mathcal{X}^\circ}) \wedge \varphi.\end{aligned}$$

PROPOSITION 2.16. *Let $\mathcal{X}$, $\boldsymbol{S}$, $\mathcal{X}^\circ$, $\boldsymbol{S}^\circ$, and ω be as above. The (p,m)-current $T^\circ = \big[\mathrm{Res}_{[\mathcal{X}^\circ;\boldsymbol{S}^\circ]}(\omega_{|\mathcal{X}^\circ})\big] \in {}'\mathcal{D}^{(p,m)}(\mathcal{X}^\circ)$ defined in (2.26) extends as the $\overline{\partial}$-closed (p,m)-current $T = R_{\boldsymbol{S}} \wedge \omega$ supported in $|\boldsymbol{S}|$. This extension is defined*[6] *in a neighborhood of each point $z \in |\boldsymbol{S}|$, independently of the choice of the reduced defining functions $s_1, \dots, s_m$ for the hypersurfaces S_j that fulfill the requirements in Definition 2.15 in an open neighborhood U of the point z, as follows:*

(2.27)

$$\begin{aligned} T_{|U} &= \Big[\overline{\partial}\Big(\frac{|s_m|^{2\lambda}}{2i\pi}\Big) \wedge \dots \wedge \overline{\partial}\Big(\frac{|s_1|^{2\lambda^m}}{2i\pi}\Big) \wedge \omega_{|U}\Big]_{\lambda=0} \\ &= \Big[\lambda^{\frac{m(m+1)}{2}} \Big(\prod_{\kappa=1}^m |s_\kappa|^{2\lambda^{m+1-\kappa}}\Big) \Big(\frac{\overline{\partial}\log|s_m|^2}{2i\pi} \wedge \dots \wedge \frac{\overline{\partial}\log|s_1|^2}{2i\pi}\Big) \wedge \omega_{|U}\Big]_{\lambda=0}. \end{aligned}$$

Moreover, when $m < N$, given any reduced holomorphic function s_{m+1} in U, which is not identically zero on any irreducible component of $\boldsymbol{S}_J$, for some specific ordering

[6]The meaning of the symbol $[\ ]_{\lambda=0}$ in (2.27) remains the same as in many places in our introductory first chapter: one considers first the current-valued function of λ as defined from the beginning for $\mathrm{Re}\,\lambda \gg 1$ in a conic sector with aperture $< \pi/(m+1)$ bisected by the real positive axis $\overline{\mathbb{R}^+}$, then takes the meromorphic continuation as a function of λ, which remains pole-free around $\overline{\mathbb{R}^+}$, and finally evaluates the value at $\lambda = 0$.

$J = \{1, \dots, m\}$, *one has also that*

(2.28) $T_{|U} =$

$$\Big[\lambda^{\frac{m(m+1)}{2}} |s_{m+1}|^{2\lambda} \Big(\prod_{\kappa=0}^{m-1} |s_\kappa|^{2\lambda^{m+1-\kappa}}\Big) \Big(\frac{\overline{\partial} \log |s_m|^2}{2i\pi} \wedge \cdots \wedge \frac{\overline{\partial} \log |s_1|^2}{2i\pi}\Big) \wedge \omega_{|U}\Big]_{\lambda=0}.$$

PROOF. Let us prove first that the definition of $T_{|U}$ is licit in a neighborhood U of $z \in |\boldsymbol{S}|$ and that it does not depend on the reduced defining functions $s_1, \dots, s_m$ in U for $S_1, \dots, S_m$ respectively, and that moreover (2.28) holds. We will assume first that $m = 2 < N$. We need at this point the following lemma.

LEMMA 2.17. *Let f, g, h, u, v be holomorphic functions in a neighborhood Ω of the origin in $\mathbb{C}^N$, $N > 2$, and let $q, r \in \mathbb{N}^*$. Suppose that u, v are invertible in Ω and that the hypersurfaces $S_f = V(f)$, $S_g = V(g)$, $S_h = V(h)$ intersect properly. Then the $'\mathcal{D}^{(0,1)}(\Omega)$-valued map*

$$\mu \longmapsto T_\mu = \frac{|ug|^{2\mu}}{g^r} \Big[\overline{\partial}\Big(\frac{|vf|^{2\lambda}}{f^q}\Big)\Big]_{\lambda=0}$$

extends from $\{\mu \in \mathbb{C} : \operatorname{Re}\mu \gg 1\}$ to the complex plane $\mathbb{C}$ as a meromorphic map with poles in $-\mathbb{Q}^+$. Its value at $\mu = 0$ is independent of u and v. Furthermore, the $'\mathcal{D}^{(0,2)}(\Omega)$-valued map $\nu \longmapsto |h|^{2\nu}\overline{\partial}T_0$ extends from $\{\nu \in \mathbb{C} : \operatorname{Re}\nu \gg 1\}$ as a meromorphic map with poles in $-\mathbb{Q}^+$, with $\big[|h|^{2\nu}\overline{\partial}T_0\big]_{\nu=0} = \overline{\partial}T_0$. In addition, both $'\mathcal{D}^{(0,2)}(\Omega)$-valued functions

$$\lambda \longmapsto \overline{\partial}\Big(\frac{|vg|^{2\lambda}}{g^r}\Big) \wedge \overline{\partial}\Big(\frac{|uf|^{2\lambda^2}}{f^q}\Big), \qquad \lambda \longmapsto |h|^{2\lambda}\, \overline{\partial}\Big(\frac{|vg|^{2\lambda^2}}{g^r}\Big) \wedge \overline{\partial}\Big(\frac{|uf|^{2\lambda^3}}{f^q}\Big)$$

extend from $\{\nu \in \mathbb{C} : \operatorname{Re}\nu \gg 1\}$ to the complex plane $\mathbb{C}$ as meromorphic maps with poles in $-\mathbb{Q}^+$. Their value at $\lambda = 0$ is independent of the invertible functions u and v and coincides there with the value of $\overline{\partial}T_0$.

PROOF. Hironaka's resolution of singularities theorem (see Theorem A.41), reduces the proof to the case where f, g, h are monomial maps. Then, all assertions are easily proved. The assertion $\big[|h|^{2\nu}\overline{\partial}T_0\big]_{\nu=0} = \overline{\partial}T_0$ is the only one that requires the fact that $\operatorname{codim}(V(f) \cap V(g) \cap V(h)) = 3$, while the current involved is a $(0,2)$-current. We will state in §2.2 a *dimension principle* that will cover this situation. □

Lemma 2.17 can be iterated and thus provides the justification needed for the definition of $T_{|U}$ regarding its independence on the choice of the reduced defining holomorphic functions s_j. Moreover, iteration of Lemma 2.17 shows that for

$$\omega_{|U} = \frac{\Psi_U}{(s_1 \cdots s_m)^q},$$

where $q \in \mathbb{N}^*$ and $\Psi_U \in \Omega^p(U)$, one has

(2.29) $T_{|U} =$

$$\Big[\overline{\partial}\Big(\frac{|s_m|^{2\lambda_m}}{2i\pi\, s_m^q}\Big) \wedge \Big[\cdots \Big[\overline{\partial}\Big(\frac{|s_2|^{2\lambda_2}}{2i\pi\, s_2^q}\Big) \wedge \Big[\overline{\partial}\Big(\frac{|s_1|^{2\lambda_1}}{2i\pi\, s_1^q}\Big)\Big]_{\lambda_1=0}\Big]_{\lambda_2=0} \cdots\Big]_{\lambda_{m-1}=0}\Big]_{\lambda_m=0} \wedge \Psi_U$$

and $\big[|s_{m+1}|^{2\lambda} T_{|U}\big]_{\lambda=0} = T_{|U}$. Let us now show that T extends T° in U. Let $z^\circ \in |\boldsymbol{S}^\circ| \cap U$ and take $(s_1)_{|U^\circ}, \dots, (s_m)_{|U^\circ}$ as defining functions for the smooth hypersurfaces S_j° around z°. The hypothesis that ω is $\boldsymbol{S}$-multilogarithmic implies that $\omega_{|U^\circ}$ is expressed as

$$\omega_{|U^\circ} = \bigwedge_{j=1}^{m} \frac{ds_j}{s_j} \wedge r^\circ[\omega] + \sigma^\circ[\omega],$$

where $r^\circ[\omega]$ is a $(p-m,0)$ holomorphic form in U° and $\sigma^\circ[\omega]$ is of the form $\sum_{\kappa=1}^m \varpi_\kappa^\circ$, where the forms ϖ_κ° have their polar sets contained in $\widehat{|\boldsymbol{S}|}_\kappa$ respectively. As a consequence

$$\Big[\overline{\partial}\Big(\frac{|s_m|^{2\lambda}}{2i\pi}\Big) \wedge \cdots \wedge \overline{\partial}\Big(\frac{|s_1|^{2\lambda^m}}{2i\pi}\Big) \wedge \sigma^\circ[\omega]\Big]_{\lambda=0} = 0.$$

Since

$$\Big[\overline{\partial}\Big(\frac{|s_m|^{2\lambda}}{2i\pi\, s_m}\Big) \wedge \cdots \wedge \overline{\partial}\Big(\frac{|s_1|^{2\lambda^m}}{2i\pi\, s_1}\Big) \wedge (ds_1 \wedge \cdots \wedge ds_m)\Big]_{\lambda=0} \wedge r^\circ[\omega]$$
$$= r^\circ[\omega] \wedge \big([\boldsymbol{S}_{\{1,\dots,m\}}]\big)_{|U^\circ} = \big[\mathrm{Res}_{[\mathcal{X}^\circ;\boldsymbol{S}^\circ]}(\omega_{|\mathcal{X}^\circ})\big]_{|U^\circ},$$

T coincides with $T^\circ = \big[\mathrm{Res}_{[\mathcal{X}^\circ;\boldsymbol{S}^\circ]}(\omega_{|\mathcal{X}^\circ})\big]$ on U°, and the proposition is proved. □

REMARK 2.18. When U is an open neighborhood of an arbitrary point in $|\boldsymbol{S}|$, the fact that $|s_{m+1}|^{2\lambda} T_{|U} = T_{|U}$ for any reduced holomorphic function in U, which is not identically zero on any irreducible component of the closed analytic subset $\boldsymbol{S}_{\{1,\dots,m\}} \cap U = S_1 \cap S_2 \cap \cdots \cap S_m \cap U$, can be interpreted as follows. The current T supported on $\boldsymbol{S}_{\{1,\dots,m\}}$ is the *standard extension* of its restriction to the $(N-m)$-dimensional complex manifold $\mathcal{X}^\circ \cap \boldsymbol{S}_{\{1,\dots,m\}}$. The current T coincides with the current induced by the iterated Leray residue $\mathrm{Res}_{[\mathcal{X}^\circ;\boldsymbol{S}^\circ]}(\omega_{|\mathcal{X}^\circ})$ on the manifold $\mathcal{X}^\circ \cap \boldsymbol{S}_{\{1,\dots,m\}}$. Actually, the action of T, as a current in $\mathcal{X}$ supported by $\boldsymbol{S}_{\{1,\dots,m\}}$, involves nontangential derivations, while the Leray residue $\mathrm{Res}_{[\mathcal{X}^\circ;\boldsymbol{S}^\circ]}(\omega_{|\mathcal{X}^\circ})$, which it extends from the point of view, is a holomorphic $(p-m,0)$ form on the complex manifold $\mathcal{X}^\circ \cap \boldsymbol{S}_{\{1,\dots,m\}}$, and is considered an open subset of $\boldsymbol{S}_{\{1,\dots,m\}}$.

2.2. Coleff–Herrera sheaves of currents

The notion of the Coleff–Herrera sheaf of currents on a complex manifold, with support in a prescribed closed analytic subset V, and its importance were pointed out by J.E. Björk in [**Bj96, Bj04**]. The original construction of sections of such a sheaf is due to N. Coleff and M. Herrera in [**CoH78**]. In this section, following the approach of J.E. Björk, M. Andersson, H. Samuelsson Kalm [**Bj96, Bj04, And09, BjS10**], we will introduce the various concepts and then revisit (with this point of view) the original theory developed in 1978 by N. Coleff and M. Herrera in [**CoH78**]. Let us fix from now on some notations that will be used throughout this section. Let $\mathcal{X}$ be an N-dimensional complex manifold and let $z \in \mathcal{X}$. Its structural sheaf will be denoted by $\mathcal{O}$. For any sheaf $\mathcal{F}$ of $\mathcal{O}$-modules on $\mathcal{X}$, we will denote by $\mathcal{F}_z$ its stalk at a point $z \in \mathcal{X}$. For a neighborhood U of z and $\mathfrak{f} \in \mathcal{F}(U)$, the germ of $\mathfrak{f}$ at z will be denoted by $\mathfrak{f}_z \in \mathcal{F}_z$. For any $0 \le p, q \le N$, we let $C^\infty_{p,q}$ be the sheaf of smooth (complex-valued) (p,q)-forms on $\mathcal{X}$ and ${}'\mathcal{D}^{(p,q)}$ be that of complex-valued currents with bidegree (p,q) on $\mathcal{X}$. We let also $\mathcal{O}_{\mathcal{X}}\langle\partial/\partial\zeta\rangle$ be the sheaf of holomorphic differential operators. It is a sheaf, whose stalk at z consists of germs of operators

$$\mathcal{Q}\Big(\zeta_1,\dots,\zeta_n,\frac{\partial}{\partial\zeta_1},\dots,\frac{\partial}{\partial\zeta_N}\Big)$$

at z, where $\mathcal{Q}$ is holomorphic in $\zeta_1,\dots,\zeta_N$ and polynomial in the $\partial/\partial\zeta_j$, $j=1,\dots,N$. As usual, a system of local coordinates in a neighborhood of $z\in\mathcal{X}$ will be denoted by $(\zeta_1,\dots,\zeta_N)$. It is said to be centered if $\zeta(z)=0$.

2.2.1. Integration currents, pseudo-meromorphicity, and holonomy. Let $V_0=\big(\{z\,:\,v_1(z)=\cdots=v_K(z)=0\}\big)_0$ be the germ at the origin of a purely dimensional closed analytic set V in a neighborhood of the origin in $\mathbb{C}^N$ with $\operatorname{codim}_0 V=M$. The local presentation of closed analytic subsets (see Proposition A.24 in §A.3.4) allows us to assume that V admits around 0 the following representation. There exist $f_1,\dots,f_M$ so that $df_1\wedge\cdots\wedge df_M$ does not vanish identically on any irreducible component of the germ of a closed analytic set $W=\bigcap_{j=1}^M V(f_j)$, where $V(f_j)$ is by definition $f_j^{-1}(\{0\})$. Furthermore, we assume that V is the union of a finite number of irreducible components $\mathfrak{W}_1,\dots,\mathfrak{W}_{L'}$ among the irreducible components $\mathfrak{W}_1,\dots,\mathfrak{W}_L$ of W. Namely, one has for some $1\le L'\le L$,

$$V=\Big(\bigcap_{j=1}^M V(f_j)\Big)\setminus(\mathfrak{W}_{L'+1}\cup\cdots\cup\mathfrak{W}_L)=\Big(\bigcap_{j=1}^M V(f_j)\Big)\setminus V(f_{M+1}), \tag{2.30}$$

where $f_{M+1}=\sum_{\kappa=1}^K\lambda_\kappa\,v_\kappa$ does not vanish identically $\mathfrak{W}_{L'+1}\setminus V,\dots,\mathfrak{W}_L\setminus V$, so that $V=W^{V(f_{M+1})}$ according to the notation in §D.1.1; see formula (D.4). It follows from Proposition 2.16 that the integration (M,M)-current $[W]$ factorizes in a neighborhood of the origin as

$$\begin{aligned}[W]&=R_{(V(f_1),\dots,V(f_M))}\wedge\Big(\bigwedge_{j=1}^M\frac{df_j}{f_j}\Big)\\ &=\Big[\overline{\partial}\Big(\frac{|f_M|^{2\lambda}}{2i\pi\,f_M}\Big)\wedge\cdots\wedge\overline{\partial}\Big(\frac{|f_1|^{2\lambda^M}}{2i\pi\,f_1}\Big)\Big]_{\lambda=0}\wedge df\\ &=\Big[\lambda^{\frac{M(M+1)}{2}}\frac{\prod_{\kappa=1}^M|f_\kappa|^{2\lambda^{M+1-\kappa}}}{f_1\cdots f_M}\bigwedge_{j=1}^M\frac{\overline{df_j}}{\overline{f_j}}\Big]_{\lambda=0}\wedge df,\end{aligned}\tag{2.31}$$

where $df=df_1\wedge\cdots\wedge df_M$. Moreover, it follows from (2.28) that

(2.32)

$$\begin{aligned}[V]&=\Big[\lambda^{\frac{M(M+1)}{2}}(1-|f_{M+1}|^{2\lambda})\frac{\prod_{\kappa=0}^{M-1}|f_\kappa|^{2\lambda^{M+1-\kappa}}}{f_1\cdots f_M}\bigwedge_{j=1}^M\frac{\overline{df_j}}{\overline{f_j}}\Big]_{\lambda=0}\wedge df\\ &=\Big[(1-|f_{M+1}|^{2\lambda_{M+1}})\Big[\overline{\partial}\Big(\frac{|f_M|^{2\lambda_M}}{f_M}\Big)\wedge\Big[\cdots\Big[\overline{\partial}\Big(\frac{|f_1|^{2\lambda_1}}{f_1}\Big)\Big]_{\lambda_1=0}\Big]\cdots\Big]_{\lambda_{M+1}=0}\Big]\wedge df.\end{aligned}$$

Let us consider such a factorization in a neighborhood Ω of the origin in $\mathbb{C}^N$ and let $\pi:\widetilde{\Omega}\to\Omega$ be a log resolution for the hypersurface $\bigcup_{j=1}^{M+1}V(f_j)$; see Definition A.40 and Theorem A.41. Consider the local centered coordinates $t=(t_1,\dots,t_N)$ in a neighborhood of a point $\tilde z$ (with $t(\tilde z)=0$) on the complex manifold $\widetilde{\Omega}$. Then, one has, for any $j = 1, \dots, M+1$, in the corresponding local chart that

$$\pi^* f_j(t) = u_j(t)\, t^{\alpha_j},\ j = 1, \dots, M+1,$$

where $t^{\alpha_j} = \prod_{\kappa=1}^{N} t_j^{\alpha_{j,\kappa}}$, $\alpha_{j,\kappa} \in \mathbb{N}$, and u_j is invertible. The last representation implies that in a neighborhood of $\tilde z$ and outside $t_1 \cdots t_N = 0$ one has that

$$\pi^*\Big(\bigwedge_{j=1}^{M} \overline{\frac{df_j}{f_j}}\Big)(t) = \sum_{1 \le j_1 < \cdots < j_M \le N} \gamma_J \bigwedge_{\kappa=1}^{M} \overline{\frac{dt_{j_\kappa}}{t_{j_\kappa}}},$$

when $\gamma_J \in \mathbb{Z}$. These computations motivate the following definition introduced by M. Andersson, H. Samuelsson, and E. Wulcan [**AndW10**, **AndS12**, **AndW18**, **AndW21**].

Definition 2.19. Let $\mathcal{X}$ be an N-dimensional complex manifold.[7] A germ of a current T_z at a point $z \in \mathcal{X}$ is said to be *pseudo-meromorphic* if and only if it is the germ at z of a finite sum of germs of currents realized in a neighborhood of z as direct images $T_\iota = (\Pi_\iota)_*(\tau_\iota)$, where Π_ι is a holomorphic map from an open subset $\mathcal{U}_\iota \subset \mathbb{C}_t^{N_\iota}$ to a neighborhood U_ι of z in $\mathcal{X}$.[8] Moreover, one requires the current τ_ι to be of the form

(2.33)

$$\begin{aligned}\tau_\iota &= \Big(\bigotimes_{1\le\kappa\le k'} \mathrm{PV}\Big[\frac{1}{t_\kappa^{\gamma_\kappa}}\Big]\Big) \otimes \Big(\bigwedge_{k'<\kappa\le k} \bar\partial\Big(\frac{1}{t_\kappa^{\gamma_\kappa}}\Big)\Big) \wedge \varpi(t),\\ &= \Big[\Big(\frac{\lambda}{2i\pi}\Big)^{k-k'} \Big(\prod_{\kappa=1}^{k} \frac{|t_\kappa^{\gamma_\kappa}|^{2\lambda}}{t_\kappa^{\gamma_\kappa}}\Big) \Big(\gamma_k \frac{\overline{dt_k}}{\overline{t_k}}\Big) \wedge \cdots \wedge \Big(\gamma_{k'+1} \frac{\overline{dt_{k'+1}}}{\overline{t_{k'+1}}}\Big) \wedge \varpi(t)\Big]_{\lambda=0},\end{aligned}$$

where ϖ is a smooth differential form with compact support[9] and $0 \le k' \le k \le N_\iota$, $\gamma_\kappa \in \mathbb{N}^*$. Here, the function

$$\lambda \longmapsto \Big(\frac{\lambda}{2i\pi}\Big)^{k-k'} \Big(\prod_{\kappa=1}^{k} \frac{|t_\kappa^{\gamma_\kappa}|^{2\lambda}}{t_\kappa^{\gamma_\kappa}}\Big) \Big(\gamma_k \frac{\overline{dt_k}}{\overline{t_k}}\Big) \wedge \cdots \wedge \Big(\gamma_{k'+1} \frac{\overline{dt_{k'+1}}}{\overline{t_{k'+1}}}\Big)$$

is considered as a meromorphic ${}'\mathcal{D}^{(0,k-k')}(\mathcal{U}_\iota)$-valued map with strictly negative rational poles.

Example 2.20. It follows from the previous discussion that germs of integration currents are pseudo-meromorphic. Let us briefly explain why, taking as a definition of pseudo-meromorphic currents the primitive one, namely, assuming that each Π_ι is a tower either of smooth modifications or open inclusions [**AndW10**]. Let V be an irreducible closed analytic subset with codimension M contained in a neighborhood $\mathcal{U}$ of the origin in $\mathbb{C}^N$. Let us make a modification $\pi : \widetilde{\mathcal{U}} \to U$ such that $\pi^{-1}(V)$ is an hypersurface in $\widetilde{\mathcal{U}}$ with normal crossings; see Theorem A.41. One of its irreducible components $Z_\iota = |\mathrm{div}(E_\iota)|$, called the *strict transform* of V, is such

[7] We refer here to the definition given in [**AndW18**, §2.1] and to [**AndW18**, Theorem 2.15 (1)].

[8] As a matter of fact, for this monograph it is enough to restrict our use to the case already introduced in [**AndW10**, §2]. There, Π_ι is a tower of either smooth modifications or open inclusions. Naturally, in the latter case $N_\iota = N$. The above comments hold for the case where $\mathcal{X}$ is a reduced complex analytic space; see Remark 2.29.

[9] Since all currents appearing as factors in (2.33) are currents in distinct complex variables t_κ, the tensor product is well defined. Note that the bidegree of currents is not specified here. The germs of currents are specified neither by their pure degree nor their bidegree.

$\pi(Z_\iota) = V$. Let $L_\iota = \mathcal{O}_{\widetilde{\mathcal{U}}}(D_\iota)$ be the line bundle attached to it (see Proposition D.17) and let $c_1(\overline{L}_\iota)$ be the first Chern form of an arbitrary smooth metrization of L_ι (see Definition B.46). The closed current $\pi_*((c_1(\overline{L}_\iota))^M \wedge [Z_\iota])$ has order 0 and is, by construction, pseudo-meromorphic according to Definition 2.19. Note that only towers of either smooth modifications or inclusions are involved here as holomorphic mappings Π. This current coincides with the integration current $[V]$ around each point in V_{reg} since one can choose π so that $\pi_{|Z_\iota}$ realizes the blowup of V outside V_{sing}. Thus, $[V]$ and $\pi_*((c_1(\overline{L}_\iota))^M \wedge [Z_\iota])$ do coincide according to the dimension principle for normal currents; see Proposition C.6. Hence, one concludes that $[V]$ is pseudo-meromorphic. Such is also the germ of the current $R_{\boldsymbol{S}} \wedge \omega$ at the point $z \in \boldsymbol{S}_{\{1,\dots,m\}}$ introduced in Proposition 2.16, where ω is a meromorphic $(0,p)$-form assumed to be $\boldsymbol{S} = (S_1, \dots, S_m)$-multilogarithmic; see Definition 2.15.

Definition 2.19 induces naturally the sheaf $\mathrm{PM}_{\mathcal{X}}$, or, for short, PM, whose stalk at the point z is the space of germs of pseudo-meromorphic currents at $z \in \mathcal{X}$. A section T of the sheaf PM in an open subset U will be called a *pseudo-meromorphic current* in U. One denotes this by $T \in \mathrm{PM}(U)$. It follows from the definition of the sheaf PM that it is closed under the action of the differential operators $\overline{\partial}$ and ∂, and it is a module over the sheaf of smooth forms. In case such a section T is E-valued, where E is a holomorphic bundle over $\mathcal{X}$, one writes $T \in \mathrm{PM}(U, E)$. The following proposition is essential for us.

PROPOSITION 2.21. *Let T_z be a germ of a pseudo-meromorphic current at a point $z \in \mathcal{X}$ and let $h_z \in \mathcal{O}_z$. Then the maps*

$$\lambda \longmapsto \Big(\frac{|h|^{2\lambda}}{h} T\Big)_z, \quad \lambda \longmapsto \Big(\overline{\partial}\Big(\frac{|h|^{2\lambda}}{h}\Big) \wedge T\Big)_z \tag{2.34}$$

extend from $\{\lambda \in \mathbb{C} : \operatorname{Re}\lambda \gg 1\}$ to the complex plane $\mathbb{C}$ as a ${}'\mathcal{D}_z$-valued meromorphic map with (possible) poles in $-\mathbb{Q}^+$. Their values at $\lambda = 0$ belong to PM_z.

PROOF. In view of Definition 2.19, it is enough to consider the case when $\mathcal{X}$ is an open neighborhood of the origin $\mathcal{U} \subset \mathbb{C}^N_t$, and the maps are

$$\lambda \longmapsto \frac{|h(t)|^{2\lambda}}{h(t)} \Big(\bigotimes_{1 \le \kappa \le k'} \mathrm{PV}\Big[\frac{1}{t_\kappa^{\gamma_\kappa}}\Big]\Big) \otimes \Big(\bigwedge_{k' < \kappa \le k} \overline{\partial}\Big(\frac{1}{t_\kappa^{\gamma_\kappa}}\Big)\Big),$$

$$\lambda \longmapsto \overline{\partial}\Big(\frac{|h|^{2\lambda}}{h}\Big)(t) \wedge \Big(\bigotimes_{1 \le \kappa \le k'} \mathrm{PV}\Big[\frac{1}{t_\kappa^{\gamma_\kappa}}\Big]\Big) \otimes \Big(\bigwedge_{k' < \kappa \le k} \overline{\partial}\Big(\frac{1}{t_\kappa^{\gamma_\kappa}}\Big)\Big),$$

where $h \in \mathcal{O}(\mathcal{U})$. One can even replace $\mathcal{U}$ by $\widetilde{\mathcal{U}}$, where $\Pi : \widetilde{\mathcal{U}} \to \mathcal{U}$ is a log resolution for the hypersurface $V(t_1 \cdots t_k\, h)$; see Definition A.40 and Theorem A.41. In particular it means that each point of $\widetilde{\mathcal{U}}$ admits a local chart in which each holomorphic function $\Pi^* t_1, \dots, \Pi^* t_k, \Pi^* h$ is expressed in centered local coordinates $\xi_1, \dots, \xi_N$ as a monomial multiplied by an invertible holomorphic function. Then, one has in such local coordinates ξ

$$\Big(\gamma_k \frac{d\overline{\Pi^* t_k}}{\overline{\Pi^* t_k}}\Big) \wedge \dots \wedge \Big(\gamma_{k'+1} \frac{d\overline{\Pi^* t_{k'+1}}}{\overline{\Pi^* t_{k'+1}}}\Big) = \sum_{|I| \le k-k'} \frac{d\overline{\xi}_I}{\overline{\xi}_I} \wedge \overline{u_I(\xi)},$$

where u_I is a holomorphic $k - k' - |I|$-form. According to the representation of elementary currents as in the second line of (2.33), the elementary current τ, considered here on $\mathcal{U}$, is expressed as a sum $\tau = \sum_\iota \Pi_* \tau_\iota$ of elementary currents in local coordinates on $\widetilde{\mathcal{U}}$. Introducing the additional log resolution Π, thus replacing $\mathcal{U}$ by $\widetilde{\mathcal{U}}$, then leads back to the situation where one considers τ elementary, as in the first line of (2.33), with in addition $h(t) = u(t)\,t^\alpha$, where $\alpha \in \mathbb{N}^N$ and u is invertible. If t^α contains as a factor t_κ with $k' < \kappa \leq k$, then

$$\frac{|h(t)|^{2\lambda}}{h(t)}\Big(\bigotimes_{1\leq\kappa\leq k'} \mathrm{PV}\Big[\frac{1}{t_\kappa^{\gamma_\kappa}}\Big]\Big)\otimes\Big(\bigwedge_{k'<\kappa\leq k}\overline{\partial}\Big(\frac{1}{t_\kappa^{\gamma_\kappa}}\Big)\Big) = 0,$$

$$\overline{\partial}\Big(\frac{|h|^{2\lambda}}{h}\Big)(t)\wedge\Big(\bigotimes_{1\leq\kappa\leq k'} \mathrm{PV}\Big[\frac{1}{t_\kappa^{\gamma_\kappa}}\Big]\Big)\otimes\Big(\bigwedge_{k'<\kappa\leq k}\overline{\partial}\Big(\frac{1}{t_\kappa^{\gamma_\kappa}}\Big)\Big) = 0$$

as currents in $\widetilde{\mathcal{U}}$ when $\mathrm{Re}\,\lambda \gg 1$. Therefore, we do not need to consider more of that case. In the situation where $h = u\,t_1^{\chi_1}\cdots t_{k'}^{\chi_{k'}} \times \prod_{\kappa>k} t^{\chi_\kappa}$, it is immediate to observe that the map

$$\lambda \longmapsto |u|^{2\lambda}\Big(\prod_{\kappa>k}\frac{|t_\kappa|^{2\chi_\kappa\lambda}}{t_\kappa^{\chi_\kappa}}\Big)\Big(\bigotimes_{1\leq\kappa\leq k'}\frac{|t_\kappa|^{2\chi_\kappa\lambda}}{t_\kappa^{\chi_\kappa}}\mathrm{PV}\Big[\frac{1}{t_\kappa^{\gamma_\kappa}}\Big]\Big)\otimes\Big(\bigwedge_{k'<\kappa\leq k}\overline{\partial}\Big(\frac{1}{t_\kappa^{\gamma_\kappa}}\Big)\Big)$$

extends to the complex plane $\mathbb{C}$ as a $\big({}'\mathcal{D}^{(0,k-k')}\big)_0$-valued meromorphic map with poles in $\mathbb{Q}^-$; see, for example, formulae (1.7). The value of this analytic continuation at $\lambda = 0$ equals the germ of a pseudo-meromorphic current

$$\Big(\bigotimes_{\kappa>k}\mathrm{PV}\Big[\frac{1}{t_\kappa^{\chi_\kappa}}\Big]\Big)\otimes\Big(\bigotimes_{1\leq\kappa\leq k'}\mathrm{PV}\Big[\frac{1}{t_\kappa^{\gamma_\kappa+\chi_\kappa}}\Big]\Big)\otimes\Big(\bigwedge_{k'<\kappa\leq k}\overline{\partial}\Big(\frac{1}{t_\kappa^{\gamma_\kappa}}\Big)\Big).$$

Since

$$\Big(\overline{\partial}\Big(\frac{|h|^{2\lambda}}{h}\Big)\wedge T\Big)_z = \overline{\partial}\Big(\frac{|h|^{2\lambda}}{h}\,T\Big)_z - \Big(\frac{|h|^{2\lambda}}{h}\,\overline{\partial}T\Big)_z$$

for $\mathrm{Re}\,\lambda \gg 1$ and $\overline{\partial}T_z$ is a germ of a pseudo-meromorphic if T_z is, the result for the second map follows immediately from the result for the first one. □

REMARK 2.22. If u is an invertible element in $\mathcal{O}_z$, then the above proof shows also that for any $h_z \in \mathcal{O}_z$ and any germ of a pseudo-meromorphic current T_z at z, one has

$$\Big[\Big(\frac{|uh|^{2\lambda}}{h}\,T\Big)_z\Big]_{\lambda=0} = \Big[\Big(\frac{|h|^{2\lambda}}{h}\,T\Big)_z\Big]_{\lambda=0}. \tag{2.35}$$

Moreover, since T_z is a germ of pseudo-meromorphic current at z, the map

$$\lambda \longmapsto |uh|^{2\lambda}\,T$$

extends from $\{\lambda \in \mathbb{C} : \mathrm{Re}\,\lambda \gg 1\}$ as a meromorphic current-valued map in a neighborhood of z with strictly negative rational poles. Its value at $\lambda = 0$ is a germ of pseudo-meromorphic current independent of the invertible element u.

REMARK 2.23. In the particular case where $T = [V]$ is an integration current, in order to prove the assertion in Proposition 2.21 (for what concerns the first map in (2.34)) one may use the fact that the distribution coefficients of integration currents are *regular holonomic*. This notion has been introduced precisely by J. E. Björk; see [**Bj93, Bj96, Bj04**] and is also presented in §C.2.5. We refer the reader to Remark C.34 for an alternative proof based on the use of the Bernstein–Sato formal equation in place of appealing to the construction of a log resolution as we did in the proof of Proposition 2.21.

Pseudo-meromorphic currents share important properties, which will help to formalize the concept of the Coleff–Herrera current in the next section.

PROPOSITION 2.24. *Let* $T \in \mathrm{PM}(U)$ *and* $h \in \mathcal{O}_{\mathcal{X}}(U)$. *If* $\operatorname{Supp} T \subset V(h)$, *then* $\overline{h}\,T = d\overline{h} \wedge T = 0$ *in* $'\mathcal{D}(U)$.

PROOF. Assume that the germ of T at $z \in U$ equals $T_z = \sum_\iota \big((\Pi_\iota)_* \tau_\iota\big)_z$, where each τ_ι is a current of the form (2.33) defined in some open subset $\mathcal{U}_\iota \subset \mathbb{C}_t^{N_\iota}$ and Π_ι is a holomorphic map. Let us repeat the reasoning previously used in the proof of Proposition 2.21. Let us perform additional log-resolutions $\widehat{\pi}_\iota : \widehat{\mathcal{X}}_\iota \to \mathcal{U}_\iota$ (see Theorem A.41) so that if each Π_ι is replaced by $\widehat{\Pi}_\iota = \Pi_\iota \circ \widehat{\pi}_\iota$, then τ_ι is replaced by a sum of elementary currents $\widehat{\tau}_{\iota,\iota'}$. The elementary currents $\widehat{\tau}_{\iota,\iota'}$ are still of the form (2.33) in an open subset of some neighborhood $\widehat{\mathcal{U}_{\iota,\iota'}}$ of the origin in $\mathbb{C}^{N_\iota}$. However, one deduces in addition that $\widehat{\Pi}_\iota^* h$ is a product of a monomial in t in $\widehat{\mathcal{U}_{\iota,\iota'}}$ and of an invertible function. The assumptions of the proposition imply that

$$
\begin{aligned}
0 = \Big[|h|^{2\lambda} T_z\Big]_{\lambda=0} &= \sum_{\{\iota,\iota' \,:\, \operatorname{Supp}\left(((\widehat{\Pi}_\iota)_* \widehat{\tau}_{\iota,\iota'})\right)_z \subset V_z\}} \Big[|h|^{2\lambda}((\widehat{\Pi}_\iota)_* \widehat{\tau}_{\iota,\iota'})_z\Big]_{\lambda=0} \\
&\quad + \sum_{\{\iota,\iota' \,:\, \operatorname{Supp}\left(((\widehat{\Pi}_\iota)_* \widehat{\tau}_{\iota,\iota'})\right)_z \not\subset V_z\}} \Big[|h|^{2\lambda}((\widehat{\Pi}_\iota)_* \widehat{\tau}_{\iota,\iota'})_z\Big]_{\lambda=0} \\
&= \Big[|h|^{2\lambda}\Big(\sum_{\{\iota,\iota' \,:\, \operatorname{Supp}\left(((\widehat{\Pi}_\iota)_* \widehat{\tau}_{\iota,\iota'})\right)_z \not\subset V_z\}} ((\widehat{\Pi}_\iota)_* \widehat{\tau}_{\iota,\iota'})_z\Big)\Big]_{\lambda=0}.
\end{aligned}
$$

As a consequence, one has

$$
\sum_{\{\iota,\iota' \,:\, \operatorname{Supp}\left(((\widehat{\Pi}_\iota)_* \widehat{\tau}_{\iota,\iota'})\right)_z \not\subset V_z\}} ((\widehat{\Pi}_\iota)_* \widehat{\tau}_{\iota,\iota'})_z = 0.
$$

Thus, we may delete from the representation of T_z all currents $(\widehat{\Pi}_\iota)_* \widehat{\tau}_{\iota,\iota'}$ such that $\operatorname{Supp}\big(((\widehat{\Pi}_\iota)_* \tau_{\iota,\iota'})\big)_z \not\subset V_z$; see [**AndW10**, proof of Proposition 2.2]. We are left to consider the case where $\tau = \widehat{\tau}_{\iota,\iota'}$ is a current of the form

$$
\tau = \Big(\bigotimes_{1 \le \kappa \le k'} \mathrm{PV}\Big[\frac{1}{t_\kappa^{\gamma_\kappa}}\Big]\Big) \otimes \Big(\bigwedge_{k' < \kappa \le k} \overline{\partial}\Big(\frac{1}{t_\kappa^{\gamma_\kappa}}\Big)\Big) \wedge \varpi(t) \tag{2.36}
$$

in a neighborhood of the origin $\mathcal{U}' \subset \mathbb{C}_t^{N'}$; moreover, it is such that if one denotes $\Pi = \widehat{\Pi}_\iota : \mathcal{U}' \subset \mathbb{C}^{N'} \to U_z \subset U$, then $(\operatorname{Supp} \Pi_* \tau)_z \subset V_z$. In the right-hand side of (2.36), the form ϖ is smooth with compact support. One has in addition that $\Pi^* h$ is a monomial in t multiplied by an invertible holomorphic function. Observe that the *essential support of* τ is, by definition, the set

$$
\operatorname{Supp}_{\mathrm{ess}}(\tau) := \{t \in \mathcal{U}' \,:\, t_{k'+1} = \cdots = t_k = 0\}. \tag{2.37}
$$

Let us assume first that $\Pi^* h$ vanishes identically on the essential support of τ. Then the conclusion $\overline{\Pi^* h}\, \tau = 0$ follows from the fact that

$$
\overline{t_\kappa}\, \overline{\partial}\Big(\frac{1}{t_\kappa^{\gamma_\kappa}}\Big) = 0
$$

holds for any $k' < \kappa \le k$. One has that $d\overline{h} \wedge T = 0$ holds for the same reasons, since $d(\overline{t_\kappa} w) = d\overline{t_\kappa} \wedge \overline{w} + \overline{t_\kappa}\, d\overline{w}$. It remains to consider the case where $\Pi^* h = g$ does not vanish identically on $\mathrm{Supp}_{\mathrm{ess}}(\tau)$. This means that g is the product of a monomial $(t')^\gamma$ and an invertible function v in t. In the monomial $(t')^\gamma$, the exponent satisfies $\gamma \in \mathbb{N}^{N'}$ and all its coordinates t' with nonzero exponent are distinct from the coordinates $t_{k'+1}, \dots, t_k$. It is readily seen now that

$$\big[|g|^{2\lambda}\,\tau\big]_{\lambda=0} = \big[|v \times (t')^\gamma|^{2\lambda}\tau\big]_{\lambda=0} = \tau.$$

Therefore, $\big(\Pi_*\big(\big[(1-|g|^{2\lambda})\,\tau\big]_{\lambda=0}\big)\big)_z = \big(\big[(1-|h|^{2\lambda})\,\Pi_*\tau\big]_{\lambda=0}\big)_z = 0$. Since one has also $\big(\big[|h|^{2\lambda}\Pi_*\tau\big]_{\lambda=0}\big)_z = 0$ taking into account the fact that $\mathrm{Supp}\,(\Pi_*\tau)_z \subset V_z$, one concludes that $\Pi_*\tau = 0$ in this case. Proposition 2.24 is then proved. □

The following result follows from the proof of Proposition 2.24; see [**AndW18**, Lemma 2.8]. We take the opportunity to state it here since it is of interest to us in what follows, view §2.2.4.

PROPOSITION 2.25. *Let $T \in \mathrm{PM}(U)$ and $h \in \mathcal{O}_{\mathcal{X}}(U)$, where U is an open subset of the N-dimensional complex manifold $\mathcal{X}$. Assume that the germ of T at $z \in U$ equals $T_z = \sum_\iota \big((\Pi_\iota)_*\tau_\iota\big)_z$, where each τ_ι is a current of the form* (2.33) *defined in some open subset $\mathcal{U}_\iota \subset \mathbb{C}_t^{N_\iota}$ and Π_ι is a holomorphic map. The map $\lambda \in \mathbb{C} \longmapsto (1-|h|^{2\lambda})T$ extends from $\{\lambda \in \mathbb{C} : \mathrm{Re}\,\lambda \gg 1\}$ to $\mathbb{C}$ as a current-valued map whose value at $\lambda = 0$ belongs to $\mathrm{PM}(U)$; see Proposition* 2.21 *and Remark* 2.22. *Moreover, one has for any $z \in U$ that*

(2.38)
$$\begin{aligned}(1-|h|^{2\lambda}T)_z = \Big(1-|h|^{2\lambda}\sum_\iota (\Pi_\iota)_*\tau_\iota\Big)_z &= \sum_{\{\iota\,:\,\mathrm{Supp}\,((\Pi_\iota)_*\tau_\iota)_z\subset V_z(h)\}} \big((\Pi_\iota)_*\tau_\iota\big)_z \\ &= \sum_{\{\iota\,:\,\mathrm{Supp}_{\mathrm{ess}}\,(\tau_\iota)\subset\Pi_\iota^{-1}(V_z(h))\}} \big((\Pi_\iota)_*\tau_\iota\big)_z.\end{aligned}$$

Pseudo-meromorphic currents obey the *dimension principle*, whose formulation is along the lines of the assertion in Proposition C.6 for normal (m,m)-currents, or that of the first assertion in Proposition D.5 for generalized cycles. In our formulas, the second entry in the bidegree of the pseudo-meromorphic current T needs to be specified and compared to the codimension of $\mathrm{Supp}\,T$.

PROPOSITION 2.26 (Dimension principle). *Let $\mathcal{X}$ be an N-dimensional complex manifold and let $U \subset \mathcal{X}$ be open. If the support of $T \in \mathrm{PM}^{(\bullet,m)}(U)$ is contained in a closed analytic subset V such that* $\mathrm{codim}(V) > m$, *then $T = 0$ is a current in U.*

PROOF. Let $(\zeta_1, \dots, \zeta_N)$ be local coordinates on $\mathcal{X}$ in a neighborhood of a point $z \in V_{\mathrm{reg}}$, such that $V = \{\zeta_1 = \cdots = \zeta_k = 0\}$. Proposition 2.24 implies that $d\overline{z}_\kappa \wedge T = 0$ around z for $\kappa = 1, \dots, k$. This means that $T = d\overline{z}_1 \wedge \cdots \wedge d\overline{z}_k \wedge T'$ for some current T'. But this is impossible since $k > m$, unless $T = 0$ around z. We repeat the same argument replacing V by V_{sing} and conclude inductively that $T = 0$. □

COROLLARY 2.27. *Let $T \in \mathrm{PM}(\mathcal{X})$ be a pseudo-meromorphic current with bidegree (p,m) so that $\mathrm{Supp}\,T \subset V$, where V is a purely dimensional, closed analytic subset of $\mathcal{X}$ of codimension m. Let S be a closed hypersurface which satisfies* $\mathrm{codim}(V \cap S) > m$. *For any $z \in V \cap S$ and any local defining function s for S around z, the map $\lambda \longmapsto (1-|s|^{2\lambda})T$ extends from $\{\lambda \in \mathbb{C} : \mathrm{Re}\,\lambda \gg 1\}$ as a* *current-valued map whose value at* $\lambda = 0$ *belongs to* $\mathrm{PM}(\mathcal{X})$*; see Proposition* 2.21 *and Remark* 2.22*. Moreover, one has in this case and for any* $z \in \mathcal{X}$ *that*

$$\Big(\big[(1 - |s|^{2\lambda})\,T\big]_{\lambda=0}\Big)_z = 0 \tag{2.39}$$

independently of the choice of the defining section s*. We rephrase* (2.39) *as*

$$\big(\mathbf{1}_S \cdot T\big)_z = 0. \tag{2.40}$$

PROOF. The germ at z of current $T - \big[|s|^{2\lambda}T\big]_{\lambda=0}$ is supported by $V \cap S$ since $|s|^0 = 1$ on $V \setminus S$. Since $\operatorname{codim}(V \cap S) > m$, it follows from Proposition 2.26 that $T_z - \big(\big[|s|^{2\lambda}T\big]_{\lambda=0}\big)_z = 0$. □

REMARK 2.28. One can restate Corollary 2.27 as follows: any $(\bullet, m)$ pseudo-meromorphic current, whose support is contained in a purely dimensional closed analytic subset of codimension m in $\mathcal{X}$, is realized as the *standard extension* $T_{V\setminus S \hookrightarrow V}$ from its *restriction* to $V \setminus S$. The codimension of the hypersurface S satisfies $\operatorname{codim}(V \cap S) > m$. For example, such was the case for the current $R_{\boldsymbol{S}} \wedge \omega$ in Proposition 2.16. Keeping its notation, consider $V = \boldsymbol{S}_{\{1,\dots,m\}}$ and choose the hypersurface S to be such that $V \setminus S = \boldsymbol{S}_{\{1,\dots,m\}} \cap \mathcal{X}^\circ$.

REMARK 2.29. The concept of a pseudo-meromorphic current on a complex manifold $\mathcal{X}$ introduced in Definition 2.19 can be carried through to a purely dimensional complex analytic space; see for example [**AndW18**, **AndW21**]. We stick here to the most general definition, as formulated in [**AndW18**, §2.1] and rephrased in [**AndW18**, Theorem 2.15(2)]. Suppose that $\widehat{\pi} : \widehat{\mathcal{X}} \to \mathcal{X}$ is a smooth modification around $z \in \mathcal{X}$; see Theorem A.41 in §A.4.3. Then a germ of a current at z is a germ of a pseudo-meromorphic current at z on $\mathcal{X}$ if it is the germ at z of a current $\widehat{\pi}_*(\widehat{\mu})$, where $\widehat{\mu}$ is a pseudo-meromorphic current on $\widehat{\mathcal{X}}$. It will be enough for our purposes in this monograph to follow a more restrictive definition. Such a restricted version corresponds to the original definition introduced in [**AndW10**, §2] and with the following formulation: a current is pseudo-meromorphic in a neighborhood of $z \in \mathcal{X}$ if it is the sum of currents $(\Pi_\iota)_*\tau_\iota$, where τ_ι is of the form (2.33) and $\Pi_\iota = \widehat{\pi}_{N_\iota} \circ \cdots \circ \widehat{\pi_1}$ is a tower of successive either smooth modifications or open inclusions. Propositions 2.21, 2.24, 2.26 (dimension principle) as well as Corollary 2.27 remain valid in these more general contexts; see [**AndW18**, §2.3] and [**AndW21**, §2.1].

2.2.2. The sheaf $\mathrm{CH}_{\mathcal{X},V}$. We introduce in this section the important notion of the *Coleff–Herrera current* for a purely dimensional, closed analytic subset in an N-dimensional complex manifold. These currents induce the *Coleff–Herrera sheaf* of currents. We start for that purpose with a preliminary definition.

DEFINITION 2.30. Let $0 \le M \le N$ and $z \in \mathcal{X}$. A germ Q_z of a holomorphic differential operator from the stalk $(C^\infty_{N,N-M})_z$ to the stalk $(C^\infty_{N-M,N-M})_z$ is a $\mathbb{C}$-linear operator between these two $\mathbb{C}$-vector spaces. Its action is described in local coordinates for the basis

$$\{d\overline{\zeta}_J \wedge (d\zeta_1 \wedge \cdots \wedge d\zeta_N) \,:\, |J| = N - M\}, \quad \{d\overline{\zeta}_J \wedge d\zeta_I \,:\, |I| = |J| = N - M\}$$

by a $\big(\binom{N}{M}^2, \binom{N}{M}\big)$ matrix with entries in $\mathcal{O}_z\langle \partial/\partial\zeta\rangle$. Its adjoint operator Q_z^* acts, at the level of corresponding stalks, from $({}'\mathcal{D}^{(M,M)})_z$ to $({}'\mathcal{D}^{(0,M)})_z$ as follows. If

T is a representative of T_z in an infinitesimal neighborhood U_z of z, where Q_z is represented by Q, then $Q_z^*(T_z) = (Q^*(T))_z$. Recall that

$$\forall \varphi \in \mathcal{D}^{(N-M,N-M)}(U), \quad \big\langle Q^*(T)\,,\,\varphi\big\rangle = \big\langle T, Q(\varphi)\big\rangle. \tag{2.41}$$

PROPOSITION 2.31. *Let $z \in \mathcal{X}$ and let $T_z \in (\mathrm{PM}^{(M,M)})_z$ be such that one has $\operatorname{Supp} T_z \subset V_z$, where V_z is the germ at z of a closed analytic subset V. We also assume for the sake of simplicity*[10] *that T_z equals the germ at z of a finite sum of germs of currents realized in a neighborhood of z as direct images $T_\iota = (\Pi_\iota)_* \tau_\iota$, where Π_ι is a tower of either smooth modifications or open inclusions and each elementary current τ_ι is of the form (2.33) (hence $N_\iota = N$). If Q_z is any germ of a holomorphic differential operator from $(C^\infty_{N-M,N})_z$ to $(C^\infty_{N-M,N-M})_z$, then one has that $\operatorname{Supp}(Q_z^*(T_z)) \subset V_z$ and $Q_z^*(T_z) \in (\mathrm{PM}^{(0,M)})_z$.*

PROOF. The first assertion is immediate: consider a small neighborhood U of z with a representative V of V_z in U, and a representative Q of Q_z. For any φ in $\mathcal{D}^{(N,N-M)}(U)$, one has that $\operatorname{Supp}(\varphi) \subset U \setminus V \Longrightarrow \operatorname{Supp}(Q(\varphi)) \subset U \setminus V$. Let us prove the second assertion. One may assume that a representative T of T_z in some neighborhood U of z is the direct image of a current τ of bidimension equal to $(N-M, N-M)$. The current τ is supposed to be defined in a neighborhood $\mathcal{U}$ of the origin in $\mathbb{C}^N$. Then one has $T = \Pi_* \tau$, where Π is a holomorphic map $\Pi : \mathcal{U} \to U$ which equals a tower of smooth modifications or open inclusions. Moreover, τ is of the form

$$\tau = \Big(\bigotimes_{1\leq\kappa\leq k'} \mathrm{PV}\Big[\frac{1}{t_\kappa^{\gamma_\kappa}}\Big]\Big) \otimes \Big(\bigwedge_{k'<\kappa\leq k} \overline{\partial}\Big(\frac{1}{t_\kappa^{\gamma_\kappa}}\Big)\Big) \wedge \varpi(t),$$

where $k \leq N-(N-M)$ and ϖ is a smooth $(N-(N-M), N-(N-M)-(k-k'))$ differential form with compact support. It follows from the chain rule that

$$\big\langle Q^*(T), \varphi\big\rangle = \big\langle T, Q(\varphi)\big\rangle = \big\langle \tau\,,\, \pi^*(Q(\varphi))\big\rangle = \big\langle \tau\,,\, \mathcal{Q}(\pi^*\varphi)\big\rangle,$$

where $\mathcal{Q}$ is a differential operator with meromorphic coefficients from $C^\infty_{N-M,N}(\mathcal{U})$ to $C^\infty_{N-M,N-M}(\mathcal{U})$. Let π be a smooth modification $\pi = \widehat{\mathcal{U}} \to \mathcal{U}$ such that, given any meromorphic coefficient χ of $\mathcal{Q}$, $\pi^*\chi$ expresses in local coordinates ζ as h/ζ^γ, where $\gamma \in \mathbb{N}^N$ and h is holomorphic. We now consider $\Pi \circ \pi$ instead of Π as a new tower of smooth modifications or open inclusions. Then, asserting the second assertion boils down to proving in the univariate case that the $(0,0)$- and $(0,1)$-currents

$$\theta\, d\overline{\zeta} \wedge d\zeta \longmapsto \Big\langle \mathrm{PV}\Big[\frac{1}{\zeta^\gamma}\Big]\,,\, \Big(\frac{\partial}{\partial\zeta}\Big)^\nu [\theta](\zeta)\, d\overline{\zeta} \wedge d\zeta\Big\rangle,$$
$$\theta\, d\zeta \longmapsto \Big\langle \overline{\partial}\Big(\frac{1}{\zeta^\gamma}\Big)\,,\, \Big(\frac{\partial}{\partial\zeta}\Big)^\nu [\theta](\zeta)\, d\zeta\Big\rangle$$

are pseudo-meromorphic for any $\gamma, \nu \in \mathbb{N}^*$. This follows immediately, after ν iterations, from the fact that according to Stokes's formula and formula (1.14) combined

[10]We refer the reader to [**AndW18**, Theorem 3.5] for what concerns the general case where the towers Π_ι are holomorphic maps, which amounts to saying that besides modifications and open inclusions, simple projections are also allowed in their realization; see [**AndW18**, Theorem 2.15].

with the relation $\overline{\zeta}\,\overline{\partial}(1/\zeta^\gamma) = 0$, one has

$$\int_{\mathbb{C}} \Big[\frac{|\zeta|^{2\lambda}}{\zeta^\gamma}\,\frac{\partial\theta}{\partial\zeta}\,d\overline{\zeta}\wedge d\zeta\Big]_{\lambda=0} = \gamma\Big[\int_{\mathbb{C}} \frac{|\zeta|^{2\lambda}}{\zeta^{\gamma+1}}\,\theta\,d\overline{\zeta}\wedge d\zeta\Big]_{\lambda=0},$$

$$\Big\langle \overline{\partial}\Big(\frac{1}{\zeta^\gamma}\Big),\,\frac{\partial\theta}{\partial\zeta}\,d\zeta\Big\rangle = \frac{1}{(\gamma-1)!}\Big(\frac{\partial}{\partial\zeta}\Big)^{\gamma}[\theta](0) = \gamma\,\Big\langle \overline{\partial}\Big(\frac{1}{\zeta^{\gamma+1}}\Big)\,\theta\,d\zeta\Big\rangle.$$

This concludes the proof of the proposition. □

As a corollary of this proposition, one gets the following result.

COROLLARY 2.32. *Let $\mathcal{X}$ be an N-dimensional complex manifold and let V be a purely dimensional closed subset with* $\operatorname{codim} V = M$. *Let also $T \in {}'\mathcal{D}^{(0,M)}(\mathcal{X})$ be such that* $\operatorname{Supp} T \subset V$, *and for any $z \in$* $\operatorname{Supp} T$ *there exists a germ Q_z of a holomorphic differential operator from $(C^\infty_{N-M,N})_z$ to $(C^\infty_{N-M,N-M})_z$, together with a germ $\delta_z \in \mathcal{O}_z$ with* $\operatorname{codim}_z(V(\delta)\cap V_z) > M$, *such that T_z admits the representation*

$$T_z = Q_z^*\Big(\Big[\frac{|\delta_z|^{2\lambda}}{\delta_z}[V_z]\Big]_{\lambda=0}\Big). \tag{2.42}$$

Then $T \in \mathrm{PM}^{(0,M)}(\mathcal{X})$. Furthermore, the current T satisfies around each point $z \in V$ the two following properties:

(1)
$$\forall\, h \in I(V_z) = \{h \in \mathcal{O}_z \,:\, V_z \subset V(h)\},\ \overline{h}\cdot T_z = 0; \tag{2.43}$$

(2) *given any germ of a hypersurface S_z at z such that* $\operatorname{codim}_z(S_z \cap V_z) > M$, *and a reduced defining germ s for it in a neighborhood of z,*

$$\big(\big[|s|^{2\lambda}T\big]_{\lambda=0}\big)_z = T_z. \tag{2.44}$$

PROOF. The first assertion follows from the fact that $[V] \in \mathrm{PM}^{(M,M)}(\mathcal{X})$; see Example 2.20. The second assertion follows from Propositions 2.31 and 2.24, together with Corollary 2.27. □

For a purely dimensional closed analytic subset V with $\operatorname{codim} V = M > 0$ in the complex manifold $\mathcal{X}$, the sheaf of $\overline{\partial}$-closed currents of bidegree $(\bullet, M)$ supported by V will play a major role. Its definition is suggested by Corollary 2.32. This sheaf is called the *Coleff–Herrera sheaf*, referring to the pioneering contribution of Nicolas R. Coleff and Miguel E. Herrera [**CoH78**]; see also [**DS85**]. Its definition was formalized by J.E. Björk in [**Bj96, Bj04**]. The variation of the definition we present here is based on the analytic continuation approach used by M. Andersson in [**And09**].

DEFINITION 2.33 (Coleff–Herrera sheaf). Let $\mathcal{X}$ be an N dimensional manifold and let V be a closed analytic subset of pure codimension $M > 0$. A *Coleff–Herrera current* T in an open subset $U \subset \mathcal{X}$, with respect to V, is a $\overline{\partial}$-closed $(\bullet, M)$-current T in U, with $\operatorname{Supp} T \subset V$ satisfying the following properties:

(1) for each $z \in U \cap V$, its germ T_z at z satisfies (2.43);
(2) for any $s \in \mathcal{O}_z$ such that $\operatorname{codim}_z(V(s) \cap V_z) > M$,[11]

$$\lambda \longmapsto |s|^{2\lambda}T_z$$

[11] This additional property is also usually reformulated as follows : T satisfies the *standard extension property* (S. E. P) locally around $V \cap S$ from its restriction to the open subset $V \setminus S$, when S is a closed hypersurface so that $\operatorname{codim}(V \cap S) > \operatorname{codim} V = M$.

extends from $\{\lambda \in \mathbb{C} : \operatorname{Re}\lambda \gg 1\}$ to the complex plane $\mathbb{C}$ as a $({}'\mathcal{D}^{(\bullet,M)})_z$-valued meromorphic function for which (possible) poles have strictly negative real part and which satisfies (2.44).

The sheaf, whose sections are (p, M) Coleff–Herrera currents, is denoted by $\mathrm{CH}^p_{\mathcal{X},V}$, with

$$\mathrm{CH}_{\mathcal{X},V} = \bigoplus_{p=0}^{N} \mathrm{CH}^p_{\mathcal{X},V}.$$

REMARK 2.34. Let V be a closed analytic subset with pure codimension M of an N-dimensional complex manifold $\mathcal{X}$. Given $z^0 \in V$, consider at the level of germs at z^0 two local embeddings $\iota' : V \cap U_{z^0} \hookrightarrow \mathrm{U}' \subset \mathbb{C}^{N_{\iota'}}$ and $\iota'' : V \cap U_{z^0} \hookrightarrow \mathrm{U}'' \subset \mathbb{C}^{N_{\iota''}}$. As in Remark A.16, one may assume that $\iota' = \iota_{\min}$ is minimal, thus $\mathrm{U}'' = \mathrm{U}' \times \mathbb{C}^\nu_w$ with $\nu = N_{\iota''} - N_{\iota_{\min}}$ and $\iota'' = \iota : z \mapsto (\iota_{\min}(z), 0)$ in a neighborhood of z^0 on V. One has

$$(2.45)\quad \mathfrak{T}_{\iota_{\min}(z^0)} \in (\mathrm{CH}^{N-M}_{\mathrm{U}',\iota_{\min}(V\cap U_{z^0})})_{\iota_{\min}(z^0)} \Longrightarrow (\mathfrak{T} \wedge [w=0])_{\iota(z^0)} \in (\mathrm{CH}^{N-M+\nu}_{\mathrm{U}'\times\mathbb{C}^\nu,\iota(V\cap U_{z^0})})_{\iota(z^0)}.$$

As a consequence, the morphism

$$(2.46)\qquad (\mathrm{CH}^{N-M+\nu}_{\mathrm{U}'',\iota(V\cap U_{z^0})})_{\iota(z^0)} \hookrightarrow \big([w=0] \wedge \mathrm{CH}^{N-M}_{\mathrm{U}',\iota_{\min}(V\cap U_{z^0})}\big)_{\iota(z^0)}$$

is injective. Such a morphism fails to induce an isomorphism of sheaves since, for example,

$$\Big(dw \wedge \bigwedge_{j=1}^{m} \overline{\partial} \frac{1}{w_j^{\gamma_j+1}} \wedge \mathrm{CH}^{N-M}_{\mathrm{U}',\iota_{\min}(V\cap U_{z^0})}\Big)_{\iota(z^0)} \subset (\mathrm{CH}^{N-M+\nu}_{\mathrm{U}'',\iota(V\cap U_{z^0})})_{\iota(z^0)}$$

while

$$(2.47)\quad \Big(dw \wedge \bigwedge_{j=1}^{\nu} \overline{\partial} \frac{1}{w_j^{\gamma_j+1}} \wedge \mathrm{CH}^{N-M}_{\mathrm{U}',\iota_{\min}(V\cap U_{z^0})}\Big)_{\iota(z^0)} \subset \big([w=0] \wedge \mathrm{CH}^{N-M}_{\mathrm{U}',\iota_{\min}(V\cap U_{z^0})}\big)_{\iota(z^0)} \Longleftrightarrow (\gamma_1, \dots, \gamma_\nu) = 0.$$

The possibility of taking derivatives in additional directions, such as those corresponding here to the w_j's, prevents as a consequence the (germs of) Coleff–Herrera sheaves $(\mathrm{CH}_{\mathcal{X},V})_z$, where $z \in V$, being independent of the local embedding $V_z \hookrightarrow \mathcal{X}_z$. Despite this fact, the Coleff–Herrera sheaf $\mathrm{CH}^{N-M}_{\mathcal{X},V} = \mathrm{CH}^{\dim V}_{\mathcal{X},V}$, once combined with the structural sheaf on V, plays a key role in the representation of the *dualizing sheaf*[12] ω_V^{dimV} as

$$(2.48)\ \omega_V^{N-M} = \omega_V^{\dim V} = \mathrm{Hom}_{\mathcal{O}_\mathcal{X}}(\mathcal{O}_\mathcal{X}/\mathcal{I}_{\mathcal{X},V}, \mathrm{CH}^{N-M}_{\mathcal{X},V}) = \mathrm{Hom}_{\mathcal{O}_\mathcal{X}}(\mathcal{O}_V, \mathrm{CH}^{N-M}_{\mathcal{X},V}).$$

The reason why the sheaf on the right-hand side of (2.48) is intrinsically defined is that

$$\mathrm{Hom}_{\mathcal{O}_{\mathrm{U}'}}(\mathcal{O}_{\mathrm{U}'}/\mathcal{I}_{\iota_{\min}(V\cap U_{z^0}),\mathrm{U}'}, \mathrm{CH}^{N-M}_{\mathrm{U}',\iota_{\min}(V\cap U_{z^0})}) \simeq \mathrm{Hom}_{\mathcal{O}_{\mathrm{U}''}}(\mathcal{O}_{\mathrm{U}''}/\mathcal{I}_{\iota(V\cap U_{z^0}),\mathrm{U}''}, \mathrm{CH}^{N-M+\nu}_{\mathrm{U}'',\iota(V\cap U_{z^0})}),$$

[12]See Definition B.34 and Remark B.35.

where $\iota' = \iota_{\min}$ is a local minimal embedding around z^0, thus $\mathrm{U}'' = \mathrm{U}' \times \mathbb{C}^\nu_w$ with $\nu = N_{\iota''} - N_{\iota_{\min}}$ and $\iota'' = \iota : z \mapsto (\iota_{\min}(z), 0)$ in a neighborhood of z^0. Instead of the structural sheaf $\mathcal{O}_V$, one may as well use the sheaves Ω_V^{N-M-p}, $p = 1, \dots, N-M$, of strongly holomorphic $(N-M-p, 0)$-forms on V which are also intrinsically defined.[13] It follows then from the isomorphisms

$$\begin{aligned}\mathrm{Hom}_{\mathcal{O}_{\mathrm{U}'}}\big(\Omega_{\mathrm{U}'}^{N-M-p}/(\mathcal{J}_{\iota_{\min}(V\cap U_{z^0})\hookrightarrow \mathrm{U}'})_{\text{Käh}}, \mathrm{CH}^{N-M}_{\mathrm{U}', \iota_{\min}(V\cap U_{z^0})}\big) \\ \simeq \mathrm{Hom}_{\mathcal{O}_{\mathrm{U}''}}\big(\Omega_{\mathrm{U}''}^{N-M-p}/(\mathcal{J}_{\iota(V\cap U_{z^0})\hookrightarrow \mathrm{U}''})_{\text{Käh}}, \mathrm{CH}^{N-M+\nu}_{\mathrm{U}'', \iota(V\cap U_{z^0})}\big)\end{aligned}$$

involving the sheaves of Kähler-Grothendieck differentials[14] under the same conditions on the local embeddings $\iota' = \iota_{\min}$ and $\iota'' = \iota$, together with the fact that Coleff–Herrera currents satisfy the standard extension property (which allows to quotient out the torsion) that for any $p = 0, \dots, N-M-1$,

$$\omega_V^p \simeq \mathrm{Hom}_{\mathcal{O}_{\mathcal{X}}}(\Omega_V^{N-M-p}, \mathrm{CH}^{N-M}_{\mathcal{X},V}), \tag{2.49}$$

where the Barlet sheaf $\omega^p_{V\subset\mathcal{X}} = \omega^p_V$ is the sheaf which sections are $(p,0)$-forms holomorphic in the sense of Barlet; see Definition B.34. We refer the reader for more details about such isomorphisms to [**Sam21**, proof of Proposition 4.3].

EXAMPLE 2.35. Proposition 2.24 shows that any pseudo-meromorphic current T in $U \subset \mathcal{X}$, whose support $\operatorname{Supp} T$ is contained in $U \cap V$, satisfies (2.43). Corollary 2.27 shows that any such pseudo-meromorphic current T, with bidegree $(\bullet, M)$, satisfies also the *standard extension property* required in Definition 2.33. As a consequence,[15] any $\overline{\partial}$-closed pseudo-meromorphic $(\bullet, M)$-current in U supported by the purely dimensional closed subset $U \cap V$ belongs to $\mathrm{CH}_{\mathcal{X},V}(U)$.

REMARK 2.36. Given a holomorphic bundle E over $\mathcal{X}$, the sheaf of E-valued Coleff–Herrera currents for a closed analytic subset V with pure codimension M is the sheaf whose sections on any open subset U are E-valued $(\bullet, M)$-currents in U supported by $U \cap V$. The coordinates of these sections expressed in a local frame in a neighborhood of any $z \in U \cap V$ have their germs belonging to the corresponding stalk $(\mathrm{CH}_{\mathcal{X},V})_z$. The sheaf is denoted by $\mathrm{CH}_{\mathcal{X},V}(\cdot, E)$ and is decomposed as

$$\mathrm{CH}_{\mathcal{X},V}(\cdot, E) = \bigoplus_{p=0}^{N} \mathrm{CH}^p_{\mathcal{X},V}(\cdot, E),$$

according to the bidegrees (p, M) of currents, for $p = 0, \dots, N$.

We state next an important *local structure theorem* for sections of $\mathrm{CH}^0_{\mathcal{X},V}$. Besides the fact that such a result (see [**Bj96, Bj04, And09, BjS10**]) shows their pseudo-meromorphicity, it also emphasizes the characteristic property that only holomorphic differential operators are involved in the action of such currents. This explains why they do play a role of an algebraic nature, despite of their analytic structure. These differential operators with analytic coefficients are, as we will see later on in this monograph, reminiscent of the so-called *Nœtherian operators* involved in the algebraic formulation of the *Ehrenpreis–Palamodov fundamental*

[13]See Definition B.32 and Remark B.33.

[14]See Remark B.33.

[15]One could view this assertion as the residue analogue to the second assertion in the dimension principle for normal currents (see Proposition C.26) or the second assertion in the dimension principle for generalized cycles (see Proposition D.5).

principle, [**Ehr, Pal, Bj74, AndW07**] or §5.1.2 and §7.6.3 in this monograph. Local structure results of this kind originally go back to the work of P. Dolbeault [**Dol93**]. The result below is due to Jan-Erik Björk [**Bj04**], who formalized for the first time the concept of the Coleff–Herrera current and showed the major role it plays.

PROPOSITION 2.37 (Structure of $\mathrm{CH}^0_{\mathcal{X},V}$-germs). *Let $\mathcal{X}$ and V be as above, where* $\operatorname{codim} V = M$, *$z \in V$, and $T_z \in (\mathrm{CH}^0_{\mathcal{X},V})_z$. There exists a germ q_z of a holomorphic differential operator from $(C^\infty_{N,N-M})_z$ to $(C^\infty_{0,N-M})_z$, together with a germ ω_z of an $(N-M,0)$ holomorphic form, and $\delta_z \in \mathcal{O}_z$ with* $\operatorname{codim}(V(\delta)\cap V_z) > M$, *such that*

$$T_z = \Big(Q^*\Big(\Big[\frac{|\delta|^{2\lambda}}{\delta}\,[V]\Big]_{\lambda=0}\Big)\Big)_z, \tag{2.50}$$

where $Q_z : (C^\infty_{N,N-M})_z \to (C^\infty_{N-M,N-M})_z$ is the germ of a holomorphic differential operator defined by

$$\forall\, \varphi \in (C^\infty_{N,N-M})_z, \quad Q_z(\varphi) = q_z(\varphi)\wedge \omega_z.$$

PROOF. We follow here the argument from [**And09**]. First, consider the case when V is smooth around z. Thus V is defined in local, centered coordinates $(\zeta_1,\dots,\zeta_N)$ in a neighborhood of this point as $V = \{\zeta_1 = \cdots = \zeta_M\} = \{\boldsymbol{\zeta} = 0\}$. We set ω to be the differential form $\omega = d\zeta_{M+1}\wedge\cdots\wedge d\zeta_N$. Let also N_T be the order of the representative T around z. For $j = 1,\dots,M$, one has $\overline{\zeta}_j T_z = 0$, because of (2.43). This implies

$$d\overline{\zeta_j}\wedge T_z = 0, \quad \forall\, j = 1,\dots,M, \tag{2.51}$$

since T is assumed to be $\overline{\partial}$-closed. From now on, we fix a small neighborhood $U = U_z$ of z, namely a polydisk $\Delta_{\boldsymbol{\zeta}}\times\Delta_{\boldsymbol{\zeta}'}$, where we separate the local coordinates in two blocks $\boldsymbol{\zeta} = (\zeta_1,\dots,\zeta_M)$ and $\boldsymbol{\zeta}' = (\zeta_{M+1},\dots,\zeta_N)$, and keep T as representative for T_z of order N_T in U. Let

$$\varphi = \Big(\sum_{|J|=N-M} \varphi_J\, d\overline{\zeta_J}\Big)\wedge (d\zeta_1\wedge\cdots\wedge d\zeta_N) \in \mathcal{D}^{(N,N-M)}(U).$$

It follows from (2.51) and the property $\overline{\zeta_j}T = 0$ for $j = 1,\dots,M$, that for any $\theta \in \mathcal{D}(U)$, which equals identically to one around z,

$$\begin{aligned}
\langle \theta T,\varphi\rangle &= \langle T, \theta\,\varphi_{\{M+1,\dots,N\}}\, d\overline{\boldsymbol{\zeta}'}\wedge d\boldsymbol{\zeta}\wedge\omega(\boldsymbol{\zeta}')\rangle = \langle T, \theta\,\psi\, d\overline{\boldsymbol{\zeta}'}\wedge d\boldsymbol{\zeta}\wedge\omega(\boldsymbol{\zeta}')\rangle\\
&= \pm\Big\langle T\,,\,\theta\Big(\sum_{\{\kappa\in\mathbb{N}^M\,:\,|\kappa|\le N_T\}}\Big(\frac{\partial}{\partial\boldsymbol{\zeta}}\Big)^{\kappa}[\psi](0,\boldsymbol{\zeta}')\,\frac{\boldsymbol{\zeta}^\kappa}{\kappa!}\Big)\,,\, d\overline{\boldsymbol{\zeta}'}\wedge\omega(\boldsymbol{\zeta}')\wedge d\boldsymbol{\zeta}\Big\rangle,
\end{aligned}$$

where $d\boldsymbol{\zeta} = d\zeta_1\wedge\cdots\wedge d\zeta_M$ and $d\boldsymbol{\zeta}' = d\zeta_{M+1}\wedge\cdots\wedge d\zeta_N$. We use here notations that will be used throughout this monograph, namely, for any $\kappa\in\mathbb{N}^M$,

$$\kappa! = \prod_{j=1}^M \kappa_j!,\quad \Big(\frac{\partial}{\partial\boldsymbol{\zeta}}\Big)^{\kappa} = \prod_{j=1}^M\Big(\frac{\partial}{\partial\zeta_j}\Big)^{\kappa_j},\quad \boldsymbol{\zeta}^\kappa = \prod_{j=1}^M \zeta_j^{\kappa_j},\quad |\kappa| = \kappa_1+\cdots+\kappa_M.$$

If π denotes the projection $(\boldsymbol{\zeta},\boldsymbol{\zeta}')\longmapsto\boldsymbol{\zeta}'$, then the direct images of $(0,0)$-currents (that is, of distributions) in $\Delta_{\boldsymbol{\zeta}'}$, defined by

$$a_\kappa(\boldsymbol{\zeta}') = \pi_*\big(\boldsymbol{\zeta}^\kappa T(\boldsymbol{\zeta},\boldsymbol{\zeta}')\wedge d\boldsymbol{\zeta}\big),\quad \kappa\in\mathbb{N}^M \text{ with } |\kappa|\le N_T,$$

are represented by holomorphic functions a_κ in $\Delta_{\boldsymbol{\zeta}'}$. This follows from the fact that T is $\overline{\partial}$-closed. Therefore, if

$$q\Big(\zeta, \frac{\partial}{\partial\zeta}\Big)(\varphi) = \pm\Big(\sum_{\{\kappa\in\mathbb{N}^M\,:\,|\kappa|\le N_T\}} \frac{a_\kappa(\boldsymbol{\zeta}')}{\kappa!}\Big(\frac{\partial}{\partial\zeta}\Big)^\kappa\Big)(\psi)\, d\overline{\boldsymbol{\zeta}'},$$

then one has

$$\langle T_z, \varphi_z\rangle = \big\langle [\boldsymbol{\zeta} = 0]\,,\, q(\varphi)\wedge\omega\big\rangle = \big\langle [V]_z\,,\, Q(\varphi_z)\big\rangle,$$

which proves the result in this case. Let us now consider the case where $z \in V_{\rm sing}$. We recall here the following fact; see Proposition A.24 in §A.3.4. In a convenient neighborhood $U = U_z$ of z, $V \cap U$ is the union of finitely many components $\mathfrak{W}_1, \dots, \mathfrak{W}_L$ of a reduced complete intersection W. More precisely $W = \{\zeta \in U : f_1(\zeta) = \cdots = f_M(\zeta) = 0\}$, where $f_j \in \mathcal{O}(U)$ for $j = 1, \dots, M$, and $df_1\wedge\cdots\wedge df_M \not\equiv 0$ on any irreducible component of W, in particular on $\mathfrak{W}_1, \dots, \mathfrak{W}_L$. Furthermore, because of the Nœther presentation of V, which follows from Proposition A.24 (see §A.3.4), one can choose centered local coordinates $(\boldsymbol{\zeta}, \boldsymbol{\zeta}') \in \Delta_{\boldsymbol{\zeta}} \times \Delta_{\boldsymbol{\zeta}'}$ in a neighborhood of z, where $\boldsymbol{\zeta} = (\zeta_1, \dots, \zeta_M)$ and $\boldsymbol{\zeta}' = (\zeta_{M+1}, \dots, \zeta_N)$ such that the projection $\pi : (\boldsymbol{\zeta}, \boldsymbol{\zeta}') \to \boldsymbol{\zeta}'$ is proper. Moreover,[16]

$$J(\boldsymbol{\zeta}, \boldsymbol{\zeta}') = \Big|\frac{\partial \boldsymbol{f}(\boldsymbol{\zeta}, \boldsymbol{\zeta}')}{\partial \boldsymbol{\zeta}}\Big| \not\equiv 0 \text{ on } (\Delta_{\boldsymbol{\zeta}} \times \Delta_{\boldsymbol{\zeta}'}) \cap \mathfrak{W}_\ell \quad \text{for } \ell = 1, \dots, L. \tag{2.52}$$

As before, let N_T be the order of a representative T of T_z in an arbitrarily small neighborhood U of z. In a neighborhood of any point $z_0 = (\boldsymbol{\zeta}_0, \boldsymbol{\zeta}'_0)$ in the open subset $\big((\Delta_{\boldsymbol{\zeta}} \times \Delta_{\boldsymbol{\zeta}'}) \cap V\big) \setminus V(J)$, one can use $(\boldsymbol{f}, \boldsymbol{\zeta}')$ as local coordinates. The chain rule implies that

$$\begin{pmatrix} \partial/\partial f_1 \\ \vdots \\ \partial/\partial f_M \end{pmatrix}(\boldsymbol{\zeta}, \boldsymbol{\zeta}') = {}^t\Big(\Big[\frac{\partial \boldsymbol{f}}{\partial \boldsymbol{\zeta}}(\boldsymbol{\zeta}, \boldsymbol{\zeta}')\Big]^{-1}\Big) \begin{pmatrix} \partial/\partial \zeta_1 \\ \vdots \\ \partial/\partial \zeta_M \end{pmatrix}(\boldsymbol{\zeta}, \boldsymbol{\zeta}'). \tag{2.53}$$

Since T is $\overline{\partial}$-closed, the direct images of $(0,0)$-currents a_κ, with $\kappa \in \mathbb{N}^M$ satisfying $|\kappa| \le N_T$, defined in $\Delta_{\boldsymbol{\zeta}'}$ by

$$a_\kappa(\boldsymbol{\zeta}') = \pi_*\big(\boldsymbol{f}^\kappa(\boldsymbol{\zeta}, \boldsymbol{\zeta}')\, d\boldsymbol{f}(\boldsymbol{\zeta}, \boldsymbol{\zeta}') \wedge T(\boldsymbol{\zeta}, \boldsymbol{\zeta}')\big)$$

(where $d\boldsymbol{f} = df_1\wedge\cdots\wedge df_M$), are $\overline{\partial}$-closed. Thus, these $(0,0)$-currents are represented by holomorphic functions a_κ in $\Delta_{\boldsymbol{\zeta}'}$. Moreover, for $\nu^\circ \in \mathbb{N}^*$ large enough,

$$q^\circ : (\boldsymbol{\zeta}, \boldsymbol{\zeta}') \longmapsto J^{\nu^\circ}(\boldsymbol{\zeta}, \boldsymbol{\zeta}') \sum_{\{\kappa\in\mathbb{N}^M\,:\,|\kappa|\le N\}} \frac{a_\kappa(\boldsymbol{\zeta}')}{\kappa!}\Big(\frac{\partial}{\partial \boldsymbol{f}}\Big)^\kappa$$

is a holomorphic differential operator in $\Delta_{\boldsymbol{\zeta}}\times\Delta_{\boldsymbol{\zeta}'}$. Observe now that any differential $(N, N-M)$ form $\varphi(\boldsymbol{\zeta}, \boldsymbol{\zeta}') = \psi\, d\overline{\boldsymbol{\zeta}'} \wedge d\zeta_1 \wedge\cdots\wedge d\zeta_N$ in $\Delta_{\boldsymbol{\zeta}} \times \Delta_{\boldsymbol{\zeta}'}$ can be rewritten as

$$\varphi(\boldsymbol{\zeta}, \boldsymbol{\zeta}') = \frac{\psi(\boldsymbol{\zeta}, \boldsymbol{\zeta}')}{J(\boldsymbol{\zeta}, \boldsymbol{\zeta}')}\, d\overline{\boldsymbol{\zeta}'} \wedge d\boldsymbol{f}(\boldsymbol{\zeta}, \boldsymbol{\zeta}') \wedge d\boldsymbol{\zeta}'.$$

[16] Note that the jacobian determinant J in (2.52) can be used as a *universal Oka denominator* for holomorphic functions on $U \cap V_{\rm reg}$ that extend as locally bounded functions on $U \cap V$; see §A.3.5.

Since, for $\nu \in \mathbb{N}^*$ large enough,

$$q : \psi \longmapsto J^\nu q^\circ\Big(\frac{\psi}{J}\Big)$$

remains a holomorphic differential operator in $\Delta_\zeta \times \Delta_{\zeta'}$, one can represent T in $(\Delta_\zeta \times \Delta_{\zeta'}) \setminus V(J)$ as

$$T = Q^*\Big(\Big[\frac{|J|^{2\lambda}}{J^{\nu^\circ+\nu}}\Big]_{\lambda=0}[V]\Big), \tag{2.54}$$

where

$$Q : \psi\, d\overline{\zeta'} \wedge (d\zeta_1 \wedge \cdots \wedge d\zeta_N) \longmapsto \pm\, q(\psi)\, d\zeta'.$$

Finally we claim that both the current T and the pseudo-meromorphic $(0, M)$-current

$$Q^*\Big(\Big[\frac{|J|^{2\lambda}}{J^{\nu^\circ+\nu}}\Big]_{\lambda=0}[V]\Big)$$

in $\Delta_\zeta \times \Delta_{\zeta'}$ inherit the standard extension property (2.44). Actually, for $\mu \in \mathbb{C}$ with $\operatorname{Re}\mu \gg 1$,

$$(|J|^{2\mu}\, T)_z = \Big(|J|^{2\mu}\, Q^*\Big(\Big[\frac{|J|^{2\lambda}}{J^{\nu^\circ+\nu}}\Big]_{\lambda=0}[V]\Big)\Big)_z$$

as a consequence of the equality of $(0, M)$-currents in $(\Delta_\zeta \times \Delta_{\zeta'}) \setminus V(J)$. The assertion of the proposition follows by applying the analytic continuation principle. The proof of Proposition 2.37 is completed. □

REMARK 2.38. One should point out that the denominator δ that appears in the local representation (2.50) in the above proof can be chosen as a power of the Oka universal denominator attached to V around the point z; see §A.3.5.

Proposition 2.37 for sections of $\mathrm{CH}^0_{\mathcal{X},V}$ can be slightly modified to show that $(\mathrm{CH}^p_{\mathcal{X},V})_z \subset \mathrm{PM}_z$ for any $z \in V$ and for any $0 \le p \le N$. In fact, one can summarize all results established up to now and conclude this subsection with the following statement.

THEOREM 2.39. *Let $\mathcal{X}$ be an N-dimensional complex manifold and let V be any purely dimensional, closed analytic subset of $\mathcal{X}$ of* $\operatorname{codim} V = M > 0$. *For each $p = 0, \dots, N$, the sheaf $\mathrm{CH}^p_{\mathcal{X},V}$ coincides with the sheaf of (p, M) pseudo-meromorphic currents supported by V and $\overline{\partial}$-closed.*

PROOF. As we have already seen (see Example 2.35), any pseudo-meromorphic (p, M)-current in $U \subset \mathcal{X}$ supported by V and $\overline{\partial}$-closed in U belongs to $\mathrm{CH}^p_{\mathcal{X},V}(U)$. It follows from Proposition 2.37, together with Example 2.20 and Proposition 2.31, that sections of $\mathrm{CH}^0_{\mathcal{X},V}$ are also sections of $\mathrm{PM}^{(0,M)}$. Thus Theorem 2.39 holds for $p = 0$. Let us consider $T_z \in \mathrm{CH}^p_{\mathcal{X},V}$ and show that it is a germ at z of pseudo-meromorphic (p, M)-current. Let us express the germ T_z as

$$T_z = \Big(\sum_{1\le i_1<\cdots<i_p\le N} (d\zeta_{i_1} \wedge \cdots \wedge d\zeta_{i_p}) \wedge T_I\Big)_z.$$

Then each $(T_I)_z$ is the germ at z of a $\overline{\partial}$-closed current since T_z is. Since properties (1) and (2) in Definition 2.33 hold for T_z, they also hold for any $(T_I)_z$ at any $z \in V$. As a consequence, one has $(T_I)_z \in (\mathrm{CH}^0_{\mathcal{X},V})_z \subset \mathrm{PM}^{(0,M)}_z$ for any $z \in V$. Since PM is a module over smooth forms as observed from Definition 2.19 before stating Proposition 2.21, one has $(d\zeta_I \wedge T_I)_z \in \mathrm{PM}_z^{(p,M)}$ for any $z \in V$ and any ordered multi-index I with length p. Hence $T \in \mathrm{PM}^{(p,M)}(U)$. Since the current T is supported by V and is $\overline{\partial}$-closed, the proof of Theorem 2.39 is complete. □

2.2.3. Coleff–Herrera currents with prescribed polar set. Let $\mathcal{X}$ be an N-dimensional complex manifold, and let $V \subset \mathcal{X}$ be a purely dimensional, closed analytic subset of $\operatorname{codim} V = M > 0$ as in §2.2.2. Let also S be a closed complex hypersurface in an open neighborhood of V such that $\operatorname{codim}(V \cap S) > M$. Recall that one defines meromorphic functions with a polar set included in a closed hypersurface $S \subset \mathcal{X}$ starting from the definition of meromorphic functions on $\mathcal{X}$ as holomorphic functions h in $\mathcal{X} \setminus S$ that are such that $[h]$ extends as a $\overline{\partial}$-closed distribution to $\mathcal{X}$. Then it is natural to use the Cauchy–Riemann operator $\overline{\partial}$ to introduce the sheaf $\mathrm{CH}^{\star S}_{\mathcal{X},V}$ of *meromorphic Coleff–Herrera currents with prescribed polar set along* S. This is done as follows.

DEFINITION 2.40 (The sheaf $\mathrm{CH}^{\star S}_{\mathcal{X},V}$). Let U be an open subset of $\mathcal{X}$. A $(\bullet, M)$-current supported by V is said to be *a meromorphic Coleff–Herrera current with prescribed polar set along* S if it satisfies both conditions (1) and (2) as in Definition 2.33, together with the condition

$$\operatorname{Supp}(\overline{\partial} T) \subset V \cap S. \tag{2.55}$$

The corresponding sheaf $\mathrm{CH}^{\star S}_{\mathcal{X},V}$ decomposes as

$$\mathrm{CH}^{\star S}_{\mathcal{X},V} = \bigoplus_{p=0}^{N} \mathrm{CH}^{\star S,p}_{\mathcal{X},V}$$

according to the bidegrees (p, M)-currents for $p = 0, \dots, N$.

REMARK 2.41. Let E be a holomorphic bundle and let U an open subset of $\mathcal{X}$. An E-valued current $T \in {}'\mathcal{D}^{(\bullet,M)}(U, E)$ is said to be a *meromorphic E-valued current with prescribed polar set along* S if its coordinates in a local frame in a neighborhood of any $z \in U \cap V$ have their germs at z lie in $(\mathrm{CH}^{\star S}_{\mathcal{X},V})_z$. This sheaf is denoted by $\mathrm{CH}^{\star}_{\mathcal{X},V}(\cdot, E)$ and splits as

$$\mathrm{CH}^{\star S}_{\mathcal{X},V}(\cdot, E) = \bigoplus_{p=0}^{N} \mathrm{CH}^{\star S,p}_{\mathcal{X},V}(\cdot, E),$$

according to the bidegrees of (p, M)-currents for $p = 0, \dots, N$.

Theorem 2.39 implies the following useful result.

THEOREM 2.42. *Let $\mathcal{X}$ be an N-dimensional complex manifold and let $V \subset \mathcal{X}$ be a purely dimensional closed analytic subset with codimension M. Assume also that S is a closed analytic hypersurface in a neighborhood of V such that one has* $\operatorname{codim}(V \cap S) > M$. *The sheaf* $\mathrm{CH}^{\star S}_{\mathcal{X},V}$ *coincides with the sheaf of* $(\bullet, M)$ *pseudo-meromorphic currents T supported by V and satisfies* $\operatorname{Supp}(\overline{\partial} T) \subset V \cap S$.

PROOF. Any pseudo-meromorphic $(\bullet, M)$-current in $U \subset \mathcal{X}$, with $\operatorname{Supp} T$ included in V, satisfies conditions (1) and (2) in Definition 2.33; see Proposition 2.24 and Corollary 2.27. Then, any pseudo-meromorphic $(\bullet, M)$-current T supported by V and which satisfies (2.55) belongs to $\mathrm{CH}^{\star S}_{\mathcal{X},V}(U)$. Let us now prove the converse inclusion. In order to do that, consider $T_z \in \mathrm{CH}^{\star S}_{\mathcal{X},V}$. That is, consider the germ

T_z of Coleff–Herrera current with prescribed polar set along S at a point in $z \in V$. Since $(\mathrm{CH}^{\star S}_{\mathcal{X},V})_z = (\mathrm{CH}_{\mathcal{X},V})_z$ for any $z \in V \setminus S$ and Theorem 2.39 holds in that case, we are led to consider the case where $z \in V \cap S$. Let s be a reduced defining holomorphic function for S around z. Then for $\nu \in \mathbb{N}^*$ strictly larger than the order of $(\overline{\partial} T)_z$ around z, one has that

$$(s^\nu T)_z \in (\mathrm{CH}_{\mathcal{X},V})_z \subset \mathrm{PM}_z,$$

where the last inclusion follows from Theorem 2.39. Furthermore, Proposition 2.21 asserts that the $('\mathcal{D})_z$ valued map

$$\lambda \longmapsto \Big(\Big[\frac{|s|^{2\nu\lambda}}{s^\nu}\,(s^\nu T)\Big]_{\lambda=0}\Big)_z$$

extends from $\{\lambda \in \mathbb{C} \,:\, \mathrm{Re}\,\lambda \gg 1\}$ to the complex plane $\mathbb{C}$ as a meromorphic map with (possible) poles in $-\mathbb{Q}^+$. Its value at $\lambda = 0$ belongs to PM_z. Since $\big(\big[|s|^{2\nu\lambda}T\big]_{\lambda=0}\big)_z = T_z$, because T satisfies condition (1) as in Definition 2.33, one gets that $T_z \in \mathrm{PM}_z$. Since T_z satisfies also that $\mathrm{Supp}\,(\overline{\partial} T)_z \subset V_z \cap S_z$, the converse inclusion is proved. This concludes the proof of the theorem. □

Sections of $\mathrm{CH}^{\star S}_{\mathcal{X},V}$ induce via the action of the $\overline{\partial}$ operator sections of $\mathrm{CH}_{\mathcal{X},V\cap S}$ as follows.

THEOREM 2.43. *Let $\mathcal{X}$ be an N-dimensional complex manifold and $V \subset \mathcal{X}$ be a purely dimensional, closed analytic subset with codimension M. Let also S be a closed analytic hypersurface in a neighborhood of V such that* $\mathrm{codim}(V \cap S) > M$. *Then the following inclusion of sheaves is valid:*

$$\overline{\partial}\,(\mathrm{CH}^{\star S}_{\mathcal{X},V})_z \subset (\mathrm{CH}_{\mathcal{X},V\cap S})_z, \ \textit{for any } z \in V. \tag{2.56}$$

PROOF. Theorem 2.42 and the obvious fact that $\overline{\partial}$ acts from $\mathrm{PM}^{(\bullet,M)}$ to $\mathrm{PM}^{(\bullet,M+1)}$ imply the inclusions

$$\overline{\partial}\,(\mathrm{CH}^{\star S}_{\mathcal{X},V})_z \subset \overline{\partial}\,(\mathrm{PM}^{(\bullet,M)})_z \subset (\mathrm{PM}^{(\bullet,M+1)})_z.$$

Given $z \in V$ and $T_z \in (\mathrm{CH}^{\star S}_{\mathcal{X},V})_z$, one has by definition

$$\mathrm{Supp}\,(\overline{\partial} T_z) \subset V_z \cap S_z.$$

Since $\mathrm{codim}_z(V \cap S) > M$, $\overline{\partial} T_z \in (\mathrm{PM}^{(\bullet,M+1)})_z$, and $\overline{\partial}(\overline{\partial} T_z) = 0$, it follows from Theorem 2.39 that $\overline{\partial} T_z \in (\mathrm{CH}_{\mathcal{X},V\cap S})_z$, which concludes the proof of the theorem. □

If $T_z \in (\mathrm{CH}_{\mathcal{X},V})_z$ is a germ of a Coleff–Herrera current at $z \in V$ and S_z is a germ of hypersurface at z with reduced defining function s around z, then it follows from Theorem 2.43 that

$$\Big(\overline{\partial}\Big(\Big[\frac{|s|^{2\lambda}}{s}\,T\Big]_{\lambda=0}\Big)\Big)_z \in (\mathrm{CH}_{\mathcal{X},V^{\mathcal{X}\setminus S}\cap S})_z. \tag{2.57}$$

Here $V^{\mathcal{X}\setminus S}$ denotes, in terms of germs of closed analytic subsets at the point z, the union of the irreducible components of V which do not lie entirely in S. When T_z belongs instead to $(\mathrm{CH}^{\star S_1}_{\mathcal{X},V})_z$, such that $(S_1)_z$ is a germ of a hypersurface with $\mathrm{codim}_z(V \cap S_1) = M + 1$, we have a slightly more involved formulation of a result, which could be iterated as well.

PROPOSITION 2.44. *Let $\mathcal{X}$ be an N-dimensional complex manifold and let $V \subset \mathcal{X}$ be a purely dimensional, closed analytic subset with* $\operatorname{codim} V = M > 0$. *Let also S_1 be a fixed closed analytic hypersurface such that* $\operatorname{codim}(V \cap S_1) > M$, $s_2 \in \mathcal{O}(\mathcal{X})$, *and $S_2 = V(s_2)$. Furthermore, assume that*

(1) *$V^{\mathcal{X}\setminus S_2}$ is the union of irreducible components $\mathfrak{V}$ of V with* $\operatorname{codim}(\mathfrak{V} \cap S_2) > M$;
(2) *$S_1^{V\setminus S_2}$ is the closed analytic subset of S_1 defined as the union of the irreducible components of S_1, whose intersection with V does not lie entirely in $V^{\mathcal{X}\setminus S_2} \cap S_2$.*

Then, for each $z \in V$ and each $T_z \in (\mathrm{CH}^{\star S_1}_{\mathcal{X},V})_z$, the $({}'\mathcal{D}^{(\bullet,M+1)})_z$-valued map

$$\lambda \longmapsto \Big(\overline{\partial}\Big(\frac{|s_2|^{2\lambda}}{s_2}\Big) \wedge T\Big)_z$$

extends from $\{\lambda \in \mathbb{C} : \operatorname{Re}\lambda \gg 1\}$ to the complex plane $\mathbb{C}$ as a meromorphic map with poles in $-\mathbb{Q}^+$. Moreover, one has for any $z \in V$ that

$$\Big[\overline{\partial}\Big(\frac{|s_2|^{2\lambda}}{s_2}\Big) \wedge \big(\mathrm{CH}^{\star S_1}_{\mathcal{X},V}\big)_z\Big]_{\lambda=0} \subset \big(\mathrm{CH}^{\star S_1^{V\setminus S_2}}_{\mathcal{X},V^{\mathcal{X}\setminus S_2}\cap S_2}\big)_z. \tag{2.58}$$

PROOF. The first assertion follows from the inclusion $(\mathrm{CH}^{\star S_1}_{\mathcal{X},V})_z \subset \mathrm{PM}_z$ for any $z \in V$; see Theorem 2.42, combined with Proposition 2.21. Proposition 2.21 implies that for any $z \in V$ and $T_z \in (\mathrm{CH}^{\star S_1}_{\mathcal{X},V})_z$,

$$\Big(\Big[\overline{\partial}\Big(\frac{|s_2|^{2\lambda}}{s_2}\Big) \wedge T\Big]_{\lambda=0}\Big)_z \in \mathrm{PM}_z.$$

Since V is purely $(N-M)$-dimensional and $\operatorname{codim}(V \cap S_1) > M$, it follows from the fact that for any $z \in V$, $(V^{\mathcal{X}\setminus S_2})_z$ is a union of some irreducible components of V_z having $\operatorname{codim}_z(V^{\mathcal{X}\setminus S_2} \cap S_1) = M+1$ at any $z \in V \cap S_1$. The principle of analytic continuation implies that one has equality between sheaves of currents

$$\Big[\overline{\partial}\Big(\frac{|s_2|^{2\lambda}}{s_2}\Big) \wedge \mathrm{CH}^{\star S_1}_{\mathcal{X},V}\Big]_{\lambda=0} = \Big[\overline{\partial}\Big(\frac{|s_2|^{2\lambda}}{s_2}\Big) \wedge \mathrm{CH}^{\star S_1}_{\mathcal{X},V^{\mathcal{X}\setminus S_2}}\Big]_{\lambda=0}.$$

Given a germ

$$T_z \in \big(\mathrm{CH}^{\star S_1}_{\mathcal{X},V^{\mathcal{X}\setminus S_2}}\big)_z,$$

one has for the same reasons as above (with V replaced by $V^{\mathcal{X}\setminus S_2}$) that for any $z \in V$

$$\mathfrak{T}_z = \Big(\Big[\overline{\partial}\Big(\frac{|s_2|^{2\lambda}}{s_2}\Big) \wedge T\Big]_{\lambda=0}\Big)_z \in \mathrm{PM}_z.$$

Moreover, the germ $\mathfrak{T}_z$ is the germ of a $(\bullet, M+1)$ pseudo-meromorphic current $\mathfrak{T}$ at z, whose support lies in $V^{\mathcal{X}\setminus S_2} \cap S_2$. This support lies in a closed analytic subset of V having codimension in $\mathcal{X}$ at least equal to $M+1$. Let us now take $z \in V^{\mathcal{X}\setminus S_2} \cap S_2$, so that we have $\operatorname{codim}_z\big(V^{\mathcal{X}\setminus S_2} \cap S_2\big) = M+1$. To prove that

$$\mathfrak{T}_z \in \big(\mathrm{CH}^{\star S_1^{V\setminus S_2}}_{\mathcal{X},V^{\mathcal{X}\setminus S_2}\cap S_2}\big)_z,$$

as the assertion (2.58) requires, it is sufficient to show that

$$\operatorname{Supp}(\overline{\partial}\,\mathfrak{T})_z \subset S_1^{V\setminus S_2}, \tag{2.59}$$

after invoking once again Theorem 2.42. It follows from Theorem 2.43 that one has $(\overline{\partial}T)_z \in (\mathrm{CH}_{\mathcal{X},V^{\mathcal{X}\setminus S_2}\cap S_1})_z$. On the other hand, one has

$$(\overline{\partial}\mathfrak{T})_z = \Big(\Big[\overline{\partial}\Big(\frac{|s_2|^{2\lambda}}{s_2}\Big)\wedge\overline{\partial}T\Big]_{\lambda=0}\Big)_z.$$

Clearly the support of $\overline{\partial}T_z$ lies in $(S_1\cap S_2)_z$ since $\mathrm{Supp}\,(\overline{\partial}T)\subset V\cap S_1$. It is therefore enough, in order to prove (2.59), to suppose that $z\in (V^{\mathcal{X}\setminus S_2}\cap S_2)\cap S_1$. Assume, in addition, that $z\notin S_1^{V\setminus S_2}\cap S_2$, in other words z does not belong to any of the irreducible components Σ of S_1 which is such that $\mathrm{codim}\,(\Sigma\cap V\cap S_2)\geq M+2$. This means that any irreducible component of $(S_1)_z$ lies entirely in $(S_2)_z$. As a consequence, one has

$$\Big(\Big[\overline{\partial}\Big(\frac{|s_2|^{2\lambda}}{s_2}\Big)\wedge\overline{\partial}T\Big]_{\lambda=0}\Big)_z = 0$$

for $\mathrm{Re}\,\lambda\gg 1$. Then, the analytic continuation principle allows us to conclude that $(\overline{\partial}\mathfrak{T})_z = 0$. This completes the proof of the proposition. □

2.2.4. Extension-restriction of pseudo-meromorphic currents. Let U be an open subset of the N-dimensional complex manifold $\mathcal{X}$ and let T be a complex-valued pseudo-meromorphic current in U. Let S be a complex hypersurface in $\mathcal{X}$, whose reduced local defining (holomorphic) function s determines the reduced defining germ $s_z\in\mathcal{O}_z$ in a neighborhood of each $z\in S$. It follows from Proposition 2.21 that for any $z\in U\cap S$, the $('\mathcal{D})_z$-valued map

$$\lambda\longmapsto (1-|s_z|^{2\lambda})\,T_z \tag{2.60}$$

extends from $\{\lambda\in\mathbb{C}\,:\,\mathrm{Re}\,\lambda\gg 1\}$ to $\mathbb{C}$ as a meromorphic map with poles in $-\mathbb{Q}^+$, whose value $\big[(1-|s_z|^{2\lambda})T_z\big]_{\lambda=0}$ at $\lambda=0$ belongs to PM_z. This value is independent of the reduced holomorphic defining germ s_z for S around z; see Remark 2.22. All such germs $\big[(1-|s_z|^{2\lambda})T_z\big]_{\lambda=0}$ globalize as a pseudo-meromorphic current in U with support contained in the closed analytic subset $S\cap U$ of U. This current is naturally denoted by $\mathbf{1}_{U\cap S}\cdot T$. On the other hand, the current $T-\mathbf{1}_{U\cap S}\cdot T$ has several alternative notations, such as

$$T-\mathbf{1}_{U\cap S}\cdot T = \mathbf{1}_{U\setminus S}\cdot T = T_{(U\setminus S)\hookrightarrow U}. \tag{2.61}$$

In the particular case when T is a $(\bullet, m)$-current supported by a closed analytic subset W with pure codimension equal to m, the dimension principle (see Proposition 2.26 and Corollary 2.27) imply that as soon as the codimension of the hypersurface S satisfies $\mathrm{codim}\,(W\cap S)>m$, then

$$\mathbf{1}_{U\cap S}\cdot T = 0,\quad\text{or alternatively,}\ \ T_{(U\setminus S)\hookrightarrow U} = T. \tag{2.62}$$

A current $T\in\mathrm{PM}(U)$, as in (2.62), is said to be a *standard extension* of its restriction to the open subset $U\setminus S$ of U. This explains the notation $T_{(U\setminus S)\hookrightarrow U}$ for $\mathbf{1}_{U\setminus S}\,T$ from above.

The procedure described previously can be iterated. Suppose that W_z is a germ at z of closed analytic subset of $\mathcal{X}$. Let $(h_{z,1},\dots,h_{z,m_z})$ be any finite sequence of elements in $\mathcal{O}_{\mathcal{X},z}$ such that $W_z = V(h_{z,1})\cap\cdots\cap V(h_{z,m_z})$. Given $T\in\mathrm{PM}(U)$, where $z\in U$, let

$$T_z = \sum_\iota (\Pi_\iota)_*\tau_\iota.$$

Each $\Pi_\iota : \mathcal{U}_\iota \subset \mathbb{C}^{N_\iota} \to U_z$ above is a holomorphic mapping[17] and τ_ι is an elementary current in $\mathcal{U}_\iota$ which is of the form (2.33). It follows by iterated use of Proposition 2.25 that, according to the notation previously introduced in (2.40),

$$(2.63) \quad \Big(\mathbf{1}_{V(h_{z,m_z})} \cdot \big(\mathbf{1}_{V(h_{z,m_z-1})} \cdot \big(\cdots \big(\mathbf{1}_{V(h_{z,1})} \cdot T\big) \cdots\big)\big)\Big)_z = \sum_{\{\iota \,:\, \mathrm{Supp}_{\mathrm{ess}} \tau_\iota \subset \Pi_\iota^{-1}(W_z)\}} \big((\Pi_\iota)_* \tau_\iota\big)_z.$$

The fact that the right-hand side of (2.63) depends only on W_z and not on the sequence $(h_{z,1}, \dots, h_{z,m_z})$, such that $W_z = V(h_{z,1}) \cap \cdots \cap V(h_{z,m_z})$, suggests the following definition.

DEFINITION 2.45. Let $T \in \mathrm{PM}(U)$, where U is an open subset of $\mathcal{X}$, and let W be a closed analytic subset of $\mathcal{X}$. The *restriction of* T *to the closed analytic subset* W is the pseudo-meromorphic current $\mathbf{1}_{U\cap W} \cdot T \in \mathrm{PM}(U)$ (with support included in $U \cap W$) whose germ at a point $z \in U \cap W$ equals

$$(2.64) \quad (\mathbf{1}_{U\cap W} \cdot T)_z := \Big(\mathbf{1}_{V(h_{z,m_z})} \cdot \big(\mathbf{1}_{V(h_{z,m_z-1})} \cdot \big(\cdots \big(\mathbf{1}_{V(h_{z,1})} \cdot T\big) \cdots\big)\big)\Big)_z,$$

where $(h_{z,1}, \dots, h_{z,m})$ is an arbitrary sequence of elements in $\mathcal{O}_z$ such that one has $W_z = V(h_{z,1}) \cap \cdots \cap V(h_{z,m_z})$ and the right-hand side of (2.64) is defined according to (2.63). The *extension of* $T \in \mathrm{PM}(U)$ *from the open subset* $U \setminus W \subset U$ is defined as the pseudo-meromorphic current

$$T - \mathbf{1}_{U\cap W} \cdot T = T_{U\setminus W \hookrightarrow U} = \mathbf{1}_{U\setminus W} \cdot T \in \mathrm{PM}(U).$$

Given W' and W'' two closed analytic subsets of $\mathcal{X}$ and $T \in \mathrm{PM}(U)$, where U is an open subset of $\mathcal{X}$, the transitivity equality $\mathbf{1}_{W''} \cdot (\mathbf{1}_{W'} \cdot T) = \mathbf{1}_{W'\cap W''} \cdot T$ in $\mathrm{PM}(U)$ follows from Definition 2.45.

EXAMPLE 2.46. Suppose that W has pure codimension equal to m. It follows from Proposition A.24 (see §A.3.4) that for any $z \in U \cap W$ there exists $m+1$ reduced germs $s_{z,1}, \dots, s_{z,m}, s_{z,m+1} \in \mathcal{O}_z$ such that $ds_{z,1} \wedge \cdots \wedge ds_{z,m} \not\equiv 0$ on any irreducible component of $V(s_{z,1}) \cap \cdots \cap V(s_{z,m})$ and

(1) the germ W_z is the finite union of $\mathfrak{V}_{z,1}, \dots, \mathfrak{V}_{z,L'}$ among the irreducible components $\mathfrak{V}_{z,1}, \dots, \mathfrak{V}_{z,L}$ of $V(s_{z,1}) \cap \cdots \cap V(s_{z,m})$, where $L \geq L'$;
(2) one has $s_{z,m+1} \equiv 0$ on W_z but $\mathrm{codim}_z(V(s_{z,m+1}) \cap (\mathfrak{V}_{z,\ell} \setminus W_z)) > m$ for any $\ell > L'$.

Given $T \in \mathrm{PM}(U)$, the sequence $(s_{z,1}, \dots, s_{z,m+1})$ can be used in order to express $(\mathbf{1}_{U\cap W} \cdot T)_z$ when $z \in U \cap V$ according to (2.64).

PROPOSITION 2.47. *Let $\mathcal{X}$ be an N-dimensional complex manifold and let W be a closed analytic subset of pure codimension m. Let also U be an open subset of $\mathcal{X}$ and $T \in \mathrm{PM}(U)$ with bidegree $(\bullet, m)$. Assume that there exists a closed analytic subset W' such that* $\mathrm{codim}\, W' > m$ *and* $\mathrm{Supp}\, T \cap W \subset W'$*. Then, one has*

$$(2.65) \quad \mathbf{1}_{U\cap W} \cdot T = 0, \quad \text{or alternatively } T_{(U\setminus W)\hookrightarrow U} = T.$$

PROOF. It is an immediate consequence of the dimension principle for pseudo-meromorphic currents; see Proposition 2.26 and Corollary 2.27. □

[17]It is enough to consider the case where each Π_ι is a tower of one of smooth modifications, open inclusions, or simple projections; see [**AndW18**, Theorem 2.15].

In particular, restriction or extension operations defined by

$$T \in \mathrm{PM}(U) \longmapsto \mathbf{1}_{U\cap W} \cdot T \in \mathrm{PM}(U),$$
$$T \in \mathrm{PM}(U) \longmapsto T_{(U\setminus W)\hookrightarrow U} \in \mathrm{PM}(U)$$

can be applied to sections of the Coleff–Herrera sheaves of currents $\mathrm{CH}_{\mathcal{X},V}$ or $\mathrm{CH}^{\star S}_{\mathcal{X},V}$ described by Definitions 2.33 and 2.40, respectively. They extend the same operations performed on integration currents (as specific examples of (m,m) positive closed currents, see Example C.13) or generalized cycles (see Definition D.4), which are current re-interpretations of *gap-sheaves construction*; see §D.1.1. We conclude this brief subsection by formalizing a principle that we already met in several examples. This principle relates to the concepts of Coleff–Herrera current or meromorphic Coleff–Herrera current with prescribed poles along a hypersurface.

DEFINITION 2.48 (Standard extension property (SEP)). Let $\mathcal{X}$ be an N-dimensional complex manifold and let U be its open subset. A pseudo-meromorphic current $T \in \mathrm{PM}(U)$, whose support is included in a closed analytic subset $V \subset U$, is said to be *a standard extension for* V or it is said to *satisfy the standard extension property* (*SEP*) *across closed hypersurfaces of* $V \supset \operatorname{Supp} T$ if and only if for any complex hypersurface S in U the following implication holds:

$$\big(\mathrm{codim}_U(V\cap S) > \mathrm{codim}_U(V)\big) \Longrightarrow T_{(U\setminus S)\hookrightarrow U} = T. \tag{2.66}$$

EXAMPLE 2.49. Given a purely dimensional closed analytic subset V of $\mathcal{X}$ and given a hypersurface S satisfying $\mathrm{codim}(V\cap S) > \mathrm{codim}\, V$, any section of the Coleff–Herrera sheaves $\mathrm{CH}_{\mathcal{X},V}$, $\mathrm{CH}^{\star S}_{\mathcal{X},V}$ or $\mathrm{CH}^{\mathcal{X},V}(\cdot,E)$, $\mathrm{CH}^{\star S}_{\mathcal{X},V}(\cdot,E)$, when E is a holomorphic bundle over $\mathcal{X}$ (see Remark 2.36 and 2.41, respectively) satisfies the SEP across any closed hypersurface of $U\cap V$. This is condition (2), formulated in the definition of the sheaves $\mathrm{CH}_{\mathcal{X},V}$ or $\mathrm{CH}^{\star S}_{\mathcal{X},V}$.

2.3. Coleff–Herrera's original construction revisited

In this section, $\mathcal{X}$ is a complex N-dimensional manifold, and $V \subset \mathcal{X}$ denotes a purely dimensional closed analytic subset with codimension M in $\mathcal{X}$. We start in §2.3.1 with a presentation of Nicolas Coleff and Miguel Herrera's construction in [**CoH78**]. We will then reformulate it in §2.3.2 in more geometric terms by introducing holomorphic sections of holomorphic line bundles equipped with hermitian metrics. Finally, in §2.3.3, we conclude Chapter 2 with an up-to-date presentation of Coleff–Herrera theory within the complete intersection setting.

2.3.1. Essential intersection and Coleff–Herrera's original construction. Let $V \subset \mathcal{X}$ be as before and let $S_1,\dots,S_m$ be complex hypersurfaces of $\mathcal{X}$, when $1 \le m \le N-M$, such that $\mathrm{codim}_{\mathcal{X}}(V\cap S_j) > M$ for $j=1,\dots,m$. We recall that, given a closed analytic subset W, together with a complex hypersurface S in $\mathcal{X}$, the closed analytic subset $W^{\mathcal{X}\setminus S} \subset W$ is defined locally as follows. For any $z \in W$ such that $\mathfrak{W}_{z,1},\dots,\mathfrak{W}_{z,L}$ are the irreducible components of W_z,

$$W_z^{\mathcal{X}\setminus S} = \bigcup_{\{1\le \ell\le L\,:\,\mathfrak{W}_{z,\ell}\not\subset S_z\}} \mathfrak{W}_{z,\ell}. \tag{2.67}$$

When $1 \le m \le N - M$, let us first recall the notion of essential intersection[18] of a purely dimensional closed analytic subset $V \subset \mathcal{X}$ with an ordered sequence of hypersurfaces $\boldsymbol{S} = (S_1, \dots, S_m)$ in $\mathcal{X}$ such that $\operatorname{codim}_{\mathcal{X}}(V \cap S_j) > M$ for any j.

DEFINITION 2.50. Let $1 \le m \le N - M = N - \operatorname{codim}_{\mathcal{X}} V$. Given an ordered sequence $\boldsymbol{S} = (S_1, \dots, S_m)$ of complex hypersurfaces $\{S_1, \dots, S_m\}$ of $\mathcal{X}$ with $\operatorname{codim}_{\mathcal{X}}(V \cap S_j) > M$, the essential intersection $\big(V \bigcap \boldsymbol{S}\big)_{\mathrm{ess}}$ is the closed analytic subset of $\mathcal{X}$, either with pure codimension $M + m$ or empty, defined through the following iterated process:

$$\begin{aligned} V^{[0]} &= V, \\ V^{[j+1]} &= V_j^{\mathcal{X} \setminus S_{j+1}} \cap S_{j+1} \text{ for } j = 0, \dots, m-1, \\ V^{[m]} &:= \big(V \bigcap \boldsymbol{S}\big)_{\mathrm{ess}}. \end{aligned} \tag{2.68}$$

REMARK 2.51. The definition of $\big(V \bigcap \boldsymbol{S}\big)_{\mathrm{ess}}$ highly depends on the ordering of $S_1, \dots, S_m$. However, if the hypersurfaces $S_1, \dots, S_m$ are in such position[19] that they satisfy $\operatorname{codim}(V \cap S_1 \cap \dots \cap S_m) = M + m$ or $+\infty$, then one has

$$\big(V \bigcap \boldsymbol{S}\big)_{\mathrm{ess}} = \big(V \bigcap \mathfrak{s}(\boldsymbol{S})\big)_{\mathrm{ess}} \tag{2.69}$$

for any permutation $\mathfrak{s}$ of $\{S_1, \dots, S_m\}$.

The reason why we mention here the concept of essential intersection, besides its intimate relation with the Coleff–Herrera construction of residue currents that we describe in this subsection, is that it appears crucially in the methods developed towards algebraic or analytic improper intersection theory ([**StV82, Tw95, AndSWY17B, AndESWY21A, AndESWY21B**]). It is therefore deeply connected with the concept of *gap sheaf*; see §D.1.1[20] or, e.g., [**Mas03**] for more details. We will come back to these important concepts later on in this monograph.

Adapting a slightly different approach to the original one, we present the construction of *residue currents* proposed by Coleff and Herrera in [**CoH78**]. These currents, also known as *Coleff–Herrera residue currents*, are attached to a closed purely $(N - M)$-dimensional analytic subset $V \subset \mathcal{X}$, to an ordered sequence of hypersurfaces $\boldsymbol{S} = (S_1, \dots, S_m)$, for $1 \le m \le N - M$, in $\mathcal{X}$ with $\operatorname{codim}(V \cap S_j) > M$ for $j = 1, \dots, m$, and to a meromorphic $(p, 0)$-form ω in an open neighborhood U of V in $\mathcal{X}$. The polar set of $(p, 0)$-form ω, $0 \le p \le N - M$, lies in $U \cap |\boldsymbol{S}|$, where $|\boldsymbol{S}| = \bigcup_{j=1}^m S_j$. For each $z \in V \cap |\boldsymbol{S}|$, Hilbert's nullstellensatz in $\mathcal{O}_z$ implies that for reduced defining germs of holomorphic sections $s_j = s_{z,j}$, $j = 1, \dots, m$, of the hypersurfaces S_j around z there exist an integer $\nu = \nu_z \in \mathbb{N}^*$ such that

$$((s_1 \cdots s_m)^\nu \omega)_z = \varpi_z \in (\Omega^p)_z. \tag{2.70}$$

Observe that the s_j's are nonidentically zero on any irreducible component of V_z since $\operatorname{codim}_{\mathcal{X}}(V \cap S_j) > M$.

[18]See also §D.1.1, in particular (D.4) and (D.5), for the related concept of a *cutting-out* operation performed on cycles. For what concerns this notion of *essential intersection* (deeply connected with our approach toward analytic residue theory), we refer the reader to [**CoH78**] and [**Sol77**].

[19]One says then they are in position to define a *complete intersection* on V; see §2.3.3.

[20]We refer in particular here to (D.4) and(D.5), where the *cutting-out* operation on cycles is described.

DEFINITION 2.52. Let V, $\boldsymbol{S}$, ω be as above. The current $T = R_{\boldsymbol{S}} \wedge \omega \wedge [V]$ is the pseudo-meromorphic $\overline{\partial}$-closed current in $\mathcal{X}$, with bidegree $(p+M, M+m)$, defined as follows. It is defined locally as $T_z = 0$, if $z \notin V_z \cap |\boldsymbol{S}|$. In case $z \in V \cap |\boldsymbol{S}|$, it is defined by the iterative procedure

$$(2.71)\qquad \begin{aligned} (T^{[0]})_z &= (\varpi \wedge [V])_z, \\ (T^{[j+1]})_z &= \Big[\overline{\partial}\Big(\frac{|s_{j+1}|^{2\nu\lambda}}{2i\pi\, s_{j+1}^{\nu}}\Big) \wedge (T^{[j]})_z\Big]_{\lambda=0} \quad \text{for } j = 0, \dots, m-1, \\ \big(R_{\boldsymbol{S}} \wedge \omega \wedge [V]\big)_z &= (T^{[m]})_z \end{aligned}$$

following Proposition 2.21. Here $s_j = s_{j,z}$ is the germ at z of a reduced defining function for S_j for every $j = 1, \dots, m$ and ϖ_z, $\nu = \nu_z$ are taken as in (2.70). This current has its support included in $V \cap |\boldsymbol{S}|$ and it depends only on $[V]$, $\boldsymbol{S}$, and ω.

REMARK 2.53. The definition of the sheaf of (p, m)-currents on the (reduced) complex analytic space $(V, \mathcal{O}/\mathcal{I}_V)$, in particular their interpretation through the identification $T \longleftrightarrow (\iota_V)_* T$, where $\iota_V : V \hookrightarrow \mathcal{X}$ (see (B.33) in §B.2.2.2), allows one to interpret $R_{\boldsymbol{S}} \wedge \omega \wedge [V]$ as the direct image through the inclusion ι of a (p, m)-current on V, denoted by $R_{\iota_V^{-1}(\boldsymbol{S})} \wedge \iota_V^* \omega$.

PROPOSITION 2.54 (Lelong–Poincaré type factorization formulae). *Let $\mathcal{X}$, V, $\boldsymbol{S}$, ω be as above. One has*

$$(2.72)\qquad R_{\boldsymbol{S}} \wedge \omega \wedge [V] \in \Big[\mathrm{CH}_{\mathcal{X},(V \bigcap \boldsymbol{S})_{\rm ess}}\Big]^{p+M}(\mathcal{X}).$$

Moreover, assume that the germ V_z at $z \in (V \bigcap \boldsymbol{S})_{\rm ess}$ is defined as the union of a finite number of some irreducible components $\mathfrak{W}_1, \dots, \mathfrak{W}_{L'}$ of the intersection germ $W_z = V(\sigma_{z,1}) \cap \cdots \cap V(\sigma_{z,M})$, where $\sigma_{z,1}, \dots, \sigma_{z,M} \in \mathcal{O}_z$ fulfill the following condition: $d\sigma_{z,1} \wedge \cdots \wedge d\sigma_{z,M} \not\equiv 0$ on any of the $L \geq L'$ irreducible components $\mathfrak{W}_\ell$ of W_z. Then, there is a germ of a hypersurface $\Sigma_{V,\boldsymbol{S},z}$ with

$$\operatorname{codim}_z\big((V \textstyle\bigcap \boldsymbol{S}')_{\rm ess} \cap \Sigma_{V,\boldsymbol{S}'',z}\big) > M + m'$$

such that [21]

$$(2.73)\qquad \big(R_{\boldsymbol{S}} \wedge \omega \wedge [V]\big)_z \subset \Big(\Big[\mathrm{CH}^{\star\Sigma_{V,\boldsymbol{S},z}}_{\mathcal{X},(V \bigcap \boldsymbol{S})_{\rm ess}}\Big]^0\Big)_z \wedge d\sigma_{1,z} \wedge \cdots \wedge d\sigma_{M,z} \wedge (\Omega^p)_z.$$

If $L = L'$, then $V_z = W_z$, and one has

$$(2.74)\qquad \big(R_{\boldsymbol{S}} \wedge \omega \wedge [V]\big)_z \subset \Big(\Big[\mathrm{CH}_{\mathcal{X},(V \bigcap \boldsymbol{S})_{\rm ess}}\Big]^0\Big)_z \wedge d\sigma_{1,z} \wedge \cdots \wedge d\sigma_{M,z} \wedge (\Omega^p)_z.$$

PROOF. Let us first prove assertion (2.72). One can see from the construction of the current $T^{[j]}$ appearing at step j in the iterative procedure (2.71) that it is supported by the closed analytic subset $V^{[j]}$. The subset $V^{[j]}$ appears at the same step j of the iterative procedure (2.68). It follows that $R_{\boldsymbol{S}} \wedge \omega \wedge [V]$ is a $(p+M, m+M)$ pseudo-meromorphic current, whose support is included in $V^{[m]} = \big(V \bigcap \boldsymbol{S}\big)_{\rm ess}$. If $V^{[m]} = \emptyset$, $R_{\boldsymbol{S}} \wedge \omega \wedge [V] = 0$. Otherwise, $V^{[m]}$ has pure codimension $M + m$ and Theorem 2.39 implies that $R_{\boldsymbol{S}} \wedge \omega \wedge [V] \in \mathrm{CH}^{p+M}_{\mathcal{X},(V \bigcap \boldsymbol{S})_{\rm ess}}(\mathcal{X})$. In order to prove the other two assertions, let us fix $z \in \big(V \bigcap \boldsymbol{S}\big)_{\rm ess}$. It means that $(V \bigcap \boldsymbol{S})_{\rm ess} \neq \emptyset$ and

[21] Multiplicative factorization formulas, such as (2.73) and (2.74), are known as factorization formulae of the *Lelong–Poincaré type*. The most classical one is the formula (2.76) for the germ of an integration current $[V]_z$.

that the defining reduced holomorphic germs $s_{z,1},\dots,s_{z,m}$ for $S_1,\dots,S_m$ around z, are not all being identically equal to zero on any irreducible component of V_z. Specifically, the defining reduced holomorphic germs $s_{z,1},\dots,s_{z,m}$ for $S_1,\dots,S_m$ around z are not all being identically equal to zero on any $\mathfrak{W}_\ell$ for $\ell=1,\dots,L'$. Let $\varpi_z\in(\Omega^p)_z$ and $\nu\in\mathbb{N}^*$ such that $(s_{z,1}\cdots s_{z,m})^\nu\omega_z=\varpi_z$. Let $\mathfrak{T}_z$ be the germ of a pseudo-meromorphic $(0,M)$-current at z defined through the iterative procedure

$$
\begin{aligned}
\mathfrak{T}^{[0]} &= [1],\\
(\mathfrak{T}^{[\kappa+1]})_z &= \Big[\overline{\partial}\Big(\frac{|\sigma_{z,\kappa+1}|^{2\lambda}}{2i\pi\,\sigma_{z,\kappa+1}}\Big)\wedge(\mathfrak{T}^{[\kappa]})_z\Big]_{\lambda=0}\quad\text{for }\kappa=0,\dots,M-1,\\
\mathfrak{T}_z &= \Big[(1-|\sigma_{z,M+1}|^{2\lambda})\,(\mathfrak{T}^{[M]})_z\Big]_{\lambda=0},
\end{aligned}\tag{2.75}
$$

where $\sigma_{z,M+1}\in I(V_z)$ is chosen such that $\operatorname{codim}_z((\mathfrak{W}_\ell\setminus V)\cap V(\sigma_{z,M+1}))=M+1$ for $\ell=L'+1,\dots,L$, when $L'<L$, or is taken to be equal to 0, if $L'=L$. It follows from Theorem 2.39 that $\mathfrak{T}_z=(\mathfrak{T}^{[M]})_z\in(\mathrm{CH}_{\mathcal{X},V})_z$. Similarly, Theorem 2.42 implies that $\mathfrak{T}_z\in\mathrm{CH}_{\mathcal{X},V}$. Moreover, it follows from Proposition 2.16 and formula (2.29) that

$$
[V]_z=\mathfrak{T}_z\wedge d\sigma_{z,1}\wedge\cdots\wedge d\sigma_{z,M}.\tag{2.76}
$$

Repeated use of Proposition 2.44 shows that the iterated procedure

$$
\begin{aligned}
(\mathfrak{U}^{[0]})_z &= \mathfrak{T}_z,\\
(\mathfrak{U}^{[j+1]})_z &= \Big[\overline{\partial}\Big(\frac{|s_{z,j+1}|^{2\nu\lambda}}{s_{z,j+1}^\nu}\Big)\wedge(\mathfrak{U}^{[j]})_z\Big]_{\lambda=0}\quad\text{for }j=0,\dots,m-1,\\
\mathfrak{U}_z &= (\mathfrak{U}^{[m]})_z
\end{aligned}\tag{2.77}
$$

leads to the construction of an element $\mathfrak{U}_z\in\big(\mathrm{CH}^{\star\Sigma_{V,\boldsymbol{S},z}}_{\mathcal{X},(V\bigcap\boldsymbol{S})_{\mathrm{ess}}}\big)_z$ for some $\Sigma_{V,\boldsymbol{S},z}$, depending on $V(\sigma_{z,M+1})$ and on the germs at z of the S_j, in case $L'<L$. In the case where $L'=L$, one has that $\sigma_{z,M+1}=0$ and thus $\mathfrak{U}_z\in\big(\mathrm{CH}_{\mathcal{X},(V\bigcap\boldsymbol{S})_{\mathrm{ess}}}\big)_z$. The factorization formulae (2.73) (when $L'<L$) or (2.74) (when $L'=L$) follow from the definition of $(R_{\boldsymbol{S}}\wedge\omega\wedge[V])_z=\big(R_{\boldsymbol{S}}\wedge[V]\wedge\varpi/(\prod_1^m\sigma_{z,j})^\nu\big)_z$, together with the factorization (2.76) for $[V]_z$. □

Following Coleff–Herrera in [**CoH78**], one may define also mixed *principal value-residue currents.*

DEFINITION 2.55. Let V, $\boldsymbol{S}=(S_1,\dots,S_{m'},S_{m'+1},\dots,S_m)=(\boldsymbol{S}',\boldsymbol{S}'')$, and let ω be as above, where $1\le m'\le m$. The current $T=R_{\boldsymbol{S}'}P_{\boldsymbol{S}''}\wedge\omega\wedge[V]=RP_{\boldsymbol{S}'\boldsymbol{S}''}\wedge\omega\wedge[V]$ is the pseudo-meromorphic current in $\mathcal{X}$, with bidegree $(p+M,M+m')$. In case $z\notin V_z\cap|\boldsymbol{S}'|$, then it is defined locally as $T_z=0$. On the other hand, if $z\in V\cap|S'|$, then it is defined by the iterative procedure

(2.78)

$$
\begin{aligned}
(T^{[0]})_z &= (\varpi\wedge[V])_z,\\
(T^{[j+1]})_z &= \Big[\overline{\partial}\Big(\frac{|s_{j+1}|^{2\nu\lambda}}{2i\pi\,s_{j+1}^\nu}\Big)\wedge(T^{[j]})_z\Big]_{\lambda=0}\quad\text{for }j=0,\dots,m'-1,\\
(T^{[j+1]})_z &= \Big[\frac{|s_{j+1}|^{2\nu\lambda}}{2i\pi\,s_{j+1}^\nu}\,(T^{[j]})_z\Big]_{\lambda=0}\quad\text{for }j=m',\dots,m-1,\\
\big(RP_{\boldsymbol{S}'\boldsymbol{S}''}\wedge\omega\wedge[V]\big)_z &= (T^{[m]})_z
\end{aligned}
$$

following Proposition 2.21, where $s_j = s_{j,z}$ is the germ of a reduced defining function for the hypersurface S_j in a neighborhood of z for every $j = 1, \dots, m$, and ϖ_z, $\nu = \nu_z$ are taken to be as in (2.70). This current has its support included in $V \cap |\boldsymbol{S}'|$ and it depends only on $[V]$, $\boldsymbol{S}$ and ω.

REMARK 2.56. The definition of the sheaf of (p, m)-currents on the (reduced) complex analytic space $(V, \mathcal{O}/\mathcal{I}_V)$ allows one to interpret, as in Remark 2.53, the mixed current $RP_{\boldsymbol{S}'\boldsymbol{S}''} \wedge \omega \wedge [V]$ as the direct image through the inclusion ι_V of a (p, m')-current on V, denoted by $RP_{\iota_V^{-1}(\boldsymbol{S}'\boldsymbol{S}'')} \wedge \iota_V^* \omega$.

PROPOSITION 2.57. *Let $\mathcal{X}$, V, $\boldsymbol{S} = (\boldsymbol{S}', \boldsymbol{S}'')$, ω be as above. Let, as usual, $((V \bigcap \boldsymbol{S}')_{\rm ess})^{\mathcal{X}\setminus|\boldsymbol{S}''|}$ be the union of the irreducible components of $(V \bigcap \boldsymbol{S}')_{\rm ess}$ which do not lie entirely in $|\boldsymbol{S}''|$. One has*

$$RP_{\boldsymbol{S}'\boldsymbol{S}''} \wedge \omega \wedge [V] \in \Big(\Big[\mathrm{CH}^{\star|\boldsymbol{S}''|}_{\mathcal{X}, ((V \bigcap \boldsymbol{S}')_{\rm ess})^{\mathcal{X}\setminus|\boldsymbol{S}''|}}\Big]^{p+M}\Big)(\mathcal{X}). \tag{2.79}$$

Moreover, assume that the germ V_z is defined as the union of a finite number of some irreducible components $\mathfrak{W}_1, \dots, \mathfrak{W}_{L'}$ of $W_z = V(\sigma_{z,1}) \cap \dots \cap V(\sigma_{z,M})$ in a neighborhood of $z \in (V \bigcap \boldsymbol{S}')_{\rm ess}$, where $\sigma_{z,1}, \dots, \sigma_{z,M} \in \mathcal{O}_z$ are such that one has $d\sigma_{z,1} \wedge \dots \wedge d\sigma_{z,M} \not\equiv 0$ on any of the $L \geq L'$ irreducible components $\mathfrak{W}_\ell$ of W_z. Then, there is a germ of a hypersurface $\Sigma_{V,\boldsymbol{S},z} \supset |\boldsymbol{S}''|_z$, whose codimension satisfies

$$\mathrm{codim}_z\big(((V \bigcap \boldsymbol{S}')_{\rm ess})^{\mathcal{X}\setminus|\boldsymbol{S}''|} \cap \Sigma_{V,\boldsymbol{S},z}\big) > M + m',$$

such that

$$\big(RP_{\boldsymbol{S}'\boldsymbol{S}''} \wedge \omega \wedge [V]\big)_z \subset \Big(\Big[\mathrm{CH}^{\star \Sigma_{V,\boldsymbol{S},z}}_{\mathcal{X}, (V \bigcap \boldsymbol{S}')_{\rm ess})^{\mathcal{X}\setminus|\boldsymbol{S}''|}}\Big]^0\Big)_z \wedge d\sigma_{1,z} \wedge \dots \wedge d\sigma_{M,z} \wedge (\Omega^p)_z. \tag{2.80}$$

If $L' = L$, then one can take $\Sigma_{V,\boldsymbol{S},z} = |\boldsymbol{S}''|_z$.

PROOF. The first assertion follows from Theorem 2.42. The proof of the second assertion is based on the same arguments present in the proof of the second assertion in Proposition 2.54. □

The currents $R_{\boldsymbol{S}} \wedge \omega \wedge [V]$ or $RP_{\boldsymbol{S}'\boldsymbol{S}''} \wedge \omega \wedge [V]$ were originally introduced in [**CoH78**] through the *integration on semi-analytic chains* and the related notion of the *admissible path of integration* with respect to $(V, \boldsymbol{S})$ in a neighborhood of $z \in V \cap S_1 \cap \dots \cap S_{m'}$. We recall the definitions of these objects below.

DEFINITION 2.58 (Semi-analytic chains $\Gamma_{V,\boldsymbol{s}}(\varepsilon)$ around $z \in V$). Let $\mathcal{X}$ be an N-dimensional complex manifold and let $V \subset \mathcal{X}$ be a purely dimensional closed analytic subset with $\mathrm{codim}\, V = M$. For $z \in V$ let $\boldsymbol{s} = (s_1, \dots, s_{m'}, s_{m'+1}, \dots, s_m)$ be an ordered sequence of m reduced elements in $\mathcal{O}_z$, which are not identically equal to 0 on any irreducible component of V_z. Assume that the elements of this ordered sequence are represented by reduced elements (also denoted by s_j) in some neighborhood U_z of z in $\mathcal{X}$. For any $\boldsymbol{\varepsilon} = (\varepsilon_1, \dots, \varepsilon_{m'}, \varepsilon_{m'+1}, \dots, \varepsilon_m)$ such that

$(\varepsilon_1, \dots, \varepsilon_{m'})$ is not a critical value for $(|s_1|^2, \dots, |s_{m'}|^2)_{|(V\cap U_z)_{\mathbb{R}}}$, the closed semi-analytic[22] subset

$$\{\zeta \in V \cap U_z \,:\, |s_j(\zeta)|^2 = \varepsilon_j \text{ for } 1 \le j \le m',\ |s_j(\zeta)|^2 > \varepsilon_j \text{ for } m'+1 \le j \le m\}$$

induces a $(2(N-M) - |m'|)$-dimensional *semi-analytic integration chain* on U_z by

$$\text{(2.81)}\quad \Gamma_{V,\boldsymbol{s}}(\boldsymbol{\varepsilon}) : \Phi \in \mathcal{D}^{2(N-M)-m'}((U_z)_{\mathbb{R}}) \longmapsto \langle [\Gamma_{V,\boldsymbol{s}}(\boldsymbol{\varepsilon})]\,,\, \Phi\rangle = \int_{\Gamma_{V,\boldsymbol{s}}(\boldsymbol{\varepsilon})} \Phi$$

$$= \lim_{\ell \to +\infty} \int_V \Big(\prod_{j>m'} \chi_\ell\Big(\frac{|s_j|^2}{\varepsilon_j}\Big)\Big)\, \overline{\partial}\Big(\chi_\ell\Big(\frac{|s_{m'}|^2}{\varepsilon_{m'}}\Big)\Big) \wedge \cdots \wedge \overline{\partial}\Big(\chi_\ell\Big(\frac{|s_1|^2}{\varepsilon_1}\Big)\Big) \wedge \Phi,$$

where $(\chi_\ell)_{\ell\in\mathbb{N}^*}$, with $\operatorname{Supp}\chi_\ell \subset [1-1/\ell, +\infty[$, is a smooth regularization by convolution with an approximation of unity of the characteristic function $\chi_{[1,+\infty[}$.

DEFINITION 2.59. Let $\mathcal{X}$ and V be as above and let $\boldsymbol{S} = (\boldsymbol{S}', \boldsymbol{S}'')$ be an ordered sequence of m hypersurfaces as in Definition 2.55. Let $\boldsymbol{s} = (s_1, \dots, s_{m'}, s_{m'+1}, \dots, s_m)$ be a sequence of reduced defining germs for the hypersurfaces $S_1, \dots, S_m$ at $z \in V$ respectively. A map $t :]0,1[\longmapsto \boldsymbol{\varepsilon}(t) \in (\mathbb{R}^+)^m$ is said to be m-admissible in a neighborhood of $z \in V \cap (S_1 \cap \cdots \cap S_{m'})$ with respect to $(V, \boldsymbol{S})$ if and only if

(1) $\operatorname{Im}\delta$ avoids the set of critical values of $(|s_1|^2, \dots, |s_{m'}|^2)_{|(V\cap U_z)_{\mathbb{R}}}$ for some sufficiently small neighborhood U_z of z;

(2) when t tends to 0_+, then one has that for all $k \in \mathbb{N}^*$,

$$\text{(2.82)}\qquad \varepsilon_j(t) = O\big((\varepsilon_{j+1}(t))^k\big) \text{ for } j = 1, \dots, m-1.$$

The presentation of germs of Coleff–Herrera currents $(R_{\boldsymbol{S}} \wedge \omega \wedge [V])_z$, when $m' = m$ or of the germs $(RP_{\boldsymbol{S}'\boldsymbol{S}''} \wedge \omega \wedge [V])_z$, when $m' < m$, at $z \in (V \cap \boldsymbol{S}')_{\text{ess}}$ for a meromorphic $(p,0)$-form ω, whose polar set is along $|\boldsymbol{S}|$ around z, is as follows (see [**CoH78**]).

THEOREM 2.60. *Suppose that $t \mapsto \boldsymbol{\varepsilon}(t)$ is m-admissible with respect to z and that U_z fulfills condition* (1) *in Definition* 2.59. *Then, when $m' = m$,*

$$\text{(2.83)}\qquad \langle R_{\boldsymbol{S}} \wedge \omega \wedge [V]\,,\, \varphi\rangle = \lim_{t\to 0_+} \Big\langle \frac{[\Gamma_{V,\boldsymbol{s}}(\boldsymbol{\varepsilon}(t))]}{(2i\pi)^m}\,,\, \omega\wedge\varphi\Big\rangle$$

for any $\varphi \in \mathcal{D}^{N-M-p,N-M-m}(U_z)$. Furthermore, if $m' < m$, then

$$\text{(2.84)}\qquad \langle RP_{\boldsymbol{S}'\boldsymbol{S}''} \wedge \omega \wedge [V]\,,\, \varphi\rangle = \lim_{t\to 0_+} \Big\langle \frac{[\Gamma_{V,\boldsymbol{s}}(\boldsymbol{\varepsilon}(t))]}{(2i\pi)^m}\,,\, \omega\wedge\varphi\Big\rangle$$

for any $\varphi \in \mathcal{D}^{N-M-p,N-M-m'}(U_z)$.

To avoid the technical condition (1) in Definition 2.59, one can fix a smooth regularization χ of $\chi_{[1,+\infty[}$ on the real line and for each $\boldsymbol{\varepsilon} \in (\mathbb{R}^+)^m$, substitute the semi-analytic integration chain $\Gamma_{V,\boldsymbol{S}}(\boldsymbol{\varepsilon})$ with bidimension $(N-M, N-M-m')$ defined in (2.81) by the $(M, M+m')$-current

$$\text{(2.85)}\quad [\Gamma^\chi_{V,\boldsymbol{s}}(\boldsymbol{\varepsilon})] = \Big(\prod_{j>m'} \chi\Big(\frac{|s_j|^2}{\varepsilon_j}\Big)\Big)\Big(\overline{\partial}\Big(\chi\Big(\frac{|s_{m'}|^2}{\varepsilon_{m'}}\Big)\Big) \wedge \cdots \wedge \overline{\partial}\Big(\chi\Big(\frac{|s_1|^2}{\varepsilon_1}\Big)\Big)\Big) \wedge [V].$$

---

[22] The terminology *semi-analytic* comes from the fact that such subset is defined in terms of either equalities $\mathfrak{s}_\iota = 0$ or inequalities $\mathfrak{s}_\iota > 0$, where the s_ι are germs at z of real analytic functions on $\mathcal{X}_{\mathbb{R}}$.

This is done in [**SamL13**, Theorems 2 and 11]. The following result also holds [**SamL13**, Theorems 2 and 11].

THEOREM 2.61. *Suppose that* $t :]0,1[\longmapsto \varepsilon(t) \in (\mathbb{R}^+)^m$ *fulfills condition* (2) *in Definition* 2.59. *Let* V, $\boldsymbol{S}$, ω *and* $z \in (V \cap \boldsymbol{S}')_{\rm ess}$ *be as in Theorem* 2.60. *Let also* $\boldsymbol{s} = (s_1, \dots, s_{m'}, s_{m'+1}, \dots, s_m) = (\boldsymbol{s}', \boldsymbol{s}'')$ *be a sequence of reduced defining germs for an ordered sequence* $(S_1, \dots, S_m) = (\boldsymbol{S}', \boldsymbol{S}'')$ *at* z. *Then one has*

$$\begin{aligned}(R_{\boldsymbol{S}} \wedge \omega \wedge [V])_z &= \lim_{t \to 0_+} \Big(\frac{[\Gamma^{\mathcal{X}}_{V,\boldsymbol{s}}(\varepsilon(t))]}{(2i\pi)^m} \wedge \omega \Big)_z \quad \text{when } m' = m, \\ (RP_{\boldsymbol{S}'\boldsymbol{S}''} \wedge \omega \wedge [V])_z &= \lim_{t \to 0_+} \Big(\frac{[\Gamma^{\mathcal{X}}_{V,\boldsymbol{s}}(\varepsilon(t))]}{(2i\pi)^m} \wedge \omega \Big)_z \quad \text{when } m' < m\end{aligned} \tag{2.86}$$

with respect to the weak convergence in $('\mathcal{D})_z$.

In this monograph we propose systematically an alternative approach for the definition of residue or mixed principal value–residue currents which illustrates in this subsection that of the currents $R_{\boldsymbol{S}} \wedge \omega \wedge [V]$ or $RP_{\boldsymbol{S}'\boldsymbol{S}''} \wedge \omega \wedge [V]$. Our point of view leads to different representations of the above objects to those on the right-hand sides of equalities in Theorem 2.60 or Theorem 2.61. The approach presented in the present text is based on the *analytic continuation principle* as a substitute for integration on semi-analytic chains along admissible paths (as in Theorem 2.60) or weak limits along admissible paths after regularization (as in Theorem 2.61). The reason for our choice is that besides the fact that it avoids appealing to the intricate condition (2) in Definition 2.59, from our point of view it makes the construction more algebraic. Therefore, it gives some hope of transposing it within an algebraic or even arithmetic frame. To invite the reader towards potential further developments, let us mention here the references [**Igu**] and [**Den92**] related to Igusa calculus as well as [**Jea74, Bar82, BarM85**] related to Mellin's inversion.

2.3.2. Coleff–Herrera currents attached to sections of line bundles. In this section, we extend the approach developed in the univariate setting in §1.2.5. Until now, the currents $R_{\boldsymbol{S}} \wedge \omega \wedge [V]$ or $RP_{\boldsymbol{S}'\boldsymbol{S}''} \wedge \omega \wedge [V]$ we introduced in §2.3.1, involve, besides geometric objects such as the purely dimensional closed analytic subset V or the ordered sequence of hypersurfaces $\boldsymbol{S}$, $(\boldsymbol{S}', \boldsymbol{S}'')$ which all intersect V properly, algebraic objects such as the meromorphic form ω in the ambient manifold $\mathcal{X}$, with poles along $|\boldsymbol{S}|$. We present in this subsection an alternative point of view, where sections of Cartier divisors on $\mathcal{X}$ play a central role. We plan to consider, besides problems of geometric nature related, for example, to proper or improper *intersection theory*, questions of algebraic nature. These are problems that appear, for example, in *division* or *interpolation* theory. In order to obtain a coherent picture of the working frame we are dealing with, all currents T, meromorphic forms ω, or test forms φ should be treated as sections of holomorphic bundles over $\mathcal{X}$. The bundle approach reveals to be useful yet for another reason. It sustains, as will be apparent later from Chapter 3, *averaging procedures* where Hilbert–Samuel multiplicity of an ideal in $\mathcal{O}_z$ is expressed as a Lelong number (Proposition C.24), or where Crofton's formula, as stated in §D.5.1, is used. We will deal in this section with holomorphic sections of (holomorphic) *line bundles* on N-dimensional complex manifold $\mathcal{X}$, in correspondence with Cartier divisors on $\mathcal{X}$; see §D.2.1. Similar constructions involving holomorphic sections of complex vector bundles of higher rank will be discussed in the forthcoming Chapter 3. The geometric point of view of the Coleff–Herrera theory, as we present it here, was introduced by M. Andersson [**And04, And05, And07**]. For the background on bundle theory we refer the reader to Appendix B.

Let E be a holomorphic bundle over $\mathcal{X}$. We start by introducing homomorphisms between sheaves of bundle-valued pseudo-meromorphic currents on $\mathcal{X}$ in compliance with the definitions of $R^s_{|\,|}$, $P^s_{|\,|}$ given in Proposition 1.27.

DEFINITION 2.62. Let $s \in \mathcal{O}_{\mathcal{X}}(\mathcal{X}, L) = \mathcal{O}(\mathcal{X}, L)$, where $L = \mathcal{O}(D)$ is the holomorphic line bundle corresponding to the Cartier divisor D on $\mathcal{X}$. Let $|\,|$ be a hermitian metric on L and let $s^* \in \mathcal{O}(\mathcal{X}, \mathcal{O}(-D))$ be its complex conjugate.[23] The operators $[R^s_{|\,|}] = [R^s_{|\,|,0} + R^s_{|\,|,1}]$, $[P^s_{|\,|}] = [P^s_{|\,|,0}]$ and $[R^s_{|\,|} + P^s_{|\,|}]$ are respectively defined as follows. For any open subset U of the N-dimensional complex manifold $\mathcal{X}$ and for any section T of $\mathrm{PM}(U, E)$, we define

(2.87)
$$\begin{aligned}
\left[R^s_{|\,|}\right](T) &= \left[\left(1 - |s|^{2\lambda} + \bar\partial |s|^{2\lambda}\, \frac{s^*}{2i\pi |s|^2}\right) \otimes T\right]_{\lambda=0} \\
&= \mathbf{1}_{V(s)}\, T + \left[\bar\partial |s|^{2\lambda}\, \frac{s^*}{2i\pi |s|^2} \otimes T\right]_{\lambda=0} \in \mathrm{PM}\big(U, (\mathbb{C} \oplus \mathcal{O}(-D)) \otimes E\big), \\
\left[P^s_{|\,|}\right](T) &= \left[\left(|s|^{2\lambda}\, \frac{s^*}{2i\pi |s|^2}\right) \otimes T\right]_{\lambda=0} \in \mathrm{PM}\big(U, \mathcal{O}(-D) \otimes E\big), \\
\left[R^s_{|\,|} + P^s_{|\,|}\right](T) &= \left[R^s_{|\,|}\right](T) + \left[P^s_{|\,|}\right](T) \in \mathrm{PM}\big(U, (\mathbb{C} \oplus \mathcal{O}(-D)) \otimes E\big).
\end{aligned}$$

REMARK 2.63. If $|\,|'$ and $|\,|''$ are two metrics of $\mathcal{O}(D)$, one has $(|s|')^2 = \rho (|s|'')^2$, where $\rho \in C^\infty(\mathcal{X}, \mathbb{R}^+)$. As a consequence $(\lambda, z) \mapsto \rho^\lambda(z)$ depends on the parameter λ as an entire function of λ, whose value at $\lambda = 0$ equals identically 1 for $z \in \mathcal{X}$. As a consequence, given $T \in \mathrm{PM}(U, E)$, the pseudo-meromorphic currents $\left[R^s_{|\,|}\right](T)$ or $\left[P^s_{|\,|}\right](T)$ do not depend on the choice of the metric $|\,|$ on $\mathcal{O}(D)$. Therefore, the currents defined by (2.87) will be denoted by $\left[R^s\right](T)$, $\left[P^s\right](T)$, and $\left[(R+P)^s\right](T)$ from now on.

PROPOSITION 2.64. *Let V be a purely dimensional closed subset of $\mathcal{X}$ and let $s \in \mathcal{O}(\mathcal{X}, \mathcal{O}(D))$ be such that* $\operatorname{codim}(V \cap V(s)) > \operatorname{codim} V$. *If $T \in \mathrm{CH}_{\mathcal{X},V}(U, E)$, one has*

$$(2.88)\qquad \begin{aligned}
\left[R^s\right](T) &= \left[\left(\bar\partial |s|^{2\lambda}\, \frac{s^*}{2i\pi |s|^2}\right) \otimes T\right]_{\lambda=0} \in \mathrm{CH}_{\mathcal{X}, V \cap V(s)}\big(U, \mathcal{O}(-D) \otimes E\big), \\
\left[P^s\right](T) &\in \mathrm{CH}^{\star V(s)}_{\mathcal{X},V}\big(U, \mathcal{O}(-D) \otimes E\big).
\end{aligned}$$

If $T \in \mathrm{CH}^{\star\Sigma}_{\mathcal{X},V}(U, E)$ for some closed hypersurface Σ in a neighborhhod of V such that $\operatorname{codim}(V \cap \Sigma) > \operatorname{codim} V$, *then one has*

$$(2.89)\qquad \begin{aligned}
\left[R^s\right](T) &= \left[\left(\bar\partial |s|^{2\lambda}\, \frac{s^*}{2i\pi |s|^2}\right) \otimes T\right]_{\lambda=0} \in \mathrm{CH}^{\star\Sigma^{V \setminus V(s)}}_{\mathcal{X}, V^{\mathcal{X}\setminus S} \cap V(s)}\big(U, \mathcal{O}(-D) \otimes E\big), \\
\left[P^s\right](T) &\in \mathrm{CH}^{\star(\Sigma \cup V(s))}_{\mathcal{X},V}\big(U, \mathcal{O}(-D) \otimes E\big),
\end{aligned}$$

[23] See §B.3.2. Note that since $L = \mathcal{O}(D)$ is a line bundle, $s^*/|s|^2$ is a meromorphic section of $\mathcal{O}(-D)$.

where $V^{\mathcal{X}\setminus V(s)}$ *is the union of the irreducible components of* S *on which* $s \not\equiv 0$ *and* $\Sigma^{V\setminus V(s)}$ *is the union of the irreducible components of* Σ*, whose intersection with* V *does not lie entirely in* $V^{\mathcal{X}\setminus V(s)} \cap V(s)$*.*

PROOF. The first assertions follow from Theorems 2.39 and 2.42, together with the fact that Coleff–Herrera currents for a closed analytic set V with pure dimension inherit the standard extension property across any proper hypersurface of V. The second assertion follows from Proposition 2.44 with $S_1 = \Sigma$ and $S_2 = V(s)$. □

PROPOSITION 2.65. *Let* ∇_s *be the operator*[24] $\nabla_s = 2i\pi \rfloor_s - \overline{\partial}$*. For any current* $T \in \mathrm{PM}(U, E)$*, one has*

$$\begin{aligned}\nabla_s\big([P_s](T)\big) &= \big(\mathrm{Id} - [R^s]\big)(T) + [P_s](\overline{\partial} T),\\ \nabla_s\big([R_s](T)\big) &= -[R^s](\overline{\partial} T).\end{aligned} \tag{2.90}$$

PROOF. For $\lambda \in \mathbb{C}$ with $\operatorname{Re}\lambda \gg 1$,

$$\begin{aligned}\nabla_s\Big(|s|^{2\lambda}\frac{s^*}{2i\pi|s|^2} \otimes T\Big) &= |s|^{2\lambda} T - \Big(\overline{\partial}|s|^{2\lambda}\frac{s^*}{2i\pi|s|^2}\Big) \otimes T + \Big(|s|^{2\lambda}\frac{s^*}{2i\pi|s|^2}\Big) \otimes (\overline{\partial} T)\\ &= T - \Big(1 - |s|^{2\lambda} + \overline{\partial}|s|^{2\lambda}\frac{s^*}{2i\pi|s|^2}\Big) \otimes T\\ &\qquad + \Big(|s|^{2\lambda}\frac{s^*}{2i\pi|s|^2}\Big) \otimes (\overline{\partial} T).\end{aligned}$$

The first equality in (2.90) follows from the analytic continuation principle. One has also

$$\begin{aligned}2i\pi\rfloor_s\Big(\Big(1 - |s|^{2\lambda} + \overline{\partial}|s|^{2\lambda}\frac{s^*}{2i\pi|s|^2}\Big) \otimes T\Big) &= -(\overline{\partial}|s|^{2\lambda})\, T,\\ \overline{\partial}\Big(\Big(1 - |s|^{2\lambda} + \overline{\partial}|s|^{2\lambda}\frac{s^*}{2i\pi|s|^2}\Big) \otimes T\Big) &= -(\overline{\partial}|s|^{2\lambda})\, T + (1 - |s|^{2\lambda})\,\overline{\partial} T\\ &\qquad + \Big((\overline{\partial}|s|^{2\lambda})\,\frac{s^*}{|s|^2}\Big) \otimes (\overline{\partial} T).\end{aligned}$$

Furthermore, subtracting the second equality from the first and then applying the analytic continuation principle leads to the second equality in (2.90). □

Observe that the action of any operator $[R^s]$ or $[P^s]$ preserves pseudo-meromorphicity. Thus, given $T \in \mathrm{PM}^{(\bullet, M)}(U, E)$ and given m Cartier divisors $D_1, \dots, D_m$ on $\mathcal{X}$ with global holomorphic sections $s_j \in \mathcal{O}(\mathcal{X}, \mathcal{O}(D_j))$, one defines the following $(\mathbb{C} \oplus \mathcal{O}(-D_1)) \otimes \cdots \otimes (\mathbb{C} \oplus \mathcal{O}(-D_m)) \otimes E$-valued pseudo-meromorphic currents in

---

[24] Such an operator was already introduced in the univariate setting; see Definition 1.26. We extend its action in the new context, respecting the following Leibniz rule: if T is a pseudo-meromorphic current and Φ is an $\mathcal{O}(-D)$-valued k differential form, then

$$\overline{\partial}(\Phi \otimes T) = \overline{\partial}\Phi \otimes T + (-1)^{k+1}\Phi \otimes \overline{\partial} T.$$

the open subset U:

$$
\begin{aligned}
R^{(s_1,\dots,s_m)}(T) &= \big([R^{s_m}]\circ\cdots\circ[R^{s_1}]\big)(T)\\
&= \sum_{r=0}^{\min(m,N-M)} \sum_{\#J=r} R_J^{(s_1,\dots,s_m)}(T).
\end{aligned}
$$

(2.91)

$$
\begin{aligned}
(R+P)^{(s_1,\dots,s_m)}(T) &= \big([(R+P)^{s_m}]\circ\cdots\circ[(R+P)^{s_1}]\big)(T)\\
&= \sum_{r=0}^{\min(m,N-M)} \sum_{\#J=r} (R+P)_J^{(s_1,\dots,s_m)}(T),
\end{aligned}
$$

where, for each subset J of $\{1,\dots,m\}$ with cardinal $|J|=r$

$$R_J^{(s_1,\dots,s_m)}(T)\in \mathrm{PM}^{(\bullet,M+r)}\big(U,\mathcal{O}(-|D|_J)\otimes E\big) \tag{2.92}$$

and

$$
\begin{aligned}
(R+P)_J^{(s_1,\dots,s_m)}(T) &= \sum_{J'\subset\{1,\dots,m\}\setminus J} (R+P)_{J,J'}^{(s_1,\dots,s_m)}(T)\\
&\in \mathrm{PM}^{(\bullet,M+r)}\Big(U,\Big(\bigoplus_{J'\subset\{1,\dots,m\}\setminus J}\mathcal{O}(-|D|_{J\cup J'})\Big)\otimes E\Big).
\end{aligned} \tag{2.93}
$$

The notation $|D|_K$ denotes the sum of the Cartier divisors D_j for $j\in K$. Moreover, given a multi-index $1\le j_1<\cdots<j_r\le m$, one has

$$\mathrm{Supp}\,\big(R_J^{(s_1,\dots,s_m)}(T)\big)\subset W_J(T;\boldsymbol{s}), \tag{2.94}$$

where $W_J(T;\boldsymbol{s}) = W_J^{[m]}(T;\boldsymbol{s})$ is the closed analytic subset of $\mathrm{Supp}\,T$ obtained iteratively by

$$
\begin{aligned}
W_J^{[0]}(T;\boldsymbol{s}) &= \mathrm{Supp}\,T,\\
W_J^{[j+1]}(T;\boldsymbol{s}) &= \begin{cases} (W_J^{[j]}(T;\boldsymbol{s}))^{V(s_{j+1})} = & \text{if } j+1\notin J\\ (W_J^{[j]}(T;\boldsymbol{s}))^{\mathcal{X}\setminus V(s_{j+1})}\cap V(s_{j+1}) & \text{if } j+1\in J\end{cases}\\
&\qquad \text{for } j=0,\dots,m-1.
\end{aligned} \tag{2.95}
$$

Given a subset $J'\subset\{1,\dots,m\}\setminus J$, one has

$$\mathrm{Supp}\,\big((R+P)_{J,J'}^{(s_1,\dots,s_m)}(T)\big)\subset W_{J,J'}(T;\boldsymbol{s}), \tag{2.96}$$

where $W_{J,J'}(T;\boldsymbol{s}) = W_{J,J'}^{[m]}(T;\boldsymbol{s})$ is the closed analytic subset of $\mathrm{Supp}\,T$ obtained iteratively by

$$
\begin{aligned}
W_{J,J'}^{[0]}(T;\boldsymbol{s}) &= \mathrm{Supp}\,T,\\
W_{J,J'}^{[j+1]}(T;\boldsymbol{s}) &= \begin{cases} (W_{J,J'}^{[j]}(T;\boldsymbol{s}))^{V(s_{j+1})} & \text{if } j+1\notin J\cup J'\\ (W_{J,J'}^{[j]}(T;\boldsymbol{s}))^{\mathcal{X}\setminus V(s_{j+1})} & \text{if } j+1\in J'\\ (W_{J,J'}^{[j]}(T;\boldsymbol{s}))^{\mathcal{X}\setminus V(s_{j+1})}\cap V(s_{j+1}) & \text{if } j+1\in J\end{cases}\\
&\qquad \text{for } j=0,\dots,m-1.
\end{aligned} \tag{2.97}
$$

PROPOSITION 2.66. *Assume that* V *is a purely dimensional closed subset of* $\mathcal{X}$ *with codimension* M. *Assume also that* $T \in \mathrm{CH}^{\star S}_{\mathcal{X},V}(U)$ *for some hypersurface* S *satisfying* $\mathrm{codim}\,(V \cap S) > M$. *If* J *and* J' *are any disjoint subsets of* $\{1,\dots,m\}$ *so that* $\mathrm{codim}\, W_{J,J'}([V];\boldsymbol{s}) \le N$, *then* $W_{J,J'}([V];\boldsymbol{s})$ *has pure codimension* $M+|J|$ *and there is a closed hypersurface* $\Sigma_{J,J'} = \Sigma_{J,J'}(V,S,\boldsymbol{s})$, *in* $\mathcal{X}$ *with*

$$\mathrm{codim}(W_{J,J'}([V];\boldsymbol{s}) \cap \Sigma_{J,J'}) > \mathrm{codim}\, W_{J,J'}([V];\boldsymbol{s})$$

satisfying

$$(R+P)^{(s_1,\dots,s_m)}_{J,J'}(T) \in \mathrm{CH}^{\star\Sigma_{J,J'}}_{\mathcal{X},W_{J,J'}([V];\boldsymbol{s})}\big(U, \mathcal{O}(-|\boldsymbol{D}|_{J\cup J'})\big).$$

PROOF. It follows from Proposition 2.44. □

DEFINITION 2.67. If $T = [V]$ and $\mathrm{codim}\,(V \cap V(s_j)) > \mathrm{codim}\, V$ for any $j = 1,\dots,m$, then the bundle-valued current $R^{(s_1,\dots,s_m)}([V])$ is called the *Coleff–Herrera residue current on* V *attached to the the sections* $(s_j)_{|V}$ *of the line bundles* $(\mathcal{O}(D_j))_{|V}$, defined as restrictions to the complex analytic space $(V, \mathcal{O}/\mathcal{I}_V)$ of the Cartier divisors $\mathcal{O}(D_j)$. The current $R^{(s_1,\dots,s_m)}([V])$ in the ambient manifold $\mathcal{X}$ is the direct image through the embedding $\iota_V : V \hookrightarrow \mathcal{X}$ of a bundle-valued current $R^{(\iota_V^* s_1,\dots,\iota_V^* s_m)}$ on the complex analytic space $(V, \mathcal{O}/\mathcal{I}_V)$.

REMARK 2.68. If V is locally defined in a neighborhood of $z \in W_J([V];\boldsymbol{s})$ as the union of a finite number of irreducible components of $V(\sigma_{z,1}) \cap \cdots \cap V(\sigma_{z,M})$ such that the differential form $d\sigma_{z,1} \wedge \cdots \wedge d\sigma_{z,M}$ does not vanish identically on any irreducible component of $V(\sigma_{z,1}) \cap \cdots \cap V(\sigma_{z,N})$, then there exists a germ of a hypersurface $\Sigma_{z,J}$ at z with $\mathrm{codim}_z(W_J(V;\boldsymbol{s}) \cap \Sigma_z) > M + r$ such that $\big(R_J^{(s_1,\dots,s_m)}([V])\big)_z$ factorizes as

$$\big(R_J^{(s_1,\dots,s_m)}([V])\big)_z \in (d\sigma_{z,1} \wedge \cdots \wedge d\sigma_{z,m}) \Big(\mathrm{CH}^{\star\Sigma_{z,J}}_{\mathcal{X},W_J(T;\boldsymbol{s})}\big(\,\cdot\,, \mathcal{O}(-|\boldsymbol{D}|_J)\big)\Big)_z. \tag{2.98}$$

This follows from Proposition 2.44 by using arguments similar to those used in the proof of the second assertion in Proposition 2.54.

Let us point out here that the construction of the current $R^{(s_1,\dots,s_m)}([V])$ through an iterated analytic continuation process, which is not so easy to describe and thus exploit, can be also realized through a direct (noniterative anymore) approach involving just one complex parameter.

PROPOSITION 2.69 ([**AndSWY17A**]). *Let* $\mathcal{X}$, $V \subset \mathcal{X}$ *with pure codimension* M *as before, and let* $s_j \in \mathcal{O}(D_j)$, $j = 1,\dots,m$, *where* $D_1,\dots,D_m$ *are Cartier divisors on* $\mathcal{X}$. *Suppose* $\mathrm{codim}\,(V \cap V(s_j)) > M$ *for* $j = 1,\dots,m$. *For any* $j = 1,\dots,m$, *let* $|\ |$ *be an arbitrary hermitian metric on* $\mathcal{O}(D_j)$. *Let also*

$$\gamma_1 > \gamma_2 > \cdots > \gamma_m \ge 1 \tag{2.99}$$

be m *strictly positive integers and let* $\eta > 0$. *The current-valued map*

(2.100)

$$\lambda \longmapsto \Big(1 - |s_m|^{2\lambda^{\gamma_m}} + \overline{\partial}|s_m|^{2\lambda^{\gamma_m}}\,\frac{s_m^*}{2i\pi|s_m|^2}\Big) \cdots \Big(1 - |s_1|^{2\lambda^{\gamma_1}} + \overline{\partial}|s_1|^{2\lambda^{\gamma_1}}\,\frac{s_1^*}{2i\pi|s_1|^2}\Big)\,[V]$$

extends from

$$\Big\{\lambda \in \mathbb{C} : |\arg_{]\pi,\pi[}(\lambda)| < \frac{\pi}{2(\gamma_1+\eta)}\,,\ |\lambda| \gg 1\Big\}$$

to $\mathbb{C}$ as a meromorphic map, whose polar set does not intersect the closed sector

$$\{0\} \cup \Big\{\lambda \in \mathbb{C}^* \,:\, |\arg_{]\pi,\pi[}(\lambda)| \leq \frac{\pi}{2\gamma_1}\Big\}. \tag{2.101}$$

The value of this map at $\lambda = 0$ equals $R^{(s_1,\dots,s_m)}([V])$, independently of the values of the positive integers γ_j, provided condition (2.99) is satisfied.

PROOF. Since the problem is a local one, one can replace $\mathcal{X}$ by an open subset $U \subset \mathcal{X}$ over which all line bundles $\mathcal{O}(D_j)$, $j = 1,\dots,m$, are trivialized. Therefore, the problem reduces to the case where s_j is a holomorphic complex-valued function on $\mathcal{X}$ and $|s_j|^2 = e^{-\varphi_j} \times |s_j|^2$, for every $j = 1,\dots,m$, where $|\,|$ is the usual absolute value in $\mathbb{C}$. Furthermore, φ_j stands for a real, C^∞ function describing the metric $|\,|$ on $\mathcal{O}(D_j)$, namely, $dd^c\varphi_j = c_1(\mathcal{O}(D_j), |\,|)$. Let $\iota_V : V \hookrightarrow \mathcal{X}$ be the inclusion embedding. Observe that each $\iota_V^* s_j$ defines an element in $\mathcal{O}_V(\mathcal{X})$. We introduce a log resolution $\mathcal{V} \overset{\pi}{\to} V$ for the closed hypersurface $\sigma_{\boldsymbol{s}} = V(\iota_V^* s_1) \cap \dots \cap V(\iota_V^* s_m)$, using Hironaka's theorem.[25] Then $\pi^{-1}(\mathcal{Y})$ becomes a closed hypersurface with normal crossings in the $(N-M)$-dimensional complex manifold $\mathcal{V}$. This implies in particular that in a neighborhood of any point $z = z^\iota \in \pi^{-1}(V \cap \sigma_{\boldsymbol{s}}) \subset \mathcal{V}$, one can choose centered local coordinates $\boldsymbol{\zeta} = (\zeta_1,\dots,\zeta_{N-M})$ such that, in a convenient local chart U in a neighborhood of the point $z \in \mathcal{V}$, one has

$$(\iota_V \circ \pi)^* s_j = \sigma_j(\zeta)\zeta^{\beta_j} = \sigma_j(\zeta) \prod_{\kappa=1}^{N-M} \zeta_\ell^{\beta_{j,\kappa}}, \quad \psi_j := (\iota_V \circ \pi)^*\varphi_j, \tag{2.102}$$

for any $j = 1,\dots,m$. Here, σ_j is an invertible holomorphic function in U for any $j = 1,\dots,m$, $\beta_{j,\kappa}$, $\kappa = 1,\dots,N-M$, are strictly positive integers, and ψ_j are real functions for any $j = 1,\dots,m$. The proof of Proposition 2.69 relies on the following lemma, whose elegant proof quoted here is due to H. Samuelsson Kalm.

LEMMA 2.70 (H. Samuelsson Kalm [**AndSWY17A**, §5]). *Let $0 \leq M < N$, $\gamma_1,\dots,\gamma_m$ be strictly positive integers satisfying (2.99), let $\mathfrak{s}$ be a permutation of $\{1,\dots,m\}$, and let $\eta > 0$. Assume also that Ω is a neighborhood of the origin in $\mathbb{C}^{N-M}_\zeta$, $\chi_1,\dots,\chi_m$ are m smooth strictly positive functions in Ω, and $\zeta^{\beta_1},\dots,\zeta^{\beta_m}$ are m monomial functions in the variables $\zeta = (\zeta_1,\dots,\zeta_{N-M})$. For any integer r such that $0 \leq r \leq \min(m, N-M)$, the holomorphic current-valued map*

$$\Theta : (\mu_1,\dots,\mu_m) \longmapsto \Big(\prod_{j>r} \frac{|\chi_j\zeta^{\beta_j}|^{2\mu_{\mathfrak{s}(j)}}}{\zeta^{\beta_j}}\Big) \bigwedge_{j\leq r} \overline{\partial}\Big(\frac{|\chi_j\zeta^{\beta_j}|^{2\mu_{\mathfrak{s}(j)}}}{\zeta^{\beta_j}}\Big) \tag{2.103}$$

in $\{\mu \in \mathbb{C}^m : \operatorname{Re}\mu_j \gg 1$ for $j = 1,\dots,m\}$ is such that the map

$$\lambda \longmapsto \Theta(\lambda^{\gamma_1},\dots,\lambda^{\gamma_m})$$

extends from

$$\Big\{\lambda \in \mathbb{C} \,:\, |\arg_{]\pi,\pi[}(\lambda)| < \frac{\pi}{2(\gamma_1+\eta)}\,,\, |\lambda| \gg 1\Big\}$$

to $\mathbb{C}$ as a meromorphic map, whose polar set does not intersect the closed sector (2.101). Its value at $\lambda = 0$ equals the pseudo-meromorphic current

$$\Big[\Theta(\lambda^{\gamma_1},\dots,\lambda^{\gamma_m})\Big]_{\lambda=0} = \Big[\Big[\cdots\Big[\Big[\Theta(\mu_1,\dots,\mu_m)\Big]_{\mu_1=0}\Big]_{\mu_2=0}\cdots\Big]_{\mu_{m-1}=0}\Big]_{\mu_m=0}. \tag{2.104}$$

[25] See §A.4.3, in particular Theorem A.41, which we apply here with $\mathcal{X} = V$ and, as $\mathcal{Y}$, the union of $\sigma_{\boldsymbol{s}}$ with any hypersurface in V which contains V_{sing}.

PROOF OF LEMMA 2.70. Take first $\chi_j \equiv 1$ for any $j = 1, \dots, m$ and $\mathfrak{s} = \mathrm{Id}$. For $\operatorname{Re}\mu_j \gg 1$, $j = 1, \dots, m$, one can rewrite $\Theta(\mu)$ as

(2.105)

$$\begin{aligned}
\Theta(\mu) &= \mu_1 \cdots \mu_r \Big(\frac{\prod_{j=1}^m |\zeta^{\beta_j}|^{2\mu_j}}{\zeta^{\sum_{j=1}^m \beta_j}}\Big) \bigwedge_{j=1}^r \frac{\overline{d\zeta^{\beta_j}}}{\overline{\zeta^{\beta_j}}} \\
&= \mu_1 \cdots \mu_r \Big(\frac{\prod_{j=1}^m |\zeta_j^{\beta_j}|^{2\mu_j}}{\zeta^{\sum_{j=1}^m \beta_j}}\Big)\Big(\sum_{1 \le \kappa_1 < \cdots < \kappa_r \le N-M} \det[\beta_{j,\kappa_\nu}]_{1 \le j,\nu \le r} \bigwedge_{\nu=1}^r \frac{\overline{d\zeta_{\kappa_\nu}}}{\overline{\zeta_{\kappa_\nu}}}\Big) \\
&= \mu_1 \cdots \mu_r \Big(\det[\beta_{j,\kappa}]_{1 \le j,\kappa \le r} \frac{\prod_{\kappa=1}^{N-M} |\zeta_\kappa|^{2\langle\!\langle \beta_{\cdot,\kappa}\,,\,\mu\rangle\!\rangle}}{\zeta^{\sum_{j=1}^m \beta_j}} \bigwedge_{\kappa=1}^r \frac{\overline{d\zeta_\kappa}}{\overline{\zeta_\kappa}} + \cdots\Big) \\
&= \det[\beta_{j,\kappa}]_{1 \le j,\kappa \le r} \Big(\prod_{\kappa=1}^r \frac{\lambda_\kappa}{\langle\!\langle \beta_{\cdot,\kappa}\,,\,\mu\rangle\!\rangle}\Big)\Big(\prod_{\kappa>r} \frac{|\zeta_\kappa|^{2\langle\!\langle \beta_{\cdot,\kappa}\,,\,\mu\rangle\!\rangle}}{\zeta_\kappa^{|\beta(\cdot,\kappa)|}}\Big) \bigwedge_{\kappa=1}^r \overline{\partial}\Big(\frac{|\zeta_\kappa|^{2\langle\!\langle \beta_{\cdot,\kappa}\,,\,\mu\rangle\!\rangle}}{\zeta_\kappa^{|\beta(\cdot,\kappa)|}}\Big) \\
&\qquad + \cdots .
\end{aligned}$$

The $'\mathcal{D}(\Omega)$-valued map

$$(\nu_1, \dots, \nu_{N-M}) \longmapsto \Big(\prod_{\kappa>r} \frac{|\zeta_\kappa|^{2\nu_\kappa}}{\zeta_\kappa^{|\beta(\cdot,\kappa)|}}\Big) \bigwedge_{\kappa=1}^r \overline{\partial}\Big(\frac{|\zeta_\kappa|^{2\nu_\kappa}}{\zeta_\kappa^{|\beta(\cdot,\kappa)|}}\Big)$$

extends from $\{\nu \in \mathbb{C}^{N-M} : \operatorname{Re}\nu_\kappa \gg 1 \text{ for } \kappa = 1, \dots, N-M\}$ to $\mathbb{C}^{N-M}$ as a meromorphic function, whose polar set does not intersect

$$\{\nu \in \mathbb{C}^{N-M} : \operatorname{Re}\nu_\kappa \ge 0 \text{ for } \kappa = 1, \dots, N-M\}.$$

Consequently, the $'\mathcal{D}(\Omega)$-valued map

$$(\mu_1, \dots, \mu_m) \longmapsto \Big(\prod_{\kappa>r} \frac{|\zeta_\kappa|^{2\langle\!\langle \beta_{\cdot,\kappa}\,,\,\mu\rangle\!\rangle}}{\zeta_\kappa^{|\beta(\cdot,\kappa)|}}\Big) \bigwedge_{\kappa=1}^r \overline{\partial}\Big(\frac{|\zeta_\kappa|^{2\langle\!\langle \beta_{\cdot,\kappa}\,,\,\mu\rangle\!\rangle}}{\zeta_\kappa^{|\beta(\cdot,\kappa)|}}\Big)$$

extends from $\{\mu \in \mathbb{C}^m : \operatorname{Re}\mu_j \gg 1 \text{ for } j = 1, \dots, m\}$ to $\mathbb{C}^m$ as a meromorphic function, whose polar set does not intersect $\{\mu \in \mathbb{C}^m : \operatorname{Re}\mu_j \ge 0 \text{ for } j = 1, \dots, m\}$. On the other hand, if $\det[\beta_{j,\kappa}]_{1 \le j,\kappa \le r} \neq 0$, then the polar set of the univariate rational function

$$\lambda \longmapsto \prod_{\kappa=1}^r \frac{\lambda^{\gamma_{\mathfrak{s}(\kappa)}}}{\sum_{j=1}^m \beta_{\kappa,j} \lambda^{\gamma_{\mathfrak{s}(j)}}}$$

does not intersect the closed sector (2.101), for any permutation $\mathfrak{s}$ of $\{1, \dots, m\}$. Moreover, its value at $\mu = 0$ equals

$$\Big[\Big[\cdots\Big[\Big[\prod_{\kappa=1}^r \frac{\mu_{\mathfrak{s}(\kappa)}}{\sum_{j=1}^m \beta_{j,\kappa}\, \mu_{\mathfrak{s}(j)}}\Big]_{\lambda_1=0}\Big]_{\lambda_2=0} \cdots\Big]_{\lambda_{m-1}=0}\Big]_{\lambda_m=0}.$$

Since this is valid for any permutation $\mathfrak{s}$ of $\{1, \dots, m\}$, with this procedure one can treat all nonzero terms in the development of $\Theta(\mu)$ in (2.105). Thus, the proof of the lemma follows in that particular case where all χ_j equals identically the constant 1. When the χ_j are arbitrary strictly positive functions, one needs to introduce m additional complex parameters $\tau_1, \dots, \tau_m$ and consider instead of Θ the $'\mathcal{D}(\Omega)$-valued map

$$\boldsymbol{\Theta} : (\mu_1, \dots, \mu_m, \tau_1, \dots, \tau_m) \longmapsto \Big(\prod_{j>r} \frac{|\chi_j|^{2\tau_j}\, |\zeta^{\beta_j}|^{2\mu_{s(j)}}}{\zeta^{\beta_j}}\Big) \bigwedge_{j \le r} \overline{\partial}\Big(\frac{|\chi_j|^{2\tau_j}\, |\zeta^{\beta_j}|^{2\mu_{s(j)}}}{\zeta^{\beta_j}}\Big).$$

From the reasoning above, the $'\mathcal{D}(\Omega)$-valued map

$$(\lambda, \tau_1, \dots, \tau_m) \longmapsto \boldsymbol{\Theta}(\lambda^{\gamma_1}, \dots, \lambda^{\gamma_m}, \tau_1, \dots, \tau_m)$$

extends from

$$\Big\{(\lambda, \tau_1, \dots, \tau_m) \in \mathbb{C}^{m+1} \,:\, |\arg_{]\pi,\pi[}(\lambda)| < \frac{\pi}{2(\gamma_1 + \eta)}\,,\ |\lambda| \gg 1\Big\}$$

to $\mathbb{C}^{m+1}$ as a meromorphic map, whose polar set does not intersect the product of the closed sector (2.101) with $\mathbb{C}^{m+1}$. Observe that we have then

$$\boldsymbol{\Theta}(\lambda^{\gamma_1}, \dots, \lambda^{\gamma_m}, 0, \dots, 0) = \Big[\Theta(\lambda^{\gamma_1}, \dots, \lambda^{\gamma_m})\Big]_{\lambda=0}.$$

One can check immediately that the right-hand side of (2.104) remains unaffected by such an operation, which completes the proof of the lemma in the case where the χ_j are arbitrary, strictly positive, smooth functions. □

We now resume the proof of Proposition 2.69. Proving the result amounts to proving it for each current $R_J^{(s_1,\dots,s_m)}([V])$, where the multi-index $J = \{j_1, \dots, j_r\}$ is in strictly increasing order and $1 \le r \le \min(m, N - M)$. Consider the $'\mathcal{D}(\mathcal{X})$-valued holomorphic map

$$\boldsymbol{\Phi}_J : \mu \longmapsto \Big(\prod_{j \notin J}(1 - |s_j|^{2\mu_j})\Big)\Big(\bigwedge_{\kappa=1}^{r} \overline{\partial}|s_{j_\kappa}|^{2\mu_{j_\kappa}}\Big)\,[V]$$

defined in $\{\mu \in \mathbb{C}^m \,:\, \operatorname{Re}\mu_1 \gg 1, \dots, \operatorname{Re}\mu_m \gg 1\}$. Let us now consider a test form $\varphi \in \mathcal{D}^{(N-M, N-M-r)}(\mathcal{X})$. One has by construction that

$$\begin{aligned}&\langle R_J^{(s_1,\dots,s_m)}([V]), \varphi\rangle\\ &\quad = (-1)^{r(r-1)/2}\Big[\Big[\cdots\Big[\Big[\langle \boldsymbol{\Phi}_J(\mu_1, \dots, \mu_m), \varphi\rangle\Big]_{\mu_1=0}\Big]_{\mu_2=0}\cdots\Big]_{\mu_{m-1}=0}\Big]_{\mu_m=0}.\end{aligned}$$

Let us introduce a partition of unity (U^ι, θ^ι) on $\mathcal{V}$, subordinated to a covering $(\iota_V \circ \pi)^{-1}(\operatorname{Supp}\varphi) \subset \bigcup_\iota U^\iota$ of $(\iota_V \circ \pi)^{-1}(\operatorname{Supp}\varphi)$. For each local chart $U = U^\iota$, where (2.102) holds with σ_j^ι, β_j^ι, φ_j^ι for $j = 1, \dots, m$, let

$$\chi_j^\iota = |\sigma_j^\iota| e^{-\psi_j^\iota/2}, \quad \beta_j = \beta_j^\iota,$$

for $1 \le j \le r$ and

$$(\chi_j^\iota, \beta_j^\iota) = (|\sigma_j^\iota| e^{-\psi_j/2}, \beta_j^\iota) \text{ or } (1, (0, \dots, 0))$$

for $j = r+1, \dots, m$. Let also $\mathfrak{s} = \mathfrak{s}_J$ be the permutation of $\{1, \dots, m\}$ such that $\mathfrak{s}(\kappa) = j_\kappa$ for $\kappa = 1, \dots, r$ and $\mathfrak{s}(r + \kappa') = j'_{\kappa'}$ for $1 \le \kappa' \le m - r$. The indexes $j'_{\kappa'}$ are defined by $\{1, \dots, m\} \setminus J = \{j'_1, \dots, j'_{m-r}\}$ and taken in strictly increasing order. With such choices of $\boldsymbol{\chi}^\iota$, $\boldsymbol{\beta}^\iota$, and $\mathfrak{s}$, one can observe that

$$\begin{aligned}&(2i\pi)^r \langle R_J^{(s_1,\dots,s_m)}([V]), \varphi\rangle\\ &= \sum_\iota \pm\Big[\Big[\cdots\Big[\Big[\Big\langle \Theta_{\mathfrak{s}}^{[\boldsymbol{\chi}^\iota, \boldsymbol{\beta}^\iota]}(\mu_1, \dots, \mu_m)\,,\, \theta_\iota (\iota_V \circ \pi)^* \varphi\Big\rangle\Big]_{\mu_1=0}\Big]_{\mu_2=0}\cdots\Big]_{\mu_{m-1}=0}\Big]_{\mu_m=0}\end{aligned}$$

with possible duplications of the index ι, where $\Theta_{\mathfrak{s}}^{[\boldsymbol{\chi}^\iota, \boldsymbol{\beta}^\iota]}$ is defined as in (2.103). Proposition 2.69 follows then immediately from Lemma 2.70. □

Principal value currents play a fundamental role in division problems. Their variation, the so-called *mixed principal value–residue currents,* namely the currents $(R+P)^{(s_1,\dots,s_m)}(T)$, where $T \in \mathrm{PM}^{(\bullet,M)}(\mathcal{X}, E)$ for a holomorphic bundle E, are also of great interest. In particular, when T is the integration current on a purely dimensional closed analytic subset, let us introduce the following notion, akin to Definition 2.67.

DEFINITION 2.71. Let $V \subset \mathcal{X}$ be a closed analytic subset with pure codimension M. Assume also that the Cartier divisors $\mathcal{O}(D_j)$, together with their holomorphic sections s_j are as in Definition 2.67. The bundle-valued current $(R+P)^{(s_1,\dots,s_m)}([V])$ is called the *Coleff–Herrera mixed principal value–residue current* on V attached to the sections $(s_j)_{|V}$ of the line bundles $(\mathcal{O}(D_j))_{|V}$ defined as restrictions to the complex analytic space $(V, \mathcal{O}/\mathcal{I}_V)$ of the Cartier divisors $\mathcal{O}(D_j)$. The current $(R+P)^{(s_1,\dots,s_m)}([V])$ in the ambient manifold $\mathcal{X}$ is the direct image through the embedding $\iota_V : V \hookrightarrow \mathcal{X}$ of a bundle-valued current $(R+P)^{(\iota_V^* s_1,\dots,\iota_V^* s_m)}$ on the complex analytic space $(V, \mathcal{O}/\mathcal{I}_V)$.

The proof of Proposition 2.69 can be carried to such mixed principal value–residue context.

PROPOSITION 2.72 ([**AndSWY17A**]). *Let $\mathcal{X}$, $V \subset \mathcal{X}$ with pure codimension M be as before, and $s_j \in \mathcal{O}(D_j)$, $j = 1,\dots,m$, where $D_1,\dots,D_m$ are Cartier divisors on $\mathcal{X}$. Assume that* $\operatorname{codim}(V \cap V(s_j)) > M$ *for $j = 1,\dots,m$. For any $j = 1,\dots,m$, let $|\ |$ be an arbitrary hermitian metric on $\mathcal{O}(D_j)$. Let also*

$$\gamma_1 > \gamma_2 > \cdots > \gamma_m \geq 1 \tag{2.106}$$

be m strictly positive integers and $\eta > 0$. The current-valued map

$$\begin{aligned}(2.107)\quad \lambda \longmapsto \Big(1 - |s_m|^{2\lambda\gamma_m} + (\mathrm{Id} + \overline{\partial})\Big(|s_m|^{2\lambda\gamma_m} \frac{s_m^*}{2i\pi|s_m|^2}\Big)\Big) \\ \cdots \Big(1 - |s_1|^{2\lambda\gamma_1} + (\mathrm{Id} + \overline{\partial})\Big(|s_1|^{2\lambda\gamma_1} \frac{s_1^*}{2i\pi|s_1|^2}\Big)\Big)\,[V]\end{aligned}$$

extends from

$$\Big\{\lambda \in \mathbb{C} : |\arg_{]\pi,\pi[}(\lambda)| < \frac{\pi}{2(\gamma_1 + \eta)}\,,\ |\lambda| \gg 1\Big\}$$

to $\mathbb{C}$ as a meromorphic map, whose polar set does not intersect the closed sector

$$\{0\} \cup \Big\{\lambda \in \mathbb{C}^* : |\arg_{]\pi,\pi[}(\lambda)| \leq \frac{\pi}{2\gamma_1}\Big\}.$$

Moreover, the value of this map at $\lambda = 0$ equals $(R+P)^{(s_1,\dots,s_m)}([V])$, independently of the values of the positive integers γ_j, provided the condition (2.106) *is satisfied.*

REMARK 2.73. In the particular case where the germs $\iota^*(s_1)_z,\dots,\iota_V^*(s_m)_z$ define a regular sequence in the local ring $\mathcal{O}_{V,z}$ for any $z \in V \cap (V(s_1) \cup \cdots \cup V(s_m))$, all currents $(R+P)_{J,J'}^{(s_1,\dots,s_m)}([V])$ vanish unless $J' = \{1,\dots,m\} \setminus J$. Proposition

2.72 implies then that

$$(R+P)^{(s_1,\dots,s_m)}([V]) = \Big[\cdots\Big[(\mathrm{Id}+\overline{\partial})\Big(|s_m|^{2\lambda_m}\frac{s_m^*}{2i\pi|s_m|^2}\Big)\cdots(\mathrm{Id}+\overline{\partial})\Big(|s_1|^{2\lambda_1}\frac{s_1^*}{2i\pi|s_1|^2}\Big)[V]\Big]_{\lambda_1=0}\cdots\Big]_{\lambda_m=0} = \Big[(\mathrm{Id}+\overline{\partial})\Big(|s_m|^{2\lambda^{\gamma_m}}\frac{s_m^*}{2i\pi|s_m|^2}\Big)\cdots(\mathrm{Id}+\overline{\partial})\Big(|s_1|^{2\lambda^{\gamma_1}}\frac{s_1^*}{2i\pi|s_1|^2}\Big)[V]\Big]_{\lambda=0}. \tag{2.108}$$

Under the sole assumption that the germs $\iota^*(s_1)_z,\dots,\iota_V^*(s_m)_z$ define a regular sequence in the local ring $\mathcal{O}_{V,z}$ for any $z\in V\cap(V(s_1)\cap\cdots\cap V(s_m))$, all closed analytic subsets $W_J([V];\boldsymbol{s})$ introduced in (2.95) are empty, except when $J=\{1,\dots,m\}$. Then, one has in this case that

$$R^{(s_1,\dots,s_m)}([V]) = \Big[\cdots\Big[\overline{\partial}\Big(|s_m|^{2\lambda_m}\frac{s_m^*}{2i\pi|s_m|^2}\Big)\cdots\overline{\partial}\Big(|s_1|^{2\lambda_1}\frac{s_1^*}{2i\pi|s_1|^2}\Big)[V]\Big]_{\lambda_1=0}\cdots\Big]_{\lambda_m=0} = \Big[\overline{\partial}\Big(|s_m|^{2\lambda^{\gamma_m}}\frac{s_m^*}{2i\pi|s_m|^2}\Big)\cdots\overline{\partial}\Big(|s_1|^{2\lambda^{\gamma_1}}\frac{s_1^*}{2i\pi|s_1|^2}\Big)[V]\Big]_{\lambda=0}. \tag{2.109}$$

To conclude this section, one should mention that Propositions 2.69 or 2.72 still hold (with a similar proof) in which case $[V]$ is replaced by a section of the sheaves $\mathrm{CH}_{\mathcal{X},V}$ or $\mathrm{CH}_{\mathcal{X},V}^{\star S}$, when V is a purely dimensional closed subset of $\mathcal{X}$ and S is a closed complex hypersurface satisfying $\mathrm{codim}\,(V\cap S)>\mathrm{codim}\,V$. Observe that in such a situation the currents $R^{(s_1,\dots,s_m)}(T)$ or $(R+P)^{(s_1,\dots,s_m)}(T)$ cannot be achieved as direct images of currents on V as was the case for $T=[V]$; see Definitions 2.67 and 2.71. The description of their action involves in general the action of holomorphic differential operators in the ambient manifold $\mathcal{X}$ transversally to V. One recovers here the fact that the definition of the Coleff-Hererra sheaves $\mathrm{CH}_{\mathcal{X},V}$ or $\mathrm{CH}_{\mathcal{X},V}^{\star S}$ relies deeply on the embedding $\iota_V : V\hookrightarrow\mathcal{X}$, as already pointed out in Remark 2.34.

2.3.3. Coleff-Herrera theory in the complete intersection setting. Let $D_1,\dots,D_m$ be Cartier divisors with global holomorphic sections $s_j\in\mathcal{O}(D_j)$ on a complex, N-dimensional manifold $\mathcal{X}$. We assume that $m\le N-M$, where M equals the pure codimension of the closed analytic subset $V\subset\mathcal{X}$. Moreover, we assume throughout this subsection[26] that

$$\forall z\in V,\quad \forall J\subset\{1,\dots,m\},\ \mathrm{codim}_z\Big(V\cap\bigcap_{j\in J}V(s_j)\Big)\ge M+|J|. \tag{2.110}$$

EXAMPLE 2.74. If $\mathrm{codim}(V\cap\big(V(s_1)\cap\cdots\cap V(s_m)\big)=M+m$, then the intersection subset $V\cap\big(V(s_1)\cap\cdots\cap V(s_m)\big)$ admits a neighborhood U in $\mathcal{X}$ such that the restrictions of the s_j to U fulfill (2.110), when $\mathcal{X}$ is replaced by U and V by $V\cap U$. Moreover, if the previous hypothesis remains fulfilled and all D_j are

[26] This amounts to saying that $\mathrm{codim}(V\cap V(s_j))>M$ for any j, and that at any point z in $V\cap\bigcup_{j=1}^m V(s_j)$, the germs in $\mathcal{O}_{V,z}$ of the holomorphic functions $\iota_V^*s_j$ which vanish at z define a regular sequence in this local ring; see §B.1.6.

equal to the same Cartier divisor D, then the holomorphic sections

$$\langle\!\langle a_j, s\rangle\!\rangle = \sum_{k=1}^{m} a_{j,k}\, s_j,\ j = 1, \dots, m$$

of $\mathcal{O}(D)$ satisfy (2.110) for any generic choice of elements $a_j \in \mathbb{C}^m$. This is a consequence of the so-called *box principle*; see §A.3.1.

The next theorem, whose proof will be the core of this subsection, reflects the robustness of the approach to the currents $R^{(s_1,\dots,s_m)}([V])$ or $(R+P)^{(s_1,\dots,s_m)}([V])$ through analytic continuation (as described previously in §2.3.2) under the regularity hypothesis (2.110) imposed in the preamble.

THEOREM 2.75 ([**Sam09**]). *Let $\mathcal{X}$, V, $D_1, \dots, D_m$ and $s_1, \dots, s_m$ be as above fulfilling* (2.110). *Let $|\ |$ be a hermitian metric on $\mathcal{O}(D_j)$ for each $j = 1, \dots, m$. Then the following $'\mathcal{D}(\mathcal{X}, \bigotimes_{j=1}^m(\mathbb{C} \oplus \mathcal{O}(-D_j)))$-valued maps*

$$(2.111)\qquad \begin{aligned} &\mu \longmapsto \Big(\overline{\partial}\Big(|s_m|^{2\mu_m}\frac{s_m^*}{2i\pi|s_m|^2}\Big)\Big)\cdots\Big(\overline{\partial}\Big(|s_1|^{2\mu_1}\frac{s_1^*}{2i\pi|s_1|^2}\Big)\Big)\,[V],\\ &\mu \longmapsto \Big((\mathrm{Id}+\overline{\partial})\Big(|s_m|^{2\mu_m}\frac{s_m^*}{2i\pi|s_m|^2}\Big)\Big)\cdots\Big((\mathrm{Id}+\overline{\partial})\Big(|s_1|^{2\mu_1}\frac{s_1^*}{2i\pi|s_1|^2}\Big)\Big)\,[V] \end{aligned}$$

extend from $\{\mu \in \mathbb{C}^m : \mathrm{Re}\,\mu_j \gg 1$ for $j = 1, \dots, m\}$ to $\mathbb{C}^m$ as meromorphic maps, whose polar set does not intersect $\{\mu \in \mathbb{C}^m : \mathrm{Re}\,\mu_j \geq 0$ for $j = 1, \dots, m\}$. Moreover, their value at $\mu = 0$ is equal to $R^{(s_1,\dots,s_m)}([V])$ and $(R+P)^{(s_1,\dots,s_m)}([V])$ respectively.

REMARK 2.76. Theorem 2.75 remains valid if one replaces $[V]$ in (2.111) by any $(\bullet, M)$ bundle valued current T belonging to the Coleff–Herrera class $\mathrm{CH}_{\mathcal{X},V}(\mathcal{X}, E)$; see [**BjS10**]. For the reasons mentioned at the end of §2.3.2, we will therefore keep considering residue currents or mixed principal value–residue currents in the ambient manifold $\mathcal{X}$.

PROOF. The proof of Theorem 2.75 is based on the following lemma, whose proof lies upon a division argument of the Whitney type for differential forms.

LEMMA 2.77 ([**Sam09**]). *Let Ω be a neighborhood of the origin $\mathbf{0}$ in $\mathbb{C}^N$ and let V be a closed analytic subset of Ω with pure codimension M such that $\mathbf{0} \in V$. For $m \in \mathbb{N}^*$ and $1 \leq r \leq \min(m, N-M)$, let $s_1, \dots, s_m$ be m holomorphic functions in Ω satisfying condition* (2.110). *Then, for any $\nu \in \mathbb{N}^*$, the $'\mathcal{D}^{(M,M+r)}$-valued map*

$$(2.112)\qquad (\mu_1, \dots, \mu_m) \longmapsto \Xi(\mu) = \Big(\prod_{j>r}\frac{|s_j|^{2\mu_j}}{s_j^\nu}\Big)\Big(\bigwedge_{j\leq r}\overline{\partial}\Big(\frac{|s_j|^{2\mu_j}}{s_j^\nu}\Big)\Big)\,[V]$$

extends from $\{\mu \in \mathbb{C}^m : \mathrm{Re}\,\mu_j \gg 1$ for $j = 1, \dots, m\}$ to $\mathbb{C}^m$ as a meromorphic function, whose polar set intersects $([-\eta, +\infty[)^m$ along a finite union of hyperplanes $\sum_{j=1}^r \alpha_{\iota,j}\mu_j = 0$. The coefficients $\alpha_\iota \in \mathbb{N}^m$ satisfy the condition

$$\mathrm{card}(\{\kappa \in \{1, \dots, r\} : \alpha_{\iota,\kappa} \in \mathbb{N}^*\}) \geq 2.$$

Moreover $\eta > 0$, as well as the polar hyperplanes $\langle\!\langle \alpha_\iota, \mu\rangle\!\rangle = 0$, are independent from the exponent ν.

PROOF OF LEMMA 2.77. If $\mu \in \mathbb{C}^m$ is such that $\mathrm{Re}\,\mu_j \gg 1$ for any j, then it is enough to apply the current $\Xi(\mu)$ on test forms of the form $\overline{\psi} \wedge \varphi$, where

$\varphi \in \mathcal{D}^{(N-M,0)}(\Omega)$ and ψ is a $(N-M-r,0)$-form with constant coefficients, for linearity reasons. Let, as usual, $\iota_V : V \hookrightarrow \Omega$ be the inclusion embedding and $\pi : \mathcal{V} = \mathcal{V}_{\mathcal{Y}} \longrightarrow V$ be a log resolution,[27] where $\mathcal{Y}$ is the hypersurface of V defined as the union of $V \cap V(s_1 \cdots s_m)$ with any hypersurface containing V_{sing}. As in the proof of Proposition 2.69, one chooses centered local coordinates $\boldsymbol{\zeta} = (\zeta_1, \dots, \zeta_{N-M})$ in a neighborhood of any point $z \in \pi^{-1}(V)$. With respect to these centered coordinates and in a convenient neighborhood U of z in $\mathcal{V}$, one has

$$(\iota_V \circ \pi)^* s_j = \sigma_j \, \zeta^{\beta_j}$$

for any $j = 1, \dots, m$, with $\sigma_j \in (\mathcal{O}(U))^*$ and $\zeta^{\beta_j} = \prod_{\kappa=1}^{N-M} \zeta_\kappa^{\beta_{j,\kappa}} \not\equiv 1$ a true monomial. Let $(\theta^\iota)_\iota$ be a partition of unity on the complex manifold $\mathcal{V}$ subordinate to a covering $(\iota_V \circ \pi)^{-1}(\operatorname{Supp} \varphi) \subset \bigcup_\iota U^\iota$ of the compact subset $(\iota_V \circ \pi)^{-1}(\operatorname{Supp} \varphi)$. Then, one has[28]

(2.113)
$$\langle \Xi, \mu \rangle = \sum_\iota \int_{U^\iota} \Big(\prod_{j>r} \frac{|\sigma_j^\iota \zeta^{\beta_j^\iota}|^{2\mu_j}}{(\sigma_j^\iota)^\nu \zeta^{\nu \beta_j^\iota}} \Big) \Big(\bigwedge_{j \le r} \overline{\partial} \Big(\frac{|\sigma_j^\iota \zeta^{\beta_j^\iota}|^{2\mu_j}}{(\sigma_j^\iota)^\nu \zeta^{\nu \beta_j^\iota}} \Big) \Big) \wedge \theta^\iota \, (\iota_V \circ \pi)^* (\overline{\psi} \wedge \varphi).$$

It is enough to consider the open charts U^ι around points $z = z^\iota$ in $\pi^{-1}(\mathcal{Y})$. Fixing such a chart, we drop the index ι for a moment to simplify the notation. Therefore, the term from the right-hand side of (2.113) will be

$$\int_U \Big(\prod_{j>r} \frac{|\sigma_j \zeta^{\beta_j}|^{2\mu_j}}{(\sigma_j)^\nu \zeta^{\nu \beta_j}} \Big) \Big(\bigwedge_{j \le r} \overline{\partial} \Big(\frac{|\sigma_j \zeta^{\beta_j}|^{2\mu_j}}{(\sigma_j)^\nu \zeta^{\nu \beta_j}} \Big) \Big) \wedge \theta \, (\iota_V \circ \pi)^* (\overline{\psi} \wedge \varphi). \tag{2.114}$$

After possible re-indexation of $s_1, \dots, s_r$, we us assume that for some $\rho \le r$ the set $\{\beta_1, \dots, \beta_\rho\}$ is a maximal subset of $\{\beta_1, \dots, \beta_r\}$, whose elements are $\mathbb{Q}$-linearly independent in $\mathbb{Q}^{N-M}$. Up to a holomorphic change of centered local coordinates $\boldsymbol{\zeta}$ in the chart U, one may assume then in addition that $\sigma_1 = \cdots = \sigma_\rho \equiv 1$ in U. Thus (2.114) becomes

$$\Big(\prod_{j=1}^r \mu_j \Big) \int_U \Big(\prod_{j>\rho} \frac{|\sigma_j|^{2\lambda_j}}{\sigma_j^\nu} \Big) \Big(\prod_{j=1}^m \frac{|\zeta^{\beta_j}|^{2\mu_j}}{\zeta^{\nu \beta_j}} \Big) \Big(\bigwedge_{j=1}^\rho \frac{d\overline{\zeta}^{\beta_j}}{\overline{\zeta}^{\beta_j}} \Big) \wedge \overline{\omega} \wedge \theta \, (\iota_V \circ \pi)^* \varphi, \tag{2.115}$$

where

$$\omega = \Big(\bigwedge_{j=\rho+1}^r \frac{d\sigma_j}{\sigma_j} \Big) \wedge (\iota_V \circ \pi)^* \psi \in \Omega^{N-M-\rho}(U).$$

Let $K \subset \{1, \dots, N-M\}$ be the set of indices of coordinates $\zeta_1, \dots, \zeta_{N-M}$ that divide the monomial $\prod_{j>r} \zeta^{\beta_j}$. Observe that $K = \emptyset$ if $r = m$. For each subset $\mathfrak{K} \subset K$, let

$$\Pi_{\mathfrak{K}} = \{ \zeta \in U \, : \, \zeta_\kappa = 0 \text{ for } \kappa \in \mathfrak{K} \} \overset{\iota_{\mathfrak{K}}}{\hookrightarrow} U.$$

Then the pullback $\omega_{\mathfrak{K}} = \iota_{\mathfrak{K}}^* \omega$ of ω extends constantly to U. If

$$\widetilde{\omega} = \sum_{k_1 \in K} \omega_{\{k_1\}} - \sum_{\{k_1 < k_2 \, : \, k_1, k_2 \in K\}} \omega_{\{k_1, k_2\}} + \cdots + (-1)^{|K|-1} \omega_K,$$

[27] See again Definition A.40, §A.4.3.

[28] For the sake of simplicity, we keep denoting by $\boldsymbol{\zeta}$ the system of local coordinates $\boldsymbol{\zeta}^\iota$ in the chart U_ι.

then it follows by construction that

$$\frac{d\zeta_\kappa}{\zeta_\kappa} \wedge (\omega - \widetilde{\omega}) \in \Omega^{N-M-\rho+1}(U). \tag{2.116}$$

We claim that condition (2.110) implies that for any $\mathfrak{K} \subset K$, one has

$$\bigwedge_{j=1}^{\rho} d\zeta^{\beta_j} \wedge \omega_{\mathfrak{K}} = 0. \tag{2.117}$$

The claim is trivially satisfied when $r = m$, since in this case $K = \emptyset$. When $r < m$, let $J_{\mathfrak{K}}$ be the subset of $\{1, \ldots, r\}$ defined by

$$j \in J_{\mathfrak{K}} \iff \gcd\Big(\prod_{\kappa \in \mathfrak{K}} \zeta_\kappa, \zeta^{\beta_j}\Big) = 1. \tag{2.118}$$

It follows from (2.118) that

$$\Pi_{\mathfrak{K}} \subset U \cap (\iota_V \circ \pi)^{-1}\big(V \cap \bigcap_{j \notin J_{\mathfrak{K}}} V(s_j)\big).$$

Since $\operatorname{codim}\big(V \cap \bigcap_{j\notin J_{\mathfrak{K}}} V(s_j)\big) \geq M + m - |J_{\mathfrak{K}}| \geq M > M + r - |J_{\mathfrak{K}}|$. Since the restriction

$$\Big(\bigwedge_{j \in J_{\mathfrak{K}}} ds_j \wedge \psi\Big)_{|V \cap \bigcap_{j \notin J_{\mathfrak{K}}} V(s_j)}$$

makes sense, one has that

$$\begin{aligned}
&\Big((\iota_V \circ \pi)^* \big(\bigwedge_{j \in J_{\mathfrak{K}}} ds_j \wedge \psi\big)\Big)_{|\Pi_{\mathfrak{K}}} \\
&= \Big(\Big(\bigwedge_{\{j \in J_{\mathfrak{K}} \,:\, 1 \leq j \leq \rho\}} d\zeta^{\beta_j}\Big) \\
&\quad \wedge \Big(\bigwedge_{\{j \in J_{\mathfrak{K}} \,:\, \rho < j \leq r\}} d(\sigma_j \zeta^{\beta_j})\Big) \wedge (\iota_V \circ \pi)^* \psi\Big)_{|\Pi_{\mathfrak{K}}} \equiv 0
\end{aligned} \tag{2.119}$$

for dimension reasons. Furthermore, the relation

$$\zeta^{\beta_j} \frac{d(\sigma_j \zeta^{\beta_j})}{\sigma_j \zeta^{\beta_j}} = \zeta^{\beta_j} \Big(\frac{d\sigma_j}{\sigma_j} + \frac{d\zeta^{\beta_j}}{\zeta^{\beta_j}}\Big)$$

for $\rho < j \leq r$ allows one to rewrite (2.119). That is, if $\boldsymbol{\beta}_{\mathfrak{K}} = \sum_{\{j \in J_{\mathfrak{K}} \,:\, \rho < j \leq r\}} \beta_j$, (2.119) is the same as

$$\begin{aligned}
\zeta^{\boldsymbol{\beta}_{\mathfrak{K}}} \Big(\bigwedge_{\{j \in J_{\mathfrak{K}} \,:\, 1 \leq j \leq \rho\}} d\zeta^{\beta_j}\Big) &\wedge \Big(\Big(\bigwedge_{\{j \in J_{\mathfrak{K}} \,:\, \rho < j \leq r\}} \frac{d\sigma_j}{\sigma_j}\Big) \wedge (\iota_V \circ \pi)^* \psi\Big)_{\mathfrak{K}} \\
&+ \Big(\bigwedge_{\{j \in J_{\mathfrak{K}} \,:\, 1 \leq j \leq \rho\}} d\zeta^{\beta_j}\Big) \wedge \Big(\sum_{\{j \in J_{\mathfrak{K}} \,:\, \rho < j \leq r\}} d\zeta^{\beta_j} \wedge \omega_j\Big) \equiv 0
\end{aligned} \tag{2.120}$$

for some holomorphic forms ω_j in U. Multiplying (2.120) by the differential form $\bigwedge_{\{j\notin J_{\mathfrak{K}}\,:\,\rho<j\leq r\}} d\sigma_j/\sigma_j$ implies

$$(2.121)\quad \Big(\bigwedge_{\{j\in J_{\mathfrak{K}}\,:\,1\leq j\leq\rho\}} d\zeta^{\beta_j}\Big)\wedge\omega_{\mathfrak{K}} + \Big(\bigwedge_{\{j\in J_{\mathfrak{K}}\,:\,1\leq j\leq\rho\}} d\zeta^{\beta_j}\Big)\wedge\Big(\sum_{\{j\in J_{\mathfrak{K}}\,:\,\rho<j\leq r\}} d\zeta^{\beta_j}\wedge\widetilde{\omega}_j\Big)\equiv 0$$

for some holomorphic forms $\widetilde{\omega}_j$ in U. If one multiplies (2.121) by the differential form $\bigwedge_{\{j\notin J_{\mathfrak{K}}\,:\,1\leq j\leq q\}} d\zeta^{\beta_j}$ and uses the fact that any β_j with $\rho<j\leq r$ is a rational linear combination of $\beta_1,\dots,\beta_\rho$, then one gets the explanation needed for our claim (2.117) since $\gcd\big(\prod_{\kappa\in\mathfrak{K}}\zeta_\kappa,\zeta^{\beta_j}\big)\neq 1$ when $j\notin J_{\mathfrak{K}}$. Thus, since (2.117) holds, one may rewrite the expression (2.115) as

$$\Big(\prod_{j=1}^{r}\mu_j\Big)\int_U\Big(\prod_{j>\rho}\frac{|\sigma_j|^{2\lambda_j}}{\sigma_j^{\nu}}\Big)\Big(\prod_{j=1}^{m}\frac{|\zeta^{\beta_j}|^{2\mu_j}}{\zeta^{\nu\beta_j}}\Big)\Big(\bigwedge_{j=1}^{\rho}\frac{d\overline{\zeta}^{\beta_j}}{\overline{\zeta}^{\beta_j}}\Big)\wedge\overline{\omega-\widetilde{\omega}}\wedge\theta\,(\iota_V\circ\pi)^*\varphi.$$

After expanding

$$\bigwedge_{j=1}^{\rho}\frac{d\overline{\zeta}^{\beta_j}}{\overline{\zeta}^{\beta_j}}=\sum_{1\leq\kappa_1<\cdots<\kappa_\rho\leq N-M}\pm\det\big[\beta_{j,\kappa_\ell}\big]_{1\leq j,\ell\leq\rho}\Big(\bigwedge_{\kappa_\ell\notin K}\frac{d\overline{\zeta_{\kappa_\ell}}}{\overline{\zeta_{\kappa_\ell}}}\Big)\wedge\Big(\bigwedge_{\kappa_\ell\in K}\frac{d\overline{\zeta_{\kappa_\ell}}}{\overline{\zeta_{\kappa_\ell}}}\Big),$$

we rewrite (2.115) as a linear combination with integer coefficients of expressions of the form

$$(2.122)\qquad \Big(\prod_{j=1}^{r}\mu_j\Big)\int_U\Big(\prod_{j=1}^{m}\frac{|\zeta^{\beta_j}|^{2\mu_j}}{\zeta^{\nu\beta_j}}\Big)\Big(\bigwedge_{\kappa_\ell\notin K}\frac{d\overline{\zeta_{\kappa_\ell}}}{\overline{\zeta_{\kappa_\ell}}}\Big)\wedge\Phi_{\boldsymbol{\kappa},\mu},$$

where $\boldsymbol{\kappa}:1\leq\kappa_1<\cdots<\kappa_\rho\leq N-M$ and $\Phi_{\boldsymbol{\kappa},\mu}$ is the smooth test form

$$\Phi_{\boldsymbol{\kappa},\mu}=\Big(\prod_{j>\rho}\frac{|\sigma|^{2\mu_j}}{\sigma_j^{\nu}}\Big)\Big(\bigwedge_{\kappa_\ell\in K}\frac{d\overline{\zeta_{\kappa_\ell}}}{\overline{\zeta_{\kappa_\ell}}}\Big)\wedge(\overline{\omega-\widetilde{\omega}})\wedge\theta\,(\iota_V\circ\pi)^*\varphi.$$

We recall here that (2.116) holds, which justifies the smoothness of $\Phi_{\boldsymbol{\kappa},mu}$. Moreover, such test form depends holomorphically on $\mu\in\mathbb{C}^m$. It remains now to use integration by parts in the univariate setting (see (1.7)) and transform each expression (2.122) by Stokes's theorem as

$$\pm\frac{\prod_{j=1}^{r}\mu_j}{\prod_{\kappa_\ell\notin K}\Big(\sum_{j=1}^{r}\beta_{j,\kappa_\ell}\,\mu_j\Big)}\int_U\Big(\prod_{j=1}^{m}\frac{|\zeta^{\beta_j}|^{2\mu_j}}{\zeta^{\nu\beta_j}}\Big)\Big(\prod_{\kappa_\ell\notin K}\overline{\partial}_{\zeta_{\kappa_\ell}}\Big)(\Phi_{\boldsymbol{\kappa},\mu}).$$

Integration by parts in the univariate identities (1.7) shows also that the map

$$\mu\longmapsto\int_U\Big(\prod_{j=1}^{m}\frac{|\zeta^{\beta_j}|^{2\mu_j}}{\zeta^{\nu\beta_j}}\Big)\Big(\prod_{\kappa_\ell\notin K}\overline{\partial}_{\zeta_{\kappa_\ell}}\Big)(\Phi_{\boldsymbol{\kappa},\mu})$$

extends from $\{\mu\in\mathbb{C}^m\,:\,\mathrm{Re}\,\mu_j\gg 1\text{ for }j=1,\dots,m\}$ to $\mathbb{C}^m$ as a meromorphic function, whose polar set does not intersect $\{\mu\in\mathbb{C}^m\,:\,\mathrm{Re}\,\mu_j\geq 0\text{ for }j=1,\dots,m\}$. More precisely, it is meromorphic in $\{\mu\in\mathbb{C}^m\,:\,\mathrm{Re}\,\mu_j>-\eta\text{ for }j=1,\dots,m\}$, where $\eta>0$ is independent of ν. This concludes the proof of Lemma 2.77. □

Proving Theorem (2.75) reduces to proving it in the case where $\mathcal{X} = \Omega$ is a neighborhood of $\mathbf{0}$ in $\mathbb{C}^N$, $\mathbf{0} \in V$, and s_j are holomorphic sections of the trivial bundle, equipped with the canonical metric in $\mathbb{C}$. Therefore, it is enough to prove it by induction on r. That is, it is enough to prove that for any r from 1 to m, for any $\nu \in \mathbb{N}^*$, for any test form $\varphi \in \mathcal{D}^{(N-M,N-M-r)}(\Omega)$, the map (2.112) extends from the domain $\{\mu \in \mathbb{C}^m : \operatorname{Re}\mu_j \gg 1 \text{ for } j = 1, \dots, m\}$ to $\mathbb{C}^m$ as a meromorphic function, whose polar set does not intersect $\{\mu \in \mathbb{C}^m\, ; \operatorname{Re}\mu_j > -\eta \text{ for } j = 1, \dots, m\}$, with $\eta > 0$ being independent of ν. Lemma 2.77 shows this indeed holds for $r = 1$. Let $r \geq 2$ and assume that the result holds when r is replaced by $r - 1$. Using Stokes's theorem, we rewrite (2.112) as

$$\langle \Xi(\mu), \varphi\rangle = (-1)^r \int_V \Big(\prod_{j>r} \frac{|s_j|^{2\mu_j}}{s_j^\nu}\Big) \bigwedge_{2\leq j\leq r} \overline{\partial}\Big(\frac{|s_j|^{2\mu_j}}{s_j^\nu}\Big) \wedge \overline{\partial}\varphi \tag{2.123}$$
$$- \sum_{j'>r} \int_V \Big(\prod_{\{j>r\,:\, j\neq j'\}} \frac{|s_j|^{2\mu_j}}{s_j^\nu}\Big) \overline{\partial}\Big(\frac{|s_{j'}|^{2\lambda_{j'}}}{s_{j'}^\nu}\Big) \wedge \Big(\bigwedge_{2\leq j\leq r} \overline{\partial}\Big(\frac{|s_j|^{2\mu_j}}{s_j^\nu}\Big)\Big) \wedge \varphi.$$

By Lemma 2.77, the function of μ in the left-hand side of (2.123) extends to a meromorphic function to $\{\mu \in \mathbb{C}^m : \operatorname{Re}\mu_j > -\eta \text{ for } j = 1, \dots, m\}$ for some $\eta > 0$, independent of ν. Moreover, its polar set is included in a union of hyperplanes with equation $\sum_{j=1}^r \alpha_{\iota,j}\, \mu_j = 0$ and is also independent of ν. Due to the inductive hypothesis (for the first term) and because of Lemma 2.77 (for the terms indexed with j'), the function of μ in the right-hand side of (2.123) also extends to a meromorphic function to $\{\mu \in \mathbb{C}^m : \operatorname{Re}\mu_j > -\widetilde{\eta} \text{ for } j = 1, \dots, m\}$ for some $\widetilde{\eta} > 0$, which is independent on ν. Its polar set is included in a union of hyperplanes, with equations of the form $\sum_{j=2}^r \widetilde{\alpha}_{\iota,j}\, \mu_j + \widetilde{\alpha}_\iota\, \mu_{j'} = 0$ for some $j' > r$. Then, the meromorphic continuation of the function (2.112) in $\{\mu \in \mathbb{C}^m ; \operatorname{Re}\mu_j > -\min(\eta, \widetilde{\eta}) \text{ for } j = 1, \dots, m\}$ has its polar set truly included in a union of hyperplanes with equation $\sum_{j=2}^r \alpha_{\iota,j}\, \mu_j = 0$. Using the same argument and restricting eventually η (still independent of ν), one can see that for any $\rho \in \{1, \dots, r\}$, this polar set lies in fact in a union of hyperplanes with equation $\sum_{1\leq j\leq r, j\neq\rho} \alpha_{\iota,j}\mu_j = 0$. Since this is true for any ρ, it implies that the function (2.112) can be analytically continued as a meromorphic function to $\{\mu \in \mathbb{C}^m : \operatorname{Re}\mu_j > -\check{\eta} \text{ for } j = 1, \dots, m\}$ for some $\check{\eta} > 0$. This concludes the proof of the inductive argument and thus the proof of Theorem 2.75. □

Theorem 2.75 suggests introducing the following important definition.

DEFINITION 2.78. Let $V \hookrightarrow \mathcal{X}$ be a closed analytic subset with pure codimension M of an N-dimensional complex manifold $\mathcal{X}$, let $D_1, \dots, D_m$, $1 \leq m \leq N - M$, be m Cartier divisors on the manifold $\mathcal{X}$, and let $|D|$ be the Cartier divisor $D_1 + \cdots + D_m$. Given a vector of holomorphic sections $\boldsymbol{s} = (s_1, \dots, s_m)$ in $\mathcal{O}(D_1) \times \cdots \times \mathcal{O}(D_m)$ such that

$$\operatorname{codim}\big(V \cap V(\boldsymbol{s})\big) = M + m, \tag{2.124}$$

where $V(\boldsymbol{s}) = |\operatorname{div}(s_1)| \cap \cdots \cap |\operatorname{div}(s_m)|$, the Coleff–Herrera $(M, M+m)$-current

$$R^{\boldsymbol{s}} \wedge [V] \in \mathrm{CH}_{\mathcal{X},V}\big(\mathcal{X}, \mathcal{O}(-|D|)\big)$$

is called the *Coleff–Herrera residue current attached to the ordered sequence* $\boldsymbol{s}$ *defining a complete intersection on* V. We will denote it from now on in this monograph by[29]

$$\overline{\partial}\Big(\frac{1}{\boldsymbol{s}}\Big)_V := R^{\boldsymbol{s}} \wedge [V]. \tag{2.125}$$

We conclude this subsection and Chapter 2 with the following major result.

THEOREM 2.79 ([**CoH78,Pas88A**]). *Let $\mathcal{X}$ be a complex, N-dimensional manifold with structural sheaf $\mathcal{O}$, let $V \subset \mathcal{X}$ be a closed analytic subset with pure codimension M, and let $\boldsymbol{s} \in \mathcal{O}(D_1) \times \cdots \times \mathcal{O}(D_m)$ satisfying (2.124) as in Definition 2.78. For any permutation $\mathfrak{s}$ of $\{s_1, \dots, s_m\}$ with signature $\varepsilon(\mathfrak{s})$, one has*

$$\overline{\partial}(1/\mathfrak{s}(\boldsymbol{s}))_V = \varepsilon(\mathfrak{s})\, \overline{\partial}(1/\boldsymbol{s})_V. \tag{2.126}$$

Moreover, in addition to the properties satisfied by any section of the Coleff–Herrera sheaf $\mathrm{CH}_{V\cap V(\boldsymbol{s})}(\cdot\,, \mathcal{O}(-|D|))$, the residue current $\overline{\partial}(1/\boldsymbol{s})_V$ also satisfies, in terms of sheaves, the annihilation property

$$\Big(\mathcal{I}_V \oplus \bigoplus_{j=1}^m \mathcal{O}s_j\Big)\, \overline{\partial}\Big(\frac{1}{\boldsymbol{s}}\Big)_V = 0, \tag{2.127}$$

where $\mathcal{I}_V$ is the radical ideal sheaf of V and $\mathcal{O}s_j$ is the subsheaf of the sheaf of sections of $\mathcal{O}(D_j)$ generated by s_j.

PROOF. The first assertion follows immediately from Theorem 2.75. This theorem, once combined with the analytic continuation principle, implies also that

$$\mathcal{I}_V\, \overline{\partial}\Big(\frac{1}{\boldsymbol{s}}\Big)_V = 0.$$

The equality between continuous differential forms

$$\begin{aligned} s_{j_0}\Big(\bigwedge_{j=1}^m \overline{\partial}\Big(|s_j|^{2\mu_j}\frac{s_j^*}{|s_j|^2}\Big)\Big) \wedge [V] \\ &= (-1)^{j_0-1}\overline{\partial}|s_{j_0}|^{2\mu_{j_0}} \wedge \Big(\bigwedge_{j\neq j_0} \overline{\partial}\Big(|s_j|^{2\mu_j}\frac{s_j^*}{|s_j|^2}\Big)\Big) \wedge [V] \\ &= (-1)^{j_0-1}\overline{\partial}\Big(|s_{j_0}|^{2\mu_{j_0}}\Big(\bigwedge_{j\neq j_0} \overline{\partial}\Big(|s_j|^{2\mu_j}\frac{s_j^*}{|s_j|^2}\Big)\Big) \wedge [V]\Big), \end{aligned}$$

when $\mathrm{Re}\,\mu_j \gg 1$ for $j = 1, \dots, m$, Theorem 2.75, Stokes's theorem, and the analytic continuation principle imply that $\mathcal{O}s_{j_0}\, \overline{\partial}(1/\boldsymbol{s})_V = 0$. □

REMARK 2.80. Assertion (2.127) constitutes the direct ($\Longrightarrow$) part of a result to be completely settled in §3.3.3. This result is known as the *duality theorem for complete intersections* (see Theorem 3.39) and it is of great importance.

[29] One has to be careful with such a notation if one interprets $\overline{\partial}(1/\boldsymbol{s})$ as $\overline{\partial}(1/s_1)\wedge\cdots\wedge\overline{\partial}(1/s_m)$. The reason is the following: according to Theorem 2.75, one has

$$\overline{\partial}(1/\boldsymbol{s})_V = (-1)^{m(m-1)/2}\Big[\Big(\bigwedge_{j=1}^m \overline{\partial}\Big(|s_j|^{2\mu_j}\frac{s_j^*}{2i\pi|s_j|^2}\Big)\Big) \wedge [V]\Big]_{\mu_1=\cdots=\mu_m=0}$$

independently of the hermitian metrics $|\ |$ which equip the line bundles $\mathcal{O}(D_j)$. One needs to take care of the correction of the sign. In other words, one should better interpret $\overline{\partial}(1/\boldsymbol{s})$ as the formal wedge product $\overline{\partial}(1/s_m)\wedge\cdots\wedge\overline{\partial}(1/s_1)$.

CHAPTER 3

Residue currents: a bundle approach

3.1. A toric complex geometry digest

Toric varieties play an important role in this monograph. They are fundamental for the construction of local log resolutions used, as we have already seen, to prove the existence of the residue currents. They play also a crucial role in the construction of related integral representations of residue symbols. Therefore, before introducing Bochner–Martinelli currents, let us review first some necessary preliminary material. Naturally, our presentation is aimed at conforming with the scope of this monograph.

3.1.1. Rational fans. We begin with some basic definitions from convex geometry. The brackets $\langle\!\langle\ ,\ \rangle\!\rangle$ denote here the duality between a $\mathbb{Z}$-lattice $\boldsymbol{N}$ (respectively a $\mathbb{R}$-vector space E such as $\mathbb{R}^N$) and its dual $\boldsymbol{N}^\star = \mathrm{Hom}_{\mathbb{Z}}(\boldsymbol{N}, \mathbb{Z})$ (respectively its $\mathbb{R}$-dual E^*). Our reference lattice $\boldsymbol{N}$ will be $\mathbb{Z}^N$, and we use the identification $(\mathbb{Z}^N)^\star \simeq \mathbb{Z}^N$.

Definition 3.1 ($\mathbb{Z}^N$-rational cones, duality, faces). A $\mathbb{Z}^N$*-rational cone* σ in $\mathbb{R}^N$ is a finite intersection of closed half-subspaces $\{x \in \mathbb{R}^N : \langle\!\langle a^\star , x\rangle\!\rangle \geq 0\}$, where $a^\star \in (\mathbb{Z}^N)^\star \simeq \mathbb{Z}^N$. Its dimension $\dim \sigma$ is that of the $\mathbb{R}$-affine subspace it spans in $\mathbb{R}^N = \mathbb{Z}^N \otimes_{\mathbb{Z}} \mathbb{R}$. The dual cone $\check{\sigma}$, defined by

$$\check{\sigma} = \{\xi \in \mathbb{R}^N : \langle\!\langle \xi , x\rangle\!\rangle \geq 0 \text{ for any } x \in \sigma\},$$

is a $\mathbb{Z}^N$-rational cone as well. The cone σ is said to be *strict* if and only if $\dim \check{\sigma} = N$, which amounts to saying that σ does not contain any $\mathbb{R}$-linear subspace of $\mathbb{R}^N$. A *face* τ of a rational cone σ is any subset of σ of the form $\tau = \{x \in \sigma : \langle\!\langle \xi , x\rangle\!\rangle = 0\}$ for some (nonunique) $\xi \in \check{\sigma}$. Any face τ of σ (one writes $\tau \prec \sigma$) is also a rational cone with dimension between 0 and $\dim \sigma$. The set of faces τ of a rational cone σ is denoted by $\tau(\sigma)$.

Remark 3.2. $\mathbb{Z}^N$-rational cones are particular examples of $\mathbb{Z}^N$-rational generalized polyhedra; see Definition 7.51. The notion of a face of a $\mathbb{Z}^N$-rational cone is a particular case of the notion of a face of a $\mathbb{Z}^N$-rational generalized polyhedron; see Definition 7.52.

Definition 3.3. A $\mathbb{Z}^N$*-rational fan* Σ in $\mathbb{R}^N$ is a finite collection of strict rational[1] cones in $\mathbb{R}^N$, so that any face $\tau \in \tau(\sigma)$ of any cone $\sigma \in \Sigma$ remains in Σ, and any intersection of two cones in Σ is a face of both of them. The fan is

[1]We mean here $\mathbb{Z}^N$-rational. More generally, one should take instead of $\mathbb{Z}^N$ a free $\mathbb{Z}$-module $\boldsymbol{N}$ with rank N (thus isomorphic to $\mathbb{Z}^N$) and consider $\boldsymbol{N}$-rational cones as cones in $\boldsymbol{N} \otimes_{\mathbb{Z}} \mathbb{R} \simeq \mathbb{R}^N$ which are of the form $\{a_1 v_1 + \cdots + a_s v_s : a_j \geq 0\}$, where $v_1, \dots, v_s$ are generators lying in $\boldsymbol{N}$. The dual lattice $\mathrm{Hom}_{\mathbb{Z}}(\boldsymbol{N}, \mathbb{Z}) = \boldsymbol{N}^\star$ remains isomorphic, but not equal, to $\mathbb{Z}^N$. For the sake

called *projectively complete* (respectively *affinely complete*) when $\bigcup_{\sigma\in\Sigma}\sigma = \mathbb{R}^N$ (respectively when $\bigcup_{\sigma\in\Sigma}\sigma = (\overline{\mathbb{R}^+})^N$).[2] A rational fan in $\mathbb{R}^N$ is called *simplicial* if any cone in the fan is generated by a set of vectors in $\mathbb{Z}^N$ that can be completed to a basis of $\mathbb{R}^N$. Any such cone is called *simplicial.* Furthermore, the fan is called *primitive* if any cone in the fan is generated by a set of *primitive* vectors in $\mathbb{Z}^N$ (that is, vectors with coprime coordinates) that can be completed to a basis of $\mathbb{Z}^N$. Such a cone is said to be *primitive.*

EXAMPLE 3.4 (Normal fan). Let $\beta_1,\ldots,\beta_M$ be a collection of points in $\mathbb{N}^N$ and

$$\Delta_{\boldsymbol{\beta}}^{+} = \text{convex hull}\Big(\bigcup_{j=1}^{M}\big(\beta_j + (\overline{\mathbb{R}^+})^N\big)\Big)$$

with set of vertices $\Delta_{\boldsymbol{\beta}}^{+}(0)$. Equivalence classes for the equivalence relation $\mathcal{R}_{\boldsymbol{\beta}}^{+}$ in $(\overline{\mathbb{R}^+})^N$, defined by

$$\begin{aligned}(3.1)\quad x\,\mathcal{R}_{\boldsymbol{\beta}}^{+}\,y \iff &\big\{u\in\Delta_{\boldsymbol{\beta}}^{+}(0)\,:\,\langle\!\langle u,x\rangle\!\rangle = \min_{v\in\Delta_{\boldsymbol{\beta}}^{+}(0)}\langle\!\langle v,x\rangle\!\rangle\big\}\\ &= \big\{u\in\Delta_{\boldsymbol{\beta}}^{+}(0)\,:\,\langle\!\langle u,y\rangle\!\rangle = \min_{v\in\Delta_{\boldsymbol{\beta}}^{+}(0)}\langle\!\langle v,y\rangle\!\rangle\big\},\end{aligned}$$

are $\mathbb{Z}^N$-rational cones forming an affinely complete $\mathbb{Z}^N$-rational fan $\Sigma_{\boldsymbol{\beta}}^{+}$. If $\beta_1,\ldots,\beta_M$ are points in $\mathbb{Z}^N$ such that the convex hull $\Delta_{\boldsymbol{\beta}}$ of $\{\beta_1,\ldots,\beta_M\}$ has a nonempty interior, then $\Delta_{\boldsymbol{\beta}}(0)$ denotes the corresponding set of vertices. The equivalence classes for the equivalence relation $\mathcal{R}_{\boldsymbol{\beta}}$ in $\mathbb{R}^N$, defined as in (3.1) but with $\Delta_{\boldsymbol{\beta}}(0)$ in place of $\Delta_{\boldsymbol{\beta}}^{+}(0)$, are $\mathbb{Z}^N$-rational cones forming a projectively complete $\mathbb{Z}^N$-rational fan $\Sigma_{\boldsymbol{\beta}}$. The fan $\Sigma_{\boldsymbol{\beta}}^{+}$ (respectively $\Sigma_{\boldsymbol{\beta}}$) is called the *normal fan* to $\Delta_{\boldsymbol{\beta}}^{+}$ (respectively to $\Delta_{\boldsymbol{\beta}}$).

Among subfamilies of cones in a $\mathbb{Z}^N$-rational fan Σ, the most important are the following.

(1) The family of the one-dimensional cones or *rays* (denoted by $\Sigma(1)$). This family is generated by the vectors $\xi_1,\ldots,\xi_d$, where $\xi_j = (\xi_{j,1},\ldots,\xi_{j,N}) \in \mathbb{Z}^N$ denotes a primitive generator of the ray $\xi_j\,\overline{\mathbb{R}^+}\in\Sigma(1)$ for every $j = 1,\ldots,d$.
(2) The family of the N-dimensional cones denoted by $\Sigma(N)$. From the family $\Sigma(N)$ it is possible to reconstruct the whole fan based on the operation of taking faces.

The notion of *convexity* for a rational fan in $\mathbb{R}^N$ plays an important role also. A rational fan Σ in $\mathbb{R}^N$ is said to be *convex* if and only if the endpoints of the primitive vectors $\xi_1,\ldots,\xi_d$ generating its rays lie on the boundary of their closed convex hull $\text{conv}(\xi_1,\ldots,\xi_d)$. Any complete convex rational fan Σ in $\mathbb{R}^N$ is dual to a reflexive

of simplicity, one takes here $\boldsymbol{N} = \mathbb{Z}^N$. In this subsection rational will always mean $\mathbb{Z}^N$-rational (unless specified).

[2]Projectively complete rational fans are crucial for studying global questions once nice geometric presentations are constructed for them (see for example Theorem 3.13), while affinely complete rational fans are, on the other hand, quite useful when constructing (toric) log resolutions of the form we will precisely need in this chapter (see Theorem 3.11 and Example 3.12).

integral N-dimensional polyhedron.[3] In particular, if the endpoints of the primitive vectors $\xi_1, \dots, \xi_d$ generating the rays of Σ are vertices of an N-dimensional reflexive polyhedron $P = P_\Sigma$ containing the origin as an interior point, then Σ is the normal fan, as defined in Example 3.4, to the dual integral polyhedron

$$\Delta_\Sigma = \check{P} := \{x^* \in (\mathbb{R}^N)^* \simeq \mathbb{R}^N \,:\, \langle\!\langle x^*, y\rangle\!\rangle \geq -1 \quad \forall y \in P\}, \tag{3.2}$$

containing the origin as an interior point also.

3.1.2. Complete simplicial fans and homogeneous coordinate rings. Let Σ be a $\mathbb{Z}^N$-rational complete simplicial fan and let $\Sigma(1) = \{\xi_1 \overline{\mathbb{R}^+}, \dots, \xi_d \overline{\mathbb{R}^+}\}$ be its set of rays generated by primitive vectors $\xi_j \in \mathbb{Z}^N$, where $d = N + r$. Let us associate to each ray $\xi_j \overline{\mathbb{R}^+}$ a homogeneous coordinate z_j, $j = 1, \dots, N + r$; see [**Cx95**] or also [**CLO**].

DEFINITION 3.5. The polynomial ring $\mathbb{C}[z_1, \dots, z_d]$ is called the *homogeneous coordinate ring* attached to the rational fan Σ. Inside this ring, the monomial ideal $I^{\mathrm{irr}}(\Sigma) \subset \mathbb{C}[z_1, \dots, z_d]$ defined by

$$I^{\mathrm{irr}}(\Sigma) = \sum_{\sigma \in \Sigma(N)} \Big(\prod_{\{1 \leq j \leq d \,:\, \xi_j \overline{\mathbb{R}^+} \not\subset \sigma\}} z_j \Big) \mathbb{C}[z_1, \dots, z_d] \tag{3.3}$$

is called the *irrelevant ideal* attached to Σ. If $\mathbb{T}_d = ((\mathbb{C}^*)^d, \times)$ denotes the d-dimensional complex torus with pointwise multiplication, then its multiplicative subgroup

$$G(\Sigma) = \Big\{ \boldsymbol{t} = (t_1, \dots, t_d) \in \mathbb{T}^d \,:\, \prod_{j=1}^{d} t_j^{\xi_{j,k}} = 1 \quad \text{for } k = 1, \dots, N \Big\} \tag{3.4}$$

is called the *toric action group* attached to Σ.

REMARK 3.6. If $\odot$ denotes the pointwise multiplication in $\mathbb{C}^d$, then one has $G(\Sigma) \odot I^{\mathrm{irr}}(\Sigma) \subset I^{\mathrm{irr}}(\Sigma)$. In the particular case where $\Sigma(N)$ contains at least one primitive cone, then $G(\Sigma)$ is isomorphic to the complex torus $\mathbb{T}^r$, where $r = d - N$. Such is the case when the fan Σ itself is primitive; see Definition 3.3. In general, the toric action group $G(\Sigma)$ factorizes as the product of a factor isomorphic to $\mathbb{T}^r$, with a group of finite order. Speaking additively instead of multiplicatively, $G(\Sigma)$ is in general the direct sum of a free group $G_{\mathrm{free}}(\Sigma)$ with rank r and of a finite abelian torsion group $G_{\mathrm{tors}}(\Sigma)$, which disappears as soon as $\Sigma(N)$ contains at least one primitive cone.

A description of the zero set of the irrelevant ideal $I^{\mathrm{irr}}(\Sigma)$ is given in [**Bat93**, §2].

PROPOSITION 3.7. *A subset of* $\Sigma(1)$ *is said to be primitive if the cone it generates in* $\mathbb{R}^N$ *does not belong to* Σ*, while it is the case for the cone generated by any of its proper subsets. Then* $V(I^{\mathrm{irr}}(\Sigma))$ *is the union of all coordinate planes* $\{z_{i_1} = \cdots = z_{i_k} = 0\}$ *such that* $\{\xi_{i_1}, \dots, \xi_{i_k}\}$ *is a primitive subset of* $\Sigma(1)$.

Having these tools within hands, one can associate to the fan Σ an N-dimensional complex analytic space; see §A.4.1.

---

[3] A polyhedron P is called ($\mathbb{Z}^N$) *integral* if its vertices are in $\mathbb{Z}^N$. An integral polyhedron P is said to be *reflexive* if its dual $\check{P} := \{y \in \mathbb{R}^N \,:\, \langle\!\langle \xi, y\rangle\!\rangle \geq -1 \text{ for any } \xi \in \Delta\}$ has its vertices in $\mathbb{Z}^N$ also.

DEFINITION 3.8 ([**Ful93, Cx95, CLO**]). The compact complex N-dimensional analytic space $\mathcal{X}_\Sigma$ is defined as the geometric quotient

$$\mathcal{X}_\Sigma = \frac{\mathbb{C}^d \setminus I^{\rm irr}(\Sigma)}{G(\Sigma)}. \tag{3.5}$$

It is a complex manifold[4] if and only if Σ is primitive. The complex analytic space $\mathcal{X}_\Sigma$ is always normal; see Definition A.32. Moreover, the torus $\mathbb{T}^N$ acts on it and at the same time embeds in it as a dense Zariski open orbit under its action.

REMARK 3.9 (Algebraic toric variety $\mathbb{X}_\Sigma$). The complex analytic space $\mathcal{X}_\Sigma$ is classically realized as the set of closed points of the algebraic toric variety $\mathbb{X}_\Sigma$ constructed from the rational fan Σ (complete or not, simplicial or not) as follows. For each cone $\sigma \in \Sigma$, let $\mathbb{X}_\sigma = \operatorname{Spec}\big(\mathbb{C}[\check{\sigma} \cap (\mathbb{Z}^N)^\star]\big)$. For any $\tau \prec \sigma$, there is a scheme embedding $\iota_{\sigma,\tau} : \mathbb{X}_\tau \hookrightarrow \mathbb{X}_\sigma$. Moreover, this embedding is compatible with the action of $\mathbb{T}^N = \operatorname{Spec}\big(\mathbb{C}[X_1^{\pm 1}, \dots, X_N^{\pm 1}]\big) = \mathbb{X}_{\{0\}}$ on each $\mathbb{X}_{\sigma'}$, $\sigma' \in \Sigma$. Starting with the disjoint union $\bigsqcup_{\sigma\in\Sigma} \mathbb{X}_\Sigma$, then identifying for each pair (σ', σ'') the algebraic varieties $\mathbb{X}_{\sigma'}$ and $\mathbb{X}_{\sigma''}$ along $\mathbb{X}_{\sigma'\cap\sigma''}$, using the embeddings $\iota_{\sigma',\sigma'\cap\sigma''}$ and $\iota_{\sigma'',\sigma'\cap\sigma''}$, one constructs an algebraic variety $\mathbb{X}_\Sigma$ (defined over $\mathbb{Q}$), together with an action on it of $\mathbb{T}^N = \mathbb{X}_{\{0\}}$; see [**Ful93**] for more details. Furthermore, each $\mathbb{X}_\sigma$ can be decomposed into a set of orbits under the action of the complex torus $\mathbb{X}_{\{0\}}$ as $\mathbb{X}_\sigma = \bigcup_{\tau\prec\sigma} \operatorname{Orb}_\tau$, where $\operatorname{Orb}_\tau \simeq \mathbb{T}^{N-\dim\tau}$. Thus, one has that $\overline{\operatorname{Orb}_\tau} = \bigcup_{\tau\prec\sigma} \operatorname{Orb}_\sigma$ for any $\tau \in \Sigma$ in terms of Zariski topology. If τ is a ray, then $\overline{\operatorname{Orb}_\tau}$ is in particular an algebraic hypersurface $\mathbb{H}$ of $\mathbb{X}_\Sigma$. The orbit $\operatorname{Orb}_{\{0\}} = \mathbb{X}_{\{0\}} = \mathbb{T}^N$ is therefore an open dense orbit in $\mathbb{X}_\Sigma$ acting on it. Observe also that $\mathbb{X}_\Sigma$ is normal. The algebraic variety $\mathbb{X}_\Sigma$ is called the toric variety attached to the fan Σ. We are interested in this monograph in the N-dimensional complex analytic space $\mathcal{X}_\Sigma$, whose points are its closed points. The complex analytic space $\mathcal{X}_\Sigma$ defined by (3.5) is a useful alternative (geometric) presentation of the set of closed points $\mathcal{X}_\Sigma$ of $\mathbb{X}_\Sigma$, when the fan Σ is complete and simplicial. See also [**CxLS11**] for an alternative presentation.

Each irreducible hypersurface $H_j = \{z_j = 0\} \subset \mathcal{X}_\Sigma$ (modulo the action of $G(\Sigma)$) corresponds from the set-theoretical point of view to the $(N-1)$-dimensional complete simplicial toric variety attached to the $\mathbb{Z}^N/\xi_j\,\mathbb{Z}$-rational complete simplicial fan $\pi_j(\Sigma)$ in $\mathbb{R}^N/\xi_j\mathbb{R}$, where $\pi_j : \mathbb{R}^N \to \mathbb{R}^N/\mathbb{R}\,\xi_j$ denotes the quotient map. It is a closed orbit under the action of the dense torus $\mathbb{T}^N$ on $\mathcal{X}_\Sigma$; see Remark 3.9. Then $\{H_j\}$ is a Weil divisor on $\mathcal{X}_\Sigma$, which is also invariant under the action of the torus $\mathbb{T}^N$ on $\mathcal{X}_\Sigma$. Hence $\{H_j\}$ is called a *toric Weil divisor*. Let

$$\nu : (m_1, \dots, m_N) \in (\mathbb{Z}^N)^\star \simeq \mathbb{Z}^N \longmapsto \big(\langle\!\langle m, \xi_1\rangle\!\rangle, \dots, \langle\!\langle m, \xi_d\rangle\!\rangle\big) \in \mathbb{Z}^d. \tag{3.6}$$

The following important result relates the toric action group $G(\Sigma)$ attached to the simplicial fan with the $(N-1)$-Chow group $A_{N-1}(\mathcal{X}_\Sigma)$ of the complex analytic space $\mathcal{X}_\Sigma$; see §D.3.2.

THEOREM 3.10 ([**Cx95**]). *One has $A_{N-1}(\mathcal{X}_\Sigma) \simeq \mathbb{Z}^d/\nu((\mathbb{Z}^N)^\star)$ and, as a consequence*

$$G(\Sigma) \simeq \operatorname{Hom}_{\mathbb{Z}}\big(A_{N-1}(\mathcal{X}_\Sigma), \mathbb{T}\big). \tag{3.7}$$

[4]Its complex structure is, in general, an *orbifold* complex structure.

In order to describe the action of the factor of $G(\Sigma)$, which is isomorphic to $\mathbb{T}^r$, it is useful to introduce a collection of $d - N = r$ linear, $\mathbb{Q}$-independent relations with integer coefficients

$$a_{j,1}\,\xi_1 + \cdots + a_{j,d}\,\xi_d = 0\,,\ j = 1, \ldots, r, \tag{3.8}$$

between the rays $\xi_1, \ldots, \xi_d$ of the fan Σ. Note that since the fan is complete, one can choose all the $a_{j,k} \geq 0$, $j = 1, \ldots, r$, $k = 1, \ldots, d$ (see [**Ky05**, Proposition 1]). Relations (3.8) describe the kernel of the map ν defined by (3.6). Thus, the factor of G isomorphic to $\mathbb{T}^r$ induces the following equivalence relation on $\mathbb{C}^d \setminus V(I^{\rm irr}(\Sigma))$:

$$z \simeq z' \iff \exists\, \boldsymbol{t} \in \mathbb{T}^r\,,\ z' = \Big(\big(\prod_{j=1}^{r} t_j^{a_{j,1}}\big) z_1, \ldots, \big(\prod_{j=1}^{r} t_j^{a_{j,d}}\big)\, z_d\Big).$$

If $\Sigma(N)$ contains at least one primitive cone, then $A_{N-1}(\mathcal{X}_\Sigma) \simeq \mathbb{Z}^r$ and $G \simeq \mathbb{T}^r$. In this case, an isomorphism between $A_{N-1}(\mathcal{X}_\Sigma)$ and $\mathbb{Z}^r$ is realized by

$$\dot{\mathfrak{m}} \in A_{N-1}(\mathcal{X}_\Sigma) = \frac{\mathbb{Z}^d}{\nu((\mathbb{Z}^N)^\star)} \longmapsto \Big(\sum_{k=1}^{d} a_{1,k}\,\mathfrak{m}_k, \ldots, \sum_{k=1}^{d} a_{r,k}\,\mathfrak{m}_k\Big) \in \mathbb{Z}^r. \tag{3.9}$$

Before concluding this subsection, let us mention two results that will be of importance to us. The first one concerns fans in general, independently if they are simplicial or not, complete or not; see [**Ful93**, §2.6]. Recall that for a $\mathbb{Z}^N$-rational fan Σ, a refinement of Σ is a $\mathbb{Z}^N$-rational fan $\widetilde{\Sigma}$ as well and such that each cone of Σ is a union of cones belonging to $\widetilde{\Sigma}$. If the fan Σ is primitive, then so is any refinement $\widetilde{\Sigma}$. In case Σ fails to be primitive, the following result holds.

THEOREM 3.11 (Resolution of singularities within the toric varieties world). *Let Σ be a $\mathbb{Z}^N$-rational fan. One can always find a primitive refinement $\widetilde{\Sigma}$ of Σ, together with a morphism $\pi : \mathbb{X}_{\widetilde{\Sigma}} \to \mathbb{X}_\Sigma$, which is birational and proper.*

As a consequence, such a proper birational morphism from $\mathbb{X}_{\widetilde{\Sigma}}$ to $\mathbb{X}_\Sigma$ induces a proper bimeromorphic[5] morphism of complex analytic spaces from the N-dimensional complex manifold $\mathcal{X}_{\widetilde{\Sigma}}$ to the N-dimensional complex analytic space $\mathcal{X}_\Sigma$, which is in general singular. This provides a resolution of singularities for $\mathcal{X}_\Sigma$; see §A.4.3.

EXAMPLE 3.12. Theorem 3.11 holds in particular for the fan $\Sigma^+_{\boldsymbol{\beta}}$ associated with a finite collection of points $\beta_1, \ldots, \beta_M$ in $\mathbb{N}^N$, as explained in Example 3.4. This fan is affinely complete, but not projectively. It is not in general simplicial. Theorem 3.11, once combined with Hironaka's theorem (see Theorem A.41) allows us for example to realize the following: given $f_1, \ldots, f_m \in \mathcal{O}_\mathcal{X}(U)$, where $\mathcal{X}$ is an N-dimensional complex analytic space and $U \subset \mathcal{X}$, one may construct a log-resolution $\pi\ :\ \widetilde{\mathcal{X}} \to U$ subordinate to a hypersurface $\mathcal{Y}$ in U containing both $U \cap \mathcal{X}_{\rm sing}$ and the $V(f_j) \cap U$ for $j = 1, \ldots, m$, in such a way that $(f_1, \ldots, f_m)(\mathcal{O}_{\widetilde{\mathcal{X}}})_{|U} := \pi^*((f_1, \ldots, f_m)\mathcal{O}_\mathcal{X})_{|U}$ is a principal sheaf of $\mathcal{O}_{\widetilde{\mathcal{X}}}$-ideals, whose zero set is a hypersurface with normal crossings on the complex analytic manifold $\widetilde{\mathcal{X}}$.

[5] In fact birational, since complex analytic spaces considered here come from algebraic varieties $\mathbb{X}_{\widetilde{\Sigma}}$ and $\mathcal{X}_\Sigma$ defined over $\mathbb{Q}$.

The second result concerns $\mathbb{Z}^N$-rational complete simplicial fans, where the set $\Sigma(N)$ contains at least one primitive cone; see [**ShcTY06**, Theorem 1]. It completes the presentation of the complex analytic space $\mathcal{X}_\Sigma$ given by (3.5) and extends the concept of perspective as introduced in the Italian Renaissance. The concept of perspective is re-interpreted in terms of the embedding $\mathbb{P}^N_{\mathbb{C}} \hookrightarrow \mathbb{P}^{N+1}_{\mathbb{C}}$ in such a way that $\mathbb{P}^{N+1}_{\mathbb{C}}$ is realized as the disjoint union of the affine space $\mathbb{C}^{N+1}$ with its hyperplane at infinity, namely $\boldsymbol{H} = \mathbb{P}^N_{\mathbb{C}}$, as already mentioned in Example 2.7.

THEOREM 3.13 (Multiperspective). *Let Σ be a complete $\mathbb{Z}^N$-rational simplicial fan with $d = N + r$ rays. If $\Sigma(N)$ contains at least one primitive cone, then there exists a d-lattice $\boldsymbol{N}$ and a complete simplicial fan $\boldsymbol{\Sigma}$ in $\boldsymbol{N}_{\mathbb{R}} = \boldsymbol{N} \otimes_{\mathbb{Z}} \mathbb{R}$ such that*

(1) *$\boldsymbol{\mathcal{X}}_{\boldsymbol{\Sigma}} = \mathbb{C}^d \sqcup (\boldsymbol{H}_1 \cup \cdots \cup \boldsymbol{H}_r)$ for some complex hypersurfaces $\boldsymbol{H}_j$, $j = 1, \ldots, r$;*
(2) *there exists an isomorphism of complex spaces $\phi :\ \boldsymbol{H}_1 \cap \cdots \cap \boldsymbol{H}_r \longleftrightarrow \mathcal{X}_\Sigma$;*
(3) *for any $z \in \mathbb{C}^d \setminus V(I^{\mathrm{irr}}(\Sigma))$, one has*

$$\overline{G(\Sigma)(z)} \cap \big(\boldsymbol{H}_1 \cap \cdots \cap \boldsymbol{H}_r\big) = \phi^{-1}\big(z \bmod G(\Sigma)\big). \tag{3.10}$$

3.1.3. Kähler cone and moment maps. Let Σ be a $\mathbb{Z}^N$-rational complete primitive fan. The compact analytic space $\mathcal{X}_\Sigma$ defined by (3.5) (or defined utilizing the algebraic construction described in Remark 3.9) is a compact N-dimensional complex manifold. Consider the set $\Sigma(1) = \{\xi_1 \overline{\mathbb{R}^+}, \ldots, \xi_d \overline{\mathbb{R}^+}\}$, where $d = N + r$. Consider also r linear relations between the d rays of Σ with coefficients $a_{j,k}$ in $\mathbb{N}$ as in (3.8), that are $\mathbb{Q}$-linearly independent. Given a set of such relations with matrix $[\boldsymbol{a}] \in \mathcal{M}_{d,r}(\mathbb{N})$, let

$$\Theta_{[\boldsymbol{a}]} : \gamma = (\gamma_1, \ldots, \gamma_d) \in \mathbb{R}^d \longmapsto \Big(\sum_{k=1}^d a_{1,k}\, \gamma_k, \ldots, \sum_{k=1}^d a_{r,k}\, \gamma_k\Big) \in \mathbb{R}^r.$$

For any $\boldsymbol{\gamma} = (\gamma_1, \ldots, \gamma_d) \in \mathbb{R}^d$, the map from $\Sigma(1)$ to $\mathbb{R}$ sending ξ_j to $-\gamma_j$ extends in a unique way as a Σ-piecewise linear map $\Phi_{\boldsymbol{\gamma}}$ from $\mathbb{R}^N$ to $\mathbb{R}$. That is, $\Phi_{\boldsymbol{\gamma}}$ is a continuous map from $\mathbb{R}^N$ to $\mathbb{R}$, whose restriction to any $\sigma \in \Sigma$ is $\mathbb{R}$-linear.

DEFINITION 3.14. The Kähler cone of the primitive fan Σ is defined by

$$K(\Sigma) = \Theta_{[\boldsymbol{a}]}\big(\{\boldsymbol{\gamma} \in \mathbb{R}^d \,:\, \Phi_{\boldsymbol{\gamma}} \text{ is strictly convex}\}\big) \tag{3.11}$$

(independently of the matrix $[\boldsymbol{a}]$).

Let us provide an alternative presentation of $K(\Sigma)$ in terms of the primitive subsets of $\Sigma(1)$; see Proposition 3.7. For any primitive subset $\{\xi\}_J = \{\xi_j \overline{\mathbb{R}^+}, j \in J\}$ of $\Sigma(1)$, the vector $|\xi|_J = \sum_{j\in J} \xi_j$ belongs to some $\sigma_J \in \Sigma(N)$, which is generated by the primitive vectors $\xi_{\ell_{J,1}}, \ldots, \xi_{\ell_{J,N}}$, that is

$$|\xi|_J = \sum_{\nu=1}^N \alpha_{J,\nu}\, \xi_{J,\ell_\nu}$$

for some rational coefficients $\alpha_{J,\nu} \geq 0$. The $\mathbb{Q}$-independent r relations (3.8) generate the $\mathbb{Q}$-vector space of $\mathbb{Q}$-linear relations between the primitive vectors $\xi_1, \ldots, \xi_d$. Thus, there exist unique rational numbers $t_1(J), \ldots, t_r(J)$ satisfying

$$|\xi|_J - \sum_{\nu=1}^N \alpha_{J,\nu}\, \xi_{J,\ell_\nu} = \sum_{\rho=1}^r t_\rho(J)\Big(\sum_{k=1}^d a_{\rho,k}\, \xi_k\Big). \tag{3.12}$$

The alternative presentation of $K(\Sigma)$, arising from the description of $V(I^{\mathrm{irr}}(\Sigma))$ (see Proposition 3.7) is then

$$K(\Sigma) = \{\gamma \in (\overline{\mathbb{R}^+})^r : \sum_{\rho=1}^{r} t_\rho(J)\,\gamma_\rho > 0 \quad \forall\, \{\xi\}_J \overset{\text{primitive}}{\subset} \Sigma(1)\}. \tag{3.13}$$

Since the symbols $|z_1|^2, \dots, |z_d|^2$ satisfy the system of $\mathbb{Q}$-linearly independent relations

$$\sum_{k=1}^{d} a_{j,k}|z_k|^2 = 0, \quad j = 1, \dots, r,$$

as well as the $\xi_k(k = 1, \dots, d)$, do, one has

$$\sum_{j\in J} |z_j|^2 - \sum_{\nu=1}^{N} \alpha_{J,\nu}\,|z_{J,\ell_\nu}|^2 = \sum_{\rho=1}^{r} t_\rho(J) \Big(\sum_{k=1}^{d} a_{\rho,k}\,|z_k|^2\Big)$$

for any $j \in J$. As a consequence, $K(\Sigma)$ is also defined by

$$K(\Sigma) = \mu(\{z \in \mathbb{C}^d : |z|_J^2 - \sum_{\nu=1}^{n} \alpha_{J,\nu}\,|z_{J,\ell_\nu}|^2 > 0 \quad \forall\, \{\xi\}_J \overset{\text{primitive}}{\subset} \Sigma(1)\}), \tag{3.14}$$

where the map

$$\mu : z \in \mathbb{C}^d \longmapsto \Big(\sum_{k=1}^{d} a_{1,k}\,|z_k|^2, \dots, \sum_{k=1}^{d} a_{r,k}\,|z_k|^2\Big) \in (\overline{\mathbb{R}^+})^r \tag{3.15}$$

is known as a *moment map* with respect to the fan Σ; see [**Shc07**, §1.4] or [**Ky05**, §1]. *The moment map* can be seen from the point of view of symplectic geometry; see §3.1.4.

3.1.4. Projective toric manifolds. Assume now that the $\mathbb{Z}^N$-rational fan Σ, where $d = N + r = |\Sigma(1)|$, is complete, primitive, and such that $\mathcal{X}_\Sigma$ is projective as an algebraic compact manifold. Being projective amounts to saying that there exists an ample[6] toric Weil divisor $\sum_{j=1}^{d} a_j\{H_j\}$, with $a_j \in \mathbb{Z}$, on $\mathcal{X}_\Sigma$. Let $D = \sum_{j=1}^{d} a_j D_j$ be the corresponding Cartier divisor. It is a well known fact that when $\mathcal{X}_\Sigma$ is a complex toric manifold, then a line bundle $\mathcal{O}(D)$ over $\mathcal{X}_\Sigma$ is ample if and only if it is very ample [**Dem70**]. If, in addition, $\mathcal{X}_\Sigma$ is projective, then Σ can be realized as the *normal fan* of an *absolutely simple*, N-dimensional, convex polyhedron $\Delta_\Sigma = \Delta_D \subset (\mathbb{R}^N)^*$; see [**Oda**, §2.4, Theorem 1.10] and Example 3.4. A convex polyhedron $\Delta_\Sigma = \Delta_D \subset (\mathbb{R}^N)^*$ is said to be *absolutely simple* if it satisfies the following properties:

(1) Any vertex of Δ_Σ belongs to exactly N edges of Δ_Σ;
(2) Primitive vectors defining the directions of N edges meeting at a vertex of Δ_Σ form a basis of the lattice $\mathbb{Z}^N$.

Note that when Σ is a convex fan, one can take as Δ_Σ the reflexive polytope defined in formula (3.2) as the dual polyhedron of the convex hull of ξ_j, $j = 1, \dots, N + r$.

---

[6]See §D.4, in particular §D.40, for the notion of ampleness of Cartier divisors on a complex algebraic variety, in particular toric divisors on a complete toric algebraic variety.

The divisor $D = D_1 + \cdots + D_{N+r}$ is very ample in this case. One can also prove that when $\mathcal{X}_\Sigma$ is smooth and projective, the subset

$$\{z \in \mathbb{C}^d \,:\, |z|_J^2 - \sum_{\nu=1}^{N} \alpha_{J,\nu}\, |z_{J,\ell_\nu}|^2 > 0 \quad \forall \{\xi\}_J \overset{\text{primitive}}{\subset} \Sigma(1)\} \subset \mathbb{C}^d,$$

whose image by the moment map

$$z \longmapsto \Big(\sum_{k=1}^{d} a_{1,k}\, |z_k|^2, \dots, \sum_{k=1}^{d} a_{r,k}\, |z_k|^2\Big)$$

equals the Kähler cone $K(\Sigma)$ (see (3.14)) does not intersect $V(I^{\text{irr}}(\Sigma))$. Detailed description of the Kähler cone $K(\Sigma)$ in terms of primitive subsets of $\Sigma(1)$ is proposed in [**Bat93**, §2]; see also [**Cx97**, Theorem 4.1]. Moreover, for any $\gamma = (\gamma_1, \dots, \gamma_r)$ in $K(\Sigma)$, the maximal compact subgroup

$$G_{\mathbb{R}}(\Sigma) = \Big\{\Big(\prod_{j=1}^{r} t_j^{a_{j,1}}, \dots, \prod_{j=1}^{r} t_j^{a_{j,d}}\Big),\ t_j \in \mathbb{S}^1,\ j = 1, \dots, r\Big\}$$

of $G(\Sigma)$ induces a group action on the real manifold $\mu^{-1}(\gamma)$ such that $\mu^{-1}(\gamma)/G_{\mathbb{R}}(\Sigma)$ is another representation of the toric projective manifold $\mathcal{X}_\Sigma$. To be more specific, the quotient $\mu^{-1}(\gamma)/G_{\mathbb{R}}(\Sigma)$ can be identified with $\mathcal{X}_\Sigma$ through a $\mathbb{R}$-diffeomorphism that preserves the symplectic form

$$\omega = (1/2i) \sum_{1}^{d} d\overline{z_j} \wedge dz_j.$$

When $\mathcal{X}_\Sigma$ is a projective toric manifold, its Kähler cone $K(\Sigma)$ coincides, modulo the identification of $\mathbb{R}^r$ with $A_{N-1}(\mathcal{X}_\Sigma) \otimes_{\mathbb{Z}} \mathbb{R}$ induced by the isomorphism (3.9) tensored with $\mathbb{R}$, with the set of cohomology classes of closed positive $(1,1)$-forms in the cohomology group $H^{1,1}(\mathcal{X}_\Sigma, \mathbb{R}) = H^2(\mathcal{X}_\Sigma, \mathbb{R}) \simeq A_{N-1}(\mathcal{X}_\Sigma) \otimes_{\mathbb{Z}} \mathbb{R}$. That is, the Kähler cone $K(\Sigma)$ coincides with the cone of *Kähler forms* inducing *Kähler metrics* on the complex Kähler manifold $\mathcal{X}_\Sigma$; see, e.g., [**Cx97**, §3]. This justifies the terminology used to denote by extension $K(\Sigma)$ as the *Kähler cone* of the projective toric manifold $\mathcal{X}_\Sigma$. The moment map (3.15) from $\mathbb{C}^d$ to $(\overline{\mathbb{R}^+})^r$ is surjective, due to the fact that rank $[a_{j,k}]_{j,k} = r$. It follows from [**Cx97**, Theorem 4.1] that when $\mathcal{X}_\Sigma$ is a projective toric manifold, then the isomorphism (3.9), with all entries $a_{j,k}$ chosen to be nonnegative, once tensored with $\mathbb{R}$, provides the representation

$$K(\Sigma) \simeq \mu\big(\mathbb{C}^d \setminus I^{\text{irr}}(\Sigma)\big) \subset (\overline{\mathbb{R}^+})^r, \tag{3.16}$$

where μ is the moment map (3.15).

3.2. Bochner–Martinelli currents attached to bundle sections

In this section, we consider an N-dimensional complex manifold $\mathcal{X}$ with structural sheaf $\mathcal{O}$ with a closed analytic subset $V \subset \mathcal{X}$ with pure codimension M. We also let E be a holomorphic bundle $E \to \mathcal{X}$ over $\mathcal{X}$ with rank $m \geq 1$, and we let $s \in \mathcal{O}(\mathcal{X}, E)$ be a global holomorphic section of E over $\mathcal{X}$. The Koszul complex attached to s is defined as in §1.2.5, namely

$$0 \longrightarrow \bigwedge^{m} E^* \xrightarrow{\ \rfloor_s\ } \bigwedge^{m-1} E^* \xrightarrow{\rfloor_s} \cdots \xrightarrow{\rfloor_s} \bigwedge^{\ell} E^* \xrightarrow{\ \rfloor_s\ } \bigwedge^{\ell-1} E^* \xrightarrow{\rfloor_s} \cdots \xrightarrow{\rfloor_s} E^* \xrightarrow{\ \rfloor_s\ } \mathbb{C}, \tag{3.17}$$

where $\rfloor_s$ denotes the interior multiplication. It is denoted by $\mathbf{K}^s_\bullet$. We introduce also the holomorphic bundle

$$\bigwedge E^* = \bigoplus_{\ell\in\mathbb{Z}} \bigwedge^\ell E^* = \mathbb{C} \oplus \bigoplus_{\ell=1}^m \bigwedge^\ell E^*. \tag{3.18}$$

Similarly to (1.83), we consider the sheaves of currents $\mathcal{L}_E^{M,\nu}$, for $\nu \in \mathbb{Z}$, defined by[7]

$$\mathcal{L}_E^{M,\nu} = \bigoplus_{\gamma\in\mathbb{Z}} \Big({}'\mathcal{D}^{(M,M+\gamma+\nu)}(U, \bigwedge^\gamma E^*)\Big)_{U \overset{\text{open}}{\subset} \mathcal{X}}. \tag{3.19}$$

As in §1.2.5, the operator $\nabla_s = 2i\pi \rfloor_s - \overline{\partial}$ acting from $\mathcal{L}_E^{M,\nu-1}$ to $\mathcal{L}_E^{M,\nu}$ for each $\nu \in \mathbb{Z}$ will play an important role. Note that, since s is holomorphic, $\rfloor_s$ and $\overline{\partial}$ anticommute, which implies $\nabla_s^2 = \rfloor_s^2 = \overline{\partial}^2 = 0$. We equip the holomorphic bundle E with a smooth hermitian metric $|\ |$ and then denote by $\overline{E} = (E, |\ |)$ the metrized bundle. The metric $|\ |$ induces the definition of a hermitian metric on the dual bundle E^* and therefore induces the definition of a hermitian metric $|\ |$ on each entry $\bigwedge^\ell E^*$ of the Koszul complex (3.17). The metric induced on each entry $\bigwedge^\ell E^*$ will still be denoted for the time being by $|\ |$, since we try to avoid overweighing notations. Recall that the conjugate section $s^* \in C^\infty(\mathcal{X}, E^*)$ is defined by

$$s^*(z)(\xi) = \langle \xi, s(z)\rangle_z \quad \forall z \in \mathcal{X},\ \forall \xi \in E_z; \tag{3.20}$$

see (B.59), §B.3.2.1. It is also the section of $E^* \to \mathcal{X}$ with pointwise minimal norm $|\ |_z$ among those satisfying $\rfloor_s s^* = |s|^2$ in $\mathcal{X}$.

3.2.1. The currents $R^s_{|\ |}$ and $P^s_{|\ |}$ for $s \in \mathcal{O}(\mathcal{X}, \overline{E})$, $\overline{E} = (E, |\ |)$. We introduce the $\mathcal{L}_E^{M,0}(\mathcal{X})$ and $\mathcal{L}_E^{M,-1}(\mathcal{X})$-valued maps defined respectively as[8]

(3.21)

$$\lambda \longmapsto R^{s,\lambda}_{|\ |}([V]) := \Big(1 - |s|^{2\lambda} + (\overline{\partial}\,|s|^{2\lambda}) \sum_{r\geq 1} \frac{s^*}{2i\pi|s|^2}\Big(\overline{\partial}\Big(\frac{s^*}{2i\pi|s|^2}\Big)\Big)^{r-1}\Big)\,[V],$$

$$\lambda \longmapsto P^{s,\lambda}_{|\ |}([V]) := \Big(|s|^{2\lambda} \sum_{r\geq 1} \frac{s^*}{2i\pi|s|^2}\Big(\overline{\partial}\Big(\frac{s^*}{2i\pi|s|^2}\Big)\Big)^{r-1}\Big)\,[V]$$

for $\{\lambda \in \mathbb{C} : \operatorname{Re}\lambda \gg 1\}$. Observe that

$$2i\pi \rfloor_s R^{s,\lambda}_{|\ |}([V]) = -\Big((\overline{\partial}\,|s|^{2\lambda}) \sum_{r\geq 0} \Big(\overline{\partial}\Big(\frac{s^*}{2i\pi|s|^2}\Big)\Big)^{r}\Big)\,[V],$$

$$\overline{\partial}\, R^{s,\lambda}_{|\ |}([V]) = \Big(-\overline{\partial}\,|s|^{2\lambda} - (\overline{\partial}\,|s|^{2\lambda}) \sum_{r\geq 1} \Big(\overline{\partial}\Big(\frac{s^*}{2i\pi|s|^2}\Big)\Big)^{r}\Big)\,[V].$$

This implies

$$\nabla_s R^{s,\lambda}_{|\ |}([V]) = 0. \tag{3.22}$$

[7]When $M = 0$, we just set $\mathcal{L}_E^{\nu}$ instead of $\mathcal{L}_E^{0,\nu}$.

[8]Multiplication is here understood as wedge multiplication. Moreover, one needs to understand a product $T'T''$ of two respectively $\bigwedge^{\ell'} E^*$ and $\bigwedge^{\ell''} E^*$-valued currents as the $\bigwedge^{\ell'+\ell''} E^*$-valued current $T' \wedge T''$.

One has also

$$2i\pi\rfloor_s P^{s,\lambda}_{|\,|}([V]) = \Big(|s|^{2\lambda}\sum_{r\geq 0}\Big(\overline{\partial}\Big(\frac{s^*}{2i\pi|s|^2}\Big)\Big)^r\Big)[V],$$

$$\overline{\partial}P^{s,\lambda}_{|\,|}([V]) = \Big(|s|^{2\lambda}\sum_{r\geq 1}\Big(\overline{\partial}\Big(\frac{s^*}{2i\pi|s|^2}\Big)\Big)^r\Big)[V] + \big(R^{s,\lambda}_{|\,|}([V]) - (1-|s|^{2\lambda})[V]\big),$$

which implies

$$\nabla_s P^{s,\lambda}_{|\,|}([V]) = \big(\mathrm{Id} - R^{s,\lambda}_{|\,|}\big)([V]). \tag{3.23}$$

Observe that (3.22) follows from (3.23) if one takes the action of ∇_s on both sides and exploits the fact that $\nabla_s^2 = 0$. One can state now a multivariate version of Proposition 1.27.

PROPOSITION 3.15. *Let $\mathcal{X}$, V, $\overline{E} = (E, |\,|)$, s be as above. The $\mathcal{L}_E^{M,0}(\mathcal{X})$ and $\mathcal{L}_E^{M,-1}(\mathcal{X})$-valued maps $\lambda \longmapsto R^{s,\lambda}_{|\,|}([V])$ and $\lambda \longmapsto P^{s,\lambda}_{|\,|}([V])$, respectively, extend from $\{\lambda \in \mathbb{C} \ \ \mathrm{Re}\,\lambda \gg 1\}$ to $\mathbb{C}$ as meromorphic maps, whose polar sets lie in $-\mathbb{Q}^+$. Moreover, one has*

$$R^s_{|\,|}([V]) = \big[R^{s,\lambda}_{|\,|}([V])\big]_{\lambda=0} \in \mathrm{PM}(\mathcal{X}, \textstyle\bigwedge E^*),$$

$$P^s_{|\,|}([V]) = \big[P^{s,\lambda}_{|\,|}([V])\big]_{\lambda=0} \in \mathrm{PM}(\mathcal{X}, \textstyle\bigwedge E^*),$$

$$\begin{aligned} &\mathrm{Supp}\,R^s_{|\,|} \subset V\cap V(s),\quad \nabla_s R^s_{|\,|} = 0,\quad \nabla_s P^s_{|\,|} = [V] - R^s_{|\,|}([V]),\\ &0 \leq r < \mathrm{codim}_V(V\cap V(s))\\ &\qquad\qquad \Longrightarrow \big(R^s_{|\,|}([V])\big)^{(M,M+r)} = \mathrm{Proj}_{\wedge^r E^*} R^s_{|\,|}([V]) = 0. \end{aligned} \tag{3.24}$$

Furthermore, for any $r \geq \mathrm{codim}_V(V\cap S)$, for any $z \in V\cap V(s)$ and for any $f \in \mathcal{O}_z$,

$$\begin{aligned} &f = 0 \text{ on } V\cap V(s) \text{ around } z \Longrightarrow \big(\overline{f}\cdot R^s_{|\,|}([V])\big)_z = \big(d\overline{f}\wedge R^s_{|\,|}([V])\big)_z = 0,\\ &|f| = O(|s|^r) \text{ on } V \text{ around } z \Longrightarrow \Big(f\cdot\big(R^s_{|\,|}([V])\big)^{(M,M+r)}\Big)_z = 0. \end{aligned} \tag{3.25}$$

PROOF. Since the integration current $[V]$ is pseudo-meromorphic, it follows from Proposition 2.21 that $\lambda \longmapsto (1-|s|^{2\lambda})[V]$ extends to $\mathbb{C}$ as a $'\mathcal{D}^{(M,M)}(\mathcal{X})$-valued meromorphic map, whose poles are in $-\mathbb{Q}^+$. Its value at $\lambda = 0$ equals the integration current $\mathbf{1}_{\mathcal{X}\setminus V(s)}\cdot[V]$. Such integration current acts over the set $V^{\mathcal{X}\setminus V(s)}$, where $V^{\mathcal{X}\setminus V(s)}$ is the union of the irreducible components of V that are not contained entirely in $V(s)$. This is a particular case of restriction of pseudo-meromorphic currents; see §2.2.4 or, for this specific situation, see §D.1.1. Then it is clear that one can replace V by $V^{\mathcal{X}\setminus S}$ and hence one can suppose that $\mathrm{codim}_V(V\cap V(s)) > 1$. Furthermore, it is also clear that it is enough to prove the result locally around $V\cap V(s)$. Thus, we fix a point $a \in V\cap V(s)$. Let $\pi : \mathcal{V} \to V\cap U_a$ be a log resolution as in Theorem A.41, applied here for $\mathcal{X} = V\cap U_a$. Let also $\mathcal{Y}$ be the union of the proper closed analytic subset $(V\cap U_a)\cap V(s)$ with a closed hypersurface of $V\cap U_a$ containing $V_{\mathrm{sing}}\cap U_a$. Then $\mathcal{V}$ is an $(N-M)$-dimensional complex manifold. Let us suppose that $z \in \mathcal{V}$ admits a neighborhood U such that $(\iota_V\circ\pi)^*s = s_0\otimes s'$, where s_0 is a section of a Cartier divisor $D = D_z$ on U and s' is a nonvanishing section of $\mathcal{O}(-D)\otimes(\iota_V\circ\pi)^*(E_{|V})$. One has

$$(\iota_V\circ\pi)^*\Big(\frac{s^*}{|s|^2}\Big) = \frac{1}{s_0}\,\sigma',$$

where σ' is a smooth section of $\mathcal{O}(D)\otimes(\iota_V\circ\pi)^*(E^*_{|V})$ in U. For any $r\geq 1$ and $\lambda\in\mathbb{C}$ such that $\operatorname{Re}\lambda\gg 1$, one has

$$(\iota_V\circ\pi)^*\Big((\overline{\partial}\,|s|^{2\lambda})\,\frac{s^*}{|s|^2}\,\Big(\overline{\partial}\Big(\frac{s^*}{|s|^2}\Big)^{r-1}\Big)\Big)=(\overline{\partial}\,\|s_0s'\|^{2\lambda})\,\frac{1}{s_0^r}\,\sigma'(\overline{\partial}\sigma')^{r-1}$$
$$=\overline{\partial}\Big(\frac{\|s_0s'\|^{2\lambda}}{s_0^r}\Big)\,\sigma'(\overline{\partial}\sigma')^{r-1},$$
$$(\iota_V\circ\pi)^*\Big(|s|^{2\lambda}\,\frac{s^*}{|s|^2}\,\Big(\overline{\partial}\Big(\frac{s^*}{|s|^2}\Big)^{r-1}\Big)\Big)=\frac{\|s_0s'\|^{2\lambda}}{s_0^r}\,\sigma'(\overline{\partial}\sigma')^{r-1}.$$

The ${}'\mathcal{D}(U,\mathcal{O}(-rD))$-valued maps

$$\lambda\longmapsto\overline{\partial}\Big(\frac{\|s_0s'\|^{2\lambda}}{s_0^r}\Big),\qquad \lambda\longmapsto\frac{\|s_0s'\|^{2\lambda}}{s_0^r}$$

extend to $\mathbb{C}$ as meromorphic maps with poles in $-\mathbb{Q}^+$ as a consequence of results established in §2.3.2; see also Proposition 2.21. Moreover, their values at $\lambda=0$ equal the pseudo-meromorphic currents $2i\pi\,\overline{\partial}(1/s_0^r)$ and $2i\pi\,P^{s_0^r}$, respectively; see §2.3.2. The direct images of these currents by $\iota\circ\pi$ define germs of pseudo-meromorphic currents at a. The first assertion of the proposition is proved, and the fact that both $R^s_{|\,|}([V])$ and $P^s_{|\,|}([V])$ are $\bigwedge E^*$-valued pseudo-meromorphic currents is established. It is immediate that $R^s_{|\,|}([V])$ is supported by $V\cap V(s)$. The final two assertions in the third line of (3.24) follow from (3.22) and (3.23) and the application of the analytic continuation principle. Observe that the current

$$\big(R^s_{|\,|}([V])\big)^{(M,M+r)}\in \mathrm{PM}^{(M,M+r)}(\mathcal{X},\bigwedge^r E^*)$$

is supported by a closed analytic subset with codimension $M+\operatorname{codim}_V(V\cap V(s))$. Then, it follows from the dimension principle for pseudo-meromorphic currents (see Proposition 2.26) that this current is the null current as soon as

$$M+r<M+\operatorname{codim}_V(V\cap V(s)),$$

that is, when $r<\operatorname{codim}_V(V\cap V(s))$. The assertion in the fourth line of (3.24) is then proved. The first assertion in (3.25) follows from the pseudo-meromorphicity of $R^s_{|\,|}([V])$, together with the fact that this current is supported by $V\cap V(s)$; see Proposition 2.24. The second assertion in (3.25) follows from the fact that

$$|f|=O(|s|^r)\text{ on }V\text{ around }a\Longrightarrow|(\iota_V\circ\pi)^*f|=O(|s_0|^r)\text{ around }z\in\mathcal{V},$$

if we refer here to our previous proof, which implies that

$$(\iota_V\circ\pi)^*f\cdot\overline{\partial}\Big(\frac{1}{s_0^r}\Big)=0.$$

This concludes the proof of Proposition 3.15. □

EXAMPLE 3.16 (The case of trivial bundles). Suppose that $s_1,\dots,s_m$ are m holomorphic functions on $\mathcal{X}$, so that $(s_1,\dots,s_m)$ is considered as a holomorphic section of the trivial m-bundle $\mathbb{C}^m$ over $\mathcal{X}$, equipped with its standard hermitian metric. Let the frame $(\epsilon_1,\dots,\epsilon_m)$ be the canonical basis of $\mathbb{C}^m$. Then, for any $r\geq 1$, the current $(R^s_{|\,|}([V]))^{M,M+r}=\mathrm{Proj}_{\wedge^r(\mathbb{C}^m)^*}R^s_{|\,|}([V])$ equals

$$\sum_{1\leq j_1<\cdots<j_r\leq m} R^s_{|\,|;j_1,\dots,j_r}([V])\,\epsilon^*_{j_1}\wedge\cdots\wedge\epsilon^*_{j_r},$$

where, for each multi-index $1 \leq i_1 < \cdots < i_r \leq m$,

(3.26) $$R^s_{|\,|;j_1,\dots,j_r}([V]) = \frac{(-1)^{r(r-1)/2}}{(2i\pi)^r}\Big[\lambda |s|^{2\lambda}\,\frac{\overline{\partial}\,|s|^2}{|s|^2} \wedge \frac{\sum_{\ell=1}^r (-1)^{\ell-1}\overline{s_{j_\ell}} \bigwedge_{\ell'\neq\ell} d\,\overline{s_{j_{\ell'}}}}{|s|^{2r}}\Big]_{\lambda=0}[V].$$

In particular, if $m \leq N - M$ and $\operatorname{codim}(V \cap V(s)) = m$, one has

(3.27) $$R^s \wedge ([V]) = \frac{(m-1)!(-1)^{m(m-1)/2}}{(2i\pi)^m}\Big(\Big[\lambda\,|s|^{2\lambda}\,\frac{\bigwedge_{j=1}^m d\,\overline{s_j}}{|s|^{2m}}\Big]_{\lambda=0}[V]\Big)\,\epsilon_1^* \wedge \cdots \wedge \epsilon_m^*.$$

The expression of $R^s_{|\,|}([V])$, in case E is a trivial bundle, involves differential $(0,r)$-forms appearing in the expression of the Bochner–Martinelli kernel that had been introduced in Example 2.7. This motivates the following definition.

DEFINITION 3.17. Let $\mathcal{X}$, V, $s \in \mathcal{O}(\mathcal{X}, E)$ be as in Proposition 3.15. Let also $\overline{E} = (E, |\,|)$ be a metrized holomorphic vector bundle on $\mathcal{X}$. The current $R^s_{|\,|}([V])$ is called the *Bochner–Martinelli residue current* induced on V by the global section s of E. Similarly, the current $P^s_{|\,|}([V])$ is called the *Bochner–Martinelli principal value current* induced on V by the global section s of E. When $\operatorname{codim}_V(V \cap V(s)) > 0$, both currents can also be interpreted as direct images through the embedding ι_V of $(\bigwedge E^*)_{|V}$-valued currents[9] on the complex analytic space V.

3.2.2. The complete intersection setting revisited. Let V be, as in the previous section, a closed analytic subset with pure codimension M in some N-dimensional complex manifold $\mathcal{X}$ with structural sheaf $\mathcal{O}$. Let $D_1, \dots, D_m$ be m Cartier divisors on the manifold $\mathcal{X}$. For the holomorphic vector bundle defined as the direct sum $E = \mathcal{O}(D_1) \oplus \cdots \oplus \mathcal{O}(D_m)$, we consider its section defined as the direct sum $s = s_1 \oplus \cdots \oplus s_m \in \mathcal{O}(\mathcal{X}, E)$, where $s_j \in \mathcal{O}(\mathcal{X}, \mathcal{O}(D_j))$ for $j = 1, \dots, m$. Proposition 3.18 enhances the geometric robustness of the Coleff–Herrera current $\big(\overline{\partial}(1/(s_1, \dots, s_m))\big)_V$ introduced in Definition 2.78 under the additional complete intersection hypothesis $\operatorname{codim}_V(V \cap V(s)) = m$. Such robustness plays an essential role in applications. Proposition 3.18 will act as motivation to extend the construction of Bochner–Martinelli residue or principal value currents $R^s_{|\,|}([V])$ or $P^s_{|\,|}([V])$, introduced in Definition 3.17, to the frame of generically exact complexes of hermitian holomorphic bundles, which will be presented in §3.3.

PROPOSITION 3.18. *Let $\mathcal{X}$, V be as above. Let also $D_1, \dots, D_m$ be Cartier divisors on $\mathcal{X}$ and let s_j be global holomorphic sections of the line bundle $\mathcal{O}(D_j)$ for $j = 1, \dots, m$, so that* $\operatorname{codim}_V(V \cap V(s)) = m$, *where $s = s_1 \oplus \cdots \oplus s_m$. Then, for any permutation $\mathfrak{s}$ of $\{1, \dots, m\}$ with signature $\varepsilon(\mathfrak{s})$ and any smooth hermitian metric $|\,|$ on the holomorphic bundle $E = \bigoplus_{j=1}^m \mathcal{O}(D_j)$, one has*

(3.28) $$R^s_{|\,|}([V]) = \overline{\partial}\Big(\frac{1}{(s_{\mathfrak{s}(1)}, \dots, s_{\mathfrak{s}(m)})}\Big)_V\,\varepsilon(\mathfrak{s}).$$

PROOF. The proof we present here is due to M. Andersson [**And04**]. Note that in the context of trivial bundles, the result was obtained for the first time in [**BGVY**] (see also [**PasTY00**]). Since the definition of $s = s_1 \oplus \cdots \oplus s_m$ does not depend on $\mathfrak{s}$, it follows from (2.126) that it is sufficient to prove (3.28) when

[9] In the sense of the theory of currents on a complex analytic space; see §B.2.2.2.

$\mathfrak{s} = \mathrm{Id}$. It is also sufficient to prove this equality of currents when $\mathcal{X}$ is replaced by a neighborhood U_a of a point $a \in V \cap V(s)$. We may also assume that in this neighborhood U_a, the supports of the divisors $\mathrm{div}(s_j)$ intersect properly on $V \cap U_a$, namely

$$\forall J \subset \{1, \dots, m\}, \quad \mathrm{codim}_{V\cap U_a}\Big(V \cap \bigcap_{j\in J} (V(s_j) \cap U_a)\Big) \geq |J|.$$

It follows from Proposition 3.15, namely from the fourth property in (3.24), that

$$R^s_{|\,|}([V]) = \big(R^s_{|\,|}([V])\big)^{(M,M+m)} \in \mathrm{PM}^{(M,M+m)}(\mathcal{X}, \det E^*) = \mathrm{PM}^{(M,M+m)}(\mathcal{X}, \mathcal{O}(-|D|)).$$

Let us first show that

$$R^s_{|\,|}([V]) = R^s_{\|\,\|}([V]), \tag{3.29}$$

when $|\ |$ and $\|\ \|$ are two distinct metrics on E. For any $\lambda \in \mathbb{C}$ with $\mathrm{Re}\,\lambda \gg 1$, consider the pseudo-meromorphic current

$$T^\lambda = |s|^{2\lambda}\, Q_{|\,|}\, Q_{\|\,\|}\, [V],$$

where

$$Q_{|\,|} = \sum_{r\geq 1} \frac{s^{*|\,|}}{2i\pi |s|^2} \Big(\overline{\partial}\Big(\frac{s^{*|\,|}}{2i\pi|s|^2}\Big)\Big)^{r-1}, \quad Q_{\|\,\|} = \sum_{r\geq 1} \frac{s^{*\|\,\|}}{2i\pi \|s\|^2} \Big(\overline{\partial}\Big(\frac{s^{*\|\,\|}}{2i\pi\|s\|^2}\Big)\Big)^{r-1},$$

and the sections $s^{*|\,|}$, $s^{*\|\,\|}$ denote the conjugates of s with respect to the metric $|\ |$, $\|\ \|$, respectively. It follows from the fact that ∇_s acts as an antiderivation,[10] and the identity $|s|^{2\lambda}\, \nabla_s Q_{|\,|} = |s|^{2\lambda}$ that

$$\begin{aligned}
\nabla_s\big(|s|^{2\lambda}\, Q_{|\,|}\, [V]\big) &= -(\overline{\partial}|s|^{2\lambda})\, Q_{|\,|}\, [V] + |s|^{2\lambda}\, [V] \\
&= \big(|s|^{2\lambda} - (\overline{\partial}|s|^{2\lambda})\, Q_{|\,|}\big)([V]),
\end{aligned}$$

$$\begin{aligned}
\nabla_s\big(|s|^{2\lambda}\, Q_{|\,|}\, Q_{\|\,\|}\, [V]\big) &= \big(\nabla_s(|s|^{2\lambda}\, Q_{|\,|}\, [V])\big)\, Q_{\|\,\|} - |s|^{2\lambda}\, Q_{|\,|}\, \nabla_s\big(Q_{\|\,\|}\, [V]\big) \\
&= |s|^{2\lambda}\, Q_{\|\,\|} - (\overline{\partial}\, |s|^{2\lambda})\, Q^s_{|\,|}\, Q^s_{\|\,\|}\, [V] + [s|^{2\lambda}\, Q_{|\,|}\, [V] \\
&= \big(|s|^{2\lambda}\, (Q_{\|\,\|} - Q_{|\,|}) - (\overline{\partial}\, |s|^{2\lambda})\, Q_{|\,|}\, Q_{\|\,\|}\big)\, [V].
\end{aligned} \tag{3.30}$$

Repeating the argument used in the proof of the first assertion in Proposition 3.15, one observes that both maps

$$\lambda \longmapsto |s|^{2\lambda}\, Q_{|\,|}\, Q_{\|\,\|}\, [V], \quad \lambda \longmapsto (\overline{\partial}\, |s|^{2\lambda})\, Q_{|\,|}\, Q_{\|\,\|}\, [V]$$

extend to $\mathbb{C}$ as meromorphic maps with poles in $-\mathbb{Q}^+$, whose value at $\lambda = 0$ are pseudo-meromorphic currents. For the second map, this pseudo-meromorphic current is supported by $V \cap V(s)$. It follows from the dimension principle for pseudo-meromorphic currents (Proposition 2.26) that all the $(\bullet, r)$ components of $\big[(\overline{\partial}\, |s|^{2\lambda})\, Q_{|\,|}\, Q_{\|\,\|}\, [V]\big]_{\lambda=0}$ are equal to 0, when $r < M + m$. Since

$$Q_{|\,|}\, Q_{\|\,\|} = \frac{s^*_{|\,|}}{2i\pi|s|^2}\, \frac{s^*_{\|\,\|}}{2i\pi\|s\|^2} \sum_{\{1\leq r', r''\leq m\,:\, r'+r''\leq m\}} \pm \Big(\overline{\partial}\Big(\frac{s^{*|\,|}}{2i\pi|s|^2}\Big)\Big)^{r'-1} \Big(\overline{\partial}\Big(\frac{s^{*\|\,\|}}{2i\pi\|s\|^2}\Big)\Big)^{r''-1},$$

---

[10] Namely, if $T' \in \mathcal{L}_E^{\nu'}$, $T'' \in \mathcal{L}_E^{\nu''}$, then $\nabla_s(T'\, T''\, [V]) = (\nabla_s T')\, T''\, [V] + (-1)^{\nu} T'\, (\nabla_s T'')\, [V]$.

all components of the current $(\overline{\partial}\,|s|^{2\lambda})\,Q_{|\,|}Q_{\|\,\|}$ have bidegree $(0,k)$ for $k \le m-1$. Therefore $\big[(\overline{\partial}\,|s|^{2\lambda})\,Q_{|\,|}Q_{\|\,\|}\,[V]\big]_{\lambda=0} = 0$. Taking the analytic continuation in λ up to $\lambda = 0$ in both sides of the final equality in (3.30), then acting on them by ∇_s, one gets

$$(\nabla_s)^2\Big(\Big[|s|^{2\lambda}\,Q_{|\,|}Q_{\|\,\|}\,[V]\Big]_{\lambda=0}\Big) = 0 = \big(\mathrm{Id} - R^s_{\|\,\|}\big)([V]) - \big(\mathrm{Id} - R^s_{|\,|}\big)([V])$$

after observing that

$$\begin{aligned}\nabla_s(|s|^{2\lambda}Q_{|\,|}\,[V]) &= \big(\mathrm{Id} - R^s_{|\,|}\big)([V]),\\ \nabla_s(|s|^{2\lambda}Q_{\|\,\|}\,[V]) &=\nabla_s(\|s\|^{2\lambda}Q_{\|\,\|}) = \big(\mathrm{Id} - R^s_{\|\,\|}\big)([V]).\end{aligned}$$

The above identities are derived taking into account the properties on the third line in (3.24). Equality (3.29) follows. Thus, to prove (3.28), it is enough to choose an arbitrary metric $|\ |_j$ on each $\mathcal{O}(D_j)$, $j = 1,\dots,m$, and take the metric $|\xi|_z = |\xi_1|_z + \cdots + |\xi_m|_z$ on each E_z. Fix now a local frame e_j for $\mathcal{O}(D_j)$ in a neighborhood of the point a. It follows from Theorem 2.75 that the $'\mathcal{D}(U_a, \bigwedge E^*)$-valued map

$$\boldsymbol{P}^{\boldsymbol{\mu}}_{|\,|} : \boldsymbol{\mu} \longmapsto \sum_{r=0}^{m-1} \frac{|s_{r+1}|^{2\mu_{r+1}}}{2i\pi s_{r+1}}\Big(\overline{\partial}\Big(\frac{|s_r|^{2\mu_r}}{2i\pi s_r}\Big)\wedge\cdots\wedge\overline{\partial}\Big(\frac{|s_1|^{2\mu_1}}{2i\pi s_1}\Big)\Big)\,[V]$$

extends from $\{\boldsymbol{\mu} \in \mathbb{C}^m : \mathrm{Re}\,\mu_j \gg 1 \text{ for } j = 1,\dots,m\}$ to $\mathbb{C}^m$ as a meromorphic map, whose polar set does not intersect $\{\boldsymbol{\mu} \in \mathbb{C}^m : \mathrm{Re}\,\mu_j \ge 0 \text{ for } j = 1,\dots,m\}$. Moreover, one has in U_a that

$$\nabla_s\big([\boldsymbol{P}^{\boldsymbol{\mu}}_{|\,|}]_{\boldsymbol{\mu}=0}\big) = [V] - \overline{\partial}\Big(\frac{1}{(s_1,\dots,s_m)}\Big)_V. \tag{3.31}$$

As in (3.30), one has that

$$\begin{aligned}\nabla_s\big(|s|^{2\lambda}\,Q_{|\,|}\,\boldsymbol{P}^{\boldsymbol{\mu}}_{|\,|}\big) &= \nabla_s\big(|s|^{2\lambda}\,Q_{|\,|}\big)\,\boldsymbol{P}^{\boldsymbol{\mu}}_{|\,|} - |s|^{2\lambda}\,Q_{|\,|}\,\nabla_s\big(\boldsymbol{P}^{\boldsymbol{\mu}}_{|\,|}\big)\\ &= \big(|s|^{2\lambda} - (\overline{\partial}|s|^{2\lambda})\,Q_{|\,|}\big)\,\boldsymbol{P}^{\boldsymbol{\mu}}_{|\,|} - |s|^{2\lambda}\,Q_{|\,|}\,\nabla_s\big(\boldsymbol{P}^{\boldsymbol{\mu}}_{|\,|}\big),\end{aligned} \tag{3.32}$$

whenever $(\lambda,\boldsymbol{\mu}) \in \mathbb{C}^{m+1}$ and $\mathrm{Re}\,\lambda \gg 1$ and $\mathrm{Re}\,\mu_j \gg 1$ for $j = 1,\dots,m$. For $\mathrm{Re}\,\lambda \gg 1$, it follows from the analytic continuation principle and from (3.31) that

$$\nabla_s\big(|s|^{2\lambda}\,Q_{|\,|}\,[\boldsymbol{P}^{\boldsymbol{\mu}}_{|\,|}]_{\boldsymbol{\mu}=0}\big) = \big(|s|^{2\lambda} - (\overline{\partial}|s|^{2\lambda})\,Q_{|\,|}\big)\,[\boldsymbol{P}^{\boldsymbol{\mu}}_{|\,|}]_{\boldsymbol{\mu}=0} - |s|^{2\lambda}\,Q_{|\,|}\,[V]. \tag{3.33}$$

Since $\big[\boldsymbol{P}^{s,\boldsymbol{\mu}}_{|\,|}\big]_{\boldsymbol{\mu}=0}$ is a pseudo-meromorphic current, one can show as in the proof of Proposition 2.21, that the current-valued maps

$$\lambda \longmapsto |s|^{2\lambda}\,[\boldsymbol{P}^{\boldsymbol{\mu}}_{|\,|}]_{\boldsymbol{\mu}=0}, \quad \lambda \longmapsto (\overline{\partial}|s|^{2\lambda})\,Q_{|\,|}\,[\boldsymbol{P}^{\boldsymbol{\mu}}_{|\,|}]_{\boldsymbol{\mu}=0}$$

extend from $\{\lambda \in \mathbb{C} : \mathrm{Re}\,\lambda \gg 1\}$ to $\mathbb{C}$ as meromorphic maps with poles in $-\mathbb{Q}^+$, whose values at $\lambda = 0$ are pseudo-meromorphic currents. The current

$$\Big[(\overline{\partial}|s|^{2\lambda}))\,Q_{|\,|}\,[\boldsymbol{P}^{\boldsymbol{\mu}}_{|\,|}]_{\boldsymbol{\mu}=0}\Big]_{\lambda=0}$$

is a pseudo-meromorphic current supported by $V \cap V(s)$. It equals 0 as a consequence of the dimension principle (Proposition 2.26) exactly for the same reason as the current $\big[(\overline{\partial}\,|s|^{2\lambda})\,Q_{|\,|}Q_{\|\,\|}\,[V]\big]_{\lambda=0}$ considered previously. Then, it follows from (3.33) and from the analytic continuation principle that

$$\nabla_s\Big(\Big[|s|^{2\lambda}\,Q_{|\,|}\,[\boldsymbol{P}^{\boldsymbol{\mu}}_{|\,|}]_{\boldsymbol{\mu}=0}\Big]_{\lambda=0}\Big) = \Big[|s|^{2\lambda}\,[\boldsymbol{P}^{\boldsymbol{\mu}}_{|\,|}]_{\boldsymbol{\mu}=0}\Big]_{\lambda=0} - P^s_{|\,|}([V]).$$

Comparing the results of the action of ∇_s on both sides of this equality leads to the equality

$$0 = \Big([V] - \overline{\partial}\Big(\frac{1}{(s_1,\dots,s_m)}\Big)_V\Big) - ([V] - R^s_{|\ |}([V])),$$

which concludes the proof of Proposition 3.18. □

3.2.3. Transformation laws from the geometric point of view. Let $\mathcal{X}$ be a complex N-dimensional manifold with structural sheaf $\mathcal{O}$ with $V \subset \mathcal{X}$ a closed analytic subset with codimension $N-M$. Let $(D_1,\dots,D_m)$ and $(\mathbf{D}_1,\dots,\mathbf{D}_m)$ be sequences of Cartier divisors on the complex manifold $\mathcal{X}$. We consider then the corresponding holomorphic vector bundles

$$E = \mathcal{O}(D_1) \oplus \cdots \oplus \mathcal{O}(D_m), \quad \mathbf{E} = \mathcal{O}(\mathbf{D}_1) \oplus \cdots \oplus \mathcal{O}(\mathbf{D}_m).$$

If $s_j \in \mathcal{O}(\mathcal{X}, O(D_j))$, $\mathbf{S}_j \in \mathcal{O}(\mathcal{X}, \mathcal{O}(\mathbf{D}_j))$, $j = 1,\dots,m$, then

$$s = s_1 \oplus \cdots \oplus s_m \in \mathcal{O}(\mathcal{X}, E) \quad \mathbf{S} = \mathbf{S}_1 \oplus \cdots \oplus \mathbf{S}_m \in \mathcal{O}(\mathcal{X}, \mathbf{E})$$

are the corresponding global sections of E and $\mathbf{E}$, respectively. We assume from now on in this subsection that

$$\operatorname{codim}_V(V \cap V(s)) = \operatorname{codim}_V(V \cap V(\mathbf{S})) = m. \tag{3.34}$$

We assume in addition that there exists $H \in \mathcal{O}\big(\mathcal{X}, \operatorname{Hom}_{\mathbb{C}}(E, \mathbf{E})\big)$ such that $\mathbf{S} = H\,s$ on V. Let

$$\Delta := \det H \in \mathcal{O}\big(\mathcal{X}, \operatorname{Hom}_{\mathbb{C}}\big(\det E, \det \mathbf{E}\big)\big) \tag{3.35}$$

and let $\Delta' \in \mathcal{O}\big(\mathcal{X}, \operatorname{Hom}_{\mathbb{C}}\big((\det \mathbf{E})^*, (\det E)^*\big)\big)$ be its transpose. We intend in this section to compare the two $\det E^*$-valued currents

$$\begin{aligned} R^s_{|\ |}([V]) &: \varphi \in \mathcal{D}^{(N-M,N-M-m)}(\mathcal{X}, \det E) \longmapsto \langle R^s_{|\ |}([V])\,,\, \varphi\rangle, \\ \Delta'\, R^{\mathbf{S}}_{\|\ \|}([V]) &: \varphi \in \mathcal{D}^{(N-M,N-M-m)}(\mathcal{X}, \det E) \longmapsto \langle R^{\mathbf{S}}_{\|\ \|}([V])\,,\, \Delta\,\varphi\rangle, \end{aligned} \tag{3.36}$$

where the smooth hermitian metrics $|\ |$ and $\|\ \|$ are arbitrarily chosen on E and $\mathbf{E}$, respectively. Observe that Proposition 3.18 implies that these currents do not depend on the choice of the metrics $|\ |$ and $\|\ \|$ on E and $\mathbf{E}$, respectively. Let us state and prove two preliminary lemmas before stating the main results of this subsection.

LEMMA 3.19. *Under the hypothesis above, the currents*

$$R^s_{|\ |}([V]),\ \Delta'\, R^{\mathbf{S}}_{\|\ \|}([V])$$

are global sections of $\mathrm{CH}_{\mathcal{X},V\cap V(s)}(\mathcal{X}, \det E^*)$.

PROOF. Proposition 3.18 and Theorem 2.79 imply that

$$R^s_{|\ |}([V]) \in \mathrm{CH}_{\mathcal{X},V\cap V(s)}(\mathcal{X}, \det E^*), \quad R^{\mathbf{S}}_{\|\ \|}([V]) \in \mathrm{CH}_{\mathcal{X},V\cap V(\mathbf{S})}(\mathcal{X}, \det \mathbf{E}^*).$$

Therefore, in order to prove that the $\det E^*$-valued current $\Delta'\, R^{\mathbf{S}}_{\|\ \|}([V])$ belongs also to $\mathrm{CH}_{\mathcal{X},V\cap V(s)}(\mathcal{X}, \det E^*)$, it is enough to show that it is also supported by $V\cap V(s) \subset V \cap V(\mathbf{S})$. The proof of this claim is as follows. If $a \in V \cap (V(\mathbf{S})\setminus V(s))$, then it follows from Cramer's rule for for sections $\mathbf{S}_j(\zeta) = g_j(\zeta)\,\mathbf{e}_j(\zeta)$ for $j = 1,\dots,m$ and $\Delta(\zeta) = \delta(\zeta)\,e(\zeta)$, expressed in local frames in a neighborhood of a, that

$$\delta_a \in \big(g_1,\dots,g_m\big)\,\mathcal{O}_a + (\mathcal{I}_V)_a.$$

Thus, Theorem 2.79 implies that $\Delta' R^{\mathbf{S}}_{\| \|}([V]) = 0$ in an open neighborhood of a. This concludes the proof of Lemma 3.19. □

LEMMA 3.20. *Let U be an open neighborhood of the origin in $\mathbb{C}^N_\zeta$, and let V be a closed analytic subset of U with pure codimension $M < N$. Let also $f_1, \dots, f_m, g_1, \dots, g_m$ be in $\mathcal{O}(U)$, satisfying $V \cap V(f) = V \cap V(g)$, $\mathrm{codim}_V(V \cap V(f)) = m$. We also suppose that there exists a $\mathbb{C}$-linear $A = [a_{j,k}]_{1\le j,k\le m}$ with entries $h_{j,k} \in \mathcal{O}(U)$ so that*

$$g_j(\zeta) = \sum_{k=1}^{m} a_{j,k}(\zeta)\, f_k(\zeta) \qquad \forall \zeta \in V \tag{3.37}$$

for $j = 1, \dots, m$. If $f = (f_1, \dots, f_m)$ and $g = (g_1, \dots, g_m)$ are sections of the trivial bundle $\mathbb{C}^m$ on U, then one has

$$R^f_{|\,|}([V]) = \det A \times R^g_{|\,|}([V]), \tag{3.38}$$

where $|\,|$ is the canonical hermitian metric on $\mathbb{C}^m$.

PROOF. The proof we propose here is very similar to the proof of Proposition 3.18 and is due to R. Lärkäng [**Lar12**, Theorem 7.1]. Let $(\epsilon_1, \dots, \epsilon_m)$ be the canonical basis of the trivial bundle $\mathbb{C}^m$ over U, $f = \sum_{j=1}^m f_j\, \epsilon_j$, $g = Af = \sum_{j=1}^m g_j \epsilon_j$. Assume also, for the time being, that the formal expressions

$$\sigma_f = \sum_{j=1}^{m} \frac{\overline{f_j}}{2i\pi |f|^2}\, \epsilon_j^*, \qquad \sigma_g = \sum_{j=1}^{m} \frac{\overline{g_j}}{2i\pi |g|^2}\, \epsilon_j^*,$$
$$Q^f = \sum_{r\ge 1} \sigma_f\, (\overline{\partial} \sigma_f)^{r-1}, \qquad Q^g = \sum_{r\ge 1} \sigma_g\, (\overline{\partial} \sigma_g)^{r-1}$$

are interpreted as smooth, bundle-valued forms in $U \setminus V(f)$. We also denote as A' the transpose of A, considered as an endomorphism of the trivial bundle $\mathbb{C}^m$. Let also

$$Q^g_A = \sum_{r\ge 1} (A'\sigma_g)\, \big(\overline{\partial}(A'\sigma_g)\big)^{r-1}.$$

Since $g = Af$ on V and $A'(\overline{\partial}\sigma_g) = \overline{\partial}(A'\sigma_g)$ on $V \setminus V(g)$ due to the holomorphy of A, one has

$$\nabla_f(Q^g_A\, [V]) = 1$$

on $V \setminus V(g)$. It follows from Leibniz's rule (after repeating the computation contained in the proof of Proposition 3.18, see (3.30) and (3.32)) that for $\lambda \in \mathbb{C}$ with $\mathrm{Re}\,\lambda \gg 1$,

$$\begin{aligned} \nabla_f\big(|g|^{2\lambda}\, Q^g_A\, Q^f\, [V]\big) &= \nabla_f\big(|g|^{2\lambda}\, Q^g_A\, [V]\big)\, Q^f - |g|^{2\lambda}\, Q^g_A\, \nabla_f\big(Q^f\, [V]\big) \\ &= \nabla_f\big(|g|^{2\lambda}\, Q^g_A\, [V]\big)\, Q^f - |g|^{2\lambda}\, Q^g_A \\ &= |g|^{2\lambda}\, (Q^f - Q^g_A)\, [V] - \overline{\partial}|g|^{2\lambda}\, Q^g_A\, Q^f\, [V]. \end{aligned} \tag{3.39}$$

Since $V \cap V(f) = V \cap V(g)$, the maps $\lambda \longmapsto |g|^{2\lambda}\, Q^f\, [V]$ and $\lambda \mapsto |f|^{2\lambda} Q^f\, [V]$ extend as meromorphic maps to $\mathbb{C}$ with polar sets in $-\mathbb{Q}^+$. The value of the meromorphic continuation of $\lambda \mapsto |f|^{2\lambda} Q^f\, [V]$ at $\lambda = 0$ equals the Bochner–Martinelli principal value current $P^f_{|\,|}([V])$; see Proposition 3.15. For the same reason, the value of the meromorphic continuation of the map $\lambda \longmapsto |g|^{2\lambda} Q_A^g [V]$ at $\lambda = 0$ equals the Bochner–Martinelli principal value current $A' P^g_{|\ |}([V])$. The map

$$\lambda \to \overline{\partial}|g|^{2\lambda} Q_A^g Q^f [V] \tag{3.40}$$

also has a meromorphic continuation to $\mathbb{C}$, with poles in $-\mathbb{Q}^+$, whose value at $\lambda = 0$ is a pseudo-meromorphic current. To see that, one needs to use a log resolution $\pi : \mathcal{V} \to V$, as in Theorem A.41, for $\mathcal{X} = V \cap U$ and for $\mathcal{Y}$ being equal to the union of $V \cap V(f)$ and a hypersurface $\mathcal{G}$ of V containing $V_{\rm sing}$. If $z_0 \in \mathcal{V}$ is in the pre-image of an open neighborhood of $a \in V \cap V(f)$ in U through $\iota_V \circ \pi$, then in a local chart around z_0 in $\mathcal{V}$, one has

$$(\iota_V \circ \pi)^* f = \mathfrak{s}_{f,0}\, \mathfrak{s}'_f, \quad (\iota_V \circ \pi)^* g = \mathfrak{s}_{g,0}\, \mathfrak{s}'_g,$$

where $\mathfrak{s}'_f$ and $\mathfrak{s}'_g$ are holomorphic nonvanishing sections of the trivial bundle $\mathbb{C}^m$ and $\mathfrak{s}_{f,0}$, $\mathfrak{s}_{g,0}$ are holomorphic functions such that $\mathfrak{s}_{f,0}$ divides $\mathfrak{s}_{g,0}$ (since one has $g_a \in \sum_{j=1}^m f_j \mathcal{O}_{U,a} + \mathcal{I}_{V,a}$). As a consequence, in the local chart around z_0, the map

$$\lambda \longmapsto (\iota_V \circ \pi)^* \big(\overline{\partial}|g|^{2\lambda} Q_A^g Q^f\big)$$

is expressed as a linear combination of maps

$$\lambda \longmapsto \overline{\partial}|\mathfrak{s}_{g,0}\mathfrak{s}'_g|^{2\lambda} \frac{\varpi_{k,\ell}}{\mathfrak{s}^k_{f,0}\, \mathfrak{s}^\ell_{g,0}} = \overline{\partial}|\mathfrak{s}_{g,0}\mathfrak{s}'_g|^{2\lambda} \Big(\frac{\mathfrak{s}_{g,0}}{\mathfrak{s}_{f,0}}\Big)^k \frac{\varpi_{k,\ell}}{\mathfrak{s}^{k+\ell}_{g,0}}, \quad k, \ell \in \mathbb{N},$$

where $\varpi_{k,\ell}$ is a smooth $(0,\bullet)$-form. Therefore, the map (3.40) extends to $\mathbb{C}$ as a meromorphic map with poles in $-\mathbb{Q}^+$. Its value at $\lambda = 0$ is a pseudo-meromorphic current supported by $V \cap V(f)$. The same reasoning that was invoked for the maps

$$\lambda \longmapsto |s|^{2\lambda} Q_{|\ |} Q_{\|\ \|} [V],$$
$$\lambda \longmapsto (\overline{\partial}|s|^{2\lambda}) Q_{|\ |} \big[\boldsymbol{P}^{\boldsymbol{\mu}}_{|\ |}\big]_{\boldsymbol{\mu}=0}$$

in the proof of Proposition 3.18 and the dimension principle for pseudo-meromorphic currents (Proposition 2.26) imply that

$$\Big[\overline{\partial}|g|^{2\lambda} Q_A^g Q^f [V]\Big]_{\lambda=0}.$$

Comparing the action of ∇_f on both sides of the identity (3.39), then using the analytic continuation principle, one deduces formula (3.38)

$$\begin{aligned} 0 = \nabla_f\big(P^f_{|\ |}([V]) - A' P^g_{|\ |}([V])\big) &= \nabla(P^f_{|\ |} - A' P^g_{|\ |}) [V] \\ &= \big((1 - R^f_{|\ |}) - (1 - A' R^g_{|\ |})\big) [V] \\ &= A' R^g_{|\ |}([V]) - R^f_{|\ |}([V]). \end{aligned}$$ □

We can now state the following very important result.[11]

THEOREM 3.21. *Let $\mathcal{X}$, V, D_j and $\mathbf{D}_j$, E, $\mathbf{E}$, s_j and $\mathbf{S}_j$, s and $\mathbf{S}$, H in $\mathcal{O}(\mathcal{X}, {\rm Hom}_{\mathbb{C}}(E, \mathbf{E}))$ satisfying $\mathbf{S} = Hs$, $\Delta = \det H$ and Δ' be as in the preamble of this subsection. Then*

$$R^s_{|\ |}([V]) = R^{\mathbf{S}}_{\|\ \|}([V]) = \overline{\partial}\Big(\frac{1}{(s_1, \dots, s_n)}\Big)_V = \Delta' \overline{\partial}\Big(\frac{1}{(\mathbf{S}_1, \dots, \mathbf{S}_n)}\Big)_V \tag{3.41}$$

for any choice of smooth metrics $|\ |$, $\|\ \|$ on E and $\mathbf{E}$, respectively.

[11]Theorem 3.21 is classically known as the *geometric transformation law*, more precisely its reformulation in terms of currents.

REMARK 3.22. When $V = \mathcal{X}$, a proof of this result can be found in [**DS96**, Proposition 2.2]. It relies on the fibered residue theorem [**CoH78**, Theorem 4.4.2]. In the case when $V = \mathcal{X}$ and $m = N$ (the so-called *absolute case*), Theorem 3.21 provides a current re-interpretation and generalization of the *transformation law for Grothendieck residues*, as formulated in algebraic geometry ([**GrH**, §7], [**Gri76**]).

PROOF. It follows from Proposition 3.18 that it is enough to prove the first equality. Lemma 3.19 allows one to assume that $V \cap V(s) = V \cap V(\mathbf{S})$. Thus, it is enough to prove the result in a local chart U around $a \in V \cap V(s)$. Above this chart U all bundles $\mathcal{O}(D_j)$ and $\mathcal{O}(\mathbf{D}_j)$ are trivial. Also, we note that the currents $R^s_{|\ |}([V])$ or $R^{\mathbf{S}}_{\|\ \|}([V])$ do not depend on the choice of the metrics on E and $\mathbf{E}$ (see Proposition 3.18). Hence, the local situation reduces to the case described precisely in Lemma 3.20. □

One can also formulate a useful generalized version of Theorem 3.21 as follows. We take $\mathcal{X}$ and V, with $\operatorname{codim} V = M$, as above, and let $m, k \in \mathbb{N}^*$ be such that $m + k \leq N - M$. In addition to the Cartier divisors $D_1, \dots, D_m, \mathbf{D}_1, \dots, \mathbf{D}_m$ filling this picture, we introduce k additional ones, $T_1, \dots, T_k$. Consider now holomorphic sections $s_1, \dots, s_m, \mathbf{S}_1, \dots, \mathbf{S}_m$, and $t_1, \dots, t_k$ of the $(2m+k)$-line bundles $\mathcal{O}(D_1), \dots, \mathcal{O}(D_m)$, $\mathcal{O}(\mathbf{D}_1), \dots, \mathcal{O}(\mathbf{D}_m)$, $\mathcal{O}(T_1), \dots, \mathcal{O}(T_k)$, respectively. We denote by $\mathfrak{E} = \bigoplus_{\kappa=1}^{k} \mathcal{O}(T_\kappa)$ the corresponding vector bundle, together with its section $t = t_1 \oplus \cdots \oplus t_k$. We assume first that

$$
\begin{aligned}
&\operatorname{codim}_V(V \cap V(s)) = \operatorname{codim}_V(V \cap V(\mathbf{S})) = m, \\
&\operatorname{codim}_V\big(V \cap V(s) \cap V(t)\big) = \operatorname{codim}_V\big(V \cap V(\mathbf{S}) \cap V(t)\big) = m + k, \\
&\operatorname{codim}_V\big(V \cap V(s) \cap V(t_\kappa)\big) > m \quad \forall \kappa = 1, \dots, k.
\end{aligned}
\tag{3.42}
$$

Furthermore, we assume that there exists a meromorphic section H of $\operatorname{Hom}_{\mathbb{C}}(E, \mathbf{E})$ such that $\mathbf{S} = Hs$, together with positive integers $q_1, \dots, q_k$, with $q = (q_1, \dots, q_k)$ such that

$$
\Delta = t_1^{q_1} t_2^{q_2} \cdots t_k^{q_k} \det H \in \mathcal{O}\big(\mathcal{X}, \mathcal{O}(\langle\!\langle q, T \rangle\!\rangle) \otimes \operatorname{Hom}_{\mathbb{C}}(\det E, \det \mathbf{E})\big).
\tag{3.43}
$$

PROPOSITION 3.23. *Let $\chi_1, \dots, \chi_k \in \mathbb{N}$. Under the previous hypothesis and the additional hypothesis $V \cap V(s) \subset V \cap V(\mathbf{S})$, one has*

$$
\begin{aligned}
&\overline{\partial}\Big(\frac{1}{\big(s_1, \dots, s_m, t_1^{1+\chi_1}, \dots, t_k^{1+\chi_k}\big)}\Big)_V \\
&\quad = \Delta' \, \overline{\partial}\Big(\frac{1}{\big(\mathbf{S}_1, \dots, \mathbf{S}_m, t_1^{1+\chi_1+q_1}, \dots, t_k^{1+\chi_k+q_k}\big)}\Big)_V.
\end{aligned}
\tag{3.44}
$$

PROOF. Choose metrics $|\ |$ on each line bundle $\mathcal{O}(D)$ involved in the setting. Each such metric induces a metric on all $\mathcal{O}(\ell D)$ for $\ell \in \mathbb{N}^*$. Given these metrics,

$\mu \in \mathbb{C}^m$, $\nu \in \mathbb{C}^k$, consider the following expressions:

$$
\begin{aligned}
\mathrm{R}^{s,\mu}_{|\,|} &= \overline{\partial}\Big(\frac{|s_m|^{2\mu_m}}{s_m}\Big) \wedge \cdots \wedge \overline{\partial}\Big(\frac{|s_1|^{2\mu_1}}{s_1}\Big),\\
\mathrm{R}^{\boldsymbol{S},\mu}_{|\,|} &= \overline{\partial}\Big(\frac{|\mathbf{S}_m|^{2\mu_m}}{\mathbf{S}_m}\Big) \wedge \cdots \wedge \overline{\partial}\Big(\frac{|\mathbf{S}_1|^{2\mu_1}}{\mathbf{S}_1}\Big),\\
\mathrm{R}^{t,\nu}_{|\,|} &= \overline{\partial}\Big(\frac{|t_k|^{2\nu_k}}{t_k^{1+\chi_k}}\Big) \wedge \cdots \wedge \overline{\partial}\Big(\frac{|t_1|^{2\nu_1}}{t_1^{1+\chi_1}}\Big),\\
\mathfrak{R}^{\nu}_{|\,|} &= \overline{\partial}\Big(\frac{|t_k|^{2\nu_k}}{t_k^{1+\chi_k+q_k}}\Big) \wedge \cdots \wedge \overline{\partial}\Big(\frac{|t_1|^{2\nu_1}}{t_1^{1+\chi_1+q_1}}\Big)\,\Delta'.
\end{aligned}
\tag{3.45}
$$

It follows from Theorem 2.75 that in a neighborhood of $V \cap V(s+t)$, the map

$$(\mu,\nu) \longmapsto \mathrm{R}^{t,\nu}_{|\,|}\,\mathrm{R}^{s,\mu}_{|\,|}\,[V]$$

extends as a meromorphic map on $\mathbb{C}^m \times \mathbb{C}^k$, whose polar set does not intersect

$$\{(\mu,\nu)\,:\, \mathrm{Re}\,\mu_j \geq 0, \mathrm{Re}\,\nu_\kappa \geq 0, 1 \leq j \leq m,\ 1 \leq \kappa \leq k\}.$$

Moreover, its value at $(0,0)$ equals the current

$$\overline{\partial}\Big(\frac{1}{(s_1,\dots,s_m,t_1^{1+\chi_1},\dots,t_k^{1+\chi_k})}\Big)_V.$$

Fix now $\nu \in \mathbb{C}^k$ with $\mathrm{Re}\,\nu_\kappa \gg 1$ for any $\kappa = 1,\dots,k$. Consider the map

$$\mu \longmapsto \mathfrak{R}^{\nu}_{|\,|}\,\mathrm{R}^{\boldsymbol{S},\mu}_{|\,|}\,[V].$$

It follows from Theorem 3.21 that its value at $\mu = 0$ equals the current

$$\mathrm{R}^{t,\nu}_{|\,|}\Big[\mathrm{R}^{s,\mu}_{|\,|}\Big]_{\mu=0}[V],$$

which is supported by $V \cap V(s)$. On the other hand, Theorem 2.75 implies that in a neighborhood of $V \cap V(\boldsymbol{S}) \cap V(t)$, containing $V \cap V(s) \cap V(t)$ by hypothesis, the map

$$(\mu,\nu) \longmapsto \mathfrak{R}^{\nu}_{|\,|}\,\mathrm{R}^{\mathbf{S},\mu}_{|\,|}\,[V]$$

extends as a meromorphic map to $\mathbb{C}^m \times \mathbb{C}^k$. The polar set of the meromorphic extension does not intersect

$$\{(\mu,\nu)\,:\, \mathrm{Re}\,\mu_j \geq 0,\ \mathrm{Re}\,\nu_\kappa \geq 0 \text{ for } 1 \leq j \leq m,\ 1 \leq \kappa \leq k\}.$$

Moreover, its value at $(0,0)$ equals the current

$$\Delta'\,\overline{\partial}\Big(\frac{1}{(\mathbf{S}_1,\dots,\mathbf{S}_m,t_1^{1+\chi_1+q_1},\dots,t_k^{1+\chi_k+q_k})}\Big)_V.$$

Then, one has, as a consequence of the principle of analytic continuation, that the equality

$$
\begin{aligned}
\overline{\partial}\Big(\frac{1}{(s_1,\dots,s_m,t_1^{1+\chi_1},\dots,t_k^{1+\chi_k})}\Big)_V &= \Big[\mathrm{R}^{t,\nu}_{|\,|}\Big[\mathrm{R}^{s,\mu}_{|\,|}\Big]_{\mu=0}[V]\Big]_{\nu=0}\\
= \Big[\mathfrak{R}^{\nu}_{|\,|}\,\mathrm{R}^{\mathbf{S},\mu}_{|\,|}\,[V]\Big]_{\mu=\nu=0} &= \Delta'\,\overline{\partial}\Big(\frac{1}{(\mathbf{S}_1,\dots,\mathbf{S}_m,t_1^{1+\chi_1+q_1},\dots,t_k^{1+\chi_k+q_k})}\Big)_V
\end{aligned}
$$

holds in a neighborhood of any point in $V \cap V(s) \cap V(t)$. On the other hand, both currents equal to 0 in a neighborhood of a point in $V \cap (V(\boldsymbol{S}) \setminus V(s)) \cap V(t)$. This concludes the proof of the proposition. $\square$

3.3. Bochner–Martinelli currents and generically exact complexes

Let $(\mathcal{X}, \mathcal{O})$ be an N-dimensional complex manifold equipped with its structure sheaf. We consider in this section, in place of a hermitian vector bundle $(E, |\ |)$, a complex $\mathbf{F}_\bullet$ of holomorphic vector bundles E_ℓ, together with morphisms F_ℓ between such bundles,

$$(3.46) \qquad 0 \longrightarrow E_L \xrightarrow{F_L} E_{L-1} \xrightarrow{F_{L-1}} \cdots \xrightarrow{F_2} E_1 \xrightarrow{F_1} E_0 \xrightarrow{F_0} E_{-1} = \frac{E_0}{\operatorname{Im} F_1} \longrightarrow 0.$$

As usual, F_0 denotes the quotient map,

$$\operatorname{Im}(F_\ell) \subset \operatorname{Ker}(F_{\ell-1})$$

for any $\ell = 0, \dots, L$, and 0 stands for the null vector bundle $\mathcal{X} \times \{0\}$. We assume that the complex $\mathbf{F}_\bullet$ of holomorphic bundles is metrized. More precisely, each entry E_ℓ, $\ell \geq 0$, is equipped with a smooth hermitian metric[12] $|\ | = |\ |_\ell$. Such a metrized complex $\overline{\mathbf{F}_\bullet}$ is paired with a closed analytic subset V with pure codimension M in $\mathcal{X}$. Under some conditions to be specified and illustrated in the next subsection, one will attach in §3.3.2 to such a complex $\mathbf{F}_\bullet$ of holomorphic bundles and to V a residue and a principal value current

$$R^{\mathbf{F}_\bullet}_{|\ |}([V]), \quad P^{\mathbf{F}_\bullet}_{|\ |}([V]),$$

of the Bochner–Martinelli type. Our reference model is the Koszul complex $\mathbf{K}^s_\bullet$ attached to a section s of a holomorphic bundle E, once it is equipped with a smooth hermitian metric. In §3.3.3, we will consider the particular case $V = \mathcal{X}$ to profit from Dolbeault's lemma there and be able to explain why the residue current $R^{\mathbf{F}_\bullet}_{|\ |} = R^{\mathbf{F}_\bullet}_{|\ |}([\mathcal{X}])$ materializes the obstruction for exactness of the complex $\mathbf{F}_\bullet$ over $\mathcal{X}$. Both construction and reasoning are inspired by the strategy outlined previously in the univariate case; see §1.2.5.

3.3.1. Generically exact Koszul and Eagon–Northcott complexes.

The concept of a *generically exact complex* and its connection to the construction of attached residue or principal value currents was introduced and described in [**And04**].

DEFINITION 3.24. A complex $\mathbf{F}_\bullet$ of holomorphic bundles (E_ℓ, F_ℓ), where r_ℓ denotes the rank of E_ℓ for $\ell \geq -1$, is said to be generically exact if and only if there exists a closed analytic subset $V(\mathbf{F}_\bullet)$, which we will always assume minimal with respect to inclusion, satisfying

$$(3.47) \qquad \forall z \in \mathcal{X} \setminus V(\mathbf{F}_\bullet),\ \forall \ell = 1, \dots, L, \quad \operatorname{rank} F_\ell(z) = \sum_{\nu \geq 0} (-1)^\nu r_{\ell+\nu}.$$

This amounts to saying that the restriction of $\mathbf{F}_\bullet$ to the open subset $\mathcal{X} \setminus V(\mathbf{F}_\bullet)$ is an exact complex of holomorphic bundles, namely $\operatorname{Im}(F_\ell) = \operatorname{Ker}(F_{\ell-1})$ over $\mathcal{X} \setminus V(\mathbf{F}_\bullet)$ for $\ell = 1, \dots, L$.

---

[12] The metric on E_{-1} is induced by that on E_0.

Given a generically exact complex $\mathbf{F}_\bullet$, one denotes by

$$
\begin{aligned}
&E(\mathbf{F}_\bullet) = E = \bigoplus_{\ell\geq -1} E_\ell, \\
&F = \bigoplus_{\ell\geq -1} F_\ell, \quad \nabla_{\mathbf{F}_\bullet} = 2i\pi\, F - \overline{\partial}.
\end{aligned}
\tag{3.48}
$$

Furthermore, we assume that the bundle E is equipped with $\mathbb{F}_2$-grading

$$
E = E^+ \oplus E^- = \Big(\bigoplus_{\ell\geq 0} E_{2\ell}\Big) \oplus \Big(\bigoplus_{\ell\geq 0} E_{2\ell-1}\Big). \tag{3.49}
$$

This grading induces a natural $\mathbb{F}_2$-grading on $\mathrm{Hom}_{\mathbb{C}}\big(E(\mathbf{F}_\bullet), E(\mathbf{F}_\bullet)\big)$. Then, the morphism F sending E^+ to E^- and E^- to E^+ is an odd map as the operator $\overline{\partial}$ is, when both morphisms are considered as acting in either of the four ways

$$
\begin{gathered}
{}'\mathcal{D}(U,E) \xrightarrow{F,\overline{\partial}} {}'\mathcal{D}(U,E) \quad \text{or} \quad {}'\mathcal{D}\big(U, \mathrm{Hom}_{\mathbb{C}}(E,E)\big) \xrightarrow{\boldsymbol{F},\boldsymbol{\overline{\partial}}} {}'\mathcal{D}\big(U, \mathrm{Hom}_{\mathbb{C}}(E,E)\big), \\
C^\infty_{\bullet,\bullet}(U,E) \xrightarrow{F,\overline{\partial}} C^\infty_{\bullet,\bullet}(U,E) \quad \text{or} \quad C^\infty_{\bullet,\bullet}\big(U, \mathrm{Hom}_{\mathbb{C}}(E,E)\big) \xrightarrow{\boldsymbol{F},\boldsymbol{\overline{\partial}}} C^\infty_{\bullet,\bullet}\big(U, \mathrm{Hom}_{\mathbb{C}}(E,E)\big)
\end{gathered}
$$

when U is an open subset of $\mathcal{X}$. Note that both morphisms are denoted either as $F, \overline{\partial}$ or $\boldsymbol{F}, \boldsymbol{\overline{\partial}}$, depending whether they are considered as acting on E-valued sections or acting on $\mathrm{Hom}_{\mathbb{C}}(E,E)$-valued ones. Since F is holomorphic, one has

$$
\begin{aligned}
&F\,\overline{\partial} = -\overline{\partial}\,F, \quad \nabla^2_{\mathbf{F}_\bullet} = (2i\pi\,F - \overline{\partial})^2 = -2i\pi\,(F\,\overline{\partial} + \overline{\partial}\,F) = F^2 = \overline{\partial}^2 = 0, \\
&\boldsymbol{F}\,\boldsymbol{\overline{\partial}} = -\boldsymbol{\overline{\partial}}\,\boldsymbol{F}, \quad \boldsymbol{\nabla}^2_{\mathbf{F}_\bullet} = (2i\pi\,\boldsymbol{F} - \boldsymbol{\overline{\partial}})^2 = -2i\pi\,(\boldsymbol{F}\,\boldsymbol{\overline{\partial}} + \boldsymbol{\overline{\partial}}\,\boldsymbol{F}) = \boldsymbol{F}^2 = \boldsymbol{\overline{\partial}}^2 = 0
\end{aligned}
\tag{3.50}
$$

for such actions. Note also that $\boldsymbol{\nabla}_{\mathbf{F}_\bullet}(\mathrm{Id}_{E(\mathbf{F}_\bullet)}) = 0$.

EXAMPLE 3.25 (Koszul complex). Given a holomorphic bundle E with rank m and $s \in \mathcal{O}(\mathcal{X}, E)$, the Koszul complex

$$
0 \to \bigwedge^m E^* \xrightarrow{(\rfloor_s)_m} \bigwedge^{m-1} E^* \cdots \xrightarrow{(\rfloor_s)_{\ell+1}} \bigwedge^{\ell+1} E^* \xrightarrow{(\rfloor_s)_\ell} \bigwedge^{\ell} E^* \xrightarrow{(\rfloor_s)_{\ell-1}} \cdots \xrightarrow{(\rfloor_s)_2} E^* \xrightarrow{(\rfloor_s)_1} \mathbb{C} \tag{3.51}
$$

is by far the most tractable among the generically exact complexes which will be considered in this monograph. It is denoted by $\mathbf{K}^s_\bullet$, and $\rfloor_s$ is the interior multiplication at each level ℓ such as introduced in (1.80). One has $V(\mathbf{K}^s_\bullet) = V(s)$. The operators $F, \boldsymbol{F}, \nabla_{\mathbf{F}_\bullet}, \boldsymbol{\nabla}_{\mathbf{F}_\bullet}$ are denoted in this case as $\rfloor_s$, $\boldsymbol{\rfloor_s}$, $\nabla_{\mathbf{K}^s_\bullet} = 2i\pi\rfloor_s - \overline{\partial}$, $\boldsymbol{\nabla}_{\mathbf{K}^s_\bullet} = 2i\pi\boldsymbol{\rfloor_s} - \boldsymbol{\overline{\partial}}$.

EXAMPLE 3.26 (Eagon–Northcott complex [**EaN62**]). Let E be a holomorphic vector bundle with rank m and $s \in \mathcal{O}(\mathcal{X}, E)$. Fix $k \in \mathbb{N}^*$. Let $(\mathbf{K}^s_\bullet)_1, \ldots, (\mathbf{K}^s_\bullet)_k$ be k copies of the Koszul complex $\mathbf{K}^s_\bullet$. Then one has for each $\kappa = 1, \ldots, k$ and for each $\ell \in \mathbb{N}$ that

$$
((\mathbf{K}^s_\bullet)_\kappa)_\ell = \Big(\bigwedge^\ell E^*\Big)_\kappa.
$$

For each $\ell \in \mathbb{N}^*$, let

$$(3.52)\quad \begin{aligned}(\mathbf{EN}^{s,k}_\bullet)_\ell &= \bigoplus_{\{\nu\in\mathbb{N}^k\,:\,|\nu|=\ell-1\}} \bigwedge_{\kappa=1}^{k}\Big(\bigwedge^{\nu_\kappa+1} E^*\Big)_\kappa \\ &= \bigoplus_{\{\nu\in\mathbb{N}^k\,:\,|\nu|=\ell-1\}} \bigwedge_{\kappa=1}^{k}((\mathbf{K}^s_\bullet)_\kappa)_{\nu_\kappa+1}.\end{aligned}$$

In particular, for $\ell = 1$,

$$(\mathbf{EN}^{s,k}_\bullet)_1 = (E^*)_1 \wedge \cdots \wedge (E^*)_k.$$

The morphism $F^{s,k}_\ell : (\mathbf{EN}^{s,k}_\bullet)_\ell \longrightarrow (\mathbf{EN}^{s,k}_\bullet)_{\ell-1}$ is defined by

$$F^{s,k}_1 = (\rfloor_s)_1 \wedge \cdots \wedge (\rfloor_s)_k \ : (\mathbf{EN}^{s,k}_\bullet)_1 \longmapsto \mathbb{C},$$

and for any $\ell \geq 2$,

$$F^{s,k}_\ell = \sum_{\kappa=1}^{k}\Big(\sum_{\nu\geq 2}((\rfloor_s)_\kappa)_\nu\Big)_{|(\mathbf{EN}^{s,k}_\bullet)_\ell}.$$

Thus, we construct a generically exact complex $\mathbf{EN}^{s,k}_\bullet$

$$(3.53)\quad \begin{aligned}&\cdots \xrightarrow{F^{s,k}_{\ell+1}} (\mathbf{EN}^{s,k}_\bullet)_\ell \xrightarrow{F^{s,k}_\ell} (\mathbf{EN}^{s,k}_\bullet)_{\ell-1} \xrightarrow{F^{s,k}_{\ell-1}} \cdots \\ &\cdots \xrightarrow{F^{s,k}_3} (\mathbf{EN}^{s,k}_\bullet)_2 \xrightarrow{F^{s,k}_2} (\mathbf{EN}^{s,k}_\bullet)_1 = \bigwedge_{\kappa=1}^{k}(E^*)_\kappa \xrightarrow{F^{s,k}_1} \mathbb{C}.\end{aligned}$$

Observe that again $V(\mathbf{EN}^{k,s}_\bullet) = V(s)$ and that one has in terms of sheaves of sections

$$\mathcal{O}(\,\cdot\,, \mathrm{Im}\,(F^{s,k}_1)) = \big(\rfloor_s\mathcal{O}(\,\cdot\,, E^*)\big)^k,$$

relating the Eagon–Northcott complex $\mathbf{EN}^{s,k}_\bullet$ to the kth power of the ideal sheaf $\mathcal{O}(\,\cdot\,, \rfloor_s E^*)$.

EXAMPLE 3.27 (Locally free resolutions of coherent sheaves of $\mathcal{O}$-modules). Let $\mathcal{F}$ be a coherent sheaf of $\mathcal{O}$-modules; see §A.2.1. For every $z \in \mathcal{X}$, one can associate to $\mathcal{F}$ (see Definition A.8) a free resolution for $\mathcal{F}$ over some neighborhood U of z. That is, one can associate to $\mathcal{F}$ a generically exact sequence of trivial vector bundles over U

(3.54)

$$0 \longrightarrow U\times\mathbb{C}^{r_L} \xrightarrow{F_L} U\times\mathbb{C}^{r_{L-1}} \xrightarrow{F_{L-1}} \cdots \xrightarrow{F_3} U\times\mathbb{C}^{r_2} \xrightarrow{F_2} U\times\mathbb{C}^{r_1} \xrightarrow{F_1} U\times\mathbb{C}^{r_0},$$

where L and the ranks r_ℓ, $\ell = 0,\dots,L$, the morphisms F_ℓ, $\ell = 1,\dots,L$ depend on z and are such that the corresponding sequence of locally free sheaves of sections

$$(3.55)\quad 0 \longrightarrow \mathcal{O}_U^{\oplus r_L} \xrightarrow{F_L} \mathcal{O}_U^{\oplus r_{L-1}} \xrightarrow{F_{L-1}} \cdots \xrightarrow{F_3} \mathcal{O}_U^{\oplus r_2} \xrightarrow{F_2} \mathcal{O}_U^{\oplus r_1} \xrightarrow{F_1} \mathcal{O}_U^{\oplus r_0},$$

where $\mathrm{coker}(F_1) \simeq \mathcal{F}_{|U}$, is exact. If the coherent sheaf $\mathcal{F}$ admits at the global level a finite, locally free resolution[13]

$$(3.56)\qquad 0 \longrightarrow \mathscr{L}^{r_L}_{\mathcal{X}} \xrightarrow{\mathfrak{f}_L} \mathscr{L}^{r_L-1}_{\mathcal{X}} \xrightarrow{\mathfrak{f}_{L-1}} \cdots \xrightarrow{\mathfrak{f}_3} \mathscr{L}^{r_2}_{\mathcal{X}} \xrightarrow{\mathfrak{f}_2} \mathscr{L}^{r_1}_{\mathcal{X}} \xrightarrow{\mathfrak{f}_1} \mathscr{L}^{r_0}_{\mathcal{X}}$$

[13]Such is the case, for example, when $U = U_\nu$ is a convenient neighborhood of a holomorphically convex compact subset taken from a sequence $(K_\nu)_{\nu\geq 1}$ of compact subsets exhausting a Stein manifold; see Proposition A.12.

with $\operatorname{coker}(\mathfrak{f}_1) \simeq \mathcal{F}$, then one can consider each locally free sheaf $\mathscr{L}_{\mathcal{X}}^{r_\ell}$ with constant rank r_ℓ as the sheaf of sections $\mathcal{O}(\cdot\,, E_\ell)$ of a holomorphic r_ℓ-bundle E_ℓ. Then, one can attach to the above finite free resolution for $\mathcal{F}$ the corresponding generically exact complex (3.46) with $V(\mathbf{F}_\bullet) = \operatorname{Supp}\mathcal{F}$. A particularly important case appears when $\mathcal{X}$ is a Stein manifold and $\mathcal{F} = \mathcal{O}^{\oplus^r}/\mathcal{J}$, where $\mathcal{J}$ is a coherent subsheaf of $\mathcal{O}^{\oplus^r}$ ideals. In this setting, if one replaces $\mathcal{X}$ by an open subset U_ν from an exhausting sequence, then one chooses the finite, locally free resolution (3.56) which obeys the conditions $\mathscr{L}_{\mathcal{X}}^{r_0} = \mathcal{O}^{\oplus^r}$ and $\operatorname{Im}\mathfrak{f}_1 = \mathcal{J}$; see Proposition A.12. Then the corresponding generically exact complex of trivial bundles is such that one has $V(\mathbf{F}_\bullet) = V(\mathcal{J}) = \operatorname{Supp}(\mathcal{O}^{\oplus^r}/\operatorname{Im}\mathfrak{f}_1)$. When $s \in \mathcal{O}(\mathcal{X}, E)$, with $\operatorname{rank} E = \operatorname{codim} V(s)$ in addition, the Koszul complex $\mathbf{K}_\bullet^s$ (see Example 3.25) provides an example a locally free resolution of $\mathcal{O}/\mathcal{I}_s$. If the sequence of locally free sheaves of sections attached to the Eagon–Northcott complex $\mathbf{EN}_\bullet^{s,k}$ (see Example 3.26) happens to be exact, it provides a locally free resolution for $\mathcal{O}/(\mathcal{I}_s)^k$, as is the case when $\operatorname{codim} V(s) = \operatorname{rank}(E)$. More generally, in polynomial algebra, Eagon–Northcott complexes are used to provide locally free resolutions for *determinantal sheaves of ideals*, when the associated sequence of locally free sheaves of sections attached to them is exact. Recall that a *determinantal sheaf of ideals* in polynomial algebra is defined by the minors of the maximal rank of a matrix with polynomial entries.

3.3.2. The currents $R^{\overline{\mathbf{F}_\bullet}}$ and $P^{\overline{\mathbf{F}_\bullet}}$ for a metrized complex $\overline{\mathbf{F}_\bullet}$. Let $\mathcal{X}$ be an N-dimensional complex manifold, let V be a closed analytic subset with pure codimension equal to $M < N$, let E be a holomorphic bundle, and let $s \in \mathcal{O}(\mathcal{X}, E)$. Let $F : \mathbb{C} \to E$ be the morphism of bundles over $\mathcal{X}$ defined by

$$F(z)(t) = t\,s(z) \in E_z \quad \forall z \in \mathcal{X},\ \forall t \in \mathbb{C}_z.$$

Let also F_0 be the quotient morphism from E to $E/\operatorname{Im} F$. Then the short complex

$$\mathbf{F}_\bullet^s : 0 \longrightarrow \mathbb{C} \xrightarrow{F} E \xrightarrow{F_0} E/\operatorname{Im} F \longrightarrow 0$$

is the most elementary model of the generically exact complex, with $V(\mathbf{F}_\bullet^s) = V(s)$. Let $|\ |$ be a smooth hermitian metric on E, which allows us to metrize the complex $\mathbf{F}_\bullet^s$ as $\overline{\mathbf{F}_\bullet^s}$. In this subsection we adapt the construction of the currents $R^s_{|\ |}([V])$ and $P^s_{|\ |}([V])$ described in Proposition 3.15 to the case of generically exact complex $(\mathbf{F}_\bullet, |\ |)$ of metrized holomorphic bundles. From the prototypical examples where $\overline{\mathbf{F}_\bullet} = \overline{\mathbf{F}_\bullet^s}$ or $\overline{\mathbf{F}_\bullet} = \overline{\mathbf{K}_\bullet^s}$, M. Andersson and E. Wulcan in [**And04**] proposed the construction of Bochner–Martinelli type residue or principal value currents attached to a metrized generically exact complex $\overline{\mathbf{F}_\bullet}$. This construction will prove to be quite helpful when translating division problems in terms of multivariate residue calculus. We present it here, following essentially their reasoning, as it was presented in [**AndW07**, §2]. Let

$$\overline{\mathbf{F}_\bullet} : 0 \longrightarrow \overline{E_L} \xrightarrow{F_L} \overline{E_{L-1}} \xrightarrow{F_{L-1}} \cdots \xrightarrow{F_2} \overline{E_1} \xrightarrow{F_1} \overline{E_0} \xrightarrow{F_0} \overline{E_{-1}} = \overline{\frac{E_0}{\operatorname{Im} F_1}} \longrightarrow 0 \tag{3.57}$$

be a metrized generalized complex. It means that each entry E_ℓ is equipped with a smooth hermitian metric $|\ |$, where the metric on E_{-1} is the quotient metric induced by the metric on E_0. Thus, $\overline{E_\ell} = (E_\ell, |\ |)$. For each $\ell \in \mathbb{N}$, we denote as ρ_ℓ

the generic rank of F_ℓ (see (3.47)) and let

$$\Delta_\ell^{\mathbf{F}_\bullet} := \det{}^{\rho_\ell} F_\ell \in \mathcal{O}\big(\mathcal{X}, \bigwedge^{\rho_\ell} \mathrm{Hom}_{\mathbb{C}}(E_\ell, E_{\ell-1})\big). \tag{3.58}$$

Observe that such a definition ensures that[14]

$$V(\mathbf{F}_\bullet) = \bigcup_{\ell \geq 1} V(\Delta_\ell^{\mathbf{F}_\bullet}), \tag{3.59}$$

where $V(\mathbf{F}_\bullet)$ is the smallest closed analytic subset of $\mathcal{X}$ such that the restriction of the complex $\mathbf{F}_\bullet$ to the open subset $\mathcal{X} \setminus V(\mathbf{F}_\bullet)$ is exact; see Definition 3.24. The fact that each bundle E_ℓ is now equipped with a metric motivates us to introduce the morphism $\sigma_\ell \in \big(\mathrm{Hom}_{\mathbb{C}}(E_{\ell-1}, E_\ell)\big)_{|\mathcal{X}\setminus V(\mathbf{F}_\bullet)}$ for each $\ell \geq 0$. This morphism is defined as the *minimal inverse* of F_ℓ for the choice of the hermitian structures $\overline{E_{\ell-1}}$ and $\overline{E_\ell}$. Namely, the minimality of F_ℓ is expressed in terms the orthogonal splitting given by $(E_{\ell-1})_z = (\mathrm{Im}\, F_\ell)_z \oplus ((\mathrm{Im}\, F_\ell)^\perp)_z$, so that, for all $z \in \mathcal{X} \setminus V(\mathbf{F}_\bullet)$,

$$\begin{aligned}\sigma_\ell(z)(\xi) \\ = \begin{cases} 0 & \text{when } \xi \in ((\mathrm{Im}\, F_\ell)^\perp)_z \\ \eta \text{ such that } |\eta| = \min_{\{\check\eta \in (E_\ell)_z \,:\, F_\ell(z)(\check\eta) = \xi\}} |\check\eta| & \text{when } \xi \in \big(\mathrm{Im}\, F_\ell\big)_z. \end{cases}\end{aligned} \tag{3.60}$$

Note that $\sigma_\ell \in C^\infty\big(\mathcal{X} \setminus V(\mathbf{F}_\bullet), \mathrm{Hom}_{\mathbb{C}}(E_{\ell-1}, E_\ell)\big)$. Let

$$\sigma := \Big(\sum_{\ell \geq 0} \sigma_\ell \,\mathrm{Proj}_{E_{\ell-1}}\Big)_{|\mathcal{X}\setminus V(\mathbf{F}_\bullet)} \in C^\infty\big(\mathcal{X} \setminus V(\mathbf{F}_\bullet), \mathrm{Hom}_{\mathbb{C}}(E(\mathbf{F}_\bullet), E(\mathbf{F}_\bullet))\big). \tag{3.61}$$

One has, by construction, that σ is an odd smooth section of $\mathrm{Hom}_{\mathbb{C}}(E(\mathbf{F}_\bullet), E(\mathbf{F}_\bullet))$ over $\mathcal{X} \setminus V(\mathbf{F}_\bullet)$ and that it satisfies

$$\begin{aligned} \boldsymbol{F}(\sigma) &= \sigma \circ F + F \circ \sigma = \mathrm{Id}_{E(\mathbf{F}_\bullet)}, \\ \overline{\boldsymbol{\partial}}\sigma := \overline{\boldsymbol{\partial}}(\sigma) &= \overline{\partial} \circ \sigma + \sigma \circ \overline{\partial}, \\ \boldsymbol{\nabla}_{\mathbf{F}_\bullet}(\sigma) &= 2i\pi\, \mathrm{Id}_{E(\mathbf{F}_\bullet)} - \overline{\boldsymbol{\partial}}\sigma, \\ \sigma\, \overline{\boldsymbol{\partial}}\sigma &= \overline{\boldsymbol{\partial}}\sigma\, \sigma \end{aligned} \tag{3.62}$$

over $\mathcal{X} \setminus V(\mathbf{F}_\bullet)$. From the third equality in (3.62), one has that

$$\begin{aligned} (\boldsymbol{\nabla}_{\mathbf{F}_\bullet}(\sigma))^{-1} &= \frac{1}{2i\pi} \sum_{r \geq 1} \Big(\overline{\boldsymbol{\partial}}\Big(\frac{\sigma}{2i\pi}\Big)\Big)^{r-1} \\ &\Longrightarrow \sum_{r \geq 1} \frac{\sigma}{2i\pi} \Big(\overline{\boldsymbol{\partial}}\Big(\frac{\sigma}{2i\pi}\Big)\Big)^{r-1} = \sigma\, \big(\boldsymbol{\nabla}_{\mathbf{F}_\bullet}(\sigma)\big)^{-1} \\ &\Longrightarrow \Big(\sum_{r \geq 1} \frac{\sigma}{2i\pi} \Big(\overline{\boldsymbol{\partial}}\Big(\frac{\sigma}{2i\pi}\Big)\Big)^{r-1}\Big)\, \boldsymbol{\nabla}_{\mathbf{F}_\bullet}(\sigma) = \sigma \end{aligned}$$

holds on $\mathcal{X} \setminus V(\mathbf{F}_\bullet)$. Next, one computes the action of $\boldsymbol{\nabla}_{\mathbf{F}_\bullet}$ on both sides of the above equality using the fact that $(\boldsymbol{\nabla}_{\mathbf{F}_\bullet})^2 = 0$. After right-multiplication by

[14]Note that $V(\Delta_0) = \emptyset$, since F_0 is surjective.

$\big(\boldsymbol{\nabla}_{\mathbf{F}_\bullet}(\sigma)\big)^{-1}$ of the results obtained, one gets the following important identity on $\mathcal{X}\setminus V(\mathbf{F}_\bullet)$:

$$\boldsymbol{\nabla}_{\mathbf{F}_\bullet}\Big(\sum_{r\geq 1}\frac{\sigma}{2i\pi}\Big(\overline{\boldsymbol{\partial}}\Big(\frac{\sigma}{2i\pi}\Big)\Big)^{r-1}\Big) = \mathrm{Id}_{E(\mathbf{F}_\bullet)}. \tag{3.63}$$

For $\lambda\in\mathbb{C}$ such that $\mathrm{Re}\,\lambda\gg 1$, let

$$|\Delta^{\mathbf{F}_\bullet}|^{2\lambda} = \Big(\prod_{\ell\geq 1}|\Delta_\ell^{\mathbf{F}_\bullet}|^{2\lambda}\Big)\,\mathrm{Id}_{E(\mathbf{F}_\bullet)}. \tag{3.64}$$

If $\mathbf{E}=\mathrm{Hom}_{\mathbb{C}}(E(\mathbf{F}_\bullet),E(\mathbf{F}_\bullet))$, then one introduces, motivated by the construction of the maps $\lambda\longmapsto R^{s,\lambda}_{|\,|}([V])$ and $\lambda\longmapsto P^{s,\lambda}_{|\,|}([V])$ in (3.21), the $\mathbf{E}$-current valued maps

(3.65)

$$R^{\mathbf{F}_\bullet,\lambda}([V]) : \lambda\longmapsto\Big((\mathrm{Id}_{E(\mathbf{F}_\bullet)}-|\Delta^{\mathbf{F}_\bullet}|^{2\lambda})+(\overline{\boldsymbol{\partial}}\,|\Delta^{\mathbf{F}_\bullet}|^{2\lambda})\sum_{r\geq 1}\frac{\sigma}{2i\pi}\Big(\overline{\boldsymbol{\partial}}\Big(\frac{\sigma}{2i\pi}\Big)\Big)^{r-1}\Big)\,[V],$$

$$P^{\mathbf{F}_\bullet,\lambda}([V]) : \lambda\longmapsto|\Delta^{\mathbf{F}_\bullet}|^{2\lambda}\Big(\sum_{r\geq 1}\frac{\sigma}{2i\pi}\Big(\overline{\boldsymbol{\partial}}\Big(\frac{\sigma}{2i\pi}\Big)\Big)^{r-1}\Big)\,[V],$$

that are well defined for $\lambda\in\mathbb{C}$ satisfying $\mathrm{Re}\,\lambda\gg 1$. The splitting[15] over $\mathcal{X}\setminus V(\mathbf{F}_\bullet)$

$$\begin{aligned}\sum_{r\geq 1}\frac{\sigma}{2i\pi}\Big(\overline{\boldsymbol{\partial}}\Big(\frac{\sigma}{2i\pi}\Big)\Big)^{r-1} &= \sum_{\ell\geq -1}\Big(\sum_{k>\ell}\frac{\sigma_k\,\overline{\boldsymbol{\partial}}\sigma_{k-1}\cdots\overline{\boldsymbol{\partial}}\sigma_{\ell+2}\,\overline{\boldsymbol{\partial}}\sigma_{\ell+1}}{(2i\pi)^{k-\ell}}\Big)\\ &= \sum_{\ell\geq -1}\Big(\sum_{k>\ell}\frac{\overline{\boldsymbol{\partial}}\sigma_k\,\overline{\boldsymbol{\partial}}\sigma_{k-1}\cdots\overline{\boldsymbol{\partial}}\sigma_{\ell+2}\,\sigma_{\ell+1}}{(2i\pi)^{k-\ell}}\Big),\end{aligned} \tag{3.66}$$

implies that the $\mathbf{E}$-current valued maps $\lambda\longmapsto R^{\mathbf{F}_\bullet,\lambda}([V])$ and $\lambda\longmapsto P^{\mathbf{F}_\bullet,\lambda}([V])$ split respectively as

$$\begin{aligned}R^{\mathbf{F}_\bullet,\lambda}([V]) &= (\mathrm{Id}_{E(\mathbf{F}_\bullet)}-|\Delta^{\mathbf{F}_\bullet}|^{2\lambda})\,[V]+\sum_{\ell\geq -1}\Big(\sum_{k>\ell}R^{\mathbf{F}_\bullet,\lambda}_{\ell,k}([V])\Big),\\ P^{\mathbf{F}_\bullet,\lambda}([V]) &= \sum_{\ell\geq -1}\Big(\sum_{k>\ell}P^{\mathbf{F}_\bullet,\lambda}_{\ell,k}([V])\Big),\end{aligned} \tag{3.67}$$

where

$$\begin{aligned}R^{\mathbf{F}_\bullet,\lambda}_{\ell,k}([V]) &= (\overline{\boldsymbol{\partial}}\,|\Delta^{\mathbf{F}_\bullet}|^{2\lambda})\,\frac{\sigma_k\,\overline{\boldsymbol{\partial}}\sigma_{k-1}\cdots\overline{\boldsymbol{\partial}}\sigma_{\ell+2}\,\overline{\boldsymbol{\partial}}\sigma_{\ell+1}}{(2i\pi)^{k-\ell}}\,[V]\\ &= (\overline{\boldsymbol{\partial}}\,|\Delta^{\mathbf{F}_\bullet}|^{2\lambda})\,\frac{\overline{\boldsymbol{\partial}}\sigma_k\,\overline{\boldsymbol{\partial}}\sigma_{k-1}\cdots\overline{\boldsymbol{\partial}}\sigma_{\ell+2}\,\sigma_{\ell+1}}{(2i\pi)^{k-\ell}}\,[V]\\ &\in {}'\mathcal{D}^{(0,k-\ell)}\big(\mathcal{X},\mathrm{Hom}_{\mathbb{C}}(E_\ell,E_k)\big)\\ P^{\mathbf{F}_\bullet,\lambda}_{\ell,k}([V]) &= |\Delta^{\mathbf{F}_\bullet}|^{2\lambda}\,\frac{\sigma_k\,\overline{\boldsymbol{\partial}}\sigma_{k-1}\cdots\overline{\boldsymbol{\partial}}\sigma_{\ell+2}\,\overline{\boldsymbol{\partial}}\sigma_{\ell+1}}{(2i\pi)^{k-\ell}}\,[V]\\ &= |\Delta^{\mathbf{F}_\bullet}|^{2\lambda}\,\frac{\overline{\boldsymbol{\partial}}\sigma_k\,\overline{\boldsymbol{\partial}}\sigma_{k-1}\cdots\overline{\boldsymbol{\partial}}\sigma_{\ell+2}\,\sigma_{\ell+1}}{(2i\pi)^{k-\ell}}\,[V]\\ &\in {}'\mathcal{D}^{(0,k-\ell-1)}\big(\mathcal{X},\mathrm{Hom}_{\mathbb{C}}(E_\ell,E_k)\big),\end{aligned} \tag{3.68}$$

[15]The equality comes from the fact that $\boldsymbol{\sigma}\,(\overline{\partial}\boldsymbol{\sigma})=(\overline{\partial}\boldsymbol{\sigma})\,\boldsymbol{\sigma}$.

for any $k > \ell \geq -1$. For $\operatorname{Re}\lambda \gg 1$, it follows from (3.63) that

(3.69)

$$\begin{aligned}\boldsymbol{\nabla}_{\mathbf{F}_\bullet}\big(P_{|\ |}^{\overline{\mathbf{F}_\bullet},\lambda}([V])\big) &= \boldsymbol{\nabla}_{\mathbf{F}_\bullet}(|\Delta^{\mathbf{F}_\bullet}|^{2\lambda})\Big(\sum_{r\geq 1}\frac{\sigma}{2i\pi}\Big(\overline{\boldsymbol{\partial}}\Big(\frac{\sigma}{2i\pi}\Big)\Big)^{r-1}\Big)[V] + |\Delta^{\mathbf{F}_\bullet}|^{2\lambda}[V] \\ &= -(\overline{\boldsymbol{\partial}}\,|\Delta^{\mathbf{F}_\bullet}|^{2\lambda})\Big(\sum_{r\geq 1}\frac{\sigma}{2i\pi}\Big(\overline{\boldsymbol{\partial}}\Big(\frac{\sigma}{2i\pi}\Big)\Big)^{r-1}\Big)[V] + |\Delta^{\mathbf{F}_\bullet}|^{2\lambda}[V] \\ &= \operatorname{Id}_{E(\mathbf{F}_\bullet)}[V] - R_{|\ |}^{\overline{\mathbf{F}_\bullet},\lambda}([V]).\end{aligned}$$

Taking the action of $\boldsymbol{\nabla}_{\mathbf{F}_\bullet}$ on both sides and using the fact that $\boldsymbol{\nabla}^2_{\mathbf{F}_\bullet} = 0$ leads to

$$\boldsymbol{\nabla}_{\mathbf{F}_\bullet}\big(R_{|\ |}^{\overline{\mathbf{F}_\bullet},\lambda}([V])\big) = 0 \tag{3.70}$$

for $\operatorname{Re}\lambda \gg 1$. In the spirit of Proposition 3.15, we obtain the following result [**AndW07**].

Proposition 3.28. *Let $\mathcal{X}$, V, $\overline{\mathbf{F}_\bullet}$, $\mathbf{E} = \operatorname{Hom}_{\mathbb{C}}(E(\mathbf{F}_\bullet), E(\mathbf{F}_\bullet))$, $\lambda \longmapsto |\Delta_{\mathbf{F}_\bullet}|^{2\lambda}$, $\boldsymbol{\sigma}$ be as above. For each $k > \ell \geq -1$, the two $\operatorname{Hom}_{\mathbb{C}}(E_\ell, E_k)$-current valued maps*

$$\lambda \longmapsto R_{\ell,k}^{\overline{\mathbf{F}_\bullet},\lambda}([V]),\quad \lambda \longmapsto P_{\ell,k}^{\overline{\mathbf{F}_\bullet},\lambda}([V])$$

extend to $\mathbb{C}$ as meromorphic maps with polar set in $-\mathbb{Q}^+$. Their values at $\lambda = 0$ are $\operatorname{Hom}_{\mathbb{C}}(E_\ell, E_k)$-valued pseudo-meromorphic currents with bidegrees $(M, M+(k-\ell))$ and $(M, M+(k-\ell-1))$, respectively. Correspondingly, we denote these currents as $R_{\ell,k}^{\overline{\mathbf{F}_\bullet}}([V])$ and $P_{\ell,k}^{\overline{\mathbf{F}_\bullet}}([V])$. Moreover, the $\mathbf{E}$-valued currents

(3.71)

$$\begin{aligned}R^{\overline{\mathbf{F}_\bullet}}([V]) &= \operatorname{Id}_{E(\mathbf{F}_\bullet)}\mathbf{1}_{V(\mathbf{F}_\bullet)}[V] + \sum_{\ell\geq -1}\Big(\sum_{k>\ell} R_{\ell,k}^{\overline{\mathbf{F}_\bullet}}([V])\Big) \\ &= \operatorname{Id}_{E(\mathbf{F}_\bullet)}\mathbf{1}_{V(\mathbf{F}_\bullet)}[V] + \Big[(\overline{\boldsymbol{\partial}}\,|\Delta^{\mathbf{F}_\bullet}|^{2\lambda})\sum_{r\geq 1}\frac{\sigma}{2i\pi}\Big(\overline{\boldsymbol{\partial}}\Big(\frac{\sigma}{2i\pi}\Big)\Big)^{r-1}\Big]_{\lambda=0}[V], \\ P^{\overline{\mathbf{F}_\bullet}}([V]) &= \sum_{\ell\geq -1}\Big(\sum_{k>\ell} P_{\ell,k}^{\overline{\mathbf{F}_\bullet}}([V])\Big) \\ &= \Big[|\Delta^{\mathbf{F}_\bullet}|^{2\lambda}\Big(\sum_{r\geq 1}\frac{\sigma}{2i\pi}\Big(\overline{\boldsymbol{\partial}}\Big(\frac{\sigma}{2i\pi}\Big)\Big)^{r-1}\Big)\Big]_{\lambda=0}[V]\end{aligned}$$

satisfy

$$\boldsymbol{\nabla}_{\mathbf{F}_\bullet}\big(R^{\overline{\mathbf{F}_\bullet}}([V])\big) = 0,\quad \boldsymbol{\nabla}_{\mathbf{F}_\bullet}\big(P^{\overline{\mathbf{F}_\bullet}}([V])\big) = \operatorname{Id}_{E(\mathbf{F}_\bullet)}[V] - R^{\overline{\mathbf{F}_\bullet}}([V]). \tag{3.72}$$

One has in addition

$$1 \leq k-\ell < \operatorname{codim}_V(V\cap V(\mathbf{F}_\bullet)) \Longrightarrow R_{\ell,k}^{\overline{\mathbf{F}_\bullet}}([V]) = 0. \tag{3.73}$$

Furthermore, for any ℓ, k with $\ell \geq -1$ and $k-\ell \geq \operatorname{codim}_V(V\cap V(\mathbf{F}_\bullet))$, for any $z \in V\cap V(\mathbf{F}_\bullet)$ and any $f \in \mathcal{O}_z$,

$$f = 0 \text{ on } V\cap V(\mathbf{F}_\bullet) \text{ around } z \Longrightarrow \big(\overline{f}\,R_{\ell,k}^{\overline{\mathbf{F}_\bullet}}\big)_z = \big(d\overline{f}\wedge R_{\ell,k}^{\overline{\mathbf{F}_\bullet}}\big)_z = 0. \tag{3.74}$$

Proof. For each $\ell \geq -1$, let $V_\ell^{\mathbf{F}_\bullet} = \bigcup_{k>\ell} V(\Delta_k^{\mathbf{F}_\bullet})$ and $\pi_\ell : \mathcal{V}_\ell \to V$ be a log resolution of V for the complex analytic subset $\mathcal{Y}$ of V defined as the union of $V\cap V_\ell^{\mathbf{F}_\bullet}$ with any hypersurface of V containing V_{sing}; see Theorem A.41. Let

$a \in V \cap V_\ell^{\mathbf{F}_\bullet}$ and $z \in \mathcal{V}_\ell$ be in the pre-image through $\iota_V \circ \pi_\ell$ of a neighborhood of a in $\mathcal{X}$. One has in a local chart $U = U_z$ around z that for any $\kappa > \ell$,

$$(\iota_V \circ \pi_\ell)^* \Delta_\kappa^{\mathbf{F}_\bullet} = s_{\kappa,0}\, S'_\kappa,$$

where $s_{\kappa,0}$ is a holomorphic section of a line bundle L_κ and S'_κ is a nonvanishing section of the holomorphic bundle $L_\kappa^{-1} \otimes (\iota_V \circ \pi_\ell)^* \big(\bigwedge^{\rho_\kappa} \mathrm{Hom}_{\mathbb{C}}(E_\kappa, E_{\kappa-1})\big)$. A linear algebra argument (see [**And06D**, Lemma 4.1]) shows that $s_{\kappa,0}\,(\iota_V \circ \pi_\ell)^* \sigma_\kappa$ is then a smooth section ψ_κ of $\mathrm{Hom}_{\mathbb{C}}\big(L_\kappa \otimes (\iota_V \circ \pi_\ell)^* E_{\kappa-1}, L_\kappa \otimes (\iota_V \circ \pi_\ell)^* E_\kappa\big)$. For each $k > \ell$ consider now the maps

$$\lambda \longmapsto (\iota_V \circ \pi)^* \Big(\big(\overline{\partial} |\Delta^{\mathbf{F}_\bullet}|^{2\lambda}\big)\, \sigma_k\, \overline{\partial} \sigma_{k-1} \cdots \overline{\partial} \sigma_{\ell+2}\, \overline{\partial} \sigma_{\ell+1} \Big),$$
$$\lambda \longmapsto (\iota_V \circ \pi)^* \Big(|\Delta^{\mathbf{F}_\bullet}|^{2\lambda}\, \sigma_k\, \overline{\partial} \sigma_{k-1} \cdots \overline{\partial} \sigma_{\ell+2}\, \overline{\partial} \sigma_{\ell+1} \Big).$$

Since $\psi_{\kappa+1}\, \psi_\kappa = 0$ for any $\kappa > \ell$, as this holds outside the union of the hypersurfaces $V(s_{0,\kappa})$ for $\kappa > \ell$, these maps equal

$$\lambda \longmapsto \frac{\overline{\partial} \big| (\iota_V \circ \pi)^* \Delta^{\mathbf{F}_\bullet} \big|^{2\lambda}}{s_{0,k} \cdots s_{0,\ell+1}}\, \psi_k\, \overline{\partial} \psi_{k-1} \cdots \overline{\partial} \psi_{\ell+2}\, \overline{\partial} \psi_{\ell+1},$$
$$\lambda \longmapsto \frac{\big| (\iota_V \circ \pi)^* \Delta^{\mathbf{F}_\bullet} \big|^{2\lambda}}{s_{0,k} \cdots s_{0,\ell+1}}\, \psi_k\, \overline{\partial} \psi_{k-1} \cdots \overline{\partial} \psi_{\ell+2}\, \overline{\partial} \psi_{\ell+1},$$

respectively. Then, it follows, as in the proof of Proposition 3.15, that these maps extend to $\mathbb{C}$ as meromorphic maps with polar set contained in $-\mathbb{Q}^+$ and whose values at $\lambda = 0$ equal pseudo-meromorphic currents. The first assertion of the proposition then follows. Formulae (3.72) follow from (3.63) and (3.64) if one invokes the analytic continuation principle. Assertion (3.65) follows from the dimension principle (Proposition 2.26) taking into account the fact that the pseudo-meromorphic current $R_{\ell,k}^{\mathbf{F}_\bullet}([V])$ has bidegree $(M, M + (k - \ell))$ and that its support equals $V \cap \bigcup_{\kappa > \ell} V_\kappa^{\mathbf{F}_\bullet} \subset V \cap V(\mathbf{F}_\bullet)$. One has the slightly more precise result

$$1 \leq k - \ell < \mathrm{codim}_V \Big(V \cap \bigcup_{\kappa > \ell} V_\kappa^{\mathbf{F}_\bullet} \Big) \Longrightarrow R_{\ell,k}^{\mathbf{F}_\bullet}([V]) = 0. \tag{3.75}$$

Assertion (3.66) follows from Proposition 2.24, since $R^{\mathbf{F}_\bullet}([V])$ is a pseudo-meromorphic current supported by $V \cap V(\mathbf{F}_\bullet)$. Proposition 3.28 is thus completely proved. □

REMARK 3.29. If $\mathrm{codim}_V(V \cap V(\mathbf{F}_\bullet)) > 0$, then the complex of holomorphic bundles

$$0 \longrightarrow (E_L)_{|V} \overset{(F_L)_{|V}}{\longrightarrow} \cdots \overset{(F_2)_{|V}}{\longrightarrow} (E_1)_{|V} \overset{(F_1)_{|V}}{\longrightarrow} (E_0)_{|V} \overset{(F_0)_{|V}}{\longrightarrow} \frac{E_{0|V}}{\mathrm{Im}\,(F_1)_{|V}} \longrightarrow 0$$

can be interpreted as a generically exact complex of holomorphic bundles over the complex analytic space $(V, \mathcal{O}_V)$. The proof of Proposition 3.28 shows that the currents $R^{\mathbf{F}_\bullet}([V])$ and $P^{\mathbf{F}_\bullet}([V])$ can be interpreted as the direct images through ι_V of pseudo-meromorphic currents on the complex analytic space $(V, \mathcal{O}_V)$, in accordance to Definition B.44 in §B.2.2.2. Therefore, the currents $R^{\mathbf{F}_{\bullet|V}}$ and $P^{\mathbf{F}_{\bullet|V}}$ satisfy

$$\boldsymbol{\nabla}_{\boldsymbol{F}}^V R^{\mathbf{F}_{\bullet|V}} = 0, \quad \boldsymbol{\nabla}_{\boldsymbol{F}}^V P^{\mathbf{F}_{\bullet|V}} = \mathrm{Id}_{E((\mathbf{F}_\bullet)_{|V})} - R^{\mathbf{F}_{\bullet|V}}, \tag{3.76}$$

where the action of the operator $\boldsymbol{\nabla}^V_{\mathbf{F}_\bullet} = 2i\pi \boldsymbol{F}_{|V} - \overline{\boldsymbol{\partial}}_V$ is now understood from ${}'\mathcal{D}\big(V, \mathrm{Hom}_{\mathbb{C}}(E((\mathbf{F}_\bullet)_{|V}), E((\mathbf{F}_\bullet)_{|V}))\big)$ into itself.

EXAMPLE 3.30 (Koszul complex). Let E be a holomorphic bundle with rank m over $\mathcal{X}$ and $s \in \mathcal{O}(\mathcal{X}, E)$. One can attach to s the generically exact Koszul complex $\mathbf{K}^s_\bullet$; see (3.51) in Example 3.25. Given a closed analytic subset V of $\mathcal{X}$ with pure dimension M, one can also attach to s and to the hermitian bundle $\overline{E} = (E, |\ |)$ the Bochner–Martinelli residue and principal value currents $R^s_{|\ |}([V])$, $P^s_{|\ |}([V])$, respectively, as in Proposition 3.15. The currents

$$
\begin{aligned}
R^s_{|\ |}([V]) &= \Big[(1 - |s|^{2\lambda}) + (\overline{\partial}|s|^{2\lambda}) \sum_{r\geq 1} \frac{s^*}{2i\pi|s|^2}\, \overline{\partial}\Big(\frac{s^*}{2i\pi|s|^2}\Big)^{r-1}\Big]_{\lambda=0} [V],\\
P^s_{|\ |}([V]) &= \Big[|s|^{2\lambda} \sum_{r\geq 1} \frac{s^*}{2i\pi|s|^2}\, \overline{\partial}\Big(\frac{s^*}{2i\pi|s|^2}\Big)^{r-1}\Big]_{\lambda=0} [V]
\end{aligned}
\tag{3.77}
$$

can be re-interpreted as follows. For each $\ell \geq 1$, the smooth section of the metrized holomorphic bundle $\big(\mathrm{Hom}_{\mathbb{C}}\big(\bigwedge^{\ell-1} E^*, \bigwedge^{\ell} E^*\big), |\ |\big)$, whose metric is induced by the metric $|\ |$ on E, defined by

$$
\sigma_\ell(z,\xi) = \Big(z, \frac{s^*(z)}{|s(z)|^2}\,\xi\Big) \quad \forall z \in \mathcal{X} \setminus V(\mathbf{K}^s_\bullet),\ \forall \xi \in \Big(\bigwedge^{\ell-1} E^*\Big)_z,
\tag{3.78}
$$

is the minimal inverse. This follows from (3.60) and the fact that

$$
(\rfloor_s)_\ell \in \mathcal{O}\big(\mathcal{X}, \mathrm{Hom}_{\mathbb{C}}\big(\bigwedge^{\ell} E^*, \bigwedge^{\ell-1} E^*\big)\big)
$$

on $\mathcal{X} \setminus V(\mathbf{K}^s_\bullet)$. The currents $R^s_{|\ |}([V])$ and $P^s_{|\ |}([V])$, considered as acting by multiplication from $E(\mathbf{K}^s_\bullet)$ into itself, define $\mathbf{E}(\mathbf{K}^s_\bullet) = \mathrm{Hom}_{\mathbb{C}}(E(\mathbf{K}^s_\bullet), E(\mathbf{K}^s_\bullet))$-valued currents. These currents coincide with $R^{\overline{\mathbf{K}^s_\bullet}}([V])$ and $P^{\overline{\mathbf{K}^s_\bullet}}([V])$, where the Koszul complex $\mathbf{K}^s_\bullet$ is equipped with the metrics induced by the original metric $|\ |$ on E. In the particular case where $\mathrm{codim}_V(V \cap V(s)) = m$, one has for any $\ell \geq 0$,

$$
\begin{aligned}
&R^{\overline{\mathbf{F}_\bullet}}_\ell = \mathrm{Hom}_{\mathbb{C}}\big(\bigwedge^{\ell} E^*, \bigwedge^{m} E^*\big)\text{-component of}\\
&\Big[(\overline{\boldsymbol{\partial}}|s|^{2\lambda})\, \frac{s^*}{2i\pi|s|^2}\, \overline{\boldsymbol{\partial}}\Big(\frac{s^*}{2i\pi|s|^2}\Big)^{m-1}\Big]_{\lambda=0} [V] = 0 \quad \text{unless } \ell = 0.
\end{aligned}
\tag{3.79}
$$

EXAMPLE 3.31 (Eagon–Northcott complex). Let $\overline{E} = (E, |\ |)$ and $s \in \mathcal{O}(\mathcal{X}, E)$ as in Example 3.30. Let $\mathbf{EN}^{s,k}_\bullet$ be the Eagon–Northcott complex constructed from k copies $(\mathbf{K}^s_\bullet)_\kappa$, where $\kappa = 1, \dots, k$, of $\mathbf{K}^s_\bullet$; see Example 3.26. For each copy $(\mathbf{K}^s_\bullet)_\kappa$, let $(\sigma)_\kappa \in C^\infty(\mathcal{X} \setminus V(s), \mathbf{E}((\mathbf{K}^s_\bullet)_\kappa))$ be the minimal inverse of $(\rfloor_s)_\kappa$, as defined in (3.78). For each $\kappa = 1, \dots, k$, let

$$
Q^{(\overline{\mathbf{K}_\bullet^s})_\kappa} = \sum_{r\geq 1} \frac{(\sigma)_\kappa}{2i\pi} \Big(\overline{\boldsymbol{\partial}}\Big(\frac{(\sigma)_\kappa}{2i\pi}\Big)\Big)^{r-1} \in C^\infty\Big(\mathcal{X} \setminus V(s), \mathbf{E}((\mathbf{K}^s_\bullet)_\kappa)\Big).
\tag{3.80}
$$

Expression (3.80) is considered in terms of its action by multiplication from $\mathcal{O}(\mathcal{X} \setminus V(s), E((\mathbf{K}^s_\bullet)_\kappa)$ into itself as in Example 3.30. It follows from simple multilinear algebra (see [**And06C**, §3]) that, if one assumes $\mathrm{codim}_V(V \cap V(s)) > 0$, then

$$
R^{\overline{\mathbf{EN}^{s,k}_\bullet}}([V]) = \Big[(\overline{\boldsymbol{\partial}}\,|s|^{2\lambda})\, Q^{(\overline{\mathbf{K}_\bullet^s})_1} \cdots Q^{(\overline{\mathbf{K}_\bullet^s})_k}\Big]_{\lambda=0} [V].
\tag{3.81}
$$

In particular, one has that

$$R_{0,r}^{\overline{\mathbf{EN}_\bullet^{s,k}}}([V]) = \mathrm{Hom}_{\mathbb{C}}(\mathbb{C}, \mathbf{EN}_r^{s,k})\text{-valued component of} \tag{3.82}$$

$$\Big[\frac{1}{(2i\pi)^r}(\overline{\partial}\,|s|^{2\lambda})\prod_{\kappa=1}^{k}(\sigma)_\kappa \sum_{\{\rho\in\mathbb{N}^k\,:\,|\rho|=r-1\}} \Big(\overline{\partial}\Big(\frac{(\sigma)_1}{2i\pi}\Big)\Big)^{\rho_1}\cdots\Big(\overline{\partial}\Big(\frac{(\sigma)_k}{2i\pi}\Big)\Big)^{\rho_k}\Big]_{\lambda=0}[V]$$

for any $r \geq 1$. It follows from assertion (3.73) in Proposition 3.28 that such a current equals 0 as soon as $r < \mathrm{codim}_V(V \cap V(s))$. Moreover, since the m coordinates of $\overline{\partial}\sigma$ (where σ denotes the minimal inverse of s) in a local frame for $E_{|V}$ in a neighborhood of $z \in V$ are linearly dependent since $\rfloor_s \sigma = 1$ outside $V(f)$, one has $R_{0,r}^{\overline{\mathbf{EN}_\bullet^{s,k}}}([V])$ as soon as $r > \min(N-M, m)$. Therefore, in the particular case where $\mathrm{codim}_V(V \cap V(s)) = m$, the only nontrivial currents from the list (3.82) are the ones where $r = m$.

3.3.3. Residual obstruction for exactness and duality theorems. We consider in this section a generically exact metrized complex of holomorphic bundles

$$\overline{\mathbf{F}_\bullet} : 0 \longrightarrow \overline{E_L} \xrightarrow{F_L} \overline{E_{L-1}} \xrightarrow{F_{L-1}} \cdots \xrightarrow{F_2} \overline{E_1} \xrightarrow{F_1} \overline{E_0} \xrightarrow{F_0} \overline{E_{-1}} = \overline{\frac{E_0}{\mathrm{Im}\,F_1}} \longrightarrow 0$$

over an N-dimensional complex manifold $\mathcal{X}$. For $\mathbf{E}(\mathbf{F}_\bullet) = \mathrm{Hom}_{\mathbb{C}}(E(\mathbf{F}_\bullet), E(\mathbf{F}_\bullet))$, let $R^{\overline{\mathbf{F}_\bullet}} = R^{\overline{\mathbf{F}_\bullet}}([\mathcal{X}])$ and $P^{\overline{\mathbf{F}_\bullet}} = P^{\overline{\mathbf{F}_\bullet}}([\mathcal{X}])$ be the $\mathbf{E}(\mathbf{F}_\bullet)$-valued currents introduced in Proposition 3.28. Their decompositions are described by

$$\begin{aligned} R^{\overline{\mathbf{F}_\bullet}} &= \sum_{\ell\geq -1}\Big(\sum_{k>\ell} R_{\ell,k}^{\overline{\mathbf{F}_\bullet}}([\mathcal{X}])\Big) = \sum_{\ell\geq -1}\Big(\sum_{k>\ell} R_{\ell,k}^{\overline{\mathbf{F}_\bullet}}\Big) = \sum_{\ell\geq -1} R_\ell^{\overline{\mathbf{F}_\bullet}},\\ P^{\overline{\mathbf{F}_\bullet}} &= \sum_{\ell\geq -1}\Big(\sum_{k>\ell} P_{\ell,k}^{\overline{\mathbf{F}_\bullet}}([\mathcal{X}])\Big) = \sum_{\ell\geq -1}\Big(\sum_{k>\ell} P_{\ell,k}^{\overline{\mathbf{F}_\bullet}}\Big) = \sum_{\ell\geq -1} P_\ell^{\overline{\mathbf{F}_\bullet}}. \end{aligned} \tag{3.83}$$

We formulate in this subsection Propositions 3.32 and 3.36 to illustrate the following important fact: the current $R^{\overline{\mathbf{F}_\bullet}}$, through its components $R_\ell^{\overline{\mathbf{F}_\bullet}}$, $\ell \geq 0$, materializes the obstruction for the associated complex of locally free sheaves of sections

$$0 \longrightarrow \mathcal{O}(\cdot, E_L) \xrightarrow{F_L} \mathcal{O}_{\mathcal{X}}(\cdot, E_{L-1}) \xrightarrow{F_{L-1}} \cdots \xrightarrow{F_2} \mathcal{O}_{\mathcal{X}}(\cdot, E_1) \xrightarrow{F_1} \mathcal{O}_{\mathcal{X}}(\cdot, E_0) \tag{3.84}$$

to be exact. Namely, one has the following proposition.

PROPOSITION 3.32. *Let $\ell \geq 0$ and $h \in \mathcal{O}(\mathcal{X}, E_\ell)$ such that either*

$$\begin{aligned} &F_\ell\, h = 0 \textit{ when } \ell \geq 1\\ &\textit{or } h_z \in \big(\mathcal{O}(\cdot, \mathrm{Im}\,F_1)\big)_z\ \forall z \in \mathcal{X} \setminus V(\mathbf{F}_\bullet) \textit{ when } \ell = 0. \end{aligned} \tag{3.85}$$

Then

$$R_\ell^{\overline{\mathbf{F}_\bullet}}\, h = 0 \Longrightarrow \big(\forall z \in \mathcal{X},\ \ h_z \in \big(\mathcal{O}(\cdot, \mathrm{Im}\,F_{\ell+1})\big)_z\big). \tag{3.86}$$

Conversely, in the case $R_{\ell+1}^{\overline{\mathbf{F}_\bullet}} = 0$,

$$\big(\forall z \in \mathcal{X},\ h_z \in \big(\mathcal{O}(\cdot, \mathrm{Im}\,F_{\ell+1})\big)_z\big) \Longrightarrow R_\ell^{\overline{\mathbf{F}_\bullet}}\, h = 0. \tag{3.87}$$

PROOF. The proof of assertion (3.86) is inspired by the approach developed to establish the same claim in the univariate setting; see Proposition 1.29. Let us first prove (3.86), when $\ell \geq 1$. Since $h \in \mathcal{O}(\mathcal{X}, E_\ell)$,

$$R_\ell^{\overline{\mathbf{F}_\bullet}}\, h = 0 \Longleftrightarrow R^{\overline{\mathbf{F}_\bullet}}\, h = 0.$$

It follows from (3.72) that

$$\begin{aligned}\nabla_{\mathbf{F}_\bullet}(P^{\overline{\mathbf{F}_\bullet}} h) &= (\boldsymbol{\nabla}_{\mathbf{F}_\bullet} P^{\overline{\mathbf{F}_\bullet}})(h) - P^{\overline{\mathbf{F}_\bullet}}(\nabla_{\mathbf{F}_\bullet} h) \\ &= (\mathrm{Id}_{E(\mathbf{F}_\bullet)} - R^{\overline{\mathbf{F}_\bullet}})(h) - P^{\overline{\mathbf{F}_\bullet}}(2i\pi F_\ell h - \overline{\partial} h) = h.\end{aligned} \tag{3.88}$$

Let $u = P^{\overline{\mathbf{F}}} h = \sum_{\kappa \geq \ell+1} u_\kappa$, with $u_\kappa \in {}'\mathcal{D}^{(0,\kappa-\ell-1)}(\mathcal{X}, E_\kappa)$. One has

$$(2i\pi F - \overline{\partial})\, u = h \Longrightarrow \begin{cases} 2i\pi F_{\ell+1}\, u_{\ell+1} = h \\ 2i\pi\, F_\kappa\, u_\kappa = \overline{\partial} u_{\kappa-1} \quad \forall \kappa > \ell+1. \end{cases}$$

Applying Dolbeault's lemma in a neighborhood of any point $z \in \mathcal{X}$, one solves locally the sequence of $\overline{\partial}$ equations

$$\overline{\partial} q_{\kappa,z} = \Big(u_\kappa + 2i\pi\, F_{\kappa+1}\, q_{\kappa+1}\Big)_z, \quad q_{\kappa,z} \in \Big({}'\mathcal{D}^{(0,\kappa-\ell-2)}(\cdot, E_\kappa)\Big)_z,$$

taking into account that $F_{\kappa+1} = 0$ for $\kappa \gg 1$ and that any $(0,N)$-form is locally $\overline{\partial}$-exact. Let

$$a_z = (u_{\ell+1})_z + 2i\pi\, F_{\ell+2}(q_{\ell+2})_z.$$

One has $(F_{\ell+1} a)_z = h_z$, since $F^2 = 0$ and

$$\begin{aligned}(\overline{\partial} a)_z &= (\overline{\partial} u_{\ell+1})_z - F_{\ell+2}\, \overline{\partial} q_{\ell+2,z} = \big(2i\pi F_{\ell+2}\,(u_{\ell+2} - \overline{\partial} q_{\ell+2})\big)_z \\ &= -(2i\pi)\,(2i\pi F_{\ell+2} F_{\ell+3} q_{\ell+3})_z = 0,\end{aligned}$$

using again the property that $F^2 = 0$. This proves the assertion (3.86) in the case where $\ell \geq 1$. Consider now the case where $\ell = 0$. Let $h \in \mathcal{O}(\mathcal{X}, E_0)$. We assume that $h_z \in \big(\mathcal{O}(\cdot, \operatorname{Im} F_1)\big)_z$ in a neighborhood of any point $z \in V(\mathbf{F}_\bullet)$. Relation (3.88) and the fact that $F_0 F_1 = 0$ (since $F^2 = 0$) imply that $\nabla_{\mathbf{F}_\bullet}(P^{\overline{\mathbf{F}_\bullet}}(h)) = h$. Therefore, the proof in the case $\ell = 0$ can be carried along the same lines as in the case $\ell \geq 1$. In order to prove assertion (3.87), it is enough to prove that under the hypothesis $R^{\overline{\mathbf{F}_\bullet}}_{\ell+1} = 0$ the implication

$$h_z \in \big(\mathcal{O}(\cdot, \operatorname{Im} F_{\ell+1})\big)_z \Longrightarrow (h\, R^{\overline{\mathbf{F}_\bullet}}_\ell)_z = 0$$

holds for any point $z \in \mathcal{X}$. If $h_z \in \big(\mathcal{O}(\cdot, \operatorname{Im} F_{\ell+1})\big)_z$, then there exists an element $a_z \in \big(\mathcal{O}(\cdot, E_{\ell+1})\big)_z$ satisfying $h_z = (\nabla_{\mathbf{F}^*_\bullet}(a))_z$. Then, one has

$$\big(R^{\overline{\mathbf{F}_\bullet}}_\ell h\big)_z = \big(R^{\overline{\mathbf{F}_\bullet}} h\big)_z = \big(R^{\overline{\mathbf{F}_\bullet}}(\nabla_{\mathbf{F}_\bullet} a)\big)_z = \big(\nabla_{\mathbf{F}_\bullet}(R^{\overline{\mathbf{F}_\bullet}}\, a)\big)_z = \big(\nabla_{\mathbf{F}_\bullet}\,(R^{\overline{\mathbf{F}_\bullet}}_{\ell+1}\, a)\big)_z = 0,$$

since $\boldsymbol{\nabla}_{\mathbf{F}_\bullet}(R^{\overline{\mathbf{F}_\bullet}}) = 0$ in view of (3.72). □

REMARK 3.33. If $\mathcal{X}$ is a Stein manifold, assertion (3.86) stands as follows:

$$R^{\overline{\mathbf{F}_\bullet}}_\ell\, h = 0 \Longrightarrow h = F_{\ell+1}\, a \text{ for some } a \in \mathcal{O}(\mathcal{X}, E_{\ell+1}). \tag{3.89}$$

EXAMPLE 3.34 (Koszul complex, Example 3.30 continued). An important particular case, where Proposition 3.32 applies, is that of the Koszul complex introduced in Example 3.30. If $\overline{E} = (E, |\ |)$ is an m-dimensional, hermitian holomorphic bundle over a complex manifold $\mathcal{X}$, whose section $s \in \mathcal{O}(\mathcal{X}, E)$ has its conjugate section equal to $s^* \in C^\infty(\mathcal{X}, E^*)$, then one has that the implication

$$\begin{aligned} h\, R^{\overline{\mathbf{F}_\bullet}}_0 = \Big[\Big((\overline{\boldsymbol{\partial}}|s|^{2\lambda})\, \frac{s^*}{2i\pi|s|^2}\, \overline{\boldsymbol{\partial}}\Big(\frac{s^*}{2i\pi|s|^2}\Big)^{m-1}\Big)_{|E_0(\mathbf{K}^s_\bullet)=\mathbb{C}}\Big]_{\lambda=0}(h) = 0 \\ \Longrightarrow h_z \in \big(\mathcal{O}(\cdot, \operatorname{Im} \rfloor_s)\big)_z \quad \forall z \in \mathcal{X},\end{aligned} \tag{3.90}$$

where $\rfloor_s : E^* \to \mathbb{C}$ is the interior multiplication by the section s, holds for any $h \in \mathcal{O}(\mathcal{X}, E_0(\mathbf{K}^s_\bullet)) = \mathcal{O}(\mathcal{X})$. Moreover, if

$$\operatorname{codim} V(s) = m = \operatorname{rank} E,$$

then it follows from (3.79) in Example 3.30, together with assertion (3.87) in Proposition 3.32, that

$$(3.91)\quad \Big(h_z \in \big(\mathcal{O}(\cdot, \operatorname{Im} \rfloor_s)\big)_z \quad \forall z \in \mathcal{X}\Big) \Longleftrightarrow \Big(h_z \in \big(\mathcal{O}(\cdot, \operatorname{Im} \rfloor_s)\big)_z \quad \forall z \in V(s)\Big)$$
$$\Longrightarrow h\, R_0^{\overline{\mathbf{F}_\bullet}} = \Big[\Big((\overline{\partial} |s|^{2\lambda}) \frac{s^*}{2i\pi |s|^2}\, \overline{\partial}\Big(\frac{s^*}{2i\pi|s|^2}\Big)^{m-1}\Big)_{|E_0(\mathbf{K}^s_\bullet) = \mathbb{C}}\Big]_{\lambda=0} (h) = 0.$$

EXAMPLE 3.35 (Eagon–Northcott complex, Example 3.31 continued). Let $\mathcal{X}$, $\overline{E} = (E, |\ |)$, and $s \in \mathcal{O}(\mathcal{X}, E)$ be as in the previous example. Let $k \in \mathbb{N}^*$. One has, for any $h \in \mathcal{O}(\mathcal{X}, E_0(\mathbf{EN}^{s,k}_\bullet)) = \mathcal{O}(\mathcal{X})$, that

$$(3.92)\quad h\, R_0^{\overline{\mathbf{EN}^{s,k}_\bullet}} = \Big[\Big((\overline{\partial}\, |s|^{2\lambda})\, Q^{(\overline{\mathbf{K}^s_\bullet})_1} \cdots Q^{(\overline{\mathbf{K}^s_\bullet})_k}\Big)_{|E_0(\mathbf{EN}^{k,s}_\bullet) = \mathbb{C}}\Big]_{\lambda=0} (h) = 0$$
$$\Longrightarrow h_z \in (\mathcal{I}_s)^k_z,$$

where $\mathcal{I}_s$ is the coherent sheaf of ideals[16] attached to s. The smooth sections $Q^{(\overline{\mathbf{K}^s_\bullet})_\kappa}$ of the vector bundle $\operatorname{Hom}_{\mathbb{C}}\big(E((\mathbf{K}^s_\bullet)_\kappa), E((\mathbf{K}^s_\bullet)_\kappa)\big)$ over $\mathcal{X} \setminus V(s)$, defined for k copies of the Koszul complex $\mathbf{K}^s_\bullet$, have been described in (3.81).

The fact that the equality $R_1^{\overline{\mathbf{K}^s_\bullet}} = 0$ holds for any section $s \in \mathcal{O}(\mathcal{X}, E)$ satisfying $\operatorname{codim} V(s) = \operatorname{rank} E$ (see Example 3.34) suggests the following proposition.

PROPOSITION 3.36. *Let*

$$\overline{\mathbf{F}_\bullet} : 0 \longrightarrow \overline{E_L} \xrightarrow{F_L} \overline{E_{L-1}} \xrightarrow{F_{L-1}} \cdots \xrightarrow{F_2} \overline{E_1} \xrightarrow{F_1} \overline{E_0} \xrightarrow{F_0} \overline{E_{-1}} = \overline{\frac{E_0}{\operatorname{Im} F_1}} \longrightarrow 0$$

be a metrized generically exact complex of holomorphic bundles over a complex N-dimensional manifold $(\mathcal{X}, \mathcal{O})$. If the associated complex of locally free sheaves of sections

$$(3.93)\qquad 0 \longrightarrow \mathcal{O}(\cdot, E_L) \xrightarrow{F_L} \mathcal{O}(\cdot, E_{L-1}) \xrightarrow{F_{L-1}} \cdots \xrightarrow{F_2} \mathcal{O}(\cdot, E_1) \xrightarrow{F_1} \mathcal{O}(\cdot, E_0)$$

is exact, then $R_1^{\overline{\mathbf{F}_\bullet}} = 0$.

PROOF. The exactness of the complex of locally free sheaves of sections (3.93) is equivalent to the so-called *Buchsbaum–Eisenbud–Fitting conditions*

$$(3.94)\qquad \forall \kappa > 0, \quad \operatorname{codim} V(\Delta_\kappa^{\mathbf{F}_\bullet}) \geq \kappa,$$

where the sections $\Delta_\kappa^{\mathbf{F}_\bullet} \in \mathcal{O}\big(\mathcal{X}, \bigwedge^{\operatorname{rank} F_\kappa} \operatorname{Hom}_{\mathbb{C}}(E_\kappa, E_{\kappa-1})\big)$ have been defined for any $\kappa \in \mathbb{N}^*$ in (3.58) (see Theorem A.18).[17] It is pointed out in Remark A.19 that, under the hypothesis that the complex (3.93) of locally free sheaves is exact, the inclusion

$$(3.95)\qquad \kappa' \geq \kappa \geq 1 \Longrightarrow V(\Delta_{\kappa'}^{\mathbf{F}_\bullet}) \subset V(\Delta_\kappa^{\mathbf{F}_\bullet}) \subset V(\Delta_1^{\mathbf{F}_\bullet}) = V(\mathbf{F}_\bullet)$$

holds. The decreasing sequence of closed analytic subsets $\big(V(\Delta_\kappa^{\mathbf{F}_\bullet})\big)_{\kappa \geq 1}$ depends only on the sheaf $\mathcal{F} = \operatorname{coker}(F_1)$, whose free resolution over $\mathcal{X}$ is given by (3.93).

[16] Namely, $h_z \in (\mathcal{I}_s)_z \Longleftrightarrow h_z = \rfloor_s a_z$ for some $a_z \in (\mathcal{O}(\cdot, E^*))_z$.

[17] The reader will also find the related references there. In any case, we invite him/her to consult §A.2.3 and §A.3.3 before reading the proof that follows.

This follows from Theorem A.9 quoted in Appendix A. Relations (3.67) and (3.68) imply that

$$R_1^{\overline{\mathbf{F}_\bullet}} = \sum_{\kappa \geq 2} R_{1,\kappa}^{\overline{\mathbf{F}}} = \sum_{\kappa \geq 2} \Big[(\overline{\partial}|\Delta_1^{\mathbf{F}_\bullet}|^{2\lambda})\,\frac{\overline{\partial}\sigma_\kappa\,\overline{\partial}\sigma_{\kappa-1}\cdots\overline{\partial}\sigma_3\,\sigma_2}{(2i\pi)^{\kappa-1}}\Big]_{\lambda=0}.$$

The current

$$R_{1,2}^{\overline{\mathbf{F}_\bullet}} = \Big[(\overline{\partial}|\Delta_1^{\mathbf{F}_\bullet}|^{2\lambda})\,\frac{\sigma_2}{2i\pi}\Big]_{\lambda=0}$$

is pseudo-meromorphic, with bidegree $(0,1)$. It is supported by $V(\Delta_2^{\mathbf{F}_\bullet})$. The dimension principle for pseudo-meromorphic currents (see Proposition 2.26) and the fact $\operatorname{codim} V(\Delta_2^{\mathbf{F}_\bullet}) \geq 2$ imply that $R_{1,2}^{\overline{\mathbf{F}_\bullet}} = 0$. Let us now consider the pseudo-meromorphic current

$$R_{1,3}^{\overline{\mathbf{F}_\bullet}} = \Big[(\overline{\partial}|\Delta_1^{\mathbf{F}_\bullet}|^{2\lambda})\,\frac{\overline{\partial}\sigma_3\,\sigma_2}{2i\pi}\Big]_{\lambda=0}.$$

In $\mathcal{X} \setminus V(\Delta_3^{\mathbf{F}_\bullet})$, one has

$$(R_{1,3}^{\overline{\mathbf{F}_\bullet}})_{\mathcal{X}\setminus V(\Delta_3^{\mathbf{F}_\bullet})} = (\overline{\partial}\sigma_3)_{|\mathcal{X}\setminus V(\Delta_3^{\mathbf{F}_\bullet})}\Big(\Big[(\overline{\partial}|\Delta_1^{\mathbf{F}_\bullet}|^{2\lambda})\,\frac{\sigma_2}{(2i\pi)^2}\Big]_{\lambda=0}\Big)_{|\mathcal{X}\setminus V(\Delta_3^{\mathbf{F}_\bullet})}.$$

The first factor $(\overline{\partial}\sigma_3)_{|\mathcal{X}\setminus V(\Delta_3^{\mathbf{F}_\bullet})}$ is a smooth form in $\mathcal{X} \setminus V(\Delta_3^{\mathbf{F}_\bullet})$. The second one, namely

$$\Big(\Big[(\overline{\partial}|\Delta_1^{\mathbf{F}_\bullet}|^{2\lambda})\,\frac{\sigma_2}{(2i\pi)^2}\Big]_{\lambda=0}\Big)_{|\mathcal{X}\setminus V(\Delta_3^{\mathbf{F}_\bullet})},$$

is also a pseudo-meromorphic current $\mathcal{X} \setminus V(\Delta_3^{\mathbf{F}_\bullet})$ with bidegree $(0,1)$. Its support is included in the closed analytic subset $(\mathcal{X}\setminus V(\Delta_3^{\mathbf{F}_\bullet}))\cap V(\Delta_2^{\mathbf{F}_\bullet})$, whose codimension is at least 2. Then, the dimension principle implies that

$$\Big(\Big[(\overline{\partial}|\Delta_1^{\mathbf{F}_\bullet}|^{2\lambda})\,\frac{\sigma_2}{(2i\pi)^2}\Big]_{\lambda=0}\Big)_{|\mathcal{X}\setminus V(\Delta_3^{\mathbf{F}_\bullet})} = 0$$

in $\mathcal{X} \setminus V(\Delta_3^{\mathbf{F}_\bullet})$. But this means that $\operatorname{Supp} R_{1,3}^{\overline{\mathbf{F}_\bullet}} \subset V(\Delta_3^{\mathbf{F}_\bullet})$. Buchsbaum–Eisenbud–Fitting conditions (3.94) imply for $\kappa = 3$ that $\operatorname{codim} V(\Delta_3^{\mathbf{F}_\bullet}) \geq 3$. Therefore, the dimension principle applies once more for the pseudo-meromorphic current $R_{1,3}^{\overline{\mathbf{F}_\bullet}}$ with bidegree $(0,2)$, and one concludes that $R_{1,3}^{\overline{\mathbf{F}_\bullet}} = 0$. The procedure iterates from step $\kappa \geq 3$ to step $\kappa + 1$, following the same reasoning, which relies on (3.94), (3.95), and the fact that all currents involved are pseudo-meromorphic, and therefore conform to the dimension principle. Thus, we have shown that $R_{1,\kappa}^{\overline{\mathbf{F}_\bullet}} = 0$ for any $\kappa \geq 2$. But, this is equivalent to $R_1^{\overline{\mathbf{F}_\bullet}} = 0$. Proposition 3.36 is thus proved. □

REMARK 3.37. If the complex of locally free sheaves (3.93) is exact, it follows from (3.95) that

$$\begin{aligned} V(\mathbf{F}_\bullet) = V(\Delta_1^{\mathbf{F}_\bullet}) &= \{z \in \mathcal{X} \,:\, \operatorname{rank} F_1(z) < \rho_1\} \\ &= \Big\{z \in \mathcal{X} \,:\, \mathcal{F} = \frac{\mathcal{O}(\cdot, E_0)}{F_1\,\mathcal{O}(\cdot, E_1)} \text{ is not locally free at } z\Big\}. \end{aligned} \tag{3.96}$$

Repeating the proof of Proposition 3.36, which relies on (3.94), (3.95), and the fact that all currents appearing in the picture are pseudo-meromorphic, one can show, using the dimension principle, that

$$R^{\overline{\mathbf{F}_\bullet}} = \sum_{\ell\geq 0} R_\ell^{\overline{\mathbf{F}_\bullet}} = R_0^{\overline{\mathbf{F}_\bullet}} = \sum_{\kappa=\operatorname{codim} V(\Delta_1^{\mathbf{F}_\bullet})}^{\min(N,L)} R_{0,\kappa}^{\overline{\mathbf{F}_\bullet}}, \tag{3.97}$$

under the hypothesis in Proposition 3.36. Observe that in the particular case where F_1 is generically surjective, one has

$$V(\mathbf{F}_\bullet) = V(\Delta_1^{\mathbf{F}_\bullet}) = \text{zero locus of the annihilator of } \mathcal{F} = \frac{\mathcal{O}(\cdot, E_0)}{F_1\,\mathcal{O}(\cdot, E_1)}. \tag{3.98}$$

We conclude the section by formulating its main result.

THEOREM 3.38 (Duality theorem). *Let*

$$\overline{\mathbf{F}_\bullet} : 0 \longrightarrow \overline{E_L} \xrightarrow{F_L} \overline{E_{L-1}} \xrightarrow{F_{L-1}} \cdots \xrightarrow{F_2} \overline{E_1} \xrightarrow{F_1} \overline{E_0} \xrightarrow{F_0} \overline{E_{-1}} = \overline{\frac{E_0}{\operatorname{Im} F_1}} \longrightarrow 0$$

be a generically exact complex of metrized holomorphic bundles over a complex N-dimensional manifold $(\mathcal{X}, \mathcal{O})$. Assume that the associated complex of locally free sheaves of sections (3.93) is exact. Then, for any $h \in \mathcal{O}(\mathcal{X}, E_0)$ satisfying

$$h_z \in \big(\mathcal{O}(\cdot, \operatorname{Im} F_1)\big)_z \quad \forall z \text{ generic in } \mathcal{X}, \tag{3.99}$$

the following are equivalent:

$$R_0^{\overline{\mathbf{F}_\bullet}} h = \Big[\Big((\overline{\partial}\,|\Delta^{\mathbf{F}_\bullet}|^{2\lambda}) \sum_{r\geq 1} \frac{\sigma}{2i\pi} \Big(\overline{\partial}\Big(\frac{\sigma}{2i\pi}\Big)\Big)^{r-1}\Big)_{|E_0}\Big]_{\lambda=0} (h) = 0 \tag{3.100}$$

$$\iff h_z \in \big(\mathcal{O}(\cdot, \operatorname{Im} F_1)\big)_z \quad \forall z \in \mathcal{X}.$$

PROOF. The direct assertion ($\Longrightarrow$) follows from assertion (3.86) for $\ell = 0$ in Proposition 3.32. Since $R_1^{\overline{\mathbf{F}_\bullet}} = 0$, by Proposition 3.36, assertion (3.87) for $\ell = 0$ applies, which proves the converse assertion ($\Longleftarrow$). This concludes the proof of Theorem 3.38. □

We point out that most important are the cases where generically exact complexes of bundles lead to *explicitly tractable* exact complexes of locally free sheaves of sections.

THEOREM 3.39 (Duality theorem for complete intersections). *Let $\overline{E} = (E, |\ |)$ be a metrized holomorphic bundle over a complex manifold $\mathcal{X}$. Let also the section $s \in \mathcal{O}(\mathcal{X}, E)$ be such that* $\operatorname{codim} V(s) = \operatorname{rank} E$. *If $\mathcal{I}_s$ is the sheaf of ideals attached to the section s, then for any $k \in \mathbb{N}^*$ and any $h \in \mathcal{O}(\mathcal{X})$ one has that*

$$\big(h_z \in (\mathcal{I}_s)_z^k \quad \forall z \in V(s)\big) \iff R_0^{\overline{\mathbf{EN}_\bullet^{s,k}}} h = 0. \tag{3.101}$$

Furthermore, if $\operatorname{rank} E = m$, $\operatorname{codim} V(s) = m$ *and there are Cartier divisors $D_1, \ldots, D_m$ on $\mathcal{X}$ such that $s = s_1 \oplus \cdots \oplus s_m$, $s_j \in \mathcal{O}(\mathcal{X}, \mathcal{O}(D_j))$, then one has that*
(3.102)

$$\big(h_z \in (\mathcal{I}_s)_z^k \quad \forall z \in V(s)\big) \iff \big(R_0^{\overline{\mathbf{K}_\bullet}^{s^\kappa}} h = 0 \quad \forall \kappa \in (\mathbb{N}^*)^m \text{ with } |\kappa| = k + m - 1\big)$$

$$\iff \Big(h\,\overline{\partial}\Big(\frac{1}{\mathbf{s}^\kappa}\Big) = 0 \quad \forall \kappa \in (\mathbb{N}^*)^m \text{ with } |\kappa| = k + m - 1\Big)$$

for any $k \in \mathbb{N}^*$, where for any multi-index $\kappa \in (\mathbb{N}^*)^m$ we used the notation

$$s^\kappa = s_1^{\kappa_1} \oplus \cdots \oplus s_m^{\kappa_m}, \quad \mathbf{s}^\kappa = (s_1^{\kappa_1}, \ldots, s_m^{\kappa_m}).$$

PROOF. Observe that $\mathcal{I}_s$ is locally an example of a determinantal ideal sheaf in a neighborhood of any $z \in V(s)$ [**EaN62**]. Thus, the generically exact complex of sheaves associated with the Eagon–Northcott complex $\mathbf{EN}_\bullet^{s,k}$, as in (3.93), is exact. Then Theorem 3.38 applies. The second equivalence in (3.102) follows from Proposition 3.18. The first equivalence in (3.102) follows from the fact that if $\operatorname{codim} V(s_1 \oplus \cdots \oplus s_m) = m$, then, for any $k \in \mathbb{N}^*$ and for any $z \in V(s_1 \oplus \cdots \oplus s_m)$, one has

$$(\mathcal{I}_s)_z^k = \bigcap_{\{\kappa \in (\mathbb{N}^*)^m \,:\, |\kappa| = k+m-1\}} (\mathcal{I}_{s^\kappa})_z.$$

This follows, for example, by the now classical algebraic argument due to Melvin Hochster found in [**LipT81**, §3]. It remains to apply assertion (3.87) in Proposition 3.36 when $\mathbf{F}_\bullet = \mathbf{K}_\bullet^{s^\kappa}$ and $\ell = 0$, which is licit in our case according to (3.79) in Example 3.30. □

Our last proposition in this section concerns the *Cohen–Macaulay situation*[18] and Coleff–Herrera sheaves of bundle-valued currents which have been introduced previously in Definition 2.33 and Remark 2.36.

PROPOSITION 3.40. *Let $\mathcal{X}$ and $\overline{\mathbf{F}_\bullet}$ be as in Theorem 3.38, where F_1 is generically surjective and $L = \operatorname{codim} V(\mathbf{F}_\bullet) = M$. Then*

$$R^{\overline{\mathbf{F}_\bullet}} = R_0^{\overline{\mathbf{F}_\bullet}} = R_{0,M}^{\overline{\mathbf{F}_\bullet}} \in \mathrm{CH}^0_{\mathcal{X}, V^{[M]}(\mathbf{F}_\bullet)}(\mathcal{X}, \mathrm{Hom}_{\mathbb{C}}(E_0(\mathbf{F}_\bullet), E_M(\mathbf{F}_\bullet))), \tag{3.103}$$

where $V^{[M]}(\mathbf{F}_\bullet)$ denotes the union of irreducible components of V with codimension $M = L$.

PROOF. The two first equalities follow from assertion (3.97) stated in Remark 3.37. The fact that $R^{\overline{\mathbf{F}_\bullet}}$ is $\bar\partial$-closed follows from the fact that $\nabla_{\mathbf{F}_\bullet} R^{\overline{\mathbf{F}_\bullet}} = 0$ (see (3.72)) combined with the fact that

$$R^{\overline{\mathbf{F}_\bullet}} = R_{0,M}^{\overline{\mathbf{F}_\bullet}} \text{ with } M = L \Longrightarrow \boldsymbol{F}\, R^{\overline{\mathbf{F}_\bullet}} = 0.$$

The current $R^{\overline{\mathbf{F}_\bullet}}$ is an $\mathbf{E}(\mathbf{F}_\bullet)$-valued current (in fact a $\mathrm{Hom}_{\mathbb{C}}(E_0(\boldsymbol{F}), E_M(\mathbf{F}_\bullet))$-valued current) with bidegree $(0, M)$. Thus $R^{\overline{\mathbf{F}_\bullet}}$ is supported by the union $V^{[M]}(\mathbf{F}_\bullet)$ of irreducible components of $V(\mathbf{F}_\bullet)$ with codimension M, according to the dimension principle for pseudo-meromorphic currents; see Proposition 2.26. It is then a Coleff–Herrera current with respect to $V^{[M]}(\mathbf{F}_\bullet)$ according to Proposition 2.24 and Corollary 2.27. □

3.4. Bochner–Martinelli residue currents: structural results

3.4.1. Decomposition of $R^s_{|\,|}([V])$ along distinguished varieties. In this section, we consider a purely n-dimensional complex analytic space $(V, \mathcal{O}_V)$, together with its embedding $\iota_V : V \hookrightarrow \mathcal{X}$ in some ambient $N = n + M$-dimensional complex manifold $\mathcal{X}$ equipped with its structural sheaf $\mathcal{O}$. The structural sheaf $\mathcal{O}_V$ equals the quotient sheaf $\mathcal{O}_V/\mathcal{I}_V$, where $(\mathcal{I}_V)_z = \{h \in \mathcal{O}_z : h = 0 \text{ on } V_z\}$.[19] More

[18] We refer the reader to Definition A.21 for the notion of the Cohen–Macaulay (coherent) sheaf, which we will revisit in the forthcoming §3.4.2, §3.4.3, and 3.4.4.

[19] Important examples for us will be the cases where $\mathcal{X} = \mathbb{C}^N$ or $\mathbb{P}^N_{\mathbb{C}}$ and V is a purely n-dimensional algebraic affine or projective subvariety. Naturally, V is defined as the set of common zeroes of a finite collection of polynomials $p \in \mathbb{C}[X_1, \dots, X_N]$ or homogeneous polynomials P in $\mathbb{C}[X_0, \dots, X_N]$, expressed in affine coordinates $z_1, \dots, z_N$ or homogeneous coordinates

specifically, we focus in this section on a metrized holomorphic bundle E over $\mathcal{X}$ and its holomorphic section $s \in \mathcal{O}(\mathcal{X}, E)$ satisfying $\mathrm{codim}_V(V \cap V(s)) > 0$. As pointed in Definition 3.17, the $\bigwedge E^*$-valued current $R^s_{|\,|}([V])$, whose definition and properties have been listed in Proposition 3.15, can be also considered a $(\bigwedge E^*_{|V})$-valued current $R^{s,V}_{|\,|}$ on the (embedded) complex analytic space $(V, \mathcal{O}_V)$. We denote by

$$\mathcal{J} = \rfloor_{s_{|V}} \mathcal{O}_V(\,\cdot\,, E^*_{|V}) \tag{3.104}$$

the coherent sheaf of $\mathcal{O}_V$ ideals on the complex analytic space $(V, \mathcal{O}_V)$ attached to $s_{|V} \in \mathcal{O}_V(V, E_{|V})$. Let us first recall the following definition; see also Definition A.35 together with Remark A.38 in §A.4.2.

DEFINITION 3.41 (Distinguished varieties for $\mathcal{J}$). Let $\pi_{\mathcal{J}} : \widehat{V} \to V$ be the normalized blowup of V along the coherent ideal sheaf $\mathcal{J}$ and let $\mathscr{E}(\mathcal{J}) = \sum_\gamma m_\gamma \{H_\gamma\}$ be the Weil (exceptional) divisor on the normal complex analytic space $\widehat{V}$, which is associated[20] to the principal ideal sheaf $\mathcal{J}\,\mathcal{O}_{\widehat{V}}$. For any $r = 0, \dots, n$, let $\mathscr{V}_r(\mathcal{J})$ (respectively $\mathscr{H}_r(\mathcal{J})$) be the set of irreducible closed analytic subsets of V (respectively irreducible hypersurfaces H of $\widehat{V}$) defined by

$$\mathscr{V}_r(\mathcal{J}) = \{\widehat{\pi}_{\mathcal{J}}(H_\gamma) \, : \, \mathrm{codim}_V(\widehat{\pi}_{\mathcal{J}}(H_\gamma)) = r\} = \{\widehat{\pi}_{\mathcal{J}}(H_\gamma) \, : \, H_\gamma \in \mathscr{H}_r(\mathcal{J})\}. \tag{3.105}$$

The family $\mathscr{V}_r(\mathcal{J})$, whose indexation by γ is highly redundant in general, is called the collection of *distinguished varieties with codimension* r for the coherent subsheaf of $\mathcal{O}_V$-ideals $\mathcal{J}$. We will denote by $V_r(\mathcal{J})$ the closed analytic subset of V with codimension r defined as

$$V_r(\mathcal{J}) = \bigcup_{V \in \mathscr{V}_r(\mathcal{J})} V. \tag{3.106}$$

The current $R^{s,V}_{|\,|}$ admits then the following decomposition.

PROPOSITION 3.42. *Let $(\mathcal{X}, \mathcal{O})$, $(V, \mathcal{O}_V)$ be as above. Assume also that the embedding $\iota_V : V \hookrightarrow \mathcal{X}$, the vector bundle $\overline{E} = (E, |\,|)$, with $|\,|$ smooth and the section $s \in \mathcal{O}(\mathcal{X}, E)$ are as above. Then, the Bochner–Martinelli residue current $R^{s,V}_{|\,|}$ is decomposed, as a bundle-valued current on V, in the following manner*

$$R^{s,V}_{|\,|} = \sum_{r=1}^{n} (S^{s,V}_{|\,|,r} + N^{s,V}_{|\,|,r}), \tag{3.107}$$

where $S^{s,V}_{|\,|,r}$ and $N^{s,V}_{|\,|,r}$ are $\bigwedge^r E^*_{|V}$-valued pseudo-meromorphic currents with bidegree $(0, r)$, whose supports satisfy

$$\mathrm{Supp}\, S^{s,V}_{|\,|,r} \subset V_r(\mathcal{J}), \quad \mathrm{Supp}\, N^{s,V}_{|\,|,r} \subset \bigcup_{1 \le \rho < r} V_\rho(\mathcal{J}). \tag{3.108}$$

The current $S^{s,V}_{|\,|,r}$ satisfies the standard extension property across any closed hypersurface of $V_r(\mathcal{J})$ and one has $\mathbf{1}_W \cdot N^{s,V}_{|\,|,r} = 0$ for any closed analytic subset W of

---

$[z_0 : \cdots : z_N]$ correspondingly. Another important example will be that of a closed analytic subset V with pure dimension N in an open neighborhood $\mathcal{X} = \Omega$ of $\mathbf{0}$ in $\mathbb{C}^N$ such that $\mathbf{0} \in V$.

[20]The fact that the complex analytic space $(\widehat{V}, \mathcal{O}_{\widehat{V}})$ is normal, hence regular in codimension 1, authorizes its construction; see §A.4.2 for more details.

V satisfying

$$\mathrm{codim}_V\Big(W \cap \bigcup_{1\le\rho<r} V_\rho(\mathscr{J})\Big) > \mathrm{codim}_V W \ge r.$$

Moreover, both $S^{s,V}_{|\ |,r}$ and $N^{s,V}_{|\ |,r}$ are linear combinations of currents of the form $\widehat{\pi}_*(\tau\wedge\varpi)$, where τ is semimeromorphic current with bidegree $(0,1)$, independent of the metric $|\ |$, and ϖ is a smooth $(0,r-1)$-form. Furthermore, the coefficients of these linear combinations are independent of the metric $|\ |$ too.

Proof. We follow the proof of Proposition 3.15. Let us first express the smooth bundle sections $\widehat{\pi}^*(s_{|V})$ and $\widehat{\pi}^*\big((s^*/|s|^2)_{|V\setminus V(s)}\big)$ as

$$\widehat{\pi}^*(s_{|V}) = e_0\,\widehat{s},\quad \widehat{\pi}^*\Big(\Big(\frac{s^*}{|s|^2}\Big)_{|V\setminus V(s)}\Big) = \frac{1}{e_0}\,\widehat{\sigma},$$

where e_0 is a section of $\mathcal{O}_{\widehat{V}}(\mathscr{E})$, $\widehat{s}$ is a smooth nonvanishing section of the holomorphic bundle $\mathcal{O}_{\widehat{V}}(-\mathscr{E})\otimes\widehat{\pi}^*(E_{|V})$ on $\widehat{V}$, and $\widehat{\sigma}$ is a smooth section of $\mathcal{O}(\mathscr{E})\otimes\widehat{\pi}^*(E^*_{|V})$. For any $r=\mathrm{codim}_V(V(\mathscr{J})),\dots,n$ and any λ with $\mathrm{Re}\,\lambda\gg 1$, we have

$$\begin{aligned}\widehat{\pi}^*\Big((\overline{\partial}\,|s|^{2\lambda})\,\frac{s^*}{|s|^2}\,\Big(\overline{\partial}\Big(\frac{s^*}{|s|^2}\Big)^{r-1}\Big)\Big) &= (\overline{\partial}\,\|e_0\widehat{s}\|^{2\lambda})\,\frac{1}{e_0^r}\,\widehat{\sigma}\,(\overline{\partial}\widehat{\sigma})^{r-1}\\ &= \frac{|e_0|^{2\lambda}}{e_0^r}\,\overline{\partial}|\widehat{\sigma}|^{2\lambda}\,\widehat{\sigma}\,(\overline{\partial}\widehat{\sigma})^{r-1} + \lambda\,\frac{|e_0|^{2\lambda}}{e_0^r}|\widehat{s}|^{2\lambda}\Big(\sum_\gamma m_\gamma\frac{\overline{\partial h_\gamma}}{\overline{h}_\gamma}\Big)\,\widehat{\sigma}\,(\overline{\partial}\widehat{\sigma})^{r-1},\end{aligned}$$

where h_γ denotes a local reduced defining section of $\mathcal{O}_{\widehat{V}}$ for the irreducible hypersurface H_γ involved in the exceptional divisor $\mathscr{E}=\mathscr{E}(\mathscr{J})$ (independently of the choice of the reduced defining equation). The map

$$\lambda\longmapsto\widehat{\pi}^*\Big((\overline{\partial}\,|s|^{2\lambda})\,\frac{s^*}{|s|^2}\,\Big(\overline{\partial}\Big(\frac{s^*}{|s|^2}\Big)^{r-1}\Big)\Big),$$

once considered as a $\widehat{\pi}^*\bigwedge^r E^*_{|V}$- current valued map on $\widehat{V}$, extends to $\mathbb{C}$ as a meromorphic map with poles in $-\mathbb{Q}^+$, whose value at $\lambda=0$ equals

$$\Big[\lambda\,\frac{|e_0|^{2\lambda}}{e_0^r}\Big(\sum_\gamma m_\gamma\frac{\overline{\partial h_\gamma}}{\overline{h}_\gamma}\Big)\,\widehat{\sigma}\,(\overline{\partial}\widehat{\sigma})^{r-1}\Big]_{\lambda=0}.$$

The direct image current

$$\pi_*\Big(\Big[\lambda\,\frac{|e_0|^{2\lambda}}{e_0^r}\Big(\sum_{\{\gamma\,:\,H_\gamma\in\bigcup_{r<\rho\le n}\mathscr{H}_\rho(\mathscr{J})\}} m_\gamma\frac{\overline{\partial h_\gamma}}{\overline{h}_\gamma}\Big)\,\widehat{\sigma}\,(\overline{\partial}\widehat{\sigma})^{r-1}\Big]_{\lambda=0}\Big)$$

is a pseudo-meromorphic current with bidegree $(0,r)$. It is supported by a closed analytic subset of V with codimension in V strictly greater than r. Then, it follows from the dimension principle (see Proposition 2.26 combined with Remark 2.29 in the setting of complex analytic spaces) that

$$\widehat{\pi}_*\Big(\Big[\lambda\,\frac{|e_0|^{2\lambda}}{e_0^r}\Big(\sum_{\{\gamma\,:\,H_\gamma\in\bigcup_{r<\rho\le n}\mathscr{H}_\rho(\mathscr{J})\}} m_\gamma\frac{\overline{\partial h_\gamma}}{\overline{h}_\gamma}\Big)\,\widehat{\sigma}\,(\overline{\partial}\widehat{\sigma})^{r-1}\Big]_{\lambda=0}\Big) = 0.$$

Therefore, we have, for each $r = \mathrm{codim}_V(V(\mathcal{J})), \dots, n$,

$$\begin{aligned}(3.109)\quad & \Big[(\overline{\partial}\,|s|^{2\lambda})\,\frac{s^*}{2i\pi|s|^2}\Big(\overline{\partial}\Big(\frac{s^*}{2i\pi|s|^2}\Big)^{r-1}\Big)\Big]_{\lambda=0} \\ &= \widehat{\pi}_*\Big(\frac{1}{(2i\pi)^r}\Big[\lambda\,\frac{|e_0|^{2\lambda}}{e_0^r}\Big(\sum_{\{\gamma\,:\,H_\gamma\in\bigcup_{\rho\le r}\mathscr{H}_\rho(\mathcal{J})\}} m_\gamma\frac{\overline{\partial}h_\gamma}{\overline{h}_\gamma}\Big)\,\widehat{\sigma}\,(\overline{\partial}\widehat{\sigma})^{r-1}\Big]_{\lambda=0}\Big).\end{aligned}$$

The pseudo-meromorphic $(0,r)$-current

$$S^{s,V}_{|\,|,r} = \widehat{\pi}_*\Big(\frac{1}{(2i\pi)^r}\Big[\lambda\,\frac{|e_0|^{2\lambda}}{e_0^r}\Big(\sum_{\{\gamma\,:\,H_\gamma\in\mathscr{H}_r(\mathcal{J})\}} m_\gamma\frac{\overline{\partial}h_\gamma}{\overline{h}_\gamma}\Big)\,\widehat{\sigma}\,(\overline{\partial}\widehat{\sigma})^{r-1}\Big]_{\lambda=0}\Big)$$

is supported by the distinguished variety $V_r(\mathcal{J})$ with pure codimension r in V. Therefore, it satisfies the standard extension property for any hypersurface of $V_r(\mathcal{J})$. The $(0,r)$ pseudo-meromorphic current

$$N^{s,V}_{|\,|,r} = \widehat{\pi}_*\Big(\frac{1}{(2i\pi)^r}\Big[\lambda\,\frac{|e_0|^{2\lambda}}{e_0^r}\Big(\sum_{\{\gamma\,:\,H_\gamma\in\bigcup_{1\le\rho<r}\mathscr{H}_\rho(\mathcal{J})\}} m_\gamma\frac{\overline{\partial}h_\gamma}{\overline{h}_\gamma}\Big)\,\widehat{\sigma}\,(\overline{\partial}\widehat{\sigma})^{r-1}\Big]_{\lambda=0}\Big)$$

is supported by the closed analytic subset $\bigcup_{1\le\rho<r} V_\rho(\mathcal{J})$. Now, let W be a closed subset of V satisfying

$$\mathrm{codim}_V\Big(W\cap\bigcup_{1\le\rho<r} V_\rho(\mathcal{J})\Big) > \mathrm{codim}_V W \ge r.$$

The dimension principle (see again Proposition 2.26 combined with Remark 2.29 in the setting of complex analytic spaces) implies that

$$\mathbf{1}_W\cdot N^{s,V}_{|\,|,r} = 0$$

as a $(0,r)$ pseudo-meromorphic current with support contained in a closed analytic subset of V with codimension strictly larger than r in V. This concludes the proof of Proposition 3.42. □

REMARK 3.43. The explicit decomposition (3.107) for residue currents of the Bochner–Martinelli type described in Proposition 3.42 can be understood as a substitute for Siu's stratification theorem ([**Siu74, De92**]) for d-closed positive currents; see also Theorem C.28. The decomposition of currents described by Proposition 3.42 will be a useful tool for us later on when revisiting King's theorem [**Ki70**] and improper intersection theory.[21]

3.4.2. $R^{\overline{\mathbf{F}_\bullet}}$ and the Buchsbaum–Eisenbud–Fitting sequence. Let $\mathcal{X}$ be an N-dimensional complex manifold, equipped with its structural sheaf $\mathcal{O}$. We focus in this section on a coherent sheaf of $\mathcal{O}$-ideals $\mathcal{I}$ satisfying

$$(3.110)\qquad V(\mathcal{I}) = \mathrm{Supp}\,\frac{\mathcal{O}}{\mathcal{I}},\quad 0<\mathrm{codim}\,V(\mathcal{I}) = M \le N.$$

[21]Note that King's theorem has been generalized in [**AndSWY17B**, Theorem 1.1] and [**AndESWY21A**, Theorem 1.1] in the local versus global setting, and from cycles to generalized cycles. See also Proposition C.25, Examples D.6, D.10, and D.16.

We will assume that there exists a generically exact sequence of holomorphic bundles

(3.111)
$$\mathbf{F}_\bullet : 0 \longrightarrow E_L \xrightarrow{F_L} E_{L-1} \xrightarrow{F_{L-1}} \cdots \xrightarrow{F_2} E_1 \xrightarrow{F_1} E_0 \xrightarrow{F_0} E_{-1} = \frac{E_0}{\operatorname{Im} F_1} \longrightarrow 0,$$

where the homomorphism F_1 is generically surjective, the complex of locally free sheaves of sections

$$0 \longrightarrow \mathcal{O}(\cdot, E_L) \xrightarrow{F_L} \mathcal{O}(\cdot, E_{L-1}) \xrightarrow{F_{L-1}} \cdots \xrightarrow{F_2} \mathcal{O}(\cdot, E_1) \xrightarrow{F_1} \mathcal{O}(\cdot, E_0) \tag{3.112}$$

is exact and, in addition,

$$\mathcal{I} = \operatorname{Im}\big(\mathcal{O}(\cdot, E_1) \xrightarrow{F_1} \mathcal{O}(\cdot, E_0)\big). \tag{3.113}$$

Condition (3.113) implies that $L \geq M$. This follows from Theorem A.9 and Proposition A.17. Moreover, Theorem A.9 implies that the decreasing sequence of closed analytic subsets $V_r(\mathbf{F}_\bullet)$, defined in (3.58),

$$V_1(\mathbf{F}_\bullet) \supset V_2(\mathbf{F}_\bullet) \supset \cdots \supset V_r(\mathbf{F}_\bullet) \supset \cdots$$

satisfies the Buchsbaum–Eisenbud conditions

$$\operatorname{codim} V_r(\mathbf{F}_\bullet) \geq r \quad \forall r \in \mathbb{N}^* \tag{3.114}$$

(recalled in §3.3.3 and used in the proof of Proposition 3.36). The decreasing sequence $\{V_k(\mathbf{F}_\bullet)\}_k$ depends only of the coherent sheaf $\mathcal{I}$. It does not depend on the exact complex of locally free sheaves (3.112). Neither does it depend on the generically exact complex (3.111), in which $\mathcal{I}$ is involved. Since F_1 is generically surjective, one has, in particular, $V_1(\mathbf{F}_\bullet) = V$; see (3.98) in Remark 3.37. Moreover (see [**Eis**, Corollary 20.12]) one has that

$$V_r(\mathbf{F}_\bullet) = V_1(\mathbf{F}_\bullet) \quad \forall r = 1, \ldots, M. \tag{3.115}$$

DEFINITION 3.44 (Buchsbaum–Eisenbud–Fitting[22] sequence). For each r in $\mathbb{N}^*$, the closed analytic subset $V_r(\mathbf{F}_\bullet)$ depends only on the coherent sheaf of $\mathcal{O}$-ideals $\mathcal{I}$. It is called the r-Buchsbaum–Eisenbud–Fitting closed analytic subset attached to the coherent sheaf of $\mathcal{O}$-ideals $\mathcal{I}$ and is denoted by $\mathbb{V}_r(\mathcal{I})$. If M denotes the codimension of $\operatorname{Supp}(\mathcal{O}/\mathcal{I})$, then the decreasing sequence

$$\mathbb{V}_1(\mathcal{I}) = \mathbb{V}_2(\mathcal{I}) = \cdots = \mathbb{V}_M(\mathcal{I}) \supset \mathbb{V}_{M+1}(\mathcal{I}) \cdots \supset \mathbb{V}_r(\mathcal{I}) \supset \cdots, \tag{3.116}$$

is called the *Buchsbaum–Eisenbud–Fitting* sequence of $\mathcal{I}$.

PROPOSITION 3.45. *Assume that the Buchsbaum–Eisenbud–Fitting conditions* (3.114) *are formulated in their strong form*

$$\mathbb{V}_1(\mathcal{I}) = \cdots = \mathbb{V}_M(\mathcal{I}) = V, \quad \operatorname{codim}(\mathbb{V}_{M+\rho}(\mathcal{I})) > M + \rho \quad \forall \rho \geq 1, \tag{3.117}$$

[22]The coherent sheaf $\mathcal{I}_r(\mathbf{F}_\bullet)$ attached to the section $\det^{\rho_r} F_r$ of $\operatorname{Hom}_{\mathbb{C}}(E_r, E_{r-1})$ satisfies $\operatorname{Supp}\big(\mathcal{O}/\mathcal{I}_r(\mathbf{F}_\bullet)\big) = V_r(\mathbf{F}_\bullet)$; see (3.58). It does not depend on the exact complex of locally free sheaves (3.112), or on the generically exact complex of holomorphic bundles (3.111). The sheaf $\mathcal{I}_r(\mathbf{F}_\bullet)$ is called the *r-Fitting ideal* of the coherent sheaf $\mathcal{I}$. This is the reason why we add the name of Hans Fitting here; see [**Fi36**] and [**Eis**, §20.2].

and that the generically exact complex $\mathbf{F}_\bullet$ *is metrized. Then the Bochner–Martinelli residue current* $R^{\overline{\mathbf{F}_\bullet}}$ *can be computed through the following iterative procedure:*
(3.118)

$$T^{[0]} = R^{\overline{\mathbf{F}_\bullet}}_{0,M} = \Big[(\overline{\partial}\,|\Delta_1^{\mathbf{F}_\bullet}|^{2\lambda})\,\frac{\overline{\partial}\sigma_M\cdots\overline{\partial}\sigma_2\,\sigma_1}{(2i\pi)^M}\Big]_{\lambda=0},$$

$$T^{[\rho+1]} = T^{[\rho]} + \Big[|\Delta^{\mathbf{F}_\bullet}_{M+\rho+1}|^{2\lambda}\,\frac{\overline{\partial}\sigma_{M+\rho+1}}{2i\pi}\,T^{[\rho]}\Big]_{\lambda=0} \quad \text{for } \rho = 0,\dots,N-M-1,$$

$$R^{\overline{\mathbf{F}_\bullet}} = T^{[N-M]}.$$

PROOF. As pointed out in Remark 3.37, the proof of Proposition 3.36 implies also that

$$R^{\overline{\mathbf{F}_\bullet}} = \sum_{\rho=0}^{N-M} R^{\overline{\mathbf{F}_\bullet}}_{0,M+\rho} = \sum_{\rho=0}^{N-M}\Big[(\overline{\partial}\,|\Delta^{\mathbf{F}_\bullet}|^{2\lambda})\,\frac{\overline{\partial}\sigma_{M+\rho}\cdots\overline{\partial}\sigma_2\,\sigma_1}{(2i\pi)^{M+\rho}}\Big]_{\lambda=0}. \tag{3.119}$$

For any $\rho\in\mathbb{N}^*$, the $(0, M+\rho)$ pseudo-meromorphic current

$$R^{\overline{\mathbf{F}_\bullet}}_{0,M+\rho} = \Big[(\overline{\partial}\,|\Delta^{\mathbf{F}_\bullet}|^{2\lambda})\,\frac{\overline{\partial}\sigma_{M+\rho}\cdots\overline{\partial}\sigma_2\,\sigma_1}{(2i\pi)^{M+\rho}}\Big]_{\lambda=0}$$

satisfies $\mathbf{1}_{\mathbb{V}_{M+\rho}(\mathcal{I})}\,R^{\overline{\mathbf{F}_\bullet}}_{0,M+\rho} = 0$, because of the dimension principle for pseudo-meromorphic currents (Proposition 2.26) and the strong Buchsbaum–Eisenbud–Fitting conditions (3.117). Consider as a starting point the case where $\rho = 1$. The $(0, M+1)$ pseudo-meromorphic currents

$$R^{\overline{\mathbf{F}_\bullet}}_{0,M+1}, \quad\text{and}\quad \Big[\Big(|\Delta^{\mathbf{F}_\bullet}_{M+1}|^{2\lambda}\,\frac{\overline{\partial}\,\sigma_{M+1}}{2i\pi}\Big)\,R^{\overline{\mathbf{F}_\bullet}}_{0,M}\Big]_{\lambda=0}$$

are such that their restrictions to $\mathbb{V}_{M+1}(\mathcal{I})$ equal 0. Moreover, their restrictions to $\mathcal{X}\setminus\mathbb{V}_{M+1}(\mathcal{I})$ coincide because of the analytic continuation principle. They are therefore equal. We now repeat the reasoning with the currents

$$R^{\overline{\mathbf{F}_\bullet}}_{0,M+2} \quad\text{and}\quad \Big[\Big(|\Delta^{\mathbf{F}_\bullet}_{M+2}|^{2\lambda'}\,\frac{\overline{\partial}\,\sigma_{M+2}}{2i\pi}\Big)\Big[\Big(|\Delta^{\mathbf{F}_\bullet}_{M+1}|^{2\lambda}\,\frac{\overline{\partial}\,\sigma_{M+1}}{2i\pi}\Big)\,R^{\overline{\mathbf{F}_\bullet}}_{0,M}\Big]_{\lambda=0}\Big]_{\lambda'=0}.$$

They are still such that their restrictions to $\mathbb{V}_{N+2}(\mathcal{I})$ equal 0 because of the dimension principle, since $\operatorname{codim}\mathbb{V}_{M+2}(\mathcal{I}) > M+2$. Furthermore, their restrictions to $\mathcal{X}\setminus\mathbb{V}_{M+2}(\mathcal{I})$, coincide in view of the analytic continuation principle. Such a process iterates. Proposition 3.45 is thus proved. □

EXAMPLE 3.46 (The case of radical sheaves of ideals). If $\mathcal{I}$ is a radical sheaf, that is $\sqrt{\mathcal{I}_z} = \mathcal{I}_z$ for all $z\in V = \operatorname{Supp}(\mathcal{O}/\mathcal{I})$, then re-inforced Buchsbaum–Eisenbud–Fitting conditions (3.117) hold when the complex of locally free sheaves (3.112) is exact; see [**Eis**, Corollary 20.14]. Moreover, it follows from the unicity of minimal free resolutions (see Theorem A.9) that $\mathbb{V}_{M+1}(\mathcal{I})\subset V_{\rm sing}$ in such a case. The same holds if $\mathcal{I}$ is a coherent sheaf of ideals such that, for each $z\in\operatorname{Supp}(\mathcal{O}/\mathcal{I})$, all associated primes of $(\mathcal{O}/\mathcal{I})_z$ are isolated with the same codimension M, independent on z.[23] In particular, if $z\in(\operatorname{Supp}(\mathcal{O}/\mathcal{I}))_{\rm reg} = V_{\rm reg}$, $(\mathcal{O}/\mathcal{I})_z$ has all its associated primes isolated, which implies that it is a free $(\mathcal{O}/\sqrt{\mathcal{I}})_z$-module, where $(\mathcal{O}/\sqrt{\mathcal{I}})_z$ admits a minimal resolution with length M since $z\in V_{\rm reg}$. Unicity of minimal resolutions imply then that $\mathbb{V}_{M+1}(\mathcal{I})\subset V_{\rm sing}$ also in this case. In case

[23]Such a coherent sheaf of ideals is said to be *purely dimensional.*

$\mathcal{X}$ is a Stein manifold, shrinking $\mathcal{X}$ if necessary,[24] one may assume that there exists a locally free resolution of minimal length L

$$0 \longrightarrow \mathscr{L}^{r_L} \xrightarrow{\mathfrak{f}_L} \mathscr{L}^{r_{L-1}} \xrightarrow{\mathfrak{f}_{L-1}} \cdots \mathscr{L}^{r_2} \xrightarrow{\mathfrak{f}_2} \mathscr{L}^{r_1} \xrightarrow{\mathfrak{f}_1} \mathscr{L}^{r_0}$$

with $\operatorname{coker}(\mathfrak{f}_1) = \mathcal{O}/\mathcal{I}_V$; see Proposition A.12. One can even take $\mathscr{L}_{\mathcal{X}}^{r_\ell} = \mathcal{O}^{\oplus r_\ell}$ for any $\ell \geq 1$, except eventually the case when $\ell = L$. The particular case $L \leq N-1$ will be studied more closely below; see §3.4.3.

If Proposition 3.45 applies, as in the case of the radical sheaf $\mathcal{I} = \mathcal{I}_V$, with $\operatorname{codim} V = M$ (see Example 3.46) then it follows from (3.118) that an extrinsic[25] structural description of the current $R^{\overline{\mathbf{F}_\bullet}}$ is subordinate to that of the current

$$R^{\overline{\mathbf{F}_\bullet}}_{0,M} = \Big[(\overline{\partial}\, |\Delta_1^{\mathbf{F}_\bullet}|^{2\lambda})\, \frac{\overline{\partial}\sigma_M \cdots \overline{\partial}\sigma_2\, \sigma_1}{(2i\pi)^M} \Big]_{\lambda=0}. \tag{3.120}$$

Let us then come back to the general situation of an ideal sheaf $\mathcal{I}$ such that there exists a generically exact complex (3.111) with F_1 being generically surjective and the locally free complex (3.112) is exact. We do not assume any additional hypothesis on $\mathcal{I}$ for the moment, besides the fact that the support V of $\mathcal{O}/\mathcal{I}$ is a closed analytic subset of $\mathcal{X}$ with codimension $M \leq N$. We have seen in Proposition 3.40 that, in the case where one could choose the generically exact complex (3.111) with $L = M$, then

$$R^{\overline{\mathbf{F}_\bullet}}_{0,M} \in \mathrm{CH}_{\mathcal{X},V}\big(\mathcal{X}, \mathrm{Hom}_{\mathbb{C}}(E_0, E_M)\big). \tag{3.121}$$

In general, the situation is much more involved. Under the sole hypothesis that $\mathcal{X}$ is Stein, one can state the following result [**AndS12**, §3].

PROPOSITION 3.47. *There exists a trivial holomorphic bundle* $(\mathbb{E}_M, \|\ \|)$ *over* $\mathcal{X}$*, equipped with its hermitian canonical metric, and a holomorphic section* G *of* $\mathrm{Hom}_{\mathbb{C}}(E_M, \mathbb{E}_M)$ *with generic rank* ρ_G*, such that if* σ_G *denotes the minimal inverse of* $G\ : \overline{E_M} \to \mathbb{E}_M$*, then*

$$\begin{aligned} R^{\overline{\mathbf{F}_\bullet}}_{0,M} &= \Big[\big|\det{}^{\rho_G} G\big|^{2\lambda}\, \sigma_G\, G R^{\overline{\mathbf{F}_\bullet}}_{0,M} \Big]_{\lambda=0} \\ G R^{\overline{\mathbf{F}_\bullet}}_{0,M} &\in \mathrm{CH}^0_{\mathcal{X},V}\big(\mathcal{X}, \mathrm{Hom}_{\mathbb{C}}(E_0, \mathbb{E}_M)\big). \end{aligned} \tag{3.122}$$

PROOF. One has

$$\begin{aligned} R^{\overline{\mathbf{F}_\bullet}}_{0,M} &= \Big[(\overline{\partial}\, |\Delta_1^{\mathbf{F}_\bullet}|^{2\lambda})\, \frac{\overline{\partial}\sigma_M \cdots \overline{\partial}\sigma_2\, \sigma_1}{(2i\pi)^M} \Big]_{\lambda=0} \\ &= \Big[(\overline{\partial}\, |\Delta_1^{\mathbf{F}_\bullet}|^{2\lambda})\, \frac{\sigma_M\, \overline{\partial}\sigma_{M-1} \cdots \overline{\partial}\sigma_1}{(2i\pi)^M} \Big]_{\lambda=0} \\ &= \Big[\big(\mathrm{Id}_{E_M} - F_{M+1}\sigma_{M+1}\big)\, (\overline{\partial}\, |\Delta_1^{\mathbf{F}_\bullet}|^{2\lambda})\, \frac{\sigma_M\, \overline{\partial}\sigma_{M-1} \cdots \overline{\partial}\sigma_1}{(2i\pi)^M} \Big]_{\lambda=0} \end{aligned}$$

since $\sigma^2 = 0$. One has

$$\sigma_{M+1}\overline{\partial}\sigma_M \cdots \overline{\partial}\sigma_2\, \sigma_1 = \sigma_{M+1}\sigma_M\, \overline{\partial}\sigma_{M-1} \cdots \overline{\partial}\sigma_1 = 0 \quad \text{outside } \mathbb{V}_{M+1}(\mathcal{I}).$$

[24]One may replace $\mathcal{X}$ by some convenient open neighborhood U_ν of a holomorphically convex compact subset K_{ν_0}, with $\nu_0 \gg 1$, taken from an exhaustion of $\mathcal{X} = \bigcup_{\nu \geq 1} K_\nu$ by compact subsets satisfying $K_\nu \subset \operatorname{int}(K_{\nu+1})$ for any ν.

[25]By *extrinsic*, we mean here a description of this $\mathbf{E}(\mathbf{F}_\bullet)$-valued current, supported by V, as a current in the ambient manifold $\mathcal{X}$. This description will be completed in §3.4.3 by an *intrinsic* one with respect to its support V in the particular case where $\mathcal{I} = \mathcal{I}_V$.

Since the pseudo-meromorphic $(0,M)$-current $R^{\overline{\mathbf{F}_\bullet}}_{0,M}$ has the standard extension property across any hypersurface of V, in particular across $\mathbb{V}_{M+1}(\mathcal{I}) \subset V$ whose codimension in V is strictly positive according to the Buchsbaum–Eisenbud–Fitting conditions, $R^{\overline{\mathbf{F}_\bullet}}_{0,M}$ is an $\operatorname{Im} F_{M+1}^{\perp}$-valued current. Consider now the dual complex of a generically exact complex of holomorphic bundles

$$(\mathbf{F}_\bullet)^* : 0 \longrightarrow E_0^* \xrightarrow{F_1^*} E_1^* \longrightarrow \cdots \longrightarrow E_{M-1}^* \xrightarrow{F_M^*} E_M^* \xrightarrow{F_{M+1}^*} E_{M+1}^* \longrightarrow \cdots$$

and its associated complex of locally free sheaves

$$0 \longrightarrow \mathcal{O}(\cdot, E_0^*) \xrightarrow{F_1^*} \cdots \longrightarrow \mathcal{O}(\cdot, E_{M-1}^*) \xrightarrow{F_M^*} \mathcal{O}(\cdot, E_M^*) \xrightarrow{F_{M+1}^*} \mathcal{O}(\cdot, E_{M+1}^*) \longrightarrow \cdots.$$

Since $\mathcal{X}$ is Stein, it follows from the coherence of the ideal sheaf

$$\operatorname{Ker}\big(\mathcal{O}(\cdot, E_M^*) \xrightarrow{F_{M+1}^*} \mathcal{O}(\cdot, E_{M+1}^*)\big)$$

(see §A.2.2) that there exists a trivial holomorphic hermitian bundle $\mathbb{E}_M$ over $\mathcal{X}$, together with a holomorphic morphism $G \in \mathcal{O}\big(\mathcal{X}, \operatorname{Hom}_{\mathbb{C}}(E_M, \mathbb{E}_M)\big)$, such that the corresponding complex of locally free sheaves of sections

$$\mathcal{O}(\cdot, \mathbb{E}_M^*) \xrightarrow{G^*} \mathcal{O}(\cdot, E_M^*) \xrightarrow{F_{M+1}^*} \mathcal{O}(\cdot, E_{M+1}^*) \tag{3.123}$$

is exact. Since the rank of F_{M+1} remains constant above $\mathcal{X} \setminus \mathbb{V}_{M+1}(\mathcal{I})$, so does the rank of F_{M+1}^*. Thus, the rank ρ_G of G remains constant too. Inspired by the construction (3.60) of the minimal inverse σ_ℓ for F_ℓ in §3.3.2, one constructs a smooth mapping

$$\sigma_G \,: (\mathbb{E}_M)_{|\mathcal{X}\setminus \mathbb{V}_{M+1}(\mathcal{I})} \longrightarrow (E_M)_{|\mathcal{X}\setminus \mathbb{V}_{M+1}(\mathcal{I})}.$$

This smooth mapping above $\mathcal{X} \setminus \mathbb{V}_{M+1}(\mathcal{I})$ satisfies the properties

$$\begin{aligned} &(\operatorname{Im} G)^{\perp} \subset \operatorname{Ker} \sigma_G, \quad \operatorname{Im} \sigma_G \perp \operatorname{Ker} G, \\ &\sigma_G \circ G = \operatorname{Id} \text{ on } (\operatorname{Ker} G)^{\perp} = (\operatorname{Im} F_{M+1})^{\perp}. \end{aligned}$$

Since G is holomorphic and the complex (3.123) is exact, one has

$$\overline{\partial}(G\, R^{\overline{\mathbf{F}_\bullet}}_{0,M}) = -G\, \overline{\partial} R^{\overline{\mathbf{F}_\bullet}}_{0,M} = -2i\pi G\, F_{M+1}\, R^{\overline{\mathbf{F}_\bullet}}_{0,M} = 0,$$

taking into account that $\nabla_{\mathbf{F}_\bullet} R^{\overline{\mathbf{F}_\bullet}} = \nabla_{\mathbf{F}_\bullet} R_0^{\overline{\mathbf{F}_\bullet}} = 0$ (see (3.72)) and that $G \circ F_{M+1} = 0$ according to the exactness of the complex (3.123). Then the pseudo-meromorphic current $GR^{\overline{\mathbf{F}_\bullet}}$, supported by V of bidegree or having a bidegree $(0,M)$, satisfies all requirements to belong to $\mathrm{CH}^0_{\mathcal{X},V}(\mathcal{X}, \operatorname{Hom}_{\mathbb{C}}(E_0, \mathbb{E}_M))$. Since $R^{\overline{\mathbf{F}_\bullet}}_{0,M}$ is $(\operatorname{Im} F_{M+1})^{\perp}$-valued and $\sigma_G \circ G = \operatorname{Id}$ on $(\operatorname{Im} F_{M+1})^{\perp}$ over $\mathcal{X} \setminus \mathbb{V}_{M+1}(\mathcal{I})$, the first equality in (3.122) follows if one invokes once more the fact that both pseudo-meromorphic $(0,M)$-currents involved satisfy the standard extension property across any hypersurface of V. In particular, both currents retain the standard extension property across the hypersurface $\mathbb{V}_{M+1}(\mathcal{I})$, whose codimension is at least $M+1$. This completes the proof of Proposition 3.47. □

3.4.3. Residue currents and structure forms. Let $(\mathcal{X}, \mathcal{O})$ be a complex N-dimensional Stein manifold and let $V \subset \mathcal{X}$ be a closed analytic subset with pure codimension M. In this subsection, we consider as coherent sheaf of $\mathcal{O}$-ideals $\mathcal{I}$ the radical sheaf $\mathcal{I} = \mathcal{I}_V$ and the (reduced) complex analytic space $(V, \mathcal{O}/\mathcal{I}_V)$, with pure dimension $N-M$, equipped with its *dualizing sheaf* ω_V^{N-M}; see Remark B.35. We assume, as in the previous section, that there exists a generically exact sequence of metrized holomorphic bundles (3.111). We assume in addition that the complex $\overline{\mathbf{F}_\bullet}$ has length[26] L equal to $N-1$, that F_1 is generically surjective, and moreover (this is the most important point) that the associated complex of locally free sheaves (3.112) is exact, with

$$\mathcal{I}_V = \operatorname{Im}\big(\mathcal{O}(\cdot, E_1) \xrightarrow{F_1} \mathcal{O}(\cdot, E_0)\big), \tag{3.124}$$

where $\operatorname{rank} E_0 = 1$, that is, $E_0 = \mathcal{O}(D)$. One recalls that for any $r \in \mathbb{N}^*$, the r-Fitting ideal sheaf $\mathcal{I}_r(\mathbf{F}_\bullet) = \mathcal{I}_{\Delta_r^{\mathbf{F}_\bullet}}$ depends only on the radical ideal sheaf $\mathcal{I}_V$ and thus will be denoted by $\operatorname{Fitt}_r(V)$. Moreover, if one denotes as

$$\mathbb{V}_r(\mathcal{I}_V) = \mathbb{V}_r = \operatorname{Supp}(\mathcal{O}/\operatorname{Fitt}_r(V)),$$

then one has

$$\begin{aligned} \mathbb{V}_1 = \cdots = \mathbb{V}_M = V \supset V_{\mathrm{sing}} \supset \mathbb{V}_{M+1} \supset \cdots \supset \mathbb{V}_{M+\rho} \supset \mathbb{V}_{M+\rho+1} \supset \cdots \\ \supset \operatorname{codim} \mathbb{V}_{M+\rho} > M+\rho \quad \forall \rho \geq 1. \end{aligned} \tag{3.125}$$

Consider the following sequence of normalized blowups:
(3.126)

$$\widehat{V}^{[N-M-1]} \longrightarrow \cdots \longrightarrow \widehat{V}^{[\rho+1]} \xrightarrow{\widehat{\pi}_{\rho+1}} \widehat{V}^{[\rho]} \xrightarrow{\widehat{\pi}_\rho} \cdots \longrightarrow \widehat{V}^{[2]} \xrightarrow{\widehat{\pi}_2} \widehat{V}^{[1]} \xrightarrow{\widehat{\pi}_1} V = \mathbb{V}_M,$$

where $\widehat{\pi}_1$ denotes the normalized blowup[27] of the complex analytic space V along the sheaf of ideals $\iota_V^* \operatorname{Fitt}_{M+1}(V)$ (with exceptional Cartier divisor $\widehat{D}_1$) and $\widehat{\pi}_{\rho+1}$ denotes the normalized blowup of $\widehat{V}^{[\rho]}$ along the sheaf of ideals

$$(\iota_V \circ \widehat{\pi}_1 \circ \cdots \circ \widehat{\pi}_\rho)^* \operatorname{Fitt}_{M+\rho+1}(V),$$

for any $\rho \geq 1$. Let also

$$\widehat{\Pi}_V = \widehat{\pi}_1 \circ \cdots \circ \widehat{\pi}_{N-M-1} : \widehat{V}^{[N-M-1]} \longrightarrow V. \tag{3.127}$$

We recall then that

$$R^{\overline{\mathbf{F}_\bullet}} = R_0^{\overline{\mathbf{F}_\bullet}} = \sum_{\rho=0}^{N-M-1} R_{0,M+\rho}^{\overline{\mathbf{F}_\bullet}}, \tag{3.128}$$

and that the current $R^{\overline{\mathbf{F}_\bullet}}$ can be recovered from $R_{0,M}^{\overline{\mathbf{F}_\bullet}}$ through the inductive process (3.118); see Proposition 3.45.

Since $\mathcal{X}$ is a Stein manifold, it can be embedded in $\mathbb{C}^{2N+1}$; see §A.2.2. Thus the canonical line bundle $T_{\mathcal{X}}^{*(N,0)}$ admits global holomorphic sections. We fix

[26] There exists always such a resolution because of reinforced Buchsbaum–Eisenbud–Fitting constraints (3.125), since a generically exact complex $\mathbf{F}_\bullet$ with length L such that the sequence of locally free sheaves of sections (3.112) is exact, exists as soon as $\mathbb{V}_r(\mathcal{I}) = \emptyset$ for $r > L$. This is the case if one takes $L = N-1$, which we will therefore do.

[27] See Definition A.4.2.2, together with Remark A.38.

such a section and denoted it by Υ. The following proposition provides a *structural intrinsic description*[28] of the $(0,M)$ pseudo-meromorphic current $R^{\overline{\mathbf{F}_\bullet}}_{0,M} \in \mathrm{PM}(\mathcal{X}, \mathrm{Hom}_{\mathbb{C}}(E_0, E_M))$.

PROPOSITION 3.48. *There exists a trivial holomorphic bundle $\overline{\mathfrak{E}} = (\mathfrak{E}, \|\ \|)$ over V, together with $\mathfrak{f} \in \omega_V^{N-M}(V, \mathrm{Hom}_{\mathbb{C}}(\iota_V^* E_0, \mathfrak{E}))$ such that*

$$R^{\overline{\mathbf{F}_\bullet}}_{0,M} \wedge \Upsilon = (\iota_V)_* \, \omega_0, \tag{3.129}$$

where $\omega_0 = \omega_0(\overline{\mathbf{F}_\bullet}, \Upsilon)$ is the direct image through $\widehat{\pi}_1$ of

$$\Big[\frac{|e_0|^{2\lambda}}{e_0}\, \mathfrak{s}_0\, \widehat{\pi}_1^* \mathfrak{f}\Big]_{\lambda=0}.$$

The holomorphic section e_0 belongs to the line bundle $\mathcal{O}_{\widehat{V}^{[1]}}(\widehat{D}_1)$, and $\mathfrak{s}_0$ is a smooth section of $\mathcal{O}_{\widehat{V}^{[1]}}(\widehat{D}_1) \otimes \widehat{\pi}_1^*(\mathrm{Hom}_{\mathbb{C}}(\mathfrak{E}, (E_M)_{|V}))$ depending on the metrics on E_M and on E_{M+1}. The hermitian metric $|\ |$ on the line bundle $\mathcal{O}_{\widehat{V}^{[1]}}(\widehat{D}_1)$ is arbitrary.

PROOF. Let $(\mathbb{E}_M, \|\ \|)$, $G \in \mathcal{O}(\mathcal{X}, \mathrm{Hom}_{\mathbb{C}}(E_M, \mathbb{E}_M))$, and its minimal inverse $\sigma_G : (\mathbb{E}_M)_{\mathcal{X}\setminus \mathbb{V}_{M+1}} \to (E_M)_{|\mathcal{X}\setminus\mathbb{V}_{M+1}}$ be as in the proof of Proposition 3.47. Let $z \in V_{\mathrm{reg}}$. It follows from the box principle, as a particular case of Bertini's theorem [**Jo83**], that there exist $s_{z,1}, \dots, s_{z,M} \in \mathcal{O}(U_z, \mathrm{Im}\, F_1)$, where $U_z \subset \mathcal{X} \setminus V_{\mathrm{sing}}$ is a local chart around z above which E_0 is trivialized, such that

$$V \cap U_z = V(s_{z,1}) \cap \cdots \cap V(s_{z,M}), \quad ds_{z,1} \wedge \cdots \wedge ds_M \neq 0 \text{ on } V \cap U_z.$$

Let $\mathfrak{E} = \iota_V^* \mathbb{E}_M$. One has in U_z that

$$\overline{\partial}\Big(\frac{1}{s_{z,1}, \dots, s_{z,M}}\Big) \wedge \Upsilon = [V \cap U_z] \wedge \xi_z \quad \text{with } \xi_z \in \Omega^{N-M}(U_z, \mathcal{O}(-MD))$$

and $GF_{M+1} = 0$ over $\mathcal{X}\setminus\mathbb{V}_{M+1} \supset \mathcal{X}\setminus V_{\mathrm{sing}}$. Besides the exact complex of locally free sheaves (3.112), one has also over U_z the (minimal) Koszul exact complex associated to the generically exact complex $\mathbf{K}_\bullet^{s_{z,1}\oplus\cdots\oplus s_{z,M}}$. Thus, uniqueness of minimal free resolutions for $(\mathcal{O}/\mathcal{I})_{|U_z}$ (see Theorem A.9 or [**AndW07**, Theorem 4.4] for the argument developed here) implies in U_z that if e denotes a local holomorphic frame for $\mathcal{O}(D)$ there, then

$$\begin{aligned} G\, R^{\overline{\mathbf{F}_\bullet}}_{0,M} \wedge \Upsilon &= G\Big(h_z\, \overline{\partial}\Big(\frac{1}{s_{z,1}, \dots, s_{z,M}}\Big)\, \frac{1}{e}\Big) \wedge \Upsilon \\ &= \Big(G\, h_z\, \overline{\partial}\Big(\frac{1}{s_{z,1}, \dots, s_{z,M}}\Big)\, \frac{1}{e}\Big) \wedge \Upsilon \\ &= G\Big(h_z\, \frac{[V \cap U_z] \wedge \xi_z}{e}\Big) = (\iota_V)_* \mathfrak{f}_z, \end{aligned} \tag{3.130}$$

where $\mathfrak{f}_z$ is a meromorphic $\mathrm{Hom}_{\mathbb{C}}(\iota_V^* E_0, \mathfrak{E})$-valued $(N-M)$-form on the complex analytic space $V \cap U_z$, depending on Υ and $\mathbf{F}_\bullet$. Since

$$G\, R^{\overline{\mathbf{F}_\bullet}}_{0,M} \wedge \Upsilon = (\iota_V)_* \mathfrak{f}_z \tag{3.131}$$

[28] We mean here intrinsic with respect to V, once Υ and the metrized complex $\overline{\mathbf{F}_\bullet}$ have been chosen.

in $U_z \cap V$ for each $z \in V_{\rm reg}$, the $\mathrm{Hom}_{\mathbb{C}}(\iota_V^* E_0, \mathfrak{E})$-valued $(N-M)$-forms $\mathfrak{f}_z$, for $z \in V_{\rm reg}$, globalize as a meromorphic $\mathrm{Hom}_{\mathbb{C}}(\iota_V^* E_0, \mathfrak{E})$-valued $(N-M)$-form $\mathfrak{f}$ on $V_{\rm reg}$. The current

$$\varphi \in {}'\mathcal{D}^{(0,N-M)}\big(\mathcal{X} \setminus V_{\rm sing}, (\mathrm{Hom}_{\mathbb{C}}(E_0, \mathbb{E}_M))^*\big) \longmapsto \int_{V_{\rm reg}} \mathfrak{f} \wedge \varphi$$

extends as a $\overline{\partial}$-closed current, because

$$G\, R_{0,M}^{\overline{\mathbf{F}_\bullet}} \in \mathrm{CH}_{\mathcal{X},V}\big(\mathcal{X}, \mathrm{Hom}_{\mathbb{C}}(E_0, \mathbb{E}_M)\big)$$

by Proposition 3.47, which implies that $\overline{\partial}(G\, R_{0,M}^{\overline{\mathbf{F}_\bullet}} \wedge \Upsilon) = 0$ in $\mathcal{X}$. Furthermore, the meromorphic $\mathrm{Hom}_{\mathbb{C}}(\iota_V^* E_0, \mathfrak{E})$-valued $(N-M)$-form $\mathfrak{f}$ on $V_{\rm reg}$ extends meromorphically to V as a section of the dualizing sheaf $\omega_V^{N-M}(V, \mathrm{Hom}_{\mathbb{C}}(\iota_V^* E_0, \mathfrak{E}))$. To justify the last claim, we invoke the following argument. Since

$$G\, R_{0,M}^{\overline{\mathbf{F}_\bullet}} \wedge \Upsilon \in \mathrm{PM}^{(N,M)}\big(\mathcal{X}, \mathrm{Hom}_{\mathbb{C}}(E_0, \mathfrak{E})\big),$$

$[\mathfrak{f}]$ (where $\mathfrak{f}$ is defined by (3.131) around $z \in V_{\rm reg}$) defines a $\mathrm{Hom}_{\mathbb{C}}(\iota_V^* E_0, \mathfrak{E})$-valued pseudo-meromorphic current with bidegree $(N-M, 0)$ on V in the sense of Remark 2.29. The restriction of such a $(N-M, 0)$ pseudo-meromorphic current to $V_{\rm reg} = V \setminus V_{\rm sing}$ is a $(N-M, 0)$-bundle valued meromorphic form, which implies the meromorphicity of $\mathfrak{f}$ on V; see [**AndS12**, Example 2.8]. Since

$$R_{0,M}^{\overline{\mathbf{F}_\bullet}} \wedge \Upsilon \in \mathrm{PM}^{(N,M)}(\mathcal{X}, \mathrm{Hom}_{\mathbb{C}}(E_0, E_M)) \quad \text{with } \mathrm{Supp}\, R_{0,M}^{\overline{\mathbf{F}_\bullet}} \subset V,$$

such a current inherits the standard extension property across any hypersurface of V, in particular $\mathbb{V}_{M+1} \subset V_{\rm sing}$, and one then has that

$$R_{0,M}^{\overline{\mathbf{F}_\bullet}} \wedge \Upsilon = \Big[|\Delta_{M+1}^{\mathbf{F}_\bullet}|^{2\lambda}\, R_{0,M}^{\overline{\mathbf{F}_\bullet}} \wedge \Upsilon\Big]_{\lambda=0}.$$

This fact and the property $\sigma_G\, G = \mathrm{Id}$ on $\mathrm{Im}\, F_{M+1}^{\perp}$ over $\mathcal{X} \setminus \mathbb{V}_{M+1}$, while $R_{0,M}^{\overline{\mathbf{F}_\bullet}}$ is $\mathrm{Hom}_{\mathbb{C}}(E_0, (\mathrm{Im}\, F_{M+1})^{\perp})$-valued there, lead to

$$\begin{aligned}
R_{0,M}^{\overline{\mathbf{F}_\bullet}} \wedge \Upsilon &= \Big[|\Delta_{M+1}^{\mathbf{F}_\bullet}|^{2\lambda} \sigma_G\, (G R_{0,M}^{\overline{\mathbf{F}_\bullet}} \wedge \Upsilon)\Big]_{\lambda=0} \\
&= \Big[|\Delta_{M+1}^{\mathbf{F}_\bullet}|^{2\lambda} \sigma_G\, (\iota_V)_* \mathfrak{f}\Big]_{\lambda=0} \\
&= (\iota_V)_* \Big((\widehat{\pi}_1)_* \Big[(\iota_V \circ \widehat{\pi}_1)^* \big(|\Delta_{M+1}^{\mathbf{F}_\bullet}|^{2\lambda}\, \sigma_G\big)\, \widehat{\pi}_1^* \mathfrak{f}\Big]_{\lambda=0}\Big).
\end{aligned}$$

The assertion of the proposition follows from the fact that

$$(\iota_V \circ \widehat{\pi}_1)^* (|\Delta_{M+1}^{\mathbf{F}_\bullet}|^{2\lambda} \sigma_G) = \frac{|e_0|^{2\lambda}}{e_0}\, \mathfrak{s}_0\, \rho^{\lambda},$$

where e_0 is a holomorphic section of the line bundle $\mathcal{O}_{\widehat{V}^{[1]}}(\widehat{D}_1)$, $\mathfrak{s}_0$ is a smooth $\mathcal{O}_{\widehat{V}^{[1]}}(\widehat{D}_1) \otimes \widehat{\pi}_1^* \mathrm{Hom}_{\mathbb{C}}(\mathfrak{E}, \iota_V^* E_M)$-valued function, and ρ is a smooth strictly positive function. □

Given Proposition 3.45, the following statement (see [**AndS12**, Proposition 3.3]) provides then a useful intrinsic structural description of the whole residue current $R^{\overline{\mathbf{F}_\bullet}} = R_0^{\overline{\mathbf{F}_\bullet}}$, once the holomorphic section Υ of the canonical bundle has been chosen.

PROPOSITION 3.49. *There exist Cartier divisors $\widehat{\boldsymbol{D}}_1, \dots, \widehat{\boldsymbol{D}}_{N-M-1}$ on the complex normal analytic space $\widehat{V}^{[N-M-1]}$, together with respective holomorphic sections $e_1, \dots, e_{N-M-1}$ and smooth $(0,1)$-forms $\mathfrak{s}_1, \dots, \mathfrak{s}_{N-M-1}$, where each form $\mathfrak{s}_\rho$ is $\mathcal{O}(\widehat{\boldsymbol{D}}_\rho) \otimes \widehat{\Pi}_V^*(\mathrm{Hom}_{\mathbb{C}}(E_{M+\rho-1}, E_{M+\rho}))$-valued, such that*

$$R^{\overline{\mathbf{F}_\bullet}} \wedge \Upsilon = (\iota_V)_* \omega, \tag{3.132}$$

where

(3.133)

$$\begin{aligned}
\omega &= \omega(\overline{\mathbf{F}_\bullet}, \Upsilon) \\
&= \Big((\widehat{\Pi}_V)_* \sum_{\rho=0}^{N-M-1} \Big[\frac{|e_\rho|^{\mu_\rho}}{e_\rho}\mathfrak{s}_\rho \Big[\cdots\Big[\frac{|e_1|^{2\mu_1}}{e_1}\mathfrak{s}_1\Big]_{\mu_1=0}\Big]\cdots\Big]_{\mu_\rho=0}\Big)\, \omega_0(\overline{\mathbf{F}_\bullet}, \Upsilon) \\
&= (\widehat{\Pi}_V)_* \sum_{\rho=0}^{N-M-1} \Big[\frac{|e_\rho|^{\mu_\rho}}{e_\rho}\mathfrak{s}_\rho \cdots \Big[\frac{|e_1|^{2\mu_1}}{e_1}\mathfrak{s}_1 \Big[\frac{|e_0'|^{2\mu_0}}{e_0'}\mathfrak{s}_0'\, \widehat{\Pi}_V^* \mathfrak{f}\Big]_{\mu_0=0}\Big]_{\mu_1=0} \cdots\Big]_{\mu_\rho=0},
\end{aligned}$$

with

$$\frac{|e_0'|^{2\mu_0}}{e_0'}\mathfrak{s}_0' = (\widehat{\pi}_2 \circ \cdots \circ \widehat{\pi}_{N-M-1})^* \Big(\frac{|e_0|^{2\mu_0}}{e_0}\mathfrak{s}_0\Big).$$

Moreover all sections $e_0', e_1, \dots, e_{N-M+1}$ vanish on $\widehat{\Pi}_V^{-1}(V_{\mathrm{sing}})$ since $V_{\mathrm{sing}} \supset \mathbb{V}_{M+\rho}$ for $\rho \geq 1$. Such a current ω is called a structure current or, as it is the direct image of a semimeromorphic form, a structure form.

PROOF. The result follows from the realization of $R^{\overline{\mathbf{F}_\bullet}} = R^{\mathbf{F}_\bullet}$ through the iteration process (3.118) described in Proposition 3.45. One needs to use the following fact: for each $\rho = 1, \dots, N-M-1$, the pullback of $\overline{\partial}\sigma_{M+\rho}$ through the normalized blowup $\widehat{\pi} : \widehat{V} \to V$ of V along the ideal sheaf $\mathrm{Fitt}_{M+\rho}(V)$ (with Cartier exceptional divisor $\widehat{D}_\rho$) is a semimeromorphic form $\check{s}_\rho/\check{e}_\rho$, where $\check{e}_\rho$ is a holomorphic section of the line bundle $\mathcal{O}_{\widehat{V}}(\widehat{D}_\rho)$ and $\check{s}_\rho$ is a $(0,1)$-smooth section of the holomorphic bundle $\mathcal{O}_{\widehat{V}}(\widehat{D}_\rho) \otimes \widehat{\pi}^*(\mathrm{Hom}_{\mathbb{C}}((\iota_V)^* E_{M+\rho-1}, (\iota_V)^* E_{M+\rho}))$. □

3.4.4. The Cohen–Macaulay case. We consider in this section a coherent ideal sheaf $\mathcal{I}$ of $\mathcal{O}$-ideals on an N-dimensional complex manifold $\mathcal{X}$, such that the closed analytic subset $V(\mathcal{I}) = \mathrm{Supp}\, \mathcal{O}/\mathcal{I}$ has pure codimension equal to M, and $(\mathcal{O}/\mathcal{I})_z$ is Cohen–Macaulay[29] in a neighborhood of any point z in $V(\mathcal{I})$. The last condition means that given any $z \in V(\mathcal{I})$ there exists a minimal locally free resolution

(3.134)

$$0 \longrightarrow \mathcal{O}_z^{\oplus r_M(z)} \xrightarrow{\mathfrak{f}_M} \mathcal{O}_z^{\oplus r_{M-1}(z)} \longrightarrow \cdots \longrightarrow \mathcal{O}_z^{\oplus r_2(z)} \xrightarrow{\mathfrak{f}_2} \mathcal{O}_z^{\oplus r_1(z)} \xrightarrow{\mathfrak{f}_1} \mathcal{O}_z^{\oplus r_0(z)},$$

where $\mathrm{coker}\, \mathfrak{f}_1 \simeq (\mathcal{O}/\mathcal{I})_z$. If $\overline{\mathbf{F}_\bullet}$ is a generically exact, metrized complex of holomorphic bundles over $\mathcal{X}$, with $F_1^{\mathbf{F}_\bullet}$ being generically surjective, such as (3.111), and if the associated sequence of locally free sheaves (3.112) is exact, with

$$\mathcal{I} = \mathrm{Im}\,\big(\mathcal{O}(\cdot, E_1^{\mathbf{F}_\bullet}) \xrightarrow{F_1^{\mathbf{F}_\bullet}} \mathcal{O}(\cdot, E_0^{\mathbf{F}_\bullet})\big),$$

[29] See Definition A.21.

then it follows then from Theorem A.9, that the locally free resolution
(3.135)

$$0 \longrightarrow \mathcal{O}_z(E_L^{\mathbf{F}_\bullet}) \overset{(F_L^{\mathbf{F}_\bullet})_z}{\longrightarrow} \mathcal{O}_z(E_{L-1}^{\mathbf{F}_\bullet}) \overset{(F_{L-1}^{\mathbf{F}_\bullet})_z}{\longrightarrow} \cdots \overset{(F_2^{\mathbf{F}_\bullet})_z}{\longrightarrow} \mathcal{O}_z(E_1^{\mathbf{F}_\bullet}) \overset{(F_1^{\mathbf{F}_\bullet})_z}{\longrightarrow} \mathcal{O}_z(E_0^{\mathbf{F}_\bullet})$$

is isomorphic to the direct sum of the complex (3.134) having length M and of a trivial complex. Thus, for the sake of simplicity, we assume that the length L of the generically exact complex $\mathbf{F}_\bullet$ is equal to M, which is the minimal value allowed; see Proposition A.17. Thus, it follows from Proposition 3.40 and Proposition 3.48, taking E_M in place of of $\mathbb{E}_M$ and assuming $G = \mathrm{Id}_{E_M}$, that the equality of residue currents $R^{\overline{\mathbf{F}_\bullet}} = R_{0,M}^{\overline{\mathbf{F}_\bullet}}$ holds. This implies that given a local holomorphic section Υ_U of the canonical bundle in a neighborhood U of $z \in V(\mathcal{I})$, one has that

$$(R^{\overline{\mathbf{F}_\bullet}})_{|U} \wedge \Upsilon_U = (\iota_{V\cap U})_* \mathfrak{f}_U, \tag{3.136}$$

where $\mathfrak{f}_U$ is a $\mathrm{Hom}_{\mathbb{C}}(\iota_{V\cap U}^* E_0^{\mathbf{F}_\bullet}, \iota_{V\cap U}^* E_M^{\mathbf{F}_\bullet})$-valued section of the dualizing sheaf $\omega_{V\cap U}^{N-M}$. Thus, the Coleff–Herrera $\mathrm{Hom}_{\mathbb{C}}(E_0^{\mathbf{F}_\bullet}, E_M^{\mathbf{F}_\bullet})$-valued current $R^{\overline{\mathbf{F}_\bullet}}$ is a current with bidegree $(0, M)$, which is independent on the choice of the metrics on $\mathbf{F}_\bullet$. Thus, we denoted it by $R^{\mathbf{F}_\bullet}$ instead of $R^{\overline{\mathbf{F}_\bullet}}$.

DEFINITION 3.50. Let $\mathcal{I}$, $V(\mathcal{I})$ and $\mathbf{F}_\bullet$ be as above. A generically exact complex $\mathbf{G}_\bullet$, with $F_1^{\mathbf{G}_\bullet}$ being generically surjective, is said to *dominate* $\mathbf{F}_\bullet$ if

$$\operatorname{codim} V(\mathbf{G}_\bullet) \geq \operatorname{codim} V(\mathbf{F}_\bullet) = \operatorname{codim} V(\mathcal{I}) = M \tag{3.137}$$

and there exists a homomorphism of $\mathcal{O}$-modules

$$\mathfrak{a}_{\mathbf{F}_\bullet}^{\mathbf{G}_\bullet} : \frac{\mathcal{O}(\cdot, E_0^{\mathbf{G}_\bullet})}{\mathrm{Im}\big(\mathcal{O}(\cdot, E_1^{\mathbf{G}_\bullet}) \overset{F_1^{\mathbf{G}_\bullet}}{\longrightarrow} \mathcal{O}(\cdot, E_0^{\mathbf{G}_\bullet})\big)} \longrightarrow \frac{\mathcal{O}(\cdot, E_0^{\mathbf{F}_\bullet})}{\mathrm{Im}\big(\mathcal{O}(\cdot, E_1^{\mathbf{F}_\bullet}) \overset{F_1^{\mathbf{F}_\bullet}}{\longrightarrow} \mathcal{O}(\cdot, E_0^{\mathbf{F}_\bullet})\big)} \tag{3.138}$$

that extends as a morphism A of generically exact complexes from $\mathbf{G}_\bullet$ to $\mathbf{F}_\bullet$, using the standard diagram chase argument:[30]

$$\begin{array}{ccccccccc} \cdots & \longrightarrow & E_M^{\mathbf{G}_\bullet} & \overset{F_M^{\mathbf{G}_\bullet}}{\longrightarrow} & \cdots & \overset{F_1^{\mathbf{G}_\bullet}}{\longrightarrow} & E_0^{\mathbf{G}_\bullet} & \overset{F_0^{\mathbf{G}_\bullet}}{\longrightarrow} & E_0^{\mathbf{G}_\bullet}/\mathrm{Im}\, F_1^{\mathbf{G}_\bullet} \\ & & A_M \big\downarrow & & & & A_0 \big\downarrow & & \\ & & E_M^{\mathbf{F}_\bullet} & \overset{F_M^{\mathbf{F}_\bullet}}{\longrightarrow} & \cdots & \overset{F_1^{\mathbf{F}_\bullet}}{\longrightarrow} & E_0^{\mathbf{F}_\bullet} & \overset{F_0^{\mathbf{F}_\bullet}}{\longrightarrow} & E_0^{\mathbf{F}_\bullet}/\mathrm{Im}\, F_1^{\mathbf{F}_\bullet}. \end{array}$$

EXAMPLE 3.51. If $A_0 : E_0^{\mathbf{G}_\bullet} \to E_0$ is a morphism of holomorphic bundles satisfying $A_0(\mathrm{Im}\, F_1^{\mathbf{G}_\bullet}) \subset \mathrm{Im}\, F_1$, then one can define the germ at $z \in \mathcal{X}$ of a morphism $\mathfrak{a}_{\mathbf{F}_\bullet}^{\mathbf{G}_\bullet}$, described in (3.138), as

$$(\mathfrak{a}_{\mathbf{F}_\bullet}^{\mathbf{G}_\bullet})_z : (g_0)_z \in \big(\mathcal{O}(\cdot, E_0^{\mathbf{G}_\bullet}/\mathrm{Im}\, F_1^{\mathbf{G}_\bullet})\big)_z \longmapsto (A_0 g_0)_z \bmod \big(\mathcal{O}(\cdot, \mathrm{Im}\, F_1^{\mathbf{F}_\bullet})\big)_z. \tag{3.139}$$

Next, let us state and prove the following result [**Lar19**, Theorem 4.1], complementing the set of results established in §3.2.3

THEOREM 3.52 (Transformation law in the Cohen–Macaulay setting). *Let* $\mathbf{F}_\bullet$ *be a generically exact complex attached to a Cohen–Macaulay sheaf of ideals* $\mathcal{I}$ *as above, and let* $\mathbf{G}_\bullet$ *be a generically exact complex of holomorphic bundles dominating the complex* $\mathbf{F}_\bullet$ *as in Definition* 3.50. *Then, one has the current identity*

$$R^{\mathbf{F}_\bullet} A_0 = A_M R_{0,M}^{\overline{\mathbf{G}_\bullet}}, \tag{3.140}$$

which is independent of the choice of metric on $\mathbf{G}_\bullet$.

[30] See for example [**Eis**, Proposition A3-13].

PROOF. The proof relies on an argument inspired by one already used in the proof of Lemma 3.20 (sustaining the proof of Theorem 3.21). Assume that the generically exact complexes $\mathbf{F}_\bullet$ and $\mathbf{G}_\bullet$ are both metrized and consider the following two formal $\mathrm{Hom}_{\mathbb{C}}(E(\mathbf{F}_\bullet)\oplus E(\mathbf{G}_\bullet))$-valued forms

$$Q^{\overline{\mathbf{F}_\bullet}} = \sum_{r\geq 1} \frac{\sigma^{\overline{\mathbf{F}_\bullet}}}{2i\pi}\Big(\overline{\boldsymbol{\partial}}\Big(\frac{\sigma^{\overline{\mathbf{F}_\bullet}}}{2i\pi}\Big)\Big)^{r-1}, \quad \boldsymbol{A}\, Q^{\overline{\mathbf{G}_\bullet}} = \boldsymbol{A}\sum_{r\geq 1} \frac{\sigma^{\overline{\mathbf{G}_\bullet}}}{2i\pi}\Big(\overline{\boldsymbol{\partial}}\Big(\frac{\sigma^{\overline{\mathbf{G}_\bullet}}}{2i\pi}\Big)\Big)^{r-1},$$

where $\boldsymbol{A}\in\mathcal{O}(\mathcal{X},\mathrm{Hom}_{\mathbb{C}}(E(\mathbf{G}_\bullet),E(\mathbf{F}_\bullet))$ is defined from the morphism A constructed from $\mathfrak{a}_{\mathbf{F}_\bullet}^{\mathbf{G}_\bullet}$ as in Definition 3.50. Then the current

$$R_{\overline{\mathbf{F}_\bullet}}^{\overline{\mathbf{G}_\bullet}} = \Big[\Big(\overline{\boldsymbol{\partial}}\Big(\prod_{\ell\geq 1}(|\Delta_\ell^{\mathbf{F}_\bullet}|\,|\Delta_\ell^{\mathbf{G}_\bullet}|)^{2\lambda}\,\mathrm{Id}_{E(\mathbf{F}_\bullet)\oplus E(\mathbf{G}_\bullet)}\Big)\Big)\, Q^{\overline{\mathbf{F}_\bullet}}\,\boldsymbol{A}\, Q^{\overline{\mathbf{G}_\bullet}}\Big]_{\lambda=0}$$

is pseudo-meromorphic. If the operator $\boldsymbol{\nabla}_{\mathbf{F}_\bullet\oplus\mathbf{G}_\bullet} = 2i\pi(\boldsymbol{F}^{\mathbf{F}_\bullet}\oplus\boldsymbol{F}^{\mathbf{G}_\bullet}) - \overline{\boldsymbol{\partial}}$ acts from the bundle $\mathrm{Hom}_{\mathbb{C}}(E(\mathbf{F}_\bullet)\oplus E(\mathbf{G}_\bullet), E(\mathbf{F}_\bullet)\oplus E(\mathbf{G}_\bullet))$ to itself, then one can show, as in the proof of Lemma 3.20 (see the proof of Theorem 3.2 in [**Lar19**] for more details) that

$$\boldsymbol{\nabla}_{\mathbf{F}_\bullet\oplus\mathbf{G}_\bullet} R_{\overline{\mathbf{F}_\bullet}}^{\overline{\mathbf{G}_\bullet}} = R^{\overline{\mathbf{F}_\bullet}}\boldsymbol{A} - \boldsymbol{A}\, R^{\overline{\mathbf{G}_\bullet}}. \tag{3.141}$$

In particular, one has, for any $r\geq 1$, that

$$R_{0,r}^{\overline{\mathbf{F}_\bullet}} A_0 - A_r\, R_{0,r}^{\overline{\mathbf{G}_\bullet}} = F_{r+1}^{\mathbf{F}_\bullet}(R_{\overline{\mathbf{F}_\bullet}}^{\overline{\mathbf{G}_\bullet}})_{0,r+1} - \overline{\partial}\big(R_{\overline{\mathbf{F}_\bullet}}^{\overline{\mathbf{G}_\bullet}}\big)_{0,r},$$

where $\big(R_{\overline{\mathbf{F}_\bullet}}^{\overline{\mathbf{G}_\bullet}}\big)_{0,r}$ denotes the $\mathrm{Hom}_{\mathbb{C}}(E_0^{\mathbf{G}_\bullet}, E_r^{\mathbf{F}_\bullet})$-component of $R_{\overline{\mathbf{F}_\bullet}}^{\overline{\mathbf{G}_\bullet}}$. Take now $r = M = \mathrm{codim}\, V(\mathbf{F}_\bullet) = \mathrm{length}(\mathbf{F}_\bullet)$. Since $F_{M+1}^{\mathbf{F}_\bullet} = 0$ and

$$\big(R_{\overline{\mathbf{F}_\bullet}}^{\overline{\mathbf{G}_\bullet}}\big)_{0,M} = 0,$$

as a consequence of the dimension principle,[31] one gets the current identity (3.140). Note that the dimension principle applies because the codimension of $V(\mathbf{F}_\bullet)$ and of $V(\mathbf{G}_\bullet)$ are assumed to be greater than or equal to $r = M$. □

EXAMPLE 3.53 (The case of Koszul complexes). Let $D_1,\dots,D_m,\mathbf{D}_1,\dots,\mathbf{D}_M$ be $m+M$ Cartier divisors on $\mathcal{X}$. Let s_j be a holomorphic section of $\mathcal{O}(D_j)$ for each $j=1,\dots,m$ and let $\mathbf{S}_k$ be a holomorphic section of $\mathcal{O}(\mathbf{D}_k)$ for each $k=1,\dots,M$. Let also

$$E^D = \mathcal{O}(D_1)\oplus\cdots\oplus\mathcal{O}(D_m), \qquad E^{\mathbf{D}} = \mathcal{O}(\mathbf{D}_1)\oplus\cdots\oplus\mathcal{O}(\mathbf{D}_M),$$
$$s = s_1\oplus\cdots\oplus s_m \in \mathcal{O}(\mathcal{X}, E^D), \qquad \mathbf{S} = \mathbf{S}_1\oplus\cdots\oplus\mathbf{S}_M\in\mathcal{O}(\mathcal{X},E^{\mathbf{D}}).$$

We assume from now on that

$$\mathrm{codim}\, V(s) = m, \quad \mathbf{S} = H\, s,\ H\in\mathcal{O}(\mathcal{X},\mathrm{Hom}_{\mathbb{C}}(E^D,E^{\mathbf{D}})). \tag{3.142}$$

Let $\mathbf{F}_\bullet = \mathbf{K}_\bullet^s$ and $\mathbf{G}_\bullet = \mathbf{K}_\bullet^{\mathbf{S}}$. Since the length of $\mathbf{F}_\bullet$ equals $m = \mathrm{codim}\, V(s)$ and the sequence of locally free sheaves (3.112) attached to $\mathbf{F}_\bullet$ is exact, $(\mathcal{O}/\mathcal{I}_s)_z$ is Cohen–Macaulay for any $z\in\mathcal{X}$. Since $E_0^{\mathbf{F}_\bullet} = E_0^{\mathbf{G}_\bullet} = \mathbb{C}$ and $\mathrm{Im}\, F_1^{\mathbf{G}_\bullet}\subset \mathrm{Im}\, F_1^{\mathbf{F}_\bullet}$, $A_0 = \mathrm{Id}_{\mathbb{C}}$ induces a morphism $\mathfrak{a}_{\mathbf{F}_\bullet}^{\mathbf{G}_\bullet}$ of $\mathcal{O}$-modules acting as (3.138); see Example 3.51. Therefore $\mathbf{G}_\bullet$ dominates $\mathbf{F}_\bullet$, in accordance with Definition 3.50. One checks

[31] We do not repeat this argument again here but refer the reader to [**Lar19**, Proposition 3.6].

easily that $A_r = \bigwedge^r H^* : \bigwedge^r (E^{\mathbf{D}})^* \longrightarrow \bigwedge^r (E^D)^*$ for any $r \geq 1$. Therefore, Theorem 3.52 implies, in view of Proposition 3.18 and Example 3.30, that

$$R^{\mathbf{K}^s_\bullet} = \overline{\partial}\Big(\frac{1}{(s_1, \dots, s_m)}\Big) = \big(\bigwedge^m H^*\big)\, R_{0,M}^{\overline{\mathbf{K}^\mathbf{S}_\bullet}} \tag{3.143}$$

independently of the metric taken on $E^{\mathbf{D}}$. This leads to the metrized generically exact complex $\overline{\mathbf{G}_\bullet}$. Such a result generalizes Theorem 3.21, proved for the case $V = \mathcal{X}$ with the corresponding assumptions being $\operatorname{codim} V(s) = \operatorname{codim} V(\mathbf{S}) = m = M$.

EXAMPLE 3.54 (Bochner–Martinelli residue currents in the Cohen–Macaulay case). Let $\mathbf{D}$ be a Cartier divisor on $\mathcal{X}$ and let $\mathbf{S}_1, \dots, \mathbf{S}_M$ be holomorphic sections of $\mathcal{O}(\mathbf{D})$. Assume that the section $\mathbf{S} = \mathbf{S}_1 \oplus \cdots \oplus \mathbf{S}_M$ is such that $V(\mathbf{S})$ is a closed analytic subset of $\mathcal{X}$ with pure codimension $\operatorname{codim} V(\mathbf{S}) = m$. Assume also that there exists a generically exact complex $\mathbf{F}_\bullet$ as (3.135) with length $L = m$ such that $E_0 = E_0^{\mathbf{F}_\bullet} = \mathcal{O}(\mathbf{D})$, $F_1 = F_1^{\mathbf{F}_\bullet}$ is generically surjective, $\mathbf{S}_1, \dots, \mathbf{S}_M$ generate

$$\operatorname{Im}\big(\mathcal{O}(\cdot, E_1^{\mathbf{F}_\bullet}) \xrightarrow{F_1^{\mathbf{F}_\bullet}} \mathcal{O}(\cdot, \mathcal{O}(\mathbf{D}))\big),$$

and the associated sequence of sheaves of sections is exact. Therefore, the coherent sheaf

$$\mathcal{F} = \frac{\mathcal{O}(\cdot, \mathcal{O}(\mathbf{D}))}{\operatorname{Im}\big(\mathcal{O}(\cdot, E_1^{\mathbf{F}_\bullet}) \xrightarrow{F_1^{\mathbf{F}_\bullet}} \mathcal{O}(\cdot, \mathcal{O}(\mathbf{D}))\big)} = \frac{\mathcal{O}(\cdot, \mathcal{O}(\mathbf{D}))}{\mathcal{O}\,\mathbf{S}_1 + \cdots + \mathcal{O}\,\mathbf{S}_M}$$

is locally Cohen–Macaulay in a neighborhood of any point in its support $V(\mathbf{S})$. Let

$$s_j = \sum_{k=1}^{M} \lambda_{j,k}\, \mathbf{S}_k, \ j = 1, \dots, m,$$

where the section $s = s_1 \oplus \cdots \oplus s_m$ is such that $\operatorname{codim}(V(s)) = m = \operatorname{codim} V(\mathbf{S})$. Let also the section $s_0 \in \mathcal{O}(\mathcal{X}, \mathcal{O}(\mathbf{D}))$ be such that $\operatorname{codim}(V(s) \cap V(s_0)) > m$. Let $\mathbf{G}_\bullet = \mathbf{K}^s_\bullet$. The following morphism of bundles $A_0 : \mathbb{C} \longrightarrow \mathcal{O}(\mathbf{D})$ defined by $A_0(z,t) = (z, t\, s_0(z))$ for any $z \in \mathcal{X}$ and $t \in \mathbb{C}$ fulfills the requirements in Example 3.51. Then, let

$$A_m : E_m^{\mathbf{K}^s_\bullet} = \det\big(\overbrace{\mathcal{O}(-\mathbf{D}) \oplus \cdots \oplus \mathcal{O}(-\mathbf{D})}^{m \text{ times}}\big) \longrightarrow E_m^{\mathbf{F}_\bullet}$$

be the morphism of bundles induced by $\mathfrak{a}^{\mathbf{G}_\bullet}_{\mathbf{F}_\bullet}$. Theorem 3.52, together with Proposition 3.18, implies that

$$R^{\mathbf{F}_\bullet} A_0 = A_m\, \overline{\partial}\Big(\frac{1}{(s_1, \dots, s_m)}\Big) \iff R^{\mathbf{F}_\bullet} = \overline{\partial}\Big(\frac{1}{(s_1, \dots, s_m)}\Big)\, \frac{1}{s_0}. \tag{3.144}$$

Since $R^{\mathbf{F}_\bullet}$ is $\overline{\partial}$-closed as a Coleff–Herrera current, it follows from Theorem 3.39 that A_M is expressed locally as

$$A_M = \sum_{j=0}^{M} u_{M,j}\, s_j, \quad \text{with } u_{M,j} \in \mathcal{O}\big(\cdot, \mathcal{O}(-\mathbf{D}) \otimes \operatorname{Hom}(E_m^{\mathbf{K}^s_\bullet}, E_m^{\mathbf{F}_\bullet})\big).$$

Theorem 2.79 implies that locally, applying (3.144),

$$R^{\mathbf{F}_\bullet} = u_{M,0}\, \overline{\partial}\Big(\frac{1}{(s_1, \dots, s_m)}\Big); \tag{3.145}$$

see [**Lar19**, Example 4.3].

EXAMPLE 3.55. Let $f = (f_1, \dots, f_m) = f_1\epsilon_1 + \cdots + f_m\epsilon_m$ be a section of the trivial m-bundle $\mathbb{C}^m$ over an open subset Ω of $\mathbb{C}^N$. Consider a metrized version of the Koszul complex $\mathbf{K}_\bullet^f$. If $V(f)$ has pure codimension M, then one can express the Bochner–Martinelli current attached to such a complex as

$$\overline{R^{\mathbf{K}_\bullet^f}} = \sum_{r=M}^{N} \sum_{1\leq i_1<\cdots<i_r\leq m} R^f_{|\ |;I}\, \epsilon^*_{i_1} \wedge \cdots \wedge \epsilon^*_{i_r}.$$

This current acts by multiplication; see Examples 3.16 and 3.30. Assume that there exists a generically exact complex $\mathbf{F}_\bullet$ as described by (3.134) and (3.135) when one assumes that $\mathcal{I} = (f_1, \dots, f_m)\,\mathcal{O}_\Omega$, with $L = M$, F_1 being generically surjective and $E_0 = \mathbb{C}$. Furthermore, the assumption that the associated sequence of sheaves of sections (3.112) is exact implies that the coherent sheaf $\mathcal{O}_\Omega / \sum_{j=1}^m f_j\mathcal{O}_\Omega$ is Cohen–Macaulay in a neighborhood of any point of its support. Let

$$\mathfrak{a}_{\mathbf{F}_\bullet}^{\mathbf{K}_\bullet^f}$$

be the morphism of $\mathcal{O}$-modules attached to Id $: \mathbb{C} \to \mathbb{C}$ as in Example 3.51. Then the complex $\mathbf{K}_\bullet^f$ dominates $\mathbf{F}_\bullet$; see Definition 3.50. Let

$$A_M : \bigwedge^M \mathbb{C}^* = E_M^{\mathbf{K}_\bullet^f} \longrightarrow E_M^{\mathbf{F}_\bullet}$$

be the morphism of bundles induced at the level M by $\mathfrak{a}_{\mathbf{F}_\bullet}^{\mathbf{K}_\bullet^f}$, as in this definition. Theorem 3.52 implies, independently of the choice of metric on the trivial bundle $\mathbb{C}^m$, that the current

$$R^{\mathbf{F}_\bullet} = \sum_{1\leq i_1<\cdots<i_M\leq m} A_M(\epsilon^*_{i_1} \wedge \cdots \wedge \epsilon^*_{i_M})\, R^f_{|\ |;I}, \tag{3.146}$$

is acting here by multiplication from $E_0^{\mathbf{F}_\bullet} = \mathbb{C}$ to $E_M^{\mathbf{F}_\bullet}$; see [**Lar19**, Example 4.5].

One should relate the results presented in §3.4.4 to factorization formulae of the Poincaré type (as in Proposition 2.54) for the integration current attached to the *fundamental cycle* $\dim_{\mathbb{C}}(\mathcal{O}_{\mathbb{C}^N,0}/\mathbb{I})\,\{0\}$ of a primary ideal, where $\mathcal{O}_0/\mathbb{I}$ admits a minimal locally free resolution (3.134) (where $z = 0$, $M = N$, $r_0 = 1$). If $D_{\mathbf{E}}$ is the connection on $\mathbf{E} = \mathrm{Hom}_{\mathbb{C}}(E(\mathbf{F}_\bullet), E(\mathbf{F}_\bullet))$ induced by the trivial connection on each trivial bundle $\mathbb{C}^{r_\ell}$, then one has[32]

(3.147)

$$\dim_{\mathbb{C}}(\mathcal{O}_{\mathbb{C}^N,0}/\mathbb{I})\,[\{0\}] = \mathrm{Tr}\Big[\frac{1}{N!} D_{\mathbf{E}}\mathfrak{f}_1 \cdots D_{\mathbf{E}}\mathfrak{f}_N\, R^{\mathbf{F}_\bullet}_{0,N}\Big] = \frac{1}{N!} D_{\mathbf{E}}\mathfrak{f}_1 \cdots D_{\mathbf{E}}\mathfrak{f}_N\, R^{\mathbf{F}_\bullet};$$

see [**Lej82**, Theorem 2.4]. A similar result generally holds when $\mathcal{I}$ is a coherent Cohen–Macaulay ideal sheaf in $\mathcal{O}_{\mathcal{X}}$ and $\mathcal{X}$ is an N-dimensional complex manifold. We refer the reader to [**LarW18**, Theorem 6.1] for further details; see also §5.5 in this monograph. By fundamental cycle of $\mathcal{I}$, we mean now the cycle defined by $Z_{\mathrm{fond}}(\mathcal{I}) = \sum_\iota \mathrm{length}\,(\mathcal{O}_{\mathcal{X}})_{Z,Z_\iota}\, Z_\iota$, where the multiplicity length $((\mathcal{O}_{\mathcal{X}})_{Z,Z_\iota})$ attached to each Z_ι is here defined as follows (see [**Ful**, I.5]) given a generic point z_ι on $|Z_\iota|$ and a complex submanifold H_{z_ι} intersecting $|Z_\iota|$ transversally in a neighborhood of z_ι, then length $(\mathcal{O}_{\mathcal{X}})_{Z,Z_\iota} = \dim_{\mathbb{C}}\big(\mathcal{O}_{\mathcal{X},z_\iota}/(\mathcal{I} + \mathcal{I}_{H_{z_\iota}})_{z_\iota}\big)$.

[32]Observe that one takes on the right-hand side the trace of a $\mathrm{Hom}_{\mathbb{C}}(E_0, E_0) \simeq \mathbb{C}$-valued current, which is in this particular case the current itself.

CHAPTER 4

Bochner–Martinelli kernels and weights

4.1. Diagonal submanifold and Koszul complex $\mathscr{K}_\bullet^{\mathscr{D}}$

Let us fix in this preamble some of the notation to be used throughout this chapter. An N-dimensional complex manifold with structural sheaf $\mathcal{O}$ is still denoted by $\mathcal{X}$, while $\mathscr{X}$ describes its duplication. Namely, $\mathscr{X}$ is the $2N$-dimensional complex manifold $\mathscr{X} = \mathcal{X} \times \mathcal{X}$ with structural sheaf $\mathscr{O} = \mathcal{O} \times \mathcal{O}$. Points in $\mathscr{X}$ are described as (z, ζ) and $\sqcap$, π denote the projections

$$\sqcap : (z,\zeta) \in \mathscr{X} \longrightarrow z \in \mathcal{X}, \quad \pi : (z,\zeta) \in \mathscr{X} \longrightarrow \zeta \in \mathcal{X}. \tag{4.1}$$

Given a holomorphic bundle E, one denotes by $\overset{\frown}{E}$ the holomorphic bundle over $\mathscr{X}$ defined by

$$\overset{\frown}{E} = \mathrm{Hom}_{\mathbb{C}}(\pi^* E, \sqcap^* E) \simeq \sqcap^* E \otimes (\pi^* E)^*. \tag{4.2}$$

More generally, let

$$\dots, E_{L+1}, E_L, E_{L-1}, \dots, E_1, E_0, E_{-1}, \dots$$

be a sequence of holomorphic bundles over $\mathcal{X}$, indexed by $\mathbb{Z}$ in decreasing order. If $\ell, r \in \mathbb{Z}$, then one denotes by $\overset{\frown}{E}_{\ell,r}$ the holomorphic bundle over $\mathscr{X}$, defined as

$$E_{\ell,r}^{\frown} = \mathrm{Hom}_{\mathbb{C}}(\pi^* E_r, \sqcap^* E_\ell) \simeq \sqcap^* E_\ell \otimes (\pi^* E_r)^*. \tag{4.3}$$

Thus, one considers the diagonal N-dimensional complex submanifold of $\mathscr{X}$:

$$\mathscr{D} = \{(z,\zeta) \in \mathscr{X} \, : \, z = \zeta\} \subset \mathscr{X}. \tag{4.4}$$

The manifold $\mathcal{X}$ embeds in its duplication $\mathscr{X}$ along the submanifold $\mathscr{D}$.

4.1.1. Koszul complex over the duplication of a good manifold.

We begin with a definition which will be essential for us all through this chapter. In the present section we will illustrate this definition using fundamental examples in this monograph.

Definition 4.1. A complex, N-dimensional manifold $\mathcal{X}$ is said to *embed as a smooth diagonal complete intersection in its $2N$-dimensional duplication* $\mathscr{X}$ if and only if there exists a holomorphic bundle $\mathscr{E}$ with rank N over $\mathscr{X}$ and $\Delta \in \mathscr{O}(\mathscr{X}, \mathscr{E})$ such that $V(\Delta) = \mathscr{D}$, with $d\Delta \neq 0$ locally on $\mathscr{D}$. Then, the Koszul complex $\mathscr{K}_\bullet^{\Delta}$ is called the $(\mathscr{E}, \Delta)$-*Koszul complex for the embedded manifold* $\mathcal{X} \hookrightarrow \mathscr{X}$ *as the diagonal* $\mathscr{D}$ and is denoted in what follows as $\mathscr{K}_\bullet^{\mathscr{D}}$ when $(\mathscr{E}, \Delta)$ are implicitly defined. A manifold $\mathcal{X}$, which embeds as a smooth diagonal complete intersection in its duplication $\mathscr{X}$, is said to be *a good complex manifold*.

PROPOSITION 4.2. *The complex affine space* $\mathbb{C}^N$ *and the complex projective space* $\mathbb{P}^N_{\mathbb{C}}$ *are examples of good complex manifolds. The Koszul complex* $\mathscr{K}_\bullet{}^{\mathscr{D}}$ *for the embedding of the manifold* $\mathbb{C}^N \hookrightarrow \mathbb{C}^N \times \mathbb{C}^N$ *as the diagonal manifold* $\mathscr{D}$ *is given by* $\mathscr{K}_\bullet{}^{\mathscr{D}} = \mathscr{K}_\bullet{}^{\Delta}$, *when* $\{\epsilon_1, \dots, \epsilon_N\}$ *denotes the canonical basis of the trivial bundle* $\mathbb{C}^N$ *and where*

(4.5)

$$\{\mathscr{E} = \mathbb{C}^N,\ \Delta = \sum_{k=1}^N (\zeta_k - z_k)\, \epsilon_k\} \quad \text{or} \quad \{\mathscr{E} = \pi^* T^{(1,0)}_{\mathbb{C}^N},\ \Delta = \sum_{k=1}^N (\zeta_k - z_k)\, \frac{\partial}{\partial \zeta_k}\}.$$

The Koszul complex $\mathscr{K}_\bullet{}^{\mathscr{D}}$ *for embedding of* $\mathbb{P}^N_{\mathbb{C}} \hookrightarrow \mathbb{P}^N_{\mathbb{C}} \times \mathbb{P}^N_{\mathbb{C}}$, *using*

$$([\zeta_0 : \dots : \zeta_N], [z_0 : \dots : z_N])$$

as homogeneous coordinates on the duplication, is given by $\mathscr{K}_\bullet{}^{\mathscr{D}} = \mathscr{K}_\bullet{}^{\Delta}$, *where*

$$\mathscr{E} = \overset{\frown}{\mathcal{O}(1)} \otimes \pi^* T^{(1,0)}_{\mathbb{P}^N_{\mathbb{C}}} = \sqcap^* \mathcal{O}(1) \otimes \pi^* \big(\mathcal{O}(-1) \otimes T^{(1,0)}_{\mathbb{P}^N_{\mathbb{C}}}\big),\ \Delta = -\sum_{\kappa=0}^N z_\kappa \frac{\partial}{\partial \zeta_\kappa}. \tag{4.6}$$

PROOF. The assertions related to $\mathbb{C}^N$ are immediate. The projective space $\mathbb{P}^N_{\mathbb{C}}$ is the archetypal example of a complete N-dimensional toric manifold. It is realized as the geometric quotient $(\mathbb{C}^{N+1} \setminus \{\mathbf{0}\})/\mathbb{C}^*$. This is a particular case of the more general situation described in Definition 3.8. Let $\Pi : \mathbb{C}^{N+1} \setminus \{\mathbf{0}\} \to \mathbb{P}^N_{\mathbb{C}}$ be the projection. Let $L = \mathcal{O}(\ell)$ be the line bundle over $\mathcal{X} = \mathbb{P}^N_{\mathbb{C}}$ whose isomorphism class is defined in Example B.9. The L-valued differential (p,q)-forms φ, $1 \le p, q \le N$, and corresponding (p,q)-currents T on $\mathbb{P}^N_{\mathbb{C}}$ are identified with $\Pi^* L$-valued (p,q)-forms Φ and (p,q)-currents $\mathfrak{T}$ defined on $\mathbb{C}^{N+1} \setminus \{\mathbf{0}\}$, respectively. The forms Φ and the currents $\mathfrak{T}$ satisfy the relations

$$\begin{aligned}
&\Big(\sum_{\kappa=0}^N \zeta_\kappa \frac{\partial}{\partial \zeta_\kappa}\Big) \rfloor \Phi = 0 \quad \Big(\text{resp. } \Big(\sum_{\kappa=0}^N \zeta_\kappa \frac{\partial}{\partial \zeta_k}\Big) \rfloor \mathfrak{T} = 0\Big),\\
&\Big(\sum_{\kappa=0}^N \overline{\zeta_\kappa} \frac{\partial}{\partial \overline{\zeta_k}}\Big) \rfloor \Phi = 0 \quad \Big(\text{resp. } \Big(\sum_{\kappa=0}^N \overline{\zeta_\kappa} \frac{\partial}{\partial \overline{\zeta_k}}\Big) \rfloor \mathfrak{T} = 0\Big).
\end{aligned} \tag{4.7}$$

The coefficients of Φ (respectively of $\mathfrak{T}$) are homogeneous in variables ζ and $\overline{\zeta}$, whose bidegree is equal to $(\ell - p, -q)$; see for example [**Ber88**, §0]. Then, one has that

$$-\sum_{\kappa=0}^N \zeta_k \frac{\partial}{\partial \zeta_\kappa}(\zeta) = \Delta(\zeta, \zeta) = 0,$$

for any $[\zeta_0 : \dots : \zeta_N] \in \mathbb{P}^N_{\mathbb{C}}$. Moreover,

$$\Delta(z, \zeta) = 0 \Longleftrightarrow z = \zeta$$

since

$$\Delta(z, \zeta)\,(\langle\!\langle u, \zeta \rangle\!\rangle) = -\langle\!\langle u, z \rangle\!\rangle,$$

for any $z = [z_0 : \dots : z_N]$ and $u = [u_0 : \dots : u_N]$ fixed in $\mathbb{P}^N_{\mathbb{C}}$. This implies also that $d\Delta \neq 0$ locally on $\mathscr{D}$. Thus, the assertion for $\mathbb{P}^N_{\mathbb{C}}$ follows. □

REMARK 4.3. Assume that $\mathcal{X} = \mathcal{X}_\Sigma$ is a toric complete manifold subordinate to a complete, primitive, $\mathbb{Z}^N$-rational fan in $\mathbb{R}^N$, having $d = N + r$ rays with primitive generators $\xi_1, \dots, \xi_{N+r}$; see §3.1.1. Let $[a_{j,\kappa}]_{1 \le j \le r, 1 \le \kappa \le d}$ be a matrix whose entries are positive integers, with rank $r = \operatorname{rank}\operatorname{Pic}(\mathcal{X}_\Sigma)$, so that the system of equations (see (3.8))

$$\sum_{\kappa=1}^{d} a_{j,\kappa}\,\xi_\kappa = 0, \quad j = 1,\dots,r,$$

is satisfied. When $\mathbb{P}^N_{\mathbb{C}}$ is replaced by such a general complete toric manifold $\mathcal{X}_\Sigma$, the role of the section $\boldsymbol{s}$, expressed in homogeneous coordinates on $\mathcal{X}_\Sigma \times \mathcal{X}_\Sigma$ (see §3.1.2), is played by

$$\boldsymbol{s}_{\mathcal{X}_\Sigma}(z,\zeta) = \Big(-\sum_{\kappa=1}^{d} a_{1,\kappa}\, z_\kappa \frac{\partial}{\partial\zeta_\kappa}, \dots, -\sum_{\kappa=1}^{d} a_{r,\kappa}\, z_\kappa \frac{\partial}{\partial\zeta_\kappa}\Big).$$

The zero set of $\boldsymbol{s}$ equals the diagonal $\mathscr{D}$, and $\boldsymbol{s}$ is no longer, in general, the section of a holomorphic bundle over $\mathscr{X} = \mathcal{X}_\Sigma \times \mathcal{X}_\Sigma$ with rank N. The projective space $\mathbb{P}^N_{\mathbb{C}}$ is indeed a very particular case among the family of smooth complete toric manifolds. It corresponds to the setting when $r = 1$ and all the entries $a_{1,\kappa}$ are equal to 1.

4.1.2. Nabla operators and $\mathcal{L}^\nu$ sheaves of currents for good manifolds. Let $\mathcal{X}$ be an N-dimensional, good complex manifold, which embeds as the diagonal submanifold $\mathscr{D}$ in its $2N$-dimensional duplication $\mathscr{X}$. Let also $\mathscr{K}_\bullet{}^{\mathscr{D}}$ be the Koszul complex for the embedding $\mathcal{X} \simeq \mathscr{D} \subset \mathscr{X}$, which is subordinate[1] to an N-bundle $\mathscr{E}$ over $\mathscr{X}$ and its holomorphic section Δ, which is such that $V(\Delta) = \mathscr{D}$ and satisfies $d\Delta \neq 0$ locally on $\mathscr{D}$. We attach to $(\mathscr{E}, \Delta)$ two distinct nabla operators:

$$\nabla_{\mathscr{D}} = 2i\pi \rfloor_\Delta - \overline{\partial}, \quad \nabla^\circ_{\mathscr{D}} = 2i\pi \rfloor_\Delta - \overline{\partial}_\zeta. \tag{4.8}$$

If $\mathscr{E}(\mathscr{K}_\bullet{}^{\mathscr{D}}) = \bigoplus_{\gamma\in\mathbb{Z}} \bigwedge^\gamma \mathscr{E}^*$ and if $\mathscr{E}'$ is an arbitrary holomorphic bundle over $\mathscr{X}$, then the above operators act from the space of $\mathscr{E}(\mathscr{K}_\bullet{}^{\mathscr{D}}) \otimes \mathscr{E}'$-valued currents into itself. More precisely, for any $\nu \in \mathbb{Z}$, we introduce the sheaves of currents

$$\begin{aligned}
\mathcal{L}^\nu_{\mathscr{D}}[\mathscr{E}'] &= \bigoplus_{\gamma\in\mathbb{Z}} {}'\mathcal{D}^{(0,\gamma+\nu)}\Big(\cdot, \Big(\bigwedge^{\gamma} \mathscr{E}^*\Big) \otimes \mathscr{E}'\Big) \\
&= \bigoplus_{\gamma=\max(0,-\nu)}^{\min(N,2N-\nu)} {}'\mathcal{D}^{(0,\gamma+\nu)}\Big(\cdot, \Big(\bigwedge^{\gamma} \mathscr{E}^*\Big) \otimes \mathscr{E}'\Big), \\
(\mathcal{L}^\nu_{\mathscr{D}}[\mathscr{E}'])^\circ &= \bigoplus_{\gamma\in\mathbb{Z}} {}'\mathcal{D}^{(0,0),(0,\gamma+\nu)}\Big(\cdot, \Big(\bigwedge^{\gamma} \mathscr{E}^*\Big) \otimes \mathscr{E}'\Big) \\
&= \bigoplus_{\gamma=\max(0,-\nu)}^{\min(N,N-\nu)} {}'\mathcal{D}^{(0,0),(0,\gamma+\nu)}\Big(\cdot, \Big(\bigwedge^{\gamma} \mathscr{E}^*\Big) \otimes \mathscr{E}'\Big),
\end{aligned} \tag{4.9}$$

where the notation $(\ )^\circ$ means that we keep only, as indicated, components with bidegree $(0,0)$ in z. Then, one has that

$$\begin{aligned}
\nabla_{\mathscr{D}} &: \mathcal{L}^\nu_{\mathscr{D}}[\mathscr{E}'] \longrightarrow \mathcal{L}^{\nu+1}_{\mathscr{D}}[\mathscr{E}'], \\
\nabla^\circ_{\mathscr{D}} &: (\mathcal{L}^\nu_{\mathscr{D}}[\mathscr{E}'])^\circ \longrightarrow (\mathcal{L}^{\nu+1}_{\mathscr{D}}[\mathscr{E}'])^\circ.
\end{aligned} \tag{4.10}$$

[1] We will call the Koszul complex $\mathscr{K}_\bullet{}^{\mathscr{D}}$ the $(\mathscr{E}, \Delta)$ Koszul complex.

Since $\rfloor_\Delta \circ \rfloor_\Delta = \overline{\partial} \circ \overline{\partial} = \overline{\partial}_\zeta \circ \overline{\partial}_\zeta = 0$, one can define two complexes of sheaves of currents

$$
\begin{aligned}
&\cdots \longrightarrow \mathcal{L}_{\mathscr{D}}^{\nu-1}[\mathscr{E}'] \xrightarrow{\nabla_{\mathscr{D}}} \mathcal{L}_{\mathscr{D}}^{\nu}[\mathscr{E}'] \xrightarrow{\nabla_{\mathscr{D}}} \mathcal{L}_{\mathscr{D}}^{\nu+1}[\mathscr{E}'] \longrightarrow \cdots \\
&\cdots \longrightarrow (\mathcal{L}_{\mathscr{D}}^{\nu-1}[\mathscr{E}'])^\circ \xrightarrow{\nabla_{\mathscr{D}}^\circ} (\mathcal{L}_{\mathscr{D}}^{\nu}[\mathscr{E}'])^\circ \xrightarrow{\nabla_{\mathscr{D}}^\circ} (\mathcal{L}_{\mathscr{D}}^{\nu+1}[\mathscr{E}'])^\circ \longrightarrow \cdots ,
\end{aligned}
\tag{4.11}
$$

where $\nabla_{\mathscr{D}}^2 = (\nabla_{\mathscr{D}}^\circ)^2 = 0$.

EXAMPLE 4.4. If $\mathcal{X} = \mathbb{C}^N$, $\mathscr{E} = T_{\mathbb{C}^N}^{(1,0)}$ and $\Delta = \sum_{k=1}^N (\zeta_k - z_k)\, \partial/\partial\zeta_k$, then one has for any holomorphic bundle $\mathscr{E}'$ over $\mathbb{C}^N \times \mathbb{C}^N$ and any $\nu \in \mathbb{Z}$ that

$$
\begin{aligned}
\mathcal{L}_{\mathscr{D}}^{\nu}[\mathscr{E}'] &= \bigoplus_{\gamma=\max(0,-\nu)}^{\min(N,2N-\nu)} \ \bigoplus_{\{(\kappa,\kappa')\in\mathbb{N}^2\,:\,\kappa+\kappa'=\gamma+\nu\}} {}'\mathcal{D}^{(0,\kappa),(\gamma,\kappa')}(\cdot, \mathscr{E}'), \\
(\mathcal{L}_{\mathscr{D}}^{\nu}[\mathscr{E}'])^\circ &= \bigoplus_{\gamma=\max(0,-\nu)}^{\min(N,N-\nu)} {}'\mathcal{D}^{(0,0),(\gamma,\gamma+\nu)}(\cdot, \mathscr{E}').
\end{aligned}
\tag{4.12}
$$

EXAMPLE 4.5. If $\mathcal{X} = \mathbb{P}_{\mathbb{C}}^N$ and $\mathscr{E}$ and Δ are as in (4.6), then one has for any holomorphic bundle $\mathscr{E}'$ over $\mathbb{P}_{\mathbb{C}}^N$ and any $\nu \in \mathbb{Z}$ that

$$
\begin{aligned}
\mathcal{L}_{\mathscr{D}}^{\nu}[\mathscr{E}'] &= \bigoplus_{\gamma=\max(0,-\nu)}^{\min(N,2N-\nu)} \ \bigoplus_{\{(\kappa,\kappa')\in\mathbb{N}^2\,:\,\kappa+\kappa'=\gamma+\nu\}} {}'\mathcal{D}^{(0,\kappa),(\gamma,\kappa')}(\cdot, \mathcal{O}(\overset{\frown}{-\gamma}) \otimes \mathscr{E}'), \\
(\mathcal{L}_{\mathscr{D}}^{\nu}[\mathscr{E}'])^\circ &= \bigoplus_{\gamma=\max(0,-\nu)}^{\min(N,N-\nu)} {}'\mathcal{D}^{(0,0),(\gamma,\gamma+\nu)}(\cdot, \mathcal{O}(\overset{\frown}{-\gamma}) \otimes \mathscr{E}').
\end{aligned}
\tag{4.13}
$$

We need also to introduce sheaves of bundle-valued currents $\boldsymbol{\mathcal{L}}_{\mathscr{D}}^{\nu}$ and $(\boldsymbol{\mathcal{L}}_{\mathscr{D}}^{\nu})^\circ$. Their definition does not involve $\mathscr{E}'$ any longer, at least for the moment,[2] but involves instead the holomorphic bundles $\mathscr{E}$ and its dual $\mathscr{E}^*$, when $\mathscr{K}_\bullet^{\mathscr{D}}$ is the $(\mathscr{E}, \Delta)$ Koszul complex. They are defined as follows. For any $\nu \in \mathbb{Z}$, let

$$
\begin{aligned}
\boldsymbol{\mathcal{L}}_{\mathscr{D}}^{\nu} &= \bigoplus_{\gamma\geq-\nu,\,\kappa\geq0} {}'\mathcal{D}^{(\kappa,\gamma+\nu)}\Big(\cdot, \Big(\bigwedge^{\kappa} \mathscr{E}\Big) \otimes \Big(\bigwedge^{\gamma} \mathscr{E}^*\Big)\Big), \\
(\boldsymbol{\mathcal{L}}_{\mathscr{D}}^{\nu})^\circ &= \bigoplus_{\gamma\geq-\nu,\,\kappa\geq0,\,\kappa'\geq0} {}'\mathcal{D}^{(\kappa,0),(\kappa',\gamma+\nu)}\Big(\cdot, \Big(\bigwedge^{\kappa+\kappa'} \mathscr{E}\Big) \otimes \Big(\bigwedge^{\gamma} \mathscr{E}^*\Big)\Big).
\end{aligned}
\tag{4.14}
$$

The operators

$$
\boldsymbol{\nabla}_{\mathscr{D}} = 2i\pi \rfloor_\Delta - \overline{\partial}, \quad \boldsymbol{\nabla}_{\mathscr{D}}^\circ = 2i\pi \rfloor_\Delta - \overline{\partial}_\zeta \tag{4.15}
$$

act respectively as

$$
\begin{aligned}
\boldsymbol{\nabla}_{\mathscr{D}} &: \boldsymbol{\mathcal{L}}_{\mathscr{D}}^{\nu} \longrightarrow \boldsymbol{\mathcal{L}}_{\mathscr{D}}^{\nu+1}, \\
\boldsymbol{\nabla}_{\mathscr{D}}^\circ &: (\boldsymbol{\mathcal{L}}_{\mathscr{D}}^{\nu})^\circ \longrightarrow (\boldsymbol{\mathcal{L}}_{\mathscr{D}}^{\nu+1})^\circ.
\end{aligned}
\tag{4.16}
$$

The action of the operators $\rfloor_\Delta$, $\overline{\partial}$, $\overline{\partial}_\zeta$ is defined as follows: let $\mathscr{U}$ be an open subset above which the vector bundle $\mathscr{E}$ is trivial and equipped with a local frame

[2] We will need to enlarge this definition a bit later on, but we postpone it to §4.5.3, when it is needed; see (4.96).

$\{e_1, \dots, e_N\}$. Let also $T \in \boldsymbol{\mathcal{L}}^{\nu}_{\mathscr{D}}(\mathscr{U})$ be expressed as

$$T = \sum_{\gamma \geq -\nu,\, \kappa \geq 0} \; \sum_{1 \leq i_1 < \cdots < i_\kappa \leq N} \; \sum_{1 \leq j_1 < \cdots < j_\gamma \leq N} \tau^{\kappa,\gamma}_{I,J}\, e_I \otimes e^*_J,$$

then

$$\rfloor_\Delta T := (-1)^\nu \sum_{1 \leq i_1 < \cdots < i_\kappa \leq N} \; \sum_{1 \leq j_1 < \cdots < j_\gamma \leq N} (-1)^\gamma\, \tau^{\kappa,\gamma}_{I,J}\, e_I \otimes \rfloor_\Delta e^*_J,$$

$$\overline{\partial} T := \sum_{1 \leq i_1 < \cdots < i_\kappa \leq N} \; \sum_{1 \leq j_1 < \cdots < j_\gamma \leq N} \overline{\partial} \tau^{\kappa,\gamma}_{I,J}\, e_I \otimes e^*_J,$$

$$\overline{\partial}_\zeta T := \sum_{1 \leq i_1 < \cdots < i_\kappa \leq N} \; \sum_{1 \leq j_1 < \cdots < j_\gamma \leq N} \overline{\partial}_\zeta \tau^{\kappa,\gamma}_{I,J}\, e_I \otimes e^*_J,$$

The same rules apply when $T \in (\boldsymbol{\mathcal{L}}^{\nu}_{\mathscr{D}})^\circ(\mathscr{U})$. Namely, if

$$T = \sum_{\gamma \geq -\nu,\, \kappa,\kappa' \geq 0} \; \sum_{1 \leq i_1 < \cdots < i_{\kappa+\kappa'} \leq N} \; \sum_{1 \leq j_1 < \cdots < j_\gamma \leq N} \tau^{\kappa,\kappa',\gamma}_{I,J}\, e_I \otimes e^*_J,$$

then one has in this case that

$$\rfloor_\Delta T := (-1)^\nu \sum_{1 \leq i_1 < \cdots < i_{\kappa+\kappa'} \leq N} \; \sum_{1 \leq j_1 < \cdots < j_\gamma \leq N} (-1)^\gamma\, \tau^{\kappa,\kappa',\gamma}_{I,J}\, e_I \otimes \rfloor_\Delta e^*_J.$$

Furthermore, one can define, as in (4.11), the two complexes of sheaves of currents

$$(4.17)\qquad \begin{array}{l} \cdots \longrightarrow \boldsymbol{\mathcal{L}}^{\nu-1}_{\mathscr{D}} \overset{\boldsymbol{\nabla}_{\mathscr{D}}}{\longrightarrow} \boldsymbol{\mathcal{L}}^{\nu}_{\mathscr{D}} \overset{\boldsymbol{\nabla}_{\mathscr{D}}}{\longrightarrow} \boldsymbol{\mathcal{L}}^{\nu+1}_{\mathscr{D}} \longrightarrow \cdots \\ \cdots \longrightarrow (\boldsymbol{\mathcal{L}}^{\nu-1}_{\mathscr{D}})^\circ \overset{\boldsymbol{\nabla}^\circ_{\mathscr{D}}}{\longrightarrow} (\boldsymbol{\mathcal{L}}^{\nu}_{\mathscr{D}})^\circ \overset{\boldsymbol{\nabla}^\circ_{\mathscr{D}}}{\longrightarrow} (\boldsymbol{\mathcal{L}}^{\nu+1}_{\mathscr{D}})^\circ \longrightarrow \cdots, \end{array}$$

where $\boldsymbol{\nabla}^2_{\mathscr{D}} = (\boldsymbol{\nabla}^\circ_{\mathscr{D}})^2 = 0$.

4.2. Affine and projective Bochner–Martinelli kernels

4.2.1. The affine case $\mathcal{X} = \mathbb{C}^N$. We begin the present section with the vector bundle on $\mathbb{C}^N \times \mathbb{C}^N$ and the corresponding holomorphic section defining the diagonal as in the second configuration of (4.5); that is, we have

$$(4.18)\qquad \mathscr{E} = \pi^* T^{(1,0)}_{\mathbb{C}^N} \quad \text{and} \quad \Delta = \sum_{k=1}^N (\zeta_k - z_k)\, \partial/\partial \zeta_k.$$

If $\|\ \|$ denotes the canonical hermitian metric on $\mathscr{E} \simeq \mathbb{C}^N$, then one has

$$(4.19)\qquad \frac{\Delta^*}{2i\pi \|\Delta\|^2} = \frac{\sum_{k=1}^N \overline{\zeta_k - z_k}\, d\zeta_k}{2i\pi \|\zeta - z\|^2} = \frac{\partial_\zeta \log \|\zeta - z\|^2}{2i\pi}.$$

Furthermore, taking into account the definition of the Bochner–Martinelli principal value current $P^\Delta_{\|\ \|}$ in (3.24) (see Proposition 3.15 with $\mathscr{X}$ in place of $\mathcal{X}$ and $V = \mathscr{X}$), one has that

$$(4.20)\qquad P^\Delta_{\|\ \|} = \Big[\|\zeta - z\|^{2\lambda} \frac{\partial_\zeta \log \|\zeta - z\|^2}{2i\pi} \wedge \sum_{r=1}^N \big(dd^c_\zeta \log \|\zeta - z\|^2 \big)^{r-1} \Big]_{\lambda = 0}.$$

Since $\operatorname{codim} V(\Delta) = N = \operatorname{rank} \mathscr{E}$, it follows from the last assertion in (3.24) from Proposition 3.15 that

$$(4.21)\quad R^\Delta_{\|\ \|} = \Big[\overline{\partial} \|\zeta - z\|^{2\lambda} \wedge \frac{\partial \log \|\zeta - z\|^2}{2i\pi} \wedge (dd^c \log \|\zeta - z\|^2)^{N-1} \Big]_{\lambda=0} = [\mathscr{D}],$$

where the last equality is a consequence of J.R. King's formula; see Remark C.31. We can therefore state the following proposition.

PROPOSITION 4.6. *The smooth differential form*

$$K_{\mathbb{C}^N}\ : (z,\zeta) \in (\mathbb{C}^N \times \mathbb{C}^N) \setminus \mathscr{D} \longmapsto \frac{\partial_\zeta \log \|\zeta - z\|^2}{2i\pi} \wedge \sum_{r=1}^{N} \left(dd^c_\zeta \log \|\zeta - z\|^2\right)^{r-1}$$

extends to $\mathbb{C}^N \times \mathbb{C}^N$ *as a locally integrable differential form.*[3] *It defines the current* $[K_{\mathbb{C}^N}]$, *which is a global section of* $(\mathcal{L}^{-1}_{\mathscr{D}}[\mathbb{C}])^\circ$ *when the defining Koszul complex for* $\mathscr{D}$ *is the* $(\mathscr{E}, \Delta)$ *Koszul complex* $\mathscr{K}_\bullet^{\mathscr{D}}$, *with* $\mathscr{E}$ *and* Δ *as in* (4.18). *Moreover, one has in* $\mathbb{C}^N \times \mathbb{C}^N$ *the current identities*

$$\text{(4.22)} \qquad \nabla_{\mathscr{D}} K_{\mathbb{C}^N} = [1_{\mathbb{C}^N \times \mathbb{C}^N}] - [\mathscr{D}], \quad \nabla^\circ_{\mathscr{D}} K_{\mathbb{C}^N} = [1_{\mathbb{C}^N \times \mathbb{C}^N}] - [\mathscr{D}]_{(0,0),(N,N)},$$

where $[\mathscr{D}]_{(0,0),(N,N)}$ *denotes the component of the integration current* $[\mathscr{D}]$ *with bidegree* $(0,0)$ *in* z *and* (N,N) *in* ζ. *On the other hand, for each* z^0 *fixed in* $\mathbb{C}^N$, *the smooth form*

$$\zeta \in \mathbb{C}^N \setminus \{z^0\} \longmapsto K_{\mathbb{C}^N}(z^0, \zeta)$$

extends to $\mathbb{C}^N$ *as a locally integrable form defining a section of* $\mathcal{L}^{-1}_E$, *if one takes*[4] $E = T^{(1,0)}_{\mathbb{C}^N}$. *Moreover, one has in* $\mathbb{C}^N$ *the current identity*

$$\text{(4.23)} \qquad \nabla_{\Delta(z^0,\cdot)} K_{\mathbb{C}^N}(z^0, \cdot) = [1_{\mathbb{C}^N}] - [\{z^0\}],$$

where $\nabla_{\Delta(z^0,\cdot)} = \rfloor_{\Delta(z^0,\cdot)} - \overline{\partial}_\zeta$.[5]

PROOF. The first assertion is immediate since, after multiplication by the factor $\|\zeta - z\|^{2N-1} = \|\zeta - z\|^{2(N-1)+1}$, one gets a bounded differential form in $\mathbb{C}^N \times \mathbb{C}^N$. The second assertion follows from the description of $(\mathcal{L}^\nu_{\mathscr{D}}[\mathbb{C}])^\circ$ in (4.12). The currential identities (4.22) follow from (4.20), (4.21), and (3.24) in Proposition 3.15. For z^0 fixed in $\mathbb{C}^N$, the form $K_{\mathbb{C}^N}(z^0, \cdot)$ extends from $\mathbb{C}^N \setminus \{z^0\}$ as a locally integrable form since the function $\zeta \mapsto \|\zeta - z^0\|^{1-2N}$ is locally integrable in $\mathbb{C}^N$. The identity (4.23) follows again from (4.21) and (3.24) in Proposition 3.15, applied this time to $\mathbb{C}^N$ (instead of $\mathbb{C}^N \times \mathbb{C}^N$) with $z = z^0$ being fixed. □

Proposition 4.6 suggests the following definition.

DEFINITION 4.7. The current $K_{\mathbb{C}^N}$ defined in $\mathbb{C}^N \times \mathbb{C}^N$ as

$$\text{(4.24)} \qquad [K_{\mathbb{C}^N}(z,\zeta)] = \left[\frac{\partial_\zeta \log \|\zeta - z\|^2}{2i\pi} \wedge \sum_{r=1}^{N} \left(dd^c_\zeta \log \|\zeta - z\|^2\right)^{r-1}\right]$$

is called the *Bochner–Martinelli kernel of the complex manifold* $\mathbb{C}^N$.

REMARK 4.8. In the particular case where $\mathcal{X} = \mathbb{C}^{N+1}$ and $z^0 = 0$, we analyzed before how $[K_{\mathbb{C}^{N+1}}(0,\zeta)]$ could have been deduced from the Cauchy kernel in $\mathbb{C}^N$ through an averaging process; see Example 2.7.

---

[3] It means that for any $(\mathscr{E}(\mathscr{K}_\bullet^{\mathscr{D}}))^*$-valued smooth test form Φ on $\mathbb{C}^N \times \mathbb{C}^N$, the integral $\int_{\mathbb{P}^N_{\mathbb{C}} \times \mathbb{P}^N_{\mathbb{C}}} K_{\mathbb{C}^N} \wedge \Phi$ is absolutely convergent.

[4] See (3.19) for the definition of the sheaves of currents $\mathcal{L}^\nu_E$ attached to a vector bundle E over $\mathcal{X} = \mathbb{C}^N$.

[5] See §3.2.

4.2.2. The projective case $\mathcal{X} = \mathbb{P}^N_{\mathbb{C}}$. In the present section, the vector bundle on $\mathbb{P}^N_{\mathbb{C}} \times \mathbb{P}^N_{\mathbb{C}}$ and the corresponding holomorphic section defining the diagonal are as in (4.6),

$$\mathscr{E} = \overset{\curvearrowleft}{\mathcal{O}}(1) \otimes \pi^* T^{(1,0)}_{\mathbb{P}^N_{\mathbb{C}}}, \quad \Delta = -\sum_{\kappa=0}^{N} z_\kappa \frac{\partial}{\partial \zeta_\kappa}, \tag{4.25}$$

where $([z_0 : \cdots : z_N], [\zeta_0 : \cdots : \zeta_N])$ denote the homogeneous coordinates in the duplication $\mathscr{X} = \mathbb{P}^N_{\mathbb{C}} \times \mathbb{P}^N_{\mathbb{C}}$. The smooth $(1,0)$-form

$$\begin{aligned} \mathcal{Q}^{\Delta}_{\|\ \|} : (z,\zeta) \in \mathbb{C}^{N+1} \times \mathbb{C}^{N+1} \setminus \{(z,\zeta) : z \wedge \zeta = 0\} \\ \longrightarrow \frac{1}{2i\pi\, \|z \wedge \zeta\|^2} \sum_{\kappa=0}^{N} \big((\zeta \bullet z)\, \overline{\zeta_\kappa} - \overline{z_\kappa}\, \|\zeta\|^2\big)\, d\zeta_\kappa, \end{aligned}$$

where $\bullet$ denotes the canonical hermitian scalar product on $\mathbb{C}^{N+1}$, and satisfies in its domain of definition

$$\begin{aligned} \Big(\sum_{\kappa=0}^{N} \zeta_\kappa \frac{\partial}{\partial \zeta_\kappa}\Big) \rfloor \mathcal{Q}^{\Delta}_{\|\ \|} &= \Big(\sum_{\kappa=0}^{N} \overline{\zeta_\kappa} \frac{\partial}{\partial \overline{\zeta_\kappa}}\Big) \rfloor \mathcal{Q}^{\Delta}_{\|\ \|} \equiv 0, \\ -2i\pi \Big(\sum_{\kappa=0}^{N} z_\kappa \frac{\partial}{\partial \zeta_\kappa}\Big) \rfloor \mathcal{Q}^{\Delta}_{\|\ \|} &= \frac{-|\zeta \bullet z|^2 + \|z\|^2\, \|\zeta\|^2}{\|z \wedge \zeta\|^2} \equiv 1. \end{aligned}$$

The coefficients of this $(1,0)$-form, once expressed in terms of $\{d\zeta_\kappa : \kappa = 0, \dots, N\}$, are $(-1,0)$ homogeneous in $(z,\overline{z})$ and $(0,0)$ homogeneous in $(\zeta, \overline{\zeta})$. Thus, the form $\mathcal{Q}^{\Delta}_{\|\ \|}$ induces a smooth section

$$q^{\Delta}_{\|\ \|} \in C^\infty_{1,0}\big(\mathbb{P}^N_{\mathbb{C}} \times \mathbb{P}^N_{\mathbb{C}} \setminus \mathscr{D}, \overset{\curvearrowleft}{\mathcal{O}}(-1) \otimes \pi^* T^{*(1,0)}_{\mathbb{P}^N_{\mathbb{C}}}\big) = C^\infty_{1,0}\left(\mathbb{P}^N_{\mathbb{C}} \times \mathbb{P}^N_{\mathbb{C}} \setminus \mathscr{D}, \mathscr{E}^*\right);$$

see [**Ber88**, §0]. Moreover, one has

$$P^{\Delta}_{|\ |} = \Big[\Big(\frac{\|z \wedge \zeta\|^2}{\|z\|^2\, \|\zeta\|^2}\Big)^{\lambda} q^{\Delta}_{\|\ \|} \sum_{r \geq 1} (\overline{\partial} q^{\Delta}_{\|\ \|})^{r-1}\Big]_{\lambda=0} \tag{4.26}$$

for some ad hoc smooth metric $|\ |$ on $\mathscr{E}$. Therefore, (4.26) is the principal value Bochner–Martinelli current attached to the section Δ of the metrized holomorphic bundle $\overline{\mathscr{E}} = (\mathscr{E}, |\ |)$; see Proposition 3.15. The corresponding Bochner–Martinelli residue current equals

$$\begin{aligned} R^{\Delta}_{|\ |} &= \Big[\overline{\partial}\Big(\frac{\|z \wedge \zeta\|^2}{\|z\|^2\, \|\zeta\|^2}\Big)^{\lambda} \wedge q^{\Delta}_{\|\ \|} \wedge \sum_{r \geq 1} (\overline{\partial} q^{\Delta}_{\|\ \|})^{r-1}\Big]_{\lambda=0} \\ &= \Big[\overline{\partial}\Big(\frac{\|z \wedge \zeta\|^2}{\|z\|^2\, \|\zeta\|^2}\Big)^{\lambda} \wedge q^{\Delta}_{\|\ \|} \wedge (\overline{\partial} q^{\Delta}_{\|\ \|})^{N-1}\Big]_{\lambda=0} \end{aligned} \tag{4.27}$$

since $\operatorname{codim} V(\Delta) = \operatorname{rank} \mathscr{E} = N$; see the last assertion in (3.24) of Proposition 3.15. Observe that Δ factorizes in a local chart $\mathscr{U}_a$ around $a \in \mathscr{D}$ as $\Delta = e \otimes \delta$, where

$$\begin{aligned} &e \in \mathscr{O}\big(\mathscr{U}_a, \overset{\curvearrowleft}{\mathcal{O}}(1)\,\big) \text{ with } e = \mathrm{Id}_{\pi^* \mathcal{O}(1)} \text{ on } U_a \cap \mathscr{D}, \\ &\delta \in \mathscr{O}\big(U_a, \pi^* T^{(1,0)}_{\mathbb{P}^N_{\mathbb{C}}}\big) \text{ with } V(\delta) = U_a \cap \mathscr{D}. \end{aligned}$$

It follows from the invariance of $R^{\Delta}_{|\,|}$ concerning the choice of the metric (see equality (3.29) in the proof of Proposition 3.18, and from J.R. King's formula, see Remark C.31) that one has

$$R^{\Delta}_{|\,|} = \overset{\curvearrowleft}{\mathcal{O}(-N)}\ [\mathscr{D}] = \mathrm{Id}_{\pi^*\mathcal{O}(-N)}\,[\mathscr{D}]. \tag{4.28}$$

Therefore, we state the following proposition [**AndG11**, Proposition 4.1].

PROPOSITION 4.9. *The smooth differential form, defined on $(\mathbb{P}^N_{\mathbb{C}}\times\mathbb{P}^N_{\mathbb{C}})\setminus\mathscr{D}$ and expressed in homogeneous coordinates $([z_0:\cdots:z_N],[\zeta_0:\cdots:\zeta_N])$ as*

$$K_{\mathbb{P}^N_{\mathbb{C}}}(z,\zeta) = \frac{\sum_{\kappa=0}^N\big((\zeta\bullet z)\,\overline{\zeta_\kappa}-\overline{z_\kappa}\,\|\zeta\|^2\big)\,d\zeta_\kappa}{2i\pi\|z\wedge\zeta\|^2}\wedge\sum_{r=1}^N\Big(\overline{\partial}\Big(\frac{\sum_{\kappa=0}^N\big((\zeta\bullet z)\,\overline{\zeta_\kappa}-\overline{z_\kappa}\,\|\zeta\|^2\big)\,d\zeta_\kappa}{2i\pi\|z\wedge\zeta\|^2}\Big)\Big)^{r-1}$$

extends to $\mathbb{P}^N_{\mathbb{C}}\times\mathbb{P}^N_{\mathbb{C}}$ as an integrable differential form.[6] *It defines the current $K_{\mathbb{P}^N_{\mathbb{C}}}$, which is a global section of the sheaf $(\mathcal{L}^{-1}_{\mathscr{D}}[\mathbb{C}])^\circ$ when the defining Koszul complex for $\mathscr{D}$ as the $(\mathscr{E},\Delta)$ Koszul complex $\mathscr{K}_\bullet^{\mathscr{D}}$, with $\mathscr{E}$ and Δ as in (4.25). Moreover, one has in $\mathbb{P}^N_{\mathbb{C}}\times\mathbb{P}^N_{\mathbb{C}}$ the current identities*

$$\begin{aligned}\nabla_{\mathscr{D}}K_{\mathbb{P}^N_{\mathbb{C}}} &= [1_{\mathbb{P}^N_{\mathbb{C}}\times\mathbb{P}^N_{\mathbb{C}}}] - \mathrm{Id}_{\pi^*\mathcal{O}(-N)}\,[\mathscr{D}],\\ \nabla^\circ_{\mathscr{D}}K_{\mathbb{P}^N_{\mathbb{C}}} &= [1_{\mathbb{P}^N_{\mathbb{C}}\times\mathbb{P}^N_{\mathbb{C}}}] - \mathrm{Id}_{\pi^*\mathcal{O}(-N)}\,[\mathscr{D}]_{(0,0),(N,N)}.\end{aligned} \tag{4.29}$$

On the other hand, for each z^0 fixed in $\mathbb{P}^N_{\mathbb{C}}$, the smooth form

$$\zeta\in\mathbb{P}^N_{\mathbb{C}}\setminus\{z^0\}\longmapsto K_{\mathbb{P}^N_{\mathbb{C}}}(z^0,\zeta)$$

extends also to $\mathbb{P}^N_{\mathbb{C}}$ as a locally integrable form defining section of the sheaf $\mathcal{L}_E^{-1}$, where $E=\mathrm{Hom}_{\mathbb{C}}(\mathcal{O}(1),(\mathcal{O}(1))_{z^0})\otimes T^{(1,0)}_{\mathbb{P}^N_{\mathbb{C}}}$. Furthermore, one has in $\mathbb{P}^N_{\mathbb{C}}$ the current identity

$$\nabla_{\Delta(z^0,\cdot)}K_{\mathbb{P}^N_{\mathbb{C}}}(z^0,\cdot) = [1_{\mathbb{P}^N_{\mathbb{C}}}] - [\{z^0\}]\,\mathrm{Id}_{(\mathcal{O}(-N))_{z^0}}. \tag{4.30}$$

PROOF. The first assertion is immediate. After multiplication of $K_{\mathbb{P}^N_{\mathbb{C}}}$ by the factor $\|z\wedge\zeta\|^{2N-1}=\|z\wedge\zeta\|^{2(N-1)+1}$, one gets a bounded bundle-valued differential form in $\mathbb{P}^N_{\mathbb{C}}\times\mathbb{P}^N_{\mathbb{C}}$. The second assertion follows from the description of $(\mathcal{L}^{\nu}_{\mathscr{D}}[\mathbb{C}])^\circ$ in (4.13). Currential identities (4.29) follow from (4.26), (4.27), and (3.24) in Proposition 3.15. For z^0 fixed in $\mathbb{P}^N_{\mathbb{C}}$, the bundle-valued form $\zeta\mapsto K_{\mathbb{P}^N_{\mathbb{C}}}(z^0,\cdot)$ extends from $\mathbb{P}^N_{\mathbb{C}}\setminus\{z^0\}$ to a bundle-valued integrable form on $\mathbb{P}^N_{\mathbb{C}}$, since the bundle-valued section $\zeta\mapsto\|z^0\wedge\zeta\|^{1-2N}$ is locally integrable in $\mathbb{P}^N_{\mathbb{C}}$. The currential identity (4.30) follows again from (4.27) and (3.24) in Proposition 3.15, applied this time in $\mathbb{P}^N_{\mathbb{C}}$ (instead of $\mathbb{P}^N_{\mathbb{C}}\times\mathbb{P}^N_{\mathbb{C}}$) with $z=z^0$ fixed. □

As in the previous subsection, one can then set the following definition.

DEFINITION 4.10. The bundle-valued current on $\mathbb{P}^N_{\mathbb{C}}\times\mathbb{P}^N_{\mathbb{C}}$ expressed by

$$\Big[\frac{\sum_{\kappa=0}^N\big((\zeta\bullet z)\,\overline{\zeta_\kappa}-\overline{z_\kappa}\,\|\zeta\|^2\big)\,d\zeta_\kappa}{2i\pi\|z\wedge\zeta\|^2}\wedge\sum_{r=1}^N\Big(\overline{\partial}\Big(\frac{\sum_{\kappa=0}^N\big((\zeta\bullet z)\,\overline{\zeta_\kappa}-\overline{z_\kappa}\,\|\zeta\|^2\big)\,d\zeta_\kappa}{2i\pi\|z\wedge\zeta\|^2}\Big)\Big)^{r-1}\Big]$$

---

[6] By this, we mean that for any $(\mathscr{E}(\mathscr{K}_\bullet^{\mathscr{D}}))^*$-valued smooth form Φ on $\mathbb{P}^N_{\mathbb{C}}\times\mathbb{P}^N_{\mathbb{C}}$, the integral $\int_{\mathbb{P}^N_{\mathbb{C}}\times\mathbb{P}^N_{\mathbb{C}}}K_{\mathbb{P}^N_{\mathbb{C}}}\wedge\Phi$ is absolutely convergent.

in homogeneous coordinates $([z_0 : \cdots : z_N], [\zeta_0 : \cdots : \zeta_N])$, is called the *Bochner–Martinelli kernel* $K_{\mathbb{P}^N_{\mathbb{C}}}$ *of the projective space* $\mathbb{P}^N_{\mathbb{C}}$ with respect to the Fubini–Study metric.

4.3. Chern connection and Koppelman's formulae on good manifolds

Let $\mathcal{X}$ be an N-dimensional, good complex manifold. That is, we assume that $\mathcal{X}$ embeds in its duplication $\mathscr{X}$ as $\mathscr{D} = V(\Delta)$, where $\Delta \in \mathscr{O}(\mathscr{X}, \mathscr{E})$ with $\operatorname{rank} \mathscr{E} = N$; see Definition 4.1. Let also $\mathscr{K}_{\bullet}^{\mathscr{D}}$ be the $(\mathscr{E}, \Delta)$ Koszul complex. We consider the bundle

$$\boldsymbol{\mathcal{E}} = \bigoplus_{\ell \geq 0} \bigwedge^{\ell} (\mathscr{E} \oplus \mathscr{E}^*) = (\det \mathscr{E} \otimes \det \mathscr{E}^*) \oplus \boldsymbol{\mathcal{E}}' = \operatorname{Hom}_{\mathbb{C}}(\det \mathscr{E}) \oplus \boldsymbol{\mathcal{E}}' \tag{4.31}$$

and observe the following:

(1) Since $\mathscr{E}$ (respectively $\mathscr{E}^*$) is a subbundle of $\boldsymbol{\mathcal{E}}$, any holomorphic section of $\mathscr{E}$ (respectively of $\mathscr{E}^*$) can be interpreted as a holomorphic section of $\boldsymbol{\mathcal{E}}$. In particular, if the section $\Delta \in \mathscr{E}$ (respectively Δ^*, when $\mathscr{E}$ is metrized), then the corresponding holomorphic section of $\boldsymbol{\mathcal{E}}$ (respectively of $\mathscr{E}^*$) is denoted by $\boldsymbol{\Delta}$ (respectively $\boldsymbol{\Delta}^*$);

(2) Since $\operatorname{Hom}_{\mathbb{C}}(\mathscr{E}, \mathscr{E}) \simeq \mathscr{E} \otimes \mathscr{E}^*$ can also be considered a subbundle of $\boldsymbol{\mathcal{E}}$ through such identification, any $\operatorname{Hom}_{\mathbb{C}}(\mathscr{E}, \mathscr{E})$-valued differential form can be interpreted as an $\boldsymbol{\mathcal{E}}$-valued differential form;

(3) Since $\det \mathscr{E}$ is a line bundle, any $\boldsymbol{\mathcal{T}} \in {}'\mathcal{D}(\mathcal{X}, \boldsymbol{\mathcal{E}})$ admits a unique splitting as

$$\boldsymbol{\mathcal{T}} = T \, \mathrm{Id}_{\det \mathscr{E}} + \boldsymbol{\mathcal{T}}' \quad \text{with } T \in {}'\mathcal{D}(\mathscr{X}), \ \boldsymbol{\mathcal{T}}' \in {}'\mathcal{D}(\mathscr{X}, \boldsymbol{\mathcal{E}}'). \tag{4.32}$$

To be more specific, let $\mathscr{U}$ be an open subset of $\mathcal{X}$ over which the bundle $\mathscr{E}$ is trivial. In the local frame $\{e_1, \dots, e_N\}$ for the bundle $\mathscr{E}$ the current $\boldsymbol{\mathcal{T}}_{|\mathscr{U}}$ is decomposed as

$$\begin{aligned} \boldsymbol{\mathcal{T}}_{|\mathscr{U}} &= \sum_{0 \leq \kappa, \gamma \leq N} \ \sum_{1 \leq i_1 < \cdots < i_\kappa \leq N} \ \sum_{1 \leq j_1 < \cdots < j_\gamma \leq N} \tau_{I,J}^{\kappa,\gamma}(\mathscr{U}) \, e_I \otimes e_J^* \\ &= \tau(\mathscr{U}) \frac{(\sum_1^N e_j \wedge e_j^*)^N}{N!} + \sum_{\min(\kappa,\gamma) < N} \ \sum_{\#I = \kappa} \ \sum_{\#J = \gamma} \tau_{I,J}^{\kappa,\gamma}(\mathscr{U}) \, e_I \otimes e_J^*. \end{aligned}$$

Hence, all locally defined currents $\tau(\mathscr{U})$ globalize as $T \in {}'\mathcal{D}(\mathscr{X}) = {}'\mathcal{D}(\mathscr{X})$. From now on, we will use the notations for T and the splitting (4.32)

$$T = \boldsymbol{\mathcal{T}}_{\mathscr{E}}, \quad \boldsymbol{\mathcal{T}} = \boldsymbol{\mathcal{T}}_{\mathscr{E}} \, \mathrm{Id}_{\det \mathscr{E}} + \boldsymbol{\mathcal{T}}' \tag{4.33}$$

respectively. We now equip the bundle $\mathscr{E}$ with a smooth hermitian metric and define an action $D_{\overline{\boldsymbol{\mathcal{E}}}}$ from the sheaf ${}'\mathcal{D}(\cdot, \boldsymbol{\mathcal{E}})$ to itself as follows.

DEFINITION 4.11. Let $\overline{\mathscr{E}} = (\mathscr{E}, |\ |)$, where $|\ |$ is a smooth hermitian metric, $D_{\overline{\mathscr{E}}}$ is its Chern connection considered as acting from the sheaf of $\mathscr{E}$-valued currents to itself. Let also $D^*_{\overline{\mathscr{E}}}$ be the dual connection acting from the sheaf of $\mathscr{E}^*$-valued currents to itself.[7] The action of $D_{\overline{\mathscr{E}}}$ and $D^*_{\overline{\mathscr{E}}}$ induces the definition of an operator $D_{\overline{\boldsymbol{\mathcal{E}}}}$, whose action from ${}'\mathcal{D}(\mathscr{U}, \boldsymbol{\mathcal{E}})$ into ${}'\mathcal{D}(\mathscr{U}, \boldsymbol{\mathcal{E}})$ is described as

$$D_{\overline{\boldsymbol{\mathcal{E}}}}(\tau \, e_I \otimes e_J^*) = d\tau \, e_I \otimes e_J^* + (-1)^{p+q} \, \tau \left(D_{\overline{\mathscr{E}}} e_I \otimes e_J^* + (-1)^{\kappa} \, e_I \otimes D^*_{\overline{\mathscr{E}}} e_J^* \right), \tag{4.34}$$

[7] We refer the reader to §B.2.1.3 and §B.3.2.2 for the notions introduced here.

where the bundle $\mathscr{E}$ is trivial over $\mathscr{U}$ with local frame $\{e_1, \dots, e_N\}$. Here, if the current $\tau \in {}'\mathcal{D}(\mathscr{U})$ has pure degree $p+q$ and

$$I : 1 \le i_1 < \cdots < i_\kappa \le N, \quad J : 1 \le j_1 < \cdots < j_\gamma \le N,$$

then, as usual,

$$D_{\overline{\mathscr{E}}} e_I = \sum_{\nu=1}^{\kappa} (-1)^{\nu-1} e_{i_1} \wedge \cdots \wedge e_{i_{\nu-1}} \wedge D_{\overline{\mathscr{E}}} e_{i_\nu} \wedge e_{i_{\nu+1}} \wedge \cdots \wedge e_{i_\kappa},$$

$$D^*_{\overline{\mathscr{E}}} e^*_J = \sum_{\nu=1}^{\gamma} (-1)^{\nu-1} e^*_{j_1} \wedge \cdots \wedge e^*_{j_{\nu-1}} \wedge D^*_{\overline{\mathscr{E}}} e^*_{i_\nu} \wedge e^*_{i_{\nu+1}} \wedge \cdots \wedge e^*_{i_\gamma}.$$

If one takes into account our observation (1), then the residue and principal value Bochner–Martinelli currents $R^{\boldsymbol{\Delta}}_{| \, |}$ and $P^{\boldsymbol{\Delta}}_{| \, |}$, respectively, are defined. Apply Proposition 3.15 with $V = \mathscr{X}$, by

$$(4.35) \qquad \begin{aligned} R^{\boldsymbol{\Delta}}_{| \, |} &= \Big[(\overline{\partial} |\Delta|^{2\lambda}) \sum_{r=1}^{N} \frac{\boldsymbol{\Delta}^*}{2i\pi |\Delta|^2} \Big(\overline{\partial} \frac{\boldsymbol{\Delta}^*}{2i\pi |\Delta|^2} \Big)^{r-1} \Big]_{\lambda=0} \\ &= \Big[(\overline{\partial} |\Delta|^{2\lambda}) \frac{\boldsymbol{\Delta}^*}{2i\pi |\Delta|^2} \Big(\overline{\partial} \frac{\boldsymbol{\Delta}^*}{2i\pi |\Delta|^2} \Big)^{N-1} \Big]_{\lambda=0}, \\ P^{\boldsymbol{\Delta}}_{| \, |} &= \Big[|\Delta|^{2\lambda} \sum_{r=1}^{N} \frac{\boldsymbol{\Delta}^*}{2i\pi |\Delta|^2} \Big(\overline{\partial} \frac{\boldsymbol{\Delta}^*}{2i\pi |\Delta|^2} \Big)^{r-1} \Big]_{\lambda=0}. \end{aligned}$$

Taking into account the definition of the sheaves of currents $\boldsymbol{\mathcal{L}}^{\nu}_{\mathscr{D}}$ for $\nu \in \mathbb{Z}$ (see (4.14)) one has

$$(4.36) \qquad R^{\boldsymbol{\Delta}}_{| \, |} \in \boldsymbol{\mathcal{L}}^0_{\mathscr{D}}(\mathscr{X}) \subset {}'\mathcal{D}(\mathscr{X}, \boldsymbol{\mathcal{E}}), \quad P^{\boldsymbol{\Delta}}_{| \, |} \in \boldsymbol{\mathcal{L}}^{-1}_{\mathscr{D}}(\mathscr{X}) \subset {}'\mathcal{D}(\mathscr{X}, \boldsymbol{\mathcal{E}}).$$

Note also that $P^{\boldsymbol{\Delta}}_{| \, |}$ is the current induced by a locally integrable differential form since $d\Delta \neq 0$ locally on $\mathscr{D}$, which means that the introduction of the complex parameter λ is revealed to be unnecessary. Taking also into account the definition of the nabla operator $\boldsymbol{\nabla}_{\mathscr{D}}$ (see 4.15 and the assertions (3.24) in Proposition 3.15) one has that

$$(4.37) \qquad \boldsymbol{\nabla}_{\mathscr{D}} P^{\boldsymbol{\Delta}}_{| \, |} = [1_{\mathscr{X}}] - R^{\boldsymbol{\Delta}}_{| \, |}.$$

We introduce also the normalized $\mathrm{Hom}_{\mathbb{C}}(\mathscr{E}, \mathscr{E})$-valued form defined as the normalized curvature tensor[8]

$$\Phi_{\overline{\mathscr{E}}} = -\frac{1}{2i\pi} \Theta_{\overline{\mathscr{E}}}$$

of the Chern connection $D_{\overline{\mathscr{E}}}$, so that

$$(4.38) \qquad \det(\mathrm{Id}_{\mathscr{E}} + \Phi_{\overline{\mathscr{E}}}) = c(\overline{\mathscr{E}}) = 1 + c_1(\overline{\mathscr{E}}) + \cdots + c_N(\overline{\mathscr{E}})$$

is the (total) Chern form of the metrized holomorphic bundle $\mathscr{E}$. Let

$$\boldsymbol{\Phi}_{\overline{\mathscr{E}}} \in \bigoplus_{k=0}^{N} C^\infty_{k,k}(\mathscr{X}, \boldsymbol{\mathcal{E}})$$

[8]See §B.3.2.2 and §B.3.3, in particular Propositions B.45 and B.48.

be the $\boldsymbol{\mathcal{E}}$-valued differential form associated to $\Phi_{\overline{\mathscr{E}}}$, taking into account our observation (2). Then observation (3) and formula (4.38) imply

$$\frac{1}{N!}\Big(\mathbf{\Phi}_{\overline{\mathscr{E}}} + D_{\overline{\boldsymbol{\mathcal{E}}}}\mathbf{\Delta}\Big)^{N} = c_N(\overline{\mathscr{E}})\,\mathrm{Id}_{\det E} + \Big(\frac{1}{N!}\Big(\mathbf{\Phi}_{\overline{\mathscr{E}}} + D_{\overline{\boldsymbol{\mathcal{E}}}}\mathbf{\Delta}\Big)^{N}\Big)'. \tag{4.39}$$

With all this material in hand, one can now state the following proposition [**Got08**, Lemma 3.3].

PROPOSITION 4.12. *One has*

$$\begin{aligned} &\frac{1}{N!}\Big(\mathbf{\Phi}_{\overline{\mathscr{E}}} + D_{\overline{\boldsymbol{\mathcal{E}}}}\mathbf{\Delta}\Big)^{N} \in \boldsymbol{\mathcal{L}}^{0}_{\mathscr{D}}(\mathscr{X}),\\ &\boldsymbol{\nabla}_{\mathscr{D}}\Big(\mathbf{\Phi}_{\overline{\mathscr{E}}} + D_{\overline{\boldsymbol{\mathcal{E}}}}\mathbf{\Delta}\Big)^{N} = \boldsymbol{\nabla}_{\mathscr{D}}\left(\mathbf{\Phi}_{\overline{\mathscr{E}}} + D_{\overline{\boldsymbol{\mathcal{E}}}}\mathbf{\Delta}\right) = 0. \end{aligned} \tag{4.40}$$

PROOF. The first assertion in (4.40) follows from the definition of the sheaf of currents $\boldsymbol{\mathcal{L}}^{0}_{\mathscr{D}}$ (see (4.14)), and from formula (B.60) (respectively (B.61)) describing the action along local frames of $D_{\overline{\mathscr{E}}}$ (respectively $\Phi_{\overline{\mathscr{E}}}$), once combined with the extensions to $\boldsymbol{\mathcal{E}}$-valued currents presented in points (2) and (3) in the beginning of this section. The first equality in the second assertion follows from the second one according to Leibniz's rule. One has

$$\boldsymbol{\nabla}_{\mathscr{D}}(D_{\overline{\boldsymbol{\mathcal{E}}}}\mathbf{\Delta}) = 2i\pi\,\rfloor_{\mathbf{\Delta}}\,(D_{\overline{\boldsymbol{\mathcal{E}}}}\mathbf{\Delta}) - \overline{\partial}\,(D_{\overline{\boldsymbol{\mathcal{E}}}}\mathbf{\Delta}) = -\overline{\partial}\,(D_{\overline{\boldsymbol{\mathcal{E}}}}\mathbf{\Delta}) \tag{4.41}$$

since no e_j^* appears in the local expression of $D_{\overline{\boldsymbol{\mathcal{E}}}}\mathbf{\Delta}$ in an open subset $\mathscr{U}$ over which $\mathscr{E}$ is trivial with $\{e_1,\dots,e_N\}$ as local holomorphic frame there. If in the open set $\mathscr{U}$ the identity

$$\Delta = \sum_{k=1}^{N} \Delta_k(\mathscr{U})\, e_k$$

holds, then one has

$$D_{\overline{\mathscr{E}}}\Delta = \sum_{k=1}^{N}\Big(\partial\left(\Delta_k(\mathscr{U})\right) + \sum_{\ell=1}^{N} a_{k,\ell}\wedge \Delta_k(\mathscr{U})\Big)\, e_k, \tag{4.42}$$

where the matrix of the connection $A = [a_{k,\ell}]_{1\leq k,\ell\leq N}$ is expressed in terms of the Gram matrix G of the holomorphic frame $\{e_1,\dots,e_N\}$ as $A = G^{-1}\,\partial G$; see Proposition B.45. Moreover, since the curvature tensor $\Theta_{\overline{\mathscr{E}}}$ is expressed in $\mathscr{U}$ as

$$\Theta_{\overline{\mathscr{E}}} = \sum_{k=1}^{N}\sum_{\ell=1}^{N} \overline{\partial} a_{k,\ell}\, e_k\otimes e_\ell^* \tag{4.43}$$

(see once more Proposition B.45), one gets that

$$\overline{\partial}(D_{\overline{\mathscr{E}}}\,\Delta) = \Theta_{\overline{\mathscr{E}}}\Delta.$$

Hence, in view of Definition 4.11 and of the definitions for $\mathbf{\Delta}$ and for $\mathbf{\Phi}_{\overline{\boldsymbol{\mathcal{E}}}}$,

$$\overline{\partial}(D_{\overline{\boldsymbol{\mathcal{E}}}}\,\mathbf{\Delta}) = -2i\pi\,\mathbf{\Phi}_{\overline{\mathscr{E}}}\,\mathbf{\Delta}. \tag{4.44}$$

On the other hand, one has in the same open subset $\mathscr{U}$ as above, with $\{e_1, \dots, e_N\}$ a local holomorphic frame for $\mathscr{E}$ there, that

$$\begin{aligned}2i\pi \rfloor_{\Delta} \Phi_{\overline{\mathscr{E}}} &= -\sum_{\nu=1}^{N} \Delta_k(\mathscr{U}) \rfloor_{e_\nu} \Big(\sum_{k=1}^{N}\sum_{\ell=1}^{N} \theta_{k,\ell}(\mathscr{U})\, e_k \otimes e_\ell^* \Big) \\ &= \sum_{k=1}^{N}\sum_{\ell=1}^{N} \theta_{k,\ell}\, \Delta_\ell(\mathscr{U})\, e_k = \Theta_{\overline{\mathscr{E}}} \Delta(\mathscr{U}) = -2i\pi\, \Phi_{\overline{\mathscr{E}}}\Delta(\mathscr{U}).\end{aligned}$$

One concludes from the definitions of both $\boldsymbol{\Delta}$ and $\boldsymbol{\Phi}_{\overline{\mathscr{E}}}$ that

$$2i\pi \rfloor_{\boldsymbol{\Delta}} \boldsymbol{\Phi}_{\overline{\mathscr{E}}} = -2i\pi\, \boldsymbol{\Phi}_{\overline{\mathscr{E}}}\, \boldsymbol{\Delta}. \tag{4.45}$$

Since $\overline{\partial}\Theta = 0$ (see (4.43)), the second equality in the second assertion (4.40) follows from (4.41), (4.44), and (4.45). □

We may now state the main result of this section [**Got08**, Theorem 3.4].

THEOREM 4.13. *Let $K^{\boldsymbol{\Delta}}_{\boldsymbol{\overline{\mathcal{E}}}}$ be the current*

$$K^{\boldsymbol{\Delta}}_{\boldsymbol{\overline{\mathcal{E}}}} = P^{\boldsymbol{\Delta}}_{|\,|} \frac{1}{N!}\Big(\boldsymbol{\Phi}_{\overline{\mathscr{E}}} + D_{\overline{\boldsymbol{\mathcal{E}}}}\boldsymbol{\Delta}\Big)^N \in \boldsymbol{\mathcal{L}}^{-1}_{\mathscr{D}}(\mathscr{X}). \tag{4.46}$$

One has that

$$\begin{aligned}\boldsymbol{\nabla}_{\mathscr{D}} K^{\boldsymbol{\Delta}}_{\boldsymbol{\overline{\mathcal{E}}}} &= \Big[\frac{1}{N!}\Big(\boldsymbol{\Phi}_{\overline{\mathscr{E}}} + D_{\overline{\boldsymbol{\mathcal{E}}}}\boldsymbol{\Delta}\Big)^N\Big] - [\mathscr{D}]\,\mathrm{Id}_{\det \mathscr{E}} \\ &= \big(c_N(\overline{\mathscr{E}}) - [\mathscr{D}]\big)\,\mathrm{Id}_{\det \mathscr{E}} + \Big[\frac{1}{N!}\Big(\boldsymbol{\Phi}_{\overline{\mathscr{E}}} + D_{\overline{\boldsymbol{\mathcal{E}}}}\boldsymbol{\Delta}\Big)^N\Big]'.\end{aligned} \tag{4.47}$$

PROOF. It follows from Leibniz's rule, Proposition 4.12, and then (4.37) that

$$\begin{aligned}\boldsymbol{\nabla}_{\mathscr{D}} K^{\boldsymbol{\Delta}}_{\boldsymbol{\overline{\mathcal{E}}}} &= \big(\boldsymbol{\nabla}_{\mathscr{D}} P^{\boldsymbol{\Delta}}_{|\,|}\big) \frac{1}{N!}\Big(\boldsymbol{\Phi}_{\overline{\mathscr{E}}} + D_{\overline{\boldsymbol{\mathcal{E}}}}\boldsymbol{\Delta}\Big)^N \\ &= \Big[\frac{1}{N!}\Big(\boldsymbol{\Phi}_{\overline{\mathscr{E}}} + D_{\overline{\boldsymbol{\mathcal{E}}}}\boldsymbol{\Delta}\Big)^N\Big] - R^{\boldsymbol{\Delta}}_{|\,|} \frac{1}{N!}\Big(\boldsymbol{\Phi}_{\overline{\mathscr{E}}} + D_{\overline{\boldsymbol{\mathcal{E}}}}\boldsymbol{\Delta}\Big)^N \\ &= \Big[\frac{1}{N!}\Big(\boldsymbol{\Phi}_{\overline{\mathscr{E}}} + D_{\overline{\boldsymbol{\mathcal{E}}}}\boldsymbol{\Delta}\Big)^N\Big] - R^{\boldsymbol{\Delta}}_{|\,|} \frac{(D_{\overline{\boldsymbol{\mathcal{E}}}}\boldsymbol{\Delta})^N}{N!}.\end{aligned}$$

Since one has (4.42) in any open subset $\mathscr{U}$ of $\mathscr{X}$ over which the bundle $\mathscr{E}$ is trivial, with local frame $\{e_1, \dots, e_N\}$, it then follows from Proposition 3.18 together with Theorem 2.79 that

$$R^{\boldsymbol{\Delta}}_{|\,|} \frac{(D_{\overline{\boldsymbol{\mathcal{E}}}}\boldsymbol{\Delta})^N}{N!} = [\mathscr{D}]\,\mathrm{Id}_{\det \mathscr{E}}.$$

The second equality in (4.47) follows from the splitting (4.33) of the current

$$\boldsymbol{\mathcal{T}} = \Big[\frac{1}{N!}\Big(\boldsymbol{\Phi}_{\overline{\mathscr{E}}} + D_{\overline{\boldsymbol{\mathcal{E}}}}\boldsymbol{\Delta}\Big)^N\Big],$$

together with the fact that $\boldsymbol{\mathcal{T}}_{\mathscr{E}} = c_N(\overline{\mathscr{E}})$ according to (4.39). □

The following corollary generalizes the Cauchy–Pompeiu integral representation formula. It belongs to the wide range of integral representation formulae for (p, q) forms (complex-valued or bundle-valued) introduced by Walter Koppelman in 1967 [**Kop67**]; see [**Got08**, Theorem 3.4].

COROLLARY 4.14. *Let* $K_{\overline{\mathcal{E}}} \in {}'\mathcal{D}(\mathscr{X}) = {}'\mathcal{D}(\mathscr{X})$ *such that one has the decomposition* $K^{\Delta}_{\overline{\mathcal{E}}} = K_{\overline{\mathcal{E}}}\,\mathrm{Id}_{\det\mathscr{E}} + (K^{\Delta}_{\overline{\mathcal{E}}})'$ *according to* (4.32). *Then*

$$\overline{\partial}\, K_{\overline{\mathcal{E}}} = [\mathscr{D}] - [c_N(\overline{\mathscr{E}})]. \tag{4.48}$$

PROOF. For any $\mathcal{T} \in {}'\mathcal{D}(\mathscr{X}, \mathcal{E})$, one observes that

$$\begin{aligned}(D_{\overline{\mathcal{E}}}\mathcal{T})_{\mathscr{E}} &= \Big(D_{\overline{\mathcal{E}}}(\mathcal{T}_{\mathscr{E}}\,\mathrm{Id}_{\det\mathscr{E}} + \mathcal{T}')\Big)_{\mathscr{E}} = \Big(D_{\overline{\mathcal{E}}}(\mathcal{T}_{\mathscr{E}}\,\mathrm{Id}_{\det\mathscr{E}})\Big)_{\mathscr{E}} \\ &= \Big(d(\mathcal{T}_{\mathscr{E}}\,\mathrm{Id}_{\det\mathscr{E}})\Big)_{\mathscr{E}} = d\mathcal{T}_{\mathscr{E}}.\end{aligned}$$

Then Corollary 4.14 follows from (4.47) provided one invokes the compatibility of $D_{\overline{\mathscr{E}}}$ (respectively of $D^*_{\overline{\mathscr{E}}}$) with the complex structure of $\mathscr{E}$ (respectively of $\mathscr{E}^*$). □

REMARK 4.15. We invite the reader to consult, for example, [**Got08**, Theorem 2.3] for the relation between Koppelman's formulae $\overline{\partial}K = [\mathscr{D}] - P$, with P smooth, on the duplication $\mathscr{X}$, and integral representation of (p,q)-forms in bounded open subsets with piecewise smooth boundary in $\mathcal{X}$. This relation is realized through Stokes's formula on $\mathcal{X}$.

In the spirit of Definitions 4.7 and 4.10, we have the following.

DEFINITION 4.16. Let $\mathcal{X}$ be an N-dimensional good complex manifold for which the Koszul complex $\mathscr{K}_{\bullet}^{\mathscr{D}}$ is the $(\mathscr{E}, \Delta)$ one. Given a hermitian metric $|\ |$ on $\mathscr{E}$, the current $K^{\Delta}_{\overline{\mathcal{E}}} \in \mathcal{L}^{-1}_{\mathscr{D}}(\mathscr{X})$, is called a *Bochner–Martinelli kernel* for $\mathcal{X}$.

4.4. Bochner–Martinelli weights, definition, and constructions

4.4.1. Global weights. Let $\mathcal{X}$ be a good N-dimensional manifold, which embeds in its duplication $\mathscr{X}$ as the smooth diagonal complete intersection denoted by $\mathscr{D} = V(\Delta)$. The section $\Delta \in \mathscr{O}(\mathscr{X}, \mathscr{E})$, where $\operatorname{rank}\mathscr{E} = N$, satisfies then $d\Delta \neq 0$ locally on $\mathscr{D}$. For any holomorphic bundle E over $\mathcal{X}$, a section of $\mathcal{L}^0_{\mathscr{D}}[\widehat{E}]$ in the open subset $\mathscr{U}$, having bidegree $(0,0)$ in z, is expressed as

$$\mathscr{W}(z,\zeta) = (\mathscr{W}(z,\zeta))_0 + \sum_{r=1}^{N} (\mathscr{W}(z,\zeta))_r,$$

where

$$\mathscr{W}_0 = \mathscr{W}_{0,0} \in {}'\mathcal{D}^{(0,0)}(\mathscr{U}, \widehat{E}),\ \mathscr{W}_r \in {}'\mathcal{D}^{(0,r)}\Big(\mathscr{U}, \Big(\bigwedge^r \mathscr{E}^*\Big)\otimes \widehat{E}\,\Big) \text{ for } r \geq 1;$$

see (4.2) and (4.9). Assume now that the singular support of $\mathscr{W}$ in $\mathscr{U}$ does not intersect $\mathscr{D}$ and that $\overline{\partial}_z \mathscr{W} = 0$. Namely, we assume that each $\mathscr{W}_r$ is holomorphic in $\mathscr{U}$ as a $(0,0)$-form with respect to the variable z. Then, one has that

$$\nabla_{\mathscr{D}}\mathscr{W} = (2i\pi \rfloor_{\Delta} - \overline{\partial})\mathscr{W} = \nabla^{\circ}_{\mathscr{D}}\mathscr{W} = (2i\pi\rfloor_{\Delta} - \overline{\partial}_{\zeta})\mathscr{W}$$

in $\mathscr{U}$. For any $K \in \mathcal{L}^{-1}_{\mathscr{D}}[\mathbb{C}](\mathscr{X})$ with singular support on $\mathscr{D}$, the definition of $\mathscr{W}K$ as a bundle-valued current in $\mathscr{U}$ is licit and one has

$$\mathscr{W}K \in \mathcal{L}^{-1}_{\mathscr{D}}[\widehat{E}](\mathscr{U}),\quad \nabla_{\mathscr{D}}(\mathscr{W}K) = (\nabla_{\mathscr{D}}\mathscr{W})\,K + \mathscr{W}\,(\nabla_{\mathscr{D}}K) \text{ in } \mathscr{U} \tag{4.49}$$

according to Leibniz's rule. In particular, one can take K to be the Bochner–Martinelli principal value current $K = P^{\Delta}_{|\ |}$ (see Proposition 3.15), where $|\ |$ is a smooth hermitian metric on $\mathscr{E}$. If one imposes the constraints

$$\nabla_{\mathscr{D}}\mathscr{W} = 0 \text{ in } \mathscr{U},\quad (\mathscr{W}(z,z))_0 = \mathrm{Id}_{(\sqcap^* E)_{(z,z)}}\ \forall\, z \in \sqcap(\mathscr{U}), \tag{4.50}$$

then the current identity (4.49) leads to the equality

$$
\begin{aligned}
\nabla_{\mathscr{D}}(\mathscr{W} K) &= \mathscr{W} - \mathscr{W}\, R^{\Delta}_{|\,|} \\
&= \mathscr{W} - \mathscr{W}\Big[\overline{\partial}|\Delta|^{2\lambda}\,\frac{\Delta^*}{2i\pi|\Delta|^2}\Big(\overline{\partial}\,\frac{\Delta^*}{2i\pi|\Delta|^2}\Big)^{N-1}\Big]_{\lambda=0} \\
&= \mathscr{W} - \mathscr{W}_0\Big[\overline{\partial}|\Delta|^{2\lambda}\,\frac{\Delta^*}{2i\pi|\Delta|^2}\Big(\overline{\partial}\,\frac{\Delta^*}{2i\pi|\Delta|^2}\Big)^{N-1}\Big]_{\lambda=0} \\
&= \mathscr{W} - \Big[\overline{\partial}|\Delta|^{2\lambda}\,\frac{\Delta^*}{2i\pi|\Delta|^2}\Big(\overline{\partial}\,\frac{\Delta^*}{2i\pi|\Delta|^2}\Big)^{N-1}\Big]_{\lambda=0}\otimes \mathrm{Id}_{\sqcap^* E}
\end{aligned}
\tag{4.51}
$$

valid in $\mathscr{U}$. This motivates our first definition.

DEFINITION 4.17. Let $\mathcal{X}$ be a good N-dimensional complex manifold, which embeds in its duplication $\mathscr{X}$ as the smooth diagonal complete intersection denoted by $\mathscr{D} = V(\Delta)$. The section $\Delta \in \mathscr{O}(\mathscr{X}, \mathscr{E})$, where $\operatorname{rank}\mathscr{E} = N$, satisfies then $d\Delta \neq 0$ locally on $\mathscr{D}$. Let E be a holomorphic bundle over $\mathcal{X}$ and $z^0 \in \mathcal{X}$. A *global E-weight in $\mathscr{U} \subset \mathscr{X}$* with respect to the $(\mathscr{E}, \Delta)$ Koszul complex $\mathscr{K}_{\bullet}^{\mathscr{D}}$ is a section $\mathscr{W}$ of the sheaf $\mathcal{L}^0_{\mathscr{D}}[\overset{\frown}{E}]$ in $\mathscr{U}$, whose singular support in $\mathscr{U}$ does not intersect $\mathscr{D}$, and such that

$$
\begin{aligned}
&\operatorname{bidegree}_z \mathscr{W} = (0,0), \quad \overline{\partial}_z \mathscr{W} = 0, \quad \nabla_{\mathscr{D}} \mathscr{W} = 0, \\
&(\mathscr{W}(z,z))_{0,0} = \mathrm{Id}_{(\sqcap^* E)_{(z,z)}} \ \forall z \in \sqcap(\mathscr{U}).
\end{aligned}
\tag{4.52}
$$

Archetypal examples of global weights are the so-called *Fubini–Study weights*,[9] when $\mathcal{X} = \mathbb{C}^N$ or $\mathcal{X} = \mathbb{P}^N_{\mathbb{C}}$.

PROPOSITION 4.18. *Let $\mathcal{X} = \mathbb{C}^N$ be a good manifold, which imbeds in its duplication $\mathbb{C}^N \times \mathbb{C}^N$ as the diagonal complete intersection $\mathscr{D} = V(\Delta)$, where $\Delta \in \mathscr{O}(\mathscr{X}, \mathscr{E})$, $\mathscr{E} = \pi^* T^{(1,0)}_{\mathbb{C}^N}$ and $\Delta = \sum_{k=1}^N (\zeta_k - z_k)\,\partial/\partial\zeta_k$. The current induced in $\mathbb{C}^N \times \mathbb{C}^N$ by the smooth form*

$$
\mathscr{W}_{\mathbb{C}^N} : (z,\zeta) \longmapsto \frac{1 + \sum_{k=1}^N z_k \overline{\zeta}_k}{1 + \|\zeta\|^2} + dd^c \log(1 + \|\zeta\|^2)
\tag{4.53}
$$

is a global $\mathbb{C}$-weight in $\mathbb{C}^N \times \mathbb{C}^N$ with respect to the $(\mathscr{E}, \Delta)$ Koszul complex $\mathscr{K}_{\bullet}^{\mathscr{D}}$. Similarly, if

$$
\mathscr{E} = \overset{\frown}{\mathcal{O}(1)} \otimes \pi^* T^{(1,0)}_{\mathbb{P}^N_{\mathbb{C}}}, \quad \Delta = -\sum_{\kappa=0}^{N} z_\kappa\, \partial/\partial\zeta_\kappa,
$$

then the current induced in $\mathbb{P}^N_{\mathbb{C}} \times \mathbb{P}^N_{\mathbb{C}}$ by the smooth bundle-valued form

$$
\mathscr{W}_{\mathbb{P}^N_{\mathbb{C}}} : (z,\zeta) \longmapsto \frac{z \bullet \zeta}{\|\zeta\|^2} + dd^c \log \|\zeta\|^2,
\tag{4.54}
$$

expressed in homogeneous coordinates $([z_0 : \cdots : z_N], [\zeta_0 : \cdots : \zeta_N])$, is a global $\mathcal{O}(1)$-weight in $\mathbb{P}^N_{\mathbb{C}} \times \mathbb{P}^N_{\mathbb{C}}$ with respect to the $(\mathscr{E}, \Delta)$ Koszul complex $\mathscr{K}_{\bullet}^{\mathscr{D}}$.

[9]The terminology used here relies on the fact that the construction of such weights is based on the Fubini–Study metric on $\mathbb{P}^N_{\mathbb{C}}$.

PROOF. One has for the first case when $\mathscr{X} = \mathbb{C}^N$,

$$\begin{aligned}
\overline{\partial}\mathscr{W}_{\mathbb{C}^N} &= \overline{\partial}\Big(\frac{1+\sum_{k=1}^N z_k\overline{\zeta}_k}{1+\|\zeta\|^2}\Big)\\
&= \frac{\sum_{k=1}^N z_k\, d\overline{\zeta}_k}{1+\|\zeta\|^2} - \Big(1+\sum_{k=1}^N z_k\overline{\zeta_k}\Big)\,\frac{\partial\|\zeta\|^2}{(1+\|\zeta\|^2)^2},\\
2i\pi\rfloor_\Delta\mathscr{W}_{\mathbb{C}^N} &= 2i\pi\rfloor_\Delta(dd^c\log(1+\|\zeta\|^2)\\
&= \frac{\sum_{k=1}^N (z_k-\zeta_k)\, d\overline{\zeta}_k}{1+\|\zeta\|^2} + \Big(\|\zeta\|^2-\sum_{k=1}^N z_k\overline{\zeta}_k\Big)\,\frac{\overline{\partial}\|\zeta\|^2}{(1+\|\zeta\|^2)^2},
\end{aligned}$$

which implies $\nabla_{\mathscr{D}}\mathscr{W}_{\mathbb{C}^N} = \nabla^\circ_{\mathscr{D}}\mathscr{W}_{\mathbb{C}^N} = 0$. The assertion for $\mathscr{W}_{\mathbb{P}^N_{\mathbb{C}}}$ follows after one comes back to affine coordinates in the local chart $\mathbb{C}^N \simeq \{z_0 \neq 0\}$. □

REMARK 4.19. The weight $\mathscr{W}_{\mathbb{P}^N_{\mathbb{C}}}$ is the direct image through the proper morphism

$$\Pi\ : (z,\zeta,u) \in \mathbb{P}^N_{\mathbb{C}}\times\mathbb{P}^N_{\mathbb{C}}\times\mathbb{P}^N_{\mathbb{C}} \mapsto (z,\zeta)\in\mathbb{P}^N_{\mathbb{C}}\times\mathbb{P}^N_{\mathbb{C}}$$

of the $(\Pi^*(\overset{\frown}{\mathscr{O}}(1)) \oplus \mathbb{C})$-valued current

$$(dd^c\log\|u\|^2)^N\langle\!\langle u,z\rangle\!\rangle\, 2i\pi P^{\langle\!\langle u,\zeta\rangle\!\rangle} + (dd^c\log\|\zeta\|^2)\ (dd^c\log\|u\|^2)^N,$$

where the principal value current $P^{\langle\!\langle u,\zeta\rangle\!\rangle}$ has been introduced in Proposition 2.64. This follows from Crofton's formula; see §D.5.1.

Besides the models of global weights described in Proposition 4.18, a strategy to generate, then combine, smooth $\mathbb{C}$-global weights was proposed (in the case where $\mathscr{X} = \Omega$ is a domain in $\mathbb{C}^N$) in the pioneering works [**AndB82, AndP88**].

PROPOSITION 4.20. *Let $q^1,\dots,q^M$ be M smooth sections of $\pi^*T^{*(1,0)}_\Omega$, which are holomorphic in z, namely*

$$q^j(z,\zeta) = \sum_{k=1}^N q^j_k(z,\zeta)\,d\zeta_k,\quad \overline{\partial}_z q^j_k = 0,\ j=1,\dots,M,\ k=1,\dots,N. \tag{4.55}$$

Let also $\Gamma : W\in\mathbb{C}^M \to \Gamma(W)$ be a holomorphic function in $\mathscr{Q}(\Omega\times\Omega)$, where

$$\mathscr{Q} : (z,\zeta) \longmapsto \big(1+\langle\!\langle \boldsymbol{q}^1, z-\zeta\rangle\!\rangle,\dots,1+\langle\!\langle \boldsymbol{q}^M, z-\zeta\rangle\!\rangle\big), \tag{4.56}$$

which satisfies $\Gamma(\mathbf{1}) = 1$. If $\mathscr{E} = \pi^*T^{(1,0)}_\Omega$ and $\Delta = \sum_{k=1}^N(\zeta_k - z_k)\partial/\partial\zeta_k$, then

$$\mathscr{W}_{\boldsymbol{q}}(z,\zeta) = \Big[\sum_{\{\kappa\in\mathbb{N}^M\,:\,|\kappa|\le N\}}\frac{1}{\kappa!}\Big(\Big(\frac{\partial}{\partial W}\Big)^\kappa\Gamma\Big)(\mathscr{Q}(z,\zeta))\prod_{j=1}^M\Big(\frac{\overline{\partial}_\zeta q^j(z,\zeta)}{2i\pi}\Big)^{\kappa_j}\Big] \tag{4.57}$$

is a global $\mathbb{C}$-weight in $\Omega\times\Omega$ with respect to the $(\mathscr{E},\Delta)$ Koszul complex $\mathscr{K}_\bullet^{\mathscr{D}}$.

PROOF. For each $j = 1,\dots,M$,

$$\mathscr{W}^j = [1] - \nabla_{\mathscr{D}}\Big(\frac{[q^j]}{2i\pi}\Big) = [1] - \nabla^\circ_{\mathscr{D}}\Big(\frac{[q^j]}{2i\pi}\Big) = \Big[\mathscr{Q}_j + \frac{\overline{\partial}_\zeta[q^j]}{2i\pi}\Big]$$

is a $\mathbb{C}$-weight, since $\nabla_{\mathscr{D}}[1] = 0$ and $\nabla^2_{\mathscr{D}} = 0$. Since all weights $\mathscr{W}^j$ are smooth, their exterior (wedge) product is licit and any polynomial function $P(\mathscr{W}^1,\dots,\mathscr{W}^M)$, satisfying $P(\mathbf{1}) = 1$, remains a global weight in $\Omega\times\Omega$ according to Leibniz's rule. The same remains true if P is replaced by a power series, provided the convergence makes sense. This is precisely what happens when P is replaced by Γ. The right-hand side of (4.57) corresponds precisely to

$$\Gamma(\mathscr{W}^1,\dots,\mathscr{W}^M) = \Gamma\Big(\mathcal{Q} + \frac{\overline{\partial}_\zeta \boldsymbol{q}}{2i\pi}\Big) = \sum_{\kappa\in\mathbb{N}^M} a_\kappa(z,\zeta) \prod_{j=1}^M \Big(\frac{\overline{\partial}_\zeta[q^j]}{2i\pi}\Big)^{\kappa_j},$$

where $\sum_{\kappa\in\mathbb{N}^M} a_\kappa(z,\zeta)(W')^\kappa$ is the Taylor series of Γ in a neighborhood of $\mathcal{Q}(z,\zeta)$. □

Positive plurisubharmonic functions $\phi : \mathbb{C}^N \to \mathbb{R}$ provide interesting global weights in $\mathbb{C}^N \times \mathbb{C}^N$ with respect to the $(\mathscr{E},\Delta)$ Koszul complex $\mathscr{K}_\bullet^{\mathscr{D}}$ if one choses $\mathscr{E}$ and Δ as $\mathscr{E} = \pi^* T^{(1,0)}_{\mathbb{C}^N}$ and $\Delta = \sum_{k=1}^N (\zeta_k - z_k)\, \partial/\partial\zeta_k$.

PROPOSITION 4.21. *Let $\phi : \mathbb{C}^N \to \mathbb{R}$ be a plurisubharmonic function and let*

$$\phi_\varepsilon = \phi * \rho_\varepsilon : z \in \mathbb{C}^N \longmapsto \int_{\mathbb{C}^N} \phi(\zeta)\, \rho_\varepsilon(z-\zeta)\, d\lambda_{2N}(\zeta)$$

be its smooth pluri-subharmonic regularization.[10] *Then*

$$\mathscr{W}_{\phi,\varepsilon} = [1] - \frac{1}{i\pi} \nabla_{\mathscr{D}} [\partial\phi_\varepsilon(\zeta)] = \Big[1 + 2\langle\!\langle \partial\phi_\varepsilon(\zeta)\,,\, z-\zeta\rangle\!\rangle + 2dd^c\phi_\varepsilon(\zeta)\Big] \tag{4.58}$$

is a smooth global $\mathbb{C}$-weight in $\mathbb{C}^N \times \mathbb{C}^N$ with respect to the $(\mathscr{E},\Delta)$ Koszul complex $\mathscr{K}_\bullet^{\mathscr{D}}$.

PROOF. An immediate computation shows that $\nabla_{\mathscr{D}} \mathscr{W}_{\phi,\varepsilon} = 0$ in $\mathbb{C}^N \times \mathbb{C}^N$. □

REMARK 4.22. It is important to observe that the exponential function involved in the realization of an entire function $\Gamma : \mathbb{C}^M \to \mathbb{C}$, as in Proposition 4.20, is such that

$$\exp(\mathscr{W}_{\phi,\varepsilon}) = e\, \exp\big(2\langle\!\langle \partial\phi_\varepsilon(\zeta)\,,\, z-\zeta\rangle\!\rangle\big) \Big(\sum_{r=0}^N \frac{(2dd^c\phi_\varepsilon(\zeta))^r}{r!}\Big), \tag{4.59}$$

where

$$\exp\big(2\langle\!\langle \partial\phi_\varepsilon(\zeta)\,,\, z-\zeta\rangle\!\rangle\big) \leq \exp\big(\phi_\varepsilon(z) - \phi_\varepsilon(\zeta)\big) \tag{4.60}$$

for any $(z,\zeta) \in \mathbb{C}^N \times \mathbb{C}^N$, because of the plurisubharmonicity of ϕ_ε. Of particular interest to applications is the case when $\phi(z) = H(\operatorname{Im} z)$, where H is a convex positive function in $\mathbb{R}^N$. This occurs, for example, if H is the so-called *support function* $H_{\mathrm{K}}(x) = \sup_{\xi\in\mathrm{K}} \langle\!\langle x,\xi\rangle\!\rangle$ of a compact convex subset $\mathrm{K} \subset \mathbb{R}^N$; see also Theorems 7.92 and 7.94.

4.4.2. Local weights. Let $\mathcal{X}$ be a good manifold and let $\mathscr{X}, \mathscr{D}, \mathscr{E}, \Delta, E$ be as in §4.4.1. Fix now $z^0 \in \mathcal{X}$ and let $\mathscr{U}^0$ be a neighborhood of $\sqcap^{-1}(\{z^0\}) = \{z^0\} \times \mathcal{X}$ in $\mathscr{X}$. Let $\mathscr{W}^{z^0} \in \mathcal{L}^0_{\mathscr{D}}[E^{\frown}](\mathscr{U}^0)$, which is assumed to be smooth in a neighborhood of $\sqcap^{-1}(\{z^0\})$ and satisfies in $\mathscr{U}^0$

$$\nabla^\circ_{\mathscr{D}} \mathscr{W}^{z^0} = (2i\pi \rfloor_\Delta - \overline{\partial}_\zeta) \mathscr{W}^{z^0} = 0,\ (\mathscr{W}^{z^0}(z^0,z^0))_{0,0} = \mathrm{Id}_{(\sqcap^* E)_{(z^0,z^0)}}. \tag{4.61}$$

Let

$$K = P^{\Delta}_{\sqcap} = \big[P^{\Delta,\lambda}_{\sqcap}\big]_{\lambda=0}$$

[10] One takes ρ to be a positive bump function with support in the unit ball $\mathbb{B}_N$ and satisfying $\int_{\mathbb{C}^N} \rho\, d\lambda_{2N} = 1$. Then $\rho_\varepsilon(\zeta) = \epsilon^{-2N} \rho(\zeta/\varepsilon)$ for $0 < \varepsilon \ll 1$. As usual, $d\lambda_{2N}$ denotes the $2N$-dimensional Lebesgue measure.

be the Bochner–Martinelli principal value current associated to $\overline{\mathscr{E}} = (\mathscr{E}, |\ |)$ and the defining section Δ for $\mathscr{D}$; see §4.4.1. Observe that the multiplication of two currents is permissible when their singular supports do not intersect. Thus, the hypothesis (4.61) on $\mathscr{W}^{z^0}$ in $\mathscr{U}^0$ imply the following current identities in $\mathcal{X}$:

(4.62)

$$\begin{aligned}
\Big[\nabla^{\circ}_{\mathscr{D}}\big(\mathscr{W}^{z^0} P^{\Delta,\lambda}_{|\ |}(z^0,\cdot)\big)\Big]_{\lambda=0} &= \mathscr{W}^{z^0}(z^0,\cdot) - \Big[\mathscr{W}^{z^0} R^{\Delta,\lambda}_{|\ |}(z^0,\cdot)\Big]_{\lambda=0} \\
&= \mathscr{W}^{z^0}(z^0,\cdot) - \Big[\Big((\mathscr{W}^{z^0})_{0,0}\,(\overline{\partial}_\zeta|\Delta|^{2\lambda})\,\frac{\Delta^*}{2i\pi|\Delta|^2}\Big(\overline{\partial}_\zeta\,\frac{\Delta^*}{2i\pi|\Delta|^2}\Big)^{N-1}\Big)(z^0,\cdot)\Big]_{\lambda=0} \\
&= \mathscr{W}^{z^0}(z^0,\cdot) - \Big[\Big((\overline{\partial}_\zeta|\Delta|^{2\lambda})\,\frac{\Delta^*}{2i\pi|\Delta|^2}\Big(\overline{\partial}_\zeta\,\frac{\Delta^*}{2i\pi|\Delta|^2}\Big)^{N-1}\Big)(z^0,\cdot)\Big]_{\lambda=0} \otimes \mathrm{Id}_{E_{z^0}}.
\end{aligned}$$

This implies that

$$\text{(4.63)}\quad \mathscr{W}^{z^0}_N(z^0,\cdot) - \Big[\Big((\overline{\partial}_\zeta|\Delta|^{2\lambda})\,\frac{\Delta^*}{2i\pi|\Delta|^2}\Big(\overline{\partial}_\zeta\,\frac{\Delta^*}{2i\pi|\Delta|^2}\Big)^{N-1}\Big)(z^0,\cdot)\Big]_{\lambda=0} \otimes \mathrm{Id}_{E_{z^0}} = \overline{\partial} T^{z^0}$$

for some current T^{z^0} in $\mathcal{X}$. As we will see in §4.5, the identity (4.63) will provide, thanks to Stokes's formula on $\mathcal{X}$, integral representation formulae for bundle-valued holomorphic functions on $\mathcal{X}$ at the point z^0. All these considerations motivate the following definition.[11]

DEFINITION 4.23. Let $\mathcal{X}$ be an N-dimensional, good complex manifold, which embeds in its duplication $\mathscr{X}$ as the diagonal complete intersection $\mathscr{D} = V(\Delta)$. As usual, $\Delta \in \mathscr{O}(\mathscr{X},\mathscr{E})$, $\operatorname{rank}\mathscr{E} = N$, and $d\Delta \neq 0$ locally on $\mathscr{D}$. Let E be a holomorphic bundle over $\mathcal{X}$ and let $z^0 \in \mathcal{X}$. A *local E-weight at z^0* with respect to the $(\mathscr{E},\Delta))$ Koszul complex $\mathscr{K}_\bullet^{\mathscr{D}}$ is a section $\mathscr{W}^{z^0}$ of $\mathcal{L}^0_{\mathscr{D}}[\widehat{E}]$ in some neighborhood $\mathscr{U}^0$ of $\sqcap^{-1}(\{z^0\})$, which is a smooth neighborhood of $\sqcap^{-1}(\{z^0\})$ and satisfies

$$\text{(4.64)}\quad \operatorname{bidegree}_z \mathscr{W}^{z^0} = (0,0), \quad \nabla^{\circ}_{\mathscr{D}}\mathscr{W}^{z^0} = 0 \text{ in } \mathscr{U}^0, \quad (\mathscr{W}^{z^0}(z^0,z^0))_{0,0} = \mathrm{Id}_{E_{z^0}}.$$

REMARK 4.24. A global E-weight $\mathscr{W}$ with respect to the Koszul complex $\mathscr{K}_\bullet^{\mathscr{D}}$ in an open subset $\mathscr{U}$ of $\mathscr{X}$ containing $\sqcap^{-1}(\{z^0\})$ defines a local E-weight $\mathscr{W}^{z^0}$ at z^0 with respect to the Koszul complex $\mathscr{K}_\bullet^{\mathscr{D}}$ as soon as it is smooth in a neighborhood of $\sqcap^{-1}(\{z^0\})$.

The following result provides an archetypal example local $\mathbb{C}$-weight $\mathscr{W}$ at a point $z^0 \in \mathcal{X}$, where $\mathcal{X}$ is an N-dimensional good complex manifold.

PROPOSITION 4.25. *Let $\mathcal{X}$, $\mathscr{X}$, $\mathscr{E}$, Δ be as in Definition 4.17, and let $z^0 \in \mathcal{X}$. Let $\chi \in \mathcal{D}(\mathcal{X})$ be such that $\chi \equiv 1$ in a neighborhood of z^0. Given a hermitian metric $|\ |$ on $\mathscr{E}$, the current*

$$\text{(4.65)}\qquad \mathscr{W}^{z^0}_\chi = [\chi(\zeta)] - \overline{\partial}\chi(\zeta)\,\Big[|\Delta|^{2\lambda}\sum_{r=1}^{N}\frac{\Delta^*}{2i\pi|\Delta|^2}\Big(\overline{\partial}_\zeta\,\frac{\Delta^*}{2i\pi|\Delta|^2}\Big)^{r-1}\Big]_{\lambda=0}$$

is smooth in some neighborhood $\mathscr{U}^0$ of $\sqcap^{-1}(\{z^0\})$ and defines in $\mathscr{U}^0$ a $\mathbb{C}$-weight at z^0 with respect to the $(\mathscr{E},\Delta)$ Koszul complex $\mathscr{K}_\bullet^{\mathscr{D}}$.

[11]The notion of a local weight at point z^0 of a good complex manifold for the Koszul complex $\mathscr{K}_\bullet^{\mathscr{D}}$ was originally introduced in [**AndB82**, **AndP88**], when $\mathcal{X} = \Omega \subset \mathbb{C}^N$. The evolved presentation given here in a more general context is due to Mats Andersson [**And03**, **And06A**].

PROOF. Since $\overline{\partial}\chi \equiv 0$ in a neighborhood of z^0 and the singular support of the current

$$\Big[|\Delta|^{2\lambda}\sum_{r=1}^{N}\frac{\Delta^*}{2i\pi|\Delta|^2}\Big(\overline{\partial}_\zeta\frac{\Delta^*}{2i\pi|\Delta|^2}\Big)^{r-1}\Big]_{\lambda=0}$$

equals the diagonal $\mathscr{D}$, the set $\sqcap^{-1}(\{z^0\})$ does not intersect the singular support of the current (4.65). Therefore, such current is smooth in some neighborhood $\mathscr{U}^0$ of $\sqcap^{-1}(\{z^0\})$. It is a smooth section of $\mathcal{L}^0_{\mathscr{D}}[\mathbb{C}]$ there. Moreover, one is in a neighborhood of $\sqcap^{-1}(\{z^0\})$ that

$$\begin{aligned}\nabla^\circ_{\mathscr{D}}[\chi(\zeta)] &= -[\overline{\partial}\chi(\zeta)]\\ &\quad+\nabla^\circ_{\mathscr{D}}\Big(-\overline{\partial}\chi(\zeta)\Big[|\Delta|^{2\lambda}\sum_{r=1}^{N}\frac{\Delta^*}{|\Delta|^2}\Big(\overline{\partial}_\zeta\frac{\Delta^*}{2i\pi|\Delta|^2}\Big)^{r-1}\Big]_{\lambda=0}\Big)\\ &=\overline{\partial}\chi(\zeta)\Big([1]-\Big[(\overline{\partial}|\Delta|^{2\lambda})\frac{\Delta^*}{2i\pi|\Delta|^2}\Big(\overline{\partial}_\zeta\frac{\Delta^*}{2i\pi|\Delta|^2}\Big)^{N-1}\Big]_{\lambda=0}\Big)=[\overline{\partial}\chi(\zeta)].\end{aligned}$$

Therefore $\nabla^\circ_{\mathscr{D}}\mathscr{W}^{z^0}_\chi = 0$ around the point $\sqcap^{-1}(\{z^0\})$. Since $\chi(z^0) = 1$, the initial condition $(\mathscr{W}^{z^0}_\chi(z^0,z^0))_0 = (\mathscr{W}^{z^0}_\chi(z^0,z^0))_{0,0} = 1$ is also fulfilled. □

When $\mathcal{X} = \Omega$ is a bounded domain of $\mathbb{C}^N$, the following result [**AndB82**, **AndP88**] provides a useful complement to Proposition 4.20. We take here as usual $\mathscr{E} = \pi^*T^{(1,0)}_\Omega$ and $\Delta = \sum_{k=1}^N(\zeta_k - z_k)\,\partial/\partial\zeta_k$.

PROPOSITION 4.26. *Let Ω be a bounded domain in $\mathbb{C}^N$. Let also $q^1,\dots,q^M$ be smooth sections of $\pi^*T^{*(1,0)}_\Omega$ holomorphic in z, $\mathcal{Q}:\Omega\times\Omega\to\mathbb{C}^M$ (see (4.55) and (4.56)) and let $\Gamma:\operatorname{Im}\mathcal{Q}\to\mathbb{C}$ be holomorphic with $\Gamma(\mathbf{1})=1$, as in Proposition 4.20. If $z^0\in\Omega$ and q^{z^0} is a smooth section of $\pi^*T^{*(1,0)}_\Omega$ in a neighborhood of $\{z^0\}\times\partial\Omega$ satisfying*

$$\rfloor_\Delta q^{z^0}(z,\zeta)=1 \text{ and } \overline{\partial}_z q^{z^0}(z,\zeta)=0 \quad \text{around } \{z^0\}\times\partial\Omega, \tag{4.66}$$

then, for any $\chi\in\mathcal{D}(\mathcal{X},[0,1])$ such that $\chi^{-1}([0,1[)$ is sufficiently close to $\partial\Omega$, the section

$$\begin{aligned}\mathscr{W}^{z^0}_{\boldsymbol{q},\chi} &= \Big(\chi(\zeta)+\frac{\overline{\partial}_\zeta\big((1-\chi(\zeta))\,q^{z^0}(z,\zeta)\big)}{2i\pi}\Big)^{N+1}\Big[\Gamma\Big(\mathcal{Q}+\frac{\overline{\partial}_\zeta\boldsymbol{q}}{2i\pi}\Big)\Big]\\ &= \Big(\chi(\zeta)+\frac{\overline{\partial}_\zeta\big((1-\chi(\zeta))\,q^{z^0}(z,\zeta)\big)}{2i\pi}\Big)^{N+1}\mathscr{W}_{\boldsymbol{q}}(z,\zeta)\end{aligned}\tag{4.67}$$

defines a $\mathbb{C}$-local weight at z^0 with respect to the $(\mathscr{E},\Delta)$ Koszul complex $\mathscr{K}_\bullet^{\mathscr{D}}$.

PROOF. Proposition 4.20 and Leibniz's rule for $\nabla^\circ_{\mathscr{D}}$ imply that it is enough to prove that the current

$$\mathscr{W}^{z^0}=\Big[\chi(\zeta)+\frac{\overline{\partial}_\zeta\big((1-\chi(\zeta))\,q^{z^0}(z,\zeta)\big)}{2i\pi}\Big]$$

defines a $\mathbb{C}$-local weight at z^0. This current is well defined and smooth in a neighborhood $\mathscr{U}^0$ of $\sqcap^{-1}(\{z^0\})$, when $\operatorname{Supp}(1-\chi)$ is sufficiently close to $\partial\Omega$, in order that $\rfloor_\Delta q^{z^0} = 1$ in a neighborhood of such support. Since $2i\pi\rfloor_\Delta$ and $\nabla^\circ_{\mathscr{D}}$ anticommute, one has

$$2i\pi\rfloor_\Delta\mathscr{W}^{z^0}=-\overline{\partial}_\zeta\circ\rfloor_\Delta\Big(\Big[\frac{(1-\chi(\zeta))q^{z^0}(z,\zeta)}{2i\pi}\Big]\Big)=-\overline{\partial}[1-\chi(\zeta)]=\overline{\partial}[\chi(\zeta)]=\overline{\partial}_\zeta\mathscr{W}^{z^0}.$$

Thus, $\nabla^{\circ}_{\mathscr{D}}\mathscr{W}^{z^0} = 0$ in $\mathscr{U}^0$. Since $(\mathscr{W}^{z^0}(z^0,z^0))_0 = (\mathscr{W}^{z^0}(z^0,z^0))_{0,0} = 1$, we are done. □

4.4.3. Global weights attached to metrized complexes. Let $\mathcal{X}$ be a good manifold and let $\mathscr{K}_\bullet^{\mathscr{D}}$ be the Koszul complex, with respect to $\mathscr{E}$, $\Delta \in \mathscr{O}(\mathscr{X},\mathscr{E})$, attached to the embedding of $\mathcal{X}$ in its duplication $\mathscr{X}$ as the diagonal complete intersection $\mathscr{D} = V(\Delta)$. We start in this section with a metrized generically exact complex of holomorphic bundles

$$\overline{\mathbf{F}_\bullet} : 0 \longrightarrow \overline{E_L} \xrightarrow{F_L} \overline{E_{L-1}} \xrightarrow{F_{L-1}} \cdots \xrightarrow{F_2} \overline{E_1} \xrightarrow{F_1} \overline{E_0} \xrightarrow{F_0} \overline{E_{-1}} = \overline{\frac{E_0}{\operatorname{Im} F_1}} \longrightarrow 0. \tag{4.68}$$

Let $V(\mathbf{F}_\bullet)$ be the smallest closed analytic set over whose complement the complex $\mathbf{F}_\bullet$ is exact. Attached to such a metrized complex are the residue and principal value currents

$$\begin{aligned}
R^{\overline{\mathbf{F}_\bullet}} &= \sum_{\ell\geq 0}\Big(\sum_{r>\ell} R^{\overline{\mathbf{F}_\bullet}}_{\ell,r}\Big)\\
&= \sum_{\ell\geq 0}\Big(\sum_{r>\ell}\Big(\Big[(\overline{\partial}\,|\Delta^{\mathbf{F}_\bullet}|^{2\lambda})\,\frac{\sigma_r\,\overline{\partial}\sigma_{r-1}\cdots\overline{\partial}\sigma_{\ell+2}\,\overline{\partial}\sigma_{\ell+1}}{(2i\pi)^{r-\ell}}\Big]_{\lambda=0}\Big)\Big),\\
P^{\overline{\mathbf{F}_\bullet}} &= \sum_{\ell\geq 0}\Big(\sum_{r>\ell} P^{\overline{\mathbf{F}_\bullet}}_{\ell,r}\Big)\\
&= \sum_{\ell\geq 0}\Big(\sum_{r>\ell}\Big[|\Delta^{\mathbf{F}_\bullet}|^{2\lambda}\,\frac{\sigma_r\,\overline{\partial}\sigma_{r-1}\cdots\overline{\partial}\sigma_{\ell+2}\,\overline{\partial}\sigma_{\ell+1}}{(2i\pi)^{r-\ell}}\Big]_{\lambda=0}\Big);
\end{aligned}\tag{4.69}$$

see (3.67), (3.68), and Proposition 3.28. The morphisms $\sigma_\ell : E_{\ell-1} \to E_\ell$ are the minimal inverse of the morphisms F_ℓ; see §3.3.2. The section $\|\Delta^{\mathbf{F}_\bullet}\|^2$ is defined in (3.58), (3.64) and satisfies $V(|\Delta^{\mathbf{F}_\bullet}|^2) = V(\mathbf{F}_\bullet)$. The currents $R^{\overline{\mathbf{F}_\bullet}}$ and $P^{\overline{\mathbf{F}_\bullet}}$ are $\operatorname{Hom}_{\mathbb{C}}(E(\mathbf{F}_\bullet), E(\mathbf{F}_\bullet))$-valued currents satisfying the current identities

$$\boldsymbol{\nabla}_{\mathbf{F}_\bullet}\big(R^{\overline{\mathbf{F}_\bullet}}\big) = 0, \quad \boldsymbol{\nabla}_{\mathbf{F}_\bullet}\big(P^{\overline{\mathbf{F}_\bullet}}\big) = \operatorname{Id}_{E(\mathbf{F}_\bullet)} - R^{\overline{\mathbf{F}_\bullet}}, \tag{4.70}$$

where the operator $\boldsymbol{\nabla} = 2i\pi\boldsymbol{F} - \overline{\boldsymbol{\partial}}$ was introduced in (3.48) and (3.50). It is acting from the sheaf of $\operatorname{Hom}_{\mathbb{C}}(E(\mathbf{F}_\bullet), E(\mathbf{F}_\bullet))$ into itself. It follows in particular from (4.70) that

$$\begin{aligned}
&2i\pi F_{r+1}R^{\overline{\mathbf{F}_\bullet}}_{0,r+1} - \overline{\partial}_\zeta R^{\overline{\mathbf{F}_\bullet}}_{0,r} = 0 \quad \forall\, r\geq 0,\\
&2i\pi F_1\, P^{\overline{\mathbf{F}_\bullet}}_{0,1} = \operatorname{Id}_{E_0}, \quad 2i\pi\, F_{r+1}\, P^{\overline{\mathbf{F}_\bullet}}_{0,r} = \overline{\partial}_\zeta P^{\overline{\mathbf{F}_\bullet}}_0 - R^{\overline{\mathbf{F}_\bullet}}_{0,r} \quad \forall\, r\geq 1.
\end{aligned}\tag{4.71}$$

Thus, one has on the duplication $\mathscr{X}$ the two complexes $\sqcap^*\mathbf{F}_\bullet$ and $\pi^*\mathbf{F}_\bullet$, namely

$$0 \longrightarrow \sqcap^*\overline{E_L} \xrightarrow{F_L(z)} \sqcap^*\overline{E_{L-1}} \xrightarrow{F_{L-1}(z)} \cdots \xrightarrow{F_2(z)} \sqcap^*\overline{E_1} \xrightarrow{F_1(z)} \sqcap^*\overline{E_0} \xrightarrow{F_0(z)} \sqcap^*\overline{E_{-1}} \longrightarrow 0,$$
$$0 \longrightarrow \pi^*\overline{E_L} \xrightarrow{F_L(\zeta)} \pi^*\overline{E_{L-1}} \xrightarrow{F_{L-1}(\zeta)} \cdots \xrightarrow{F_2(\zeta)} \pi^*\overline{E_1} \xrightarrow{F_1(\zeta)} \pi^*\overline{E_0} \xrightarrow{F_0(\zeta)} \pi^*\overline{E_{-1}} \longrightarrow 0.$$

The corresponding Koszul complex $\mathscr{K}_\bullet^{\mathscr{D}}$ is

$$0 \to \bigwedge^N \mathscr{E}^* \xrightarrow{\rfloor_\Delta} \bigwedge^{N-1}\mathscr{E}^* \cdots \xrightarrow{\rfloor_\Delta} \bigwedge^{\ell+1}\mathscr{E}^* \xrightarrow{\rfloor_\Delta} \bigwedge^{\ell}\mathscr{E}^* \longrightarrow \cdots \xrightarrow{\rfloor_\Delta} \mathscr{E}^* \xrightarrow{\rfloor_\Delta} \mathbb{C},$$

while the nabla operators are $\nabla^\circ_{\mathscr{D}} = 2i\pi\rfloor_\Delta - \overline{\partial}$ and $\nabla^\circ_{\mathscr{D}} = 2i\pi\rfloor_\Delta - \overline{\partial}_\zeta$. For each $r \geq -1$ and $\ell \geq 0$, the operator $\nabla^\circ_{\mathscr{D}}$ acts as

$$\begin{aligned}\nabla^\circ_{\mathscr{D}} : \big(\mathcal{L}^{\ell-r}_{\mathscr{D}}[E^{\frown}_{\ell,r}]\big)^\circ(\mathscr{X}) \\ &= \big(\mathcal{L}^{\ell-r}_{\mathscr{D}}[\mathrm{Hom}_{\mathbb{C}}(\pi^* E_r, \sqcap^* E_\ell)]\big)^\circ(\mathscr{X}) \longrightarrow \big(\mathcal{L}^{\ell-r+1}_{\mathscr{D}}[E^{\frown}_{\ell,r}]\big)^\circ(\mathscr{X});\end{aligned}$$

see (4.3) and (4.10). Consider the family

$$\mathbf{H} = \{H_{\ell,r} \in \big(\mathcal{L}^{\ell-r}_{\mathscr{D}}[E^{\frown}_{\ell,r}]\big)^\circ(\mathscr{X}) \,:\, \ell \geq 0,\ r \geq -1\}.$$

For any $\ell \geq 0$, let

$$H_\ell = \sum_{r\geq -1} H_{\ell,r} \in \bigoplus_{r\geq -1} \big(\mathcal{L}^{\ell-r}_{\mathscr{D}}[E^{\frown}_{\ell,r}]\big)^\circ(\mathscr{X}) \subset {}'\mathcal{D}(\mathscr{X}, \mathrm{Hom}_{\mathbb{C}}(E(\pi^*\mathbf{F}_\bullet), \sqcap^* E_\ell)) \tag{4.72}$$

and

$$\begin{aligned} H = \sum_{\ell\geq 0} H_\ell = \sum_{\ell \geq 0}\Big(\sum_{r\geq -1} H_{\ell,r}\Big) \in \bigoplus_{\ell\geq 0}\bigoplus_{r\geq -1} \big(\mathcal{L}^{\ell-r}_{\mathscr{D}}[E^{\frown}_{\ell,r}]\big)^\circ(\mathscr{X}) \\ \subset {}'\mathcal{D}(\mathscr{X}, \mathrm{Hom}_{\mathbb{C}}(E(\pi^*\mathbf{F}_\bullet), E(\sqcap^*\mathbf{F}_\bullet))).\end{aligned} \tag{4.73}$$

Inspired by its analogy with the Hefer division problem in $\Omega \times \Omega$, where Ω is a pseudo-convex domain in $\mathbb{C}^N$ (see the important lemma formulated by Hans Hefer in [**Hef50**], as well as §A.2.2),[12] we introduce the following definition [**And06A**, §5].

DEFINITION 4.27. Consider the generically exact complex $\mathbf{F}_\bullet$ as in[13] (4.68) on an N-dimensional good complex manifold $\mathcal{X}$ duplicated as $\mathscr{X}$. Suppose that the Koszul complex $\mathscr{K}_\bullet^{\mathscr{D}}$ is the $(\mathscr{E}, \Delta)$ one, where $\operatorname{rank}\mathscr{E} = N$ and the section $\Delta \in \mathscr{O}(\mathscr{X}, \mathscr{E})$ is such that $V(\Delta) = \mathscr{D}$ and satisfies $d\Delta \neq 0$ locally on $\mathscr{D}$. A family

$$\mathbf{H} = \{H_{\ell,r} \in \big(\mathcal{L}^{\ell-r}_{\mathscr{D}}[E^{\frown}_{\ell,r}]\big)^\circ(\mathcal{X}) \,:\, \ell \geq 0,\ r \geq -1\},$$

where H is defined by (4.73), is said to be a solution of the $\mathbf{F}_\bullet$-*Hefer problem* with respect to $\mathscr{K}_\bullet^{\mathscr{D}}$ on $\mathscr{X}$ if and only if

$$\begin{aligned} &H \text{ is smooth satisfying } \overline{\partial}_z H = 0,\\ &(H_{\ell,\ell})_{0,0}(z,z) = \mathrm{Id}_{(\pi^* E_\ell)_{(z,z)}} \quad \forall \ell \geq 0,\ \forall z \in \mathcal{X},\\ &H_{\ell,r} = 0 \text{ unless } r \geq \ell,\\ &\nabla_{\mathscr{D}} H = \nabla^\circ_{\mathscr{D}} H = H \circ (2i\pi F(\zeta)) - (2i\pi F(z)) \circ H.\end{aligned} \tag{4.74}$$

A solution $\mathbf{H}$ to the $\mathbf{F}_\bullet$-Hefer problem is called a *Hefer morphism* for $\mathbf{F}_\bullet$.

REMARK 4.28. In terms of the components $H_{\ell,r}$ involved in the splitting (4.73) of H, the two last conditions in (4.74) can be reformulated as

$$\begin{aligned} &H_{\ell,r} = 0 \text{ unless } r \geq \ell,\\ &\nabla^\circ_{\mathscr{D}} H_{\ell,r} = H_{\ell,r-1} \circ (2i\pi F_r(\zeta)) - (2i\pi F_{\ell+1}(z)) \circ H_{\ell+1,r} \quad \forall r \geq \ell \geq 0.\end{aligned} \tag{4.75}$$

Therefore, the following proposition[14] is the key result of this subsection.

---

[12]The Hefer problem is also known as the Gleason problem (given the contribution of Andrew Gleason, see [**Gl64**]) within the community of operator theory or functional analysis.

[13]One forgets for the time being that the complex is metrized.

[14]This proposition appeared in [**And06A**, §5]. The proof we give here is inspired by assertion (5.4) in [**And06A**]. The reader could also refer to [**AndG11**, Proposition 4.2], where an alternative proof leading to a more general result (not to be used in this monograph) is proposed.

PROPOSITION 4.29. *Let* $\overline{\mathbf{F}_\bullet}$ *be a metrized, generically exact complex on an* N*-dimensional good complex manifold* $\mathcal{X}$ *as in* (4.68). *Let* $\mathbf{H}$ *be a solution to the* $\mathbf{F}_\bullet$*-Hefer problem on its duplication* $\mathscr{X}$ *with respect to the* $(\mathscr{E},\Delta)$ *Koszul complex* $\mathscr{K}_\bullet^{\mathscr{D}}$. *Then*

$$\mathscr{W}^{\overline{\mathbf{F}_\bullet}}(z,\zeta) = \Big(2i\pi F_1(z)\,H_1(z,\zeta)\,P_0^{\overline{\mathbf{F}_\bullet}}(\zeta) + H_0(z,\zeta)\,R_0^{\overline{\mathbf{F}_\bullet}}(\zeta)\Big)_{|\mathscr{U}} \tag{4.76}$$

is a global E_0*-weight in* $\mathscr{U} = \mathscr{X}\setminus V(\sqcap^*\mathbf{F}_\bullet)$ *with respect to* $\mathscr{K}_\bullet^{\mathscr{D}}$.

PROOF. Since H is smooth and the singular supports of $P^{\overline{\mathbf{F}_\bullet}}$ and $R^{\overline{\mathbf{F}_\bullet}}$ lie in $V(\mathbf{F}_\bullet)$, the singular support of $\mathscr{W}^{\overline{\mathbf{F}_\bullet}}$ does not intersect $\mathscr{D}\cap\mathscr{U}$, taking into account the definition of $\mathscr{U}$. Moreover, the bidegree of $\mathscr{W}^{\overline{\mathbf{F}_\bullet}}$ equals $(0,0)$ as the bidegree in z of H is. Observe that for each $r\geq 1$, one has the equality of bidegrees of $\mathrm{Hom}_{\mathbb{C}}(E_0,E_r)$-valued currents

$$\mathrm{bidegree}_z\, R_{0,r}^{\overline{\mathbf{F}_\bullet}} = \mathrm{bidegree}_z\, P_{0,r}^{\overline{\mathbf{F}_\bullet}} = (0,0),$$
$$\mathrm{bidegree}_\zeta\, R_{0,r}^{\overline{\mathbf{F}_\bullet}} = (0,r),\ \ \mathrm{bidegree}_\zeta P_{0,r}^{\overline{\mathbf{F}_\bullet}} = (0,r-1).$$

On the other hand

$$H_{0,r}\in(\mathcal{L}_{\mathscr{D}}^{-r}[\widehat{E_{0,r}}])^\circ(\mathscr{X}),\quad H_{1,r}\in(\mathcal{L}_{\mathscr{D}}^{1-r}[\widehat{E_{1,r}}])^\circ(\mathscr{X}).$$

Therefore, it is clear that

$$\mathscr{W}^{\overline{\mathbf{F}_\bullet}}\in\mathcal{L}_{\mathscr{D}}^0[\widehat{E_0}](\mathscr{U}).$$

One has also $\overline{\partial}_z\mathscr{W}^{\overline{\mathbf{F}_\bullet}}=0$ since $\overline{\partial}_z H=0$. The equality

$$\mathscr{W}_0(z,z) = \mathrm{Id}_{(\sqcap^* E)_{(z,z)}}\ \forall z\in\sqcap(\mathscr{U})$$

follows from

$$(H_{1,1})_{0,0}(z,z) = \mathrm{Id}_{(\sqcap^* E_1)_{(z,z)]}}\ \forall z\in\sqcap(\mathscr{U}),$$

once it combined with the first current identity on $\mathscr{X}$ in the second line of (4.71). It remains to prove

$$\nabla_{\mathscr{D}}^\circ\mathscr{W}^{\overline{\mathbf{F}_\bullet}} = 0 \text{ in } \mathscr{U}.$$

Using the Hefer identities (4.75) and respecting the $\mathbb{F}_2$-grading of the holomorphic bundle $\mathrm{Hom}_{\mathbb{C}}(E(\pi^*\mathbf{F}_\bullet),E(\sqcap^*\mathbf{F}_\bullet))$ (see §3.3.1) while

$$H_{1,r}\in(\mathcal{L}_{\mathscr{D}}^{1-r}[\widehat{E_{1,r}}])^\circ(\mathscr{X})$$

for any $r\geq 1$, one deduces that

$$\begin{aligned}\nabla_{\mathscr{D}}^\circ H_1\,P_0^{\overline{\mathbf{F}_\bullet}} &= 2i\pi\,\rfloor_\Delta(H_1P_0^{\overline{\mathbf{F}_\bullet}}) - \overline{\partial}_\zeta(H_1P_0^{\overline{\mathbf{F}_\bullet}}) = 2i\pi\,\rfloor_\Delta(H_1P_0^{\overline{\mathbf{F}_\bullet}}) - H_1\,\overline{\partial}_\zeta P_0^{\overline{\mathbf{F}_\bullet}}\\ &= (\nabla_{\mathscr{D}}^\circ H_1)\,P_0^{\overline{\mathbf{F}_\bullet}} - H_1\,\overline{\partial}_\zeta P_0^{\overline{\mathbf{F}_\bullet}} = \textstyle\sum_{r\geq1}(\nabla_{\mathscr{D}}^\circ H_{1,r})\,P_0^{\overline{\mathbf{F}_\bullet}} - H_1\,\overline{\partial}_\zeta P_0^{\overline{\mathbf{F}_\bullet}}\\ &= 2i\pi\textstyle\sum_{r\geq1}(H_{1,r-1}F_r(\zeta)-F_2(z)H_{2,r})\,P_0^{\overline{\mathbf{F}_\bullet}} - H_1\,\overline{\partial}_\zeta P_0^{\overline{\mathbf{F}_\bullet}}.\end{aligned}$$

Computing the action of $\nabla_{\mathscr{D}}^\circ$ on $H_{0,r}\in(\mathcal{L}_{\mathscr{D}}^{-r}[\widehat{E_{0,r}}])^\circ(\mathscr{X})$ for any $r\geq 1$, one gets

$$\begin{aligned}\nabla_{\mathscr{D}}^\circ H_0\,R_0^{\overline{\mathbf{F}_\bullet}} &= 2i\pi\,\rfloor_\Delta(H_0R_0^{\overline{\mathbf{F}_\bullet}}) - \overline{\partial}_\zeta(H_0R_0^{\overline{\mathbf{F}_\bullet}}) = 2i\pi\,\rfloor_\Delta(H_0R_0^{\overline{\mathbf{F}_\bullet}}) - H_0\,\overline{\partial}_\zeta R_0^{\overline{\mathbf{F}_\bullet}}\\ &= (\nabla_{\mathscr{D}}^\circ H_0)\,R_0^{\overline{\mathbf{F}_\bullet}} - H_0\,\overline{\partial}_\zeta R_0^{\overline{\mathbf{F}_\bullet}} = \textstyle\sum_{r\geq1}(\nabla_{\mathscr{D}}^\circ H_{0,r})\,R_0^{\overline{\mathbf{F}_\bullet}} - H_0\,\overline{\partial}_\zeta R_0^{\overline{\mathbf{F}_\bullet}}\\ &= 2i\pi\textstyle\sum_{r\geq1}(H_{0,r-1}F_r(\zeta)-F_1(z)H_{1,r})\,R_0^{\overline{\mathbf{F}_\bullet}} - H_0\,\overline{\partial}_\zeta R_0^{\overline{\mathbf{F}_\bullet}}.\end{aligned}\tag{4.77}$$

If one respects the $\mathbb{F}_2$-grading of $\mathrm{Hom}_{\mathbb{C}}(E(\mathbf{F}_\bullet), E(\mathbf{F}_\bullet))$ and uses the second identity in the second line of (4.71), then it follows from (4.77) and from $F^2 = 0$ that

$$\begin{aligned}
&\nabla^\circ_{\mathscr{D}}(2i\pi\, F_1(z)\, H_1\, P_0^{\overline{\mathbf{F}_\bullet}}) \\
&\quad = -2i\pi F_1(z) \textstyle\sum_{r\geq 2} H_{1,r-1}\, 2i\pi F_r(\zeta)\, P^{\overline{\mathbf{F}_\bullet}} + 2i\pi\, F_1(z) H_1 \overline{\partial}_\zeta P_0^{\overline{\mathbf{F}_\bullet}} \\
&\quad = -2i\pi F_1(z) \textstyle\sum_{r\geq 1} H_{1,r}(\overline{\partial}_\zeta P_0^{\overline{\mathbf{F}_\bullet}} - R_0^{\overline{\mathbf{F}_\bullet}}) + 2i\pi\, F_1(z) H_1 \overline{\partial}_\zeta P_0^{\overline{\mathbf{F}_\bullet}} \\
&\quad = 2i\pi\, F_1(z) H_1 R_0^{\overline{\mathbf{F}_\bullet}}.
\end{aligned} \tag{4.78}$$

Adding (4.78) to (4.77) leads to

$$\nabla^\circ_{\mathscr{D}} \mathscr{W}^{\overline{\mathbf{F}_\bullet}} = \sum_{r\geq 0} H_{0,r}\left(2i\pi F_{r+1}(\zeta) R^{\overline{\mathbf{F}_\bullet}}_{0,r+1} - \overline{\partial}_\zeta R^{\overline{\mathbf{F}_\bullet}}_{0,r}\right) = 0$$

in view of the current identities on $\mathcal{X}$ in the first line of (4.71). This concludes the proof of the proposition. □

4.5. Bochner–Martinelli weighted integral representation formulae

As we already mentioned in §4.4.1 and §4.4.2, the reason for the introduction of the concepts of global or local weights in §4.4 was to understand the essence of multivariate analogues of the Cauchy–Pompeiu integral representation formula for smooth functions. More precisely, we intend to imitate the strategy used in the univariate setting, when we transformed the representation formula (1.54) in its ponderated[15] form (1.55), as a weighted integral representation formula for holomorphic functions. We will illustrate this strategy first in the important particular cases when $\mathcal{X}$ is a subdomain of $\mathbb{C}^N$ (locally and semilocally) and then in the case $\mathcal{X} = \mathbb{P}^N_{\mathbb{C}}$ (this time globally). Finally, we describe it briefly in the general case when $\mathcal{X}$ is an N-dimensional good complex manifold.

4.5.1. The affine case $\mathcal{X} = \Omega \subset \mathbb{C}^N$. In this subsection, Ω is a subdomain of $\mathbb{C}^n$. It is considered as an N-dimensional good complex manifold and the Koszul complex $\mathscr{K}_\bullet^{\mathscr{D}}$ on $\Omega \times \Omega$, that we refer to, is the $(\mathscr{E}, \Delta)$ one, where $\mathscr{E} = \pi^* T_\Omega^{(1,0)}$ and $\Delta = \sum_{k=1}^N (\zeta_k - z_k)\, \partial/\partial\zeta_k$. Let us start with a purely local result.

PROPOSITION 4.30. *Let E be a holomorphic bundle over Ω, $z^0 \in \Omega$, and let $\mathscr{W}^{z^0} \in \mathcal{L}^0_{\mathscr{D}}[E^\frown](\mathscr{U}^0)$ be a local E-weight at z^0 with respect to the Koszul complex $\mathscr{K}_\bullet^{\mathscr{D}}$. Let also $\mathscr{U}^0$ be a neighborhood of $\sqcap^{-1}(\{z^0\})$, while $\chi^0 \in \mathcal{D}(\Omega)$ satisfies $\chi^0 \equiv 1$ in a neighborhood of the point z^0. Then, for any $s \in \mathcal{O}(\Omega, E)$ one has that*

$$s(z^0) = \int_\Omega (\chi^0)\,(\mathscr{W}^{z^0}(z^0,\cdot))_{N,N}\, s(\cdot) - \int_\Omega (\overline{\partial}\chi^0)\left(K_{\mathbb{C}^N}(z^0,\cdot)\mathscr{W}^{z^0}(z^0,\cdot)\right)_{N,N-1} s(\cdot). \tag{4.79}$$

PROOF. We already know (see §4.2.1) that in the affine setting the identity

$$\left[|\Delta|^{2\lambda} \sum_{r=1}^N \frac{\Delta^*}{2i\pi|\Delta|^2}\left(\overline{\partial}_\zeta \frac{\Delta^*}{2i\pi|\Delta|^2}\right)^{r-1}\right]_{\lambda=0} = (K_{\mathbb{C}^N})_{|\Omega\times\Omega}$$

holds, where the Bochner–Martinelli kernel $K_{\mathbb{C}^N}$ on $\mathbb{C}^N \times \mathbb{C}^N$ has been defined in Definition 4.7. It follows from Proposition 4.25 that

$$[\chi^0(\zeta)] - \overline{\partial}\chi^0(\zeta)\, K_{\mathbb{C}^N}(z,\zeta)$$

[15]One should better say here *weighted* form, to keep track of the terminology we are using.

is a $\mathbb{C}$-local weight at z^0 with respect to $\mathscr{K}_\bullet^{\mathscr{D}}$. The current

$$\check{\mathscr{W}}^{z^0} = \big(\chi^0(\zeta) - \overline{\partial}\chi^0(\zeta)\, K_{\mathbb{C}^N}(z,\zeta)\big)\, \mathscr{W}^{z^0}(z,\zeta)$$

is well defined and smooth in a neighborhood $\mathscr{U}$ of $\sqcap^{-1}(\{z^0\})$ and defines a E-local weight at z^0 since

$$\nabla^{\circ}_{\mathscr{D}} \check{\mathscr{W}}^{z^0} = 0$$

in a neighborhood of $\sqcap^{-1}(\{z^0\})$ according to Leibniz's rule. Observe that the projection $\pi(\mathrm{Supp}\,(\check{\mathscr{W}}^{z^0}))$ is a compact subset of Ω. One has in Ω that

$$\begin{aligned}
\nabla^{\circ}_{\mathscr{D}}\big(\check{\mathscr{W}}^{z^0}(z^0,\cdot)\, K_{\mathbb{C}^N}(z^0,\cdot)\big) &= \nabla^{\circ}_{\mathscr{D}}\check{\mathscr{W}}^{z^0}(z^0,\cdot)\, K_{\mathbb{C}^N}(z^0,\cdot) + \check{\mathscr{W}}^{z^0}(z^0,\cdot)\, \nabla^{\circ}_{\mathscr{D}} K_{\mathbb{C}^N}(z^0,\cdot)\\
&= \check{\mathscr{W}}^{z^0}(z^0,\cdot)\, \nabla^{\circ}_{\mathscr{D}} K_{\mathbb{C}^N}(z^0,\cdot)\\
&= \check{\mathscr{W}}^{z^0}(z^0,\cdot)\, \nabla_{\Delta(z^0,\cdot)} K_{\mathbb{C}^N}(z^0,\cdot)\\
&= \check{\mathscr{W}}^{z^0}(z^0,\cdot) - \check{\mathscr{W}}^{z^0}(z^0,\cdot)[\{z^0\}]\\
&= \check{\mathscr{W}}^{z^0}(z^0,\cdot) - (\check{\mathscr{W}}^{z^0}(z^0,\cdot))_{0,0}[\{z^0\}]\\
&= \check{\mathscr{W}}^{0}(z^0,\cdot) - [\{z^0\}]\,\mathrm{Id}_{E_{z^0}},
\end{aligned}$$

in view of (4.23) (see Proposition 4.6) and of identity $(\check{\mathscr{W}}^{z^0}(z^0,z^0))_{0,0} = \mathrm{Id}_{E_{z^0}}$. Comparing the bidegrees in both sides of the last identity, one gets

$$-\overline{\partial}\big(\check{\mathscr{W}}^{z^0}(z^0,\cdot)\, K_{\mathbb{C}^N}(z^0,\cdot)\big)_{N,N-1} = (\check{\mathscr{W}}^{z^0}(z^0,\cdot))_{N,N} - [\{z^0\}]\,\mathrm{Id}_{E_{z^0}}.$$

Since $\check{\mathscr{W}}^{z^0}(z^0,\cdot)$ has compact support in Ω, it follows from Stokes's theorem that for any $s \in \mathcal{O}(\Omega, E)$,

$$s(z^0) = \int_\Omega (\check{\mathscr{W}}^{z^0}(z^0,\zeta))_{N,N}\, s(\zeta),$$

which is formula (4.79). □

REMARK 4.31. If the E-local weight $\mathscr{W}^0$ at z^0 is in $\mathscr{U}^0$ the direct image through $\iota_{\sqcap^{-1}(\{z_0\})} : \sqcap^{-1}(\{z_0\}) \hookrightarrow \mathscr{U}^0$ of a $\mathrm{Hom}_{\mathbb{C}}(E, E_{z^0})$-valued current $\mathscr{W}^0(z^0,\cdot)$, which is smooth around the point z^0, then the integral representation formula (4.79) still holds.

Proposition 4.30 leads to the following semilocal result.

PROPOSITION 4.32. *Let E be a holomorphic bundle over Ω, and let ω be a relatively compact subset of Ω. Assume also that $\mathscr{W}$ is a smooth global E-weight with respect to the Koszul complex $\mathscr{K}_\bullet^{\mathscr{D}}$ in $\omega \times \Omega$ and that $\chi \in \mathcal{D}(\Omega)$ satisfies $\chi \equiv 1$ in a neighborhood of $\overline{\omega}$. Any $s \in \mathcal{O}(\Omega, E)$ admits then in ω the integral representation formula,*

$$s(z) = \int_\Omega (\chi\, \mathscr{W}(z,\cdot))_{N,N}\, s(\cdot) - \int_\Omega \big(\overline{\partial}\chi\, K_{\mathbb{C}^N}(z,\cdot)\mathscr{W}(z,\cdot)\big)_{N,N-1}\, s(\cdot). \tag{4.80}$$

PROOF. For any $z^0 \in \omega$ the mapping

$$(z,\zeta) \longmapsto [\chi(\zeta)] - \overline{\partial}\chi(\zeta)\, K_{\mathbb{C}^N}(z,\zeta)$$

is a $\mathbb{C}$-local weight at z^0. One has $\chi(z^0) = 1$, and the above mapping, as a current-valued map, is smooth and $\nabla^{\circ}_{\mathscr{D}}$-closed in a neighborhood of $\sqcap^{-1}(\{z^0\})$. Since $\mathscr{W}$ is a smooth global E-weight in $\omega \times \Omega \supset \sqcap^{-1}(\{z^0\})$, it defines a local E-weight at any point in ω, in particular at z^0. Proposition 4.30 therefore applies at this point. It remains true for any $z^0 \in \omega$. □

Alternatively, we have also the following.

PROPOSITION 4.33. *Let E, ω, $\mathscr{W}$, and χ be as in Proposition 4.32. Let $q_{\omega,\chi}$ be a smooth section of $\pi^*T^{*(0,1)}_\Omega$ in a neighborhood of* $\operatorname{Supp}\chi \times \operatorname{Supp}(1-\chi)$, *which is also holomorphic in z, and satisfies*

$$\rfloor_\Delta q_{\omega,\chi}(z,\zeta) = 1 \text{ in a neighborhood of } \overline{\omega} \times \operatorname{Supp}(\overline{\partial}\chi). \tag{4.81}$$

Then any $s \in \mathcal{O}(\Omega, E)$ admits in ω the following integral representation formula

$$s(z) = \int_\Omega \Big(\Big(\chi + \frac{\overline{\partial}_\zeta\big((1-\chi)\, q_{\omega,\chi}(z,\cdot)\big)}{2i\pi}\Big)^{N+1} \mathscr{W}(z,\cdot)\Big)_{N,N} s(\cdot). \tag{4.82}$$

PROOF. Let $z^0 \in \omega$. As in Proposition 4.26, one observes that the hypothesis on $q_{\omega,\chi}$ implies that

$$(z,\zeta) \longmapsto \Big[\chi(\zeta) + \frac{\overline{\partial}_\zeta\big((1-\chi(\zeta))\, q_{\omega,\chi}(z,\zeta)\big)}{2i\pi}\Big]$$

is a $\mathbb{C}$-local weight at z^0. Since $\mathscr{W}$ is a smooth, global E-weight in $\omega \times \Omega$, the mapping

$$(z,\zeta) \longmapsto \Big(\chi(\zeta) + \frac{\overline{\partial}_\zeta\big((1-\chi(\zeta))\, q_{\omega,\chi}(z,\zeta)\big)}{2i\pi}\Big)^{N+1} \mathscr{W}(z,\zeta)$$

defines a smooth, local E-weight at z^0 according to Leibniz's rule. Moreover,

$$\Big[\Big(\chi + \frac{\overline{\partial}_\zeta\big((1-\chi)\, q_{\omega,\chi}(z^0,\cdot)\big)}{2i\pi}\Big)^{N+1} \mathscr{W}(z^0,\cdot)\Big]$$

is compactly supported as a $\operatorname{Hom}_{\mathbb{C}}(E, E_{z^0})$-valued current in Ω since

$$\Big(\frac{\overline{\partial}_\zeta\big((1-\chi)\, q_{\omega,\chi}(z^0,\cdot)\big)}{2i\pi}\Big)^{N+1} = 0.$$

The same proof as that of Proposition 4.30 implies the integral representation formula

$$s(z^0) = \int_\Omega \Big(\Big(\chi(\zeta) + \frac{\overline{\partial}_\zeta\big((1-\chi(\zeta))\, q_{\omega,\chi}(z^0,\zeta)\big)}{2i\pi}\Big)^{N+1} \mathscr{W}(z^0,\zeta)\Big)_{N,N} s(\zeta)$$

holds for any $s \in \mathcal{O}(\Omega, E)$. This remains true for any $z^0 \in \omega$, which concludes the proof of Proposition 4.33. □

Consider the particular case where Ω is bounded. Proposition 4.33 leads then to the following integral representation formula with boundary terms.

PROPOSITION 4.34. *Assume that Ω is bounded and has piecewise smooth boundary, and let E be a holomorphic bundle over a neighborhood U of $\overline{\Omega}$ in $\mathbb{C}^N$. Let ω be a relatively compact subset of Ω, and let $\mathscr{W}$ be a smooth, global E-weight in $\omega \times \Omega$ represented by a continuous differential bundle-valued form, extending continuously to $\omega \times \overline{\Omega}$. Let also $U \subset \mathbb{C}^N \setminus \overline{\omega}$ be a neighborhood of $\partial\Omega$ in $\mathbb{C}^N$ and $\sigma_{\omega,U}$ be a smooth section of $\pi^*T^{*(1,0)}_\Omega$ in $\omega \times (\Omega \cap U)$ admitting a C^1 extension to $\omega \times (\overline{\Omega} \cap U)$ that satisfies*

$$\overline{\partial}_z \sigma_{\omega,U}(z,\zeta) = 0,\ \rfloor_\Delta \sigma_{\omega,U}(z,\zeta) = 1 \quad \forall\, (z,\zeta) \in \omega \times (\overline{\Omega} \cap U). \tag{4.83}$$

Then, any $s \in \mathcal{O}(\Omega, E)$, extending continuously to $\overline{\Omega}$, admits in ω the integral representation

$$(4.84)\quad s(z) = \int_{\Omega} (\mathscr{W}(z,\cdot))_{N,N}\, s(\cdot) + \int_{\partial\Omega} \frac{\sigma_{\omega,U}(z,\cdot)}{2i\pi} \Big(\sum_{r=0}^{N-1} \Big(\overline{\partial}_\zeta \frac{\sigma_{\omega,U}(z,\cdot)}{2i\pi}\Big)^{N-1-r} (\mathscr{W}(z,\cdot))_{r,r}\Big)\, s(\cdot).$$

PROOF. Let $(\chi_\nu)_{\nu\geq 1}$ be a sequence of cut-off functions identically equal to 1 on a neighborhood of $\overline{\omega}$ satisfying $\lim_{\nu\longrightarrow\infty} \chi_\nu = \chi_{\overline{\Omega}}$, where the convergence is in the sense of distributions in $\mathbb{C}^N$ and $\chi_{\overline{\Omega}}$ is the characteristic function of the bounded open subset with piecewise boundary Ω. It follows from Proposition 4.33 that for any $\nu \geq 1$ and any $z \in \omega$,

$$(4.85)\quad s(z) = \int_{\Omega} \Big(\Big(\chi_\nu(\zeta) + \frac{\overline{\partial}_\zeta((1-\chi_\nu(\zeta))\,\sigma_{\omega,U}(z,\zeta))}{2i\pi}\Big)^{N+1} \mathscr{W}(z,\zeta)\Big)_{N,N} s(\zeta).$$

Observe now that one has in Ω for any such ν and any such z that

$$\Big(\chi_\nu + \frac{\overline{\partial}_\zeta((1-\chi_\nu)\,\sigma_{\omega,U}(z,\cdot))}{2i\pi}\Big)^{N+1} = \Big(\chi_\nu + (1-\chi_\nu)\frac{\overline{\partial}_\zeta\sigma_{\omega,U}(z,\cdot)}{2i\pi}\Big)^{N+1} - (N+1)\,\overline{\partial}\chi_\nu \frac{\sigma_{\omega,U}(z,\cdot)}{2i\pi}\Big(\chi_\nu + (1-\chi_\nu)\frac{\overline{\partial}_\zeta\sigma_{\omega,U}(z,\cdot)}{2i\pi}\Big)^{N}.$$

For each (N,N) (respectively $(N,N-1)$) form θ (respectively σ) on Ω, which extends continuously to the boundary, one has that

(4.86)

$$\lim_{\nu\to+\infty} \int_\Omega \chi_\nu\,\theta = \int_\Omega \theta,\quad \lim_{\nu\to+\infty}\int_\Omega (1-\chi_\nu)^r\theta = 0 \quad \forall r>0,$$
$$\lim_{\nu\to+\infty} \binom{N}{r}\int_\Omega (\chi_\nu)^r(1-\chi_\nu)^{N-r}\,(\overline{\partial}\chi_\nu)\,\sigma = -\frac{1}{N+1}\int_{\partial\Omega}\sigma \quad \forall r = 0,\dots,N,$$

thanks to the jump formula; see for example [**BGVY**, Lemma 2.5]. As a consequence

$$\lim_{\nu\to+\infty}\int_\Omega \Big(\Big(\chi_\nu + (1-\chi_\nu)\frac{\overline{\partial}_\zeta\sigma_{\omega,U}(z,\cdot)}{2i\pi}\Big)^{N+1}\mathscr{W}(z,\cdot)\Big)_{N,N} s = \int_\Omega (\mathscr{W}(z,\cdot))_{N,N}\, s$$

and

$$-(N+1)\lim_{\nu\to+\infty}\int_\Omega \Big(\overline{\partial}\chi_\nu\frac{\sigma_{\omega,U}(z,\cdot)}{2i\pi}\Big(\chi_\nu + (1-\chi_\nu)\frac{\overline{\partial}_\zeta\sigma_{\omega,U}(z,\cdot)}{2i\pi}\Big)^{N}\mathscr{W}(z,\cdot)\Big)_{N,N} s = \int_{\partial\Omega}\frac{\sigma_{\omega,U}(z,\cdot)}{2i\pi}\Big(\sum_{r=0}^{N-1}\Big(\overline{\partial}_\zeta\frac{\sigma_{\omega,U}(z,\cdot)}{2i\pi}\Big)^{N-1-r}(\mathscr{W}(z,\zeta))_{r,r}\Big)\, s.$$

Formula (4.84) for $z \in \omega$ follows from (4.85) if $\nu \longrightarrow +\infty$. □

EXAMPLE 4.35. Important examples where Propositions 4.32 to 4.34 apply are those where the global weight $\mathscr{W}$ is realized utilizing the construction described in Proposition 4.20. Weighted integral formulae of the form just mentioned were obtained by Mats Andersson and Bo Berndtsson in [**AndB82**]; see also [**AndP88**] or [**BGVY**, Theorem 2.7]. In these formulas one has for any $0 \le r \le N$ that

$$\mathscr{W}_{r,r}(z,\zeta) = \Big[\sum_{\{\kappa \in \mathbb{N}^M \,:\, |\kappa| = r\}} \frac{1}{\kappa!} \Big(\Big(\frac{\partial}{\partial W} \Big)^{\kappa} \Gamma \Big) (\mathcal{Q}(z,\zeta)) \prod_{j=1}^{M} \Big(\frac{\overline{\partial}_\zeta q^j(z,\zeta)}{2i\pi} \Big)^{\kappa_j} \Big], \tag{4.87}$$

where Γ is holomorphic on $\operatorname{Im} \mathcal{Q}$ with $\Gamma(\mathbf{1}) = 1$; see (4.57) and (4.56) for the definition of $\mathcal{Q}$. They generalize in the multivariate setting the weighted Cauchy formulae previously introduced in the univariate setting in §1.2.2 and §1.2.3.

Let now $\mathbf{F}_\bullet$ be a generically exact complex of holomorphic bundles over Ω so that the related $\mathbf{F}_\bullet$-Hefer problem for the Koszul complex $\mathscr{K}_\bullet^{\mathscr{D}}$ admits a solution $\mathbf{H}$ in $\Omega \times \Omega$; see Definition 4.27.[16] One can take outside the set $\sqcap^{-1}(V(\mathbf{F}_\bullet))$, instead of the global weight $\mathscr{W}$ described in Propositions 4.32 and 4.33, the global $(E_0 \otimes E)$-weight

$$\mathscr{W}(z,\zeta) = \mathscr{W}(z,\zeta)\, \mathscr{W}^{\overline{\mathbf{F}_\bullet}}(z,\zeta),$$

where

$$\mathscr{W}^{\overline{\mathbf{F}_\bullet}}(z,\zeta) = \Big(2i\pi F_1^{\overline{\mathbf{F}_\bullet}}(z)\, H_1(z,\zeta)\, P_0^{\overline{\mathbf{F}_\bullet}}(\zeta) + H_0(z,\zeta)\, R_0^{\overline{\mathbf{F}_\bullet}}(\zeta) \Big)_{|(\Omega \times \Omega) \setminus \sqcap^{-1}(V(\mathbf{F}_\bullet))};$$

see Proposition 4.29 and formula (4.72) for the definition of H_0 and H_1 in terms of $\mathbf{H}$, attached to a metrized version $\overline{\mathbf{F}_\bullet}$ of $\mathbf{F}_\bullet$. Then integral representation formulae lead to explicit solutions for division problems. Let us state, for example, the following proposition derived from Proposition 4.33. For the sake of simplicity, we take $E = \mathbb{C}$.

PROPOSITION 4.36. *Let $\omega, \chi, q_{\omega,\chi}$ be as in Proposition 4.33 and let $\mathscr{W}$ be a global $\mathbb{C}$-weight with respect to the Koszul complex $\mathscr{K}_\bullet^{\mathscr{D}}$ in $\omega \times \Omega$. Let $\mathbf{F}_\bullet$ be a generically exact complex of holomorphic bundles over Ω such that the $\mathbf{F}_\bullet$-Hefer problem with respect to the Koszul complex $\mathscr{K}_\bullet^{\mathscr{D}}$ has a solution $\mathbf{H}$ on $\Omega \times \Omega$. Then any $s \in \mathcal{O}(\Omega, E_0^{\mathbf{F}_\bullet})$ splits in ω as*

$$s(z) = 2i\pi F_1^{\overline{\mathbf{F}_\bullet}}(z)\, Q(z) + R(z), \tag{4.88}$$

where

$$\begin{aligned} Q(z) &= \int_\Omega \Big(\Big(\chi + \frac{\overline{\partial}_\zeta \big((1-\chi)\, q_{\omega,\chi}(z,\cdot) \big)}{2i\pi} \Big)^{N+1} \mathscr{W}(z,\zeta)\, H_1(z,\cdot)\, P_0^{\overline{\mathbf{F}_\bullet}} \Big)_{N,N} s(\cdot), \\ R(z) &= \int_\Omega \Big(\Big(\chi + \frac{\overline{\partial}_\zeta \big((1-\chi)\, q_{\omega,\chi}(z,\cdot) \big)}{2i\pi} \Big)^{N+1} \mathscr{W}(z,\zeta)\, H_0(z,\cdot)\, R_0^{\overline{\mathbf{F}_\bullet}} \Big)_{N,N} s(\cdot). \end{aligned} \tag{4.89}$$

Moreover, one has

$$\big(R_0^{\overline{\mathbf{F}_\bullet}} s = 0 \text{ in } \Omega \big) \Longrightarrow \big(s(z) = 2i\pi F_1^{\overline{\mathbf{F}_\bullet}}(z)\, Q(z) \text{ in } \omega \big). \tag{4.90}$$

PROOF. Repeating the proof of Proposition 4.33 with $\mathscr{W}\mathscr{W}^{\mathbf{F}_\bullet}$ instead of $\mathscr{W}$, one gets (4.88) for any $z \in \omega \setminus V(\mathbf{F}_\bullet)$. Note that $\mathscr{W}^{\mathbf{F}_\bullet}$ fails to be smooth around $\{z\} \times \Omega$, but we are in the situation where $\mathscr{W}^{\mathbf{F}_\bullet}(z,\cdot)$ defines a current in Ω; see Remark 4.31. Then the result follows using analytic continuation across the closed hypersurface $\omega \cap V(\mathbf{F}_\bullet)$. □

[16] We will show in §4.6 that such is the case when Ω is pseudo-convex. That is, $(\Omega, \mathcal{O}(\Omega))$ is a Stein manifold and the bundles $E_\ell = E_\ell^{\mathbf{F}_\bullet}$ are trivial; see Proposition 4.43.

REMARK 4.37. One cannot use Proposition 4.34 instead of Proposition 4.33 since the singular supports of the currents $P^{\overline{\mathbf{F}_\bullet}}$ and $R^{\overline{\mathbf{F}_\bullet}}$ intersect the piecewise smooth boundary $\partial\Omega$, unless the analytic set $V(\mathbf{F}_\bullet)$ is discrete.

4.5.2. The projective case $\mathcal{X} = \mathbb{P}^N_{\mathbb{C}}$. The projective space $\mathbb{P}^N_{\mathbb{C}}$ is an N-dimensional, compact, good manifold. It makes Stokes's theorem easier to handle. The Koszul complex $\mathscr{K}_\bullet^{\mathscr{D}}$ on $\mathbb{P}^N_{\mathbb{C}} \times \mathbb{P}^N_{\mathbb{C}}$ that we refer to in this section is the $(\mathscr{E}, \Delta)$ one, where

$$\mathscr{E} = \overset{\curvearrowleft}{\mathcal{O}(1)} \otimes \pi^* T^{(1,0)}_{\mathbb{P}^N_{\mathbb{C}}}, \quad \Delta = -\sum_{\kappa=0}^{N} (z_\kappa)\, \partial/\partial\zeta_k,$$

and $([z_0 : \cdots : z_N], [\zeta_0 : \cdots : \zeta_N])$ denote the homogeneous coordinates on the duplication $\mathbb{P}^N_{\mathbb{C}} \times \mathbb{P}^N_{\mathbb{C}}$; see (4.6) in Proposition 4.2. Let us begin with a purely local result [**AndG11**, Proposition 4.1].

PROPOSITION 4.38. *Let E be a holomorphic bundle over $\mathbb{P}^N_{\mathbb{C}}$, $z^0 \in \mathbb{P}^{\mathrm{N}}_{\mathbb{C}}$, and let $\mathscr{U}^0$ be a neighborhood of $\sqcap^{-1}(\{z^0\})$. If $\mathscr{W}^{z^0} \in \mathcal{L}^0_{\mathscr{D}}[E^{\curvearrowleft}](\mathscr{U}^0)$ is a local E-weight at z^0 with respect to the Koszul complex $\mathscr{K}_\bullet^{\mathscr{D}}$, then for any $s \in \mathcal{O}(\mathcal{O}(-N) \otimes E)$ one has that*

$$s(z^0) = \int_{\mathbb{P}^N_{\mathbb{C}}} (\mathscr{W}^{z^0}(z^0, \cdot))_{N,N}\, s(\cdot). \tag{4.91}$$

PROOF. We already know (see §4.2.2) that in the projective setting the equality

$$\Big[|\Delta|^{2\lambda} \sum_{r=1}^{N} \frac{\Delta^*}{2i\pi|\Delta|^2} \Big(\overline{\partial}_\zeta \frac{\Delta^*}{2i\pi|\Delta|^2}\Big)^{r-1}\Big]_{\lambda=0} = K_{\mathbb{P}^N_{\mathbb{C}}}$$

holds, where the Bochner–Martinelli kernel $K_{\mathbb{P}^N_{\mathbb{C}}}$ on $\mathbb{P}^N_{\mathbb{C}} \times \mathbb{P}^N_{\mathbb{C}}$ has been defined in Definition 4.10. One has in $\mathbb{P}^N_{\mathbb{C}}$ that

$$\begin{aligned}
\nabla^\circ_{\mathscr{D}}(\mathscr{W}^{z^0}(z^0,\cdot)\, K_{\mathbb{P}^N_{\mathbb{C}}}(z^0,\cdot)) &= \nabla^\circ_{\mathscr{D}}\mathscr{W}^{z^0}(z^0,\cdot)\, K_{\mathbb{P}^N_{\mathbb{C}}}(z^0,\cdot) + \mathscr{W}^{z^0}(z^0,\cdot)\, \nabla^\circ_{\mathscr{D}} K_{\mathbb{P}^N_{\mathbb{C}}}(z^0,\cdot) \\
&= \mathscr{W}^{z^0}(z^0,\cdot)\, \nabla^\circ_{\mathscr{D}} K_{\mathbb{P}^N_{\mathbb{C}}}(z^0,\cdot) \\
&= \mathscr{W}^{z^0}(z^0,\cdot)\, \nabla_{\Delta(z^0,\cdot)} K_{\mathbb{P}^N_{\mathbb{C}}}(z^0,\cdot) \\
&= \mathscr{W}^{z^0}(z^0,\cdot) - [\{z^0\}]\, (\mathscr{W}^{z^0}(z^0,z^0))_{0,0}\, \mathrm{Id}_{(\mathcal{O}(-N))_{z^0}} \\
&= \mathscr{W}^{z^0}(z^0,\cdot) - [\{z^0\}]\, \mathrm{Id}_{(\mathcal{O}(-N)\otimes E)_{z^0}},
\end{aligned}$$

in view of the identity (4.30) (see Proposition 4.9) and because of the initial condition $(\mathscr{W}^{z^0}(z^0,z^0))_{0,0} = \mathrm{Id}_{E_{z^0}}$. Comparing the bidegrees in both sides of the last identity, one gets

$$-\overline{\partial}\big(\mathscr{W}^{z^0}(z^0,\cdot)\, K_{\mathbb{P}^N_{\mathbb{C}}}(z^0,\cdot)\big)_{N,N-1} = (\mathscr{W}^{z^0}(z^0,\cdot))_{N,N} - [\{z^0\}]\, \mathrm{Id}_{(\mathcal{O}(-N)\otimes E)_{z^0}}.$$

Since $\mathbb{P}^N_{\mathbb{C}}$ is compact without boundary, it follows from Stokes's theorem that for any section $s \in \mathcal{O}(\mathbb{P}^N_{\mathbb{C}}, \mathcal{O}(-N) \otimes E)$ the integral representation formula

$$s(z^0) = \int_{\mathbb{P}^N_{\mathbb{C}}} (\mathscr{W}^{z^0}(z^0,\zeta))_{N,N}\, s(\zeta)$$

holds. But, this is the desired formula (4.91). □

An immediate corollary of this local result is the following global one.

COROLLARY 4.39. *Let E be a holomorphic bundle and let $\mathscr{W}$ be a smooth global E-weight in $(\mathbb{P}^N(\mathbb{C}) \times \mathbb{P}^N(\mathbb{C})) \setminus \sqcap^{-1}(V)$, where V is a closed algebraic subvariety of $\mathbb{P}^N_{\mathbb{C}}$. Any holomorphic section $s \in \mathcal{O}(\mathbb{P}^N_{\mathbb{C}}, \mathcal{O}(-N) \otimes E)$ admits in $\mathbb{P}^N_{\mathbb{C}} \setminus V$ the integral representation*

$$s(z) = \int_{\mathbb{P}^N_{\mathbb{C}}} (\mathscr{W}(z, \cdot))_{N,N}\, s(\cdot). \tag{4.92}$$

PROOF. For any $z^0 \notin V$, $\mathscr{W}$ defines a local E-weight at z^0. Then, Proposition 4.38 applies, which proves (4.92) for any $z \in \mathbb{P}^N_{\mathbb{C}} \setminus V$. □

Corollary 4.39 implies the important proposition.

PROPOSITION 4.40. *Let $\mathbf{F}_\bullet$ be a generically exact complex of holomorphic bundles over $\mathbb{P}^N_{\mathbb{C}}$ such that the $\mathbf{F}_\bullet$-Hefer problem with respect to $\mathscr{K}_\bullet^{\mathscr{D}}$ admits a solution $\mathbf{H}$. Then, given a metrized version $\overline{\mathbf{F}_\bullet}$ of $\mathbf{F}_\bullet$, any holomorphic section $s \in \mathcal{O}(\mathbb{P}^N_{\mathbb{C}}, \mathcal{O}(-N) \otimes E_0^{\mathbf{F}_\bullet})$ splits as*

$$s(z) = 2i\pi\, F_1^{\mathbf{F}_\bullet}(z)\, Q(z) + R(z), \tag{4.93}$$

where

$$Q(z) = \int_{\mathbb{P}^N_{\mathbb{C}}} \big(H_1(z, \cdot) P_0^{\overline{\mathbf{F}_\bullet}}\big)_{N,N}\, s(\cdot), \quad R(z) = \int_{\mathbb{P}^N_{\mathbb{C}}} \big(H_0(z, \cdot)\, R_0^{\overline{\mathbf{F}_\bullet}}\big)_{N,N}\, s(\cdot). \tag{4.94}$$

One has, as a consequence, that

$$\big(R_0^{\overline{\mathbf{F}_\bullet}} s = 0 \text{ in } \mathbb{P}^N_{\mathbb{C}}\big) \Longrightarrow \big(s(z) = 2i\pi F_1^{\mathbf{F}_\bullet}(z)\, Q(z) \text{ in } \mathbb{P}^N_{\mathbb{C}}\big). \tag{4.95}$$

PROOF. It follows from Proposition 4.29 and Corollary 4.39 that the equality $s(z) = 2i\pi F_1^{\mathbf{F}_\bullet}(z)\, Q(z) + R(z)$ holds in $\mathbb{P}^N_{\mathbb{C}} \setminus V(\mathbf{F}_\bullet)$. Note that $\mathscr{W}^{\mathbf{F}_\bullet}$ fails to be smooth in a neighborhood of $\{z\} \times \Omega$, but we are in the situation where $\mathscr{W}^{\mathbf{F}_\bullet}(z, \cdot)$ defines a current in Ω; see Remark 4.31. The principle of analytic continuation across the analytic set $V(\mathbf{F}_\bullet)$ implies that the same equality holds everywhere in $\mathbb{P}^N_{\mathbb{C}}$. It proves the result. □

4.5.3. Weighted Koppelman's formulae in the general setting. Let $\mathcal{X}$ be an N-dimensional good complex manifold. Recall that it means that $\mathcal{X}$ embeds in its duplication $\mathscr{X}$ as $\mathscr{D} = V(\Delta)$, where $\Delta \in \mathscr{O}(\mathscr{X}, \mathscr{E})$ and $\operatorname{rank} \mathscr{E} = N$; see Definition 4.1. Let also $\mathscr{K}_\bullet^{\mathscr{D}}$ be the $(\mathscr{E}, \Delta)$ Koszul complex. We refer in this short subsection to the material introduced at the end of §4.1.2. This subsection can be viewed as a complement to §4.3 from which it exploits some material. In particular, we consider the holomorphic bundle

$$\boldsymbol{\mathcal{E}} = \bigoplus_{\ell \geq 0} \bigwedge^{\ell} (\mathscr{E} \oplus \mathscr{E}^*) = (\det \mathscr{E} \otimes \det \mathscr{E}^*) \oplus \boldsymbol{\mathcal{E}}' = \operatorname{Hom}_{\mathbb{C}}(\det \mathscr{E}) \oplus \boldsymbol{\mathcal{E}}'$$

over the duplication $\mathscr{X}$ of the good complex manifold $\mathcal{X}$. We refer the reader to §4.3 for the definition of the holomorphic section $\boldsymbol{\Delta}$ of $\boldsymbol{\mathcal{E}}$ induced by Δ and to the end of §4.1.2 for the corresponding definition of the sheaves of currents $\boldsymbol{\mathcal{L}}^{\nu}_{\mathscr{D}}$ (see (4.14)) together with the operator $\boldsymbol{\nabla}_{\mathscr{D}}$ between them (see (4.15)). We need to extend slightly the definition of the sheaves $\boldsymbol{\mathcal{L}}^{\nu}_{\mathscr{D}}$ as follows. Given a holomorphic bundle E over $\mathcal{X}$, one sets

$$\boldsymbol{\mathcal{L}}^{\nu}_{\mathscr{D}}[\widehat{E}] = \bigoplus_{\gamma \geq -\nu,\, \kappa \geq 0} {}'\mathcal{D}^{(\kappa, \gamma+\nu)}\Big(\cdot, \Big(\bigwedge^{\kappa} \mathscr{E}\Big) \otimes \Big(\bigwedge^{\gamma} \mathscr{E}^*\Big) \otimes \widehat{E}\Big), \tag{4.96}$$

for any $\nu \in \mathbb{Z}$. We will also need the residue and principal value currents $R^{\boldsymbol{\Delta}}_{|\,|}$ and $P^{\boldsymbol{\Delta}}_{|\,|}$, defined in (4.35), once the bundle $\mathscr{E}$ is metrized as $\overline{\mathscr{E}} = (\mathscr{E}, |\ |)$. Furthermore, we will need the Bochner–Martinelli kernel

$$K^{\boldsymbol{\Delta}}_{\overline{\boldsymbol{\mathcal{E}}}} = P^{\boldsymbol{\Delta}}_{|\,|}\,\frac{1}{N!}\Big(\boldsymbol{\Phi}_{\overline{\mathscr{E}}} + D_{\overline{\boldsymbol{\mathcal{E}}}}\boldsymbol{\Delta}\Big)^N \in \boldsymbol{\mathcal{L}}^{-1}_{\mathscr{D}}(\mathscr{X})$$

introduced in (4.46). We recall that the smooth form involved here,

$$\frac{1}{N!}\Big(\boldsymbol{\Phi}_{\overline{\mathscr{E}}} + D_{\overline{\boldsymbol{\mathcal{E}}}}\boldsymbol{\Delta}\Big)^N$$

satisfies the relation

$$\frac{1}{N!}\Big(\boldsymbol{\Phi}_{\overline{\mathscr{E}}} + D_{\overline{\boldsymbol{\mathcal{E}}}}\boldsymbol{\Delta}\Big)^N = c_N(\overline{\mathscr{E}})\,\mathrm{Id}_{\det E} + \Big(\frac{1}{N!}\Big(\boldsymbol{\Phi}_{\overline{\mathscr{E}}} + D_{\overline{\boldsymbol{\mathcal{E}}}}\boldsymbol{\Delta}\Big)^N\Big)',$$

where $c_N(\overline{\mathscr{E}})$ is the N-Chern form of the metrized bundle $\overline{\mathscr{E}}$; see (4.39).

Let us first transpose in terms of the material presented in §4.5.3 the notion of global E-weight introduced in Definition 4.17 and extensively used in §4.5.1 and §4.5.2.

Definition 4.41. Given a holomorphic bundle E over $\mathcal{X}$, a $\boldsymbol{\mathcal{E}}$-valued global E-weight in $\mathscr{U} \subset \mathscr{X}$ with respect to $\boldsymbol{\Delta}$ is a section $\mathbb{W}$ of $\boldsymbol{\mathcal{L}}^0_{\mathscr{D}}[E^{\frown}]$ in $\mathscr{U}$, whose singular support does not intersect $\mathscr{U} \cap \mathscr{D}$, such that

$$\text{(4.97)}\qquad \begin{aligned} &\mathrm{bidegree}_z \mathbb{W} = (0,0), \quad \overline{\partial}_z \mathbb{W} = 0, \quad \boldsymbol{\nabla}_{\mathscr{D}} \mathbb{W} = 0,\\ &(\mathbb{W}(z,z))_{0,0} = \mathrm{Id}_{(\sqcap^* E)_{(z,z)}}\ \forall\, z \in \sqcap(\mathscr{U}). \end{aligned}$$

If $\sqcap^{-1}(\{z^0\}) \subset \mathscr{U}$ and $\mathbb{W}$ is smooth in a neighborhood of $\sqcap^{-1}(\{z^0\})$, then $\mathbb{W}$ is said to define an $\boldsymbol{\mathcal{E}}$-valued local E-weight at z^0 with respect to $\boldsymbol{\Delta}$. Then one can formulate a weighted version of Corollary 4.14; see [**Got08**, Theorem 3.4 and (17)].

Proposition 4.42. *Let E be a holomorphic bundle over the good complex manifold $\mathscr{X}$. Let also $\mathbb{W}$ be a $\boldsymbol{\mathcal{E}}$-valued global E-weight in $\mathscr{U} \subset \mathscr{X}$ with respect to $\boldsymbol{\Delta}$. If*

$$\text{(4.98)}\qquad \begin{aligned} \mathbb{W}K^{\boldsymbol{\Delta}}_{\overline{\boldsymbol{\mathcal{E}}}} &= K_{\overline{\boldsymbol{\mathcal{E}}},\mathbb{W}}\,\mathrm{Id}_{\det\mathscr{E}} + \big(\mathbb{W}K^{\boldsymbol{\Delta}}_{\overline{\boldsymbol{\mathcal{E}}}}\big)',\\ \frac{\mathbb{W}}{N!}\Big(\boldsymbol{\Phi}_{\overline{\mathscr{E}}} + D_{\overline{\boldsymbol{\mathcal{E}}}}\boldsymbol{\Delta}\Big)^N &= \Phi_{\overline{\boldsymbol{\mathcal{E}}},\mathbb{W}}\,\mathrm{Id}_{\det\mathscr{E}} + \Big(\frac{\mathbb{W}}{N!}\Big(\boldsymbol{\Phi}_{\overline{\mathscr{E}}} + D_{\overline{\boldsymbol{\mathcal{E}}}}\boldsymbol{\Delta}\Big)^N\Big)' \end{aligned}$$

are the decompositions (4.33) *for the sections*

$$\mathbb{W}\,K^{\boldsymbol{\Delta}}_{\overline{\boldsymbol{\mathcal{E}}}} \in \boldsymbol{\mathcal{L}}^{-1}_{\mathscr{D}}[\widehat{E}](\mathscr{U}), \quad \frac{\mathbb{W}}{N!}\Big(\boldsymbol{\Phi}_{\overline{\mathscr{E}}} + D_{\overline{\boldsymbol{\mathcal{E}}}}\boldsymbol{\Delta}\Big)^N \in \boldsymbol{\mathcal{L}}^0_{\mathscr{D}}[\widehat{E}](\mathscr{U}),$$

then for any $z^0 \in \mathcal{X}$ satisfying $\sqcap^{-1}(\{z^0\}) \subset \mathscr{U}$ the current weighted Koppelman-type identity

$$\text{(4.99)}\qquad -\overline{\partial}\,K_{\overline{\boldsymbol{\mathcal{E}}},\mathbb{W}} = \Phi_{\overline{\boldsymbol{\mathcal{E}}},\mathbb{W}} - [\mathscr{D}]\,\mathrm{Id}_{\pi^* E}$$

holds in a neighborhood of $\sqcap^{-1}(\{z^0\})$.

Proof. The multiplication of $\mathbb{W}$ and $K^{\boldsymbol{\Delta}}_{\overline{\boldsymbol{\mathcal{E}}}}$ in $\mathscr{U}$ is well defined since the singular supports of these two currents are disjoint. It follows from Leibniz's rule, the fact that $\boldsymbol{\nabla}_{\mathscr{D}}\mathbb{W} = 0$, combined with the current identity (4.47) (see Theorem 4.13), and the initial condition on $\mathscr{D}$ imposed by (4.97), that

$$\begin{aligned}\boldsymbol{\nabla}_{\mathscr{D}}(\mathbb{W}\, K_{\overline{\boldsymbol{\mathcal{E}}}}^{\boldsymbol{\Delta}}) &= (\boldsymbol{\nabla}_{\mathscr{D}}\mathbb{W})\, K_{\overline{\boldsymbol{\mathcal{E}}}}^{\boldsymbol{\Delta}} + \mathbb{W}\,\boldsymbol{\nabla}_{\mathscr{D}} K_{\overline{\boldsymbol{\mathcal{E}}}}^{\boldsymbol{\Delta}} = \mathbb{W}\,\boldsymbol{\nabla}_{\mathscr{D}} K_{\overline{\boldsymbol{\mathcal{E}}}}^{\boldsymbol{\Delta}} \\ &= \frac{\mathbb{W}}{N!}\Big(\boldsymbol{\Phi}_{\overline{\mathscr{E}}} + D_{\overline{\boldsymbol{\mathcal{E}}}}\boldsymbol{\Delta}\Big)^{N} - \mathbb{W}\,[\mathscr{D}]\,\mathrm{Id}_{\det \mathscr{E}} \\ &= \frac{\mathbb{W}}{N!}\Big(\boldsymbol{\Phi}_{\overline{\mathscr{E}}} + D_{\overline{\boldsymbol{\mathcal{E}}}}\boldsymbol{\Delta}\Big)^{N} - \mathbb{W}_{0,0}\,[\mathscr{D}]\,\mathrm{Id}_{\det \mathscr{E}} \\ &= \frac{\mathbb{W}}{N!}\Big(\boldsymbol{\Phi}_{\overline{\mathscr{E}}} + D_{\overline{\boldsymbol{\mathcal{E}}}}\boldsymbol{\Delta}\Big)^{N} - [\mathscr{D}]\,\mathrm{Id}_{\pi^* E \otimes \det \mathscr{E}}.\end{aligned}$$

Formula (4.99) follows from the identification of the $\mathbb{C}$-valued coordinate currents along $\mathrm{Hom}_{\mathbb{C}}(\det \mathscr{E})$ in decomposition (4.33). □

4.6. Solving the $\mathbf{F}_\bullet$-Hefer problem in specific situations

This section illustrates some results from §4.3. Namely, it provides important examples of configurations where, given a generically exact complex $\mathbf{F}_\bullet$ over a good complex manifold with Koszul complex $\mathscr{K}_\bullet^{\mathscr{D}}$, one can solve explicitly the $\boldsymbol{F}$-Hefer–Gleason problem; see Definition 4.27. Thus, when $\mathcal{X} = \Omega \subset \mathbb{C}^N$ (respectively when $\mathcal{X} = \mathbb{P}^N_{\mathbb{C}}$), Proposition 4.36 (respectively Proposition 4.40) will apply to such configurations. Then integral representation formulae will lead to division/interpolation formulae of the Lagrange–Kronecker type. This section complements §4.5.

4.6.1. Complexes of trivial bundles over Stein manifolds. In this subsection, $\mathcal{X}$ is a good N-dimensional Stein complex manifold. We denote by $(\mathscr{E}, \Delta)$ the configuration to which the Koszul complex $\mathscr{K}_\bullet^{\mathscr{D}}$ over the duplication $\mathscr{X}$ is subordinate, namely $\operatorname{rank}\mathscr{E} = N$, $\Delta \in \mathscr{O}(\mathscr{X}, \mathscr{E})$ with $\mathscr{D} = V(\Delta)$ and $d\Delta \neq 0$ locally on the diagonal $\mathscr{D}$.

PROPOSITION 4.43. *Let $\mathcal{X}$ be a good N-dimensional Stein complex manifold with $(\mathscr{E}, \Delta)$ Koszul complex $\mathscr{K}_\bullet^{\mathscr{D}}$ on its duplication $\mathscr{X}$. Given any generically exact complex of trivial holomorphic bundles*

$\mathbf{F}_\bullet$:

$$0 \longrightarrow E_L \simeq \mathbb{C}^{M_L} \xrightarrow{F_L} E_{L-1} \simeq \mathbb{C}^{M_{L-1}} \xrightarrow{F_{L-1}} \cdots \longrightarrow E_1 \simeq \mathbb{C}^{M_1} \xrightarrow{F_1} E_0 \simeq \mathbb{C}^{M_0} \longrightarrow 0,$$

the $\mathbf{F}_\bullet$-Hefer problem admits a solution.

PROOF. For any $r \geq \ell \geq 0$, one has $E^{\frown}_{\ell,r} \simeq \mathbb{C}^{M_r M_\ell}$. If

$$\{\epsilon_1^{[r]*}, \ldots, \epsilon_{M_r}^{[r]*}\},\ \{\epsilon_1^{[\ell]*}, \ldots, \epsilon_{M_\ell}^{[\ell]*}\}$$

are global frames for E_r and E_ℓ, respectively, then

$$\{\epsilon_\mu^{[\ell]} \otimes \epsilon_\nu^{[r]*} \,:\, \mu = 1, \ldots, M_\ell, \nu = 1, \ldots, M_r\}$$

is a global frame for $E^{\frown}_{\ell,r}$. One needs to express each $H_{\ell,r}$ along a global frame. One takes for any $r \geq 0$,

$$H_{r,r}(z,\zeta) = \sum_{\nu=1}^{M_r} \epsilon_\nu^{[r]} \otimes \epsilon_\nu^{[r*]} = \mathrm{Id}_{E_r}.$$

Suppose now that $r-\ell=1$ and consider the section

$$(z,\zeta)\longmapsto F_{\ell+1}(\zeta)-F_{\ell+1}(z)=\sum_{\mu=1}^{M_\ell}\sum_{\nu=1}^{M_r}(u_{\ell,\ell+1})_{\mu,\nu}\,\epsilon_\mu^{[\ell]}\otimes\epsilon_\nu^{[\ell+1]*}$$

of $E_{\ell,\ell+1}^{\frown}$. Each coordinate function $(u_{\ell,\ell+1})_{\mu,\nu}$ annihilates the Bochner–Martinelli current $R^\Delta_{\lceil\,\rceil}=R^\Delta$ since $d\Delta\neq 0$ on $\mathscr{D}$. It follows then from Proposition 3.18 and Theorem 3.39 that the equation $\nabla_{\mathscr{D}}H_{\ell,\ell+1}(z,\zeta)=2i\pi(F_{\ell+1}(\zeta)-F_{\ell+1}(z))$ (with unknown $H_{\ell,\ell+1}\in\mathscr{O}(\mathscr{X},\mathscr{E}^*\otimes E_{\ell,\ell+1}^{\frown}))$ can be solved locally in a neighborhood of each point in $\mathscr{X}$. Since $\mathscr{X}$ is Stein as $\mathcal{X}$ is, it follows from Henri Cartan's Theorem A (see Theorem A.6) that all such local solutions globalize into a global one $H_{\ell,\ell+1}\in\mathscr{O}(\mathscr{X},\mathscr{E}^*\otimes E_{\ell,\ell+1}^{\frown})$. We now process by induction, assuming that all the

$$H_{\ell,r}\in\mathscr{O}\big(\mathscr{X},(\bigwedge^{r-\ell}\mathscr{E}^*)\otimes E_{\ell,r}^{\frown}\big)$$

have been defined for any $r>\ell$ with $r-\ell\leq k$ in order to satisfy the Hefer rules (4.75). Fix now $\ell\geq 0$ and $r=\ell+k+1$. One has, according to the already satisfied Hefer rules and to $F^2=0$, that

$$\begin{aligned}\nabla_{\mathscr{D}}\big(H_{\ell,r-1}F_r(\zeta)-F_{\ell+1}(z)H_{\ell+1,r}\big)&=2i\pi\,\Delta\,\big(H_{\ell,r-1}F_r(\zeta)-F_{\ell+1}(z)H_{\ell+1,r}\big)\\&=(\nabla_{\mathscr{D}}H_{\ell,r-1})F_r(\zeta)+F_{\ell+1}(z)\,\nabla_{\mathscr{D}}H_{\ell+1,r}\\&=-2i\pi\big(F_{\ell+1}(z)H_{\ell+1,r-1}F_r(\zeta)-F_{\ell+1}(z)\,H_{\ell+1,r-1}F_r(\zeta)\big)=0.\end{aligned}$$

Since the Koszul complex $\mathscr{K}_\bullet^{\mathscr{D}}$ is exact because $\operatorname{rank}\mathscr{E}=N=\operatorname{codim}V(\Delta)$, there exists $H_{\ell,r}\in\mathscr{O}\big(\mathscr{X},(\bigwedge^{k+1}\mathscr{E}^*)\otimes E_{\ell,r}^{\frown}\big)$ such that

$$\nabla_{\mathscr{D}}H_{\ell,r}(z,\zeta)=H_{\ell,r-1}F_r(\zeta)-F_{\ell+1}(z)H_{\ell+1,r}.$$

It concludes the argument for this choice of ℓ and r. The proof of Proposition 4.43 is complete using induction on k. □

In the particular case where $\mathcal{X}=\Omega$ is a pseudo-convex subset of $\mathbb{C}^N$, with, as usual

$$\mathscr{E}=\pi^*T_\Omega^{(1,0)},\qquad \Delta=\sum_{k=1}^N(\zeta_k-z_k)\,\partial/\partial\zeta_k, \tag{4.100}$$

one obtains from Proposition 4.36 the following division/interpolation theorem of Lagrange–Kronecker type.

THEOREM 4.44. *Let $\omega,\chi,q_{\omega,\chi}$ be as in Proposition 4.33 and let $\mathscr{W}$ be a global $\mathbb{C}$-weight with respect to the $(\mathscr{E},\Delta)$ Koszul complex $\mathscr{K}_\bullet^{\mathscr{D}}$ in $\omega\times\Omega$. Let $\mathbf{F}_\bullet$ be a generically exact complex of holomorphic trivial bundles over Ω and $\{H_{\ell,r}\}_{\ell\geq0,r\geq-1}$ which solves the $\mathbf{F}_\bullet$-Hefer problem according to Proposition 4.43. Let $H_0=\sum_{r\geq0}H_{0,r}$ and $H_1=\sum_{r\geq1}H_{1,r}$. Then any holomorphic section $S\in\mathcal{O}(\Omega,E_0^{\mathbf{F}_\bullet}\simeq\mathbb{C}^{M_0})$ splits in ω as*

$$S(z)=2i\pi F_1^{\mathbf{F}_\bullet}(z)\,Q(z)+R(z), \tag{4.101}$$

where

(4.102)
$$Q(z) = \int_\Omega \Big(\Big(\chi + \frac{\overline{\partial}_\zeta((1-\chi)\, q_{\omega,\chi}(z,\cdot))}{2i\pi}\Big)^{N+1} \mathscr{W}(z,\zeta)\, H_1(z,\cdot)\, P_0^{\mathbf{F}_\bullet}\Big)_{N,N} S(\cdot),$$
$$R(z) = \int_\Omega \Big(\Big(\chi + \frac{\overline{\partial}_\zeta((1-\chi)\, q_{\omega,\chi}(z,\cdot))}{2i\pi}\Big)^{N+1} \mathscr{W}(z,\zeta)\, H_0(z,\cdot)\, R_0^{\mathbf{F}_\bullet}\Big)_{N,N} S(\cdot).$$

Moreover, one has

$$\big(R_0^{\mathbf{F}_\bullet} S = 0 \text{ in } \Omega\big) \Longrightarrow \big(S(z) = 2i\pi F_1^{\mathbf{F}_\bullet}(z)\, Q(z) \text{ in } \omega\big). \tag{4.103}$$

REMARK 4.45. In the particular case where $\mathcal{X} = \mathbb{C}^N$ and all morphisms F_ℓ in $\mathbf{F}_\bullet$ are polynomial with coefficients in a subfield $\mathbb{K}$ of $\mathbb{C}$, the proof of Proposition 4.43 shows that the $H_{\ell,r}$ can also be taken as such polynomial morphisms.

The proof of Proposition 4.43 also leads to the concept of *Hefer form* attached to a section of a trivial bundle.

DEFINITION 4.46. Let s be a holomorphic section of a trivial bundle E with rank m over a Stein good N-dimensional manifold with its $(\mathscr{E},\Delta)$ Koszul complex $\mathscr{K}_\bullet^{\mathscr{D}}$ considered as a section of $\mathrm{Hom}_{\mathbb{C}}(E^*,\mathbb{C})$. A *Hefer form* H_s for s is any holomorphic section of

$$\mathscr{E}^* \otimes \mathrm{Hom}_{\mathbb{C}}(\pi^* E^*, \sqcap^* \mathbb{C}) = \mathscr{E}^* \otimes \mathrm{Hom}_{\mathbb{C}}(\pi^* E^*, \mathbb{C}) = \mathscr{E}^* \otimes \pi^* E$$

such that

$$\nabla_{\mathscr{D}} H_s(z,\zeta) = \nabla^\circ_{\mathscr{D}} H_s(z,\zeta) = 2i\pi(s(\zeta) - s(z)). \tag{4.104}$$

Consider now the Koszul complex $\mathbf{K}_\bullet^s$ subordinate to s.

$$0 \to \bigwedge^m E^* \xrightarrow{\rfloor_s} \bigwedge^{m-1} E^* \xrightarrow{\rfloor_s} \cdots \xrightarrow{\rfloor_s} \bigwedge^{\ell+1} E^* \xrightarrow{\rfloor_s} \bigwedge^{\ell} E^* \xrightarrow{\rfloor_s} \cdots \xrightarrow{\rfloor_s} E^* \xrightarrow{\rfloor_s} \mathbb{C}.$$

Let also $\bigwedge^{-\rho} E^* = 0$ for any $\rho \in \mathbb{N}^*$ and let $\{\epsilon_1, \dots, \epsilon_m\}$ be a global holomorphic frame for E over $\mathcal{X}$. For any $\ell \geq 0$ and $r \geq -1$, let $H_{\ell,r}$ be the section of $\bigwedge^{r-\ell} \mathscr{E}^* \otimes (\mathbf{K}^s)^{\frown}_{\ell,r}$ defined as $H_{\ell,r} = 0$ whenever $r < \ell$ and, when $r \geq \ell$, as

$$H_{\ell,r} = \frac{\mathbb{I}^\ell}{\ell!}\, \frac{\mathbb{H}_s^{r-\ell}}{(r-\ell)!}, \tag{4.105}$$

where the operators $\mathbb{I}$ (respectively $\mathbb{H}_s$) act from $\bigwedge^{\rho'} \mathscr{E}^* \otimes \bigwedge^\rho E^*$ to itself (respectively from $\bigwedge^{\rho'} \mathscr{E}^* \otimes \bigwedge^\rho E^*$ to $\bigwedge^{\rho'+1} \mathscr{E}^* \otimes \bigwedge^{\rho-1} E^*$) as follows: for any smooth $\bigwedge^{\rho'} \mathscr{E}^*$-valued current T with pure degree and any $1 \leq j_1 < \cdots < j_\rho \leq m$,

$$\mathbb{I}\,(T\, \epsilon_J^*) = \rho\, T\, \epsilon_J^*, \quad \mathbb{H}_s(T\, \epsilon_J^*) = (-1)^{\deg T + \rho'}\, T \wedge H_s(\epsilon_J^*). \tag{4.106}$$

Observe that $\mathbb{I} = \sum_1^m \epsilon_j^* \otimes \epsilon_j$. One can state the following proposition in the specific case of Koszul complexes.

PROPOSITION 4.47. *Let s be a holomorphic section of a trivial bundle E over a good Stein N-dimensional complex manifold, with $(\mathscr{E},\Delta)$ Koszul complex $\mathscr{K}_\bullet^{\mathscr{D}}$, and let H_s be a Hefer form for s with respect to $\mathscr{K}_\bullet^{\mathscr{D}}$. Then the collection*

$$\mathbf{H} = \{H_{\ell,r} \,:\, \ell \geq 0, r \geq -1\}$$

with $H_{\ell,r}$ defined as (4.105) solves the $\mathbf{K}_\bullet^s$-Hefer problem with respect to $\mathscr{K}_\bullet^{\mathscr{D}}$.

PROOF. It is clear by construction that $H_{\ell,r}$ is a section of $\big(\mathcal{L}_{\mathscr{D}}[(\mathbf{K}^s_\bullet)^{\frown}_{\ell,r}]\big)^\circ$ when $r \geq \ell$ and that $H_{\ell,r} = 0$ whenever $r < \ell$. Moreover $H_{\ell,\ell}(z,z) = \mathrm{Id}_{((\mathbf{K}^s_\bullet)^{\frown}_{\ell,\ell})(z,z)}$ for any $\ell \geq 0$. Fix $r \geq 0$ and consider the polynomial in the formal variable X

$$\begin{aligned}\mathbf{T}(X) &= 2i\pi \sum_{\ell=0}^{r} X^\ell \big(H_{\ell,r-1} \circ \rfloor_{s(\zeta)} - \rfloor_{s(z)} \circ H_{\ell+1,r}\big)\\ &= 2i\pi \sum_{\ell=0}^{r-1} X^\ell \big(H_{\ell,r-1} \circ \rfloor_{s(\zeta)} - \rfloor_{s(z)} \circ H_{\ell+1,r}\big)\\ &\qquad\in \Big(\bigoplus_{\ell=0}^{r-1}\bigoplus_{\nu} \big(\mathcal{L}^\nu_{\mathscr{D}}[(\mathbf{K}^s_\bullet)^{\frown}_{\ell,r}]\big)^\circ\Big)[X].\end{aligned}$$

One has

$$\mathbf{T}(X) = 2i\pi\Big(\Big(\sum_{\ell=0}^{r-1} X^\ell \frac{\mathbb{I}^\ell}{\ell!}\,\frac{\mathbb{H}_s^{r-1-\ell}}{(r-1-\ell)!}\Big)\rfloor_{s(\zeta)} - \rfloor_{s(z)}\Big(\sum_{\ell=0}^{r-1} X^\ell\,\frac{\mathbb{I}^{\ell+1}}{(\ell+1)!}\,\frac{\mathbb{H}_s^{r-1-\ell}}{(r-1-\ell)!}\Big)\Big).$$

Given the definition of $\mathbb{I}$ and the $\mathbb{F}_2$ grading, the polynomial expression is rewritten as

$$\mathbf{T}(X) = \sum_{\ell=0}^{r-1} X^\ell \,(\rfloor_{s(\zeta)} - \rfloor_{s(z)})\,\frac{\mathbb{I}^\ell}{\ell!}\,\frac{\mathbb{H}_s^{r-1-\ell}}{(r-1-\ell)!} = (\rfloor_{s(\zeta)} - \rfloor_{s(z)})\,\frac{(X\,\mathbb{I} + \mathbb{H}_s)^{r-1}}{(r-1)!}.$$

Since $\nabla_{\mathscr{D}}\mathbb{H}_s = \rfloor_{s(\zeta)} - \rfloor_{s(z)}$ as $\nabla_{\mathscr{D}} H_s = s(\zeta) - s(z)$, it follows from Leibniz's rule and the definition of the $H_{\ell,r}$ for $\ell = 0,\dots,r$ that

$$\mathbf{T}(X) = \nabla_{\mathscr{D}}\Big(\frac{(X\,\mathbb{I} + \mathbb{H}_s)^r}{r!}\Big) = \nabla_{\mathscr{D}}\Big(\sum_{\ell=0}^{r} H_{\ell,r}\,X^\ell\Big). \tag{4.107}$$

The identities

$$\nabla_{\mathscr{D}} H_{\ell,r} = H_{\ell,r-1} \circ \rfloor_{s(\zeta)} - \rfloor_{s(z)} \circ H_{\ell+1,r}$$

for $\ell \geq 0$ and $r \geq -1$ are obtained by comparing the coefficients in X in the two sides of (4.107), which concludes the proof of Proposition 4.47. □

Let us conclude this subsection with an alternative approach. Proposition 4.20 provides another construction of a global weight attached (in a different way than through Propositions 4.43 and 4.29) to the Koszul complex $\mathbf{K}^f_\bullet$ or, more precisely, to the Eagon–Northcott complex $\mathbf{EN}^{f,k}_\bullet$ for $k \in \mathbb{N}^*$ (see Example 3.26) when $f = f_1\epsilon_1 + \cdots + f_m\epsilon_m$ is a section of the bundle $\mathbb{C}^m$ over a pseudo-convex domain Ω in $\mathbb{C}^N$. We need first to construct a Hefer form

$$H_f = \sum_{j=1}^{m} h_{f,j}\,\epsilon_j \text{ with } h_{f,j} \in \mathscr{O}(\Omega\times\Omega, \pi^* T_\Omega^{(1,0)}) \text{ for } j = 1,\dots,m,$$

for f (see Definition 4.46) and let $\|\ \|$ be the canonical metric on $\mathbb{C}^m$. Then take a complex parameter λ such that $\operatorname{Re}\lambda \gg 1$ and define $q^\lambda \in \mathcal{O}(\Omega\times\Omega, \pi^* T_\Omega^{(1,0)})$ by

$$q^\lambda(z,\zeta) = H_f\,\|f\|^{2\lambda}\,\frac{f^*}{\|f\|^2} = \sum_{k=1}^{N} q_k^\lambda(z,\zeta)\,d\zeta_k, \tag{4.108}$$

so that

$$\frac{\overline{\partial} q^\lambda}{2i\pi} = H_f\,\Big(|f|^{2\lambda}\,\overline{\partial}\big(\frac{f^*}{2i\pi\|f\|^2}\big) + (\overline{\partial}\|f\|^{2\lambda})\,\frac{f^*}{2i\pi\|f\|^2}\Big)$$

and

$$\mathcal{Q}^\lambda(z,\zeta) = 1 + \langle\!\langle \boldsymbol{q}^\lambda, z-\zeta\rangle\!\rangle = 1 - \|f\|^{2\lambda} + \|f\|^{2\lambda}\, 2i\pi f(z)\frac{f^*}{2i\pi\|f\|^2}. \tag{4.109}$$

For any $r \geq 1$, one has

$$\begin{aligned}\Big(\frac{\overline{\partial} q^\lambda}{2i\pi}\Big)^r &= \|f\|^{2r\lambda}\Big(\overline{\partial}\big(H_f\,\frac{f^*}{2i\pi\|f\|^2}\big)\Big)^r \\ &\quad + (\overline{\partial}\|f\|^{2r\lambda})\,\big(H_f\,\frac{f^*}{2i\pi\|f\|^2}\big)\Big(\overline{\partial}\big(H_f\,\frac{f^*}{2i\pi\|f\|^2}\big)\Big)^{r-1}.\end{aligned} \tag{4.110}$$

It follows from Proposition 4.20 that for any $k \in \mathbb{N}^*$,

$$\mathscr{W}^\lambda_{f,k} = \sum_{r=0}^{N}\binom{N+k}{r}(\mathcal{Q}^\lambda)^{N+k-r}\Big(\frac{\overline{\partial} q^\lambda}{2i\pi}\Big)^r \tag{4.111}$$

is a global weight in $(\Omega\times\Omega)\setminus\sqcap^{-1}(V(f))$. Moreover, one can easily show, invoking a log resolution as in the proof of Proposition 3.15, that

$$\lambda \longrightarrow [\mathscr{W}^\lambda_{f,k}],$$

considered as a current-valued map, extends to $\mathbb{C}$ as a meromorphic function whose poles lie in $-\mathbb{Q}^+$. Moreover, the value at $\lambda = 0$ is a pseudo-meromorphic current which remains a global $\mathbb{C}$-weight $\mathscr{W}_{f,k}$. One has in particular that

(4.112)
$$\Big[\sum_{r=0}^{N}\binom{N+k}{r}(\mathcal{Q}^\lambda)^{N+k-r}\,\|f\|^{2r\lambda}\Big(\overline{\partial}\big(H_f\,\frac{f^*}{2i\pi\|f\|^2}\big)\Big)^r\Big]_{\lambda=0} = 2i\pi f(z)\,\sigma_{f,k},$$

where

$$\sigma_{f,k} = \sum_{r=1}^{N+1}\binom{N+k}{r-1}\Big[\|f\|^{2\lambda}\Big(\frac{f(z)f^*}{\|f\|^2}\Big)^{N+k-r}\mathscr{P}_{f,r}\Big]_{\lambda=0}, \tag{4.113}$$

with, for any $r \geq 1$,

$$\mathscr{P}_{f,r} = \frac{f^*}{2i\pi\|f\|^2}\Big(\overline{\partial}\big(H_f\,\frac{f^*}{2i\pi\|f\|^2}\big)\Big)^{r-1}. \tag{4.114}$$

According to the binomial current identities

$$\begin{aligned}&\big[(1-\|f\|^{2\lambda})^{N+k-r}\,(\overline{\partial}\|f\|^{2r\lambda})\,(\cdot)\big]_{\lambda=0}\\ &\quad = r\big[(1-\|f\|^{2\lambda})^{(N+k-1)-(r-1)}\,\|f\|^{2(r-1)\lambda}\,(\overline{\partial}\|f\|^{2\lambda})\,(\cdot)\big]_{\lambda=0}\\ &\quad = \frac{r}{(N+k)\binom{N+k-1}{r-1}}\,\big[(\overline{\partial}\|f\|^{2\lambda})\,(\cdot)\big]_{\lambda=0} = \frac{1}{\binom{N+k}{r}}\,\big[(\overline{\partial}\|f\|^{2\lambda})\,(\cdot)\big]_{\lambda=0},\end{aligned}$$

which rely on the same computations as those that led to (4.86), the remaining contribution

$$\Big[\sum_{r=1}^{N}\binom{N+k}{r}(\mathcal{Q}^\lambda)^{N+k-r}\|f\|^{2r\lambda}(\overline{\partial}\|f\|^{2r\lambda})\big(H_f\frac{f^*}{2i\pi\|f\|^2}\big)\Big(\overline{\partial}\big(H_f\frac{f^*}{2i\pi\|f\|^2}\big)\Big)^{r-1}\Big]_{\lambda=0}$$

to $\mathscr{W}_{f,k}$ splits as $f(z)\,\widetilde{\sigma}_{f,k} + \rho_{f,k}$, where, if one sets

$$\mathscr{R}_{f,r} = H_f\;\mathscr{P}_{f,r} = \big(H_f\,\frac{f^*}{2i\pi\|f\|^2}\big)\Big(\overline{\partial}\big(H_f\,\frac{f^*}{2i\pi\|f\|^2}\big)\Big)^{r-1}, \tag{4.115}$$

$$
\begin{aligned}
\rho_{f,k} &= \Big[\sum_{r=1}^{N} (\overline{\partial}\|f\|^{2\lambda})\,\mathscr{R}_{f,r}\Big]_{\lambda=0}, \\
\widetilde{\sigma}_{f,k} &= \Big[\frac{f^*}{\|f\|^2}\Big(\sum_{r=1}^{N}\sum_{r'=r+1}^{N+k}\binom{r'}{r}\Big(\frac{f(z)f^*}{\|f\|^2}\Big)^{(r'-r)-1}(\overline{\partial}\|f\|^{2\lambda})\,\mathscr{R}_{f,r}\Big)\Big]_{\lambda=0}.
\end{aligned}
\tag{4.116}
$$

Observe that the pseudo-meromorphic current $\rho_{f,k}$ is supported by $\pi^{-1}(V(f))$, thus plays the role of a residue current, while the pseudo-meromorphic currents $\sigma_{f,k}$ and $\widetilde{\sigma}_{f,k}$ have support $\Omega\times\Omega$ and singular support contained in $\pi^*(V(f))$, thus play the role of principal value currents. As a consequence of Proposition 4.36, the splitting $\mathscr{W}_{f,k} = f(z)(2i\pi\,\sigma_{f,k} + \widetilde{\sigma}_{f,k}) + \rho_{f,k}$ leads to an alternative version of the division/interpolation theorem stated as Theorem 4.44. Observe that, in the univariate case, such a result is described in §1.2.2 and 1.2.3; see Propositions 1.16 and 1.18. Such a weight $\mathscr{W}_{f,k}$ is, in some sense, a mean value (through Crofton's averaging formula, see §D.5.2) of weights subordinate to exact[17] Eagon–Northcott complexes $\mathbf{EN}_\bullet^{g,k}$; see [**VidY22**, §2] where the case where $\operatorname{codim} V(f) = n$ is studied in detail.

4.6.2. Projective Hefer forms and Koszul complex over $\mathbb{P}^N_{\mathbb{C}}$. The N-dimensional complex projective plane $\mathbb{P}^N_{\mathbb{C}}$ is considered in this subsection as a good complex manifold, which embeds in $\mathbb{P}^N_{\mathbb{C}}\times\mathbb{P}^N_{\mathbb{C}}$ as the diagonal complete intersection $\mathscr{D} = V(\Delta)$, where, in homogeneous coordinates on$([z_0:\cdots:z_N],[\zeta_0:\cdots:\zeta_N])$,

$$
\Delta = -\sum_{\kappa=0}^{N} z_\kappa \frac{\partial}{\partial\zeta_\kappa} \in \mathscr{O}(\mathbb{P}^N_{\mathbb{C}}\times\mathbb{P}^N_{\mathbb{C}},\mathscr{E}) \text{ with } \mathscr{E} = \mathcal{O}(1)\otimes\pi^* T^{(1,0)}_{\mathbb{P}^N_{\mathbb{C}}}.
$$

We will refer now to the Koszul complex $\mathscr{K}_\bullet^{\mathscr{D}}$ in such a projective setting, with its related sheaves of currents $\mathcal{L}^\nu_{\mathscr{E}}[\,]$, $(\mathcal{L}^\nu_{\mathscr{E}}[\,])^\circ$ and nabla operators $\nabla_{\mathscr{D}}$ or $\nabla^\circ_{\mathscr{D}}$. We need first to adapt the notion of Hefer form presented in Definition 4.46 for sections of the trivial bundle within the affine setting and formalize a similar notion for global sections of the line bundles $\mathcal{O}(d)$, $d\in\mathbb{N}^*$, over $\mathbb{P}^N_{\mathbb{C}}$, namely, according to the GAGA (Géométrie Algébrique et Géométrie Analytique) principle, homogeneous polynomial functions $[\zeta_0:\cdots:\zeta_N]\longmapsto P(\zeta)$, where $P\in\mathbb{C}[X_0,\dots,X_N]$ with $\deg P = d$. Thus the main thing to do to is to restore homogeneity, which we do as follows.

DEFINITION 4.48. Let $d\in\mathbb{N}^*$ and F be a holomorphic section of the line bundle $\mathcal{O}(d)$ over $\mathbb{P}^N_{\mathbb{C}}$. A Hefer form H_F for F corresponding to the Fubini–Study metric on $\mathcal{O}(1)$ is a smooth $\sqcap^*\mathcal{O}(d)$-valued section of $\mathscr{E}^*$, therefore represents an element of $\mathcal{L}^{-1}_{\mathscr{E}}[\sqcap^*\mathcal{O}(d)](\mathbb{P}^N_{\mathbb{C}}\times\mathbb{P}^N_{\mathbb{C}})$, which is holomorphic in z and satisfies

$$
\begin{aligned}
\nabla_{\mathscr{D}} H_F(z,\zeta) &= -\Big(2i\pi\sum_{\kappa=0}^{N} z_\kappa\frac{\partial}{\partial\zeta_\kappa} + \overline{\partial}_\zeta\Big) H_F(z,\zeta) \\
&= 2i\pi\,\big((\mathscr{W}_{\mathbb{P}^N_{\mathbb{C}}}(z,\zeta))^d F(\zeta) - F(z)\big).
\end{aligned}
\tag{4.117}
$$

The following lemma shows how to realize such Hefer forms.

---

[17]By "exact", we mean here Eagon–Northcott complexes $\mathbf{EN}_\bullet^{g,k}$ attached to a holomorphic section g of $\mathbb{C}^m$ such that $\operatorname{codim} V(g) = m$.

LEMMA 4.49. *Let $d \in \mathbb{N}^*$, let F be an holomorphic section of $\mathcal{O}(d)$, and let $A_0, \dots, A_N$ be $N+1$ homogeneous polynomials with degree $d-1$ such that $\sum_{\kappa=0}^{N} A_\kappa(z,\zeta)\, d\zeta_\kappa$ is a Hefer form for the section $\zeta \mapsto F(\zeta_0, \dots, \zeta_N)$ of $\mathbb{C}$ over $\mathbb{C}^{N+1}$ according to Definition 4.46. Then the smooth $\sqcap^*\mathcal{O}(d-1) \otimes \pi^*\mathcal{O}(1)$-valued $(1,0)$-form*

$$H = \sum_{\kappa=0}^{N} A_\kappa\big(z, \mathscr{W}_{\mathbb{P}^N_{\mathbb{C}}}(z,\zeta)\,\zeta\big)\, D\zeta_\kappa, \tag{4.118}$$

where D is the Chern connection of the metrized bundle $\mathcal{O}(1)$ with the Fubini–Study metric, is a Hefer form for F corresponding to the Fubini–Study metric.

PROOF. Since $\mathscr{W}_{\mathbb{P}^N_{\mathbb{C}}}(z,\zeta)$ is a section of $\mathrm{Hom}_{\mathbb{C}}(\pi^*\mathcal{O}(1), \sqcap^*\mathcal{O}(1))$ and each A_κ is homogeneous with degree $d-1$ while D acts from the sheaf ${}'\mathcal{D}(\cdot, \pi^*\mathcal{O}(1))$ to the sheaf ${}'\mathcal{D}^{(1,0)}(\cdot, \pi^*\mathcal{O}(1))$, one has

$$[H] \in {}'\mathcal{D}^{(1,0)}\big(\mathbb{P}^N_{\mathbb{C}} \times \mathbb{P}^N_{\mathbb{C}}, \sqcap^*\mathcal{O}(d-1) \otimes \pi^*\mathcal{O}(1)\big) = {}'\mathcal{D}\big(\mathbb{P}^N_{\mathbb{C}} \times \mathbb{P}^N_{\mathbb{C}}, \mathscr{E}^* \otimes \sqcap^*\mathcal{O}(d)\big),$$

which implies that the smooth form H, which is holomorphic in z, represents an element in $\mathcal{L}_{\mathscr{E}}^{-1}[\sqcap^*\mathcal{O}(d)](\mathbb{P}^N_{\mathbb{C}} \times \mathbb{P}^N_{\mathbb{C}})$ such that $\overline{\partial}_z[H] = 0$. Since

$$D\zeta_\kappa = d\zeta_\kappa - \zeta_\kappa \frac{\partial \|\zeta\|^2}{\|\zeta\|^2} \tag{4.119}$$

(see Proposition B.45), one has

(4.120)

$$\begin{aligned}
\nabla^\circ_{\mathscr{D}}(D\zeta_\kappa)\,(z,\zeta) &= -\Big(2i\pi \sum_{\nu=0}^{N} z_\nu \frac{\partial}{\partial \zeta_\nu} + \overline{\partial}_\zeta\Big)\Big(d\zeta_\kappa - \zeta_\kappa \frac{\sum_{\nu'=0}^{N} \overline{\zeta}_{\nu'}\, d\zeta_{\nu'}}{\|\zeta\|^2}\Big) \\
&= 2i\pi\Big(\zeta_\kappa\Big(\frac{z \bullet \zeta}{\|\zeta\|^2} + dd^c \log \|\zeta\|^2\Big) - z_\kappa\Big) = 2i\pi\big(\mathscr{W}_{\mathbb{P}^N_{\mathbb{C}}}(z,\zeta)\,\zeta_\kappa - z_\kappa\big).
\end{aligned}$$

Since $\mathscr{W}_{\mathbb{P}^N_{\mathbb{C}}}$ is a $\mathbb{C}$-global weight, it follows from Leibniz's rule and from the homogeneity of F that

$$\begin{aligned}
\nabla_{\mathscr{D}} H(z,\zeta) = \nabla^\circ_{\mathscr{D}} H(z,\zeta) &= 2i\pi \sum_{\kappa=0}^{N} A_\kappa\big(z, \mathscr{W}_{\mathbb{P}^N_{\mathbb{C}}}(z,\zeta)\,\zeta\big)\big(\mathscr{W}_{\mathbb{P}^N_{\mathbb{C}}}(z,\zeta)\,\zeta_\kappa - z_\kappa\big) \\
&= 2i\pi\,\big(F\big(\mathscr{W}_{\mathbb{P}^N_{\mathbb{C}}}(z,\zeta)\,\zeta\big) - F(z)\big) \\
&= 2i\pi\,\big((\mathscr{W}_{\mathbb{P}^N_{\mathbb{C}}}(z,\zeta))^d P(\zeta) - F(z)\big),
\end{aligned}$$

which concludes the proof of Lemma 4.49. □

Given $P_1, \dots, P_m \in \mathbb{C}[X_0, \dots, X_N]$, homogeneous with respective strictly positive degrees $d_1, \dots, d_N$, $E = \bigoplus_{j=1}^{m} \mathcal{O}(d_j)$, $P = P_1 \oplus \cdots \oplus P_m$, and $M \in \mathbb{N}^*$, one considers the tensorized Koszul complex

$\mathbf{K}^P_{\bullet,M} = \mathcal{O}(M) \otimes \mathbf{K}^P_\bullet$:

$$0 \to \mathcal{O}(M) \otimes \bigwedge^m E^* \xrightarrow{\lrcorner_P} \cdots \xrightarrow{\lrcorner_P} \mathcal{O}(M) \otimes \bigwedge^\ell E^* \xrightarrow{\lrcorner_P} \cdots \xrightarrow{\lrcorner_P} \mathcal{O}(M) \otimes E^* \xrightarrow{\lrcorner_P} \mathcal{O}(M).$$

We also adopt the convention that $\mathcal{O}(M)\otimes\bigwedge^{-1}E^*=0$. For each $\ell\geq 0$, one has, if $\{\epsilon_1,\dots,\epsilon_m\}$ denotes the canonical basis of the trivial bundle $\mathbb{C}^m$ over $\mathbb{P}^N_{\mathbb{C}}$,

(4.121)

$$\mathcal{O}(M)\otimes\bigwedge^{\ell}E^*=\mathcal{O}(M)\otimes\bigwedge^{\ell}\bigoplus_{j=1}^{m}\mathcal{O}(d_j)=\sum_{1\leq j_1<\cdots<j_\ell\leq m}\mathcal{O}\Big(M-\sum_{1}^{\ell}d_{j_\nu}\Big)\bigwedge_{\nu=1}^{\ell}\epsilon^*_{j_\nu}$$
$$=\sum_{1\leq j_1<\cdots<j_\ell\leq m}\mathcal{O}(M-|d_J|)\,\epsilon^*_J.$$

For each $j=1,\dots,m$, let H_{P_j} be a Hefer form for P_j corresponding to the Fubini–Study metric; see Definition 4.48 and Lemma 4.49 for the construction of such Hefer form. Consider now the formal $\bigoplus_{\ell\geq 0}\bigwedge^{\ell}(E\oplus E^*)$-valued expressions

$$\mathbb{I}_d=\sum_{j=1}^{m}\frac{\epsilon^*_j\otimes\epsilon_j}{\mathscr{W}^{d_j}_{\mathbb{P}^N_{\mathbb{C}}}},\qquad \mathbb{H}_P=\sum_{j=1}^{m}\frac{H_{P_j}}{\mathscr{W}^{d_j}_{\mathbb{P}^N_{\mathbb{C}}}}\,\epsilon_j. \tag{4.122}$$

They act formally on the sheaves of current-valued sections of

$$E(\mathbf{K}^P_{\bullet,M})\simeq\sum_{\ell\geq 0}\ \sum_{1\leq j_1<\cdots<j_\ell\leq m}\mathcal{O}(M-|d_J|)\,\epsilon^*_J \tag{4.123}$$

as follows: for any ordered subset J (in increasing order) with cardinal ρ and any $\mathcal{O}(M-|d_J|)$-valued current T with pure degree,

(4.124)

$$\mathbb{I}_d(T\,\epsilon^*_J)=T\sum_{\nu=1}^{\rho}\epsilon^*_{j_1}\wedge\cdots\wedge\epsilon^*_{j_{\nu-1}}\wedge\frac{\epsilon^*_{j_\nu}}{\mathscr{W}^{d_{j_\nu}}_{\mathbb{P}^N_{\mathbb{C}}}}\wedge\epsilon^*_{j_{\nu+1}}\wedge\cdots\wedge\epsilon^*_{j_\rho},$$

$$\mathbb{H}_P(T\,\epsilon^*_J)=(-1)^{\deg T}\,T\sum_{\nu=1}^{\rho}(-1)^{\nu-1}\frac{H_{P_{j_\nu}}}{\mathscr{W}^{d_{j_\nu}}_{\mathbb{P}^N_{\mathbb{C}}}}\,\epsilon^*_{j_1}\wedge\cdots\wedge\epsilon^*_{j_{\nu-1}}\wedge\epsilon^*_{j_{\nu+1}}\wedge\cdots\wedge\epsilon^*_{j_\rho}.$$

Once all these preliminaries are settled, one can now state the following important proposition [**AndG11**, Proposition 4.4 and §5].

PROPOSITION 4.50. *Let $M\geq|d_J|$ for any subset J of $\{1,\dots,m\}$ with cardinal* $\min(m,N)$. *For each $\ell\geq 0$ and $r\geq -1$, the action of the operator*

$$(W_{\mathbb{P}^N_{\mathbb{C}}}(z,\zeta))^M\,\frac{(\mathbb{I}_d(z,\zeta))^\ell}{\ell!}\circ\frac{(\mathbb{H}_P(z,\zeta))^{r-\ell}}{(r-\ell)!} \tag{4.125}$$

defines an element

$$H_{\ell,r}\in\mathcal{L}^{-(r-\ell)}_{\mathscr{E}}[(E^{\mathbf{K}^P_{\bullet,M}})^{\frown}_{\ell,r}](\mathbb{P}^N_{\mathbb{C}}\times\mathbb{P}^N_{\mathbb{C}}) \tag{4.126}$$

and the collection $\mathbf{H}=\{H_{\ell,r}\,:\,\ell\geq 0,\ r\geq -1\}$ is a solution[18] *of the $\mathbf{K}^P_{\bullet,M}$-Hefer problem; see Definition* 4.27.

PROOF. One has $H_{\ell,r}=0$ if $\ell>r$. In other cases, the hypothesis of M ensures that either $H_{\ell,r}=0$ (when $r>\min(m,N)$) or $H_{\ell,r}$ is well defined as a $\mathrm{Hom}_{\mathbb{C}}(\pi^*E(\mathbf{K}^P_{\bullet,M}),\sqcap^*E(\mathbf{K}^P_{\bullet,M}))$-valued current since the factor $\mathscr{W}_{\mathbb{P}^N_{\mathbb{C}}}$ absorbs the denominators involved in $\mathbb{I}_d$ or $\mathbb{H}_P$. For any ordered subset J with length $r\geq\ell\geq 0$,

[18] This solution depends on the choice of M. For the sake of simplicity, we did not mention this dependency within our notations.

one has

$$\mathbb{H}_P^{r-\ell}(\epsilon_J^*) = \sum_{0\leq j_1'<\cdots<j_\ell'\leq m,\ J'\subset J} \tau_{J,J'}\epsilon_{J'}^*,$$

where $\tau_{J,J'}$ is a $(r-\ell,0)$ which is $\pi_*\mathcal{O}(|d_{J\setminus J'}|)$-valued, that is, it defines a global section of $\mathcal{L}_{\mathscr{E}}^{-(r-\ell)}[\mathrm{Hom}_{\mathbb{C}}(\pi^*\mathcal{O}(-|d_{J\setminus J'}|),\mathbb{C})]$. For any $J'\subset J$ with cardinal ℓ,

$$\mathscr{W}_{\mathbb{P}_{\mathbb{C}}^N}^M \frac{\mathbb{I}_d^\ell}{\ell!}(\epsilon_{J'}^*) = \mathscr{W}^{M-|d_{J'}|}\,\epsilon_{J'}^*.$$

Since for such J',

$$\begin{aligned}\mathscr{W}_{\mathbb{P}_{\mathbb{C}}^N}^{M-|d_{J'}|}\,\mathcal{L}_{\mathscr{E}}^{-(r-\ell)}&[\mathrm{Hom}_{\mathbb{C}}(\pi^*\mathcal{O}(-|d_{J\setminus J'}|),\mathbb{C})]\\ &\subset \mathcal{L}_{\mathscr{E}}^{-(r-\ell)}\big[\mathrm{Hom}_{\mathbb{C}}(\pi^*\mathcal{O}(M-|d_J|)\,,\sqcap^*\mathcal{O}(M-|d_{J'}|)\big],\end{aligned}$$

it follows from the identification (4.123) that $H_{\ell,r}$, through such action, can be understood as a global section of the sheaf

$$\mathcal{L}_{\mathscr{E}}^{-(r-\ell)}[\mathcal{O}(M)\otimes \overset{\curvearrowleft}{E^{\mathbf{K}_\bullet^P}}_{\ell,r}],$$

which is our first assertion (4.126). For any $r\geq 0$, consider the two univariate polynomials in the variable X

(4.127)

$$\begin{aligned}\mathbf{T}' &= \sum_{\ell=0}^r X^\ell\, H_{\ell,r} = \mathscr{W}_{\mathbb{P}_{\mathbb{C}}^N}^M \frac{(X\mathbb{I}_d+\mathbb{H}_P)^r}{r!},\\ \mathbf{T}'' &= 2i\pi\sum_{\ell=0}^r X^\ell\Big(H_{\ell,r-1}\,(P(\zeta))_{|\pi^*(\mathcal{O}(M)\otimes E_r)} - (P(z))_{|\sqcap^*(\mathcal{O}(M)\otimes E_r)}\,H_{\ell+1,r}\Big)\\ &\in {}'\mathcal{D}\big(\mathbb{P}_{\mathbb{C}}^N\times\mathbb{P}_{\mathbb{C}}^N,(E^{\mathbf{K}_{\bullet,M}^P})^{\curvearrowleft}_{\ell,r}\big)\,[X].\end{aligned}$$

It follows from the fact that $\mathscr{W}_{\mathbb{P}_{\mathbb{C}}^N}$ is a global weight and from the definition of $\mathbb{H}_P$ combined with the current identities (4.117) for $F=P_1,\dots,P_m$, that

$$\begin{aligned}\nabla_{\mathscr{D}}\mathbf{T}' = \nabla_{\mathscr{D}}^\circ\mathbf{T}' &= \mathscr{W}_{\mathbb{P}_{\mathbb{C}}^N}^M\,\nabla_{\mathscr{D}}\mathbb{H}_P\,\frac{(X\mathbb{I}_d+\mathbb{H}_P)^{r-1}}{(r-1)!}\\ &= 2i\pi\,\mathscr{W}_{\mathbb{P}_{\mathbb{C}}^N}^M\Big(\sum_{j=1}^m P_j(\zeta) - \frac{P_j(z)}{\mathscr{W}_{\mathbb{P}_{\mathbb{C}}^N}^{d_j}}\Big)\,\epsilon_j\,\frac{(X\mathbb{I}_d+\mathbb{H}_P)^{r-1}}{(r-1)!}\\ &= 2i\pi\,\mathscr{W}_{\mathbb{P}_{\mathbb{C}}^N}^M\Big(\frac{(X\mathbb{I}_d+\mathbb{H}_P)^{r-1}}{(r-1)!}\,P(\zeta) - \Big(\sum_{j=1}^m\frac{P_j(z)}{\mathscr{W}_{\mathbb{P}_{\mathbb{C}}^N}^{d_j}}\,\epsilon_j\Big)\frac{(X\mathbb{I}_d+\mathbb{H}_P)^{r-1}}{(r-1)!}\Big) = \mathbf{T}''\end{aligned}$$

since

$$\mathscr{W}_{\mathbb{P}_{\mathbb{C}}^N}^M\,\epsilon_j\sum_{\ell=0}^{r-1}X^\ell\,\frac{\mathbb{I}_d^{\ell+1}}{(\ell+1)!}\,\frac{\mathbb{H}_P^{r-1-\ell}}{(r-1-\ell)!} = \mathscr{W}_{\mathbb{P}_{\mathbb{C}}^N}^{M-d_j}\,\epsilon_j\,\Big(\frac{(X\mathbb{I}_d+\mathbb{H}_P)^{r-1}}{(r-1)!}\Big)$$

for $j=1,\dots,m$ according to the definition of $\mathbb{I}_d$. Current identities $\nabla_{\mathscr{D}}\mathbf{T}'=\mathbf{T}''$ for any $r\geq 0$ imply the identity $\nabla_{\mathscr{D}}H=\nabla_{\mathscr{D}}^\circ H = H\circ(2i\pi P(\zeta)) - (2i\pi P(z))\circ H$, where $H=\sum_{\ell\geq 0}\sum_{r\geq\ell}H_{\ell,r}$. One has immediately that

$$(H_{\ell,\ell}(z,z))_{0,0} = \mathrm{Id}_{(\sqcap^*(E_\ell^{\mathbf{K}_{\bullet,M}^P}))_{(z,z)}}$$

since $(\mathscr{W}_{\mathbb{P}_{\mathbb{C}}^N}(z,z))_{0,0}=\mathrm{Id}_{(\sqcap^*\mathcal{O}(1))_{(z,z)}}$. Proposition 4.50 is therefore proved. □

Let us now equip each $\mathcal{O}(d_j)$, $j=1,\dots,m$, with the Fubini–Study metric $|\ |_j$ and the direct sum $\bigoplus_{j=1}^m \mathcal{O}(d_j)$ with the metric $|\ | = \sqrt{|\ |_1^2+\cdots+|\ |_m^2}$. Then one has

(4.128)

$$|P| = \sqrt{\sum_{j=1}^m \frac{|P_j|^2}{\|\zeta\|^{2d_j}}},\quad P^* = \sum_{j=1}^m \frac{\overline{P_j}}{\|\zeta\|^{2d_j}}\,\epsilon_j^*,$$

$$R_0^{\overline{\mathbf{K}_\bullet^P}} = \Big[(\overline{\partial}|P|^{2\lambda}) \sum_{r=1}^{\min(m,N)} \frac{P^*}{2i\pi|P|^2}\Big(\overline{\partial}\frac{P^*}{2i\pi|P|^2}\Big)^{r-1}\Big]_{\lambda=0} = \sum_{r=1}^{\min(m,N)} R_{0,r}^{\overline{\mathbf{K}_\bullet^P}},$$

$$P_0^{\overline{\mathbf{K}_\bullet^P}} = \Big[|P|^{2\lambda} \sum_{r=1}^{\min(m,N+1)} \frac{P^*}{2i\pi|P|^2}\Big(\overline{\partial}\frac{P^*}{2i\pi|P|^2}\Big)^{r-1}\Big]_{\lambda=0} = \sum_{r=1}^{\min(m,N+1)} P_{0,r}^{\overline{\mathbf{K}_\bullet^P}},$$

where $R_0^{\overline{\mathbf{K}_\bullet^P}}$ and $P_0^{\overline{\mathbf{K}_\bullet^P}}$ are considered, for any $M\in\mathbb{N}$, as sections of the holomorphic bundle $\mathrm{Hom}_{\mathbb{C}}(\mathcal{O}(M), E(\mathbf{K}_{\bullet,M}^P))$ acting by multiplication; see Example 3.30. A major consequence of Proposition 4.50, once combined with Proposition 4.40, is the following division/interpolation theorem of Lagrange–Kronecker type; see [**AndG11**, §5].

THEOREM 4.51. *Let $P = P_1\oplus\cdots\oplus P_m$, $|P|$, P^*, $R_{0,r}^{\overline{\mathbf{K}_\bullet^P}}$, $P_{0,r}^{\overline{\mathbf{K}_\bullet^P}}$ for $r\geq 1$ as in* (4.128), *$\deg P_j = d_j \geq 1$, $M\geq |d_J|$ for any subset J of $\{1,\dots,m\}$ with cardinal $\min(m,N)$, $\mathbb{I}_d$, $\mathbb{H}_P$ with the description of their action as operators as in* (4.122), (4.124). *Any global holomorphic section s of $\mathcal{O}(M-N)$ splits in $\mathbb{P}^N_{\mathbb{C}}$ as*

$$s(z) = 2i\pi \rfloor_{P(z)} Q(z) + R(z), \tag{4.129}$$

where

(4.130)

$$Q(z) = \int_{\mathbb{P}^N_{\mathbb{C}}} \Big(\Big((\mathscr{W}_{\mathbb{P}^N_{\mathbb{C}}}(z,\cdot))^M \sum_{r=1}^{\min(m,N+1)} \mathbb{I}_d(z,\cdot)\,\frac{(\mathbb{H}_P(z,\zeta))^{r-1}}{(r-1)!}\Big) P_{0,r}^{\overline{\mathbf{K}_\bullet^P}}\Big)_{N,N} s,$$

$$R(z) = \int_{\mathbb{P}^N_{\mathbb{C}}} \Big(\Big((\mathscr{W}_{\mathbb{P}^N_{\mathbb{C}}}(z,\cdot))^M \sum_{r=1}^{\min(m,N)} \frac{(\mathbb{H}_P(z,\zeta))^{r}}{r!}\Big) R_{0,r}^{\overline{\mathbf{K}_\bullet^P}}\Big)_{N,N} s.$$

If moreover $R_0^{\overline{\mathbf{K}_\bullet^P}} s = 0$, then $s(z) = 2i\pi\rfloor_{P(z)} Q(z)$ in $\mathbb{P}^N_{\mathbb{C}}$.

4.6.3. Hefer problem, syzygies for homogeneous polynomial ideals. Let $\mathbb{S} = \mathbb{C}[X_0,\dots,X_N]$ be the graded homogeneous ring of the projective space $\mathbb{P}^N_{\mathbb{C}}$ considered as the archetypal example of a smooth complete toric manifold (see §3.1.2) and, for each $d\in\mathbb{Z}$, let $\mathbb{S}_{-d}$ be its shift

$$\mathbb{S}_{-d} = \sum_{\{\nu\in\mathbb{N}^{N+1}\,:\,|\nu|=d\}} \Big(\prod_{\kappa=0}^N X_\kappa^{-\nu_\kappa}\Big)\mathbb{S} = \sum_{\{\nu\in\mathbb{Z}^{N+1}\,:\,|\nu|=d\}} X^{-\nu}\mathbb{S}. \tag{4.131}$$

A *polynomial morphism of graded free $\mathbb{S}$-modules*

$$\mathbb{F} : \mathbb{M}_1 := \bigoplus_{\nu=1}^{M_1} \mathbb{S}_{-d_\nu^{[1]}} \longrightarrow \mathbb{M}_0 := \bigoplus_{\mu=0}^{M_0} \mathbb{S}_{-d_\mu^{[0]}}$$

is given by an (M_0, M_1) matrix for which each polynomial entry $\mathbb{F}_{\mu,\nu}$, $1 \leq \mu \leq M_0$, $1 \leq \nu \leq M_1$, is a homogeneous polynomial with degree

$$\deg \mathbb{F}_{\mu,\nu} = d_\nu^{[1]} - d_\mu^{[0]} \geq 0.$$

An example of such a polynomial morphism is that which is associated with a system of generators $\{P_1, \dots, P_m\}$, with $\deg P_j = d_j \in \mathbb{N}^*$, of a homogeneous ideal $\mathbb{I}(P)$ in $\mathbb{S}$, namely

$$\mathbb{F}_P : Q_1 \oplus \cdots \oplus Q_m \in \bigoplus_{j=1}^m \mathbb{S}_{-d_j} \longmapsto Q_1P_1 + \cdots + Q_mP_m \in \mathbb{S}.$$

We consider in this section an exact complex of graded free $\mathbb{S}$-modules
(4.132)

$$\mathbb{M}_\bullet^P : 0 \longmapsto \mathbb{M}_L \xrightarrow{\mathbb{F}_L} \cdots \xrightarrow{\mathbb{F}_{\ell+1}} \mathbb{M}_\ell \xrightarrow{\mathbb{F}_\ell} \mathbb{M}_{\ell-1} \xrightarrow{\mathbb{F}_{\ell-1}} \cdots \xrightarrow{\mathbb{F}_2} \mathbb{M}_1 \xrightarrow{\mathbb{F}_P} \mathbb{S} \xrightarrow{\Pi} \mathbb{M}_{-1} \longrightarrow 0,$$

where $\mathbb{M}_{-1}$ is the finitely generated $\mathbb{S}$-module $\mathbb{M}_{-1} = \mathbb{S}/\mathrm{Im}\,\mathbb{F}_P$ and Π denotes the quotient map. Given $(P_1, \dots, P_m)$, such an exact complex of graded free $\mathbb{S}$-modules always exists,[19] with $L \leq N+1$, as it follows from Hilbert's syzygy theorem [**Hil1890**]; see also [**Eis05**, Chapter 1A]. One can attach to such an exact complex of graded free $\mathbb{S}$-modules $\mathbb{M}_\bullet^P$ a generically exact complex $\mathbf{F}_\bullet(\mathbb{M}^P)$ as follows. For each $\ell \geq 0$, let $\{\epsilon_\nu^{[\ell]} : \nu = 1, \dots, M_\ell\}$ be a copy of the canonical basis of the trivial bundle $\mathbb{C}^{M_\ell}$ with $\{\epsilon_\nu^{[\ell]*} : \nu = 1, \dots, M_\ell\}$ as a dual basis. Let us associate a holomorphic bundle E_ℓ with rank M_ℓ to the graded free $\mathbb{S}$-module $\mathbb{M}_\ell$ as follows:

$$\mathbb{M}_\ell^P = \bigoplus_{\nu=1}^{M_\ell} \mathbb{S}_{-d_\nu^{[\ell]}} \longleftrightarrow E_\ell = \bigoplus_{\nu=1}^{M_\ell} \mathcal{O}(-d_\nu^{[\ell]})\, \epsilon_\nu^{[\ell]*}. \tag{4.133}$$

In the same vein, let us consider the following correspondence at the level of morphisms:

$$\begin{array}{ccc} \displaystyle\bigoplus_{\nu=1}^{M_\ell} \mathbb{S}_{-d_\nu^{[\ell]}} & \xrightarrow{\mathbb{F}_\ell = [(\mathbb{F}_\ell)_{\mu,\nu}]} & \displaystyle\bigoplus_{\nu=1}^{M_{\ell-1}} \mathbb{S}_{-d_\mu^{[\ell-1]}} \\ & \updownarrow & \\ \displaystyle\bigoplus_{\nu=1}^{M_\ell} \mathcal{O}(-d_\nu^{[\ell]})\, \epsilon_\nu^{[\ell]*} & \xrightarrow{F_\ell = [(\mathbb{F}_\ell)_{\mu,\nu}]} & \displaystyle\bigoplus_{\mu=1}^{M_{\ell-1}} \mathcal{O}(-d_\mu^{[\ell-1]})\, \epsilon_\mu^{[\ell]*}, \end{array} \tag{4.134}$$

with the identifications

$$(F_\ell)_{\mu,\nu} \in \mathbb{S}_{d_\nu^{[\ell]} - d_\mu^{[\ell-1]}} \simeq \mathcal{O}\big(\mathbb{P}^N_{\mathbb{C}}, \mathcal{O}(d_\nu^{[\ell]} - d_\mu^{[\ell-1]})\big).$$

Then one associates in such a way the generically exact complex
(4.135)

$$\mathbf{F}_\bullet(\mathbb{M}^P) :\ 0 \longrightarrow E_L \xrightarrow{F_L} E_{L-1} \xrightarrow{F_{L-1}} \cdots \xrightarrow{F_2} E_1 \xrightarrow{F_P} \mathbb{C} \xrightarrow{F_0} E_{-1} = \frac{\mathbb{C}}{\mathrm{Im}\, F_P} \longrightarrow 0$$

[19]It is even unique up to isomorphism (see [**Eis05**, §1A and §1B]) if one takes it *minimal*, that is

$$\mathrm{Im}\,\mathbb{F}_\ell \subset \sum_{\kappa=0}^{N} \sum_{\nu=1}^{M_{\ell-1}} X_\kappa\, \mathbb{S}_{-d_\mu^{[\ell-1]}}$$

for any $\ell \geq 1$. A major stumbling block that remains is the fact that such free resolution, or also *syzygy*, fails, in general, to be easily tractable.

to the exact complex of free graded $\mathbb{S}$-modules (4.132). The objective of this subsection is to solve the $\mathbf{F}_\bullet(\mathbb{M}^P)$-Hefer problem. To do that, one first considers the auxiliary lifted generically exact complex

(4.136)

$$(\mathbf{F}_\bullet(\mathbb{M}^P))^\dagger :$$

$$0 \longrightarrow \mathbb{C}^{M_L} \xrightarrow{F_L^\dagger} \mathbb{C}^{M_{L-1}} \xrightarrow{F_{L-1}^\dagger} \cdots \xrightarrow{F_2^\dagger} \mathbb{C}^m \xrightarrow{(F_P)^\dagger} \mathbb{C} \xrightarrow{F_0^\dagger} E_{-1} = \frac{\mathbb{C}}{\mathrm{Im}\,(F_P)^\dagger} \longrightarrow 0$$

as a generically exact complex over the Stein manifold $\mathbb{C}^{N+1}$, where

$$F_\ell^\dagger = \sum_{\mu=1}^{M_{\ell-1}} \sum_{\nu=1}^{M_\ell} (F_\ell)_{\mu,\nu}\, \epsilon^{[\ell-1]*} \otimes \epsilon_\nu^{[\ell]}.$$

We will need the following lemma.

LEMMA 4.52. *The $(\mathbf{F}_\bullet(\mathbb{M}^P))^\dagger$-Hefer problem (see Definition 4.27) with respect to the $(\mathscr{E}^\dagger, \Delta^\dagger)$ Koszul complex $\mathscr{K}_\bullet^{\mathscr{D}_{N+1}}$, where*

$$\mathscr{E}^\dagger = \pi^* T^{(1,0)}_{\mathbb{C}^{N+1}}, \quad \Delta^\dagger = \sum_{\kappa=0}^{N} (\zeta_\kappa - z_\kappa) \frac{\partial}{\partial \zeta_\kappa}$$

admits a solution

$$\mathbf{H}^\dagger = \{H^\dagger_{\ell,r} \,:\, \ell \geq 0,\ r \geq -1\}$$

such that for each $r \geq \ell \geq 0$,

$$H^\dagger_{\ell,r}(z,\zeta) = \sum_{\mu=0}^{M_\ell} \sum_{\nu=0}^{M_r} \Big(\sum_{0 \leq \kappa_1 < \cdots < \kappa_{r-\ell} \leq N} (H^\dagger_{\ell,r})^\kappa_{\mu,\nu}(z,\zeta) \bigwedge_{\rho=1}^{r-\ell} d\zeta_{\kappa_\rho} \Big)\, \epsilon_\mu^{[\ell]*} \otimes \epsilon_\nu^{[r]}, \tag{4.137}$$

where, for any μ, ν, κ,

$$(H_{\ell,r})^\kappa_{\mu,\nu}(X,Y) \in \big(\mathbb{C}[X_0,\ldots,X_N,Y_0,\ldots,Y_N]\big)_{d_\nu^{[r]} - d_\mu^{[\ell]} - (r-\ell)}. \tag{4.138}$$

PROOF. The proof relies on that of Proposition 4.43 when $\mathcal{X} = \mathbb{C}^{N+1}$. We just need to take care of homogeneity when solving in H the $\nabla_{\mathscr{D}}$-problems $\nabla_{\mathscr{D}} H = \Phi$ which occur at each step; namely, keep only the adequate homogeneous contribution in H according to the right-hand side Φ of the equation. $\square$

In the same vein as Proposition 4.50, one has Proposition 4.53; see [**AndG11**, Proposition 4.4].

PROPOSITION 4.53. *Let M be an integer such that*

$$M \geq \max\{d_\nu^{[\ell]} \,:\, \ell \geq 0,\ 1 \leq \nu \leq M_\ell\}. \tag{4.139}$$

Then, the $\big(\mathbf{F}_\bullet(\mathbb{M}^P) \otimes \mathcal{O}(M)\big)$-Hefer problem with respect to the $(\mathscr{E}, \Delta)$ Koszul complex $\mathscr{K}_\bullet^{\mathscr{D}}$, where

$$\mathscr{E} = (\overset{\frown}{\mathcal{O}(1)}) \otimes \pi^* T^{(1,0)}_{\mathbb{P}^N_{\mathbb{C}}}, \quad \Delta = -\sum_{\kappa=0}^{N} z_\kappa \frac{\partial}{\partial \zeta_\kappa}$$

admits a solution[20]

$$\mathbf{H} = \{H_{\ell,r} \,:\, \ell \geq 0,\ r \geq -1\},$$

[20]This solution depends on the choice of M. For the sake of simplicity, we did not mention this dependency within our notation.

where, for any $r \geq \ell \geq 0$,

$$H_{\ell,r}(z,\zeta) = \sum_{\mu=0}^{M_\ell} \sum_{\nu=0}^{M_r} \Big(\sum_{0 \leq \kappa_1 < \cdots < \kappa_{r-\ell} \leq N} (H_{\ell,r})^{\kappa}_{\mu,\nu}(z,\zeta) \bigwedge_{\rho=1}^{r-\ell} D\zeta_{\kappa_\rho} \Big) \epsilon^{[\ell]*}_{\mu} \otimes \epsilon^{[r]}_{\nu}, \tag{4.140}$$

D being the connection of the line bundle $\mathcal{O}(1)$ equipped with the Fubini–Study metric (see (4.119)) with, for any μ, ν and $0 \leq \kappa_1 < \cdots < \kappa_{r-\ell} \leq N$,

$$(H_{\ell,r})^{\kappa}_{\mu,\nu}(z,\zeta) = \big(\mathscr{W}_{\mathbb{P}^N_{\mathbb{C}}}(z,\zeta)\big)^{M-d^{[r]}_\nu} (H^{\dagger}_{\ell,r})^{\kappa}_{\mu,\nu}\big(z, \mathscr{W}_{\mathbb{P}^N_{\mathbb{C}}}(z,\zeta)\,\zeta\big). \tag{4.141}$$

PROOF. One has for any $r \geq 0$ and $\nu = 1, \dots, M_r$ that

$$\mathscr{W}^{M-d^{[r]}_\nu}_{\mathbb{P}^N_{\mathbb{C}}} \in \mathcal{L}^0_{\mathscr{E}}\big[(\mathcal{O}(M - d^{[r]}_\nu))^{\frown}\big]\,(\mathbb{P}^N_{\mathbb{C}} \times \mathbb{P}^N_{\mathbb{C}}).$$

On the other hand, it follows from (4.138) that for any $r \geq \ell \geq 0$, for any μ, ν, κ,

$$(H_{\ell,r})^{\kappa}_{\mu,\nu} \bigwedge_{\rho=1}^{r-\ell} d\zeta_{\kappa_\rho} \in C^{\infty}_{r-\ell,0}\big(\mathbb{P}^N_{\mathbb{C}} \times \mathbb{P}^{\mathbb{N}}_{\mathbb{C}}, \sqcap^* \mathcal{O}(d^{[r]}_\nu - d^{[\ell]}_\mu - (r-\ell)) \otimes \pi^* \mathcal{O}(-(r-\ell))\big).$$

As a consequence

$$\mathscr{W}^{M-d^{[r]}_\nu}_{\mathbb{P}^N_{\mathbb{C}}} (H_{\ell,r})^{\kappa}_{\mu,\nu} \bigwedge_{\rho=1}^{r-\ell} d\zeta_{\kappa_\rho}$$
$$\in C^{\infty}_{0,r-\ell}\Big(\mathbb{P}^N_{\mathbb{C}} \times \mathbb{P}^{\mathbb{N}}_{\mathbb{C}},\, \big(\mathcal{O}(M-(r-\ell))^{\frown}\big) \otimes \mathrm{Hom}_{\mathbb{C}}\big(\pi^*\mathcal{O}(-d^{[r]}_\nu), \sqcap^*\mathcal{O}(-d^{[\ell]}_\mu)\big)\Big).$$

According to the definition of the generically exact complex $\mathbf{F}_\bullet(\mathbb{M}^P)$ through the correspondences (4.133) and (4.134), one has that

$$H_{\ell,r} \in \mathcal{L}^{-(r-\ell)}_{\mathscr{E}}\big[(\mathcal{O}(M) \otimes \mathbf{F}_\bullet(\mathbb{M}^P))^{\frown}_{\ell,r}\big]\,(\mathbb{P}^N_{\mathbb{C}} \times \mathbb{P}^N_{\mathbb{C}}).$$

It is clear that $H_{\ell,r} = 0$ when $\ell > r$. Let now $r \geq \ell$. Leibniz's rule, together with the fact that $\mathscr{W}_{\mathbb{P}^N_{\mathbb{C}}}$ is a global weight, implies that

$$\nabla_{\mathscr{D}}(H_{\ell,r})_{\mu,\nu} = \nabla^{\circ}_{\mathscr{D}}(H_{\ell,r})_{\mu,\nu} = \mathscr{W}^{M-d^{[r]}_\nu}_{\mathbb{P}^N_{\mathbb{C}}} \nabla_{\mathscr{D}}\Big(\sum_{\#\kappa = r-\ell} (H^{\dagger}_{\ell,r})^{\kappa}_{\mu,\nu}(z, \mathscr{W}_{\mathbb{P}^N_{\mathbb{C}}}\,\zeta)\, D\zeta_{\kappa}\Big),$$

where $D\zeta_\kappa = D\zeta_{\kappa_1} \wedge \cdots \wedge D\zeta_{\kappa_{r-\ell}}$. If $\nabla^{\dagger}$ denotes the nabla operator attached to the $(\mathscr{E}^{\dagger}, \Delta^{\dagger})$ Koszul complex over $\mathbb{C}^{N+1} \times \mathbb{C}^{N+1}$, we know from Lemma 4.52 that for any μ, ν with $1 \leq \mu \leq M_\ell$, $1 \leq \nu \leq M_r$,

(4.142)

$$(\nabla^{\dagger} H^{\dagger}_{\ell,r})_{\mu,\nu} = \nabla^{\dagger}\big((H^{\dagger}_{\ell,r})_{\mu,\nu}\big) = 2i\pi\Big(\big(H^{\dagger}_{\ell,r-1} F^{\dagger}_r(\zeta)\big)_{\mu,\nu} - \big(F^{\dagger}_{\ell+1}(z)\, H^{\dagger}_{\ell+1,r}\big)_{\mu,\nu}\Big)$$
$$= 2i\pi\Big(\sum_{\nu'=1}^{M_{r-1}} \big(H^{\dagger}_{\ell,r-1}\big)_{\mu,\nu'} (F^{\dagger}_r(\zeta))_{\nu',\nu} - \sum_{\mu'=0}^{M_{\ell+1}} (F^{\dagger}_{\ell+1}(z))_{\mu,\mu'} \big(H^{\dagger}_{\ell+1,r}\big)_{\mu',\nu}\Big).$$

Let us now perform on the right-hand side of (4.142) all substitutions

$$\begin{aligned} \zeta &\longleftarrow \mathscr{W}_{\mathbb{P}^N_{\mathbb{C}}}(z,\zeta)\,\zeta \quad \big(\text{in } \big(H^{\dagger}_{\ell,r-1}\big)^{\kappa}_{\mu,\nu'} \text{ or } \big(H^{\dagger}_{\ell+1,r}\big)^{\kappa}_{\mu',\nu}\big), \\ d\zeta_{\kappa_\rho} &\longleftarrow D\zeta_{\kappa_\rho} \quad \big(\text{in any } (r-\ell-1,0)\text{-form } d\zeta_\kappa \text{ involved}\big), \\ (F^{\dagger}_r(\zeta))_{\nu',\nu} &\longleftarrow (\mathscr{W}(z,\zeta))^{d^{[r]}_\nu - d^{[r-1]}_{\nu'}} (F_r(\zeta))_{\nu',\nu}. \end{aligned} \tag{4.143}$$

Let us call $(\Upsilon(z,\zeta))_{\mu,\nu}$ the expression obtained for this right-hand side once all substitutions have been performed. It follows from Leibniz's rule, combined with the fact that $\mathscr{W}_{\mathbb{P}^N_{\mathbb{C}}}$ is a global weight (hence $\nabla_{\mathscr{D}}\mathscr{W}_{\mathbb{P}^N_{\mathbb{C}}} = 0$) and with the identities (4.120), that (4.142) becomes

$$\nabla_{\mathscr{D}}\Big(\sum_{\#\kappa=r-\ell} (H^\dagger_{\ell,r})^\kappa_{\mu,\nu}(z,\mathscr{W}_{\mathbb{P}^N_{\mathbb{C}}}\,\zeta)\,D\zeta_\kappa\Big) = (\Upsilon(z,\zeta))_{\mu,\nu}.$$

Moreover, it follows from the definition of $(H_{\ell,r-1})_{\mu,\nu'}$ and $(H_{\ell+1,r})_{\mu',\nu}$ (see (4.140) and (4.141)) that (4.142) also implies for any μ,ν, once substitutions (4.142) have been performed, that

$$(\nabla_{\mathscr{D}} H_{\ell,r})_{\mu,\nu} = \mathscr{W}^{N-d^{[r]}_\nu}\Upsilon_{\mu,\nu} = 2i\pi\Big(\big(H_{\ell,r-1}\,F_r(\zeta)\big)_{\mu,\nu} - \big(F_{\ell+1}(z)\,H_{\ell+1,r}\big)_{\mu,\nu}\Big).$$

Since $(H_{\ell,\ell}(z,z))_{0,0} = \mathrm{Id}_{(\sqcap^*(\mathcal{O}(M)\otimes\mathbf{F}_\bullet(\mathbb{M}^P))_\ell)_{(z,z)}}$ because $\mathbf{H}^\dagger$ solves the $(\mathbf{F}_\bullet(\mathbb{M}^P))^\dagger$-Hefer problem and $\mathscr{W}_{\mathbb{P}^N_{\mathbb{C}}}$ is a global $\mathcal{O}(1)$-weight, Proposition 4.53 is proved. □

Proposition 4.53, once combined with Proposition 4.40, leads to the following division/interpolation theorem of Lagrange–Kronecker type.

THEOREM 4.54. *For $r \geq 1$, let the currents $R^{\overline{\mathbf{F}_\bullet(\mathbb{M}^P)}}_{0,r}$, $P^{\overline{\mathbf{F}_\bullet(\mathbb{M}^P)}}_{0,r}$ be as in Proposition 3.28 and formulae (3.83), with the Fubini–Study metric equipping each $\mathcal{O}(-d)$. Let also $\mathbf{H} = \{H_{\ell,r} : \ell \geq 0, r \geq -1\}$ be as in Proposition 4.53. If $M \in \mathbb{N}^*$ is such that the condition (4.139), depending on the complex of free $\mathbb{S}$-graded modules $\mathbb{M}^P$, is satisfied, then any global holomorphic section s of $\mathcal{O}(M-N)$ decomposes in $\mathbb{P}^N_{\mathbb{C}}$ as*

$$s(z) = 2i\pi F_P(z)\,Q(z) + R(z), \tag{4.144}$$

where

$$\begin{aligned} Q(z) &= \int_{\mathbb{P}^N_{\mathbb{C}}}\Big(\sum_{r=1}^{\min(m,N+1)} H_{1,r}(z,\cdot)\,P^{\overline{\mathbf{F}_\bullet(\mathbb{M}^P)}}_{0,r}\Big)_{N,N}\, s,\\ R(z) &= \int_{\mathbb{P}^N_{\mathbb{C}}}\Big(\sum_{r=1}^{\min(m,N)} H_{0,r}(z,\cdot)\,R^{\overline{\mathbf{F}_\bullet(\mathbb{M}^P)}}_{0,r}\Big)_{N,N}\, s. \end{aligned} \tag{4.145}$$

Moreover, if $R^{\overline{\mathbf{F}_\bullet(\mathbb{M}^P)}}_{0}s = 0$, then $s(z) = 2i\pi F_P(z)\,Q(z)$ in $\mathbb{P}^N_{\mathbb{C}}$.

CHAPTER 5

Integral closure, Briançon–Skoda type theorems

5.1. Integral closure of ideals and valuative criterion

5.1.1. Integral closure of ideals, the analytic reduced case. Let $\mathbb{A}$ be a nœtherian commutative ring and let $\mathbb{I} \subset \mathbb{A}$ be an ideal. The objective of this section is to associate to $\mathbb{I}$ an ideal $\overline{\mathbb{I}}$ lying between $\mathbb{I}$ and its radical $\sqrt{\mathbb{I}}$, in order to satisfy the following imperatives:

(1) there exists an analytically tractable test for the membership to $\overline{\mathbb{I}}$, once an explicit collection $\{a_1, \dots, a_m\}$ of generators for $\mathbb{I}$ is given;

(2) though the ideal $\overline{\mathbb{I}}$ does not reflect the complete algebraic information carried by $\mathbb{I}$, it reflects much more than the sole geometric information carried by $\sqrt{\mathbb{I}}$. Namely, there exists a uniform integer $\mu_{\mathbb{A}} \in \mathbb{N}^*$ such that for any ideal $\mathbb{I} \subset \mathbb{A}$ one has that

$$\forall\, k \in \mathbb{N}^*, \quad \overline{\mathbb{I}^{\mu_{\mathbb{A}}+k-1}} \subset \mathbb{I}^k. \tag{5.1}$$

When $\mathbb{A}$ is additionally assumed to be reduced (namely $\sqrt{\{0\}} = \{0\}$ as ideals in $\mathbb{A}$), the first candidate for such an association $\mathbb{I} \longrightarrow \overline{\mathbb{I}}$ is that of *integral closure of ideals* (see also §A.4.2.3). We encountered this notion in Proposition 3.15, where it was playing a role in the annihilation property of Bochner–Martinelli residue currents $R^s_{|\mathbb{I}|}([V])$.

Definition 5.1. Let $\mathbb{I}$ be an ideal in the commutative nœtherian ring $\mathbb{A}$. The integral closure of $\mathbb{I}$ is defined to be the ideal[1] of $\mathbb{A}$ consisting of all $h \in \mathbb{A}$ that satisfy a homogeneous integral dependency relation of the form

$$h^N + \sum_{k=1}^{N} u_k h^{N-k} = 0, \quad N \in \mathbb{N}^*, \quad u_k \in \mathbb{I}^k \quad \forall\, k = 1, \dots, N. \tag{5.2}$$

It is apparent that Definition 5.1 does not provide us in general with a suitable tractable criterion to test the membership to $\overline{\mathbb{I}}$. However, in some special reduced nœtherian rings in the analytic context, such a criterion does exist. Consider the fundamental examples of *reduced* local rings appearing in local classical complex analytic geometry. Namely, consider the local ring $\mathcal{O}_z$, when $(\mathcal{X}, \mathcal{O})$ is an N-dimensional complex manifold or a purely N-dimensional reduced complex analytic space, and $z \in \mathcal{X}$. Such a local ring is regular if and only if $z \in \mathcal{X}_{\rm reg}$. If we assume that $\mathcal{X}$ is an open subset Ω of $\mathbb{C}^N$ containing the origin 0, then the case of the local (then regular) ring $\mathbb{A}$ of interest to us will be the ring $\mathcal{O}_{\Omega,0} = \mathcal{O}_0$. On the other hand, if we assume that $\mathcal{X}$ is a purely dimensional closed analytic subset $V \subset \Omega \subset \mathbb{C}^N$ with $\operatorname{codim} V = M$, $0 \in V$, considered as an $(N-M)$-purely

[1] It is not immediate from the definition that it is an ideal; see §A.4.2.3 for an equivalent presentation clarifying this point.

dimensional reduced analytic space, its structural sheaf $\mathcal{O}_{\mathcal{X}} = \mathcal{O}_V$ equals $\mathcal{O}_\Omega/\mathcal{I}_V$, where the radical sheaf of ideals $\mathcal{I}_V = \mathcal{I}_{V,\Omega}$ is defined as the annihilator of the integration (M, M)-current $[V]$. In this setting the local ring $\mathbb{A}$ (regular or not) of interest to us will be $\mathbb{A} = (\mathcal{O}_\Omega/\mathcal{I}_V)_0 = \mathcal{O}_0/(\mathcal{I}_V)_0$. In each of these two cases, one has for any ideal $\mathbb{I}$ with a set of generators $\{a_1, \dots, a_m\}$ that

$$h \in \overline{\mathbb{I}} \Longleftrightarrow |h| \leq C|a| \quad \text{in a neighborhood of } 0$$

(5.3) $$\text{or}$$

$$h \bmod \mathcal{I}_{V,0} \in \overline{\mathbb{I}} \Longleftrightarrow |h| \leq C|a| \quad \text{on } V \text{ in a neighborhood of } 0,$$

respectively for some positive constant C and where $|\ |$ denotes any norm on $\mathbb{C}$ or $\mathbb{C}^m$. This membership criterion is called *analytic valuative criterion in the reduced case.* It can be reformulated in terms of the normalized blowup of Ω (respectively of V) along the ideal sheaf $\bigoplus_{j=1}^m a_j\, \mathcal{O}_\Omega$ (respectively the ideal sheaf $\bigoplus_{j=1}^m a_j\, (\mathcal{O}_\Omega/\mathcal{I}_V)$); see Proposition A.37. For such examples, one will show in §5.2 that (5.1) is satisfied with $\mu = \min(m, N) \leq N = \mu_{\mathbb{A}}$, when $\mathbb{A} = \mathcal{O}_0$, and therefore with

$$\mu = \min(N - M, m) \leq N - M = \dim V = \mu_{\mathbb{A}}$$

when $\mathbb{A} = (\mathcal{O}/\mathcal{I}_V)_0$ and $0 \in V_{\text{reg}}$. In the case where $0 \in V_{\text{sing}}$, such a $\mu_{\mathbb{A}}$ for which inclusions (5.1) hold does exist (see §5.3) but it depends on a locally free resolution for the quotient sheaf $\mathcal{O}/\mathcal{I}_V$ at the origin. If one considers instead of the local rings $\mathcal{O}_0$ or $(\mathcal{O}/\mathcal{I}_V)_0$ the (nonlocal) nœtherian rings

$$\mathbb{A}(U) = \mathcal{O}_\Omega(U), \quad \mathbb{A}(U) = (\mathcal{O}_\Omega/\mathcal{I}_V)(U),$$

where U is a relatively compact open subset of Ω, then the same results hold with $\mu_{\mathbb{A}} = N$ in the first case or with $\mu_{\mathbb{A}} = N - M$ in the second case provided that $U \cap V_{\text{sing}} = \emptyset$. If $V \cap V_{\text{sing}} \neq \emptyset$, then it still holds with some $\mu = \mu_{\mathbb{A},U}$.

5.1.2. The nonreduced case, valuative criterion, nœtherian operators. In the case where $\mathbb{A}$ fails to be reduced, an evaluation test similar to (5.3) is not valid anymore. For example, if $\mathbb{A} = \mathcal{O}_{\mathbb{C}^2,0}/z_2^3\, \mathcal{O}_{\mathbb{C}^2,0}$ and $\mathbb{I} = \mathbb{A}\, z_2^2$, then it would follow from (5.3) that $z_2 \in \overline{\mathbb{I}}$. But this is not the case, since z_2^N cannot belong to $\mathbb{A} z_2^{N+1}$, as it would follow from (5.3) applied in case $z_2 \in \overline{\mathbb{I}}$. The test for the membership of a given element $h \in \mathbb{A}$ to the integral closure $\overline{\mathbb{I}}$ of a given ideal $\mathbb{I}$ can be achieved in terms of the following *algebraic valuative criterion* (see, e.g., [**SwH06**, Theorem 6.8.3]).

PROPOSITION 5.2 (Valuative criterion for the membership to the integral closure of an ideal). *Let $\mathbb{I}$ be an ideal in the commutative nœtherian ring $\mathbb{A}$. An element $h \in \mathbb{A}$ belongs to $\overline{\mathbb{I}}$ if and only if, for any minimal prime ideal $\mathfrak{P}$ in $\mathbb{A}$, for any discrete valuation ring Γ_v with rank 1 lying between $\mathbb{A}/\mathfrak{P}$ and its fraction field, one has $h \in \mathbb{I}\,\Gamma_v$.*

Let us come back to the analytic setting, where $0 \subset V \subset \Omega \subset \mathbb{C}^N$ and V is a closed analytic subset with codimension M. Instead of $\mathcal{I}_V$, we take now a coherent sheaf of $\mathcal{O} = \mathcal{O}_\Omega$ ideals $\mathcal{I}$ satisfying $V = \operatorname{Supp}(\mathcal{O}/\mathcal{I})$. Let $\mathbb{D}$ be the unit disk. The morphisms $\gamma : \mathbb{D}_0 \longrightarrow (\mathcal{O}/\mathcal{I})_0$ play in the analytic context the role of discrete valuation rings with rank 1. The algebraic valuative criterion stated as Proposition 5.2 can be reformulated in $\mathbb{A} = (\mathcal{O}/\mathcal{I})_0$ as

(5.4) $$h \in \overline{\mathbb{I}} \Longleftrightarrow \forall \gamma : \mathbb{D}_0 \longrightarrow (\mathcal{O}/\mathcal{I})_0,\ \gamma^* h \in \gamma^* \mathbb{I};$$

see [**LejT08**]. A particularly interesting case occurs when $(\mathcal{O}/\mathcal{I})_0$ is purely dimensional. It means here, besides the fact that its support is purely dimensional with codimension being equal to M, that all associated primes are isolated; see Example 3.46. In other words, it means that no embedded component occurs. In such a case, germs of Coleff–Herrera currents for the support V of $\mathcal{O}/\mathcal{I}$ can be used to test the membership of a germ $h \in \mathcal{O}_0$ to $\mathcal{I}_0$, as the following result states [**And11**, Theorem 1.2 and (1.6)].

THEOREM 5.3. *Let $0 \in \Omega \subset \mathbb{C}^N$ be a neighborhood and let $\mathcal{O} = \mathcal{O}_{\Omega,0}$ be the ring of germs of holomorphic functions at the origin. Let also $\mathcal{I} \subset \mathcal{O}$ be a coherent sheaf of ideals. Assume that $(\mathcal{O}/\mathcal{I})_0$ has pure codimension equal to M and denote it by $V = \mathrm{Supp}(\mathcal{O}/\mathcal{I})$. If $h \in \mathcal{O}_0$, then one has*

$$h \in \mathcal{I}_0 \Longleftrightarrow \big(\forall T_0 \in (\mathrm{CH}^0_{\Omega,V})_0,\ \mathcal{I}_0 \cdot T_0 = 0 \Longrightarrow h\,T_0 = 0\big). \tag{5.5}$$

PROOF. Let $0 < \eta \ll 1$ be such that there exists a generically exact complex of trivial bundles

$$\mathbf{F}_\bullet : 0 \longrightarrow E_L = \mathbb{C}^{r_L} \xrightarrow{F_L} E_{L-1} = \mathbb{C}^{r_{L-1}} \xrightarrow{F_{L-1}} \cdots \xrightarrow{F_2} E_1 = \mathbb{C}^{r_1} \xrightarrow{F_1} E_0 = \mathbb{C}$$

over $\eta\,\mathbb{B}_N$, where $\mathbb{B}_N$ is the euclidean unit ball in $\mathbb{C}^N$, with the morphism F_1 being generically surjective, such that the corresponding sequence of sheaves

$$0 \longrightarrow \mathcal{O}^{\oplus r_L}_{\eta\,\mathbb{B}_N} \xrightarrow{F_L} \mathcal{O}^{\oplus r_{L-1}}_{\eta\,\mathbb{B}_N} \xrightarrow{F_{L-1}} \cdots \xrightarrow{F_2} \mathcal{O}^{\oplus r_1}_{\eta\,\mathbb{B}_N} \xrightarrow{F_1} \mathcal{O}_{\eta\,\mathbb{B}_N}$$

is exact with $\operatorname{coker} F_1 = (\mathcal{O}_\Omega/\mathcal{I})_{|\eta\mathbb{B}_N}$; see §A.2.3. Moreover, one has necessarily $L \geq M$; see Theorem A.9 and Proposition A.17. Let $\overline{\mathbf{F}_\bullet}$ be a metrized version of the generically exact complex $\mathbf{F}_\bullet$ and let

$$R^{\overline{\mathbf{F}_\bullet}} = R^{\overline{\mathbf{F}_\bullet}}_0 = \sum_{r=M}^{r=\min(L,M)} R^{\overline{\mathbf{F}_\bullet}}_{0,r}$$

be the pseudo-meromorphic Bochner–Martinelli current attached to $\overline{\mathbf{F}_\bullet}$ as in §3.3.2. For any holomorphic section s of E^*_M such that $F^*_{M+1}s = 0$, one has that

$$\overline{\partial}(sR^{\overline{\mathbf{F}_\bullet}}_{0,M}) = \pm s\,\overline{\partial}R^{\overline{\mathbf{F}_\bullet}}_{0,M} = \pm s\,F_{M+1}R^{\overline{\mathbf{F}_\bullet}}_{0,M} = \pm F^*_{M+1}s\,R^{\overline{\mathbf{F}_\bullet}}_{0,M} = 0.$$

The derivation of the above equality takes into account the $\mathbb{F}_2$-grading (3.49), the definition of the $\boldsymbol{\nabla}_{\mathbf{F}_\bullet}$ operator (see (3.50)) and the fact that $\boldsymbol{\nabla} R^{\overline{\mathbf{F}_\bullet}} = 0$ (Proposition 3.28). This implies that the $\mathbb{C}$-valued current $sR^{\overline{\mathbf{F}_\bullet}}_{0,M}$ is $\overline{\partial}$-closed. Since it is pseudo-meromorphic, with bidegree $(0, M)$ and has support included in the closed analytic subset V with codimension M, one has that $sR^{\overline{\mathbf{F}_\bullet}}_{0,M} \in \mathrm{CH}^0_{\Omega,V}(\eta\,\mathbb{B}_N)$; see Theorem 2.39. Let $h \in \mathcal{O}(\eta\,\mathbb{B}_N)$, which satisfies

$$\forall\, s \in \mathcal{O}(\eta\,\mathbb{B}_N, E^*_M) \text{ such that } F^*_{M+1}s = 0, \quad h(sR^{\overline{\mathbf{F}_\bullet}}_{0,M}) = 0. \tag{5.6}$$

Let us observe that the sheaf $\mathcal{O}/\mathcal{I}$ is Cohen–Macaulay at any generic point z which belongs to $V_{\mathrm{reg}} \cap \eta\,\mathbb{B}_N$. Actually, at such a generic point z the ring $(\mathcal{O}/\mathcal{I})_z$ is a free $(\mathcal{O}/\sqrt{\mathcal{I}})_z$-module $((\mathcal{O}/\sqrt{\mathcal{I}})_z)^{\oplus r_z}$, with $(\mathcal{O}/\sqrt{\mathcal{I}})_z$ being Cohen–Macaulay.[2] Take any such regular point $z \in V_{\mathrm{reg}} \cap \eta\,\mathbb{B}_N$. Since $F^*_{M+1} = 0$ there, the condition on the left-hand side of (5.6) is fulfilled for any $s \in \mathcal{O}(\eta\,\mathbb{B}_N, E^*_M)$. Then the

[2]Such a generic point of V_{reg} is called a *regular* point of the quotient sheaf $\mathcal{O}/\mathcal{I}$. The fact that $\mathcal{O}/\mathcal{I}$ has pure dimension is irrelevant here since one can take the generic point outside the support of each embedded component.

relation (5.6) implies $R^{\overline{\mathbf{F}_\bullet}}_{0,M}h = R^{\overline{\mathbf{F}_\bullet}}h = 0$ in a neighborhood of z; see Proposition 3.40. This means that $h_z \in \mathcal{I}_z$ as a consequence of the duality theorem (Theorem 3.38). Since the current $R^{\overline{\mathbf{F}_\bullet}}_{0,M}h$ is pseudo-meromorphic with bidegree $(0,M)$ and is supported by the closed analytic set V with pure codimension M, it inherits the standard extension property across any closed hypersurface of V. Thus, one has that $R^{\overline{\mathbf{F}_\bullet}}_{0,M}h = 0$ in $\eta\,\mathbb{B}_N$. Let $\mathcal{P} \in \mathrm{Ass}(\mathcal{O}_\Omega/\mathcal{I})(\eta\,\mathbb{B}_N)$ and

$$R^{\overline{\mathbf{F}_\bullet}}_{|V(\mathcal{P})} = \mathbf{1}_{V(\mathcal{P})\setminus\bigcup_{\mathcal{P}'\supset\mathcal{P}}V(\mathcal{P}')}\cdot R^{\overline{\mathbf{F}_\bullet}} = \mathbf{1}_{\eta\,\mathbb{B}_N\setminus\bigcup_{\mathcal{P}'\supset\mathcal{P}}V(\mathcal{P}')}\cdot\big(\mathbf{1}_{V(\mathcal{P})}\cdot R^{\overline{\mathbf{F}_\bullet}}\big).$$

Then (see [**AndW10**, Theorem 5.1] for details) one has, in terms of annihilators of pseudo-meromorphic currents, that

$$\bigcap_{\mathcal{P}\in\mathrm{Ass}(\mathcal{O}_\Omega/\mathcal{I})(\eta\,\mathbb{B}_N)} \mathrm{Ann}\,(R^{\overline{\mathbf{F}_\bullet}}_{|V(\mathcal{P})})_0,$$

gives a primary decomposition of $(\mathcal{O}/\mathcal{I})_0$. Moreover, for each associated prime $\mathcal{P}$ (here with codimension M by hypothesis), the annihilator of the germs of pseudo-meromorphic currents $(R^{\overline{\mathbf{F}_\bullet}}_{0,M})_{|\mathcal{P}}$ and $(R^{\overline{\mathbf{F}_\bullet}}_0)_{|\mathcal{P}}$ do coincide; see [**AndW10**, Lemma 5.1]. It follows from these observations that the annihilators of the germs of currents $(R^{\overline{\mathbf{F}_\bullet}})_0 = (R^{\overline{\mathbf{F}_\bullet}}_0)_0$ and $(R^{\overline{\mathbf{F}_\bullet}}_{0,M})_0$ do coincide in the case where $\mathcal{O}/\mathcal{I}$ is purely dimensional. We thus have proved that if conditions (5.6) hold, then $(h\,R^{\overline{\mathbf{F}_\bullet}})_0 = 0$. This implies that $h \in \mathcal{I}_0$ thanks to the duality theorem, Theorem 3.38. Moreover, if $h \in \mathcal{I}_0$ and $T_0 \in (\mathrm{CH}^0_{\Omega,V})_0$ is such that $\mathcal{I}_0 T_0 = 0$, then one has in particular that $h\,T_0 = 0$. This ends the proof of Theorem 5.3. □

Remark 5.4. Since $(\mathrm{CH}^0_{\Omega,V})_0$ is finitely generated as an $(\mathcal{O}_\Omega/\mathcal{I})_0$-module, one can use a finite set of generators μ_ι to test the membership of any $h \in \mathcal{O}_0$ to $\mathcal{I}_0$; see (2.37).

Theorem 5.3 implies the fundamental result, known as the Ehrenpreis–Palamodov (local) fundamental principle; see [**Pal**] or [**Bj74**].[3]

Theorem 5.5. *Let $\mathcal{I}$ be a coherent sheaf of $\mathcal{O} = \mathcal{O}_\Omega$ ideals in a neighborhood $\Omega \subset \mathbb{C}^N$ of the origin satisfying*

$$\mathcal{I}_0 = \bigcap_\iota (\mathcal{J}_\iota)_0.$$

Here $\mathcal{J}_\iota$ are coherent sheaves of ideals in $\eta\,\mathbb{B}_N$, where $0 < \eta \ll 1$ is such that each $\mathcal{O}_0$-module $(\mathcal{O}/\mathcal{J}_\iota)_0$ has pure codimension M_ι and $M_\iota \neq M_{\iota'}$ as soon as $\iota \neq \iota'$. For each index ι, there exists a finite collection of germs of holomorphic differential operators

$$\mathcal{N}_{\iota,\nu}\big(\zeta, \frac{\partial}{\partial\zeta}\big) = \sum_{\{\kappa\in\mathbb{N}^N\,:\,|\kappa|\leq d_{\iota,\nu}\}} a^{\kappa}_{\iota,\nu}(\zeta)\left(\frac{\partial}{\partial\zeta}\right)^{\kappa}$$

at the origin such that for any $h \in \mathcal{O}_0$ the following is valid:

$$h \in \mathcal{I}_0 \iff \big(\forall\,\iota,\ \forall\,\nu,\ \mathcal{N}_{\iota,\nu}(h) = 0 \text{ on } \big(\mathrm{Supp}\,\mathcal{O}/\mathcal{J}_\iota\big)_0\big). \tag{5.7}$$

Proof. For each index ι, let $V_\iota = \mathrm{Supp}\,\mathcal{O}/\mathcal{J}_\iota$. Since $h \in \mathcal{I}_0$ if and only if $h \in (\mathcal{J}_\iota)_0$ for all different ι and each $\mathcal{O}_0$-module $(\mathcal{O}/\mathcal{J}_\iota)_0$ is purely dimensional, one

[3] We plan to discuss semilocal as well as global versions of such principles in §7.6.3.

can apply Theorem 5.3 and Remark 5.4 with $\mathcal{I} = \mathcal{J}_\iota$ for each ι. Thus, one gets for each ι a finite collection $\{(T_{\iota,\ell_\iota})_0\}_{\iota,\ell_\iota}$ of elements in $(\mathrm{CH}^0_{\eta\mathbb{B}_N,V_\iota})_0$ such that

$$h \in \mathcal{I}_0 \Longleftrightarrow \big(\forall\,\iota,\ \forall\,\ell_\iota,\quad h(T_{\iota,\ell_\iota})_0 = 0\big).$$

We conclude the proof by invoking the local structure of Coleff–Herrera currents; see Proposition 2.37. An alternative proof based on Hilbert's nullstellensatz can be found in [**And06B**, Theorem 2.2]. □

Classically, the local version of the fundamental principle is formulated as follows: for each associated prime $\mathcal{P}$ of the $\mathcal{O}_0$-module $(\mathcal{O}/\mathcal{I})_0$ there exists a finite collection

$$\{\mathcal{N}_{\mathcal{P},\nu_\mathcal{P}}(\zeta,\partial/\partial\zeta)\}_{\mathcal{P}\in\mathrm{Ass}\,(\mathcal{O}/\mathcal{I})_0}$$

of germs of holomorphic differential operators at the origin such that for any $h \in \mathcal{O}_0$,

$$h \in \mathcal{I}_0 \Longleftrightarrow \big(\forall\,\mathcal{P} \in \mathrm{Ass}((\mathcal{O}/\mathcal{I})_0),\ \forall\,\nu_\mathcal{P},\ \mathcal{N}_{\mathcal{P},\nu_\mathcal{P}}(h) = 0 \text{ on } (V(\mathcal{P}))_0\big). \tag{5.8}$$

DEFINITION 5.6. A finite collection $\mathcal{N} = \{\mathcal{N}_{\mathcal{P},\nu_\mathcal{P}}(\zeta,\partial/\partial\zeta)\,:\,\mathcal{P} \in \mathrm{Ass}\,(\mathcal{O}/\mathcal{I})_0\}$ such that (5.8) holds is called *a collection of local nœtherian operators* for the $\mathcal{O}_0$-module $(\mathcal{O}/\mathcal{I})_0$.

Given Theorem 5.3, it is natural to address the following question. In case the $\mathcal{O}_0$-module $\mathbb{A} = (\mathcal{O}/\mathcal{I})_0$ is purely dimensional with $V = \mathrm{Supp}\,\mathcal{O}/\mathcal{I}$ in a neighborhood of 0, does there exist a finite collection $\mathcal{N}^\mathbb{A} = \{\mathcal{N}^\mathbb{A}_\nu(\zeta,\partial/\partial\zeta)\}_\nu$ of germs of differential operators with holomorphic coefficients at the origin such that

(1) the assertion

$$\big(\forall\,\nu,\ \mathcal{N}^\mathbb{A}_\nu(h) = 0 \text{ on } \big(\mathrm{Supp}\,\mathcal{O}/\mathcal{I}\big)_0\big) \Longrightarrow h \in \mathcal{I}_0 \tag{5.9}$$

holds;

(2) the construction of $\mathcal{N}^\mathbb{A}$ is tractable analytically, starting from a generating system $\{f_1,\dots,f_r\}$ for $\mathcal{I}_0$ in $\mathcal{O}_0$;

(3) there exists a uniform integer $\mu_\mathbb{A} \in \mathbb{N}^*$, such that, for any ideal $\mathbb{I}$ of $\mathbb{A}$ with $\{a_1,\dots,a_m\}$ mod $\mathcal{I}_0$ as a set of generators and for any $k \in \mathbb{N}^*$, one has for $(h)_0 \in \mathcal{O}_0$

$$\big(\max_\nu |\mathcal{N}^\mathbb{A}_\nu(h)| \le C|a|^{\mu_\mathbb{A}+k-1} \text{ around 0 on } V\big) \Longrightarrow (h)_0 \in \mathbb{I}^k\,? \tag{5.10}$$

We will sketch an approach to this question in §5.4.

5.2. Briançon–Skoda theorem in $\mathbb{A} = \mathcal{O}_0$

In this section, we focus our interest on the case where $\mathbb{A} = \mathcal{O}_0 := \mathcal{O}_{\mathbb{C}^N,0}$, since in this case one can take $\mu_\mathbb{A} = N = \dim\mathcal{O}_0$.

THEOREM 5.7 (Briançon–Skoda theorem).[4] *If $\mathbb{I} \subset \mathbb{A} = \mathcal{O}_0$ is an ideal, then one has that $\overline{\mathbb{I}^{N+k-1}} \subset \mathbb{I}^k$ for any $k \in \mathbb{N}^*$.*

[4]This is a fundamental result concerning applications. In its sharp form, as given in Theorem 5.8, it was proved by Joël Briançon and Henri Skoda [**BriS74**]. The original proof of this result, which goes back to 1974, is based on L^2 methods introduced by Lars Hörmander (see [**Hor73**]) in multivariate complex analysis.

PROOF. The proof that we propose here relies on the approach towards the construction of a weight attached to the Eagon–Northcott complex $\mathbf{EN}_{\bullet}^{a,k}$, described in §4.6.1, following Proposition 4.47 and its proof. Let $\mathbb{B}_N$ be the unit euclidean ball in $\mathbb{C}^N$ and $0 < \eta \ll 1$. Let $\mathscr{W}_{a,k}$ be the global weight in $(\eta\,\mathbb{B}_N \setminus V(a)) \times \eta\,\mathbb{B}_N$ defined for a complex parameter λ with sufficiently large real part $\operatorname{Re}\lambda \gg 1$ by

$$\mathscr{W}_{a,k} = \big[\mathscr{W}_{a,k}^{\lambda}\big]_{\lambda=0},$$

where the current valued map $\mathscr{W}_{a,k}^{\lambda}$ is defined by (4.109), (4.110), and (4.111) when assuming that $f = a$. Let $\{\epsilon_1, \dots, \epsilon_m\}$ be the canonical basis for the trivial bundle $\mathbb{C}^m$ over $\eta\,\mathbb{B}_N$. If $a = \sum_{j=1}^m a_j \epsilon_j$, then let H_a be corresponding Hefer form H_a in the pseudo-convex domain $\eta\,\mathbb{B}_N \times \eta\,\mathbb{B}_N$; see Definition 4.46. Let also $a^* = \sum_{j=1}^m \overline{a_j}\,\epsilon_j^*$ be the conjugate section of a, and let $a^\kappa = a_1^{\kappa_1} \cdots a_m^{\kappa_m}$ for any $\kappa \in \mathbb{N}^m$. Then one has that

$$\mathscr{W}_{a,k}(z,\cdot) = \rho_{a,k}(z,\cdot) + \sum_{\{\kappa \in \mathbb{N}^m \,:\, |\kappa| = k\}} a^\kappa(z)\, T_{a,\kappa}^{\kappa}(z,\cdot). \tag{5.11}$$

In the above relation

(5.12)

$$\rho_{a,k}(z,\cdot) = \big[\rho_{a,k}^{\lambda}(z,\cdot)\big]_{\lambda=0} = \Big[\sum_{r=1}^{N} \sum_{r'=r}^{r+k-1} \binom{r'}{r} \Big(\frac{a(z)\,a^*}{\|a\|^2}\Big)^{r'-r} (\overline{\partial}\|a\|^{2\lambda})\, \mathscr{R}_{a,r}\Big]_{\lambda=0},$$

$$\mathscr{R}_{a,r}(z,\cdot) = \Big(H_a \frac{a^*}{2i\pi\|a\|^2}\Big) \Big(\overline{\partial}\Big(H_a \frac{a^*}{2i\pi\|a\|^2}\Big)\Big)^{r-1} \quad \text{for } r \geq 1,$$

and each $T_{a,\kappa}(z,\cdot)$ for $z \in \eta\,\mathbb{B}_N$, $\kappa \in \mathbb{N}^m$ with $|\kappa| = k$ is a pseudo-meromorphic current in $\eta\,\mathbb{B}_N$ depending holomorphically on z and supported by $V(a)$. Let also $\chi_\eta \in \mathcal{D}(\eta\,\mathbb{B}_N)$ be identically equal to 1 around $(\eta/2)\,\overline{\mathbb{B}_N}$ and let $q_\eta = q_{(\eta/2)\,\mathbb{B}_N, \chi_\eta}$ be as in Proposition 4.33. It follows Proposition 4.33, repeating the argument used in the proof of Proposition 4.36 and presented in the computations next to the proof of Proposition 4.47, that the following integral representation formula holds for h in the open ball $(\eta/2)\,\mathbb{B}_N$:

(5.13)

$$\begin{aligned} h(z) &= \int_\Omega \Big(\Big(\chi_\eta + \frac{\overline{\partial}_\zeta((1-\chi_\eta)\,q_\eta(z,\cdot))}{2i\pi}\Big)^{N+1} \mathscr{W}_{a,k}(z,\cdot)\Big)_{N,N} h \\ &= \int_{\eta\,\mathbb{B}_N} \Big(\Big(\chi_\eta + \frac{\overline{\partial}_\zeta((1-\chi_\eta)\,q_\eta(z,\cdot))}{2i\pi}\Big)^{N+1} \rho_{a,k}(z,\cdot)\Big)_{N,N} h \\ &\qquad\qquad \mod \Big(\sum_{j=1}^m a_j\, H((\eta/2)\,\mathbb{B}_N)\Big)^k. \end{aligned}$$

As is pointed out, the validity of the (5.13) is established for $z \in (\eta/2)\,\mathbb{B}_N \setminus V(a)$ and then for all $z \in (\eta/2)\,\mathbb{B}_N$, thanks to the analytic continuation principle. Let $\widehat{\pi} : \widehat{\mathscr{X}} \to \eta\,\mathbb{B}_N$ be the normalized blowup of $\eta\,\mathbb{B}_N$ along the $\mathcal{O}$-ideal sheaf $\sum_1^m \mathcal{O}\,a_j$. Let also $\widehat{L} = \mathcal{O}(\widehat{D})$ be the line bundle over the complex normal analytic space $\widehat{\mathscr{X}}$, which corresponds to its exceptional divisor $\widehat{D}$. One has that

$$\widehat{\pi}_\zeta^*\Big(H_a \frac{a^*}{2i\pi\|a\|^2}\Big) = \frac{\sigma(z,\cdot)}{e}$$

on $(\eta/2)\,\mathbb{B}_N \times (\widehat{\mathcal{X}} \setminus |\widehat{D}|)$, where e is a section of $\widehat{L}$ and σ is a smooth section of the line bundle $\pi^*\widehat{L}$. Note that the projection $\pi\ :\ (z,\widehat{\zeta}) \in (\eta/2)\,\mathbb{B}_N \times \widehat{\mathcal{X}} \to \widehat{\zeta} \in \widehat{\mathcal{X}}$ is holomorphic in z. Thus, for any $r = 1,\dots,N$ and $\lambda \in \mathbb{C}$ satisfying $\operatorname{Re}\lambda \gg 1$, one has that

$$(5.14)\qquad \begin{aligned} \widehat{\pi}_\zeta^*\Big((\overline{\partial}\|a\|^{2\lambda})\,\mathscr{R}_{a,r}(z,\cdot)\Big) &= (\overline{\partial}\|e\,u\|^{2\lambda})\,\frac{\sigma(z,\cdot)\,\big(\overline{\partial}\sigma(z,\cdot)\big)^{r-1}}{e^r} \\ &= \lambda\,\|eu\|^{2\lambda}\,(\overline{\partial}\log\|eu\|^2)\,\frac{\sigma(z,\cdot)\,\big(\overline{\partial}\sigma(z,\cdot)\big)^{r-1}}{e^r}, \end{aligned}$$

where u is a nonvanishing section of $\widehat{L}^{-1}$. For any $r' = r+1,\dots,r+k-1$ (when $k \ge 2$), it holds in $(\eta/2)\,\mathbb{B}_N \times (\widehat{\mathcal{X}} \setminus |\widehat{D}|)$ that

$$(5.15)\qquad \widehat{\pi}_\zeta^*\Big(\frac{a(z)\,a^*}{\|a\|^2}\Big)^{r'-r} = \frac{\tau_{r,r'}(z,\cdot)}{e^{r'-r}},$$

where $\tau_{r,r'}$ is a smooth section of $\pi^*\widehat{L}^{r'-r}$ that is holomorphic in z. Combining formulae (5.14) with (5.15), one gets for any $r = 1,\dots,N$ and any $r' = r,\dots,r+k-1$ that

$$(5.16)\qquad \pi_\zeta^*\Big(\Big(\frac{a(z)\,a^*}{\|a\|^2}\Big)^{r'-r}(\overline{\partial}\|a\|^{2\lambda})\,\mathscr{R}_{a,r}(z,\cdot)\Big) = \lambda\,\|eu\|^{2\lambda}\,(\overline{\partial}\log\|eu\|^2)\,\frac{\sigma_{r,r'}(z,\cdot)}{e^{r'}},$$

where $\sigma_{r,r'}$ is a smooth section of $\pi^*\widehat{L}^{r'}$, which is holomorphic in z. Let $\widehat{z}^0 \in |\widehat{D}|$, and let $\mathfrak{e}$, $\mathfrak{u}$ be the coordinates of e (correspondingly of u) with respect to the local frame ϵ^* (correspondingly, with respect to its dual frame ϵ) for $\widehat{L}$ around $\widehat{z}^0$, so that

$$(5.17)\qquad \overline{\partial}\log\|eu\|^2 = \check{\theta} + \frac{\overline{\partial\mathfrak{e}}}{\overline{\mathfrak{e}}} + \frac{\overline{\partial\mathfrak{u}}}{\overline{\mathfrak{u}}} = \theta + \frac{\overline{\partial\mathfrak{e}}}{\overline{\mathfrak{e}}}.$$

Here θ is a smooth $(0,1)$-form in a neighborhood of $\widehat{z}^0$. If $h \in \mathbb{I}^{N+k-1}$ admits a representative $\tilde{h}$ in $\eta\,\mathbb{B}_N$ such that $|\tilde{h}| \le C|a|^{N+k-1}$, then $\widehat{\pi}^*h/\mathfrak{e}^{r'}$ defines at $\widehat{z}^0$ an element in $\mathcal{O}_{\widehat{\mathcal{X}},\widehat{z}^0}$ since $\widehat{\mathcal{X}}$ is normal for any $r' = r,\dots,r+k-1$, where $r = 1,\dots,N$. Then, it follows from (5.16), together with (5.17), that

$$\widehat{\pi}^*h\,\Big[\pi_\zeta^*\Big(\Big(\frac{a(z)\,a^*}{\|a\|^2}\Big)^{r'-r}(\overline{\partial}\|a\|^{2\lambda})\,\mathscr{R}_{a,r}(z,\cdot)\Big)\Big]_{\lambda=0} = 0$$

in a neighborhood of $\{z\} \times \{\widehat{z}^0\}$ in $(\eta/2)\,\mathbb{B}_N \times \widehat{\mathcal{X}}$. This can be seen by taking, for example, a log resolution above a neighborhood of $\widehat{z}^0$ for a hypersurface containing $\widehat{\mathcal{X}}_{\text{sing}}$ and $|\widehat{D}|$. Since this is true for any $\widehat{z}^0$ in $|\widehat{D}|$, one has that

$$\Big[\int_{\eta\,\mathbb{B}_N}\Big(\Big(\chi_\eta + \frac{\overline{\partial}_\zeta\big((1-\chi_\eta)\,q_\eta(z,\cdot)\big)}{2i\pi}\Big)^{N+1}\rho_{a,k}^\lambda(z,\cdot)\Big)_{N,N}h\Big]_{\lambda=0} = 0.$$

Then (5.13) implies $(h)_0 \in \mathbb{I}^k$. This concludes the proof of Theorem 5.7. □

In the proof of Theorem 5.7, one could have taken into account the number of generators $a_1,\dots,a_m$ of the ideal $\mathbb{I}$. Namely, one has for any $r \ge m$

$$(5.18)\qquad (\overline{\partial}\|a\|^{2\lambda})\,\mathcal{R}_{a,r}(z,\cdot) = \lambda\,\|a\|^{2\lambda}\,\frac{\overline{\partial}\|a\|^2}{\|a\|^2}\Big(\frac{H_a(z,\cdot)\,a^*}{2i\pi\|a\|^2}\Big)\,\frac{\big(H_a(z,\cdot)\,\overline{\partial}a^*\big)^{r-1}}{\|a\|^{2(r-1)}} = 0.$$

Therefore, one has in Theorem 5.7 that $\overline{\mathbb{I}^{\min(m,N)+k-1}} \subset \mathbb{I}^k$. Let us present a variation of the above proof still based on Propositions 4.29 and 4.43, but using in place of the weight $\mathscr{W}_{a,k}$, the weight

$$\mathscr{W}_{\mathbf{K}^a_\bullet,k} = 2i\pi\,\big((\imath_{a(z)})_1 \wedge \cdots \wedge (\imath_{a(z)})_k\big) H_1^{\mathbf{EN}^{a,k}_\bullet}(z,\cdot) P_0^{\overline{\mathbf{EN}^{a,k}_\bullet}} + H_0^{\mathbf{EN}^{a,k}_\bullet}(z,\cdot)\, R_0^{\overline{\mathbf{EN}^{a,k}_\bullet}}.$$

We recall that the Eagon–Northcott complex $\mathbf{EN}^{a,k}_\bullet$ over $\eta\,\mathbb{B}_N$, for $0 < \eta \ll 1$, appearing here is build from k copies $\{(\mathbf{K}^a_\bullet, (\lrcorner_a)_\kappa),\ \kappa = 1,\dots,k\}$ of the Koszul complex $\mathbf{K}^a_\bullet$ as described in Example 3.26. We will denote it for the sake of simplicity by $\mathbf{E}^{a,k}_\bullet$. It is metrized as $\overline{\mathbf{E}}^{a,k}_\bullet := \overline{\mathbf{E}^{a,k}_\bullet}$ by taking the canonical metric on any trivial bundle over $\eta\,\mathbb{B}_N$, and

$$\mathbf{H}^{\mathbf{E}^{a,k}_\bullet} = \{H_{\ell,r}^{\mathbf{E}^{a,k}_\bullet} : \ell \geq 0,\ r \geq -1\}$$

is the solution of the $\mathbf{E}^{a,k}_\bullet$-Hefer problem deduced by using plain linear algebra from the solution we gave for the $\mathbf{K}^a_\bullet$-Hefer problem (that is when $k = 1$) in Proposition 4.47. Thus, we are ready to state the following more precise result [**BriS74, And06C**]:

THEOREM 5.8. *Let $k \in \mathbb{N}^*$ and $(h)_0 \in \overline{\mathbb{I}^{\min(m,N)+k-1}}$, where $\mathbb{I} \subset \mathbb{A} = \mathcal{O}_0$ is an ideal, whose minimal number of generators equals m. Then $(h)_0 \in \mathbb{I}^k$. More precisely, for $0 < \eta \ll 1$, h admits in $(\eta/2)\,\mathbb{B}_N$ as representative of the mapping*

$$z \mapsto 2i\pi\big((\lrcorner_{a(z)})_1 \wedge \cdots \wedge (\lrcorner_{a(z)})_k\big) q(z), \tag{5.19}$$

where

$$q(z) = \int_{\eta\,\mathbb{B}_N} \Big(\Big(\chi_\eta + \frac{\overline{\partial}_\zeta\big((1-\chi_\eta)\,q_\eta(z,\cdot)\big)}{2i\pi}\Big)^{N+1} H_1^{\mathbf{E}^{a,k}_\bullet}(z,\cdot) P^{\overline{\mathbf{E}}^{a,k}_\bullet}\Big)_{N,N} h. \tag{5.20}$$

In (5.20) *the support function $\chi_\eta \in \mathcal{D}(\eta\,\mathbb{B}_N)$ equals identically* 1 *around $(\eta/2)\,\overline{\mathbb{B}_N}$ and the* $(1,0)$*-form q_η is defined as $q_{(\eta/2)\,\mathbb{B}_N,\chi_\eta}$ in Proposition* 4.33, *when taking $\omega_\eta = (\eta/2)\,\mathbb{B}_N$.*

PROOF. For $0 < \eta \ll 1$, let $\{a_1,\dots,a_m\}$ be representatives in $\mathcal{O}(\eta\,\mathbb{B}_N)$ of elements generating minimally the ideal $\mathbb{I}$. It follows from Propositions 4.29, 4.33, and 4.43 that proving this result boils down to proving that

$$(h)_0 \in \overline{\mathbb{I}^{\min(N,m)+k-1}} \Longrightarrow R_0^{\overline{\mathbf{E}}^{a,k}_\bullet} h = 0.$$

For $\kappa = 1,\dots,k$, let $\{(\epsilon_1)_\kappa,\dots,(\epsilon_m)_\kappa\}$ be the copy of the canonical basis of $\mathbb{C}^m$ and let

$$(\sigma)_\kappa = \sum_{j=1}^m \frac{\overline{a}_j}{\|a\|^2}\,(\epsilon_j^*)_\kappa \in \mathcal{O}(\eta\,\mathbb{B}_N, (\mathbb{C}^m)^*),\ \kappa = 1,\dots,k.$$

We recall (see Example 3.31) that

(5.21)

$$\begin{aligned}
R_0^{\overline{\mathbf{E}}_\bullet^{a,k}} &= \sum_{r=1}^{N} R_{0,r}^{\overline{\mathbf{E}}_\bullet^{a,k}}\\
&= \sum_{r=1}^{\min(N,m)} \frac{1}{(2i\pi)^r}\Big[(\overline{\partial}|a|^{2\lambda})\,(\sigma)_1\cdots(\sigma)_k \sum_{\{\rho\in\mathbb{N}^k\,:\,|\rho|=r-1\}} \bigwedge_{\kappa=1}^{k}\Big(\overline{\partial}\Big(\frac{(\sigma)_\kappa}{2i\pi}\Big)\Big)^{\rho_\kappa}\Big]_{\lambda=0},\\
P_0^{\overline{\mathbf{E}}_\bullet^{a,k}} &= \sum_{r=1}^{N+1} P_{0,r}^{\overline{\mathbf{E}}_\bullet^{a,k}}\\
&= \sum_{r=1}^{\min(N+1,m)} \frac{1}{(2i\pi)^r}\Big[|a|^{2\lambda}\,(\sigma)_1\cdots(\sigma)_k \sum_{\{\rho\in\mathbb{N}^k\,:\,|\rho|=r-1\}} \bigwedge_{\kappa=1}^{k}\Big(\overline{\partial}\Big(\frac{(\sigma)_\kappa}{2i\pi}\Big)\Big)^{\rho_\kappa}\Big]_{\lambda=0}.
\end{aligned}$$

Let, as in the proof of Theorem 5.7, $\widehat{\pi} : \widehat{\mathcal{X}} \to \eta\,\mathbb{B}_N$ be the normalized blowup of $\eta\,\mathbb{B}_N$ along the $\mathcal{O}$ ideal sheaf $\sum_1^m \mathcal{O}\,a_j$ and let $\widehat{L} = \mathcal{O}(\widehat{D})$ be the line bundle over $\widehat{\mathcal{X}}$ corresponding to its exceptional divisor $\widehat{D}$. For any $r = 1,\dots,\min(N,m)$, one has in $\widehat{\mathcal{X}} \setminus |\widehat{D}|$ that

$$\widehat{\pi}^*\Big(\frac{(\sigma)_1\cdots(\sigma)_k}{(2i\pi)^k} \sum_{\{\rho\in\mathbb{N}^k\,:\,|\rho|=r-1\}} \bigwedge_{\kappa=1}^{k}\Big(\overline{\partial}\Big(\frac{(\sigma)_\kappa}{2i\pi}\Big)\Big)^{\rho_\kappa}\Big) = \frac{\Sigma_r}{e^{\min(N,m)+k-1}}, \tag{5.22}$$

where Σ_r is a smooth $\widehat{L}^{\min(N,m)+k-1} \otimes \widehat{\pi}^*\big(\bigwedge^r(\mathbb{C}^m)^*\big)$-valued form. At this stage, we repeat the argument from (5.16) used in the proof of Theorem 5.7. Since h admits a representative in $\eta\,\mathbb{B}_N$ satisfying

$$|\widehat{\pi}^* h| \le C|e|^{\min(N,m)+k-1},$$

we conclude, if $\mathfrak{e}$ is the local coordinate of e in a local frame in a neighborhood of a point $\widehat{z}^0 \in |\widehat{D}|$, that the bounded meromorphic function $\widehat{\pi}^* h/\mathfrak{e}^{\min(N,m)+k-1}$ is regular in a neighborhood of $\widehat{z}^0$ on the normal complex analytic space $\mathcal{X}$, and that

$$\widehat{\pi}^* h\,\Big[(\overline{\partial}|\widehat{\pi}^* a|^{2\lambda})\,\frac{\Sigma_r}{e^{\min(N,m)+k-1}}\Big]_{\lambda=0} = 0$$

is a current in a neighborhood of $\widehat{z}^0$, for the same reason as in the proof of Theorem 5.7. It follows that

$$R_{0,r}^{\overline{\mathbf{E}}_\bullet^{a,k}}\, h = 0$$

for $r = 0,\dots,\min(N,m)$, which concludes the proof of Theorem 5.8. □

Let us consider the case where $\mathbb{I}$ admits a finite system of generators, whose cardinal equals its codimension. It means that if $\mathbb{I}$ is generated by a regular sequence $(a_1,\dots,a_m)$ in the local ring $\mathcal{O}_0$, then the generically exact Eagon–Northcott complex $\mathbf{E}_\bullet^{a,k}$ induces, at the level of locally free sheaves of bundle sections, a free resolution for the Cohen–Macaulay $\mathcal{O}_0$-module $(\mathcal{O}/(\sum_1^m \mathcal{O}\,a_j)^k)_0$. This occurs for each integer $k \in \mathbb{N}^*$. Then, it follows from the uniqueness of a minimal, finite, free resolution for $\mathcal{F}_z$ (see Theorem A.9) together with the results presented in §3.4.4 (see in particular Theorem 3.52) that

$$R^{\overline{\mathbf{E}}_\bullet^{a,k}} = R_0^{\overline{\mathbf{E}}_\bullet^{a,k}} = R_0^{\mathbf{E}_\bullet^{a,k}}\, R_{0,m}^{\mathbf{E}_\bullet^{a,k}} = R^{\mathbf{E}_\bullet^{a,k}} \in (\mathrm{CH}^0_{\mathbb{C}^N,V(a)})_0 \tag{5.23}$$

is independent of the way the Eagon–Northcott complex $\mathbf{E}^{a,k}_\bullet = \mathbf{EN}^{a,k}_\bullet$ is metrized. Let us describe in the global setting the relation of the residue current (5.23) to the Coleff–Herrera currents

$$\overline{\partial}\Big(\frac{1}{a_1,\dots,a_m}\Big)_V$$

introduced on a complex manifold $\mathcal{X}$ in Definition 2.78 and whose role is of crucial importance when $V=\mathcal{X}$; see Theorem 3.39. First, we fix some notation:

$$[1:m]=\{1,\dots,m\},\ [1:k]=\{1,\dots,k\},\ [2:k]=\{2,\dots,k\},$$
$$\mathfrak{T}=\{\mathfrak{t}:[2:k]\longrightarrow[1:m]\},\quad \nu(\mathfrak{t})=\big(|\mathfrak{t}^{-1}(\{1\})|,\dots,|\mathfrak{t}^{-1}(\{m\})|\big),$$
$$N_{m,k}(\mathfrak{t})=\binom{(m-1)+(k-1)}{m-1,\nu_1(\mathfrak{t}),\dots,\nu_m(\mathfrak{t})}=\frac{m-1+|\nu(\mathfrak{t})|!}{(m-1)!\prod_{j=1}^m(\nu_j(\mathfrak{t}))!}\quad \forall\,\mathfrak{t}\in\mathfrak{T}.$$

PROPOSITION 5.9. *Let $\mathcal{X}$ be a complex N-dimensional manifold, let V be a closed purely dimensional analytic subset with* $\operatorname{codim}V=M$, *and let $D_1,\dots,D_m$, where $m\le N-M$, be Cartier divisors on $\mathcal{X}$, with holomorphic sections $a_1,\dots,a_m$ respectively, such that* $\operatorname{codim}(V\cap V(a))=M+m$ *and* $\operatorname{codim}_{V\cap V(a)}(V_{\mathrm{sing}}\cap V(a))>0$. *Let*

$$(E)_\kappa=(\mathcal{O}(D_1)\oplus\cdots\oplus\mathcal{O}(D_m))_\kappa,\quad \kappa=1,\dots,k,$$

be k copies of $E=\mathcal{O}(D_1)\oplus\cdots\oplus\mathcal{O}(D_m)$ and let

$$(a)_\kappa=(a_1\oplus\cdots\oplus a_m)_\kappa=\big((a_1)_\kappa\oplus\cdots\oplus(a_m)_\kappa\big),\qquad \kappa=1,\dots,k,$$

be the copy of a as a section of $(E)_\kappa$. Then, one has formally

(5.24)

$$R^{\mathbf{EN}^{a,k}_\bullet}([V])=\sum_{\rho\in[1:k]^m}\overline{\partial}\Big(\frac{1}{(a_1)_{\rho_1},\dots,(a_m)_{\rho_m}}\Big)_V\sum_{\mathfrak{t}\in\mathfrak{T}}\frac{1}{N_{m,k}(\mathfrak{t})}\bigwedge_{\kappa=2}^k\Big[\frac{1}{(a_{\mathfrak{t}(\kappa)})_\kappa}\Big].$$

PROOF. Fix metrics on each $\mathcal{O}(D_j)$, $j=1,\dots,m$. Let $\|\ \|$ be the metric on E defined by $\|\ \|=(\sum_{j=1}^m|\ |_j^2)^{1/2}$. Such choice of metrics is irrelevant, since the Eagon–Northcott residue is independent of the choice of metrics, as we just recalled; see (5.23). One can choose (locally) holomorphic frames $e_1,\dots,e_m$ for $\mathcal{O}(D_1)$, ..., $\mathcal{O}(D_m)$ respectively, such that the minimal inverse of $a=\sum_1^m\alpha_j\,e_j$ is expressed locally as $\sigma=\sum_1^m(\overline{\alpha_j}/\|\alpha\|^2)\,e_j^*$. For $j=1,\dots,m$, let $\boldsymbol{e}_j=\sum_{\kappa=1}^k(e_j)_\kappa$. Let also

$$\boldsymbol{\sigma}=\sum_{\kappa=1}^m(\sigma)_\kappa=\sum_{j=1}^m\frac{\overline{\alpha_j}}{\|\alpha\|^2}\,\boldsymbol{e}_j^*.$$

In such local frames—let's say over an open subset U of $\mathcal{X}$ around a point in V—the current $R^{\mathbf{EN}^{a,k}_\bullet}([V])$ is expressed locally as

(5.25)

$$\begin{aligned}R^{\mathbf{EN}^{a,k}_\bullet}([V])&=\frac{1}{(2i\pi)^m}\Big[(\overline{\partial}|a|^{2\lambda})\bigwedge_{\kappa=1}^k(\sigma)_\kappa\sum_{\{\rho\in\mathbb{N}^k\,:\,|\rho|=m-1\}}\bigwedge_{\kappa=1}^k\Big(\overline{\partial}\Big(\frac{(\sigma)_\kappa}{2i\pi}\Big)\Big)^{\rho_\kappa}[V]\Big]_{\lambda=0}\\&=\frac{1}{(2i\pi)^m}\Big[(\overline{\partial}|a|^{2\lambda})\,(\boldsymbol{\sigma}\wedge(\overline{\partial}\boldsymbol{\sigma})^{m-1})\,\Big(\bigwedge_{\kappa=2}^k(\sigma)_\kappa\Big)\,[V]\Big]_{\lambda=0}\\&=\frac{(m-1)!(-1)^{\frac{m(m-1)}{2}}}{(2i\pi)^m}\Big[\lambda\|\alpha\|^{2\lambda}\,\frac{\overline{d\alpha}}{\|\alpha\|^{2m}}\,\Big(\bigwedge_{j=1}^m\boldsymbol{e}_j^*\Big)\,\Big(\bigwedge_{\kappa=2}^k(\sigma)_\kappa\Big)\,[V]\Big]_{\lambda=0},\end{aligned}$$

where $d\alpha = d\alpha_1 \wedge \cdots \wedge d\alpha_m$; see (3.82) and the final observation in Example 3.31. One has

$$\begin{aligned}(5.26)\quad & \Big[\lambda\|\alpha\|^{2\lambda} \frac{\overline{d\alpha}}{\|\alpha\|^{2m}} \Big(\bigwedge_{j=1}^{m} e_j^*\Big) \Big(\bigwedge_{\kappa=2}^{k} (\sigma)_\kappa\Big) [V]\Big]_{\lambda=0} \\ &= \sum_{\mathfrak{t}\in\mathfrak{T}} \Big[\lambda\|\alpha\|^{2\lambda} \Big(\frac{\overline{\alpha}}{\|\alpha\|^2}\Big)^{\nu(\mathfrak{t})} \frac{\overline{d\alpha}}{\|\alpha\|^{2m}} \Big(\bigwedge_{j=1}^{m} e_j^*\Big) \wedge \Big(\bigwedge_{\kappa=2}^{k} (e^*_{\mathfrak{t}(\kappa)})_\kappa\Big) [V]\Big]_{\lambda=0}.\end{aligned}$$

For each $\mathfrak{t} \in \mathfrak{T}$, the current defined as

$$\frac{(-1)^{\frac{m(m-1)}{2}}}{(2i\pi)^m} \frac{(m-1+|\nu(\mathfrak{t})|)!}{\prod_{j=1}^m (\nu_j(\mathfrak{t}))!} \Big[\lambda\|\alpha\|^{2\lambda} \Big(\frac{\overline{\alpha}}{\|\alpha\|^2}\Big)^{\nu(\mathfrak{t})} \frac{\overline{d\alpha}}{\|\alpha\|^{2m}} [V]\Big]_{\lambda=0}$$

in U belongs to $\mathrm{CH}^0_{U, U\cap(V\cap V(a))}(U)$ according to Theorem 2.39. Actually, it is a pseudo-meromorphic current with bidegree $(0, M+m)$, which is supported by the closed analytic subset $U \cap (V \cap V(a))$, whose pure codimension equals $M+m$, and moreover it is $\overline{\partial}$-closed in U, as one can check easily. Thus, it shares with the restriction to U of the current

$$\overline{\partial}\Big(\frac{1}{a_1^{\nu_1(\mathfrak{t})+1}, \dots, a_m^{\nu_m(\mathfrak{t})+1}}\Big)_V$$

the standard extension property across any hypersurface of $U \cap (V \cap V(a))$. Integral representation formulae (due to Aldo Andreotti and François Norguet [**AN64**, **Nor74**], see also [**BGVY**, Proposition 3.16]) show that these two currents do coincide around each point $z^0 \in U \cap (V_{\mathrm{reg}} \cap V(a))$. The equality of Coleff–Herrera currents

$$\begin{aligned}(5.27)\quad & \frac{(-1)^{\frac{m(m-1)}{2}}}{(2i\pi)^m} \frac{(m+|\nu(\mathfrak{t})|-1)!}{\prod_{j=1}^m (\nu_j(\mathfrak{t}))!} \Big[\lambda\|\alpha\|^{2\lambda} \Big(\frac{\overline{\alpha}}{\|\alpha\|^2}\Big)^{\nu(\mathfrak{t})} \frac{\overline{d\alpha}}{\|\alpha\|^{2m}} [V]\Big]_{\lambda=0} \\ &= \overline{\partial}\Big(\frac{1}{a_1^{\nu_1(\mathfrak{t})+1}, \dots, a_m^{\nu_m(\mathfrak{t})+1}}\Big)_V\end{aligned}$$

in U follows then by standard extension. Since $|\nu(\mathfrak{t})| = k-1$, relations (5.25) to (5.27) and the definition of $N(\mathfrak{t})$ for $\mathfrak{t} \in \mathfrak{T}$ imply the equality of currents (5.24) in U. Since U is an arbitrary local chart above which each bundle $\mathcal{O}(D_j)$ is trivial, we have the desired equality in $\mathcal{X}$. □

5.3. Briançon–Skoda theorem in $\mathbb{A} = (\mathcal{O}/\mathcal{I}_V)_0$

Let $\Omega \subset \mathbb{C}^N$ be an open set containing the origin 0. We consider in this section the local nœtherian ring $\mathbb{A} = (\mathcal{O}/\mathcal{I}_V)_0$, where $\mathcal{O} = \mathcal{O}_\Omega$ and $V \subset \Omega$ is a purely dimensional closed analytic subset with $\operatorname{codim} V = M$ and $0 \in V_{\mathrm{sing}}$. For $0 < \eta \ll 1$ small enough, let $\mathbf{F}_\bullet$ be a generically exact complex of trivial holomorphic bundles in $\eta\,\mathbb{B}_N$,

$$\mathbf{F}_\bullet : 0 \to E_L \simeq \mathbb{C}^{M_L} \xrightarrow{F_L} E_{L-1} \simeq \mathbb{C}^{M_{L-1}} \xrightarrow{F_{L-1}} \cdots \longrightarrow E_1 \simeq \mathbb{C}^{M_1} \xrightarrow{F_1} E_0 = \mathbb{C} \longrightarrow 0,$$

where $L \leq N-1$. Furthermore, we assume for the above complex that the morphism F_1 is generically surjective and that the corresponding sequence of sheaves of sections

$$0 \longrightarrow \mathcal{O}_{\eta\,\mathbb{B}_N}^{\oplus r_L} \xrightarrow{F_L} \mathcal{O}_{\eta\,\mathbb{B}_N}^{\oplus r_{L-1}} \xrightarrow{F_{L-1}} \cdots \xrightarrow{F_2} \mathcal{O}_{\eta\,\mathbb{B}_N}^{\oplus r_1} \xrightarrow{F_1} \mathcal{O}_{\eta\,\mathbb{B}_N}$$

is exact over $\eta\,\mathbb{B}_N$ with coker $F_1 = (\mathcal{O}_\Omega/\mathcal{I}_V)_{|\eta\,\mathbb{B}_N}$; see §A.2.3. Since $\mathcal{I}_V$ is radical, such a complex $\mathbf{F}_\bullet$ always exists; see §3.4.3. For a metrized version $\overline{\mathbf{F}_\bullet}$ of the complex $\mathbf{F}_\bullet$ we let

$$R^{\overline{\mathbf{F}_\bullet}} = R_0^{\overline{\mathbf{F}_\bullet}} = \sum_{\rho=0}^{N-M-1} R_{0,M+\rho}^{\overline{\mathbf{F}_\bullet}}, \quad P^{\overline{\mathbf{F}_\bullet}} = \sum_{\ell\geq 0} P_\ell^{\overline{\mathbf{F}_\bullet}} = \sum_{\ell\geq 0}\Big(\sum_{r=\ell+1}^{N-1} P_{\ell,r}^{\overline{\mathbf{F}_\bullet}}\Big)$$

be the residue current and the principal value current attached to $\overline{\mathbf{F}_\bullet}$, respectively, as in Proposition 3.28. If Υ denotes the N-form $dz = dz_1\wedge\cdots\wedge dz_N$, then Proposition 3.49 implies that in a small enough ball $\eta\,\mathbb{B}_N$, $0<\eta\ll 1$, the equality

$$R^{\overline{\mathbf{F}_\bullet}}\wedge dz = R_0^{\overline{\mathbf{F}_\bullet}}\wedge dz = (\iota_V)_*\,\omega \tag{5.28}$$

holds, where

(5.29)

$$\begin{aligned}\omega &= (\widehat{\Pi}_V)_* \sum_{\rho=0}^{N-M-1}\bigg[\frac{|e_\rho|^{\mu_\rho}}{e_\rho}\mathfrak{s}_\rho\Big[\cdots\Big[\frac{|e_1|^{2\mu_1}}{e_1}\mathfrak{s}_1\Big[\frac{|e_0'|^{2\mu_0}}{e_0'}\mathfrak{s}_0'\,\widehat{\Pi}_V^*\mathfrak{f}\Big]_{\mu_0=0}\Big]\cdots\Big]_{\mu_{\rho-1}=0}\bigg]_{\mu_\rho=0}\\ &= (\widehat{\Pi}_V)_* \sum_{\rho=0}^{N-M-1}\bigg[\Big(\frac{|e_\rho|^{2\lambda}}{e_\rho}\mathfrak{s}_\rho\Big)\cdots\Big(\frac{|e_1|^{2\lambda^\rho}}{e_1}\mathfrak{s}_1\Big)\Big(\frac{|e_0'|^{2\lambda^{\rho+1}}}{e_0'}\mathfrak{s}_0'\Big)\widehat{\Pi}_V^*\mathfrak{f}\bigg]_{\lambda=0}\\ &= \sum_{\rho=0}^{N-M-1}\omega_\rho.\end{aligned}$$

Note that the second equality follows from Proposition 2.72. In the expression on the right-hand side of (5.29), $e_0', e_1, \ldots, e_\rho$ are holomorphic sections of Cartier divisors on the normal complex analytic space $\widehat{V}^{N-M-1}$. These sections are constructed in (3.126) and (3.127) as the ending point of a sequence of iterated normalized blowups of V along the ideals $(\iota_V)^*\mathrm{Fitt}_{M+\rho}(V)$, $\rho\geq 1$. The normalized blowups start with $\widehat{\pi}_1$ and culminate as $\widehat{\Pi}_V$. In addition, the $\mathfrak{s}_j$ are bundle-valued, smooth $(0,1)$-forms and $\mathfrak{f}$ is a bundle-valued section of the dualizing sheaf ω_V^{N-M}. Cartier divisors $\widehat{D}_0,\ldots,\widehat{D}_{N-M-1}$, as well as their sections $e_0', e_1,\ldots,e_{N-M-1}$, depend on the coherent radical sheaf ideal $\mathcal{I}_V$ through the sequence of its Fitting ideals, while the smooth forms $\mathfrak{s}_j$ depend on the metrized complex $\overline{\mathbf{F}_\bullet}$. The construction of the section $\mathfrak{f}$ of the dualizing sheaf relies on the complex $\mathbf{F}_\bullet$; see Proposition 3.48. We also know from Proposition 4.43 that the $\mathbf{F}_\bullet$-Hefer problem with respect to $(\mathscr{E},\Delta)$, as in (4.100), admits a solution $\mathbf{H} = \{H_{\ell,r} : \ell\geq 0,\ r\geq -1\}$ in $\eta\,\mathbb{B}_N\times\eta\,\mathbb{B}_N$. As a consequence of Proposition 4.29,

$$\mathscr{W}^{\mathbf{F}_\bullet}(z,\cdot) = 2i\pi F_1(z)H_1(z,\cdot)P_0^{\overline{\mathbf{F}_\bullet}} + H_0(z,\cdot)\,R_0^{\overline{\mathbf{F}_\bullet}} \tag{5.30}$$

is a global weight in $(\eta\,\mathbb{B}_N\times\eta\,\mathbb{B}_N)\setminus\sqcap^{-1}(V)$ with respect to the $(\mathscr{E},\Delta)$ Koszul complex $\mathscr{K}_\bullet^{\mathscr{D}}$. One can now state the main theorem of this section [**AndSS10**, Theorem 1.1].

THEOREM 5.10 (Briançon–Skoda–Huneke theorem).[5] *Let $\mathbb{I}$ be any ideal in the local ring $\mathbb{A} = (\mathcal{O}/\mathcal{I}_V)_0$. Then, one has that $\overline{\mathbb{I}^{\mu'_{\mathbb{A}}+(N-M)+k-1}} \subset \mathbb{I}^k$ for some positive integer $\mu'_{\mathbb{A}}$, which is independent of $\mathbb{I}$, and for any $k \in \mathbb{N}^*$. The smallest integer $\mu_{\mathbb{A}}$ satisfying $\overline{\mathbb{I}^{\mu_{\mathbb{A}}+k-1}} \subset \mathbb{I}^k$, for any $k \in \mathbb{N}^*$ and any ideal $\mathbb{I} \subset \mathbb{A}$, is called the Briançon–Skoda exponent of the local ring $\mathbb{A}$.*

PROOF. Let $a_1, \dots, a_m$ be representatives of generators of $\mathbb{I}$ modulo $\mathcal{I}_V(\eta\,\mathbb{B}_N)$, where $\eta\,\mathbb{B}_N$ is a sufficiently small ball centered at the origin. Let λ' be a complex parameter satisfying Re $\lambda' \gg 1$ and let $\mathscr{W}^{\lambda'}_{a,k}$ be the global weight in $\eta\,\mathbb{B}_N \times \eta\,\mathbb{B}_N$ defined by (4.109), (4.110), and (4.111) by taking a in place of f and λ' instead of λ. Then $\mathscr{W}^{\overline{\mathbf{F}_\bullet}}\,\mathscr{W}^{\lambda'}_{a,k}$ is considered a global weight[6] in $(\eta\,\mathbb{B}_N \times \eta\,\mathbb{B}_N) \setminus \sqcap^{-1}(V \cap \eta\,\mathbb{B}_N)$ provided one takes

$$\text{Re } \lambda' \gg \max\big(\text{orders in a neighborhood of } \eta\,\overline{\mathbb{B}_N} \text{ of the currents } P_0^{\overline{\mathbf{F}_\bullet}}, R_0^{\overline{\mathbf{F}_\bullet}}\big).$$

Let $h \in \mathbb{A}$, whose representative in $\eta\,\mathbb{B}_N$ satisfies

$$|h| \leq C|a|^{\mu'+(N-M)+k-1} \quad \text{on } V \cap \eta\,\mathbb{B}_N, \tag{5.31}$$

where the exponent $\mu' \in \mathbb{N}$ is to be specified later on. Let also $\chi_\eta \in \mathcal{D}(\eta\,\mathbb{B}_N)$ be the support function identically equal to 1 around $(\eta/2)\,\overline{\mathbb{B}_N}$ and let $q_\eta = q_{(\eta/2)\,\mathbb{B}_N, \chi_\eta}$ be as in Proposition 4.33. It follows from Proposition 4.33, which uses the same argument relying on the proof of Proposition 4.36 and on the computations after the proof of Proposition 4.47, that the following integral representation formula holds for h on $(\eta/2)\,\mathbb{B}_N$:

$$\begin{aligned} h(z) = &\int_\Omega \Big(\Big(\chi_\eta + \frac{\overline{\partial}_\zeta\big((1-\chi_\eta)\,q_\eta(z,\cdot)\big)}{2i\pi}\Big)^{N+1} H_0(z,\cdot)\,R_0^{\overline{\mathbf{F}_\bullet}}\Big)_{N,N} \mathscr{W}^{\lambda'}_{a,k} h \\ &+ 2i\pi F_1(z)\Big(\int_\Omega \Big(\Big(\chi_\eta + \frac{\overline{\partial}_\zeta\big((1-\chi_\eta)\,q_\eta(z,\cdot)\big)}{2i\pi}\Big)^{N+1} H_1(z,\cdot)\,P_0^{\overline{\mathbf{F}_\bullet}}\,\mathscr{W}^{\lambda'}_{a,k}\Big)_{N,N} h\Big). \end{aligned} \tag{5.32}$$

That is, the above formula is derived in two steps. Namely, one establishes first the formula for $z \in (\eta/2)\,\mathbb{B}_N \setminus V$ and then for all $z \in (\eta/2)\,\mathbb{B}_N$ using the analytic continuation principle. As a consequence, one has that the representation formula

$$h(z) = \int_{\eta\,\mathbb{B}_N} \Big(\Big(\chi_\eta + \frac{\overline{\partial}_\zeta\big((1-\chi_\eta)\,q_\eta(z,\cdot)\big)}{2i\pi}\Big)^{N+1} H_0(z,\cdot)\,R_0^{\overline{\mathbf{F}_\bullet}}\,\mathscr{W}^{\lambda'}_{a,k}\Big)_{N,N} h \tag{5.33}$$

---

[5] Theorem 5.10 as shown here stands as a particular case of Briançon–Skoda's theorem for local rings, which are essentially of finite type over an excellent nœtherian local ring, or analytically unramified nœtherian local rings containing $\mathbb{Q}$; [**Hu92**], see also [**HoH90, Smi01**], [**SwH06**, Chapter 13]. The analytic proof presented here is inspired and adapted from the proof of Theorem 1.1 in [**AndSS10**].

[6] One could have taken instead of $\mathscr{W}^{\lambda'}_{a,k}$

$$\mathscr{W}^{\lambda'}_{\mathbf{K}^a_\bullet,k} = 2i\pi\,\big((\mathbb{J}_{a(z)})_1 \wedge \cdots \wedge (\mathbb{J}_{a(z)})_k\big) H_1^{\overline{\mathbf{EN}^{a,k}_\bullet}}(z,\cdot) P_0^{\overline{\mathbf{EN}^{a,k}_\bullet},\lambda'} + H_0^{\overline{\mathbf{EN}^{a,k}_\bullet}}(z,\cdot)\,R_0^{\overline{\mathbf{EN}^{a,k}_\bullet},\lambda'}$$

as in the proof described in [**AndSS10**, Theorem 1.1], which would have led to a slightly refined choice of the absolute integer $\mu'_{\mathbb{A}}$ (in particular taking into account the cardinal m of a set of generators for the ideal $\mathbb{I}$). Since such refinement is irrelevant here, we keep to the choice of the weight $\mathscr{W}^{\overline{\mathbf{F}_\bullet}}\,\mathscr{W}^{\lambda'}_{a,k}$.

holds for any $z \in V \cap (\eta/2)\,\mathbb{B}_N$. For any $r = 1,\dots,\min(r, N-M)$, $\lambda' \in \mathbb{C}$ such that $\mathrm{Re}\,\lambda' \gg 1$, and $z \in (\eta/2)\,\mathbb{B}_N$, let

$$\rho^{\lambda'}_{a,k,r}(z,\cdot) = \Big(\sum_{r'=r}^{r+k-1} \binom{r'}{r} \Big(\frac{a(z)\,a^*}{\|a\|^2}\Big)^{r'-r}\Big)\,(\overline{\partial}\|a\|^{2\lambda})\,\mathscr{R}_{a,r}(z,\cdot)$$

(see (5.12)) and

$$\mathscr{W}^{\dagger,\lambda'}_{a,k}(z,\cdot) = \mathscr{W}^{\lambda'}_{a,k}(z,\cdot) - \sum_{r=1}^{N} \rho^{\lambda'}_{a,k,r}(z,\cdot).$$

It follows from the pseudo-meromorphicity of $R^{\overline{\mathbf{F}_\bullet}}$ and Proposition 2.21 that the maps

$$\lambda' \longmapsto \int_{\eta\,\mathbb{B}_N} \Big(\Big(\chi_\eta + \frac{\overline{\partial}_\zeta\big((1-\chi_\eta)\,q_\eta(z,\cdot)\big)}{2i\pi}\Big)^{N+1} H_0(z,\cdot)\,R_0^{\overline{\mathbf{F}_\bullet}}\,\mathscr{W}^{\dagger,\lambda'}_{a,k}\Big)_{N,N} h,$$

$$\lambda' \longmapsto \int_{\eta\,\mathbb{B}_N} \Big(\Big(\chi_\eta + \frac{\overline{\partial}_\zeta\big((1-\chi_\eta)\,q_\eta(z,\cdot)\big)}{2i\pi}\Big)^{N+1} H_0(z,\cdot)\,R_0^{\overline{\mathbf{F}_\bullet}}\,\rho^{\lambda'}_{a,k,r}(z,\zeta)\Big)_{N,N} h,$$
$$r = 1,\dots,N,$$

extend from $\{\lambda' \in \mathbb{C} : \mathrm{Re}\,\lambda' \gg 1\}$ to $\mathbb{C}$ as meromorphic maps with poles in $\mathbb{Q}^+$. Moreover, for any $z \in (\eta/2)\,\mathbb{B}_N \setminus V$, the computations leading to the current identity

$$[\mathscr{W}^{\lambda'}_{a,k}(z,\cdot)]_{\lambda'=0} = \Big[\sum_{r=1}^{N} \rho^{\lambda'}_{a,k,r}(z,\cdot)\Big]_{\lambda'=0} + \sum_{\{\kappa\in\mathbb{N}^m\,:\,|\kappa|=k\}} \boldsymbol{a}^\kappa(z)\,T^\kappa_{a,k}(z,\cdot) \tag{5.34}$$

in $\eta\,\mathbb{B}_N$ (see the end of §4.6.1 and the proof of Theorem 5.7) show that the map

$$z \in (\eta/2)\,\mathbb{B}_N$$
$$\longmapsto \Big[\int_{\eta\,\mathbb{B}_N} \Big(\Big(\chi_\eta + \frac{\overline{\partial}_\zeta\big((1-\chi_\eta)\,q_\eta(z,\cdot)\big)}{2i\pi}\Big)^{N+1} H_0(z,\cdot)\,R_0^{\overline{\mathbf{F}_\bullet}}\,\mathscr{W}^{\dagger,\lambda'}_{a,k}\Big)_{N,N} h\Big]_{\lambda'=0}$$

is in the kth power of the ideal generated by $a_1,\dots,a_m$ in $\mathcal{O}((\eta/2)\,\mathbb{B}_N)$. For any $0 \le \rho \le N-M-1$, let us introduce the bundle-valued $(N-M,\rho)$-current on the complex analytic space $\widehat{V}^{N-M-1}$ defined as

$$\omega_\rho = [\omega^\lambda_\rho]_{\lambda=0} = \Big[\Big(\frac{|e_\rho|^{2\lambda}}{e_\rho}\mathfrak{s}_\rho\Big)\cdots\Big(\frac{|e_1|^{2\lambda^\rho}}{e_1}\mathfrak{s}_1\Big)\Big(\frac{|e_0'|^{2\lambda^{\rho+1}}}{e_0'}\mathfrak{s}_0'\Big)\widehat{\Pi}_V^*\mathfrak{f}\Big]_{\lambda=0}, \tag{5.35}$$

where the polar set of $\widehat{\Pi}_V^*\mathfrak{f}$ lies in $\widehat{\Pi}_V^{-1}(V_{\mathrm{sing}} \cap \eta\,\mathbb{B}_N)$ and hence in the hypersurface $V(e_0')$; see (5.29). In order to prove that $h \in \mathbb{I}^k$, it is enough to show that for any ρ, r satisfying $r + \rho \le N-M$ with $r \le \min(m, N-M)$, the relation

$$h\Big[R^{\overline{\mathbf{F}_\bullet}}_{0,M+\rho} \wedge \rho^{\lambda'}_{a,k,r}\Big]_{\lambda'=0} = 0 \tag{5.36}$$

holds. The expression of $\mathscr{R}_{a,r}$ (see (5.12)) says that (5.36) will follow from the fact that for all such ρ, r, for any subset $\mathcal{J} \subset \{1,\dots,m\}$ with cardinal r and any map χ from $\{1,\dots,k\}$ to $\{1,\dots,m\}$, for any $j = 1,\dots,m$,

$$\widehat{H}\Big[\lambda'\,\omega_\rho\,\|\widehat{A}\|^{2\lambda'}\,\frac{\widehat{A}_j\,\overline{\widehat{A}_\chi}\,\overline{d\widehat{A}_{\mathcal{J}}}}{\|\widehat{A}\|^{2(k+r+1)}}\Big]_{\lambda'=0} = 0 \tag{5.37}$$

holds as a bundle-valued current on the complex analytic space $\widehat{V}^{N-M-1}$, where

$$\widehat{H} = (\iota_V \circ \widehat{\Pi}_V)^* H, \quad \widehat{A}_j = (\iota_V \circ \widehat{\Pi}_V)^* a_j, \ j = 1, \dots, m,$$

$$d\widehat{A}_{\mathcal{J}} = \bigwedge_{j \in \mathcal{J}} d\widehat{A}_j, \quad \widehat{A}_\chi = \prod_{\kappa=1}^{k} \widehat{A}_{\chi(\kappa)}.$$

Let Z^0 be a point in $\widehat{V}^{N-M+1} \cap \widehat{\Pi}_V^{-1}(V(a) \cap V)$ and let $\Pi = \Pi^{Z^0}$ be a log resolution $\Pi : \mathcal{V} \to U^0$ of a neighborhood U^0 of Z^0 in $\widehat{V}^{N-M-1}$ for an arbitrary hypersurface, which contains both the set $\widehat{V}^{N-M-1}_{\text{sing}} \cap U^0$, and the sets $V(e_0') \cap U^0$, $V(e_j) \cap U^0$ for $j = 1, \dots, \rho$; see Theorem A.41. Let $v^0 \in (\pi^0)^{-1}(\{Z^0\})$. For any $(0, N - M - \rho)$ test form $\varphi = \sum_J \varphi_J \overline{dw_J}$ on $\mathcal{V}$ with support in a neighborhood of v^0, one can express, in view of (5.35) and of the formal univariate integration by parts (1.7), the action of the current $[\Pi^* \omega_\rho^\lambda]_{\lambda=0}$ on φ as

$$(5.38) \quad \big\langle [\Pi^* \omega_\rho^\lambda]_{\lambda=0}, \varphi \big\rangle = \kappa(n(v^0)) \sum_{\#J = N-M-\rho} \int_{\mathcal{V}} \Big(\Big(\frac{\partial}{\partial w}\Big)^{n(v^0)} (\varphi_J \sigma_J) \Big) \frac{dw}{w} \wedge \overline{dw},$$

where $n(v^0) \in \mathbb{N}^{N-M}$ with $|n(v^0)| = \boldsymbol{N}$, $\kappa(n(v^0)) \in \mathbb{Q}^*$, $\sigma_J = \sigma_J^{v^0}$ are smooth forms in a neighborhood of z^0, and $w_1, \dots, w_{N-M}$ are local coordinates in a neighborhood of this point. Let us suppose that μ' in (5.31) satisfies $\mu' \geq \boldsymbol{N} + 2(N-M) + 1$.[7] If $H = \Pi^* \widehat{H}$ and $A = \Pi^* \widehat{A}$, then one has that $|H| \leq C |A|^{\mu' + N - M + k - 1}$ in a neighborhood of v^0. Thus, since $\mathcal{V}$ is a complex $(N-M)$-dimensional manifold, it follows from Theorem 5.7 (with $\mathbb{A} = \mathcal{O}_{\mathcal{V}, v^0}$ instead of $\mathcal{O}_0$) that for any $\alpha^\circ \in \mathbb{N}^{N-M}$ such that $|\alpha^\circ| \leq \boldsymbol{N}$ and any $j = 1, \dots, m$,

$$(5.39) \qquad \Big(\frac{\partial}{\partial \zeta}\Big)^{\alpha^\circ} (A_j H) \in \Big(\sum_{j=1}^m A_j \, \mathcal{O}_{\mathcal{V}, v^0}\Big)^{(\boldsymbol{N} - |\alpha^\circ|) + k + 2(N-M+1)}.$$

On the other hand, for any $\beta^\circ \in \mathbb{N}^{N-M}$ with $|\alpha^\circ| + |\beta^\circ| = \boldsymbol{N}$, the equality

$$(5.40) \quad \lambda' \Big(\frac{\partial}{\partial \zeta}\Big)^{\beta^\circ} \|A\|^{2(\lambda' - r - k - 1)} = \lambda' \Big(\prod_{\ell = r+k+1}^{r+k+|\beta^\circ|} (\lambda' - \ell) \Big) \frac{\Psi_{\beta^\circ}(\zeta, \overline{A})}{\|A\|^{r+k+1+|\beta^\circ|}} \|A\|^{2\lambda'}$$

holds, where Ψ_{β° is homogeneous with degree $|\beta^\circ|$ in $A_1, \dots, A_m$. Let $\widehat{\pi}_A : \widehat{\mathcal{V}}_A \to \mathcal{V}_{v^0}$ be the normalized blowup of a neighborhood of v^0 in $\mathcal{V}$ along the ideal sheaf $\sum_{j=1}^m A_j \, \mathcal{O}_{\mathcal{V}}$. Then, for any such β° and any map $\chi : \{1, \dots, k\} \to \{1, \dots, m\}$ as in (5.37),

$$\widehat{\pi}_A^* \Big(\lambda' \Big(\frac{\partial}{\partial \zeta}\Big)^{\beta^\circ} \overline{\textstyle\prod_1^k A_{\chi(\kappa)}} \, \|A\|^{2(\lambda' - r - k - 1)} \Big)$$
$$= \lambda' \Big(\prod_{\ell = r+k+1}^{r+k+|\beta^\circ|} (\lambda' - \ell) \Big) \frac{\|\widehat{\pi}_A^* A\|^{2\lambda'}}{\mathfrak{e}^{|\beta^\circ| + k}} \frac{\varpi_{\chi, \nu'}}{\|\mathfrak{e}\|^{2(r+1)}},$$

where $\mathfrak{e}$ is a holomorphic section of the line bundle corresponding to the exceptional divisor and $\varpi_{\chi, \nu'}$ is a smooth form. Since

$$(\boldsymbol{N} - |\alpha^\circ| + k) - (|\beta^\circ| + k) = 0,$$

[7] A more careful analysis taking into account weaker sufficient conditions than the rough ones (5.37), proposed here, would show that the choice $\mu' \geq |\alpha| + (N-M)$ is in fact enough for the rest of the book; see [**AndSS10**, §4].

it follows from (5.38) and from (5.39) that for any $j \in \{1, \dots, m\}$ and any subset $\mathcal{J} \subset \{1, \dots, m\}$ with cardinal r the product

$$\widehat{\pi}_A^* H \Big[\big[(\Pi \circ \widehat{\pi}_A)^* \omega_\rho^\lambda \big]_{\lambda=0} \pi_A^* \Big(\lambda' \, \|A\|^{2\lambda'} \, \frac{A_j \, \overline{A}_\chi \, \overline{dA_{\mathcal{I}}}}{\|A\|^{2(k+r+1)}} \Big) \Big]_{\lambda'=0} \tag{5.41}$$

is well defined as a current on $\widehat{\mathcal{V}}_A$. Since $\widehat{\Pi}_V$ and Π are proper morphisms, a compactness argument ensures that the number of conditions

$$\mu' \geq \boldsymbol{N} + 2(N-M) + 1$$

imposed, which depends only on the ω_ρ, hence on the sequence of Fitting ideals attached to the locally free resolution of $(\mathcal{O}/\mathcal{I}_V)_0 = \mathbb{A}$, is indeed finite. This ensures the existence of $\mu'_{\mathbb{A}}$ such that whenever h represents an element of $\overline{\mathbb{I}^{\mu'_{\mathbb{A}}+(N-M)+1}}$, then conditions (5.36) hold. But this implies that the representation

$$\begin{aligned} h(z) &= \Big[\int_\Omega \Big(\Big(\chi_\eta + \frac{\overline{\partial}_\zeta \big((1-\chi_\eta) \, q_\eta(z, \cdot) \big)}{2i\pi} \Big)^{N+1} H_0(z, \cdot) \, R_0^{\mathbf{F}_\bullet} \, \mathscr{W}_{a,k}^{\dagger, \lambda'} \Big)_{N,N} h \Big]_{\lambda=0} \\ &= \sum_{\{\kappa \in \mathbb{N}^m \, : \, |\kappa| = k\}} a^\kappa(z) \, Q_\kappa(z) \end{aligned}$$

holds for any $z \in (\eta/2) \, \mathbb{B}_N$. Here

$$Q_\kappa(z) = \Big[\int_\Omega \Big(\Big(\chi_\eta + \frac{\overline{\partial}_\zeta \big((1-\chi_\eta) \, q_\eta(z, \cdot) \big)}{2i\pi} \Big)^{N+1} H_0(z, \cdot) R_0^{\mathbf{F}_\bullet} \, T_{a,k}^{\kappa, \lambda'}(z, \cdot) \Big)_{N,N} h \Big]_{\lambda'=0};$$

see (5.34). This concludes the proof of Theorem 5.10. □

In some particular cases, the cardinality of the generating system of $\mathbb{I}$ can be taken into account, which was not the case in the proof above. To deal with that question we will use, as in the proof of Theorem 5.8, the Eagon–Northcott complex. The construction described below is general, but we consider here just the particular case we are concerned with. Take $a \in \mathcal{O}_0(\cdot, \mathbb{C}^m)$ and $k \in \mathbb{N}^*$. Let us form the complex $\mathbf{EN}_\bullet^{a,k} \oplus \mathbf{F}_\bullet$ as follows. For any $\ell \in \mathbb{Z}$ take

$$(\mathbf{EN}_\bullet^{a,k} \oplus \mathbf{F}_\bullet)_\ell = \bigoplus_{\ell' + \ell'' = \ell} \mathbf{EN}_{\ell'}^{a,k} \otimes E_{\ell''}. \tag{5.42}$$

From now on for the sake of simplicity, let us denote $\mathbf{EN}_\bullet^{a,k}$ by $\mathbf{E}_\bullet^{a,k}$, as we already did in §5.2. The rules for extending the morphisms

$$F^{a,k} = \sum_{\ell'} F_{\ell'}^{a,k} : \; E(\mathbf{E}_\bullet^{a,k}) \longrightarrow E(\mathbf{E}_\bullet^{a,k}),$$

$$F = \sum_\ell F_{\ell''} : E(\mathbf{F}_\bullet) \longrightarrow E(\mathbf{F}_\bullet)$$

to actions from $('\mathcal{D}(\cdot, E(\mathbf{E}_\bullet^{a,k} \oplus \mathbf{F}_\bullet)))_0$ into itself are given by

$$\begin{aligned} F^{a,k} &: \xi' \otimes \xi'' \longrightarrow (F^{a,k} \xi') \otimes \xi'', \\ F &: \xi' \otimes \xi'' \longrightarrow (-1)^{\deg \xi'} \, \xi' \otimes (F \xi''), \end{aligned} \tag{5.43}$$

when ξ'' has a pure degree. The actions

$$\boldsymbol{F}^{a,k}, \; \boldsymbol{F} : \boldsymbol{F}^{a,k}, \; \boldsymbol{F} : \big('\mathcal{D}(\cdot, \mathbf{E}(\mathbf{E}_\bullet^{a,k} \oplus \mathbf{F}_\bullet)) \big)_0 \longrightarrow \big('\mathcal{D}(\cdot, \mathbf{E}(\mathbf{E}_\bullet^{a,k} \oplus \mathbf{F}_\bullet)) \big)_0, \tag{5.44}$$

are similarly defined. If the complexes of trivial bundles $\mathbf{E}_\bullet^{a,k}$ and $\mathbf{F}_\bullet$ are metrized, say with the canonical hermitian metric on any $\mathbf{E}_{\ell'}^{a,k}$ or $E_{\ell''}$, then the fact that $R^{\overline{\mathbf{F}_\bullet}}$ is pseudo-meromorphic makes licit the construction of the $E(\mathbf{E}_\bullet^{a,k} \oplus \mathbf{F}_\bullet)$-valued currents

$$\left[P^{\overline{\mathbf{E}_\bullet^{a,k}},\lambda'}\, R^{\overline{\mathbf{F}_\bullet}}\right]_{\lambda'=0}, \quad \left[R^{\overline{\mathbf{E}_\bullet^{a,k}},\lambda'}\, R^{\overline{\mathbf{F}_\bullet}}\right]_{\lambda'=0};$$

see Proposition 2.21. It follows from Proposition 3.28 that

$$\begin{aligned}\boldsymbol{\nabla}_{\mathbf{E}_\bullet^{a,k}\oplus\mathbf{F}_\bullet}\left[P^{\overline{\mathbf{E}_\bullet^{a,k}},\lambda'}\, R^{\overline{\mathbf{F}_\bullet}}\right]_{\lambda'=0} &= \left(\boldsymbol{\nabla}_{\mathbf{E}_\bullet^{a,k}}\left[P^{\overline{\mathbf{E}_\bullet^{a,k}},\lambda'}\right]_{\lambda'=0}\right) R^{\overline{\mathbf{F}_\bullet}}\\ &= R^{\overline{\mathbf{F}_\bullet}} - \left[R^{\overline{\mathbf{E}_\bullet^{a,k}},\lambda'}\right]_{\lambda'=0} R^{\overline{\mathbf{F}_\bullet}}\\ &= R^{\overline{\mathbf{F}_\bullet}} - R^{\overline{\mathbf{E}_\bullet^{a,k}}}\, R^{\overline{\mathbf{F}_\bullet}}.\end{aligned}$$

This implies

$$\boldsymbol{\nabla}_{\mathbf{E}_\bullet^{a,k}\oplus\mathbf{F}_\bullet}\left(P^{\overline{\mathbf{F}_\bullet}} + \left[P^{\overline{\mathbf{E}_\bullet^{a,k}},\lambda'}\, R^{\overline{\mathbf{F}_\bullet}}\right]_{\lambda'=0}\right) = \mathrm{Id}_{\mathbf{E}(\mathbf{E}_\bullet^{a,k}\oplus\mathbf{F}_\bullet)} - R^{\overline{\mathbf{E}_\bullet^{a,k}}}\, R^{\overline{\mathbf{F}_\bullet}}. \tag{5.45}$$

Under the reinforced Buchsbaum–Eisenbud–Fitting conditions (3.117), we can now state (see [**AndSS10**, Theorem 1.1(ii)]) the following result.

THEOREM 5.11. *Assume that* $0 \leq m \leq N - M$ *is such that*

$$\begin{aligned}&\mathrm{codim}_V(V_{\mathrm{sing}} \cap V(a)) \geq 1 + m,\\ &\mathrm{codim}_V\,(\mathbb{V}_{M+\rho}(\mathcal{I}_V) \cap V(a)) \geq (1+\rho) + m \quad \forall\,\rho = 1, \ldots, N-M+1.\end{aligned} \tag{5.46}$$

If $\mathbb{I}$ *is any ideal in* $\mathbb{A} = (\mathcal{O}/\mathcal{I})_0$, *whose minimal number of generators is less than or equal to* m, *then one has*

$$\overline{\mathbb{I}^{m+k-1}} \subset \mathbb{I}^k \tag{5.47}$$

for any $k = 1, 2, \ldots$.

PROOF. Let us assume that $\mathbb{I}$ is generated by m elements represented by $a_1, \ldots, a_m$ (modulo $\mathcal{I}_V$) in $\mathcal{O}_0$. One has

$$\left[R_0^{\overline{\mathbf{E}_\bullet^{a,k}},\lambda'}\, R^{\overline{\mathbf{F}_\bullet}}\right]_{\lambda'=0} = \left[\Phi_{a,k}^{\lambda'}\, R^{\overline{\mathbf{F}_\bullet}}\right]_{\lambda'=0},$$

where

$$\Phi_{a,k}^{\lambda'} = \sum_{r=1}^{\min(N,m)} \frac{1}{(2i\pi)^r}(\overline{\partial}|a|^{2\lambda'})\,(\sigma)_1\cdots(\sigma)_k \sum_{\{\rho\in\mathbb{N}^k\,:\,|\rho|=r-1\}} \bigwedge_{\kappa=1}^{k}\left(\overline{\partial}\Big(\frac{(\sigma)_\kappa}{2i\pi}\Big)\right)^{\rho_\kappa}.$$

Let h be a representative of an element in $\overline{\mathbb{I}^{m+k-1}}$ defined in a sufficiently small ball $\eta\,\mathbb{B}_N$, with $0 < \eta \ll 1$. The pseudo-meromorphic current

$$h\left[\Phi_{a,k}^{\lambda'}\, R_{0,M}^{\overline{\mathbf{F}_\bullet}}\right]_{\lambda'=0}$$

has maximal bidegree $(N, M+m)$, and its support is contained in $V_{\mathrm{sing}} \cap V(a)$. This follows from Theorem 5.8, when it is applied in the regular $(N-M)$-dimensional local ring $\mathcal{O}_{V,z}$ with $z \in V_{\mathrm{reg}} \in \eta\,\mathbb{B}_N$ for sufficiently small $0 < \eta \ll 1$ instead of $\mathcal{O}_0$. Namely, one has

$$\left[\iota_V^*(h\Phi_{a,k}^{\lambda'})\right]_{\lambda=0} = \left[(\iota_V^* h)\,\Phi_{\iota_V^* a,k}^{\lambda'}\right]_{\lambda'=0} = 0.$$

Since $\operatorname{codim}(V_{\text{sing}} \cap V(a)) \geq M+m+1$ given (5.46), it follows from the dimension principle (see Proposition 2.26) that

$$h\left[\Phi_{a,k}^{\lambda'} R_{0,M}^{\overline{\mathbf{F}_\bullet}}\right]_{\lambda'=0} = 0.$$

It follows from Proposition 3.45 that outside $\mathbb{V}_{M+1}(\mathcal{I}_V)$ and for some smooth bundle section σ_{M+1} there, one has that

$$\begin{aligned} h\Big[\Phi_{a,k}^{\lambda'} R_{0,M+1}^{\overline{\mathbf{F}_\bullet}}\Big]_{\lambda'=0} &= h\Big[\Phi_{a,k}^{\lambda'} (\overline{\partial}\sigma_{M+1})\, R_{0,M}^{\overline{\mathbf{F}_\bullet}}\Big]_{\lambda'=0} \\ &= (\overline{\partial}\sigma_{M+1}) \Big(h\Big[\Phi_{a,k}^{\lambda'} R_{0,M}^{\overline{\mathbf{F}_\bullet}}\Big]_{\lambda'=0}\Big) = 0. \end{aligned} \tag{5.48}$$

Hence

$$\operatorname{Supp}\left(h\left[\Phi_{a,k}^{\lambda'} R_{0,M+1}^{\overline{\mathbf{F}_\bullet}}\right]_{\lambda'=0}\right) \subset \mathbb{V}_{M+1}(\mathcal{I}_V) \cap V(a).$$

Since $\mathbb{V}_{M+1}(\mathcal{I}_V) \cap V(a)$ has codimension that equals at least $M+m+2$, in view of (5.46), the dimension principle applies once again, and one has

$$h\left[\Phi_{a,k}^{\lambda'} R_{0,M+1}^{\overline{\mathbf{F}_\bullet}}\right]_{\lambda'=0} = 0.$$

This procedure iterates given the hypothesis (5.46) and one concludes that

$$h\Big[\Phi_{a,k}^{\lambda'} R^{\overline{\mathbf{F}_\bullet}}\Big]_{\lambda'=0} = 0.$$

Thus, (5.45) implies that

$$\boldsymbol{\nabla}_{\mathbf{E}_\bullet^{a,k} \oplus \mathbf{F}_\bullet}\Big(hP^{\overline{\mathbf{F}_\bullet}} + h\Big[P^{\overline{\mathbf{E}_\bullet^{a,k}},\lambda'} R^{\mathbf{F}_\bullet}\Big]_{\lambda'=0}\Big) = h.$$

Using the argument developed in the proof of Proposition 3.32, namely the proof of (3.86) in the case $\ell = 0$, one concludes that $h \in \mathbb{I}^k$. This ends the proof of the theorem. □

Under some additional constraints on the ideal $\mathbb{I}$, one has that

$$\overline{\mathbb{I}^{(N-M)+k-1}} \subset \mathbb{I}^k,$$

as in the case where $\mathbb{A} = (\mathcal{O}/\mathcal{I}_V)_0$ is regular. The following result illustrates such a situation.

PROPOSITION 5.12. *Assume that the ideal* $\mathbb{I} \subset \mathbb{A} = (\mathcal{O}/\mathcal{I}_V)_0$ *is such that*

$$\forall \mathcal{P} \in \operatorname{Ass}(\mathbb{A}/\mathbb{I}), \quad V(\mathcal{P}) \not\subset V_{\text{sing}}. \tag{5.49}$$

Then $\overline{\mathbb{I}^{(N-M)+k-1}} \subset \mathbb{I}^k$ *for any* $k \in \mathbb{N}^*$.

PROOF. Consider a nonunique, primary decomposition of the ideal $\mathbb{I}$, namely $\mathbb{I} = \mathcal{Q}_1 \cap \cdots \cap \mathcal{Q}_K$, where $\sqrt{\mathcal{Q}_\kappa} = \mathcal{P}_\kappa \in \operatorname{Ass}(\mathbb{A}/\mathbb{I})$ (with possible redundancy). For each $\kappa = 1, \ldots, K$, let $a_{[\kappa],1}, \ldots, a_{[\kappa],m_\kappa} \in \mathcal{O}_0$ be such that their classes modulo $\mathcal{I}_V$ generate $\mathcal{Q}_\kappa$. Let h be the representative in $\mathcal{O}_0$ of an element in $\overline{\mathbb{I}^{(N-M)+k-1}}$. Since

$$\mathbb{I} \subset \mathcal{Q}_\kappa \Longrightarrow \overline{\mathbb{I}^{(N-M)+k-1}} \subset \overline{\mathcal{Q}_\kappa^{(N-M)+k-1}} \quad \forall \kappa = 1, \ldots, K,$$

there exists $C > 0$ and $0 < \eta \ll 1$ such that one has

$$\forall \kappa = 1, \ldots, K, \quad |h| \leq C|a_{[\kappa]}|^{N-M+k-1} \quad \text{on } V \cap \eta\mathbb{B}_N.$$

Then, it follows from Theorem 5.7 that

$$(h)_z \in \mathcal{I}_{V,z} + \Big(\sum_{\mu=1}^{m_\kappa} a_{[\kappa],\mu}\, \mathcal{O}_z\Big)^k$$

for any generic z in $V(\mathcal{P}_\kappa) \cap V_{\rm reg}$. Thus, it remains true for any $z \in V(\mathcal{P}_\kappa)$. Hence, one has that

$$(h)_0 \in \mathcal{I}_{V,0} + \Big(\sum_{\mu=1}^{m_\kappa} a_{[\kappa],\mu}\, \mathcal{O}_0\Big)^k.$$

Then the class $(h)_0$ modulo $\mathcal{I}_V$ belongs to $\mathcal{Q}_\kappa^k$ for any $\kappa = 1, \dots, K$. Hence $(h)_0$ belongs to $\mathbb{I}^k$. □

EXAMPLE 5.13 (The case $M = N - 1$ [**Sz14**]). Let V be a closed analytic curve in a neighborhood $\eta\,\mathbb{B}_N$ of the origin in $\mathbb{C}^N$ $(0 < \eta \ll 1$, $\mathbb{B}_N$ being the unit euclidean ball) and let $\widetilde{\mathcal{O}}_{V,0}$ be the integral closure of $\mathcal{O}_{V,z}$ in $\mathcal{M}_{V,z}$; see §A.4.1.[8] As observed in [**Sz14**, Lemma 2.1], when $M = N - 1$, namely, when V is a curve in a neighborhood of 0, the Briançon–Skoda exponent $\mu_{\mathbb{A}}$ of $\mathbb{A} = (\mathcal{O}/\mathcal{I}_V)_0$ is defined as the least integer $\mu \in \mathbb{N}^*$ such that, for any a in the maximal ideal $\mathfrak{M}_{V,0}$ of $\mathcal{O}_{V,0}$,

(5.50)

$$\big\{(h)_0 \in \widetilde{\mathcal{O}}_{V,0} \,:\, |h| \le C\,|a|^{\mu-1} \text{ in a neighborhood of 0 on } V\big\} \subset \mathcal{O}_{V,0} = (\mathcal{O}/\mathcal{I}_V)_0.$$

Suppose that V is additionally defined in $\eta\,\mathbb{B}_N$ as a reduced complete intersection. That is, assume that there are $f_1, \dots, f_{N-1}$ in $\mathcal{O}(\eta\,\mathbb{B}_N)$ such that

$$V = \{(z, w_1, \dots, w_{N-1}) \in \eta\,\mathbb{B}_N \,:\, f_1(z,w) = \cdots = f_{N-1}(z,w) = 0\}, \quad df \not\equiv 0 \text{ on } V.$$

We also assume that $(z,w) \in V \cap \eta\,\mathbb{B}_N \to z$ is proper. It follows from the duality theorem, Theorem 3.39, for complete intersections[9] that the germ $(h)_0 \in \widetilde{\mathcal{O}}_{V,0}$ belongs to $\mathcal{O}_{V,0}$ if and only if the semimeromorphic current

$$h\,\overline{\partial}\Big(\frac{1}{f_1, \dots, f_{N-1}}\Big) = \Big[|\delta|^{2\lambda}\,\frac{g}{\delta}\,\overline{\partial}\Big(\frac{1}{f_1, \dots, f_{N-1}}\Big)\Big]_{\lambda=0},$$

where δ is a universal Oka denominator (see Theorem A.27), satisfies

(5.51)
$$\begin{aligned}\overline{\partial}\Big(h\,\overline{\partial}\Big(\frac{1}{f_1, \dots, f_{N-1}}\Big)\Big) &= g\,\Big[\overline{\partial}\Big(\frac{|\delta|^{2\lambda}}{\delta}\Big) \wedge \overline{\partial}\Big(\frac{1}{f_1, \dots, f_{N-1}}\Big)\Big]_{\lambda=0} \\ &= g\,\overline{\partial}\Big(\frac{1}{\delta, f_1, \dots, f_{N-1}}\Big) = 0.\end{aligned}$$

Thus, one can rephrase (5.50) in this context as follows: given a in $\mathfrak{M}_{V,0}$, for any germ of a weakly holomorphic function h on V at the origin,

(5.52)
$$\big(|h| \le C|a|^{\mu-1} \text{ in a neighborhood of 0 on } V\big) \Longrightarrow \Big(\forall \varphi \in \big(\mathcal{D}(\Omega)\big)_0,\ \int_{V_{\rm reg}} \Big[\frac{h\,d\zeta}{df}\Big]\,\overline{\partial}\varphi = 0\Big),$$

where $[\cdot/\cdot]$ means the division of $(N,0)$-forms by nonvanishing $(N-1,0)$-forms. Let J be the jacobian determinant $\det\big[\partial f_j/\partial w_k\big]_{1\le j,k\le N-1}$ and let $\widehat{\Pi} : \mathbb{C} \to V$ be the normalized blowup of V along the maximal ideal $\mathfrak{M}_{V,0}$ of $\mathbb{A}$. Let also $\mu_0(V)$ be the

[8] The sheaf $\widetilde{\mathcal{O}}_V$ is the sheaf of weakly holomorphic functions on $V \hookrightarrow \Omega$; see Definition A.25.

[9] This was pointed out in [**Ts**] and generalized in [**TsY04**, Theorem 5.2]. Another generalization of a strong holomorphicity criterion is proposed in [**And10**].

multiplicity of V at the origin. That is, $\mu_0(V)$ is the Lelong number of $[V]$ at 0; see §A.3.4 and Example C.16. Then, one may assume, according to Puiseux's theorem [**Lef53, Wh**], up to a change of coordinates, that

$$\widehat{\Pi}\ : t \mapsto \big(t^{\mu_0(V)}, \textstyle\sum_{k\geq \mu_0(V)} \gamma_k t^k\big),$$

with $\gamma_k \in \mathbb{C}^{N-1}$. Then such normalization provides a local parametrization of the germ of curve V_0, and (5.52) can be reformulated on the normalization $\mathbb{C}_0$ as

$$(5.53)\quad \forall\, a \in \mathfrak{M}_{V,0},\ \forall \varphi \in (\mathcal{D}(\Omega))_0,\ \forall \mathfrak{h} \in \mathcal{O}_{\mathbb{C},0},$$

$$\int_{\mathbb{C}} \mathfrak{h}(t)\, \frac{t^{(\mu-1)\nu_0(\widehat{\Pi}^* a)}}{t^{\nu_0(\widehat{\Pi}^* J)}}\, d\, t^{\mu_0(V)} \wedge (\widehat{\Pi}^*\overline{\partial}\varphi)(t) = 0.$$

This is also equivalent to

$$(5.54)\qquad \forall\, a \in \mathfrak{M}_{V,0},\quad (\mu-1)\,\nu_0(\widehat{\Pi}^* a) \geq \nu_0(\widehat{\Pi}^* J) - (\mu_0(V)-1).$$

Since the possible minimal value for $\nu_0(\widehat{\Pi}^* a)$ equals $\mu_0(V)$ (and is achieved when $a = z$ modulo $\mathcal{I}_{V,0}$), one gets the upper estimate

$$(5.55)\qquad \mu_{\mathbb{A}} \leq \operatorname{ceil}\Big(\frac{1+\nu_0(\widehat{\Pi}^* J)}{\mu_0(V)}\Big).$$

In fact (5.55) is an equality; see [**Sz14**, Lemma 2.4]. In the particular case where $N = 2$ and V is a germ at 0 of an irreducible hypersurface $\{f(z_1, z_2) = 0\}$, then $\mu_0(V)$ coincides with the Milnor number of f at the isolated singularity $(0,0)$. Namely, $\mu_0(V)$ coincides with the dimension of the vector space $\mathcal{O}_0(\mathbb{C}^2)/(\partial_{z_1} f, \partial_{z_2} f)$. We refer the reader to [**Sz14**], Theorem 1.1, for more details. Note that such a Briançon–Skoda exponent $\mu_{\mathbb{A}}$ can also be interpreted in that case using Puiseux theory (see [**Lef53, Wh**]) in terms of Puiseux characteristics.

5.4. Briançon–Skoda theorem for purely dimensional $\mathbb{A} = (\mathcal{O}/\mathcal{I})_0$

Let $\Omega \subset \mathbb{C}^N$ be a domain containing the origin and let $\mathcal{I} \subset \mathcal{O}_\Omega$ be a coherent sheaf of ideals such that $\mathbb{A} = (\mathcal{O}/\mathcal{I})_0$ is purely dimensional. Let also $0 < \eta \ll 1$ be sufficiently small, and let the complex

$$\mathbf{F}_\bullet : 0 \to E_L \simeq \mathbb{C}^{M_L} \xrightarrow{F_L} E_{L-1} \simeq \mathbb{C}^{M_{L-1}} \xrightarrow{F_{L-1}} \cdots \longrightarrow E_1 \simeq \mathbb{C}^{M_1} \xrightarrow{F_1} E_0 = \mathbb{C} \longrightarrow 0,$$

with F_1 generically surjective, be such that the corresponding sequence of sheaves of sections

$$0 \longrightarrow \mathcal{O}^{\oplus r_L}_{\eta\,\mathbb{B}_N} \xrightarrow{F_L} \mathcal{O}^{\oplus r_{L-1}}_{\eta\,\mathbb{B}_N} \xrightarrow{F_{L-1}} \cdots \xrightarrow{F_2} \mathcal{O}^{\oplus r_1}_{\eta\,\mathbb{B}_N} \xrightarrow{F_1} \mathcal{O}_{\eta\,\mathbb{B}_N}$$

is exact over $\eta\,\mathbb{B}_N$ with $\operatorname{coker} F_1 = (\mathcal{O}/\mathcal{I})_{|\eta\,\mathbb{B}_N}$. It follows from the fact that $\mathcal{O}/\mathcal{I}$ is purely dimensional (see Example 3.46) that the strong Buchsbaum–Eisenbud–Fitting conditions

$$\mathbb{V}_1(\mathcal{I}) = \cdots = \mathbb{V}_M(\mathcal{I}) = V, \quad \operatorname{codim}(\mathbb{V}_{M+\rho}(\mathcal{I})) > M+\rho \quad \forall \rho \geq 1$$

remain fulfilled. In addition, $\mathbb{V}_{M+1}(\mathcal{I}) \subset V_{\text{sing}}$ holds, where $V = \operatorname{Supp}(\mathcal{O}/\mathcal{I})_{|\eta\,\mathbb{B}_N}$. Moreover, if $\overline{\mathbf{F}_\bullet}$ is a metrized version of $\mathbf{F}_\bullet$, then the following extrinsic description of the residue current $R^{\overline{\mathbf{F}_\bullet}} = R^{\mathbf{F}_\bullet}$ is given by the conjunction of Propositions 3.45 and 3.47. Actually, without loss of generality, when all bundles happen to be trivial over $\eta\,\mathbb{B}_N$ and are assumed to be equipped with the canonical hermitian metric, there exists a Coleff–Herrera current

$$T = \sum_{\iota=1}^{r(T)} T_\iota\,\epsilon_\iota \in \mathrm{CH}^0_{\eta\,\mathbb{B}_N, V\cap\eta\,\mathbb{B}_N}(\eta\,\mathbb{B}_N, \mathbb{C}^{r(T)}) \tag{5.56}$$

and a collection of vector-valued differential forms $\Sigma_0, \dots, \Sigma_{N-M}$ defined almost everywhere in $\eta\,\mathbb{B}_N$ that have the following properties. The above forms are expressed by

$$\begin{aligned} \Sigma_0 &= \sigma_G = \sum_{\iota=1}^{r(\Sigma_0)} \frac{\overline{\delta_{0,\iota}}}{\|\delta_0\|^2}\,\epsilon_\iota^{[0]}, \\ \Sigma_\rho &= \overline{\partial}\Big(\frac{\sigma_{M+\rho}}{2i\pi}\Big) = \sum_{\iota=1}^{r(\Sigma_\rho)} \overline{\partial}\Big(\frac{\overline{\delta_{\rho,\iota}}}{\|\delta_\rho\|^2}\Big)\,\epsilon_\iota^{[\rho]}, \quad \rho = 1, \dots, N-M, \end{aligned} \tag{5.57}$$

where by $\delta_{\rho,\iota}$ we denote holomorphic functions in $\eta\,\mathbb{B}_N$. The residue current

$$R^{\overline{\mathbf{F}_\bullet}} = R_0^{\overline{\mathbf{F}_\bullet}} = \sum_{\rho=0}^{N-M} R^{\overline{\mathbf{F}_\bullet}}_{0,M+\rho} \tag{5.58}$$

is such that for $\rho = 0, \dots, N-M$,

$$\begin{aligned} R^{\overline{\mathbf{F}_\bullet}}_{0,M+\rho} &= \Big[\cdots\Big[\Big(\prod_{\gamma=0}^{\rho} \|\delta_{\rho-\gamma}\|^{2\mu_{\rho-\gamma}}\Sigma_{\rho-\gamma}\Big)\,T\Big]_{\mu_0=0}\cdots\Big]_{\mu_\rho=0} \\ &= \Big[\cdots\Big[\Omega_\rho^{\mu}\,T\Big]_{\mu_0=0}\cdots\Big]_{\mu_\rho=0}. \end{aligned} \tag{5.59}$$

Any coordinate T_ι of the vector-valued Coleff–Herrera current T inherits the local structure of Coleff–Herrera currents; see Proposition 2.37. More precisely, given $h \in \mathcal{O}(\eta\,\mathbb{B}_N)$ and $\varphi \in \mathcal{D}^{(N,N-M)}(\eta\,\mathbb{B}_N)$, one has[10]

$$\langle h\,T_\iota, \varphi\rangle = \sum_\nu \int_{V\cap\eta\,\mathbb{B}_N} \mathcal{N}_{\iota,\nu}(\zeta, \partial/\partial\zeta)(h)\,\frac{\boldsymbol{\mathcal{Q}}_{\iota,\nu}\,(\theta_\iota\rfloor\varphi)}{\Delta}, \tag{5.60}$$

where

$$\begin{aligned} \mathcal{N}_{\iota,\nu} &= \mathcal{N}_{\iota,\nu}(\zeta, \partial/\partial\zeta) : \mathcal{O}(\eta\,\mathbb{B}_N) \longrightarrow \mathcal{O}(\eta\,\mathbb{B}_N), \\ \boldsymbol{\mathcal{Q}}_{\iota,\nu} &= \boldsymbol{\mathcal{Q}}_{\iota,\nu}(\zeta, \partial/\partial\zeta) : C^\infty_{N-M,N-M}(\eta\,\mathbb{B}_N) \longrightarrow C^\infty_{N-M,N-M}(\eta\,\mathbb{B}_N) \end{aligned} \tag{5.61}$$

are holomorphic differential operators, θ_ι is a smooth section of $\bigwedge^M T^{(1,0)}_{\eta\,\mathbb{B}_N}$ acting from the left by interior contraction $C^\infty_{N,N-M}(\eta\,\mathbb{B}_N) \longrightarrow C^\infty_{N-M,N-M}(\eta\,\mathbb{B}_N)$, and the denominator $\Delta \in \mathcal{O}(\eta\,\mathbb{B}_N)$ is a holomorphic function in $\eta\,\mathbb{B}_N$ satisfying

$$(V \cap \eta\,\mathbb{B}_N) \cap V(\Delta) \subset V_{\mathrm{sing}} \cap \eta\,\mathbb{B}_N. \tag{5.62}$$

We remark that Δ is a power of the Oka universal denominator for V in a neighborhood of the origin; see §A.3.5. Moreover, one has

$$h \in \operatorname{ann} T_\iota \iff \big((\mathcal{N}_{\iota,\nu}(\zeta, \partial/\partial\zeta)\,h)_{|V\cap\eta\,\mathbb{B}_N} = 0\big). \tag{5.63}$$

---

[10] See [**Sz18**, Theorem 2.2] for such reformulation of Jan-Erik Björk's local structure theorem as stated previously in Proposition 2.37.

The collection of differential operators $\mathcal{N}_{\iota,\nu}(\zeta,\partial/\partial\zeta)$, determined by all possible values of ι,ν, defines the annihilator of the Coleff–Herrera current T. Since

$$\operatorname{ann} T \subset \operatorname{ann} R^{\overline{\mathbf{F}_\bullet}}_{0,M} = \operatorname{ann} R_0^{\mathbf{F}_\bullet} = \operatorname{ann} R^{\mathbf{F}_\bullet},$$

because $(\mathcal{O}/\mathcal{I})_0$ is purely dimensional (see the proof of Theorem 5.3) it follows from Theorem 3.38 that

$$\Big(\big(\mathcal{N}_{\iota,\nu}(\zeta,\partial/\partial\zeta)\, h\big)_{|V\cap\eta\,\mathbb{B}_N} = 0 \quad \forall\, \iota,\nu\Big) \Longrightarrow h \in \mathcal{I}(\eta\,\mathbb{B}_N). \tag{5.64}$$

With all preliminary material settled, one can state the following version of the Briançon–Skoda's theorem in this particular nonreduced case. The result is due to Jacob Sznajdman [**Sz18**, Theorem 1.2].

THEOREM 5.14. *Let $\mathbb{A} = (\mathcal{O}/\mathcal{I})_0$ be purely dimensional and let $\mathbf{F}_\bullet$ be a generically exact complex of trivial holomorphic bundles inducing a locally free resolution of $\mathcal{O}/\mathcal{I}$ over $\eta\,\mathbb{B}_N$ for $0<\eta\ll 1$. Assume also that T is a bundle-valued Coleff–Herrera current in $\eta\,\mathbb{B}_N$ such that the residue current $R^{\mathbf{F}_\bullet}$ attached to some metrized version of $\mathbf{F}_\bullet$ splits there as (5.58), (5.59). Let also $\mathcal{N}_\nu^{\mathbb{A}} = \{\mathcal{N}_{\iota,\nu}\}_{\iota,\nu}$ be the collection of holomorphic differential operators involved in Leibniz's relations (5.60) for all coordinates T_ι of T and defining the annihilator of the germ of Coleff–Herrera current T. Then there exists a positive integer $\mu'_{\mathbb{A}}$ such that for any ideal $\mathbb{I}$ in $\mathbb{A}$, for any $(h)_0 \in \mathbb{I}$, where $a_1,\dots,a_m, h$ are representing generators of $\mathbb{I}$ and of $(h)_0$ modulo $\mathcal{I}(\eta\,\mathbb{B}_N)$ in $\eta\,\mathbb{B}_N$, respectively,*

$$\Big(\max_{\nu} |\mathcal{N}_\nu^{\mathbb{A}}(\zeta,\partial/\partial\zeta)(h)| \le C\,|a|^{\mu'_{\mathbb{A}}+N-M+k-1} \ \text{on } V\cap\eta\,\mathbb{B}_N\Big) \Longrightarrow (h)_0 \in \mathbb{I}^k, \tag{5.65}$$

for any $k\in\mathbb{N}^*$.

PROOF. Let $h\in\mathbb{A}$, with its representative also denoted by h in $\eta\,\mathbb{B}_N$, and let $a_1,\dots,a_m$ be representatives of generators of $\mathbb{I}$ modulo $\mathcal{I}(\eta\,\mathbb{B}_N)$ such that

$$|\mathcal{N}_{\iota,\nu} h| \le C|a|^{\mu'+(N-M)+k-1} \quad \text{on } V\cap\eta\,\mathbb{B}_M, \tag{5.66}$$

for $\mu'\gg 1$. The value of μ' will be specified later on. Keeping the notations of the proof of Theorem 5.10 with the same initial reasoning, it is sufficient, in order to prove that $h\in\mathbb{I}^k$, to show that for any $\rho, r\le m$ such that $\rho+r\le N-M$

$$\begin{aligned} h\bigg[\Big[\cdots\big[\rho^{\lambda'}_{a,k,r}\,\Omega^\mu_\rho\, T\big]_{\mu_0=0}\cdots\Big]_{\mu_\rho=0}\bigg]_{\lambda'=0} \\ = h\bigg[\Big[\cdots\big[\rho^{\lambda'}_{a,k,r}\,\Omega^\mu_\rho\, h\,T\big]_{\mu_0=0}\cdots\Big]_{\mu_\rho=0}\bigg]_{\lambda'=0} = 0. \end{aligned} \tag{5.67}$$

First, we take first both $\operatorname{Re}\mu_\gamma\gg 1$ for $\gamma=0,\dots,\rho$ and $\operatorname{Re}\lambda'\gg 1$. Since $\rho^{\lambda'}_{a,k,r}\,\Omega^\mu_\rho$ is a differential $(r, r+\rho)$-form, which is smooth up to an arbitrarily high order, the current defined in the left-hand side of (5.67) is an $(N-M+r, N-M+r+\rho)$-current. This current is to be tested against $\mathbb{C}$-valued $(M-r, M-(r+\rho))$-forms ψ in $\eta\,\mathbb{B}_N$. It is enough to prove for each such test form ψ in $\eta\,\mathbb{B}_N$ and each ι that

$$\bigg[\Big[\cdots\big[\langle h\,T_\iota\,,\,\rho^{\lambda'}_{a,k,r}\Omega^\mu_\rho\,\psi\rangle\big]_{\mu_0=0}\cdots\Big]_{\mu_\rho=0}\bigg]_{\lambda'=0} = 0. \tag{5.68}$$

We may even assume that each differential form $\Sigma_{\rho-\gamma}$, $\gamma = 0, \dots, \rho$, involved in the definition of Ω^λ_ρ (see (5.59)) is replaced by one of its $\mathbb{C}$-valued coordinates. These coordinates are of the form specified in (5.57). Let then

$$\breve{\omega}^\mu_\rho = \Big(\prod_{\gamma=0}^{\rho-1} \|\delta_{\rho-\gamma}\|^{2\mu_{\rho-\gamma}}\, \overline{\partial}\Big(\frac{\overline{\delta_{\rho-\gamma,1}}}{\|\delta_{\rho-\gamma}\|^2}\Big)\Big)\, \|\delta_0\|^{2\mu_0}\, \frac{\overline{\delta_{0,1}}}{\|\delta_0\|^2}$$

be the contribution corresponding to the use of the first coordinate. Proving (5.68) boils down to proving that

$$\Big[\Big[\cdots\Big[[\langle h\,|\Delta|^{2\lambda}\, T_\iota\,,\, \rho^{\lambda'}_{a,k,r}\,\breve{\omega}^\mu_\rho\, \chi\rangle]_{\lambda=0}\Big]_{\mu_0=0}\cdots\Big]_{\mu_\rho=0}\Big]_{\lambda'=0} = 0. \tag{5.69}$$

Let $\alpha = (\alpha_1, \dots, \alpha_N) \in \mathbb{N}^N$. One has that

$$\Big(\frac{\partial}{\partial\zeta}\Big)^\alpha \Big(\|\delta_0\|^{2\mu_0}\Big(\frac{\overline{\delta_{0,1}}}{\|\delta_0\|^2}\Big)\Big) = \Big(\prod_{\kappa=1}^{|\alpha|}(\mu_0-\kappa)\Big)\, \frac{\Phi_{0,\alpha}(\zeta, \overline{\delta_{0,1}}, \dots, \overline{\delta_{\gamma,r(\Sigma_0)}})}{\|\delta_0\|^{2(1+|\alpha|)}}, \tag{5.70}$$

where $\Phi_{0,\alpha}$ is holomorphic in the first variables ζ and homogeneous with degree $1+|\alpha|$ in the block formed by the last $r(\Sigma_0)$ ones. Similarly, for each $\gamma = 1, \dots, \rho$, one has

$$\begin{aligned}\Big(\frac{\partial}{\partial\zeta}\Big)^\alpha \Big(\overline{\partial}\Big(\|\delta_\gamma\|^{2\mu_\gamma}\frac{\overline{\delta_{\gamma,1}}}{\|\delta_\gamma\|^2}\Big)\Big) &= \overline{\partial}\Big(\Big(\frac{\partial}{\partial\zeta}\Big)^\alpha \Big(\|\delta_\gamma\|^{2\mu_\gamma}\frac{\overline{\delta_{\gamma,1}}}{\|\delta_\gamma\|^2}\Big)\Big) \\ &= \Big(\prod_{\kappa=1}^{|\alpha|}(\mu_\gamma-\kappa)\Big)\,\overline{\partial}\Big(\frac{\Phi_{\gamma,\alpha}(\zeta, \overline{\delta_{\gamma,1}}, \dots, \overline{\delta_{\gamma,r(\Sigma_\gamma)}})}{\|\delta_\gamma\|^{2(1+|\alpha|)}}\Big),\end{aligned} \tag{5.71}$$

where $\Phi_{\gamma,\alpha}$ is holomorphic in the first variables ζ and homogeneous with degree $1+|\alpha|$ in the block formed by the last $r(\Sigma_\gamma)$ ones. Furthermore, for $\beta \in \mathbb{N}^N$ and $\mathrm{Re}\,\lambda' \gg 1$ we consider the forms, defined using the following test function $\chi : [1:k] \to [1:m]$ and the increasing multi-index $J : 1 \le j_1 < \cdots < j_r \le m$, as

$$\Big(\frac{\partial}{\partial\zeta}\Big)^\beta \Big(\lambda'\,\|a\|^{2\lambda'}\frac{\overline{a}_\chi}{\|a\|^{2(r+1+k)}}\,\overline{da_J}\Big) = \Big(\Big(\frac{\partial}{\partial\zeta}\Big)^\beta \Big(\lambda'\,\|a\|^{2\lambda'}\frac{\overline{a}_\chi}{\|a\|^{2(r+1+k)}}\Big)\Big)\,\overline{da_J},$$

where $a_\chi = \prod_{\kappa=1}^k a_{\chi(\kappa)}$ and $\overline{da_{\mathcal{J}}} = \overline{da_{j_1}} \wedge \cdots \wedge \overline{da_{j_r}}$. One has

$$\begin{aligned}&\Big(\frac{\partial}{\partial\zeta}\Big)^\beta \Big(\lambda'\,\|a\|^{2\lambda'}\frac{\overline{a}_\chi}{\|a\|^{2(r+1+k)}}\,\overline{da_{\mathcal{J}}}\Big) \\ &\qquad = \lambda'\Big(\prod_{\kappa=r+k+1}^{r+k+|\beta|}(\lambda'-\kappa)\Big)\|a\|^{2\lambda'}\,\frac{\Psi_{\chi,\beta}(\zeta, \overline{a_1}, \dots, \overline{a_m})}{\|a\|^{2(r+1+k+|\beta|)}}\,\overline{da_{\mathcal{J}}},\end{aligned} \tag{5.72}$$

where $\Psi_{\chi,\beta}(\zeta, \overline{a}_1, \dots, \overline{a}_m)$ is holomorphic in ζ and homogeneous with degree $k+|\beta|$ in the block of variables $(\overline{a}_1, \dots, \overline{a}_m)$. Consider a log resolution[11]

$$\Pi\, : \mathcal{V} \longrightarrow V \cap \eta\,\mathbb{B}_N$$

such that $\Pi^{-1}(V(\Delta) \cap V \cap \eta\,\mathbb{B}_N)$ is a hypersurface with normal crossings and the inverse image $\mathrm{Fitt}_{M+\gamma}(\mathcal{I})\,\mathcal{O}_{\mathcal{V}}$ of the $(M+\gamma)$-Fitting coherent sheaf of ideals through the log resolution Π is principal with zero locus also a hypersurface with normal

[11]The existence of such a log resolution follows from Hironaka's theorem, Theorem A.41, when combined with the combinatorial resolution of singularities of toric varieties; see Theorem 3.11 and Example 3.12.

crossings for any $\gamma = 1, \ldots, \rho$. Leibniz's theorem and relations (5.70), (5.71) imply that one has for any $\alpha \in \mathbb{N}^N$, in local coordinates $w_1, \ldots, w_{N-M}$ in a neighborhood of a point v^0 on $\mathcal{V}$, that

$$\text{(5.73)} \quad (\iota_V \circ \Pi)^* \Big(\frac{|\Delta|^{2\lambda}}{\Delta} \Big(\frac{\partial}{\partial \zeta}\Big)^{\alpha} (\check{\omega}^{\mu}_{\rho}) \Big) = \sum_{\ell} Q^{\ell}_{\alpha}(\mu)\, |u^{\ell}|^{2\lambda} \prod_{\gamma=0}^{\rho} |u^{\ell}_{\gamma}|^{2\mu_\ell} \Big(\prod_{j=1}^{N-M} \frac{|w_j|^{2(\nu^{\ell}_j \lambda + \langle\!\langle m^{\ell}_j, \mu \rangle\!\rangle)}}{w_j^{N^{\ell}_{\alpha,j}}} \Big)\, \varpi^{\ell}_{\alpha},$$

where

$$\begin{aligned} &Q^{\ell}_{\alpha} \in \mathbb{C}[\mu_0, \ldots, \mu_\rho],\ \varpi^{\ell}_{\alpha} \in (C^{\infty}_{0,\rho})_{v^0},\ N^{\ell}_{\alpha,j} \in \mathbb{N}, \\ &\nu^{\ell}_j \in \mathbb{N},\ m^{\ell}_j \in \mathbb{N}^{\rho+1}, \text{ with } N^{\ell}_{\alpha,j} \geq 1 \Longrightarrow (\nu^{\ell}_j, \mu^{\ell}_j) \neq (0, \ldots, 0), \\ &u^{\ell},\ u^{\ell}_{\gamma} \in \mathcal{O}^*_{\mathcal{X}, v^0}. \end{aligned}$$

Let $M(\mathcal{Q})$ be the maximal order of the differential operators $\mathcal{Q}_{\iota,\nu}$ in (5.60). Choose μ' in (5.66) such that

$$\text{(5.74)} \qquad \mu' \geq M(\mathcal{Q}) + \boldsymbol{N}(v^0) + 2(N - M + 1),$$

where

$$\boldsymbol{N}(v^0) = \max_{|\alpha| \leq M(\mathcal{Q})} \max_{\ell} \Big(\sum_{\{j \,:\, N^{\ell}_{\alpha,j} \geq 1\}} n^{\ell}_{\alpha,j} \Big),$$

with $n^{\ell}_{\alpha,j} = N^{\ell}_{\alpha,j} - 1$. Let, for any ι, ν,

$$\text{(5.75)} \qquad \begin{aligned} H_{\iota,\nu} &= (\iota_V \circ \Pi)^* \big(\mathcal{N}_{\iota,\nu}(\zeta, \partial/\partial\zeta)(h) \big), \\ A &= (\iota_V \circ \Pi)^*(a). \end{aligned}$$

Since there is a constant C such that, for any ι, ν,

$$|H_{\iota,\nu}| \leq C |A|^{\mu' + (N-M) + k - 1}$$

in a neighborhood of v^0 on $\mathcal{V}$ according to (5.66), it follows from Theorem 5.8 (with $\mathbb{A} = \mathcal{O}_{\mathcal{V}, v^0}$ instead of $\mathcal{O}_0$) that for all $\alpha^{\circ} \in \mathbb{N}^{N-M}$ with

$$|\alpha^{\circ}| \leq \max_{|\alpha| \leq M(\mathcal{Q})} \max_{\ell} \Big(\sum_{\{j \,:\, N^{\ell}_{\alpha,j} \geq 1\}} n^{\ell}_{\alpha,j} \Big),$$

one has, for any pair of indexes ι, ν, that

$$\text{(5.76)} \qquad \Big(\Big(\frac{\partial}{\partial w}\Big)^{\alpha^{\circ}} (H_{\iota,\nu}) \Big)_{v^0} \in \Big(\sum_{j=1}^{m} A_j\, \mathcal{O}_{\mathcal{V}, v^0} \Big)^{M(\mathcal{Q}) + (\boldsymbol{N}(v^0) - |\alpha^{\circ}|) + k + 2(N-M+1)}.$$

For $\iota, \nu, \rho, r, \psi$ fixed and $\operatorname{Re} \lambda' \gg 1$ sufficiently large, we consider the vector-valued integral

$$\text{(5.77)} \qquad \int_{V \cap \eta \mathbb{B}_N} \mathcal{N}_{\iota,\nu}(\zeta, \partial/\partial\zeta)(h)\, \frac{\mathcal{Q}_{\iota,\nu}\, (\theta_\iota \rfloor \Omega^{\mu}_{\rho}\, \rho^{\lambda'}_{a,k,r}\, \psi)}{\Delta}$$

involved in the splitting of

$$\langle h\, T_\iota\,,\, \rho^{\lambda'}_{a,k,r} \Omega^{\mu}_{\rho}\, \chi \rangle$$

as given by (5.60). Given Leibniz's formula, each of the coordinates of the integral (5.77) decomposes as a linear combination of integrals of the form

(5.78)
$$\int_{V\cap\eta\,\mathbb{B}_N} \frac{\mathcal{N}_{\iota,\nu}(\zeta,\partial/\partial\zeta)(h)}{\Delta}\Big(\Big(\frac{\partial}{\partial\zeta}\Big)^{\alpha}\check{\omega}_\rho^\mu\Big)\Big(\Big(\Big(\frac{\partial}{\partial\zeta}\Big)^{\beta}\Big(\lambda'\,\|a\|^{2\lambda'}\frac{\overline{a}_\chi}{\|a\|^{2(r+1+k)}}\Big)\Big)\overline{da_J}\Big)\check{\psi}_z,$$

where multi-orders of differentiation $\alpha,\beta\in\mathbb{N}^N$ satisfy $|\alpha|+|\beta|\le M(\boldsymbol{\mathcal{Q}})$. As usual, in (5.78) χ maps $[1:k]$ to $[1:m]$, the multi-index $J: 1\le j_1<\cdots<j_r\le m$ is increasing, and $\check{\psi}_z$ is a test form with bidegree $(N-M, N-M-(\rho+r))$, depending holomorphically of the parameter $z\in(\eta/2)\,\mathbb{B}_N$. We now express the lifted integral (5.78) over $\eta\,\mathbb{B}_N$, after introducing the multiplicative correction factor $|\Delta|^{2\lambda}$ within the integral, keeping for the moment all complex parameters $\lambda,\mu_0,\dots,\mu_\rho,\lambda'$ with arbitrarily large real parts, as an integral over the smooth $N-M$ complex manifold $\mathcal{V}$. We first use (5.73) and consider the contribution arising from a neighborhood of a point v^0 on the complex manifold $\mathcal{V}$. Consider the analytic continuation of the lifted integral (5.78) in (λ,μ) up to $\lambda=0$, $\mu=0$, when the integrand is multiplied by a test function θ supported by an open sufficiently small neighborhood of v^0. The test function originates from the partition of unity subordinate to $\Pi^{-1}(\operatorname{Supp}\psi)$. After univariate integrations by parts corresponding to the holomorphic differential operators $\partial/\partial w_j$, $j=0,\dots,N-M$ (see (1.7)), we are lead to a linear combination of expressions

(5.79)
$$\int_{\mathcal{V}}\Big(\frac{\partial}{\partial w}\Big)^{n_\alpha^\ell(v^0)}\Big(H_{\iota,\nu}\,\Pi^*\Big(\Big(\frac{\partial}{\partial\zeta}\Big)^{\beta}\Big(\lambda'\,\|a\|^{2\lambda'}\frac{\overline{a}_\chi}{\|a\|^{2(r+1+k)}}\Big)\Big)\Big)\,\frac{dw}{w}\wedge\overline{dA_J}\wedge\xi_z^{v^0,\ell}(w)\,\theta,$$

where $\xi_z^{v^0,\ell}$ is a smooth $(0,N-M-(\rho+r))$-form depending holomorphically on z. Using Leibniz's rule, we expand each

$$\Big(\frac{\partial}{\partial w}\Big)^{n_\alpha^\ell(v^0)}\Big(H_{\iota,\nu}\,\Pi^*\Big(\Big(\frac{\partial}{\partial\zeta}\Big)^{\beta}\Big(\lambda'\,\|A\|^{2\lambda'}\frac{\overline{A}_\chi}{\|A\|^{2(r+1+k)}}\Big)\Big)\Big)$$
$$= \sum_{|\alpha^\circ|+|\beta^\circ|\le|n_\alpha^\ell(v^0)|}\kappa_{\alpha^\circ}^{\beta^\circ}$$
$$\times\Big(\Big(\frac{\partial}{\partial w}\Big)^{\alpha^\circ}H_{\iota,\nu}\Big)\Big(\frac{\partial}{\partial w}\Big)^{\beta^0}\Big(\Pi^*\Big(\Big(\frac{\partial}{\partial\zeta}\Big)^{\beta}\Big(\lambda'\|a\|^{2\lambda'}\frac{\overline{a}_\chi}{\|a\|^{2(r+1+k)}}\Big)\Big)\Big).$$

Consider now the normalized blowup $\widehat{\mathcal{V}}_A$ of a neighborhood of v^0 in $\mathcal{V}$ along the ideal sheaf $\sum_{j=1}^m A_j\mathcal{O}_{\mathcal{V}}$. If one takes into account the membership realizations (5.76) together with formulae (5.72), the definition of $M(\boldsymbol{\mathcal{Q}})$, and the fact that

$$\boldsymbol{N}(v^0)\ge|n_\alpha^\ell(v^0)|=|\alpha^\circ|+|\beta^\circ|$$

for any $\alpha\in\mathbb{N}^{N-M}$ with $|\alpha|\le M(\boldsymbol{\mathcal{Q}})$, then one concludes, after lifting all integrals (5.79) to $\widehat{\mathcal{V}}_A$, that each coordinate of

$$\Big[\int_{V\cap\eta\,\mathbb{B}_N}\mathcal{N}_{\iota,\nu}(\zeta,\partial/\partial\zeta)(h)\,\frac{\boldsymbol{\mathcal{Q}}_{\iota,\nu}\,(\theta_\iota\rfloor\Omega_\rho^\mu\,\rho_{a,k,r}^{\lambda'}\,\psi)}{\Delta}\Big]_{\lambda'=0}$$

equals 0, provided that

$$\mu'\ge M(\boldsymbol{\mathcal{Q}})+\boldsymbol{N}(v^0)+2(M-N+1)$$

for any v^0 in $\Pi^{-1}(\operatorname{Supp}\psi)$. Since only a finite number of such v^0 needs to be considered because of the compactness argument, the proof of Theorem 5.14 is thus completed. We refer the reader to [**Sz18**, §4]. There, the proof of the above theorem involves λ' regularization of the weight attached to the Eagon–Northcott complex and it is described with more detail,[12] leading to the refined condition $\mu'_{\mathbb{A}} \geq M(\boldsymbol{Q})+\boldsymbol{N}+k+\max(m, N-M)$ in place of $\mu'_{\mathbb{A}} \geq M(\boldsymbol{Q})+\boldsymbol{N}+k+2(N-M+1)$, which we required in the proof presented here. □

5.5. Primary ideals in $\mathcal{O}_0$

Let us now come back to the case where $\mathbb{A} = \mathcal{O}_0$ (see §5.2) and consider the particular case of $\mathfrak{M}_0$-primary ideals $\mathbb{I}$, where $\mathfrak{M}_0$ denotes the maximal ideal of $\mathbb{A}$. We let $a_1, \dots, a_m$ be representatives of generators for $\mathbb{I}$ in $\eta\,\mathbb{B}_N$, where $0 < \eta \ll 1$ is sufficiently small. Since $\mathbb{I}$ is $\mathfrak{M}_0$-primary, one may assume that there exists a generically exact complex of trivial bundles $\mathbf{F}^a_\bullet$ having length $L = N = \operatorname{codim} V(a)$, $E_0(\mathbf{F}^a_\bullet) = \mathbb{C}$, and the morphism F^a_1 being generically surjective. It induces, at the level of sheaves of sections in $\eta\,\mathbb{B}_N$, a minimal free resolution of the Cohen–Macaulay quotient sheaf $(\mathcal{O}/\sum_{j=1}^m a_j\mathcal{O})_{\eta\,\mathbb{B}_N}$. We denote as R^a the vector-valued Coleff–Herrera current

$$R^a = R^{\mathbf{F}^a_\bullet} = R^{\overline{\mathbf{F}^a_\bullet}}_{0,N}. \tag{5.80}$$

Let

$$H_a(z,\cdot) = \sum_{j=1}^m h_{a,j}(z,\cdot)\,\epsilon_j \in \mathscr{O}(\eta\,\mathbb{B}_N \times \eta\,\mathbb{B}_N, \pi^* T^{(1,0)*}_{\eta\,\mathbb{B}_N})$$

be a Hefer form for $a = a_1\,\epsilon_1 + \cdots + a_m\,\epsilon_m$ (see §4.6.1) where

$$h_{a,j}(z,\cdot) = \sum_{k=1}^N h_{a,j,k}(z,\cdot)\,d\zeta_k.$$

We denote by $1\otimes a - a\otimes 1$ the holomorphic section of $\mathscr{O}(\eta\,\mathbb{B}_N \times \eta\,\mathbb{B}_N, \mathbb{C}^m)$ defined by

(5.81)

$$(1 \otimes a - a \otimes 1)(z,\zeta) = \sum_{j=1}^m (a_j(\zeta) - a_j(z))\,\epsilon_j = \sum_{k=1}^N (\zeta_k - z_k)\Big(\sum_{j=1}^m h_{a,j,k}(z,\zeta)\,\epsilon_j\Big).$$

Let also $H \in \mathscr{O}\big(\eta\,\mathbb{B}_N \times \eta\,\mathbb{B}_N, \operatorname{Hom}_{\mathbb{C}}(\mathbb{C}^N, \mathbb{C}^m)\big)$ be expressed in the canonical bases $\{\check{\epsilon}_1, \dots, \check{\epsilon}_N\}$ of $\mathbb{C}^N$ and $\{\epsilon_1, \dots, \epsilon_m\}$ of $\mathbb{C}^m$, respectively, by

$$H(\check{\epsilon}_k) = \sum_{j=1}^m h_{a,k,j}\epsilon_j \quad \text{for } k = 1, \dots, N, \tag{5.82}$$

$$\Longrightarrow H^*(\epsilon_j^*) = \sum_{k=1}^N h_{a,j,k}\,\check{\epsilon}_k^* \quad \text{for } j = 1, \dots, m.$$

[12] The proof in [**Sz18**, §4] uses extensively the notion of *almost* semimeromorphic current. We did not insist on it here, since we preferred to focus on the specific examples we were dealing with. For what concerns almost semimeromorphic currents, namely direct images through log-resolutions of currents defined by semimeromorphic forms and the properties they share and on which such proof relies, see [**AndW18**, §4], in particular Theorem 4.25 with its subsequent comments.

Observe that $\operatorname{codim} V(1\otimes a - a\otimes 1) \overset{\text{pure}}{=} N \overset{\text{pure}}{=} \operatorname{codim} V(\zeta - z)$ and that the ring $\mathscr{O}/(\zeta - z)\,\mathscr{O}$ is Cohen–Macaulay over $\eta\,\mathbb{B}_N \times \eta\,\mathbb{B}_N$, since $(\zeta_1 - z_1, \dots, \zeta_N - z_N)$ is regular in $\mathscr{O}(\mathbb{C}^N \times \mathbb{C}^N)$. It follows from Theorem 3.52 (see Example 3.53) that

(5.83)

$$\begin{aligned}
&\overline{\partial}\Big(\frac{1}{\zeta_1 - z_1, \dots, \zeta_N - z_N}\Big) \\
&= \Big(\bigwedge^N H^*\Big) R_{0,N}^{\overline{\mathscr{K}_\bullet^{1\otimes a - a\otimes 1}}} = \Big(\bigwedge^N H^*\Big) \\
&\quad \times \Big[(\overline{\partial}|1\otimes a - a\otimes 1|^{2\lambda})\frac{(1\otimes a - a\otimes 1)^*}{2i\pi\big|1\otimes a - a\otimes 1\big|^2} \bigwedge^{N-1} \overline{\partial}\Big(\frac{(1\otimes a - a\otimes 1)^*}{2i\pi\big|1\otimes a - a\otimes 1\big|^2}\Big)\Big]_{\lambda=0}
\end{aligned}$$

for any metric $|\ |$ on the trivial bundle $\mathbb{C}^m$ over $\eta\,\mathbb{B}_N \times \eta\,\mathbb{B}_N$ and the metrization it induces on the Koszul complex $\mathscr{K}_\bullet^{1\otimes a - a\otimes 1}$. If one takes in particular the canonical metric $\|\ \|$ on $\mathbb{C}^m$, then (5.83) becomes, if $\mathfrak{a} = 1\otimes a - a\otimes 1 = \sum_{j=1}^m \mathfrak{a}_j\,\epsilon_j$,

(5.84)

$$\begin{aligned}
&\overline{\partial}\Big(\frac{1}{\zeta_1 - z_1, \dots, \zeta_N - z_N}\Big) \\
&= \frac{(-1)^{\frac{N(N-1)}{2}}}{(2i\pi)^N} \times \Big[\lambda\,\frac{\|\mathfrak{a}\|^{2\lambda}}{\|\mathfrak{a}\|^{2(N+1)}}\,\overline{\partial}\|\mathfrak{a}\|^2 \\
&\qquad \times \sum_{1\le j_1<\cdots<j_N\le m} \Big(\sum_{\ell=1}^N (-1)^{\ell-1}\overline{\mathfrak{a}_{i_\ell}} \bigwedge_{\ell'\neq \ell} \overline{d\,\mathfrak{a}_{j'_\ell}}\Big)\Big(\bigwedge_{\ell=1}^N H^*(\epsilon^*_{j_\ell})\Big)\Big]_{\lambda=0};
\end{aligned}$$

see Example 3.16.

EXAMPLE 5.15. If $m = N$ and $(a_1, \dots, a_N)$ defines the origin as a complete intersection in $\eta\,\mathbb{B}_N$, then the current identity (5.83) becomes the equality between Coleff–Herrera currents

(5.85)

$$\overline{\partial}\Big(\frac{1}{\zeta_1 - z_1, \dots, \zeta_N - z_N}\Big) = (\det H)\,\overline{\partial}\Big(\frac{1}{a_1(\zeta) - a_1(z), \dots, a_N(\zeta) - a_N(z)}\Big).$$

This equality is independent of the choice of the metric on the trivial bundle $\mathbb{C}^N$ over $\eta\,\mathbb{B}_N$.

Since $a_j R^a = \overline{\mathfrak{M}_0}\,R^a = 0$ for $j = 1, \dots, m$, it follows from (5.83) that

$$\begin{aligned}
&\overline{\partial}\Big(\frac{1}{\zeta_1 - z_1, \dots, \zeta_N - z_N}\Big) \wedge R^a(z) = \Big[\Big(\bigwedge^N H^*\Big) R_{0,N}^{\overline{\mathscr{K}_\bullet^{1\otimes a - a\otimes 1}},\lambda} \wedge R^a(z)\Big]_{\lambda=0} \\
&= \Big[\lambda|\mathfrak{a}|^{2\lambda}\Big(\bigwedge^N H^*\Big)\frac{\overline{\partial}_\zeta\,|\mathfrak{a}|^2}{|\,\mathfrak{a}\,|^2}\,\frac{\mathfrak{a}^*}{2i\pi\,|\mathfrak{a}|^2}\Big(\bigwedge^{N-1}\overline{\partial}_\zeta\Big(\frac{\mathfrak{a}^*}{2i\pi\,|\,\mathfrak{a}\,|^2}\Big)\Big) \wedge R^a(z)\Big]_{\lambda=0} \\
&= \Big[\lambda|\mathfrak{a}|^{2\lambda}\Big(\bigwedge^N H^*\Big)\frac{\overline{\partial}_\zeta\,|a|^2}{|a|^2}\,\frac{a^*}{2i\pi\,|a|^2}\Big(\bigwedge^{N-1}\overline{\partial}_\zeta\Big(\frac{a}{2i\pi\,|a|^2}\Big)\Big) \wedge R^a(z)\Big]_{\lambda=0} \\
&= \Big[\lambda|a|^{2\lambda}\Big(\bigwedge^N H^*(z,\cdot)\Big)\frac{\overline{\partial}_\zeta\,|a|^2}{|a|^2}\,\frac{a^*}{2i\pi\,|a|^2}\Big(\bigwedge^{N-1}\overline{\partial}_\zeta\Big(\frac{a^*}{2i\pi\,|a|^2}\Big)\Big)\Big]_{\lambda=0} R^a(z),
\end{aligned} \tag{5.86}$$

where $a^* = \mathfrak{a}^*(0,\cdot)$ is now the conjugate of the section a of the trivial bundle $\mathbb{C}^m$ over $\eta\,\mathbb{B}_N$. Formula (5.86) leads to the following *trace formula* in $(\mathcal{O}/\mathbb{I})_0$.

PROPOSITION 5.16. *Let $\mathbb{I}$ be a $\mathfrak{M}_0$-primary ideal in $\mathbb{A} = \mathcal{O}_0$ whose generators admit representatives $a_1, \dots, a_m$ in $\mathcal{O}(\eta\,\mathbb{B}_N)$ for $0 < \eta \ll 1$ sufficiently small. Let also $|\ |$ be a hermitian metric on the trivial bundle $\mathbb{C}^m$ over $\eta\,\mathbb{B}_N$. Then any $h \in \mathcal{O}(\eta\,\mathbb{B}_N)$ is such that $(h)_0$ admits modulo $\mathbb{I}$ in $\eta\,\mathbb{B}_N$ the representative*

$$(5.87)\quad \mathrm{Tr}_{\mathbb{I},|\ |}:$$

$$z \longmapsto \Big\langle \Big[\lambda |a|^{2\lambda} \Big(\bigwedge^{N} H^*(z,\cdot)\Big) \frac{\overline{\partial}_\zeta |a|^2}{|a|^2}\, \frac{a^*}{2i\pi |a|^2} \Big(\bigwedge^{N-1} \overline{\partial}_\zeta \Big(\frac{a^*}{2i\pi |a|^2}\Big)\Big)\Big]_{\lambda=0}, h\,\theta\, d\zeta \Big\rangle,$$

where the test function $\theta \in \mathcal{D}(\eta\,\mathbb{B}_N)$ equals identically one in a neighborhood of the origin and $d\zeta = d\zeta_1 \wedge \cdots \wedge d\zeta_N$.

PROOF. Testing the current identity (5.86) against $(z,\zeta)\theta(\zeta)\,\theta(z) d\zeta \wedge dz$ leads to $(h - \mathrm{Tr}_{\mathbb{I},|\ |})\, R^a = 0$. One then concludes the proof by applying Theorem 3.38. □

EXAMPLE 5.17. If $m = N$, then $(a_1, \dots, a_N)$ is a regular sequence in $\mathcal{O}_0$. Thus, the trace function $\mathrm{Tr}_{\mathbb{I}}[h]$ is independent on the choice of the hermitian metric on $\mathbb{C}^N$ and equals

$$(5.88)\qquad \mathrm{Tr}_{\mathbb{I}}[h] : z \longmapsto \Big\langle (\det H(z,\cdot))\, \overline{\partial}\Big(\frac{1}{a_1, \dots, a_N}\Big),\ h\,\theta\, d\zeta \Big\rangle.$$

REMARK 5.18. Formulae (5.83), (5.84), and (5.85) in the important case $m = N$ appear as (current) local versions of a well known integral formula due to André Weil [**We35**] (see also [**Gl63**]). We already mentioned it in the one-variable case; see §1.3.1. We point out that such a formula in its global version plays a central role concerning effectivity problems in polynomial algebraic geometry. In parallel with such a transcription of André Weil type integral formulas, the congruencies

$$(5.89)\qquad h \equiv \mathrm{Tr}_{\mathbb{I},|\ |}[h] \bmod \mathbb{I} \quad \text{or} \quad h \equiv \mathrm{Tr}_{\mathbb{I}}[h] \bmod \mathbb{I} \ (\text{when } m = N)$$

(see (5.87) or (5.88)), which are called *trace formulae*, extend to the (local) multivariate setting (for $\mathfrak{M}_0$-primary ideals) the formulae (1.109) relating univariate residue calculus with euclidean division; see Proposition 1.38. Similarly, *trace formulae* (5.89) extend within the multivariate settings of Propositions 1.59 and 1.61 in the case $m = N = 1$ or Remark (1.17), together with (1.168) and (1.169), when $m > N = 1$ in the context of Gorenstein algebras.

Proposition (5.16) can be iterated, as it was already the case for the euclidean division algorithm in the univariate setting (if we refer to Proposition 1.38). Let $a_1, \dots, a_m$ be representatives of generators of the $\mathfrak{M}_0$-ideal $\mathbb{I}$ in $\mathcal{O}(\eta\,\mathbb{B}_N)$ and let $|\ |$ be any hermitian metric on the trivial bundle $\mathbb{C}^m$. Then, for each $k \in \mathbb{N}^*$ and each $h \in \mathcal{O}(\eta\,\mathbb{B}_N)$, for sufficiently small $0 < \eta \ll 1$, we let

$$(5.90)\quad \mathrm{Tr}^k_{\mathbb{I},|\ |}[h] : z \longmapsto$$

$$\Big\langle \Big[\lambda |a|^{2\lambda} \Big(\frac{a(z) a^*}{|a|^2}\Big)^k \Big(\bigwedge^{N} H^*(z,\cdot)\Big) \frac{\overline{\partial}_\zeta |a|^2}{|a|^2}\, \frac{a^*}{2i\pi |a|^2} \Big(\bigwedge^{N-1} \overline{\partial}_\zeta \Big(\frac{a^*}{2i\pi |a|^2}\Big)\Big)\Big]_{\lambda=0}, h\theta d\zeta \Big\rangle,$$

where $\theta \in \mathcal{D}(\eta\,\mathbb{B}_N)$ is identically equal to one in a neighborhood of the origin. Note that one has $\mathrm{Tr}^k_{\mathbb{I},|\ |}[h] = \mathrm{Tr}_{\mathbb{I},|\ |}[h]$ for $k = 1$.

PROPOSITION 5.19. *For any $k \in \mathbb{N}^*$, one has that*

$$(h_0) \equiv \sum_{\kappa=0}^{k-1} \Big(\binom{N+\kappa}{N} \operatorname{Tr}^{\kappa}_{\mathbb{I},|\;|}[h] \Big)_0 \quad \mathrm{mod} \quad \mathbb{I}^k. \tag{5.91}$$

Hence

$$(h)_0 = \sum_{\kappa=0}^{\infty} \binom{N+\kappa}{N} \big(\operatorname{Tr}^{\kappa}_{\mathbb{I},|\;|}[h] \big)_0 \tag{5.92}$$

provides an $\mathbb{I}$-adic expansion of $(h)_0$ in the ring $\mathcal{O}_0[[z]]$ of formal power series. The development (5.92) *is called the Bergman–Weil $\mathbb{I}$-adic expansion of the Bochner–Martinelli type.*

PROOF. Assertion (5.91) was proved at the end of §4.6.1 (following the proof of Proposition 4.47) using the canonical metric $\|\;\|$ as a metric on the trivial bundle $\mathbb{C}^m$ and f in place of a. We observe that changing the canonical metric $\|\;\|$ into an arbitrary one $|\;|$ does not make any difference in the proof. Let us repeat briefly this proof here for the sake of completeness with the canonical metric $\|\;\|$ on $\mathbb{C}^m$. We follow the notations used in §4.6.1 (this time a is in place of f). Thus, we introduce the global weight $\mathscr{W}^{\lambda}_{a,k}$ in $(\eta\,\mathbb{B}_N \times \eta\,\mathbb{B}_N) \setminus \sqcap^{-1}(\{0\})$, where λ is a complex parameter with Re $\lambda \gg 1$, as in (4.111). It follows from Proposition 4.30 that one has that

$$h(z) = \int_{\eta\,\mathbb{B}_N} \theta\,(\mathscr{W}^{\lambda}_{a,k}(z,\cdot))_{N,N}\, h - \int_{\eta\,\mathbb{B}_N} (\overline{\partial}\theta)\,\big(K_{\mathbb{C}^N}(z,\cdot)\mathscr{W}^{\lambda}_{a,k}(z,\cdot)\big)_{N,N-1}\, h,$$

for any $z \in \eta\,\mathbb{B}_N \setminus \{0\}$. Furthermore, we define $\sigma_{a,k}$, $\rho_{a,k,N}$ and $\widetilde{\sigma}_{a,k,N}$ (see (4.113), (4.115), and (4.116)) as

$$\begin{aligned}
\sigma_{a,k}(z,\cdot) &= \sum_{r=1}^{N+1} \binom{N+k}{r-1} \Big[\|f\|^{2\lambda} \Big(\frac{f(z)f^*}{\|f\|^2} \Big)^{N+k-r} \mathscr{P}_{a,r}(z,\cdot) \Big]_{\lambda=0}, \\
\rho_{a,k,N}(z,\cdot) &= \big[(\overline{\partial}\|a\|^{2\lambda})\, \mathscr{R}_{a,N}(z,\cdot) \big]_{\lambda=0}, \\
\widetilde{\sigma}_{a,k,N}(z,\cdot) &= \Big[\frac{a^*}{\|a\|^2} \sum_{r'=N+1}^{N+k} \binom{r'}{N} \Big(\frac{a(z)a^*}{\|a\|^2} \Big)^{r'-N-1} (\overline{\partial}\|a\|^{2\lambda})\, \mathscr{R}_{a,N}(z,\cdot) \Big]_{\lambda=0} \\
&= \Big[\frac{a^*}{\|a\|^2} \sum_{\kappa=1}^{k} \binom{N+\kappa}{N} \Big(\frac{a(z)a^*}{\|a\|^2} \Big)^{\kappa-1} (\overline{\partial}\|a\|^{2\lambda})\, \mathscr{R}_{a,N}(z,\cdot) \Big]_{\lambda=0},
\end{aligned}$$

where

$$\begin{aligned}
\mathscr{P}_{a,r}(z,\cdot) &= \frac{a^*}{2i\pi\|a\|^2} \Big(\overline{\partial}\big(H_a(z,\cdot)\, \frac{a^*}{2i\pi\|a\|^2}\big) \Big)^{r-1} \quad \text{for } r \geq 1, \\
\mathscr{R}_{a,N}(z,\cdot) &= H_a(z,\cdot)\ \mathscr{P}_{a,N} = \big(H_a(z,\cdot)\, \frac{a^*}{2i\pi\|a\|^2} \big) \Big(\overline{\partial}\big(H_a(z,\cdot)\, \frac{a^*}{2i\pi\|a\|^2}\big) \Big)^{N-1}.
\end{aligned}$$

One has

$$\begin{aligned}\Big[\int_{\eta\,\mathbb{B}_N} \theta\,(\mathscr{W}^\lambda_{a,k}(z,\cdot))_{N,N}\,h\Big]_{\lambda=0} &= \int_{\eta\,\mathbb{B}_N} \theta\,h\,\big(a(z)\,\rho_{a,k}(z,\cdot)+\widetilde{\sigma}_{a,k,N}(z,\cdot)\big)+2i\pi a(z)\int_{\eta\,\mathbb{B}_N}\theta\,h\,\sigma_{a,r}(z,\cdot)\\ &\equiv \sum_{\kappa=0}^{k-1}\binom{N+\kappa}{N}\,\mathrm{Tr}^{\kappa}_{\mathbb{I},\|\ \|}[h] \mod \mathbb{I}^k.\end{aligned}$$

Since $a(z)\,\sigma_{a,k}$ is a smooth form with holomorphic coefficients in $\mathbb{I}^k$, one has in a neighborhood of 0, which does not intersect $\mathrm{Supp}\,\overline{\partial}\theta$, that

$$\int_{\eta\,\mathbb{B}_N}(\overline{\partial}\theta)\,\big(K_{\mathbb{C}^N}(z,\cdot)\mathscr{W}^\lambda_{a,k}(z,\cdot)\big)_{N,N-1}\,h=\sum_{\#I=k} a^I(z)\,\xi_I(z),$$

where ξ_I are real analytic functions there. Note that $z\to K_{\mathbb{C}^N}(z,\cdot)$ is a real analytic function in this neighborhood too. Assertion (5.91) follows by a flatness argument. Thus, Proposition 5.19 is proved. □

REMARK 5.20. Using the normalized blowup of $\eta\,\mathbb{B}_N$ along the ideal sheaf generated by $a_1,\dots,a_m$, one sees easily that for any $\kappa\in\mathbb{N}$ with $\kappa\le k-1$ and any $z\in\eta\,\mathbb{B}_N$, the germ at the origin of the current

$$\Big[\lambda|a|^{2\lambda}\Big(\frac{a(z)a^*}{|a|^2}\Big)^{\kappa}\Big(\bigwedge^N H^*(z,\cdot)\Big)\frac{\overline{\partial}_\zeta|a|^2}{|a|^2}\,\frac{a^*}{2i\pi|a|^2}\Big(\bigwedge^{N-1}\overline{\partial}_\zeta\Big(\frac{a^*}{2i\pi|a|^2}\Big)\Big)\Big]_{\lambda=0}$$

is annihilated by $\overline{\mathbb{I}^{N+k-1}}$. It follows that we recover the fact $h\in\mathbb{I}^k$, using (5.91). Thus, the $\mathbb{I}$-adic development (5.92) in Proposition 5.19 fits at all stages k with the Briançon–Skoda theorem in $\mathbb{A}=\mathcal{O}_0$; Theorem 5.7.

REMARK 5.21. Let $\mathbf{F}^{a,k}_\bullet$ be a generically exact complex of trivial holomorphic bundles over $\eta\,\mathbb{B}_N$, with length $L=N=\mathrm{codim}\,V(a)$, $E_0(\mathbf{F}^{a,k}_\bullet)=\mathbb{C}$. We also assume that the morphism $F^{a,k}_1$ in the above complex is generically surjective and that $\mathbf{F}^{a,k}_\bullet$ induces a minimal free resolution of the Cohen–Macaulay quotient sheaf $(\mathcal{O}/(\sum_{j=1}^m a_j\mathcal{O})_{\eta\,\mathbb{B}_N})^k$ at the level of sheaves of sections in $\eta\,\mathbb{B}_N$. Let $R^{a,k}$ be the Coleff–Herrera residue realized from an arbitrary metrization of the complex $\mathbf{F}^{a,k}_\bullet$. One can start with the current identity

$$\begin{aligned}\overline{\partial}\Big(\frac{1}{\zeta_1-z_1,\dots,\zeta_N-z_N}\Big)\wedge R^{a,k}(z) &= \Big[\Big(\bigwedge^N H^*\Big)R_{0,N}^{\overline{\mathscr{K}^{1\otimes a-a\otimes 1}_\bullet},\lambda}\wedge R^{a,k}(z)\Big]_{\lambda=0}\\ &= \Big[\lambda|\mathfrak{a}|^{2\lambda}\Big(\bigwedge^N H^*\Big)\frac{\overline{\partial}_\zeta|\mathfrak{a}|^2}{|\mathfrak{a}|^2}\,\frac{\mathfrak{a}^*}{2i\pi\,|\mathfrak{a}|^2}\Big(\bigwedge^{N-1}\overline{\partial}_\zeta\Big(\frac{\mathfrak{a}^*}{2i\pi\,|\mathfrak{a}|^2}\Big)\Big)\wedge R^{a,k}(z)\Big]_{\lambda=0}\end{aligned}$$

and take into account that

$$\begin{aligned}\Big(\frac{1}{|\mathfrak{a}(z,\cdot)|^2}\Big)^{N+1} &\equiv \Big(\frac{1}{\|a\|^2}\Big)^{N+1}\Big(\frac{1}{1-(a(z)a^*)/|a|^2}\Big)^{N+1}\\ &\equiv \Big(\frac{1}{\|a\|^2}\Big)^{N+1}\sum_{\kappa=0}^{k-1}\binom{N+\kappa}{\kappa}\Big(\frac{a(z)\,a^*}{|a|^2}\Big)^{\kappa} \mod \sqcap^*\,(\mathbb{I}^k+\overline{\mathscr{M}_0}),\end{aligned}$$

while $\sqcap^*(\mathbb{I}^k+\overline{\mathfrak{M}_0})$ annihilates $R^{a,k}(z)$, to derive from the same reasoning used to prove Proposition 5.16 an alternative proof of Proposition 5.19; see [**VidY22**, §5].

EXAMPLE 5.22. In the particular case where $m = N$, the $\mathbb{I}$-adic development (5.92) is independent of the metric $|\ |$. If the test function $\theta \in \mathscr{D}(\eta\,\mathbb{B}_N)$ is equal to 1 in a neighborhood of the origin 0, then (5.92) becomes the convergent expansion in $\mathcal{O}_0$ given by
(5.93)

$$(h)_0 = \sum_{\kappa\in\mathbb{N}^N} \Big(\Big\langle \overline{\partial}\Big(\frac{1}{a_1^{\kappa_1+1},\dots,a_N^{\kappa_N+1}}\Big)(\zeta)\,,\,(\det H(\cdot,\zeta))\,\theta\,h\,d\zeta\Big\rangle\Big)_0\, a_1^{\kappa_1}\cdots a_N^{\kappa_N}\,.$$

The content of this convergence will be specified in §6.1.2.

Let us conclude this section and this chapter with the following observation. Let $\mathscr{F}_\bullet^{1\otimes a-a\otimes 1}$ be a generically exact complex of trivial bundles over $\eta\,\mathbb{B}_N\times\eta\,\mathbb{B}_N$, with $E_0(\mathscr{F}_\bullet^{1\otimes a-a\otimes 1}) = \mathbb{C}$ and let $F_1^{1\otimes a-a\otimes 1}$ be generically surjective such that $\mathscr{F}_\bullet^{1\otimes a-a\otimes 1}$ induces a minimal free resolution of $(\mathscr{O}/\sum_{j=1}^m (a_j(\zeta)-a_j(z))\mathscr{O})_{\eta\,\mathbb{B}_N\times\eta\,\mathbb{B}_N}$ at the level of sheaves of sections in $\eta\,\mathbb{B}_N\times\eta\,\mathbb{B}_N$.[13] Theorem 3.52 still implies that there exists $\Delta_N \in \mathscr{O}(\eta\,\mathbb{B}_N\times\eta\,\mathbb{B}_N, (E_N(\mathscr{F}_\bullet^{1\otimes a-a\otimes 1}))^*)$ such that the current identity
(5.94)

$$\overline{\partial}\Big(\frac{1}{\zeta_1-z_1,\dots,\zeta_N-z_N}\Big)\wedge R_a(z) = \Big[\Delta_N(z,\cdot)\,R_{0,N}^{\overline{\mathscr{F}_\bullet^{1\otimes a-a\otimes 1}},\lambda}(z,\zeta)\wedge R_a(z)\Big]_{\lambda=0}$$

holds in $\eta\,\mathbb{B}_N\times\eta\,\mathbb{B}_N$. Since $(1\otimes a - a\otimes 1, a(z))$ and $(a(\zeta), a(z))$ define the same ideal sheaf in $\mathcal{I}$ in $\mathscr{O}_{\eta\,\mathbb{B}_N\times\eta\,\mathbb{B}_N}$ and the ring $\mathscr{O}_{\eta\,B_N\times\eta\,\mathbb{B}_N}/\mathcal{I}$ is Cohen–Macaulay, there exists

$$\delta \in \mathscr{O}\big(\eta\,\mathbb{B}_N\times\eta\,\mathbb{B}_N, \mathrm{Hom}_{\mathbb{C}}\big(E_N(\sqcap^*\mathbf{F}_\bullet^a)\oplus E_N(\pi^*\mathbf{F}_\bullet^a), E_N(\mathscr{F}_\bullet^{1\otimes a-a\otimes 1})\big)\big)$$

such that

$$\overline{\partial}\Big(\frac{1}{\zeta_1-z_1,\dots,\zeta_N-z_N}\Big)\wedge R^a(z) = (\Delta_N\delta)(z,\zeta)\,R^a(\zeta)\wedge R^a(z). \tag{5.95}$$

Current identities (5.94) or (5.95) lead to matrix-valued versions of Cauchy–Weil type formulae.

[13]Here is a question which looks to us interesting, but we suspect hard to answer: decide under which conditions on a such a quotient sheaf is Cohen–Macaulay.

CHAPTER 6

Residue calculus and trace formulae

We will present in this chapter the important concepts of trace and trace formulae. We will start with the $\mathbb{C}$-algebra $\mathcal{O}_{\mathbb{C}^n}(\Delta)$, where Δ is an analytic Weil polyhedron in $\mathbb{C}^n$ in §6.1, and continue with the presentation of the *algebraic residue symbol* in a more abstract algebraic context in §6.2 and §6.3. Our main illustrative examples will nevertheless remain the $\mathbb{K}$-algebras

$$(6.1)\quad \mathbb{K}[X_1,\dots,X_n],\quad \mathbb{K}_{\mathbb{J}}[X_1,\dots,X_n,Y_1,\dots,Y_{N-n}] = K_{\mathbb{J}}[X,Y] := \frac{\mathbb{K}[X,Y]}{\mathbb{J}},$$

where $\mathbb{K}$ is a field (with arbitrary characteristic, but mostly with characteristic 0) and $\mathbb{J}$ is a radical ideal in $\mathbb{K}[X,Y]$ such that $\dim(\operatorname{Spec}(\mathbb{K}_{\mathbb{J}}[X,Y])) = n$. It is the reason for denoting all through this chapter by N not the dimension of the surrounding complex manifold as was the case in Chapters 2 through 5, but the total number $N = n+(N-n)$ of variables involved in $\mathbb{K}_{\mathbb{J}}[X,Y]$. The $\mathbb{K}$-algebra $\mathbb{K}_{\mathbb{J}}[X,Y]$ is our most general algebraic model. The algebraic notion of trace, together with the properties it inherits, will be introduced for both examples described by (6.1) in §6.4. Then, we continue the presentation of trace and its properties in two particular situations. Namely, we first study it more closely when $\operatorname{char}\mathbb{K} = 0$ and then focus on the case when $\mathbb{K} = \mathbb{Q}$. This material is is the subject of §6.5 and §6.6, respectively.

6.1. Trace in analytic polyhedra

6.1.1. Analytic Weil polyhedron, nondegeneracy, skeleton. The concept of *analytic Weil polyhedron* is important in presenting the results in this section, whenever they are within the frame of multidimensional complex analysis. Throughout this subsection, Ω denotes an open pseudo-convex domain (or, which is equivalent, a domain of holomorphy) in $\mathbb{C}^n$. Thus, Ω is equipped with the complex structure inherited from the ambient space $\mathbb{C}^n$ and is assumed to be a Stein manifold. Let also $f : \Omega \to \mathbb{C}^m$ (with $m \geq n$) be a holomorphic map and let $r = (r_1,\dots,r_m)$, $r_j > 0$ for $j = 1,\dots,m$, be a multiradius.

DEFINITION 6.1 (Analytic Weil polyhedron subordinate to (Ω, f, r)). Any open subset $\Delta \subset \Omega$ of the form

$$\Delta = \{\zeta \in \Omega \,:\, |f_j(\zeta)| < r_j \text{ for } j = 1,\dots,m\},$$

where $r_j > 0$ for $j = 1,\dots,m$, is called an *analytic Weil polyhedron* subordinate to (Ω, f, r)[1] if and only if it is nonempty and relatively compact in Ω. It is convenient to extend slightly this definition,[2] so that *any finite union of connected components*

[1]The reader will find (for example in [**AY**]) a more general definition of the analytic Weil polyhedron involving a bounded domain in $\mathbb{C}$ instead of $U_j = D(o, r_j)$.

[2]Which we will do here.

of such an analytic Weil polyhedron is considered as an analytic Weil polyhedron subordinate to (Ω, f, r) as well.

EXAMPLE 6.2.

(1) Let $f : \mathbb{C}^n \to \mathbb{C}^m$ be a holomorphic map satisfying $|f(\zeta)| \geq \kappa$ for $|\zeta| \geq C$ for some $\kappa > 0$, $C \geq 0$ and arbitrary norms on $\mathbb{C}$ and $\mathbb{C}^m$. If we take $r = (r_1, \dots, r_m)$ such that $\max_{1 \leq j \leq m} r_j$ is sufficiently small (depending on the norms and on κ), then the open subset

$$\{\zeta \in \mathbb{C}^n : |f_j(\zeta)| < r_j \text{ for } j = 1, \dots, m\}$$

is an analytic Weil polyhedron subordinate to $(\mathbb{C}^n, f, r)$, whenever it is nonempty. In particular, this holds when the entries f_j are polynomials and the map $f : \mathbb{C}^n$ to $\mathbb{C}^m$ is proper.

(2) The polydisc $\mathbb{D}^n = \{\zeta \in \mathbb{C}^n : |\zeta_j| < 1 \text{ for } j = 1, \dots, n\}$ is the archetypal example of an analytic Weil polyhedron subordinate to $(\mathbb{C}^n, z, (1, \dots, 1))$.

(3) On the other hand, let $f_1 = p_1/q_1, \dots, f_n = p_n/q_n$ be n rational inner functions[3] in the Hardy space $H^2(\mathbb{D}^n)$. Such rational functions are also assumed to be pole-free on $\overline{\mathbb{D}^n}$ and such that the subset $\mathbb{D}^n \cap \bigcap_{j=1}^n f_j^{-1}(\{0\})$ is a zero-dimensional subset. Thus, the subset $\mathbb{D}^n \cap \bigcap_{j=1}^n f_j^{-1}(\{0\})$ is finite, since the f_j are rational. The polydisc $\mathbb{D}^n$ is not in general a Weil polyhedron for the configuration $(((1+\eta)\mathbb{D})^n, f, (1, \dots, 1))$, where $\eta > 0$ is such that $((1+\eta)\mathbb{D})^n \cap (q_1 \cdots q_n)^{-1}(\{0\}) = \emptyset$, since it may fail to be relatively compact in $((1+\eta)\mathbb{D})^n$. The reader will find connections between the material presented in this section and interpolation questions in the Hardy space of the polydisc $\mathbb{D}^n$ involving precisely rational inner functions.

REMARK 6.3. Observe that f does not vanish on the boundary of any analytic Weil polyhedron $\Delta \Subset \Omega$ subordinate to (Ω, f, r), since $|f_j(\zeta)| = r_j > 0$ for at least one $j \in \{1, \dots, m\}$, whenever $\zeta \in \partial\Delta$. Thus, one concludes that the closed analytic subset $f^{-1}(\{0\}) = \{\zeta \in \Omega : f(\zeta) = 0\}$ contains only zero-dimensional (hence discrete) components in an open neighborhood of $\overline{\Delta}$. This implies that the set $f^{-1}(\{0\})$ is a finite subset of Δ.

DEFINITION 6.4 (Nondegenerate, strongly nondegenerate analytic Weil polyhedron [**Boo97, AAC99**]). An analytic Weil polyhedron Δ subordinate to (Ω, f, r) is said to be *nondegenerate* if for each ordered subset $1 \leq j_1 < \cdots < j_p$ of $\{1, \dots, m\}$, with arbitrary cardinal $p \leq n$, the differential form $\bigwedge_{\ell=1}^p df_{j_\ell}$ does not vanish on $\{\zeta \in \partial\Delta : |f_{j_\ell}| = r_{j_\ell} \text{ for } \ell = 1, \dots, p\}$. It is said to be *strongly nondegenerate* if, in addition,

$$\{\zeta \in \partial\Delta : |f_{j_\ell}| = r_{j_\ell} \text{ for } \ell = 1, \dots, p\} = \emptyset$$

whenever $p > n$ and $1 \leq j_1 < \cdots < j_p \leq m$.

REMARK 6.5. As pointed out in [**AAC99**], it is equivalent to say that an analytic Weil polyhedron is strongly nondegenerate and that it is locally biholomorphic to some part of the polydisc $\mathbb{D}^n$.

---

[3]Rational inner functions, pole-free on $\overline{\mathbb{D}^n}$, can be realized as follows (see [**Rud**]): choose $q \in \mathbb{C}[\zeta_1, \dots, \zeta_n]$ such that $\overline{\mathbb{D}^n} \cap q^{-1}(\{0\}) = \emptyset$. Then, we set $p = \zeta^\chi \overline{q(1/\bar{\zeta})}$, where $\chi \in \mathbb{N}^n$, so that $p \in \mathbb{C}[\zeta_1, \dots, \zeta_n]$. Then $f = p/q$ is the prototype of a rational inner function. Clearly $|f(\zeta)| = 1$ whenever $|\zeta_1| = \cdots = |\zeta_n| = 1$. That is, $|f(\zeta)| = 1$ on the Shilov boundary of the polydisc $\mathbb{D}^n$.

DEFINITION 6.6 (Skeleton of a strongly nondegenerate analytic Weil polyhedron). Let Δ be a strongly nondegenerate analytic Weil polyhedron subordinate to (Ω, f, r). Its *skeleton* $\Gamma(\Delta)$ is the real analytic n-cycle, whose support is the n-dimensional real submanifold of $\mathbb{C}^n_{\mathbb{R}} = \mathbb{R}^{2n}$ defined as the disjoint union

$$\mathrm{Supp}(\Gamma(\Delta)) = \bigsqcup_{1\leq j_1<\cdots<j_n\leq m} \overbrace{\{\zeta \in \overline{\Delta}\,;\, |f_{j_\ell}| = r_{j_\ell} \text{ for } \ell = 1,\dots,n\}}^{\sigma_{\Delta,J}},$$

oriented so that for any ordered n-index $1 \leq j_1 < \cdots < j_n \leq m$, the restriction to $\sigma_{\Delta,J}$ of the n-form $i^{-n}\bigwedge_{\ell=1}^n df_{j_\ell}/f_{j_\ell}$, that is the real n-form $\big(\bigwedge_{\ell=1}^n d(\arg(f_{j_\ell}))\big)_{|\sigma_{\Delta,J}}$, is a volume form on $\sigma_{\Delta,J}$. $\Gamma(\Delta)$ is also known as the *Shilov boundary* of the analytic Weil polyhedron Δ.

REMARK 6.7. If Δ is simply nondegenerate and $|J| = n$, then all $\sigma_{\Delta,J}$ are real n-dimensional manifolds. However, their union, considered as the support of the skeleton, is a piecewise n-dimensional real manifold. Its pieces intersect transversally.

6.1.2. The case when $m = n$. In the particular case when $m = n$, the support of the Shilov boundary of an analytic Weil polyhedron subordinate to (U, f, r) is the n-dimensional real manifold $\sigma_\Delta := \sigma_{\Delta,[1,\dots,n]}$ and the positive measure ν_Δ on it, attached to the volume form $\big(\bigwedge_{j=1}^n d(\arg(f_j))\big)_{|\sigma_\Delta}/(2\pi)^n$, is such that for any continuous function, φ on σ_Δ, one has that

$$\int_{\sigma_\Delta} \varphi(\zeta)\, d\nu_\Delta(\zeta) = \frac{1}{(2i\pi)^n}\int_{\Gamma(\Delta)} \varphi(\zeta)\,\frac{df_1}{f_1}\wedge\cdots\wedge\frac{df_n}{f_n}. \tag{6.2}$$

Let us state the following important proposition connecting multivariate residue calculus, as introduced previously in Chapter 2, with the topics discussed in this section. with the topics discussed in this section. From now on in this subsection let Δ be an analytic Weil polyhedron subordinate to (Ω, f, r), where f is a holomorphic map from $\mathbb{C}^n$ to itself. We already mentioned that Sard's lemma [**Sa42**] implies that for almost any multi-radius $(\rho_1, \dots, \rho_n)$ in $]0, r_1] \times \cdots \times]0, r_n]$, the open set

$$\Delta_\rho = \{\zeta \in \Delta \,:\, |f_j(\zeta)| < \rho_j \text{ for } j = 1,\dots,n\}$$

is an analytic, nondegenerate Weil polyhedron with respect to (Ω, f, ρ). Let $d\nu_{\Delta,\rho}$ be the positive measure on the support of its skeleton $\Gamma(\Delta_\rho)$ as in (6.2). Let also $J : \Omega \to \mathbb{C}$ be the determinant of the jacobian matrix of f in the ζ coordinates. We assume from now on that the functions $f_1, \dots, f_n$ have common zeroes inside Δ, and we denote by $\mu_\Delta \in \mathbb{N}^*$ the total mass of the atomic integration current defined as $[\mathrm{div}((f_1)_{|\Delta}) \wedge \cdots \wedge \mathrm{div}((f_m)_{|\Delta})]$ in Δ. Let $V_\Delta(f) = \bigcap_1^n \{\zeta \in \Delta \,:\, f_j(\zeta) = 0\}$.

PROPOSITION 6.8. *Let h be any holomorphic function in Δ. Let also ρ in $]0, r_1[\, \times \cdots \times\,]0, r_n[$ be such that Δ_ρ is nondegenerate (recall that this happens for almost any $\rho \in\,]0, \rho_1[\, \times \cdots \times\,]0, \rho_n[$). Then, one has that for every test function θ in $\mathcal{D}(\Delta)$, which is identically equal to 1 in a neighborhood of the finite set $V_\Delta(f)$,* *the following equality holds:*

$$(6.3)\quad \forall \kappa \in \mathbb{N}^n, \quad \frac{1}{\rho_1^{2\kappa_1}\cdots\rho_n^{2\kappa_n}}\int_{\sigma\Delta_\rho}\frac{h(\zeta)}{J(\zeta)}\,\overline{f^\kappa(\zeta)}\,d\nu_{\Delta_\rho}(\zeta)$$

$$= \frac{1}{(2i\pi)^n}\int_{\Gamma(\Delta_\rho)} h(\zeta)\bigwedge_{j=1}^{n}\frac{d\zeta_j}{(f_j(\zeta))^{\kappa_j+1}} = \Big\langle \bar\partial\Big(\frac{1}{f_1^{\kappa_1+1},\dots,f_n^{\kappa_n+1}}\Big),\, \theta\, h\, d\zeta\Big\rangle,$$

where $d\zeta = d\zeta_1\wedge\cdots\wedge d\zeta_n$. Each Coleff–Herrera current on the right-hand side is attached[4] *to the sections $f_j^{\kappa_j+1}$, $j=1,\dots,n$, of the trivial bundle on Ω defining $V_\Delta(f)$ as a complete intersection in a neighborhood of $\overline{\Delta}$.*

REMARK 6.9. Proposition 2.37 implies that the complex numbers $a_{\Delta,\kappa}[h\,d\zeta]$, with $d\zeta := d\zeta_1\wedge\cdots\wedge d\zeta_n$, defined by the equalities (6.3) (in particular by the last one) are of the form

$$a_{\Delta,\kappa}[h\,d\zeta] = \sum_{\alpha\in V_\Delta(f)}\Big(\mathcal{Q}^\alpha_{\Delta,f,\kappa}(\partial/\partial\zeta)(h)\Big)_{\zeta=\alpha},$$

where $\mathcal{Q}^\alpha_{\Delta,f,\kappa}(\partial/\partial\zeta)$ is a differential operator with constant coefficients and degree at most $\mu_{\kappa,\alpha}-1$ (thanks to Theorem 2.79; see also [**GrH**, pp. 667–670]). Recall that

$$\mu_{\kappa,\alpha} := \dim_{\mathbb{C}}\frac{\mathcal{O}_{\mathbb{C}^n,\alpha}}{(f_1^{\kappa_1+1},\dots,f_n^{\kappa_n+1})\mathcal{O}_{\mathbb{C}^n,\alpha}} = (\kappa_1+1)\cdots(\kappa_n+1)\,\mu_{0,\alpha}.$$

PROOF. The first equality in (6.3) follows from the definition of the positive measure ν_{Δ_ρ} attached to a nondegenerate Weil polyhedron subordinate to (Ω, f,ρ). It follows from Stokes's formula that the integral expression

$$\frac{1}{(2i\pi)^n}\int_{\Gamma(\Delta_\rho)} h(\zeta)\bigwedge_{j=1}^{n}\frac{d\zeta_j}{(f_j(\zeta))^{\kappa_j+1}}$$

does not depend on the choice of ρ, for almost all ρ in $]0,r_1[\,\times\cdots\times\,]0,r_n[$ such that Δ_ρ is nondegenerate. Actually, first fix ρ and ρ' such that Δ_ρ, $\Delta_{\rho'}$ are both nondegenerate and ρ, ρ' differ by exactly one coordinate. Then, Stokes's theorem implies the equality of the corresponding integral expressions in the left-hand side of (6.3). Next, proceed by induction on the number of coordinates at which ρ and ρ' differ. Let us prove the last equality in (6.3) for $\rho\in\,]0,\rho_1[\,\times\cdots\times\,]0,\rho_n[$ generic. Fix $\kappa\in\mathbb{N}$ and let

$$g_j = f_j^{\kappa_j+1},\ \rho_j^{[\kappa]} = \rho_j^{\kappa_j+1}\quad \text{for } 1\le j\le n,\ 0<\rho_j\le r_j.$$

Then, one can consider Δ an analytic Weil polyhedron subordinate to $(\Omega, g, r^{[\kappa]})$, while $\Delta_{g,\rho}$, for almost any $\rho\in\,]0,r_1[\,\times\cdots\times\,]0,r_n[$, is considered as a nondegenerate Weil polyhedron $\Delta_{g;\rho}$ subordinate to $(\Omega, g, \rho^{[\kappa]})$. Fix such ρ and observe that

$$\|g\|^2 = \sum_{j=1}^{n}|g_j|^2 \ge \big(\min_{1\le j\le n}\rho_j^{[\kappa]}\big)^2 \quad \text{on } \partial\Delta_{g;\rho}.$$

Let $\eta < \min_{1\le j\le n}\rho_j^{[\kappa]}$ be such that the set $B_\eta = \{\zeta\in\Delta_{g;\rho} : \|g(\zeta)\|^2<\eta^2\}$ is a relatively compact open subset of $\Delta_{g;\rho}$ whose boundary ∂B_η is smooth. Averaging

---

[4] We refer here to §2.3.3; see Theorem 2.79. See [**KyS10**] for a similar averaging approach.

on the $(n-1)$-dimensional simplex

$$\{u \in \mathbb{R}^n \,:\, u_j \geq 0 \quad \text{for } j = 1, \dots, n; \sum_1^n u_j = \eta^2\},$$

the almost everywhere constant function

$$u \longmapsto \frac{1}{(2i\pi)^n} \int_{\Gamma(\Delta_{g;\sqrt{u_1},\dots,\sqrt{u_n}})} h(\zeta) \bigwedge_{j=1}^n \frac{d\zeta_j}{g_j(\zeta)} = \frac{1}{(2i\pi)^n} \int_{\Gamma(\Delta_{g;\rho})} h(\zeta) \bigwedge_{j=1}^n \frac{d\zeta_j}{g_j(\zeta)}$$

allows us to represent its almost everywhere constant value as

$$\begin{aligned}
&(-1)^{n(n-1)/2} \frac{(n-1)!}{(2i\pi)^n} \int_{\partial B_\eta} \frac{\sum_{\ell=1}^n (-1)^{\ell-1} \overline{g_\ell} \bigwedge_{\ell' \neq \ell} \overline{dg_{\ell'}}}{\|g\|^{2n}} \wedge h\, d\zeta \\
&= \Big[(-1)^{n(n-1)/2} \frac{(n-1)!}{(2i\pi)^n} \int_{\partial B_\eta} \|g\|^{2\lambda} \frac{\sum_{\ell=1}^n (-1)^{\ell-1} \overline{g_\ell} \bigwedge_{\ell' \neq \ell} \overline{dg_{\ell'}}}{\|g\|^{2n}} \wedge h\, d\zeta\Big]_{\lambda=0} \\
&= \Big[(-1)^{n(n-1)/2} \frac{n!}{(2i\pi)^n} \int_{B_\eta} \|g\|^{2(\lambda-n)} \Big(\bigwedge_{\ell=1}^n \overline{dg_\ell}\Big) \wedge h\, d\zeta\Big]_{\lambda=0}.
\end{aligned}$$

We recognize here the action of the Bochner–Martinelli current for sections of the trivial bundle (see Example 3.16) and conclude the proof by using Proposition 3.18. □

COROLLARY 6.10. *Let $\kappa, \kappa' \in \mathbb{N}^n$ and let $\rho \in \,]0, r_1[\, \times \cdots \times \,]0, r_n[$ be such that Δ_ρ is nondegenerate (recall that this occurs for almost all $\rho \in \,]0, \rho_1[\, \times \cdots \times \,]0, \rho_n[$). Then one has that*

$$\frac{1}{\rho_1^{2\kappa_1} \cdots \rho_n^{2\kappa_n}} \int_{\sigma_\rho} f^\kappa(\zeta)\, \overline{f^{\kappa'}(\zeta)}\, \frac{d\nu_{\Delta_\rho}(\zeta)}{\mu_\Delta} = \delta_\kappa^{\kappa'}. \tag{6.4}$$

PROOF. One profits from the equalities in (6.3). When $\kappa \neq \kappa'$, the result follows immediately from the fact that the residue current $\overline{\partial}(1/(g_1, \dots, g_n))$ attached to a system $(g_1, \dots, g_n)$ in a position of a complete intersection is annihilated (as a current) by each of the g_j (see Theorem 2.79). As for the remaining case where $\kappa = \kappa'$, it follows from the fact that the Coleff–Herrera current $\overline{\partial}(1/(f_1, \dots, f_n))$, when factorized in terms of a wedge product with $df_1 \wedge \cdots \wedge df_n$, leads to the integration current $[\mathrm{div}((f_1)_{|\Delta})] \wedge \cdots \wedge [\mathrm{div}((f_n)_{|\Delta})]$ with total mass precisely μ. □

When $\rho_1, \dots, \rho_n$ are sufficiently small, the analytic Weil polyhedron Δ_ρ is the disjoint union of $\nu = |V_\Delta(f)| \leq \mu$ connected components, where $|V_\Delta(f)|$ denotes the cardinal number of $V_\Delta(f)$. Each of these components is still a Weil polyhedron subordinate to (U, f, ρ), as we assumed in Definition 6.1, and contains a single common zero of $f_1, \dots, f_n$. From now on, we will focus on one of such connected components of a nondegenerate analytic Weil polyhedron Δ_ρ and denote it as Δ_ρ^α, where α stands for the unique common zero of $f_1, \dots, f_n$ it contains. Moreover, one can assume, shrinking if necessary the value of ρ, that $\zeta \mapsto f_{|\Delta_\rho^\alpha}(\zeta)$ realizes a (in general ramified) covering with $\mu_\alpha = \mu_{0,\alpha} := \dim_{\mathbb{C}} \mathcal{O}_\alpha/(f_1, \dots, f_n)\mathcal{O}_\alpha$ sheets from Δ_ρ^α onto a neighborhood of the closure of the polydisc

$\mathbb{D}_\rho = D(0,\rho_1)\times\cdots\times D(0,\rho_n)\subset\mathbb{C}^n$. One can then rewrite, for any $\kappa\in\mathbb{N}^n$,

$$\frac{1}{(2i\pi)^n}\int_{\Gamma(\Delta_\rho^\alpha)} h(\zeta)\bigwedge_{j=1}^n \frac{d\zeta_j}{(f_j(\zeta))^{\kappa_j+1}} = \frac{1}{(2i\pi)^n}\int_{\Gamma(\Delta_\rho^\alpha)} \frac{h(\zeta)}{J(\zeta)}\bigwedge_{j=1}^n \frac{df_j(\zeta)}{(f_j(\zeta))^{\kappa_j+1}}$$
$$= \frac{1}{(2i\pi)^n}\int_{\Gamma(\mathbb{D}_\rho)} \Big(\sum_{\xi\in f^{-1}_{|\Delta_\rho^\alpha}(t)} \frac{h(\xi)}{J(\xi)}\Big)\bigwedge_{j=1}^n \frac{dt_j}{t_j^{\kappa_j+1}},$$

where the *trace function*

$$t\in\mathbb{D}_\rho \longmapsto \sum_{\xi\in f^{-1}_{|\Delta_\rho^\alpha}(t)} \frac{h(\xi)}{J(\xi)}$$

is holomorphic in $\mathbb{D}_\rho$, since h is holomorphic in Δ^α; see, e.g., [**AVGZ**] or [**Ts**, II.6].[5] For reasons of simplicity, one assumes that $\Delta=\Delta_\rho$. Then one can conclude this discussion with the following statement.

PROPOSITION 6.11. *Let $f:\Omega\longrightarrow\mathbb{C}^m$ be a holomorphic map and let Δ be an analytic Weil polyhedron subordinate to (Ω,f,r). For a holomorphic function h in Δ, let $(a_{\Delta,\kappa}[h\,d\zeta])_{\kappa\in\mathbb{N}^n}$ be the multi-indexed list of complex numbers defined as in (6.3). Then the series $\sum_{\kappa\in\mathbb{N}^n} a_{\Delta,\kappa}[h\,d\zeta]\,t^\kappa$ can be interpreted as the Taylor series[6] at the origin of the holomorphic trace function*

$$(6.5)\quad t\in\mathbb{D}_\rho\longmapsto \sum_{\alpha\in V_\Delta(f)}\Big(\sum_{\xi\in f^{-1}_{|\Delta^\alpha}(t)} \frac{h\,d\zeta}{\bigwedge_{j=1}^n df_j}(\xi)\Big) = \sum_{\alpha\in V_\Delta(f)}\Big(\sum_{\xi\in f^{-1}_{|\Delta^\alpha}(t)} \frac{h(\xi)}{J(\xi)}\Big),$$

where $\bigcup\Delta^\alpha\subset\Delta$ is a nondegenerate analytic Weil polyhedron subordinate to (Ω,f,r) containing all common zeroes α of $f_1,\dots,f_n$ in Δ. Furthermore, α is the only zero in the connected component Δ^α and $\mathbb{D}_\rho$ is a polydisc centered at the origin such that for any α, $f_{|\Delta^\alpha}$ realizes a covering with $\mu_\alpha=\dim_{\mathbb{C}}\mathcal{O}_\alpha/(f_1,\dots,f_n)\mathcal{O}_\alpha$ sheets from Δ^α onto a neighborhood of $\overline{\mathbb{D}}_\rho$.

Assume that Δ is a nondegenerate analytic Weil polyhedron subordinate to the configuration (Ω,f,r). Since Ω is pseudo-convex, there exists an (n,n)-matrix H with holomorphic entries $h_{j,k}$ in $\Omega\times\Omega$ for every $1\le j,k\le n$, such that

$$(6.6)\qquad \forall\,\zeta,z\in\Omega\times\Omega,\quad f_j(\zeta)-f_j(z)=\sum_{k=1}^n h_{j,k}(z,\zeta)\,(\zeta_k-z_k).$$

Let $\mathcal{O}(\Delta)\cap C(\overline{\Delta})$ be the space of functions $h:\overline{\Delta}\to\mathbb{C}$ that are holomorphic in Δ and continuous in $\overline{\Delta}$. The following result is a particular case of a more general integral representation formula due to André Weil [**We35**], which already appeared in this monograph with a current formulation; see Remark 5.18, (5.94) or (5.95). Such a formula will be reformulated later on (according to the approach proposed here) in a formal algebraic context as Proposition 6.13.

[5] Although here h/J is a meromorphic function, its trace function happens to be holomorphic in this particular case. It follows from the proof in [**GrH**, p. 668], or one can use the ad hoc argument as in [**Ts**, II.6].

[6] As a matter of fact, Proposition 6.11 is, in some sense, a substitute for Taylor expansion along shifted coordinate functions when $z_1-a_1,\dots,z_n-a_n$ are replaced by $f_1,\dots,f_n$; see also Remark 6.22 (which is in an algebraic setting that the present situation illustrates).

PROPOSITION 6.12 (Weil's integral representation formula[7] in the case $m = n$). *Let Ω, f, r, Δ, and H be as above. Let also $h \in \mathcal{O}(\Delta) \cap C(\overline{\Delta})$. Then one has for any $z \in \Delta$ that*

$$h(z) = \frac{1}{(2i\pi)^n} \int_{\Gamma(\Delta)} h \det H(z,\cdot) \bigwedge_{j=1}^{n} \frac{d\zeta_j}{f_j - f_j(z)}$$

$$(6.7) \qquad = \sum_{\kappa \in \mathbb{N}^n} \Big(\int_{\sigma_\Delta} h \frac{\det H(z,\cdot)}{\det H(\cdot,\cdot)} \prod_{j=1}^{n} \Big(\frac{\overline{f_j}}{r_j}\Big)^{\kappa_j} d\nu_\Delta \Big) \prod_{j=1}^{n} \Big(\frac{f_j}{r_j}\Big)^{\kappa_j}(z),$$

where the series on the right-hand side converges uniformly on each compact subset of Δ.

PROOF. The function

$$z \in \Delta \longmapsto \frac{1}{(2i\pi)^n} \int_{\Gamma(\Delta)} h \det H(z,\cdot) \bigwedge_{j=1}^{n} \frac{d\zeta_j}{f_j - f_j(z)}$$

is holomorphic in Δ and expands there as

$$\sum_{\kappa \in \mathbb{N}^n} \Big(\frac{1}{(2i\pi)^n} \int_{\Gamma(\Delta)} h \det H(z,\cdot) \bigwedge_{j=1}^{n} \frac{d\zeta_j}{f_j^{\kappa_j+1}} \Big) f^\kappa(z)$$
$$= \sum_{\kappa \in \mathbb{N}^n} a_{\Delta,\kappa}\big[h \det A(z,\cdot)\, d\zeta\big] f^\kappa(z).$$

Let $z \in \Delta$. If one compares the Taylor development in a neighborhood of the origin $u = 0$ of the holomorphic function $u \longmapsto f_j(z+u) - f_j(z)$ with that of the holomorphic function $u \longmapsto \sum_{k=1}^{n} a_{j,k}(z+u, z) u_k$ (such developments coincide for any $j = 1, \dots, n$ due to (6.6)), then one concludes that $\det H(z,z) = J(z)$ for any $z \in \Delta$. Take $z \in \Delta$ such that $f(z) \in \mathbb{D}_\rho$, where the polydisc $\mathbb{D}_\rho$ has been introduced in the statement of Proposition 6.11 and is paired with the Weil polyhedron $\bigcup \Delta^\alpha \subset \Delta$. Since $f_{|\Delta}$ is necessarily an open mapping, then for generic $z \in \Delta$ with $f(z) \in \mathbb{D}_\rho$, one has that

$$\prod_{\xi \in f^{-1}_{|\bigcup \Delta^\alpha}(\{f(z)\})} J(\xi) \neq 0.$$

Fix one such z and take ξ to be in a pre-image of $f(z)$ for the holomorphic map $f_{|\bigcup \Delta^\alpha}$. If $\xi = z$, then $h(\xi) \det H(z,\xi)/J(\xi) = h(z) J(z)/J(z) = h(z)$. In a neighborhood of each $\xi \neq z$, which is a pre-image of $f(z)$ in $\bigcup \Delta^\alpha$, it follows from Cramer's rule that $(z', \zeta') \mapsto \det H(z', \zeta')$ is in the ideal generated by the functions $(z', \zeta') \longmapsto f_j(z') - f_j(\zeta')$, whenever $(z', \zeta') \in \Delta \times \Delta$ belongs to a neighborhood $U_{z,\xi}$. Since $f_j(\xi) = f_j(z)$, then for such ξ one has that $h(\xi) \det H(z,\xi)/J(\xi) = 0$. Finally

$$\sum_{\xi \in f^{-1}_{|\bigcup \Delta^\alpha}(\{f(z)\})} \frac{h(\xi) \det H(z,\xi)}{J(\xi)} = h(z) + 0 + \cdots + 0 = h(z).$$

[7]Such a representation formula is also known as the Cauchy–Weil formula since it inherits the intrinsic algebraic nature of Cauchy's formula; see [**AlY18**, **AlY21**] for applications.

Hence Proposition 6.11 implies that

$$\sum_{\kappa\in\mathbb{N}^n} a_{\Delta,\kappa}\big[h\,\det H(z,\cdot)\,d\zeta\big]\,f^{\kappa}(z) = h(z)$$

for such generic z close to $V_\Delta(f)$. Then Proposition 6.12 follows by analytic continuation. □

6.1.3. The case when $m > n$. Let Δ be a nondegenerate analytic Weil polyhedron subordinate to (Ω, f, r), where $f = (f_1, \ldots, f_m)$, with $m > n$, is a vector of holomorphic functions defined in the open pseudo-convex subset $\Omega \subset \mathbb{C}^n$ and $r \in (\mathbb{R}^{+*})^m$. The Shilov boundary $\Gamma(\Delta)$ splits now as $\sum_J \Gamma_J(\Delta)$, where $\Gamma_J(\Delta)$ is an oriented n-dimensional analytic chain with boundary, whose support is the n-dimensional real manifold

$$\sigma_{\Delta,J} = \{\zeta \in \mathrm{Supp}(\Gamma(\Delta)) \,:\, |f_{j_\ell}(\zeta)| = r_j, \quad j = 1, \ldots, n\}$$

with nonempty boundary $\partial\sigma_{\Delta,J}$. Such $\sigma_{\Delta,J}$ carries the positive measure defined as $\nu_{\Delta,J} = \big(\bigwedge_{\ell=1}^n d(\arg(f_{j_\ell}))\big)_{|\sigma_{\Delta,J}}/(2\pi)^n$ attached to the volume form defined. Let also $\{\epsilon_1, \ldots, \epsilon_m\}$ be the canonical basis of the trivial bundle $\mathbb{C}^m$ over Ω, so that $f = \sum_{j=1}^m f_j\,\epsilon_j$ and

$$df = \sum_{j=1}^m df_j\,\epsilon_j, \quad H_f(z,\cdot) = \sum_{j=1}^m h_{f,j}(z,\cdot)\,\epsilon_j,$$

where H_f is a Hefer form for f; see Definition 4.46. Then

$$\frac{(df)^n}{n!} = \sum_{1\le j_1<\cdots<j_n\le m} df_J \bigwedge_{\ell=1}^n \epsilon_{j_\ell}, \quad \frac{(H_f(z,\cdot))^n}{n!} = \sum_{1\le j_1<\cdots<j_n\le m} H^n_{f,J}(z,\cdot) \bigwedge_{\ell=1}^n \epsilon_{j_\ell}.$$

Given any $\omega \in \mathcal{O}(\Delta, \bigwedge^n(T_\Omega^{(1,0)*} \otimes \mathbb{C}^m))$, namely

$$\omega = \sum_{1\le j_1<\cdots<j_n\le m} \omega_J \bigwedge_{\ell=1}^n \epsilon_{j_\ell} = \sum_{1\le j_1<\cdots<j_n\le m} h_J\,d\zeta \bigwedge_{\ell=1}^n \epsilon_{j_\ell},$$

which extends continuously to $\overline{\Delta}$, the function

$$\begin{aligned} t \in \mathbb{D}_r \longmapsto \begin{bmatrix} \omega \\ f - t \end{bmatrix} &:= \sum_{1\le j_1<\cdots<j_n\le m} \frac{1}{(2i\pi)^n} \int_{\Gamma_J(\Delta)} \frac{\omega_J}{\prod_{\ell=1}^n (f_{j_\ell} - t_{j_\ell})} \\ &= \sum_{1\le j_1<\cdots<j_n\le m} \frac{1}{(2i\pi)^n} \int_{\Gamma_J(\Delta)} \frac{\omega_J}{(df)_J} \bigwedge_{\ell=1}^n \frac{df_{j_\ell}}{f_{j_\ell} - t_{j_\ell}} \end{aligned}$$

is well defined due to the nondegeneracy of the analytic Weil polyhedron Δ. Moreover, it is holomorphic in the polydisc $\mathbb{D}_r$ and admits there the following absolutely convergent Taylor development:

$$\begin{bmatrix} \omega \\ f_1 - t_1, \ldots, f_m - t_m \end{bmatrix}_\Delta = \sum_{\kappa\in\mathbb{N}^n} \sum_{1\le j_1<\cdots<j_n\le m} a_{\Delta,J,\kappa}[\omega]\, t_{j_1}^{\kappa_1} \cdots t_{j_n}^{\kappa_n}, \tag{6.8}$$

where

$$a_{\Delta,J,\kappa}[\omega] = \frac{1}{\prod_{\ell=1}^n r_{j_\ell}^{2\kappa_\ell}} \int_{\sigma_{\Delta,J}} \frac{\omega_J}{(df)_J} \prod_1^n \overline{f_{j_\ell}^{\kappa_\ell}}\,d\nu_{\Delta,J}. \tag{6.9}$$

A key difference with the case where $m = n$ is that the function

$$t \in \mathbb{D}_r \longmapsto \sum_{\kappa\in\mathbb{N}^n} \sum_{1\leq j_1<\cdots<j_n\leq m} a_{\Delta,J,\kappa}[\omega]\, t_{j_1}^{\kappa_1}\cdots t_{j_n}^{\kappa_n}$$

cannot be interpreted anymore as a trace function of the form

$$t \longmapsto \sum_{1\leq j_1<\cdots<j_n\leq m} \gamma_J \sum_{\xi\in f_{|\Delta}^{-1}(\{t\})} \frac{\omega_J}{(df)_J}(\xi)$$

(for some combinatorial positive weights γ_J with sum 1) as in Proposition 6.11, since no statement substitutes for Proposition 6.8 is in this case.

Nevertheless, a substitute for Proposition 6.12 remains valid in such context and keeps carrying the concept of trace. For any $t \in \mathbb{D}_r$, the closed analytic subset $f_{|\Delta}^{-1}(\{t\}) = f_{|\overline{\Delta}}^{-1}(\{t\})$ is a finite subset of Δ, possibly empty. In the particular case where $t = f(z) \in f(\Delta)$, the set $f_{|\Delta}^{-1}(\{f(z)\})$ is the disjoint union of $\{z\}$ with the (possibly empty) subset $E_z = f_{\Delta\setminus\{z\}}^{-1}(\{f(z)\})$. As a consequence of Cramer's rule, one has that for any $z \in \Delta$ and any $z' \in E_z$, $H_f^n(z,\cdot)(z') = 0$. On the other hand, $(H_f(\zeta,\zeta))^n(\zeta) = (df)^n(\zeta)$ for any $\zeta \in \Omega$. As Proposition 6.12 does in the case where $m = n$, Proposition 6.13 reflects the following trace property, namely that for any ordered index $1 \leq j_1 < \cdots < j_n \leq m$, for any $z \in \Delta$ and any holomorphic function h in Δ, then, at least formally

$$\sum_{\xi\in f_{|\Delta}^{-1}(\{f(z)\})} h(\xi)\,\frac{H_{f,J}^n(z,\xi)}{(df)_J(\xi)} = h(z) + 0 + \cdots + 0 = h(z).$$

PROPOSITION 6.13. *Let Ω, f, r, Δ, and $H_f(z,\cdot)$ be as above and let h be a holomorphic function in Δ which extends continuously to $\overline{\Delta}$. Then one has for any $z \in \Delta$,*

$$\begin{aligned} h(z) &= \left[\frac{h\, H_f^n(z,\cdot)/n!}{f_1 - w_1,\ldots,f_m - w_m}\right]_\Delta \\ &= \frac{1}{(2i\pi)^n} \sum_{1\leq j_1<\cdots<j_n\leq m} \int_{\Gamma_J(\Delta)} h\,\frac{H_{f,J}^n(z,\cdot)}{\prod_{\ell=1}^n (f_{j_\ell} - f_{j_\ell}(z))} \qquad (6.10)\\ &= \sum_{\substack{1\leq j_1<\cdots<j_n\leq m\\ \kappa\in\mathbb{N}^n}} \Big(\int_{\sigma_{\Delta,J}} h\,\frac{H_{f,J}^n(z,\cdot)}{H_{f,J}^n(\cdot,\cdot)} \prod_{\ell=1}^n \Big(\frac{\overline{f_{j_\ell}}}{r_{j_\ell}}\Big)^{\kappa_\ell} d\nu_{\Delta,J}\Big) \prod_1^n \Big(\frac{f_{j_\ell}}{r_{j_\ell}}\Big)^{\kappa_\ell}(z), \end{aligned}$$

where the series on the right-hand side converges uniformly on each compact subset of Δ.

PROOF. The proof we give here is taken from [**And92**]; see also [**AAC99**, §2]. It does not rely on a trace argument, which makes it quite different from that of Proposition 6.12. We use instead a weighted integral representation formula involving a global weight in $\Delta \times \Delta$ constructed according to Proposition 4.20. For any $j = 1,\ldots,m$, let q^j be the $(1,0)$-differential form in $\Delta \times \overline{\Delta}$ defined as

$$q^j(z,\zeta) = \frac{\overline{f_j(\zeta)}}{r_j^2 - \overline{f_j(\zeta)}f_j(z)} H_{f,j}.$$

Note that the forms q^j depend holomorphically on the variables z. Moreover, for each $j = 1, \dots, m$, the $(1,1)$-form

$$\bar{\partial}_\zeta((q^j)_{|\Delta\times\Delta})(z,\zeta) = \frac{r_j^2}{\left(r_j^2 - \overline{f_j(\zeta)} f_j(z)\right)^2} \overline{df_j(\zeta)} \wedge H_{f,j}(z,\zeta) \tag{6.11}$$

extends continuously from Δ^2 to $\Delta \times \overline{\Delta}$. Let $\lambda \in \mathbb{C}$ be a complex parameter with $\operatorname{Re}\lambda \gg 1$ and

$$\Gamma^\lambda : (W_1, \dots, W_{2m}) \in \{W \in \mathbb{C}^{2m} : \operatorname{Re} W_j > 0 \text{ for } j = 1, \dots, m\} \longmapsto \prod_{j=1}^{2m} W_j^\lambda.$$

Since

$$\Theta_j(z,\zeta) := 1 + \langle\!\langle q^j(z,\zeta), z - \zeta \rangle\!\rangle = \frac{r_j^2 - |f_j(\zeta)|^2}{r_j^2 - \overline{f_j(\zeta)}\, f_j(z)}$$

for any $z \in \Delta$, $\zeta \in \overline{\Delta}$, and $j = 1, \dots, m$, one has for such z, ζ, and j that

$$\operatorname{Re}\, \Theta_j(z,\zeta) > 0, \quad \left|\Theta_j(z,\zeta)\right| \leq \frac{r_j + |f_j(\zeta)|}{r_j} \leq 2. \tag{6.12}$$

Let $\boldsymbol{q} = (q^1, \dots, q^m, q^1, \dots, q^m)$ and let $\mathscr{W}_{\boldsymbol{q}}$ be defined as in (4.57). Note that for any $j = 1, \dots, m$, $z \in \Delta$, $\zeta \in \overline{\Delta}$, and Re $\lambda \gg 1$,
(6.13)

$$\lambda(\Theta_j(z,\zeta))^{\lambda-1}\,\overline{\partial}_\zeta q^j(z,\zeta) = -\frac{\overline{\partial}\left(1 - |f_j(\zeta)|^2/r_j^2\right)^\lambda}{f_j(\zeta)} \wedge \frac{h_{f,j}(z,\zeta)}{\left(1 - \overline{f_j(\zeta)} f_j(z)/r_j^2\right)^{\lambda+1}}.$$

Since $(\Theta(z,\zeta))^\lambda = \prod_{j=1}^m (\Theta_j(z,\zeta))^\lambda = 0$ for $z \in \Delta$ and $\zeta \in \partial\Delta$, it follows from Proposition 4.34 (namely from the integral representation formula (4.84)) that for any $z \in \Delta$ and any λ with Re $\lambda \gg 1$,

$$h(z) = \frac{1}{(2i\pi)^n} \int_\Delta h\,(\mathscr{W}_{\boldsymbol{q}}(z,\cdot))_{n,n} = \sum_{1 \leq j_1 < \cdots < j_n \leq m} \tag{6.14}$$

$$\times \frac{1}{(2i\pi)^n} \int_\Delta h(\Theta(z,\cdot))^\lambda \Big(\prod_{j \notin J} \Theta_j(z,\cdot)\Big)^\lambda \bigwedge_{\ell=1}^n \left(\lambda(\Theta_{j_\ell}(z,\zeta))^{\lambda-1}\overline{\partial}_\zeta q^{j_\ell}(z,\cdot)\right).$$

We now substitute in the integrands on the right-hand side of the integral representation formula (6.14) the identities (6.13) for $j = j_\ell$, $\ell = 1, \dots, n$. One has for any $j = 1, \dots, m$, $z \in \Delta$, and $\zeta \in \overline{\Delta}$ that

$$\Big[\frac{h_{f,j}(z,\zeta)}{\left(1 - \overline{f_j(\zeta)} f_j(z)/r_j^2\right)^{\lambda+1}}\Big]_{\lambda=0} = r_j^2\, \frac{h_{f,j}(z,\zeta)}{r_j^2 - \overline{f_j(\zeta)} f_j(z)}.$$

On the other hand, for any ordered subset $J : 1 \leq j_1 < \cdots < j_n \leq m$ and for any $(1,0)$-smooth forms ϖ_ℓ in a neighborhood of $\overline{\Delta}$ one has, as a consequence of the jump formula in distribution theory, that

$$\Big[\int_\Delta \bigwedge_{\ell=1}^n \left(\left(-\overline{\partial}(1 - |f_j(\zeta)|^2/r_j^2)^\lambda\right) \wedge \varpi_\ell\right)\Big]_{\lambda=0} = \int_{\Gamma_J(\Delta)} \bigwedge_{\ell=1}^n \varpi_\ell.$$

The jump formula applies here since Δ is assumed to be nondegenerate. One can then profit from estimates (6.12), in particular from the first one, to conclude that formula (6.10) follows asymptotically from (6.14), when λ tends to 0^+ in $\mathbb{R}^+$. □

REMARK 6.14. What stands as the main stumbling block in the $\mathbb{I}$-adic convergent development (6.10) in the case where $m > n$ (compared to the case where $m = n$) is that the coefficients

$$z \longmapsto \int_{\sigma_{\Delta,J}} h \frac{H^n_{f,J}(z,\cdot)}{H^n_{f,J}(\cdot,\cdot)} \prod_{\ell=1}^{n} \Big(\frac{\overline{f_{j_\ell}}}{r_{j_\ell}}\Big)^{\kappa_\ell} d\nu_{\Delta,J}$$

depend drastically on Δ. One should compare such convergent development in $\mathcal{O}(\Delta)$ to the formal developments in $\mathcal{O}[[a_1,\dots,a_m]]$ given, for example, by Proposition 5.19. The coefficients of these formal developments depend on the Taylor series of h at the origin. We refer the reader to [**VidY22**, §3 and §4] for suggestions about how to combine convergent $\mathbb{I}$-dic developments, such as (6.10), and the formal ones such as (5.90), (5.92).

6.1.4. Hardy spaces of strongly nondegenerate analytic Weil polyhedra. Given a strongly nondegenerate analytic Weil polyhedron Δ subordinate to (Ω, f, r), its open subset $\Delta_{r-\varepsilon} = \{\zeta \in \Delta : |f_j(\zeta)| < r_j - \varepsilon_j \text{ for } j = 1,\dots,m\}$ remains a strongly nondegenerate Weil polyhedron subordinate to $(\Delta, f_{|\Delta}, r-\varepsilon)$ provided each $\varepsilon_j \in]0, r_j[$ is sufficiently close to 0. According to such observation, let us attach to Δ the following Banach spaces [**Boo97, AAC99**].

DEFINITION 6.15 (Hardy space $H^q(\Delta)$). Let Δ be a strongly nondegenerate analytic Weil polyhedron. For any $q \in [1,\infty[$, the Hardy space $H^q(\Delta)$ is defined to be the space of holomorphic functions h in Δ such that

$$\|h\|_{H^q(\Delta)} := \limsup_{\varepsilon\to 0} \Big(\sum_{1\le j_1<\cdots<j_n\le m} \int_{\sigma_{\Delta_{r-\varepsilon},J}} |h|^q \, d\nu_{\Delta_{r-\varepsilon},J}\Big)^{1/q} < +\infty$$

for $1 \le q < \infty$. It inherits a Banach space structure when equipped with the norm as above. The Banach space $H^\infty(\Delta)$ is defined as the space of holomorphic bounded functions h in Δ, equipped with the norm $\|h\|_{H^\infty(\Delta)} = \sup_\Delta |h|$.

Given a strongly nondegenerate Weil polyhedron Δ subordinate to (Ω, f, r), one has that $\mathrm{Supp}(\Gamma(\Delta)) = \bigsqcup_{1\le j_1<\cdots<j_n\le m} \sigma_{\Delta,J}$. For any $q \in [1,\infty]$, the Cauchy–Weil operator

$$\mathrm{CW} : L^2(\sigma_\Delta, d\nu_\Delta) := \bigoplus_{1\le j_1<\cdots<j_n\le m} L^q(\sigma_{\Delta,J}, d\nu_{\Delta,J}) \bigoplus_{\ell=1}^{n} \epsilon_{j_\ell} \longmapsto \mathcal{O}(\Delta),$$

where

$$(6.15)\quad \big(\mathrm{CW}\big(\sum_J \mathfrak{h}_J\, \epsilon_J\big)\big)(z) = \sum_{\substack{1\le j_1<\cdots<j_n\le m\\ \kappa\in\mathbb{N}^n}} \Big(\int_{\sigma_{\Delta,J}} \mathfrak{h}_J \frac{H^n_{f,J}(z,\cdot)}{H^n_{f,J}(\cdot,\cdot)} \prod_{\ell=1}^{n} \Big(\frac{\overline{f_{j_\ell}}}{r_{j_\ell}}\Big)^{\kappa_\ell} d\nu_{\Delta,J}\Big) \prod_{1}^{n} \Big(\frac{f_{j_\ell}}{r_{j_\ell}}\Big)^{\kappa_\ell}(z)$$

acts continuously provided $\mathcal{O}(\Delta)$ is equipped with the topology of uniform convergence on any compact subset. Even more, one has the following precise result; see [**Boo97**, Remark 2].

PROPOSITION 6.16. *For any $q \in [1,\infty]$, the Cauchy–Weil operator* (6.15) *acts continuously from $L^q(\sigma_\Delta, d\nu_\Delta)$ to the Hardy space $H^q(\Delta)$.*

PROOF. As pointed out in [**AAC99**, §1], the analytic Weil polyhedron Δ subordinate to (Ω, f, r) is strongly nondegenerate if and only if it is locally biholomorphic to an open subset of the polydisc $\mathbb{D}^n$. More precisely, each point ξ in $\mathrm{Supp}(\Gamma(\Delta))$ admits an open neighborhood U_ξ in Ω, which is biholomorphically equivalent (via the biholomorphim ι_ξ) to a neighborhood of some point $\iota_\xi(\xi) \in \Gamma(\mathbb{D}^n)$ in such a way that $\iota_\xi(U_\xi \cap \Delta) = \iota_\xi(U_\xi) \cap \mathbb{D}^n$. Moreover, the biholomorphism ι_ξ is given by $\iota_\xi \ : \zeta \mapsto f_{J(\xi)}(\zeta)$, where the ordered index $1 \leq j_1(\xi) < \cdots < j_n(\xi) \leq m$ is such that one has $\sigma_\Delta \cap U_\xi = \sigma_{\Delta, J(\xi)} \cap U_\xi$. Then the result follows from the fact that the Cauchy–Weil operator attached to the polydisc $\mathbb{D}^n$ (subordinate to $(\mathbb{C}^n, z, 1)$) acts continuously from $L^q(\Gamma(\mathbb{D}^n), d\nu_{\Gamma(\mathbb{T}^n)}) = L^q((\mathbb{R}/(2\pi\mathbb{Z}))^n, d\theta)$ into the Hardy space $H^q(\mathbb{D}^n)$; see also the proof of Weil's formula in the strongly nondegenerate case combined with [**Boo97**, §1, Remark 2]. □

6.2. Algebraic approach to residue symbols

6.2.1. Quasi-regular sequences in a commutative $\mathbb{K}$-algebra $\mathbb{A}$.

In this subsection, $\mathbb{A}$ denotes a commutative[8] ring. The notion of *quasi-regularity* plays a role in the algebraic approach towards multivariate residue calculus. *Quasi-regularity* is intrinsically associated with the exactness of the *Koszul complex* introduced in Chapter 3 (see Examples 3.25 and 3.30, or also §1.2.5). The material presented in this subsection is from [**Lip87**, §3].

DEFINITION 6.17 (Quasi-regular sequence). A sequence[9] $(f_1, \dots, f_m)$ of m elements in $\mathbb{A}$ is said to be *quasi-regular* if and only if for any integer $N \in \mathbb{N}$ and any family $\{u_\kappa \ : \ \kappa \in \mathbb{N}^m, \ |\kappa| = N\}$ of elements in $\mathbb{A}$, one has
(6.16)

$$\sum_{\{\kappa \in \mathbb{N}^m \,:\, |\kappa| = N\}} u_\kappa \prod_{\ell=1}^m f_\ell^{\kappa_\ell} \in (f_1\mathbb{A} + \cdots + f_m\mathbb{A})^{N+1} \Longrightarrow u_\kappa \in f_1\mathbb{A} + \cdots + f_m\mathbb{A} \quad \forall \kappa.$$

Let $f = (f_1, \dots, f_m)$ be a finite sequence in $\mathbb{A}$ and let $\mathbb{I}_f$ be the finitely generated ideal $\mathbb{I}_f = f_1\mathbb{A} + \cdots + f_m\mathbb{A}$. We recall here that the $\mathbb{I}_f$*-dic completion of* $\mathbb{A}$, namely

$$\widehat{\mathbb{A}}_f = \lim_{N \geq 0} \mathbb{A}/\mathbb{I}_f^N,$$

is the $\mathbb{A}$-module, whose elements are identified with equivalence classes of Cauchy sequences $(a_N)_{N\geq 0}$ of elements of $\mathbb{A}$ in the $\mathbb{I}_f$-dic topology[10] modulo the equivalence relation:

$$(a'_N)_{N\geq 0} \sim (a''_N)_{N\geq 0} \Longleftrightarrow \lim_{N\to+\infty} (a'_N - a''_N) = 0.$$

[8]The *commutative* setting here is fundamental, even though some attempts towards developing residue calculus make sense outside such frame, for example in the *quaternionic* context; see §7.7.1 and §7.7.2.

[9]Observe that the ordering of the terms of the sequence is irrelevant in such a definition. Thus, one could speak about a *quasi-regular family* $\{f_1, \dots, f_m\}$ instead of a *quasi-regular sequence* $(f_1, \dots, f_m)$. Nevertheless, since later on the ordering will play an important role in the definition of algebraic residue symbols, we will always speak in this monograph about *quasi-regular sequences*.

[10]This means that for any $N_0 \in \mathbb{N}$, all elements a_N for $N \gg 1$ (depending on N_0) differ by an element of $\mathbb{I}_f^{N_0}$.

Let also $\widehat{\mathbb{I}}_f = f_1\widehat{\mathbb{A}}_f + \cdots + f_m\widehat{\mathbb{A}}_f \subset \widehat{\mathbb{A}}_f$ and let $\widehat{\pi}_f$ be the canonical projection from $\widehat{\mathbb{A}}_f$ to its quotient $\widehat{\mathbb{A}}_f/\widehat{\mathbb{I}}_f = \mathbb{A}/\mathbb{I}_f$; see [**Lip87**, Lemma 3.3.2]. Given a map[11]

$$T : \mathbb{P}_f = \mathbb{A}/\mathbb{I}_f \to \widehat{\mathbb{A}}_f, \text{ with } \widehat{\pi}_f \circ T = \mathrm{Id}_{\mathbb{P}_f} \text{ and } T(0) = 0, \tag{6.17}$$

elements of $\widehat{\mathbb{A}}_f$ can be identified with power series $\sum_{\kappa\in\mathbb{N}^m} T(\xi_\kappa)\, f^\kappa$, where $\xi_\kappa \in \mathbb{P}_f$. Thus, the following four assertions are equivalent (see [**Lip87**, Lemma 3.3.2]):

(1) The sequence $(f_1, \ldots, f_m)$ is quasi-regular in $\mathbb{A}$.

(2) For any $\widehat{h} \in \widehat{\mathbb{A}}_f$, the list $[\xi_\kappa]_{\kappa\in\mathbb{N}^m}$ such that $\widehat{h}$ can be represented in $\widehat{\mathbb{A}}_f$ as

$$\widehat{h} = \sum_{\kappa\in\mathbb{N}^m} T(\xi_\kappa)\, f^\kappa$$

is uniquely determined in terms of $\widehat{h}$ once the section T is prescribed,

(3) The Koszul complex[12] $\mathbf{K}_\bullet^f$ determined by f over $\mathbb{A}$ is exact except in degree zero.

(4) For any maximal ideal $\mathfrak{M}$ containing $\mathbb{I}_f$, the images $\mathfrak{f}_j$ of the f_j in the localization $\mathbb{A}_{\mathfrak{M}}$ form a *regular sequence* in this local ring.[13]

From now on, we assume that $f = \{f_1, \ldots, f_m\}$ is a quasi-regular sequence in $\mathbb{A}$. Let us now be more specific about the commutative ring $\mathbb{A}$ and assume that it also has the structure of $\mathbb{K}$-algebra: the commutative ring $\mathbb{K}$ (called the *ring of constants* or *basis ring*) is paired with a ring homomorphism $\iota : \mathbb{K} \to \mathbb{A}$. Then the $\mathbb{I}_f$-dic completion $\widehat{\mathbb{A}}_f$ inherits a structure of $\mathbb{K}[[X_1, \ldots, X_m]] = \mathbb{K}[[X]]$-algebra[14] through the ring homomorphism

$$\widehat{\iota}_f\ :\ \sum_{\kappa\in\mathbb{N}^m} \gamma_\kappa X^\kappa \in \mathbb{K}[[X]] \longmapsto \sum_{\kappa\in\mathbb{N}^m} \iota(\gamma_\kappa) f^\kappa \in \widehat{\mathbb{A}}_f. \tag{6.18}$$

To any element $\widehat{h}$ in $\widehat{\mathbb{A}}_f$, one can associate the element $\widehat{H}$ of $\mathrm{Hom}_{\mathbb{K}[[X]]}(\widehat{\mathbb{A}}_f, \widehat{\mathbb{A}}_f)$[15] acting by multiplication as follows

$$\widehat{H} : \widehat{a} \in \widehat{\mathbb{A}}_f \longrightarrow \widehat{h} \times \widehat{a} \in \widehat{\mathbb{A}}_f. \tag{6.19}$$

When the section T of $\widehat{\pi}_f$ introduced in (6.17) is, in addition, $\mathbb{K}$-linear, which we will assume from now on, one has that

$$\xi_\kappa(\widehat{H}(\widehat{a})) = \phi_\kappa(\widehat{H})(\xi_\kappa(\widehat{a}))$$

for some $\phi_\kappa(\widehat{H}) \in \mathrm{Hom}_{\mathbb{K}}(\mathbb{P}_f, \mathbb{P}_f)$ and for any $\kappa \in \mathbb{N}^m$. More generally, one may define in the same way $\phi_\kappa(U) \in \mathrm{Hom}_{\mathbb{K}}(\mathbb{P}_f, \mathbb{P}_f)$ for $\kappa \in \mathbb{N}^m$ and any U in

---

[11] Such a map is called a *section* of the projection morphism $\widehat{\pi}_f$.

[12] Such a complex is built in the algebraic setting of the commutative ring $\mathbb{A}$ using the interior contraction operators $\rfloor_{f,k} : \bigwedge^k(\sum_{j=1}^m \mathbb{A}\,\epsilon_j^*) \longmapsto \bigwedge^{k-1}(\sum_{j=1}^m \mathbb{A}\,\epsilon_j^*)$, where $f = f_1\epsilon_1 + \cdots + f_m\,\epsilon_m$ and $k \in \mathbb{N}^*$, $\{\epsilon_1, \ldots, \epsilon_m\}$ being here a nonsense basis, defined as in Definition 1.25.

[13] That is, $\mathfrak{f}_1 \neq 0$ and for any $k = 2, \ldots, m$, $\mathfrak{f}_j$ is not a zero-divisor in $\mathbb{A}_{\mathfrak{M}}/(\mathfrak{f}_1, \ldots, \mathfrak{f}_{k-1})$ (since $\mathbb{A}_{\mathfrak{M}}$ is a local ring, the ordering of the sequence f is irrelevant).

[14] In our context, given two commutative rings $\mathbb{K}_0$ and $\mathbb{K}_1$, $\mathbb{K}_1$ *is said to inherit a structure of* $\mathbb{K}_0$*-algebra* if and only if there is a ring homomorphism $\iota_{0,1} : \mathbb{K}_0 \to \mathbb{K}_1$. This point of view slightly enlarges the classical situation where $\mathbb{K}_1$ contains $\mathbb{K}_0$ as a subring and $\iota_{0,1}$ equals the natural inclusion map. For example, the ring $\mathbb{A}$ inherits the structure of $\mathbb{K}$-algebra because of the existence of the ring homomorphism $\iota\ : \mathbb{K} \to \mathbb{A}$.

[15] One keeps track here of the structure of $\mathbb{K}[[X]]$-module induced by that of $\mathbb{K}[[X]]$-algebra on the $\mathbb{I}_f$-dic completion $\widehat{\mathbb{A}}_f$. The quotient $\mathbb{K}$-algebra $\mathbb{P}_f = \mathbb{A}/\mathbb{I}_f = \widehat{\mathbb{A}}_f/\widehat{\mathbb{I}}_f$ inherits in the same way a structure of $\mathbb{K}$-module.

$\mathrm{Hom}_{\mathbb{K}[[X]]}(\widehat{\mathbb{A}}_f, \widehat{\mathbb{A}}_f)$ (besides just those $U = \widehat{H}$ corresponding to multiplication operators by elements $\widehat{h}$ in $\widehat{\mathbb{A}}_f$). Finally, if we introduce the (noncommutative) $\mathbb{K}[[X]]$-algebra

$$\widehat{\mathbb{E}}_f = \mathrm{Hom}_{\mathbb{K}}(\mathbb{P}_f, \mathbb{P}_f)[[X]], \tag{6.20}$$

we may attach to the $\mathbb{K}$-linear section $T : \mathbb{A}/\mathbb{I}_f \longrightarrow \widehat{\mathbb{A}}_f$ (see (6.17)) the following $\mathbb{K}[[X]]$-linear map

$$T^\sharp : U \in \mathrm{Hom}_{\mathbb{K}[[X]]}(\widehat{\mathbb{A}}_f, \widehat{\mathbb{A}}_f) \longmapsto T^\sharp(U) = \sum_{\kappa \in \mathbb{N}^m} \phi_\kappa(U)\, X^m \in \widehat{\mathbb{E}}_f. \tag{6.21}$$

Most interesting is the particular case where $\mathbb{P}_f = \mathbb{A}/\mathbb{I}_f$ is a *finitely presented* $\mathbb{K}$*-module.*[16] If such is the case, then the $\mathbb{K}[[X]]$-linear map

$$T^* : \xi \otimes \Big(\sum_{\kappa \in \mathbb{N}^m} \gamma_\kappa\, X^\kappa\Big) \in \mathbb{P}_f \otimes_{\mathbb{K}} \mathbb{K}[[X]] \longmapsto T(\xi) \times \widehat{\iota}_f\Big(\sum_{\kappa \in \mathbb{N}^m} \gamma_\kappa\, X^\kappa\Big) \in \widehat{\mathbb{A}}_f \tag{6.22}$$

realizes a bijection between the $\mathbb{I}_f$-dic completion $\hat{\mathbb{A}}_f$ and $\mathbb{P}_f \otimes \mathbb{K}[[X]]$, where $\widehat{\iota}_f$ is as in (6.18); see [**Lip87**, Lemma 3.3.2]. Thus $T^\sharp$, as defined in (6.21), realizes an isomorphism of $\mathbb{K}[[X]]$-modules between $\mathrm{Hom}_{\mathbb{K}[[X]]}(\widehat{\mathbb{A}}_f, \widehat{\mathbb{A}}_f)$ and $\widehat{\mathbb{E}}_f$. Observe in particular that through such an isomorphism $X_j \in \widehat{\mathbb{E}}_f$ corresponds to the multiplication operator by $\widehat{f_j}$, namely $\widehat{F}_j$, considered as acting from the $\mathbb{K}[[X]]$-module $\widehat{\mathbb{A}}_f$ to itself as in (6.19). The operator $\widehat{F}_j$ identifies with f_j. Moreover, one can easily see that $(T^\sharp)^{-1}$ maps isomorphically the $\mathbb{K}$-algebra $\mathrm{Hom}_{\mathbb{K}}(\mathbb{P}_f, \mathbb{P}_f) \subset \widehat{\mathbb{E}}_f$ onto a $\mathbb{K}$-subalgebra of $\mathrm{Hom}_{\mathbb{K}[[X]]}(\widehat{\mathbb{A}}_f, \widehat{\mathbb{A}}_f)$. Once $\mathrm{Hom}_{\mathbb{K}}(\mathbb{P}_f, \mathbb{P}_f)$ identifies with such $\mathbb{K}$-subalgebra $(T^\sharp)^{-1}(\mathrm{Hom}_{\mathbb{K}}(\mathbb{P}_f, \mathbb{P}_f))$, by taking into account the correspondence already mentioned between $X_j \in \widehat{\mathbb{E}}_f[[X]]$ and $\widehat{F_j} \simeq f_j$, one gets the identification

$$\mathrm{Hom}_{\mathbb{K}[[X]]}(\widehat{\mathbb{A}}_f, \widehat{\mathbb{A}}_f) \simeq_T \mathrm{Hom}_{\mathbb{K}}(\mathbb{P}_f, \mathbb{P}_f)[[f]]. \tag{6.23}$$

The identification above depends, as materialized with the notation $\simeq_T$, on the choice of the $\mathbb{K}$-linear section T for $\widehat{\pi}_f$.

EXAMPLE 6.18. The case $\mathbb{A} = \mathbb{K}[X_1, \dots, X_n] = \mathbb{K}[X]$, where $\mathbb{K}$ is a field, is the archetypal example of the situation described above. A sequence $p = (p_1, \dots, p_m)$ of elements in $\mathbb{K}[X]$ such that $1 \notin \mathbb{I}_p$ is quasi-regular in $\mathbb{K}[X]$ if and only if $m \leq n$ and the algebraic variety $\mathrm{Spec}\,(\mathbb{K}[X]/\mathbb{I}_p)$ has dimension $n - m$. Thus, the quotient algebra $\mathbb{P}_p = \mathbb{K}[X]/\mathbb{I}_p$ is finitely presented as a $\mathbb{K}$-module when $m = n$, in which case it inherits a structure of finitely dimensional $\mathbb{K}$-vector space. A $\mathbb{K}$-linear section T of $\widehat{\pi}_p$ is, for example, provided as follows: given $\xi \in \mathbb{P}_p$ with representative q, take $T(\xi) = T(q)$ as the remainder in the Gröbner division algorithm $\mathscr{G}_{\mathscr{B}}$ by the ideal $\mathbb{I}_p$ once a standard basis $\mathscr{B}$ for $\mathbb{I}_p$ has been chosen. When $\mathbb{K} = \mathbb{C}$ and H_p is a Hefer

[16]This means that there exists an exact sequence of $\mathbb{K}$-modules

$$\mathbb{K}^{r_1} \xrightarrow{F_1} \mathbb{K}^{r_0} \xrightarrow{F_0} \mathbb{P}_f \longrightarrow 0 \quad (r_0, r_1 \in \mathbb{N}^*).$$

In other words, $\mathbb{P}_f$ equals the quotient of some $\mathbb{K}^{r_0}$ by a finitely generated $\mathbb{K}$-submodule. When $\mathbb{K}$ is nœtherian, this is equivalent to saying that $\mathbb{P}_f$ is finitely generated. Observe that any projective finitely generated $\mathbb{K}$-module is a direct summand of some $\mathbb{K}^{r_0}$ as a consequence of being projective, and therefore is finitely presented.

form $\sum_{j=1}^n h_{p,j}\,\epsilon_j$, where $h_{p,j} \in \Omega_{\mathbb{C}[X]/\mathbb{C}}$ is a Kähler differential for $j = 1, \dots, n$, then, since $\#V(p) < \infty$,

$$T : \bar{q} \in \mathbb{K}[X]/\mathbb{I}_p \longmapsto \Big\langle \overline{\partial}\Big(\frac{1}{p_1, \dots, p_n}\Big)\, ,\, q\, H_p^n(X, \cdot)\Big\rangle \in \mathbb{K}[X]$$

is also an important example of section of $\widehat{\pi}_p$; see Proposition 6.8 and Theorem 3.39.

EXAMPLE 6.19. Let

$$\mathbb{A} = \mathbb{K}_{\mathbb{J}}[X, Y] = \frac{\mathbb{K}[X_1, \dots, X_n, Y_1, \dots, Y_{N-n}]}{\mathbb{J}}, \tag{6.24}$$

where $\mathbb{J}$ is a purely n-dimensional radical ideal in $\mathbb{K}[X, Y]$. A sequence of polynomials $p = (p_1, \dots, p_m)$ in $\mathbb{K}[X, Y]$ that satisfies $1 \notin \mathbb{I}_p + \mathbb{J}$ is such that the classes $\bar{p}_j$ of the p_j modulo $\mathbb{J}$ define a quasi-regular sequence in $\mathbb{A}$ if and only $m \leq n$ and the algebraic variety $\mathrm{Spec}(\mathbb{K}_{\mathbb{J}}[X, Y]/\mathbb{I}_{\bar{p}})$ is purely $n - m$-dimensional. In this case, the quotient algebra $\mathbb{P}_{\bar{p}}$ is finitely presented if and only if $m = n$, in which case $\mathbb{K}_{\mathbb{J}}[X, Y]/\mathbb{I}_{\bar{p}}$ inherits a structure of $\mathbb{K}$-vector space. A $\mathbb{K}$-linear section of $\widehat{\pi}_{\bar{p}}$ is given as follows: consider on $\mathbb{K}[X, Y]$ the Gröbner division algorithm $\mathscr{G}_{\mathscr{B}}$ by the ideal $\mathbb{I}_p + \mathbb{J}$, once a standard basis[17] $\mathscr{B}$ for $\mathbb{I}_p + \mathbb{J}$ has been chosen. Then the correspondence

$$\xi \in \mathbb{P}_{\bar{p}} \longmapsto \mathrm{remainder}_{\mathscr{G}_{\mathscr{B}}}(\Xi),$$

where $\Xi \in \mathbb{K}[X, Y]$ represents ξ modulo $\mathbb{I}_p + \mathbb{J}$, provides the required $\mathbb{K}$-linear section.

6.2.2. Algebraic residue symbols as traces. Let $\mathbb{A}$ be a commutative ring equipped with a structure of $\mathbb{K}$-algebra over the commutative basis ring $\mathbb{K}$. Let also $f = (f_1, \dots, f_m)$ be a quasi-regular sequence in $\mathbb{A}$ such that $\mathbb{P}_f = \mathbb{A}/\mathbb{I}_f$ is a finitely generated projective module. Since $\mathbb{P}_f$ is then finitely presented, the $\mathbb{K}[[X]]$-modules $\widehat{\mathbb{A}}_f$ and $\mathbb{P}_f \otimes_{\mathbb{K}} \mathbb{K}[[X]]$ are isomorphic. In order to materialize such a $\mathbb{K}[[X]]$-isomorphism, one chooses a $\mathbb{K}$-linear section T for $\widehat{\pi}_f$ and then uses T^*, defined by (6.22), as its materialization; see §6.2.1. Since $\mathbb{P}_f$ is finitely generated and projective over $\mathbb{K}$, so is $\widehat{\mathbb{A}}_f \simeq \mathbb{P}_f \otimes_{\mathbb{K}} \mathbb{K}[[X]]$ over $\mathbb{K}[[X]]$. Therefore, one can define *trace maps*:

$$\begin{aligned} &\mathrm{Tr}_{\,\mathbb{P}_f\,:\,\mathbb{K}} : \mathrm{Hom}_{\mathbb{K}}(\mathbb{P}_f, \mathbb{P}_f) \longrightarrow \mathbb{K}, \\ &\mathrm{Tr}_{\,\widehat{\mathbb{A}}_f\,:\,\mathbb{K}[[X]]} : \mathrm{Hom}_{\mathbb{K}[[X]]}(\widehat{\mathbb{A}}_f, \widehat{\mathbb{A}}_f) \simeq_T \mathrm{Hom}_{\mathbb{K}}(\mathbb{P}_f, \mathbb{P}_f)[[f]] \longrightarrow \mathbb{K}[[X]]. \end{aligned} \tag{6.25}$$

The identification $\mathrm{Hom}_{\mathbb{K}[[X]]}(\widehat{\mathbb{A}}_f, \widehat{\mathbb{A}}_f) \simeq_T \mathrm{Hom}_{\mathbb{K}}(\mathbb{P}_f, \mathbb{P}_f)[[f]]$ allows us to profit from the m natural $\mathrm{Hom}_{\mathbb{K}}(\mathbb{P}_f, \mathbb{P}_f)$-derivations on $\mathrm{Hom}_{\mathbb{K}}(\mathbb{P}_f, \mathbb{P}_f)[[f]]$ acting as

$$\frac{\partial}{\partial f_j} : \sum_{\kappa \in \mathbb{N}^m} \xi_\kappa f^\kappa \longmapsto \sum_{\{\kappa \in \mathbb{N}^m\,:\,\kappa_j \neq 0\}} \kappa_j\, \xi_\kappa\, f_1^{\kappa_1} \cdots f_j^{\kappa_j - 1} \cdots f_m^{\kappa_m} \tag{6.26}$$

for $j = 1, \dots, m$. Let $\Omega_{\mathbb{A}/\mathbb{K}}$ be the $\mathbb{A}$-module of Kähler differentials attached to the structure of $\mathbb{A}$ as a $\mathbb{K}$-algebra and $d : \mathbb{A} \to \Omega_{\mathbb{A}/\mathbb{K}}$ be its associated universal derivation. Recall that such universal derivation is defined imposing the conditions $d\gamma = 0$ for any $\gamma \in \mathbb{K}$, $d(a' + a'') = da' + da''$, $d(a'a'') = a'\, da'' + a''\, da'$ (namely

[17]Note that one needs to know generators for $\mathbb{J}$ in $\mathbb{K}[X, Y]$ to settle such Gröbner division algorithm.

Leibniz's rule) for all elements $a', a'' \in \mathbb{A}$. Given $h, a_1, \dots, a_m \in \mathbb{A}$, consider the element defined as

$$\omega_a(h) = h\, da_1 \wedge \cdots \wedge da_m \in \bigwedge^m \Omega_{\mathbb{A}/\mathbb{K}}. \tag{6.27}$$

Given the $\mathbb{K}$ linear section T, which governs the identification (6.23), let us explain how to attach to $h, a_1, \dots, a_m$ an element $(\omega_a^{\#}(h))_T \in \mathrm{Hom}_{\mathbb{K}}(\mathbb{P}_f, \mathbb{P}_f)[[f]]$. We proceed in three steps to detail such a construction.

(1) We consider first each a_j, $j = 1, \dots, m$ as an element $\widehat{a_j} \in \widehat{\mathbb{A}}_f$ and let

$$\widehat{A_j} \in \mathrm{Hom}_{\mathbb{K}[[X]]}(\widehat{\mathbb{A}}_f, \widehat{\mathbb{A}}_f)$$

be the multiplication operator by $\widehat{a_j}$ defined by (6.19). The corresponding element[18]

$$\widehat{A}_j \longleftrightarrow \sum_{\kappa \in \mathbb{N}^m} \phi_\kappa(\widehat{A_j})\, f^\kappa \in \widehat{\mathbb{E}}_f = \mathrm{Hom}_{\mathbb{K}}(\mathbb{P}_f, \mathbb{P}_f)\,[[f]]$$

is attached to $\widehat{A}_j$ according to the isomorphism (6.23).

(2) We form the jacobian determinant

$$\begin{vmatrix} \dfrac{\partial}{\partial f_1}\Big(\displaystyle\sum_{\kappa \in \mathbb{N}^m} \phi_\kappa(\widehat{A_1})\, f^\kappa\Big) & \cdots & \dfrac{\partial}{\partial f_1}\Big(\displaystyle\sum_{\kappa \in \mathbb{N}^m} \phi_\kappa(\widehat{A_m})\, f^\kappa\Big) \\ \vdots & \vdots & \vdots \\ \dfrac{\partial}{\partial f_m}\Big(\displaystyle\sum_{\kappa \in \mathbb{N}^m} \phi_\kappa(\widehat{A_1})\, f^\kappa\Big) & \cdots & \dfrac{\partial}{\partial f_m}\Big(\displaystyle\sum_{\kappa \in \mathbb{N}^m} \phi_\kappa(\widehat{A_m})\, f^\kappa\Big) \end{vmatrix}.$$

Then we expand it, respecting the noncommutativity of $\mathrm{Hom}_{\mathbb{K}}(\mathbb{P}_f, \mathbb{P}_f)$ and the rules

$$\Big(\sum_{\kappa'} \phi_{\kappa'} f^{\kappa'}\Big)\Big(\sum_{\kappa''} \psi_{\kappa''} f^{\kappa''}\Big) = \sum_{\kappa'}\sum_{\kappa''} (\phi_{\kappa'} \circ \psi_{\kappa''})\, f^{\kappa' + \kappa''} \tag{6.28}$$

as

$$\sum_{\kappa \in \mathbb{N}^m} \phi_\kappa\Big(\frac{da_1 \wedge \cdots \wedge da_m}{df_1 \wedge \cdots \wedge df_m}\Big)\, f^\kappa \in \mathrm{Hom}_{\mathbb{K}}(\mathbb{P}_f, \mathbb{P}_f)[[f]].$$

(3) Finally, define

$$\begin{aligned}(6.29)\quad & (\omega_a^{\sharp}(h))_T \\ &= \Big(\sum_{\kappa \in \mathbb{N}^m} \phi_\kappa(\widehat{h})\, f^\kappa\Big) \circ \Big(\sum_{\kappa \in \mathbb{N}^n} \phi_\kappa\Big(\frac{da_1 \wedge \cdots \wedge da_m}{df_1 \wedge \cdots \wedge df_m}\Big)\, f^\kappa\Big) = \sum_{\kappa \in \mathbb{N}^m} (\omega^{\sharp}_{\mathbf{a},\kappa}(h))_T\, f^\kappa\end{aligned}$$

respecting once more the noncommutativity of $\mathrm{Hom}_{\mathbb{K}}(\mathbb{P}_f, \mathbb{P}_f)$ and the multiplication rules (6.28).

It follows from [**Lip87**, Theorem 3.6 and Corollary 3.7] that one may define the *algebraic residue symbols* in this abstract algebraic setting as follows.

---

[18] As soon as the $\mathbb{K}$-linear section T is fixed, the $\mathbb{K}$-homomorphisms $\phi_\kappa(\widehat{A_j}) \in \mathrm{Hom}_{\mathbb{K}}(\mathbb{P}_f, \mathbb{P}_f)$, $\kappa \in \mathbb{N}^m$, which appears there can thus be computed, in principle, algorithmically.

DEFINITION 6.20. Let $\mathbb{A}$, $\mathbb{K}$, f, with $\mathbb{P}_f = \mathbb{A}/\mathbb{I}_f$ being finitely generated and projective as a $\mathbb{K}$-module, be as before. The *algebraic residue symbols*

$$\operatorname{Res}_{\mathbb{A}/\mathbb{K}} \begin{bmatrix} h\, du_1 \wedge \cdots \wedge du_m \\ f_1^{\kappa_1+1}, \ldots, f_m^{\kappa_m+1} \end{bmatrix} \in \mathbb{K} \tag{6.30}$$

are defined through the relations

$$\operatorname{Res}_{\mathbb{A}/\mathbb{K}} \begin{bmatrix} h\, da_1 \wedge \cdots \wedge da_m \\ f_1^{\kappa_1+1}, \ldots, f_m^{\kappa_m+1} \end{bmatrix} = \operatorname{Tr}_{\mathbb{P}_f : \mathbb{K}}\big((\omega^\sharp_{a,\kappa}(h))_T\big) \tag{6.31}$$

for any $\kappa \in \mathbb{N}^m$.

REMARK 6.21. Although the $\mathbb{K}$-homomorphisms $(\omega^\sharp_{a,\kappa}(h))_T$ from $\mathbb{P}_f$ into itself depend on the identification (6.23), and hence they depend on the choice of the $\mathbb{K}$-linear section T for $\widehat{\pi}_f$ sustaining the algorithmic procedure described above, their traces in $\mathbb{K}$, as endomorphisms of the finitely generated projective $\mathbb{K}$-module $\mathbb{P}_f$, do not. We will admit here this crucial fact[19] and refer the reader to [**Lip87**, Theorem 3.6 and Corollary 3.7], and the construction of the residue homomorphism together with its functional properties as described in [**Lip87**, §1].

REMARK 6.22 (Taylor formula in the algebraic setting). It is fundamental to observe that the residue symbols (6.30) are all given together as a family indexed by $\kappa \in \mathbb{N}^m$. Once the algorithmic process initiates, h and $a = (a_1, \ldots, a_m)$ all given and a $\mathbb{K}$-linear section T for $\widehat{\pi}_f$ is chosen, the whole list of symbols (6.30) are generated. Assuming $\boldsymbol{a}$ to be fixed and h to be kept free, this is indeed reminiscent of the Taylor formula. In fact, as pointed out in [**Lip87**, 3.4.2], one has the trace formula

$$\operatorname{Tr}_{\widehat{\mathbb{A}}_f : \mathbb{K}[[X]]}(U) = \sum_{\kappa \in \mathbb{N}^m} \operatorname{Tr}_{\mathbb{P}_f : \mathbb{K}}(\phi_\kappa(U))\, X^\kappa$$

for any $U \in \operatorname{Hom}_{\mathbb{K}[[X]]}(\widehat{\mathbb{A}}_f, \widehat{\mathbb{A}}_f)$. Once such U identifies with $\sum_{\kappa \in \mathbb{N}^n} \phi_\kappa(U)\, f^\kappa$ through identification (6.23), the choice of T is, in fact, irrelevant to the computation of traces, as we pointed out before. Therefore

$$\operatorname{Tr}_{\widehat{\mathbb{A}}_f : \mathbb{K}[[X]]}\big((\omega^\sharp_a(h))_T\big) = \sum_{\kappa \in \mathbb{N}^m} \operatorname{Res}_{\mathbb{A}/\mathbb{K}} \begin{bmatrix} h\, da_1 \wedge \cdots \wedge da_m \\ f_1^{\kappa_1+1}, \ldots, f_m^{\kappa_m+1} \end{bmatrix} X^\kappa \tag{6.32}$$

appears (in a reformulated form) as a generalization of the Taylor formula. One should compare it to the result stated in the analytic context of an analytic Weil polyhedron Δ subordinate to (Ω, f, r), where $\Omega \subset \mathbb{C}^n$ and the number of the components f_i of f is n, as in Proposition 6.11. If one takes the ring of holomorphic functions in Δ to play the role of the ring $\mathbb{A}$ and $\mathbb{C}$ to play that of $\mathbb{K}$, and as a_j is the coordinate function z_j for $j = 1, \ldots, n$, then the left-hand side of (6.32) is, when considered as an element in $\mathcal{O}_0$, the trace germ $X \mapsto \sum_{\xi \in f^{-1}(\{X\})} h(\xi)/J_f(\xi)$.

[19] Since the general spirit of our monograph is analytic, such considerations would lead us too far from our goal. But we wanted at least to sketch here how the algebraic procedure leading concretely to the realization of the "division by the jacobian" carries in algebraic terms (see (6.29)) towards the definition (6.31) (or realization, if one follows the presentation in [**Lip87**]) of algebraic residue symbols.

6.2.3. Wiebe's theorem and the algebraic transformation law. Let $f = \{f_1, \ldots, f_m\}$ and $g = \{g_1, \ldots, g_m\}$ be two quasi-regular sequences with the same length m in a commutative ring $\mathbb{A}$. Moreover, assume that there exists a matrix $A = [a_{j,k}]_{1\leq j,k\leq m}$ with entries in $\mathbb{A}$ such that

$$g_j = \sum_{k=1}^{m} a_{j,k}\, f_k, \quad j = 1, \ldots, m. \tag{6.33}$$

Note that we keep here notations already used in Lemma 3.20, which illustrates the results presented in this subsection. Let us start by recalling an important theorem due to H. Wiebe [**Wi69**], whose proof relies essentially upon the application of Cramer's rule.

THEOREM 6.23 (Wiebe theorem). *The class of* $\det A$ *in* $\mathbb{P}_g$ *is independent of the choice of the matrix* A *provided the entries of* A *satisfy* (6.33). *Moreover, if* $f\,\mathbb{P}_g$ *is the ideal of* $\mathbb{P}_g$ *generated by the classes of the* f_j *modulo* $\mathbb{I}_g$*, then the one has*

$$\begin{aligned} \operatorname{ann}_{\mathbb{P}_g}((\det A)\,\mathbb{P}_g) &= \big\{\eta \in \mathbb{P}_g \,:\, \eta\,(\det A)\,\mathbb{P}_g = 0\big\} = f\,\mathbb{P}_g,\\ \operatorname{ann}_{\mathbb{P}_g}(f\,\mathbb{P}_g) &= \big\{\eta \in \mathbb{P}_g \,:\, \eta\cdot(f\,\mathbb{P}_g) = 0\big\} = (\det A)\,\mathbb{P}_g. \end{aligned} \tag{6.34}$$

PROOF. One can take instead of $\mathbb{A}$ the local ring $\mathbb{A}_{\mathfrak{M}}$, which is defined as the localization of $\mathbb{A}$ at an arbitrary maximal ideal $\mathfrak{M}$ containing the ideal generated by the g_j. Thus, f and g can be assumed to be regular sequences, independent of the ordering chosen for each of the families $\{f_1, \ldots, f_m\}$ and $\{g_1, \ldots, g_m\}$. The main ingredient used in the proof of Wiebe's theorem is Cramer's rule. We refer the reader to [**ElM06**, §8.4] for more details. □

Let us now come back to the situation where $\mathbb{A}$ has additionally a structure of $\mathbb{K}$-algebra over the commutative ring of constants $\mathbb{K}$. In the case where the $\mathbb{K}$-modules $\mathbb{P}_f$ and $\mathbb{P}_g$ are both finitely generated and projective, comparing the computation of algebraic residue symbols (for the choice of f or g) introduced as two all-together families indexed by $\kappa \in \mathbb{N}^m$ in [**Lip87**, §2] (which are both interpreted within this monograph as traces given the generalized Taylor formula (6.32), see Definition 6.20 above) leads to the so-called *transition formula* or *algebraic transformation law*, whose geometric counterpart has been presented in §3.2.3 [**Ky89**, **BY91**, **BoyH97**, **BY99**].

THEOREM 6.24. *Let* $f = (f_1, \ldots, f_m)$ *and* $g = (g_1, \ldots, g_m)$ *be two quasi-regular sequences in* $\mathbb{A}$ *such that* (6.33) *holds for some* (m,m)*-matrix* A *with entries in the commutative ring* $\mathbb{A}$. *Moreover, assume that the* $\mathbb{K}$*-modules* $\mathbb{P}_f$ *and* $\mathbb{P}_g$ *are both finitely generated and projective. Then for any* $h, a_1, \ldots, a_m \in \mathbb{A}$, *for any multi-index* $\kappa \in \mathbb{N}^m$ *with length* $|\kappa| = \kappa_1 + \cdots + \kappa_m$, *one has*

$$\operatorname{Res}_{\mathbb{A}/\mathbb{K}} \begin{bmatrix} h\, da_1 \wedge \cdots \wedge da_m \\ f_1^{\kappa_1+1}, \ldots, f_m^{\kappa_m+1} \end{bmatrix} = \operatorname{Res}_{\mathbb{A}/\mathbb{K}} \begin{bmatrix} (\det A)\, A_\kappa^{|\kappa|}[g]\, h\, da_1 \wedge \cdots \wedge da_m \\ g_1^{|\kappa|+1}, \ldots, g_m^{|\kappa|+1} \end{bmatrix}, \tag{6.35}$$

where $A_\kappa^{|\kappa|}[g]$ *is the coefficient of* X^κ *in the expansion*

$$\prod_{j=1}^{m} \Big(\sum_{\ell=0}^{|\kappa|} g_j^{|\kappa|-\ell} \Big(\sum_{k=1}^{m} a_{j,k} X_k \Big)^{\ell} \Big) = \sum_{\{\nu\in\mathbb{N}^m \,:\, |\nu|\leq m|\kappa|\}} A_\nu^{|\kappa|}[g]\, X^\nu \tag{6.36}$$

in the $\mathbb{K}$*-algebra* $\mathbb{A}[X]$*. One can alternatively reformulate* (6.35) *as*

$$\text{(6.37)}\quad \operatorname{Res}_{\mathbb{A}/\mathbb{K}}\begin{bmatrix} h\,da_1\wedge\cdots\wedge da_m \\ f_1^{\kappa_1+1},\ldots,f_m^{\kappa_m+1}\end{bmatrix} = \sum_{\{\nu^1,\ldots,\nu^m\in\mathbb{N}^m\,:\,\sum_j \nu^j=\kappa\}} \prod_{j=1}^m \binom{|\nu^j|}{\tau^j} \operatorname{Res}_{\mathbb{A}/\mathbb{K}}\begin{bmatrix} (\det A)\Big(\prod_{1\le j,k\le m} a_{j,k}^{\nu_k^j}\Big)\,h\,da_1\wedge\cdots\wedge da_m \\ g_1^{|\nu^1|+1},\ldots,g_m^{|\nu^m|+1}\end{bmatrix}.$$

In particular, when $\kappa=(0,\ldots,0)$,

$$\text{(6.38)}\qquad \operatorname{Res}_{\mathbb{A}/\mathbb{K}}\begin{bmatrix} h\,da_1\wedge\cdots\wedge da_m \\ f_1,\ldots,f_m\end{bmatrix} = \operatorname{Res}_{\mathbb{A}/\mathbb{K}}\begin{bmatrix} (\det A)\,h\,da_1\wedge\cdots\wedge da_m \\ g_1,\ldots,g_m\end{bmatrix}.$$

PROOF. We refer to [**Lip87**, §2, Corollary §2.8] for the statement and proof of formula (6.38) in the particular case when $\kappa=0$ in the most general algebraic context as the one settled here. Let us only mention that Cramer's rule and Wiebe's Theorem 6.23 play a crucial role in the proof. Note that in §6.5.1, we will present a detailed proof in the case of most interest to us. Namely, we will present the particular case of *residue symbols in the affine algebraic setting*, where $\mathbb{A}=\mathbb{K}[X_1,\ldots,X_n]$ with $\mathbb{K}$ a subfield of $\mathbb{C}$. Since the residue symbols are defined as an all-together family indexed by $\kappa\in\mathbb{N}^m$, it is natural for a transition formula (6.38) to stand as a particular case of a more general transition formula (6.35) for $\kappa=(0,\ldots,0)$. We explain here how a more general formula (6.35) is deduced from (6.38). Let $\omega=h\,da_1\wedge\cdots\wedge da_m$. We fix the multi-index $\kappa\in\mathbb{N}^m$. Consider the two sequences

$$\text{(6.39)}\qquad \begin{aligned} F&=(f_1^{\kappa_1+1},\ldots,f_m^{\kappa_m+1},X_1,\ldots,X_m),\\ G&=(f_1-X_1,\ldots,f_m-X_m,X_1^{\kappa_1+1},\ldots,X_m^{\kappa_m+1})\end{aligned}$$

in $\mathbb{A}[X]$. Both sequences are quasi-regular in the $\mathbb{K}$-algebra $\mathbb{A}[X]$, and the corresponding $\mathbb{K}$-modules $\mathbb{A}[X]/\mathbb{I}_F$ and $\mathbb{A}[X]/\mathbb{I}_G$ are finitely generated and projective over $\mathbb{K}$ as $\mathbb{P}_f$ is. Then it follows from (6.38) applied to F and G in the $\mathbb{K}$-algebra $\mathbb{A}[X]$ instead of being applied to f and g in the $\mathbb{K}$-algebra $\mathbb{A}$ that

$$\text{(6.40)}\qquad \begin{aligned} &\operatorname{Res}_{\mathbb{A}/\mathbb{K}}\begin{bmatrix} \omega \\ f_1^{\kappa_1+1},\ldots,f_m^{\kappa_m+1}\end{bmatrix}\\ &= \operatorname{Res}_{\mathbb{A}/\mathbb{K}}\begin{bmatrix} \omega \\ f_1^{\kappa_1+1},\ldots,f_m^{\kappa_m+1}\end{bmatrix} \operatorname{Res}_{\mathbb{K}[X_1,\ldots,X_m]/\mathbb{K}}\begin{bmatrix} dX_1\wedge\cdots\wedge dX_m \\ X_1,\ldots,X_m\end{bmatrix}\\ &= \operatorname{Res}_{\mathbb{A}[X]/\mathbb{K}}\begin{bmatrix} \omega\wedge dX_1\wedge\cdots\wedge dX_m \\ f_1-X_1,\ldots,f_m-X_m,X_1^{\kappa_1+1},\ldots,X_m^{\kappa_m+1}\end{bmatrix}.\end{aligned}$$

Let $(A\,X)_j:=\sum_{k=1}^m a_{j,k}X_k$ for $j=1,\ldots,m$. We redeclare now F and G as

$$\begin{aligned} F&=(f_1-X_1,\ldots,f_m-X_m,X_1^{\kappa_1+1},\ldots,X_m^{\kappa_m+1}),\\ G&=\big(g_1-(A\,X)_1,\ldots,g_m-(A\,X)_m,X_1^{\kappa_1+1},\ldots,X_m^{\kappa_m+1}\big).\end{aligned}$$

Both sequences are again quasi-regular in $\mathbb{A}[X]$ as f and g are in $\mathbb{A}$. Moreover, the $\mathbb{K}$-modules $\mathbb{A}[X]/\mathbb{I}_F$ and $\mathbb{A}[X]/\mathbb{I}_G$ are finitely generated and projective over $\mathbb{K}$ as $\mathbb{P}_f$ and $\mathbb{P}_g$ are. The transition formula (6.38) applied once more to F and G in place of f and g implies now that

$$\text{(6.41)}\quad \operatorname{Res}_{\mathbb{A}[X]/\mathbb{K}}\begin{bmatrix} \omega\wedge dX_1\wedge\cdots\wedge dX_m \\ f_1-X_1,\dots,f_m-X_m,X_1^{\kappa_1+1},\dots,X_m^{\kappa_m+1}\end{bmatrix} = \operatorname{Res}_{\mathbb{A}[X]/\mathbb{K}}\begin{bmatrix} (\det A)\,\omega\wedge dX_1\wedge\cdots\wedge dX_m \\ g_1-(A\,X)_1,\dots,g_m-(A\,X)_m,X_1^{\kappa_1+1},\dots,X_m^{\kappa_m+1}\end{bmatrix}.$$

Again, we redeclare

$$F=(g_1-(A\,X)_1,\dots,g_m-(A\,X)_m,X_1^{\kappa_1+1},\dots,X_m^{\kappa_m+1}),$$
$$G=\big(g_1^{|\kappa|+1}-(A\,X)_1^{|\kappa|+1},\dots,g_m^{|\kappa|+1}-(A\,X)_m^{|\kappa|+1},X_1^{\kappa_1+1},\dots,X_m^{\kappa_m+1}\big).$$

Both sequences are quasi-regular in $\mathbb{A}[X]$ as g is in $\mathbb{A}$, and the $\mathbb{K}$-modules $\mathbb{A}[X]/\mathbb{I}_F$ and $\mathbb{A}[X]/\mathbb{I}_G$ are finitely generated and projective as the $\mathbb{K}$-module $\mathbb{P}_g$ is. The right-hand side of (6.41) is then transformed through (6.38) (with F and G in place of f and g, respectively) into

$$\text{(6.42)}\quad \operatorname{Res}_{\mathbb{A}[X]/\mathbb{K}}\begin{bmatrix} (\det A)\prod\limits_{j=1}^{m}\Big(\sum\limits_{\ell=0}^{|\kappa|} g_j^{|\kappa|-\ell}((A\,X)_j)^{\ell}\Big)\,\omega\wedge dX_1\wedge\cdots\wedge dX_m \\ g_1^{|\kappa|+1}-(A\,X)_1^{|\kappa|+1},\dots,g_m^{|\kappa|+1}-(A\,X)_m^{|\kappa|+1},X_1^{\kappa_1+1},\dots,X_m^{\kappa_m+1}\end{bmatrix}.$$

Finally, we redeclare again

$$F=\big(\tilde f_1^{|\kappa|+1}-(A\,X)_1^{|\kappa|+1},\dots,\tilde f_m^{|\kappa|+1}-(A\,X)_m^{|\kappa|+1},X_1^{\kappa_1+1},\dots,X_m^{\kappa_m+1}\big),$$
$$G=\big(g_1^{|\kappa|+1},\dots,g_m^{|\kappa|+1},X_1^{\kappa_1+1},\dots,X_m^{\kappa_m+1}\big).$$

These two sequences are quasi-regular in $\mathbb{A}[X]$, and the $\mathbb{K}$-modules $\mathbb{A}[X]/\mathbb{I}_F$ and $\mathbb{A}[X]/\mathbb{I}_G$ are finitely generated and projective over $\mathbb{K}$ as $\mathbb{P}_f$ is. The transition formula (6.38), applied as before with f and g replaced once more by F and G, respectively, implies this time that (6.42) equals

$$\text{(6.43)}\qquad \operatorname{Res}_{\mathbb{A}[X]/\mathbb{K}}\begin{bmatrix} (\det A)\prod\limits_{j=1}^{m}\Big(\sum\limits_{\ell=0}^{|\kappa|} g_j^{|\kappa|-\ell}((A\,X)_j)^{\ell}\Big)\,\omega\wedge dX_1\wedge\cdots\wedge dX_m \\ g_1^{|\kappa|+1},\dots,g_m^{|\kappa|+1},X_1^{\kappa_1+1},\dots,X_m^{\kappa_m+1}\end{bmatrix},$$

that is, the right-hand side of (6.35). Since

$$\text{(6.44)}\quad \prod_{j=1}^{m}\Big(\sum_{\ell=0}^{|\kappa|} g_j^{|\kappa|-\ell}\Big(\sum_{k=1}^{m} a_{j,k}X_k\Big)^{\ell}\Big) = \sum_{\{\nu\in\mathbb{N}^m\,:\,\nu\prec|\kappa|\underline{1}\}} g^{|\kappa|\underline{1}-\nu}\prod_{j=1}^{m}\Big(\sum_{k=1}^{m} a_{j,k}X_k\Big)^{\nu_j},$$

one has

$$\text{(6.45)}\quad A_\kappa^{|\kappa|}[g] = \sum_{\{\nu^1,\dots,\nu^m\in\mathbb{N}^m\,:\,\sum_j \tau^j=\kappa\}} g^{|\kappa|\underline{1}-(|\nu^1|,\dots,|\nu^m|)}\prod_{j=1}^{m}\binom{|\nu^j|}{\nu^j}\prod_{1\le j,k\le m} a_{j,k}^{\nu_k^j}.$$

Then plotting (6.45) in the residue symbol on the right-hand side of (6.35) and using that

$$\operatorname{Res}_{\mathbb{A}/\boldsymbol{b}}\begin{bmatrix} g^{\nu'}\varpi \\ g_1^{\nu_1'+\nu_1''+1},\dots,g_m^{\nu_m'+\nu_m''+1}\end{bmatrix} = \operatorname{Res}_{\mathbb{A}/\boldsymbol{b}}\begin{bmatrix} \varpi \\ g_1^{\nu_1''+1},\dots,g_m^{\nu_m''+1}\end{bmatrix}$$

for any $\nu',\nu''\in\mathbb{N}^m$ and any $\varpi\in\bigwedge^m\Omega_{\mathbb{A}/\mathbb{K}}$, one can reformulate (6.35) as (6.37). □

REMARK 6.25 (Leibniz-type transition laws). Suppose first that $\mathbb{K}$ is an integral domain with characteristic 0. Let $\omega = h\, da_1 \wedge \cdots \wedge da_m$. The right-hand side of (6.37) becomes after a re-indexation

$$\sum_{\{\sigma^1,\dots,\sigma^m\in\mathbb{N}^m\,;\,|\sigma^j|=\kappa_j,\,1\le j\le m\}} \prod_{k=1}^{m} \binom{(\sum_j \sigma^j)_k}{\sigma^1_k,\dots,\sigma^m_k} \operatorname{Res}_{\mathbb{A}/\mathbb{K}} \left[\begin{matrix} (\det A)\Big(\prod\limits_{1\le j,k\le m} a_{j,k}^{\sigma^j_k}\Big)\,\omega \\ g_1^{(\sum_j \sigma^j)_1+1},\dots,g_m^{(\sum_j \sigma^j)_m+1}\end{matrix}\right]$$

$$= \sum_{\{\sigma^1,\dots,\sigma^m\in\mathbb{N}^m\,;\,|\sigma^j|=\kappa_j,\,1\le j\le m\}} \prod_{k=1}^{m} \binom{\kappa_k}{\sigma^k_1,\dots,\sigma^k_m}$$

$$\times \Big(\prod_{k=1}^{m} \frac{\big(\sum_j \sigma^j\big)_k!}{\kappa_k!}\Big) \operatorname{Res}_{\mathbb{A}/\mathbb{K}} \left[\begin{matrix} (\det A)\Big(\prod\limits_{1\le j,k\le m} a_{j,k}^{\sigma^j_k}\Big)\,\omega \\ g_1^{(\sum_j \sigma^j)_1+1},\dots,g_m^{(\sum_j \sigma^j)_m+1}\end{matrix}\right]$$

One can then rewrite (6.37) in the fraction field of $\mathbb{K}$ as

(6.46)

$$\operatorname{Res}_{\mathbb{A}/\mathbb{K}} \left[\begin{matrix} \omega \\ f_1^{\kappa_1+1},\dots,f_m^{\kappa_m+1}\end{matrix}\right] = \sum_{\{\nu\in\mathbb{N}^m\,:\,|\nu|=|\kappa|\}} \frac{\nu!}{\kappa!} \operatorname{Res}_{\mathbb{A}/\mathbb{K}} \left[\begin{matrix} (\det A)\, A^\kappa_\nu\, \omega \\ g_1^{\nu_1+1},\dots,g_m^{\nu_m+1}\end{matrix}\right],$$

where $\kappa! = \kappa_1!\cdots\kappa_m!$, $\nu! = \nu_1!\cdots\nu_m!$ and A^κ_ν is the coefficient of X^ν in the expansion

$$\prod_{k=1}^{m} \Big(\sum_{j=1}^{m} a_{j,k} X_j\Big)^{\kappa_k} = \sum_{\{\nu\in\mathbb{N}^m\,:\,|\nu|=|\kappa|\}} A^\kappa_\nu X^\nu. \tag{6.47}$$

Let us drop the hypothesis on $\mathbb{K}$. Given the well known relations (1.14), it is natural to introduce, given an element $\omega \in \bigwedge^m \Omega_{\mathbb{A}/\mathbb{K}}$, the following formal partial derivations symbols

$$\Big(\frac{\partial}{\partial f}\Big)^\kappa \omega := \kappa!\, \operatorname{Res}_{\mathbb{A}/\mathfrak{b}} \left[\begin{matrix} \omega \\ f_1^{\kappa_1+1},\dots,f_m^{\kappa_m+1}\end{matrix}\right].$$

One has as a consequence of (6.35) (which implies (6.46) with both sides multiplied by $\kappa!$) the following set of Leibniz-type transition laws:

$$\Big(\frac{\partial}{\partial f}\Big)^\kappa \omega = \sum_{\{\nu\in\mathbb{N}^m\,:\,|\nu|=|\kappa|\}} \Big(\frac{\partial}{\partial g}\Big)^\nu \big((\det A)\, A^\kappa_\nu\, \omega\big) \quad (\kappa \in \mathbb{N}^m). \tag{6.48}$$

Such Leibniz-type transition laws support our interpretation of (6.32) as a generalization of the Taylor formula.

6.3. Residue symbols in $\mathbb{K}$-polynomial algebras

All through this section, the commutative ring $\mathbb{K}$ will be a field of any characteristic, unless otherwise specified. We denote as $\overline{\mathbb{K}}$ an algebraic closure[20] of $\mathbb{K}$ and as $\mathbb{K}^{\rm sep}$ the separable closure of $\mathbb{K}$ in $\overline{\mathbb{K}}$. $\mathbb{K}^{\rm sep}$ is the unique separable extension of $\mathbb{K}$

[20] We recall that two algebraic closures of a given field $\mathbb{K}$ are isomorphic through an isomorphism which fixes elements of $\mathbb{K}$.

containing all separable extensions[21] of $\mathbb{K}$ within $\overline{\mathbb{K}}$. Observe that $\overline{\mathbb{K}}$ is necessarily an infinite field.

6.3.1. Residue symbols $\mathrm{Res}_{\mathbb{K}[X]/\mathbb{K}}$. Let $\mathbb{A}$ be $\mathbb{K}$-algebra $\mathbb{A} = \mathbb{K}[X_1, \dots, X_n]$ (also denoted in short by $\mathbb{K}[X]$). Let also $p = (p_1, \dots, p_m) \in \mathbb{A}$ be such that $(p_1, \dots, p_m)$ is quasi-regular, $\mathbb{I}_p \neq \mathbb{A}$, and $\mathbb{P}_p$ is a finitely dimensional $\mathbb{K}$-vector space. This amounts to saying (see Example 6.18) that $m = n$ and that

$$V_{\overline{\mathbb{K}}}(p) = \{x \in \overline{\mathbb{K}}^n : p(x) = 0\}$$

is a finite subset in $\overline{\mathbb{K}}^n$. Given $q \in \mathbb{A}$ and $\omega = q\, dX_1 \wedge \cdots \wedge dX_n \in \Omega^n_{\mathbb{A}/\mathbb{K}}$, the residue symbols

$$\mathrm{Res}_{\mathbb{K}[X]/\mathbb{K}} \begin{bmatrix} \omega \\ p_1^{\kappa_1+1}, \dots, p_n^{\kappa_n+1} \end{bmatrix} \tag{6.49}$$

are defined for any $\kappa \in \mathbb{N}^n$ according to Definition 6.20. Given $p_0 \in \mathbb{K}[X]$ such that $p_0\mathbb{K}[X] + \mathbb{I}_p = \mathbb{K}[X]$, there exists $a_0 \in \mathbb{K}[X]$ satisfying

$$1 - a_0\, p_0 \in \mathbb{I}_p, \quad \deg a_0 p_0 \leq \prod_{j=0}^{n} \deg p_j; \tag{6.50}$$

see [**Jel05**, Theorem 1.1 and Remark 3.7].[22] For each $\kappa \in \mathbb{N}^n$ and $\ell \in \mathbb{N}$, let

$$\begin{aligned} \left(1 - (a_0 p_0)^{\ell+1}\right)^{|\kappa|+1} &= 1 - p_0^{\ell+1} \sum_{\nu=1}^{|\kappa|+1} (-1)^{\nu-1} \binom{|\kappa|+1}{\nu} a_0^{(\ell+1)\nu}\, p_0^{(\ell+1)(\nu-1)} \\ &= 1 - p_0^{\ell+1} D_\ell^{|\kappa|}[p_0]. \end{aligned} \tag{6.51}$$

It will be useful in what follows to enlarge $\mathbb{A} = \mathbb{K}[X]$, where the family of residue symbols (6.49), already introduced in Definition 6.20, exist.

Definition 6.26. Let $(p_0, p_1, \dots, p_n)$ be a sequence of elements in $\mathbb{A} = \mathbb{K}[X]$ such that $(p_1, \dots, p_n)$ is a quasi-regular sequence defining a proper ideal $\mathbb{I}_p$ and $p_0\, \mathbb{K}[X] + \mathbb{I}_p = \mathbb{K}[X]$. For any $\omega \in \Omega^n_{\mathbb{K}[X]/\mathbb{K}}$, any $\ell \in \mathbb{N}$ and $\kappa \in \mathbb{N}^n$,

$$\mathrm{Res}_{\mathbb{K}[X]/\mathbb{K}} \begin{bmatrix} \omega/p_0^{\ell+1} \\ p_1^{\kappa_1+1}, \dots, p_n^{\kappa_n+1} \end{bmatrix} := \mathrm{Res}_{\mathbb{K}[X]/\mathbb{K}} \begin{bmatrix} D_\ell^{|\kappa|}[p_0]\, \omega \\ p_1^{\kappa_1+1}, \dots, p_n^{\kappa_n+1} \end{bmatrix} \tag{6.52}$$

is an element of $\mathbb{K}$, which is independent of the choice of a_0. The algebraic expression $D_\ell^{|\kappa|}[p_0]$ is defined then as in (6.51). The whole list of elements (6.52) for all $\ell \in \mathbb{N}$, $\kappa \in \mathbb{N}^n$ is called the list of $\mathbb{K}[X]/\mathbb{K}$ *algebraic residue symbols* attached to $(p_0, p_1, \dots, p_n)$.

It is crucial to justify the independence of algebraic residue symbols from the choice of a_0 in the above definition. It is the content of the next proposition.

---

[21]An algebraic extension $\mathbb{K}^\circ$ of $\mathbb{K}$ within an algebraic closure $\overline{\mathbb{K}}$ is said to be a *separable extension* if and only if, for any $\gamma \in \mathbb{K}^\circ$, the minimal polynomial Φ of γ over $\mathbb{K}$ is separable; that is, Φ has exactly $\deg \Phi$ distinct roots in $\overline{\mathbb{K}}$.

[22]This follows from the sharp geometric effective nullstellensatz, which we will discuss in §7.1.5.

INDEPENDENCE ON THE CHOICE OF a_0. If $a_{0,1}$ and $a_{0,2}$ are alternative choices for a_0, one has

$$p_0^{\ell+1}(D_{\ell,1}^{|\kappa|}[p_0] - D_{\ell,2}^{|\kappa|}[p_0]) \in \sum_{j=1}^{n} p_j^{\kappa_j+1}\,\mathbb{K}[X]. \tag{6.53}$$

Since $D_{\ell,1}^{|\kappa|}[p_0]\,p_0^{\ell+1} \equiv 1$ modulo $\sum_{j=0}^{n} p_j^{\kappa_j+1}\mathbb{K}[X]$, multiplying (6.53) by $D_{\ell,1}^{|\kappa|}[p_0]$ leads to

$$D_{\ell,1}^{|\kappa|}[p_0] - D_{\ell,2}^{\kappa}[p_0] \in \sum_{j=1}^{n} p_j^{\kappa_j+1}\,\mathbb{K}[X]$$

as well. Since one has $(\omega_{a,\kappa}^{\sharp}(h))_T = 0$, when $h \in \sum_{j=1}^{m} f_j^{\kappa_j+1}\mathbb{A}$ in the formal algebraic construction (6.29) leading to the realization of algebraic residue symbols as traces, one has for any $j = 1, \dots, n$ and any $u_j \in \mathbb{K}[X]$ that

$$\mathrm{Res}_{\mathbb{K}[X]/\mathbb{K}}\begin{bmatrix} p_j^{\kappa_j+1}\quad u_j\,\omega \\ p_1^{\kappa_1+1}, \dots, p_n^{\kappa_n+1}\end{bmatrix} = 0,$$

after placing $\mathbb{A} = \mathbb{K}[X]$, $f = (p_1, \dots, p_n)$, $a = (X_1, \dots, X_n)$. □

Theorem 6.24 implies when $\mathbb{A} = \mathbb{K}[X]$ the following result.

PROPOSITION 6.27. *Let $p = (p_1, \dots, p_n)$ and $\phi = (\phi_1, \dots, \phi_n)$ be two quasi-regular sequences in $\mathbb{K}[X]$ such that there exists an $n \times n$ matrix A, whose entries $a_{j,k} \in \mathbb{K}[X]$ satisfy $\phi_j = \sum_{k=1}^{n} a_{j,k}\,p_k$ for $j = 1, \dots, n$. Let $p_0 \in \mathbb{K}[X]$ be such that $p_0\,\mathbb{K}[X] + \mathbb{I}_p = \mathbb{K}[X]$. Then, for any $\omega \in \bigwedge^n \Omega_{\mathbb{K}[X]/\mathbb{K}}$ and any $\kappa \in \mathbb{N}^n$,*

$$\mathrm{Res}_{\mathbb{K}[X]/\mathbb{K}}\begin{bmatrix} \omega/p_0^{\ell+1} \\ p_1^{\kappa_1+1}, \dots, p_n^{\kappa_n+1}\end{bmatrix} = \mathrm{Res}_{\mathbb{K}[X]/\mathbb{K}}\begin{bmatrix} (\det A)\,D_\ell^{|\kappa|}[p_0]\,A_\kappa^{|\kappa|}[\phi]\,\omega \\ \phi^{|\kappa|+1}, \dots, \phi^{|\kappa|+1}\end{bmatrix}, \tag{6.54}$$

where $A_\kappa^{|\kappa|}[\phi]$ is defined as in (6.36) with ϕ replacing g.

What makes the case $\mathbb{A} = \mathbb{K}[X_1, \dots, X_n]$, or more generally the $\mathbb{K}[t][X]$, where t is a finite collection of algebraically independent parameters t_ι transcendental over $\mathbb{K}$, of great interest is that one can formulate in such cases an algebraic counterpart of the Cauchy–Weil analytic formula previously stated for a Weil polyhedron such as $\Delta = \Delta(\Omega, f, r)$ in Proposition 6.12; see [**BoyH99**, §4, Example 1].

THEOREM 6.28 (Algebraic Cauchy–Weil formula). *Let $\mathfrak{R}$ be a commutative nœtherian ring*[23] *and let $p = (p_1, \dots, p_n)$ be a quasi-regular sequence in the $\mathfrak{R}$ polynomial algebra $\mathbb{A} = \mathfrak{R}[X_1, \dots, X_n]$ such that the $\mathfrak{R}$-module $\mathbb{A}/\mathbb{I}_p$ is finitely generated and projective. Let $H = [h_{j,k}]_{1 \le j,k \le n}$ be an $n \times n$ matrix of elements in the polynomial algebra $\mathfrak{R}[X_1, \dots, X_n, Y_1, \dots, Y_n]$ satisfying*

$$p_j(Y) - p_j(X) = \sum_{k=1}^{n} h_{j,k}(X,Y)(Y_k - X_k) \tag{6.55}$$

for $j = 1, \dots, n$. Any element $q \in \mathbb{A}$ is reproduced as

$$\begin{aligned} q(X) &= \mathrm{Res}_{\mathfrak{R}[X][Y]/\mathfrak{R}[X]}\begin{bmatrix} q(Y)\,dY_1 \wedge \cdots \wedge dY_n \\ Y - X\end{bmatrix} \\ &= \mathrm{Res}_{\mathfrak{R}[X][Y]/\mathfrak{R}[X]}\begin{bmatrix} (\det H(X,Y))\,q(Y)\,dY_1 \wedge \cdots \wedge dY_n \\ p(Y) - p(X)\end{bmatrix}\end{aligned} \tag{6.56}$$

[23]In particular, the commutative field $\mathbb{K}$ or the nœtherian ring $\mathbb{K}[t]$.

and admits then in the $\mathbb{I}_p$-dic completion $\widehat{\mathbb{A}}_p$ the expansion

$$(6.57)\quad q(X) = \sum_{\kappa\in\mathbb{N}^n} \operatorname{Res}_{\mathfrak{R}[X][Y]/\mathfrak{R}[X]} \begin{bmatrix} (\det H(X,Y))\, q(Y)\, dY_1 \wedge \cdots \wedge dY_n \\ (p_1(Y))^{\kappa_1+1}, \dots, (p_n(Y))^{\kappa_n+1} \end{bmatrix} p^{\kappa}(X).$$

PROOF. The proof relies on the algebraic transformation law stated as the transition formula (6.38) in Theorem 6.24, combined with the compatibility of the construction of algebraic residue symbols with the change of basis, as formulated in [**Lip87**, §2, Corollary 2.4]. We refer the reader to the detailed proof presented in [**BoyH99**, §4, Theorem 4.4 and Example 1]. We will present later on a proof with more analytic flavor in the case when $\mathfrak{R} = \mathbb{K}$ or when $\mathfrak{R} = \mathbb{K}[t]$, where $\mathbb{K}$ is a subfield of $\mathbb{C}$ and t is a finite set of parameters t_ι; see §6.5.1, together with Example 5.15. □

Let $(p_1, \dots, p_n)$ be a quasi-regular sequence in $\mathbb{K}[X]$ such that $\mathbb{I}_p \neq \mathbb{K}[X]$. Then the elimination theorem[24] [**Jel05**, Theorem 1.6] implies that for each $j = 1, \dots, n$, there exists a univariate polynomial ϕ_j with coefficients in $\mathbb{K}$ such that

$$(6.58)\qquad \phi_j(X_j) = \sum_{k=1}^{n} a_{j,k}(X)\, p_j(X), \quad \max_{1\leq j\leq n} \deg(a_{j,k}\, p_k) \leq \prod_{j=1}^{n} \deg p_j.$$

If $p_0 \in \mathbb{K}[X]$ is such $p_0\, \mathbb{K}[X] + \mathbb{I}_p = \mathbb{K}[X]$, it follows then from Proposition 6.27 that for any $\omega \in \bigwedge^n \Omega_{\mathbb{K}[X]/\mathbb{K}}$, for any $\ell \in \mathbb{N}$ and any $\kappa \in \mathbb{N}^n$,

$$(6.59)\quad \operatorname{Res}_{\mathbb{K}[X]/\mathbb{K}} \begin{bmatrix} \omega/p_0^{\ell+1} \\ p_1^{\kappa_1+1}, \dots, p_n^{\kappa_n+1} \end{bmatrix} = \operatorname{Res}_{\mathbb{K}[X]/\mathbb{K}} \begin{bmatrix} (\det A)\, D_\ell^{|\kappa|}[p_0]\, A_\kappa^{|\kappa|}[\phi]\, \omega \\ \phi_1^{|\kappa|+1}(X_1), \dots, \phi_n^{|\kappa|+1}(X_n) \end{bmatrix}.$$

Formula (6.59) for $\kappa \in \mathbb{N}^n$ trivializes the computation of the family of $\mathbb{K}[X]/\mathbb{K}$-algebraic residue symbols attached to the family to $(p_0, \dots, p_n)$ as in (6.52). It occurs because these formulas reduce such computation to the univariate case $n = 1$ and therefore to Euclid's division algorithm thanks to Proposition 6.29 below, which is a reformulation of Proposition 1.38 (corresponding to the particular case where $\mathbb{K} = \mathbb{C}$ and $n = 1$), together with Corollary 6.30.

PROPOSITION 6.29. *Let $p = \gamma_0 + \gamma_1 X + \cdots + \gamma_d X^d$ be a univariate polynomial with degree $d \in \mathbb{N}^*$ in the polynomial algebra $\mathbb{K}[X]$. Consider the iterated Euclid's algorithm initiated with X^k, where $k \in \mathbb{N}$:*

(6.60)

$$X^k = \sum_{\nu=0}^{d-1} \gamma_{0,d-1-\nu}^k\, X^{d-1-\nu} + p\, q_0^k, \quad q_{\kappa-1}^k = \sum_{\nu=0}^{d-1} \gamma_{\kappa,d-1-\nu}^k\, X^{d-1-\nu} + p\, q_\kappa^k \quad \forall \kappa \geq 1.$$

Then one has for any $k, \kappa \in \mathbb{N}$,

$$(6.61)\qquad \operatorname{Res}_{\mathbb{K}[X]/\mathbb{K}} \begin{bmatrix} X^k\, dX \\ p^{\kappa+1} \end{bmatrix} = \frac{\gamma_{\kappa,d-1}^k}{\gamma_d}.$$

---

[24] Jelonek's result holds in the case where $\mathbb{K}$ is algebraically closed. But if $\overline{\mathbb{K}}$ denotes an algebraic closure of $\mathbb{K}$, one can check that the polynomial $\Phi_j \in \overline{\mathbb{K}}[X]$ in (6.58) can be taken in $\mathbb{K}[X]$, and so the $a_{j,k}$. It follows from the fact that the finite subset $X_j(\{x \in \overline{\mathbb{K}}^n : p(x) = 0\})$ is invariant under the action of the Galois group $\mathrm{Gal}(\mathbb{K}^{\mathrm{sep}}/\mathbb{K})$, where $\mathbb{K}^{\mathrm{sep}}$. See, for example, [**SomY21**, Theorem 5.1], in the particular case where $\mathbb{K} = \mathbb{Q}$.

PROOF. The argument relies on the following observation: the $\mathbb{I}_p$-dic development

$$X^k = \sum_{\kappa=0}^{\infty} \Big(\sum_{\nu=0}^{d-1} \gamma^k_{\kappa,d-1-\nu}\, X^{d-1-\nu}\Big)\, p^\kappa(X)$$

obtained through the iterated euclidean division algorithm coincides with the Bergman–Weil development

$$X^k = \sum_{\kappa=0}^{\infty} \mathrm{Res}_{\mathbb{K}[X][Y]/\mathbb{K}[Y]} \begin{bmatrix} Y^k \, \dfrac{p(Y)-p(X)}{Y-X}\, dY \\ p^{\kappa+1}(Y) \end{bmatrix} p^\kappa(X),$$

which is given as (6.56) in Theorem 6.28, since we are in the univariate case. Formula (6.61) follows from such identification. □

COROLLARY 6.30. *Let $p \in \mathbb{K}[X]$ be a univariate polynomial with degree $d \in \mathbb{N}^*$ as in Proposition* 6.29. *Then, for any $k \in \mathbb{N}$, one has*

$$k+1 < (\kappa+1)d \Longrightarrow \mathrm{Res}_{\mathbb{K}[X]/\mathbb{K}} \begin{bmatrix} Y^k\, dY \\ (p(Y))^{\kappa+1} \end{bmatrix} = 0.$$

PROOF. It follows from the fact that $\sum_{\kappa' \leq \kappa} \big(\sum_{\nu=0}^{d-1} \gamma^k_{\kappa',\nu} X^\nu\big) p^{\kappa'}$ is the remainder in the euclidean division of X^k by $p^{\kappa+1}$. If $k \leq (\kappa+1)d - 1$, such a remainder coincides with X^k. In particular, if $k < (\kappa+1)d-1$, one has necessarily $\gamma^k_{\kappa,d-1} = 0$. The result then follows from (6.61). □

The following corollary to Proposition 6.29 will also prove to be extremely useful for us later on; see, for example, Proposition 6.41.

COROLLARY 6.31. *Let $p = (p_1,\dots,p_n)$ be a quasi-regular sequence in the $\mathbb{K}$-algebra $\mathbb{A} = \mathbb{K}[X_1,\dots,X_n]$. Then, for any $r \in \mathbb{N}^*$, $(p_1(X^r),\dots,p_n(X^r))$ remains a quasi-regular sequence in $\mathbb{A}$, and one has*

(6.62)
$$\forall\, q \in \mathbb{K}[X],\ \mathrm{Res}_{\mathbb{K}[X]/\mathbb{K}} \begin{bmatrix} q\, dX \\ p_1,\dots,p_n \end{bmatrix} = \mathrm{Res}_{\mathbb{K}[X]/\mathbb{K}} \begin{bmatrix} q(X^r)\,(X_1\cdots X_n)^{r-1}\, dX \\ p_1(X^r),\dots,p_n(X^r) \end{bmatrix},$$

where $dX = dX_1 \wedge \cdots \wedge dX_n$ and $X^r = (X_1^r,\dots,X_n^r)$.

PROOF. If one substitutes X^r to X in the set of conditions (6.16), then one sees immediately that the sequence $(p_1(X^r),\dots,p_n(X^r))$ is quasi-regular in $\mathbb{A}$ as the sequence p is. For the same reason, given the polynomial identities involving univariate polynomials $\Phi_j(X_j)$, for $j = 1,\dots,n$, as in (6.58), the polynomial identities $\Phi_j(X_j^r) = \sum_{k=1}^n a_{j,k}(X^r)\, p_j(X^r)$ also hold for any $j = 1,\dots,n$ and $r \in \mathbb{N}^*$. It follows from Theorem 6.24, that is, in the polynomial context Proposition 6.27, that proving (6.62) reduces to the case where $p_1,\dots,p_n$ are univariate polynomials respectively in $X_1,\dots,X_n$, namely the polynomials $\Phi_1(X_1),\dots,\Phi_n(X_n)$. Then proving (6.58) boils down to the univariate case. But it is immediate to observe that if p is a distinguished univariate polynomial with degree d in X and γ_κ is the coefficient of X^{d-1} in the remainder of the euclidean division of X^κ by p, then γ_κ is also the coefficient of $X^{rd} - 1$ in the euclidean division of $X^{r\kappa}$ by $p(X^r)$. Then Corollary 6.31 is an immediate consequence of Proposition 6.29, namely (6.61) with $\kappa = 0$. □

6.3.2. Residue symbols $\mathrm{Res}_{\mathbb{K}_{\mathbb{J}}[X,Y]/\mathbb{K}}$. Let $\mathbb{A} = \mathbb{K}_{\mathbb{J}}[X,Y] = \mathbb{K}[X,Y]/\mathbb{J}$, with $X = (X_1, \dots, X_n)$, $Y = (Y_1, \dots, Y_{N-n})$, and $\mathbb{J}$ is a purely n-dimensional radical ideal in $\mathbb{K}[X,Y]$ as in Example 6.19, namely

$$\mathbb{X} = \mathrm{Spec}\,(\mathbb{K}_{\mathbb{J}}[X,Y])$$

is an n-dimensional affine algebraic variety defined over $\mathbb{K}$. Let $p_1, \dots, p_n$ be n elements in $\mathbb{K}[X,Y]$ such that $\mathbb{X}_{\overline{\mathbb{K}}} \cap V(p)$ is a nonempty, zero-dimensional algebraic subvariety of $\mathbb{X}_{\overline{\mathbb{K}}} \subset \overline{\mathbb{K}}^N$. Let $\mathfrak{M}$ be a maximal ideal of $\overline{\mathbb{K}}_{\mathbb{J}}[X,Y]$ containing $\mathbb{I}_{\bar{p}}\,\overline{\mathbb{K}}_{\mathbb{J}}[X,Y]$, where $\bar{p} = p$ modulo $\mathbb{J}$. It follows from the fact that $\mathbb{J}$ is purely n-dimensional and radical that for any such maximal ideal $\mathfrak{M}$, the images of $\bar{p}_1, \dots, \bar{p}_n$ in the localization $(\overline{\mathbb{K}}_{\mathbb{J}}[X,Y])_{\mathfrak{M}}$ define a regular sequence there. This implies, according to (4) in the characterization of quasi-regularity (following formula (6.17) in §6.2.1) that $(\bar{p}_1, \dots, \bar{p}_n)$ is a regular sequence in $\mathbb{K}_{\mathbb{J}}[X,Y]$. Let $p_0 \in \mathbb{K}[X,Y]$ be such that

$$\sum_{j=0}^{n} p_j\, \mathbb{K}[X,Y] + \mathbb{J} = \mathbb{K}[X,Y], \tag{6.63}$$

$a_0 \in \mathbb{K}[X,Y]$ be such that[25]

$$1 - a_0 p_0 \in \sum_{j=1}^{n} p_j\, \mathbb{K}[X,Y] + \mathbb{J}, \quad \deg(a_0\, p_0) \le \deg \mathbb{X}_{\overline{\mathbb{K}}} \prod_{j=1}^{n} \deg p_j \tag{6.64}$$

(see [**Jel05**, Theorem 1.1]), and for any $\ell \in \mathbb{N}$, $\kappa \in \mathbb{N}^n$, let $D_\ell^{|\kappa|}[p_0] \in \mathbb{K}[X,Y]$ be as in (6.51). Its class in $\mathbb{K}_{\mathbb{J}}[X,Y]$ is denoted by $D_\ell^{|\kappa|}[\bar{p}_0]$.

DEFINITION 6.32. Let $\mathbb{J}$, $\mathbb{X}$, $p_0, \dots, p_n \in \mathbb{K}[X,Y]$ be as above. The $\mathbb{K}_{\mathbb{J}}[X,Y]/\mathbb{K}$ *algebraic residue symbols* attached to $(\bar{p}_0, \dots, \bar{p}_n)$ are defined as follows. For any $\varpi \in \Omega_{\mathbb{K}_{\mathbb{J}}[X,Y]/\mathbb{K}}$, any $\ell \in \mathbb{N}$, and any $\kappa \in \mathbb{N}^n$,

$$\mathrm{Res}_{\mathbb{K}_{\mathbb{J}}[X,Y]/\mathbb{K}} \begin{bmatrix} \varpi/\bar{p}_0^{\ell+1} \\ \bar{p}_1^{\kappa_1+1}, \dots, \bar{p}_n^{\kappa_n+1} \end{bmatrix} := \mathrm{Res}_{\mathbb{K}_{\mathbb{J}}[X,Y]/\mathbb{K}} \begin{bmatrix} D_\ell^{|\kappa|}[\bar{p}_0]\,\varpi \\ \bar{p}_1^{\kappa_1+1}, \dots, \bar{p}_n^{\kappa_n+1} \end{bmatrix}. \tag{6.65}$$

Such algebraic residue symbols do not depend[26] on the choice of a_0.

Since $\mathbb{X}_{\overline{\mathbb{K}}} \cap V(p)$ is a finite subset of $\overline{\mathbb{K}}^N$, sharp effective elimination (see [**Jel05**, Theorem 3.8]) implies there exists univariate polynomials $\Phi_1, \dots, \Phi_n$ such that for $j = 1, \dots, n$,

$$\phi_j(X_j) = \sum_{k=1}^{n} a_{j,k}\, p_k \quad \text{modulo } \mathbb{J}, \quad \max_{1\le k\le n} \deg(a_{j,k}\, p_k) \le \deg \mathbb{X}_{\overline{\mathbb{K}}} \prod_{j=1}^{n} \deg p_j. \tag{6.66}$$

Then, Theorem 6.24 implies the following result when $\mathbb{A} = \mathbb{K}_{\mathbb{J}}[X,Y]$ and $p_0, \dots, p_n$ are as in Definition 6.32: for any $\varpi \in \Omega^n_{\mathbb{K}_{\mathbb{J}}[X,Y]/\mathbb{K}}$, for any $\ell \in \mathbb{N}$, and for any $\kappa \in \mathbb{N}^n$,
(6.67)

$$\mathrm{Res}_{\mathbb{K}_{\mathbb{J}}[X,Y]/\mathbb{K}} \begin{bmatrix} \varpi/\bar{p}_0^{\ell+1} \\ \bar{p}_1^{\kappa_1+1}, \dots, \bar{p}_n^{\kappa_n+1} \end{bmatrix} = \mathrm{Res}_{\mathbb{K}_{\mathbb{J}}[X,Y]/\mathbb{K}} \begin{bmatrix} D_\ell^{|\kappa|}[\bar{p}_0]\, A_\kappa^{|\kappa|}[\bar{\phi}]\,\varpi \\ \phi_1^{|\kappa|+1}(\bar{X}_1), \dots, \phi_n^{|\kappa|+1}(\bar{X}_n) \end{bmatrix}.$$

---

[25] The degree of $\mathbb{X}_{\overline{\mathbb{K}}}$ does not depend on the choice of the algebraic closure since any two algebraic closures of $\mathbb{K}$ are isomorphic. The same remark made above (about the fact that Jelonek's result holds in [**Jel05**] for algebraically closed fields) still applies here.

[26] One needs to repeat here the argument already used to justify the definition of the $\mathbb{K}[X]/\mathbb{K}$ algebraic residue symbols in Definition 6.26.

Under the geometric hypothesis, known as Nœther presentation of $\mathbb{X}_{\overline{\mathbb{K}}}$,

$$(6.68)\qquad \left|\{(x,y)\in \mathbb{X}_{\overline{\mathbb{K}}}\subset \overline{\mathbb{K}}^N : x_1=\cdots=x_n=0\}\right| = \deg \mathbb{X}_{\overline{\mathbb{K}}} = \deg \mathbb{X},$$

and subsequently, when

$$(6.69)\qquad \varpi = \bar{q}\, d\bar{X}_1\wedge\cdots\wedge d\bar{X}_n,$$

where $q\in\mathbb{K}[X,Y]$, we will specify in §6.4.2, in terms of trace calculus, the right-hand side of (6.67); see Proposition 6.45. The results obtained there will be, as the results in §6.3.1, in concurrence with the iterated Euclid division algorithm in the univariate setting.

6.3.3. Residues symbols in $\mathbb{K}[X_1^{\pm 1},\dots,X_n^{\pm 1}]$. One considers now in place of the algebraic variety $\mathbb{K}^n=\mathrm{Spec}(\mathbb{K}[X_1,\dots,X_n])$ (see §6.3.1) or more generally, in place of the algebraic variety $\mathbb{X}=\mathrm{Spec}(\mathbb{K}_{\mathbb{J}}[X_1,\dots,X_n,Y])$, where $\mathbb{J}$ is a purely n-dimensional radical ideal (see §6.3.2) another particular important example. In this example, one considers the case where $\mathbb{X}$ is a Zariski open subvariety of the affine variety $\mathbb{K}^n$, namely, $\mathbb{X}=\mathbb{T}^n_{\mathbb{K}}=\mathrm{Spec}(\mathbb{K}[X_1^{\pm 1},\dots,X_n^{\pm 1}])$. Assume throughout this section that $F_0,\dots,F_n$ are $n+1$ elements in $\mathbb{K}[X^{\pm 1}]:=\mathbb{K}[X_1^{\pm 1},\dots,X_n^{\pm 1}]$[27] such that $(F_0,\dots,F_n)$ generate $\mathbb{K}[X^{\pm 1}]$ as a $\mathbb{K}[X^{\pm 1}]$-module, while $F=(F_1,\dots,F_n)$ is a quasi-regular sequence in $\mathbb{K}[X^{\pm 1}]$. Such a hypothesis implies the existence of univariate polynomials $\Phi_j(X_j^{\pm 1})$ such that $\Phi_j(X_j^{\pm 1})\in\sum_{k=1}^n \mathbb{K}[X_1^{\pm 1},\dots,X_n^{\pm 1}]\,F_k$ for $j=1,\dots,n$. Choose also a Laurent polynomial $a_0\in\mathbb{K}[X^{\pm 1}]$ satisfying $a_0F_0\equiv 1$ modulo $\sum_{j=1}^n F_j\mathbb{K}[X^{\pm 1}]$. Furthermore, for any $\ell\in\mathbb{N}$ and any $\kappa\in\mathbb{N}^n$, let also $U_\ell^{|\kappa|}\in\mathbb{K}[X^{\pm 1}]$ be such that

$$\left(1-(a_0\,F_0)^{\ell+1}\right)^{|\kappa|+1} = 1-F_0^{\ell+1}D_\ell^{|\kappa|}[F_0]\equiv 1 \text{ modulo } \sum_{j=1}^n F_j^{\kappa_j+1}\mathbb{K}[X^{\pm 1}].$$

Given $\omega\in\bigwedge^n\Omega_{\mathbb{K}[X^{\pm 1}]/\mathbb{K}}$, the list of *toric residue symbols* attached to $(F_0,\dots,F_n)$ in the toric setting $\mathbb{T}^n_{\mathbb{K}}$ is the list, indexed by $\ell\in\mathbb{N}$ and $\kappa\in\mathbb{N}^n$, whose entries are the elements in $\mathbb{K}$ defined by

$$(6.70)\quad \mathrm{Res}\begin{bmatrix}\omega/F_0^{\ell+1}\\ F_1^{\kappa_1+1},\dots,F_n^{\kappa_n+1}\end{bmatrix}_{\mathbb{T}^n_{\mathbb{K}}} := \mathrm{Res}_{\mathbb{K}[X^{\pm 1}]/\mathbb{K}}\begin{bmatrix}\omega/F_0^{\ell+1}\\ X_1\,F_1^{\kappa_1+1},\dots,X_n\,F_n^{\kappa_n+1}\end{bmatrix}$$
$$= \mathrm{Res}_{\mathbb{K}[X^{\pm 1}]/\mathbb{K}}\begin{bmatrix}D_\ell^{|\kappa|}[F_0]\,\omega\\ X_1\,F_1^{\kappa_1+1},\dots,X_n\,F_n^{\kappa_n+1}\end{bmatrix}.$$

The above list of *toric symbols* is again independent of the choice of a_0.

REMARK 6.33. If $X=Y^U$, where $U\in\mathrm{GL}(n,\mathbb{Z})$ with $\det U=\pm 1$, so that $\mathbb{K}[X^{\pm 1}]=\mathbb{K}[Y^{\pm 1}]$, then one has

$$\bigwedge_1^n dX_j/X_j = \det U\bigwedge_1^n dY_j/Y_j = \pm 1\bigwedge_1^n dY_j/Y_j,$$

which motivates the presence of the factor X_j in front of $F_j^{\kappa_j+1}$ in each lower entry of the toric residue symbol (6.70).

[27] Elements in this $\mathbb{K}$-algebra are called Laurent polynomials.

For more references about toric residues, we refer the reader to [**CaD97**, **CaCD97**, **CaDS96**, **Cx96**] or also [**Wei08**] concerning interpolation problems on toric varieties.

6.4. Trace in $\mathbb{K}[X]$ or $\mathbb{K}_{\mathbb{J}}[X,Y]$

6.4.1. The algebraic trace function in the $\mathbb{K}[X]$-setting. Throughout this section, we assume that $p = (p_1, \dots, p_n)$ is a quasi-regular sequence in the polynomial algebra $\mathbb{A} = \mathbb{K}[X_1, \dots, X_n] = \mathbb{K}[X]$, where $\mathbb{K}$ is a (commutative) field with arbitrary characteristic, such that $\mathbb{I}_p$ is a proper ideal in $\mathbb{K}[X]$. Let also

$$d_j := \deg p_j \text{ for } j = 1, \dots, n, \quad D := d_1 \cdots d_n. \tag{6.71}$$

The quasi-regularity constraints (6.16) imply that the family $\{p^\kappa : \kappa \in \mathbb{N}^n\}$ is $\mathbb{K}$-linearly independent in $\mathbb{K}[X]$. On the other hand, a fundamental result by Oskar Perron ([**Per**, Satz 57, S. 129], see also [**Pl05**, Theorem 1.1 and §2] for an actualized proof[28]) implies that for each $j = 1, \dots, n$, one can find a polynomial E_j in $\mathbb{K}[t_1, \dots, t_n, \tau] = \mathbb{K}[t, \tau]$ such that

$$E_j(p_1, \dots, p_n, X_j) = 0 \quad \text{and} \quad \deg\big(E_j(t_1^{d_1}, \dots, t_n^{d_n}, t)\big) \le D. \tag{6.72}$$

Since $\{p^\kappa : \kappa \in \mathbb{N}^n\}$ is a $\mathbb{K}$-linearly independent family in $\mathbb{K}[X]$, one has that

$$\deg_\tau E_j := \nu_j \ge 1 \text{ for } j = 1, \dots, m, \quad \nu := \min_{1 \le j \le n} \nu_j \ge 1.$$

Then one can express for any $j = 1, \dots, n$ the polynomial $E_j(t, X_j) \in \mathbb{K}[t, X_j]$ with degree ν_j in X_j as

$$\begin{aligned} E_j(t, X_j) &= E_j(p_1 - (p_1 - t_1), \dots, p_n - (p_n - t_n), X_j) \\ &= \sum_{k=1}^n a_{j,k}(t, X)(p_j(X) - t_j). \end{aligned} \tag{6.73}$$

The polynomials $a_{j,k} \in \mathbb{K}[t, X]$ are such that $\max_{j,k} \deg_{t,X}((p_j - t_j)a_{j,k}) \le D$. Let us denote as A the $n \times n$ matrix $[a_{j,k}]_{1 \le j,k \le n}$.

As a consequence of (6.73), the family $(p_1 - t_1, \dots, p_n - t_n)$ is quasi-regular in $\mathbb{K}(t)[X]$. On the other hand, if $p_0 \in \mathbb{K}[X]$ with $\deg p_0 = d_0$ is such that $(p_0, \dots, p_n)$ generate $\mathbb{K}[X]$ as a $\mathbb{K}[X]$-module, then the polynomials $(p_0, p_1 - t_1, \dots, p_n - t_n)$ generate $\mathbb{K}(t)[X]$ as a $\mathbb{K}(t)$-module. Thus Theorem 1.1 from [**Jel05**] implies that there exists a polynomial $D[p_0] \in \mathbb{K}(t)[X]$ with

$$1 - D[p_0](t, \cdot)\, p_0 \in \sum_{j=1}^n (p_j - t_j)\, \mathbb{K}(t)[X], \quad \deg_X(D[p_0](t, \cdot)\, p_0) \le d_0\, D. \tag{6.74}$$

Therefore, we are ready to state the following proposition.

[28] The proof of the Perron theorem relies on a plain linear algebra elementary argument, which boils down to a comparison of dimensions over $\mathbb{K}$. It admits a generalization to the case where the polynomial algebra $\mathbb{K}[X_1, \dots, X_n]$ is replaced by $\mathbb{K}_{\mathbb{J}}[X_1, \dots, X_n]$, where $\mathbb{J}$ is a purely n-dimensional radical ideal, as in Example 6.19; see [**Jel05**, Theorem 3.3]. The main stumbling block in the Perron theorem or its generalization is that they fail to rely on an algorithmic process or a closed formula. Hence, they are ineffective.

PROPOSITION 6.34. *Let $(p_0,\dots,p_n)$ and A_0 be as above. Given $\omega\in\bigwedge^n\Omega_{\mathbb{K}[X]/\mathbb{K}}$,*

$$
(6.75)\qquad \mathrm{Res}_{\mathbb{K}(t)[X]/\mathbb{K}(t)}\begin{bmatrix}\omega/p_0\\ p-t\end{bmatrix} := \mathrm{Res}_{\mathbb{K}(t)[X]/\mathbb{K}(t)}\begin{bmatrix}D[p_0](t,\cdot)\,\omega\\ p_1-t_1,\dots,p_n-t_n\end{bmatrix}
$$
$$
= \mathrm{Res}_{\mathbb{K}(t)[X]/\mathbb{K}(t)}\begin{bmatrix}(\det A(t,\cdot))\,D[p_0](t,\cdot)\,\omega(X)\\ E_1(t,\cdot),\dots,E_n(t,\cdot)\end{bmatrix}
$$

is an element of $\mathbb{K}(t)$, whose definition does not depend on $D[p_0]\in\mathbb{K}(t)[X]$, provided (6.74) *holds. Furthermore,* (6.75) *does not depend on the choice of the polynomials $E_1,\dots,E_n$ and of the matrix A either, provided* (6.73) *holds.*

PROOF. The first equality in (6.75) fits with Definition 6.26, where $\mathbb{K}$ is replaced by $\mathbb{K}(t)$ and p by $p-t$, $\ell=|\kappa|=0$, $D[p_0]=D_0^0[p_0]$. It also justifies the independence from the choice of $D[p_0]$. The second equality in (6.75) follows from Theorem 6.24 and holds whenever the E_j and A fulfill (6.73). Once $D[p_0]$, the E_j's and the matrix A are specified, the explicit computation of the residue symbol (6.75) boils down to iterated euclidean division in $\mathbb{K}(t)[X]$ thanks to Proposition 6.29. Hence, Proposition 6.34 follows directly. □

Therefore, the following definition comes as the algebraic companion of the definition motivated by Proposition 6.11 in the analytic context.

DEFINITION 6.35. The $\mathbb{K}$-rational function

$$
(6.76)\qquad \begin{bmatrix}\omega/p_0\\ p-t\end{bmatrix} : t\longmapsto \mathrm{Res}_{\mathbb{K}(t)[X]/\mathbb{K}(t)}\begin{bmatrix}\omega/p_0\\ p-t\end{bmatrix}
$$

is called *the $\mathbb{K}$-rational trace of ω/p_0* for the the quasi-regular sequence $(p_1,\dots,p_n)$ in $\mathbb{K}[X]$.

Let $\overline{\mathbb{K}}$ be an algebraic closure of $\mathbb{K}$. We recall that $\overline{\mathbb{K}}$ is necessarily an infinite field. Among sequences $(p_1,\dots,p_n)$ of elements in $\mathbb{K}[X]$ defining a quasi-regular sequence, those for which $x\in\overline{\mathbb{K}}^n\longmapsto p(x)\in\overline{\mathbb{K}}^n$ is a *proper polynomial morphism* from the affine algebraic variety $\overline{\mathbb{K}}^n$ defined over $\mathbb{K}$ to itself, will play an important role. Recall that the properness property is independent of the choice of the integral closure since it amounts to saying that $\mathbb{K}[X]$ is a finitely generated $\mathbb{K}[p]$ module.[29] Furthermore, any submodule $\mathbb{K}[p][X_j]$ is a finitely generated $\mathbb{K}[p]$-module for every $j=1,\dots,n$, as a submodule of a finitely generated module over the nœtherian ring $\mathbb{K}[p]$. The fact that

$$
x\in\overline{\mathbb{K}}^n\longmapsto p(x)\in\overline{\mathbb{K}}^n
$$

is a proper polynomial morphism is therefore equivalent to the following: for each $j=1,\dots,n$ the related polynomial $E_j(t,\tau)$ in (6.72) can be taken of the form

$$
(6.77)\quad E_j(t,\tau)=\tau^{\nu_j}+\sum_{\rho=1}^{\nu_j}u_{j,\rho}(t)\,\tau^{\nu_j-\rho},\ \deg_t u_{j,\rho}+\nu_j-\rho\le D \text{ for } \rho=1,\dots,\nu_j.
$$

In this context, we can formulate a sharp version of Proposition 6.34 in the particular case when $p_0=1$.

PROPOSITION 6.36. *Let $\mathbf{p}=(p_1,\dots,p_n)$ be a quasi-regular sequence in $\mathbb{K}[X]$ such that $x\in\overline{\mathbb{K}}^n\mapsto p(x)\in\overline{\mathbb{K}}^n$ is a proper polynomial morphism. Then, for any*

[29] Observe that this implies that $p_1,\dots,p_n$ are necessarily $\mathbb{K}$-algebraically independent.

$\omega = q\, dX_1 \wedge \cdots \wedge dX_n \in \bigwedge^n \Omega_{\mathbb{K}[X]/\mathbb{K}}$, *the trace of* ω *with respect to* p *is an element of* $\mathbb{K}[t]$ *which satisfies*

$$\deg \begin{bmatrix} \omega \\ p-t \end{bmatrix} \leq \frac{D}{\nu}\, d_\omega + \frac{D(n(D-1)-|d|)}{\nu}, \tag{6.78}$$

where $\nu = \min_{1\leq j\leq n} \deg_\tau E_j$, $d_\omega = \deg q$, *and* $|d| = d_1 + \cdots + d_n$. *Moreover, one has*

$$\begin{bmatrix} \omega \\ p-t \end{bmatrix} = \sum_{\nu|\kappa|\leq D(d_\omega+n(D-1)-|d|)} \operatorname{Res}_{\mathbb{K}[X]/\mathbb{K}} \begin{bmatrix} \omega \\ p_1^{\kappa_1+1}, \ldots, p_n^{\kappa_n+1} \end{bmatrix} t^\kappa. \tag{6.79}$$

PROOF. One has $\det_{t,X}((\det A)\, q) \leq d_\omega + n(D-1) - |d|$ taking into account (6.77) and (6.73). It follows from Proposition 6.29 that

$$\operatorname{Res}_{\mathbb{K}(t)[X]/\mathbb{K}(t)} \begin{bmatrix} X^\kappa\, dX_1 \wedge \cdots \wedge dX_n \\ E_1(t,X_1), \ldots, E_n(t,X_n) \end{bmatrix} \in \mathbb{K}_{|\kappa|D/\nu}[t],$$

for any $\kappa \in \mathbb{N}^n$. The first assertion of Proposition 6.36, with the degree estimate (6.78), comes as a consequence. Theorem 6.24 implies that, for any $\kappa \in \mathbb{N}^n$,

$$\begin{aligned} \operatorname{Res}_{\mathbb{K}[X]/\mathbb{K}} \begin{bmatrix} \omega \\ p_1^{\kappa_1+1}, \ldots, p_n^{\kappa_n+1} \end{bmatrix} &= \operatorname{Res}_{\mathbb{K}[t,X]/\mathbb{K}} \begin{bmatrix} \omega \wedge dt \\ p-t, t_1^{\kappa_1+1}, \ldots, t_n^{\kappa_n+1} \end{bmatrix} \\ &= \operatorname{Res}_{\mathbb{K}[t,X]/\mathbb{K}} \begin{bmatrix} (\det A)\, \omega \wedge dt \\ E(t,X), t_1^{\kappa_1+1}, \ldots, t_n^{\kappa_n+1} \end{bmatrix}. \end{aligned}$$

Here, $dt = dt_1 \wedge \cdots \wedge dt_n$ and $E(t,X) = (E_1(t,X_1), \ldots, E_n(t,X_n))$. It corresponds to the coefficient of t^κ in the development of the polynomial

$$\begin{bmatrix} \omega \\ p-t \end{bmatrix} = \operatorname{Res}_{\mathbb{K}(t)[X]/\mathbb{K}(t)} \begin{bmatrix} (\det A(t,\cdot))\, \omega \\ E(\cdot, X) \end{bmatrix}$$

along powers of t. This proves (6.79). □

Proposition 6.36 admits the following reciprocal result.

PROPOSITION 6.37. *Let* $p = (p_1, \ldots, p_n)$ *be a quasi-regular sequence in the polynomial algebra* $\mathbb{A} = \mathbb{K}[X_1, \ldots, X_n]$ *such that* $\mathbb{I}_p \neq \mathbb{A}$, *and let* $d_j = \deg p_j$ *for* $j = 1, \ldots, n$. *A sufficient condition for* $x \in \overline{\mathbb{K}}^n \to p(x) \in \overline{\mathbb{K}}^n$ *to be a proper polynomial morphism is that*

(6.80)

$$\forall \kappa \in \mathbb{N}^n,\ |\kappa| \leq |d| - n \Longrightarrow \operatorname{Res}_{\mathbb{K}(t)[X]/\mathbb{K}(t)} \begin{bmatrix} X^\kappa X_j^{|d|-n+1}\, dX_1 \wedge \cdots \wedge dX_n \\ p-t \end{bmatrix} \in \mathbb{K}[t],$$

for all $j = 1, \ldots, n$.

PROOF. Assume that (6.80) holds. It follows from Theorem 6.28 that

$$X_j^{|d|-n+1} = \operatorname{Res}_{\mathbb{K}[X][Y]/\mathbb{K}[X]} \begin{bmatrix} (\det H(X,Y))\, Y_j^{|d|-n+1}\, dY_1 \wedge \cdots \wedge dY_n \\ p_1 - p_1(X), \ldots, p_n - p_n(X) \end{bmatrix}$$

holds for every $j = 1, \ldots, n$ and where $\deg_{X,Y}(\det H(X,Y)) \leq |d| - n$; see (6.55). Let

$$\det H(X,Y) = \sum_{|\kappa|+|\kappa'|\leq |d|-n} \gamma_{\kappa,\kappa'} Y^\kappa X^{\kappa'}.$$

Then, one has that

$$\begin{aligned} X_j^{|d|-n+1} &= \sum_{|\kappa|+|\kappa'|\le|d|-n} \gamma_{\kappa,\kappa'} X^{\kappa'} \begin{bmatrix} Y^\kappa Y_j^{|d|-n+1}\, dY_1\wedge\cdots\wedge dY_n \\ p-t \end{bmatrix}_{t=p(X)} \\ &= \sum_{|\kappa|\le|d|-n} u_{j,\kappa}(p_1(X),\dots,p_n(X))X^\kappa = 0 \end{aligned}$$

for each $j = 1,\dots,n$ and where $u_{j,\kappa} \in \mathbb{K}[t_1,\dots,t_n]$. As a consequence $\mathbb{K}[X]$ is a finitely generated $\mathbb{K}[p]$-module, which means that $x \in \overline{\mathbb{K}}^n \longmapsto p(x) \in \overline{\mathbb{K}}^n$ is a proper morphism in the algebraic sense. □

We now fix a nontrivial archimedean[30] absolute value on $\overline{\mathbb{K}}$. That is, we fix a map $|\ | : \overline{\mathbb{K}} \to [0,+\infty[$ such that

$$|\gamma| = 0 \Longleftrightarrow \gamma = 0,\ |\gamma'\gamma''| = |\gamma'|\,|\gamma''|,\ |\gamma'+\gamma''| \le |\gamma'| + |\gamma''|$$

for any $\gamma,\gamma',\gamma'' \in \overline{\mathbb{K}}$, and $\gamma \not\equiv 1$ on $\overline{\mathbb{K}}\setminus\{0\}$. Such an absolute value is a unique extension of a nontrivial absolute value on $\mathbb{K}$. Let $|x| := \max_{1\le k\le n}|x_k|$ for any $x \in \overline{\mathbb{K}}^n$. We aim to quantify the notion of properness for a polynomial morphism $p = (p_1,\dots,p_n)$ from $\overline{\mathbb{K}}^n$ to itself. For that purpose, we introduce the concept of δ-properness, where $\delta = (\delta_1,\dots,\delta_n) \in \mathbb{Q}^n$ with $0 \le \delta_j < d_j = \deg p_j$ for $j = 1,\dots,n$. Such δ is called *p-admissible* or a *multi-index of properness.* Definition 6.38 below does not rely on the choice of nontrivial archimedean absolute values $|\ |$ chosen on $\overline{\mathbb{K}}$ or $\overline{\mathbb{K}}^n$.

DEFINITION 6.38. Let $p = (p_1,\dots,p_n)$ be a quasi-regular sequence in the polynomial algebra $\mathbb{A} = \mathbb{K}[X_1,\dots,X_n]$. The polynomial morphism $x \in \overline{\mathbb{K}}^n \longmapsto p(x) \in \overline{\mathbb{K}}^n$ is said to be *δ-proper* if and only if for any $\varepsilon = (\varepsilon_1,\dots,\varepsilon_n)$, where $0 < \varepsilon_j \ll 1$ for $j = 1,\dots,n$, there exist strictly positive constants C_ε and c_ε such that

$$|x| \ge C_\epsilon \Longrightarrow \max_{1\le j\le n} \frac{|p_j(x)|}{|x|^{d_j}}\,|x|^{\delta_j+\varepsilon_j} \ge c_\varepsilon. \tag{6.81}$$

The following proposition shows that the concept of δ-properness quantifies that of properness for a polynomial morphism from $\overline{\mathbb{K}}^n$ to itself.

PROPOSITION 6.39. *A polynomial morphism $p : \overline{\mathbb{K}}^n \to \mathbb{K}^n$ defined over $\mathbb{K}$ is proper if and only if it is δ-proper for some p-admissible δ such that*

$$\forall\,\kappa \in \mathbb{N}^n,\ \langle\!\langle \kappa, d\rangle\!\rangle \le D \Longrightarrow \langle\!\langle \kappa\,,\, d-\delta\rangle\!\rangle < 1. \tag{6.82}$$

Moreover, if it is the case, then p is δ-proper for any p-admissible δ which satisfies (6.82).

PROOF. Suppose that a polynomial morphism $p : \overline{\mathbb{K}}^n \to \mathbb{K}^n$ defined over $\mathbb{K}$ is proper. Let us express for each $j = 1,\dots,n$ the polynomial $E_j(t,\tau)$ in (6.77) as

$$E_j(t,\tau) = \tau^{\nu_j} + \sum_{\rho=1}^{\nu_j}\Big(\sum_{|\kappa|\le\rho+D-\nu_j} \gamma_{j,\rho,\kappa}\, t^\kappa\Big)\,\tau_j^{\nu_j-\rho} = 0,$$

[30]When the stronger inequality $|\gamma'+\gamma''| \le \max(|\gamma'|,|\gamma''|)$ replaces the triangle inequality involved in this notion, the absolute value $|\ |$ is said to be *ultrametric*. It is also called a (general) *valuation*.

where $\gamma_{j,\rho,\kappa} \in \mathbb{K}$. Let $\delta = (\delta_1, \dots, \delta_n)$ be p-admissible. One has for any $x \in \overline{\mathbb{K}}^n \setminus \{0\}$ and any $j = 1, \dots, n$ that

$$x_j^{\nu_j} = -\sum_{\rho=1}^{\nu_j} \Big(\sum_{|\kappa| \le \rho + D - \nu_j} \gamma_{j,\rho,\kappa} \, |x|^{\langle\!\langle \kappa, d-\delta \rangle\!\rangle} \prod_{j=1}^{n} \Big(\frac{p_j(x)}{|x|^{d_j}} \, |x|^{\delta_j} \Big) \Big) \, x_j^{\nu_j - \rho}.$$

For $0 < \eta \ll 1$, one has then

$$\max_{1 \le j \le n} \Big(|x|^{\delta_j} \frac{|p_j(x)|}{|x|^{d_j}} \Big) < \eta \Longrightarrow |x_j|^{\nu_j} \le \sum_{\rho=1}^{\nu_j} |x|^{\mu(\rho + D - \nu_j)} \, |x_j|^{\nu_j - \rho}, \tag{6.83}$$

where

$$\mu(\rho) = \max_{\{\kappa \in \mathbb{N}^n \,:\, \langle\!\langle \kappa, d \rangle\!\rangle \le \rho\}} \langle\!\langle \kappa, d - \delta \rangle\!\rangle.$$

Therefore, if $\mu(D) < 1$, then $\mu(\rho + D - \nu_j) + \nu_j - \rho < \nu_j - \rho + 1 < \nu_j$ for $1 \le \rho \le \nu_j$ and it follows from (6.83) that

$$\max_{1 \le j \le n} \Big(|x|^{\delta_j} \frac{|p_j(x)|}{|x|^{d_j}} \Big) < \eta \Longrightarrow |x| \le C(\eta) < +\infty.$$

Therefore the condition $\mu(D) < 1$ implies that p is δ-proper. On the other hand, it is a classical fact that any polynomial morphism $p : \overline{\mathbb{K}}^n \longrightarrow \overline{\mathbb{K}}^n$, which is δ-proper for some p-admissible δ is proper from $\overline{\mathbb{K}}^n \longrightarrow \overline{\mathbb{K}}^n$; see for example [**BY99**, Proposition 3.1] for a proof of this well known result. It concludes the proof of Proposition 6.39. □

For the notion of δ-properness described above, it is natural to introduce the complete algebraic variety

$$\mathbb{P}_{\overline{\mathbb{K}}} = \mathrm{Spec}\big(\mathrm{Proj}\,(\overline{\mathbb{K}}[X_0, \dots, X_n])\big)$$

with its structure sheaf $\mathbb{O}$. Let us denote from now on as $x_0, \dots, x_n$ the homogeneous coordinates on $\mathbb{P}_{\overline{\mathbb{K}}}$ and as $x = [x_0 : \cdots : x_n]$ the representation of $x \in \mathbb{P}_{\overline{\mathbb{K}}}$ in terms of such coordinates. Let $P_1, \dots, P_n \in \mathbb{K}[X_0, \dots, X_n]$ be the homogenizations of $p_1, \dots, p_n$ with respective degrees $d_1, \dots, d_n$. We consider first the case of $(0, \dots, 0)$-properness. One can state in this particular situation the following proposition.

Proposition 6.40. *Let p be a quasi-regular sequence in $\mathbb{K}[X]$. Then, the polynomial morphism $x \in \overline{\mathbb{K}}^n \mapsto p(x) \in \overline{\mathbb{K}}^n$ is $(0, \dots, 0)$-proper if and only if $\{x \in \mathbb{P}_{\overline{\mathbb{K}}} : x_0 = P_1(x) = \cdots = P_n(x) = 0\} = \emptyset$.*

Proof. Suppose that $x \mapsto p(x)$ is $(0, \dots, 0)$-proper. Then, it follows from (6.81) that for any $N \in \mathbb{N}^*$ sufficiently large, there exists $C(N) > 0$ and $c(N) > 0$ such that, for any $x = [x_0 : x'] \in \mathbb{P}_{\overline{\mathbb{K}}}$,

$$|x'| \ge C(N)\, |x_0| \Longrightarrow |x_0|^{N+n-1} \le \frac{1}{c(N)} \max_{1 \le j \le n} \Big(\frac{|P_j(x)|}{|x'|^{d_j - 1/(N+n-1)}} \Big). \tag{6.84}$$

Assume that the set $\{x \in \mathbb{P}_{\overline{\mathbb{K}}} : x_0 = P_1(x) = \cdots = P_n(x) = 0\}$ is nonempty. Then, let $\alpha = [0 : \alpha_1 : \cdots : \alpha_n]$ be a point of this set, where $\alpha_1 \neq 0$. It follows from (6.84) that the germ at α of x_0/x_1 lies in the integral closure of the $(N+n-1)$th power of the ideal generated by the germs at α of $P_1/x_1^{d_1}, \dots, P_n/x_1^{d_n}$. The Lipman–Teissier theorem [**LipS81, LipT81**] applied in the regular n-dimensional local nœtherian ring $\mathbb{O}_\alpha$ implies then that the germ of x_0/x_1 at α belongs to the Nth power of the ideal generated by the germs at α of the $P_j/x_1^{d_j}$. Since this is true for any $N \in \mathbb{N}^*$ and $\bigcap_{N\in\mathbb{N}^*}\big((P_1/z_1^{d_1},\dots,P_n/z_n^{d_n})_\alpha\big)^N$ is the null ideal in $\mathbb{O}_\alpha$, one is led to a contradiction. It follows that $\{x \in \mathbb{P}_{\overline{\mathbb{K}}} : x_0 = P_1(x) = \cdots = P_n(x) = 0\} = \emptyset$. Conversely, if the above hypothesis is satisfied, the functions

$$x \in \mathbb{P}_{\overline{\mathbb{K}}} \longmapsto \frac{|P_j(x)|}{|x|^{d_j}}$$

do not share any common zero in $\{x \in \mathbb{P}_{\overline{\mathbb{K}}} : |x'| \geq C|x_0|\}$ for some $C > 0$ sufficiently large. A continuity argument asserts then that in this subset,

$$\max_{1\leq j\leq n} \frac{|P_j(x)|}{|x|^{d_j}} \geq c$$

for some $c > 0$. This says that the polynomial morphism $x \mapsto p(x)$ is $(0,\dots,0)$-proper. □

We now study δ-properness in the case when δ is sufficiently close to $(0,\dots,0)$. Namely, we study the case when

$$(n-1)|\delta|_\infty < \min_{1\leq j\leq n} |d_j - \delta_j|, \tag{6.85}$$

for $|\delta|_\infty = \max_{1\leq j\leq n}\delta_j$.

PROPOSITION 6.41. *Let p be a quasi-regular sequence in $\mathbb{K}[X]$ such that the polynomial morphism $x \in \overline{\mathbb{K}}^n \mapsto p(x) \in \overline{\mathbb{K}}^n$ is δ-proper for some p-admissible multi-exponent δ satisfying* (6.85). *For any $\omega = q\,dX \in \Omega^n_{\mathbb{K}[X]/\mathbb{K}}$ with degree defined by $d_\omega = \deg q$, one has*

$$\begin{bmatrix} \omega \\ p-t \end{bmatrix} = \sum_{\kappa\in E^d_\delta[\omega]} \operatorname{Res}_{\mathbb{K}[X]/\mathbb{K}} \begin{bmatrix} \omega \\ p_1^{\kappa_1+1},\dots,p_n^{\kappa_n+1} \end{bmatrix} t^\kappa, \tag{6.86}$$

where

$$E^d_\delta[\omega] = \big\{\kappa \in \mathbb{N}^n : \langle\!\langle \kappa, d-\delta\rangle\!\rangle + |d-\delta| \leq (n-1)(|\kappa|+n)|\delta|_\infty + d_\omega + n\big\},$$

$d-\delta = (d_1-\delta_1,\dots,d_n-\delta_n)$ and $|d-\delta| = \sum_{j=1}^n (d_j-\delta_j)$.

PROOF. Observe first that condition (6.85) implies that the subset

$$\{\kappa \in \mathbb{N}^n : \langle\!\langle \kappa, d-\delta\rangle\!\rangle + |d-\delta| \leq (n-1)(|\kappa|+n)|\delta|_\infty + d_\omega + n\} \subset \mathbb{N}^n$$

is finite. Let us first consider the case when $d_j \gg 1$ and $\delta \in \mathbb{N}^n$, which amounts to saying that $\delta_j \in \{0,\dots,d_j-1\}$ for $j = 1,\dots,n$. Let $\phi_1(X_1),\dots,\phi_n(X_n)$ be univariate polynomials with common degree $\chi \geq \max_j d_j$ satisfying the algebraic identities $\phi_j(X_j) = \sum_{k=1}^n a_{j,k}(X)\,p_k(X)$. Such polynomials do exist, having, in addition, the property $\max_k \deg(a_{j,k}p_k) \leq 2D$; see [**Jel05**, Theorem 1.6].[31] Let $P_j \in \mathbb{K}[X_0,\dots,X_n]$ and $\Phi_j \in \mathbb{K}[X_0,X_j]$ be the homogenizations of p_j,ϕ_j for each $j = 1,\dots,n$. It is equivalent to saying that p is δ-proper and that for any $N \in \mathbb{N}^*$ sufficiently large, there exists $C(N) > 0$ and $c(N) > 0$ such that, for any point $x = [x_0 : x'] \in \mathbb{P}_{\overline{\mathbb{K}}}$,

$$\begin{aligned} |x'| \geq C(N)\,|x_0| \Longrightarrow |x_0|^{(N+n)|\delta|_\infty+1} \\ \leq \frac{1}{c(N)} \max_{1\leq j\leq n}\Big(|x'|^{(N+n)(\delta_j-d_j)+1}\,\big|P_j(x)\,x_0^{|\delta|_\infty-\delta_j}\big|^{N+n}\Big) \end{aligned} \tag{6.87}$$

[31] In fact, we will not here make use of any estimate on the $\deg(a_{j,k}p_k)$ for $1 \leq j,k \leq n$.

holds. On the other hand,

$$(6.88)\quad |x'| < C(N)\,|x_0| \Longrightarrow |\Phi_j(x)|^{N+n} \leq K(N) \max_{1\leq 1\leq n} \left(|x_0|^{(N+n)(\chi-d_j)}|P_j^{N+n}(x)|\right)$$

for some $K(N) > 0$. Consider in the local ring $\mathbb{K}[[X_0,\dots,X_n]]$ the ideal

$$\mathbb{I} = \left(X_0^{|\delta|_\infty-\delta_1}p_1,\dots,X_0^{|\delta|_\infty-\delta_n}p_n\right)\mathbb{K}[[X_0,\dots,X_n]].$$

It follows from (6.87), (6.88) and the valuative criterion in the local nœtherian ring $\mathbb{K}[[X_0,\dots,X_n]]$ (see Proposition 5.2 or also [**Ko99**, Theorem 3.3]) that each $X_0^{(N+n)|\delta|_\infty+1}\Phi_j^{N+n}$ considered here as an element in $\mathbb{K}[[X_0,\dots,X_n]]$ belongs to the integral closure of the $(N+n)$-power of the ideal $\mathbb{I}$. Since $\mathbb{I}$ is generated by n elements, it follows from Lipman–Teissier theorem [**LipS81, LipT81**] that

$$\begin{aligned}X_0^{(N+n)|\delta|_\infty+1}\,\Phi_j^{N+n}(X) &\in \mathbb{I}^{N+1}\\ &\subset \bigcap_{\{\kappa\in\mathbb{N}^n\,:\,|\kappa|=N\}} \left((X_0^{|\delta|_\infty-\delta_1}p_1)^{\kappa_1+1},\dots,(X_0^{|\delta|_\infty-\delta_n}p_n)^{\kappa_n+1}\right)\mathbb{K}[[X_0,\dots,X_n]]\end{aligned}$$

for any $j = 1,\dots,n$. Then, the homogeneity argument implies that one has a polynomial equality

$$\phi_j^{N+n} = \sum_{k=1}^{N} h_{j,k}^{N,\kappa}\, p_k^{\kappa_k+1}$$

for any $j = 1,\dots,n$ and any $\kappa \in \mathbb{N}^n$ such that $|\kappa| = N$, where

$$\deg(h_{j,k}^{N,\kappa}) \leq (N+n)(\chi+|\delta|_\infty) - (\kappa_k+1)(d_k+|\delta|_\infty-\delta_k)$$

for any $1 \leq j,k \leq n$. Furthermore, one has that

$$(6.89)\quad \deg(\det H^{N,\kappa}) \leq n(N+n)\,\chi + \left((n-1)(|\kappa|+n)\,|\delta|_\infty - \langle\kappa, d-\delta\rangle - |d-\delta|\right).$$

Proposition 6.27 implies

$$\operatorname{Res}_{\mathbb{K}[X]/\mathbb{K}}\begin{bmatrix}\omega\\ p_1^{\kappa_1+1},\dots,p_n^{\kappa_n+1}\end{bmatrix} = \operatorname{Res}_{\mathbb{K}[X]/\mathbb{K}}\begin{bmatrix}\det H^{N,\kappa}\,\omega\\ \phi_1^{N+n}(X_1),\dots,\phi_n^{N+n}(X_n)\end{bmatrix}.$$

On the other hand, Proposition 6.29 and Corollary 6.30 show that

$$\deg(\det H^{N,\kappa}) + d_\omega < n(N+n)\chi - n \Longrightarrow \operatorname{Res}_{\mathbb{K}[X]/\mathbb{K}}\begin{bmatrix}\omega\\ p_1^{\kappa_1+1},\dots,p_n^{\kappa_n+1}\end{bmatrix} = 0.$$

Hence, if one takes into account (6.89), then

$$\begin{aligned}&\langle\!\langle\kappa, d-\delta\rangle\!\rangle + |d-\delta| > (n-1)(|\kappa|+n)\,|\delta|_\infty + d_\omega + n\\ &\qquad\Longrightarrow \deg(\det H^{N,\kappa}) + d_\omega < n(N+n)\chi - n \Longrightarrow \begin{bmatrix}\omega\\ p_1^{\kappa_1+1},\dots,p_n^{\kappa_n+1}\end{bmatrix} = 0.\end{aligned}$$

Then equality (6.86) follows in the particular case $\delta \in \mathbb{N}^n$ from formula (6.79), in Proposition 6.36. In the general case, let r be a common denominator for δ_j, $j = 1,\dots,n$. Take now instead of $p = (p_1,\dots,p_n)$ the polynomial map

$$p^{[r]} = (p_1(X_1^r),\dots,p_n(X_1^r)),$$

which is $r\delta$-proper. The same reasoning as above leads to

$$(6.90)\quad r\,(n-1)(|\kappa|+n)\,|\delta|_\infty - r\,\langle\!\langle \kappa, d-\delta\rangle\!\rangle - r|d-\delta| < -d_{\omega^{[r]}} - n$$
$$\Longrightarrow \operatorname{Res}_{\mathbb{K}[X]/\mathbb{K}}\begin{bmatrix} \omega^{[r]} \\ p_1^{\kappa_1+1}(X^r),\ldots,p_n^{\kappa_n+1}(X^r)\end{bmatrix} = 0,$$

where $\omega^{[r]} = q(X^r)(X_1\cdots X_n)^{r-1}\,dX_1\wedge\cdots\wedge dX_n$, so that $d_{\omega^{[r]}} = r\,d_\omega + (r-1)n$. It follows then from (6.90) and Corollary 6.31 that

(6.91)
$$(n-1)(|\kappa|+n)|\delta|_\infty - \langle\!\langle \kappa, d-\delta\rangle\!\rangle - |d-\delta| < -\frac{rd_\omega + (r-1)n)}{r} - \frac{n}{r} = -d_\omega - n$$
$$\Longrightarrow \operatorname{Res}_{\mathbb{K}[X]/\mathbb{K}}\begin{bmatrix} \omega \\ p_1^{\kappa_1+1},\ldots,p_n^{\kappa_n+1}\end{bmatrix} = 0.$$

Thus, formula (6.86) also holds in this general case. The proof of Proposition 6.41 is now complete. □

REMARK 6.42. When p is $(0,\ldots,0)$-proper, it follows from Proposition 6.41 that

$$(6.92)\qquad d_\omega < |d| - n \Longrightarrow \operatorname{Res}_{\mathbb{K}[X]/\mathbb{K}}\begin{bmatrix} \omega \\ p_1,\ldots,p_n\end{bmatrix} = 0.$$

This Euler–Jacobi formula is in [**KK87**]. When the characteristic of $\mathbb{K}$ is zero, and its cardinality is at most that of the continuum, then $\mathbb{K}$ is known to be (ring) isomorphic to a subfield of $\mathbb{C}$. In such a situation, one can profit from the powerful machinery of Bochner–Martinelli kernels providing explicitness for the algebraic residue symbols (without any need to invoke Proposition 6.27) to refine (6.86) as

$$(6.93)\quad \begin{bmatrix} \omega \\ p-t\end{bmatrix} = \sum_{\{\kappa\in\mathbb{N}^n\,:\,\langle\!\langle \kappa,d-\delta\rangle\!\rangle + |d-\delta| \le d_\omega + n\}} \operatorname{Res}_{\mathbb{K}[X]/\mathbb{K}}\begin{bmatrix} \omega \\ p_1^{\kappa_1+1},\ldots,p_n^{\kappa_n+1}\end{bmatrix} t^\kappa.$$

Such a formula holds for any p-admissible δ; see Theorem 6.56. It unknown whether such a result holds independently of the characteristic of the field $\mathbb{K}$.

6.4.2. The algebraic trace function in the $\mathbb{K}_{\mathbb{J}}[X,Y]$ setting. Let us return in this section to the framework considered in §6.3.2 and use the same notations. Throughout this section, we assume that the geometric assumption (6.68) holds, which is essential. Such a hypothesis implies that the classes $\bar{X}_j$ of X_j modulo $\mathbb{J}$ are such that $(\bar{X}_1,\ldots,\bar{X}_n)$ is a quasi-regular sequence in $\mathbb{A} = \mathbb{K}_{\mathbb{J}}[X,Y]$. Moreover, the sequence $(\bar{X}_1 - t_1,\ldots,\bar{X}_n - t_n)$ is also a quasi-regular sequence in the $\mathbb{K}[t]$-algebra $\mathbb{A}[t]$, and it is such that the quotient

$$\frac{\mathbb{A}[t]}{(\bar{X}_1 - t_1,\ldots,\bar{X}_n - t_n)\,\mathbb{A}[t]}$$

is finitely generated and projective as a $\mathbb{K}[t]$-module. This happens because for each $k = 1,\ldots,N-n$ there exists a monic relation of integral dependency

$$Y_k^{\chi_k} - \sum_{\rho=1}^{\chi_k} y_{k,\rho}(X)\,Y_k^{\chi_k-\rho} \in \mathbb{J},$$

where the coefficients $y_{k,\rho}$ belong to $\mathbb{K}[X_1,\dots,X_n]$. It follows from the construction of algebraic residue symbols, as described in §6.2.2 (see [**Lip87**, Corollary 3.7]), that

(6.94)
$$\operatorname{Res}_{\mathbb{A}[t]/\mathbb{K}[t]}\begin{bmatrix}\bar{X}^\kappa\,\bar{Y}^\chi\,d\bar{X}_1\wedge\cdots\wedge d\bar{X}_n\\ \bar{X}_1-t_1,\dots,\bar{X}_n-t_n\end{bmatrix}=t^\kappa\operatorname{Res}_{\mathbb{A}[t]/\mathbb{K}[t]}\begin{bmatrix}\bar{Y}^\chi\,d\bar{X}_1\wedge\cdots\wedge d\bar{X}_n\\ \bar{X}_1-t_1,\dots,\bar{X}_n-t_n\end{bmatrix}$$
$$=\sum_{\{\kappa'\in\mathbb{N}^n\,:\,|\kappa'|\le d(\chi)\}}\operatorname{Res}_{\mathbb{A}/\mathbb{K}}\begin{bmatrix}\bar{Y}^\chi\,d\bar{X}_1\wedge\cdots\wedge d\bar{X}_n\\ \bar{X}_1^{\kappa_1+1},\dots,\bar{X}_n^{\kappa_n+1}\end{bmatrix}t^{\kappa+\kappa'},$$

where $\kappa\in\mathbb{N}^n,\chi\in\mathbb{N}^{N-n}$, equals the trace of the multiplication operator from $\mathbb{A}[t]/(\bar{X}-t)\mathbb{A}[t]$ to itself by $\bar{X}^\kappa\bar{Y}^\chi$. We will denote such element of $\mathbb{K}[t]$ as

$$\begin{bmatrix}\bar{X}^\kappa\bar{Y}^\chi\,d\bar{X}\\ \bar{X}-t\end{bmatrix}_{\mathbb{X}}. \tag{6.95}$$

We recall at this point that $\mathbb{X}$ denotes the affine algebraic variety defined over $\mathbb{K}$ as $\mathbb{X}=\operatorname{Spec}\mathbb{K}_{\mathbb{J}}[X,Y]$. Let us point out the following result concerning the degree estimates.

PROPOSITION 6.43. *Let $\kappa\in\mathbb{N}^n,\chi\in\mathbb{N}^{N-n}$. Then one has*

$$\deg_t\begin{bmatrix}\bar{X}^\kappa\bar{Y}^\chi\,d\bar{X}\\ \bar{X}-t\end{bmatrix}_{\mathbb{X}}\le|\kappa|+|\chi|\deg\mathbb{X}, \tag{6.96}$$

where $\deg\mathbb{X}$ *is defined as* $\deg\mathbb{X}=\deg\mathbb{X}_{\overline{\mathbb{K}}}$, *independently of the algebraic closure* $\overline{\mathbb{K}}$.

PROOF. It follows from (6.94) that it is enough to prove the result when $\kappa=0$, which we assume from now on. It follows from the generalized Perron theorem as stated in [**Jel05**, Theorem 3.3] that, since the polynomial morphism $X\;:\mathbb{X}\to\mathbb{K}^n$ is generically finite (in fact here finite), there exists a polynomial $E_\chi\in\mathbb{K}[t_1,\dots,t_n,\tau]$ such that $E_\chi(X_1,\dots,X_n,Y^\chi)\in\mathbb{J}$ and

$$\deg_t E_\chi(t^{\deg X_1},\dots,t^{\deg X_n},t^{\deg Y^\chi})=\deg_t E_\chi(t,\dots,t,t^{|\chi|})\le\deg\mathbb{X}\,|\chi|.$$

Since X is a finite morphism, the minimal polynomial $E_{\chi,\min}(t,\cdot)$ among polynomials which satisfy $E_{\chi,\min}(X,Y^\mu)\in\mathbb{J}$ is monic in τ (which means that the coefficient of the leading monomial in τ equals 1) and divides E_χ in $\mathbb{K}[t,\tau]$. It implies that $\deg_t E_{\chi,\min}(t,\dots,t,t^{|\chi|})\le\deg\mathbb{X}\,|\chi|$. Then, one has that

$$E_{\chi,\min}(X_1,\dots,X_n,Y^\chi)=(Y^\chi)^\nu+\sum_{\rho=1}^{\nu}u_{\chi,\rho}(X)\,(Y^\chi)^{\nu-j}\in\mathbb{J}$$

with $\nu=\nu_\chi\in\mathbb{N}^*$ and $\deg u_{\chi,\rho}\le|\chi|\,(\rho+\deg\mathbb{X}-\nu)$ for $\rho=1,\dots,\nu$. Moreover,

$$(\bar{Y}^\chi)^\nu+\sum_{\rho=1}^{\nu}u_{\chi,\rho}(t)\,(\bar{Y}^\chi)^{\nu-\rho}\in(\bar{X}-t)\,\mathbb{K}_{\mathbb{J}}[X,Y][t].$$

In particular $\deg u_{\chi,1}\le|\chi|\deg\mathbb{X}$. This estimate leads to the conclusion of the lemma, since $-u_{\chi,1}(t)$ is the trace of the multiplication operator by $\bar{Y}^\chi$ from the $\mathbb{K}[t]$-projective finitely generated module $\mathbb{K}_{\mathbb{J}}[X,Y][t]/(\bar{X}-t)\mathbb{K}_{\mathbb{J}}[X,Y][t]$ into itself. □

If one takes the univariate polynomials $\phi_1(\bar{X}_1),\dots,\phi_n(\bar{X}_n)$ with respective strictly positive degrees $\nu_1,\dots,\nu_n$, in place of $\bar{X}_1,\dots,\bar{X}_n$, then one can define in the same way, for any $\kappa\in\mathbb{N}^n$ and $\chi\in\mathbb{N}^{N-n}$ an element of $\mathbb{K}[\mathfrak{t}]=\mathbb{K}[\mathfrak{t}_1,\dots,\mathfrak{t}_n]$ by

(6.97)

$$\begin{aligned}\begin{bmatrix} X^\kappa\, Y^\chi\, d\bar{X} \\ \phi(\bar{X})-\mathfrak{t}\end{bmatrix}_{\mathbb{X}} &:= \operatorname{Res}_{\mathbb{K}_{\mathbb{J}}[X,Y][\mathfrak{t}]/\mathbb{K}[\mathfrak{t}]}\begin{bmatrix} \bar{X}^\kappa\, \bar{Y}^\chi\, d\bar{X}_1\wedge\cdots\wedge d\bar{X}_n \\ \phi_1(\bar{X}_1)-\mathfrak{t}_1,\dots,\phi_n(\bar{X}_n)-\mathfrak{t}_n\end{bmatrix} \\ &= \sum_{\{\kappa'\in\mathbb{N}^n\,:\,|\kappa'|\le d_\phi(\kappa,\chi)\}} \operatorname{Res}_{\mathbb{K}_{\mathbb{J}}[X,Y]/\mathbb{K}}\begin{bmatrix} \bar{X}^\kappa\bar{Y}^\chi\, d\bar{X}_1\wedge\cdots\wedge d\bar{X}_n \\ \phi_1^{\kappa_1'+1}(\bar{X}_1),\dots,\phi_n^{\kappa_n'+1}(\bar{X}_n)\end{bmatrix}\mathfrak{t}^{\kappa'}.\end{aligned}$$

The above definition extends by $\mathbb{K}$-linearity to any element of $\mathbb{K}_{\mathbb{J}}[X,Y][t]$ in place of $X^\kappa Y^\chi$. In particular, it follows from Theorem 6.24 that if

$$b_\phi(t,\bar{X})=\prod_{j=1}^n \frac{\phi_j(\bar{X}_j)-\phi_j(t)}{\bar{X}_j-t_j},$$

then one has

$$\begin{aligned}\begin{bmatrix} \bar{X}^\kappa\bar{Y}^\chi\, d\bar{X} \\ \bar{X}-t\end{bmatrix}_{\mathbb{X}} &= \operatorname{Res}_{\mathbb{K}_{\mathbb{J}}[X,Y][t]/\mathbb{K}[t]}\begin{bmatrix} b_\phi(t,\bar{X})\,\bar{X}^\kappa\bar{Y}^\chi\, d\bar{X}_1\wedge\cdots\wedge d\bar{X}_n \\ \phi_1(\bar{X}_1)-\phi_1(t),\dots,\phi_n(\bar{X}_n)-\phi_n(t)\end{bmatrix} \\ &= \left(\begin{bmatrix} b_\phi(t,\bar{X})\,\bar{X}^\tau\,\bar{Y}^\chi\, d\bar{X} \\ \phi(\bar{X})-\mathfrak{t}\end{bmatrix}_{\mathbb{X}}\right)_{\mathfrak{t}=\phi(t)}.\end{aligned} \tag{6.98}$$

The following transitivity property also holds.

PROPOSITION 6.44. *Let $\kappa\in\mathbb{N}^n$ and $\chi\in\mathbb{N}^{N-n}$. Then*

$$\begin{bmatrix} \bar{X}^\kappa\bar{Y}^\chi d\bar{X} \\ \phi(\bar{X})-\mathfrak{t}\end{bmatrix}_{\mathbb{X}} = \sum_{\kappa'\in\mathbb{N}^n} \operatorname{Res}_{\mathbb{K}[t]/\mathbb{K}}\begin{bmatrix} \begin{bmatrix} \bar{X}^\kappa\bar{Y}^\chi\, d\bar{X} \\ \bar{X}-t\end{bmatrix}_{\mathbb{X}} dt \\ \phi_1^{\kappa_1'+1}(t_1),\dots,\phi_n^{\kappa_n'+1}(t_n)\end{bmatrix}\mathfrak{t}^{\kappa'}, \tag{6.99}$$

where $dt=dt_1\wedge\cdots\wedge dt_n$.

PROOF. Clearly, it is enough to prove that

$$\operatorname{Res}_{\mathbb{K}_{\mathbb{J}}[X,Y]/\mathbb{K}}\begin{bmatrix} \bar{X}^\kappa\bar{Y}^\chi\, d\bar{X} \\ \phi(\bar{X})\end{bmatrix} = \operatorname{Res}_{\mathbb{K}[t]/\mathbb{K}}\begin{bmatrix} \begin{bmatrix} \bar{X}^\kappa\bar{Y}^\chi\, d\bar{X} \\ \bar{X}-t\end{bmatrix}_{\mathbb{X}} dt \\ \phi\end{bmatrix}. \tag{6.100}$$

It follows from (6.98) that

$$\begin{aligned}\operatorname{Res}_{\mathbb{K}[t]/\mathbb{K}}&\begin{bmatrix} \begin{bmatrix} \bar{X}^\kappa\bar{Y}^\chi\, d\bar{X} \\ \bar{X}-t\end{bmatrix}_{\mathbb{X}} dt \\ \phi\end{bmatrix} \\ &= \operatorname{Res}_{\mathbb{K}[t]/\mathbb{K}}\begin{bmatrix} \operatorname{Res}_{\mathbb{K}_{\mathbb{J}}[X,Y][t]/\mathbb{K}[t]}\begin{bmatrix} b_\phi(t,\bar{X})\,X^\kappa Y^\chi\, d\bar{X} \\ \phi(\bar{X})\end{bmatrix} \\ \phi\end{bmatrix}.\end{aligned}$$

Then, equality (6.100) follows immediately from Proposition 6.29. □

Proposition 6.44 provides us with an explicit description of the algebraic residue symbols defined in (6.67). Namely,

$$\operatorname{Res}_{\mathbb{K}_{\mathbb{J}}[X,Y]/\mathbb{K}}\begin{bmatrix} \varpi/\bar{p}_0^{\ell+1} \\ \bar{p}_1^{\kappa_1+1},\dots,\bar{p}_n^{\kappa_n+1}\end{bmatrix} = \operatorname{Res}_{\mathbb{K}_{\mathbb{J}}[X,Y]/\mathbb{K}}\begin{bmatrix} D_\ell^{|\kappa|}[\bar{p}_0]\,A_\kappa^{|\kappa|}[\bar{\phi}]\,\varpi \\ \phi_1^{|\kappa|+1}(\bar{X}_1),\dots,\phi_n^{|\kappa|+1}(\bar{X}_n)\end{bmatrix},$$

provided $\varpi=\bar{q}\,d\bar{X}_1\wedge\cdots\wedge d\bar{X}_n$.

PROPOSITION 6.45. *Let $p = (p_1, \ldots, p_n)$ be such that $\bar{p}$ defines a quasi-regular sequence in $\mathbb{K}_{\mathbb{J}}[X,Y]$ and let p_0 be such that $(\bar{p}_0, \ldots, \bar{p}_n)\mathbb{K}_{\mathbb{J}}[X,Y] = \mathbb{K}_{\mathbb{J}}[X,Y]$. For each $\ell \in \mathbb{N}$ and each $\kappa \in \mathbb{N}^n$, let $D_\ell^{|\kappa|}[\bar{p}_0]$ be the element in $\mathbb{K}_{\mathbb{J}}[X,Y]$ defined by (6.64) and (6.51). Let $\phi_1, \ldots, \phi_n$ be univariate polynomials in variables $X_1, \ldots, X_n$ respectively, with degrees $\nu_1, \ldots, \nu_n$ and leading coefficients γ_{j,ν_j}, as in (6.66). That is,*

$$\phi_j(X_j) - \sum_{k=1}^{n} a_{j,k}\, p_k \in \mathbb{J}$$

for $j = 1, \ldots, n$, with $a_{j,k}$ in $\mathbb{K}[X,Y]$. Let the $A_\kappa^{|\kappa|}[\bar{\phi}] \in \mathbb{K}_{\mathbb{J}}[X,Y]$ be defined as in (6.36) with $\bar{\phi}$ replacing g. Let $\varpi = \bar{q}\, d\bar{X}_1 \wedge \cdots \wedge d\bar{X}_n$. For $\ell \in \mathbb{N}$ and $\kappa \in \mathbb{N}^n$, let

$$\left[\frac{((\det A)\, A_\kappa^{[\kappa]}[\bar{\phi}]\, D_\ell^{|\kappa|}[\bar{p}_0]\, q)(\bar{X},\bar{Y})\, d\bar{X}}{\bar{X} - t}\right]_{\mathbb{X}} = \sum_{\kappa' \in \mathbb{N}^n} q_{\kappa'}^{[\ell,\kappa]}(t)\, \phi^{\kappa'}(t), \tag{6.101}$$

where

$$\deg_t \Big(q_{\kappa'}^{[\ell,\kappa]} - \gamma^{[\ell,\kappa]}(\kappa') \prod_{j=1}^{n} t_j^{\nu_j - 1}\Big) < |\nu| - n.$$

Then

$$\operatorname{Res}_{\mathbb{K}_{\mathbb{J}}[X,Y]/\mathbb{K}} \begin{bmatrix} \varpi / \bar{p}_0^{\ell+1} \\ \bar{p}_1^{\kappa_1+1}, \ldots, \bar{p}_n^{\kappa_n+1} \end{bmatrix} = \frac{\gamma^{[\ell,\kappa]}(|\kappa|+1, \ldots, |\kappa|+1)}{\gamma_\phi}, \tag{6.102}$$

where γ_ϕ denotes the product of the leading coefficients of the polynomials ϕ_j.

REMARK 6.46. One can assume that the ϕ_j's are monic, but we keep them here as nonmonic since we focus later on questions where the field $\mathbb{K}$ is replaced by an integral domain; see §6.6. An archetypal example of such a situation is the case when one considers $\mathbb{K} = \mathbb{Z}$ instead of its fraction field $\mathbb{Q}$.

PROOF. The residue symbol on the left-hand side of (6.102) transforms as

$$\operatorname{Res}_{\mathbb{K}_{\mathbb{J}}[X,Y]/\mathbb{K}} \begin{bmatrix} D_\ell^{|\kappa|}[\bar{p}_0]\, A_\kappa^{|\kappa|}[\bar{\phi}]\, \varpi \\ \phi_1^{|\kappa|+1}(\bar{X}_1), \ldots, \phi_n^{|\kappa|+1}(\bar{X}_n) \end{bmatrix}$$

according to Theorem 6.24; see §6.3.2. The development (6.101) is unique and is obtained through the iterated Euclid division algorithm by $\phi_j(t_j)$ in each variable t_j. Then the result follows from Proposition 6.44 combined with Proposition 6.29. □

Proposition 6.44 also implies the following result, which generalizes Corollary 6.31, valid when $\mathbb{A} = \mathbb{K}[X]$, to the setting when $\mathbb{A} = \mathbb{K}_{\mathbb{J}}[X,Y]$. For any $r \in \mathbb{N}^*$, let $\mathbb{J}_r$ be the purely n-dimensional radical ideal defined as the radical of the ideal generated by $J_\iota(X^r, Y^r)$, where $\mathbb{J} = \sum_\iota J_\iota \mathbb{K}[X,Y]$. Let $\mathbb{X}_r = \operatorname{Spec} \mathbb{K}_{\mathbb{J}_r}$.

PROPOSITION 6.47. *Let $p = (p_1, \ldots, p_n)$ be a sequence of elements in $\mathbb{K}[X,Y]$ such that $(\bar{p}_1, \ldots, \bar{p}_n)$ is quasi-regular in $\mathbb{A} = \mathbb{K}_{\mathbb{J}}[X,Y]$ and $\mathbb{I}_{\bar{p}} \neq \mathbb{A}$. Let $r \in \mathbb{N}^*$ be such that r is not a multiple of the characteristic of $\mathbb{K}$. Then, for any $q \in \mathbb{K}[X,Y]$*

$$r^{N-n} \operatorname{Res}_{\mathbb{K}_{\mathbb{J}}[X,Y]/\mathbb{K}} \begin{bmatrix} \bar{q}\, d\bar{X} \\ \bar{p}_1, \ldots, \bar{p}_n \end{bmatrix} = \operatorname{Res}_{\mathbb{K}_{\mathbb{J}_r}[X,Y]/\mathbb{K}} \begin{bmatrix} \bar{q}^{[r]}\, (\bar{X}_1 \cdots \bar{X}_n)^{r-1}\, d\bar{X} \\ \bar{p}_1^{[r]}, \ldots, \bar{p}_n^{[r]} \end{bmatrix}, \tag{6.103}$$

where $q^{[r]} = q(X^r, Y^r)$, $p_j^{[r]} = p_j(X^r, Y^r)$ for $j = 1, \ldots, n$, and $X^r = (X_1^r, \ldots, X_n^r)$, $Y^r = (Y_1^r, \ldots, Y_{N-n}^r)$.

PROOF. Let $\phi_1,\dots,\phi_n$ be the univariate polynomials in variables $X_1,\dots,X_r$, respectively, having correspondingly positive degrees $\nu_1,\dots,\nu_n$ and such that for any $j=1,\dots,n$,

$$\phi_j - \sum_{k=1}^{n} a_{j,k}\, p_k \in \mathbb{J}$$

for some $a_{j,k}\in\mathbb{K}[X,Y]$. Therefore, if $\phi_j^{[r]}=\phi_j(X_j^r)$ and $a_{j,k}^{[r]}=a_{j,k}(X^r,Y^r)$, then one has

$$\phi_j^{[r]} - \sum_{k=1}^{n} a_{j,k}^{[r]}\, p_k^{[r]} \in \mathbb{J}_r.$$

It follows then from Theorem 6.24 that

$$(6.104)\qquad \operatorname{Res}_{\mathbb{K}_{\mathbb{J}_r}[X,Y]/\mathbb{K}}\begin{bmatrix}\bar{q}^{[r]}\,(\bar{X}_1\cdots\bar{X}_n)^{r-1}\,d\bar{X}\\ \bar{p}_1^{[r]},\dots,\bar{p}_n^{[r]}\end{bmatrix} = \operatorname{Res}_{\mathbb{K}_{\mathbb{J}_r}[X,Y]/\mathbb{K}}\begin{bmatrix}\bar{U}^{[r]}\,(\bar{X}_1\cdots\bar{X}_n)^{r-1}\,d\bar{X}\\ \bar{\phi}_1^{[r]},\dots,\bar{\phi}_n^{[r]}\end{bmatrix},$$

where $U=(\det A)\,q$. One has

$$\begin{bmatrix}\bar{U}^{[r]}(\bar{X}_1\cdots\bar{X}_n)^{r-1}\,d\bar{X}\\ \bar{X}-t\end{bmatrix}_{\mathbb{X}_r} = (t_1\cdots t_n)^{r-1}\sum_{\{\kappa\in\mathbb{N}^n\,;\ |\kappa|\le d(U,r)\}}\gamma_\kappa\, t^\kappa.$$

It follows from (6.104) and Proposition 6.44 that

$$(6.105)\qquad \operatorname{Res}_{\mathbb{K}_{\mathbb{J}_r}[X,Y]/\mathbb{K}}\begin{bmatrix}\bar{U}^{[r]}\,(\bar{X}_1\cdots\bar{X}_n)^{r-1}\,d\bar{X}\\ \bar{p}_1^{[r]},\dots,\bar{p}_n^{[r]}\end{bmatrix} = \sum_{\{\kappa\in\mathbb{N}^n\,;\ |\kappa|\le d(U,r)\}}\gamma_\kappa \operatorname{Res}_{\mathbb{K}[t]/\mathbb{K}}\begin{bmatrix}(t_1\cdots t_n)^{r-1}\,t^\kappa\,dt\\ \phi_1(t_1^r),\dots,\phi_n(t_n^r)\end{bmatrix}.$$

The same argument as in the proof of Corollary 6.31 shows that

$$(6.106)\qquad \sum_{\{\kappa\in\mathbb{N}^n\,;\ |\kappa|\le d(U,r)\}}\gamma_\kappa \operatorname{Res}_{\mathbb{K}[t]/\mathbb{K}}\begin{bmatrix}(t_1\cdots t_n)^{r-1}\,t^\kappa\,dt\\ \phi_1(t_1^r),\dots,\phi_n(t_n^r)\end{bmatrix} = \sum_{\{\kappa'\in\mathbb{N}^n\,;\ r|\kappa'|\le d(U,r)\}}\gamma_{r\kappa'} \operatorname{Res}_{\mathbb{K}[t]/\mathbb{K}}\begin{bmatrix}t^{\kappa'}\,dt\\ \phi_1(t_1),\dots,\phi_n(t_n)\end{bmatrix}.$$

Furthermore, for any $\kappa'\in\mathbb{N}^n$ one has that

$$\gamma_{r\kappa'} = \operatorname{Res}_{\mathbb{K}_{\mathbb{J}_r}[X,Y]/\mathbb{K}}\begin{bmatrix}\bar{U}^{[r]}\,d\bar{X}\\ \bar{X}_1^{r\kappa_1'+1},\dots,\bar{X}_n^{r\kappa_n'+1}\end{bmatrix} = \operatorname{Res}_{\mathbb{K}_{\mathbb{J}_r}[X,Y]/\mathbb{K}}\begin{bmatrix}\bar{U}^{[r]}\,\bar{X}_1^{r-1}\,d\bar{X}_1\wedge\cdots\wedge\bar{X}_n^{r-1}\,d\bar{X}_n\\ \bar{X}_1^{r(\kappa_1'+1)},\dots,\bar{X}_n^{r(\kappa_n'+1)}\end{bmatrix}.$$

Therefore

$$
\begin{aligned}
r^n \sum_{\kappa' \in \mathbb{N}^n} \gamma_{r\kappa'} t^{\kappa'} &= \operatorname{Res}_{\mathbb{K}_{\mathbb{J}_r}[X,Y][t]/\mathbb{K}[t]} \begin{bmatrix} U(\bar{X}^r, \bar{Y}^r)\, d\bar{X}_1^r \wedge \cdots \wedge d\bar{X}_n^r \\ \bar{X}_1^r - t_1, \dots, \bar{X}_n^r - t_n \end{bmatrix} \\
&= r^n \operatorname{Res}_{\mathbb{K}_{\mathbb{J}_r^Y}[X,Y][t]/\mathbb{K}[t]} \begin{bmatrix} U(\bar{X}, \bar{Y}^r)\, d\bar{X}_1 \wedge \cdots \wedge d\bar{X}_n \\ \bar{X}_1 - t_1, \dots, \bar{X}_n - t_n \end{bmatrix} \\
&= r^N \operatorname{Res}_{\mathbb{K}_{\mathbb{J}_r}[X,Y][t]/\mathbb{K}[t]} \begin{bmatrix} U(\bar{X}, \bar{Y})\, d\bar{X}_1 \wedge \cdots \wedge d\bar{X}_n \\ \bar{X}_1 - t_1, \dots, \bar{X}_n - t_n \end{bmatrix},
\end{aligned} \tag{6.107}
$$

where $\mathbb{J}_r^Y = \sum_\iota J_\iota(X, Y^r)\, \mathbb{K}[X,Y]$, provided that J_ι generate $\mathbb{J}$, as a consequence of [**Lip87**, Corollary 3.7]. When the characteristic of the field char$\,(\mathbb{K})$ and r are coprime, then $r^n = 1 + \cdots + 1$ (r^n times) is nonzero. Thus, one can divide both sides of (6.107) by r^n. Then, it follows from Proposition 6.44 that the right-hand side of (6.106) equals

$$
r^{N-n} \operatorname{Res}_{\mathbb{K}_{\mathbb{J}}[X,Y]/\mathbb{K}} \begin{bmatrix} U(\bar{X}, \bar{Y})\, d\bar{X} \wedge \cdots \wedge d\bar{X}_n \\ \phi(\bar{X}_1), \dots, \phi_n(\bar{X}_n) \end{bmatrix}.
$$

That is, the right-hand side of (6.106) equals r^{N-n} times the residue symbol in the left-hand side of (6.103) if one invokes once more Theorem 6.24. □

Let $p_1, \dots, p_n, q \in \mathbb{K}[X,Y]$. We now come to the definition of a trace

$$
t \longmapsto \begin{bmatrix} \bar{q}\, d\bar{X}_1 \wedge \cdots \wedge d\bar{X}_n \\ p_1 - t_1, \dots, p_n - t_n \end{bmatrix}_{\mathbb{X}} = \begin{bmatrix} \bar{q}\, d\bar{X} \\ p - t \end{bmatrix}_{\mathbb{X}}
$$

under the sole hypothesis that $\bar{p} = (\bar{p}_1, \dots, \bar{p}_n)$ defines a quasi-regular sequence in $\mathbb{A} = \mathbb{K}_{\mathbb{J}}[X,Y]$ such that $\mathbb{I}_{\bar{p}} \neq \mathbb{K}_{\mathbb{J}}[X,Y]$. Since $\mathbb{I}_{\bar{p}} \neq \mathbb{K}_{\mathbb{J}}[X,Y]$, the polynomials $p_1, \dots, p_n$ are $\mathbb{K}$-algebraically independent modulo $\mathbb{J}$. Thus, it follows from the generalized version of the Perron theorem as formulated in [**Jel05**, Theorem 3.3] that for each $j = 1, \dots, n$ there exists a polynomial $E_j \in \mathbb{K}[t_1, \dots, t_n, \tau]$ such that (6.108)

$$
E_j(p_1, \dots, p_n, X_j) \in \mathbb{J} \quad \text{and} \quad \deg\left(E_j(t_1^{\deg p_1}, \dots, t_n^{\deg p_n}, t)\right) \leq \deg \mathbb{X} \prod_{j=1}^n \deg p_j.
$$

Moreover, $\deg_\tau E_j \geq 1$ and $E_j(t, 0) \not\equiv 0$. Let us rewrite (6.108) as

$$
E_j(t, X_j) - \sum_{k=1}^n a_{j,k}(t, X, Y)\, (p_j - t_j) \in \mathbb{J},
$$

with $\max_{j,k} \deg_{t,X,Y}((p_j - t_j)a_{j,k}) \leq \deg \mathbb{X} \prod_{j=1}^n \deg p_j$. This implies that the sequence $(\bar{p}_1 - t_1, \dots, \bar{p}_n - t_n)$ is quasi-regular in $\mathbb{K}(t)[X,Y]/\mathbb{J}\,\mathbb{K}(t)[X,Y]$, which we will denote by $(\mathbb{K}(t))_{\mathbb{J}}[X,Y]$. Observe that the quotient

$$
\frac{(\mathbb{K}(t))_{\mathbb{J}}[X,Y]}{(\bar{p} - t)(\mathbb{K}(t))_{\mathbb{J}}[X,Y]} = \frac{\mathbb{K}(t)[X,Y]}{(p - t, \mathbb{J})\, \mathbb{K}(t)[X,Y]}
$$

is a $\mathbb{K}(t)$-finitely dimensional vector space. Let also $p_0 \in \mathbb{K}[X,Y]$ be such that $(\bar{p}_0, \dots, \bar{p}_n)\,\mathbb{K}_{\mathbb{J}}[X,Y] = \mathbb{K}_{\mathbb{J}}[X,Y]$ and let $D_{\mathbb{X}}[p_0] \in \mathbb{K}(t)[X,Y]$ be such that

$$(6.109)\quad \begin{aligned} &1 - D_{\mathbb{X}}[p_0](t,\cdot)\,p_0 \in \sum_{j=1}^{n}(p_j - t_j)\,\mathbb{K}(t)[X,Y] + \mathbb{J}\,\mathbb{K}(t)[X,Y],\\ &\deg_{X,Y}(D_{\mathbb{X}}[p_0](t,\cdot)\,p_0) \leq \deg \mathbb{X} \prod_{j=0}^{n} d_j; \end{aligned}$$

see [**Jel05**, Theorem 1.1]. Then Theorem 6.24, combined with Proposition 6.44, implies the following result whose proof follows along the same lines as that of Proposition 6.34.

PROPOSITION 6.48. *For any* $q \in \mathbb{K}[X,Y]$,

$$(6.110)\quad \begin{bmatrix} \bar{q}\,d\bar{X}/\bar{p}_0 \\ \bar{p} - t \end{bmatrix}_{\mathbb{X}} := \operatorname{Res}_{(\mathbb{K}(t))_{\mathbb{J}}[X,Y]/\mathbb{K}(t)} \begin{bmatrix} \big((\det A)\,D_{\mathbb{X}}[p_0]\,q\big)(t,\bar{X},\bar{Y})\,d\bar{X}_1 \wedge \cdots \wedge d\bar{X}_n \\ \bar{p}_1 - t_1, \dots, \bar{p}_n - t_n \end{bmatrix}$$

is an element of $\mathbb{K}(t)$, *which does not depend on the choice of* $D_{\mathbb{X}}[p_0]$.

Moreover, in case the dominant coefficient of each E_j, as a polynomial in τ with coefficients in $\mathbb{K}[t]$, can be taken equal to 1, which is equivalent to saying that $\bar{p}$ defines a proper morphism from $\mathbb{X}$ to $\mathbb{K}^n$, then one has that for any $q \in \mathbb{K}[X,Y]$

$$(6.111)\quad \begin{bmatrix} \bar{q}\,d\bar{X} \\ \bar{p} - t \end{bmatrix}_{\mathbb{X}} \in \mathbb{K}[t].$$

We will determine precisely in §6.5.2 (see Theorem 6.56) the degree of the polynomial (6.111) in t, for a given q is given, in the particular case where $\mathbb{K}$ is isomorphic to a subfield of $\mathbb{C}$, hence has characteristic zero. It will be necessary to introduce, as we will do in Definition 6.53, a notion of δ-properness that extends Definition 6.38 naturally.

6.5. The characteristic zero case

We assume in this section that $\mathbb{K}$ denotes a subfield of $\mathbb{C}$, whose algebraic closure $\overline{\mathbb{K}}$ is within the field of complex numbers.[32] Since we work within the frame of complex analytic geometry, we are now in a situation to profit extensively from the involution $z \mapsto \overline{z}$ in the complex setting.[33] The following observation is then essential, namely that antiholomorphic coordinates in $\mathbb{C}^N$ behave as constant functions do under the action of holomorphic differential operators since any such operator annihilates them. In this section, we place ourselves in the setting of §6.3.2 and §6.4.2. Let then $X = (X_1, \dots, X_n)$, $Y = (Y_1, \dots, Y_{N-n})$ with $n, N \in \mathbb{N}^*$ be such that $N \geq n$. Let $\mathbb{J}$ be a purely n-dimensional radical ideal in $\mathbb{K}[X,Y]$, $\mathbb{X} = \operatorname{Spec}(\mathbb{K}_{\mathbb{J}}[X,Y])$ be the corresponding algebraic variety defined over $\mathbb{K} \subset \mathbb{C}$, and let $\mathbb{X}_{\mathbb{C}}$ be its complex analytification.[34] Generators of $\mathbb{J}$ in $\mathbb{K}[X,Y]$, as well as

[32]Note that any field with characteristic zero and cardinality at most that of the continuum is (ring) isomorphic to a subfield of $\mathbb{C}$. So the results in this section cover the case where $\mathbb{K}$ has characteristic zero and cardinality at most that of the continuum.

[33]One could suggest that the Frobenius operator $\gamma \mapsto \gamma^{\mathrm{p}}$ is a substitute candidate for the complex conjugation in the characteristic $\mathrm{p} > 0$ case, where p is a prime number, but this is indeed very far from being understood for the time being.

[34]One needs to be careful to distinguish the notation $\overline{z}$ from the notation used in §6.3.2 and §6.4.2 for the class $\bar{q}$ of $q \in \mathbb{K}[X,Y]$ modulo the n-dimensional radical ideal $\mathbb{J}$.

any element of $\mathbb{K}[X,Y]$, can be as well considered as elements of $\mathbb{C}[X,Y]$. We will denote as $(z_1,\dots,z_n)$ and $(w_1,\dots,w_{N-n})$ the coordinates in the N-dimensional affine space $\mathbb{C}^N$ and we will also denote as $V(\mathbb{J})$ the closed analytic subset with pure codimension $M=N-n$ of $\mathbb{C}^N$ corresponding to $\mathbb{X}_{\mathbb{C}} \hookrightarrow \mathbb{C}^N$. We assume

$$\#\{(z,w)\in V(\mathbb{J}) \,:\, z_1=\cdots=z_n=0\} = \deg \mathbb{X}_{\mathbb{C}} = \deg \mathbb{X}, \tag{6.112}$$

which corresponds to the geometric assumption (6.68) already introduced at the end of §6.3.2 and all through §6.4.2.

6.5.1. Algebraic residue symbols realized as currents. Let $(p_1,\dots,p_n)$ be a sequence of elements in $\mathbb{K}[X,Y]$ which classes $\bar{p}_j$ modulo $\mathbb{J}$ define a quasi-regular sequence $\bar{p}=(\bar{p}_1,\dots,\bar{p}_n)$ in $\mathbb{A}=\mathbb{K}_{\mathbb{J}}[X,Y]$ such that $\mathbb{I}_{\bar{p}} \neq \mathbb{A}$. Let also $p_0\in\mathbb{K}[X,Y]$ such that $(\bar{p}_0,\dots,\bar{p}_n)\,\mathbb{A}=\mathbb{A}$. Let $a_0\in\mathbb{K}[X,Y]$ be such that

$$1-a_0p_0 \in \sum_{j=1}^n p_j\,\mathbb{K}[X,Y]+\mathbb{J}, \quad \deg(a_0p_0)\le \deg\mathbb{X}\,d_0\,D, \tag{6.113}$$

where $d_0=\deg p_0$ and $D=\prod_1^n \deg p_j$; see [**Jel05**, Theorem 1.1]. For any $\ell\in\mathbb{N}$ and $\kappa\in\mathbb{N}^n$, let $D_\ell^{|\kappa|}[p_0]$ be the polynomial in $\mathbb{K}[X,Y]$ defined by

$$\big(1-(a_0p_0)^{\ell+1}\big)^{|\kappa|+1} = 1-p_0^{\ell+1}D_\ell^{|\kappa|}[p_0]$$

and $D_\ell^{|\kappa|}[\bar{p}_0]\in\mathbb{K}_{\mathbb{J}}[X,Y]$ its class modulo $\mathbb{J}$. Let also $\boldsymbol{\sigma}$ be the minimal inverse of

$$(z,w)\in\mathbb{C}^N \longmapsto \sum_{j=1}^n p_j(z,w)\,\epsilon_j$$

for the canonical hermitian metric on $\mathbb{C}^N$, namely

$$\boldsymbol{\sigma} : (z,w)\in\mathbb{C}^N\setminus V(p)\longmapsto \sum_{j=1}^n \frac{\overline{p_j(z,w)}}{\|p(z,w)\|^2}\,\epsilon_j^* = \big(\sigma_1(z,w),\dots,\sigma_n(z,w)\big). \tag{6.114}$$

Algebraic residue symbols

$$\mathrm{Res}_{\mathbb{K}_{\mathbb{J}}[X,Y]/\mathbb{K}}\begin{bmatrix} \varpi/\bar{p}_0^{\ell+1} \\ \bar{p}_1^{\kappa_1+1},\dots,\bar{p}_n^{\kappa_n+1}\end{bmatrix} := \mathrm{Res}_{\mathbb{K}_{\mathbb{J}}[X,Y]/\mathbb{K}}\begin{bmatrix} D_\ell^{|\kappa|}[\bar{p}_0]\,\varpi \\ \bar{p}_1^{\kappa_1+1},\dots,\bar{p}_n^{\kappa_n+1}\end{bmatrix},$$

where $\varpi=\bar{q}\,d\bar{X}_1\wedge\cdots\wedge d\bar{X}_n=\bar{q}\,d\bar{X}$ such as those introduced in Definition 6.32 are realized in case where $\mathbb{K}\subset\mathbb{C}$ in terms of the action of Coleff–Herrera currents or, which turns out to be more efficient when $\bar{p}$ induces on analytifications a proper polynomial morphism from $\mathbb{X}_{\mathbb{C}}$ to $\mathbb{C}^n$, of residue currents of the Bochner–Martinelli type. We formulate the fundamental result of this subsection.

THEOREM 6.49. *Let $\mathbb{J}$, $\mathbb{X}$, $p=(p_1,\dots,p_n)$, $\boldsymbol{\sigma}$, $\sigma_1,\dots,\sigma_n$, p_0, be as above. For any $q\in\mathbb{K}[X,Y]$, $\ell\in\mathbb{N}$, and $\kappa\in\mathbb{N}^n$,*

(6.115)
$$\mathrm{Res}_{\mathbb{K}_{\mathbb{J}}[X,Y]/\mathbb{K}}\begin{bmatrix} \bar{q}\,d\bar{X}/\bar{p}_0^{\ell+1} \\ \bar{p}_1^{\kappa_1+1},\dots,\bar{p}_n^{\kappa_n+1}\end{bmatrix} = \Big\langle \bar{\partial}\Big(\frac{1}{p_1^{\kappa_1+1},\dots,p_n^{\kappa_n+1}}\Big)_{V(\mathbb{J})},\,(D_\ell^{|\kappa|}[p_0]q)\,d\zeta\Big\rangle$$
$$= \binom{n-1+|\kappa|}{n-1,\kappa}\Big\langle \sigma^\kappa\,\frac{\boldsymbol{\sigma}}{2i\pi}\Big(\bar{\partial}\,\frac{\boldsymbol{\sigma}}{2i\pi}\Big)^{n-1}([V(\mathbb{J})])\,,\,(D_\ell^{|\kappa|}[p_0]q)\,(\bar{\partial}\chi\wedge d\zeta)\bigwedge_{j=1}^n\epsilon_j\Big\rangle.$$

where $d\zeta = d\zeta_1 \wedge \cdots \wedge d\zeta_n$, $\sigma^\kappa = \prod_{j=1}^n \sigma_j^{\kappa_j}$, *and* $\chi \in \mathcal{D}(\mathbb{C}^N, \mathbb{C})$ *is identically equal to* 1 *in a neighborhood of* $V(\mathbb{J}) \cap V(p)$.

PROOF. Let $R > 0$ such that $\operatorname{Supp}\chi \subset R\,\mathbb{B}_N$, where $\mathbb{B}_N$ is the unit euclidean ball in $\mathbb{C}^N$. Since $\|p\| > 0$ on $V(\mathbb{J}) \cap \partial(R\,\mathbb{B}_N)$, there exists $\eta > 0$ such that for any $t \in \eta\,\mathbb{B}_n$,

$$V(\mathbb{J}) \cap V(p-t) \cap B(0,R) \subset \text{interior}(\operatorname{Supp}\chi).$$

Moreover, it follows from Stokes's theorem and Proposition 3.18 that for any holomorphic function h in $R\,\mathbb{B}_N$, for any $t \in \eta\,\mathbb{B}_N$ with $0 < \eta \ll 1$,

$$\Big\langle \overline{\partial}\Big(\frac{1}{p-t}\Big)_{V(\mathbb{J})}, \chi\, h\, d\zeta \Big\rangle = \Big\langle \frac{\boldsymbol{\sigma}_t}{2i\pi}\Big(\overline{\partial}\,\frac{\boldsymbol{\sigma}_t}{2i\pi}\Big)^{n-1}([V(\mathbb{J})])\,,\, \overline{\partial}\chi \wedge h d\zeta \bigwedge_{j=1}^n \epsilon_j \Big\rangle, \tag{6.116}$$

where

$$\boldsymbol{\sigma}_t : (z,w) \in \mathbb{C}^N \setminus V(p-t) \longmapsto \sum_{j=1}^n \frac{\overline{p_j(z,w) - t_j}}{\|p(z,w) - t\|^2}\,\epsilon_j^* = \big(\sigma_{t,1}(z,w), \dots, \sigma_{t,n}(z,w)\big).$$

Let also, for (z,w) in a neighborhood of $\operatorname{Supp}(\overline{\partial}\chi) \cap V(\mathbb{J})$,

$$\begin{aligned} \check{\boldsymbol{\sigma}}_t(z,w) &= \sum_{j=1}^n \frac{\overline{p_j(z,w)}}{\langle\!\langle \overline{p(z,w)}, p(z,w) - t \rangle\!\rangle}\,\epsilon_j^* = \big(\check{\sigma}_{t,1}(z,w), \dots, \check{\sigma}_{t,n}(z,w)\big) \\ &= \Big(\sum_{k=0}^\infty \langle\!\langle \frac{\overline{p(z,w)}}{\|p(z,w)\|^2}, t \rangle\!\rangle^k\Big) \sum_{j=1}^n \frac{\overline{p_j(z,w)}}{\|p(z,w)\|^2}\,\epsilon_j^*. \end{aligned} \tag{6.117}$$

Since

$$\boldsymbol{\sigma}_t(z,w)\Big(\sum_{j=1}^n (p_j(z,w) - t)\,\epsilon_j\Big) = \check{\boldsymbol{\sigma}}_t(z,w)\Big(\sum_{j=1}^n (p_j(z,w) - t)\,\epsilon_j\Big) \equiv 1$$

for (z,w) in a neighborhood of $\operatorname{Supp}(\overline{\partial}\chi) \cap V(\mathbb{J})$ and $t \in \eta\,\mathbb{B}_n$, a classical cohomological deformation argument based on Stokes's formula on the complex analytic space $V(\mathbb{J}) \hookrightarrow \mathbb{C}^N$ ensures that one also has, as well as (6.116), that

$$\Big\langle \overline{\partial}\Big(\frac{1}{p-t}\Big)_{V(\mathbb{J})}, \chi\, h\, d\zeta \Big\rangle = \Big\langle \frac{\check{\boldsymbol{\sigma}}_t}{2i\pi}\Big(\overline{\partial}\,\frac{\check{\boldsymbol{\sigma}}_t}{2i\pi}\Big)^{n-1}([V(\mathbb{J})])\,,\, \overline{\partial}\chi \wedge h d\zeta \bigwedge_{j=1}^n \epsilon_j \Big\rangle. \tag{6.118}$$

After some easy combinatorial computations, one has for $t \in \eta\,\mathbb{B}_n$ that

$$\begin{aligned} &\Big\langle \overline{\partial}\Big(\frac{1}{p-t}\Big)_{V(\mathbb{J})}, \chi\, h\, d\zeta \Big\rangle \\ &\quad = \sum_{\kappa \in \mathbb{N}^n} \binom{n-1+|\kappa|}{n-1, \kappa} \\ &\qquad \times \Big\langle \sigma^\kappa \frac{\boldsymbol{\sigma}}{2i\pi}\Big(\overline{\partial}\,\frac{\boldsymbol{\sigma}}{2i\pi}\Big)^{n-1}([V(\mathbb{J})])\,,\, \overline{\partial}\chi \wedge h\, d\zeta \bigwedge_{j=1}^n \epsilon_j \Big\rangle\, t^\kappa \\ &\quad = \sum_{\kappa \in \mathbb{N}^n} a_{\chi,\kappa}[h\, d\zeta]\, t^\kappa \end{aligned} \tag{6.119}$$

as a convergent power series in t in $\mathcal{O}_{\mathbb{C}^n}(\eta\,\mathbb{B}_n)$. Let $\boldsymbol{V} = V(\mathbb{J}) \times \mathbb{C}^n \subset \mathbb{C}^{\mathbb{N}}_{z,w} \times \mathbb{C}^n_t$ and $\theta \in \mathcal{D}(\mathbb{C}^n_t, \mathbb{C})$ identically equal to one in a neighborhood of $t = 0$. It follows from the geometric transformation law (see Theorem 3.21) together with Theorem 2.75 that for each $\kappa \in \mathbb{N}^n$,

$$
\begin{aligned}
a_{\chi,\kappa}[h\,d\zeta] &= \Big\langle \overline{\partial}\Big(\frac{1}{p-t, t_1^{\kappa_1+1}, \dots, t_n^{\kappa_n+1}}\Big)_{\mathbf{V}}, \, h\,\chi\,\theta\, d\zeta \wedge dt\Big\rangle \\
&= \Big\langle \overline{\partial}\Big(\frac{1}{p^{\kappa_1+1}, \dots, p_n^{\kappa_n+1}, t_1^{\kappa_1+1}, \dots, t_n^{\kappa_n+1}}\Big)_{\mathbf{V}}, \, h\,t^\kappa\,\chi\,\theta\, d\zeta \wedge dt\Big\rangle \qquad (6.120)\\
&= \Big\langle \overline{\partial}\Big(\frac{1}{p^{\kappa_1+1}, \dots, p_n^{\kappa_n+1}}\Big)_{\mathbf{V}}, \, h\, d\zeta \wedge dt\Big\rangle.
\end{aligned}
$$

The second equality in (6.115) then holds for any κ, with $h = (D_\ell^{|\kappa|}[p_0]q)_{|R\mathbb{B}_N}$. To prove the first equality, let us introduce univariate polynomials $\phi_1, \dots, \phi_n$ with coefficients in $\mathbb{K}$ such that

$$
\phi_j(X_j) - \sum_{k=1}^{n} a_{j,k}\, p_k \in \mathbb{J}, \quad a_{j,k} \in \mathbb{K}[X,Y], \ \deg(a_{j,k}p_k) \le (\deg \mathbb{X})\, D;
$$

see [**Jel05**, Theorem 1.6]. Let us start with

$$
\mathrm{Res}_{\mathbb{K}_{\mathbb{J}}[X,Y]/\mathbb{K}}\left[\frac{\bar{q}\, d\bar{X}/\bar{p}_0^{\ell+1}}{\bar{p}_1^{\kappa_1+1}, \dots, \bar{p}_n^{\kappa_n+1}}\right] = \mathrm{Res}_{\mathbb{K}_{\mathbb{J}}[X,Y]/\mathbb{K}}\left[\frac{\varpi_{\ell,\kappa}}{\bar{p}_1^{\kappa_1+1}, \dots, \bar{p}_n^{\kappa_n+1}}\right],
$$

where $\varpi_{\ell,\kappa} = \bar{q}\, D_\ell^{|\kappa|}[\bar{p}_0]\, d\bar{X}$; see Definition 6.32. It follows from Theorem 6.24 and Remark 6.25 (see (6.46)) that

$$
\begin{aligned}
&\kappa!\, \mathrm{Res}_{\mathbb{K}_{\mathbb{J}}[X,Y]/\mathbb{K}}\left[\frac{\varpi_{\ell,\kappa}}{\bar{p}_1^{\kappa_1+1}, \dots, \bar{p}_n^{\kappa_n+1}}\right] \qquad (6.121)\\
&\quad = \sum_{\{\nu \in \mathbb{N}^n \,:\, |\nu| = |\kappa|\}} \nu!\, \mathrm{Res}_{\mathbb{K}_{\mathbb{J}}[X,Y]/\mathbb{K}}\left[\frac{(\det \bar{A})\, A_\nu^\kappa\, \varpi_{\ell,\kappa}}{\phi_1^{\nu_1+1}(\bar{X}_1), \dots, \phi_n^{\nu_n+1}(\bar{X}_n)}\right],
\end{aligned}
$$

where the A_ν^κ are the classes modulo $\mathbb{J}$ of the polynomials A_ν^κ defined as in (6.47). Let us prove first that one also has in parallel that

$$
\begin{aligned}
&\kappa!\, \Big\langle \overline{\partial}\Big(\frac{1}{p_1^{\kappa_1+1}, \dots, p_n^{\kappa_n+1}}\Big)_{V(\mathbb{J})}, \, \omega_{\ell,\kappa}\Big\rangle \qquad (6.122)\\
&\quad = \sum_{\{\nu \in \mathbb{N}^n \,:\, |\nu| = |\kappa|\}} \nu!\, \Big\langle \overline{\partial}\Big(\frac{1}{\phi_1^{\nu_1+1}(\zeta_1), \dots, \phi_n^{\nu_n+1}(\zeta_n)}\Big)_{V(\mathbb{J})}, \, (\det A)\, A_\nu^\kappa\, \omega_{\ell,\kappa}\Big\rangle,
\end{aligned}
$$

where $\omega_{\ell,\kappa}(z,w) = q D_\ell^{|\kappa|}[p_0](z,w)\, dz$. In order to prove (6.121), we introduce

$$
\begin{aligned}
&\mathbf{\Sigma} : (z,w) \in \mathbb{C}^N \setminus V(\phi) \qquad (6.123)\\
&\quad \longmapsto \sum_{j=1}^{n} \frac{\sum_{k=1}^{n} a_{j,k}(z,w)\, \overline{\phi_k(z_k)}}{\|\phi(z)\|^2}\, \epsilon_j^* = (\Sigma_1(z,w), \dots, \Sigma_n(z,w)),
\end{aligned}
$$

so that, for any $(z,w) \in V(\mathbb{J}) \setminus \phi^{-1}(\{0\})$,

$$
\mathbf{\Sigma}(z,w)\Big(\sum_{j=1}^{n} p_j(z,w)\, \epsilon_j\Big) = \boldsymbol{\sigma}(z,w)\Big(\sum_{j=1}^{n} p_j(z,w)\, \epsilon_j\Big) \equiv 1. \qquad (6.124)
$$

Let χ_0 and χ_1 in $\mathcal{D}(\mathbb{C}^N, \mathbb{R})$ be both identically equal to one in a neighborhood of $V(\mathbb{J}) \cap \phi^{-1}(\{0\})$, such that $\mathrm{Supp}\,(\overline{\partial}\chi_0) \cap \mathrm{Supp}\,(\overline{\partial}\chi_1) = \emptyset$. For any $\psi \in \mathcal{D}(\mathbb{C}^N, \mathbb{C})$

which is identically equal to 1 in a neighborhood of Supp $(\overline{\partial}\chi_0)$ and identically equal to 0 in a neighborhood of Supp $(\overline{\partial}\chi_1)$, the $(n, n-1)$ form

$$\Theta = \Big(\prod_{j=1}^{n} (\psi \sigma_j + (1-\psi)\Sigma_j)^{\kappa_j}\Big) \frac{\psi \boldsymbol{\sigma} + (1-\psi)\,\boldsymbol{\Sigma}}{2i\pi} \Big(\overline{\partial}\Big(\frac{\psi\boldsymbol{\sigma} + (1-\psi)\boldsymbol{\Sigma}}{2i\pi}\Big)\Big)^{n-1}$$

induces a closed form on the open subset $V(\mathbb{J}) \setminus \{0\}$ of the n-dimensional complex analytic space $V(\mathbb{J}) \hookrightarrow \mathbb{C}^N$. Stokes's theorem on $V(\mathbb{J})$ therefore implies

$$\int_{V(\mathbb{J})} \overline{\partial}\chi_0 \wedge \Theta = \int_{V(\mathbb{J}} \overline{\partial}\chi_1 \wedge \Theta. \tag{6.125}$$

Since the left-hand side of (6.122) is independent on χ, if follows from the second equality in (6.115) and from (6.125) that

$$\begin{aligned}\kappa! \Big\langle \overline{\partial}\Big(\frac{1}{p_1^{\kappa_1+1}, \dots, p_n^{\kappa_n+1}}\Big)_{V(\mathbb{J})}, \omega_{\ell,\kappa}\Big\rangle \\ = \frac{(n-1+k)!}{(n-1)!} \Big\langle \Sigma^{\kappa} \frac{\boldsymbol{\Sigma}}{2i\pi} \Big(\overline{\partial}\,\frac{\boldsymbol{\Sigma}}{2i\pi}\Big)^{n-1} ([V(\mathbb{J})])\,, \overline{\partial}\chi \wedge \omega_{\ell,\kappa} \bigwedge_{j=1}^{n} \epsilon_j\Big\rangle.\end{aligned} \tag{6.126}$$

Since

$$\Sigma^{\kappa}(z,w) = \prod_{j=1}^{n} \Big(\sum_{j=1}^{n} a_{j,k}(z,w)\, \frac{\overline{\phi_j(z_j)}}{\|\phi(z)\|^2}\Big)^{\kappa_j} = \sum_{\{\nu \in \mathbb{N}^n\,:\,|\nu| = |\kappa|\}} A_\nu^{\kappa} \prod_{j=1}^{n} \Big(\frac{\overline{\phi_j(z_j)}}{\|\phi(z)\|^2}\Big)^{\nu_j}$$

according to the definition of the A_ν^κ (see (6.47)), equality (6.122) follows. Given the similarity of computation rules (6.121) and (6.122), proving the first equality in (6.115) amounts to proving that for any polynomials $u(X_1), \dots, u(X_n), v(X,Y)$ in $\mathbb{K}[X,Y]$,[35] one has

(6.127)

$$\mathrm{Res}_{\mathbb{K}_{\mathbb{J}}[X,Y]/\mathbb{K}} \begin{bmatrix} \bar{v}\, d\bar{X} \\ u_1(\bar{X}_1), \dots, u_n(\bar{X}_n)\end{bmatrix} = \Big\langle \overline{\partial}\Big(\frac{1}{u_1(\zeta_1), \dots, u_n(\zeta_n)}\Big)_{V(\mathbb{J})}, v(\zeta)\, d\zeta\Big\rangle.$$

Let $R = (R_1, \dots, R_n) \in (\mathbb{R}^+)^n$, such that the polydisc $\Delta = R_1\mathbb{D} \times \dots \times R_n\mathbb{D}$ of $\mathbb{C}^n$ contains $\{\alpha \in \mathbb{C}^n \,:\, u_1(\alpha_1) = \dots = u_n(\alpha_n) = 0\}$ and is such that the support of its Shilov boundary $\Gamma(\Delta)$ does not intersect the discriminant locus of the proper projection $\pi\,:\,(z,w) \in V(\mathbb{J}) \longrightarrow z \in \mathbb{C}^n$. It follows from Stokes's formula on $V(\mathbb{J})$, as in the proof of Proposition 6.8, that

$$\Big\langle \overline{\partial}\Big(\frac{1}{u_1(\zeta_1), \dots, u_n(\zeta_n)}\Big)_{V(\mathbb{J})}, v(\zeta)\, d\zeta\Big\rangle = \frac{1}{(2i\pi)^n} \int_{\pi^*(\Gamma(\Delta))} v \bigwedge_{j=1}^{n} \frac{d\zeta_j}{u_j(\zeta_j)}. \tag{6.128}$$

One has in $\mathbb{K}[t]$ (after specification of t on an element of $\mathbb{C}^n$) the equality

$$\begin{bmatrix} \bar{v}\, d\bar{X} \\ \bar{X} - t\end{bmatrix}_{\mathbb{X}} = \Big\langle \overline{\partial}\Big(\frac{1}{\zeta_1 - t_1, \dots, \zeta_n - t_n}\Big)_{V(\mathbb{J})}, v\, d\zeta\Big\rangle = \sum_{\xi \in \pi^{-1}(\{t\})} v(\xi).$$

[35] Let us say, for example, given $\nu \in \mathbb{N}^n$ such that $|\nu| = |\kappa|$,

$$u_j(X_j) = \phi_j^{\nu_j+1}(X_j) \text{ for } j = 1, \dots, n, \quad v = q A_\nu^\kappa D_\ell^{|\kappa|}[p_0](X,Y).$$

Then, one has for any $\kappa \in \mathbb{N}^n$ that

$$\Big\langle \overline{\partial}\Big(\frac{1}{\zeta_1^{\kappa_1+1},\dots,\zeta_n^{\kappa_n+1}}\Big)_{V(\mathbb{J})}, v\, d\zeta\Big\rangle = \mathrm{Res}_{\mathbb{K}_{\mathbb{J}}[X,Y]/\mathbb{K}}\left[\begin{matrix}\bar v\, d\bar X\\ \bar X_1^{\kappa_1+1},\dots,\bar X_n^{\kappa_n+1}\end{matrix}\right]. \tag{6.129}$$

Let

$$\frac{1}{\prod_{j=1}^n u_j(\zeta_j)} = \frac{1}{\prod_{j=1}^n \zeta_j^{\deg u_j}} \sum_{\kappa\in\mathbb{N}^n} c_\kappa \zeta^{-\kappa}$$

be the convergent Laurent development of $1/\prod_{j=1}^n u_j(\zeta_j)$ on the support of the cycle $\pi^*(\Gamma(\Delta))$. One has then the (finite, since v is a polynomial function) expansion

$$\begin{aligned}\frac{1}{(2i\pi)^n}\int_{\pi^*(\Gamma(\Delta))} v \bigwedge_{j=1}^n \frac{d\zeta_j}{u_j(\zeta_j)} &= \sum_{\kappa\in\mathbb{N}^n} c_\kappa \Big\langle \overline{\partial}\Big(\frac{1}{\zeta_1^{\deg u_1+\kappa_1},\dots,\zeta_n^{\deg u_n+\kappa_n}}\Big)_{V(\mathbb{J})}, v\, d\zeta\Big\rangle\\ &= \sum_{\kappa\in\mathbb{N}^n} c_\kappa\, \mathrm{Res}_{\mathbb{K}_{\mathbb{J}}[X,Y]/\mathbb{K}}\left[\begin{matrix}\bar v\, d\bar X\\ \bar X_1^{\deg u_1+\kappa_1},\dots,\bar X_n^{\deg u_n+\kappa_n}\end{matrix}\right],\end{aligned}$$

where the second equality follows from (6.129). Since

$$\begin{aligned}\sum_{\kappa\in\mathbb{N}^n} c_\kappa\, \mathrm{Res}_{\mathbb{K}_{\mathbb{J}}[X,Y]/\mathbb{K}}&\left[\begin{matrix}\bar v\, d\bar X\\ \bar X_1^{\deg u_1+\kappa_1},\dots,\bar X_n^{\deg u_n+\kappa_n}\end{matrix}\right]\\ &= \mathrm{Res}_{\mathbb{K}_{\mathbb{J}}[X,Y]/\mathbb{K}}\left[\begin{matrix}\bar v\, d\bar X\\ u_1(\bar X_1),\dots,u_n(\bar X_n)\end{matrix}\right]\end{aligned}$$

as a consequence (for example) of Proposition 6.44, equality (6.127) is proved, which concludes the proof of Theorem 6.49. □

REMARK 6.50. Since

$$\frac{\boldsymbol{\sigma}}{2i\pi}\Big(\overline{\partial}\,\frac{\boldsymbol{\sigma}}{2i\pi}\Big)^{n-1} = \frac{(n-1)!}{(2i\pi)^n}\Big(\sum_{j=1}^n(-1)^{j-1}\sigma_j \bigwedge_{j'\neq j}\overline{\partial}\sigma_{j'}\Big)\bigwedge_{j=1}^n \epsilon_j^*,$$

one can also rewrite the second equality in (6.115) as

$$\mathrm{Res}_{\mathbb{K}_{\mathbb{J}}[X,Y]/\mathbb{K}}\left[\begin{matrix}\bar q\, d\bar X/\bar p_0^{\ell+1}\\ \bar p_1^{\kappa_1+1},\dots,\bar p_n^{\kappa_n+1}\end{matrix}\right] = -\binom{n-1+|\kappa|}{n-1,\kappa}\int_{V(\mathbb{J})}\overline{\partial}\chi\wedge\sigma^\kappa\,\Omega\wedge(D_\ell^{|\kappa|}[p_0]q)\, d\zeta, \tag{6.130}$$

where

$$\Omega = \frac{(-1)^{n(n-1)/2}(n-1)!}{(2i\pi)^n}\Big(\sum_{j=1}^n(-1)^{j-1}\sigma_j\bigwedge_{j'\neq j}\overline{\partial}\sigma_{j'}\Big). \tag{6.131}$$

Let us mention here an important corollary of this result, also known as the Euler–Jacobi residue formula, already mentioned in the context of multivariate algebraic residue calculus over $\mathbb{K}[X]$ instead of $\mathbb{K}_{\mathbb{J}}[X,Y]$; see Remark 6.42 (see also [**KK87**]).

COROLLARY 6.51. *Let $\mathbb{K}\subset\mathbb{C}$, $\mathbb{J}\subset\mathbb{K}[X,Y]$, $\mathbb{X} = \mathrm{Spec}\ \mathbb{K}_{\mathbb{J}}[X,Y]$. Let also $p_1,\dots,p_n\in\mathbb{K}[X,Y]$ be such that $\bar p = (\bar p_1,\dots,\bar p_n)$ is a quasi-regular sequence in $\mathbb{A} = \mathbb{K}_{\mathbb{J}}[X,Y]$, with $\mathbb{I}_{\bar p}\neq\mathbb{A}$. Let $P_1,\dots,P_n$ be the homogenizations of the p_j's in $\mathbb{K}[X_0,X,Y]$ and let $\overline{\mathbb{X}_{\mathbb{C}}}$ be the Zariski closure of $\mathbb{X}_{\mathbb{C}}$ in $\mathbb{P}^N_{\mathbb{C}}$. Suppose that*

$$\overline{\mathbb{X}_{\mathbb{C}}}\cap V(P_1)\cap\cdots\cap V(P_n) = \emptyset. \tag{6.132}$$

Then, for any $q \in \mathbb{K}[X,Y]$ *and any subsets* $I \subset \{1,\dots,n\}$, $J \subset \{1,\dots,N-n\}$ *whose sum of cardinals equals* n,

$$\deg q < \sum_{j=1}^{n} \deg p_j - n \Longrightarrow \operatorname{Res}_{\mathbb{K}_{\mathbb{J}}[X,Y]/\mathbb{K}} \begin{bmatrix} q(\bar{X},\bar{Y})\, d\bar{X}_I \wedge d\bar{Y}_J \\ \bar{p}_1,\dots,\bar{p}_n \end{bmatrix} = 0. \tag{6.133}$$

PROOF. Let $d_j = \deg p_j$. The homogenizations $P_1,\dots,P_n$ of the polynomials $p_1,\dots,p_n$ induce holomorphic sections respectively of the holomorphic bundles $\mathcal{O}_{\mathbb{P}^N_{\mathbb{C}}}(d_j)$ over $\mathbb{P}^N_{\mathbb{C}}$. Consider the homogeneous coordinates

$$[z_0 : Z_1 : \cdots : Z_N],\ Z_j = z_0 z_j \text{ for } j=1,\dots,n,\ Z_{n+j} = z_0 w_j \text{ for } j = 1,\dots,N-n,$$

on $\mathbb{P}^N_{\mathbb{C}}$. Since $P_1,\dots,P_n$ do not share any common zero on $\overline{V(\mathbb{J})} \cap \{z_0 = 0\}$, one has

$$\operatorname{Supp}\Big(\bar\partial\Big(\frac{1}{P_1,\dots,P_n}\Big)_{V(\mathbb{J})}\Big) \subset V(\mathbb{J}) \cap V(p),$$

where the Coleff–Herrera current on the left-hand side was introduced in Definition 2.78. The condition (6.133) on the degrees ensures that the rational n-form

$$q(Z_1/z_0,\dots,Z_0/z_0) \bigwedge_{j=1}^{n} \frac{d(Z_j/z_0)}{p_j(Z_1/z_0,\dots,Z_N/z_0)}$$

does not have the hyperplane $\{z_0 = 0\}$ as polar hyperplane in $\mathbb{P}^N_{\mathbb{C}}$. Then the assertion results from Stokes's theorem on the complex analytic space $\overline{V(\mathbb{J})}$, together with the first equality in (6.115), since one can find a system of coordinates (z', w') in $\mathbb{C}^N$ such that for any subset $K \subset \{1,\dots,N\}$ with cardinal n, the cardinal of the intersection of $\mathbb{X}_{\mathbb{C}}$ with $\bigcap_{j\in K}\{z'_k = 0\}$ equals $\deg \mathbb{X}$. □

REMARK 6.52. Suppose that the geometric condition (6.112) is fulfilled. Take $q = X^{\kappa} Y^{\chi}$, $I = \{1,\dots,n\}$, $J = \emptyset$, where $\kappa \in \mathbb{N}^n$ and $\chi \in \mathbb{N}^{N-n}$. If the conditions in Corollary 6.51 are satisfied, then one has

$$\deg \begin{bmatrix} \bar{X}^{\kappa}\bar{Y}^{\chi}\, d\bar{X} \\ \bar{X} - t \end{bmatrix}_{\mathbb{X}} \le |\kappa| + |\chi| \tag{6.134}$$

instead of the upper estimate by $|\kappa| + |\chi| \deg \mathbb{X}$ which was provided by Proposition 6.43.

6.5.2. Properness and algebraic residue symbols in characteristic 0. Let $(p_1,\dots,p_n)$ be a sequence of polynomials in $\mathbb{K}[X,Y]$ whose classes $\bar{p}_j$ modulo the purely n-dimensional radical ideal $\mathbb{J}$ define a quasi-regular sequence $\bar{p}$ in $\mathbb{K}_{\mathbb{J}}[X,Y]$ such that $\mathbb{I}_{\bar{p}} \neq \mathbb{K}_{\mathbb{J}}[X,Y]$. Following the ideas introduced by Stanisłas Łojasiewicz [**Lo59, Lo, JKS92**] (see also [**Boc75**]), we introduce the following definition.

DEFINITION 6.53. Let δ_j, $j = 1,\dots,n$ be nonnegative rational numbers. The sequence p is said *to admit* δ *as Łojasiewicz multi-exponent for the algebraic variety* $\mathbb{X} = \operatorname{Spec} \mathbb{K}_{\mathbb{J}}[X,Y]$ if and only if for any $\varepsilon = (\varepsilon_1,\dots,\varepsilon_n)$, where $0 < \varepsilon_j \ll 1$ for $j = 1,\dots,n$, there exist strictly positive constants C_ε and c_ε such that

$$(z,w) \in \mathbb{X}_{\mathbb{C}} \text{ and } \max(|z|,|w|) \ge C_\epsilon \tag{6.135}$$
$$\Longrightarrow \max_{1\le j\le n} \frac{|p_j(z,w)|}{(\max(|z|,|w|))^{\deg p_j}} (\max(|z|,|w|))^{\delta_j + \varepsilon_j} \ge c_\varepsilon,$$

where $|z| = \max(|z_1|, \ldots, |z_n|)$, $|w| = \max(|w_1|, \ldots, |w_{N-n}|)$. Such a Łojasiewicz multi-exponent is called a *multi-index of properness for* $\mathbb{X}$ in the particular case when $\delta_j < \deg p_j$ for $j = 1, \ldots, n$.

REMARK 6.54. Definition 6.53 holds when $\mathbb{K}$ is an arbitrary field of any characteristic, $\mathbb{C}$ being replaced by an algebraic closure $\overline{\mathbb{K}}$, $\mathbb{X}_{\mathbb{C}}$ by $\mathbb{X}_{\overline{\mathbb{K}}}$, and $|\ |$ by a nontrivial archimidean absolute value on $\overline{\mathbb{K}}$, as in §6.4.1. The notion of a p-admissible Łojasiewicz multi-exponent with respect to $\mathbb{X}$ extends then the notion of a p-admissible $(\delta_1, \ldots, \delta_n)$ properness multi-exponent such as introduced in Definition 6.38.

PROPOSITION 6.55. *Let $\mathbb{K}$ be a field with arbitrary characteristic, let $\mathbb{J} \subset \mathbb{K}[X, Y]$ be purely n-dimensional and radical, and let*

$$\mathbb{X} = \operatorname{Spec} \mathbb{K}_{\mathbb{J}}[X, Y], p_1, \ldots, p_n \in \mathbb{K}[X, Y]$$

as above. Let $\deg \mathbb{X} = \deg \mathbb{X}_{\overline{\mathbb{K}}}$ and $D = \prod_{j=1}^{n} \deg p_j$. Then

$$(\delta_1, \ldots, \delta_n)^\circ = \big(D \deg \mathbb{X}, \ldots, D \deg \mathbb{X}\big) \tag{6.136}$$

is a Łojasiewicz multi-exponent for p with respect to the algebraic variety $\mathbb{X}$. Moreover p induces a proper polynomial morphism from $\mathbb{X}_{\overline{\mathbb{K}}}$ to $\overline{\mathbb{K}}^n$ if and only if it admits any $(\delta_1, \ldots, \delta_n)$ with $\delta_j < \deg p_j$ for $j = 1, \ldots, n$, such that

$$\langle\!\langle \kappa, d \rangle\!\rangle \leq D \deg \mathbb{X} \Longrightarrow \langle\!\langle \kappa, d - \delta \rangle\!\rangle < 1 \tag{6.137}$$

as a multi-index exponent of properness for $\mathbb{X}$.

PROOF. Up to a $\mathbb{K}$-linear change of coordinates, one may assume that the geometric hypothesis (6.68) holds. The first assertion follows from the fact that there exist univariate polynomials $\phi_1(X_1), \ldots, \phi_n(X_n)$ such that, for $j = 1, \ldots, n$,

$$\phi_j(X_j) - \sum_{k=1}^{n} a_{j,k}\, p_k \in \mathbb{J} \quad \text{with} \quad a_{j,k} \in \mathbb{K}[X, Y], \quad \max_k \deg(a_{j,k}\, p_k) \leq D \deg \mathbb{X};$$

see [**Jel05**, Theorem 3.8]. Therefore, one has

$$\begin{aligned} &(x, y) \in \mathbb{X}_{\overline{\mathbb{K}}} \quad \text{and} \quad |x| \geq C \\ &\qquad \Longrightarrow c|\phi(x)| \leq \max_{1 \leq j \leq n} \big(1 + \max(|x|, |y|)\big)^{D \deg \mathbb{X} - \deg p_j} |p_j(x, y)| \end{aligned}$$

for some $C > 0$ and $c > 0$, which implies that $(\delta_1, \ldots, \delta_n)^\circ$ is a Łojasiewicz exponent. The second assertion follows from the fact that $\bar{p}_1, \ldots, \bar{p}_n$ are $\mathbb{K}$-algebraically independent in $\mathbb{K}_{\mathbb{J}}[X, Y]$ (since $\bar{p}$ is quasi-regular such that $\mathbb{I}_{\bar{p}} \neq \mathbb{K}_{\mathbb{J}}[X, Y]$). Then, if p realizes a proper polynomial morphism from $\mathbb{X}_{\overline{\mathbb{K}}}$ to $\overline{\mathbb{K}}^n$, then there exists for any $j = 1, \ldots, n$, a polynomial $E_j(t, \tau) = \tau^{\nu_j} + \sum_{\rho=1}^{\nu_j} u_{j,\rho}(t)\, \tau^{\nu_j - \rho}$ with

$$\deg_t E_j(t^{\deg p_1}, \ldots, t^{\deg p_n}, t) \leq D \deg \mathbb{X}, \quad E_j(p_1, \ldots, p_n, X_j) \in \mathbb{J};$$

see [**Jel05**, Theorem 3.3]. Then the second claim of the proposition is proved similarly to Proposition 6.39. □

If $\mathbb{K} \subset \mathbb{C}$, the following important result generalizes Corollary 6.51; see [**BVY05**, Proposition 4.1].

THEOREM 6.56. *Let $\mathbb{K} \subset \mathbb{C}$, $\mathbb{J} \subset \mathbb{K}[X, Y]$, $p_1, \ldots, p_n \in \mathbb{K}[X, Y]$ such that $\bar{p}$ is a quasi-regular sequence in $\mathbb{K}_{\mathbb{J}}[X, Y]$ with $\mathbb{I}_{\bar{p}} \neq \mathbb{K}_{\mathbb{J}}[X, Y]$. Suppose that p admits a multi-exponent of properness $(\delta_1, \ldots, \delta_n)$ with respect to $\mathbb{X}$. Then, for any q in*

$\mathbb{K}[X,Y]$ and any subsets $I \subset \{1,\dots,n\}$, $J \subset \{1,\dots,N-n\}$ whose sum of cardinals equals n,

$$\deg q < \sum_{j=1}^{n}(d_j - \delta_j) - n \Longrightarrow \mathrm{Res}_{\mathbb{K}_{\mathbb{J}}[X,Y]/\mathbb{K}} \begin{bmatrix} q(\bar{X},\bar{Y})\, d\bar{X}_I \wedge d\bar{Y}_J \\ \bar{p}_1,\dots,\bar{p}_n \end{bmatrix} = 0. \tag{6.138}$$

Moreover the element in $\mathbb{K}[t]$ defined as in Proposition 6.48 and (6.111) equals

$$\begin{bmatrix} \bar{q}\, d\bar{X} \\ \bar{p} - t \end{bmatrix}_{\mathbb{X}} = \sum_{\kappa \in E_\delta^d[\bar{q}\, d\bar{X}]} \mathrm{Res}_{\mathbb{K}_{\mathbb{J}}[X,Y]/\mathbb{K}} \begin{bmatrix} \bar{q}\, d\bar{X} \\ \bar{p}_1^{\kappa_1+1},\dots,\bar{p}_n^{\kappa_n+1} \end{bmatrix} t^\kappa, \tag{6.139}$$

where, if $d = (\deg p_1,\dots,\deg p_n)$ and $|d-\delta| = |d| - |\delta|$,

$$E_\delta^d[\bar{q}\, d\bar{X}] = \big\{\kappa \in \mathbb{N}^n \,:\, \langle\!\langle \kappa\,,\, d-\delta\rangle\!\rangle + |d-\delta| \le \deg q + n\big\}. \tag{6.140}$$

PROOF. Let us explain why the first assertion (6.138) implies the second one (6.139). Let $\chi \in \mathcal{D}(\mathbb{C}^n,\mathbb{C})$ be identically equal to one in a neighborhood of $\mathbb{X}_{\mathbb{C}} \cap V(p)$. Since p defines a proper polynomial morphism from $\mathbb{X}_{\mathbb{C}}$ to $\mathbb{C}^n$, it follows from Theorem 6.49 and specification of parameters t as complex numbers in $\eta\,\mathbb{B}_n$ with $0 < \eta \ll 1$ that one has in $\mathbb{C}[[t_1,\dots,t_n]]$

$$\begin{aligned} \begin{bmatrix} \bar{q}\, d\bar{X} \\ \bar{p} - t \end{bmatrix}_{\mathbb{X}} = \begin{bmatrix} \bar{q}\, d\bar{X} \\ \bar{p} - t \end{bmatrix}_{\mathbb{X}_{\mathbb{C}}} &= \Big\langle \bar{\partial}\Big(\frac{1}{p_1 - t_1,\dots,p_n - t_n}\Big)_{V(\mathbb{J})}, q\,\chi\, d\zeta\Big\rangle \\ &= \sum_{\kappa\in\mathbb{N}^n} \Big\langle \bar{\partial}\Big(\frac{1}{p_1^{\kappa_1+1},\dots,p_n^{\kappa_n+1}}\Big)_{V(\mathbb{J})}, q\,\chi\, d\zeta\Big\rangle t^\kappa \\ &= \sum_{\kappa\in\mathbb{N}^n} \mathrm{Res}_{\mathbb{C}_{\mathbb{J}}[X,Y]/\mathbb{C}} \begin{bmatrix} \bar{q}\, d\bar{X} \\ \bar{p}_1^{\kappa_1+1},\dots,\bar{p}_n^{\kappa_n+1} \end{bmatrix} t^\kappa \\ &= \sum_{\kappa\in\mathbb{N}^n} \mathrm{Res}_{\mathbb{K}_{\mathbb{J}}[X,Y]/\mathbb{K}} \begin{bmatrix} \bar{q}\, d\bar{X} \\ \bar{p}_1^{\kappa_1+1},\dots,\bar{p}_n^{\kappa_n+1} \end{bmatrix} t^\kappa. \end{aligned} \tag{6.141}$$

It follows from (6.138) that the development on the right-hand side of (6.141) truncates as (6.139), with $E_\delta^d[\bar{q}\, d\bar{X}]$ defined as (6.140). Proving (6.138) amounts to proving it when $I = \{1,\dots,n\}$, $J = \emptyset$ and condition (6.112) is satisfied, since, up to a $\mathbb{K}$-linear change or coordinates, one can always assume that

$$\big|\{Z = (Z_1,\dots,Z_n) \in \mathbb{X}_{\mathbb{C}} \subset \mathbb{K}^N \,:\, Z_{i_1} = \cdots = Z_{i_n} = 0\}\big| = \deg \mathbb{X}$$

for any $1 \le i_1 < \cdots < i_n \le N$. It is also clear that one may replace δ_j by $\delta_j + 1/\nu$ with $\nu \in \mathbb{N}^*$ arbitrarily large and then suppose that for some $C > 0$ and $c > 0$, one has

$$(z,w) \in V(\mathbb{J}) \text{ and } \max(|z|,|w|) \ge C \Longrightarrow \max_{1\le j\le n} \frac{|p_j(z,w)|}{(\max(|z|,|w|))^{\deg p_j - \delta_j}} \ge c. \tag{6.142}$$

The proof we give here is inspired by that presented in [**Z05**, Theorem 1] when $n = N$. Let, for $j = 1,\dots,n$,

$$\prod_{j'\ne j} (\deg p_j - \delta_j) = \frac{m_j}{m}, \quad m_j, m \in \mathbb{N}^* \quad \text{with } m_j \ge 3.$$

Denote

$$\boldsymbol{m} = (m_1,\dots,m_n),\ M = m \prod_{j=1}^{n} (\deg p_j - \delta_j)$$

and let $\rho_1, \dots, \rho_n$ the C^1 functions in $\mathbb{C}^N$ defined by

$$(z, w) \longmapsto \rho_j(z, w) = \sqrt{|p_j^{m_j}(z, w)|}$$

for $j = 1, \dots, n$. Let $\boldsymbol{\Sigma_m}$ be the C^1 $\mathbb{C}^n$-valued map defined as follows

$$\begin{aligned}\boldsymbol{\Sigma_m} : (z, w) \in \mathbb{C}^N \setminus V(p) \\ \longmapsto \sum_{j=1}^n \frac{\rho_j^2(z, w)}{\|\rho(x, w)\|^2} \frac{\epsilon_j^*}{p_j(z, w)} = \big(\Sigma_{\boldsymbol{m},1}(z, w), \dots, \Sigma_{\boldsymbol{m},n}(z, w)\big).\end{aligned}$$

Since

$$\boldsymbol{\Sigma_m}(z, w)\Big(\sum_{j=1}^n p_j(z, w)\, \epsilon_j\Big) = \boldsymbol{\sigma}(z, w)\Big(\sum_{j=1}^n p_j(z, w)\, \epsilon_j\Big) \equiv 1,$$

where $\boldsymbol{\sigma}$ is defined as in (6.114), the same cohomological argument as that used in the proof of Theorem 6.49, with $(\boldsymbol{\sigma}, \boldsymbol{\Sigma_m})$ instead of $(\boldsymbol{\sigma}, \boldsymbol{\Sigma})$ as in (6.123) and $\kappa = 0$ there, together with the second equality in (6.115), shows that, if $\chi \in \mathcal{D}(\mathbb{C}^N, \mathbb{C})$ is identically equal to one in a neighborhood of $V(\mathbb{J}) \cap V(p)$,

(6.143)

$$\begin{aligned}\operatorname{Res}_{\mathbb{K}_{\mathbb{J}}[X,Y]/\mathbb{K}} \begin{bmatrix} \bar{q}\, d\bar{X} \\ \bar{p}_1, \dots, \bar{p}_n \end{bmatrix} &= \Big\langle \frac{\boldsymbol{\sigma}}{2i\pi} \Big(\overline{\partial} \frac{\boldsymbol{\sigma}}{2i\pi}\Big)^{n-1} ([V(\mathbb{J})])\, ,\, q\, (\overline{\partial}\chi \wedge d\zeta) \bigwedge_{j=1}^n \epsilon_j \Big\rangle \\ &= \Big\langle \frac{\boldsymbol{\Sigma_m}}{2i\pi} \Big(\overline{\partial} \frac{\boldsymbol{\Sigma_m}}{2i\pi}\Big)^{n-1} ([V(\mathbb{J})])\, ,\, q\, (\overline{\partial}\chi \wedge d\zeta) \bigwedge_{j=1}^n \epsilon_j \Big\rangle.\end{aligned}$$

It follows from (6.142) that for some $c_m > 0$

(6.144)

$$(z, w) \in V(\mathbb{J}) \text{ and } \max(|z|, |w|) \geq C \Longrightarrow \|\rho(z, w)\|^2 \geq c_m\, (\max(|z|, |w|))^M.$$

Let θ be an even function such that

$$(6.145) \qquad \theta \in \mathcal{D}\Big(\Big] -\frac{2c_m}{3}, \frac{2c_m}{3}\Big[, [0, 1]\Big), \quad \theta \equiv 1 \text{ on } \Big[\frac{c_m}{3}, \frac{c_m}{3}\Big].$$

Let $R > C$ and $\chi_R(z, w) = \theta(\rho^2(z, w)/R^M)$. Since

$$p_j \Sigma_{\boldsymbol{m},j} \equiv (\rho_j/\|\rho\|)^2$$

for $j = 1, \dots, n$ on $\mathbb{C}^N \setminus V(p)$, one can rewrite (6.143) when $\chi = \chi_R$ as (see also (6.130))

$$(6.146) \quad \operatorname{Res}_{\mathbb{K}_{\mathbb{J}}[X,Y]/\mathbb{K}} \begin{bmatrix} \bar{q}\, d\bar{X} \\ \bar{p}_1, \dots, \bar{p}_n \end{bmatrix} = -\int_{V(\mathbb{J})} \overline{\partial}\chi_R \wedge \Big(\prod_{j=1}^n \frac{|p_j|}{p_j} \rho_j^{1-2/m_j}\Big) \Omega_{\boldsymbol{m}} \wedge q\, d\zeta,$$

where

$$\Omega_{\boldsymbol{m}} = \frac{(-1)^{n(n-1)/2}(n-1)!}{(2i\pi)^n} \Big(\frac{2^{n-1}\big(\sum_{j=1}^n (-1)^{j-1} \rho_j \bigwedge_{j' \neq j} \overline{\partial}\rho_{j'}\big)}{\|\rho\|^{2n}}\Big).$$

Since $\overline{\partial}\chi_R = R^{-M}\, \theta'(\|\rho\|^2/R^M)\, \overline{\partial}\|\rho\|^2$, one gets from (6.146) that

$$\begin{aligned}(6.147) \quad &\operatorname{Res}_{\mathbb{K}_{\mathbb{J}}[X,Y]/\mathbb{K}} \begin{bmatrix} \bar{q}\, d\bar{X} \\ \bar{p}_1, \dots, \bar{p}_n \end{bmatrix} \\ &= -\frac{2^{n-1}}{R^M} \int_{V(\mathbb{J})} \Big(q\, \theta'\Big(\frac{\|\rho\|^2}{R^M}\Big) \Big(\prod_{j=1}^n \frac{|p_j|}{p_j}\Big)\Big) \Big(\Big(\prod_{j=1}^n \rho_j^{1-2/m_j}\Big) \frac{\bigwedge_{j=1}^n d\rho_j}{\|\rho\|^{2(n-1)}}\Big) \wedge d\zeta.\end{aligned}$$

For any $(z,w)\in V(\mathbb{J})$,

$$\max(|z|,|w|)\geq R>C\Longrightarrow \rho^2(z,w)\geq c_m(\max(|z|,|w|))^M\geq c_m\,R^M$$

according to (6.144). It follows then from the choice of θ (see (6.145)) that

$$\operatorname{Supp}\Big(\theta'\Big(\frac{\|\rho\|^2}{R^M}\Big)\Big)\cap V(\mathbb{J})\subset\{(z,w)\in\mathbb{C}^N\ :\ \max(|z|,|w|)\leq R\}.$$

There is a positive constant $C_\theta(q)$ such that
(6.148)

$$\forall\,(z,w)\text{ almost everywhere }\in V(\mathbb{J}),\ \Big|q\,\theta'\Big(\frac{\|\rho\|^2}{R^M}\Big)\,\Big(\prod_{j=1}^n\frac{|p_j|}{p_j}\Big)\Big|\leq C_\theta(q)\,R^{\deg q}.$$

Given $I,I'\subset\{1,\dots,n\}$ whose sum of cardinals equals n and $\zeta_j=x_j+i\,y_j$ for $j=1,\dots,n$, let us study the asymptotic behavior when $R\to+\infty$ of

$$R\longmapsto\Theta_{I,I'}(R):=\int_{V(\mathbb{J})}\Theta_R\,\frac{\bigwedge_{j=1}^n\rho_j^{1-2/m_j}\,d\rho_j}{\|\rho\|^{2(n-1)}}\wedge dx_I\wedge dy_{I'},$$

where

$$\Theta_R\overset{\text{a.e. on }V(\mathbb{J})}{=}q\,\theta'\Big(\frac{\|\rho\|^2}{R^M}\Big)\,\Big(\prod_{j=1}^n\frac{|p_j|}{p_j}\Big).$$

If W_π denotes the algebraic hypersurface of $\mathbb{C}^n$ defined as the discriminant locus of the proper projection $\pi\ :(z,w)\in V(\mathbb{J})\to z\in\mathbb{C}^n$, one has

$$\int_{V(\mathbb{J})}\Theta_R\,\frac{\bigwedge_{j=1}^n\rho_j^{1-2/m_j}\,d\rho_j}{\|\rho\|^{2(n-1)}}\wedge dx_I\wedge dy_{I'}$$
$$=\int_{V(\mathbb{J})\setminus\pi^{-1}(W_\pi)}\Theta_R\,\frac{\bigwedge_{j=1}^n\rho_j^{1-2/m_j}\,d\rho_j}{\|\rho\|^{2(n-1)}}\wedge dx_I\wedge dy_{I'}.$$

The $2n$-dimensional real manifold $\big(V(\mathbb{J})\setminus\pi^{-1}(W_\pi)\big)_{\mathbb{R}}$ defined as the underlying real manifold of the n-dimensional complex manifold $V(\mathbb{J})\setminus\pi^{-1}(W_\pi)$ is a union of $\deg\mathbb{X}$ sheets $\mathscr{V}_\iota$. Each such sheet is a $2n$-differentiable manifold on which each map

$$\pi_{I,I',\iota}:(\operatorname{Re}z,\operatorname{Im}z,\operatorname{Re}w,\operatorname{Im}w)\in\mathscr{V}_\iota\longrightarrow(\rho_1,\dots,\rho_n,x_I,y_{I'})\in\mathbb{R}^{2n}$$

is proper with degree $\deg\pi_{I,I',\iota}=d_\iota(I,I')$. Let $|d(I,I')|=\sum_\iota d_\iota(I,I')$. Since

$$dv_n^{[\boldsymbol{m}]}=\frac{\bigwedge_{j=1}^n\rho_j^{1-2/m_j}\,d\rho_j}{\|\rho\|^{2(n-1)}}$$

is a volume form on $(\mathbb{R}^+)^n$, one has then, taking also into account estimates (6.148), that

$$\Big|\int_{V(\mathbb{J})\setminus\pi^{-1}(W_\pi)}\Theta_R\,\frac{\bigwedge_{j=1}^n\rho_j^{1-2/m_j}\,d\rho_j}{\|\rho\|^{2(n-1)}}\wedge dx_I\wedge dy_{I'}\Big|$$
$$\leq C_\theta(q)\,R^{\deg q+n-M}\,|d(I,I')|\int_{\frac{c_m}{3}\leq\frac{\|\rho\|^2}{R^M}\leq\frac{2c_m}{3}}dv_n^{[\boldsymbol{m}]}=C'_\theta(q)\frac{R^{\deg q+n}}{R^{\sum_{j=1}^n M/m_j}},$$

where

$$C'_\theta(q)=\frac{C_\theta(q)}{(c_m/3)^{n-1}}\int_{\frac{c_m}{3}\leq\|\rho\|^2\leq\frac{2c_m}{3}}dv_n^{[\boldsymbol{m}]}.$$

When $\deg q + n < \sum_{j=1}^n M/m_j = \sum_{j=1}^n (\deg p_j - \delta_j)$, one has

$$\lim_{R\to+\infty} \Big| \int_{V(\mathbb{J})\setminus \pi^{-1}(W_\pi)} \Theta_R \frac{\bigwedge_{j=1}^n \rho_j^{1-2/m_j}\, d\rho_j}{\|\rho\|^{2(n-1)}} \wedge dx_I \wedge dy_{I'} \Big| = 0$$

for any I, I', which implies that

$$\mathrm{Res}_{\mathbb{K}_{\mathbb{J}}[X,Y]/\mathbb{K}} \begin{bmatrix} \bar{q}\, d\bar{X} \\ \bar{p}_1, \dots, \bar{p}_n \end{bmatrix} = 0$$

since (6.147) holds independently on the choice of $R \gg 1$. It completes the proof of the first assertion (6.138), hence of Theorem 6.56. $\square$

Let us present another approach to the first assertion (6.138) in Theorem 6.56 which involves the closure of the closed analytic subset $V(\mathbb{J})$ (namely the support of the complex affine algebraic variety $\mathbb{X}_{\mathbb{C}}$) in the projective space $\mathbb{P}^N_{\mathbb{C}}$. Such an approach fits with one already proposed when $\mathbb{X} = \mathbb{A}^N_{\mathbb{K}}$, that is $N = n$, in Propositions 6.40 and 6.41. The smooth strictly positive function

$$(z, w) \longmapsto \rho(z, w) = \frac{1}{\sqrt{1 + \|z\|^2 + \|w\|^2}} \tag{6.149}$$

plays there an important role. Let $p_1, \dots, p_n \in \mathbb{K}[X, Y]$ be such that $\bar{p}$ defines a quasi-regular sequence in $\mathbb{K}_{\mathbb{J}}[X, Y]$. Assume, in addition, that $\mathbb{I}_{\bar{p}} \neq \mathbb{K}_{\mathbb{J}}[X, Y]$. We do not assume for the moment any properness assumption. If $\delta = (\delta_1, \dots, \delta_n)$ with $|\delta|_\infty = \max_{1\le j\le n} \delta_j$ is a vector of n nonnegative real numbers, let

$$\rho_{\delta,j} = \rho^{\deg p_j + |\delta|_\infty - \delta_j} \quad \text{for } j = 1, \dots, n.$$

If $\mathbf{\Sigma}^\delta : \mathbb{C}^N \setminus V(p) \to \mathbb{C}^n$ is the smooth map defined by

$$\mathbf{\Sigma}^\delta = \sum_{j=1}^m \frac{\rho_{\delta,j}^2}{\sum_{\iota=1}^n \rho_{\delta,\iota}^2 |p_\iota|^2}\, \overline{p_j}\, \epsilon_j^* = (\Sigma_1^\delta, \dots, \Sigma_n^\delta),$$

it follows from Theorem 6.49[36] and from the fact that

$$\mathbf{\Sigma}^\delta \Big(\sum_{j=1}^n p_j\, \epsilon_j \Big) \equiv 1$$

on $\mathbb{C}^N \setminus V(p)$, together with Stokes's theorem on $V(\mathbb{J})$, that for any $q \in \mathbb{K}[X, Y]$

(6.150)

$$\begin{aligned} \mathrm{Res}_{\mathbb{K}_{\mathbb{J}}[X,Y]/\mathbb{K}} \begin{bmatrix} \bar{q}\, d\bar{X} \\ \bar{p}_1, \dots, \bar{p}_n \end{bmatrix} &= \Big\langle \frac{\mathbf{\Sigma}^\delta}{2i\pi} \Big(\bar\partial \frac{\mathbf{\Sigma}^\delta}{2i\pi} \Big)^{n-1} ([V(\mathbb{J})])\,,\, q\,(\bar\partial\chi \wedge d\zeta) \bigwedge_{j=1}^n \epsilon_j \Big\rangle \\ &= -\Big[\int_{V(\mathbb{J})} \bar\partial\chi \wedge \Omega^{\delta,\lambda} \wedge q\, d\zeta \Big]_{\lambda=0}, \end{aligned}$$

[36]With its reformulation as in (6.130) and (6.131); see also the proofs of Theorems 6.49 and 6.56 where the cohomological deformation argument was already used.

where χ is any test function in $\mathbb{C}^N$ identically equal to one in a neighborhood of $V(\mathbb{J}) \cap V(p)$,

$$\begin{aligned}\Omega^{\delta,\lambda} &= \frac{(-1)^{n(n-1)/2}(n-1)!}{(2i\pi)^n} \|p\|_{\rho_\delta}^{2\lambda} \frac{\sum_{j=1}^n (-1)^{j-1}\rho_{\delta,j}^2 \overline{p_j} \bigwedge_{j'\neq j} \overline{\partial}(\rho_{\delta,j'}^2 \overline{p_{j'}})}{\|p\|_{\rho_\delta}^{2n}} \\ &= \frac{(-1)^{n(n-1)/2}(n-1)!}{(2i\pi)^n} U^{\delta,\lambda}\end{aligned} \tag{6.151}$$

with

$$\|p\|_{\rho_\delta}^2 = \sum_{j=1}^n \rho_{\delta,j}^2 \, |p_j|^2$$

and λ a complex parameter starting from $\{\lambda \in \mathbb{C} : \operatorname{Re} \lambda \gg 1\}$ to follow analytic continuation. For $\operatorname{Re} \lambda \gg 1$, Stokes's theorem applied on the closed analytic subset $\overline{V(\mathbb{J})} \subset \mathbb{P}_{\mathbb{C}}^N$ implies that

(6.152)

$$\begin{aligned}\Big[- \int_{V(\mathbb{J})} \overline{\partial}\chi \wedge U^{\delta,\lambda} \wedge q\, d\zeta \Big]_{\lambda=0} &= \Big[\int_{\overline{V(\mathbb{J})}} \overline{\partial}(1-\chi) \wedge U^{\delta,\lambda} \wedge q\, d\zeta \Big]_{\lambda=0} \\ &= - \int_{\overline{V(\mathbb{J})}} (1-\chi)\, \overline{\partial} U^{\delta,\lambda} \wedge q\, d\zeta \\ &= -\lambda \int_{\overline{V(\mathbb{J})}} (1-\chi)\, \|p\|_{\rho_\delta}^{2\lambda} \frac{\bigwedge_{j=1}^n \overline{\partial}(\rho_{\delta,j}^2 \overline{p_j})}{\|p\|_{\rho_\delta}^{2n}} \wedge q\, d\zeta.\end{aligned}$$

We can then reformulate (6.150) as

$$\begin{aligned}&\operatorname{Res}_{\mathbb{K}_{\mathbb{J}}[X,Y]/\mathbb{K}} \begin{bmatrix} \bar{q}\, d\bar{X} \\ \bar{p}_1, \dots, \bar{p}_n \end{bmatrix} \\ &= - \frac{(-1)^{n(n-1)/2}(n-1)!}{(2i\pi)^n} \Big[\lambda \int_{\overline{V(\mathbb{J})}} (1-\chi)\, \|p\|_{\rho_\delta}^{2\lambda} \frac{\bigwedge_{j=1}^n \overline{\partial}(\rho_{\delta,j}^2 \overline{p_j})}{\|p\|_{\rho_\delta}^{2n}} \wedge q\, d\zeta \Big]_{\lambda=0}.\end{aligned} \tag{6.153}$$

Let $Z = (z_0, Z)$ be the $n+1$ homogeneous coordinates on $\mathbb{P}_{\mathbb{C}}^N$, so that

$$z_j = \frac{Z_j}{z_0} \quad \text{for } j = 1, \dots, n, \quad w_j = \frac{Z_{n+j}}{z_0} \quad \text{for } j = 1, \dots, N-n.$$

Observe that if $P_1, \dots, P_n, Q$ denote the homogenizations of $p_1, \dots, p_n, q$ in $\mathbb{K}[X_0, X, Y]$, respectively, then one has, when expressed in homogeneous coordinates,

$$\begin{aligned}\|p(z,w)\|_{\rho_\delta}^2 &= \sum_{j=1}^n \Big(\frac{|P_j(Z)|}{\|Z\|^{\deg p_j}} \Big)^2 \Big(\frac{|z_0|}{\|Z\|} \Big)^{2(|\delta|_\infty - \delta_j)}, \\ \bigwedge_{j=1}^n \overline{\partial}(\rho_{\delta,j}^2 \overline{p_j}) &= z_0^{n|\delta|_\infty + \sum_{j=1}^n (\deg p_j - \delta_j)} \bigwedge_{j=1}^n \overline{\partial} \Big(\overline{\frac{P_j(Z) z_0^{|\delta|_\infty - \delta_j}}{\|Z\|^{2(|\delta|_\infty - \delta_j + \deg p_j)}}} \Big), \\ q(z,w)\, d\zeta &= z_0^{-\deg q - n - 1}\, Q(Z)\, \Omega(Z),\end{aligned} \tag{6.154}$$

with Ω the Euler form in $(z_0, Z_1, \dots, Z_N)$. Consider on the analytic space $\overline{V(\mathbb{J})} \subset \mathbb{P}^N_{\mathbb{C}}$ with $\iota : \overline{X(\mathbb{J})} \hookrightarrow \mathbb{P}^N_{\mathbb{C}}$ the n-holomorphic bundle

$$E = \bigoplus_{j=1}^{n} \iota^* \mathcal{O}_{\mathbb{P}^N_{\mathbb{C}}}(|\delta|_\infty + \deg p_j - \delta_j)$$

and its holomorphic section

$$s_\delta : Z \longmapsto \big(z_0^{|\delta|_\infty - \delta_1} P_1(Z), \dots, z_0^{|\delta|_\infty - \delta_n} P_n(Z)\big).$$

Any point $Z^0 \in (\overline{V(\mathbb{J})} \setminus V(\mathbb{J})) \cap V(s_\delta)$ admits a neighborhood U with a log resolution $\pi_U : \mathscr{U} \to U$ such that $\pi_U^* s_\delta$ factorizes on $\mathscr{U}$ as

$$\pi_U^* s_\delta = e_{\delta,\mathscr{U}} \otimes \mathfrak{s}_{\delta,\mathscr{U}},$$

where $e_{\delta,\mathscr{U}}$ is a holomorphic section of a line bundle $L_{\delta,\mathscr{U}}$ over $\mathscr{U}$ such that $V(e)$ is a hypersurface with normal crossings and $\mathfrak{s}_{\delta,\mathscr{U}}$ is a nonvanishing section of $L_{\delta,\mathscr{U}}^{-1} \otimes \mathbb{C}^n$. This follows from Theorem A.41 combined with Theorem 3.11; see Example 3.12. It follows from (6.154) that the differential form

$$\lambda \, \|p\|_{\rho_\delta}^{2\lambda} \frac{\bigwedge_{j=1}^n \overline{\partial}(\rho_{\delta,j}^2 \overline{p_j})}{\|p\|_{\rho_\delta}^{2n}} = \Big(\overline{\partial} \|p\|_{\rho_\delta}^{2\lambda} \wedge \mathbf{\Sigma}_\delta \wedge (\overline{\partial} \mathbf{\Sigma}_\delta)^{n-1}\Big) \Big(\frac{\epsilon_n \wedge \cdots \wedge \epsilon_1}{(n-1)!}\Big)$$

is such that

$$(\iota \circ \pi_{\mathscr{U}})^* \Big(\lambda \, \|p\|_{\rho_\delta}^{2\lambda} \frac{\bigwedge_{j=1}^n \overline{\partial}(\rho_{\delta,j}^2 \overline{p_j})}{\|p\|_{\rho_\delta}^{2n}}\Big) \tag{6.155}$$

expresses in centered local coordinates $(v_1, \dots, v_n)$ around $\mathscr{Z}^0 \in \pi_{\mathscr{U}}^{-1}(\{Z^0\})$ as a linear combination of expressions

$$\lambda \, (\iota \circ \pi)^* \big(z_0^{n|\delta|_\infty + \sum_{j=1}^n (\deg p_j - \delta_j)}\big) \, \frac{|e_{\mathscr{U}}|^{2\lambda}}{e_{\mathscr{U}}^n} \, u_\iota^\lambda \Big(\frac{\overline{dv_\iota}}{\overline{v}_\iota} \wedge \xi_\iota + \eta_\iota\Big),$$

where u_ι is a strictly positive smooth function and ξ_ι, η_ι are smooth $(0, n-1)$ and $(0,n)$, respectively, forms. We have all elements in hand to state the following result.

PROPOSITION 6.57. *Let $\mathbb{K} \subset \mathbb{C}$, $\mathbb{J}$ be a purely n-dimensional radical ideal in $\mathbb{K}[X,Y]$, $p_1, \dots, p_n \in \mathbb{K}[X,Y]$ such that $\bar{p}$ is a quasi-regular sequence in $\mathbb{K}_{\mathbb{J}}[X,Y]$ and $\mathbb{I}_{\bar{p}} \neq \mathbb{K}_{\mathbb{J}}[X,Y]$. Suppose that $p = (p_1, \dots, p_n)$ admits $(\delta_1, \dots, \delta_n)$ as a Lojasiewicz multi-exponent for $\mathbb{X} = \operatorname{Spec} \mathbb{K}_{\mathbb{J}}[X,Y]$, such that*

$$\sum_{j=1}^n (\deg p_j - \delta_j) - n > 0. \tag{6.156}$$

Then one has, for $q \in \mathbb{K}[X,Y]$, for all subsets $I \subset \{1, \dots, n\}$, $J \subset \{1, \dots, N-n\}$ whose sum of cardinals equals n,

$$\deg q < \sum_{j=1}^n (\deg p_j - \delta_j) - n \Longrightarrow \operatorname{Res}_{\mathbb{K}_{\mathbb{J}}[X,Y]/\mathbb{K}} \begin{bmatrix} \bar{q} \, d\bar{X}_I \wedge d\bar{Y}_J \\ \bar{p}_1, \dots, \bar{p}_n \end{bmatrix} = 0. \tag{6.157}$$

PROOF. We first observe (if one replaces each δ_j by $\delta_j + 1/\nu$ with $\nu \gg 1$) that one may assume that there exists $C, c > 0$ such that (6.142) holds. Up to a change of coordinates, one may assume that $I = \{1, \dots, n\}$, $J = \emptyset$, with the geometric assumption (6.112) fulfilled. If p admits $(\delta_1,\dots,\delta_n)$ as a Łojasiewicz multi-exponent with respect to $\mathbb{X}$, then, for any $r\in\mathbb{N}^*$,

$$p^{[r]}=\big(p_1(X^r,Y^r),\dots,p(X^r,Y^r)\big)\quad \big(X^r=(X_1^r,\dots,X_n^r),\ Y^r=(Y_1^r,\dots,Y_{N-n}^r)\big),$$

admits $(r\delta_1,\dots,r\delta_n)$ as a Łojasiewicz multi-exponent with respect to the affine algebraic variety $\mathbb{X}_r=\mathrm{Spec}\,(\mathbb{K}_{\mathbb{J}_r}[X,Y])$, $\mathbb{J}_r$ being the radical ideal in $\mathbb{K}[X,Y]$ which zero set in $\mathbb{C}^N$ is $\{(z,w)\in\mathbb{C}^N : (z^r,w^r)\in\mathbb{X}_{\mathbb{C}}\}$. Since

$$r^{N-n}\,\mathrm{Res}_{\mathbb{K}_{\mathbb{J}}[X,Y]/\mathbb{K}}\begin{bmatrix}\bar q\,d\bar X\\ \bar p_1,\dots,\bar p_n\end{bmatrix}=\mathrm{Res}_{\mathbb{K}_{\mathbb{J}_r}[X,Y]/\mathbb{K}}\begin{bmatrix}\bar q^{[r]}\,(\bar X_1\cdots\bar X_n)^{r-1}\,d\bar X\\ \bar p_1^{[r]},\dots,\bar p_n^{[r]}\end{bmatrix},$$

with $\deg q_r=r\,\deg q+(n-1)r$, proving the result for $p=(p_1,\dots,p_n)$ amounts to proving it when p is replaced by $p^{[r]}$, thus one may assume that $\delta_1,\dots,\delta_n\in\mathbb{N}^*$. Up to a change of coordinates, one may assume that $I=\{1,\dots,n\}$, $J=\emptyset$. Let us take $q\in\mathbb{K}[X,Y]$ and start with the representation (6.153) for the residue symbol, where $\chi\in\mathcal{D}(\mathbb{P}^N_{\mathbb{C}},[0,1])$ is identically equal to one in a neighborhood of $V(\mathbb{J})\cap V(p)$ and identically equal to 0 in a neighborhood of $\overline{V(\mathbb{J})}\setminus V(\mathbb{J})$. As in the proof of Proposition 6.41, we observe that hypothesis (6.142) implies that

$$\big|z_0^{|\delta|_\infty}\big|\le C_{Z^0}\max_{1\le j\le n}\big|P_j(Z)\,z_0^{|\delta|_\infty-\delta_j}\big|$$

in a neighborhood of any $Z^0\in\overline{V(\mathbb{J})}\setminus V(\mathbb{J})$ for some $C_{Z^0}>0$. As a consequence $e^n_{\mathscr{U}}$ divides $(\iota\circ\pi_{\mathscr{U}})^*(z_0^{|\delta|_\infty})$ in any expression such as (6.155) in a neighborhood of a point $\mathscr{Z}^0\in\pi_{\mathscr{U}}^{-1}(Z^0)$ in the complex manifold $\mathscr{U}$ such that $\pi_{\mathscr{U}}:\mathscr{U}\to U$ is a convenient toric log resolution for $\mathcal{I}_{s_\delta}$ in a neighborhood of Z^0. Since

$$(\iota\circ\pi_{\mathscr{U}})^*\big(z_0^{-\deg q-n-1}\,Q(Z)\,\Omega(Z)\big)=(\iota\circ\pi_{\mathscr{U}})^*\big(z_0^{-\deg q-n-1}\big)\,\varpi,$$

where ϖ is a smooth $(n,0)$ holomorphic form in a neighborhood of $\mathscr{Z}^0$, one concludes that whenever $\sum_{j=1}^n(\deg p_j-\delta_j)-n-1-\deg q\ge 0$, then any (n,n)-current such as

(6.158)
$$\Big[\lambda\,(\iota\circ\pi)^*\big(z_0^{n|\delta|_\infty+\sum_{j=1}^n(\deg p_j-\delta_j)-n-1-\deg q}\big)\,\frac{|e_{\mathscr{U}}|^{2\lambda}}{e^n_{\mathscr{U}}}\,u_\iota^\lambda\Big(\frac{\overline{dv_\iota}}{\overline{v_\iota}}\wedge\xi_\iota+\eta_\iota\Big)\wedge\varpi\Big]_{\lambda=0}$$

in a neighborhood of $\mathscr{Z}^0$ is the null current. As a consequence of the preliminary discussion above, the conclusion of Proposition 6.57 follows. □

REMARK 6.58. When δ is p-admissible, Proposition 6.57 provides an alternative proof of Theorem 6.56. If one takes $\delta_1=\cdots=\delta_n=\deg\mathbb{X}\prod_{j=1}^n\deg p_j$, which is always a Łojasiewicz multi-exponent for p according to Proposition 6.55, then

$$z_0^{n\deg\mathbb{X}\prod_{j=1}^n\deg p_j-\sum_{j=1}^n\deg p_j+\deg q+n+1}$$

annihilates each (n,n)-current (6.158).

6.5.3. Properness and toric residue symbols (characteristic 0). Let $\Delta_1,\dots,\Delta_n$ be n-dimensional $\mathbb{Z}^n$-rational polyhedra[37] in $\mathbb{R}^n=\mathbb{Z}^n\otimes_{\mathbb{Z}}\mathbb{R}$ such that the Minkowski sum $\sum_1^n\Delta_j$ is n-dimensional. One can attach to such a configuration (see, e.g., [**Kh77**]) a primitive $\mathbb{Z}^n$-rational fan Σ compatible with $\sum_1^n\Delta_j$. It means that Σ is a primitive refinement of the normal fan of the n-dimensional $\mathbb{Z}^n$-rational

[37] A convex polyhedron in $\mathbb{R}^n$ is said to be $\mathbb{Z}^n$- (respectively $\mathbb{Q}^n$-) rational if its vertices lie in the lattice $\mathbb{Z}^n$ (respectively, in $\mathbb{Z}^n/N$ for some $N\in\mathbb{N}^*$).

polyhedron $\sum_1^n \Delta_j$ which is also the $\mathbb{Z}^n$-rational fan *dual* to Δ; see Example 3.4. Any Δ_j is interpreted as the stabilizer

$$\Delta_j = \{\xi \in \mathbb{R}^n \,:\, \langle \xi\,,\cdot\rangle \geq \Psi_{D_j} \text{ on } \mathbb{R}^n\}$$

of the concave virtual function Ψ_{D_j} of a toric Cartier divisor D_j on $\mathbb{X}_\Sigma$ with toric meromorphic distinguished section s_{D_j}; see §D.2.2, Example D.40 and §D.5.2. If $\xi_1, \dots, \xi_d$, where $d = n + r$, denote the rays in Σ, it is convenient to attach to the toric complete complex manifold $(\mathbb{X}_\Sigma)_\mathbb{C}$ its Cox ring of homogeneous coordinates $\mathbb{C}[z_1, \dots, z_d]$; see §3.1.2.

For each $j = 1, \dots, n$, let F_j be a Laurent polynomial in $\mathbb{K}[X_1^{\pm 1}, \dots, X_n^{\pm 1}]$ whose support $\mathscr{A}_j$ is such that $\operatorname{conv}(\mathscr{A}_j) = \Delta_j$ for $j = 1, \dots, n$, which means that

$$F_j = \sum_{\alpha \in \mathscr{A}_j} \gamma_{j,\alpha}\, X^\alpha \quad \text{with } \gamma_{j,\alpha} \in \mathbb{K}^*, \quad j = 1, \dots, n, \quad \mathscr{A}_j \subset \Delta_j \cap \mathbb{Z}^n.$$

We assume that $(F_1, \dots, F_n)$ defines a quasi-regular sequence in $\mathbb{I} = \mathbb{K}[X_1^{\pm 1}, \dots, X_n^{\pm 1}]$ with $\mathbb{I}_F \neq \mathbb{K}[X_1^{\pm 1}, \dots, X_n^{\pm 1}]$. In the same vein as Definition 6.53, let us introduce the notion of *multi-Łojasiewicz polyhedron* with respect to the complete (nonsingular) toric variety $\mathbb{X}_\Sigma$.

DEFINITION 6.59. Let $(\delta_1, \dots, \delta_n)$ be a collection of $\mathbb{Q}^n$-rational convex polyhedra with $\{0\} \subset \delta_j$ for $j = 1, \dots, n$. One says that $F = (F_1, \dots, F_n)$ admits $(\delta_1, \dots, \delta_n)$ as *multi-Lojasiewicz polyhedron* with respect to the complete toric variety $\mathbb{X}_\Sigma$ (defined over $\mathbb{Q}$) if and only if for each $(\epsilon_1, \dots, \epsilon_n)$ with $0 < \epsilon_j \ll 1$, there exist strictly positive constants C_ϵ and c_ϵ such that

$$\text{(6.159)} \quad \forall\, z \in \mathbb{C}^n \text{ and } \operatorname{Re} z \geq C_\epsilon$$

$$\Longrightarrow \max_{1 \leq j \leq n} \Big(\frac{F_j(\exp(z_1), \dots, \exp(z_n))}{\exp\big(\max_{\xi \in \Delta_j} \langle\!\langle \xi\,,\, \operatorname{Re}\, z\rangle\!\rangle\big)} \Big) \exp\big(\max_{\xi \in \delta_j + \epsilon_j \mathbb{B}_n} \langle\!\langle \xi\,,\, \operatorname{Re}\, z\rangle\!\rangle\big) \geq c_\epsilon.$$

Such a multi-Łojasiewicz polyedron $(\delta_1, \dots, \delta_n)$ is said to be a *multi-polyhedron of properness* if and only if it satisfies in addition that

$$\sum_{j=1}^n \{\xi \in \mathbb{R}^n \,:\, \xi + \delta_j \subset \Delta_j\} = \Delta_1' + \cdots + \Delta_n' \subset \Delta_1 + \cdots + \Delta_n$$

is an n-dimensional $\mathbb{Q}$-rational polyhedron. Any multipolyhedron of properness $(\delta_1, \dots, \delta_n)$ is also called an *F-properness multi-exponent* since the algebraic morphism $z \in \mathbb{T}_\mathbb{C}^n \to F(z) \in \mathbb{C}^n$ is then proper.

EXAMPLE 6.60. It is equivalent to say that the extreme allowed multi-Łojasiewicz polyhedron $(\{0\}, \dots, \{0\})$ is a true F-multiproperness exponent and that the closures $\overline{V(F_j)}$ in $(\mathbb{X}_\Sigma)_\mathbb{C}$ of the closed analytic subsets $\{z \in \mathbb{T}_\mathbb{C}^n \,:\, F_j(z) = 0\}$ of $\mathbb{T}_\mathbb{C}^n$ intersect only in $\mathbb{T}_\mathbb{C}^n$. It can also be reformulated as

$$\text{(6.160)} \quad \forall u \in \mathbb{S}^{n-1}, \quad \bigcap_{j=1}^n \Big\{ z \in \mathbb{T}_\mathbb{C}^n\,; \sum_{\{\alpha \in \mathscr{A}_j \,:\, \langle\!\langle \alpha, u\rangle\!\rangle = \min_{\xi \in \Delta_j} \langle\!\langle \xi, u\rangle\!\rangle\}} \gamma_{j,\alpha}\, z^\alpha = 0 \Big\} = \emptyset,$$

which are the nondegeneracy conditions introduced by A. Khovanski [**Kh78**]. The proof of such equivalence relies on using the Cox ring of homogeneous coordinates and the representation of $(\mathbb{X}_\Sigma)_\mathbb{C}$ as in (3.5). As in the proof of Proposition 6.40, we may invoke the Briançon–Skoda theorem (Theorem 5.7) in the local ring $\mathcal{O}_{(\mathbb{X}_\Sigma)_\mathbb{C}, \alpha}$, where $\alpha \in ((\mathbb{X}_\Sigma)_{\mathbb{C}} \setminus \mathbb{T}^n) \cap \bigcap_{j=1}^n \overline{V(F_j)}$, to show that such subset is empty as soon as $(\{0\}, \dots, \{0\})$ is a properness multi-exponent for F. Under such conditions, it was proved by D. Bernstein in [**Be75**] (see also [**Kou75**], [**Kh78**], or [**PasR04**, Theorems 4 and 5] for a re-interpretation in terms of Monge–Ampère real measures of Ronkin functions (see §D.5.2), which provides, in fact, a local version of the so-called *Bernstein–Kouchnirenko theorem*) that the degree of the proper intersection 0-cycle $\operatorname{div}(F_1) \cap \cdots \cap \operatorname{div}(F_n)$ in $\mathbb{T}^n_{\mathbb{C}}$ (or here $(\mathbb{X}_\Sigma)_{\mathbb{C}}$ according to (6.160)) equals $n!$ times the mixed Minkowski volume

$$\operatorname{vol}(\Delta_1, \dots, \Delta_n) = \frac{1}{n!} \sum_{k=1}^n \sum_{1 \le j_1 < \cdots < j_k \le n} (-1)^{n-k} \operatorname{vol}_n(\Delta_{j_1} + \cdots + \Delta_{j_k}).$$

It is *Bézout's toric geometric intersection theorem* in the proper intersection case.

One can state in the toric context the following toric version of Euler–Jacobi's formula [**Kh78**], which is the transposition in the toric context of the general result [**KK87**] formulated in Remark 6.42 as a consequence of Proposition 6.41; see also Corollary 6.51 for the particular case where $\mathbb{K} \subset \mathbb{C}$ as here.

THEOREM 6.61. *Let $F_1, \dots, F_n \in \mathbb{K}[X_1^{\pm 1}, \dots, X_n^{\pm 1}]$, with $\mathbb{K} \subset \mathbb{C}$, be n Laurent polynomials such that $\dim(\sum_{j=1}^n \Delta_j) = n$, where $\Delta_j = \operatorname{conv}(\operatorname{Supp}(F_j))$ for $j = 1, \dots, n$, and the nondegeneracy conditions* (6.160) *are fulfilled. Then the sequence $F = (F_1, \dots, F_n)$ is quasi-regular with $V_{\mathbb{T}^n_{\mathbb{C}}}(F) \neq \emptyset$ and, for any $G \in \mathbb{K}[X_1^{\pm 1}, \dots, X_n^{\pm 1}]$,*

(6.161)

$$\operatorname{Supp} G \subset \operatorname{int}(\Delta_1 + \cdots + \Delta_n) \Longrightarrow \operatorname{Res}_{\mathbb{K}[X_1^{\pm 1}, \dots, X_n^{\pm 1}]/\mathbb{K}} \begin{bmatrix} G\, dX_1 \wedge \cdots \wedge dX_n \\ F_1, \dots, F_n \end{bmatrix} = 0.$$

PROOF. It relies on the same arguments as the proof of Corollary 6.51 (take $\mathbb{K} = \mathbb{C}$, $N = n$). The Cox ring of homogeneous coordinates $\mathbb{K}[X_0, \dots, X_n]$ is now replaced by $\mathbb{C}[z_1, \dots, z_d]$ and homogenization is a toric homogenization with respect to the complete primitive fan Σ; see [**CaCD97, CaD97, Cx96**]. □

The following result, which is the transposition in the toric setting of Theorem 6.56, also holds; see [**VidY01**, Proposition 2.2].

THEOREM 6.62. *Let $F_1, \dots, F_n \in \mathbb{K}[X_1^{\pm 1}, \dots, X_n^{\pm 1}]$, with $\mathbb{K} \subset \mathbb{C}$, be n Laurent polynomials as in Theorem* 6.61. *Suppose that $(\delta_1, \dots, \delta_n)$ is an F-properness multi-exponent with respect to $\mathbb{X}_\Sigma$. Since the map F is proper from $\mathbb{X}_\Sigma$ to $\mathbb{C}^n$, hence from $\mathbb{T}^n_{\mathbb{C}}$ to $\mathbb{C}^n$, then $(F_1, \dots, F_n)$ is quasi-regular in $\mathbb{K}[X_1^{\pm 1}, \dots, X_n^{\pm 1}]$ with $V_{\mathbb{T}^n_{\mathbb{C}}}(F) \neq \emptyset$. Moreover, if*

$$\Delta_j' = \{\xi \in \mathbb{R}^n : \xi + \delta_j \subset \Delta_j\} \subset \Delta_j$$

for $j = 1, \dots, n$, one has, for any $G \in \mathbb{K}[X_1^{\pm 1}, \dots, X_n^{\pm 1}]$, that

(6.162)

$$\operatorname{Supp} G \subset \operatorname{int}(\Delta_1' + \cdots + \Delta_n') \Longrightarrow \operatorname{Res}_{\mathbb{K}[X_1^{\pm 1}, \dots, X_n^{\pm 1}]/\mathbb{K}} \begin{bmatrix} G\, dX_1 \wedge \cdots \wedge dX_n \\ F_1, \dots, F_n \end{bmatrix} = 0.$$

PROOF. A proof involving the toric compactification $\mathbb{X}_\Sigma$, more precisely the Cox ring of homogeneous coordinates $\overline{\mathbb{C}}[z_1, \dots, z_d]$ attached to $(\mathbb{X}_\Sigma)_{\mathbb{C}}$, is proposed in [**VidY01**, §3.2]. It follows along the same lines as the proof of Proposition 6.57 in the previous subsection. An alternative proof, entirely affine, which does not use any toric compactification of $\mathbb{T}^n_{\mathbb{C}}$, but relies instead on Bochner–Martinelli integral representation formulas for total sums of residues, is proposed in [**Z05**, Theorem 2 and Corollary 3]. This proof follows along the same argument as the proof of Theorem 6.56. Since both proofs follow similar arguments to those respectively of Proposition 6.57 and Theorem 6.56 quoted above, we do not repeat them here and refer the reader to [**VidY01**, **Z05**] instead, or [**KhG96**, **KhG02**] for related results. □

6.6. Residue symbols and arithmetics

The basis field $\mathbb{K}$ throughout this section will be the field $\mathbb{Q}$ of rational numbers, whose invertible elements $\gamma = \pm\boldsymbol{\gamma}'/\boldsymbol{\gamma}$, with $\boldsymbol{\gamma}, \boldsymbol{\gamma}' \in \mathbb{N}^*$ being coprime, inherit the logarithmic size $[\gamma] = \max(\log\boldsymbol{\gamma}, \log\boldsymbol{\gamma}')$. The n-dimensional radical ideal $\mathbb{J} \subset \mathbb{Q}[X,Y]$ in Example 6.19 induces the definition of the algebraic n-dimensional algebraic variety $\mathbb{X} = \mathrm{Spec}\,(\mathbb{Q}_{\mathbb{J}}[X,Y])$ with degree $\mathbf{d} := \deg \mathbb{X}_{\mathbb{C}}$ and canonical logarithmic height $\mathbf{h}$. We will recall these definitions in §6.6.2. We assume unless specified otherwise, that the geometric assumption

$$\big|\{(z,w) \in \mathbb{X}_{\mathbb{C}} \hookrightarrow \mathbb{C}^n \times \mathbb{C}^{N-n} \,:\, z_1 = \cdots = z_n = 0\}\big| = \mathbf{d} \tag{6.163}$$

is fulfilled. Given n polynomials $p_1, \dots, p_n \in \mathbb{Z}[X,Y]$ with

$$p_j(X,Y) = \sum_{\kappa\in\mathbb{N}^n} \sum_{\chi\in\mathbb{N}^{N-n}} \pm\boldsymbol{\gamma}_{j,\kappa,\chi}\, X^\kappa Y^\chi \quad (\boldsymbol{\gamma}_{j,\kappa,\chi} \in \mathbb{N}),$$

we introduce the following respective markers for degrees and logarithmic sizes:

$$D(p) = \prod_{j=1}^n \deg p_j, \quad \mathrm{h}(p) = \max_{1\le j\le n} \mathrm{h}(p_j) := \max_{1\le j\le n} \log\max(1, |\boldsymbol{\gamma}_{j,\kappa,\chi}|). \tag{6.164}$$

Furthermore, given the differential form

$$\begin{aligned}\omega &= q(X,Y)\, dX_1 \wedge \cdots \wedge dX_n \\ &= \Big(\sum_{\kappa\in\mathbb{N}^n} \sum_{\chi\in\mathbb{N}^{N-n}} \pm\boldsymbol{\gamma}'_{\kappa,\chi}\, X^\kappa Y^\chi\Big)\, dX_1 \wedge \cdots \wedge dX_n \in \Omega^n_{\mathbb{Z}[X,Y]/\mathbb{Z}} \quad (\boldsymbol{\gamma}'_{\kappa,\chi} \in \mathbb{N}),\end{aligned}$$

we let

$$d_\omega = \deg q, \quad \mathrm{h}(\omega) = \mathrm{h}_1(q) := \log\|\mathrm{coeff}(q)\|_1. \tag{6.165}$$

The marker $\mathrm{h}_1(q)$ is called the *length* of the polynomial q. A distinctive property of the length is its submultiplicativity, that is

$$\mathrm{h}_1(q_1q_2) \le \mathrm{h}_1(q_1) + \mathrm{h}_2(q_2)$$

for $q_1, q_2 \in \mathbb{Z}[X,Y]$; the logarithmic size and the length can be compared thanks to the inequalities

$$\mathrm{h}(q) \le \mathrm{h}_1(q) \le \mathrm{h}(q) + (N+1)\deg q. \tag{6.166}$$

As we have seen in §6.3.1 (see Proposition 6.27 combined with Proposition 6.29), when $N = n$, $\mathbb{J} = \{0\}$ and $\mathbb{X} = \mathbb{Q}^n$, Euclid's algorithm governs the computation of coefficients in the formal pseudo-Taylor series

$$\sum_{\kappa\in\mathbb{N}^n} \mathrm{Res}_{\mathbb{Q}_{\mathbb{J}}[X,Y]/\mathbb{Q}} \begin{bmatrix} \omega(\bar{X},\bar{Y})/\bar{p}_0^{\ell+1} \\ \bar{p}_1^{\kappa_1+1}, \dots, \bar{p}_n^{\kappa_n+1} \end{bmatrix} t^\kappa \in \mathbb{Q}[[t]], \tag{6.167}$$

where $p = (p_1, \dots, p_n) \in \mathbb{Z}[X,Y]$ is such that its class $\bar{p}$ modulo $\mathbb{J}$ is quasi-regular in $\mathbb{Q}_{\mathbb{J}}[X,Y]$ and $p_0 \in \mathbb{Z}[X,Y]$ is such that $(\bar{p}_0, \bar{p})\,\mathbb{Q}_{\mathbb{J}}[X,Y] = \mathbb{Q}_{\mathbb{J}}[X,Y]$, $\ell \in \mathbb{N}$. For the general case of (6.167) see Proposition 6.45 in §6.4.2. We begin by recalling the arithmetic complexity of Euclid's iterated algorithm, hence that of algebraic residue computations over $\mathbb{Q}$ in §6.6.1. Once respective markers for the geometric and arithmetic complexities are settled (this is the content of §6.6.2), we will give in §6.6.3 explicit upper estimates about the logarithmic size

$$\left[\operatorname{Res}_{\mathbb{Q}_{\mathbb{J}}[X,Y]/\mathbb{Q}}\begin{bmatrix} \omega(\bar{X},\bar{Y})/\bar{p}_0^{\ell+1} \\ \bar{p}_1^{\kappa_1+1}, \dots, \bar{p}_n^{\kappa_n+1}\end{bmatrix}\right]$$

in terms of $|\kappa|$, ℓ, $(\mathbf{d},\mathbf{h})$, $(D(p),\mathrm{h}(p))$, $(\deg p_0, \mathrm{h}(p_0))$, $(d_\omega, \mathrm{h}(\omega))$. Thus we derive estimates for the arithmetic complexity of the formal power series (6.167). A dynamical approach relying on the algebraic trace is presented in §6.6.4. It will provide us with an alternative presentation to such arithmetic complexity estimates.

6.6.1. The univariate case. Given the crucial role played by Euclid's division algorithm in the univariate setting, it seems essential to begin this section by quantifying precisely Euclid's algorithm, when the input data belong to the polynomial nœtherian ring $\mathbb{Z}[X]$ in the single variable X over the integral domain $\mathbb{Z}$. Surprisingly, such a result is not readily available in the literature. Let us formulate it in the language of univariate residue theory; see §1.3.1 or (and) Proposition 6.29.

LEMMA 6.63. *Let $p(X) = \gamma\, X^D + \sum_{\kappa=1}^{D} \gamma_{D-j}\, X^{D-\kappa} \in \mathbb{Z}[X]$, where $\gamma \in \mathbb{N}^*$. Let $\omega = q\, dX \in \Omega_{\mathbb{Z}[X]/\mathbb{Z}}$. Then, one has that*

$$\gamma^{d_\omega+1}\left(\left(\frac{1}{\gamma}\right)^{(D-1)(\kappa+1)} \operatorname{Res}_{\mathbb{Q}[X]/\mathbb{Q}}\begin{bmatrix}\omega \\ p^{\kappa+1}\end{bmatrix}\right) \in \mathbb{Z} \tag{6.168}$$

for any $\kappa \in \mathbb{N}$. The integer (6.168) *satisfies the estimate*

$$\left|\gamma^{d_\omega+1}\left(\left(\frac{1}{\gamma}\right)^{(D-1)(\kappa+1)} \operatorname{Res}_{\mathbb{Q}[X]/\mathbb{Q}}\begin{bmatrix}\omega \\ p^{\kappa+1}\end{bmatrix}\right)\right| \le \mathrm{h}(\omega) + \big(d_\omega + 1 - (\kappa+1)D\big)\, h(p) + (d_\omega + 1 - D)\log 2. \tag{6.169}$$

Moreover, if $d_\omega + 1 - (\kappa+1)D < 0$, then

$$\gamma^{d_\omega+1}\left(\left(\frac{1}{\gamma}\right)^{(D-1)(\kappa+1)} \operatorname{Res}_{\mathbb{Q}[X]/\mathbb{Q}}\begin{bmatrix}\omega \\ p^{\kappa+1}\end{bmatrix}\right) = 0.$$

PROOF. We refer to [**SomY21**, Theorem 3.2] for a detailed presentation of this result using elementary reasoning. Nevertheless, the proof is interesting since it controls almost optimally the effectiveness of Euclid's iterated algorithm regarding the computation of algebraic residue symbols in the univariate setting. ☐

REMARK 6.64. The above estimates are nearly optimal. Let D, $\gamma \le \gamma_{D-1}$ be strictly positive integers and let $p(X) = \gamma X^D - \gamma_{D-1} X^{D-1}$ be a polynomial as in Lemma 6.63. For the form $\omega = \gamma'\, X^{d_\omega}\, dX$, where $d_\omega \in \mathbb{N}^*$ and $\kappa \in \mathbb{N}$ are such that

$d_\omega - (\kappa+1)D \geq 0$, then

$$\begin{aligned}
\operatorname{Res}_{\mathbb{Q}[X]/\mathbb{Q}}\begin{bmatrix}\omega \\ p^{\kappa+1}\end{bmatrix} &= \gamma' \binom{d_\omega - (\kappa+1)D}{\kappa} \frac{\gamma_{D-1}^{d_\omega+1-(\kappa+1)D}}{\gamma^{d_\omega+1-(\kappa+1)(D-1)}} \\
&= \binom{d_\omega + 1 - (\kappa+1)D}{\kappa} \frac{\exp\big(\mathrm{h}(\omega) + (d_\omega+1-(\kappa+1)D)\,\mathrm{h}(p)\big)}{\gamma^{d_\omega+1-(\kappa+1)(D-1)}},
\end{aligned}$$

where the binomial coefficient is estimated by $2^{d_\omega+1-D}$, which fits with upper estimates (6.169).

When $p_0 \in \mathbb{Z}[X]$ is coprime with p, one can derive from Lemma 6.63 estimates for the logarithmic sizes

$$\left\lceil \operatorname{Res}_{\mathbb{Q}[X]/\mathbb{Q}}\begin{bmatrix}\omega/p_0 \\ p^{\kappa+1}\end{bmatrix} \right\rceil \tag{6.170}$$

in terms of $\kappa \in \mathbb{N}$, $(D, \mathrm{h}(p))$, $(d_\omega, \mathrm{h}(\omega))$, $(\deg(p_0), \mathrm{h}(p_0))$. To do that, observe that the Sylvester resultant determinant $\operatorname{res}(p_0, p)$ of p and p_0 is a nonzero integer satisfying

$$\log|\operatorname{res}(p_0,p)| \leq \deg(p_0)\,\mathrm{h}(p) + D\,\mathrm{h}(p_0). \tag{6.171}$$

Moreover, one has that $(|\operatorname{res}(p_0,p)|)^{\kappa+1} \equiv \boldsymbol{D}_0^\kappa[p_0]\,p_0$ modulo $p^{\kappa+1}\,\mathbb{Z}[X]$, where $\deg \boldsymbol{D}_0^\kappa[p_0] = (\kappa+1)D - 1$ and

$$\mathrm{h}\big(\boldsymbol{D}_0^\kappa[p_0]\,dX\big) \leq \big((\kappa+1)D - 1\big)\,\mathrm{h}(p_0) + (\kappa+1)\,\deg p_0\,\mathrm{h}(p). \tag{6.172}$$

Since

$$\operatorname{Res}_{\mathbb{Q}[X]/\mathbb{Q}}\begin{bmatrix}\omega/p_0 \\ p^{\kappa+1}\end{bmatrix} = \frac{1}{(\operatorname{res}(p_0,p))^{\kappa+1}} \operatorname{Res}_{\mathbb{Q}[X]/\mathbb{Q}}\begin{bmatrix}\boldsymbol{D}_0^\kappa[p_0]\,\omega \\ p^{\kappa+1}\end{bmatrix}$$

for any $\kappa \in \mathbb{N}$, it follows from Lemma 6.63 and the relations

$$\begin{aligned}
d_{\boldsymbol{D}_0^\kappa[p_0]\,\omega} &= d_\omega + (\kappa+1)D - 1, \\
\mathrm{h}\big(\boldsymbol{D}_0^\kappa[p_0]\,\omega\big) &\leq \mathrm{h}(\omega) + \mathrm{h}\big(\boldsymbol{D}_0^\kappa[p_0]\,dX\big) \\
&\leq \mathrm{h}(\omega) + (\kappa+1)\big(D\,\mathrm{h}(p_0) + \deg(p_0)\,\mathrm{h}(p)\big) - \mathrm{h}(p_0)
\end{aligned}$$

that

$$(\operatorname{res}(p_0,p))^{\kappa+1}\,\gamma^{d_\omega+(\kappa+1)} \operatorname{Res}_{\mathbb{Q}[X]/\mathbb{Q}}\begin{bmatrix}\omega/p_0 \\ p^{\kappa+1}\end{bmatrix} \in \mathbb{Z}, \tag{6.173}$$

with

$$\begin{aligned}
\left|(\operatorname{res}(p_0,p))^{\kappa+1}\,\gamma^{d_\omega+(\kappa+1)} \operatorname{Res}_{\mathbb{Q}[X]/\mathbb{Q}}\begin{bmatrix}\omega/p_0 \\ p^{\kappa+1}\end{bmatrix}\right| &\leq \mathrm{h}(\omega) \\
+ \big(d_\omega + (\kappa+1)\deg(p_0)\big)\,\mathrm{h}(p) + \big((\kappa+1)D - 1\big)\,\mathrm{h}(p_0) &+ (d_\omega + \kappa\,D)\log 2.
\end{aligned} \tag{6.174}$$

6.6.2. Geometric and arithmetic markers for complexity. We recall that the degree $\deg \mathbb{X}$ of $\mathbb{X}_\mathbb{C}$ is defined as the degree of its Zariski closure $\overline{\mathbb{X}_\mathbb{C}}$ in $\mathbb{P}^N_\mathbb{C}$. Namely, it is defined as

$$\deg \mathbb{X} = \int_{\overline{\mathbb{X}_\mathbb{C}}} \big(c_1(L^\circ, |\ |)\big)^n \qquad \big(L^\circ = \mathcal{O}_{\mathbb{P}^N_\mathbb{C}}(1)\big), \tag{6.175}$$

where $|\ |$ is a continuous metric on L° and $c_1(L^\circ, |\ |)$ denotes the first Chern current of the metrized bundle thus obtained. This is the Wirtinger formula. If $|\ |$ is the Fubini–Study metric $\|\ \|$, then $c_1 = dd^c \log\|Z\|^2$, where $Z = [z_0 : Z_1 : \cdots : Z_N]$

denotes the homogeneous coordinates on $\mathbb{P}^N_{\mathbb{C}}$, with $z_j z_0 = Z_j$ for $j = 1, \dots, n$ and $w_j z_0 = Z_j$ for $j = 1, \dots, N - n$. Furthermore, the expression (6.175) for $\deg \mathbb{X}$ is equivalent to

(6.176)

$$\deg \mathbb{X} = \int_{\left(\prod_{\ell=1}^{n} (\mathbb{P}^N_{\mathbb{C}})_\ell\right) \times \overline{\mathbb{X}_{\mathbb{C}}}} \Big(\prod_{\ell=1}^{n} c_1((L^\circ)_\ell, \| \ \|)(u_\ell) \Big) \, [\langle\!\langle u_1, Z \rangle\!\rangle = \cdots = \langle\!\langle u_n, Z \rangle\!\rangle = 0],$$

as a consequence of Crofton's formula; see [**De**], [**AndSWY17B**, Theorem 6.2] or §D.5.1. In the case where the metric $| \ |$ is not smooth but just continuous, the nth power of $c_1(L^\circ, | \ |)$ in (6.175) is defined along the Bedford–Taylor recursive process [**BedT76**]; see also (C.7), §C.1.2. Since $\mathbb{X}$ remains fixed throughout this section, one will use from now the notation

$$\mathbf{d} := \deg \mathbb{X} = \deg \overline{\mathbb{X}}_{\mathbb{C}}. \tag{6.177}$$

In particular $\mathbf{d} = 1$ in the setting where $N = n$, $\mathbb{J} = \{0\}$, $X = (X_1, \dots, X_n)$, $Y = \{ \ \}$, $\mathbb{X} = \mathbb{A}^N_{\mathbb{Q}} = \mathbb{A}^n_{\mathbb{Q}}$. Given holomorphic sections $P_j \in \mathcal{O}(\mathbb{P}^N_{\mathbb{C}}, (L^\circ)^{d_j})$, $j = 1, \dots, n$, $d_j \in \mathbb{N}^*$, whose divisors $\mathrm{div}(P_j)$ have their supports intersecting properly with $\overline{\mathbb{X}_{\mathbb{C}}}$ in $\mathbb{P}^N_{\mathbb{C}}$, one has

$$\begin{aligned} \int_{\overline{\mathbb{X}_{\mathbb{C}}}} [P_1 = 0] \wedge \cdots \wedge [P_n = 0] &= \deg \Big(\mathrm{Spec} \Big(\frac{\mathrm{Proj}\, \mathbb{Q}_{\mathbb{J}}[z_0, Z_1, \dots, Z_n]}{(P_1, \dots, P_n)} \Big) \Big) \\ &= \deg \mathbb{X} \prod_{j=1}^{n} d_j = \mathbf{d} \prod_{j=1}^{n} d_j \end{aligned} \tag{6.178}$$

in accordance with *Bézout's geometric intersection theorem.* If $p_1, \dots, p_n \in \mathbb{Z}[X, Y]$ are such that $(\bar{p}_1, \dots, \bar{p}_n)$ is a quasi-regular sequence in $\mathbb{Q}_{\mathbb{J}}[X, Y]$, then it follows from Bézout's geometric intersection theorem within the frame of improper intersection theory in $\mathbb{P}^N_{\mathbb{C}}$, that

$$\deg \Big(\mathrm{Spec} \Big(\frac{\mathbb{Q}_{\mathbb{J}}[X, Y]}{(\bar{p}_1, \dots, \bar{p}_n)\, \mathbb{Q}_{\mathbb{J}}[X, Y]} \Big) \Big) \leq \mathbf{d} \prod_{j=1}^{n} \deg p_j = \mathbf{d}\, D(p); \tag{6.179}$$

see [**Ful84, Ful**], [**AndESWY21B**, Theorem 1.1] or Proposition D.59. It follows from the situation described in (6.179) that $\mathbf{d}\, D(p)$ is a marker from above for the geometric complexity of the intersection 0-cycle realized as the proper intersection

$$\mathbb{X}_{\mathbb{C}} \,\dot{\cap}\, \mathrm{div}_{\mathbb{C}^N}(p_1) \,\dot{\cap} \cdots \dot{\cap}\, \mathrm{div}_{\mathbb{C}^N}(p_n).$$

From now on we use the notations (compatible with Bézout's geometric intersection theorem)

$$\mathbf{d}\,(p) = \deg \mathbb{X} \times D(p) = \deg \mathbb{X} \times \prod_{j=1}^{n} \deg p_j, \qquad \mathbf{d}\,(p, a) = \mathbf{d}\,(p)\, \deg a, \tag{6.180}$$

whenever the sequence $\bar{p} = (\bar{p}_1, \dots, \bar{p}_n)$, $p_j \in \mathbb{Z}[X, Y]$, is quasi-regular in $\mathbb{Q}_{\mathbb{J}}[X, Y]$ and $a \in \mathbb{Z}[X, Y]$ with $\deg a > 0$.

Similarly, to quantify the arithmetic complexity of $\mathbb{X}$, we quantify the arithmetic complexity of the closure $\overline{\mathbb{X}}$ of $\iota(\mathbb{X})$, where

$$\iota : \mathbb{A}^N_{\mathbb{Q}} \hookrightarrow \mathbb{P}^N_{\mathbb{Q}} = \mathrm{Spec}\big(\mathrm{Proj}\, \mathbb{Q}[z_0, Z_1, \dots, Z_N]\big).$$

Namely, we consider on $\overline{\mathbb{X}}$ the algebraic line bundle $\mathbb{L}^\circ = \mathbb{O}_{\mathbb{P}^N_{\mathbb{Q}}}(1)$. Let $\mathrm{s}_0, \dots, \mathrm{s}_{n+1}$ be $n+1$ generic global sections of $\mathbb{L}^\circ$, that is,

$$\mathrm{s}_j = \gamma_{j,0}\, z_0 + \sum_{k=1}^{N} \gamma_{j,k} Z_k, \quad (\gamma_{j,0}, \dots, \gamma_{j,n}) \in \mathbb{Z}^{N+1} \setminus \{(0,\dots,0)\}. \tag{6.181}$$

Assume first that $|\ |$ is the canonical metric on $\mathbb{L}^\circ_{\mathbb{C}} = L^\circ$, defined on the fiber L°_Z by

$$|t|_Z = \frac{|t|}{\max(|z_0|, |Z_1|, \dots, |Z_N|)} \quad \forall t \in L^\circ_Z.$$

Its first Chern current equals

$$c_1(L^\circ, |\ |) = dd^c \log \max(|z_0|, |Z_1|, \dots, |Z_N|) \overset{\text{weakly}}{=} \lim_{r \to +\infty} \frac{dd^c \log \|Z^r\|}{r}, \tag{6.182}$$

where $Z^r = (z_0^r, Z_1^r, \dots, Z_N^r)$. Since the complex hyperplanes $|(\mathrm{div}(\mathrm{s}_j))_{\mathbb{C}}|$ intersect properly with $\overline{\mathbb{X}}_{\mathbb{C}}$ in $\mathbb{P}^N_{\mathbb{C}}$, when $(\gamma_{j,0}, \dots, \gamma_{j,N})$, $j = 0, \dots, n$, are taken generically, the corresponding (n,n)-current

$$G_{\mathrm{can},\mathrm{s}} = -\sum_{j=0}^{n} \log|\mathrm{s}_{n-j}| \Big(\prod_{\ell < n-j} c_1(L^\circ, |\ |) \Big) \bigwedge_{\ell > n-j} [(\mathrm{div}(\mathrm{s}_\ell))_{\mathbb{C}}] \Big) \tag{6.183}$$

is well defined (see for example [**De**, III, Theorem 4.5]) and satisfies the Green–Lelong–Poincaré equation

$$dd^c G_{\mathrm{can},\mathrm{s}} + [(\mathrm{div}(\mathrm{s}_0))_{\mathbb{C}}] \wedge \cdots \wedge [(\mathrm{div}(\mathrm{s}_n))_{\mathbb{C}}] = \prod_{\ell=0}^{n} c_1(L^\circ, |\ |). \tag{6.184}$$

In addition (6.183) is such that

$$\begin{aligned} h^\infty_{\mathrm{can},\mathrm{s}}(\overline{\mathbb{X}}) &:= \int_{\overline{\mathbb{X}}_{\mathbb{C}}} G_{\mathrm{can},\mathrm{s}} \\ &= \lim_{r\to+\infty} \frac{1}{r} \int_{\overline{\mathbb{X}}_{\mathbb{C}}} \Big(-\sum_{j=0}^{n} \log|\mathrm{s}_{n-j}| \Big(\prod_{\ell<n-j} c_1(L^\circ, \|\ \|)(Z^r) \Big) \bigwedge_{\ell>n-j} [(\mathrm{div}(\mathrm{s}_\ell))_{\mathbb{C}}] \Big) \end{aligned} \tag{6.185}$$

is also well defined as a real number. The second equality in (6.185) follows from (6.182), as a consequence of the monotone continuity theorem due to E. Bedford and B. A. Taylor [**BedT82**] (see also [**De**, III, Theorem 3.7]) since the sequence of smooth plurisubharmonic (psh) functions $\big((\log\|Z^r\|)/r\big)_{r\geq 1}$ decreases towards the psh function $\log(|z_0|, |Z_1|, \dots, |Z_N|)$ on $\mathbb{C}^{N+1} \setminus \{(0,\dots,0)\}$.

Now, for each prime integer p let v_{p} be the equivalence class (*place*) of the p-adic (ultrametric) absolute value $|\ |_{\mathrm{p}}$. Such absolute values are jointly defined by

$$\pm\frac{\gamma}{\gamma'} = \pm \prod_{\{\mathrm{p\ prime}\}} \mathrm{p}^{-\nu_{\mathrm{p}}(\gamma'/\gamma)} = \pm \prod_{\{\mathrm{p\ prime}\}} \Big|\frac{\gamma'}{\gamma}\Big|_{\mathrm{p}} \tag{6.186}$$

for coprime numbers $\gamma, \gamma' \in \mathbb{N}^*$, so that the product formula holds logarithmically

$$\sum_{\mathrm{p\ prime}} \log\Big|\frac{\gamma'}{\gamma}\Big|_{\mathrm{p}} = -\log\Big|\frac{\gamma'}{\gamma}\Big| = -\log\Big|\frac{\gamma'}{\gamma}\Big|_\infty \tag{6.187}$$

for any such γ, γ'. The usual archimedean absolute value $|\ |$ on $\mathbb{Q}$ is naturally denoted by $|\ |_\infty$, with its associated place being denoted by v_∞. The completion of $\mathbb{Q}$ for such archimedean absolute value $|\ |_\infty$ is $\mathbb{R}$ and its integral closure is $\mathbb{C}$, while the completion of $\mathbb{Q}$ with respect to $|\ |_{\rm p}$ is the field of p-adic numbers $\mathbb{Q}_{\rm p}$, with $\mathbb{C}_{\rm p}$ as an integral closure. To any such nonarchimedean place $v_{\rm p}$, one can associate the *Berkovich analytification* $(\mathbb{P}^N)^{\rm an}_{v_{\rm p}}$ with respect to the nonarchimedean place $v_{\rm p}$ of the complete algebraic variety $\mathbb{P}^N_{\mathbb{Q}}$. One may refer, for example, to [**BGPS14**, §1.2–§1.4] for a self-contained presentation of the construction of such compact topological space $(\mathbb{P}^N)^{\rm an}_{v_p}$ in the more general setting, where $\mathbb{P}^N_{\mathbb{Q}}$ is replaced by the complete toric variety $\mathbb{X}_\Sigma$ attached to a $\mathbb{Z}$-rational complete fan in the real vector space $\mathbb{R}^N = \mathbb{Z}^N \otimes_{\mathbb{Z}} \mathbb{R}$; see §3.1. From the point of view of complex analysis, it is highly interesting that one can, following an idea which was introduced by A. Lagerberg [**Lag12**], construct sheaves of (k,k')-*smooth forms*, together with the dual sheaves of $(N-k, N-k')$-*currents* on the analytification $(\mathbb{P}^N)^{\rm an}_{v_{\rm p}}$ for any $0 \le k, k' \le N$ with $k+k' \le N$. In addition, one can construct two operators $\boldsymbol{d}'$, respectively $\boldsymbol{d}''$, carrying (k,k')-forms (or $(N-k, N-k')$-currents) to $(k+1,k')$ forms (or $(N-(k-1), N-k')$-currents), respectively to $(k, k'+1)$ forms (or $(N-k, N-(k'-1))$-currents). One can see directly a similarity with the role played by ∂ and $\bar\partial$ on the complex analytification $\mathbb{P}^N_{\mathbb{C}} = (\mathbb{P}^N)^{\rm an}_{v_\infty}$. Thus the operator $\boldsymbol{d}'\boldsymbol{d}''$ substitutes naturally dd^c in such context. For such construction, we refer the reader to the seminal introductive paper [**ChD12**] as well as to [**GubK17**]. The algebraic line bundle $\mathbb{L}^\circ$ induces an *analytified line bundle* $(\mathbb{L}^\circ)^{\rm an}_{v_{\rm p}}$ over the analytification $(\mathbb{P}^N)^{\rm an}_{v_{\rm p}}$. The analytified line bundle $(\mathbb{L}^\circ)^{\rm an}_{v_{\rm p}}$ can be metrized as well, with its so-called continuous canonical metric $|\ |_{{\rm can}, v_p}$ being as described in [**BGPS14**, Example 1.3.11]. As a special case one has

$$\big((\mathbb{L}^\circ)^{\rm an}_{v_{\rm p}}, |\ |_{{\rm can},\infty}\big) = (L^\circ, |\ |)$$

when $v = v_\infty$ is the archimedean place. Fix now a prime number p. Given a global section

$$\mathrm{s} = \gamma_0 z_0 + \gamma_1 Z_1 + \cdots + \gamma_N Z_N \qquad ((\gamma_0, \dots, \gamma_n) \in \mathbb{Z}^{N+1} \setminus \{(0,\dots,0)\})$$

with divisor $\mathrm{div}_{\mathbb{P}^N_{\mathbb{Q}}}(\mathrm{s})$ on $\mathbb{P}^N_{\mathbb{Q}}$, there corresponds to it a $(\boldsymbol{d}', \boldsymbol{d}'')$-closed (integration) current

$$[\mathrm{div}(\mathrm{s})]_{v_{\rm p}} \in {}'\mathcal{D}^{(1,1)}((\mathbb{P}^N)^{\rm an}_{v_{\rm p}}, \mathbb{R}).$$

As in the complex setting, the first Chern current $c_1((\mathbb{L}^\circ)^{\rm an}_{v_{\rm p}}, |\ |_{{\rm can}, v_{\rm p}})$ is defined independently of the choice of the section s of $\mathbb{L}^\circ$ through the Lelong–Poincaré equation [**ChD12**, §6.4],

$$\boldsymbol{d}'\boldsymbol{d}''(-\log|\,\mathrm{s}\,|_{{\rm can}, v_{\rm p}}) + [\mathrm{div}(\mathrm{s})]_{v_{\rm p}} = c_1((\mathbb{L}^\circ)^{\rm an}_{v_{\rm p}}, |\ |_{{\rm can}, v_{\rm p}}). \tag{6.188}$$

The algebraic n-dimensional projective variety $\overline{\mathbb{X}}$ is such that

$$(\overline{\mathbb{X}})^{\rm an}_{v_{\rm p}} \hookrightarrow (\mathbb{P}^N)^{\rm an}_{v_{\rm p}},$$

and one can attach to it a $(\boldsymbol{d}', \boldsymbol{d}'')$-closed (integration) current

$$\big[(\overline{\mathbb{X}})^{\rm an}_{v_{\rm p}}\big] \in {}'\mathcal{D}^{(N-n,N-n)}((\mathbb{P}^N)^{\rm an}_{v_{\rm p}}, \mathbb{R}).$$

If $\mathrm{s}_0, \dots, \mathrm{s}_n$ are $n+1$ sections of $\mathbb{L}^\circ$ with vectors of generic coefficients $(\gamma_{j,0}, \dots, \gamma_{j,n})$ in $\mathbb{Z}^{N+1}\setminus\{(0,\dots,0)\}$, then it follows from the fact that the canonical metric $|\ |_{{\rm can}, v_{\rm p}}$

on $(\mathbb{L}^\circ)^{\rm an}_{v_{\rm p}}$ is semipositive (see [**BGPS14**, §1.3 and Definition 1.4.1]) and from the fact that the rational sections s_j intersect (jointly) properly with $\mathbb{X}$, that the integral

$$
\begin{aligned}
(6.189)\quad & h^{v_{\rm p}}_{{\rm can},{\rm s}}(\overline{\mathbb{X}}) : \\
&= \int_{(\mathbb{X})^{\rm an}_{v_{\rm p}}} \Big(-\sum_{j=0}^{n} \log|{\rm s}_{n-j}|_{{\rm can},v_{\rm p}} \Big(\prod_{\ell<n-j} c_1((\mathbb{L}^\circ)^{\rm an}_{v_{\rm p}}, |\ |_{{\rm can},v_{\rm p}})\Big) \bigwedge_{\ell>n-j} [{\rm div}({\rm s}_\ell)]_{v_{\rm p}}\Big) \\
&= \int_{(\mathbb{P}^N)^{\rm an}_{v_{\rm p}}} \Big(-\sum_{j=0}^{n} \log|{\rm s}_{n-j}|_{{\rm can},v_{\rm p}} \Big(\prod_{\ell<n-j} c_1((\mathbb{L}^\circ)^{\rm an}_{v_{\rm p}}, |\ |_{{\rm can},v_{\rm p}})\Big) \bigwedge_{\ell>n-j} [{\rm div}({\rm s}_\ell)]_{v_{\rm p}}\Big) \\
&\qquad\qquad\qquad\qquad\qquad\qquad\qquad\qquad\qquad\qquad\qquad\qquad\qquad [(\overline{\mathbb{X}})^{\rm an}_{v_{\rm p}}]
\end{aligned}
$$

is well defined as a real number; see [**BGPS14**, Proposition 1.4.5 and Theorem 1.4.10]. The following proposition is important because it substantiates the definition of the canonical height $\mathbf{h}$ of the algebraic affine variety $\mathbb{X}$ which is defined over $\mathbb{Q}$ as $\mathbb{X} = {\rm Spec}\,(\mathbb{Q}_{\mathbb{J}}[X_1,\dots,X_n,Y_1,\dots,Y_{N-n}])$.

PROPOSITION 6.65. *For all primes* p, *but a finite number primes, one has that* $h^{v_{\rm p}}_{{\rm can},{\rm s}}(\overline{\mathbb{X}}) = 0$. *The real number*

$$
\mathbf{h} = h^{\infty}_{{\rm can},{\rm s}}(\overline{\mathbb{X}}) + \sum_{\rm p\ prime} h^{v_{\rm p}}_{{\rm can},{\rm s}}(\overline{\mathbb{X}}) \tag{6.190}
$$

does not depend on the choice of the generic sections ${\rm s}_0,\dots,{\rm s}_n$ *of* $\mathbb{L}^\circ$. *It is nonnegative and equals* 0 *when* $N = n$ *and thus* $\mathbb{X} = \mathbb{A}^n_{\mathbb{Q}}$. *This nonnegative real number is called the global logarithmic canonical height of* $\mathbb{X}$ *and realizes a marker of the arithmetic complexity of the algebraic variety* $\mathbb{X}$.

PROOF. We refer here the reader to the seminal monograph [**BGPS14**] or to [**Gua18**, §3B and §3C] for a presentation of the concepts of local or global logarithmic heights for toric divisors equipped with toric metrics in an N-dimensional complete toric variety attached to a rational fan in $\mathbb{R}^N = \mathbb{Z}^N \otimes_{\mathbb{Z}} \mathbb{R}$. For the first assertion, see [**Gub02**, Proposition 5.3.7 (d)]. The independence of the right-hand side of (6.190) concerning ${\rm s}_0,\dots,{\rm s}_n$ follows from the product formula (6.187). Since $\mathbf{h} = h^{\infty}_{{\rm can},Z}$ when $\mathbb{X} = \mathbb{P}^N$, one has $\mathbf{h} \geq 0$ according to the semipositivity of the metrics $|\ |$ and $|\ |_{{\rm can},v_{\rm p}}$ and $|\ |$ respectively on L° and $(\mathbb{L}^\circ)^{\rm an}_{v_{\rm p}}$. □

REMARK 6.66. Instead of the canonical metric on L°, one takes the Fubini–Study metric (it exists on $(L^\circ)^{\rm an}_{v_{\rm p}}$; see [**ChM20**, §2.2.3]). The nonnegative real number

$$
\mathbf{h}_{\rm proj} = h^{\infty}_{\|\ \|,{\rm s}}(\overline{\mathbb{X}}) + \sum_{\rm p\ prime} h^{v_{\rm p}}_{{\rm can},{\rm s}}(\overline{\mathbb{X}}) \tag{6.191}
$$

is again independent of the choice of the sections ${\rm s}_0,\dots,{\rm s}_n$ of $\mathbb{L}^\circ$ and is called the *projective global logarithmic height of the algebraic variety* $\mathbb{X}$. One has the comparison inequality

$$
\big|\mathbf{h} - \mathbf{h}_{\rm proj}\big| \leq \frac{7}{2}\log(N+1)\,\mathbf{d}; \tag{6.192}
$$

see [**DP99**, Proposition 2.1 (v)]. One has also (according to (6.185) and its transposition on each $(\mathbb{P}^N)^{\mathrm{an}}_{v_{\mathrm{p}}}$ for prime p) asymptotically that

$$\frac{\mathbf{h}}{\mathbf{d}} = \frac{1}{r_0}\,\frac{\mathbf{h}^{[r_0]}}{\mathbf{d}^{[r_0]}} = \lim_{r\to+\infty}\frac{1}{r}\,\frac{\mathbf{h}^{[r]}_{\mathrm{proj}}}{\mathbf{d}^{[r]}} \tag{6.193}$$

for any $r_0 \in \mathbb{N}^*$. In (6.193) notations $\mathbf{d}^{[r]}$, $\mathbf{h}^{[r]}$, $\mathbf{h}^{[r]}_{\mathrm{proj}}$ denote the degree, the canonical and projective global logarithmic heights, respectively, of

$$\mathbb{X}_r = \mathrm{Spec}\,(\mathbb{Q}_{\mathbb{J}_r}[X,Y]),$$

where $\mathbb{J}_r = \pi_r^*\mathbb{J}$, $\pi_r : (X,Y) \to (X^r, Y^r)$, as in Proposition 6.47; see [**PhS08**, §1.2]. The projective logarithmic height is developed in the univariate setting (when $N = 1$ and $n = 0$) in §1.3.3 using the Chow arithmetic product; see Definition 1.54. The same construction carries through in the multivariate setting. We refer here to [**GS90, BGS94**] or to the presentation involving multivariate residue calculus proposed in [**BY98**, §1] in the particular case where $\overline{\mathbb{X}}_{\mathbb{C}}$ is defined as a complete intersection in $\mathbb{P}^N_{\mathbb{C}}$ and then extended in [**BY00**, §3] to the general case.

EXAMPLE 6.67 (Codimension 1 case). Let $N = n+1$ and let $\mathbb{J}$ be the ideal generated by a polynomial $\mathbf{p} \in \mathbb{Z}[X_1,\dots,X_n,Y]$ with jointly coprime coefficients in $\mathbb{Z}$. Then one has a global canonical logarithmic height for $\mathbb{X} = \mathrm{Spec}\,(\mathbb{Q}_{\mathbb{J}}[X,Y])$ the real nonnegative number

$$\mathbf{h} = \frac{1}{(2\pi)^{n+1}}\int_{(\mathbb{R}/\mathbb{Z})^{n+1}} \log|\mathbf{p}(e^{i\theta_1},\dots,e^{i\theta_{n+1}})|\,d\theta_1\cdots d\theta_{n+1}. \tag{6.194}$$

Thus the number (6.194) (in general transcendental) is the logarithm of the Mahler measure of the irreducible polynomial $\mathbf{p}$; see also §D.5.2. In the univariate setting, for $n = 0$ and $N = 1$, (6.194) is a re-interpretation of Jensen's formula.

Bézout's geometric intersection theorem in $\mathbb{P}^N_{\mathbb{Q}}$ asserts that whenever

$$\mathbb{X} = \sum_\iota \nu_\iota\,\mathbb{X}_\iota, \quad \mathbb{X}' = \sum_{\iota'} \nu'_\iota\,\mathbb{X}'_\iota$$

are two purely dimensional effective cycles on $\mathbb{P}^N$ with corresponding dimensions $n(\mathbb{X})$ and $n(\mathbb{X}')$, that satisfy $n(\mathbb{X}) + n(\mathbb{X}') \geq N$ and

$$\mathrm{codim}\,\big((\overline{\mathbb{X}_\iota})_{\mathbb{C}} \cap (\overline{\mathbb{X}'_{\iota'}})_{\mathbb{C}}\big) = n(\mathbb{X}) + n(\mathbb{X}') - N$$

for any ι, ι', then one has

$$\deg\big(\mathbb{X}\,\dot{\cap}\,\mathbb{X}'\big) = \deg\mathbb{X} \times \deg\mathbb{X}'. \tag{6.195}$$

Bézout's arithmetic intersection theorem (see [**BGS94**, Corollary 4.2.4]), combined with (6.192), asserts that for some positive constants

$$c_{\mathrm{proj}}(n(\mathbb{X}), n(\mathbb{X}'), N),\ c(n(\mathbb{X}), n(\mathbb{X}'), N),$$

the following estimates

$$\begin{aligned}(\mathbf{h}_{\mathrm{proj}})_{\mathbb{X}\dot{\cap}\mathbb{X}'} &\leq \mathbf{h}_{\mathrm{proj}}\,\mathbf{d}' + \mathbf{h}'_{\mathrm{proj}}\,\mathbf{d} + c_{\mathrm{proj}}(n(\mathbb{X}), n(\mathbb{X}'), N)\,\mathbf{d}\,\mathbf{d}',\\ \mathbf{h}_{\mathbb{X}\dot{\cap}\mathbb{X}'} &\leq \mathbf{h}\,\mathbf{d}' + \mathbf{h}'\,\mathbf{d} + c(n(\mathbb{X}), n(\mathbb{X}'), N)\,\mathbf{d}\,\mathbf{d}'\end{aligned} \tag{6.196}$$

hold, in terms of the degrees $\mathbf{d}$, $\mathbf{d}'$, the projective and canonical logarithmic heights $(\mathbf{h}_{\mathrm{proj}}, \mathbf{h}'_{\mathrm{proj}})$, $(\mathbf{h}, \mathbf{h}')$ of $\mathbb{X}$ and $\mathbb{X}'$, respectively. Consider now the sequence of polynomials $(p_1,\dots,p_n) \in \mathbb{Z}[X,Y]$ such that $(\bar{p}_1,\dots,\bar{p}_n)$ is quasi-regular in $\mathbb{Q}_{\mathbb{J}}[X,Y]$.

Then we introduce as markers for arithmetic complexity, in parallel with the marker $\mathbf{d}(p)$ introduced in (6.180), the numbers

$$\begin{aligned}\mathbf{h}_c(p) &= \mathbf{d}(p) \times \Big(\frac{\mathbf{h}}{\mathbf{d}} + \sum_{j=1}^{n} \frac{\mathrm{h}(p_j)}{\deg(p_j)} + c\Big) \\ &= \mathbf{h}\, D(p) + \sum_{j=1}^{n} \mathrm{h}(p_j)\, \mathbf{d}(p_1, \dots, \widehat{p_j}, \dots, p_n) + c\, \mathbf{d}(p),\end{aligned} \tag{6.197}$$

where c is a real positive parameter. We also use the notations (compatible with Bézout's arithmetic intersection theorem)

$$\begin{aligned}\mathbf{h}_c(a\, dX) &= \mathrm{h}(a\, dX) + (\mathbf{h} + c\, \mathbf{d}) \deg a, \\ \mathbf{h}_c(p, a) &= \mathrm{h}(a)\, \mathbf{d}(p) + \mathbf{h}_c(p) \deg(a)\end{aligned} \tag{6.198}$$

for any $a \in \mathbb{Z}[X, Y]$ with $\deg a > 0$ and $c \in \mathbb{R}$. These markers control arithmetic elimination and arithmetic nullstellensatz, as the two following theorems, both of major interest to us, illustrate. The first one concerns the arithmetic elimination theory.

THEOREM 6.68 (Arithmetic elimination (AE)). *Let $\mathbb{J}$ be a purely n-dimensional radical ideal in the polynomial ring $\mathbb{Q}[X_1, \dots, X_n, Y_1, \dots, Y_{N-n}] = \mathbb{Q}[X, Y]$. Assume also that the sequence of polynomials $p_1, \dots, p_n \in \mathbb{Z}[X, Y]$ is such that the classes $(\bar{p}_1, \dots, \bar{p}_n)$ modulo $\mathbb{J}$ define a quasi-regular sequence in $\mathbb{Q}_{\mathbb{J}}[X, Y]$. Suppose moreover that the algebraic variety $\mathbb{X} = \operatorname{Spec}(\mathbb{Q}_{\mathbb{J}}[X, Y])$ satisfies the condition (6.163). Then, there are univariate polynomials*

$$\phi_j = \gamma_j X_j^{\nu_j} + \sum_{\rho=1}^{\nu_j} \gamma_{j,\rho} X_j^{\nu_j - \rho} \in \mathbb{Z}[X_j] \quad (\nu_j, \gamma_j \in \mathbb{N}^*)$$

in the variables $X_1, \dots, X_n$, respectively, such that, for any $j = 1, \dots, n$,

$$\phi_j - \sum_{k=1}^{n} a_{j,k} p_k \in \mathbb{J}. \tag{6.199}$$

Furthermore, the following estimates hold

$$\begin{aligned}&\max_{1 \le j \le n} \big(\nu_j, \deg(a_{j,1} p_1), \dots, \deg(a_{j,n} p_n)\big) \le \mathbf{d}(p), \\ &\max_{1 \le j \le n} \big(\mathrm{h}(\phi_j), \mathrm{h}(a_{j,1}) + \mathrm{h}(p_1), \dots, \mathrm{h}(a_{j,n}) + \mathrm{h}(p_n)\big) \le \mathbf{h}_{c_{\mathrm{ae}}(n,N)}(p),\end{aligned} \tag{6.200}$$

where

$$c_{\mathrm{ae}}(n, N) = (n+1) \log \big(2(n+2)(N+1)^2\big).$$

PROOF. This result is proved in [**SomY21**, Theorem 5.1 and Corollary 5.6]. The reasoning of Z. Jelonek in [**Jel05**] inspires the proof. Among his arguments, Perron's theorem [**Jel05**, Theorem 3.3] needs to be transposed in the arithmetic frame and then combined with deep results from arithmetic intersection theory stated in [**AKS13**]. Some of these are related to the discussion above about logarithmic heights and Bézout's arithmetic intersection theorem. Note that the generalized Perron theorem [**Jel05**, Theorem 3.3] transposed in the arithmetic setting is formulated (in the form needed by us) in [**AKS13**, Theorem 3.15 and Corollary 3.23]. We present it later as Theorem 7.17 for completeness. The arithmetic frame is a product of projective spaces $\mathbb{P}_{\mathbb{Q}}^{N_\iota}$ instead of just $\mathbb{P}_{\mathbb{Q}}^{N}$. Since such proof will lead us far beyond the scope of this monograph, we refer the reader to [**SomY21**, §5] for details. □

REMARK 6.69. It follows from (6.199) and (6.200) that there exists C and $c > 0$ such that

$$(6.201)\quad (z,w) \in \mathbb{X}_{\mathbb{C}} \text{ and } \max(|z|,|w|) \geq C \Longrightarrow \max_{1\leq j\leq n} \frac{|p_j(z,w)|}{\|(z,w)\|^{\deg p_j}}\, \|(z,w)\|^{\mathbf{d}(p) - \min_{1\leq \ell\leq n}\nu_\ell} \geq c.$$

Given any smooth metric on the holomorphic bundle $\bigoplus_{j=1}^n \mathcal{O}(\deg p_j)$ over $\mathbb{P}^N_{\mathbb{C}}$, if $P = (P_1,\dots,P_n)$ is the homogenization of $p_1,\dots,p_n$, then one has that

$$(6.202)\qquad \Big(\frac{z_0}{\|Z\|}\Big)^{n(\mathbf{d}(p) - \min_{1\leq\ell\leq n}\nu_\ell)} \mathbf{1}_{\{z_0=0\}} \cdot R^P_{\|\,\|}([\mathbb{X}_{\mathbb{C}}]) = 0.$$

The restrictive clause

$$(6.203)\qquad 1 \leq j,\ell \leq n \Longrightarrow \mathbf{d}(p) \leq d_j + \nu_\ell$$

ensures that the restriction to $\mathbb{X}_{\mathbb{C}}$ of the polynomial $(p_1,\dots,p_n)$ is topologically proper. More precisely, it admits

$$(\mathbf{d}(p) - \min_{1\leq\ell\leq n}\nu_\ell, \dots, \mathbf{d}(p) - \min_{1\leq\ell\leq n}\nu_\ell)$$

as a multi-index of properness; see Definition 6.53.

The following result is an arithmetic version of Hilbert's nullstellensatz. We will revisit it in §7.1.6.

THEOREM 6.70 (Arithmetic nullstellensatz (AN)). *Let $\mathbb{J}$, $\mathbb{X}$, $p = (p_1,\dots,p_n)$ be as in Theorem 6.68, except that condition (6.163) on $\mathbb{X}_{\mathbb{C}}$ is not anymore required. Let $p_0 \in \mathbb{Z}[X,Y]$ be such that $\mathbb{J} + \sum_{j=0}^n p_j\mathbb{Q}[X,Y] = \mathbb{Q}[X,Y]$. Then, there exists $\gamma \in \mathbb{N}^*$ and polynomials $q_0,\dots,q_n$ in $\mathbb{Z}[X,Y]$ such that $\gamma - \sum_{j=0}^n q_jp_j \in \mathbb{J}$ and*

$$(6.204)\qquad \begin{aligned} &\max_{0\leq j\leq n} \deg(p_jq_j) \leq \mathbf{d}(p,p_0),\\ &\max_{0\leq j\leq n} \big(\log\gamma, \mathrm{h}(q_j) + \mathrm{h}(p_j)\big) \leq \mathbf{h}_{c_{\mathrm{an}}(n,N)}(p,p_0), \end{aligned}$$

with $c_{\mathrm{an}}(n,N) = (4n+8)\log(N+3)$.

PROOF. See [**AKS13**, Theorem 0.1]. The proof consists of deep arguments developed for the generalized Perron Theorem 3.3 from [**Jel05**] but formulated the arithmetic setting here. We will state later on the parametric version of the Perron generalized theorem formulated as Theorem 7.17; see [**AKS13**, Theorem 3.15] for its original version. This result is the cornerstone of this approach. Its presentation will follow after we detail Z. Jenonek's view toward Hilbert's nullstellensatz in §7.1.5. □

Another consequence of the arithmetic generalized Perron theorem formulated in [**AKS13**, Theorem 3.15 and Lemma 3.16], and which rules the arithmetic complexity of trace calculus, plays an important role in the following subsection. We only mention it here and refer to [**SomY21**, Lemma 4.6] for a proof.

PROPOSITION 6.71 (Arithmetic complexity of trace (ACT)). *Let* $\mathbb{J}$ *and* $\mathbb{X}$ *be as in Theorem* 6.68, *with condition* (6.163) *fulfilled. For each* $a \in \mathbb{Z}[X,Y]$ *with* $\deg a > 0$, *there exists* $\gamma(a) \in \mathbb{N}^*$ *such that*

$$\gamma(a) \begin{bmatrix} a(\bar{X},\bar{Y})\, d\bar{X} \\ \bar{X} - t \end{bmatrix}_{\mathbb{X}} \in \mathbb{Z}[t] \qquad (t = (t_1,\dots,t_n)), \tag{6.205}$$

where

(6.206)

$$\log \gamma(a) \le (\mathbf{h} + c_{\text{act}}(N)\,\mathbf{d})\, \deg a$$

$$\mathrm{h}\left(\gamma(a) \begin{bmatrix} a(\bar{X},\bar{Y})\, d\bar{X} \\ \bar{X} - t \end{bmatrix}_{\mathbb{X}}\right) \le \mathrm{h}(a\,dX) + (\,\mathbf{h} + c_{\text{act}}(N)\,\mathbf{d}\,)\,\deg a = \mathrm{h}_{c_{\text{act}}}(a\,dX)$$

for $c_{\text{act}}(N) = (N+2)\log(2N+3)$.

6.6.3. Size estimates of rational algebraic residue symbols. Global residues are interconnected to duality for algebraic varieties [**Ku08**], Jacobi–Lagrange–Kronecker-type interpolation formulae and methods, Hilbert's nullstellensatz and more generally division theorems of Briançon–Skoda–Huneke type; see §7.2 and §7.3 below. Moreover, they appear in several fields of mathematics and theoretical physics, such as scattering amplitude computations beyond Feynman diagrams; see for example [**SoZ16**]. For all these reasons, it seemed important to us to make explicit, in terms of the geometric and arithmetic complexity of the input data,[38] the arithmetic complexity of some Taylor series, when understood as traces. We begin with Remark 6.22, where one takes $W = (W_1,\dots,W_n)$ as a formal set of variables instead of X, and we consider the formal power series

$$\sum_{\kappa \in \mathbb{N}^n} \mathrm{Res}_{\mathbb{Q}_{\mathbb{J}}[X,Y]/\mathbb{Q}} \begin{bmatrix} \varpi \\ \bar{p}_1^{\kappa_1+1},\dots,\bar{p}_n^{\kappa_n+1} \end{bmatrix} W^\kappa \in \mathbb{Q}[[W]].$$

Then, we will deduce the arithmetic complexity of a more general Taylor series (still understood as traces)

$$\sum_{\kappa \in \mathbb{N}^n} \mathrm{Res}_{\mathbb{Q}_{\mathbb{J}}[X,Y]/\mathbb{Q}} \begin{bmatrix} \varpi/\bar{p}_0^{\ell+1} \\ \bar{p}_1^{\kappa_1+1},\dots,\bar{p}_n^{\kappa_n+1} \end{bmatrix} W^\kappa$$
$$= \sum_{\kappa \in \mathbb{N}^n} \mathrm{Res}_{\mathbb{Q}_{\mathbb{J}}[X,Y]/\mathbb{Q}} \begin{bmatrix} D_\ell^{|\kappa|}[\bar{p}_0]\,\varpi \\ \bar{p}_1^{\kappa_1+1},\dots,\bar{p}_n^{\kappa_n+1} \end{bmatrix} W^\kappa,$$

---

[38] Such input data are the following:

- The purely n-dimensional affine algebraic variety $\mathbb{X}$ defined over $\mathbb{Q}$ as $\mathrm{Spec}(\mathbb{Q}_{\mathbb{J}}[X,Y])$, where $X = (X_1,\dots,X_n)$, $Y = (Y_1,\dots,Y_{N-n})$, and $\mathbb{J}$ is a purely n-dimensional radical ideal in $\mathbb{Q}[X,Y]$ such that the Nœther presentation (6.112) for $\mathbb{X}_{\mathbb{C}} = V(\mathbb{J})$ holds;
- The polynomials $p_j \in \mathbb{Z}[X,Y]$ such that the sequence $(\bar{p}_1,\dots,\bar{p}_n)$, where $\bar{p}_j = p_j$ modulo $\mathbb{J}$, is quasi-regular in $\mathbb{Q}_{\mathbb{J}}[X,Y]$, but fails to generate such quotient algebra;
- The differential form

$$\varpi = q(\bar{X},\bar{Y})\, d\bar{X}_1 \wedge \cdots \wedge d\bar{X}_n \in \bigwedge^n \Omega_{\mathbb{Q}_{\mathbb{J}}[X,Y]/\mathbb{Q}},$$

where $q \in \mathbb{Z}[X,Y]$, which induces through pullback along the embedding $\mathbb{X}_{\mathbb{C}} \hookrightarrow \mathbb{C}^N$ a n-form on the complex algebraic space $\mathbb{X}_{\mathbb{C}}$.

where $p_0 \in \mathbb{Z}[X,Y]$ is such that $\sum_{j=0}^n p_j Q[X,Y] + \mathbb{J} = \mathbb{Q}[X,Y]$, $\ell \in \mathbb{N}$ and for any $\kappa \in \mathbb{N}^n$

$$D_\ell^{|\kappa|}[\bar{p}_0] = \sum_{\nu=1}^{|\kappa|+1} (-1)^{\nu-1} \binom{|\kappa|+1}{\nu} \bar{a}_0^{(\ell+1)\nu}\, \bar{p}_0^{(\ell+1)(\nu-1)}.$$

Here $a_0 \in \mathbb{Q}[X,Y]$ and is such that $1 - a_0 p_0 \in \mathbb{J} + \sum_{j=1}^n p_j \mathbb{Q}[X,Y]$; see (6.51). It follows from Theorem 6.70 that one can chose $a_0 = q_0/\boldsymbol{\gamma}$, where $\boldsymbol{\gamma} \in \mathbb{N}^*$ and $q_0 \in \mathbb{Z}[X,Y]$ to be such that the degree and logarithmic size estimates

$$\deg(p_0 q_0) \leq \mathbf{d}\,(p,p_0),\ \max(\log \boldsymbol{\gamma}, \mathrm{h}(q_0) + \mathrm{h}(p_0)) \leq \mathbf{h}_{c_{\mathrm{an}}(n,N)}(p,p_0) \tag{6.207}$$

hold. With such choice

(6.208)

$$\begin{aligned} D_\ell^{|\kappa|}[\bar{p}_0] &= \frac{1}{\boldsymbol{\gamma}^{(\ell+1)(|\kappa|+1)}} \sum_{\nu=1}^{|\kappa|+1} (-1)^{\nu-1} \binom{|\kappa|+1}{\nu} \boldsymbol{\gamma}^{(\ell+1)(|\kappa|+1-\nu)}\, \bar{q}_0^{(\ell+1)\nu}\, \bar{p}_0^{(\ell+1)(\nu-1)} \\ &= \frac{\boldsymbol{D}_\ell^{|\kappa|}[\bar{p}_0]}{\boldsymbol{\gamma}^{(\ell+1)(|\kappa|+1)}} \in \frac{\mathbb{Z}[\bar{X},\bar{Y}]}{\boldsymbol{\gamma}^{(\ell+1)(|\kappa|+1)}}. \end{aligned}$$

Let us consider first the case where $N = n$ and $\mathbb{X} = \mathbb{Q}^n$. In this particular case

$$\mathbf{d}\,(p) = \prod_{j=1}^n d_j, \quad \mathbf{h}_c(p) = \Big(\prod_{j=1}^n d_j\Big)\Big(\sum_{j=1}^n \frac{\mathrm{h}(p_j)}{d_j} + c\Big) \tag{6.209}$$

for any $c \geq 0$, since $\mathbf{d} = 1$ and $\mathbf{h} = 0$ in this case. Let $\phi = (\phi_1(X_1), \dots, \phi_n(X_n))$ be as in Theorem 6.68. Besides the markers $\mathbf{d}\,(p)$ and $\mathbf{h}_c(p)$ defined in (6.209), it is natural, given Remark 6.69 and Remark 6.73, to introduce the (geometric) marker $\mu_\phi^\infty(p)$. This marker is defined by

$$\begin{aligned} \mu_\phi^\infty(p) &= n\,\mathbf{d}\,(p) - |\nu(\phi)| = n\,\mathbf{d}(p) - (\nu_1(\phi) + \cdots + \nu_n(\phi)) \\ &\leq n\,\big(\mathbf{d}\,(p) - \min_{1\leq \ell\leq n} \nu_\ell(\phi)\big). \end{aligned} \tag{6.210}$$

We also introduce the arithmetic marker

$$\mathfrak{d}_\phi = \prod_{j=1}^n |\gamma_j(\phi)|, \tag{6.211}$$

where $\gamma_j(\phi)$ is the leading coefficient of ϕ_j. We state in this context the following result [**SomY21**, Theorem 6.3].

THEOREM 6.72. *For any $\kappa \in \mathbb{N}^n$, one has*

$$\mathfrak{d}_\phi^{\deg q + (|\kappa|+1)(\mu_\phi^\infty(p)+1)}\, \mathrm{Res}_{\mathbb{Q}[X]/\mathbb{Q}} \begin{bmatrix} q(X)\,dX \\ p_1^{\kappa_1+1}, \dots, p_n^{\kappa_n+1} \end{bmatrix} = m_{p,\phi,q}(\kappa) \in \mathbb{Z}. \tag{6.212}$$

Moreover, for $c = c(n) = 3(n+2)\log(n+2)$ the estimates

$$\begin{aligned} &-\infty \leq \log|m_{p,\phi,q}(\kappa)| \leq n\big((|\kappa|+1)\mu_\phi^\infty(p) + \deg q\big)\,\mathbf{h}_c(p) + \mathrm{h}(q(X)\,dX), \\ &\log \mathfrak{d}_\phi \leq n\,\mathbf{h}_{c_{\mathrm{ae}}(n,0)}(p) \leq n\,\mathbf{h}_c(p) \end{aligned} \tag{6.213}$$

hold.

PROOF. Let Z_p be the zero-dimensional algebraic cycle in $\mathbb{C}^n$ defined as

$$Z_p = \sum_{\alpha \in V(p)} \dim_{\mathbb{C}} \Big(\frac{\mathcal{O}_{\mathbb{C}^n,\alpha}}{\sum_{j=1}^n p_j \mathcal{O}_{\mathbb{C}^n,\alpha}} \Big) \{\alpha\}.$$

For any $j = 1, \dots, n$, one has $e_j(p) = \deg z_j(Z_p) \le \deg \phi_j = \nu_j$. It follows from (6.59) that

$$\text{(6.214)} \quad \operatorname{Res}_{\mathbb{Q}[X]/\mathbb{Q}} \left[\begin{matrix} q(X)\, dX \\ p_1^{\kappa_1+1}, \dots, p_n^{\kappa_n+1} \end{matrix} \right] = \operatorname{Res}_{\mathbb{Q}[X]/\mathbb{Q}} \left[\begin{matrix} (\det A)\, A_\kappa^{|\kappa|}[\phi]\, q(X)\, dX \\ \phi_1^{|\kappa|+1}(X_1), \dots, \phi_n^{|\kappa|+1}(X_n) \end{matrix} \right]$$

for any $\kappa \in \mathbb{N}^n$. The terms $A_\kappa^{|\kappa|}[\phi]$ are defined as in (6.36) and A is the $n \times n$ matrix with entries $a_{j,k} \in \mathbb{Z}[X,Y]$ involved in the algebraic identities (6.199) with $\mathbb{J} = \mathbf{0}$. Let

$$(\det A)\, A_\kappa^{|\kappa|}[\phi]\, q(X)\, dX = \Theta_\kappa(X)\, dX.$$

One has, in view of degree and size estimates (6.200) of the submultiplicativity of the length combined with the comparison inequalities (6.166), that

$$\text{(6.215)} \quad \begin{aligned} \deg \Theta_\kappa &\le \deg q - |d| + n\,(|\kappa|+1)\, \mathbf{d}\,(p), \\ \mathrm{h}(\Theta_\kappa\, dX) &\le \mathrm{h}(q(X)\, dX) + n\,(|\kappa|+1)\, \mathbf{h}_c(p), \end{aligned}$$

where $c = 3(n+2)\log(n+2) \ge c_{\mathrm{ae}}(n,0)$. Since

$$\text{(6.216)} \quad \deg \Theta_\kappa + n - (|\kappa|+1)(|\nu(\phi)|-1) \le \deg q + (|\kappa|+1)(\mu_\phi^\infty(p)+1),$$

the assertion (6.212) follows from (6.168) in Lemma 6.63. As a matter of fact, (6.168) implies that

$$\mathfrak{d}_\phi^{\deg \Theta_\kappa + n - (|\kappa|+1)(|\nu(\phi)|-1)} \operatorname{Res}_{\mathbb{Q}[X]/\mathbb{Q}} \left[\begin{matrix} q(X)\, dX \\ p_1^{\kappa_1+1}, \dots, p_n^{\kappa_n+1} \end{matrix} \right] = m'_{p,\phi,q}(\kappa) \in \mathbb{Z}.$$

An estimate for $\log|m'_{p,\phi,q}(\kappa)|$ is provided by (6.169) in Lemma 6.63, once this lemma is transposed to the multivariate tensorial case. Then, one gets

$$\begin{aligned} \log|m'_{p,\phi,q}(\kappa)| \le \ & \mathrm{h}(\Theta_\kappa\, dX) + \big(\deg \Theta_\kappa + n - (|\kappa|+1)|\nu(\phi)|\big)|\,\mathrm{h}(\phi)| \\ & + \big(\deg \Theta_\kappa - |\nu(\phi)| + n\big)\, \log 2, \end{aligned}$$

where $|\,\mathrm{h}(\phi)| := \sum_{j=1}^n \mathrm{h}(\phi_j)$ and $|\nu(\phi)| = \sum_{j=1}^n \nu_j(\phi)$. Then, the upper estimate for $\log|m_{p,\phi,q}(\kappa)|$ follows from the fact that

$$m_{p,\phi,q}(\kappa) = m'_{p,\phi,q}(\kappa)\, \mathfrak{d}_\phi^{\deg q + (|\kappa|+1)(\mu_\phi^\infty(p) + |\nu(\phi)|) - n - \deg \Theta_\kappa}. \qquad \square$$

REMARK 6.73. The geometric marker $\mu_\phi^\infty(p)$ is bounded from above by the marker $n\,\mathbf{d}\,(p) - \sum_{j=1}^n \deg(z_j(Z_p)) = n\,\mathbf{d}\,(p) - \sum_{j=1}^n e_j(p)$. It is indeed a rough estimate. It is worthwhile to point out that a sharper bound for $\mu_\phi^\infty(p)$ from above is the positive integer $\tilde{\mu}_\phi(p) = n\big(\mathbf{d}\,(p) - \min_{1 \le \ell \le n} \nu_\ell\big)$, satisfying

$$\text{(6.217)} \quad (\xi_0^{\tilde{\mu}_\phi(p)})_x \in \sum_{j=1}^n P_j \mathcal{O}_{\mathbb{P}^n_{\mathbb{C}},x}$$

for any $x \in \mathbb{P}^n_{\mathbb{C}} \setminus \mathbb{C}^n$. The universality of the constant n involved in such markers as $\tilde{\mu}_\phi(p)$ follows from the estimates (6.201), together with the Briançon–Skoda theorem in the regular n-dimensional local ring $\mathcal{O}_{\mathbb{P}^n_{\mathbb{C}},x}$; see Theorem 5.7. Therefore, it is tempting to conjecture that one could replace the geometric marker $\mu_\phi^\infty(p)$ in (6.212) and (6.213) by the smallest positive integer $\mu^\infty(p)$ among those satisfying the same property (6.217) as $\tilde{\mu}^\infty(p)$ does in a neighborhood of any point at infinity with respect to $P_1\mathcal{O}_{\mathbb{P}^n_{\mathbb{C}}} + \cdots + P_n\mathcal{O}_{\mathbb{P}^n_{\mathbb{C}}}$. Note also the factor n is present twice. It appears firstly inside $\mu^\infty_\phi(p)$ and secondly in front of the multiplication by $\mathbf{h}_c(p)$ in the right-hand side of (6.213). This is evidence of a deep relation between algebraic residue symbols and the Briançon–Skoda theorem, even at the level of underlying computations. In the particular case where $\deg Z_p = \mathbf{d}(p)$, that is, in the case where the homogeneous polynomials P_j do not share any common zero in $\mathbb{P}^n_{\mathbb{C}} \setminus \mathbb{C}^n$, one has

$$\mathrm{Res}_{\mathbb{Q}[X]/\mathbb{Q}}\begin{bmatrix} q(X)\,dX \\ p_1^{\kappa_1+1}, \dots, p_n^{\kappa_n+1}\end{bmatrix} \neq 0 \Longrightarrow |\kappa| \leq \langle \kappa, d\rangle \leq \deg q.$$

Applying a linear change of coordinates $X = AX'$, where A is an invertible $n\times n$ matrix with integer entries, whose size is indeed easy to control, one gets that $e_j(p(AX')) = \mathbf{d}(p)$ for any $j = 1,\dots,n$. If so, $\nu_j(\phi) = \mathbf{d}(p)$ for any j and hence $\mu^\infty_\phi(p) = 0$ in (6.212) and (6.213). It corroborates the conjecture stated above about the integer that would be a candidate to replace $\mu^\infty_\phi(p)$, namely $\tilde{\mu}^\infty(p) = 0$ in this particular case.

EXAMPLE 6.74. Given $q \in \mathbb{Z}[X]$, the arithmetic complexity of Bergman–Weil formal pseudo-$(p_1,\dots,p_n)$-adic developments

(6.218)
$$q(X) = \sum_{\kappa\in\mathbb{N}^n} \mathrm{Res}_{\mathbb{Q}[X,Y]/\mathbb{Q}}\begin{bmatrix} \det H(X,Y)\,q(Y)\,dY_1\wedge\cdots\wedge dY_n \\ p_1^{\kappa_1+1}(Y),\dots,p_n^{\kappa_n+1}(Y)\end{bmatrix} p^\kappa \in \mathbb{Q}[X][[\boldsymbol{p}]]$$
$$\forall\, h \in \mathbb{Z}[X,Y],$$

where $p_j(Y) - p_j(X) = \sum_{k=1}^n h_{j,k}(X,Y)(Y_k - X_k)$ for $j = 1,\dots,n$, is governed as a particular case by Theorem 6.72. The polynomials $h_{j,k} \in \mathbb{Z}[X,Y]$ and the matrix $H = [h_{j,k}]_{1\leq j,k\leq n}$ are those given in Theorem 6.28.

One can estimate the arithmetic complexity of residue symbols
(6.219)
$$\mathrm{Res}_{\mathbb{Q}[X]/\mathbb{Q}}\begin{bmatrix} q(X)\,dX/p_0^{\ell+1} \\ p_1^{\kappa_1+1},\dots,p_n^{\kappa_n+1}\end{bmatrix} = \frac{1}{\gamma^{(\ell+1)(|\kappa|+1)}}\mathrm{Res}_{\mathbb{Q}[X]/\mathbb{Q}}\begin{bmatrix} \boldsymbol{D}_\ell^{|\kappa|}[p_0]\,q(X)\,dX \\ p_1^{\kappa_1+1},\dots,p_n^{\kappa_n+1}\end{bmatrix},$$

when γ and $\boldsymbol{D}_\ell^{|\kappa|}[p_0]$ are attached to $(p_0,\dots,p_n)$ as well. It is understood according to Theorem 6.70 as in (6.208), with degree and logarithmic size estimates (6.207). In order to estimate the arithmetic complexity of residue symbols (6.219), one needs to replace $q(X)$ by $\boldsymbol{D}_\ell^{|\kappa|}[p_0](X)\,q(X)$ in (6.212) and (6.213). Estimates for $\deg\Theta_\kappa$ as well as $\mathrm{h}(\Theta_\kappa\,dX)$ depend then in addition on $d_0 = \deg p_0$, $\mathrm{h}(p_0\,dX)$, and of course, on the exponent ℓ. We refer the reader to [**SomY21**, Theorem 6.7] for precise degree and logarithmic size estimates (in terms of $\mathbf{d}(p,p_0)$, $\mathrm{h}(p_0)$ and $\mathbf{h}_{c(n)}(p,p_0)$, κ, ℓ) for numerators and denominators of the residue symbols (6.219).

The case where $\mathbb{J} \neq \mathbf{0}$ is slightly more involved. One still needs to introduce $\phi = (\phi_1,\dots,\phi_n)$ attached to $\mathbb{X}$ and $(p_1,\dots,p_n)$ as in Theorem 6.68 and to consider

$$\Theta_\kappa(\bar{X},\bar{Y}) = (\det A)\,A_\kappa^{|\kappa|}[\bar{\phi}]\,q(\bar{X},\bar{Y}) \tag{6.220}$$

so that, according to Theorem 6.24, once combined with Proposition 6.44, they will be able to deduce that

(6.221)
$$\mathrm{Res}_{\mathbb{Q}_{\mathbb{J}}[X,Y]/\mathbb{Q}}\left[\frac{q(\bar X,\bar Y)\,d\bar X}{\bar p_1^{\kappa_1+1},\dots,\bar p_n^{\kappa_n+1}}\right] = \mathrm{Res}_{\mathbb{Q}_{\mathbb{J}}[X,Y]/\mathbb{Q}}\left[\frac{\Theta_\kappa(\bar X,\bar Y)\,d\bar X}{\phi_1^{|\kappa|+1}(\bar X_1),\dots,\phi_n^{|\kappa|+1}(\bar X_n)}\right] = \mathrm{Res}_{\,\mathbb{Q}[t]/\mathbb{Q}}\left[\frac{\Psi_\kappa(t)\,dt_1\wedge\cdots\wedge dt_n}{\phi_1^{|\kappa|+1}(t_1),\dots,\phi_n^{|\kappa|+1}(t_n)}\right].$$

Note that

$$\Psi_\kappa(t) = \left[\frac{\Theta_\kappa(\bar X,\bar Y)\,d\bar X}{\bar X - t}\right]_{\mathbb{X}} \in \mathbb{Q}[t_1,\dots,t_n] \tag{6.222}$$

is the trace polynomial function of $\Theta_\kappa \in \mathbb{Z}[X,Y]$. In the place of the arithmetic marker $\mathfrak{d}_\phi$ (as defined by (6.211) for the particular case where $\mathbb{J} = \mathbf{0}$, independently of the multi-exponent $\kappa \in \mathbb{N}^n$), one needs to invoke Proposition 6.71 with $a = \Theta_\kappa$ and introduce $\mathfrak{d}_\phi(\kappa) \in \mathbb{N}^*$ for each such $\kappa \in \mathbb{N}^*$. Then the number $\mathfrak{d}_\phi(\kappa)$ is the least positive integer such that

$$\begin{gathered}\mathfrak{d}_\phi(\kappa)\left[\frac{\Theta_\kappa(\bar X,\bar Y)\,d\bar X}{\bar X - t}\right]_{\mathbb{X}} \in \mathbb{Z}[t],\\ \log \mathfrak{d}_\phi(\kappa) \leq (\mathbf{h} + c_{\mathrm{act}}(N)\,\mathbf{d})\,\deg\Theta_\kappa,\\ \mathrm{h}\left(\mathfrak{d}_\phi(\kappa)\left[\frac{\Theta_\kappa(\bar X,\bar Y)\,d\bar X}{\bar X - t}\right]_{\mathbb{X}}\right) \leq \mathrm{h}(\Theta_\kappa\,dX) + (\,\mathbf{h} + c_{\mathrm{act}}(N)\,\mathbf{d}\,)\,\deg\Theta_\kappa.\end{gathered} \tag{6.223}$$

As for the geometric marker $\mu_\phi^\infty(p)$, we keep its definition as in (6.210). The analogue of Theorem 6.72, in the general context of an n-dimensional radical ideal $\mathbb{J}$ instead of $\mathbb{J} = \mathbf{0}$ when $N = n$, is the following [**SomY21**, Theorem 6.2].

THEOREM 6.75. *For any $\kappa \in \mathbb{N}^n$, one has that*

(6.224)
$$\mathfrak{d}_\phi(\kappa)\,\mathfrak{d}_\phi^{\deg q + (|\kappa|+1)(\mu_\phi^\infty(p)+1)}\,\mathrm{Res}_{\mathbb{Q}_{\mathbb{J}}[X,Y]/\mathbb{Q}}\left[\frac{q(\bar X,\bar Y)\,d\bar X}{\bar p_1^{\kappa_1+1},\dots,\bar p_n^{\kappa_n+1}}\right] = m_{p,\phi,q}(\kappa) \in \mathbb{Z}.$$

Furthermore, if $c = c(N) = 3(N+2)\log(N+2)$, then the estimates

$$\log \mathfrak{d}_\phi(\kappa) \leq (\,\deg q - |d| + n(|\kappa|+1)\,\mathbf{d}\,(p))\,(\mathbf{h} + c_{\mathrm{act}}\,\mathbf{d}) \tag{6.225}$$

and

$$\begin{aligned}\max\big(\log|m_{p,\phi,q}(\kappa)|, \log\big(\mathfrak{d}_\phi(\kappa)\,\mathfrak{d}_\phi^{\deg q+(|\kappa|+1)(\mu_\phi^\infty(p)+1)}\big)\big)&\\ \leq n(|\kappa|+1)\big((\mu_\phi^\infty(p)+1)\,\mathbf{h}_c(p) + n\,\mathbf{d}\,(p)\,(\mathbf{h} + c_{\mathrm{act}}\,\mathbf{d})\big)&\\ + \deg q\big(n\,\mathbf{h}_c(p) + \mathbf{h} + c_{\mathrm{act}}\,\mathbf{d} + \log 2) + \mathrm{h}(q(X)\,dX)&\end{aligned} \tag{6.226}$$

hold.

PROOF. One has here

$$\begin{gathered}\deg\Theta_\kappa \leq \deg q - |d| + n\,(|\kappa|+1)\,\mathbf{d}\,(p),\\ \mathrm{h}(\Theta_\kappa\,dX) \leq \mathrm{h}(q(X)\,dX) + n\,(|\kappa|+1)\,\mathbf{h}_c(p).\end{gathered} \tag{6.227}$$

It follows from (6.221), the definition of $\mathfrak{d}_\phi(\kappa)$ and the assertion (6.168) in Lemma 6.63, that

$$\mathfrak{d}_\phi(\kappa)\,\mathfrak{d}_\phi^{\deg(\Theta_\kappa)+n-(|\kappa|+1)(|\nu(\phi)|-1)}\,\mathrm{Res}_{\mathbb{Q}_{\mathbb{J}}[X,Y]/\mathbb{Q}}\left[\frac{q(\bar X,\bar Y)\,d\bar X}{\bar p_1^{\kappa_1+1},\dots,\bar p_n^{\kappa_n+1}}\right] = m'_{p,\phi,q}(\kappa) \in \mathbb{Z}.$$

One deduces from assertion (6.169) in Lemma 6.63 (once this lemma has been transferred in the multivariate tensorial setting) that

$$\begin{aligned}\log|m'_{p,\phi,q}(\kappa)| &\le \mathrm{h}(\mathfrak{d}_\phi(\kappa)\,\Psi_\kappa(t)\,dt) + \big(\deg\Theta_\kappa - (|\kappa|+1)|\nu(\phi)| + n\big)\,|\mathrm{h}(\phi)| \\ &\quad + (\deg\Theta_\kappa - |\nu(\phi)| + n)\log 2 \\ &\le \mathrm{h}(\mathfrak{d}_\phi(\kappa)\,\Psi_\kappa(t)\,dt) + \big(\deg q + (|\kappa|+1)\mu_\phi^\infty(p)\big)\,|\mathrm{h}(\phi)| \\ &\quad + \big(\deg q + n(|\kappa|+1)\,\mathbf{d}\,(p)\big)\log 2.\end{aligned}$$

Observe that $m_{p,\phi,q}(\kappa)$ is an integer because it is the product of $m'_{p,\phi,q}(\kappa)$ by a positive integer. The reasoning here is the same as that used in the proof of Theorem 6.72. Since

$$\begin{aligned}\mathrm{h}(\mathfrak{d}_\phi(\kappa)\,\Psi_\kappa(t)\,dt) &\le \mathrm{h}(q\,dX) + n(|\kappa|+1)\,\mathbf{h}_c(p) + \deg\Theta_\kappa\,(\mathbf{h} + c_{\mathrm{act}}\,\mathbf{d}) \\ &\le \mathrm{h}(q\,dX) + n(|\kappa|+1)\,\big(\mathbf{h}_c(p) + (\mathbf{h} + c_{\mathrm{act}}\,\mathbf{d})\,\mathbf{d}\,(p)\big) + \deg q\,(\mathbf{h} + c_{\mathrm{act}}\,\mathbf{d}),\end{aligned}$$

the upper estimate for $\log|m_{p,\phi,q}(\kappa)|$ in (6.226) follows. It is immediate to check that an identical upper estimate holds for the logarithm of the integer

$$\mathfrak{d}_\phi(\kappa)\,\mathfrak{d}_\phi^{\deg q + (|\kappa|+1)(\mu_\phi^\infty(p)+1)}.$$

This step concludes the proof of Theorem 6.75. □

The fact that the positive integer $\mathfrak{d}_\phi(\kappa)$ involved in the denominator of the residue symbol (6.221) depends on κ looks like a stumbling block. In order to find a way around it, following [**AKS13**, Corollary 3.23], let us introduce for each integer $k = 1, \dots, N-n$ a polynomial $E_k \in \mathbb{Z}[X, Y_k]$ with degree at most $\mathbf{d}$ and logarithmic size

$$\mathrm{h}(E_k) \le \mathbf{h} + (n+2)\log(2N+8)\,\mathbf{d}$$

such that

$$E_k(X, Y_k) = \chi_k Y_k^{e_k} + \sum_{\rho=1}^{e_k} u_{k,\rho}(X)\,Y_k^{e_k-\rho} \equiv 0 \quad \text{modulo } \mathbb{J} \text{ in } \mathbb{Q}[X,Y].$$

Given a duplication $\{W_1, \dots, W_{N-n}\}$ of the set of coordinates $\{Y_1, \dots, Y_{N-n}\}$, let

$$\begin{aligned}b_E(X,Y,W) &= \prod_{k=1}^{N-n} \frac{E_k(X,Y_k) - E_k(X,W_k)}{Y_k - W_k} \\ &= \sum_{\{\rho\in\mathbb{N}^{N-n}\,:\,\rho\prec(e_1-1,\dots,e_{N-n}-1)\}} b_{E,\rho}(X,Y)\,W^\rho.\end{aligned} \tag{6.228}$$

Let $\rho \in \mathbb{N}^{N-n}$ be such that $\rho \prec (e_1 - 1, \dots, e_{N-n} - 1)$, that is, $0 \le \rho_j \le e_j - 1$ for $j = 1, \dots, N-n$. It follows from Proposition 6.71 that there exists a natural number $\mathfrak{d}_{E,\rho} \in \mathbb{N}^*$ such that

$$\begin{aligned}&\mathfrak{d}_{E,\rho}\begin{bmatrix} b_{E,\rho}(\bar X, \bar Y)\,d\bar X \\ \bar X - t\end{bmatrix}_{\mathbb{X}} \in \mathbb{Z}[t], \\ &\log\mathfrak{d}_{E,\rho} \le (\mathbf{h} + c_{\mathrm{act}}(N)\,\mathbf{d})\,\deg b_{E,\rho} \le (N-n)\,(\mathbf{h} + c_{\mathrm{act}}(N)\,\mathbf{d})\,\mathbf{d}.\end{aligned} \tag{6.229}$$

Given any $\Theta \in \mathbb{Z}[X,Y]$ (in particular $\Theta = \Theta_\kappa$ for $\kappa \in \mathbb{N}^n$ as in (6.220)), Theorem 6.28 implies that in $\mathbb{Q}[X,Y]$,

$$\Theta(X,Y) \equiv \mathrm{Res}_{\mathbb{Q}[X,Y][W]/\mathbb{Q}[X,Y]}\begin{bmatrix} \Theta(X,W)\,b_E(X,Y,W)\,dW \\ E_1(X,W_1), \dots, E_{N-n}(X,W_{N-n})\end{bmatrix}$$

modulo the ideal $\sum_{k=1}^{N-n} E_k(X,Y_k)\,\mathbb{Q}[X,Y]$. As a consequence

$$\Psi(t) := \begin{bmatrix} \Theta(\bar X,\bar Y)\,d\bar X \\ \bar X - t \end{bmatrix}_{\mathbb{X}}$$

$$(6.230)\qquad = \begin{bmatrix} \operatorname{Res}_{\mathbb{Q}[t,Y][W]/\mathbb{Q}[t,Y]} \begin{bmatrix} \Theta(t,W)\,b_E(t,Y,W)\,dW \\ E_1(t,W_1),\dots,E_{N-n}(t,W_{N-n}) \end{bmatrix} d\bar X \\ \\ \bar X - t \end{bmatrix}_{\mathbb{X}}$$

since $E_k(X,Y_k) \in \mathbb{J}$ for $k = 1,\dots,N-n$. The realization of algebraic residue symbols through currents (see Theorem 6.49) shows that the right-hand side of (6.230) equals, up to a sign, to

$$\sum_{\{\rho\in\mathbb{N}^{N-n}\,:\,\rho\prec(e_1-1,\dots,e_{N-n}-1)\}} \operatorname{Res}_{\mathbb{Q}[t][W]/\mathbb{Q}[t]} \begin{bmatrix} \begin{bmatrix} b_{E,\rho}(\bar X,\bar Y)\,d\bar X \\ \bar X - t \end{bmatrix}_{\mathbb{X}} W^\rho\,\Theta(t,W)\,dW \\ \\ E_1(t,W_1),\dots,E_{N-n}(t,W_{N-n}) \end{bmatrix}.$$

As a consequence, there exists $\mathfrak{d}(\Theta) \in \mathbb{N}^*$ such that

$$(6.231)\qquad \begin{aligned} &\mathfrak{d}(\Theta) \begin{bmatrix} \Theta(\bar X,\bar Y)\,d\bar X \\ \bar X - t \end{bmatrix}_{\mathbb{X}} \in \mathbb{Z}, \\ &\operatorname{Supp}(\{\mathfrak{d}(\Theta)\}) \subset \operatorname{Supp}\Big(\Big(\prod_{\{\rho\in\mathbb{N}^{N-n}\,:\,\rho\prec(e_1-1,\dots,e_{N-n}-1)\}} \mathfrak{d}_{E,\rho}\Big)\Big(\prod_{k=1}^{N-n}\chi_k\Big)\Big). \end{aligned}$$

In particular, taking $\Theta = \Theta_\kappa$ for $\kappa \in \mathbb{N}^n$, the denominator $\mathfrak{d}_\phi(\kappa)$ involved in (6.224) can be chosen with support lying in the finite subset of primes

$$\operatorname{Supp}\Big(\Big(\prod_{\{\rho\in\mathbb{N}^{N-n}\,:\,\rho\prec(e_1-1,\dots,e_{N-n}-1)\}} \mathfrak{d}_{E,\rho}\Big)\Big(\prod_{k=1}^{N-n}\chi_k\Big)\Big),$$

which is independent of the multi-index κ.

One can estimate as well the arithmetic complexity of residue symbols

$$(6.232)\qquad \operatorname{Res}_{\mathbb{Q}_{\mathbb{J}}[X]/\mathbb{Q}} \begin{bmatrix} q(\bar X,\bar Y)\,d\bar X/\bar p_0^{\ell+1} \\ \bar p_1^{\kappa_1+1},\dots,\bar p_n^{\kappa_n+1} \end{bmatrix} = \frac{1}{\gamma^{(\ell+1)(|\kappa|+1)}} \operatorname{Res}_{\mathbb{Q}_{\mathbb{J}}[X,Y]/\mathbb{Q}} \begin{bmatrix} \boldsymbol{D}_\ell^{|\kappa|}[\bar p_0]q(\bar X,\bar Y)d\bar X \\ \bar p_1^{\kappa_1+1},\dots,\bar p_n^{\kappa_n+1} \end{bmatrix}$$

when γ and $\boldsymbol{D}_\ell^{|\kappa|}[\bar p_0]$ are attached to $(p_0,\dots,p_n)$ according to Theorem 6.70 as in (6.208), with degree and logarithmic size estimates (6.207). In order to estimate the arithmetic complexity of residue symbols (6.232), one needs to replace $q(X,Y)$ by $\boldsymbol{D}_\ell^{|\kappa|}[p_0](X,Y)\,q(X,Y)$ in (6.225) and (6.226). Then the estimates for $\deg\Theta_\kappa$ and $\mathrm{h}(\Theta_\kappa\,dX)$ depend in addition on $d_0 = \deg p_0$, $\mathrm{h}(p_0\,dX)$ and, of course, on the exponent ℓ. We refer the reader to [**SomY21**, Theorem 6.6] for precise degree and logarithmic size estimates (in terms of $\mathbf{d}\,(p,p_0)$, $\mathrm{h}(p_0)$ and $\mathbf{h}\,(p,p_0)$, κ and ℓ) for numerators and denominators of the residue symbols (6.232).

6.6.4. Rational algebraic residue symbols: a dynamic approach. We describe here an alternative approach toward complexity estimates for rational algebraic residue symbols (see [**ElY07**] or [**ElM06**, §10.5]), whose nature is intrinsically more *dynamical* than algebraic, even though it also profits from arithmetic intersection theory through the Perron arithmetic generalized theorem, as formulated in

[**AKS13**, Corollary 3.23]. The dynamic nature of reasoning developed here relies upon the generalized transformation law as stated in terms of geometric currents in Proposition 3.23. As in the previous subsection, let $\mathbb{J}$ be an n-dimensional radical ideal in $\mathbb{Q}[X,Y]$, where, as usual, $X=(X_1,\dots,X_n)$, $Y=(Y_1,\dots,Y_{N-n})$. We also assume that Nœther's presentation (6.112) for $\mathbb{X}_{\mathbb{C}}=V(\mathbb{J})$ holds. If t is an additional parameter, then the ideal $\mathbb{J}_t=\mathbb{J}\,\mathbb{Q}[X,Y,t]\subset\mathbb{Q}[X,Y,t]$ is an $(n+1)$-dimensional radical ideal in $\mathbb{Q}[t,X,Y]$. One denotes by $\mathbb{X}_t=\operatorname{Spec}(\mathbb{Q}[t,X,Y]/\mathbb{J}_t)=\mathbb{X}\times\operatorname{Spec}(\mathbb{Q}[t])$ the corresponding $(n+1)$-dimensional algebraic subvariety. Let $\eta=(\eta_1,\dots,\eta_n)$ be a generic element in $\mathbb{Z}^n$. Given a sequence of polynomials $p=(p_1,\dots,p_n)$ in $\mathbb{Z}[X,Y]$, it is equivalent to say that $\mathbb{X}_{\mathbb{C}}\cap V(p)$ is a zero-dimensional algebraic subvariety of $\mathbb{C}^N$ or that

$$(\mathbb{X}_t)_{\mathbb{C}}\cap\{(t,z,w)\in\mathbb{C}^{N+1}\,:\,p_1(z,w)-\eta_1\,t=\cdots=p_n(z,w)-\eta_n\,t=t=0\}$$

is a zero-dimensional algebraic subvariety of $\mathbb{C}^{N+1}$. We assume in this subsection that $\deg p_j\geq 2$ for $j=1,\dots,n$. If not, one could reduce by 1 the number of variables n and N. It follows from Theorem 6.24[39] that if a polynomial $q\in\mathbb{Z}[X,Y]$ is given, then for any $k\in\mathbb{N}$,

(6.233)

$$\begin{aligned}
&\operatorname{Res}_{\mathbb{Q}_{\mathbb{J}_t}[X,Y,t]/\mathbb{Q}}\left[\begin{matrix}q(\bar X,\bar Y)\,d\bar X\wedge d\bar t\\ \bar p_1-\eta_1\bar t,\dots,\bar p_n-\eta_n\bar t,\bar t^{\,k+1}\end{matrix}\right]\\
&\quad=\operatorname{Res}_{\mathbb{Q}_{\mathbb{J}_t}[t,X,Y]/\mathbb{Q}}\left[\begin{matrix}\Big(\prod_{j=1}^n\Big(\sum_{\ell=0}^k\bar p_j^{\,k-\ell}(\eta_j\bar t)^\ell\Big)\Big)\,q(\bar X,\bar Y)\,d\bar X\wedge d\bar t\\ \bar p_1^{\,k+1},\dots,\bar p_n^{\,k+1},\bar t^{\,k+1}\end{matrix}\right]\\
&\quad=\sum_{\{\kappa\in\mathbb{N}^n\,:\,|\kappa|=k\}}\operatorname{Res}_{\mathbb{Q}_{\mathbb{J}}[X,Y]/\mathbb{Q}}\left[\begin{matrix}q(\bar X,\bar Y)\,d\bar X\\ \bar p_1^{\,\kappa_1+1},\dots,\bar p_n^{\,\kappa_n+1}\end{matrix}\right]\eta^\kappa.
\end{aligned}$$

One can alternatively reinterpret (6.233) as a polynomial identity in $\mathbb{Q}[\eta]$, where $\eta=(\eta_1,\dots,\eta_n)$ is considered as a set of n parameters. One needs to understand the right-hand side of this formula as

$$\operatorname{Res}_{(\mathbb{Q}(\eta))_{\mathbb{J}_t}[t,X,Y]/\mathbb{Q}(\eta)}\left[\begin{matrix}q(\bar X,\bar Y)\,d\bar X\wedge d\bar t\\ \bar p_1-\eta_1\bar t,\dots,\bar p_n-\eta_n\bar t,\bar t^{\,k+1}\end{matrix}\right].\tag{6.234}$$

We already mentioned the Perron geometric generalized theorem in §6.4.1 (when $N=n$ and $\mathbb{J}=\mathbf{0}$) and within the proof of Proposition 6.43 (in the general case). Our formulation of this result (geometric in nature) can be found in Z. Jelonek [**Jel05**, Theorem 3.3]. We need something more here since we are dealing with the arithmetic (instead of geometric) setting. As we will see in the next chapter (see §7.1.6) a *parametric version* of such a (geometric) generalized Perron theorem is the cornerstone leading to the arithmetic nullstellsensatz already stated in Theorem 6.70. We postpone to Chapter 7 the precise formulation of the parametric version of the generalized Perron theorem; see Theorem 7.17. Let us say here only that [**AKS13**, Corollary 3.23], following from such a parametric version of the generalized Perron theorem, and which we will invoke in the present section, is the arithmetic companion to [**Jel05**, Theorem 3.3]. It is also crucial for the arithmetic elimination theorem, already stated in Theorem 6.68. One has that $\bar p_1,\dots,\bar p_n$ are

[39]See also Theorem 6.49 for a proof in the particular setting where algebraic residue symbols are viewed through the action of multivariate residual currents, which is the case here since $\mathbb{K}=\mathbb{Q}$.

$\mathbb{Q}$-algebraically independent in $\mathbb{Q}_{\mathbb{J}}[X,Y]$ since the sequence $(\bar{p}_1,\ldots,\bar{p}_n)$ is quasi-regular in $\mathbb{Q}_{\mathbb{J}}[X,Y]$ and $\mathbb{X}_{\mathbb{C}} \cap V(p) \neq \emptyset$. Let $\{\varepsilon_1,\ldots,\varepsilon_n\}$ be the canonical basis of $\mathbb{Z}^n$. Then [**AKS13**, Corollary 3.23] asserts that for any $j = 1,\ldots,n$, there exists a polynomial

$$\begin{aligned}\boldsymbol{E}_j = E_{\varepsilon_j} \in \mathbb{Z}[W_1,\ldots,W_n,X^{\varepsilon_j}] &= \sum_{\rho=0}^{\boldsymbol{\nu}_j} \boldsymbol{E}_{j,\rho}(W)\, X_j^{\boldsymbol{\nu}_j-\rho} \\ &= \sum_{\rho=0}^{\boldsymbol{\nu}_j} \Big(\sum_{\kappa\in\mathbb{N}^n} \gamma_{j,\rho,\kappa} W^{\kappa}\Big)\, X_j^{\boldsymbol{\nu}_j-\rho_j} \in \mathbb{Z}[W,X_j]\end{aligned}$$

satisfying

$$\begin{aligned}&\boldsymbol{E}_j(p(X),X_j) \in \mathbb{J}, \\ &\deg_{w_1,\ldots,w_n,X_j}\big(\boldsymbol{E}_j(w_1^{\deg p_1},\ldots,w_n^{\deg p_n},X_j)\big) \leq \mathbf{d}\,(p), \\ &\max_{0\leq\rho\leq\boldsymbol{\nu}_j,\kappa\in\mathbb{N}^n}\big(\log|\gamma_{j,\rho,\kappa}| + \langle\kappa,\mathrm{h}(p)\rangle\big) \leq \mathbf{h}_{\,c_{\mathrm{aP}}(n,N)}\,(p),\end{aligned} \tag{6.235}$$

whenever $c_{\text{arithmetic Perron}}(n,N) = c_{\mathrm{aP}}(n,N) = (n+2)\log(2N+8)$. If v_j denotes the valuation in W of $\boldsymbol{E}_j(W,X_j)$, then the polynomial $\boldsymbol{E}_j$ splits as

$$\boldsymbol{E}_j = \boldsymbol{E}'_j + \boldsymbol{E}''_j, \tag{6.236}$$

where

$$\begin{aligned}\boldsymbol{E}'_j &= \sum_{\rho=0}^{\boldsymbol{\nu}_j}\Big(\sum_{\{\kappa\in\mathbb{N}^n\,:\,|\kappa|=v_j\}} \gamma_{j,\rho,\kappa}W^{\kappa}\Big)\, X_j^{\boldsymbol{\nu}_j-\rho}, \\ \boldsymbol{E}''_j &= \sum_{\rho=0}^{\boldsymbol{\nu}_j}\Big(\sum_{\{\kappa\in\mathbb{N}^n\,:\,|\kappa|>v_j\}} \gamma_{j,\rho,\kappa}W^{\kappa}\Big)\, X_j^{\boldsymbol{\nu}_j-\rho}.\end{aligned} \tag{6.237}$$

Furthermore, if $(\eta,t) = (\eta_1,\ldots,\eta_n,t)$ denotes a set of $n+1$ formal variables, the polynomials

(6.238)

$$\begin{aligned}R_j(\eta,X_j) &= \boldsymbol{E}'_j(\eta\,t,X_j)/t^{v_j} = \mathfrak{D}_{\boldsymbol{E},j}(\eta)\,X^{r_j} + \sum_{\rho=1}^{r_j} r_{j,\rho}(\eta)X_j^{r_j-\rho_j} \in \mathbb{Z}[\eta,X_j], \\ S_j(\eta,X_j,t) &= \boldsymbol{E}''_j(\eta\,t,X_j)/t^{v_j+1} \in \mathbb{Z}[\eta,t,X_j]\end{aligned}$$

are attached to the splitting (6.236). These polynomials will play an important role in what follows. It is worthwhile to observe that $\mathfrak{D}_{\boldsymbol{E},j}$ and $r_{j,\rho}$ for $j = 1,\ldots,n$, with $1 \leq \rho \leq r_j$, are homogeneous polynomials in η, whose degree is the valuation v_j of $\boldsymbol{E}_j$ in η. The Taylor formula implies, for any $j = 1,\ldots,n$, the equality

(6.239)

$$\begin{aligned}\boldsymbol{E}_j(p,X_j) = \boldsymbol{E}_j(\eta t,X_j) + \sum_{\kappa\in\mathbb{N}^n\setminus\{0\}} \frac{1}{\kappa!}\Big[\Big(\frac{\partial}{\partial W}\Big)^{\kappa}(\boldsymbol{E}_j(W,X_j))\Big]_{W=\eta t}(p-\eta t)^{\kappa} \\ \in \mathbb{J}_t\,\mathbb{Q}[\eta,t,X,Y].\end{aligned}$$

As a consequence, if one takes classes in $\mathbb{Q}[t,X,Y]/\mathbb{J}_t$, then

$$
\begin{aligned}
(6.240)\qquad & \bar{t}^{v_j}\big(R_j(\eta,\bar{X}_j)+\bar{t}S_j(\eta,\bar{t},\bar{X}_j)\big)\\
&= -\sum_{\kappa\in\mathbb{N}^n\setminus\{0\}} \frac{1}{\kappa!}\Big[\Big(\frac{\partial}{\partial W}\Big)^{\kappa}(\boldsymbol{E}_j(W,\bar{X}_j))\Big]_{W=\eta\bar{t}}(\bar{p}-\eta\bar{t})^{\kappa}\\
&\equiv -\sum_{k=1}^{n}\Big[\Big(\frac{\partial}{\partial W_k}\Big)(\boldsymbol{E}_j(W,\bar{X}_j))\Big]_{W=\eta\bar{t}}(\bar{p}_k-\eta_k\bar{t})
\end{aligned}
$$

modulo $(\bar{p}_1-\eta_1\bar{t},\dots,\bar{p}_n-\eta_n\bar{t})^2$. Let us now invoke (6.233) and the generalized transformation law stated in the geometric context as Proposition 3.23. Both elements combined with the current interpretation of algebraic residue symbols, when the base field $\mathbb{K}$, namely $\mathbb{K}=\mathbb{Q}$ here, is embedded in $\mathbb{C}$ (see Theorem 6.49) lead to the algebraic identity

$$
\begin{aligned}
&\sum_{\{\kappa\in\mathbb{N}^n\,:\,|\kappa|=k\}} \mathrm{Res}_{\mathbb{Q}_{\mathbb{J}}[X,Y]/\mathbb{Q}}\begin{bmatrix} q(\bar{X},\bar{Y})\,d\bar{X}\\ \bar{p}_1^{\kappa_1+1},\dots,\bar{p}_n^{\kappa_n+1}\end{bmatrix}\eta^{\kappa}\\
(6.241)\qquad &= \mathrm{Res}_{(\mathbb{Q}(\eta))_{\mathbb{J}_t}[t,X,Y]/\mathbb{Q}(\eta)}\begin{bmatrix} q(\bar{X},\bar{Y})\,d\bar{X}\wedge d\bar{t}\\ \bar{p}_1-\eta_1\bar{t},\dots,\bar{p}_n-\eta_n\bar{t},\bar{t}^{k+1}\end{bmatrix}\\
&= (-1)^n\mathrm{Res}_{(\mathbb{Q}(\eta))_{\mathbb{J}_t}[t,X,Y]/\mathbb{Q}(\eta)}\begin{bmatrix} J(\eta,t,\bar{X})\,q(\bar{X},\bar{Y})\,d\bar{X}\wedge d\bar{t}\\ R(\eta,\bar{X})+\bar{t}\,S(\eta,t,\bar{X}),\bar{t}^{k+|v|+1}\end{bmatrix}
\end{aligned}
$$

holding in $\mathbb{Q}(\eta)$ for any $q\in\mathbb{Z}[X,Y]$ and $k\in\mathbb{N}$. Here

(6.242)

$$
\begin{aligned}
J(\eta,t,X) &= \det\Big[\Big[\Big(\frac{\partial}{\partial W_k}\Big)(\boldsymbol{E}_j(W,X_j))\Big]_{W=\eta t}\Big]_{1\le j,k\le n},\\
R(\eta,X)+t\,S(\eta,t,X) &= \big(R_1(\eta,X_1)+t\,S_1(\eta,t,X_1),\dots,R_n(\eta,X_n)+t\,S_n(\eta,t,X_n)\big).
\end{aligned}
$$

For any $c\ge 0$, let us attach to $\mathbb{X}$, p, $\boldsymbol{E}=(\boldsymbol{E}_1,\dots,\boldsymbol{E}_n)$, the following three arithmetic markers:

$$
\begin{aligned}
&\mathbf{k}_c(\boldsymbol{E},p) = n\big(n\,\mathbf{d}\,(p)-|r|+1\big)\,\mathbf{h}_c(p),\\
(6.243)\qquad &\mathbf{k}_c^{'}(p) = n\,\mathbf{h}_c(p)+(\mathbf{h}+c\,\mathbf{d}),\\
&\mathbf{k}_c^{''}(p) = n^2\,\mathbf{h}_c(p),
\end{aligned}
$$

where $|r|=\sum_{j=1}^n \deg R_j$. The geometric (respectively arithmetic) markers $\mathbf{d},\mathbf{d}\,(p)$ (respectively $\mathbf{h}$, $\mathbf{h}_c(p)$) attached to $\mathbb{X},p$ are those that have been introduced in (6.177) and (6.180) (respectively (6.190) and (6.197)). We also attach to a polynomial $q\in\mathbb{Z}[X,Y]$ (following Proposition 6.71) an integer $\mathfrak{d}[q]\in\mathbb{N}^*$ satisfying

$$
\begin{aligned}
&\mathfrak{d}[q]\begin{bmatrix} q(\bar{X},\bar{Y})\,d\bar{X}\\ \bar{X}-\mathrm{t}\end{bmatrix}_{\mathbb{X}}\in\mathbb{Z}[\mathrm{t}],\\
(6.244)\qquad &\log\mathfrak{d}[q]\le(\mathbf{h}+c_{\mathrm{act}}(N)\,\mathbf{d})\deg q,\\
&\mathrm{h}\Big(\mathfrak{d}[q]\begin{bmatrix} q(\bar{X},\bar{Y})\,d\bar{X}\\ \bar{X}-\mathrm{t}\end{bmatrix}_{\mathbb{X}}\Big)\le \mathrm{h}(q(X,Y)\,dX)+(\,\mathbf{h}+c_{\mathrm{act}}(N)\,\mathbf{d}\,)\deg q.
\end{aligned}
$$

We now have in our hands all the necessary elements to state the main result of this subsection.

THEOREM 6.76. *For any $q \in \mathbb{Z}[X,Y]$ and $k \in \mathbb{N}$, one has in $\mathbb{Q}(\eta)$ the algebraic identity*

$$\sum_{\{\kappa\in\mathbb{N}^n\,:\,|\kappa|=k\}} \operatorname{Res}_{\mathbb{Q}_J[X,Y]/\mathbb{Q}} \begin{bmatrix} q(\bar X,\bar Y)\,d\bar X \\ \bar p_1^{\kappa_1+1},\dots,\bar p_n^{\kappa_n+1}\end{bmatrix} \eta^\kappa = \frac{1}{\mathfrak{d}[q]}\,\frac{\mathfrak{N}_{\boldsymbol{E},k}(\eta)}{\left(\prod_{j=1}^n \mathfrak{D}_{\boldsymbol{E},j}\right)^{\deg q+n+(k+|v|+1)(n\,\mathbf{d}\,(p)-|r|+1)}}, \tag{6.245}$$

where the polynomial $\mathfrak{N}_{\boldsymbol{E},k} \in \mathbb{Z}[\eta]$ is homogeneous and its degree equals

$$\deg \mathfrak{N}_{\boldsymbol{E}}^{[k]} = k + n|v|\big(\deg q + n + (k+|v|+1)(\mathbf{d}\,(p)+|r|-1)\big). \tag{6.246}$$

Furthermore, the homogeneous polynomial $\mathfrak{D}_{\boldsymbol{E},j} \in \mathbb{Z}[\eta]$ with degree v_j is the dominant coefficient of $R_j \in \mathbb{Z}[\eta][X_j]$ for each $j=1,\dots,n$. Moreover, one has the logarithmic size estimates

$$\begin{aligned} &\log \mathfrak{d}[q] + \mathrm{h}\left(\Big(\prod_{j=1}^n \mathfrak{D}_{\boldsymbol{E},j}\Big)^{\deg q+n+(k+|v|+1)(n\,\mathbf{d}\,(p)-|r|+1)}\right) \le \mathbf{k}_c(\boldsymbol{E},p,k),\\ &\mathrm{h}(\mathfrak{N}_{\boldsymbol{E}}^{[k]}) \le \mathbf{k}_c(\boldsymbol{E},p,k) + \mathrm{h}\,(q(X,Y)\,dX), \end{aligned} \tag{6.247}$$

where $\mathbf{k}_c(\boldsymbol{E},p,k) := \mathbf{k}_c\,(\boldsymbol{E},p)(k+|v|+1) + \mathbf{k}_c'\,(p)\deg q + \mathbf{k}_c''\,(p)$ with

$$c = c(n,N) = \max\big((n+5)\,\log(2N+8),(N+2)\,\log(2N+3)\big).$$

PROOF. Let us start with the algebraic identity (6.241) in $\mathbb{Q}(\eta)$ and denote by $A_{k+|v|}(\eta,X)$ the coefficient of $t^{k+|v|}$ in the development of the polynomial

$$J(\eta,t,X)\prod_{j=1}^n\Big(\sum_{\ell=0}^{k+|v|} R_j^{k+|v|-\ell}(\eta,X_j)(-tS_j(\eta,t,X_j))^\ell\Big).$$

Then one can reformulate the identity (6.241) as

$$\begin{aligned} &\sum_{\{\kappa\in\mathbb{N}^n\,:\,|\kappa|=k\}} \operatorname{Res}_{\mathbb{Q}_J[X,Y]/\mathbb{Q}} \begin{bmatrix} q(\bar X,\bar Y)\,d\bar X \\ \bar p_1^{\kappa_1+1},\dots,\bar p_n^{\kappa_n+1}\end{bmatrix} \eta^\kappa \\ &\quad = (-1)^n \operatorname{Res}_{(\mathbb{Q}(\eta))_J[X,Y]/\mathbb{Q}(\eta)} \begin{bmatrix} A_{k+|v|}(\eta,\bar X)\,q(\bar X,\bar Y)\,d\bar X \\ R_1^{k+|v|+1}(\eta,\bar X_1),\dots,R_n^{k+|v|+1}(\eta,\bar X_n)\end{bmatrix}. \end{aligned} \tag{6.248}$$

If one specializes η as a generic point in $\mathbb{C}^n$, then one can see that Proposition 6.44 implies that (6.248) can be reformulated as

$$\begin{aligned} &\sum_{\{\kappa\in\mathbb{N}^n\,:\,|\kappa|=k\}} \operatorname{Res}_{\mathbb{Q}_J[X,Y]/\mathbb{Q}} \begin{bmatrix} q(\bar X,\bar Y)\,d\bar X \\ \bar p_1^{\kappa_1+1},\dots,\bar p_n^{\kappa_n+1}\end{bmatrix} \eta^\kappa \\ &\quad = \frac{(-1)^n}{\mathfrak{d}[q]} \operatorname{Res}_{(\mathbb{Q}(\eta))[\mathfrak{t}]/\mathbb{Q}(\eta)} \begin{bmatrix} A_{k+|v|}(\eta,\mathfrak{t})\left(\mathfrak{d}[q]\begin{bmatrix} q(\bar X,\bar Y)\,d\bar X \\ \bar X-\mathfrak{t}\end{bmatrix}_{\mathbb{X}}\right) d\mathfrak{t} \\ R^{k+|v|+1}(\eta,\mathfrak{t})\end{bmatrix}, \end{aligned} \tag{6.249}$$

where $\mathfrak{t} = (\mathfrak{t}_1,\dots,\mathfrak{t}_n)$ and $R^{k+|v|+1}(\eta,\mathfrak{t}) = (R_1^{k+|v|+1}(\eta,\mathfrak{t}_1),\dots,R_n^{k+|v|+1}(\eta,\mathfrak{t}_n))$. Let us now specify the degree and the logarithmic size estimates. If $a \in \mathbb{Z}[u]$, where here $u=(u_1,\dots,u_M)$ is a finite set of variables, the length $\mathrm{h}_1(a)$ is defined as the arithmetic marker $\log\|\{\mathrm{coeff}\}(a)\|_1$, that is, $\mathrm{h}_1(a) = \mathrm{h}(a(u)\,du_1\wedge\cdots\wedge du_M)$ according to the conventions (6.165). An important point about the length is its submultiplicativity. It is also useful to recall the comparison inequalities between length and logarithmic size:

$$\mathrm{h}(a) \le \mathrm{h}_1(a) \le \mathrm{h}(a) + \log(M+1)\,\deg a. \tag{6.250}$$

It follows from the degree and logarithmic size estimates (6.235) for each $\boldsymbol{E}_j$, the definition of R_j, S_j in (6.238), and the fact that $d_j \ge 2$ for $j = 1, \dots, n$, that

$$\begin{aligned} &\max(\deg R_j, \deg S_j) \le \mathbf{d}\,(p), \\ &\max(\mathrm{h}(R_j), \mathrm{h}(S_j)) \le \mathbf{h}_{\,c_{\mathrm{aP}}(n,N)}\,(p) \\ &\qquad \Longrightarrow \max\big(\mathrm{h}_1(R_j), h_1(S_j)\big) \le \mathbf{h}_{\,(n+3)\log(2N+8)}\,(p) \end{aligned} \tag{6.251}$$

since R_j and S_j depend on at most $n+2$ variables. On the other hand, one has for any $1 \le j,k \le n$ that

$$\mathrm{h}_1\Big(\frac{\partial \boldsymbol{E}_j}{\partial W_k}\Big) \le \mathrm{h}(\boldsymbol{E}_j) + \log \mathbf{d}\,(p) + \mathbf{d}\,(p)\,\log 2. \tag{6.252}$$

Furthermore,

(6.253)

$$\begin{aligned} \deg_\eta\Big(A_{k+|v|}(\eta,\mathfrak{t})\Big(\mathfrak{d}[q]\left[\frac{q(\bar X,\bar Y)\,d\bar X}{\bar X-\mathfrak{t}}\right]_{\mathbb{X}}\Big)\Big) &\le n(k+|v|+1)\,\mathbf{d}\,(p), \\ \deg_{\mathfrak{t}}\Big(A_{k+|v|}(\eta,\mathfrak{t})\Big(\mathfrak{d}[q]\left[\frac{q(\bar X,\bar Y)\,d\bar X}{\bar X-\mathfrak{t}}\right]_{\mathbb{X}}\Big)\Big) = D_{\boldsymbol{E}}^{[k]} &\le \deg q + n(k+|v|+1)\,\mathbf{d}\,(p). \end{aligned}$$

It follows from the submultiplicativity of the length, the inequalities (6.235), (6.244), (6.251), (6.252), and the definition of c, that

$$\begin{aligned} &\mathrm{h}_1\Big(A_{k+|v|}(\eta,\mathfrak{t})\Big(\mathfrak{d}[q]\left[\frac{q(\bar X,\bar Y)\,d\bar X}{\bar X-\mathfrak{t}}\right]_{\mathbb{X}}\Big)\Big) \\ &\qquad \le h(q(X)\,dX) + \deg q\,(\mathrm{h} + c\,\mathbf{d}) + n(k+|v|+1)\,\mathbf{h}_{\,c}\,(p). \end{aligned} \tag{6.254}$$

As in the proof of Theorem 6.72 or of Theorem 6.75, one has that

$$\mathfrak{d}[q]\Big(\prod_{j=1}^{n} \mathfrak{D}_{E,j}(\eta)\Big)^{D_{\boldsymbol{E}}^{[k]}+n-(k+|v|+1)(|r|-1)} = \widetilde{\mathfrak{D}}_{\boldsymbol{E}}^{[k]}(\eta) \in \mathbb{Z}[\eta] \tag{6.255}$$

is a homogeneous polynomial, whose degree is equal to

$$\begin{aligned} \deg \widetilde{\mathfrak{D}}_{\boldsymbol{E}}^{[k]} &= |v|\big(D_{\boldsymbol{E}}^{[k]} + n - (k+|v|+1)(|r|-1)\big) \\ &\le |v|\big(\deg q + n + (k+|v|+1)\big(n\,\mathbf{d}\,(p) - |r| + 1\big)\big). \end{aligned} \tag{6.256}$$

The degree (6.256) is such that

$$\widetilde{\mathfrak{D}}_{\boldsymbol{E}}^{[k]}(\eta)\,\mathrm{Res}_{(\mathbb{Q}(\eta)[\mathfrak{t}]/\mathbb{Q}(\eta)}\left[\frac{A_{k+|v|}(\eta,\mathfrak{t})\Big(\mathfrak{d}[q]\left[\frac{q(\bar X,\bar Y)\,d\bar X}{\bar X-\mathfrak{t}}\right]_{\mathbb{X}}\Big)\,d\mathfrak{t}}{R^{k+|v|+1}(\eta,\mathfrak{t})}\right] \in \mathbb{Z}[\eta].$$

Therefore, if one let

$$\mathfrak{D}_{\boldsymbol{E}}^{[k]}(\eta) := \mathfrak{d}[q]\Big(\prod_{j=1}^{n} \mathfrak{D}_{\boldsymbol{E},j}(\eta)\Big)^{\deg q+n+(k+|v|+1)(n\,\mathbf{d}\,(p)-|r|+1)}, \tag{6.257}$$

then one also has

$$(6.258)\quad \mathfrak{N}_{\boldsymbol{E}}^{[k]}(\eta)$$
$$:= \mathfrak{D}_{\boldsymbol{E}}^{[k]}(\eta)\Big((-1)^n \mathrm{Res}_{(\mathbb{Q}(\eta)[\mathrm{t}]/\mathbb{Q}(\eta)}\left[\begin{matrix} A_{k+|v|}(\eta,\mathrm{t})\left(\mathfrak{d}[q]\begin{bmatrix} q(\bar{X},\bar{Y})\,d\bar{X} \\ \bar{X}-\mathrm{t}\end{bmatrix}_{\mathbb{X}}\right)d\mathrm{t} \\ R^{k+|v|+1}(\eta,\mathrm{t})\end{matrix}\right]\Big) \in \mathbb{Z}[\eta].$$

In view of the algebraic identity (6.235), the polynomial (6.258) is a homogeneous polynomial in η with integer coefficients and degree equal to

$$\deg_\eta \mathfrak{N}_{\boldsymbol{E}}^{[k]} = k + n|v|\big(\deg q + n + (k+|v|+1)(\mathbf{d}\,(p)+|r|-1)\big).$$

It remains to estimate the lengths of the polynomials $\mathfrak{D}_{\boldsymbol{E}}^{[k]}$ and $\mathfrak{N}_{\boldsymbol{E}}^{[k]}$. The estimate (6.247) for

$$\mathrm{h}\,(\mathfrak{D}_{\boldsymbol{E}}^{[k]}) \le \mathrm{h}_1\,(\mathfrak{D}_{\boldsymbol{E}}^{[k]}) = \log \mathfrak{d}[q] + \mathrm{h}_1\left(\Big(\prod_{j=1}^n \mathfrak{D}_{\boldsymbol{E},j}(\eta)\Big)^{\deg q+n+(k+|v|+1)(n\,\mathbf{d}\,(p)-|r|+1)}\right)$$

follows from the estimates (6.244) and (6.251), together with the submultiplicativity of the length. Let now

$$A_{k+|v|}(\eta,\mathrm{t})\left(\mathfrak{d}[q]\begin{bmatrix} q(\bar{X},\bar{Y})\,d\bar{X} \\ \bar{X}-\mathrm{t}\end{bmatrix}_{\mathbb{X}}\right) = \sum_{\{\kappa\in\mathbb{N}^n\,:\,|\kappa|\le D_{\boldsymbol{E}}^{[k]}\}} a_\kappa(\eta)\,\mathrm{t}^\kappa.$$

The coefficients $a_\kappa \in \mathbb{Z}[\eta]$ satisfy

$$\deg a_\kappa \le n(k+|v|+1)\,\mathbf{d}\,(p),$$
$$(6.259)\quad \mathrm{h}_1(a_\kappa) \le \mathrm{h}_1\left(A_{k+|v|}(\eta,\mathrm{t})\left(\mathfrak{d}[q]\begin{bmatrix} q(\bar{X},\bar{Y})\,d\bar{X} \\ \bar{X}-\mathrm{t}\end{bmatrix}_{\mathbb{X}}\right)\right)$$
$$\le \mathrm{h}(q(X)\,dX)) + \deg q\,(\mathbf{h}+c\,\mathbf{d}) + n(k+|v|+1)\,\mathbf{h}_c\,(p)$$

as a consequence of the estimates (6.253) and (6.254). For any $\kappa \in \mathbb{N}^n$, let

$$w_\kappa(\eta) = \mathfrak{D}_{\boldsymbol{E}}^{[k]}\,\mathrm{Res}_{\,\mathbb{Q}(\eta)[\mathrm{t}]/\mathbb{Q}(\eta)}\begin{bmatrix} \mathrm{t}^\kappa\,d\mathrm{t} \\ R^{k+|v|+1}(\eta,\mathrm{t})\end{bmatrix} \in \mathbb{Z}[\eta].$$

It follows from Euclid's division algorithm, namely Lemma 6.63 transposed from the univariate setting to that of tensorial products,[40] and from the submultiplicativity of the length, that $\mathrm{h}_1(w_\kappa) \le \big(\deg q + n + (k+|v|-1)\,(n\,\mathbf{d}\,(p)-|r|)\big)\,|\,\mathrm{h}_1(R)|$, where $|\,\mathrm{h}_1(R)| := \sum_{j=1}^n \mathrm{h}_1(R_j) \le n\,\mathbf{h}_c(p)$. Since

$$\mathrm{h}\left(\mathfrak{D}_{\boldsymbol{E}}^{[k]}\mathrm{Res}_{(\mathbb{Q}(\eta)[\mathrm{t}]/\mathbb{Q}(\eta)}\left[\begin{matrix} A_{k+|v|}(\eta,\mathrm{t})\left(\mathfrak{d}[q]\begin{bmatrix} q(\bar{X},\bar{Y})\,d\bar{X} \\ \bar{X}-\mathrm{t}\end{bmatrix}_{\mathbb{X}}\right)d\mathrm{t} \\ R^{k+|v|+1}(\eta,\mathrm{t})\end{matrix}\right]\right)$$
$$\le \max_{\{\kappa\in\mathbb{N}^n\,:\,|\kappa|\le D_{\boldsymbol{E}}^{[\kappa]}\}} (\mathrm{h}_1(a_\kappa)+\mathrm{h}_1(w_\kappa)) + \big(\deg q + n(k+|v|+1)\,\mathbf{h}_c\,(p)\big)\,\log(n+1),$$

one gets, according to the choice of c, the estimate for $\mathrm{h}(\mathfrak{N}_{\boldsymbol{E}}^{[k]})$ stated in (6.247). This concludes the proof of Theorem 6.76. □

[40] We refer the reader to [**SomY21**, §3]; more precisely, its Theorem 3.2 and Propositions 3.3 to 3.5.

CHAPTER 7

Miscellaneous applications: intersection, division

7.1. Hilbert's nullstellensatz in $\mathbb{K}[X_1, \dots, X_n]$ and residue calculus

7.1.1. Rabinowitsch's trick and Grete Hermann's algorithmic procedure. We recall here Hilbert's nullstellensatz [**Hil1893**, p. 320] within the polynomial context.

THEOREM 7.1 (Nullstellensatz in the polynomial context and Nœther exponent). *Let $m, n \in \mathbb{N}^*$, $\mathbb{K}$ be an algebraically closed field and polynomials $p_1, \dots, p_m, q$ be in the polynomial algebra $\mathbb{K}[X_1, \dots, X_n] = \mathbb{K}[X]$, such that $V(q) \supset \bigcap_{j=1}^m V(p_j)$. Then q lies in the radical $\sqrt{(p)}$ of $(p) = \sum_{j=1}^m p_j \mathbb{K}[X]$, with*

$$q^e = \sum_{j=1}^m q_j p_j \in (p) \tag{7.1}$$

for some minimal exponent $e((p))$ depending only of the ideal (p). Such a minimal exponent $e((p))$ is called the Nœther exponent of the ideal (p) in the polynomial algebra $\mathbb{K}[X]$.

REMARK 7.2. Concerning the independence of $e((q))$ from q, which sustains the notion of the Nœther exponent as defined in [**Pl90**], we refer to [**Las1905**].

Hilbert's nullstellensatz in the polynomial context is closely related to the *effective realization of Bézout's identity.*

DEFINITION 7.3 (Effective realization of Bézout identity). Let $m, n \in \mathbb{N}^*$ and $\mathbb{K}$ be a field. Assume also that the polynomials $p_0, p_1, \dots, p_m$ in the polynomial algebra $\mathbb{K}[X_1, \dots, X_n] = \mathbb{K}[X]$ are such that $(p_0, \dots, p_m) = \mathbb{K}[X]$. Bézout identity *for the set of generators $\{p_0, \dots, p_m\}$ for* $\mathbb{K}[X]$ is said to be *N-solvable in* $\mathbb{K}[X]$ for some $N \in \mathbb{N}^*$ if and only if one can find polynomials $q_0, \dots, q_m \in \mathbb{K}[X]$ such that

$$1 = \sum_{j=0}^m q_j p_j \quad \text{with} \quad \max_{0 \le j \le m} \deg q_j p_j \le N. \tag{7.2}$$

When $\mathbb{K}$ is algebraically closed, a famous trick, introduced by the Russian physicist G. Y. Rainich (under his former name J. L. Rabinowitsch [**Ra30**]) and subtly refined in [**Bro87B**], provides a path from the effective realization of Bézout identity (see Definition 7.3) to an upper estimation for the Nœther exponent $e((p))$ of an ideal in $\mathbb{K}[X]$, explicitly described in terms of a set of generators $\{p_1, \dots, p_m\}$.

PROPOSITION 7.4 (Rabinowitsch trick). *Let $m, n \in \mathbb{N}^*$ and let $\mathbb{K}$ be an algebraically closed field. Let also the polynomials $p_1, \dots, p_m$ in $\mathbb{K}[X_1, \dots, X_n] = \mathbb{K}[X]$ and $q \in \sqrt{((p))}$ with $\deg q > 0$. If $D \ge \deg q$ and $N_D(\{p_1, \dots, p_m\}, q)$ is the minimal positive integer N such that Bézout identity with respect to the set of generators*

$\{1 - t^D q(X), p_1(X), \dots, p_m(X)\}$ *is* N*-solvable in* $\mathbb{K}[t, X_1, \dots, X_n]$*, then*

$$q^\nu = \sum_{j=1}^m q_j p_j \in (p) \tag{7.3}$$

with

$$\begin{aligned} \nu &\le \frac{N_D(\{p_1, \dots, p_m\}, q)}{D}, \\ \max_{1 \le j \le m} \deg q_j &\le \deg q \Big(\nu + \frac{N_D(\{p_1, \dots, p_m\}, q)}{D}\Big). \end{aligned} \tag{7.4}$$

PROOF. It follows from the N_D-effective solvability in $\mathbb{K}[X]$ of Bézout identity with respect to $\{1 - t^D q(X), p_1(X), \dots, p_m(X)\}$ that there exist polynomials $Q_0, Q_1, \dots, Q_m$ in the polynomial algebra $\mathbb{K}[t, X]$ satisfying

$$1 = Q_0(t, X)(1 - t^D q(X)) + \sum_{j=1}^m Q_j(t, X) p_j(X), \tag{7.5}$$

where

$$\max\big(\deg_{t,X} Q_0 + D + \deg_X q, \max_{1 \le j \le m} (\deg_{t,X} Q_j + \deg_X p_j)\big) \le N_D(\{p_1, \dots, p_m\}, q). \tag{7.6}$$

For every $j = 0, \dots, m$, we consider the polynomials

$$Q_j(t, X) = \sum_{\ell=0}^{D-1} t^\ell Q_{j,\ell}(t^D, X).$$

It follows from (7.5) that one has also the polynomial identity

$$1 = Q_{0,0}(T, X)(1 - T\, q(X)) + \sum_{j=1}^m Q_{j,0}(T, X)\, p_j(X), \tag{7.7}$$

where

$$\max_{0 \le j \le m} \deg_T Q_{j,0} \le \frac{N_D(\{p_1, \dots, p_m\}, q)}{D},$$

and T denotes another parameter. Let us now replace T by $1/q$ in (7.7) and eliminate denominators. Then one gets the polynomial identity

$$q^\nu = \sum_{j=1}^m q_j p_j$$

in $\mathbb{K}[X]$, where

$$\nu \le \frac{N_D(\{p_1, \dots, p_m\})}{D}, \quad \max_{1 \le j \le m} \deg q_j \le \deg q \Big(\nu + \frac{N_D(\{p_1, \dots, p_m\}, q)}{D}\Big).$$

This concludes the proof of Proposition 7.4. □

Given a field $\mathbb{K}$ and a set of generators $\{p_0, \dots, p_m\}$ for $\mathbb{K}[X_1, \dots, X_n]$, an explicit resolution of Bézout identity $1 = \sum_{j=0}^m p_j q_j$ was proposed by Grete Hermann, a student of Emmy Nœther and Edmund Landau, in [**He1926**]. Hermann's algorithmic approach relies on elimination theory, that is, it relies on the iterated construction of Sylvester determinants in $\mathbb{K}(X_1, \dots, X_j)[X_{j+1}]$, $j = 0, \dots, n-1$;

see also [**L, VdW**]. Such an algorithmic procedure leads to the construction of polynomials q_j, $j=0,\dots,m$, satisfying $1=\sum_{j=0}^m q_jp_j$, whose degrees satisfy

$$\deg(q_jp_j) \le \deg p_j + 2\big(2 \max_{0\le j\le m} \deg p_j\big)^{2^{n-1}}. \tag{7.8}$$

7.1.2. Lipman–Teissier theorem and N-solvability of Bézout identity.

There is an intimate connection between elimination theory and the theory of determinants as developed by A. Cayley and F. Macaulay. As a consequence, multivariate residue calculus connects to Hilbert's nullstellensatz in $\mathbb{K}[X_1,\dots,X_n]$ when $\operatorname{char}\mathbb{K}=0$. At the core of such a connection is the link with the Briançon–Skoda theorem [**BriS74**], whose effective realization thrives on multivariate residue calculus (Theorem 5.7 and Theorem 5.8). Another essential ingredient is geometric intersection theory, in particular, the geometric Bézout's theorem; see §6.6.2 or (and) §D.5. The fact that the Nœther exponent $e((p))$ can be estimated through geometric intersection theory was suggested without proof already in the monograph by Eugen Netto [**Ne**].[1] It says that Bézout's geometric theorem governs the upper estimation of N (in term of the p_j's) such that Bézout identity (7.2) is N-solvable. To quantify the realization of (7.2), one needs to revisit some considerations for the notion of algebraic properness already sketched in §6.4. Such considerations describe precisely how Bézout's geometric theorem controls effective solvability of (7.2) with respect to the degrees of the q_j's. Given an algebraically closed field $\mathbb{K}$ and a polynomial map p from $\mathbb{K}^n$ to $\mathbb{K}^{m+1}$, let

$$\mathbb{P}^n_{\mathbb{K}} = \operatorname{Spec}\big(\operatorname{Proj}(\mathbb{K}[X_0,\dots,X_n])\big)$$

be the complete algebraic scheme with its structural sheaf $\mathcal{O}_{\mathbb{P}^n_{\mathbb{K}}}$. Let also

$$\widehat{\pi}_P : \widehat{\mathbb{P}^n_{\mathbb{K}}} \to \mathbb{P}^n_{\mathbb{K}}$$

be its normalized blowup[2] along

$$(P) = \sum_j P_j \mathcal{O}_{\mathbb{P}^n_{\mathbb{K}}},$$

where $P_j \in \mathbb{K}[X_0,\dots,X_n]$ is the homogenization of the entry p_j of p. Furthermore, let

$$H = \sum_\gamma m_\gamma H_\gamma$$

be the effective Weil divisor (on the normal complete algebraic variety $\mathbb{P}^n_{\mathbb{K}}$) attached to the exceptional divisor $(P)\,\mathcal{O}_{\widehat{\mathbb{P}^n_{\mathbb{K}}}} := \widehat{\pi}_P^*\big((P)\big)$. Every local ring

$$\mathcal{O}_{\widehat{\mathbb{P}^n_{\mathbb{K}}},H_\gamma} \quad (\text{when } m_\gamma \in \mathbb{N}^* \text{ and } \widehat{\pi}_P(H_\gamma) \subset |\operatorname{div}(X_0)| = \mathbb{P}^n_{\mathbb{K}} \setminus \mathbb{K}^n)$$

is a discrete valuation ring, whose maximal ideal contains

$$(X_0)\,\mathcal{O}_{\widehat{\mathbb{P}^n_{\mathbb{K}}}} := \widehat{\pi}_P((X_0)),$$

since $\widehat{\mathbb{P}^n_{\mathbb{K}}}$ is normal in codimension 1. Such considerations lead us to the following definition.

---

[1] We are grateful to Arkadiuz Płoski for mentioning this fascinating monograph to us.

[2] See §A.4.2.2 for the construction of the normalized blowup within the context of algebraic complex geometry. Remark A.38 describes the normalized blowup within the context of algebraic schemes instead of complex analytic spaces.

DEFINITION 7.5. Consider a polynomial map p from $\mathbb{K}^n$ to $\mathbb{K}^{m+1}$. The strictly positive rational number

$$\mu^\infty(p) := \max_{\{\gamma\,:\, m_\gamma \in \mathbb{N}^*,\ \widehat{\pi}_P(H_\gamma)\subset |\mathrm{div}(X_0)|\}} \Big(\frac{m_\gamma}{\mathrm{length}\big(\mathbb{O}_{\widehat{\mathbb{P}^n_{\mathbb{K}}},H_\gamma}/(X_0)\,\mathbb{O}_{\widehat{\mathbb{P}^n_{\mathbb{K}}},H_\gamma}\big)}\Big) \tag{7.9}$$

is called *Lojasiewicz exponent of* p *along* $|\mathrm{div}(X_0)| = \mathbb{P}^n_{\mathbb{K}} \setminus \mathbb{K}^n$.

EXAMPLE 7.6. Suppose that $(p) = \mathbb{K}[X]$. It means that $V(p) = \emptyset$ according to Theorem 7.1. Fix a nontrivial absolute value $|\ |$ on $\mathbb{K}$. Then, for any $\varepsilon > 0$, there exist positive constants $c_\varepsilon, C_\varepsilon$ such that

$$x \in \mathbb{K}^n \text{ with } |x| = |x|_\infty \geq C_\epsilon \Longrightarrow \max_j \frac{|p_j(x)|}{|x|^{\deg p_j}}\,|x|^{\mu^\infty(p)+\varepsilon} \geq c_\varepsilon. \tag{7.10}$$

In other terms, the polynomial map p admits $(\mu^\infty(p), \dots, \mu^\infty(p))$ as Łojasiewicz multi-exponent with respect to the affine algebraic variety $\mathbb{K}^n$, according to Definition 6.53. The minimal N such that Bézout identity with respect to $\{p_j\}_j$ is N-solvable (see Definition 7.3) is then bounded from below by $\max_j(\mu^\infty(p), \deg p_j)$ as observed by M. Hickel in [**Hi01**, Theorem 5.1(1)].

Bézout's geometric intersection theory governs the Łojasiewicz exponent along the hyperplane at infinity $\mathbb{P}^n_{\mathbb{K}} \setminus \mathbb{K}^n$, according to the result from [**Hi01**, §4, pp. 728–731].

PROPOSITION 7.7. *Suppose that the* $m+1$ *entries* p_j *of the polynomial map* p *are ordered in increasing size by their degrees* d_j. *Then*

$$\mu^\infty(p) \leq \min\Big(|d(p)|^n_\infty, \frac{\prod_j d_j}{(\min_j d_j)^{(m+1-n)_+}}\Big), \tag{7.11}$$

where $|d(p)|_\infty := \max_j \deg_j$.

PROOF. We refer the reader to [**Hi01**, §4] for a proof relying on refined Bézout's geometric theorem in $\mathbb{P}^n_{\mathbb{K}}$ as stated in [**Ful**, Theorem 12.3 and Example 12.3.3]. When $\mathbb{K} = \mathbb{C}$, we encourage the reader to see also, for what concerns Fulton's results mentioned here, Theorems D.54 and D.56, as well as Proposition D.59 in §D.5.4. □

In the local $(n+1)$-dimensional regular local ring $\mathbb{K}[[X_0, \dots, X_n]]$ with maximal ideal $\mathfrak{M}_{n+1} = (X_0, \dots, X_n)\mathbb{K}[[X_0, \dots, X_n]])$, let $\mathbb{I}_P = \sum_j P_j\, \mathbb{K}[[X_0, \dots, X_n]]$ and

$$\ell(\mathbb{I}_P) = \dim\Big(\bigoplus_{\nu=0}^\infty \frac{\mathbb{I}_P^\nu}{\mathfrak{M}_{n+1}\,\mathbb{I}_P^\nu}\Big) \leq \min(m+1, n+1) \tag{7.12}$$

be the *analytic spread* of the homogeneous ideal $\mathbb{I}_P$ in $\mathbb{K}[[X_0, \dots, X_n]]$. Namely, the minimal number of generators for a *reduction* of $\mathbb{I}_P$ in $\mathbb{K}[[X_0, \dots, X_n]]$, that is, an ideal $(\mathbb{I}_P)_{\mathrm{red}} \subset \mathbb{I}_P$ in $\mathbb{K}[[X_0, \dots, X_n]]$ sharing the same integral closure as $\mathbb{I}_P$ in this $(n+1)$-dimensional regular local ring; see [**NR54**]. Since $\mathbb{K}[[X_0, \dots, X_n]]$ is a regular local ring, the Lipman–Teissier theorem, which is the algebraic counterpart of the Briançon–Skoda theorem (in the analytic geometry setting of the regular local ring $\mathcal{O}_{n+1}$, see Theorem 5.8), holds [**LipS81, LipT81**]. It implies, in particular, that

$$\overline{(\mathbb{I}_P)^{\ell(\mathbb{I}_P)}} \subset \mathbb{I}_P. \tag{7.13}$$

Let us now state how the Briançon–Skoda theorem (more precisely, its algebraic counterpart, namely the Lipman–Teissier theorem), once combined with Bézout's geometric intersection theorem (more precisely Proposition 7.7), governs the effective realization of Bézout identity for a set of generators $\{p_j\}_j$ of $\mathbb{K}[X_1, \dots, X_n]$. The following result [**Hi01**, Theorem 5.1 (1,2,3)] holds and is a particular case of a more general result stated as Theorem 7.49 in §7.3.

THEOREM 7.8. *Let $m, n \in \mathbb{N}^*$ and $\mathbb{K}$ be an algebraically closed field. Assume that $\{p_0, \dots, p_m\}$ is a set of generators for $\mathbb{K}[X_1, \dots, X_n]$ with respective degrees such that $\deg p_0 = d_0 \leq \cdots \leq \deg p_m = d_m$. Then Bézout identity with respect to the p_j's is N-solvable, with*

$$(7.14)\qquad \begin{aligned} N &\leq \ell(\mathbb{I}_P) \max\big(\mu^\infty(p), d_m\big) \\ &\leq \min(m+1, n+1) \max\Big(\min\Big(d_m^n, \frac{\prod_j d_j}{d_0^{(m+1-n)_+}}\Big), d_m\Big). \end{aligned}$$

Moreover, the minimal integer $N_{\min}$, such that Bézout identity with respect to the p_j's is $N_{\min}$-solvable, satisfies

$$(7.15)\qquad N_{\min} \geq \max(\mu^\infty(p), d_m).$$

PROOF. The second assertion has been stated already as part of Example 7.6. On the other hand, it follows from inequalities (7.10) (when $|\ |$ is a nontrivial absolute value on $\mathbb{K}$) that[3]

$$X_0^{\max(\mu^\infty(p), d_m)} \in \overline{\mathbb{I}_P} = \overline{(\mathbb{I}_P)_{\rm red}}$$

for some ideal $(\mathbb{I}_P)_{\rm red} \subset \mathbb{I}_P$ generated by $\ell(\mathbb{I}_P) \leq \min(m+1, n+1)$ elements in the $(n+1)$-dimensional regular local ring $\mathbb{K}[[X_0, \dots, X_n]]$. Under these conditions, $X_0^{\ell(\mathbb{I}_P)\max(\mu^\infty(p), d_m)}$ belongs to $\mathbb{I}_P$ according to the Lipman–Teissier theorem in the regular local ring $\mathbb{K}[[X_0, \dots, X_n]]$ and the first assertion in the statement of the theorem follows by dehomogenization. The second inequality in (7.14) follows from Proposition 7.7. □

REMARK 7.9. Let $\mathbb{K}$ be an arbitrary field with $p_0, \dots, p_m \in \mathbb{K}[X_1, \dots, X_n]$ generating $\mathbb{K}[X_1, \dots, X_n]$. Given $N = N(m, n, d_0, \dots, d_m) \in \mathbb{N}^*$, it is equivalent saying that Bézout identity with respect to $\{p_0, \dots, p_m\}$ is N-solvable and that an affine system of equations in $M(N(n, m, d_0, \dots, d_m))$ variables with coefficients in $\mathbb{K}$ is compatible. It amounts to saying that such an affine system of equations admits a solution in $\overline{\mathbb{K}}^{M(N(m,n,d_0,\dots,d_m))}$, where $\overline{\mathbb{K}}$ is an arbitrary algebraic closure of $\mathbb{K}$. It remains therefore true that Bézout identity for $p_0, \dots, p_m$ is N-solvable with some

$$N \leq \min(m+1, n+1) \max\Big(\min\Big(d_m^n, \frac{\prod_j d_j}{d_0^{(m+1-n)_+}}\Big), d_m\Big).$$

The first result in this direction was obtained in the pioneering work by Dale W. Brownawell [**Bro87A**]. Estimates for the Nœther exponent $e(p_1, \dots, p_n)$ of an ideal

[3] Since $\mu^\infty(p) = a/b$ with $a \in \mathbb{N}$ and $b \in \mathbb{N}^*$, one should consider the polynomials $p_j(X_1^b, \dots, X_n^b)$ to make the argument below correct. We do not enter here in such considerations and, for the sake of simplicity, assume that $\mu^\infty(p)$ is replaced by $[\mu^\infty(p)] + 1$. For an exact proof, we refer the reader to [**Hi01**, §4]. We also refer the reader to the proof of Theorem 7.15 below, where we took care of the true rationality of $\mu^\infty(p) = a/b$ in the case where $m < n$.

of $\mathbb{K}[X]$ generated by polynomials $p_1, \dots, p_m$ follow from the Rabinowitsch trick; see Proposition 7.4, as explained in [**Bro87B**]. Nullstellen inequalities in the same vein as in (7.10) (from which the statement related to the N-solvability of Bézout identity with respect to $\{p_0, \dots, p_m\}$ follows as in Theorem 7.8 up to the use of the Lipman–Teissier theorem) came originally from techniques introduced by Yuri Nesterenko, in the particular case where $\mathbb{K}$ is a number field, in a subsequent collection of seminal papers. We refer here to [**Bro88**] for an exhaustive bibliography. A major ingredient there is the concept of *Chow form* that we already introduced in §1.3.3 for the $\mathbb{P}^1(\mathbb{C})$ setting. Remark D.13 introduces the concept of a coherent sheaf of Chow ideals of an analytic cycle in (D.19).

7.1.3. N-solvability of Bézout identity in $\mathbb{C}[X]$ and Bochner–Martinelli currents. When $\mathbb{K} = \mathbb{C}$, Bézout identity with respect to $\{p_0, \dots, p_m\}$ is N-solvable for some convenient $N = N(\{p_0, \dots, p_m\})$ in terms of the action of Bochner–Martinelli principal value currents on the complex projective space $\mathbb{P}^n_{\mathbb{C}}$. We detail this realization of Bézout identity in the present subsection. Assume that p is a polynomial map on $\mathbb{C}^n$ with entries $p_0, \dots, p_m \in \mathbb{C}[X_1, \dots, X_n]$ not sharing any common zero. Let $P_j \in \mathbb{C}[X_0, \dots, X_n]$ be the homogenization of the polynomial entry p_j for each j, with $d_j = \deg p_j$, and let ϵ_j be, as usual, the element of canonical basis. Let $M \in \mathbb{N}$. We consider, as in (4.128) for n instead of N and $m+1$ instead of m,

(7.16)

$$|P| = \sqrt{\sum_j \frac{|P_j|^2}{\|\zeta\|^{2d_j}}}, \quad P^* = \sum_j \frac{\overline{P_j}}{\|\zeta\|^{2d_j}} \epsilon_j^*,$$

$$R_0^{\overline{\mathbf{K}_\bullet^P}} = \Big[(\bar\partial |P|^{2\lambda}) \sum_{r=1}^{\min(m+1,n)} \frac{P^*}{2i\pi |P|^2} \Big(\bar\partial \frac{P^*}{2i\pi|P|^2}\Big)^{r-1}\Big]_{\lambda=0} = \sum_{r=1}^{\min(m+1,n)} R_{0,r}^{\overline{\mathbf{K}_\bullet^P}},$$

$$P_0^{\overline{\mathbf{K}_\bullet^P}} = \Big[|P|^{2\lambda} \sum_{r=1}^{\min(m+1,n+1)} \frac{P^*}{2i\pi |P|^2} \Big(\bar\partial \frac{P^*}{2i\pi|P|^2}\Big)^{r-1}\Big]_{\lambda=0} = \sum_{r=1}^{\min(m+1,n+1)} P_{0,r}^{\overline{\mathbf{K}_\bullet^P}},$$

where $R_0^{\mathbf{K}_\bullet^P}$ and $P_0^{\mathbf{K}_\bullet^P}$ are viewed as current-valued sections (respectively residue and principal value) of the bundle $\mathrm{Hom}_{\mathbb{C}}(\mathcal{O}(M), E(\mathbf{K}^P_{\bullet,M}))$ acting by multiplication. Here, $\mathbf{K}^P_{\bullet,M}$ denotes the exact tensorial complex of holomorphic bundles over $\mathbb{P}^n_{\mathbb{C}}$ defined by

$$\mathbf{K}^P_{\bullet,M} = \mathcal{O}(M) \otimes \mathbf{K}^P_\bullet :$$

$$0 \to \mathcal{O}(M) \otimes \bigwedge^m E^* \overset{\rfloor_P}{\to} \cdots \overset{\rfloor_P}{\to} \mathcal{O}(M) \otimes \bigwedge^\ell E^* \overset{\rfloor_P}{\to} \cdots \overset{\rfloor_P}{\to} \mathcal{O}(M) \otimes E^* \overset{\rfloor_P}{\to} \mathcal{O}(M),$$

where $E = \bigoplus_j \mathcal{O}(-d_j)$. Let $\mu^\infty(p)$ be the Łojasiewicz exponent of p along the hyperplane at infinity, as defined by (7.9); see Definition 7.5. We first state a fundamental property of current nature satisfied by the Łojasiewicz exponent $\mu^\infty(p)$ of a $\mathbb{C}^{m+1}$-valued polynomial map $p : \mathbb{C}^n \to \mathbb{C}^{m+1}$, whether one considers the situation where $V(p) = \{z \in \mathbb{C}^n : p(z) = 0\} = \emptyset$ or not.

PROPOSITION 7.10. *Let $N \in \mathbb{N}^*$ and let $q : \mathbb{C}^n \to \mathbb{C}$ be a polynomial map such that $q/\|p\|^{\min(m+1,n)}$ belongs to $L^\infty_{\mathrm{loc}}(\mathbb{C}^n)$. If Q denotes the homogenization of q,* *then the implication*

$$N \geq [\min(m+1,n)\,\mu^\infty(p)] + 1 \Longrightarrow R_0^{\overline{\mathbf{K}_\bullet^P}}\,\zeta_0^N Q = 0 \tag{7.17}$$

holds.

PROOF. One has

$$\mathbf{1}_{\mathbb{P}^n_{\mathbb{C}}\setminus\{\zeta_0=0\}} \cdot R_0^{\overline{\mathbf{K}_\bullet^P}}\,\zeta_0^N Q = 0,$$

since $q/\|p\|^{\min(m+1,n)}$ is locally bounded in $\mathbb{C}^n$. It is checked locally in a neighborhood of each $z \in \mathbb{C}^n$. Firstly, one introduces the normalized blowup of a neighborhood U_z of z along the ideal sheaf $(p)\,\mathcal{O}_z$ and then invokes assertion (3.25) in Proposition 3.15. To show that

$$\mathbf{1}_{\{\zeta_0=0\}} \cdot R_0^{\overline{\mathbf{K}_\bullet^P}}\,\zeta_0^N Q = 0,$$

one uses the same reasoning in a neighborhood of a point $z \in \mathbb{P}^n_{\mathbb{C}} \setminus \mathbb{C}^n$. The main difference is that one has to distinguish the components H_γ of the exceptional divisor of $\mathbb{P}^n_{\mathbb{C}}$ along the ideal $(P)\,\mathcal{O}$. Among these components there are those whose image $\widehat{\pi}_P(H_\gamma)$ lies entirely in $|\mathrm{div}(\zeta_0)|$ in a neighborhood of z and those whose image $\widehat{\pi}_P(H_\gamma)$ hits $\mathbb{C}^n$ in a neighborhood of z; see for example [**AndG11**, proof of Theorem 1.2, p. 355]. □

EXAMPLE 7.11. If $p = (p_0,\dots,p_m)$, where the polynomials p_j do not share any common zero in $\mathbb{C}^n$, which is the case in the present subsection, then one can take $q = 1$ in Proposition 7.10. Hence,

$$N \geq [\min(m+1,n)\,\mu^\infty(p)] + 1 \Longrightarrow R_0^{\overline{\mathbf{K}_\bullet^P}}\,\zeta_0^N = 0, \tag{7.18}$$

in this case.

More precisely, for a subfield $\mathbb{K}$ of $\mathbb{C}$, we consider $m+1$ polynomials $p_0,\dots,p_m$ in $\mathbb{K}[X_1,\dots,X_n]$ with respective degrees $d_0,\dots,d_m$, such that $\{p_0,\dots,p_m\}$ is a set of generators for $\mathbb{K}[X_1,\dots,X_n]$. This last assumption implies that $V(p) = \emptyset$. For each entry p_j, let H_{P_j} be a Hefer form for the homogenization P_j for the Fubini–Study metric; see Definition 4.49. The homogeneous polynomials $A_{j,\kappa}$ in the homogeneous coordinates $X_0,\dots,X_n,Y_0,\dots,Y_n$, with total degree $d_j - 1$, as in (4.118), can be then chosen with coefficients in $\mathbb{K}$. Let $d = (d_0,\dots,d_m)$ and assume that the maps $(z,\zeta) \longmapsto \mathbb{I}_d(z,\zeta)$, $(z,\zeta) \longmapsto \mathbb{H}_P(z,\zeta)$ are constructed formally from d and the H_{P_j} as in (4.122). Then, the following current realization of Bézout identity in terms of Bochner–Martinelli principal value currents holds in $\mathbb{C}^n$; see [**AndG11**, Theorem 1.2 (*ii*)].[4]

THEOREM 7.12. *Let p_j, P_j, d_j for $j = 0,\dots,m$, $\mathbb{I}_d, \mathbb{H}_d$ be as above, and let*

$$N = \max\big(\big[\min(m+1,n)\,\mu^\infty(p)\big] + 1, |d(p)|_{1,\min(m+1,n+1)} - n\big),$$

where $|d(p)|_{1,\min(m+1,n+1)} := \max_{\{J\subset\{0,\dots,m\}\,:\,|J|=\min(m+1,n+1)\}} \sum_\nu d_{j_\nu}$. *The integral representation formula*

(7.19)

$$z_0^N = 2i\pi \rfloor_{P(z)} \int_{\mathbb{P}^n_{\mathbb{C}}} \Big(\Big(\big(\mathscr{W}_{\mathbb{P}^n_{\mathbb{C}}}(z,\cdot)\big)^{N+n} \sum_{r=1}^{\min(m,n)+1} \mathbb{I}_d(z,\cdot)\,\frac{(\mathbb{H}_P(z,\cdot))^{r-1}}{(r-1)!}\Big)\,P_{0,r}^{\overline{\mathbf{K}_\bullet^P}}\Big)_{n,n}\,\zeta_0^N$$

[4] Such result is a particular case of a more general statement, namely Theorem 7.47.

(*where*

$$\mathscr{W}_{\mathbb{P}^n_{\mathbb{C}}} : (z,\zeta) \in \mathbb{P}^n_{\mathbb{C}} \times \mathbb{P}^n_{\mathbb{C}} \longmapsto \frac{z \bullet \zeta}{\|\zeta\|^2} + dd^c \log \|\zeta\|^2$$

is the Fubini–Study global weight) provides, after dehomogenization, an explicit solution to the N*-solvability of Bézout identity for* $\{p_0, \dots, p_m\}$.

PROOF. The result is an immediate consequence of Theorem 4.51, once combined with assertion (7.18) derived from Proposition 7.10 in the case where entries of the polynomial map p generate $\mathbb{K}[X_1, \dots, X_n]$. The first constraint, namely $N \geq \big[\min(m+1,n)\mu^\infty(p)\big] + 1$, justifies that Proposition 7.10 applies. As for the second constraint

$$N \geq |d(p)|_{1,\min(m+1,n+1)} - n,$$

it ensures that the formal expressions under the integral in the right-hand side of (7.19) make sense instead of just being formal. □

REMARK 7.13. According to Remark 4.19, the Fubini–Study global weight $\mathscr{W}_{\mathbb{P}^n_{\mathbb{C}}}$ can be interpreted as an averaged current in view of Crofton's formula presented in §D.5.1. Such is the case, when $d_0 = \cdots = d_m$, for each current

$$\Big[|P|^{2\lambda} \frac{P^*}{|P|^2} \Big(\overline{\partial} \frac{P^*}{|P|^2}\Big)^{r-1}\Big]_{\lambda=0}$$

for $r = 1, \dots, \min(m+1, n+1) = \min(m,n) + 1$, which is the direct image by

$$\Pi_r : \mathbb{P}^n_{\mathbb{C}} \times \prod_{\rho=1}^{r} \mathbb{P}^m_{u_\rho} : (\zeta, u_1, \dots, u_r) \longmapsto \zeta \in \mathbb{P}^n_{\mathbb{C}}$$

of the current

$$\Bigg[|\langle u_r, P\rangle|^{2\lambda^r} \frac{\langle u_r, \varepsilon^*\rangle}{\langle u_r, P\rangle} \Big(\bigwedge_{\rho=1}^{r-1} \overline{\partial}\Big(|\langle u_\rho, P\rangle|^{2\lambda^\rho} \frac{\langle u_\rho, \varepsilon^*\rangle}{\langle u_\rho, P\rangle}\Big)\Big) \Big(\prod_{\rho=1}^{r} dd^c \log \|u_\rho\|^2\Big)^m\Bigg]_{\lambda=0};$$

see Remark D.44. One can interpret then in the case where all d_j's are equal the dehomogenization of the integral representation formula (7.19) as the result of an averaging process. It is precisely along such an averaging process that the arithmetic information (the fact that the p_j's have entries in a subfield $\mathbb{K}$ of $\mathbb{C}$ such as $\mathbb{Q}$ or a number field) is lost.

REMARK 7.14. One could also have used instead of Theorem 4.51, for $R > 0$ and $z \in \mathbb{C}^n$ with $\|z\| < R$, the integral representation formula

$$1 = \int_{\mathbb{B}_n(0,R)} (\mathscr{W}(z,\cdot))_{n,n} + \int_{\partial(\mathbb{B}_n(0,R))} \frac{\sigma(z,\cdot)}{2i\pi} \Big(\sum_{r=0}^{n-1} \Big(\overline{\partial}_\zeta \frac{\sigma(z,\cdot)}{2i\pi}\Big)^{n-1-r} (\mathscr{W}(z,\cdot))_{r,r}\Big),$$

where $\mathbb{B}_n(0,R)$ denotes the euclidean ball of radius R centered at the origin, $\sigma(z,\zeta) = \partial\|\zeta\|^2/\langle\overline{\zeta}, \zeta - z\rangle$, and $\mathscr{W}$ is a global weight in $\mathbb{C}^n \times \mathbb{C}^n$. This integral representation formula is a particular case of (4.84); see Proposition 4.34. One takes here $\mathscr{W} = \mathscr{W}_{(q^{[1]},q^{[2]})}$ as in Proposition 4.20 with $M = 2$, with

$$q^{[1]}(\zeta,z) = \sum_{j=0}^{m} \frac{\overline{p_j(\zeta)}}{\|p(\zeta)\|^2} H_{p_j}(z,\zeta), \quad q^{[2]}(\zeta,z) = \partial \log(1+\|\zeta\|^2),$$
$$\Gamma(W_1,W_2) = W_1^{n+1} W_2^N, \tag{7.20}$$

where $N \in \mathbb{N}^*$. Furthermore, here, for any $j = 0,\dots,m$, $H_{p_j} = \sum_{k=1}^n h_{j,k}(z,\zeta)\, d\zeta_k$ is an algebraic Hefer form for p_j, which means that the $h_{j,k}$ belong to $\mathbb{K}[X,Y]$ and satisfy $\sum_{k=1}^n h_{j,k}(X,Y)(Y_k - X_k) = p_j(Y) - p_j(X)$. When N is chosen large enough in terms of the value of the Łojasiewicz exponent $\mu^\infty(p)$ along $\mathbb{P}^n_{\mathbb{C}} \setminus \mathbb{C}^n$, the integral

$$\int_{\mathbb{B}_n(0,R)} (\mathscr{W}_{q^{[1]},q^{[2]}}(z,\cdot))_{n,n}$$

converges absolutely to an expression

$$\int_{\mathbb{C}^n} (\mathscr{W}_{(q^{[1]},q^{[2]})}(z,\cdot))_{n,n} = \sum_{j=0}^{m} q_{N,j}(z) p_j(z)$$

for some explicit $q_{N,j} \in \mathbb{C}[X_1,\dots,X_n]$, while

$$\int_{\partial(\mathbb{B}_n(0,R))} \frac{\sigma(z,\cdot)}{2i\pi} \Big(\sum_{r=0}^{n-1} \Big(\overline{\partial}_\zeta \frac{\sigma(z,\cdot)}{2i\pi}\Big)^{n-1-r} (\mathscr{W}_{(q^{[1]},q^{[2]})}(z,\cdot))_{r,r} \Big)$$

tends to 0 uniformly in z kept in a compact subset of $\mathbb{C}^n$. It provides an explicit N'-solvability for Bézout identity with respect to $\{p_0,\dots,p_m\}$. This approach is presented in the univariate setting in [**BGVY**, §I.5]. It leads in the multivariate setting, if one takes $N = [(2n+1)\,\mu^\infty(p)] + n\,\max_j d_j - (n-1)$, to the realization of Bézout identity as $1 = \sum_{j=0}^m q_j p_j$ in $\mathbb{C}[X]$, with degree estimates

$$\max_{0\le j\le m} \deg q_j p_j \le [(2n+1)\,\mu^\infty(p)] + (2n+1)\max_j d_j - (n-1),$$

of the same kind, but worse, than those provided by Theorem 7.12. Thus, we decided to skip its detailed presentation in this monograph.

7.1.4. N-solvability in $\mathbb{C}[X]$ of Bézout identity via a diagram chase. In this section, we once more consider a collection of generators $\{p_0,\dots,p_m\}$ for $\mathbb{C}[X_1,\dots,X_n]$. Since the N-solvability of Bézout identity we propose for the p_j's in this subsection is not effective, the assumption that the coefficients of the p_j's lie in a specified subfield of $\mathbb{C}$ becomes irrelevant. We consider the Koszul complex $\mathbf{K}^P_\bullet$ as in §7.1.3. Vanishing results for Dolbeault cohomology allow us to make global instead of local the iterated applications of Dolbeault's lemma occurring within the proof of Proposition 3.32(a), where the generically exact complex of hermitian holomorphic bundles implied is $\mathbf{K}^P_\bullet$. Despite $\mathbb{P}^n_{\mathbb{C}}$ failing to be a Stein manifold, important vanishing results for Dolbeault cohomology of line bundles $\mathcal{O}(\delta)$ over $\mathbb{P}^n_{\mathbb{C}}$, where $n \in \mathbb{N}^*$, hold, as recalled in §B.2.3, Example B.31(2). One has in particular that

$$1 \le r \le n-1 \text{ and } \delta \in \mathbb{Z} \Longrightarrow H^{0,r}(\mathbb{P}^n_{\mathbb{C}}, \mathcal{O}(\delta)) = 0,$$
$$\delta + n \in \mathbb{N} \Longrightarrow H^{0,n}(\mathbb{P}^n_{\mathbb{C}}, \mathcal{O}(\delta)) = 0. \tag{7.21}$$

A diagram chase argument as in the proof of Proposition 3.32(a), with $\mathbf{K}^P_\bullet$ as generically exact complex, leads to the following result about the N-solvability of Bézout's identity for $\{p_0, \dots, p_m\}$. Let $\mu^\infty(p)$ be the Łojasiewicz exponent of $p = (p_0, \dots, p_m) : \mathbb{C}^n \to \mathbb{C}^{m+1}$ along the hyperplane at infinity $\mathbb{P}^n_{\mathbb{C}} \setminus \mathbb{C}^n$, as defined in Definition 7.5 and estimated from above as in Proposition 7.7 assuming that the polynomial entries p_j's are ordered along increasing degrees. Then, one can state in the context of Hilbert's geometric nullstellensatz the following result, which is a particular case of [**AndG11**, Theorem 1.2 (i)].[5]

THEOREM 7.15. *Bézout's identity for a collection of generators* $\{p_0, \dots, p_m\}$ *is* N*-solvable in* $\mathbb{C}[X_1, \dots, X_n]$*, where*

$$N \leq (m+1)\mu^\infty(p), \quad \text{when } m < n, \tag{7.22}$$
$$N \leq \max\big([n\mu^\infty(p)] + 1, |d(p)|_{1,n+1} - n\big), \quad \text{when } m \geq n,$$

where $|d(p)|_{1,n+1} := \max_{\{J \subset \{0,\dots,m\} : \#J = n+1\}} \sum d_{j_\nu}$.

PROOF. Consider first the case where $m < n$. Let $(m+1)\,\mu^\infty(p) = a_m/b_m$, where $a_m, , b_m \in \mathbb{N}^*$. For $j = 0, \dots, m$, let

$$P'_j(X_0, \dots, X_n) = P_j(X_0^{b_m}, \dots, X_n^{b_m}).$$

It follows from Proposition 3.15 that, as in (7.18), the equality

$$R_0^{\overline{\mathbf{K}_\bullet^{P'}}} \zeta_0^{a_m} = 0$$

holds. The proof of Proposition 3.32(a) carries through globally instead of locally, since $m < n$, due to the vanishing of the Dolbeault cohomology groups $H^{0,r}(\mathbb{P}^n_{\mathbb{C}}, \mathcal{O}(\delta))$ for any $r = 1, \dots, n-1$ and any $\delta \in \mathbb{Z}$; see (7.21). Therefore, one concludes in this case that

$$\zeta_0^{a_m} = \sum_{j=0}^m Q'_j(\zeta)\, P'_j(\zeta) = \sum_{j=0}^m Q'_j(\zeta)\, P_j(\zeta_0^{b_m}, \dots, \zeta_n^{b_m}), \tag{7.23}$$

where Q'_j is either 0 or a homogeneous polynomial function with degree $a_m - d_j b_m$. Observe also here that the fact that the p_j's do not share any common zero in $\mathbb{P}^n_{\mathbb{C}}$ prevents p, hence also $p' : \zeta \mapsto p(\zeta_1^{b_m}, \dots, \zeta_n^{b_m})$, from being a proper map from $\mathbb{C}^n$ to $\mathbb{C}^{m+1}$. As a consequence a_m cannot be strictly smaller than $b_m \min_{0 \leq j \leq m} d_j$. A flatness argument applied in $\mathbb{C}[[X_0, \dots, X_n]]$ implies that one can take the Q'_j's in $\mathbb{C}[X_0, X_1^{b_m}, \dots, X_n^{b_m}]$ as observed in the proof of [**Hi01**, Theorem 2.1 (*i*)], namely, for $j = 0, \dots, m$,

$$Q'_j(X_0, \dots, X_n) = Q_j(X_0, X_1^{b_m}, \dots, X_n^{b_m})$$

for some $Q_j \in \mathbb{C}[X_0, Y_1, \dots, Y_n]$. After dehomogenization, one obtains from (7.23) Bézout identity

$$1 = \sum_{j=0}^m q_j(X_1^{b_m}, \dots, X_n^{b_m})\, p_j(X_1^{b_m}, \dots, X_n^{b_m})$$

where $b_m \max_{0 \leq j \leq m} \deg(q_j p_j) \leq a_m$. It corresponds to the assertion in (7.22) in the case where $m < n$. If $m \geq n$, let

$$N = \max\big([n\mu^\infty(p)] + 1, |d(p)|_{1,n+1} - n\big).$$

It follows again from Proposition 3.15 that

$$R_0^{\overline{\mathbf{K}_\bullet^{P}}} \zeta_0^N = 0$$

[5]The general statement is in §7.3 as Theorem 7.45.

since $N \geq n\mu^\infty(p)$. As for the additional constraint

$$N \geq \max\left(|d(p)|_{1,n+1} - n\right),$$

it ensures that the argument in the proof of Proposition 3.32(a) can be carried through at all steps (including this time the step $\ell = n$) for *global* application of Dolbeault's lemma instead of *local* ones. It is feasible in this case (see (7.21)), though this time taking into account the vanishing of Dolbeault cohomology groups $H^{0,n}(\mathbb{P}^n_{\mathbb{C}}, \mathcal{O}(\delta))$ for $\delta \geq -n$. One concludes again in this second case that

$$\zeta_0^N = \sum_{j=0}^m Q_j(\zeta)P_j(\zeta),$$

where Q_j is a global section of $\mathcal{O}(N - d_j)$. That is, Q_j is either 0 or a homogeneous polynomial function in $\zeta_0, \dots, \zeta_n$ with degree $N - d_j$. The fact that the assertion (7.22) holds in the second case follows once again by dehomogenization. □

7.1.5. Sharp geometric nullstellensatz. Let $\mathbb{K}$ be an algebraically closed field. We consider in this section, instead of $\mathbb{K}^n = \operatorname{Spec}(\mathbb{K}[X_1,\dots,X_n])$, a purely n-dimensional algebraic subvariety $\mathbb{X}$ of $\mathbb{K}^N = \operatorname{Spec}(\mathbb{K}[X_1,\dots,X_n,Y_1,\dots,Y_{N-n}])$. Geometrically speaking, one interprets $\mathbb{X}$ as the set of common zeroes in $\mathbb{K}^N$ of a radical ideal $\mathbb{J}$ of the nœtherian ring $\mathbb{K}[X,Y]$, when the Krull dimension of the local ring $(\mathbb{K}[X,Y]/\mathbb{J})_{(x,y)}$ equals n for any point $(x,y) \in \mathbb{X}$.[6] We recall that the geometric marker $\mathbf{d} = \deg \mathbb{X}$ is defined as the cardinal number of the intersection of the Zariski closure $\overline{\mathbb{X}}$ of $\mathbb{X}$ in the projective algebraic variety $\mathbb{P}^N_{\mathbb{K}} = \operatorname{Spec}(\operatorname{Proj}(\mathbb{K}[X_0,\dots,X_n,Y_1,\dots,Y_n]))$ with a generic projective $(N-n)$-dimensional linear subspace of $\mathbb{P}^N_{\mathbb{K}}$. Throughout this subsection, one considers $m+1$ polynomials $p_0,\dots,p_m$ in $\mathbb{K}[X,Y]$ with respective (total) degrees $d_0,\dots,d_m$ in (X,Y) in increasing order and assumes that in $\mathbb{K}^N$

$$\{(x,y) \in \mathbb{X} : p_0(x,y) = \cdots = p_m(x,y) = 0\} = \emptyset. \tag{7.24}$$

We need to extend the definition of the geometric marker $\mathbf{d}(p, p_0)$, as introduced in (6.180), when $m = n$ and thus $p = (p_1,\dots,p_n)$, as follows. According to J. Kollár in [**Ko88**] and Z. Jelonek in [**Jel05**], we let $\mathbf{d}(p)$ be defined by

$$\mathbf{d}(p) = \mathbf{d} \times \begin{cases} d_m & \text{if } n = 1, \\ \prod_{j=0}^m d_j & \text{if } n > 1 \text{ and } m < n, \\ \left(\prod_{j=0}^{n-2} d_{m-j}\right) d_0 & \text{if } n > 1 \text{ and } m \geq n > 1. \end{cases} \tag{7.25}$$

There exists a significant difference between the geometric marker $\mathbf{d}(p)$ and the marker

$$N(\{p_0,\dots,p_m\}) = \min(m+1, n+1) \max\left(\min\left(d_m^n, \frac{\prod_j d_j}{d_0^{(m+1-n)_+}}\right), d_m\right), \tag{7.26}$$

which dominates the N-solvability of Bézout identity with respect to $\{p_0,\dots,p_m\}$ on the algebraic variety $\mathbb{X} = \mathbb{K}^n$ according to Theorem 7.8. To be more specific, the difference lies in that the marker (7.26) involves the factor $\min(m+1, n+1)$ reflecting the role played by the Lipman–Teissier theorem, while the definition of $\mathbf{d}(p)$ reflects only Bézout's geometric intersection theorem; see §6.6.2. One

---

[6] Note that this setting was already considered in §6.3.2, §6.4.2 or §6.5.2 and in §6.6.2, §6.6.3, and §6.6.4 as well. In the two last cases, the context was the arithmetic one, where $\mathbb{K} = \overline{\mathbb{Q}}$ was the field of algebraic numbers.

may state the following sharp version of Hilbert's geometric nullstellensatz ([**Ko88**, Theorem 1.5] and [**Som99**, Theorem 3.19] in the case where $N = n$ and $\mathbb{X} = \mathbb{K}^n$, [**Jel05**, Theorem 1.1] in the general case).

THEOREM 7.16. *Let $\mathbb{K}[X,Y]$, $\mathbb{X}$, $\mathbb{J}$, $\{p_0,\dots,p_m\}$ be as above. There are polynomials $q_0,\dots,q_m \in \mathbb{K}[X,Y]$ such that*

$$(7.27)\quad 1-\sum_{j=0}^{m} q_j p_j \in \mathbb{J} \quad \textit{with} \quad \max_{0\le j\le m} \deg q_j p_j \le \begin{cases} \mathbf{d}\,(p) \ \textit{when}\ m<n, \\ 2\,\mathbf{d}\,(p)-1 \ \textit{when}\ m\ge n. \end{cases}$$

Moreover, if m, n and the d_j's are prescribed with in addition $m<n$, then

$$(7.28)\qquad 1-\sum_{j=0}^{m} q_j p_j \in \mathbb{J} \Longrightarrow \max_{0\le j\le n} \deg q_j p_j \ge \mathbf{d}\,(p).$$

When $\mathbb{X} = \mathbb{K}^n$ and $m \ge n$, the upper bound $2\,\mathbf{d}\,(p)-1$ in (7.27) can be (possibly) refined as $(3/2)^{\iota_d}\,\mathbf{d}\,(p)$, where $\iota_d = \mathrm{card}(\{j : d_j = 2\})$. In the particular case where $\mathbb{X}$ and $(m, d_0,\dots,d_n)$ are prescribed, then assertion (7.28) holds in the three following situations: $n=1$, $m<n$, $m\ge n$ with $\iota_d = 0$.

PROOF. The proof of this result, as presented in [**Jel05**], is based on considerations that we introduced in §6.4 when exploring the concept of an algebraic trace in the polynomial setting of $\mathbb{K}[X]$ or, more generally, of the quotient setting of $\mathbb{K}_{\mathbb{J}}[X,Y] = \mathbb{K}[X,Y]/\mathbb{J}$.

Consider first the case when $m<n$. We sketch here Jelonek's proof of [**Jel05**, Theorem 3.6]. It follows from Hilbert's nullstellensatz (see Theorem 7.1) that there exist $q_0^\circ,\dots,q_m^\circ \in \mathbb{K}[X,Y]$ such that

$$(7.29)\qquad 1-\sum_{j=0}^{m} q_j^\circ p_j \in \mathbb{J}.$$

Observe that from the very beginning this initial step prevents the construction of $q_0,\dots,q_m \in \mathbb{Z}[X,Y]$ in (7.27) from being explicit in terms of the input data $\mathbb{X}, p_0,\dots,p_m$. Let t be an additional parameter besides the variables X, Y. The polynomial map $\Phi : (x,y,t) \in \mathbb{X}\times\mathbb{K} \longmapsto \big(x,y,t\,p_0(x,y),\dots,t\,p_m(x,y)\big) \in \mathbb{K}^{N+m+1}$ has a closed image in $\mathbb{X}\times\mathbb{K}^{m+1} \subset \mathbb{K}^{N+m+1}$ since $t = \sum_0^m q_j^\circ(X,Y)(tp_j(X,Y))$ in $\mathbb{K}[X,Y][t]$ (with respect to the Zariski topology) according to (7.29). Moreover, Φ embeds $\mathbb{X}\times\mathbb{K}$ as an affine, purely $(n+1)$-dimensional algebraic subvariety $\Phi(\mathbb{X}\times\mathbb{K})$ of the affine algebraic variety $\mathbb{X}\times\mathbb{K}^{m+1} \subset \mathbb{K}^{N+m+1}$. Let us complete the sequence of $m+1$ polynomials $(p_m,\dots,p_0)$ as the sequence with length n defined as $\mathfrak{p} = (\mathfrak{p}_0,\dots,\mathfrak{p}_{n-1}) := (p_m,\dots,p_0,0,\dots,0)$, which is also ordered in such a way. Let $\pi : \Phi(\mathbb{X}\times\mathbb{K}) \to \mathbb{K}^{n+1}$ be a generic projection. It amounts to saying that the extension $(\pi\circ\Phi)^* : \mathbb{K}[w_0,\dots,w_n] \hookrightarrow \mathbb{K}[X,Y,t]/\mathbb{J}(\mathbb{X}\times\mathbb{A}^1_{\mathbb{K}})$ is integral. Since $\mathbb{K}$ is algebraically closed, hence infinite, it follows from Carl Siegel's box principle (see for example [**MW83**, §4, Lemma 1] or [**Jel05**, Lemma 3.5]) that one can choose π to satisfy $\pi(\Phi(x,y,t)) = \big(u_0(x,y,t),\dots,u_n(x,y,t)\big)$, with

$$(7.30)\qquad \begin{aligned} &u_j(x,y,t) = \ell_j(x,y) + t\sum_{\kappa=j}^{n-1} a_{j,\kappa}\,\mathfrak{p}_\kappa(x,y) \quad \text{for} \quad j=0,\dots,n-1, \\ &u_n(x,y,t) = \ell_n(x,y), \end{aligned}$$

where $\ell_0,\dots,\ell_n$ are $n+1$ generic linear forms in $\mathbb{K}[X,Y]$ and the coefficients $a_{j,\kappa}$ belong to $\mathbb{K}$. We now consider $\mathbb{X}$ as defined over $\mathbb{K}(t)$ instead of $\mathbb{K}$. Finiteness of the map $u_t = (u_0,\dots,u_n) : \mathbb{X} \to (\mathbb{K}(t))^{n+1}$ and the generalized version of the Perron theorem [**Per**, Satz 57, S. 129][7] imply that there exists a polynomial $E \in \mathbb{K}(t)[W_0,\dots,W_n]$ such that

$$\begin{aligned} &E(u_t(X,Y)) \in \mathbb{K}(t)[X,Y]\,\mathbb{J},\\ &\deg_{w_0,\dots,w_n} E\big(w_0^{\deg_{X,Y} u_{t,0}},\dots,w_n^{\deg_{X,Y} u_{t,n}}\big) \le \mathbf{d}\,(p). \end{aligned}$$

That is,

$$\begin{aligned} &(x,y)\in\mathbb{X} \Longrightarrow E(u(x,y,t))=0,\\ &\deg_{w_0,\dots,w_n} E\big(w_0^{d_m},\dots,w_m^{d_0},w_{m+1},\dots,w_n\big) \le \mathbf{d}\,(p). \end{aligned} \tag{7.31}$$

Since Φ is an embedding and π is a generic projection, $(x,y,t)\mapsto u(x,y,t)$ realizes a finite morphism from $\mathbb{X}\times\mathbb{K}$ to $\mathbb{K}^{n+1}$. It follows from the fact that $(\pi\circ\Phi)^*$ is an integral extension, that t admits for this extension a monic minimal polynomial satisfying

$$t^\nu + \sum_{\rho=1}^{\nu} \gamma_\rho(u(X,Y,t))\,t^{\nu-\rho} \in \mathbb{K}(t)[X,Y]\,\mathbb{J}.$$

It follows from the minimality of such a monic minimal polynomial and (7.31) that $t^\nu + \sum_{\rho=1}^{\nu} \gamma(u(X,Y,t))\,t^{\nu-\rho}$ divides $E(u(X,Y,t))$ in $\mathbb{K}(t)[X,Y]\,\mathbb{J}$. Then

$$\deg_{w_0,\dots,w_n} \gamma_\rho(w_0^{d_m},\dots,w_m^{d_0},w_{m+1},\dots,w_n) \le \mathbf{d}\,(p) \quad \text{for } \rho=1,\dots,\nu. \tag{7.32}$$

If one adds all terms of the form $t^\nu q(X,Y)$ for some $q\in\mathbb{K}[X,Y]$, which occur in the polynomial expression $t^\nu + \sum_{\rho=1}^{\nu} \gamma_\rho(u(X,Y,t))\,t^{\nu-\rho}$, then one gets in view of (7.32) the relation $1-\sum_{j=0}^m q_j(X,Y)p_j(X,Y)\in\mathbb{J}$ with $\max_{0\le j\le m}\deg(q_jp_j)\le\mathbf{d}\,(p)$. It is the result we were looking for.

As a subordinate step, let us consider the extreme case when $m+1=n$, but where $\dim\mathbb{X}\cap V(p)=0$ instead of $-\infty$. That is, $\mathbb{X}\cap V(p)=\{\alpha^1,\dots,\alpha^\mu\}$, where the α^ℓ are distinct points[8] in $\mathbb{X}$. One repeats the construction of the polynomial map $\Phi:\mathbb{X}\times\mathbb{K}\longrightarrow\mathbb{X}\times\mathbb{K}^n\subset\mathbb{K}^N\times\mathbb{K}^n$, which now becomes a nonclosed embedding outside $\{\alpha^1,\dots,\alpha^\mu\}$. Let $(x,y)\longleftrightarrow\xi=(\xi_1,\dots,\xi_N)$ be a generic linear change of coordinates in $\mathbb{K}^N$. For each coordinate ξ, let π^ξ be a generic projection from the Zariski closure of $\Phi(\mathbb{X}\times\mathbb{K})$ in $\mathbb{K}^N\times\mathbb{K}^n$ onto $\mathbb{K}^{n+1}$ such that, if $u^\xi=\pi^\xi\circ\Phi$, then

$$\begin{aligned} &u_j^\xi(\xi,t) = \ell_j^\xi(\xi) + t\sum_{\kappa=j}^{n-1} a_{j,\kappa}^\xi\, p_{n-1-j}(\xi) \quad \text{for} \quad 0\le j\le n-1,\\ &u_n^\xi(\xi,t)=\xi. \end{aligned}$$

[7]See also [**Pl05**, Theorem 1.1 and §2] and [**Jel05**, Theorem 3.3] for the generalized formulation already quoted and used in §6.4.1 as well as in §6.4.2; see, e.g., Proposition 6.43 there. It is [**Jel05**, Theorem 3.3] that we invoke again here with $\mathbb{K}(t)$ being a reference field instead of $\mathbb{K}$. It would be enough for u_t to be in such context a generically finite morphism from $\mathbb{X}$ (considered as being defined over $\mathbb{K}(t)$) to $(\mathbb{K}(t))^{n+1}$. The important point to observe is that the Perron theorem is an immediate consequence of of two facts: Bézout's geometric intersection theorem on one side (which is the most important side), and on the other side the fact that any substitution of variables $(w_0,\dots,w_n)\longmapsto(t_0^{\delta_0}+t_0,\dots,t_n^{\delta_n}+t_n)$ with $\delta_j\in\mathbb{N}^*$ preserves the reducibility of elements E in $\mathbb{K}[w]$ provided $E^{-1}(\{0\})$ does not contain hyperplanes of the form $w_j=a_j$ with $a_j\in\mathbb{K}$; see [**Jel05**, Corollary 3.2].

[8]We do not care here about multiplicities.

Here $\ell_0^\xi, \dots, \ell_{n-1}^\xi$ are generic linear forms and the coefficients $a_{j,\kappa}^\tau$ belong to $\mathbb{K}$. Perron's theorem [**Jel05**, Theorem 3.3] still applies since $(u_0^\xi(\cdot,t), \dots, u_{n-1}^\xi(\cdot,t))$ can be considered as a generically finite morphism from $\mathbb{X}$ (viewed now as being defined over $\mathbb{K}(t)$) to $(\mathbb{K}(t))^{n+1}$. Let

$$E^\xi(u_0^\xi(\cdot,t), \dots, u_{n-1}^\xi(\cdot,t)) \in \mathbb{K}(t)[\xi]\,\mathbb{J}$$

be the Perron algebraic dependency relation with a weighted degree in $(w_0, \dots, w_n)$ less than $\mathbf{d}\,(p)$ as in (7.31). Since $(\xi,t) \in \mathbb{X} \times \mathbb{K} \longmapsto u^\xi(\xi,t) \in \mathbb{K}^{n+1}$ admits the hypersurface $\bigcup_{\ell=1}^{\mu} V(w_n - \xi(\alpha^\ell))$ as the set of nonproperness, the minimal equation of t, considered as an element of $\mathbb{K}[\mathbb{X}][t]$, with respect to u^ξ, is not monic any more, but of the form

$$\gamma_0^\xi(u^\xi(X,Y,t))\, t^{\nu^\xi} + \sum_{\rho=1}^{\nu} \gamma_\rho^\xi(u^\xi(X,Y,t))\, t^{\nu^\xi-\rho} \in \mathbb{K}(t)[X,Y]\mathbb{J},$$

with $\gamma_\rho^\xi \in \mathbb{K}[w_0, \dots, w_n]$ for $\rho = 0, \dots, \nu^\xi$. Moreover,

$$V(\gamma_0^\xi) \subset \{w \in \mathbb{K}^{n+1} \,:\, \prod_{\ell=1}^{\mu}(w_n - \xi(\alpha^\ell)) = 0\}.$$

We now repeat the same argument as in the previous paragraph. In place of relation of the form $1 - \sum_{j=0}^{m} q_j p_j \in \mathbb{J}$, with the degree estimates

$$\max_{0\le j\le m} \deg(q_j p_j) \le \mathbf{d}\,(p)$$

deduced in the case studied previously, one derives now the relation

$$\prod_{\ell=1}^{\mu}(\xi - \xi(\alpha^\ell)) - \sum_{j=0}^{n-1} q_j^\xi p_j \in \mathbb{J},$$

where $\max_{0\le j\le n-1} \deg(q_j^\xi p_j) \le \mathbf{d}\,(p)$. This result is stated in [**Jel05**, Theorem 3.8] and is presented there as an *elimination theorem*, which is natural since it provides a cartesian product that contains $\{\alpha^1, \dots, \alpha^\mu\} = \mathbb{X} \cap V(p)$. Observe that one can interpret such relation as a sharp version of Max Nœther's $AF + BG$ division theorem [**No1873**] concerning the degree estimates.

We can finally consider the case $m \ge n$, with the hypothesis $\mathbb{X} \cap V(p) = \emptyset$. To begin with observe that it is enough to consider the case where $m = n$. Actually, if one re-organizes the sequence $(p_0, \dots, p_m)$ as $(\tilde{p}_0, \dots, \tilde{p}_m)$, where

$$\tilde{d}_1 \ge \tilde{d}_2 \ge \cdots \ge \tilde{d}_m \ge \tilde{d}_0$$

when $\tilde{d}_j = \deg \tilde{p}_j$, then one can again invoke Carl Siegel's box principle (see [**MW83**, §4, Lemma 1] or [**Jel05**, Lemma 3.5]) to consider $n+1$ generic combinations

$$\check{p}_0 = \tilde{p}_0, \quad \check{p}_j = \sum_{\kappa \ge j} \gamma_{j,\kappa}\, \tilde{p}_\kappa \quad \text{for } j = 1, \dots, n$$

which do not share any common on $\mathbb{X}$ (as the original p_j's). Then we replace the sequence $(p_0, \dots, p_m)$ by the sequence obtained from the re-organization of the sequence of polynomials $(\check{p}_0, \dots, \check{p}_n)$ along their increasing degrees. The proof of assertion (7.27) for the remaining case $m = n$ goes as follows; see [**Jel05**, Theorem 3.10]. Extract two distinct n-valued polynomial maps $p^{[0]}$ and $p^{[1]}$ from the $(n+1)$-valued polynomial map p. Then $\mathbb{X} \cap V(p^{[0]})$ and $\mathbb{X} \cap V(p^{[1]})$ are both of dimension less than or equal to 0. If one of these algebraic subsets is empty, then there is nothing to prove because of the first case considered in this proof. Accordingly, one can assume both of them are truly zero dimensional. A generic linear change of coordinates $(x,y) \leftrightarrow (x',y')$ in $\mathbb{K}^N$, if necessary, allows us to chose a coordinate ξ among the new coordinates x',y' such that the images by ξ of $\mathbb{X} \cap V(p^{[0]})$ and of $\mathbb{X} \cap V(p^{[1]})$ are disjoint. We now conclude with the elimination result, whose proof is in our second step, together with the fact that if $\phi^{[0]}$ and $\phi^{[1]}$ are two univariate polynomials with respective degrees δ_0 and δ_1 and without common zeroes, then Euclid's algorithm provides an identity $1 = \psi^{[0]}\phi^{[0]} + \psi^{[1]}\phi^{[1]}$ with $\max \deg(\psi^{[j]}\phi^{[j]}) \leq \delta_0 + \delta_1 - 1$.

As for the proof of the last assertion of the theorem in case $\mathbb{X} = \mathbb{K}^n$, which is due to Janos Kollár, we refer the reader to [**Ko88**, Theorem 1.5] or to the instructional presentation by Bernard Teissier [**T90**] in the Bourbaki seminar. We refer to [**Jel05**] and [**Ko88**] for the proofs of the assertions relative to the sharpness (optimality). It is also worth comparing the optimality of the present results to the optimality assertion in Theorem 7.8 (stated for the case $\mathbb{X} = \mathbb{K}^n$). □

7.1.6. An arithmetic Perron theorem in its parametric form. Let $\mathbb{K}$ be an algebraically closed field and let $\mathbb{X}$ be a purely n-dimensional affine algebraic variety defined over $\mathbb{K}$ and embedded in $\mathbb{K}^N = \mathrm{Spec}\,(\mathbb{K}[X,Y])$. As we observed in §7.1.5, the Perron theorem, in its generalized form [**Jel05**, Theorem 3.3], appears to be the cornerstone leading to the formulation of an almost sharp version of Hilbert's nullstellensatz in $\mathbb{K}_{\mathbb{J}(\mathbb{X})}[X,Y]$, in terms of $\mathbf{d} = \deg \mathbb{X}$ and the degrees of the polynomial entries $p_0,\dots,p_m$; see Theorem 7.16. This approach is realized through $\mathbf{d}\,(p)$ or $(2\,\mathbf{d}\,(p)-1)$-solvability of Bézout identity for $\{p_0,\dots,p_m\}$, combined with the Rabinowitsch trick, as stated in Proposition 7.4. We return in this section to the arithmetic frame of §6.6.3 and §6.6.4. We assume that $\mathbb{K} = \overline{\mathbb{Q}}$ and that the algebraic variety $\mathbb{X} \subset \mathbb{K}^N = \mathrm{Spec}\,(\mathbb{K}[X,Y])$ is defined over $\mathbb{Q}$. Namely, we assume that elements of $\mathbb{Z}[X,Y]$ generate the radical ideal $\mathbb{J} = \mathbb{J}(\mathbb{X})$. Then the arithmetic marker for the arithmetic complexity of $\mathbb{X}$ is its canonical global logarithmic height $\mathbf{h}$, as defined in Proposition 6.65.

To analyze how the ideas introduced in the proof of Theorem 7.16 can be transposed to the arithmetic setting and to see why the Perron theorem plays a pivotal role, one needs to review the proof of Theorem 7.16 from its beginning, which we do now. Let $(p_0,\dots,p_n)$ be a sequence of elements in $\mathbb{Z}[X,Y]$ such that

$$\{(x,y) \in \mathbb{X} :,\ p_0(x,y) = \cdots = p_n(x,y) = 0\} = \emptyset$$

in $\overline{\mathbb{Q}}^N$. It follows from Hilbert's nullstellensatz in $\overline{\mathbb{Q}}[X,Y]$ and from the fact that $\mathbb{X}$ is defined over $\mathbb{Q}$ (see also Remark 7.9) that there exist $\gamma^\circ \in \mathbb{N}^*$ and $q_1^\circ,\dots,q_n^\circ$ in $\mathbb{Z}[X,Y]$ such that $\gamma^\circ - \sum_{j=0}^n q_j^\circ p_j \in \mathbb{J}$. As a consequence, one can repeat the initial argument in the proof of Theorem 7.16 and claim that the mapping

$$\Phi : (x,y,t) \in \mathbb{X} \times \overline{\mathbb{Q}} \longmapsto \big(x,y,t\,p_0(x,y),\dots,t\,p_n(x,y)\big) \in \overline{\mathbb{Q}}^{N+n+1}$$

realizes a closed embedding of the $(n+1)$-dimensional algebraic variety $\mathbb{X} \times \overline{\mathbb{Q}}$ as an $(n+1)$-dimensional algebraic subvariety

$$\Phi(\mathbb{X} \times \overline{\mathbb{Q}}) \subset \mathbb{X} \times \overline{\mathbb{Q}}^{n+1}.$$

The left inverse $\Phi_{\text{left}}^{-1} : \Phi(\mathbb{X} \times \overline{\mathbb{Q}}) \to \mathbb{X} \times \overline{\mathbb{Q}}$ is defined by

$$\Phi_{\text{left}}^{-1}((x,y), w_0, \ldots, w_n) = \Big(x, y, \frac{q_0^\circ(x,y)w_0 + \cdots + q_n^\circ(x,y)w_n}{\gamma^\circ}\Big).$$

Let $\tau_j = (\tau_{j,1}, \ldots, \tau_{j,N})$, for $j = 0, \ldots, n$, be $n+1$ sets of parameters, whose cardinal is N. Specifying the parameters as generic integers, which means that $(\tau_0, \ldots, \tau_n)$ is a generic element in $\mathbb{Z}^{(n+1)N}$, the projection

$$\pi^{[\tau]} : ((x,y), z_0, \ldots, z_n) \in \Phi(\mathbb{X}\times\overline{\mathbb{Q}}) \mapsto \big(z_0+\langle\tau_0,(x,y)\rangle, \ldots, z_n+\langle\tau_n,(x,y)\rangle\big) \in \overline{\mathbb{Q}}^{n+1}$$

is a finite projection.[9] Then

(7.33)

$$\begin{aligned}(x,y,t) \longmapsto u^{[\tau]}(x,y,t) = u^{[\tau]}(x,y,t) &:= (\pi^{[\tau]} \circ \Phi)(x,y,t) \\ &= \big(t\,p_0(x,y) + \langle\tau_0,(x,y)\rangle, \cdots, t\,p_n(x,y) + \langle\tau_n,(x,y)\rangle\big)\end{aligned}$$

realizes a finite morphism from $\mathbb{X}_{\mathbb{Q}(\tau)} \times \mathbb{A}^1_{\mathbb{Q}(\tau)} = \mathbb{X}_{\mathbb{Q}(\tau)} \times \mathbb{Q}(\tau)$ to $\mathbb{A}^{n+1}_{\mathbb{Q}} = (\mathbb{Q}(\tau))^{n+1}$. It means that the inclusion of algebras

$$(\pi^{[\tau]} \circ \Phi)^* : \mathbb{Q}(\tau)[w_0, \ldots, w_n] \hookrightarrow \mathbb{Q}(\tau)[X,Y,t]/\mathbb{J}(\mathbb{X}_{\mathbb{Q}(\tau)} \times \mathbb{A}^1_{\mathbb{Q}}) \tag{7.34}$$

is an integral extension. Let $P(\tau, w, T) \in \mathbb{Z}[\tau, W][T]$ be the minimal polynomial of t with respect to this extension. It is primitive with respect to $\mathbb{Z}[\tau]$, is necessarily square-free, and is also primitive for $\mathbb{Z}[\tau, T]$. It is of the form

$$P(\tau, w, T) = \gamma_0(\tau)T^\nu + \sum_{\rho=1}^{\nu} \Big(\sum_{\kappa\in\mathbb{N}^{n+1}} \gamma_{\rho,\kappa}(\tau)w^\kappa\Big) T^{\nu-\rho}, \tag{7.35}$$

where γ_0 and $\gamma_{\rho,\kappa}$ belong to $\mathbb{Z}[\tau] = \mathbb{Z}[\tau_0, \ldots, \tau_n]$, and one also has that

$$\gamma_0(\tau)t^\nu + \sum_{\rho=1}^{\nu}\Big(\sum_{\kappa\in\mathbb{N}^{n+1}} \gamma_{\rho,\kappa}(\tau) \prod_{j=0}^{n} \big(t\,p_j(X,Y) + \langle\tau_j,(x,y)\rangle\big)^{\kappa_j}\Big) t^{\nu-\rho} \in \mathbb{Q}(\tau)[X,Y]\,\mathbb{J}. \tag{7.36}$$

Moreover,

$$(x,y) \longmapsto u_t^{[\tau]}(x,y) = \big(t\,p_0(x,y) + \langle\tau_0,(x,y)\rangle, \ldots, t\,p_n(x,y) + \langle\tau_n,(x,y)\rangle\big)$$

is a generically finite morphism from $\mathbb{X}_{\mathbb{Q}(\tau,t)}$; that is, the variety $\mathbb{X}$ is considered over $\mathbb{Q}(\tau,t)$ instead of $\mathbb{Q}$, onto its image in $\mathbb{A}^{n+1}_{\mathbb{Q}(\tau,t)} = (\mathbb{Q}(t,\tau))^{n+1}$. Then, one has, taking the Zariski closure of such image in $(\overline{\mathbb{Q}(\tau,t)})^{n+1}$, that

$$\overline{\operatorname{Im} u_t^{[\tau]}} = \{\varpi \in (\overline{\mathbb{Q}(\tau,t)})^{n+1} \ :\ P(\tau,\varpi,t) = 0\}; \tag{7.37}$$

see [**AKS13**, Lemma 4.8]. The coefficient of t^ν on the left-hand side of (7.36) is of the form

$$\gamma_0(\tau) - \sum_{j=0}^{n} q_j(\tau,X,Y)\,p_j(X,Y),$$

[9] The notion of a "Chow ideal" of a cycle (see Remark D.13) as well as that of a "Chow form" already introduced in the univariate setting (see Definition 1.49) remain constantly present in the approach presented here. Deep machinery relying on the concept of multi-Chow form (attached to multi-projective cycles) is developed extensively within the frame of arithmetic intersection theory by Yuri V. Nesterenko, Dale W. Brownawell, Patrice Philippon; see, e.g., [**Bro88**], [**PhS08**] and [**AKS13**, §2.3] for up-to-date bibliography about it. This machinery is involved in the approach developed in [**AKS13**], sustaining the presentation we sketch here.

where $q_j(\tau, X, Y) \in \mathbb{Z}[\tau][X, Y]$. Take any $\alpha \in \mathbb{N}^{(n+1)N}$ in the support of the polynomial in $(n+1)N$ variables $\gamma_0(\tau_0, \dots, \tau_n)$ and let $\gamma_{0,\alpha} \in \mathbb{Z}^*$ be its coefficient. Then, the coefficient of τ^α in the development as a polynomial in $(\tau_0, \dots, \tau_n)$ of $\sum_{j=0}^n q_j(\tau, X, Y)\, p_j(X, Y)$ provides, when subtracted from $\gamma_{0,\alpha}$, Bézout identity

$$\gamma - \sum_{j=0}^{m} q_j(X, Y) p_j(X, Y) \in \mathbb{J}. \tag{7.38}$$

One needs now to concentrate on upper bounds for the degrees of the q_j's and the logarithmic sizes of absolute values of their coefficients, including the estimate of $\log |\gamma_{0,\alpha}|$. To deal with such estimates, one needs the following parametric version of the Perron generalized theorem [**AKS13**, Theorem 3.15].[10]

THEOREM 7.17. *Let $\mathbb{X}$ be as above, and let $\tau = \{\tau_\iota\}_\iota$ be a finite collection of parameters realizing a transcendental basis of $\mathbb{Q}(\tau)$. Let also $U_0, \dots, U_n$ be $n+1$ elements in $U_j \in \mathbb{Z}[\tau, X, Y]$ such that the mapping*

$$(x, y) \longmapsto U^\tau(x, y) = \big(U_0(\tau, x, y), \dots, U_n(\tau, x, y)\big)$$

realizes a generically finite morphism from $\mathbb{X}_{\mathbb{Q}(\tau)}$ onto its image in the affine space $\mathbb{A}^{n+1}_{\mathbb{Q}(\tau)}$. Let $E \in \mathbb{Z}[\tau, w] = \sum_{\kappa \in \mathbb{N}^{n+1}} E_\kappa(\tau) w^\kappa$, reduced in $\mathbb{Q}(\tau)[X, Y]$ and primitive in $\mathbb{Z}[\tau]$, such that in terms of Zariski topology in $(\mathbb{A}_{\overline{\mathbb{Q}(\tau)}})^{n+1}$, one has

$$\overline{\operatorname{Im} U} = \big\{\varpi \in (\overline{\mathbb{Q}(\tau)})^{n+1} \,:\, E(\tau, \varpi) = 0\big\}.$$

Then, the following degree and logarithmic size estimates hold:

$$\begin{aligned}
&\deg_{w_0, \dots, w_n} E\big(\tau, w_0^{\deg_{X,Y} U_0}, \dots, w_n^{\deg_{X,Y} U_n}\big) \leq \mathbf{d}\,(\deg_{X,Y} U), \\
&\deg_{\tau_\iota} E\big(\tau, U(\tau, X, Y)\big) \leq \mathbf{d}\,(\deg_{X,Y} U) \sum_{j=0}^{m} \frac{\deg_{\tau_\iota} U_j}{\deg_{X,Y} U_j},
\end{aligned} \tag{7.39}$$

and, for any $\kappa \in \mathbb{N}^{n+1}$,

$$\begin{aligned}
h(E_\kappa) + \langle \kappa, h(U) \rangle &\leq \mathbf{d}\,(\deg_{X,Y} U)\Big(\frac{\mathbf{h}}{\mathbf{d}} + \log(n+2) \\
&+ \sum_{j=0}^{n} \frac{1}{\deg_{X,Y} U_j} \Big(h(U_j) + \log\big(\operatorname{card}(\operatorname{Supp}(U_j)) + 2\big) + \sum_\iota \log(|\tau_\iota| + 1) \deg_{\tau_\iota} U_j\Big)\Big),
\end{aligned} \tag{7.40}$$

where $\deg_{X,Y} U := (\deg_{X,Y} U_0, \dots, \deg_{X,Y} U_n)$ and $h(U) := (h(U_0), \dots, h(U_n))$.

REMARK 7.18. Observe that estimates for the degrees in X, Y (respectively for the degrees in the parameters τ_ι or logarithmic sizes estimates) in the first line of (7.39) (respectively in the second line of (7.39) or in (7.40)) fit with Bézout markers (6.180) for geometric intersection theory (respectively with Bézout markers (6.197) or (6.198) for arithmetic intersection theory) as introduced in §6.6.2.

Theorem 7.17 is pivotal for the formulation of the sharpest version for the arithmetic nullstellensatz known up to now, namely Theorem 6.70, following the reasoning that we described previously in this subsection. Technical estimates needed to justify degree and logarithmic size estimates for $\log |\gamma|$ and the q_j in

[10] We admit this result here since proving it would require the whole machinery of Chow forms of multi-projective cycles, together with control on their logarithmic heights in intersection theory; see [**AKS13**, §1,2]. This would take us too far from the purpose of the present monograph.

(7.38) become routine, once one invokes Theorem 7.17 with $\{\tau, t\}$ as parameters, since then $U(\tau,t,X,Y) = u^{[\tau]}(X,Y,t)$ as in (7.33); see [**AKS13**, §4.3]. As for the proof of the arithmetic elimination theorem (Theorem 6.70) quoted previously in this monograph from [**SomY21**, Theorem 5.1 and Corollary 5.6], it follows essentially along the same lines as those carried through in the argument developed for the second step of the proof of Theorem 7.16 to get an effective sharp geometric elimination (or Max Nœther's $AF + BG$ theorem). This theorem provides then a nearly optimal effective arithmetic elimination theorem. We refer to [**SomY21**, §5] for more details.

REMARK 7.19. As we have already seen in §6.6.3 and §6.6.4, the arithmetic Perron theorem, either in its parametric version stated as Theorem 7.17 or in its version without parameters τ, remains crucial for estimating the arithmetic complexity of multivariate residue calculus over $\mathbb{Q}$.

7.1.7. Hilbert's nullstellensatz and $P = NP$. Let $\mathbf{A}$ be a commutative ring. Within this subsection, $\mathbf{A}$ is one of the rings $\mathbb{Z}$, $\mathbb{R}$, $\mathbb{Q}$, $\overline{\mathbb{Q}}$, $\mathbb{C}$ or one of their transcendental extensions $\mathbb{Z}[\tau] = \mathbb{Z}[\tau_0, \dots, \tau_m]$, $\mathbb{R}[\tau]$, $\mathbb{Q}[\tau]$, $\overline{\mathbb{Q}}[\tau]$, $\mathbb{C}[\tau]$,[11] all with characteristic 0, as well as their p-adic companions $\mathbb{Z}_{\mathrm{p}}$, $\mathbb{R}_{\mathrm{p}}$, $\mathbb{C}_{\mathrm{p}}$ or $\mathbb{Z}_{\mathrm{p}}[\tau]$, $\mathbb{R}_{\mathrm{p}}[\tau]$, $\mathbb{C}_{\mathrm{p}}[\tau]$ with p a prime number. Within the characteristic 0 setting, one could consider as well $\mathbb{F}_{\mathrm{p}}$ or $\mathbb{F}_{\mathrm{p}}[\tau]$. The concept of *logarithmic height* on such ring $\mathbf{A}$ is the pendant of the notion of *entropy*. It is assigned to play the same role as that of a chaos (or disorder) quantifier in physics or dynamics. For example, when $\mathbf{A} = \mathbb{Z}$, the logarithmic height defined by $\mathrm{h}(a) = \log(|a|+1)$ quantifies the number of necessary bits required to encode $a \in \mathbb{Z}^*$, besides its sign. Such is the case also for various notions of *naive logarithmic height*, *Mahler measure*, etc., that we introduced in §6.6.1 or §6.6.2. When $\mathbf{A} = \mathbb{R}$ or $\mathbf{A} = \mathbb{C}$, it is natural to consider $\mathrm{h}(a) = 1$ for any $a \in \mathbf{A}$ to take into account that the cost of the multiplication by a does not depend on the size of the real or complex number a.

Let $\mathbf{A}$ be a ring, which is also a domain, equipped with a logarithmic size h. Let also $p_0, \dots, p_m \in \mathbf{A}[X_1, \dots, X_n]$ be $m+1$ polynomials with coefficients in $\mathbf{A}$. The ability to decide whether or not the polynomial functions $x \mapsto p_j(x)$, $j = 0, \dots, m$, share a common zero in an integral closure (in fact any) of the fraction field $\mathrm{Frac}(\mathbf{A})$ of $\mathbf{A}$ is a typical example of a *decision problem*. This is the case for the following more general problem: given the polynomials $p_0, \dots, p_m$ as before and $q \in \mathbf{A}[X_1, \dots, X_n]$, decide whether or not q belongs to the ideal $\sum_{j=0}^m p_j \mathrm{Frac}(\mathbf{A})[X_1, \dots, X_n]$ or at least to the integral closure of such ideal in $\mathrm{Frac}(\mathbf{A})[X]$. The first problem mentioned above is the *nulltellensatz decision problem* over $\mathbf{A}$, while the second one is known as the *membership decision problem* over $\mathbf{A}$. Let us describe the problems a little more conceptually, following, for example, the pioneering paper [**BlSS89**] devoted specifically to the case where $\mathbf{A} = \mathbb{R}$. A *decision problem* over $\mathbf{A}$ is a pair of subsets $(\boldsymbol{D}, \boldsymbol{D}_{\mathrm{yes}})$, with $\boldsymbol{D}_{\mathrm{yes}} \subset \boldsymbol{D}$, of the space $\mathbf{A}^\ell$ (here the integer ℓ may be finite or infinite). The space $\mathbf{A}^\ell$ plays the role of the space of *inputs* for some *machine*[12] $\mathbb{M}$ over $\mathbf{A}$. An *algorithm solving the decision*

[11] The field $\mathbb{C}$ is here visualized as its subjacent $\mathbb{R}$-vector space $\mathbb{R}^2$ in order to profit from the fact that $\mathbb{R}$ is ordered, as $\mathbb{Z}$ and $\mathbb{Q}$ are, which $\mathbb{C}$ is not.

[12] A machine over $\mathbf{A}$ consists of inputs space $\mathbf{A}^{\mathtt{init}}$, with $\mathtt{init} \in \mathbb{N}^*$, possibly infinite, a space of outputs $\mathbf{A}^{\mathtt{fin}}$, and a space of transient states $\mathbf{A}^{\mathtt{trans}}$, with $\mathtt{fin}, \mathtt{trans} \in \mathbb{N}^*$, together with a graph $\mathbb{G}_{\mathbb{M}}$ having a finite number of nodes N are classified into *computation* nodes, *branching* nodes, and *exit* nodes. We refer the reader to [**BlSS89**, §2] for a didactic presentation of the

problem is a machine $\mathbb{M}$ initiated from any input $e \in \boldsymbol{D}$, which eventually stops, then providing an output $s_{\mathbb{M}}(e)$ such that $s_{\mathbb{M}}(e) = 1 \Longleftrightarrow e \in \boldsymbol{D}_{\text{yes}}$. More precisely, given an element e in $\mathbf{A}^{\texttt{init}}$, the machine is said to *stop when initiated at* e when, as soon as it starts having e as *input*, then some *exit* node is reached after some minimal time $t_{\mathbb{M}}(e)$ (the time is being indexed here with the number of steps the machine proceeds). Whenever e is such an entry, the *cost* of the machine (when initiated in e) is by definition the quantity

$$c_{\mathbb{M}}(e) = t_{\mathbb{M}}(e) \times \mathrm{h}_{\mathbb{M}}(e)\,. \tag{7.41}$$

The quantity $\mathrm{h}_{\mathbb{M}}(e)$ denotes the maximal logarithmic height of all elements in the ring $\mathbf{A}$ that have been involved in all transient states, which appear as intermediate states before the machine reaches the *exit* node that will lead to the output $s_{\mathbb{M}}(e)$ (in the output space) within the time $t_{\mathbb{M}}(e)$. The machine $\mathbb{M}$ is said to *operate in polynomial time* if its cost function satisfies for any $e \in \mathbf{A}^{\texttt{init}}$

$$c_{\mathbb{M}}(e) \leq C_{\mathbb{M}} \left(|e| + \mathrm{h}(e)\right)^{N_{\mathbb{M}}} \tag{7.42}$$

for some $N_{\mathbb{M}} \in \mathbb{N}^*$, $C_{\mathbb{M}} > 0$. Here $|e|$ denotes the length of the support of e. That is, it denotes the number of nonzero entries of e as an element in $\mathbf{A}^{\texttt{init}_{\mathbb{M}}}$.

The nullstellensatz decision problem over $\mathbf{A} = \mathbb{C}$ is known to be NP-complete; see [**BlSS89**, §6, Main theorem]. The whole content of this paper is an introduction to the NP complexity class of decision problems over $\mathbf{A} = \mathbb{R}$. The terminology NP stands for *decidable in Nondeterminist Polynomial time.* Let us recall what it means, at least heuristically. A decision problem $(\boldsymbol{D}, \boldsymbol{D}_{\text{yes}})$ with $\boldsymbol{D}_{\text{yes}} \subset \boldsymbol{D} \subset \mathbf{A}^{\ell}$ ($\ell \in \mathbb{N}^*$, possibly infinite) over the ring $\mathbf{A}$ is said to be in the NP *class* if there exist two integers $\ell' \in \mathbb{N}^*$, possibly infinite, a machine $\mathbb{M}$ over $\mathbf{A}$ with input space $\mathbf{A}_{\mathbb{M}}^{\texttt{init}}$ such that

$$\boldsymbol{D} \times \mathbf{A}^{\ell'} \subset \mathbf{A}_{\mathbb{M}}^{\texttt{init}} \subset \mathbf{A}^{\ell} \times \mathbf{A}^{\ell'},$$

where $\mathbf{A}^{\ell'}$ plays the role of a probability space, such that

- the machine eventually stops when it is initiated at any (e,ω) in $\boldsymbol{D} \times \mathbf{A}^{\ell'}$;
- one has $s_{\mathbb{M}}(e,\omega) \in \{0,1\}$ for any $(e,\omega) \in \boldsymbol{D} \times \mathbf{A}^{\ell'}$;
- $s_{\mathbb{M}}(e,\omega) = 1 \Longleftrightarrow e \in \boldsymbol{D}_{\text{yes}}$;
- most importantly, there exists $N_{\mathbb{M}} \in \mathbb{N}$ and $C_{\mathbb{M}} > 0$ such that

$$\begin{aligned} &\forall\, e \in \boldsymbol{D}_{\text{yes}},\ \exists \omega \in \mathbf{A}^{\ell'} \quad \text{such that } s_{\mathbb{M}}(e,\omega) = 1 \\ &\qquad\qquad\qquad \text{with } c_{\mathbb{M}}(e,\omega) \leq C_{\mathbb{M}} \left(|e| + \mathrm{h}(e)\right)^{N_{\mathbb{M}}}. \end{aligned} \tag{7.43}$$

Given such a decision problem $(\boldsymbol{D}, \boldsymbol{D}_{\text{yes}})$ belonging to the NP class over $\mathbf{A}$, it is said to be NP-*complete* if it satisfies some universal property within the category of decision problems in the NP class over $\mathbf{A}$. This means the following: given any decision problem $(\tilde{\boldsymbol{D}}, \tilde{\boldsymbol{D}}_{\text{yes}})$ over $\mathbf{A}$, there exists a map $\Psi\ : \boldsymbol{D} \to \tilde{\boldsymbol{D}}$ such that $\Psi(\tilde{e}) \in \boldsymbol{D}_{\text{yes}} \Longleftrightarrow \tilde{e} \in \tilde{\boldsymbol{D}}_{\text{yes}}$ and Ψ is the restriction to $\tilde{\boldsymbol{D}}$ of the output $\tilde{e} \mapsto s_{\mathbb{M}}(\tilde{e})$ of a machine $\mathbb{M}$ (with $\tilde{\boldsymbol{D}} \subset \mathbf{A}^{\texttt{init}_{\mathbb{M}}}$) that operates over $\mathbf{A}$ in polynomial time.

---

role devoted to such nodes within the class to which it belongs. Each node comes either with one edge to the next mode and no incoming edge (*input* node), with incoming edges and exactly one outgoing edge (*computation* node), incoming edges and two outgoing edges together with a polynomial map from $\mathbf{A}^{\texttt{trans}}$ to itself governing the switching of the machine towards one of the two alternative nodes (*branching* node), or no outgoing edge and thus a unique linear map from $\mathbf{A}^{\texttt{trans}}$ to $\mathbf{A}^{\texttt{fin}}$ (*exit* node).

A decision problem $(\boldsymbol{D}, \boldsymbol{D}_{\text{yes}})$ over the ring $\mathbf{A}$ is said to be in *the class* P if there is a machine $\mathbb{M}$ over $\mathbf{A}$ that solves it in polynomial time. Any decision problem in the P class over $\mathbf{A}$ is in the NP class over $\mathbf{A}$. It is still unknown and it remains a highly challenging question whether the P class over any of the classical commutative rings $\mathbf{A}$ is strictly included or equal to the NP class over $\mathbf{A}$ [**Sm98**]. Given what we discuss within the present monograph,[13] let us quote here the following results [**BlCSS96**, §7.1, Theorems 1 and 2] or [**ShuS96**, Main Theorem] about the conjecture P = NP (or P $\neq$ NP) over the rings $\mathbb{C}$ (where the logarithmic size is taken as constant equal to 1) or $\overline{\mathbb{Q}}$ (where the logarithmic size derives from the logarithmic size $a \mapsto \mathrm{h}(a) = \log(|a|+1)$ on $\mathbb{Z}$).

THEOREM 7.20. *If* P = NP *is true over* $\mathbb{C}$*, then* P = NP *is true over* $\overline{\mathbb{Q}}$*. The converse is true as well. Moreover, if one makes the hypothesis that the sequence* $(k!)_{k\geq 0}$ *is ultimately hard to compute,*[14] *then the* NP *nullstellensatz decision problem over* $\mathbb{C}$ *fails to be in the* P *class over* $\mathbb{C}$*. This implies that* P $\neq$ NP *over* $\mathbb{C}$ *or, which is equivalent,* P $\neq$ NP *over* $\overline{\mathbb{Q}}$*.*

Sharp versions of the arithmetic Hilbert's nullstellensatz formulated as Bézout identity described in Theorem 6.70, with (nearly) sharp estimates both in degree and logarithmic size, are obtained (as we have seen already in §7.1.6 and will make more precise in §7.1.7) through a nonalgorithmic procedure relying on the use of a parametric version of the Perron theorem; see Theorem 7.17. The statement requires starting with an *a priori* solution of Bézout identity. Consequently, such proofs do not provide proof of Hilbert's nullstellensatz since they rely on it from the beginning; see, for example, the proof of Theorem 7.16. On the other hand, surprisingly, multivariate residue calculus, inspired originally by a more analytic point of view than algebraic, will provide (as we will see in §7.2.1, Theorem 7.27), a closed formula of the Lagrange–Kronecker type solving *explicitly* with reasonable bounds in terms of degrees and logarithmic sizes (fitting with geometric and arithmetic Bézout type control in intersection theory) the Bézout arithmetic identity, that is, Hilbert's nullstellensatz over $\overline{\mathbb{Q}}$. The sequence $(k!)_{k\geq 0}$, whose complexity is involved in Theorem 7.20, is also constantly present there as the sequence of denominators that appear when Leibniz's differential calculus (governed by Leibniz's formula) is reformulated in terms of multivariate residue calculus (dominated by the transformation law or its generalizations presented in §3.2.3); see Remark 6.25. For these reasons, a dynamical approach to the arithmetic complexity of multivariate residue symbols defined over $\mathbb{Q}$ or $\overline{\mathbb{Q}}$, as sketched in §6.6.4, would need to be pursued. A deformation argument of Arnold's type (see [**AVGZ**]) could be invoked instead of appealing to the somehow unexplicit Perron reasoning. It is hard to see at this point whether some algorithmic procedure could lead to the closed formula

[13]It should be added here that a large part of this heuristical section arose from correspondence with Lenor Blum. The starting point for the correspondence was the content of the informal notes of an introductory lecture A.Yger gave in Bordeaux in 2001. Here we take the opportunity to refer the reader to the seminal Emmy Nœther lecture [**Bl04**].

[14]Computation with length ℓ of $k \in \mathbb{Z}$ is a sequence of integers $[\nu_0\, \nu_1, \dots, \nu_\ell]$, initiating at $\nu_0 = 1$, ending at $\nu_\ell = k$, so that any intermediate ν_ι is expressed as $\nu_{\iota'} \bullet \nu_{\iota''}$ with $\iota' < \iota$, $\iota'' < \iota$ and $\bullet$ denotes the addition, the substraction, or the multiplication, in which case, the minimum length is denoted as $\ell_{\min}(k)$. Given a sequence $(u_k)_{k\geq 0}$ of integers, such a sequence is said to be *ultimately easy to compute* if there exists a sequence of nonzero integers $(m_k)_{k\geq 0}$ and $N \in \mathbb{N}^*$ such that $\ell(m_k u_k) \leq (\log k)^N$ for any $k \geq 2$. Otherwise, the sequence $(u_k)_{k\geq 0}$ is said to be *ultimately hard to compute*.

that we shall propose in the following subsection, but this remains a challenge about the key test role of the NP-complete nullstellensatz decision problem over $\mathbb{C}$ or $\overline{\mathbb{Q}}$ concerning the P=NP (or P$\neq$ NP) conjecture over one of these two rings. Even though a closed formula is far from sustaining an algorithmic procedure, it seems significant to us to make such heuristic observations here.

Let us go on with some additional heuristical considerations. Another trick one can use to encode the complexity of a system of polynomial inputs in $\mathbb{Z}[X_1,\dots,X_n]$ is to encode the construction process itself. Then continue by encoding the process of evaluation of the different entries of the system. For example, the univariate polynomial

$$x^{2^{2^n}} - 1 \in \mathbb{Z}[x]$$

is quite easy to encode this way (iterating 2^n times the process $x \mapsto x^2$), although it has a doubly exponential degree. This idea arose in the 1970s through the works of J. Heintz, J. Morgenstern, C. P. Schnorr, and others. Its development has been swift since that time. For example, one refers to the works [**KP94, GHMMPM97, GHMMP98**] to find both a presentation and a prospective outlook about the role this approach could have for diophantine approximation questions in $\overline{\mathbb{Q}}[X_1,\dots,X_n]$. The key notion here is the concept of *straight line program* on $\mathbb{Z}[X_1,\dots,X_n] = \mathbb{Z}[X]$; see e.g [**GHMMPM97**, Definition 18]. This concept is inspired by the notion of simplicity (where "simplicity" means "easy to compute" algorithmically) of a numerical sequence already mentioned when asking whether the factorial sequence $(k!)_{k\geq 0}$ is (or is not) ultimately hard to compute; see Theorem 7.20. Such a program $\mathbb{SL}$ consists of the following data: a directed acyclic graph[15] $\mathbb{G}_{\mathbb{SL}}$ paired with a list of instructions $\mathbb{I}_{\mathbb{SL}}$, all of them defining some kind of protocol, one for each different entry gate of the graph. Together with $(\mathbb{G}_{\mathbb{SL}}, \mathbb{I}_{\mathbb{SL}})$, a finite set $\tau_{\mathbb{SL}}$ of integer parameters is also given. Moreover, the graph $\mathbb{G}_{\mathbb{SL}}$ presents exactly $n+1$ entry gates labeled as $X_1,\dots,X_n$ and 1. The *depth* of a gate g of $\mathbb{G}_{\mathbb{SL}}$ denotes the length of the longest path from g until one of the entry gates. One can label the gates of $\mathbb{G}_{\mathbb{SL}}$ by pairs (k,ℓ), where $k \in \mathbb{N}$ means the depth of the gate and $\ell \in \mathbb{N}$ denotes another index. This index is used to label the gates $\mathbb{G}_{\mathbb{SL}}$ with fixed depth arbitrarily. Gates are then classified along the lexicographic order $\prec$ on $\mathbb{N}\times\mathbb{N}$; that is, all gates $g_{k,\ell}$ of $\mathbb{G}_{\mathbb{SL}}$ sharing k as depth are ordered along lexicographic order with respect to l. To each such gate $g = g_{k,\ell}$ of $\mathbb{G}_{\mathbb{SL}}$, one may associate

(7.44)

$$q_{k,\ell} = \Big(\sum_{0\leq\kappa'<k,\,\lambda'\geq 0} \tau_{k,\ell}^{\kappa',\lambda'} q_{\kappa',\lambda'}\Big) \times \Big(\sum_{0\leq\kappa''<k\,,\lambda''\geq 0} \widetilde{\tau}_{k,\ell}^{\kappa'',\lambda''}\, q_{\kappa'',\lambda''}\Big) \in \mathbb{Z}[\tau_{\mathbb{SL}}][X]$$

$$\text{with} \quad \tau_{k,\ell}^{\kappa',\lambda'}, \widetilde{\tau}_{k,\ell}^{\kappa'',\lambda''} \in \tau_{\mathbb{SL}} \quad \text{and} \quad q_{\kappa',\lambda'}, q_{\kappa'',\lambda''} \in \mathbb{Z}[\tau_{\mathbb{SL}}][X],$$

where $q_{\kappa',\lambda'}$, $q_{\kappa'',\ell''}$ are precalculated at the gate (k,ℓ) since $\max(\kappa',\kappa'') < \kappa$ (depending on the gate (k,ℓ)). Then, a given polynomial $p \in \mathbb{Z}[X]$ is said to be evaluated by the straight line program $\mathbb{SL}$ if there is at least one gate $g_{k,\ell}$ of $\mathbb{G}_{\mathbb{SL}}$ such that p equals $q_{k,\ell}$. One may define the size $|\mathbb{SL}|$ of the straight line program

[15]That is, a graph consisting of a finite number $|\mathbb{G}|$ (the size of $\mathbb{G}$) of vertices g (or nodes) and a finite number of edges (or arrows). Every edge directed from one vertex to another and realizing closed loops is excluded (acyclicity). Entry gates are gates with in-degree 0. That is, entry gates are those not supporting any incoming edge.

$\mathbb{SL}$ as

$$|\mathbb{SL}| := |\mathbb{G}_{\mathbb{SL}}|$$

and its depth as

$$D_{\mathbb{SL}} = \operatorname{depth} \mathbb{G}_{\mathbb{SL}}.$$

That is, the depth is the length of the longest path in the directed acyclic graph $\mathbb{G}_{\mathbb{SL}}$. Most interesting is the fact that one can also define the logarithmic height $h_{\mathbb{SL}}$ of the straight line program $\mathbb{SL}$ as

$$\mathrm{h}_{\mathbb{SL}} := \max \{\log(|\tau_\iota| + 1) \ : \ \tau_\iota \in \tau_{\mathbb{SL}}\}.$$

Observe that the role assigned to $D_{\mathbb{SL}}$ could be compared to that of the *degree* in geometric intersection theory, while the role assigned to $h_{\mathbb{SL}}$ could be compared to that of the *canonical logarithmic height*; see Definition 6.65 in arithmetic intersection theory. With such a concept in hand, one can state the following algorithmic version of Hilbert's arithmetic nullstellensatz [**KP94**, Theorem 1].

THEOREM 7.21. *Let $p_0, \dots, p_m \in \mathbb{Z}[X_1, \dots, X_n] = \mathbb{Z}[X]$ be $m+1$ polynomials, whose degrees satisfy $\deg p_i < d$, $d \geq n$ and whose naive logarithmic sizes are less than h. Suppose that $\bigcap_{j=0}^m V(p_j) = \emptyset$ in $\mathbb{C}^n$ or in $\overline{\mathbb{Q}}^n$. There is a straight line program $\mathbb{SL}(p)$ with parameters in $\mathbb{Z}[X]$ satisfying*

(7.45)

$$|\,\mathbb{SL}\,(p)| \leq (m+1)\, d^{O(n)}, \quad D_{\mathbb{SL}\,(p)} \leq O(n \log_2 d), \quad \mathrm{h}_{\,\mathbb{SL}(p)} \leq \max(\mathrm{h}, d^{O(n)}).$$

Moreover, such a straight line program evaluates an integer $a \in \mathbb{Z}^*$ as well as polynomials $q_0, \dots, q_m$ in $\mathbb{Z}[X]$ such that $a = \sum_{j=0}^m q_j(X)\, p_j(X)$.

REMARK 7.22. Let $m = n$. One could compare Theorem 7.21, which carries an algorithmic process materialized by the straight line program $\mathbb{SL}\,(p)$ on $\mathbb{Z}[X]$, with the (nearly sharp) effective arithmetic nullstellensatz stated as Theorem 6.70. Such effective arithmetic nullstellensatz provides much sharper bounds (in fact nearly sharp), but its proof relies on the parametric version of the Perron theorem (Theorem 7.17). Hence, it is not accompanied (as far as we can see for the time being) by any constructive algorithmic procedure.

One could also view the problem differently. Suppose that the polynomials $p_0, \dots, p_m \in \mathbb{Z}[X]$ are obtained through evaluation by straight line programs with parameters $\mathbb{SL}\,(p_j)$, $j = 0, \dots, m$, where the quantities $|\,\mathbb{SL}\,(p_j)|$, $D_{\,\mathbb{SL}\,(p_j)}$ and $\mathrm{h}_{\,\mathbb{SL}\,(p_j)}$, for $j = 0, \dots, m$, are respectively bounded by $|\mathbf{S}|, \mathbf{D}, \mathbf{h}$. In terms of $|\mathbf{S}|$, $\mathbf{D}$, $\mathbf{h}$, and, if necessary, other geometric parameters such that the degrees of the p_j, the size, depth, and height of a some straight line program $\mathbb{SL}$ with parameters on $\mathbb{Z}[X]$ that would be able to evaluate some $a \in \mathbb{Z}^*$ and $q_1, \dots, q_n \in \mathbb{Z}[X]$ with $a = \sum_{j=0}^m q_j(X)\, p_j(X)$, can one estimate when such a Bézout identity is realizable? If it were feasible, this would provide a test to decide (for p) Hilbert's nulstellensatz decision problem over the field of algebraic numbers: do the p_j's, yes or no, share a common zero with coordinates in an integral closure of $\mathbb{Q}$, for example $\overline{\mathbb{Q}}$? Sometimes, this the case if one adds some geometric constraint. For example, if $m = n - 1$ and the n polynomials $p_0, \dots, p_{n-1}$ are known to define a zero-dimensional algebraic variety, then for any $j = 1, \dots, n$ there exists a straight line program $\mathbb{SL}_j$ with

$$|\,\mathbb{SL}_j| \leq \big(n\,\mathbf{D}\ \deg(V(p))\,|\mathbf{S}|\big)^{O(1)},$$
$$h_{\,\mathbb{SL}_j} \leq O\big(n\,\mathbf{h}\,\big(\log(n\mathbf{D}) + d\big)\ \log\deg(V(p))\big)$$

evaluating some univariate polynomial $q \in \mathbb{Z}[X_j]$ vanishing on $V(p)$. It is interesting to compare such an algorithmic result with the arithmetic elimination, nearly sharp, result stated as Theorem 6.68. Since the proof of Theorem 6.68 relies again on the parametric version of the Perron theorem (Theorem 7.17), such a result fails *a priori* (as the sharp arithmetic nullstellensatz stated as Theorem 6.70 also does) to be accompanied with any algorithmic process.

We should mention that any elimination procedure with natural universal properties inherits, in algorithmic terms we just mentioned, a nonpolynomial complexity character [**CGHMP03**, Theorem 4]. Besides this fact, any known elimination procedure leading to optimality in the algorithmic approach appears up to now to rely on the approach towards elimination based on Jacobi–Lagrange–Kronecker method, which multivariate residue calculus governs, even though it may use intensively other duality ideas such as polarization, as for example in [**BGHM01**]. Closed formulae for N-solvability of Bézout identity for $\{p_0, \dots, p_n\}$ in $\mathbb{C}[X_1, \dots, X_n]$ or $\mathbb{Q}[X_1, \dots, X_n]$ do exist and rely on Jacobi–Lagrange–Kronecker division/interpolation; see §7.2.1. Is there some hint towards an algorithmic formulation of the statement P $\neq$ NP over $\mathbb{C}$? We hope that such heuristical considerations may convince the reader that it is time to think about complexity questions inherent to multivariate residue calculus within the arithmetic setting in terms of programming language or pure logic instead of just logarithmic size control. Estimates for the logarithmic size control of multidimensional residues within the arithmetic setting, even though unfortunately probably not sharp, were presented in §6.6.3 and in §6.6.4 (motivated by some dynamic approach). Nevertheless, such lack of sharpness suggests that the ball should now be in the court of formal calculus specialists. One could hope that new ideas from such a world will lead to new progress for some key questions in diophantine analysis remaining unsolved, such as the well-known Schanuel's conjecture. Solving Schanuel's conjecture would imply in the simplest cases of dimension 2, the algebraic independence over $\overline{\mathbb{Q}}$ of e and π, or $\log 2$ and $\log 3$. Here again, one could hope that analysis or geometry would play the same stimulating role they played for the effectivity or complexity problems that lie behind questions related to decision problems such as Hilbert's nullstellensatz over $\mathbb{C}$ or $\overline{\mathbb{Q}}$. We should also mention that the classical P $=$ NP conjecture is the conjecture over $\mathbb{F}_2$. Its positive solution amounts proving that the NP-complete nullstellensatz decision problem over $\mathbb{F}_2$ lies in the P class.

Let us conclude this section with a few words about a major decision problem over $\mathbb{C}$ or $\overline{\mathbb{Q}}$, which does not share the so-important NP-completeness property with Hilbert's nullstellensatz. Concerning the ideal membership decision problem over $\mathbb{C}$, a breakthrough arose in 1988 when E. Mayr and A. Meyer exhibited in [**MaM82**] the following example. For any $d \geq 5$ and any $k \geq 2$, one can find $10k+1$ polynomials $p_0, \dots, p_{10k}$ in $\mathbb{Z}[X_1, \dots, X_{10k}]$ (which happen to be binomials) such that X_1 lies in the ideal generated by the p_j's in $\mathbb{Q}[X]$, but

$$X_1 = \sum_{j=0}^{10k} q_j(X)\, p_j(X) \Longrightarrow \max_{0 \leq j \leq 10k} \deg q_j \geq (d-2)^{2^{k-1}}.$$

The ideal membership decision problem over $\mathbb{C}$ is known to be EXPSPACE-complete since the pioneering work of E. Mayr and A. Meyer. In other words, it is in the

EXPSPACE class of decision problems over $\mathbb{C}$ that are decidable with a determinist Turing machine in exponential time. In addition, it satisfies a universality property within this class (which means EXPSPACE-completeness). We refer the reader to [**As04**, Theorem A] for a discussion about the arithmetic counterpart of such a result, with $\mathbb{C}$ replaced by $\mathbb{Q}$, and how effective its realizations could be in terms of degree and logarithmic size estimates of the input entries. Under some additional constraints on the input p_j's, one can get degree bounds fitting with Bézout's geometric intersection theorem as well as height estimates fitting with Bézout's arithmetic intersection theorem, when $\mathbb{C}$ is replaced by $\overline{\mathbb{Q}}$. Among such peculiar favorable situations, the most important one is the case where $m+1 \leq n$ and $(p_0, \dots, p_m)$ is a quasi-regular sequence in $\mathbb{C}[X]$. Then the analogue of Theorem 7.21 holds; see [**KP94**, Theorem 2]. We shall also produce in §7.2.2 a realization of the membership $q = \sum_{j=0}^{m} q_j p_j$ in $\mathbb{C}[X]$ or $\overline{\mathbb{Q}}[X]$ in terms of a closed formula involving multivariate residue calculus that works over $\mathbb{C}$ as well as over the field of algebraic numbers. This demonstrates again in such a situation the importance of the Jacobi–Lagrange–Kronecker approach .

7.2. Jacobi–Lagrange–Kronecker (JLK) parametric identities

Let $\mathbf{A}$ be an integral domain whose fraction field $\mathbb{K}$ is a subfield of $\mathbb{C}$. For $m \leq n$, let $p = (p_0, \dots, p_n)$ or $p = (p_1, \dots, p_m)$ be sequences of elements in $\mathbf{A}[X_1, \dots, X_n]$ with length $\leq n+1$. We assume all through this section that

$$0 \leq j \leq j' \leq n \Longrightarrow 1 \leq d_j = \deg p_j \leq d_{j'} = \deg p_{j'}. \tag{7.46}$$

Since the univariate setting is described in §1.3 already, we assume from now on that $n > 1$. In this case the geometric marker $\mathbf{d}\,(p)$ introduced in (7.25) equals

$$\mathbf{d}\,(p) = \begin{cases} \prod_{j \neq 1} d_j & \text{when } p = (p_0, \dots, p_n), \\ d_1 \cdots d_m & \text{when } p = (p_1, \dots, p_m) \text{ with } m \leq n. \end{cases} \tag{7.47}$$

When $p = (p_0, \dots, p_n)$, we make the following properness assumption about the intersection of the supports of the divisors $\mathrm{div}_{\mathbb{C}^n}(p_j)$ in $\mathbb{C}^n$ for $j = 0, \dots, n$:

$$\begin{gathered} |\mathrm{div}_{\mathbb{C}^n}(p_0)| \cap \cdots \cap |\mathrm{div}_{\mathbb{C}^n}(p_n)| = \emptyset, \\ \forall\, J \subset \{0, 1, \dots, n\} \text{ with } |J| \leq n,\ \dim \Big(\bigcap_{j \in J} |\mathrm{div}_{\mathbb{C}^n}(p_j)| \Big) = n - |J|. \end{gathered} \tag{7.48}$$

Instead, when $p = (p_1, \dots, p_m)$ and $m \leq n$, we make the following properness assumption about the intersection of the supports of the divisors $\mathrm{div}_{\mathbb{C}^n}(p_j)$ this time for $j = 1, \dots, m$:

$$\forall\, J \subset \{1, \dots, m\},\ \dim_{\mathbb{C}^n} \Big(\bigcap_{j \in J} |\mathrm{div}_{\mathbb{C}^n}(p_j)| \Big) = n - |J|. \tag{7.49}$$

Let $\{\mathrm{t}\} = \{\mathrm{t}_1, \dots, \mathrm{t}_n\}$ be a set of n parameters realizing a transcendental basis of $\mathbb{K}(\mathrm{t})$. Given $\nu \in \mathbb{N}^*$, for each entry p_j of p let

$$\phi_{j,\nu}(\mathrm{t}, X) = \big(1 + \langle \mathrm{t}, X \rangle\big)^{\nu} p_j(X) \in \mathbf{A}[\mathrm{t}, X]. \tag{7.50}$$

For each entry p_j of the sequence p with degree d_j, let also

$$H_{p_j}(z, \zeta) = \sum_{k=1}^{n} H_{p_j,k}(z, \zeta)\, d\zeta_k, \tag{7.51}$$

where $H_{p_j,k} \in \mathbf{A}[X,Y]$ with $\deg H_{p_j,k} = d_j - 1$ for $1 \le k \le n$, be an algebraic Hefer form for the section p_j of the trivial bundle over $\mathbb{C}^n$ and

$$H_{\phi_{j,\nu}}(\mathfrak{t}, z, \zeta) = \big(1 + \langle \mathfrak{t}, z\rangle\big)^\nu H_{p_j}(z,\zeta) + p_j(\zeta)\Big(\sum_{\rho=0}^{\nu-1} \big(1+\langle \mathfrak{t}, z\rangle\big)^\nu \big(1+\langle \mathfrak{t},\zeta\rangle\big)^{\nu-1-\rho}\Big)\, \langle \mathfrak{t}, d\zeta\rangle \tag{7.52}$$

be the corresponding algebraic Hefer form for the section $\Phi_{j,\nu}(\mathfrak{t},\cdot)$.

7.2.1. Parametric Bézout identity of the JLK form. We consider in this subsection the case where $p = (p_0, \dots, p_n)$. Since the supports of the divisors $\mathrm{div}_{\mathbb{C}^n}(p_j)$ intersect properly in the affine space $\mathbb{C}^n$ (see (7.48)) E. Nœther's preparation theorem (see for example [**L**, **VdW**]) admits the following reformulation.

PROPOSITION 7.23. *Let*

$$\tau = \Big\{\mathfrak{t}_k^{[j]} \,:\, j = 0,2,\dots,n,\ k = 1,\dots,n\Big\} \tag{7.53}$$

be a set of n^2 parameters realizing a transcendental basis of $\mathbb{K}(\tau)$, which is organized in $n+1$ packets $\mathfrak{t}^{[j]} = (\mathfrak{t}_1^{[j]}, \dots, \mathfrak{t}_n^{[j]})$ for $j = 0,2,\dots,n$. There exists a polynomial $\Phi \in \mathbb{C}[\tau]$, with

$$\max_{j=0,2,\dots,n} \deg_{\mathfrak{t}^{[j]}} \Phi \le 2^n\, \mathbf{d}(p)\, (d_1/d_0 + 2\,\mathbf{d}(p)), \tag{7.54}$$

such that if τ is specified as an element in $\mathbb{C}^{n^2}$, then

$$\Phi(\tau) \neq 0 \Longrightarrow \bigcap_{j \in J \cup\{1\}} V(p_j) \cap \Big(\bigcap_{j' \notin J\cup\{1\}} \{\zeta \in \mathbb{C}^n \,:\, 1 + \langle \mathfrak{t}^{[j']}, \zeta\rangle = 0\}\Big) = \emptyset \tag{7.55}$$

and

$$\Big|\Big(\bigcap_{j\in J} V(p_j) \cap \Big(\bigcap_{j' \notin J\cup\{1\}} \{\zeta \in \mathbb{C}^n \,:\, 1 + \langle \mathfrak{t}^{[j']}, \zeta\rangle = 0\}\Big)\Big)\Big| = \deg\Big(\bigcap_{j\in J} V(p_j)\Big).$$

Note that the multiplicities are taken into account in the cardinality of the above intersection for any subset J of $\{0,2,\dots,n\}$.

PROOF. For each $J \subset \{0,2,\dots,n\}$ with cardinal $0 \le |J| < n$, the complex algebraic subvariety $V_{1,J} = \bigcap_{j\in J\cup\{1\}} V(p_j)$ of $\mathbb{C}^n$ is purely $(n-|J|-1)$-dimensional according to (7.48). Such is its Zariski closure $\overline{V_{1,J}}$ in $\mathbb{P}^n_{\mathbb{C}}$. Then, the Chow form $\Phi_{1,J}$ of $\overline{V_{1,J}}$ is a symmetric form in $n - |J|$ blocks of variables $(T_{\iota,0}, \dots, T_{\iota,n})$ that is homogeneous, with degree at most $d_1 \prod_{j\in J} d_j \le (d_1/d_0)\,\mathbf{d}(p)$, in each block. If τ is specified as an element in the complex space $\mathbb{C}^{n^2}$, then one has that

$$\Phi_{1,J}\big(\mathfrak{t}^{[j']} \,:\, j' \notin J \cup \{1\}\big) \neq 0 \Longrightarrow V_{1,J} \cap \Big(\bigcap_{j'\notin J\cup\{1\}} \{\zeta \in \mathbb{C}^n \,:\, 1 + \langle \mathfrak{t}^{[j']}, \zeta\rangle = 0\}\Big) = \emptyset. \tag{7.56}$$

It follows from [**KPS01**, Proposition 4.5] (see also [**SomY21**, Proposition 5.7]) that for any subset $J \subset \{0,2,\dots,n\}$, with cardinal $0 \le |J| \le n$, there exists a polynomial

$$\Psi_J \in \mathbb{C}\big[\mathfrak{t}_k^{[j']} \,:\, j' \in \{0,2,\dots,n\} \setminus J,\ k = 1,\dots,n\}\big]$$

satisfying

$$\max_{j'\in\{0,2,\dots,n\}\setminus J} \deg_{\mathrm{t}^{[j']}} \Psi_J \leq 2\,(\mathbf{d}(p))^2$$

and such that if τ is again specified as an element in the complex space $\mathbb{C}^{n^2}$, then one has

$$\begin{aligned}(7.57)\qquad &\Psi_J\big(\mathrm{t}^{[j']}\,:\, j'\notin J\cup\{1\}\big)\neq 0\\ &\Longrightarrow \operatorname{card}\Big(\bigcap_{j\in J} V(p_j)\cap\Big(\bigcap_{j'\notin J\cup\{1\}}\{\zeta\in\mathbb{C}^n\,:\,1+\langle \mathrm{t}^{[j']},\zeta\rangle=0\}\Big)\Big)\\ &=\deg\Big(\bigcap_{j\in J} V(p_j)\Big).\end{aligned}$$

The assertion of the proposition follows if one takes

$$\Phi=\Big(\prod_{\{J\subset\{0,2,\dots,n\}\,:\,\operatorname{card} J<n\}}\Phi_{1,J}\Big)\times\Big(\prod_{J\subset\{0,2,\dots,n\}}\Psi_J\Big). \qquad\square$$

REMARK 7.24. It follows from the estimates (7.55) for the degree of Φ in each block $\mathrm{t}^{[j]}$, $j=0,2,\dots,n$, that one can specify τ as an element in $\mathbb{Z}^{n^2}$ with a control of logarithmic size. The logarithmic sizes of the entries of at least one $\tau\in\mathbb{Z}^{n^2}$ such that $\Phi(\tau)\neq 0$ are bounded by $n\log 2+\log(\mathbf{d}\,(p)\,(d_1/d_0+2\,\mathbf{d}\,(p))$. Since such specified τ satisfies $\Phi(\tau)\neq 0$, both assertions on the right-hand side of the implication (7.55) are satisfied.

COROLLARY 7.25. *Let Φ be as in Proposition* 7.23. *If τ is specified as an element of $\mathbb{C}^{n^2}$ such that $\Phi(\tau)\neq 0$ and the $\phi_j(\mathrm{t}^{[j]},X)\in\mathbf{A}[\mathrm{t},X]$ are defined as in* (7.50) *for $j=0,2,\dots,n$, then*

$$(7.58)\quad \{\zeta\in\mathbb{C}^n\,:\,p_1(\zeta)=\phi_{0,\nu}(\mathrm{t}^{[0]},\zeta)=\phi_{2,\nu}(\mathrm{t}^{[2]},\zeta)=\cdots=\phi_{n,\nu}(\mathrm{t}^{[n]},\zeta)=0\}=\emptyset,$$

for any $\nu\in\mathbb{N}^*$. Moreover, there exist strictly positive constants $C(\tau)$ and $c(\tau)$ such that, for any disjoint decomposition $J\cup J'=\{0,2,\dots,n\}$ one has that

$$(7.59)\qquad \zeta\in\bigcap_{j\in J}V(p_j)\ \text{and}\ \|\zeta\|\geq C(\tau)\Longrightarrow \max_{j'\in J'}\big|1+\langle\mathrm{t}^{[j']},\zeta\rangle\big|\geq c(\tau)\,\|\zeta\|.$$

PROOF. The first assertion follows from the first statement on the right-hand side of the implication (7.55). The second assertion follows from the second statement on the right-hand side of (7.55) since the Zariski closure of $\bigcap_{j\in J}V(p_j)$ in $\mathbb{P}^n_{\mathbb{C}}$ does not intersect in such a case the linear subspace

$$\Big\{[\zeta_0:\cdots:\zeta_n]\in\mathbb{P}^n_{\mathbb{C}}\,:\,\zeta_0=0,\ \zeta_0+\sum_{k=1}^n \mathrm{t}_k^{[j']}\,\zeta_k=0\quad\text{for } j\in J'\Big\}. \qquad\square$$

Corollary 7.25 puts us in a position to state what will be a key lemma in this subsection. Its proof relies on the Łojasievicz inequalities, also known in this algebraic context as *nullstellen inequalities.* Nullstellen inequalities were originally introduced by Dale Brownawell in [**Bro88**] from the local diophantine point of view. These inequalities have been studied extensively since, in view of their relationship with separation problems in analytic or algebraic geometry; see in particular [**JKS92, CyKT99, Ko88, Cy98, Cy05**].

PROPOSITION 7.26. *Let Φ be as in Proposition 7.23 and let τ be a set of parameters as in (7.53). This set of parameters is specified here as an element of $\mathbb{C}^{n^2}$ satisfying $\Phi(\tau) \neq 0$. For any integer ν, satisfying $\nu > \mathbf{d}(p)$, the polynomial map*

$$\zeta \in \mathbb{C}^n \longmapsto \big(\phi_{0,\nu}(\mathfrak{t}^{[0]}, \zeta), \phi_{2,\nu}(\mathfrak{t}^{[2]}, \zeta), \dots, \phi_{n,\nu}(\mathfrak{t}^{[n]}, \zeta)\big) \in \mathbb{C}^n$$

admits $\big(d_0 + \mathbf{d}(p), d_2 + \mathbf{d}(p), \dots, d_n + \mathbf{d}(p)\big)$ as a multi-exponent of properness with respect to the affine space $\mathbb{A}^n_{\mathbb{C}}$.[16]

PROOF. From now on,[17] we fix $0 < \varepsilon < \nu - \mathbf{d}(p)$. Sharp nullstellen inequalities, as formulated in [**Cy05**, Theorem 5.1], show that there exists $\mathfrak{c} \in\,]0,1]$ such that for any nonempty subset J of $\{0, 2, \dots, n\}$, one has that

$$\max_{j \in J} |p_j(z)| \geq \mathfrak{c} \left(\frac{\operatorname{dist}\big(z, \bigcap_{j\in J} V(p_j)\big)}{\max(1, \|z\|^2)}\right)^{\prod_{j\in J} d_j}. \tag{7.60}$$

In particular, if one takes $J = \{0, 2, \dots, n\}$, then it follows that there exists $\mathfrak{C} \geq 1$ such that

$$\forall\, z \in \mathbb{C}^n,\ \|z\| \geq \mathfrak{C} \Longrightarrow \max_{j\neq 1} |p_j(z)|\, \|z\|^{\mathbf{d}(p)+\varepsilon} \geq \mathfrak{c}.$$

Let $z \in \mathbb{C}^n$ satisfying $\|z\| \geq \mathfrak{C}$. Note that later on, we will need to re-enforce this constraint. Let us order in increasing size the n positive numbers $|p_j(z)|\, \|z\|^{\mathbf{d}(p)+\varepsilon}$ for $j = 1, \dots, n$, namely

$$|p_{j_1(z)}(z)|\, \|z\|^{\mathbf{d}(p)+\varepsilon} \leq |p_{j_2(z)}(z)|\, \|z\|^{\mathbf{d}(p)+\varepsilon} \leq \cdots \leq |p_{j_n(z)}(z)|\, \|z\|^{\mathbf{d}(p)+\varepsilon}. \tag{7.61}$$

If $|p_{j_1(z)}(z)|\, \|z\|^{\mathbf{d}(p)+\varepsilon} \geq \mathfrak{c}$, then one has

$$\max_{j\neq 1} |\phi_{j,\nu}(\mathfrak{t}^{[j]}, z)| \geq \mathfrak{c}(\tau)\, \|z\|^{\nu - \mathbf{d}(p) - \varepsilon}$$

for some $\mathfrak{c}(\tau) > 0$. If not, let

$$r(z) = \max\big\{\rho \in \{1, \dots, n\} \,:\, |p_{j_\rho}(z)|\, \|z\|^{\mathbf{d}(p)+\varepsilon} < \mathfrak{c}\big\}.$$

It follows from (7.60) that

$$\frac{\operatorname{dist}\big(z, \bigcap_{\rho=1}^{r(z)} V(p_{j_\rho})\big)}{\|z\|} \leq \|z\|^{-\varepsilon / \prod_{\rho=1}^{r(z)} d_{j_\rho(z)}}. \tag{7.62}$$

Let $\zeta_z \in \bigcap_{\rho=1}^{r(z)} V(p_{j_\rho})$ such that

$$\frac{\|z - \zeta_z\|}{\|z\|} \leq \|z\|^{-\varepsilon / \prod_{\rho=1}^{r(z)} d_{j_\rho(z)}}.$$

It follows from (7.59) that

$$\max_{r(z)+1 \leq \rho \leq n} \frac{\big|1 + \langle \mathfrak{t}^{[j_\rho]}, \zeta_z \rangle\big|}{\|z\|} \geq c(\tau) \frac{\|\zeta_z\|}{\|z\|} \geq c(\tau) - c(\tau) \frac{\|z - \zeta_z\|}{\|z\|} \geq \frac{3}{4}\, c(\tau),$$

provided $\|z\|$ is assumed to be large enough, though depending on $c(\tau)$, $C(\tau)$, $\mathfrak{c}$, the d_j's, and ε. Without loss of generality, such an assumption could have been

---

[16] See Definition 6.53 for what it means, for a $\mathbb{C}^n$-valued polynomial map on an n-dimensional algebraic subvariety $\mathbb{X}_{\mathbb{C}}$ of $\mathbb{C}^N$, to admit $(\delta_1, \dots, \delta_n)$ as a multi-exponent of properness with respect to $\mathbb{X}_{\mathbb{C}}$.

[17] Although the proof of this result is closely related to that of [**BY91**, Lemma 5.3] and relies, as in [**BY91**, §5], on the use of Nœther's preparation theorem, we present below a different and more direct approach to such types of results.

made from the beginning of our discussion. For any ρ between $r(z)+1$ and n, one also has that

$$\begin{aligned}&\big|\big(1+\langle \mathrm{t}^{[j_\rho]}, z\rangle\big) - \big(1+\langle \mathrm{t}^{[j_\rho]}, \zeta_z\rangle\big)\big| \\ &\qquad\qquad \leq \|\tau\|\,\|z-\zeta_z\| \leq \|\tau\|\,\|z\|^{-\varepsilon/\prod_{\rho=1}^{r(z)} d_{j_\rho(z)}}\,\|z\| \leq \frac{c(\tau)}{4}\,\|z\|,\end{aligned}$$

again assuming $\|z\|$ to be large enough and depending on $c(\tau)$, $\|\tau\|$, the d_j's as well as ε. Thus, one has for such z that

$$\max_{r(z)+1\leq\rho\leq n} \big|1+\langle \mathrm{t}^{[j_\rho]}, \zeta_z\rangle\big| \geq \frac{c(\tau)}{2}\,\|z\|.$$

Therefore

$$\max_{r(z)+1\leq\rho\leq n} |\phi_{j_\rho,\nu}(\mathrm{t}^{[j_\rho]}, z)| \geq c'(\tau,\varepsilon)\,\|z\|^{\nu-\mathbf{d}\,(p)-\varepsilon}.$$

Assuming that $\varepsilon>0$ is arbitrarily close to 0, this concludes the proof of the proposition, taking into account the definition of δ-properness for polynomial maps from $\mathbb{C}^n$ to $\mathbb{C}^n$ as formulated in Definition 6.53. □

Now, let $\Delta_{\nu_p}\in \mathbf{A}[\tau,X,Y]$ be the $(n+1)\times(n+1)$ determinant defined by

(7.63) $\Delta_{\nu_p}(\tau,X,Y) :=$

$$\begin{vmatrix} \phi_{0,\nu_p}(\mathrm{t}^{[0]},X) & p_1(X) & \phi_{2,\nu_p}(\mathrm{t}^{[2]},X) & \cdots & \phi_{n,\nu_p}(\mathrm{t}^{[n]},X) \\ H_{\phi_{0,\nu_p},1}(\mathrm{t}^{[0]},X,Y) & H_{p_1,1}(Y) & H_{\phi_{2,\nu_p},1}(\mathrm{t}^{[2]},X,Y) & \cdots & H_{\phi_{n,\nu_p},1}(\mathrm{t}^{[n]},X,Y) \\ \vdots & \vdots & \vdots & \vdots & \vdots \\ \vdots & \vdots & \vdots & \vdots & \vdots \\ \vdots & \vdots & \vdots & \vdots & \vdots \\ H_{\phi_{0,\nu_p},n}(\mathrm{t}^{[0]},X,Y) & H_{p_1,n}(Y) & H_{\phi_{2,\nu_p},n}(\mathrm{t}^{[2]},X,Y) & \cdots & H_{\phi_{n,\nu_p},n}(\mathrm{t}^{[n]},X,Y) \end{vmatrix},$$

where

$$\nu_p := (n+1)\,\mathbf{d}\,(p) + |d| \tag{7.64}$$

and the Hefer forms H_{0,ν_p}, H_{p_1}, H_{2,ν_p}, ..., H_{n,ν_p} have been defined in (7.51), (7.52). The following result is a realization of Bézout identity using Jacobi–Lagrange–Kronecker interpolation. It extends to the multivariate setting the result formulated in Corollary 1.39 in the univariate setting. When $\mathbf{A}=\mathbb{Z}$, such a result was stated in a nonparametric form as [**BY91**, Theorem 5.1].

THEOREM 7.27. *Let* $\mathbf{A}$, $\mathbb{K}$, $p=(p_0,\dots,p_n)$, $\tau=\{\mathrm{t}^{[j]}\}_{1\leq j,k\leq n}$, ν_p, *and* Δ_{ν_p} *in* $\mathbf{A}[\tau,X,Y]$ *be as above. Then, one has in* $\mathbb{K}(\tau)[X]$ *the parametric Bézout identity*

$$\begin{aligned} 1 &= -\mathrm{Res}_{\mathbb{K}(\tau)[Y]/\mathbb{K}(\tau)}\begin{bmatrix} \dfrac{1}{p_1(Y)}\,\Delta_{\nu_p}(\tau,X,Y)\,dY_1\wedge\cdots\wedge dY_n \\ \phi_{0,\nu_p}(\mathrm{t}^{[0]},Y), \phi_{2,\nu_p}(\mathrm{t}^{[2]},Y),\dots,\phi_{n,\nu_p}(\mathrm{t}^{[n]},Y)\end{bmatrix} \\ &= q_1(\tau,X)p_1(X) + \sum_{j\neq 1} q_j(\tau,X)\,\phi_{j,\nu_p}(\mathrm{t}^{[j]},X), \end{aligned} \tag{7.65}$$

where

$$\max_{j\neq 1}\big(\deg_X(q_1(\tau,\cdot)p_1), \deg_X(q_j(\tau,\cdot)\phi_{j,\nu_p}(\mathrm{t}^{[j]},\cdot))\big) \leq |d| + n(\nu_p-1). \tag{7.66}$$

PROOF. Let $\nu > \mathbf{d}(p)$ and $b_\nu(\tau, X, Y) \in \mathbf{A}[\tau, X, Y]$ be such that

$$H_{\phi_{0,\nu}}(\mathfrak{t}^{[0]}, z, \zeta) \wedge H_{\phi_{2,\nu}}(\mathfrak{t}^{[2]}, z, \zeta) \wedge \cdots \wedge H_{\phi_{n,\nu}}(\mathfrak{t}^{[n]}, z, \zeta) = b_\nu(\tau, z, \zeta) \bigwedge_{j=1}^{n} d\zeta_j.$$

Since $\deg_\zeta b_\nu \le \sum_{j\neq 1} d_j + n(\nu - 1)$, it follows from Proposition 7.26 and Theorem 6.56 (once τ has been specified on an element in $\mathbb{C}^{n^2} \setminus V(\Phi)$) that

(7.67)

$$\begin{aligned} &|\kappa|(\nu - \mathbf{d}(p)) - n\,\mathbf{d}(p) \\ &\quad > \sum_{j\neq 1} d_j \Longrightarrow \mathrm{Res}_{\mathbb{K}(\tau)[Y]/\mathbb{K}(\tau)} \left[\begin{matrix} b_\nu(\tau, X, Y)\, dY_1 \wedge \cdots \wedge dY_n \\ \phi_{0,\nu}^{\kappa_1+1}(\mathfrak{t}^{[0]}, Y), \phi_{2,\nu}^{\kappa_2+1}(\mathfrak{t}^{[2]}, Y), \ldots, \phi_{n,\nu}^{\kappa_n+1}(\mathfrak{t}^{[n]}, Y) \end{matrix} \right] \\ &\quad = 0 \text{ in } \mathbb{K}(\tau)[X]. \end{aligned}$$

If $\nu = \nu_p$, then one has in particular that

(7.68)

$$\begin{aligned} \kappa \neq (0, \ldots, 0) &\Longrightarrow |\kappa|(\nu_p - \mathbf{d}(p)) - n\,\mathbf{d}(p) \ge \nu_p - (n+1)\,\mathbf{d}(p) = |d| > \sum_{j\neq 1} d_j \\ &\Longrightarrow \mathrm{Res}_{\mathbb{K}(\tau)[Y]/\mathbb{K}(\tau)} \left[\begin{matrix} b_{\nu_p}(\tau, X, Y)\, dY_1 \wedge \cdots \wedge dY_n \\ \phi_{0,\nu_p}^{\kappa_1+1}(\mathfrak{t}^{[0]}, Y), \phi_{2,\nu}^{\kappa_2+1}(\mathfrak{t}^{[2]}, Y), \ldots, \phi_{n,\nu_p}^{\kappa_n+1}(\mathfrak{t}^{[n]}, Y) \end{matrix} \right] = 0 \end{aligned}$$

in $\mathbb{K}(\tau)[X]$, according to (7.67). It then follows from Theorem 6.28 that one has in $\mathbb{K}(t)[X]$ the equality of polynomials

$$1 = \mathrm{Res}_{\mathbb{K}(\tau)[Y]/\mathbb{K}(\tau)} \left[\begin{matrix} b_{\nu_p}(\tau, X, Y)\, dY_1 \wedge \cdots \wedge dY_n \\ \phi_{0,\nu_p}(\mathfrak{t}^{[0]}, Y), \phi_{2,\nu_p}(\mathfrak{t}^{[2]}, Y), \ldots, \phi_{n,\nu_p}(\mathfrak{t}^{[n]}, Y) \end{matrix} \right]. \tag{7.69}$$

It follows from (7.58) that one can rewrite (7.69) as

$$1 = \mathrm{Res}_{\mathbb{K}(\tau)[Y]/\mathbb{K}(\tau)} \left[\begin{matrix} \dfrac{1}{p_1(Y)} b_{\nu_p}(\tau, X, Y)\, p_1(Y)\, dY_1 \wedge \cdots \wedge dY_n \\ \phi_{0,\nu_p}(\mathfrak{t}^{[0]}, Y), \phi_{2,\nu_p}(\mathfrak{t}^{[2]}, Y), \ldots, \phi_{n,\nu_p}(\mathfrak{t}^{[n]}, Y) \end{matrix} \right]. \tag{7.70}$$

Since the $(n+1)\times(n+1)$ determinant $\Delta_{\nu_p}(\tau, X, Y)$ remains invariant after adding to the first line the linear combination of the n last rows with coefficients respectively $Y_1 - X_1, \ldots, Y_n - X_n$, it follows from the formal construction of local residue symbols in $\mathscr{O}_{\mathbb{C}^n, z}$ (where $z \in \bigcap_{j\neq 1} V(\phi_{\nu,j}(\mathfrak{t}^{[j]}, \cdot))$ as traces,[18] combined with the rule of Sarrus, that one can rewrite the algebraic identity (7.70) as (7.65) with degree estimates (7.66). □

EXAMPLE 7.28. Suppose that $\mathbf{A} = \mathbb{Z}$. If one specifies the set of parameters τ on an element of $\mathbb{Z}^{n^2} \setminus V(\Phi)$, where logarithmic sizes of the entries are controlled according to Remark 7.9, then the algebraic identity (7.65) provides, after lifting denominators, Bézout identity $\gamma = \sum_{j=0}^{n} \boldsymbol{q}_j p_j$ with $\gamma \in \mathbb{N}^*$, $\boldsymbol{q}_j \in \mathbb{Z}[X]$, and $\max_j \deg(\boldsymbol{q}_j p_j) \le |d| + n(\nu_p - 1)$. Since some (probably not sharp) control on the

[18] One needs to invoke here the fact that $(\omega^{\sharp}_{a,(0,\ldots,0)}(h))_T = 0$, when h belongs to $\sum_{j=1}^{n} f_j\, \mathbb{A}$ in the formal algebraic construction (6.29).

arithmetic complexity of residue symbols

$$\text{(7.71)} \quad \operatorname{Res}_{\mathbb{Q}[X]/\mathbb{Q}} \begin{bmatrix} \dfrac{q}{p_1}\, dY \\ L_0^{\nu_p} p_0, L_2^{\nu_p} p_2, \dots, L_n^{\nu_p} p_n \end{bmatrix}$$

$$= \operatorname{Res}_{\mathbb{Q}[Y]/\mathbb{Q}} \begin{bmatrix} \dfrac{q\Big(\prod_{j\neq 1} L_j\Big)^{\nu_p}}{p_1}\, dY \\ (L_0 p_0)^{\nu_p}, (L_2 p_2)^{\nu_p}, \dots, (L_n p_n)^{\nu_p} \end{bmatrix}$$

is provided according to the transformation law and the arithmetic nullstellensatz Theorem 6.70 or to the Perron generalized theorem, Theorem 7.17 (see §6.6.3 and §6.6.4) the approach described in the present subsection leads to an effective arithmetic nullstellensatz. One should mention that a primitive result [**BY91**, Theorem 5.1] conjectured the fact that Bézout's arithmetic intersection theorem controls the N-solvability of Bézout's identity in $\mathbb{Q}[X_1, \dots, X_n]$. We remark though that the logarithmic size estimates obtained there were far from being sharp[19] when compared to, for example, those specified in Theorem 6.70. The positive point here remains that the identity (7.65), when τ is specified in $\mathbb{Z}^{n^2}$ with $\Phi(\tau) \neq 0$, provides a formulation of Hilbert's nullstellensatz as a *closed formula in terms of algebraic residue symbols* $\operatorname{Res}_{\mathbb{Q}[X]/\mathbb{Q}}[\ ]$. One should compare this approach with the result stated as Theorem 7.21, although this last result does not seem to be able to be supported by an algorithmic process. It seems reasonable to guess that A. Cayley, C. Jacobi, L. Kronecker, and F. Macaulay could have been already aware of a kind of division-interpolation path towards an explicit solution of Bézout identity; see for example, Eugen Netto's monographs [**Ne, Ne1910**].

Let us conclude this subsection with a particular case, confirming the intimate connection between Macaulay determinants and resultant theory [**Mac**] using multivariate residue calculus. Proposition 7.29 is the most immediate extension of Corollary 1.39 from the univariate to the multivariate setting.

PROPOSITION 7.29. *Suppose that the homogenization $P_0, P_2, \dots, P_n$ of the polynomials p_j's define a regular sequence in the local ring $\mathbb{C}[[X_0, \dots, X_n]]$.*[20] *Moreover, assume that*

$$\text{(7.72)} \qquad \{\zeta_0 = 0\} \cap \bigcap_{j\neq 1} |\operatorname{div}_{\mathbb{P}^n_{\mathbb{C}}}(P_j)| = \emptyset.$$

[19]They were nevertheless sharp enough to conjecture that Bézout's geometric and arithmetic intersection theorems rule jointly the effective solvability of Bézout identity. A rough upper estimate for $\max(\log \gamma, h(\boldsymbol{q}_j))$ is expressed in terms of $\kappa(n)(\mathbf{d}\,(p))^2\, \mathbf{h}_{c(n)}(p)$, where the arithmetic markers $\mathbf{h}_{c(n)}(p)$ were introduced in (6.197). This follows from the size estimates in [**SomY21**, Theorem 6.7] in a straightforward way. Note that the logarithmic size estimates originally given in [**BY91**, Theorem 5.1] are improved substantially in this parametric presentation.

[20]This amounts to saying that $R_{n+1\,:\,d_0,\dots,d_n}[P_0, \dots, P_n] \in \mathbf{A}^*$, where the P_j's are the homogenization of the p_j's in $\mathbf{A}[X_0, \dots, X_n]$ and $R_{n+1\,:\,d_0,\dots,d_n}$ denotes the Macaulay resultant of $n+1$ homogeneous forms in $n+1$ variables with respective degrees $d_0, \dots, d_n$; see [**Mac1902, Mac**], [**GKZ**, Chapter 13, §1], or [**L**].

Then one has Bézout identity

(7.73)

$$1 = -\mathrm{Res}_{\mathbb{K}[X]/\mathbb{K}}\left[\begin{matrix}\dfrac{1}{p_1(Y)}\begin{vmatrix} p_0(X) & p_1(X) & \cdots & p_n(X) \\ H_{p_0,1}(X,Y) & H_{p_1,1}(X,Y) & \cdots & H_{p_n,1}(X,Y) \\ \vdots & \vdots & \vdots & \vdots \\ H_{p_0,n}(X,Y) & H_{p_1,n}(X,Y) & \cdots & H_{p_n,n}(X,Y)\end{vmatrix} dY \\ \\ p_0(Y), p_2(Y), \dots, p_n(Y)\end{matrix}\right]$$

$$= \sum_{j=0}^{n} q_j(X) p_j(X),$$

and the following degree estimates, expressed in terms of the so-called Macaulay exponent $|d| - (n+1) + 1 = |d| - n$,

$$\max_{0 \le j \le n} \deg q_j p_j \le |d| - n \tag{7.74}$$

hold.

PROOF. Proposition 6.40 implies that the map $(p_0, p_2, \dots, p_n)$ admits $(0, \dots, 0)$ as a multiproperness exponent. It follows from Theorem 6.28 and Theorem 6.56 that the identity

$$\begin{aligned} 1 &= \mathrm{Res}_{\mathbb{K}[Y]/\mathbb{K}}\left[\begin{matrix} b_0(X,Y)\,dX \\ p_0(Y), p_2(Y), \dots, p_n(Y)\end{matrix}\right] \\ &= \mathrm{Res}_{\mathbb{K}[Y]/\mathbb{K}}\left[\begin{matrix} \dfrac{1}{p_1(Y)}\, b_0(X,Y) p_1(Y)\,dX \\ p_0(Y), p_2(Y), \dots, p_n(Y)\end{matrix}\right], \end{aligned} \tag{7.75}$$

where $(H_{p_0} \wedge H_{p_2} \wedge \cdots \wedge H_{p_n})(z,\zeta) = b_0(z,\zeta)\,d\zeta$ holds in $\mathbb{K}[X]$. The reformulation of the identity (7.75) as (7.73) relies on the same argument as that used to deduce the identity (7.70) from (7.65), with straightforward degree estimates (7.74). □

7.2.2. Parametric membership identities of the JLK form. We consider in this subsection the case of the sequence $p = (p_1, \dots, p_m)$ when $m \le n$. We still let $|d| = \sum_{j=1}^m \deg p_j$. As before, one introduces the multiparameter

$$\tau = \left\{ \mathrm{t}_k^{[j]} \,:\, j = 1, \dots, n,\ k = 1, \dots, n \right\}, \tag{7.76}$$

whose length equals n^2, and we let $\mathrm{t}^{[j]} = (\mathrm{t}_1^{[j]}, \dots, \mathrm{t}_n^{[j]})$. Given $\nu \in \mathbb{N}^*$, one extends the definition of the $\phi_{j,\nu}(\mathrm{t}, X) \in \mathbf{A}[\mathrm{t}, X]$, where $\mathrm{t} = \{\mathrm{t}_1, \dots, \mathrm{t}_n\}$, as given in (7.50) when $1 \le j \le m$, by letting

$$\phi_{j,\nu}(\mathrm{t}, X) = \begin{cases} \left(1 + \langle \mathrm{t}, X\rangle^{\nu} p_j(X)\right. & \text{for } j = 1, \dots, m, \\ \left(1 + \langle \mathrm{t}, X\rangle^{\nu}\right. & \text{for } m < j \le n. \end{cases} \tag{7.77}$$

Then we extend the definition of the Hefer forms for $m < j \le n$ by

$$H_{\phi_{j,\nu}}(\mathrm{t}, z, \zeta) = \Big(\sum_{\rho=0}^{\nu-1} \left(1 + \langle \mathrm{t}, z\rangle\right)^{\nu} \left(1 + \langle \mathrm{t}, \zeta\rangle\right)^{\nu-1-\rho} \Big) \langle \mathrm{t}, d\zeta\rangle. \tag{7.78}$$

For $j=1,\dots,m$ we consider the Hefer forms $H_{\phi_{\nu,j}}(\mathrm{t},z,\zeta)$ introduced in (7.52). Let also

$$\mathfrak{D}_\nu(\tau,X) := \Big(\prod_{j=1}^{m}(1+\langle \mathrm{t}^{[j]},X\rangle)\Big)^{\nu} \tag{7.79}$$

and let

$$b_\nu(\tau,X,Y) = \sum_{\{\ell\in\mathbb{N}^n\,:\,|\kappa|\le|d|+(n-1)\nu\}} Y^{\ell} b_{\nu,\ell}(\nu,X) \tag{7.80}$$

be the polynomial with degree in X,Y

$$\deg_{X,Y} b_\nu(\tau,X,Y) = |d| + n\,\nu - n$$

defined using the product of Hefer forms

$$H_{\phi_{1,\nu}}(\mathrm{t}^{[1]},z,\zeta)\wedge\cdots\wedge H_{\phi_{n,\nu}}(\mathrm{t}^{[n]},z,\zeta) = b_\nu(\tau,z,\zeta)\bigwedge_{j=1}^{n} d\zeta_j. \tag{7.81}$$

Since the supports of the divisors $\mathrm{div}_{\mathbb{C}^n}(p_j)$ intersect properly in the affine space $\mathbb{C}^n$ (see (7.49)) E. Nœther's preparation theorem (see for example [**L, VdW**]) admits now the following reformulation.

PROPOSITION 7.30. *Let τ be as in (7.76). There exists a polynomial $\Psi\in\mathbb{C}[\tau]$, whose degree satisfies*

$$\max_{j=0,2,\dots,n} \deg_{\mathrm{t}^{[j]}} \Psi \le 2^{m+1}(\mathbf{d}(p))^2, \tag{7.82}$$

such that, if τ is specified as an element in $\mathbb{C}^{n^2}$, then

$$\begin{aligned}
&\Psi(\tau)\neq 0\\
&\quad\Longrightarrow \Big|\Big(\bigcap_{j\in J} V(p_j)\cap\Big(\bigcap_{j'\in\{1,\dots,n\}\setminus J}\{\zeta\in\mathbb{C}^n\,:\,1+\langle \mathrm{t}^{[j']},\zeta\rangle=0\}\Big)\Big)\Big|\\
&\quad= \deg\Big(\bigcap_{j\in J} V(p_j)\Big)
\end{aligned} \tag{7.83}$$

for any $J\subset\{1,\dots,m\}$, with multiplicities taken into account.

PROOF. The argument of the proof relies on [**KPS01**, Proposition 4.5], as the construction of the polynomial $\prod_{J\subset\{0,2,\dots,n\}}\Psi_J$ in the proof of Proposition 7.23 does. □

REMARK 7.31. It follows from the estimates (7.82) for the degree of Ψ in each block $\mathrm{t}^{[j]}$ for $j=1,\dots,n$, that one can specify τ as an element in $\mathbb{Z}^{n^2}$ with a control of logarithmic size. The logarithmic sizes of the entries of at least one $\tau\in\mathbb{Z}^{n^2}$ such that $\Psi(\tau)\neq 0$ are bounded by $(m+1)\log 2 + 2\log \mathbf{d}(p)$. Both assertions on the right-hand side of the implication (7.55) are satisfied for such specialization of τ such that $\Psi(\tau)\neq 0$.

COROLLARY 7.32. *Let the parameter τ be specified on an element of $\mathbb{C}^{n^2}$ such that $\Psi(\tau)\neq 0$. There exist strictly positive constants $C(\tau)$ and $c(\tau)$ such that for any subset $J\subset\{1,\dots,m\}$,*

$$\zeta\in\bigcap_{j\in J}V(p_j) \text{ and } \|\zeta\|\ge C(\tau) \Longrightarrow \max_{j'\in\{1,\dots,n\}\setminus J}\big|1+\langle \mathrm{t}^{[j']},\zeta\rangle\big| \ge c(\tau)\,\|\zeta\|. \tag{7.84}$$

PROOF. It relies on the same argument as the proof of assertion (7.59) in Corollary 7.25 does. □

One can transpose in such a new context Proposition 7.26 as follows.

PROPOSITION 7.33. *Let the multiparameter τ, as in (7.76), be specified as an element of $\mathbb{C}^{n^2}$ satisfying $\Psi(\tau) \neq 0$, where Ψ is given by Proposition 7.30. For any integer ν such that $\nu > \mathbf{d}(p)$, the polynomial map*

$$\zeta \in \mathbb{C}^n \longmapsto \big(\phi_{1,\nu}(\mathfrak{t}^{[1]}, \zeta), \dots, \phi_{n,\nu}(\mathfrak{t}^{[n]}, \zeta)\big) \in \mathbb{C}^n$$

admits $\big(d_1 + \mathbf{d}(p), \dots, d_m + \mathbf{d}(p), \mathbf{d}(p), \dots, \mathbf{d}(p)\big)$ as a multi-exponent of properness with respect to the affine space $\mathbb{A}^n_{\mathbb{C}}$.

PROOF. Fix from now on[21] $0 < \varepsilon < \nu - \mathbf{d}(p)$. Sharp nullstellen inequalities, as formulated in [**Cy05**, Theorem 5.1], show that there exists $\mathfrak{c} \in]0,1]$ such that one has

$$\max_{j \in J} |p_j(z)| \geq \mathfrak{c} \Big(\frac{\operatorname{dist}\big(z, \bigcap_{j\in J} V(p_j)\big)}{\max(1, \|z\|^2)}\Big)^{\prod_{j\in J} d_j} \tag{7.85}$$

for any subset $J \subset \{1, \dots, n\}$. Moreover, for any $z \in \mathbb{C}^n$ the relations

$$\|z\| \geq \mathfrak{C} \Longrightarrow \max\big(\max_{1\leq j\leq m} |p_j(z)|\, \|z\|^{\mathbf{d}(p)+\varepsilon}, \max_{j>m} \big|1 + \langle \mathfrak{t}^{[j]}, z\rangle\big|\, \|z\|^{\mathbf{d}(p)+\varepsilon}\big) \geq \mathfrak{c} \tag{7.86}$$

also hold. Let $z \in \mathbb{C}^n$ such that $\|z\| \geq \mathfrak{C}$. Note that we will need to re-inforce this constraint later. We now distinguish two cases.

Suppose first that

$$\max_{j>m} \big|1 + \langle \mathfrak{t}^{[j]}, z\rangle\big|\, \|z\|^{\mathbf{d}(p)+\varepsilon} \geq \mathfrak{c}. \tag{7.87}$$

Either $\max_{1\leq j\leq m} |p_j(z)|\, \|z\|^{\mathbf{d}(p)+\varepsilon} < \mathfrak{c}$ or $\max_{1\leq j\leq m} |p_j(z)|\, \|z\|^{\mathbf{d}(p)+\varepsilon} \geq \mathfrak{c}$.

- If $\max_{1\leq j\leq m} |p_j(z)|\, \|z\|^{\mathbf{d}(p)+\varepsilon} < \mathfrak{c}$, then it follows from the estimate (7.85) with $J = \{1, \dots, m\}$ that

$$\frac{\operatorname{dist}\big(z, \bigcap_{j=1}^m V(p_j)\big)}{\|z\|} \leq \|z\|^{-\varepsilon/\mathbf{d}(p)}. \tag{7.88}$$

  Let $\zeta_z \in \bigcap_{j=1}^m V(p_j)$ be such that $\|z - \zeta_z\| = \operatorname{dist}\big(z, \bigcap_{j=1}^m V(p_j)\big)$. Since

$$\max_{j>m} \big|1 + \langle \mathfrak{t}^{[j]}, \zeta_z\rangle\big| \geq c(\tau)\, \|\zeta_z\|$$

  according to (7.84), the inequality

$$\max_{j>m} \big|1 + \langle \mathfrak{t}^{[j]}, z\rangle\big| \geq \frac{c(\tau)}{2}\, \|\zeta_z\|$$

  also holds, in view of (7.88), provided the hypothesis $\|z\| \geq \mathfrak{C}$ is re-inforced as $|z| \geq \mathfrak{C}_{\tau,\varepsilon} \geq \mathfrak{C}$ sufficiently large. Therefore, one has

$$\max_{j>m} \big|\phi_{\nu,j}(\mathfrak{t}^{[j]}, z)\big| \geq (\mathfrak{c}(\tau)\, \mathfrak{c})^{\nu} \|z\|^{\nu - \mathbf{d}(p) - \varepsilon} = \mathfrak{c}'(\tau)\, \|z\|^{\nu - \mathbf{d}(p)-\varepsilon} \tag{7.89}$$

  according to (7.87).

[21] Although the proof of the proposition is closely related to that of [**El93**, Lemma 2.2] (see also [**BGVY**, Proposition 5.8]) and relies (as in [**BY91**, §5]) on the use of Nœther's preparation theorem, we present a different and more direct approach.

- If $\max_{1\le j\le m}|p_j(z)|\,\|z\|^{\mathbf{d}(p)+\varepsilon}\ge \mathfrak{c}$, then let us order in increasing size the m positive numbers $|p_j(z)|\,\|z\|^{\mathbf{d}(p)+\varepsilon}$, namely

$$|p_{j_1(z)}(z)|\,\|z\|^{\mathbf{d}(p)+\varepsilon}\le |p_{j_2(z)}(z)|\,\|z\|^{\mathbf{d}(p)+\varepsilon}\le\cdots\le |p_{j_m(z)}(z)|\,\|z\|^{\mathbf{d}(p)+\varepsilon}. \tag{7.90}$$

One repeats now the argument, already used in the proof of Proposition 7.26, starting from the reordering (7.90) instead of (7.61). Let

$$r(z)=\max\big\{\rho\in\{1,\dots,m\}\,:\,|p_{j_\rho}(z)|\,\|z\|^{\mathbf{d}(p)+\varepsilon}<\mathfrak{c}\big\}.$$

It follows from (7.85) that

$$\frac{\operatorname{dist}\big(z,\bigcap_{\rho=1}^{r(z)}V(p_{j_\rho})\big)}{\|z\|}\le \|z\|^{-\varepsilon/\prod_{\rho=1}^{r(z)}d_{j_\rho}}\le \|z\|^{-\varepsilon/\mathbf{d}(p)}. \tag{7.91}$$

Let again $\zeta_z\in\bigcap_{\rho=1}^{r}V(p_{j_\rho})$ be such that $\|z-\zeta_z\|=\operatorname{dist}\big(z,\bigcap_{\rho=1}^{r}V(p_{j_\rho})\big)$. Since

$$\max_{j'\in\{1,\dots,n\}\setminus\{j_1,\dots,j_{r(z)}\}}\big|1+\langle \mathrm{t}^{[j']},\zeta_z\rangle\big|\ge c(\tau)\,\|\zeta_z\| \tag{7.92}$$

according once again to (7.84), it follows from (7.91) that the inequality

$$\max_{j'\in\{1,\dots,n\}\setminus\{j_1,\dots,j_r\}}\big|1+\langle \mathrm{t}^{[j']},z\rangle\big|\ge \frac{c(\tau)}{2}\,\|z\|$$

also holds, provided, once again, that the initial constraint $\|z\|\ge\mathfrak{C}$ is re-inforced as $|z|\ge \mathfrak{C}_{\tau,\varepsilon}\ge\mathfrak{C}$ sufficiently large. Then, taking into account the definition of $r(z)$, one has that

$$\max_{j'\in\{1,\dots,n\}\setminus\{j_1,\dots,j_{r(z)}\}}\big|\phi_{\nu,j'}(\mathrm{t}^{[j']},z)\big|\ge \mathfrak{c}'(\tau)\,\|z\|^{\nu-\mathbf{d}(p)-\varepsilon}. \tag{7.93}$$

Suppose now that we are in the second case, namely, we assume that

$$\max_{j>m}\big|1+\langle \mathrm{t}^{j},z\rangle\big|\,\|z\|^{\mathbf{d}(p)+\varepsilon}<\mathfrak{c}$$

holds in place of (7.87). This implies according to (7.86) that

$$\max_{1\le j\le m}|p_j(z)|\,\|z\|^{\mathbf{d}(p)+\varepsilon}\ge\mathfrak{c}.$$

From now on, we repeat once more the argument used to treat the situation in the second subcase above or that we invoked when proving Proposition 7.26. It leads this time, provided the constraint on $|z|\gg 1$ is re-inforced, to the lower estimate

$$\max_{1\le j\le m}|\phi_{j,\nu}(\mathrm{t}^{[j]},z)|\ge c''(\tau)\,\|z\|^{\nu-\mathbf{d}(p)-\varepsilon}. \tag{7.94}$$

Since the inequalities (7.89), (7.93), or (7.94) are satisfied by z for $\|z\|\ge\mathfrak{C}_{\tau,\varepsilon}$ (depending on the situation among the three we analyzed) and since ε is arbitrarily small, Proposition 7.33 is proved. □

We may now state, as a consequence of Proposition 7.33, a parametric realization with denominators of the membership of $q\in\mathbf{A}[X]$ to the polynomial ideal $(p_1,\dots,p_m)\,\mathbb{K}[X]$.[22]

[22] A nonparametric version of such result (with $M=3$) was first proved by Mohamed Elkadi [**El93**, Theorem 3.1] within the arithmetic setting where $\mathbf{A}=\mathbb{Z}$.

THEOREM 7.34. *Let $m \leq n$ and $p_1, \dots, p_m \in \mathbf{A}[X] = \mathbf{A}[X_1, \dots, X_n]$ be such that the supports of the divisors $\mathrm{div}_{\mathbb{C}^n}(p_j)$ intersect properly in $\mathbb{C}^n$, according to (7.49). Assume also that $q \in \mathbf{A}[X] \cap (p_1, \dots, p_m)\,\mathbb{K}[X]$. Fix $M \in \mathbb{N}^*$ with $M \geq 2$ and let*

(7.95)

$$\nu := \nu_p(M) = M\,\mathbf{d}\,(p), \quad K = K_{p,q}(M) := \frac{\deg q + |d| + n\,\mathbf{d}\,(p)}{M-1} + m\frac{M}{M-1}\,\mathbf{d}\,(p),$$

and let τ be the set of n^2 parameters as in (7.76). Then the following algebraic identity holds in $\mathbb{K}(\tau, X)$:

(7.96)

$$\begin{aligned} q(X) &= \frac{1}{\mathfrak{D}_\nu(\tau, X)} \sum_{\{\kappa \in \mathbb{N}^n \,:\, |\kappa|\,\mathbf{d}\,(p) \leq K_{p,q},\ \min_{1\leq j\leq m} \kappa_j \geq 1\}} \\ &\qquad \mathrm{Res}_{\mathbb{K}(\tau)[Y]/\mathbb{K}(\tau)} \left[\begin{matrix} q(Y)\,\mathfrak{D}_\nu(\tau, Y)\, b_\nu(\tau, X, Y)\, dY \\ \phi_{1,\nu}^{\kappa_1+1}(\mathfrak{t}^{[1]}, Y), \dots, \phi_{n,\nu}^{\kappa_n+1}(\mathfrak{t}^{[n]}, Y) \end{matrix} \right] \prod_{j=1}^n \phi_{j,\nu}^{\kappa_j}(\mathfrak{t}^{[j]}, X) \\ &= \frac{1}{\mathfrak{D}_\nu(\tau, X)} \sum_{j=1}^m q_j(\tau, X)\, p_j(X). \end{aligned}$$

The polynomials q_j in (7.96) belong to $\mathbb{K}(\tau)[X]$ and satisfy the degree estimates

$$\max_{1\leq j\leq m} \deg_X (q_j(\tau, X) p_j(X)) \leq \frac{M+1}{M-1}\Big(\deg q + |d| + (mM + n)\,\mathbf{d}\,(p)\Big). \tag{7.97}$$

PROOF. It follows Proposition 7.33 and Theorem 6.56, applied once τ is specified as an element in $\mathbb{C}^{n^2} \setminus V(\Psi)$, that

$$\begin{aligned} & |\kappa|\,\mathbf{d}\,(p) > K_{p,q} \\ & \iff (M-1)(|\kappa| + n)\,\mathbf{d}\,(p) > \deg q + |d| + M\,(m+n)\,\mathbf{d}\,(p) \\ & \implies (M-1)(|\kappa| + n)\,\mathbf{d}\,(p) > \deg_Y \Big(q(Y)\mathfrak{D}_\nu(\tau, Y)\, b_\nu(\tau, X, Y)\Big) + n \\ & \implies \mathrm{Res}_{\mathbb{K}(\tau)[Y]/\mathbb{K}(\tau)} \left[\begin{matrix} q(Y)\,\mathfrak{D}_\nu(\tau, Y)\, b_\nu(\tau, X, Y)\, dY \\ \phi_{1,\nu}^{\kappa_1+1}(\mathfrak{t}^{[1]}, Y), \dots, \phi_{n,\nu}^{\kappa_n+1}(\mathfrak{t}^{[n]}, Y) \end{matrix} \right] = 0, \end{aligned} \tag{7.98}$$

where $dY = \bigwedge_{k=1}^n dY_k$. Furthermore, Theorem 6.28 implies that one has the polynomial identity

$$\begin{aligned} q(X)\,\mathfrak{D}_\nu(\tau, X) &= \sum_{\{\kappa \in \mathbb{N}^n \,:\, |\kappa|\,\mathbf{d}\,(p) \leq K_{p,q}\}} \\ &\qquad \mathrm{Res}_{\mathbb{K}(\tau)[Y]/\mathbb{K}(\tau)} \left[\begin{matrix} q(Y)\,\mathfrak{D}_\nu(\tau, Y)\, b_\nu(\tau, X, Y)\, dY \\ \phi_{1,\nu}^{\kappa_1+1}(\mathfrak{t}^{[1]}, Y), \dots, \phi_{n,\nu}^{\kappa_n+1}(\mathfrak{t}^{[n]}, Y) \end{matrix} \right] \prod_{j=1}^n \phi_{j,\nu}^{\kappa_j}(\mathfrak{t}^{[j]}, X) \end{aligned} \tag{7.99}$$

in $\mathbb{K}(\tau)[X]$. The formal construction of algebraic symbols as traces in (6.29), namely the fact that $(\omega^\sharp_{a,(0,\dots,0)}(h))_T = 0$, when h belongs to $\sum_{j=1}^n f_j\, \mathbb{A}$, implies that for any $\kappa \in \mathbb{N}^n$ such that $\kappa_1 = \cdots = \kappa_m = 0$,

$$\mathrm{Res}_{\mathbb{K}(t)[Y]/\mathbb{K}(t)} \left[\begin{matrix} q(Y)\,\mathfrak{D}_\nu(\tau, Y)\, b_\nu(\tau, X, Y)\, dY \\ \phi_{1,\nu}^{\kappa_1+1}(\mathfrak{t}^{[1]}, Y), \dots, \phi_{n,\nu}^{\kappa_n+1}(\mathfrak{t}^{[n]}, Y) \end{matrix} \right] = 0,$$

since the polynomial $q(X)\mathfrak{D}_\nu(\tau, X)$ lies in the ideal generated by the polynomials $\phi_{1,\nu}(\tau, X), \dots, \phi_{m,\nu}(\tau, X)$ in $\mathbb{K}(\tau)[X]$. The algebraic identity (7.96) then follows.

Let us split $b_\nu(\tau, X, Y)$ as in (7.80) and consider for $\kappa \in \mathbb{N}^n$ with $\min_{1\le j\le m} \kappa_j \ge 1$ the contribution
(7.100)
$$b_{\nu,\ell}(\tau, X)\mathrm{Res}_{\mathbb{K}(t)[Y]/\mathbb{K}(t)}\left[\frac{q(Y)\,\mathfrak{D}_\nu(\tau,Y)\,Y^\ell\,dY}{\phi_{1,\nu}^{\kappa_1+1}(\mathrm{t}^{[1]},Y),\dots,\phi_{n,\nu}^{\kappa_n+1}(\mathrm{t}^{[n]},Y)}\right]\prod_{j=1}^{n}\phi_{j,\nu}^{\kappa_j}(\mathrm{t}^{[j]},X)$$

in the numerator of the right-hand side of the algebraic identity (7.96). Since $\min_{1\le j\le m}\kappa_j \ge 1$, such a contribution splits as $\sum_{j=1}^m q_{j,\ell}(\tau,X)p_j(X)$. It remains then to estimate $\max_{1\le j\le m}\deg_X(q_{j,\ell}(\tau,X)p_j(X))$. Observe first that such a contribution (7.100) equals 0 unless

$$|\kappa|\,\mathbf{d}\,(p) \le \frac{\deg q + n\,\mathbf{d}\,(p) + |\ell| + M(m-n)\,\mathbf{d}\,(p) + n}{M-1}, \tag{7.101}$$

according to Theorem 6.56, which we assume from now on. Then relation (7.101) implies

(7.102)
$$\begin{aligned}\deg_X \prod_{j=1}^{n}\phi_{j,\nu}^{\kappa_j}(\mathrm{t}^{[j]},X) &\le |\kappa|\,(M\,\mathbf{d}\,(p)+d_m) \le |\kappa|\,\mathbf{d}\,(p)\Big(M+\frac{1}{\prod_1^{m-1}d_j}\Big)\\ &\le \frac{M+1}{M-1}\big(\deg q + n\,\mathbf{d}\,(p) + |\ell| + M(m-n)\,\mathbf{d}\,(p) + n\big).\end{aligned}$$

On the other hand

$$\deg_X b_{\nu,\ell}(\tau,X) \le |d| + n\big(M\,\mathbf{d}\,(p) - 1\big) - |\ell|. \tag{7.103}$$

It follows from the estimates (7.102) and (7.103) that

$$\begin{aligned}&\max_{1\le j\le n}\deg_X\big(q_{j,\ell}(\tau,X)p_j(X)\big)\\ &\quad\le |d| + \frac{M+1}{M-1}\deg q + \frac{2}{M-1}(|\ell|+n) + \Big(m\,M\,\frac{M+1}{M-1} - n\Big)\,\mathbf{d}\,(p)\\ &\quad\le \frac{M+1}{M-1}\Big(\deg q + |d| + (mM+n)\,\mathbf{d}\,(p)\Big),\end{aligned}$$

since $|\ell|+n \le |d|+nM\,\mathbf{d}\,(p)$. This provides the required estimates (7.97), and thus concludes the proof of Theorem 7.34. □

Consider now $n+1$ sets of parameters $\tau_0,\dots,\tau_n$ of the form

$$\tau_\rho = \Big\{\mathrm{t}_{\rho,k}^{[j]} : j=1,\dots,n,\ k=1,\dots,n\Big\},\quad \rho=0,\dots,n, \tag{7.104}$$

such that $T=\{\tau_0,\dots,\tau_n\}$ is a transcendental basis of $\mathbb{K}(T)$.

Let $R_{n+1\,:\,m,\dots,m}$ be the Macaulay resultant of $n+1$ homogeneous forms in $n+1$ variables with respective degrees $m,\dots,m$. Let also $R_{n\,:\,m,\dots,m}$ be the Macaulay resultant of n homogeneous forms in n variables with corresponding degrees $m,\dots,m$. We recall that $R_{n+1\,:m,\dots,m}$ (respectively $R_{n\,:\,m,\dots,m}$) is a multi-homogeneous polynomial with integer coefficients in $(n+1)$ (respectively n) blocks of variables. All blocks of variables have length $\binom{n+m}{m}$ (respectively $\binom{n-1+m}{m}$). The polynomials

$R_{n+1\,:\,m,\dots,m}$ (respectively $R_{n\,:\,m,\dots,m}$) are homogeneous with degree m^n (respectively m^{n-1}) in each block. For each $\rho = 0,\dots,n$, let

$$\mathcal{H}_{\rho,\nu}(T,X,Y) = \sum_{k=1}^{n} \mathcal{H}_{\rho,\nu,k}(T,X,Y)\,dY_k = \sum_{k=1}^{n} \mathcal{H}_{\rho,k}(\tau_\rho,X,Y)\,dY_k$$

be the Kähler 1-form with coefficients in $\mathbb{Z}[T,X,Y]$ corresponding to a Hefer algebraic form for

$$\zeta \longmapsto \mathcal{D}_\nu(\tau_\rho,\zeta).$$

It follows from Proposition 7.29 that Bézout identity

(7.105)

$$\begin{aligned} 1 &= -\mathrm{Res}_{\mathbb{Q}(T)[X]/\mathbb{Q}(T)} \\ &\left[\begin{matrix} \dfrac{1}{\mathfrak{D}_\nu(\tau_1,Y)} \begin{vmatrix} \mathfrak{D}_\nu(\tau_0,X) & \mathfrak{D}_\nu(\tau_1,X) & \cdots & \mathfrak{D}_\nu(\tau_n,X) \\ \mathcal{H}_{0,\nu,1}(T,X,Y) & \mathcal{H}_{1,\nu,1}(T,X,Y) & \cdots & \mathcal{H}_{n,\nu,1}(T,X,Y) \\ \vdots & \vdots & \vdots & \vdots \\ \mathcal{H}_{0,\nu,n}(T,X,Y) & \mathcal{H}_{1,\nu,n}(T,X,Y) & \cdots & \mathcal{H}_{n,\nu,n}(T,X,Y) \end{vmatrix} dY \\ \mathfrak{D}_\nu(\tau_0,Y), \mathfrak{D}_\nu(\tau_2,Y),\dots,\mathfrak{D}_\nu(\tau_n,Y) \end{matrix}\right] \\ &= \sum_{\rho=0}^{n} \mathcal{Q}_{\nu,\rho}(T,X)\,\mathfrak{D}_\nu(\tau_\rho,X), \end{aligned}$$

holds in $\mathbb{Q}(T)[X]$ for any $\nu \in \mathbb{N}^*$. Furthermore, Macaulay degree estimates

$$\max_{0\le\rho\le n} \left(\deg_X \left(\mathcal{Q}_{\nu,\rho}(T,X)\,\mathfrak{D}_\nu(\tau_\rho,X)\right)\right) \le (n+1)m\,\nu - n, \tag{7.106}$$

hold in $\mathbb{Q}(T)[X]$ for any $\nu \in \mathbb{N}^*$ also. Moreover, (7.105) provides a Bézout identity in $\mathbb{Q}(X)$, provided T is specified on an element of $\mathbb{Z}^{n^2(n+1)}$ such that

$$\begin{aligned} &R_{n\,:\,m,\dots,m}\left[(\mathfrak{D}_1(\tau_0,\cdot))_m, (\mathfrak{D}_1(\tau_2,\cdot))_m,\dots,(\mathfrak{D}_1(\tau_n,\cdot))_m\right] \neq 0, \\ &R_{n+1\,:\,m,\dots,m}[\mathfrak{D}_1^{\mathrm{hom}}(\tau_0,\cdot),\mathfrak{D}_1^{\mathrm{hom}}(\tau_1,\cdot),\dots,\mathfrak{D}_1^{\mathrm{hom}}(\tau_n,\cdot)] \neq 0, \end{aligned} \tag{7.107}$$

where the $(\mathfrak{D}_1(\tau_\rho,\cdot))_m$'s denote the homogeneous parts of higher degree m for the polynomials $\mathfrak{D}_1(\tau_\rho,X)$ considered as polynomials in X, and the $\mathfrak{D}_1^{\mathrm{hom}}(\tau_\rho,X)$'s denote their homogenization. We may now conclude this subsection with the following main result.

THEOREM 7.35. *Let $m \le n$ and let the polynomials $p_1,\dots,p_m \in \mathbf{A}[X]$ be such that the supports of the divisors $\mathrm{div}_{\mathbb{C}^n}(p_j)$ intersect properly in $\mathbb{C}^n$ as in (7.49). Let also $q \in \mathbf{A}[X] \cap (p_1,\dots,p_m)\,\mathbb{K}[X]$. Fix $M \in \{2,3,\dots,\}$ and let*

(7.108)

$$\nu := \nu_p(M) = M\,\mathbf{d}(p), \quad K = K_{p,q}(M) := \frac{\deg q + |d| + n\,\mathbf{d}(p)}{M-1} + m\frac{M}{M-1}\mathbf{d}(p).$$

If $T = \{\tau_0, \dots, \tau_n\}$ is a set of $n^2(n+1)$ parameters as in (7.104), then the following membership parametric identity holds in $\mathbb{K}(T)[X]$:

(7.109)

$$q(X) = \sum_{\rho=0}^{n} \mathcal{Q}_{\nu,\rho}(T,X) \sum_{\{\kappa \in \mathbb{N}^n \,:\, |\kappa|\, \mathbf{d}(p) \le K_{p,q},\ \min_{1\le j\le m} \kappa_j \ge 1\}} \mathrm{Res}_{\mathbb{K}(\tau)[Y]/\mathbb{K}(\tau)} \begin{bmatrix} q(Y)\,\mathfrak{D}_\nu(\tau_\rho, Y)\, b_\nu(\tau_\rho, X, Y)\, dY \\ \phi_{1,\nu}^{\kappa_1+1}(\mathrm{t}_{\rho,\cdot}^{[1]}, Y), \dots, \phi_{n,\nu}^{\kappa_n+1}(\mathrm{t}_{\rho,\cdot}^{[n]}, Y) \end{bmatrix} \prod_{j=1}^{n} \phi_{j,\nu}^{\kappa_j}(\mathrm{t}_{\rho,\cdot}^{[j]}, X)$$

$$= \sum_{\rho=0}^{n} \mathcal{Q}_{\nu,\rho}(T,X) \Big(\sum_{j=1}^{m} q_j(\tau_\rho, X)\, p_j(X) \Big) = \sum_{j=1}^{m} Q_j(T,X)\, p_j(X).$$

The polynomials $q_j(\tau_\rho, X) \in \mathbf{A}[\tau_\rho, X]$ and $\mathcal{Q}_{\nu,\rho}(T,X) \in \mathbb{Z}[T,X]$ are defined for $\rho = 0, \dots, n$ and $j = 1, \dots, n$ by (7.96) and (7.105), respectively. Moreover, the polynomials $Q_j(T,X) \in \mathbb{K}(T)[X]$ satisfy the degree estimates

$$\text{(7.110)} \quad \max_{1\le j\le m} \deg_X (Q_j(T,X) p_j(X)) \le \frac{M+1}{M-1} \Big(\deg q + |d| + (mM+n)\, \mathbf{d}(p) \Big) + (n+1)\, M\, \mathbf{d}(p) - n.$$

PROOF. The results follow from Bézout identity (7.105) in $\mathbb{Q}(T)[X]$, once combined with the parametric realization with denominators of the membership of q with respect to $(p_1, \dots, p_m)\, \mathbf{A}[X]$ given in Theorem 7.34 by (7.96). □

EXAMPLE 7.36. Let $\mathbf{A} = \mathbb{Z}$. The parameters T introduced in Theorem 7.35 can be specified on an element $T^\circ = (\tau_0, \dots, \tau_n) \in \mathbb{Z}^{n^2(n+1)}$ such that

$$\text{(7.111)} \quad \begin{aligned} &\Big(\prod_{\rho=0}^{n} \Psi(\tau_\rho) \Big) \neq 0 \\ &R_{n\,:\,m,\dots,m} \left[(\mathfrak{D}_1(\tau_0, \cdot))_m, (\mathfrak{D}_1(\tau_2, \cdot))_m, \dots, (\mathfrak{D}_1(\tau_n, \cdot))_m \right] \neq 0 \\ &R_{n+1\,:\,m,\dots,m} [\mathfrak{D}_1^{\mathrm{hom}}(\tau_0, \cdot), \mathfrak{D}_1^{\mathrm{hom}}(\tau_1, \cdot), \dots, \mathfrak{D}_1^{\mathrm{hom}}(\tau_n, \cdot)] \neq 0. \end{aligned}$$

Moreover, one deduces from the estimates for $\deg \Psi$ (see (7.82)) and from the fact that the degrees of $R_{n+1\,:\,m,\dots,m}$ and $R_{n\,:\,m,\dots,m}$ in each block of variables equal respectively m^n and m^{n-1}, an upper estimate for the logarithmic size of such specialization of T; see for example Remark 7.31. With T specialized as such an element in $\mathbb{Z}^{n^2(n+1)}$, (7.96) provides a formula

$$\gamma\, q(X) = \sum_{j=1}^{m} \boldsymbol{q}_j(X)\, p_j(X),$$

after lifting denominators, where $\gamma \in \mathbb{N}^*$ and $\boldsymbol{q}_j \in \mathbb{Z}[X]$ for $j = 1, \dots, m$. Thus, one could compare Theorem 7.35 to [**KP94**, Theorem 2] (in the case where $m = n$), which carries an algorithmic process materialized by the straight line program $\mathbb{SL}((p_1, \dots, p_m))$ on $\mathbb{Z}[X]$. Although Theorem 7.35, which produces after specialization of the parameters a closed formula for the membership problem in $\mathbb{Q}(X)$ in the case of proper intersections, is not accompanied by such an algorithmic process, it is worthwhile to observe that multivariate residue calculus (that is, the Jacobi–Lagrange–Kronecker approach) is deeply involved in the realization of the closed formula it produces. Estimates for $\log \gamma$ and $\mathrm{h}(\boldsymbol{q}_j)$ in terms of the geometric and arithmetic markers $\mathbf{d}\,(p)$ and $\mathbf{h}_c(p)$ follow from Theorem 6.72 or Theorem 6.76 (in the case where $N = n$ and $\mathbb{X} = \mathbb{A}^n_{\mathbb{Q}}$). One needs to use for that the observation (7.71). The quantities $\log \boldsymbol{\gamma}$ and the $\mathrm{h}\,(\boldsymbol{q}_j)$ are roughly estimated from above by $\kappa(n)\,(\mathbf{d}\,(p))^3\,\mathbf{h}_{c(m,n)}$. Such identities, together with logarithmic size estimates, are due to Mohamed Elkadi [**El93**, Theorem 3.1]. Bounds obtained in [**El93**, Theorem 3.1] can be improved in view of the results established since then. Namely, the estimates can be made sharper using the Perron parametric theorem (Theorem 7.17) and its arithmetic implications, once combined with the transformation law or its generalizations (Theorem 3.21 or Proposition 3.23). Nevertheless, they still remain probably far from optimal. The main point to insist on is the fact that Theorem 7.35, as Theorem 7.27 does for Bézout identity, leads to closed formulae while (almost) respecting jointly geometric and arithmetic intersection theory.

7.3. Effective geometric Briançon–Skoda–Huneke type theorems

7.3.1. Global algebraic Briançon–Skoda–Huneke exponents $\mu(\mathbb{X}:\mathbb{P}^N_{\mathbb{C}})$. It is assumed throughout this section that $\mathbb{X}$ is a purely n-dimensional algebraic subvariety of $\mathbb{C}^N$. Let $\mathbb{J} = \mathbb{J}(\mathbb{X})$ be its radical ideal in $\mathbb{C}[X_1,\dots,X_N]$ and let $\overline{\mathbb{X}}$ be its n-dimensional Zariski closure in $\mathbb{P}^N_{\mathbb{C}}$. Let also $\mathbb{I} = \mathbb{I}(\overline{\mathbb{X}})$ be the homogeneous radical ideal in the graded algebra $\mathbb{S} = \mathbb{C}[X_0,\dots,X_N] = \mathbb{C}[\boldsymbol{X}]$ satisfying

$$\Phi \in \mathbb{I} \iff \Phi(z) = 0 \quad \forall\, z = [z_0 : \cdots : z_N] \in \overline{\mathbb{X}}. \tag{7.112}$$

Following the presentation in §4.6.3, we consider a finite collection of homogeneous polynomials $\{\Phi_\iota\}_\iota$ in the graded algebra $\mathbb{S} = \mathbb{C}[X_0,\dots,X_N]$, with corresponding degrees D_ι and such that

$$\mathbb{I} = \sum_\iota \Phi_\iota\, \mathbb{C}[\boldsymbol{X}]. \tag{7.113}$$

Consider the mapping

$$\mathbb{F}_\Phi : \bigoplus_\iota \mathcal{Q}_\iota \in \bigoplus_\iota \mathbb{S}_{-D_\iota} \longmapsto \sum_\iota \mathcal{Q}_\iota\, \Phi_\iota \in \mathbb{S},$$

where the graded modules $\mathbb{S}_{-\delta}$ for $\delta \in \mathbb{N}$ are defined in (4.131). The $\mathbb{S}$-module $\mathbb{M}_{-1} = \mathbb{S}/\mathrm{Im}\,\mathbb{F}_\Phi$ is finitely generated and the complex of graded free $\mathbb{S}$-modules

(7.114)

$$\mathbb{M}^\Phi_\bullet : 0 \longmapsto \mathbb{M}_L \xrightarrow{\mathbb{F}_L} \cdots \xrightarrow{\mathbb{F}_{\ell+1}} \mathbb{M}_\ell \xrightarrow{\mathbb{F}_\ell} \mathbb{M}_{\ell-1} \xrightarrow{\mathbb{F}_{\ell-1}} \cdots \xrightarrow{\mathbb{F}_2} \mathbb{M}_1 \xrightarrow{\mathbb{F}_\Phi} \mathbb{S} \xrightarrow{\Pi} \mathbb{M}_{-1} \longrightarrow 0,$$

where

$$\mathbb{M}^\ell = \bigoplus_\iota \mathbb{S}_{-d_\iota^{[\ell]}}, \quad \ell = 1, 2, \dots, \tag{7.115}$$

and Π denotes the quotient map, is exact. Hilbert's syzygy theorem [**Hil1890**] (see also [**Eis05**, Chapter 1A]) implies that this complex has length $L \le N + 1$. Moreover, the complex (7.114) becomes unique to an isomorphism (see [**Eis05**, §1A and §1B]) if one takes it *minimal*, that is

$$\mathrm{Im}\,\mathbb{F}_\ell \subset \sum_{k=0}^N \sum_\iota X_k\, \mathbb{S}_{-d_\iota^{[\ell-1]}}$$

for any $\ell \geq 1$.[23] From now on the resolution (7.114) will be taken as minimal. Since $\mathbb{X}$ is purely dimensional, the depth of the quotient ring satisfies $\mathrm{depth}(\mathbb{C}[\boldsymbol{X}]/\mathbb{I}) \geq 1$ and the length L of such a minimal resolution in (7.114) is bounded from above by $N = (N+1) - 1$, improving the upper bound $N+1$ provided by Hilbert's syzygy theorem. Besides the marker $\deg \mathbb{X} = \deg \overline{\mathbb{X}} = \mathbf{d}$ for the *geometric complexity* of $\mathbb{X}$ (see §6.6.2) defined intrinsically and independently of the embedding $\mathbb{X} \hookrightarrow \mathbb{P}^N_{\mathbb{C}}$, one needs to introduce the *Castelnuovo–Mumford regularity* $r(\mathbb{X})$, more precisely the Castelnuovo–Mumford regularity of $\mathbb{X} \hookrightarrow \mathbb{P}^N_{\mathbb{C}}$. It is a geometric invariant of $\mathbb{X} \hookrightarrow \mathbb{P}^N_{\mathbb{C}}$, which depends on the embedding. On the other hand, $r(\mathbb{X})$ is a marker for the *algebraic complexity* of $\mathbb{X}$ as embedded in $\mathbb{P}^N_{\mathbb{C}}$. Let us recall here the definition of this geometric invariant of $\mathbb{X} \hookrightarrow \mathbb{P}^N_{\mathbb{C}}$.

DEFINITION 7.37.[24] The *Castelnuovo–Mumford regularity* $r(\mathbb{X})$ of $\mathbb{X} \subset \overline{\mathbb{X}} \hookrightarrow \mathbb{P}^N_{\mathbb{C}}$ is defined analogously to the regularity of the quotient of $\mathbb{S}$ by the homogeneous ideal $\mathbb{I}$. That is, given any set $\Phi = \{\Phi_\iota\}_\iota$ of generators for $\mathbb{I}$ in $\mathbb{C}[\boldsymbol{X}]$ and any minimal exact complex of graded free $\mathbb{S}$-modules $\mathbb{M} = \mathbb{M}^\Phi_\bullet$ for $\mathbb{C}[X_0, \dots, X_N]/\mathbb{I}$ as in (7.114),

$$\mathrm{reg}(\mathbb{X}) = \mathrm{reg}(\overline{\mathbb{X}}) := \mathrm{reg}\big(\mathbb{C}[X_0, \dots, X_N]/\mathbb{I}\big) + 1 = \max_{\ell,\iota}(d_\iota^{[\ell]} - \ell) + 1. \tag{7.116}$$

The degrees $d_\iota^{[\ell]}$ in (7.115) for $\ell = 1, 2, \dots$ are called the *graded Betti numbers* of $\mathbb{I}$ (or of $\mathbb{S}/\mathbb{I}$) and are independent of the minimal exact complex $\mathbb{M}$.

Both markers $\mathbf{d}$ and $\mathrm{reg}(\mathbb{X})$, defined for the geometric complexity of $\mathbb{X}$ (intrinsically defined) and for its algebraic complexity (when considered as embedded in $\mathbb{P}^N_{\mathbb{C}}$) respectively, will play a key role throughout §7.3. It is now time to come back to multidimensional residue theory. We proceed as follows. Consider the correspondences

$$\mathbb{M}_\ell = \bigoplus_\iota \mathbb{S}_{-d_\iota^{[\ell]}} \longleftrightarrow E_\ell = \bigoplus_\iota \mathcal{O}(-d_\iota^{[\ell]})\, \epsilon_\iota^{[\ell]*}, \quad \ell = 1, 2, \dots,$$

where the notation $\{\epsilon_\iota^{[\ell]}\}_\iota$ denotes, as usual, a copy of the canonical basis of $\mathbb{Z}^{\mathrm{rank}\,\mathbb{M}_\ell}$, indexed by ℓ. We construct, as in §4.6.3, from $\mathbb{M}^\Phi_\bullet$ a generically exact complex of holomorphic bundles

(7.117)

$$\mathbf{F}_\bullet : \ 0 \longrightarrow E_L \xrightarrow{F_L} E_{L-1} \xrightarrow{F_{L-1}} \cdots \xrightarrow{F_2} E_1 \xrightarrow{F_1 = F_\Phi} \mathbb{C} \xrightarrow{F_0} E_{-1} = \frac{\mathbb{C}}{\mathrm{Im}\, F_\Phi} \longrightarrow 0.$$

Let us metrize $\mathbf{F}_\bullet$ as $\overline{\mathbf{F}_\bullet}$ by equipping with the Fubini–Study metric each $\mathcal{O}(-d_\iota^{[\ell]})$ for $\ell = 1, \dots, L$ and any ι. Let us also fix, for the time being, a smooth nonvanishing section of the inverse of the canonical bundle $K_{\mathbb{P}^N_{\mathbb{C}}} = \bigwedge^n T^{1,0}_{\mathbb{P}^N_{\mathbb{C}}}$ on $\mathbb{P}^N_{\mathbb{C}}$. Namely, we consider

$$\Upsilon = \frac{\sum_{k=0}^N (-1)^k z_k \bigwedge_{0 \leq k' \leq n\,:\, k' \neq k} dz_{k'}}{\chi(z_0, \dots, z_n)} \in C^\infty_{N,0}(\mathbb{P}^N(\mathbb{C})).$$

Then the equality of currents

$$R^{\overline{\mathbf{F}_\bullet}} \wedge \Upsilon = R_0^{\overline{\mathbf{F}_\bullet}} \wedge \Upsilon = (\iota_{\overline{\mathbb{X}} \hookrightarrow \mathbb{P}^N_{\mathbb{C}}})_* \, \omega_\Upsilon \tag{7.118}$$

---

[23] We refer the reader for more details and relevant bibliography to §4.6.3, where these concepts are described.

[24] For more details about the assertions contained in the following definition, we refer the reader to [**Eis05**, §4A and §4E, Exercice 4.3].

defines in the local setting the *structure current*, as described in §3.4.3. The *structure current* is also known as *structure form* since it is represented by an almost semimeromorphic form on the complex analytic space $\overline{\mathbb{X}}$. The *structure form* is attached to the complex $\overline{\mathbf{F}_\bullet}$ over the ambient manifold $\mathbb{P}^N_{\mathbb{C}}$, to its Fubini–Study metrization, and to the choice of the smooth section Υ. Moreover, one has locally on $\overline{\mathbb{X}}$ (see Proposition 3.49) that

(7.119)

$$\begin{aligned}\omega = \omega_{\mathbf{F}_\bullet,\Upsilon} &= \sum_{r=0}^{n-1}\omega_r\\ &= (\widehat{\Pi}_{\overline{\mathbb{X}}})_* \sum_{\rho=0}^{n-1}\Big[\frac{|e_\rho|^{\mu_\rho}}{e_\rho}\mathfrak{s}_\rho\Big[\cdots\Big[\frac{|e_1|^{2\mu_1}}{e_1}\mathfrak{s}_1\Big[\frac{|e_0'|^{2\mu_0}}{e_0'}\mathfrak{s}_0'\,\widehat{\Pi}^*_{\overline{\mathbb{X}}}\mathfrak{f}\Big]_{\mu_0=0}\Big]\cdots\Big]_{\mu_{\rho-1}=0}\Big]_{\mu_\rho=0}\end{aligned}$$

is realized as the pushforward of semimeromorphic forms through iterated normalized blowups, as specified in (3.133). Actually, the e_ρ's in (7.119) are holomorphic sections of holomorphic line bundles, while the $\mathfrak{s}_\rho$'s are smooth bundle-valued $(0,1)$ forms, $\mathfrak{s}_0'$ is a smooth (bundle-valued) function, and $\mathfrak{f}$ is a regular (bundle valued) differential n-form on $\overline{\mathbb{X}}$ in the sense of Barlet (this concept is defined in §B.2.5). Moreover, the realization (7.119) of ω is uniquely determined, since Υ does not vanish. Observe that ω is smooth on $\overline{\mathbb{X}}_{\rm reg}$ and that the holomorphic sections $e_0', e_1, \dots, e_{n-1}$ of line bundles over $\widehat{\Pi}_{\overline{\mathbb{X}}}$ in (7.119) depend only on the Fitting ideal sheaves of $\mathcal{I}_{\overline{\mathbb{X}}}$, as introduced in Definition 3.44; see §3.4.3. Hence the holomorphic sections $e_0', e_1, \dots, e_{n-1}$ of line bundles in (7.119) depend only on the radical ideal sheaf $\mathcal{I}_{\overline{\mathbb{X}}}$, that is, only on the homogeneous ideal $\mathbb{I}$ of $\mathbb{S}$. One concludes from the preceding discussion that the singularities of the structure form ω, appearing in its realization as the pushforward in (7.119), depend only on the embedding $\overline{\mathbb{X}} \hookrightarrow \mathbb{P}^N_{\mathbb{C}}$.

Let $P_1, \dots, P_m \in \mathbb{C}[\boldsymbol{X}]$ with respective degrees $d_1, \dots, d_m$. The corresponding Koszul complex $\overline{\mathbf{K}^P_\bullet}$ over $\mathbb{P}^N_{\mathbb{C}}$ is attached to the section $P = \bigoplus_{j=1}^m P_j\,\varepsilon_j$ of the holomorphic m-bundle $\bigoplus_{j=1}^m \mathcal{O}_{\mathbb{P}^N_{\mathbb{C}}}(d_j)$, where each $\mathcal{O}_{\mathbb{P}^N_{\mathbb{C}}}(d_j)$ is equipped with a Fubini–Study metric. The principal value and residue currents attached to such a metrized Koszul complex are of the form

$$\begin{aligned} P^{\overline{\mathbf{K}^P_\bullet}} &= \Big[|P|^{2\lambda}\sum_{r=1}^{\min(m,N+1)}\frac{\boldsymbol{P}^*}{2i\pi|P|^2}\,\overline{\partial}\Big(\frac{\boldsymbol{P}^*}{2i\pi|P|^2}\Big)^{r-1}\Big]_{\lambda=0}\\ &= \Big[|P|^{2\lambda}\sum_{r=1}^{\min(m,N+1)}\boldsymbol{Q}^P_r\Big]_{\lambda=0}, \\ R^{\overline{\mathbf{K}^P_\bullet}} &= \Big[\overline{\partial}|P|^{2\lambda}\sum_{r=1}^{\min(m,N)}\boldsymbol{Q}^P_r\Big]_{\lambda=0}, \end{aligned} \tag{7.120}$$

described in Proposition 3.28 and Example 3.30. Then, given the structure form denoted by $\omega = \omega_\Upsilon$ attached by (7.118) to the residue current

$$R^{\overline{\mathbf{F}_\bullet}} = R_0^{\overline{\mathbf{F}_\bullet}},$$

the current

$$R^{\overline{\mathbf{K}^P_\bullet}}_{\overline{\mathbb{X}}}\,\omega := \Big[\iota^*_{\overline{\mathbb{X}}\hookrightarrow\mathbb{P}^N_{\mathbb{C}}}\Big(\overline{\partial}|P|^{2\lambda}\sum_{r=1}^{\min(m,N)}\boldsymbol{Q}^P_r\Big)\wedge\omega\Big]_{\lambda=0} \tag{7.121}$$

is well defined as a pseudo-meromorphic bundle-valued current on the complex analytic space $\overline{\mathbb{X}}$ by (7.119) and Proposition 2.21; see also Remark 2.29. We already used such a construction locally in the more general situation where the metrized Eagon–Northcott complex replaced the metrized Koszul complex; see §5.3. In fact, while proving Theorem 5.10 we established a local result that can be made global on the compact complex analytic space $\overline{\mathbb{X}}$. It is the content of the following theorem.

THEOREM 7.38. *Let $\mathbb{X} \subset \overline{\mathbb{X}} \subset \mathbb{P}^N_{\mathbb{C}}$ and $\mathbb{I}$ be as above. There is a uniform minimal exponent $\mu(\mathbb{X} : \mathbb{P}^N_{\mathbb{C}}) \in \mathbb{N}$ such that for any $\overline{\mathbf{F}_\bullet}$ constructed from a minimal resolution of $\mathbb{C}[\boldsymbol{X}]/\mathbb{I}$, any Υ, any $P = (P_1, \dots, P_m)$ as above, and any $\Theta \in \mathcal{O}_{\mathbb{P}^N_{\mathbb{C}}}(D)$ with $D \in \mathbb{N}^*$,*

$$\sup_{z\in\overline{\mathbb{X}}\setminus V(P)} \frac{|\Theta(z)|}{|P(z)|^{\mu(\mathbb{X}:\mathbb{P}^N_{\mathbb{C}})+n}} < +\infty \Longrightarrow \big(R^{\overline{\mathbf{K}^P_\bullet}}_{\overline{\mathbb{X}}}\,\omega\big)\,\iota_{\overline{\mathbb{X}}\to\mathbb{P}^N_{\mathbb{C}}}\Theta \equiv 0 \tag{7.122}$$

as a bundle-valued current on the complex analytic space $\overline{\mathbb{X}}$. Moreover, one has $\mu(\mathbb{X} : \mathbb{P}^N_{\mathbb{C}}) = 0$ if $\overline{\mathbb{X}}$ is a complex n-dimensional submanifold of $\mathbb{P}^N_{\mathbb{C}}$.

PROOF. Such a uniform integer was defined locally as

$$\mu'_{\mathbb{A}(z)} \text{ with } \mathbb{A}(z) = \big(\mathcal{O}_{\mathbb{P}^N_{\mathbb{C}}}/\mathcal{I}_{\overline{\mathbb{X}}}\big)_z$$

for any $z \in \overline{\mathbb{X}}$, while proving Theorem 5.10 with $k = 1$. The globalization of all $\mu'_{\mathbb{A}(a)}$'s as $\mu(\mathbb{X} : \mathbb{P}^N_{\mathbb{C}})$ follows from a compactness argument. Such a global exponent $\mu(\mathbb{X} : \mathbb{P}^N_{\mathbb{C}}) \in \mathbb{N}$ is finally chosen to be minimal. The fact that $\mu(\mathbb{X} : \mathbb{P}^N_{\mathbb{C}}) = 0$ when $\overline{\mathbb{X}}$ is smooth follows from Briançon–Skoda's theorem (Theorem 5.7 with $k = 1$) in the local regular N-dimensional ring $\mathcal{O}_{\mathbb{P}^N_{\mathbb{C}},z}$ for any $z \in \overline{\mathbb{X}}$. □

Theorem 5.7 is due to Joël Briançon and Henri Skoda [**BriS74**], while Theorem 5.10 is due to Craig Huneke [**Hu92**]. From the point of multivariate residue theory, the following definition makes sense.

DEFINITION 7.39. The exponent $\mu(\mathbb{X} : \mathbb{P}^N_{\mathbb{C}})$ is called the *Briançon–Skoda–Huneke exponent of $\mathbb{X}$ with respect to the embedding $\mathbb{X} \subset \mathbb{P}^N_{\mathbb{C}}$.*

We introduce now the direct sum $\mathbf{K}^P_\bullet \oplus \mathbf{F}_\bullet$ of the complex $\mathbf{F}_\bullet$ with the Koszul complex $\mathbf{K}^P_\bullet = \mathbf{EN}^{P,1}_\bullet$ as in (5.42). Thus, we extend the action of the morphisms $(\daleth^P)_{\ell'}$ and F_ℓ as in (5.43) and define the induced actions[25]

$$'\mathcal{D}\big(\mathbb{P}^N_{\mathbb{C}}, \mathbf{E}(\mathbf{K}^P_\bullet \oplus \mathbf{F}_\bullet)\big) \longrightarrow {'\mathcal{D}}\big(\mathbb{P}^N_{\mathbb{C}}, \mathbf{E}(\mathbf{K}^P_\bullet \oplus \mathbf{F}_\bullet)\big).$$

In particular, which is important here for us, it induces the action of the operator $\boldsymbol{\nabla}_{\mathbf{K}^P_\bullet\oplus\mathbf{F}_\bullet}$ from the space of $\mathbf{E}(\mathbf{K}^P_\bullet \oplus \mathbf{F}_\bullet) = \mathrm{Hom}_{\mathbb{C}}(E(\mathbf{K}^P_\bullet \oplus \mathbf{F}_\bullet), E(\mathbf{K}^P_\bullet \oplus \mathbf{F}_\bullet))$-valued currents on $\mathbb{P}^N_{\mathbb{C}}$ to itself. Let us recall the current identity:

$$\boldsymbol{\nabla}_{\mathbf{K}^P_\bullet\oplus\mathbf{F}_\bullet}\Big(P^{\overline{\mathbf{F}_\bullet}} + \Big[P^{\overline{\mathbf{K}^P_\bullet},\lambda}\, R_0^{\overline{\mathbf{F}_\bullet}}\Big]_{\lambda=0}\Big) = \mathrm{Id}_{E(\mathbf{K}^P_\bullet\oplus\mathbf{F}_\bullet)} - \Big[R^{\overline{\mathbf{K}^P_\bullet},\lambda}\, R_0^{\overline{\mathbf{F}_\bullet}}\Big]_{\lambda=0} \tag{7.123}$$

(see (5.45) with $a = P$ and $k = 1$, combined with $R^{\overline{\mathbf{F}_\bullet}} = R_0^{\overline{\mathbf{F}_\bullet}}$).

[25] We refer the reader to §3.3.1 and §3.3.2 for the notation used here and the definition of the ∇ operator.

COROLLARY 7.40. *For any* $\mathbb{X}$, $\mathbf{F}_\bullet$, $P = (P_1, \dots, P_m)$, D, *and* $\Theta \in \mathcal{O}(D)$ *as in Theorem* 7.38, *one has that*

$$\sup_{z \in \overline{\mathbb{X}} \setminus V(P)} \frac{|\Theta(z)|}{|P(z)|^{\mu(\mathbb{X}\,:\,\mathbb{P}^N_{\mathbb{C}})+n}} < +\infty \tag{7.124}$$
$$\Longrightarrow \Theta = \nabla_{\mathbf{K}^P_\bullet \oplus \mathbf{F}_\bullet}\Big(\Big(P^{\overline{\mathbf{F}_\bullet}} + \Big[P^{\overline{\mathbf{K}^P_\bullet},\lambda} R_0^{\mathbf{F}_\bullet}\Big]_{\lambda=0}\Big)\Theta\Big)$$

in the sense of currents on $\mathbb{P}^N_{\mathbb{C}}$.

PROOF. Since

$$\Big[R^{\overline{\mathbf{K}^P_\bullet},\lambda} R_0^{\overline{\mathbf{F}_\bullet}}\Big]_{\lambda=0} \Theta = 0 \Longleftrightarrow \big(R^{\overline{\mathbf{K}^P_\bullet}}_{\overline{\mathbb{X}}}\, \omega\big)\iota_{\overline{\mathbb{X}} \to \mathbb{P}^N_{\mathbb{C}}} \Theta = 0,$$

by (7.118) and (7.121), the result follows from Theorem 7.38. □

REMARK 7.41. In fact, the integer $\mu(\mathbb{X} : \mathbb{P}^N_{\mathbb{C}})$ could be chosen as an integer $\mu(\overline{\mathbb{X}})$ depending intrinsically on the projective algebraic variety $\overline{\mathbb{X}}$, since the singularities of any structure form $\omega_{\mathbf{F}_\bullet, \Upsilon}$ (a priori attached to some embedding such as $\overline{\mathbb{X}} \hookrightarrow \mathbb{P}^N_{\mathbb{C}}$, as is the case here) do not depend on such an embedding; see [**AndW15**, Proposition 2.5 and §3]. We refer the reader to these results for more details. We will not enter into such considerations here and will always remain in the more restrictive embedded situation, which had been our point of view throughout the present monograph.

7.3.2. An effective Briançon–Skoda–Huneke type theorem on $\mathbb{X} \subset \mathbb{C}^N$. We present in this subsection an effective global version of Briançon–Skoda–Huneke's theorem formulated in Theorem 5.10 for the local ring $(\mathcal{O}_{\mathbb{C}^N}/\mathcal{I}_V)_0$. In such a global setting, $\mathbb{X}$ is, as in the previous subsection, a purely n-dimensional algebraic subvariety of $\mathbb{C}^N$. We attach to its closure $\overline{\mathbb{X}}$ in $\mathbb{P}^N_{\mathbb{C}}$ the Fubini–Study metrized complex $\overline{\mathbf{F}_\bullet}$ with length $L \leq N$ as in (7.117). Let

$$p = (p_\iota)_\iota \,:\, \mathbb{C}^N \longrightarrow \mathbb{C}^m$$

be a polynomial map and let

$$|d(p)|_\infty := \max_\iota \deg p_\iota. \tag{7.125}$$

Recall Definition 7.5 as formulated within the context of a general algebraically closed field $\mathbb{K}$ instead of $\mathbb{C}$ but with $n = N$ and the affine space $\mathbb{A}^N_{\mathbb{K}} = \mathbb{K}^N$ in place of $\mathbb{X}$. Instead, we consider here the subvariety $\mathbb{X}$ as above and introduce the notion of *Łojasiewicz exponent on* $\mathbb{X}$ *along* $\overline{\mathbb{X}} \cap (\mathbb{P}^N_{\mathbb{C}} \setminus \mathbb{C}^N)$. We consider for that purpose the normalized blowup

$$\widehat{\pi}_P : \widehat{\mathbb{X}} \longrightarrow \overline{\mathbb{X}} \tag{7.126}$$

of the purely n-dimensional complex analytic space $\overline{\mathbb{X}}$ with respect to the coherent ideal sheaf generated by the pullbacks via the embedding of $\overline{\mathbb{X}}$ into $\mathbb{P}^N_{\mathbb{C}}$, of the sections $z \mapsto P_\iota(z)$, where the P_ι's is the homogenization of the p_ι's. Let $\sum_\gamma m_\gamma H_\gamma$ be its exceptional divisor. Normality in codimension 1 implies that the localizations

$$\mathcal{O}_{\widehat{\mathbb{X}}, H_\gamma}, \quad \text{where} \quad m_\gamma \in \mathbb{N}^* \quad \text{and} \quad \widehat{\pi}_P(H_\gamma) \subset \overline{\mathbb{X}} \cap (\mathbb{P}^N_{\mathbb{C}} \setminus \mathbb{C}^N),$$

are discrete valuation rings. Then we can reformulate Definition 7.5 in this new context as follows.

DEFINITION 7.42. The strictly positive rational number

$$(7.127)\quad \mu^\infty(\mathbb{X}, p) := \max_{\{\gamma\,:\, m_\gamma\in\mathbb{N}^*,\ \widehat{\pi}_P(H_\gamma)\subset\overline{\mathbb{X}}\cap(\mathbb{P}^N_{\mathbb{C}}\setminus\mathbb{C}^N)\}} \left(\frac{m_\gamma}{\text{length}\big(\mathcal{O}_{\widehat{\mathbb{P}^N_{\mathbb{C}}}, H_\gamma}/(z_0)\,\mathcal{O}_{\widehat{\mathbb{P}^N_{\mathbb{C}}}, H_\gamma}\big)}\right)$$

is called the *Lojasiewicz exponent of p on* $\mathbb{X}$ *along* $\overline{\mathbb{X}}\cap(\mathbb{P}^N_{\mathbb{C}}\setminus\mathbb{C}^N)$.

It follows from the fact that

$$\sum_\gamma m_\gamma \deg_{\mathcal{O}_{\mathbb{P}^N_{\mathbb{C}}}(|d(p)|_\infty)} \widehat{\pi}(H_\gamma) \le \deg_{\mathcal{O}_{\mathbb{P}^N_{\mathbb{C}}}(|d(p)|_\infty)} \overline{\mathbb{X}},$$

see [**EinL99**], [**Laz**, (5.20)], [**AndW15**, p. 641], and from Wirtinger's formula (6.175), that one has the important inequality

$$(7.128)\qquad [\mu^\infty(\mathbb{X}, p)] + 1 \le \mathbf{d}^\infty(p) = \mathbf{d}\,|d(p)|_\infty^{r^\infty_{\mathbb{X}}(p)} \le \mathbf{d}\,|d(p)|_\infty^n,$$

where

$$(7.129)\quad r^\infty_{\mathbb{X}}(p) := \max\{\text{codim}_{\overline{\mathbb{X}}}\overline{\mathbb{X}}\cap(\mathbb{P}^N_{\mathbb{C}}\setminus\mathbb{C}^N)\ :\ \gamma \text{ such that } \widehat{\pi}_P(H_\gamma)\subset\overline{\mathbb{X}}\cap(\mathbb{P}^N_{\mathbb{C}}\setminus\mathbb{C}^N)\} \le n.$$

We can now state the main result in this section; see [**AndW15**, Theorem A].

THEOREM 7.43. *Let* $\mathbb{X}$ *be a purely n-dimensional algebraic subvariety of* $\mathbb{C}^N$ *with degree* $\mathbf{d}$ *and Castelnuovo–Mumford regularity* $\text{reg}(\mathbb{X})$, *when embedded into* $\mathbb{P}^N_{\mathbb{C}}$. *Let* $\mathbb{J}$ *be its defining ideal in* $\mathbb{C}[X]$ *and let* $\mu(\mathbb{X} : \mathbb{P}^N(\mathbb{C}))$ *be the Briançon–Skoda–Huneke exponent as described in Definition 7.39. Let also p be a polynomial map from* $\mathbb{C}^N$ *to* $\mathbb{C}^m$, *with* $|d(p)|_\infty$ *and let* $r^\infty_{\mathbb{X}}(p)$ *be defined by (7.125) and (7.129), respectively. Suppose that* $q\in\mathbb{C}[X]$ *is such that the almost everywhere defined function*

$$(7.130)\qquad \phi : z\in\mathbb{X}\longmapsto \frac{|q(z)|}{\|p(z)\|^{\mu(\mathbb{X}:\mathbb{P}^N_{\mathbb{C}})+n}}$$

is locally bounded on $\mathbb{X}$. *Then* $\bar q\in\sum_\iota \bar p_\iota$ *in* $\mathbb{C}[X]/\mathbb{J}\,\mathbb{C}[X]$, *and one can realize this membership as*

$$(7.131)\qquad q(X) = \sum_\iota q_\iota p_\iota \quad \text{modulo}\quad \mathbb{J}$$

with the degree estimates

$$(7.132)\quad \max_\iota \deg q_\iota p_\iota \le \max\Big(\deg q + \big(\mu(\mathbb{X} : \mathbb{P}^N_{\mathbb{C}}) + n\big)\,\mathbf{d}\,|d|_\infty^{r^\infty_{\mathbb{X}}(p)},\ (n+1)(|d(p)|_\infty - 1) + \text{reg}(\mathbb{X})\Big),$$

PROOF. Let $Q\in\mathbb{C}[X_0,\dots,X_N]$ be the homogenization of q,

$$M = \max\Big(\deg q + \big(\mu(\mathbb{X} : \mathbb{P}^N_{\mathbb{C}}) + n\big)\,\mathbf{d}\,|d|_\infty^{r^\infty_{\mathbb{X}}(p)},\ (n+1)(|d(p)|_\infty - 1) + \text{reg}(\mathbb{X})\Big)$$

and let $\Theta(X_0,\dots,X_N) = Q(X_0,\dots,X_N)X_0^{M-\deg q}$. Let $\widehat{\pi}_P : \widehat{\mathbb{X}} \longrightarrow \overline{\mathbb{X}}$ be the normalized blowup of the n-dimensional complex analytic space $\overline{\mathbb{X}}$ along the coherent sheaf of ideals generated in $\mathcal{O}_{\overline{\mathbb{X}}}$ by the sections

$$\iota^*_{\overline{\mathbb{X}}\hookrightarrow\mathbb{P}^N_{\mathbb{C}}}(z \mapsto P_\iota(z))$$

as in (7.126). We repeat at this point the argument already used in the proof of our previous Proposition 7.10. Let $\sum_\gamma m_\gamma\, H_\gamma$ be the corresponding exceptional divisor on the normal complex analytic space $\widehat{\mathbb{X}}$. It follows from the definition of $\mu^\infty(\mathbb{X}\,,p)$ with its upper estimate (7.128) that

$$\big(\mu(\mathbb{X}\,:\,\mathbb{P}^N_{\mathbb{C}}) + n\big)\,\mathbf{d}\,|d|_\infty^{r^\infty_{\mathbb{X}}(p)}\, m_{\gamma,\infty} \geq \big(\mu(\mathbb{X}\,:\,\mathbb{P}^N_{\mathbb{C}}) + n\big)\, m_\gamma$$

for all the irreducible hypersurfaces H_γ such that $\widehat{\pi}_P(H_\gamma)$ lies entirely in the algebraic hypersurface $\overline{\mathbb{X}} \cap (\mathbb{P}^N_{\mathbb{C}} \setminus \mathbb{C}^N)$ of $\overline{\mathbb{X}}$. On the other hand, it follows from the fact that ϕ is locally bounded on $\mathbb{X}$ that Q vanishes with some order greater than or equal to $\big(\mu(\mathbb{X}\,:\,\mathbb{P}^N_{\mathbb{C}}) + n\big)\, m_\gamma$ on any hypersurface H_γ such that $\widehat{\pi}(H_\gamma)$ hits the open subset $\mathbb{X}$ of $\overline{\mathbb{X}}$. As a consequence, the condition on the left-hand side of (7.124) is satisfied with Θ, and we conclude from Corollary 7.40 that

$$\Theta = \nabla_{\mathbf{K}^P_\bullet \oplus \mathbf{F}_\bullet}\Big(\Big(P^{\overline{\mathbf{F}_\bullet}} + \Big[P^{\overline{\mathbf{K}^P_\bullet},\lambda}\, R_0^{\mathbf{F}_\bullet}\Big]_{\lambda=0}\Big)\Theta\Big) \tag{7.133}$$

in the sense of currents on $\mathbb{P}^N_{\mathbb{C}}$. From this point, the diagram chase argument used to prove the first assertion in Proposition 3.32 carries through. It concludes that given any $z \in \overline{\mathbb{X}} \cap V(P)$, one has in $\mathcal{O}_{\mathbb{P}^N_{\mathbb{C}},z}$ that

$$\Theta_z = \sum_\iota \theta_{\iota,z}\,(P_\iota)_z + \sum_\iota \varpi_\iota\,(\Phi_\iota)_z,$$

where $\theta_{\iota,z} \in \big(\mathcal{O}_{\mathbb{P}^N_{\mathbb{C}}}(\deg\Theta - \deg p_\iota)\big)_z$ and $\varpi_{\iota,z} \in \big(\mathcal{O}_{\mathbb{P}^N_{\mathbb{C}}}(\deg\Theta - \deg\Phi_\iota)\big)_z$.[26] It remains to add the constraint

$$\deg\Theta \geq (n+1)(|d(p)|_\infty - 1) + \mathrm{reg}\,(\mathbb{X}) \tag{7.134}$$

to ensure that the argument carries through at all steps, including, in particular the final one, corresponding to $\ell = N$, with respect to a global application of Dolbeault's lemma in the ambient manifold $\mathbb{P}^N_{\mathbb{C}}$ instead of a local application of such a lemma in a Stein neighborhood of each point in $\overline{\mathbb{X}} \cap V(P)$. Since $H^{0,r}(\mathbb{P}^N_{\mathbb{C}}, \mathcal{O}_{\mathbb{P}^N_{\mathbb{C}}}(\delta)) = 0$ for any $\delta \in \mathbb{Z}$ as soon as $r < N$ (see §B.2.3, Example B.31(2)) one needs to concentrate only on the step $\ell = N$. The only $\overline{\partial}$-global obstruction that remains to be solved globally on $\mathbb{P}^N_{\mathbb{C}}$ is then

$$\overline{\partial}\, Q = \Big(P^{\overline{\mathbf{F}_\bullet}} + \Big[P^{\overline{\mathbf{K}^P_\bullet},\lambda}\, R_0^{\mathbf{F}_\bullet}\Big]_{\lambda=0}\Big)_{N+1}\Theta, \tag{7.135}$$

where

$$\Big(P^{\overline{\mathbf{F}_\bullet}} + \Big[P^{\overline{\mathbf{K}^P_\bullet},\lambda}\, R_0^{\mathbf{F}_\bullet}\Big]_{\lambda=0}\Big)_{N+1}$$

---

[26]Observe that this argument has already been used in the proof of Theorem 5.11 and could have been used in order to prove Theorem 5.10 as well, while our proof of Theorem 5.10 relied on the use of integral representation formulae. The original proof of Theorem 5.10 by Mats Andersson [AndSS, Theorem 1.1] was based on the chase diagram argument we keep following this time in the present approach.

is the component with bidegree $(0,N)$ of the current

$$P^{\overline{\mathbf{F}_\bullet}} + \Big[P^{\overline{\mathbf{K}^P_\bullet},\lambda}\, R_0^{\mathbf{F}_\bullet}\Big]_{\lambda=0}.$$

Since $R_{0,r}^{\overline{\mathbf{F}_\bullet}} = 0$ for $r < N-n$ according to the dimension principle for PM currents (see Proposition 2.26) and $L \leq N$, where L denotes the length of $\mathbf{F}_\bullet$, one has

$$\begin{aligned}\Big(P^{\overline{\mathbf{F}_\bullet}} + \Big[P^{\overline{\mathbf{K}^P_\bullet},\lambda}\, R_0^{\mathbf{F}_\bullet}\Big]_{\lambda=0}\Big)_{N+1} &= \Big[\sum_{r=1}^{n+1} |P|^{2\lambda} \boldsymbol{Q}_r\, R_{0,N+1-r}^{\overline{\mathbf{F}_\bullet}}\Big]_{\lambda=0} + \Big([P^{\overline{\mathbf{F}_\bullet}}]\Big)_{N+1}\\ &= \Big[\sum_{r=1}^{n+1} |P|^{2\lambda} \boldsymbol{Q}_r\, R_{0,N+1-r}^{\overline{\mathbf{F}_\bullet}}\Big]_{\lambda=0}.\end{aligned}$$

For each $r \leq n+1$, $R_{0,N+1-r}^{\overline{\mathbf{F}_\bullet}}$ is $E_{N+1-r} = \bigoplus_\iota \mathcal{O}_{\mathbb{P}^N_{\mathbb{C}}}(-d_\iota^{[N+1-r]})$-valued while $\boldsymbol{Q}_r$ is a section of $\mathrm{Hom}_{\mathbb{C}}\big(E(\mathbf{K}^P_\bullet), \bigwedge^r(\sum_\iota \mathcal{O}_{\mathbb{P}^N_{\mathbb{C}}}(-\deg p_\iota))\big)$. Therefore, if

$$\deg\Theta - r|d(p)|_\infty - d_\iota^{[N+1-r]} \geq -N \tag{7.136}$$

for any $1 \leq r \leq n+1$ and any ι, the $\overline{\partial}$-problem (7.135) is globally solvable in $\mathbb{P}^N_{\mathbb{C}}$ since $H^{0,N}(\mathbb{P}^N_{\mathbb{C}}, \mathcal{O}_{\mathbb{P}^N_{\mathbb{C}}}(\delta)) = 0$ for any $\delta \geq -N$ (see §B.2.3, Example B.31(2)). Since

$$\max(d_\iota^{[N+1-r]} - (N+1-r)) \leq \mathrm{reg}(\mathbb{X}) - 1$$

for any $r = 1,\dots,n+1$ according to Definition 7.37 for $r(\mathbb{X})$, condition (7.136) is satisfied as soon as

$$\deg\Theta \geq (n+1)(|d(p)|_\infty - 1) + \mathrm{reg}(\mathbb{X}).$$

One concludes from the chase diagram argument that $\Theta \in \sum_\iota P_\iota \mathbb{C}[\boldsymbol{X}]$ modulo $\mathbb{J}$, which implies the existence of the Θ_ι's with Θ_ι an homogeneous polynomial with degree $M - \deg p_\iota$ such that $\Theta = \sum_\iota \Theta_\iota P_\iota$ modulo $\mathbb{J}$. The existence of the q_ι's such that (7.131) holds with the degree estimates (7.132) follows by dehomogenization. □

REMARK 7.44. One can always refine $\mu(\mathbb{X} : \mathbb{P}^N_{\mathbb{C}}) + n)\, \mathbf{d}\, |d|_\infty^{r^\infty(p)}$ as the smallest integer $\big[\mu(\mathbb{X} : \mathbb{P}^N_{\mathbb{C}}) + n)\, \mu^\infty(\mathbb{X}, p)\big] + 1$ satisfying the inequality (7.128) and then get

$$\begin{aligned}&\max_\iota \deg q_\iota p_\iota \\ &\leq \max\big(\deg q + \big[(\mu(\mathbb{X} : \mathbb{P}^N_{\mathbb{C}}) + n)\, \mu^\infty(\mathbb{X}, p)\big] + 1,\ (n+1)(|d(p)|_\infty - 1) + \mathrm{reg}(\mathbb{X})\big)\end{aligned} \tag{7.137}$$

instead of (7.132). Since $|d(p)|_\infty$ was our marker for the geometric complexity of p, we did not account for this refinement but only mention it here. We will be more careful in the next subsection, where $n = N$ and $\mathbb{X} = \mathbb{C}^N$. In the particular case where $\overline{\mathbb{X}}$ is smooth (in which case $\mu(\mathbb{X} : \mathbb{P}^N_{\mathbb{C}}) = 0$ according to the last assertion in Theorem 7.38) and $m \leq n$, it follows from Theorem 5.8 when $k = 1$ that the function ϕ in (7.130) can be replaced by

$$z \in \mathbb{X} \longmapsto \frac{|q(z)|}{\|p(z)\|^m}$$

and the estimate (7.132) be replaced by

$$\max_{\iota} \deg q_\iota p_\iota \leq \max\big(\deg q + \big[m\,\mu^\infty(\mathbb{X}, p)\big] + 1\,,\, m(|d(p)|_\infty - 1) + \operatorname{reg}(\mathbb{X})\big). \tag{7.138}$$

7.3.3. Global Briançon–Skoda or Lipman–Teissier theorems in $\mathbb{K}^n$. We begin this subsection with the particular case of Theorem 7.43, where one has $N = n$. It means that $\mathbb{X} = \mathbb{A}^n_{\mathbb{C}} = \mathbb{C}^n$ and that the degree $\deg \mathbb{X} = \mathbf{d}$ is equal to one. Furthermore, the Castelnuovo–Mumford regularity $\operatorname{reg}(\mathbb{X})$ equals also 1 for $\mathbb{X} = \mathbb{C}^n$ embedded into $\mathbb{P}^n_{\mathbb{C}}$, with Zariski closure equal to $\mathbb{P}^n_{\mathbb{C}}$. It follows from Theorem 7.43 together with Remark 7.44 that if $p = (p_\iota)_\iota$ is a polynomial map from $\mathbb{C}^n$ to $\mathbb{C}^m$ and $q \in \mathbb{C}[X]$ is such that the function

$$z \in \mathbb{X} \longmapsto \frac{|q(z)|}{\|p(z)\|^{\min(m,n)}}$$

is locally bounded on $\mathbb{C}^n$, then one can realize in an effective way the membership of q to the ideal $\sum_\iota p_\iota(X)\,\mathbb{C}[X]$ as $p = \sum_\iota q_\iota p_\iota$ with

$$\max_{\iota} \deg q_\iota p_\iota \leq \max\big(\deg q + [\min(m,n)\,\mu^\infty(p)] + 1, \min(m, n+1)(|d(p)|_\infty - 1) + 1\big), \tag{7.139}$$

where $\mu^\infty(p) = \mu^\infty(\mathbb{C}^n, p)$ (see Definition 7.5 generalized as Definition 7.42). In fact, one may refine the degree estimates above thanks to the two following observations.

- When $m \leq n$, the additional constraint

  $$M \geq \min(m, n+1)(|d(p)|_\infty - 1) + \operatorname{reg}(\mathbb{X}) = m(|d(p)|_\infty - 1) + 1$$

  required for the choice on $M = \deg \Theta$ in the proof of Theorem 7.43 in order to solve the $\overline{\partial}$-problem at any step until the final one ($\ell = N = n$) is not necessary since $H^{0,r}(\mathbb{P}^n_{\mathbb{C}}, \mathcal{O}_{\mathbb{P}^N_{\mathbb{C}}}(\delta)) = 0$ for any $\delta \in \mathbb{Z}$. Hence we drop it.
- When $m > n$, the fact that $\overline{\mathbb{X}} = \mathbb{P}^n_{\mathbb{C}}$ implies that the initiating current identity for the diagram chase (a cornerstone in the proof of Theorem 7.43) is

  $$\Theta = \nabla_{\mathbf{K}^P_\bullet}\Big(\Big(\big[P^{\overline{\mathbf{K}^P_\bullet},\lambda}\big]_{\lambda=0}\Big)\Theta\Big)$$

  instead of (7.133). The precise constraint on $\deg \Theta = M$ required in order for the $\overline{\partial}$-problem at the final step ($\ell = n$) in the diagram chase to be globally solvable is

  $$M \geq |d(p)|_{1,n+1} - (n+1) + 1 = |d(p)|_{1,n+1} - n,$$

  where

  $$|d(p)|_{1,n+1} = \max_{\iota_0 < \cdots < \iota_n} \sum d_{\iota_\nu} \tag{7.140}$$

  instead of (7.136).

As a consequence, we may state the following result [**AndG11**, Theorem 2.1(i)], whose particular case Theorem 7.15 is.

THEOREM 7.45. *Let $p : \mathbb{C}^n \to \mathbb{C}^m$ be a polynomial map and $q \in \mathbb{C}[X_1, \dots, X_n]$ such that the almost everywhere defined function*

$$\phi \ : z \in \mathbb{C}^n \longmapsto \frac{|q(z)|}{\|p(z)\|^{\min(m,n)}}$$

is locally bounded in $\mathbb{C}^n$. Then, there is a polynomial map $(q_\iota)_\iota \ : \mathbb{C}^n \to \mathbb{C}^m$ such that $q = \sum_\iota q_\iota p_\iota$ with

$$(7.141) \quad \max_\iota \deg q_\iota p_\iota \leq \begin{cases} \deg q + \big[m\, \mu^\infty(p)\big] + 1 & \text{if } m \leq n \\ \max\big(\deg q + \big[\min(m,n)\, \mu^\infty(p)\big] + 1, |d(p)|_{1,n+1} - n\big) & \text{if } m > n, \end{cases}$$

where the Lojasiewicz exponent at infinity of p is defined by (7.9) *and $|d(p)|_{1,n+1}$ by* (7.140).

REMARK 7.46. When $m \leq n$ and $m\, \mu^\infty(p) = a_m/b_m$, with $a_m, b_m \in \mathbb{N}^*$, one can replace each P_ι by $P'_\iota(\boldsymbol{X}) = P_\iota(X_0^{b_m}, \dots, X_n^{b_m})$, as was already done in the proof of Theorem 7.15, which shows that the upper estimate (7.141) for $\max_\iota \deg q_\iota p_\iota$ can be refined as the upper estimate

$$\max_\iota \deg q_\iota p_\iota \leq \deg q + m\, \mu^\infty(p) \leq \deg q + m\, \mathbf{d}\,(p),$$

where $\mathbf{d}\,(p) = \prod_\iota \deg p_\iota$.

Since $\mathbb{P}^n_{\mathbb{C}}$ is a good complex manifold according to Definition 4.1, integral representation formulae on $\mathbb{P}^n_{\mathbb{C}}$ provide, after dehomogenization, an explicit solution of $q = \sum_\iota q_\iota p_\iota$ provided the almost everywhere defined function

$$z \in \mathbb{C}^n \longmapsto q(z)/\|p(z)\|^{\min(m,n)},$$

where m denotes the length of the sequence p, is locally bounded. We refer from this point to the notations introduced prior to the formulation of Theorem 7.12; see §4.6.2. The following result [**AndG11**, Theorem 1.2(ii)] admits Theorem 7.12 as a particular case (where $V(p) = \emptyset$).

THEOREM 7.47. *Let p be a polynomial map from $\mathbb{C}^n$ to $\mathbb{C}^m$ and let P be its homogenization. Let also $\mathbb{I}_{d(p)}, \mathbb{H}_{d(p)}$ be the (formal) bundle sections on $\mathbb{P}^n_{\mathbb{C}} \times \mathbb{P}^n_{\mathbb{C}}$ attached to the metrized Koszul complex $\overline{\mathbf{K}^P_\bullet}$ (where the metric on each $\mathcal{O}_{\mathbb{P}^n_{\mathbb{C}}}(\delta)$ is the Fubini–Study metric) as in* (4.122). *Let $q \in \mathbb{C}[X]$ be such that the almost everywhere defined function*

$$\phi \ : z \in \mathbb{C}^n \longmapsto \frac{|q(z)|}{\|p(z)\|^{\min(m,n)}}$$

is locally bounded. Let

$$M(p,q) := \deg q + \max\big([\min(m,n)\, \mu^\infty(p)] + 1, |d(p)|_{1,\min(m,n+1)} - n\big)$$

and Q be the homogenization of q. The dehomogenization of the polynomial identity in $\mathbb{C}[z_0, \dots, z_n]$, realized as the integral representation formula

$$(7.142) \quad z_0^{M(p,q)-\deg q} Q = 2i\pi \rfloor_{P(z)}$$

$$\int_{\mathbb{P}^n_{\mathbb{C}}} \Big((\mathscr{W}_{\mathbb{P}^n_{\mathbb{C}}}(z,\cdot))^{M(p,q)+n} \sum_{r=1}^{\min(m,n)+1} \mathbb{I}_d(z,\cdot)\, \frac{(\mathbb{H}_P(z,\cdot))^{r-1}}{(r-1)!} \Big) P_{0,r}^{\overline{\mathbf{K}^P_\bullet}} \Big)_{n,n} \zeta_0^{M(p,q)},$$

where

$$\mathscr{W}_{\mathbb{P}^n_{\mathbb{C}}} : (z,\zeta) \in \mathbb{P}^n_{\mathbb{C}} \times \mathbb{P}^n_{\mathbb{C}} \longmapsto \frac{z \bullet \zeta}{\|\zeta\|^2} + dd^c \log \|\zeta\|^2,$$

is the Fubini–Study global weight, which provides an explicit realization $q = \sum_\iota q_\iota p_\iota$ *for the membership problem with degree estimates*

$$\max \deg q_\iota p_\iota \leq \max \big(\deg q + [\min(m,n)\, \mu^\infty(p)] + 1, |d|_{1,\min(m,n+1)} - n \big). \tag{7.143}$$

PROOF. The result is an immediate consequence of Theorem 4.51, once it is combined with the fact that

$$R^{\overline{\mathbf{K}^p_\bullet}} z_0^{M(p,q)} Q = 0. \tag{7.144}$$

Since $M(p,q) \geq \big[\min(m,n)\, \mu^\infty(p)\big] + 1$ and $q/\|p\|^{\min(m,n)}$ is locally bounded in $\mathbb{C}^n$, assertion (7.144) follows from Proposition 7.10 and from the proof of Briançon–Skoda's theorem in its specific form (Theorem 5.8). The second constraint

$$M(p,q) \geq |d(p)|_{1,\min(m,n+1)} - n$$

ensures that the formal expressions under the integral in the right-hand side of (7.142) are mathematically sound. □

REMARK 7.48. When the polynomial entries of p have the same degree, the observations made in Remark 7.13 about the averaging role played by Crofton's formula (see §D.5.1) remain accurate in the present more general situation, which is outside the frame of Hilbert's nullstellensatz as in Theorem 7.12. The search for explicit division formulae in *division theory* seems to be reminiscent of that occurring in (improper) *intersection theory*, which leads to the introduction of the concept of *generalized cycle* (or that of *class of generalized cycle*) in order, for example, to conciliate local with global geometric improper intersection in the projective space $\mathbb{P}^n_{\mathbb{C}}$; see, e.g., [**AndSWY17B**], [**AndESWY21A**, §3 and §4], [**AndESWY21B**, Theorem 1.1], or also Theorem D.56 and Proposition D.59, in §D.5.4. It remains an open question[27] whether there exists some global arithmetic realization of Briançon-Skoda's theorem

$$\gamma a = \sum_\iota \boldsymbol{q}_\iota p_\iota, \qquad \gamma \in \mathbb{N}^*,\ \boldsymbol{q}_\iota \in \mathbb{Z}[X],$$

which is such that the degrees of the $\boldsymbol{q}_\iota$ and their logarithmic size, together with that of $\boldsymbol{\gamma}$, fit with geometric and arithmetic intersection theory, as it occurs in the particular case where $V(p) = \emptyset$; see Theorem 6.70 with $N = n$ and $\mathbb{X} = \mathbb{A}^n_{\overline{\mathbb{Q}}}$.

Theorem 7.45, with a slightly different formulation, holds in the general setting where $\mathbb{K}$ is an algebraically closed field with any characteristic. In place of the particular version of Briançon–Skoda's theorem in the regular n-dimensional local rings $\mathcal{O}_{\mathbb{P}^n_{\mathbb{C}},z}$, $z \in \mathbb{P}^n_{\mathbb{C}}$, namely Theorem 5.8, one needs to invoke for that purpose the Lipman–Teissier theorem [**LipS81, LipT81**] in each localization $(\mathbb{K}[X])_{X-x}$ of $\mathbb{K}[X]$ at any maximal ideal $(X-x)\,\mathbb{K}[X]$, where $x \in \mathbb{K}^n$. One needs to invoke the same theorem in the $(n+1)$-dimensional regular local ring $\mathbb{K}[[\boldsymbol{X}]] = \mathbb{K}[[X_0,\dots,X_n]]$. The result, conjectured in [**BGVY**, §5.2], see [**Am96**], is as follows[**Hi01**, Theorem 2.1(i) and (ii)].

[27] See also the conclusion of §7.4.3.

THEOREM 7.49. *Let $\mathbb{K}$ be an algebraically closed field and let $p = (p_\iota)_\iota$ be a polynomial map from $\mathbb{K}^n$ to $\mathbb{K}^m$, where $P_\iota \in \mathbb{K}[X_0, \dots, X_n] = \mathbb{K}[\boldsymbol{X}]$ is the homogenization of $p_\iota \in \mathbb{K}[X] = \mathbb{K}[X_1, \dots, X_n]$. Let $q \in \mathbb{K}[X]$ be such that for any $x \in \mathbb{K}^n$,*

$$q_x \in \overline{\sum_\iota p_{\iota,x}(\mathbb{K}[X])_{X-x}}, \tag{7.145}$$

where the bar denotes the integral closure.[28] *If $\ell(\mathbb{I}_P) \leq \min(m, n+1)$ denotes the analytic spread of the ideal $\mathbb{I}_P = \sum_\iota P_\iota \mathbb{K}[[\boldsymbol{X}]]$ in the $(n+1)$-regular local ring $\mathbb{K}[[\boldsymbol{X}]]$, then the effective global membership $q^{\ell(\mathbb{I}_P)} = \sum_\iota q_\iota p_\iota$ of $q^{\ell(\mathbb{I}_P)}$ with respect to the ideal $\sum_\iota p_\iota \mathbb{K}[X]$*[29] *can be realized with some q_ι such that*

$$\max_\iota \deg q_\iota p_\iota \leq \ell(\mathbb{I}_P) \max(\deg q + \mu^\infty(p)\,,\, |d(p)|_\infty). \tag{7.146}$$

In (7.146) $\mu^\infty(p)$ is the Łojasiewicz exponent along $\mathbb{P}^n_{\mathbb{K}} \setminus \mathbb{K}^n$ introduced in Definition 7.5 and $|d(p)|_\infty = \max_\iota \deg p_\iota$.

PROOF. We refer the reader to [**Hi01**, §4] for a proof of this result. It relies, as the proof of Theorem 7.8 which the above statement generalizes, on the Lipman–Teissier theorem in $\mathbb{K}[\boldsymbol{X}]$ and on the definition of the Łojasiewicz exponent $\mu^\infty(p)$. We omit here the details of the proof since $\mathbb{K} = \mathbb{C}$ was our main field of interest all through this section. □

REMARK 7.50. We recall that $\mu_\infty(p) \leq \mathbf{d}'(p)$, where the geometric marker $\mathbf{d}'(p)$ is defined by

$$\mathbf{d}'(p) = \max\Big(\min\Big(|d(p)|_\infty^n, \frac{\prod_j d_j}{(\min_j d_j)^{(m-n)_+}}\Big)\,,\, |d(p)|_\infty\Big), \tag{7.147}$$

when the p_j's are ordered by degree $d_1 \leq \cdots \leq d_m$. One can compare in the case when $m > n$ the results produced by Theorems 7.45 or 7.47 with q^n instead of q. Namely, one can compare the estimates

$$q^n = \sum_\iota q_\iota p_\iota, \text{ with } \max_\iota \deg q_\iota p_\iota \leq \max\big(\deg q + \lceil n\,\mu^\infty(p)\rceil + 1, |d(p)|_{1,n+1} - n\big)$$

produced by Theorems 7.45 or 7.47 to the estimates (formulated slightly differently)

$$q^{n+1} = \sum_\iota q_\iota p_\iota \text{ with } \max \deg q_\iota p_\iota \leq (n+1)\max(\deg q + \mu^\infty(p), |d(p)|_\infty)$$

derived by Theorem 7.49.

7.4. Algebraic residues and tropical considerations

7.4.1. Rational polyhedral complexes and tropical cycles. Let $\boldsymbol{N}$ be an n-dimensional lattice and let $\boldsymbol{N}_\mathbb{R} = \boldsymbol{N} \otimes_\mathbb{Z} \mathbb{R}$ be the corresponding n-dimensional vector space. The archetypal example for this setting is the lattice $\boldsymbol{N} = \mathbb{Z}^n$ with $\boldsymbol{N}_\mathbb{R} = \mathbb{R}^n$. We refer to [**MiR18**, §2.1.2] for the following definition, see also [**BabH17**, §2.4], when $\boldsymbol{N} \subset \mathbb{Q}^n$ and hence $\boldsymbol{N}_\mathbb{R} \subset \mathbb{R}^n$.

[28] Condition (7.145) is equivalent to the fact that q belongs to the integral closure of the ideal $\sum_\iota p_\iota \mathbb{K}[X]$ in $\mathbb{K}[X]$, following Definition A.36.

[29] As a matter of fact $q^{\min(m,n)}$ is already known to lie in $\sum_\iota p_\iota \mathbb{K}[X]$ in view of the Lipman–Teissier theorem in each n-dimensional regular local ring $(\mathbb{K}[X])_{X-x}$, $x \in \mathbb{K}^n$.

DEFINITION 7.51. A subset $\Delta \subset \boldsymbol{N}_{\mathbb{R}}$, defined as a finite intersection of half-spaces $\{x \in \boldsymbol{N}_{\mathbb{R}} : \langle a^\star, x\rangle \geq c\}$, where $a^\star \in \boldsymbol{N}^\star = \mathrm{Hom}_{\mathbb{Z}}(\boldsymbol{N}, \mathbb{Z})$ and $c \in \mathbb{R}$, is called a generalized[30] $\boldsymbol{N}$-rational polyhedron. Its dimension is defined to be the dimension of the $\mathbb{R}$-affine subspace A_Δ (whose direction[31] is denoted by V_Δ) that Δ generates in $\boldsymbol{N}_{\mathbb{R}}$ and it satisfies $0 \leq \dim \Delta \leq n = \mathrm{rank}\, \boldsymbol{N}$. Its relative interior Δ° is its interior in the $\mathbb{R}$-affine subspace A_Δ.

Let us state a few basic definitions introduced already for $\boldsymbol{N}$-rational cones (see §3.1.1), which are $\boldsymbol{N}$-rational generalized polyhedra stable under the homotheties $x \mapsto \lambda x$ of $\boldsymbol{N}_{\mathbb{R}}$.

DEFINITION 7.52. Let Δ be a generalized $\boldsymbol{N}$-rational polyhedron. A subset $\delta \subset \Delta$ is said to be a face of Δ, which we denote by $\delta \prec \Delta$, if and only if there exist $a^\star \in \boldsymbol{N}^\star$ and $c_\delta \in \mathbb{R}$ (not necessarily unique) so that $\delta = \{x \in \Delta : \langle a^\star\,, x\rangle = c_\delta\}$.

Any generalized $\boldsymbol{N}$-rational cone Δ of dimension m generates a collection of generalized $\boldsymbol{N}$-rational polyhedra, namely its faces, whose dimensions vary between 0 and m. Faces of dimension 0 are called *vertices* of Δ, faces of dimension 1 are called *rays* of Δ, and faces having dimension $m-1$ are called *facets* of Δ. Thus, we are ready to introduce the following notion.

DEFINITION 7.53. A generalized $\boldsymbol{N}$-rational polyhedron Δ is said to be $\boldsymbol{N}$-*entire* if and only if its vertices belong to the lattice $\boldsymbol{N}$.

EXAMPLE 7.54. Given an $\boldsymbol{N}$-rational polyhedron with exactly two distinct vertices $\boldsymbol{a}, \boldsymbol{b} \in \boldsymbol{N}$, its pairing $\Delta^{\rightleftharpoons}$ with an ordering $\rightleftharpoons$ of $\{\boldsymbol{a}, \boldsymbol{b}\}$ is called an *oriented ray* of the lattice $\boldsymbol{N}$. The $\boldsymbol{N}$-dimension of such an oriented ray is defined as the number of points of the lattice $\boldsymbol{N}$ on the oriented ray $[\boldsymbol{a}, \boldsymbol{b}]$ minus one. That is,

$$d_{\boldsymbol{N}}(\Delta^{\rightleftharpoons}) := |([\boldsymbol{a}, \boldsymbol{b}] \cap \boldsymbol{N}) - 1| \geq 1. \tag{7.148}$$

DEFINITION 7.55. Given a generalized $\boldsymbol{N}$-rational polyhedron Δ with a face $\delta \prec \Delta$, the *$\boldsymbol{N}$-rational cone of Δ from its face δ* is defined by

$$\sigma_{\delta\prec\Delta} := \{\lambda\,(x-y) \,:\, x \in \Delta,\ y \in \delta,\ \lambda \geq 0\}. \tag{7.149}$$

EXAMPLE 7.56. Let $\delta \prec \Delta$ be a facet of the generalized $\boldsymbol{N}$-polyhedron Δ with dimension $\dim \Delta = m \geq 1$. Let also V_δ be the $(m-1)$-dimensional $\mathbb{R}$-subspace directing the affine subspace A_δ generated by δ in $\boldsymbol{N}_{\mathbb{R}}$. Let us consider the $(n-(m-1))$-lattice $\boldsymbol{N}/\boldsymbol{N}_{V_\delta}$ defined as the quotient lattice of $\boldsymbol{N}$ by the $(m-1)$-*trace lattice* $\boldsymbol{N}_{V_\delta} := \boldsymbol{N} \cap V_\delta$. Then

$$\big(\sigma_{\delta\prec\Delta} \bmod V_\delta\big) \subset \big(V_\Delta \bmod V_\delta\big) \subset \big(\boldsymbol{N}_{\mathbb{R}} \bmod V_\delta\big) = \big(\boldsymbol{N}/\boldsymbol{N}_{V_\delta}\big)_{\mathbb{R}} \tag{7.150}$$

defines a $\boldsymbol{N}/\boldsymbol{N}_{V_\delta}$-rational (oriented) one-dimensional cone (that is a ray) in the $\mathbb{R}$ quotient vector space $\big(\boldsymbol{N}/\boldsymbol{N}_{V_\delta}\big)_{\mathbb{R}}$. Let

$$\bar{\boldsymbol{u}} = \boldsymbol{u}(\sigma_{\delta\prec\Delta} \bmod V_\delta) \in \boldsymbol{N}/\boldsymbol{N}_{V_\delta}$$

[30]By "generalized", we mean that Δ is not assumed to be bounded. In other words, the polyhedron under consideration is not defined necessarily as the convex envelope of a finite subset in $\boldsymbol{N}_{\mathbb{R}}$. For example, in the case where $\boldsymbol{N} = \mathbb{Z}^n$, Δ is not assumed to be the Newton polyhedron of an element in $\mathbb{K}[X_1^{\pm 1}, \ldots, X_n^{\pm 1}]$ (when $\mathbb{K}$ is a commutative field) as it was the case for the Δ_j's in §6.5.3.

[31]This means the underlying $\dim A_\Delta$-dimensional $\mathbb{R}$-vector space.

be a $\boldsymbol{N}/\boldsymbol{N}_{V_\delta}$-primitive vector of such ray,[32] which one can lift as a unique $\boldsymbol{N}$-primitive element in $\boldsymbol{N}$ denoted as $\boldsymbol{u}_{\tau\prec\Delta}$.

Generalized $\boldsymbol{N}$-rational polyhedra will play the role of elementary building blocks in the construction of so-called *tropical cycles* or *tropical subspaces* in $\mathbb{R}^n$ (see [**MiR18**, §2.1.3 and §4.2]) as closed irreducible analytic subsets of irreducible algebraic subvarieties analogously do in analytic or algebraic geometry; see §D.1. Let us introduce a notion similar to a cycle $\sum_\iota \{V_\iota\}$ with no multiplicities.

DEFINITION 7.57. An $\boldsymbol{N}$-rational polyhedral complex $\mathscr{L}$ in $\boldsymbol{N}_{\mathbb{R}}$ is a finite collection $\mathscr{P} = \{\Delta_\iota\}_\iota$ of generalized $\boldsymbol{N}$-rational polyhedra in $\boldsymbol{N}_{\mathbb{R}}$ satisfying

$$\begin{aligned} &\delta \prec \Delta_\iota \text{ for some } \iota \Longrightarrow \delta \in \mathscr{P} \\ &\Delta_\iota \text{ and } \Delta_{\iota'} \in \mathscr{P} \Longrightarrow \Delta_\iota \cap \Delta_{\iota'} \prec \Delta_\iota \text{ and } \Delta_\iota \cap \Delta_{\iota'} \prec \Delta_{\iota'}. \end{aligned} \tag{7.151}$$

The dimension of $\mathscr{P}$ is defined to be

$$\dim \mathscr{P} = \max_\iota \dim \Delta_\iota. \tag{7.152}$$

$\mathscr{P}$ is said to be purely dimensional if all the maximal polyhedra Δ_ι (also known as maximal cells) ordered by inclusion share the same dimension in $\boldsymbol{N}_{\mathbb{R}}$. Moreover,

$$\begin{aligned} &\forall r \in \{0, \dots, \dim \mathscr{P}\}, \quad \mathscr{P}(r) := \{\Delta_\iota \in \mathscr{P} : \dim \Delta_\iota = r\}, \\ &|\mathscr{P}| = \operatorname{Supp} \mathscr{P} := \bigcup_\iota \Delta_\iota. \end{aligned} \tag{7.153}$$

EXAMPLE 7.58. $\boldsymbol{N}$-rational fans, as introduced in Definition 3.3, provide typical examples of $\boldsymbol{N}$-rational polyhedral complexes, whose cells Δ are all $\boldsymbol{N}$-rational cones σ.

Let us now introduce the multiplicities.

DEFINITION 7.59. A purely dimensional, weighted $\boldsymbol{N}$-rational polyhedral complex is a formal combination

$$\mathscr{L} = \sum_\iota \nu(\Delta_\iota)\,\{\Delta_\iota\} = (\mathscr{P}, \nu),$$

where $\{\Delta_\iota\}_\iota = \mathscr{P}$ is a $\boldsymbol{N}$-rational polyhedral complex and $\nu\ : \Delta \in \mathscr{P} \mapsto \mathbb{Z}$ is a weight function. A weighted, purely dimensional $\boldsymbol{N}$-rational polyhedral complex is said to be *balanced* if and only if it satisfies the *balancing conditions*

$$\delta \in \mathscr{P}(\dim \mathscr{P} - 1) \Longrightarrow \sum_{\{\Delta \in \mathscr{P}(\dim \mathscr{P}) \,:\, \delta \prec \Delta\}} \nu(\Delta)\, \boldsymbol{u}_{\delta\prec\Delta} = 0, \tag{7.154}$$

where the primitive elements of $\boldsymbol{u}_{\delta\prec\Delta}$ of $\boldsymbol{N}$ are defined as in Example 7.56.

Next, we introduce an equivalence relation between weighted, purely dimensional $\boldsymbol{N}$-rational polyhedral complexes. Two weighted purely $(n-m)$-dimensional weighted $\boldsymbol{N}$ - rational polyhedral complexes $(\mathscr{P}, \nu)$ and $(\mathscr{P}', \nu')$ are equivalent, that is, they represent the same object as an equivalence class, if and only if the implications

$$\Delta \in \mathscr{P}(n-m),\ \Delta' \in \mathscr{P}'(n-m),\ \Delta^\circ \cap (\Delta')^\circ \neq \emptyset \Longrightarrow \nu(\Delta) = \nu'(\Delta') \tag{7.155}$$

hold. This amounts to saying that $(\mathscr{P}, \nu)$ and $(\mathscr{P}', \nu')$ inherit a common weighted refinement $(\mathscr{P}'', \nu'')$; see [**MiR18**, Definition 4.2.1 and Exercice 4.2.10] or also

[32] See Definition 3.3 with $\mathbb{Z}^n$ replaced by the lattice $\boldsymbol{N}/\boldsymbol{N}_{V_\delta}$.

[**BabH17**, Definition 2.6]. Then, let us propose the following definition when $\boldsymbol{N} = \mathbb{Z}^n$.

DEFINITION 7.60. Let $0 \leq m \leq n$. A tropical $(n-m)$*-cycle on* $\mathbb{R}^n$ (or also a tropical $(n-m)$-subvariety of $\mathbb{R}^n$, according to the terminology in [**BabH17**]), is the equivalence class $\overline{\mathscr{X}}$ (modulo the equivalence relation above) of a weighted, balanced, purely $(n-m)$-dimensional $\mathbb{Z}^n$-rational polyhedral complex

$$\mathscr{X} = \sum_\iota \nu(\Delta_\iota)\,\{\Delta_\iota\}.$$

Such a tropical $(n-m)$-cycle $\overline{\mathscr{X}}$ is said to be *effective* if it admits an effective representative. That is if it admits as a representative a weighted, balanced $\mathbb{Z}^n$-rational polyhedral complex such as $\mathscr{X} = \sum_\iota \nu(\Delta_\iota)\,\{\Delta_\iota\}$ with $\nu(\Delta_\iota) > 0$ for any ι.

Before discussing the particular example of $(n-1)$-tropical cycles in $\mathbb{R}^n$, one needs to recall the definition of the semiring $(\text{Trop}, \boxplus, \boxtimes)$. On $[-\infty, \infty[$, let us define two operations, namely the *tropical addition* and the *tropical multiplication*:

$$\begin{aligned} \forall\, a, b \in [-\infty, \infty[, \quad a \boxplus b &:= \max(a, b), \\ \forall\, a, b \in [-\infty, \infty[, \quad a \boxtimes b &:= a + b. \end{aligned} \tag{7.156}$$

The role of zero is now played by $-\infty$ since $a \boxplus (-\infty) = (-\infty) \boxplus a = a$ for any $a \in [-\infty, \infty[$, while the role of the unit element is played by 0 since

$$a \boxtimes 0 = 0 \boxtimes a = a + 0 = 0 + a = a \qquad \forall\, a \in [-\infty, \infty[.$$

Tropical addition is *idempotent*, that is

$$a \boxplus a = \max(a, a) = a \qquad \forall\, a \in [-\infty, \infty[. \tag{7.157}$$

When $a \in \mathbb{R}$, then there cannot be any $b \in [-\infty, \infty[$ such that $a \boxplus b = -\infty$, which means that no element (besides $-\infty$) admits an *inverse* for tropical addition. Hence it does not make sense to speak about *tropical subtraction* either: the idempotency relation (7.157) is a substitute for the lack of subtraction. Nevertheless, we note that one can define *tropical division*, namely $a/_{\text{trop}} b := a - b$, when $a \in [-\infty, \infty[$ and $b \in \mathbb{R}$. The tropical semiring[33] thus constructed is denoted as $(\text{Trop}, \boxplus, \boxtimes)$. We will from now on interpret $\mathbb{R}^n$ as $(\text{Trop} \setminus \{-\infty\})^n$.

EXAMPLE 7.61. The following result was established in [**Mi04**, Proposition 2.4]; see also [**MiR18**, §2, Theorem 2.4.10]. Given an effective $(n-1)$-tropical cycle $\overline{\mathscr{X}}$, there is a tropical Laurent polynomial[34]

$$\mathfrak{p}(X) = \boxplus_{\alpha \in \mathcal{A}} (-v(\alpha)) \boxtimes X_1^{\boxtimes \alpha_1} \boxtimes \cdots \boxtimes X_n^{\boxtimes \alpha_n}, \tag{7.158}$$

where $\mathcal{A}$ is a finite subset of $\mathbb{Z}^n$, whose closed convex envelope $\boldsymbol{\Delta}$ is an n-dimensional $\mathbb{Z}^n$-entire polyhedron, and $v : \mathcal{A} \to \mathbb{R} = \text{Trop} \setminus \{-\infty\}$ is such that $\overline{\mathscr{X}}$ admits as

[33] One may call it sometimes "semifield" since any nonzero element a admits an inverse $(-a)$ for tropical multiplication. Note anyway that 0 is different from the neutral element $(-\infty)$ with respect to tropical addition $\boxplus$.

[34] Which is unique, up to the tropical multiplication by a tropical monomial.

particular representative $(\mathscr{P},\nu)$. Namely, the support $|\mathscr{P}|$ of $\mathscr{P}$ is the *corner locus*[35] of the evaluation function

$$x \in \mathbb{R}^n \longmapsto p(x) = \max_{\alpha\in\mathcal{A}} \big(\langle \alpha, x\rangle - v(\alpha)\big) \tag{7.159}$$

of $\mathfrak{p}$ in $\mathbb{R}^n = (\text{Trop} \setminus \{-\infty\})^n$. The $(n-1)$-dimensional $\mathbb{Z}^n$-rational polyhedral complex $\mathscr{P}$ itself is realized from $\mathfrak{p}$; see [**PasR04**, Proposition 1]. Let

$$\Phi\ : (x,\xi) \in \mathbb{R}^n \times \mathbf{\Delta} \longmapsto p(x) + \check{p}(\xi) - \langle \xi, x\rangle,$$

where $\check{p}$ is the Legendre–Fenchel transform[36] of p, so that Φ is convex in $\mathbb{R}\times\mathbf{\Delta}$. One has

$$\mathscr{P} = \big\{\Delta^\xi := \{x\in\mathbb{R}^n\ :\ \Phi(x,\xi) = 0\}\,,\ \xi\in\mathbf{\Delta} \text{ such that } \dim \Delta^\xi \le n-1\big\}. \tag{7.160}$$

As for the weight function $\nu : \mathscr{P} \to \mathbb{N}$, it is realized from the tropical polynomial (7.158): for any $\Delta \in \mathscr{P}(n-1)$,

$$\Delta^\star = \bigcap_{x\in\Delta} \Delta^\star_x, \tag{7.161}$$

where $\Delta^\star_x = \{\xi \in \mathbf{\Delta}\ :\ \Phi(x,\xi) = 0\}$, is a one-dimensional $\mathbb{Z}^n$-integer polyhedron lying in $\mathbf{\Delta}$, whose $\mathbb{Z}^n$-dimension $d_{\mathbb{Z}^n}((\Delta^\star_x)^{\rightleftharpoons})$, defined in (7.148), equals $\nu(\Delta)$. The two families of generalized $\mathbb{Z}^n$-rational polyhedra

$$\{\Delta^\xi\ :\ \xi\in\mathbf{\Delta}\}\ ,\ \{\Delta^\star_x\ :\ x\in\mathbb{R}^n\}$$

define two n-dimensional $\mathbb{Z}^n$-rational polyhedral complexes $\mathscr{P}$ and $\mathscr{P}^\star$ of $\mathbb{R}^n_x$ and $\mathbb{R}^n_\xi$, respectively. Their corresponding supports $\mathbb{R}^n$ and $\mathbf{\Delta}$ are *dual to each other*; see [**PasR04**, Proposition 1] for details. It is natural to denote the effective $(n-1)$-dimensional tropical cycle which is represented by such $(\mathscr{P},\nu)$ by

$$\overline{(\mathscr{P},\nu)} = \mathrm{div}_{(\text{Trop}\setminus\{-\infty\})^n}(\mathfrak{p}). \tag{7.162}$$

7.4.2. The tropical current $\text{trop}^*\overline{\mathscr{Z}}$**.** For $n\in\mathbb{N}^*$, let trop be the tropicalization morphism, namely

$$\begin{aligned} \text{trop}\ : z = (z_1,\dots,z_n) \in (\mathbb{C}^*)^n = \mathrm{Hom}_{\mathbb{Z}}((\mathbb{Z}^n)^\star,\mathbb{C}^*) \\ \longmapsto \text{trop}(z) = -(\log|z_1|,\dots,\log|z_n|) = -\mathrm{Log}\, z \in \mathbb{R}^n = \mathrm{Hom}_{\mathbb{Z}}((\mathbb{Z}^n)^\star,\mathbb{R}) \end{aligned} \tag{7.163}$$

(see §D.5.2) and let arg be its companion morphism

$$\begin{aligned} \arg : z = (z_1,\dots,z_n) \in (\mathbb{C}^*)^n = \mathrm{Hom}_{\mathbb{Z}}((\mathbb{Z}^n)^\star,\mathbb{C}^*) \\ \longmapsto \arg(z) = (\arg(z_1),\dots,\arg(z_n)) \in (\mathbb{S}^1)^n = \mathrm{Hom}_{\mathbb{Z}}((\mathbb{Z}^n)^\star,\mathbb{S}^1). \end{aligned} \tag{7.164}$$

We recall the exact sequence of $\mathbb{Z}$-modules

$$\begin{aligned} \mathbb{V}_{\mathbb{Z}^n} : 0 \longrightarrow \mathrm{Hom}_{\mathbb{Z}}((\mathbb{Z}^n)^\star,\mathbb{S}^1) = (\mathbb{S}^1)^n \overset{\iota_{\mathbb{S}^1\hookrightarrow\mathbb{C}^*}}{\longrightarrow} \mathrm{Hom}_{\mathbb{Z}}((\mathbb{Z}^n)^\star,\mathbb{C}^*) \\ \overset{\text{trop}}{\longrightarrow} \mathrm{Hom}_{\mathbb{Z}}((\mathbb{Z}^n)^\star,\mathbb{R}) = \mathbb{R}^n \longrightarrow 0. \end{aligned} \tag{7.165}$$

[35] The corner locus of a real-valued continuous function is the subset of points in $\mathbb{R}^n$ in a neighborhood of which it fails to be smooth.

[36] That is, $\check{p}$ is the convex function defined on a dual copy $\mathbb{R}^n_\xi$ of $\mathbb{R}^n_x$ by

$$\check{p}(\xi) = \sup_{y\in\mathbb{R}^n} (\langle\xi,y\rangle - p(y)).$$

The polyhedron $\mathbf{\Delta} = \overline{\mathrm{conv}(\mathscr{A})}$ is the stability domain of $\check{p}$ equal to $\check{p}^{-1}(\mathbb{R})$.

One has similar exact sequences $\mathbb{V}_{\boldsymbol{N}}$, when one replaces $\mathbb{Z}^n$ by any n-lattice, namely
(7.166)

$$\mathbb{V}_{\boldsymbol{N}} : 0 \longrightarrow \mathrm{Hom}_{\mathbb{Z}}(\boldsymbol{N}^\star, \mathbb{S}^1) \overset{\iota_{\mathbb{S}^1 \hookrightarrow \mathbb{C}^*}}{\longrightarrow} \mathrm{Hom}_{\mathbb{Z}}(\boldsymbol{N}^\star, \mathbb{C}^*) \overset{\mathrm{trop}}{\longrightarrow} \mathrm{Hom}_{\mathbb{Z}}(\boldsymbol{N}^\star, \mathbb{R}) \longrightarrow 0.$$

For any $(n-m)$-dimensional subspace V of $\mathbb{R}^n$, let $\mathbb{H}_{\mathbb{Z}^n}(V)$ be the exact sequence of $\mathbb{Z}$-modules

$$\begin{aligned}(7.167)\quad \mathbb{H}_{\mathbb{Z}^n}(V) : 0 \longrightarrow (\mathbb{Z}^n)_V \otimes_{\mathbb{Z}} \mathbb{R} = V &\overset{\iota_{V \hookrightarrow \mathbb{R}^n}}{\longrightarrow} \mathbb{Z}^n \otimes_{\mathbb{Z}} \mathbb{R} = \mathbb{R}^n \\ &\overset{\pi_V}{\longrightarrow} (\mathbb{Z}^n/(\mathbb{Z}^n)_V) \otimes_{\mathbb{Z}} \mathbb{R} = \mathbb{R}^n/V \longrightarrow 0,\end{aligned}$$

following the notations introduced in Example 7.56. Then, consider the diagram of complexes with the exact sequence $\mathbb{H}_{\mathbb{Z}^n}(V)$ represented as horizontal and the exact sequences

$$\mathbb{V}_{(\mathbb{Z}^n)_V},\ \mathbb{V}_{\mathbb{Z}^n},\ \mathbb{V}_{\mathbb{Z}^n/(\mathbb{Z}^n)_V}$$

(in this order) represented as vertical. Keeping track of the commutativity of diagrams, one may raise for such a configuration the exact sequence $\mathbb{H}_{\mathbb{Z}^n}(V)$ to the exact sequence of $\mathbb{Z}$-modules

$$\begin{aligned}(7.168)\quad 0 \longrightarrow \mathrm{Hom}_{\mathbb{Z}}(((\mathbb{Z}^n)_V)^\star, \mathbb{S}^1) &\longrightarrow \mathrm{Hom}_{\mathbb{Z}}(((\mathbb{Z}^n))^\star, \mathbb{S}^1) = \mathbb{S}_1^n \\ &\overset{\Gamma_V}{\longrightarrow} \mathrm{Hom}_{\mathbb{Z}}((\mathbb{Z}^n/(\mathbb{Z}^n)_V)^\star, \mathbb{S}^1) \longrightarrow 0.\end{aligned}$$

Thus, we have the following technical result.

PROPOSITION 7.62. *For any affine subset A of $\mathbb{R}^n$ with direction V and any $a = (a_1, \dots, a_n) \in A$, the map*
(7.169)

$$\Pi_A :\ z \in \mathrm{trop}^{-1}(A) \longmapsto \Gamma_V\big(\arg(e^{a_1} z_1, \dots, e^{a_n} z_n)\big) \in \mathrm{Hom}_{\mathbb{Z}}((\mathbb{Z}^n/(\mathbb{Z}^n)_V)^\star, \mathbb{S}^1)$$

does not depend on the choice of a and realizes a submersion from $\mathrm{trop}^{-1}(A)$ to the multiplicative compact group $\mathrm{Hom}_{\mathbb{Z}}((\mathbb{Z}^n/(\mathbb{Z}^n)_V)^\star, \mathbb{S}^1)$. Moreover, each fiber $\Pi_A^{-1}(\{\theta\})$, where $\theta \in \mathrm{Hom}_{\mathbb{Z}}((\mathbb{Z}^n/(\mathbb{Z}^n)_V)^\star, \mathbb{S}^1)$, is an $(n-m)$-dimensional complex torus in $(\mathbb{C}^*)^n$. It is invariant under the action of the group $\mathrm{Hom}_{\mathbb{Z}}(((\mathbb{Z}^n)_V)^\star, \mathbb{C}^*)$ and is the common zero set of a finite family of binomials.

PROOF. One can see that the kernel of the map

$$(7.170)\qquad z \in \mathrm{trop}^{-1}(V) \longmapsto \Gamma_V\big(\arg(z)\big) \in \mathrm{Hom}_{\mathbb{Z}}((\mathbb{Z}^n/(\mathbb{Z}^n)_V)^\star, \mathbb{S}^1)$$

is the $(n-m)$-dimensional torus $\mathrm{Hom}_{\mathbb{Z}}(((\mathbb{Z}^n)_V)^\star, \mathbb{C}^*)$. Moreover, any of the fibers of (7.170) is of the form $e^{i\xi} \bullet \mathrm{Hom}_{\mathbb{Z}}(((\mathbb{Z}^n)_V)^\star, \mathbb{C}^*)$, where $\xi \in \mathbb{S}_1^n$ and $\mathbb{S}_1^n$ acts on $(\mathbb{C}^*)^n$ by pointwise multiplication. Such pointwise action is denoted here by $\bullet$. A straightforward argument leads then to the result. □

EXAMPLE 7.63. Let Δ be a generalized $(n-m)$-dimensional $\mathbb{Z}^n$-rational polyhedron with relative interior Δ° in the $(n-m)$-dimensional subspace A_Δ (with direction V_Δ) which it spans in $\mathbb{R}^n$. The restriction of Π_{A_Δ} to $\mathrm{trop}^{-1}(\Delta^\circ)$ still realizes a submersion from $\mathrm{trop}^{-1}(\Delta^\circ)$ to the compact group $\mathrm{Hom}_{\mathbb{Z}}((\mathbb{Z}^n/(\mathbb{Z}^n)_V)^\star, \mathbb{S}^1)$. Each fiber of $(\Pi_{A_\Delta})_{|\mathrm{trop}^{-1}(\Delta^\circ)}$ is the intersection of a $(n-m)$-dimensional complex torus with $\mathrm{trop}^{-1}(\Delta^\circ)$. The closure

$$(7.171)\qquad \overline{(\Pi_{A_\Delta})^{-1}_{|\mathrm{trop}^{-1}(\Delta^\circ)}(\{\theta\})}$$

in $\mathbb{T}^n$ of such an orbit, where $\theta \in \mathrm{Hom}_{\mathbb{Z}}((\mathbb{Z}^n/(\mathbb{Z}^n)_V)^\star, \mathbb{S}^1)$, is an $(n-m)$-dimensional complex manifold with piecewise smooth boundary. To the complex manifold (7.171) one attaches the (m,m)-integration current

$$\Big[\overline{(\Pi_{A_\Delta})^{-1}_{|\mathrm{trop}^{-1}(\Delta^\circ)}(\{\theta\})}\Big].$$

This current is positive but fails to be closed since (7.171) is a complex manifold with a boundary.

For $1 \le m < n$, let $\overline{\mathscr{L}}$ be an $(n-m)$-tropical effective cycle in $\mathbb{R}^n = \mathbb{Z}^n \otimes_{\mathbb{Z}} \mathbb{R}$ according to Definition 7.60. This cycle admits as representative the weighted, balanced, effective and purely $(n-m)$-dimensional $\mathbb{Z}^n$-rational complex $\mathscr{L} = (\mathscr{P}, \nu)$. Following Farhad Babaee and June Huh [**BabH17**, §2 and §3], we may attach to $\mathscr{L}$, hence to $\overline{\mathscr{L}}$, a positive closed current $\mathrm{trop}^*\overline{\mathscr{L}} \in {}'\mathcal{D}^{(m,m)}((\mathbb{C}^*)^n)$, in such a way that our construction will not depend on the choice of its representative $\mathscr{L} = (\mathscr{P}, \nu)$. Let us make this precise with the following statement.

THEOREM 7.64. *Let the $(n-m)$-tropical effective cycle $\overline{\mathscr{L}}$ in $\mathbb{R}^n$ be as above with representative $\mathscr{L} = (\mathscr{P}, \nu)$, a purely $(n-m)$-dimensional $\mathbb{Z}^n$-rational complex both balanced and effective. Then the (m,m)-current defined on $(\mathbb{C}^*)^n$ by*
(7.172)

$$\mathrm{trop}^*\overline{\mathscr{L}} = \sum_{\Delta \in \mathscr{P}(n-m)} \nu(\Delta) \int_{\mathrm{Hom}_{\mathbb{Z}}((\mathbb{Z}^n/(\mathbb{Z}^n)_V)^\star, \mathbb{S}^1)} \Big[\overline{(\Pi_{A_\Delta})^{-1}_{|\mathrm{trop}^{-1}(\Delta^\circ)}(\{\theta\})}\Big] d\mu_{\mathrm{haar}},$$

where μ_{haar} denotes the normalized Haar measure on $\mathrm{Hom}_{\mathbb{Z}}((\mathbb{Z}^n/(\mathbb{Z}^n)_V)^\star, \mathbb{S}^1)$, is well defined since it is independent of the representative $(\mathscr{P}, \nu)$ of $\overline{\mathscr{L}}$. Moreover, the current (7.64) *is d-closed.*

PROOF. The proof relies on Fourier analysis techniques.[37] We assume this result here and refer instead the reader to [**BabH17**, §2 and §3]. □

DEFINITION 7.65. The (m,m) closed positive current attached to the effective $(n-m)$-tropical cycle $\overline{\mathscr{L}}$ by (7.172) is called the $(n-m)$*-tropical current of* $\overline{\mathscr{H}}$.

Theorem 7.64 bridges tropical geometry, whose intrinsic, combinatorial nature has as effective objects $(n-m)$-tropical cycles, to the world of closed, (m,m)-positive closed currents of specified averaged type, which is of intrinsic complex analytic nature.

EXAMPLE 7.66. If $\overline{\mathscr{H}}$ is the effective $(n-1)$-tropical cycle represented by the tropical Laurent polynomial $\mathfrak{p}$ (with evaluation function p) as in Example 7.61, then the corresponding $(1,1)$-tropical current is the positive closed $(1,1)$-current defined as $dd^c(p \circ \mathrm{trop})$; see [**Bab14**, Theorem 5.2].

7.4.3. Tropicalization and (p,q)-supercurrents on $\mathbb{R}^n$. Our approach displays in many places the key role played by the pairing of local holomorphic coordinates $\zeta_1, \dots, \zeta_N$ to their "shadow" companions $\overline{\zeta}_1, \dots, \overline{\zeta}_N$ in the current presentation of multidimensional residue theory on a complex N-dimensional manifold $\mathcal{X}$ (for example, $\mathbb{C}^N$ or $\mathbb{P}^N_{\mathbb{C}}$) or an embedded, purely n-dimensional complex

[37] We shall present briefly in the next subsection Fourier analysis within the context of spaces of distributions or currents on the complex torus $\mathbb{T}^n$.

analytic space $V \subset \mathcal{X}$ (for example, an affine or projective algebraic subvariety $\mathbb{X} \subset \mathbb{C}^N = \mathbb{A}^N_{\mathbb{C}}$ or $\mathbb{X} \subset \mathbb{P}^N_{\mathbb{C}}$). In particular, observe that

$$\frac{\partial}{\partial \overline{\zeta}_k} \zeta_j = 0 \tag{7.173}$$

for $1 \leq j, k \leq N$ and that the sum $\|\zeta\|^2 = \sum_{k=1}^N \overline{\zeta}_k \zeta_k \geq 0$ is heuristically considered as a linear form in ζ with *constant* coefficients. Our object of interest in this subsection is the differential manifold $\mathbb{R}^n = \mathbb{Z}^n \otimes_{\mathbb{Z}} \mathbb{R}$, which prevents us from profiting from such combined roles between local coordinates and their shadow companions. Nevertheless, we consider instead the n-dimensional complex manifold

$$\mathrm{Hom}_{\mathbb{Z}}((\mathbb{Z}^n)^\star, \mathbb{C}^*) = (\mathbb{C}^*)^n = -\mathrm{trop}^{-1}(\mathbb{R}^n) = \mathrm{Log}^{-1}(\mathbb{R}^n), \tag{7.174}$$

where $\mathrm{Log}\, z = (\log|z_1|, \dots, \log|z_n|)$ for $z \in (\mathbb{C}^*)^n$, following the definition of the tropicalization morphism; see (7.163). Since Fourier analysis plays a major role in the proof of Theorem 7.64 (see [**BabH17**, §3]), let us describe it in the context of the space $'\mathcal{D}^{(p,q)}((\mathbb{C}^*)^n)$ for $0 \leq p, q \leq n$. We follow for that purpose the presentation in [**Os19**, §2] and introduce first the concept of (p,q) *supercurrent on* $\mathbb{R}^n$ *in the tropical sense.*[38]

DEFINITION 7.67. A $\mathbb{C}$-valued (p,q)-current T on $(\mathbb{C}^*)^n$ is said to induce *a* $\mathbb{C}$*-valued* (p,q)*-supercurrent on* $\mathbb{R}^n = \mathrm{trop}\,((\mathbb{C}^*)^n)$ if and only if T is invariant under the action of $(\mathbb{S}^1)^n$ on $(\mathbb{C}^*)^n$ by pointwise multiplication. That is, for each $\theta \in (\mathbb{S}^1)^n$,

$$\left(z \longmapsto e^{i\theta} z\right)_* T = T. \tag{7.175}$$

We denote by $'\mathscr{D}^{(p,q)}_{\mathrm{trop}}(\mathbb{R}^n)$ the space of such $\mathbb{C}$-valued (p,q)-currents on $(\mathbb{C}^*)^n$.

EXAMPLE 7.68. Let $0 \leq m < n$. If $\overline{\mathscr{Z}}$ is an $(n-m)$-tropical effective cycle on $\mathbb{R}^n = \mathbb{Z}^n \otimes_{\mathbb{Z}} \mathbb{R}$, then its associated closed, positive, (m,m)-tropical current $\mathrm{trop}^* \overline{\mathscr{Z}}$ (introduced in Definition 7.65) is the archetypal example of an (m,m)-current on $(\mathbb{C}^*)^n$ inducing an (m,m)-supercurrent on $\mathbb{R}^n$. One has

$$\mathrm{Supp}\left(\mathrm{trop}^* \overline{\mathscr{Z}}\right) = \mathrm{trop}^{-1}(|\mathscr{P}|), \tag{7.176}$$

where the weighted $\mathbb{Z}^n$-rational polyhedral complex $(\mathscr{P}, \nu)$ represents $\overline{\mathscr{Z}}$.

Fourier analysis allows us to attach to any current $T \in {'\mathcal{D}}((\mathbb{C}^*)^n)$ a multi-sequence

$$\widehat{T} = \left(\widehat{T}_\kappa\right)_{\kappa \in \mathbb{Z}^n} \in \left({'\mathscr{D}}^{(p,q)}_{\mathrm{trop}}(\mathbb{R}^n)\right)^{\mathbb{Z}^n},$$

which will be naturally called the *spectrum* of T.

DEFINITION 7.69. Let

$$T = \sum_{|J|=p} \sum_{|K|=q} T^{J,K} \frac{dz_J}{z_J} \wedge \frac{d\overline{z}_K}{\overline{z}_K},$$

where J and K are ordered subsets of $\{1, \dots, n\}$ in increasing order and the $T^{J,K}$'s are $\mathbb{C}$-valued distributions on $(\mathbb{C}^*)^n$. The spectrum of T is the multisequence of

[38] The (formal) definition of (p,q)-supercurrents on $\mathbb{R}^n$ was introduced in [**Lag12**]. It became a fundamental tool in nonarchimedean analytic geometry; see [**ChD12**, §1.2]. The definition suggested below is a concrete realization of such a concept in the tropical setting.

elements $\widehat{T}_\kappa \in {}'\mathcal{D}^{(p,q)}_{\rm trop}(\mathbb{R}^n)$ indexed by $\kappa \in \mathbb{Z}^n$ defined as follows: for every $\kappa \in \mathbb{Z}^n$ the corresponding term of the spectrum is given by

$$(7.177)\quad \widehat{T}_\kappa :=$$

$$\exp_*\Big(\sum_{|J|=p}\sum_{|K|=q}\Big(\mathrm{Log}_*\prod_{j=1}^n\Big(\frac{\overline{z_j}}{|z_j|}\Big)^{\kappa_j} T^{J,K}\Big)(x)\,(dx+id\theta)_J\wedge(dx-id\theta)_K\Big)$$

$$=\sum_{|J|=p}\sum_{|K|=q}\exp_*\Big(\Big(\mathrm{Log}_*\Big(\prod_{j=1}^n\Big(\frac{\overline{z_j}}{|z_j|}\Big)^{\kappa_j} T^{J,K}\Big)(x)\Big)\Big)\,\frac{dz_J}{z_J}\wedge\frac{d\overline{z}_K}{\overline{z}_K},$$

where $\exp : (x,\theta)\in\mathbb{R}^n\times(\mathbb{S}^1)^n \longmapsto e^{x+i\theta}\in(\mathbb{C}^*)^n$.

REMARK 7.70. Any $T\in {}'\mathcal{D}^{(p,q)}((\mathbb{C}^*)^n)$ splits as

$$T=\sum_{\kappa\in\mathbb{Z}^n}\Big(\prod_{j=1}^n\Big(\frac{z_j}{|z_j|}\Big)^{\kappa_j}\Big)\widehat{T}_\kappa, \tag{7.178}$$

justifying the terminology used in Definition 7.69.

Let $T\in {}'\mathcal{D}((\mathbb{C}^n)^*)$. Its 0-Fourier coefficient $\widehat{T}_0$, whose action on any smooth $(n-p,n-q)$-form φ is given by

$$\langle \widehat{T}_0,\varphi\rangle=\int_{(\mathbb{S}^1)^n}\langle T,(z\longmapsto e^{i\theta}z)^*\varphi\rangle\,\frac{d\theta}{(2\pi)^n} \tag{7.179}$$

belongs to ${}'\mathcal{D}^{(p,q)}_{\rm trop}(\mathbb{R}^n)$. The construction of the $(n-m)$-tropical current $\mathrm{trop}^*\overline{\mathscr{H}}$ attached to an effective $(n-m)$-tropical cycle (see Definition 7.65 and Example 7.68) suggest other constructions of (p,q)-supercurrents on $\mathbb{R}^n$ that could be of interest. Suppose, for example, that the effective $(n-m)$-tropical cycle $\overline{\mathscr{H}}$, with $0<m<n$, is represented by a weighted balanced $\mathbb{Z}^n$-polyhedral complex $(\mathscr{P},\nu)$. Moreover, assume that for any $\Delta\in\mathscr{P}(n-m)$ the $(n-m)$-dimensional complex torus $\mathrm{Hom}_{\mathbb{Z}}(((\mathbb{Z}^n)_{V_\Delta})^\star,\mathbb{C}^*)$ is defined as a smooth complete intersection in $(\mathbb{C}^*)^n$. Namely, assume that there are exactly m binomials $z^{\boldsymbol{a}_{\Delta,j}}-z^{\boldsymbol{b}_{\Delta,j}}$, where $\boldsymbol{a}_{\Delta,j},\boldsymbol{b}_{\Delta,j}\in(\mathbb{N}^*)^n$ satisfying[39]

$$\mathrm{Hom}_{\mathbb{Z}}(((\mathbb{Z}^n)_{V_\Delta})^\star,\mathbb{C}^*)=\big\{z\in(\mathbb{C}^*)^n\,:\,z^{\boldsymbol{a}_{\Delta,j}}-z^{\boldsymbol{b}_{\Delta,j}}=0\text{ for }j=1,\dots,m\big\} \tag{7.180}$$

and such that

$$z\in\mathrm{Hom}_{\mathbb{Z}}(((\mathbb{Z}^n)_{V_\Delta})^\star,\mathbb{C}^*)\Longrightarrow\bigwedge_{j=1}^m d\big(z^{\boldsymbol{a}_{\Delta,j}}-z^{\boldsymbol{b}_{\Delta,j}}\big)\neq 0. \tag{7.181}$$

Then every fiber of any map[40]

$$z\in\mathrm{trop}^{-1}(V_\Delta)\longmapsto\Gamma_{V_\Delta}(\arg(z))\in\mathrm{Hom}_{\mathbb{Z}}((\mathbb{Z}^n/(\mathbb{Z}^n)_{V_\Delta})^\star,\mathbb{S}^1)$$

is also defined as the smooth complete intersection

$$\big\{z\in(\mathbb{C}^*)^n\,:\,(e^{i\xi}\bullet z)^{\boldsymbol{a}_{\Delta,j}}-(e^{i\xi}\bullet z)^{\boldsymbol{b}_{\Delta,j}}=0\text{ for }j=1,\dots,m\big\}$$

[39]Such is the case when $m=1$.

[40]We refer here to (7.168) and Proposition 7.62 for the notations used here.

for some $\xi \in (\mathbb{S}^1)^n$, where $\bullet$ denotes the pointwise multiplication. Let a be an arbitrary point in A_Δ. It follows Proposition 7.62 that for any θ in the compact group $\mathrm{Hom}_{\mathbb{Z}}((\mathbb{Z}^n/(\mathbb{Z}^n)_V)^\star, \mathbb{S}^1)$,

$$
\begin{aligned}
(7.182)\quad & \Pi_{A_\Delta}^{-1}(\{\theta\}) \\
&= \big\{z \in (\mathbb{C}^*)^n : (e^{a+i\,\xi_\theta} \bullet z)^{\boldsymbol{a}_{\Delta,j}} - (e^{a+i\,\xi_\theta} \bullet z)^{\boldsymbol{b}_{\Delta,j}} = 0 \text{ for } j = 1, \dots, m\big\}
\end{aligned}
$$

for some $\xi_\theta \in (\mathbb{S}^1)^n$, in addition to

$$
(7.183)\quad z \in \Pi_{A_\Delta}^{-1}(\{\theta\}) \Longrightarrow \omega_{\Delta,\theta}(z) = \bigwedge_{j=1}^{m} d\big((e^{a+i\,\xi_\theta} \bullet z)^{\boldsymbol{a}_{\Delta,j}} - (e^{a+i\,\xi_\theta} \bullet z)^{\boldsymbol{b}_{\Delta,j}}\big) \neq 0.
$$

One can state under the hypothesis specified above Proposition 7.71.

PROPOSITION 7.71. *For $0 \le m < n$, let $\overline{\mathscr{Z}}$ be an effective $(n-m)$-tropical cycle represented by an effective, balanced weighted complex $(\mathscr{P}, \nu)$ such that the hypotheses above are fulfilled. Let T be the $(0,m)$-current on $(\mathbb{C}^*)^n$, whose action on test forms $\varphi \in \mathcal{D}^{(n,n-m)}((\mathbb{C}^*)^n)$ is described by*

$$
\begin{aligned}
(7.184)\quad \langle T, \varphi\rangle = & \sum_{\Delta \in \mathscr{P}(n-m)} \\
& \nu(\Delta) \int_{\mathrm{Hom}_{\mathbb{Z}}((\mathbb{Z}^n/(\mathbb{Z}^n)_V)^\star, \mathbb{S}^1)} \Big\langle \Big[\overline{(\Pi_{A_\Delta})^{-1}_{|\mathrm{trop}^{-1}(\Delta^\circ)}(\{\theta\})}\Big], \frac{\varphi}{\omega_{\Delta,\theta}}\Big\rangle\, d\nu_{\mathrm{haar}}(\theta),
\end{aligned}
$$

where $\varphi/\omega_{\Delta,\theta}$ is the $(n-m, n-m)$-form obtained in a neighborhood of $\Pi_{A_\Delta}^{-1}(\{\theta\})$ when dividing the $(n, n-m)$-form φ by the nonvanishing form $(m,0)$-form $\omega_{\Delta,\theta}$. Then the current $T \in {}'\mathcal{D}^{(0,m)}_{\mathrm{trop}}(\mathbb{R}^n)$.

PROOF. The result is an immediate consequence of the averaging process involved in the construction itself, as was already the case for the current $\mathrm{trop}^*\overline{\mathscr{Z}}$, whose realization inspired the construction of T. □

REMARK 7.72. When $f_1, \dots, f_m$ define a complete intersection in the complex manifold $(\mathbb{C}^*)^n$, one has

$$
\begin{aligned}
(7.185)\quad & \\
\overline{\partial}\Big(\frac{1}{f}\Big) \wedge df_1 \wedge \cdots \wedge df_m &= \Big[\bigwedge_{j=1}^{m} \overline{\partial}\Big(\frac{|f_{m+1-j}|^{2\lambda_j}}{2i\pi\, f_{m+1-j}}\Big) \wedge df_1 \wedge \cdots \wedge df_m\Big]_{\lambda_1 = \cdots = \lambda_m = 0} \\
&= R^f_{|\ |} \wedge df_1 \wedge \cdots \wedge df_m = \mathbf{1}_{\{f=0\}} \cdot \lim_{\varepsilon \to 0^+} dd^c \log(\|f\|^2 + \epsilon) \\
&= [\mathrm{div}(f_1)] \wedge \cdots \wedge [\mathrm{div}(f_m)];
\end{aligned}
$$

see Proposition 2.54 and the Lelong—Poincaré factorization formula (2.76). Observe that this is trivial when $f = 0 \Longrightarrow df_1 \wedge \cdots \wedge df_m \neq 0$, which corresponds to the situation we have here. Therefore, the $(0,m)$-supercurrent T, whose action on test forms is described by (7.184), is the natural candidate for a *residue supercurrent* attached to the tropical cycle $\overline{\mathscr{Z}}$. The property a priori missing for such residue supercurrent is its $\overline{\partial}$-closedness. To prove that balancing conditions imply that the residue current the T attached to the $(n-m)$-tropical cycle $\overline{(\mathscr{P}, \nu)}$, as in (7.184), is $\overline{\partial}$-closed looks like an interesting question. The particular case in which $m = 1$ seems to be easier to handle, since the current $\mathrm{trop}^*\overline{(\mathscr{P}, \nu)}$ admits in this case another representation; see Example 7.66.

Treating compact algebraic situations requires extending the concept of supercurrent on $\mathbb{R}^n$, considered from the tropical point of view as the tropical n-dimensional torus $(\mathrm{Trop}\setminus\{-\infty\})^n)$ (as introduced in Definition 7.67) to that of *supercurrent on the compactification with corners* $\mathbb{X}_\Sigma(\mathbb{R}_{\geq 0})$ *of* $\mathbb{R}^n$ *with respect to a complete* $\mathbb{Z}^n$*-rational fan* Σ *in* $\mathbb{R}^n = \mathbb{Z}^n \otimes_{\mathbb{Z}} \mathbb{R}$. We refer to [**AKS13**, §4.1] for the use of notations and for exact references, as well as for a sketchy presentation of the construction of such an object $\mathbb{X}_\Sigma(\mathbb{R}_{\geq 0})$ inspired by the work of W. Fulton. We refer in particular to the realization of the proper continuous mapping

$$\rho_\Sigma : (\mathbb{X}_\Sigma)_\mathbb{C} \longrightarrow \mathbb{X}_\Sigma(\mathbb{R}_{\geq 0})$$

extending

$$\mathrm{Log}\ : (\mathbb{C}^*)^n = \mathrm{Hom}_{\mathbb{Z}}((\mathbb{Z}^n)^\star, \mathbb{C}^*) \longrightarrow \mathbb{R}^n = \mathrm{Hom}_{\mathbb{Z}}((\mathbb{Z}^n)^\star, \mathbb{R}) = (\mathrm{Trop}\setminus\{-\infty\})^n,$$

where $(\mathbb{X}_\Sigma)_\mathbb{C}$ is the complete toric complex variety attached to the fan Σ, with $\mathrm{Orb}_{\{0\}} \simeq \mathbb{T}^N = (\mathbb{C}^*)^n$ as dense orbit acting on it; see Remark 3.9. Denote by $\bullet$ the action of the dense orbit $\mathrm{Orb}_{\{0\}} \simeq (\mathbb{C}^*)^n$ on $(\mathbb{X}_\Sigma)_\mathbb{C}$. This construction suggests the following definition, a companion to Definition 7.67.

DEFINITION 7.73. A $\mathbb{C}$-valued (p,q)-current T on the complete toric complex variety $(\mathbb{X}_\Sigma)_\mathbb{C}$ is said to induce *a* $\mathbb{C}$*-valued* (p,q)*-supercurrent on the compactification with corners* $\mathbb{X}_\Sigma(\mathbb{R}_{\geq 0}) = \rho_\Sigma((\mathbb{X}_\Sigma)_\mathbb{C})$ *in the tropical sense* if and only if T is invariant under the action of the real torus $(\mathrm{Orb}_{\{0\}})_\mathbb{R} \simeq (\mathbb{S}^1)^n$ of the dense orbit $\mathrm{Orb}_{\{0\}} \simeq (\mathbb{C}^*)^n$on $(\mathbb{X}_\Sigma)_\mathbb{C}$. Namely, T induces a supercurrent on $\mathbb{X}_\Sigma(\mathbb{R}_{\geq 0})$ if and only if for each $u \in (\mathrm{Orb}_{\{0\}})_\mathbb{R}$,

$$\big(x \longmapsto u \bullet x\big)_* T = T. \tag{7.186}$$

We denote by ${}'\mathscr{D}^{(p,q)}_{\mathrm{trop}}(\mathbb{X}_\Sigma(\mathbb{R}_{\geq 0}))$ the space of such $\mathbb{C}$-valued currents T on $(\mathbb{X}_\Sigma)_\mathbb{C}$.

From now on, let us focus on the algebraic situation, where

$$F \in \mathbb{K}[X_1^{\pm 1}, \dots, X_n^{\pm 1}] = \sum_{\boldsymbol{a} \in (\mathbb{Z}^n)^\star} c_{\boldsymbol{a}} X^{\boldsymbol{a}}$$

is a Laurent polynomial with coefficients in a subfield $\mathbb{K}$ of $\mathbb{C}$, whose Newton polyhedron Δ, namely the convex hull of $\{\boldsymbol{a} \in (\mathbb{Z}^n)^\star : c_{\boldsymbol{a}} \in \mathbb{K}^*\}$, is n-dimensional. Lev Ronkin introduced and studied an important example of a $(0,0)$-supercurrent on $\mathbb{R}^n$ in his last paper [**Ro00**]. Namely, he considered the current

$$\widehat{([\log|F|])_0} = \Big[\mathrm{Log}^*\Big(x \longmapsto \int_{(\mathbb{S}^1)^n} \mathrm{Log}\,|F(e^{x+i\theta})|\,\frac{d\theta}{(2\pi)^n}\Big)\Big]. \tag{7.187}$$

The convex real function

$$R_F : x \in \mathbb{R}^n \longmapsto \int_{(\mathbb{S}^1)^n} \log|F(e^{x+i\theta})|\,\frac{d\theta}{(2\pi)^n} \tag{7.188}$$

is called the *Ronkin function of the algebraic* $(n-1)$*-cycle* $\mathrm{div}_{(\mathbb{C}^*)^n}(F)$ in the affine complex variety $(\mathbb{C}^*)^n$. One can alternatively consider the real concave function

$$\rho_F : x \in \mathbb{R}^n \longmapsto -\int_{(\mathbb{S}^1)^n} \log|F(e^{-x+i\theta})|\,\frac{d\theta}{(2\pi)^n}, \tag{7.189}$$

(see §D.5.2) and the corresponding $(0,0)$-supercurrent on $\mathbb{R}^n$, if one prefers to use in the constructions the tropicalization morphism $\mathrm{trop} = -\mathrm{Log}$ instead of Log from $\mathrm{Hom}_{\mathbb{Z}}((\mathbb{Z}^n)^\star, \mathbb{C}^*) = (\mathbb{C}^*)^n$ to $\mathrm{Hom}_{\mathbb{Z}}((\mathbb{Z}^n)^\star, \mathbb{R}) = \mathbb{R}^n$. The convex (respectively concave) functions R_F (respectively ρ_F) play an important role in the so-called *croftonization process*; see Proposition D.49.

In the particular arithmetic setting, where $\mathbb{K} = \mathbb{Q}$, Roberto Gualdi pointed in [**Gua18**, Remark 2.8] that the Ronkin function R_F admits ultrametric companions corresponding to nonarchimedean places on $\mathbb{Q}$. For each such place $v = v_{\mathrm{p}}$, where $\mathrm{p} = 2, 3, 5, \ldots$ is a prime integer, such an ultrametric companion of R_F is the tropical polynomial

$$\mathfrak{R}_{F,\mathrm{p}}(X) = \boxplus_{\{\boldsymbol{a}\in(\mathbb{Z}^n)^* \,:\, c_{\boldsymbol{a}}\in\mathbb{Q}^*\}}(-v_{\mathrm{p}}(c_{\boldsymbol{a}})) \boxtimes X_1^{\boxtimes a_1} \boxtimes \cdots \boxtimes X_n^{\boxtimes a_n}, \tag{7.190}$$

where $v_{\mathrm{p}}(c_{\boldsymbol{a}})$ is the exponent of p in the prime decomposition of the strictly positive rational number $|c_{\boldsymbol{a}}|$. That is, $-v_{\mathrm{p}}(c_{\boldsymbol{a}}) = \log|c_{\boldsymbol{a}}|_{\mathrm{p}}$, where $|\ |_{\mathrm{p}}$ is the p-adic ultrametric absolute value on $\mathbb{Q}$. The evaluation function

$$\begin{aligned} R_{F,\mathrm{p}} : x \in \mathbb{R}^n &= (\mathrm{Trop} \setminus \{-\infty\})^n \\ &\longmapsto \max_{\{\boldsymbol{a}\in((\mathbb{Z}^n))^* \,:\, c_{\boldsymbol{a}}\in\mathbb{Q}^*\}} \big(\log|c_{\boldsymbol{a}}|_{\mathrm{p}} + \langle a\,,\, x\rangle\big) \in \mathbb{R} = \mathrm{Trop} \setminus \{-\infty\} \end{aligned}$$

is such that the $(0,0)$-supercurrent $[\mathrm{trop}^* R_{F,\mathrm{p}}]$ satisfies in $(\mathbb{C}^*)^n$ the Green equation

$$-dd^c[\mathrm{trop}^* R_{F,\mathrm{p}}] + \mathrm{trop}^* \overline{\mathscr{Z}_{\mathrm{p}}(F)} = 0. \tag{7.191}$$

In (7.191), the $(n-1)$-tropical current $\mathrm{trop}^* \overline{\mathscr{Z}_{\mathrm{p}}(F)}$ is attached (as described in Theorem 7.64) to the effective $(n-1)$-dimensional tropical cycle

$$\overline{\mathscr{Z}_{\mathrm{p}}(F)} = \mathrm{div}_{(\mathrm{Trop}\setminus\{-\infty\})^n}(\mathfrak{R}_{F,\mathrm{p}});$$

see Example 7.61 and Example 7.66. On the other hand, one has, given (7.187), that

$$[\mathrm{trop}^* R_F](z) = (\widehat{[\log|F|]})_0(1/z),$$

since $\mathrm{trop} = -\mathrm{Log}$. From these observations, we conclude that the following Green equations

$$\begin{aligned} -dd^c[\mathrm{trop}^* R_{F,\mathrm{p}}] + [\mathrm{trop}^* \mathrm{div}_{(\mathrm{Trop}\setminus\{-\infty\})^n}(\mathfrak{R}_{F,\mathrm{p}})] &= 0 \text{ for p prime integer,} \\ -dd^c[\mathrm{trop}^* R_F] + \widehat{[\mathrm{div}_{(\mathbb{C}^n)^*}(F(1/z))]}_0 &= 0 \end{aligned} \tag{7.192}$$

hold jointly in $(\mathbb{C}^*)^n$.

Now, let Σ be a complete, $\mathbb{Z}^n$-rational fan in $\mathbb{R}^n$, which is compatible with the n-dimensional Newton polyhedron $\Delta \subset (\mathbb{Z}^n)^* \otimes_{\mathbb{Z}} \mathbb{R} = (\mathbb{R}^n)^*$ of F. It means that the Laurent polynomial F defines a global holomorphic section of a toric Cartier divisor $\mathbb{D}$ with a concave virtual function $\Psi_{\mathbb{D}}$; see Proposition D.21 and Example D.40. Suppose first that F has rational coefficients and that it is irreducible in the polynomial ring $\mathbb{Q}[X_1^{\pm 1}, \ldots, X_n^{\pm n}]$. Given the product formula

$$\forall\, c \in \mathbb{Q}_{>0}, \quad \log|c| + \sum_{\mathrm{p\ prime}} \log|c|_{\mathrm{p}} = 0,$$

the pairing of the convex function R_F with its ultrametric companions $\mathfrak{R}_{F,\mathrm{p}}$, once combined with potential theory within the archimedean and nonarchimedean setting (as sketched in §6.6.2), leads to closed expressions for the total logarithmic height of the Zariski closure $\overline{F^{-1}(\{0\})}$ in $\mathbb{X}_\Sigma$ with respect to metrized toric divisors

[**Gua18**, Theorem 1]. The current Green identities (7.192), involving jointly the algebraic hypersurface $(\widehat{[\mathrm{div}_{(\mathbb{C}^*)^n}(F(1/z))]})_0$ and its ultrametric companions, namely the tropical currents attached to the $(n-1)$-tropical cycles

$$\mathrm{div}_{(\mathrm{Trop}\setminus\{-\infty\})^n}(\mathfrak{R}_{F,\mathrm{p}}),\ \mathrm{p}\ \text{prime},$$

reveal themselves to be of high importance. Exploring the possibilities of bringing multivariate residue theory into the picture (as we modestly tried to do in Proposition 7.71) deserves to be considered as concerns arithmetic division instead of intersection questions. Let us now return to the general setting, where the coefficients of the Laurent polynomial F considered here are only assumed to be complex. If

$$\Delta_{\mathbb{D}} = \mathrm{Stab}(\Psi_{\mathbb{D}}) = \{\xi \in (\mathbb{R}^n)^\star \,:\, \langle \xi\,, \cdot\rangle - \Psi_{\mathbb{D}} \geq 0\}$$

is the $\mathbb{Z}^n$-integral polyhedron of stability for $\Psi_{\mathbb{D}}$, then let

$$p_{\mathbb{D}} : x \in \mathbb{R}^n \longmapsto \max_{\boldsymbol{a}\in\Delta_{\mathbb{D}}\cap(\mathbb{Z}^n)^\star} \langle \boldsymbol{a}, x\rangle = \max_{\boldsymbol{a}\ \text{vertex of}\ \Delta_{\mathbb{D}}} \langle \boldsymbol{a}, x\rangle. \tag{7.193}$$

The $(0,0)$-supercurrent on $\mathbb{R}^n$

$$(\widehat{[\log|F|]})_0 - \big[\mathrm{Log}^* p_{\mathbb{D}}\big] = \big[\mathrm{Log}^* R_{F,\mathrm{p}}\big] - \big[\mathrm{Log}^* p_{\mathbb{D}}\big],$$

as well as the currents

$$\big[\mathrm{Log}^* R_{F,\mathrm{p}}\big] - \big[\mathrm{Log}^* p_{\mathbb{D}}\big]$$

for p prime in the arithmetic setting where F has rational coefficients, extend as $(0,0)$-supercurrents on the compactification with corners $\mathbb{X}_\Sigma(\mathbb{R}_{\geq 0})$ of $\mathbb{R}^n$, thus providing archetypal examples for such objects. For further considerations about similar constructions, including the concept of *Ronkin current* instead of Ronkin function, attached this time to a sequence $(F_1,\dots,F_m)$ of elements in $\mathbb{C}[X_1^{\pm 1},\dots,X_n^{\pm 1}]$, where $m>1$, we refer the reader to [**Os19**, §6].

Let us conclude this subsection by observing that the results presented here are far from being satisfactory at this stage. Our modest purpose was to address the following point. As we pointed out in Remark 7.13, Theorem 7.47, and its subsequent Remark 7.48, multivariate residue calculus is a most powerful tool for proving the effective Briançon–Skoda's theorem in $\mathbb{C}[X_1,\dots,X_n]$, possibly combining then with the use of Crofton's averaging formula. In addition, it allows us to formulate and prove an effective version of Hilbert's nullstellensatz by solving Bézout identity in the arithmetic setting, where the polynomial entries have integer coefficients; see Theorem 7.27 with Example 7.28. The main point is that one views Hilbert's nullstellensatz as a particular case of Briançon–Skoda's theorem; see §7.1.2–§7.1.4. Unfortunately, as we already mentioned in Remark 7.48, nothing is known presently concerning the possibility of an arithmetic version of Briançon–Skoda's theorem in $\mathbb{Z}[X_1,\dots,X_n]$. By *arithmetic Briançon–Skoda's theorem* we mean a reformulation of Theorem 7.47, when the polynomial entries p_j lie in $\mathbb{Z}[X_1,\dots,X_n]$. But a major difference is that the solution should be provided this time in $\mathbb{Q}[X_1,\dots,X_n]$ with (possibly not sharp) estimates fitting with Bézout's arithmetic intersection theorem both for numerators and denominators.[41] For these reasons, it is tempting to transpose the concept of residue current and multivariate algebraic residue calculus to the setting of nonarchimedean geometry. We only attempted to do that here in

[41] This is the case in the discrete setting, that is, with notations as in Theorem 7.47, when the p_j's, $j=1,\dots,m$ define a zero-dimensional variety. M. Elkadi proved this result in [**El94**].

a very naive way within the tropical setting using Babaee and Huh's realization of tropical currents [**BabH17**]. We aim to suggest from the closed expression for such currents the construction (still very primitive) of tropical residues as in Proposition 7.71, under the additional hypothesis which includes, as we observed, the hypersurface case $m = 1$. Another interesting direction to explore, given the role devoted to analytic continuation throughout this monograph, would be that of Igusa calculus [**Igu**]. All these questions remain indeed a challenge for the future.

7.5. Jacobian and socle, radical and top radical

Let $\mathbb{I}$ be a proper ideal in the local ring $\mathcal{O}_{\mathbb{C}^n,0} = \mathcal{O}_0$ and let $\{\mathcal{P}_\iota\}_\iota$ be the uniquely determined finite list of distinct prime ideals $\mathcal{P}_\iota$ in $\mathcal{O}_0$ so that

$$\sqrt{\mathbb{I}} = \bigcap_\iota \mathcal{P}_\iota; \tag{7.194}$$

see decomposition (A.12) in §A.3.1. Let $m = m(\mathbb{I})$ be the minimal cardinal of a possible set for generators of $\mathbb{I}$ in $\mathcal{O}_0$. We set $\mathbb{I} = \mathbb{I}_{f_1,\dots,f_m} = \mathbb{I}_f$, whenever f is a set of generators for $\mathbb{I}$ whose cardinal number $|f|$ equals the minimal number $m(\mathbb{I})$. The positive integer

$$\mathrm{dev}(\mathbb{I}_f) = |f| - \mathrm{codim}_0\, V(\mathbb{I}_f) \tag{7.195}$$

is called the *deviation* of $\mathbb{I}_f$ in $\mathcal{O}_0$. For any $r = 1,\dots,n$, let

$$(\sqrt{\mathbb{I}})_r := \bigcap_{\{\iota\,:\,\mathrm{codim}\,\mathcal{O}_0/\mathcal{P}_\iota = r\}} \mathcal{P}_\iota, \tag{7.196}$$

where the distinct prime ideals $\mathcal{P}_\iota$ are involved in the unique (up to the ordering) representation (7.194). Let us recall the terminology introduced by Wolmer Vercosa Vasconcelos in [**Va92**].[42]

DEFINITION 7.74. The top radical of $\mathbb{I}_f$, whenever $|f| = m$, is defined as the radical ideal

$$\mathrm{top}\sqrt{\mathbb{I}_f} := (\sqrt{\mathbb{I}_f})_{|f|} = (\sqrt{\mathbb{I}_f})_{\mathrm{codim}_0\, V(\mathbb{I}_f) + \mathrm{dev}\,(\mathbb{I}_f)}. \tag{7.197}$$

REMARK 7.75. If $\mathbb{I}_f$ is such that $|f| > n$, then one has immediately that $\mathrm{top}\sqrt{\mathbb{I}_f} = \{0\}$. When $\mathrm{codim}\,\mathbb{I}_f = |f|$, which is equivalent to saying that f is a regular sequence in $\mathcal{O}_0$, then one has the equality $\mathrm{top}\sqrt{\mathbb{I}_f} = \sqrt{\mathbb{I}_f}$.

In this section we are interested in the ideals $\mathbb{I}_f$ whose generators are few, i.e., $|f| \le n$. In this case we also introduce the nonzero ideal $\mathscr{J}(\mathbb{I}_f\,;\,f)$ generated by the $\binom{n}{m}$ coefficients $\mathscr{J}_K$ of the germ of the $(m,0)$ holomorphic differential form

$$\bigwedge_{j=1}^m df_j(z) = \sum_{1\le k_1<\cdots<k_m\le n} \mathscr{J}_K(\mathbb{I}_f\,;\,f)(z)\, dz_K. \tag{7.198}$$

Namely, we consider the ideal

$$\mathscr{J}(\mathbb{I}_f\,;\,f) = \sum_{1\le k_1<\cdots<k_m\le n} \mathscr{J}_K(\mathbb{I}_f\,;\,f)\, \mathcal{O}_0. \tag{7.199}$$

---

[42] W.V. Vasconcelos introduced in [**Va92**, §2] the notion of *top radical* in a polynomial ring $\mathbb{K}[X_1,\dots,X_n]$, where $\mathbb{K}$ is a field of characteristic 0, instead of the local ring $\mathcal{O}_0$. The minimality of $m(\mathbb{I})$ in such a global situation is then understood as follows: the supremum of the least numbers of generators of $\mathbb{I}$ in any localization of $\mathbb{K}[X_1,\dots,X_n]$ equals m.

The ideal $\mathscr{J}(\mathbb{I}_f\,;f)$ is clearly independent of the choice of the generating system f with the minimal cardinal. It is called the *jacobian ideal of* $\mathbb{I}_f$ and it is denoted by $\mathscr{J}(\mathbb{I}_f)$. When $m = n$, $\mathscr{J}(\mathbb{I}_f)$ is the principal ideal generated by the jacobian determinant $J(f) := \det\left[\partial f_j/\partial z_k\right]_{1\le j,k\le n}$ of $f_1,\dots,f_n$. We know that

$$\operatorname{codim}_0 V(\mathbb{I}_f) = |f| - \operatorname{dev}(\mathbb{I}_f)$$

provides a lower bound for the length $n - \operatorname{depth}(\mathcal{O}_0/\mathbb{I}_f)$ of a minimal free resolution

$$0 \longrightarrow \mathcal{O}_0^{\oplus r_L} \xrightarrow{\mathfrak{f}_L} \mathcal{O}_0^{\oplus r_{L-1}} \cdots \xrightarrow{\mathfrak{f}_2} \mathcal{O}_0^{\oplus |f|} \xrightarrow{\mathfrak{f}} \mathcal{O}_0 \tag{7.200}$$

for $\mathcal{O}_0/\mathbb{I}_f$, where $\mathfrak{f}(h_1\oplus\cdots\oplus h_m) = \sum_{j=1}^m h_j f_j$; see Definitions A.8 and A.10, together with Proposition A.17. We also know from Hilbert's syzygy theorem (see Example A.13) that there always exists such a minimal resolution (7.200) for $\mathcal{O}_0/\mathbb{I}_f$ with length $L \le n$. Multidimensional residue theory provides, using currents, a proof of the following result [**Lar19**, Theorem 1.4].

THEOREM 7.76. *Suppose that* $\operatorname{depth}\mathbb{I}_f \ge n - |f|$. *It means that* $\mathcal{O}_0/\mathbb{I}_f$ *admits a minimal free resolution* (7.200) *with length* $L \le |f|$. *Then one has*

$$\operatorname{top}\sqrt{\mathbb{I}_f} = \{h \in \mathcal{O}_0 \,:\, h\,\mathscr{J}(\mathbb{I}_f) \subset \mathbb{I}_f\}. \tag{7.201}$$

In other words, the top radical of $\mathbb{I}_f$ *is the carrier* $[\mathbb{I}_f : \mathscr{J}(\mathbb{I}_f)]$ *of the jacobian ideal* $\mathscr{J}(\mathbb{I}_f)$ *into* $\mathbb{I}_f$ *itself.*

PROOF. Let $\sqrt{\mathbb{I}_f}$ be decomposed as in (7.194). Let us first prove the inclusion

$$[\mathbb{I}_f : \mathscr{J}(\mathbb{I}_f)] \subset \operatorname{top}\sqrt{\mathbb{I}_f}. \tag{7.202}$$

We consider for that purpose $h \in [\mathbb{I}_f : \mathscr{J}(\mathbb{I}_f)]$ and an irreducible component $W = V(\mathcal{P}_\iota)$ with codimension $|f|$ of $V(\mathbb{I}_f)$. This makes sense only when $|f| \le n$, otherwise, $\operatorname{top}\sqrt{\mathbb{I}_f} = \mathcal{O}_0$. Hence $h \in \operatorname{top}\sqrt{\mathbb{I}_f}$. Given a generic point z on a complex analytic space $\mathcal{W}$, representing the germ W in a neighborhood of the origin, f defines a regular sequence in $\mathcal{O}_z$. Then, one has (see for example Proposition 3.18 and Proposition C.32) that

$$\Big(h\,\overline{\partial}\Big(\frac{1}{f}\Big) \wedge df_1 \wedge\cdots\wedge df_m\Big)_z = \nu\big(h\,[\mathcal{W}]\big)_z,$$

where $\nu = \nu(W)$ is an appropriate Hilbert–Samuel multiplicity. It follows from the fact that $h\,\mathscr{J}(\mathbb{I}_f) \subset \mathbb{I}_f$ and from Theorem 2.79 that $h_z = 0$ on $\mathcal{W}_z$. Since z is generic on $\mathcal{W}$, we conclude that $h \in \mathcal{P}_\iota$. It is true for any ι satisfying $\operatorname{codim}_0 V(\mathcal{P}_\iota) = m$, which means that $h \in \operatorname{top}\sqrt{\mathbb{I}_f}$ and concludes the proof of the inclusion (7.202).

It remains to prove the converse inclusion

$$\operatorname{top}\sqrt{\mathbb{I}_f} \subset [\mathbb{I}_f : \mathscr{J}(\mathbb{I}_f)]. \tag{7.203}$$

Let $\overline{\mathbf{F}_\bullet}$ be a metrized version of the generically exact complex of trivial holomorphic bundles (metrized with the canonical metric on all $\mathbb{C}^{r_\ell}$) over a sufficiently small neighborhood $\Omega \subset \mathbb{C}^n$. Its associated complex of sheaves of sections is the minimal free resolution (7.200) with length $L = n - \operatorname{depth}(\mathbb{I}_f) \le |f|$. Proving (7.203) amounts to proving, according to Theorem 3.38, that for any $h \in \operatorname{top}\sqrt{\mathbb{I}_f}$ the equality

$$h\,(df_1\wedge\cdots\wedge df_m)\,R_0^{\overline{\mathbf{F}_\bullet}} = 0 \tag{7.204}$$

holds. Let

$$R^f_{|\,|} = \sum_{r=\mathrm{codim}_0 (V(\mathbb{I}_f))}^{\min(m,n)} (R^f_{|\,|})^{0,r}$$

be the Bochner–Martinelli current attached to the section f of the trivial bundle $\mathbb{C}^m$; see Example 3.16. At this point, we need a generalization of the transformation law in the Cohen–Macaulay setting as formulated in Theorem 3.52, where $\mathbf{F}_\bullet$ is as defined above and $\mathbf{G}_\bullet = \mathbf{K}^f_\bullet$. More precisely, we are in the situation described in Example 3.55, except that our complex $\mathbf{F}_\bullet$ is not associated with a Cohen–Macaulay resolution. Hence, the metrization of such a complex does now enter into the picture. For any $r \geq 1$, the correction[43]

$$\boldsymbol{\delta}_r(\overline{\mathbf{F}_\bullet} : \overline{\mathbf{K}^f_\bullet}) = R^{\overline{\mathbf{F}_\bullet}}_{0,r} \boldsymbol{A}_0 - \boldsymbol{A}_r R^{\overline{\mathbf{K}^f_\bullet}}_{0,r} \tag{7.205}$$

in Theorem 3.52 has to be computed. For example, when $r = M$, $\mathbf{G}_\bullet = \mathbf{K}^f_\bullet$, $\boldsymbol{A}_0 = \mathrm{Id}_{\mathbb{C}}$, and $\mathbf{F}_\bullet$ is attached to a Cohen–Macaulay resolution, it equals 0; see (3.140). When $\mathbf{F}_\bullet$ is not attached to a Cohen–Macaulay resolution and $r \geq 1$, we refer the reader to [**Lar19**, Theorem 1.2 and Theorem 3.2], where the corrections (7.205) are computed. Let us only sketch the main steps of the argument leading to (7.204). One has that

$$r < m \Longrightarrow (df_1 \wedge \cdots \wedge df_m)(R^f_{|\,|})^{0,r} = 0, \tag{7.206}$$

which can be easily read on the normalized blowup of Ω along the ideal sheaf $\sum_{j=1}^m f_j \mathcal{O}_\Omega$; see the first assertion in [**Lar19**, Lemma 1.6]. On the other hand,

$$h\, df_1 \wedge \cdots \wedge df_m\, \boldsymbol{A}_m\, R^{\overline{\mathbf{K}^f_\bullet}}_{0,m} = 0, \tag{7.207}$$

since $\boldsymbol{A}_m$ vanishes on all irreducible components of $V(\mathbb{I}_f)$ with codimension strictly less than m and $h \in \mathrm{top}\sqrt{\mathbb{I}_f}$; see [**Lar19**, Lemma 6.2]. It follows from (7.206) and (7.207) that

$$h\, df_1 \wedge \cdots \wedge df_m\, \boldsymbol{A}_r\, R^{\overline{\mathbf{K}^f_\bullet}}_{0,r} = 0 \tag{7.208}$$

for all $r \geq 1$. An inductive argument on r, based on the dimension principle (Proposition 2.26) and on the computation of the corrections (7.205), shows that

$$h\, df_1 \wedge \cdots \wedge df_m\, \boldsymbol{\delta}_r(\overline{\mathbf{F}_\bullet} : \overline{\mathbf{K}^f_\bullet}) = 0$$

for any $r \geq 1$. This and (7.208), holding for any $r \geq 1$, imply (7.204), which concludes the proof of Theorem 7.76. □

REMARK 7.77. In the global setting $\mathcal{O}_0$ is replaced by $\mathbb{K}[X_1, \dots, X_n]$, where $\mathbb{K}$ is a field with characteristic 0. Then, the corresponding result formulated by W. V. Vasconcelos in [**Va92**, Theorem 2.4] is the following: given an ideal $\mathbb{I} = \mathbb{I}_p = \sum_1^m p_j\, \mathbb{K}[X_1, \dots, X_n]$ with $m = |p|$ minimal,[44] the equality

$$\mathrm{top}\sqrt{\mathbb{I}_p} = \left[\mathbb{I}_p : \mathscr{J}(\mathbb{I}_p)\right]$$

[43] We refer the reader to Definition 3.50 for what concerns the construction through a chase diagram argument of the morphisms $\boldsymbol{A}_r$ from the complex $\mathbf{K}^f_\bullet$ to the complex $\mathbf{F}_\bullet$.

[44] The minimality is understood in the context of footnote [42].

holds under the hypothesis that every embedded prime of $\mathbb{I}_p$ has codimension at most equal to $m = |p|$. A different formulation of Theorem 7.76, in the local setting, namely in $\mathcal{O}_0$, was given by H. Wiebe in [**Wi69**].

An essential consequence of Theorem 7.76 is Theorem 7.78, proved by Michel Hickel [**Hi08**].

THEOREM 7.78. *Given exactly n elements $f_1, \dots, f_n$ in the n-dimensional regular local ring $\mathcal{O}_0$ with $f(0) = 0$, the following alternative holds: either the sequence f is regular, namely $\sum_{j=1}^n f_j\mathcal{O}_0 = \mathbb{I}_f$ and $V(\mathbb{I}_f) = \{0\}$, or $J(f) \in \sum_{j=1}^n f_j\mathcal{O}_0$, where $J(f)$ denotes the jacobian determinant of f.*

PROOF. If f is a regular sequence in $\mathcal{O}_0$, then one has

$$\Big(\bar\partial\Big(\frac{1}{f}\Big) \wedge df_1 \wedge \cdots \wedge df_n\Big)_0 = \dim_{\mathbb{C}} \Big(\mathcal{O}_0/\sum_{j=1}^n f_j\mathcal{O}_0\Big) \big([(dd^c \log \|z\|^2)^n]\big)_0;$$

see, for example, Proposition 3.18 and Proposition C.32. Then, Theorem 2.79 implies that $J(f) \notin \sum_{j=1}^n f_j\mathcal{O}_0$ since $\dim_{\mathbb{C}}(\mathcal{O}_0/\sum_{j=1}^n f_j\mathcal{O}_0) \in \mathbb{N}^*$. Suppose now that the sequence f is not regular. One has $\sum_{j=1}^n f_j\mathcal{O}_0 = \mathbb{I}_g$ with $|g| \leq n$. If $|g| = m < n$, one has $df_j = \sum_{k=1}^m u_{j,k}\, dg_k$ modulo $\mathbb{I}_g\, dz$, thus $J(f) \in \mathbb{I}_g$. It remains to consider the case where $|g| = n$, in which case Theorem 7.76 applies as $n - \operatorname{depth}(\sum_{j=1}^n f_j\mathcal{O}_0) \leq n = |g|$. Since $\operatorname{codim}_0 V(\sum_{j=1}^n f_j\mathcal{O}_0) < n$ in the case we consider, then one has

$$\operatorname{top}\sqrt{\textstyle\sum_{j=1}^n f_j\mathcal{O}_0} = \mathcal{O}_0 \Longrightarrow J(f) \in \sum_{j=1}^n f_j\mathcal{O}_0,$$

according to (7.201). Theorem 7.78 is then proved. □

The proof of Theorem 7.78 by Michel Hickel [**Hi08**] was based on Krull's identity

$$\mathbb{I}_f = \bigcap_{k\in\mathbb{N}^*} (\mathbb{I}_f + \mathfrak{M}^k),$$

where $\mathfrak{M}$ denotes the maximal ideal of $\mathcal{O}_0$. M. Hickel deduced from Lipman's theory[45] more than that. He gave a formulation of Theorem 7.78 in the algebraic setting of the ring of power series $\mathbb{A} = \mathbb{K}[[X_1, \dots, X_n]]$, where $\mathbb{K}$ is a field with characteristic 0: given an ideal $\mathbb{I}_F$ with $|F| = n$ in $\mathbb{A}$, it is equivalent to say that the Krull dimension of $\mathbb{A}/\mathbb{I}_F$ is strictly positive or that $J(F) \in \mathbb{I}_F$ [**Hi08**, Theorem 1.1]. Proving that one has the implication $\dim_{\mathrm{Krull}}(\mathbb{A}/\mathbb{I}_F) > 0 \Longrightarrow J(F) \in \mathbb{I}_F$ does not appeal there to multidimensional residue theory. Instead, it is an algebraic proof. In the alternative case, where $\dim_{\mathrm{Krull}}(\mathbb{A}/\mathbb{I}_F) = 0$, which amounts to saying that $\mathbb{A}/\mathbb{I}_F$ is a finitely dimensional $\mathbb{K}$-vector space, one has in addition, despite the fact that $J(F)$ fails to lie in $\mathbb{I}_F$, the inclusion

$$J(F)\,\mathfrak{M} \subset \mathbb{I}_F, \tag{7.209}$$

where $\mathfrak{M}$ is the maximal ideal of $\mathbb{A}$. The inclusion (7.209) gives us the opportunity to recall the definition of the socle of an $\mathbb{A}$-module, where $\mathbb{A}$ is a commutative ring.

[45]We refer for a presentation of this theory the reader to [**Lip87**] or to §6.2.2 of the present monograph, where it is sketched.

DEFINITION 7.79. The *socle* of a $\mathbb{A}$-module $\mathbb{M}$ is the intersection of all essential $\mathbb{A}$-submodules. That is, the socle of $\mathbb{M}$ is the intersection of $\mathbb{A}$-submodules having a nontrivial intersection with any nontrivial $\mathbb{A}$-submodule. When $\mathbb{M}$ is Artinian, the socle is an essential submodule also.

Recalling here the notion of socle allows us the opportunity to point out, besides the work of Hartmut Wiebe [**Wi69**] we already mentioned, the pioneering paper of Günther Scheja and Uwe Störch [**ScS75**], whose content, unfortunately, we had no opportunity to discuss within the core of our monograph, except to mention it here. Instead we refer the reader to [**ScS75**] (together with the references it contains) for what concerns, besides Lipman's approach that we sketched briefly in §6.2.2, the theory of multidimensional residues from the algebraic point of view. In the setting we consider here, the inclusion (7.209) amounts to saying that the class $\bar{J}(F)$ of $J(F)$ modulo $\mathbb{I}_F$ generates the socle of the Artinian $\mathbb{K}[X_1, \dots, X_n]$-module $\mathbb{K}[[X_1, \dots, X_n]]/\sum_{j=1}^m F_j\mathbb{K}[[X_1, \dots, X_n]]$.

Before concluding this section, let us make the following observation. In the case where one has $\text{codim}_0(\mathcal{O}_0/\sum_{j=1}^n f_j\mathcal{O}_0) < n$ in Theorem 7.78, then $J(f) \in \sum_{j=1}^n f_j\mathcal{O}_0$. Thus, the construction of weights attached to Hefer morphisms in the affine setting (see Proposition 4.29 and Proposition 4.43) once combined with the realization of weighted Bochner–Martinelli type integral representation formulae (see Proposition 4.20 and Proposition 4.33) allows the possibility to realize the membership $J(f) = \sum_{j=1}^n g_j f_j$ in a neighborhood of the origin in $\mathbb{C}^n$. The realization of such explicit division formulae rely on the same strategy we already invoked to prove Briançon–Skoda's theorem in the quotient ring $\mathbb{A} = (\mathcal{O}/\mathcal{I}_V)_0$, namely Theorem 5.10. The metrized complex $\overline{\mathbf{F}_\bullet}$ satisfying $(df_1 \wedge \dots \wedge df_n)\, R_0^{\overline{\mathbf{F}_\bullet}} = 0$ (see (7.204) with $h = 1$) and the Koszul complex $\mathbf{K}_\bullet^f$ with the related global weights $\mathscr{W}_{f,1}^\lambda$, defined by (4.111), are the initial tools for such constructions. We note that some attempts in this direction already appeared in [**BY00**, Proposition 2.1, Theorems 2.2 and 2.3]. In the very particular case where $P_0, \dots, P_n$ are homogeneous forms in $\mathbb{C}[X_0, \dots, X_n]$ with complex coefficients such that the corresponding sections of holomorphic bundles $\mathcal{O}_{\mathbb{P}^n_{\mathbb{C}}}(\deg P_j)$ define a nonempty zero set in $\mathbb{P}^n_{\mathbb{C}}$, the membership of $J(P)$ to $\sum_{j=0}^n P_j\, \mathbb{C}[X_0, \dots, X_n]$ was stated without proof in the pioneering book of Eugen Netto [**Ne**, p. 142]. It was proved later and made effective by S. Spodzieja [**Sp89**][46] using Cramer's rule and Euler's formula for homogeneous forms. We have up to now no idea for a simple way (similar to that proposed by S. Spodzieja in [**Sp89**]) to make effective (without the heavy machinery of weighted Bochner–Martinelli currents, involving a noneffective free resolution for $\mathcal{O}_0/\sum_{j=1}^n f_j\mathcal{O}_0$) the membership $J(f) = \sum_{j=1}^n g_j f_j$ in the affine local setting, when $\text{codim}_0(V(f)) < n$.

7.6. Multivariate residue calculus and the exponential function

One considers in this section as a complex manifold the affine space $\mathbb{C}^n$ with its structure of Stein manifold. Let $r \in \mathbb{N}^*$. Throughout this section, $\mathbb{C}^r$-valued currents on $\mathbb{C}^n$ will be considered as linear combinations with $\mathbb{C}^r$-valued distribution

[46] Arkadius Płoski generously mentioned this problem to M. Hickel and A. Yger during a conference at the University of Calabria in 1998. It became the starting point of the reflections we tried to summarize in this section.

coefficients of the differential forms that constitute the basis

$$\mathscr{B} := \{dz_J \wedge d\overline{z}_K \;:\; J, K \text{ in increasing order, } 0 \leq |J|, |K| \leq n\}. \tag{7.210}$$

We recall that the $\mathbb{C}$-vector space of $\mathbb{C}$-*valued tempered distributions in* $\mathbb{C}^n$, that is the Schwartz space $\mathscr{S}'(\mathbb{R}^{2n})$, or also the $\mathbb{C}$-vector space of $\mathbb{C}$-*valued spherical distributions*, coincides with the $\mathbb{C}$-vector space $\big(\mathcal{D}'(\mathbb{P}^n_{\mathbb{C}})\big)_{|\mathbb{C}^n}$. We will nevertheless constantly work in the affine space $\mathbb{C}^n$ instead of $\mathbb{P}^n_{\mathbb{C}}$, since the transcendental exponential function will play a major role throughout this section.

7.6.1. Weighted algebras of entire functions, tempered currents. Recall first the definition of a *plurisubharmonic (psh) weight function* ϕ in $\mathbb{C}^n$; see for example [**Ehr, BT79, BT80**].

DEFINITION 7.80. A positive plurisubharmonic function $\phi : \mathbb{C}^n \to \mathbb{R}^+$ is called a *psh weight function*[47] if it satisfies (i) $\log(1+|z|) = O(\phi(z))$ and (ii) there are positive constants c, c', C, C' such that for any $(z,\zeta) \in \mathbb{C}^n \times \mathbb{C}^n$,

$$|\zeta - z| < \exp(-c\,\phi(z) - c') \Longrightarrow \phi(\zeta) \leq C\,\phi(z) + C'. \tag{7.211}$$

Given such a psh weight function, $\mathbf{A}_\phi(\mathbb{C}^n)$ denotes the algebra of entire functions f satisfying the growth conditions:

$$f \in \mathbf{A}_\phi(\mathbb{C}^n) \Longrightarrow |f(z)| \leq A e^{B\phi(z)} \text{ for some } A \geq 0,\; B \geq 0. \tag{7.212}$$

EXAMPLE 7.81. The function $z \in \mathbb{C}^n \mapsto \phi(z) = \|z\|^\gamma$, with $\gamma \in [1, +\infty[$, is a psh weight. The corresponding algebra $\mathbf{A}_\phi(\mathbb{C}^n)$ is also denoted by $\mathbf{A}_\gamma(\mathbb{C}^n)$. The most important among such weights are those where $\gamma = 1$ and $\gamma = 2$. When $\gamma = 1$, the algebra $\mathbf{A}_1(\mathbb{C}^n)$ is the algebra $\mathrm{Exp}(\mathbb{C}^n)$, whose elements are Fourier–Borel transforms $z \mapsto T(e^{\langle z, \cdot\rangle})$ of analytic functionals T in $\mathbb{C}^n$. When $\gamma = 2$, $\mathbf{A}_2(\mathbb{C}^n)$ inherits some interesting properties, since in this case $\phi = \|z\|^2 = \sum_{j=1}^n z_j \overline{z}_j$ is the canonical hermitian norm in $\mathbb{C}^n$. All the psh weights $z \mapsto \|z\|^\gamma$ with $\gamma \in [1, +\infty[$ are convex and radial. The psh positive smooth weight $\phi : z \mapsto \log(1 + \|z\|^2)/2$ is also radial, and the corresponding algebra $\mathbf{A}_\phi(\mathbb{C}^n)$ is the algebra of polynomial functions with complex coefficients in n variables. A major example of nonradial weight (concerning harmonic analysis) is

$$\phi(z) = \|\mathrm{Im}\, z\| + \log(1 + \|z\|^2)/2.$$

In this case, $\mathbf{A}_\phi(\mathbb{C}^n)$ is the Paley–Wiener algebra, whose elements are analytic continuations to $\mathbb{C}^n$ of Fourier transforms $\omega \in (\mathbb{R}^n)^\star \longmapsto \int_{\mathbb{R}^n} e^{-i\langle \omega\,, t\rangle}\, du(t)$ of $\mathbb{C}$-valued distributions u with compact support in $\mathbb{R}^n$.

Given a psh weight function ϕ, often, it will be necessary to introduce, as in Proposition 4.21, its smooth regularization $\phi_{\rho,\varepsilon} = \phi_\varepsilon * \rho$. Any choice of a smooth, usually radial, positive bump function ρ supported by the euclidean ball $\mathbb{B}_{\mathbb{C}^n}(0,1)$ and whose L^1-norm corresponding to $2n$-dimensional Lebesgue measure equal to 1 induces such a regularization ϕ_ε. For such a psh positive weight function ϕ on $\mathbb{C}^n$, let us now introduce the $\mathbb{C}$-subspace of $'\mathcal{D}(\mathbb{C}^n)$, whose elements are ϕ-tempered currents.

---

[47] Despite the terminology used, this concept is disjoint from that of weight in an integral representation formula, introduced and discussed in §4.4, see Definitions 4.17 and 4.23. We differentiate the two concepts by using here the terminology *psh weight function*.

DEFINITION 7.82. A $\mathbb{C}$-valued current on $\mathbb{C}^n$ is said to be ϕ-tempered if and only if each of its coordinate distributions τ along $\mathscr{B}$ is ϕ-tempered. That is, if for each of its coordinate distributions τ, there exist $\alpha, \beta \in \mathbb{N}^n$, $B \geq 0$, and a complex measure $d\mu$, depending on the coordinate distribution τ, such that

$$\tau = \Big(\frac{\partial}{\partial \zeta}\Big)^{\alpha}\Big(\frac{\partial}{\partial \overline{\zeta}}\Big)^{\beta}(d\mu) \quad \text{and} \quad \int_{\mathbb{C}^n}(1+\|\zeta\|)^{|\alpha|+|\beta|}|d\mu(\zeta)|\, e^{-B\phi(\zeta)} < +\infty. \tag{7.213}$$

Given a ϕ-tempered, $\mathbb{C}$-valued current T, let $\mathscr{B}(T)$ be the list of its coordinate distribution coefficients along the basis $\mathscr{B}$ and

$$\begin{aligned} B_\phi(T) := \inf \big\{ & B \geq 0 : \forall \tau \in \mathscr{B}(T), \\ & \tau = \Big(\frac{\partial}{\partial z}\Big)^{\alpha}\Big(\frac{\partial}{\partial \overline{z}}\Big)^{\beta}(d\mu)\,, \int_{\mathbb{C}^n}(1+\|\zeta\|)^{|\alpha|+|\beta|}\, e^{-B\phi(\zeta)}\,|d\mu(\zeta)| < \infty\big\}. \end{aligned} \tag{7.214}$$

If a $\mathbb{C}^r$-valued current T is ϕ-tempered, then $B_\phi(T)$ denotes the maximum of all $B_\phi(T_\iota)$ for the r components T_ι of T.

EXAMPLE 7.83. Let $\phi(z) = \|\mathrm{Im}\, z\| + \log(1+\|z\|^2)/2$ or $\phi(z) = \|z\|^\gamma$ with $\gamma \geq 1$. Since

$$\begin{aligned} & \|\zeta - z\| \leq 1 \\ & \Longrightarrow \big\|\, \|\mathrm{Im}\, z\| - \|\mathrm{Im}\, \zeta\|\, \big\| \leq 1 \text{ and } \big\|\, \|z\|^\gamma - \|\zeta\|^\gamma\, \big\| \leq \gamma(1+\|z\|)^{\gamma-1} = o((1+\|z\|)^\gamma, \end{aligned}$$

a $\mathbb{C}^r$-valued current T is ϕ-tempered if and only if it is $(\phi * \rho)$-tempered, where the regularizing function $\rho : \mathbb{C}^n \to \mathbb{R}^+$ is any smooth bump function supported by the euclidean ball $\overline{\mathbb{B}_{\mathbb{C}^n}(0,1)}$, whose total mass equals 1 with respect to the $2n$-dimensional Lebesgue measure. Moreover, one has

$$B_{\phi*\rho}(T) = B_\phi(T). \tag{7.215}$$

If E is a trivial bundle over $\mathbb{C}^n$, then an E-valued current T is said to be ϕ-tempered if and only all its coordinates along the canonical basis of E are ϕ-tempered. With such a definition in hand, one can state the following result.

PROPOSITION 7.84. *Let $\mathbf{F}_\bullet$ be a generically exact complex of trivial bundles over $\mathbb{C}^n$ such that each morphism $F_\ell : E_\ell \to E_{\ell-1}$, $\ell \in \mathbb{N}^*$, is polynomial. Let $\overline{\mathbf{F}_\bullet}$ be its metrization with the canonical metric on $E(\mathbf{F}_\bullet)$. If $\phi(z) = \log(1+\|z\|^2)/2$, then the currents $P^{\overline{\mathbf{F}_\bullet}}$ and $R^{\overline{\mathbf{F}_\bullet}}$ are ϕ-tempered.*

PROOF. Proposition 7.84 follows from the following observation. Let $p = \{p_\iota\}_\iota$ be a finite collection of polynomials in $\mathbb{C}[X_1, \dots, X_n]$ and let $\mathbb{K} = \mathbb{Q}(\mathrm{coeff}(p)) \subset \mathbb{C}$ be the algebraic extension of $\mathbb{Q}$ with the coefficients of the polynomial entries p_ι and their conjugates. There are polynomials

$$\mathcal{Q}(T, X, Y, U, V) \in \mathbb{K}[T, X_1, \dots, X_n, Y_1, \dots, Y_n, U_1, \dots, U_n, V_1, \dots, V_n], \quad b \in \mathbb{Q}[T],$$

in $4n+1$ and 1 variables respectively such that the formal relation[48]

$$\Big(\sum_\iota p_\iota(X)\overline{p_\iota}(Y)\Big)^\lambda = \frac{1}{b(\lambda)}\mathcal{Q}\Big(\lambda, X, Y, \frac{\partial}{\partial X}, \frac{\partial}{\partial Y}\Big)\Big(\sum_\iota p_\iota(X)\overline{p_\iota}(Y)\Big)^{\lambda+1} \tag{7.216}$$

[48]Formal algebraic relations (7.216) are called *global Bernstein–Sato identities.* We mentioned several times the existence of such identities either in the local or global setting; see (1.6), (1.9), (2.13), or also §C.2.5. Note that multi-Bernstein–Sato formal equations for a collection of polynomials $(Q_1, \dots, Q_m)$ instead of just one, as $Q = \sum_\iota p_\iota(X)\overline{p_\iota}(Y)$ here, also exist; see [**Lic88**] or also [**Sab87, Gy93**] for local results. The polynomial b, once chosen as minimal in $\mathbb{Q}[T]$ is the *Bernstein–Sato polynomial* attached to $\sum_\iota p_\iota(X)p_\iota(Y) \in \mathbb{K}(p)[X, Y]$.

holds; see [**Be72**]. If one specifies X as $(z_1, \dots, z_n)$ and Y as $(\overline{z}_1, \dots, \overline{z}_n)$ and invokes the fact that the z_j and $\overline{z}_k$ behave formally as independent variables (see (7.173)), one deduces from the formal identity (7.216) a true identity

$$\|p(z)\|^{2\lambda} = \frac{1}{b(\lambda)} \mathcal{Q}\Big(\lambda, z, \overline{z}, \frac{\partial}{\partial z}, \frac{\partial}{\partial \overline{z}}\Big) \|p(z)\|^{2(\lambda+1)}, \tag{7.217}$$

provided λ is a complex parameter with $\mathrm{Re}\,\lambda \gg 1$. Combining the identities (7.217) with the analytic continuation principle leads to a concretization in terms of integration by parts of the coordinates along the basis $\mathscr{B}$ of the vector-valued currents $P^{\overline{\mathbf{F}_\bullet}}$ and $R^{\overline{\mathbf{F}_\bullet}}$. From these particular concrete expressions, it follows that the coordinates are ϕ-tempered. □

In the same vein, one has a similar result.

PROPOSITION 7.85. *Let Γ be a free subgroup of $(\mathbb{C}^n, +)$ with rank less than or equal to n. Let also $\mathbf{F}_\bullet$ be a generically exact complex of trivial bundles over $\mathbb{C}^n$ such that each morphism $F_\ell : E_\ell \to E_{\ell-1}$, $\ell \in \mathbb{N}^*$, is given by a matrix whose entries are finite linear combinations of exponential monomial functions*

$$z \in \mathbb{C}^n \longmapsto \exp\langle \gamma, z\rangle \quad (\gamma \in \Gamma).$$

Then the currents $P^{\overline{\mathbf{F}_\bullet}}$ and $R^{\overline{\mathbf{F}_\bullet}}$ are $\|z\|$-tempered. In the particular case where $\Gamma \subset i\,\mathbb{R}^n$, both currents are $\big(\|\mathrm{Im}\, z\| + \log(1 + \|z\|^2)/2\big)$-tempered.

PROOF. We repeat the proof of Proposition 7.84 except that we use formal identities

$$\Big(\sum_\iota p_\iota(e^X)\overline{p_\iota}(e^Y)\Big)^\lambda = \frac{1}{b(\lambda)} \mathcal{Q}\Big(\lambda, e^{\pm X}, e^{\pm Y}, \frac{\partial}{\partial X}, \frac{\partial}{\partial Y}\Big) \Big(\sum_\iota p_\iota(X)\overline{p_\iota}(Y)\Big)^{\lambda+1} \tag{7.218}$$

instead of (7.216) as an initial step in the argument. □

REMARK 7.86. The study of exponential polynomials with coefficients in a subfield $\mathbb{K}$ of $\mathbb{C}$ and exponentiation multi-exponents γ in $\mathbb{K}^n$ (that is, the study of linear combinations of exponential-polynomial monomial functions

$$z \in \mathbb{C}^n \longmapsto z^\alpha \exp\langle \gamma, z\rangle \quad (\alpha \in \mathbb{N}^n,\ \gamma \in \mathbb{K}^n),$$

whose coefficients are in $\mathbb{K}$) leads to deep questions of arithmetic and diophantine nature, since it appeals to Ax–Schanuel's theorem [**Ax71, Col80**]. A conjecture arose from [**BY86, BY95**]: does Proposition 7.85 remain true under the hypothesis that the entries of each F_ℓ in $\mathbf{F}_\bullet$ are exponential polynomials with coefficients in $\overline{\mathbb{Q}}$ and exponentiation multi-exponents in $\overline{\mathbb{Q}}^n$? We refer the reader to [**Y13**] for related references about this question that go back to Leon Ehrenpreis's pioneering monograph [**Ehr**].

7.6.2. Trace formulae in weighted algebras of entire functions. In the present subsection ϕ denotes either the so-called Paley–Wiener psh weight function $\phi(z) = \|\mathrm{Im}(z)\| + \log(1 + \|z\|^2)/2$ or the radial weight $\phi(z) = \|z\|^\gamma$ with $\gamma \geq 1$. Because of Example 7.83, we focus now on a smooth, positive bump function $\rho : \mathbb{C}^n \to \mathbb{R}^+$ supported by the euclidean ball $\overline{\mathbb{B}_{\mathbb{C}^n}(0,1)}$, whose mass equals 1 for $2n$-dimensional Lebesgue measure. The bump function ρ will allow us to overcome the possible nonsmoothness of ϕ; see Example 7.83. We also need to fix once and for all a cut-off function $\chi \in \mathcal{D}(\mathbb{C}^n, [0,1])$ which is identically equal to 1 in a neighborhood of $\overline{\mathbb{B}_{\mathbb{C}^n}(0,1)}$. The role of χ will be essential in realizing the Bochner–Martinelli weighted integral representation formula playing a pivotal role throughout this subsection. We keep such ρ and χ fixed until the end of this subsection.

Now, let $f = (f_1, \dots, f_m)$ be a sequence of elements in $\mathbf{A}_\phi(\mathbb{C}^n)$ satisfying three crucial assumptions:

$$1 \leq m \leq n, \quad \operatorname{codim}_{\mathbb{C}^n} V(f) = m, \quad \overline{\partial}(1/f) \text{ is } \phi - \text{tempered} \tag{7.219}$$

in the sense of Definition 7.82, where the $(0,m)$ $\mathbb{C}$-valued Coleff–Herrera residue current $\overline{\partial}(1/f)$ was introduced in Definition 2.78. As usual $\{\epsilon_1, \dots, \epsilon_m\}$ denotes the canonical basis of $\mathbb{C}^m$, so that $f = \sum_1^m f_j \epsilon_j$. The generically exact complex attached to f is the Koszul complex $\mathbf{K}_\bullet^f$. It follows from conditions (7.211) for the weight function (see for example [**BT80**, Corollary 2.5]) that one can construct a Hefer form $\mathbb{H}_f$ for f, as in Definition 4.46 and (4.105) with $\ell = 0$, $r = 1$. It is of the form

$$\mathbb{H}_f(z,\zeta) = \sum_{j=1}^m h_{f,j}(z,\zeta)\, \epsilon_j, \tag{7.220}$$

where each $h_{f,j}(z,\zeta) = \sum_{k=1}^n h_{f,j,k}(z,\zeta)\, d\zeta_k$ satisfies

$$f_j(\zeta) - f_j(z) = \sum_{k=1}^n h_{f,j,k}(z,\zeta)(\zeta_k - z_k), \tag{7.221}$$

for $j = 1, \dots, m$ and

$$\max_{1 \leq j \leq m, 1 \leq k \leq n} |h_{f,j,k}(z,\zeta)| \leq A_f \exp\big(B_f(\phi(z) + \phi(\zeta))\big) \tag{7.222}$$

for some positive constants A_f and B_f. The Hefer form $\mathbb{H}_f$, together with

$$\mathbb{I} = \sum_{j=1}^m \epsilon_j^* \otimes \epsilon_j, \tag{7.223}$$

generate a solution for the $\mathbf{K}_\bullet^f$-Hefer problem with respect to the Koszul complex $\mathscr{K}_\bullet^{\mathscr{D}}$ attached to $(\mathscr{E}, \Delta)$ as in (4.100) with $\Omega = \mathbb{C}^n$; see Proposition 4.47. Proposition 4.29 applied on $\mathbb{C}^n \times \mathbb{C}^n \setminus \{(z,\zeta) : f(z) = 0\}$ implies that the global Bochner–Martinelli weight[49] attached to such a solution of the Hefer problem is described by

$$\begin{aligned} \mathscr{W}^{\overline{\mathbf{K}_\bullet^f}}(z,\zeta) &= 2i\pi \rfloor_{f(z)} H_1(z,\zeta)\, P_0^{\overline{\mathbf{K}_\bullet^f}}(\zeta) + H_0(z,\zeta)\, R_0^{\overline{\mathbf{K}_\bullet^f}}(\zeta) \\ &= 2i\pi \rfloor_{f(z)} H_1(z,\zeta)\, P_0^{\overline{\mathbf{K}_\bullet^f}}(\zeta) + \frac{\mathbb{H}_f^m(z,\zeta)}{m!}\, \overline{\partial}\Big(\frac{1}{f}\Big)\, \epsilon_1^* \wedge \cdots \wedge \epsilon_m^*, \end{aligned} \tag{7.224}$$

where the second equality follows from the fact that $\operatorname{codim} V(f) = m$; see Proposition 3.18 and Example 3.30.

Global Bochner–Martinelli weights attached to the transcendental exponential function will also play a crucial role in this subsection. These are the global Bochner–Martinelli weights $\mathscr{W}_{|\mathrm{Im}|}$ and $\mathscr{W}_{|\ |^\gamma}$ attached to the psh positive functions

[49] Be careful here. We consider the notion of weight is within the context of §4.4. We will talk about Bochner–Martinelli weight to avoid confusion.

$z \longmapsto \|\mathrm{Im}\, z\|$ and $z \longmapsto \|z\|^\gamma$ respectively, where $\gamma \geq 1$. These global weights are described in Proposition 4.21, namely

$$\begin{aligned} \mathscr{W}_{\|\mathrm{Im}\|}(z,\zeta) &= \Big[1 + 2\sum_{k=1}^n \frac{\partial\,(\|\mathrm{Im}\| * \rho)}{\partial \zeta_k}(\zeta)\,(z_k - \zeta_k) + 2\,dd^c(\|\mathrm{Im}\| * \rho)\,(\zeta)\Big], \\ \mathscr{W}_{\|\ \|^\gamma}(z,\zeta) &= \Big[1 + 2\sum_{k=1}^n \frac{\partial\,(\|\ \|^\gamma * \rho)}{\partial \zeta_k}(\zeta)\,(z_k - \zeta_k) + 2\,dd^c(\|\ \|^\gamma * \rho)\,(\zeta)\Big]. \end{aligned} \tag{7.225}$$

Similarly, their exponentials are defined by

$$\begin{aligned} e^{C(\mathscr{W}_{\|\mathrm{Im}\|}-1)} &= \exp\Big(2\,C\,\Big\langle \frac{\partial\,(\|\mathrm{Im}\| * \rho)}{\partial \zeta}\,,\, z - \zeta\Big\rangle\Big)\,\Big(\sum_{r=0}^n \frac{\big(2C\,dd^c(\|\mathrm{Im}\| * \rho)\big)^r(\zeta)}{r!}\Big), \\ e^{C(\mathscr{W}_{\|\ \|^\gamma}-1)} &= \exp\Big(2\,C\,\Big\langle \frac{\partial\,(\|\ \|^\gamma * \rho)}{\partial \zeta}\,,\, z - \zeta\Big\rangle\Big)\,\Big(\sum_{r=0}^n \frac{\big(2C\,dd^c(\|\ \|^\gamma * \rho)\big)^r(\zeta)}{r!}\Big) \end{aligned} \tag{7.226}$$

for $C > 0$. Finally, the global Bochner–Martinelli weight

$$\mathscr{W}_{\mathbb{C}^n} : (z,\zeta) \longmapsto \frac{1 + \sum_{k=1}^n z_k \overline{\zeta}_k}{1 + \|\zeta\|^2} + dd^c \log(1 + \|\zeta\|^2) \tag{7.227}$$

(see Proposition 4.18) completes the list of Bochner–Martinelli weights naturally. We will use them in this subsection to realize trace formulae through integral representation ones. The introduction of weights (7.226) (respectively (7.227)) attached to exponential (respectively polynomial) growth in $\mathbb{C}^n$ is justified by the definition of ϕ-tempered distributions on $\mathbb{C}^n$ in terms of tempered ones; see Definition 7.82. For any $C > 0$, $N \in \mathbb{N}^*$, we introduce the Bochner–Martinelli weights

$$\begin{aligned} \mathscr{W}_{\|\mathrm{Im}\|,C,N}[f] &:= \mathscr{W}_{\mathbb{C}^n}^N\, e^{C(\mathscr{W}_{\|\mathrm{Im}\|}-1)}\, \mathscr{W}^{\overline{\mathbf{K}_\bullet^f}}, \\ \mathscr{W}_{\|\ \|^\gamma,C,N}[f] &:= \mathscr{W}_{\mathbb{C}^n}^N\, e^{C(\mathscr{W}_{\|\ \|^\gamma}-1)}\, \mathscr{W}^{\overline{\mathbf{K}_\bullet^f}} \quad (\gamma \geq 1). \end{aligned} \tag{7.228}$$

Let also

$$q\ : (z,\zeta) \in \{\mathbb{C}^n \times \mathbb{C}^n\ :\ \|z\| < \|\zeta\|\} \longmapsto \frac{\partial \|\zeta\|^2}{\langle \overline{\zeta}\,,\, \zeta - z\rangle}, \tag{7.229}$$

so that

$$\|z\| < \|\zeta\| \Longrightarrow \rfloor_{\sum_{k=1}^n (\zeta_k - z_k)\partial/\partial\zeta_k} q(z,\zeta) = \rfloor_{\Delta(z,\zeta)} q(z,\zeta) = 1. \tag{7.230}$$

The integral representation we invoke in this subsection is derived from Proposition 4.33. For any $\ell \in \mathbb{N}^*$, any $h \in \mathbf{A}_\phi(\mathbb{C}^n)$ admits in $\mathbb{B}_{\mathbb{C}^n}(0,\ell) \setminus V(f)$ the integral representation formulae

$$\begin{aligned} h(z) &= \int_{\mathbb{C}^n} \Big(\Big(\chi(\zeta/\ell) + \frac{\overline{\partial}_\zeta\big((1 - \chi(\zeta/\ell))\, q(z,\zeta)\big)}{2i\pi}\Big)^{n+1} \mathscr{W}_{\|\mathrm{Im}\|,C,N}[f](z,\zeta)\Big)_{n,n} h(\zeta) \\ &= \int_{\mathbb{C}^n} \Big(\Big(\chi(\zeta/\ell) + \frac{\overline{\partial}_\zeta\big((1 - \chi(\zeta/\ell))\, q(z,\zeta)\big)}{2i\pi}\Big)^{n+1} \mathscr{W}_{\|\ \|^\gamma,C,N}[f](z,\zeta)\Big)_{n,n} h(\zeta). \end{aligned} \tag{7.231}$$

The following theorem describes how such integral representation formulae (for any $\ell \in \mathbb{N}^*$, then letting ℓ go to infinity) produce explicit trace formulae in $\mathbf{A}_\phi(\mathbb{C}^n)$ in the case of the Paley–Wiener weight; see also [**Vid19**].

THEOREM 7.87. *Let $\mu_1, \dots, \mu_m$ be m compactly supported distributions in $\mathbb{R}^n$ satisfying $\mathrm{Supp}(\mu_j) \subset \mathbb{B}_{\mathbb{R}^n}(0, R_j)$. Assume that $\mathrm{codim}_{\mathbb{C}^n} V(\widehat{\mu}) = m$ and that the $(0,m)$-current $\overline{\partial}(1/\widehat{\mu}) = \overline{\partial}(1/(\widehat{\mu}_1, \dots, \widehat{\mu}_m))$ is ϕ-tempered, where ϕ is the Paley–Wiener weight $\phi(z) = \|\mathrm{Im}(z)\| + \log(1 + \|z\|^2)/2$. For each $0 < \varepsilon < 1$, $R > 0$, $M \in \mathbb{N}^*$, there exist a Hefer form $\mathbb{H}_{\widehat{\mu},\varepsilon}$ and an explicit smooth kernel $K_{\varepsilon,R,M}(z,\zeta)$, depending on the supports and the orders of the $\widehat{\mu}_j$'s, with bidegree $(0,0)$ in z and $(n-m, n-m)$ in ζ, holomorphic in z, such that the following assertion holds: any compactly supported distribution u in $\mathbb{R}^n$ with $\mathrm{Supp}\, u \subset \overline{\mathbb{B}_{\mathbb{R}^n}(0, R)}$ and order less than M is such that the function*

(7.232)

$$z \longmapsto \mathrm{Tr}_{\varepsilon,R,M}[\widehat{u} : \widehat{\mu}](z) = \int_{\mathbb{C}^n} \widehat{u}(\zeta)\, K_{\varepsilon,R,M}(z,\zeta)\, \frac{(\mathbb{H}_{\widehat{\mu},\varepsilon}(z,\zeta))^m}{m!}\, \overline{\partial}\Big(\frac{1}{\overline{\widehat{\mu}}}\Big)\, \epsilon_1^* \wedge \cdots \wedge \epsilon_m^*$$

is a well defined entire function on $\mathbb{C}^n$ belonging to $\mathbf{A}_\phi(\mathbb{C}^n)$. Moreover, for any such u, this entire function satisfies the growth estimates

$$\sup_{z \in \mathbb{C}^n} \frac{\big|\mathrm{Tr}_{\varepsilon,R,M}[\widehat{u} : \widehat{\mu}](z)\big|}{(1 + \|z\|)^{[B_\phi(\overline{\partial}(1/\widehat{\mu}))] + N(\mu,M)}}\, e^{-(B_\phi(\overline{\partial}(1/\widehat{\mu})) + B(\mu,R) + \varepsilon)\|\mathrm{Im}(z)\|} < +\infty, \tag{7.233}$$

where $N(\mu, M) := M + 2(\sum_1^m \mathrm{ord}(\mu_j)) + 1$, $B(\mu, R) = R + \sum_1^m R_j$. The exponent $B_\phi(\overline{\partial}(1/\widehat{\mu}))$ is attached to the ϕ-tempered residue current $\overline{\partial}(1/\widehat{\mu})$ as in (7.214). Furthermore, most important is the fact that one has the following trace formula:

$$\widehat{u} - \mathrm{Tr}_{\varepsilon,R,M}[\widehat{u} : \widehat{\mu}] \in \sum_{j=1}^m \widehat{\mu}_j\, \mathcal{O}(\mathbb{C}^n). \tag{7.234}$$

PROOF. For any $0 < \varepsilon < 1$, one can represent each distribution μ_j with support in $\overline{\mathbb{B}_{\mathbb{R}^n}(0, R_j)}$ as a sum of derivatives of measures $\mu_{j,\varepsilon,\iota}$ supported by the euclidean ball $\overline{\mathbb{B}_{\mathbb{R}^n}(0, R_j + \varepsilon/(3m))}$. The order of these derivatives is at most $\mathrm{ord}(\mu_j)$. Given such measure $\mu_{j,\varepsilon,\iota}$, one has

$$\begin{aligned}\widehat{\mu}_{j,\varepsilon,\iota}(\zeta) - \widehat{\mu}_{j,\varepsilon,\iota}(z) &= \int (e^{-i\langle \zeta, t\rangle} - e^{-i\langle z, t\rangle})\, d\mu_{j,\varepsilon,\iota}(t) \\ &= \sum_{k=1}^n \Big(\int_{\mathbb{R}^n} \Big(\int_0^1 e^{-i\langle z + \varpi(\zeta - z)\,,\, t\rangle}\, d\varpi \Big) (-it_k)\, d\mu_{j,\varepsilon,\iota}(t) \Big) (\zeta_k - z_k).\end{aligned}$$

As a consequence, one can construct a Hefer form $(z,\zeta) \longmapsto \mathbb{H}_{\widehat{\mu},\varepsilon}(z,\zeta)$ for $\widehat{\mu}$ as in (7.220) satisfying

(7.235)

$$\sup_{z,\zeta} \frac{\|\mathbb{H}_{\widehat{\mu},\varepsilon}(z,\zeta)\|^m}{(1 + \|z\| + \|\zeta\|)^{\sum_1^m \mathrm{ord}(\mu_j)}}\, e^{-(\sum_{j=1}^m R_j + \varepsilon/3)\,((\|\mathrm{Im}\| * \rho)(z) + (\|\mathrm{Im}\| * \rho)(\zeta))} < +\infty.$$

Moreover, if

$$\widehat{u}(\mathbb{H}_{\widehat{\mu},\varepsilon}(z,\zeta))^m = \sum_{\# K = m} \Big(\widehat{u}(\mathbb{H}_{\widehat{\mu},\varepsilon}(z,\zeta))^m \Big)_{d\overline{z}_K} d\overline{z}_K,$$

then the same argument shows that for any multi-index $K \subset \{1,\dots,n\}$ such that $|K| = m$ and for any multi-index $\alpha \in \mathbb{N}^n$ for any $(\zeta, z) \in \mathbb{C}^n \times \mathbb{C}^n$,

$$\sup_{z,\zeta\in\mathbb{C}^n} \frac{\big|(\partial/\partial\zeta)^\alpha \big(\widehat{u}(\mathbb{H}_{\widehat{\mu},\varepsilon}(z,\zeta))^m\big)_{d\overline{z}_K}\big|}{(1+\|\zeta\|)^M\,(1+\|\zeta\|+\|z\|)^{\sum_1^m \mathrm{ord}(\mu_j)}} \times e^{-(R+\sum_{j=1}^m R_j+2\,\varepsilon/3)((\|\mathrm{Im}\|*\rho)(z)+(\|\mathrm{Im}\|*\rho)(\zeta)))} < +\infty. \tag{7.236}$$

Let us decompose the Bochner–Martinelli weight $\mathscr{W}^{\overline{\mathbf{K}^f_\bullet}}$ from (7.224). Fix for the time being $N \in \mathbb{N}^*$ and $C > 0$. Then represent the restriction of $\widehat{u}$ to $\mathbb{B}_{\mathbb{C}^n}(0,\ell)$, where $\ell \in \mathbb{N}^*$, using the first integral representation formula in (7.231) as

$$\begin{aligned}\widehat{u}(z) = \int_{\mathbb{C}^n} \widehat{u}(\zeta)\,\Big(\Big(\chi(\zeta/\ell) + \frac{\overline{\partial}_\zeta\big((1-\chi(\zeta/\ell))\,q(z,\zeta)\big)}{2i\pi}\Big)^{n+1} \Omega_{C,N}(z,\zeta)\Big)_{n-m,n-m} \\ \times \frac{(\mathbb{H}_{\widehat{\mu},\varepsilon}(z,\zeta))^m}{m!}\,\overline{\partial}\Big(\frac{1}{\widehat{\mu}}\Big)\,\epsilon_1^*\wedge\cdots\wedge\epsilon_m^* \quad \text{modulo} \sum_{j=1}^m \widehat{\mu_j}\,\mathcal{O}(\mathbb{B}_{\mathbb{C}^n}(0,\ell)).\end{aligned} \tag{7.237}$$

In (7.237) one has that

$$\begin{aligned}\Omega_{C,N}(z,\zeta) = \exp\Big(2\,C\Big\langle\frac{\partial\,(\|\mathrm{Im}\|*\rho)}{\partial\zeta}, z-\zeta\Big\rangle\Big)\,\big(\mathscr{W}_{\mathbb{C}^n}(z,\zeta)\big)^N \\ \times\Big(\sum_{r=0}^n \frac{\big(2\,C\,dd^c(\|\mathrm{Im}\|*\rho)\big)^r(\zeta)}{r!}\Big),\end{aligned} \tag{7.238}$$

using (7.228) and (7.226). As pointed out in Remark 4.22, the plurisubharmonicity of $\|\mathrm{Im}\|*\rho$ implies the convexity inequality

$$\begin{aligned}\Big|\exp\Big(2\,C\Big\langle\frac{\partial\,(\|\mathrm{Im}\|*\rho)}{\partial\zeta}, z-\zeta\Big\rangle\Big)\Big| &= \exp\Big(2\,C\ \mathrm{Re}\Big(\Big\langle\frac{\partial\,(\|\mathrm{Im}\|*\rho)}{\partial\zeta}, z-\zeta\Big\rangle\Big)\Big) \\ &\le \frac{\exp\big(C((\|\mathrm{Im}\|*\rho)(z)\big)}{\exp\big(C(\|\mathrm{Im}\|*\rho)(\zeta)\big)} \quad \text{for any} \quad (z,\zeta)\in\mathbb{C}^n\times\mathbb{C}^n.\end{aligned} \tag{7.239}$$

Also, the fact that $(\overline{\partial}(1/\widehat{\mu})$ is $(\|\mathrm{Im}\,z\| + \log(1+\|z\|^2)/2)$-tempered, the definition of $B_\phi(\overline{\partial}(1/\widehat{\mu}))$ (see (7.214)) once combined with (7.215), imply that for any $\varepsilon > 0$ any distribution coordinate coefficient τ of the $(0,m)$-current $\overline{\partial}(1/\widehat{\mu})$ along the basis of differential forms $\{d\overline{z}_K : |K| = m\}$ is of the form

$$\tau = \Big(\frac{\partial}{\partial\zeta}\Big)^\alpha \Big(\frac{\partial}{\partial\overline{\zeta}}\Big)^\beta (d\mu), \tag{7.240}$$

where $\alpha, \beta \in \mathbb{N}^n$ and $\theta \in C^0(\mathbb{C}^n)$ is such that

$$\int_{\mathbb{C}^n} \frac{|d\mu(\zeta)|}{(1+\|\zeta\|)^{B_\phi(\overline{\partial}(1/\widehat{\mu}))-|\alpha|-|\beta|+\varepsilon/3}} e^{-(B_\phi(\overline{\partial}(1/\widehat{\mu}))+\epsilon/3)\,(\|\mathrm{Im}\|*\rho)(\zeta))} < +\infty. \tag{7.241}$$

Let us now specify $N \in \mathbb{N}^*$ and $C = C_\varepsilon > 0$ as

$$
\begin{aligned}
N &= M + \sum_1^m \mathrm{ord}(\mu_j) + \big[B_\phi\big(\overline{\partial}(1/\widehat{\mu})\big)\big] + 1, \\
C &= C_\varepsilon = R + \sum_{j=1}^m R_j + B_\phi\big(\overline{\partial}(1/\widehat{\mu})\big) + \varepsilon.
\end{aligned}
\tag{7.242}
$$

Using the explicit representation (7.240) of all distributions $\tau \in \mathscr{B}(\overline{\partial}(1/\widehat{\mu}))$, together with the growth estimates (7.236) and (7.241), the expressions (7.238) of the smooth form $\Omega_{C_\varepsilon,N_\varepsilon}$, combined with that for the Bochner–Martinelli weight $\mathscr{W}_{\mathbb{C}^n}$ (7.227), the convexity inequality (7.239), and, finally, the fact that the derivatives up to a prescribed order M of $\|\mathrm{Im}\| * \rho$ are uniformly bounded in $\mathbb{C}^n$ by $C(M)$, allow us to show that the integral

$$
\int_{\mathbb{C}^n} \widehat{u}(\zeta)\big(\Omega_{C_\epsilon,N}(z,\zeta)\big)_{n-m,n-m} \frac{(\mathbb{H}_{\widehat{\mu},\varepsilon}(z,\zeta))^m}{m!}\, \overline{\partial}\Big(\frac{1}{\widehat{\mu}}\Big)\, \epsilon_1^* \wedge \cdots \wedge \epsilon_m^*
$$

is absolutely convergent. Moreover, it defines, as a function of z, an entire function $\mathrm{Tr}_{\varepsilon,R,M}[\widehat{u} : \widehat{\mu}]$ belonging to $\mathbf{A}_\phi(\mathbb{C}^n)$. The function $\mathrm{Tr}[\widehat{u} : \widehat{\mu}]$ satisfies the growth estimates

$$
\begin{aligned}
\Big| \int_{\mathbb{C}^n} & \widehat{u}(\zeta)\big(\Omega_{C_\epsilon,N}(z,\zeta)\big)_{n-m,n-m} \frac{(\mathbb{H}_{\widehat{\mu},\varepsilon}(z,\zeta))^m}{m!}\, \overline{\partial}\Big(\frac{1}{\widehat{\mu}}\Big)\, \epsilon_1^* \wedge \cdots \wedge \epsilon_m^* \Big| \\
&\le A_\varepsilon (1+\|z\|)^{[B_\phi(\overline{\partial}(1/\widehat{\mu}))] + N(\mu,M)} e^{(B_\phi(\overline{\partial}(1/\widehat{\mu})) + B(\mu,R) + \varepsilon)\|\mathrm{Im}(z)\|)}.
\end{aligned}
$$

When $\|z\| < \ell_0$, the sequence of holomorphic functions

$$
\begin{aligned}
& z \in \mathbb{B}_{\mathbb{C}^n}(0,\ell_0) \longmapsto \\
& \int_{\mathbb{C}^n} \Big(\chi\Big(\frac{\zeta}{\ell}\Big)\Big)^{n+1} \widehat{u}(\zeta)\big(\Omega_{C_\epsilon,N}(z,\zeta)\big)_{n-m,n-m} \frac{(\mathbb{H}_{\widehat{\mu},\varepsilon}(z,\zeta))^m}{m!}\, \overline{\partial}\Big(\frac{1}{\widehat{\mu}}\Big)\, \epsilon_1^* \wedge \cdots \wedge \epsilon_m^*,
\end{aligned}
$$

for $\ell \ge \ell_0$ converges uniformly on any compact subset of $B_{\mathbb{C}^n}(0,\ell_0)$ towards the restriction of $\mathrm{Tr}_{\varepsilon,R,M}[\widehat{u} : \widehat{\mu}]$ to $B_{\mathbb{C}^n}(0,\ell_0)$. Moreover, the same reasoning shows that the sequence of functions

$$
\begin{aligned}
& z \in \mathbb{B}_{\mathbb{C}^n}(0,\ell_0) \longmapsto \\
& \int_{\mathbb{C}^n} \widehat{u}(\zeta)\Big(\Big(\chi(\zeta/\ell) + \frac{\overline{\partial}_\zeta\big((1-\chi(\zeta/\ell))\, q(z,\zeta)\big)}{2i\pi}\Big)^{n+1} \Omega_{C_\varepsilon,N}(z,\zeta)\Big)_{n-m,n-m} \\
& \qquad \times \frac{(\mathbb{H}_{\widehat{\mu},\varepsilon}(z,\zeta))^m}{m!}\, \overline{\partial}\Big(\frac{1}{\widehat{\mu}}\Big)\, \epsilon_1^* \wedge \cdots \wedge \epsilon_m^* \\
& - \int_{\mathbb{C}^n} \Big(\chi\Big(\frac{\zeta}{\ell}\Big)\Big)^{n+1} \widehat{u}(\zeta)\big(\Omega_{C_\epsilon,N}(z,\zeta)\big)_{n-m,n-m} \frac{(\mathbb{H}_{\widehat{\mu},\varepsilon}(z,\zeta))^m}{m!}\, \overline{\partial}\Big(\frac{1}{\widehat{\mu}}\Big)\, \epsilon_1^* \wedge \cdots \wedge \epsilon_m^*
\end{aligned}
$$

converges uniformly towards 0 on any compact subset of $\mathbb{B}_{\mathbb{C}^n}(0,\ell_0)$. The fact that the entire function $\widehat{u}$ admits in each $B_{\mathbb{C}^n}(0,\ell)$, where $\ell \in \mathbb{N}^*$, the integral representation formula (7.237) (with N and $C = C_\varepsilon$ as in (7.242)) modulo the ideal $\sum_{j=1}^m \widehat{\mu_j}\, \mathcal{O}(\mathbb{B}_{\mathbb{C}^n}(0,\ell))$ then concludes the proof of the theorem. □

REMARK 7.88. Let the principal value current $P_0^{\mathbf{K}\bullet^{\widehat{\mu}}}$ be also ϕ-tempered, where $\phi(z) = \|\mathrm{Im} z\| + \log(1+\|z\|^2)/2$. Then for $C > 0$, $N \in \mathbb{N}^*$ and $0 < \varepsilon < 1$, one could also consider, by analogy to (7.228), the $(\mathbb{C}^m)^*$ valued current

$$\begin{aligned}\mathscr{W}^{\circ}_{\|\mathrm{Im}\|,C,N}[\widehat{\mu}] &= \mathscr{W}^N_{\mathbb{C}^n}\, e^{C(\mathscr{W}_{\|\mathrm{Im}\|}-1)}\, H_1(z,\zeta)\, P_0^{\overline{\mathbf{K}_\bullet}^{\widehat{\mu}}}(\zeta)\\ &= \mathscr{W}^N_{\mathbb{C}^n}\, e^{C(\mathscr{W}_{\|\mathrm{Im}\|}-1)}\Big(\sum_{r=1}^m \mathbb{I}\,\frac{\mathbb{H}^{r-1}_{\widehat{\mu},\varepsilon}}{(r-1)!}\,P_{0,r}^{\overline{\mathbf{K}_\bullet}^{\widehat{\mu}}}\Big),\end{aligned} \tag{7.243}$$

where $\mathbb{I}$ is from (7.223). One can reinforce the choice of C,N in terms of $\widehat{\mu}$, R, M and the associated constants $B_\phi(\overline{\partial}(1/\widehat{\mu}))$ and $B_\phi(P_0^{\overline{\mathbf{K}_\bullet}^{\widehat{\mu}}})$ in order to construct for each $r=1,\dots,m$ a smooth form $K^{(1:r)}_{\varepsilon,R,M}$ with bidegree $(0,0)$ in the variables z and $(n-(r-1),n-(r-1))$ in ζ, holomorphic in z, and a smooth form $K^{(0:m)}_{\varepsilon,R,M}$ with bidegree $(0,0)$ in the variables z and $(n-m,n-m)$ in ζ such that the complete division formula

$$\begin{aligned}\widehat{u}(z) = 2i\pi \,\rfloor_{\widehat{\mu}(z)} \int_{\mathbb{C}^n} \widehat{u}(\zeta)\Big(\sum_{r=1}^m K^{(1:r)}_{\varepsilon,R,M}(z,\zeta)\,\mathbb{I}\,\frac{\mathbb{H}^{r-1}_{\widehat{\mu},\varepsilon}(z,\zeta)}{(r-1)!}\,P_{0,r}^{\overline{\mathbf{K}_\bullet}^{\widehat{\mu}}}(\zeta)\Big)\\ + \int_{\mathbb{C}^n} \widehat{u}(\zeta)\,K^{(0:m)}_{\varepsilon,R,M}(z,\zeta)\,\frac{\mathbb{H}^m_{\widehat{\mu},\varepsilon}(z,\zeta)}{m!}\,\overline{\partial}\Big(\frac{1}{\widehat{\mu}}\Big)\,\epsilon_1^*\wedge\cdots\wedge\epsilon_m^*\end{aligned} \tag{7.244}$$

holds in $\mathbf{A}_\phi(\mathbb{C}^n)$. Note also that the smooth forms $K^{(1:r)}_{\varepsilon,R,M}$ and $K^{(0:m)}_{\varepsilon,R,M}$ depend on $\widehat{\mu}$. It leads to an explicit division formula

$$\begin{aligned}\widehat{u}(z) &= \int_{\mathbb{C}^n} \widehat{u}(\zeta)\,K^{(0:m)}_{\varepsilon,R,M}(z,\zeta)\,\frac{\mathbb{H}^m_{\widehat{\mu},\varepsilon}(z,\zeta)}{m!}\,\overline{\partial}\Big(\frac{1}{\widehat{\mu}}\Big)\,\epsilon_1^*\wedge\cdots\wedge\epsilon_m^* + \sum_{j=1}^m \widehat{u}_j(z)\,\widehat{\mu}_j(z)\\ &= \mathrm{Tr}^{\circ}_{\varepsilon,R,M}[\widehat{u}:\widehat{\mu}](z) + \sum_{j=1}^m \widehat{u}_j(z)\,\widehat{\mu}_j(z),\end{aligned} \tag{7.245}$$

but this time with estimates

$$\begin{aligned}\sup_{z\in\mathbb{C}^n}\Big(\frac{\max_{1\le j\le m}\big(\big|\mathrm{Tr}^{\circ}_{\varepsilon,R,M}[\widehat{u}:\widehat{\mu}](z)\big|,|\widehat{u}_j(z)|\big)}{(1+\|z\|)^{[B_\phi(\overline{\partial}(1/\widehat{\mu}),P_0^{\overline{\mathbf{K}_\bullet}^{\widehat{\mu}}}))]+N(\mu,M)}}\\ \times e^{-(B_\phi((\overline{\partial}(1/\widehat{\mu}),P_0^{\overline{\mathbf{K}_\bullet}^{\widehat{\mu}}}))+B(\mu,R)+\varepsilon)\|\mathrm{Im}(z)\|}\Big) < +\infty.\end{aligned} \tag{7.246}$$

The following important result follows from Theorem 7.87 and concerns mean periodicity.[50]

COROLLARY 7.89. *Let ϕ, $\mu_1,\dots,\mu_m$ be as in Theorem 7.87, with in addition* $\mathrm{ord}(\mu) = \max_{1\le j\le m}\mathrm{ord}(\mu_j)$ *and* $R(\mu) = \max_{1\le j\le m} R_j$. *Furthermore, assume that*[51]

$$\mathbf{A}_\phi(\mathbb{C}^n)\cap\Big(\sum_{j=1}^m \widehat{\mu}_j\,\mathcal{O}(\mathbb{C}^n)\Big) = \sum_{j=1}^m \widehat{\mu_j}\,\mathbf{A}_\phi(\mathbb{C}^n). \tag{7.247}$$

---

[50] The notation used in the statement, as well as in the proof, refers to one introduced in Theorem 7.87 or in Remark 7.88.

[51] Observe that this condition is satisfied under the additional assumption suggested in the previous Remark 7.88.

Let $\theta \in C^\infty(\mathbb{R}^n)$ be mean periodic for the μ_j's acting by convolution. Namely, assume that the equations

$$\int_{\mathbb{R}^n} \theta(t-\xi)\, d\mu_j(\xi) = (\mu_j * \theta)(t) = 0 \text{ for any } t \in \mathbb{R}^n \tag{7.248}$$

hold for any $j = 1, \dots, m$. Let ε, R, R' be such that

$$0 < \varepsilon < 1, \quad R' \geq R + R(\mu)$$

and $\eta_{\varepsilon,R'} \in \mathcal{D}(\mathbb{R}^n, [0,1])$ be a radial cut-off function identically equal to 1 in a neighborhood of the euclidean ball

$$\big\{t \in \mathbb{R}^n \,:\, \|t\| \leq B_\phi(\overline{\partial}(1/\widehat{\mu})) + R' + R(\mu) + \varepsilon\big\}. \tag{7.249}$$

If $M \geq \operatorname{ord}(\mu)$, then the mean-periodic function θ is reproduced in $\mathbb{B}_{\mathbb{R}^n}(0,R)$ by

$$\theta(t) = \tag{7.250}$$
$$\int_{\mathbb{C}^n} e^{i\langle\zeta,t\rangle} \Big(\frac{1}{(2\pi)^n}\int_{\mathbb{R}^m} \widehat{\eta_{\varepsilon,R'}\,\theta}(\omega)\, K_{\varepsilon,R',M}(\omega,\zeta)\, \frac{(\mathbb{H}_{\widehat{\mu},\varepsilon}(\omega,\zeta))^m}{m!} d\omega\Big)\, \overline{\partial}\Big(\frac{1}{\widehat{\mu}}\Big)\, \epsilon^*,$$

where $\epsilon^* = \epsilon_1^* \wedge \cdots \wedge \epsilon_m^*$ is dual of the canonical basis $\{\epsilon_1, \dots, \epsilon_m\}$ for $\mathbb{C}^m$.

Proof. It follows from Theorem 7.87 and the Paley–Wiener theorem [**Hor**, Theorem 7.3.1] that, for any $t \in \mathbb{B}_{\mathbb{R}^n}(0,R)$, the entire function

$$\operatorname{Tr}_{\varepsilon,R',M}[\widehat{\delta_t} \,:\, \widehat{\mu}] :$$
$$z \longmapsto \int_{\mathbb{C}^n} e^{-i\langle\zeta\,,\,t\rangle}\, K_{\varepsilon,R',M}(z,\zeta)\, \frac{(\mathbb{H}_{\widehat{\mu},\varepsilon}(z,\zeta))^m}{m!}\, \overline{\partial}\Big(\frac{1}{\widehat{\mu}}\Big)\, \epsilon_1^* \wedge \cdots \wedge \epsilon_m^*$$

is the analytic continuation to the complex plane of the Fourier transform of a distribution $u_{\varepsilon,R',M}[t]$ on $\mathbb{R}^n$ with compact support in the euclidean ball (7.249). The hypothesis (7.247), once combined with (7.234), implies that there are distributions $u_{\varepsilon,R',M,j}[t]$, $j = 1, \dots, m$, in $\mathbb{R}^n$, with compact support, such that

$$\delta_t - u_{\varepsilon,R',M}[t] = \sum_{j=1}^m u_{\varepsilon,R',M,j}[t] * \mu_j.$$

The above fact implies $(\delta_t * \theta)(\xi) = \theta(\xi - t) = (u_{\varepsilon,R',M}[t] * \theta)(\xi)$ for all $\xi \in \mathbb{R}^n$, since θ is mean periodic (7.248). Applying this at $\xi = 0$, then changing t into $-t$, one gets, thanks to the hypothesis on the cut-off function $\eta_{\varepsilon,R'}$ and to Fourier inversion formula, that

$$\begin{aligned}
\theta(t) &= \big\langle u_{\varepsilon,R',M}[-t](\xi)\,,\, \theta(-\xi)\big\rangle = \big\langle u_{\varepsilon,R',M}[-t](\xi)\,,\, \eta_{\varepsilon,R'}(\xi)\theta(-\xi)\big\rangle \\
&= \frac{1}{(2\pi)^n}\Big\langle u_{\varepsilon,R',M}[-t](\xi)\,,\, \int_{\mathbb{R}^n} \widehat{\eta_{\varepsilon,R'}\,\theta}(\varpi) e^{-i\langle\varpi,\xi\rangle}\, d\varpi\Big\rangle \\
&= \frac{1}{(2\pi)^n}\big\langle\, \widehat{u_{\varepsilon,R',M}[-t]}\,,\, \widehat{\eta_{\varepsilon,R'}\,\theta}\,\big\rangle \\
&= \frac{1}{(2\pi)^n}\int_{\mathbb{R}^n}\Big(\int_{\mathbb{C}^n} e^{i\langle\zeta,t\rangle}\, K_{\varepsilon,R',M}(\omega,\zeta)\, \frac{(\mathbb{H}_{\widehat{\mu},\varepsilon}(\omega,\zeta))^m}{m!}\, \overline{\partial}\Big(\frac{1}{\widehat{\mu}}\Big)\, \epsilon^*\Big)\, \widehat{\eta_{\varepsilon,R'}\,\theta}(\omega)\, d\omega \\
&= \int_{\mathbb{C}^n} e^{i\langle\zeta,t\rangle}\Big(\frac{1}{(2\pi)^n}\int_{\mathbb{R}^m} \widehat{\eta_{\varepsilon,R'}\,\theta}(\omega)\, K_{\varepsilon,R',M}(\omega,\zeta)\, \frac{(\mathbb{H}_{\widehat{\mu},\varepsilon}(\omega,\zeta))^m}{m!} d\omega\Big)\, \overline{\partial}\Big(\frac{1}{\widehat{\mu}}\Big)\, \epsilon^*,
\end{aligned}$$

where $\epsilon^* = \epsilon_1^* \wedge \cdots \wedge \epsilon_m^*$. It is precisely the desired integral representation formula. □

REMARK 7.90. One could have chosen $R' = R$ and $M = 0$ according to Theorem 7.87. However, since

$$\int_{\mathbb{R}^n} e^{i\langle \zeta, t-\xi\rangle}\, d\mu_j(\xi) = \widehat{\mu_j}(\zeta)\, e^{i\langle \zeta\,,\, t\rangle}$$

for $t \in \mathbb{R}^n$ and since $\widehat{\mu_j}\,\overline{\partial}(1/\widehat{\mu}) = 0$ for $j = 1, \dots, m$ in the sense of currents (see Theorem 2.79) the choices of $R' = R + R(\mu)$ in place of R and M in place of 0 are justified by the fact that we want the representation formula (7.250) to reproduce the mean-periodicity of θ with respect to the μ_j's acting by convolution in $\mathbb{B}_{\mathbb{R}^n}(0, R)$. In any case, it follows from the local structure of the Coleff–Herrera current $\bar{\partial}(1/\widehat{\mu})$ (namely Proposition 2.37, see also §5.1.2) that the right-hand side of (7.250) can be understood as follows. Locally, it is a superposition

$$\sum_{\iota} \lim_{\varepsilon\to 0} \int_{V(\widehat{\mu})\,\cap\,\{|h_\iota|\geq \varepsilon\}} q_\iota(\zeta, \partial/\partial\zeta)(e^{i\langle \zeta, t\rangle})\, \frac{\omega_\iota(\zeta)}{h_\iota(\zeta)}, \tag{7.251}$$

where ω_ι is an $(n-m, n-m)$-smooth differential form in ζ, q_ι is a holomorphic differential operator in ζ, and h_ι is a holomorphic function in ζ, whose zero set intersects properly $V(\widehat{\mu})$; see §2.2.2 or §5.1.2. Moreover, the superposition (7.251) is such that it realizes, as a function of t, a mean-periodic function for the μ_j's acting by convolution.

Let us comment on the results presented above by recalling some facts about the origin of the problem. In their pioneering paper [**BT80**, Definition 5.1], Carlos Berenstein and Alan Taylor introduced, given $f_1, \dots, f_m$ in some weighted algebra $\mathbf{A}_\phi(\mathbb{C}^n)$ such that the f_j's define a quasi-regular sequence, the concept of *slowly decreasing family of entire functions with respect to a family of affine* $(n-m)$-*dimensional subspaces of* $\mathbb{C}^n$. More specifically, let $\mathfrak{L}$ be a family of $(n-m)$-dimensional affine subspaces of $\mathbb{C}^n$, whose union covers $V(f)$. One says that the family $f_1, \dots, f_m$ belonging to $\mathbf{A}_\phi(\mathbb{C}^n)$ is slowly decreasing with respect to $\mathfrak{L}$ if and only if there exist positive constants c_f, c'_f, C_f, C'_f, *independent* of any affine subspace $\mathfrak{l} \in \mathfrak{L}$ so that for any $\mathfrak{l} \in \mathfrak{L}$, the connected components of

$$\mathcal{O}(\mathfrak{l}, c_f, c'_f) := \left\{ z \in \mathfrak{l} \,:\, \|f(z)\| \leq e^{-c_f\phi(z) - c'_f} \right\} \tag{7.252}$$

are all bounded and such that $\phi(\zeta) \leq C_f\phi(z) + C'_f$ when both $\zeta, z \in \mathfrak{l}$ belong to the same bounded connected component of (7.252). Going further, the authors introduced a concept of *almost parallelism* that the family of linear subspaces $\mathfrak{L}$ is required to satisfy. Under such a hypothesis, they were able to show, using Jacobi–Lagrange interpolation through Cauchy–Weil's formula, that the ideal generated by the f_j's in $\mathbf{A}_\phi$ is closed since it coincides then with $\mathbf{A}_\phi(\mathbb{C}^n) \cap (f_1, \dots, f_m)\, \mathcal{O}(\mathbb{C}^n)$; see [**BT80**, Theorem 5.6]. So the additional hypothesis (7.247) needed to ensure that Corollary 7.89 is valid holds under the slowly decreasing (SD) (for an almost parallel family $\mathfrak{L}$) requirements. Under the further assumption that the Coleff–Herrera residue current $\bar{\partial}(1/\widehat{\mu})$ is $\mathbf{A}_\phi$-tempered, we obtain, when ϕ is the Paley–Wiener weight, a more explicit version, in terms of currents, of [**BT80**, Theorem 10.1]. Corollary 7.89 also holds when the conditions in Remark 7.88 are fulfilled. We refer also the reader to [**BY95, Mat18**] for further examples involving PDEs with commensurate time lags in a fixed direction.

REMARK 7.91. In this subsection we present the trace results with the Paley–Wiener weight since it appeals to questions in harmonic analysis. The analogue of

Theorem 7.87 can be formulated, with almost the same proof, for any weight $\| \ \|^\gamma$ with $\gamma \geq 1$. The cases $\gamma = 1$ and $\gamma = 2$ are special. The case $\gamma = 1$ corresponds, through the Fourier–Borel transform, to the setting of analytic functionals, while the case $\gamma = 2$, where the weight is smooth, profits from the fact that ϕ is attached to a scalar product.

7.6.3. Ehrenpreis–Palamodov's fundamental principle revisited. Here is a well known result going back to Leonhard Euler. Let $p \in \mathbb{C}[X]$ be a polynomial with $\deg p > 0$ and distinct complex roots $\alpha_1, \dots, \alpha_\ell$. Then, the $\deg p$-dimensional $\mathbb{C}$-vector space of solutions of the homogeneous differential equation with constant coefficients

$$p\Big(i\frac{d}{dt}\Big)(y) = 0 \tag{7.253}$$

in $C^\infty([a,b])$, where $-\infty \leq a < b \leq +\infty$, admits as a basis the elementary solutions

$$t \longmapsto t^\kappa e^{-i\alpha_\nu t}, \ \nu = 1, \dots, \ell, \ \kappa = 0, \dots, \mu_{\alpha_\nu}(p) - 1,$$

where $\mu_\alpha(p)$ denotes the order of α as a zero of p. In this subsection, we consider the semilocal n-dimensional analogue of the above result, where U, in place of $]a,b[$, is *an open convex subset of* $\mathbb{R}^n$. For technical reasons, we will assume first that U is bounded and that its closure K (playing the role of $[a,b]$) is a *strictly convex* compact subset of $\mathbb{R}^n$ with *smooth boundary* containing the origin. Then, the positive support function[52]

$$H_{\mathrm{K}} : \omega \in (\mathbb{R}^n)^\star \longmapsto \sup_{t \in \mathrm{K}} \langle \omega, t \rangle \tag{7.254}$$

is smooth in $\mathbb{R}^n \setminus \{0\}$ and its gradient

$$\nabla H_{\mathrm{K}} : \omega \in \mathbb{R}^n \setminus \{0\} \longmapsto \Big(\frac{\partial H_{\mathrm{K}}}{\partial \omega_1}(\omega), \dots, \frac{\partial H_{\mathrm{K}}}{\partial \omega_n}(\omega)\Big) \in \partial \mathrm{K} \tag{7.255}$$

is surjective. We let also

$$C^\infty(\mathrm{K}) := \varinjlim_{\{\mathrm{U} \text{ open in } \mathbb{R}^n \,:\, \mathrm{U} \supset \mathrm{K}\}} C^\infty(\mathrm{U}). \tag{7.256}$$

In order to overcome the nonsmoothness of H_{K} at the origin, we will frequently regularize H_{K} by convolution $H_{\mathrm{K}} * \rho_{\mathbb{R}}$ with a smooth positive bump function $\rho_{\mathbb{R}}$, whose support is contained in the closed ball $\overline{\mathbb{B}_{\mathbb{R}^n}(0,1)}$ and whose total mass equals 1 with respect to the n-dimensional Lebesgue measure $d\omega$. Our choice for $\rho_{\mathbb{R}}$ is irrelevant throughout this subsection, as the choice of ρ was in the previous one. Observe that one has $(\nabla(H_{\mathrm{K}} * \rho_{\mathbb{R}}))(\mathbb{R}^n) \subset \mathrm{K}$, since $\rho_{\mathbb{R}}$ is positive with mass 1 and K is convex. Instead of the space of solutions of (7.253), we consider the $\mathbb{C}$-vector subspace of solutions $y \in C^\infty(\mathrm{K})$ of a system of m homogeneous PDEs with constant coefficients. Namely, we consider the $\mathbb{C}$-vector space

(7.257)

$$\begin{aligned}\mathrm{Ker}_{C^\infty(\mathrm{K})}\Big(p_1\Big(i\frac{\partial}{\partial t}\Big), \dots, p_m\Big(i\frac{\partial}{\partial t}\Big)\Big) &= \mathrm{Ker}_{C^\infty(\mathrm{K})}\, p\Big(i\frac{\partial}{\partial t}\Big) \\ &= \Big\{y \in C^\infty(\mathrm{K}) \,:\, p\Big(i\frac{\partial}{\partial t}\Big)(y) \equiv 0 \quad \text{on } \mathrm{K}\Big\},\end{aligned}$$

where $p_j \in \mathbb{C}[X_1, \dots, X_n]$ for $j = 1, \dots, m$.

[52] Since Fourier analysis is very much present in this subsection (as it was in the previous one), we make a distinction between the time domain $\mathbb{R}^n = \mathbb{R}^n_t$ (with variables denoted as usual by $t = (t_1, \dots, t_n)$) and its dual copy $(\mathbb{R}^n)^\star$, namely the frequency domain, where the variables will be $\omega = (\omega_1, \dots, \omega_n)$.

It follows from Hilbert's syzygy theorem [**Hil1890**] for $\mathbb{C}[X_1,\dots,X_n]$-finitely generated modules that there exists a generically exact complex, with length $L \le n$ and polynomial morphisms,

$$(7.258)\quad (\mathbf{F}^p_\bullet) : \ 0 \longrightarrow \mathbb{C}^{m_L} \xrightarrow{F_L} \mathbb{C}^{m_{L-1}} \xrightarrow{F_{L-1}} \cdots \xrightarrow{F_2} \mathbb{C}^m = \sum_1^m \mathbb{C}\,\epsilon_j^* \xrightarrow{\rfloor_{\sum_1^m p_j \epsilon_j} = \rfloor_{p(z)}} \mathbb{C}$$

such that the associated complex of sheaves of sections

$$0 \longrightarrow \mathcal{O}^{\oplus^{m_L}} \longrightarrow \mathcal{O}^{\oplus^{m_{L-1}}} \longrightarrow \cdots \longrightarrow \mathcal{O}^{\oplus^m} \xrightarrow{(h_1,\dots,h_m)\mapsto \sum_1^n h_j p_j} \mathcal{O} \longrightarrow \frac{\mathcal{O}}{\sum_1^m p_j \mathcal{O}} \longrightarrow 0$$

is exact. It follows from Proposition 7.84 that the currents $P_0^{\overline{\mathbf{F}^p_\bullet}}$ and $R_0^{\overline{\mathbf{F}^p_\bullet}}$ attached to the metrization of $\mathbf{F}^p_\bullet$ (with the canonical metric on each trivial bundle $\mathbb{C}^{m_\ell}$) by Proposition 3.28, are $\log(1+\|z\|^2)/2$-tempered. We let

$$(7.259)\qquad B_{\overline{\mathbf{F}^p_\bullet}} = B_{\log(1+\|z\|^2)/2}\big((P_0^{\overline{\mathbf{F}^p_\bullet}}, R_0^{\overline{\mathbf{F}^p_\bullet}})\big)$$

according to (7.214). Now, Proposition 4.43 and Remark 4.45 imply that the $\mathbf{F}^p_\bullet$-Hefer problem (with respect to the Koszul complex $\mathscr{K}_\bullet^{\mathscr{D}}$ attached to $(\mathscr{E},\Delta)$ as in (4.100)) admits a solution $\{H^p_{\ell,r}\}_{\ell\ge 0, r\ge -1}$, where the $H^p_{\ell,r}$ are polynomial. Proposition 4.29, applied on $\mathbb{C}^n\times\mathbb{C}^n \setminus \{(z,\zeta) : p(z)=0\}$, shows that the global Bochner–Martinelli weight attached to such a solution of the Hefer problem is described by

$$(7.260)\quad \begin{aligned} \mathscr{W}^{\overline{\mathbf{F}^p_\bullet}}(z,\zeta) &= 2i\pi \rfloor_{p(z)} H_1^p(z,\zeta)\, P_0^{\overline{\mathbf{F}^p_\bullet}}(\zeta) + H_0^p(z,\zeta)\, R_0^{\overline{\mathbf{F}^p_\bullet}}(\zeta) \\ &= 2i\pi \rfloor_{\sum_1^m p_j(z)\epsilon_j} \sum_{r=1}^{\min(m,n+1)} H^p_{1,r}(z,\zeta)\, P_{0,r}^{\overline{\mathbf{F}^p_\bullet}}(\zeta) + \sum_{r=1}^{\min(m,n)} H^p_{0,r}(z,\zeta)\, R_{0,r}^{\overline{\mathbf{F}^p_\bullet}}(\zeta). \end{aligned}$$

We define the degree of the solution $\{H^p_{\ell,r}\}_{\ell\ge 0,r\ge -1}$ to the $\mathbf{F}^p_\bullet$-Hefer problem by

$$(7.261)\qquad \deg_{\ell\le 1} \mathbf{H}^p = \max_{\ell\le 1}\max_{r\ge 1} \deg_{z,\zeta} H^p_{\ell,r}(z,\zeta),$$

where the degree of a polynomial map $(z,\zeta) \longmapsto \Phi(z,\zeta)$ from $\mathbb{C}^n\times\mathbb{C}^{m'}$ to $\mathbb{C}^n\times\mathbb{C}^{m''}$ denotes the maximum of the degrees of its polynomial entries. As in (7.228), we introduce for $N\in\mathbb{N}^*$ the Bochner–Martinelli global weight on the open subset $\{(z,\zeta)\in\mathbb{C}^n\times\mathbb{C}^n : p(z)\neq 0\}$ defined by

$$(7.262)\qquad \mathscr{W}_{\mathrm{K},N}[p] = \mathscr{W}_{\mathbb{C}^n}^N\, e^{\mathscr{W}_{H_{\mathrm{K}}*\rho_{\mathbb{R}}}-1}\, \mathscr{W}^{\overline{\mathbf{F}^p_\bullet}},$$

where

$$\begin{aligned} &\mathscr{W}_{H_{\mathrm{K}}*\rho_{\mathbb{R}}}(z,\zeta) \\ &\quad = \Big[1 + 2\sum_{k=1}^n \frac{\partial\big((H_{\mathrm{K}}*\rho_{\mathbb{R}})(\operatorname{Im}\zeta)\big)}{\partial\zeta_k}(z_k-\zeta_k) + 2\,dd^c\big((H_{\mathrm{K}}*\rho_{\mathbb{R}})(\operatorname{Im}\zeta)\big)\Big], \end{aligned}$$

so that

$$\begin{aligned}
&e^{\mathscr{W}_{H_{\mathrm{K}}*\rho_{\mathbb{R}}}-1}(z,\zeta)\\
&=\exp\Big(2\Big\langle\frac{\partial\big((H_{\mathrm{K}}*\rho_{\mathbb{R}})(\operatorname{Im}\zeta)\big)}{\partial\zeta}\,,\,z-\zeta\Big\rangle\Big)\Big(\sum_{r=0}^{n}\frac{\big(2\,dd^c\big((H_{\mathrm{K}}*\rho_{\mathbb{R}})(\operatorname{Im}\zeta)\big)\big)^r}{r!}\Big)\\
&=e^{-i\langle z\,,\,(\nabla\,(H_{\mathrm{K}}*\rho_{\mathbb{R}}))(\operatorname{Im}\zeta)\rangle}\\
&\quad\times\Big(e^{i\langle\zeta\,,\,(\nabla\,(H_{\mathrm{K}}*\rho_{\mathbb{R}}))(\operatorname{Im}\zeta)\rangle}\Big(\sum_{r=0}^{n}\frac{\big(2\,dd^c\big((H_{\mathrm{K}}*\rho_{\mathbb{R}})(\operatorname{Im}\zeta)\big)\big)^r}{r!}\Big)\Big).
\end{aligned}$$

Then the convexity inequality

(7.263)

$$\begin{aligned}
\big|e^{-i\langle z\,,\,(\nabla\,(H_{\mathrm{K}}*\rho_{\mathbb{R}}))(\operatorname{Im}\zeta)\rangle}\,e^{i\langle\zeta\,,\,(\nabla\,(H_{\mathrm{K}}*\rho_{\mathbb{R}}))(\operatorname{Im}\zeta)\rangle}\big| &\le \frac{\exp\big((H_{\mathrm{K}}*\rho_{\mathbb{R}})(\operatorname{Im}z)\big)}{\exp\big((H_{\mathrm{K}}*\rho_{\mathbb{R}})(\operatorname{Im}\zeta)\big)}\\
&\le e^{2\sup_{\mathrm{K}}\|\zeta\|}\,\frac{\exp\big(H_{\mathrm{K}}(\operatorname{Im}z)\big)}{\exp\big(H_{\mathrm{K}}(\operatorname{Im}\zeta)\big)},
\end{aligned}$$

valid for all $(z,\zeta)\in\mathbb{C}^n\times\mathbb{C}^n$, plays in this subsection the role played by the convexity inequality (7.239). We can now state the following multivariate, semilocal analogue of Euler's result; see [**AndW07**, Theorem 8.2] or also [**BerP89**].

THEOREM 7.92. *Let the p_j's, K, H_{K}, $\rho_{\mathbb{R}}$ be as above. Let $\deg p=\max_j\deg p_j$. There is an integer $N\in\mathbb{N}^*$, independent of K, and for each $\nu\in\mathbb{N}^n$ such that $|\nu|\le N$, there exists a distribution $T_{N,\nu}\in\mathcal{D}'(\mathbb{C}^n)$, also independent on K, which is a sum of distributions of the form $(\partial/\partial\zeta)^\alpha(\partial/\partial\overline{\zeta})^\beta[d\mu]$, where $\alpha,\beta\in\mathbb{N}^n$ and $(1+\|\zeta\|)^{\deg p}\,|d\mu|$ is a positive measure with finite total mass on $\mathbb{C}^n$, such that $T_{N,\nu}$ is annihilated by each p_j. Furthermore, any*

$$y\in\operatorname{Ker}_{C^\infty(\mathrm{K})}\Big(p_1\Big(i\frac{\partial}{\partial t}\Big),\dots,p_m\Big(i\frac{\partial}{\partial t}\Big)\Big)$$

is reproduced in K *as*

(7.264)

$$y(t)=\sum_{|\nu|\le N}\Big\langle T_{N,\nu}(\zeta)\,,\,e^{i\langle\zeta,\nabla(H_{\mathrm{K}}*\rho_{\mathbb{R}})(\operatorname{Im}\zeta)-t\rangle}\Big(\frac{\partial}{\partial\tau}\Big)^{\nu}[y]\big(\nabla(H_{\mathrm{K}}*\rho_{\mathbb{R}})(\operatorname{Im}\zeta)\big)\Big\rangle.$$

PROOF. Let $\ell\in\mathbb{N}^*$. We set

$$N'=B_{\overline{\mathbf{F}^p_\bullet}}+\deg_{\ell\le 1}\mathbf{H}^p+\deg p+1. \tag{7.265}$$

Let also $(z,\zeta)\longmapsto q(z,\zeta)$ be as defined in (7.229) and $\chi\in\mathcal{D}(\mathbb{C}^n,[0,1])$ be a cut-off function identically equal to one in a neighborhood of $\overline{\mathbb{B}_{\mathbb{C}^n}(0,1)}$. For any $t\in\mathrm{K}$ and $z\in\mathbb{B}_{\mathbb{C}^n}(0,\ell)$, one has, according to Proposition 4.33, that

$$\begin{aligned}
(7.266)\quad e^{-i\langle z,t\rangle}&=\widehat{\delta_t}(z)=\int_{\mathbb{C}^n}e^{-i\langle\zeta,t\rangle}\\
&\times\Big(\Big(\chi(\zeta/\ell)+\frac{\overline{\partial}_\zeta\big((1-\chi(\zeta/\ell))q(z,\zeta)\big)}{2i\pi}\Big)^{n+1}\big(\mathscr{W}^{N'}_{\mathbb{C}^n}e^{\mathscr{W}_{H_{\mathrm{K}}*\rho_{\mathbb{R}}}-1}\mathscr{W}^{\overline{\mathbf{F}^p_\bullet}}\big)(z,\zeta)\Big)_{n,n}.
\end{aligned}$$

It follows from the choice of N' in (7.265) (according to the definition of $B_{\overline{\mathbf{F}^p_\bullet}}$, see (7.214)), from the convexity inequality (7.263), together with the fact that all derivatives up to a finite order of χ and $H_{\mathrm{K}}*\rho_{\mathbb{R}}$ are uniformly bounded, that the following assertion holds:[53] for each integer $r' = 1, \dots, \min(m,n)$, and for each integer $r'' = 1, \dots, \min(m, n+1)$, for each $\nu \in \mathbb{N}^n$ with $|\nu| \leq N = N' + \deg_{\ell\leq 1} \mathbf{H}^p$, one can find smooth integral kernels

$$\Omega^{(0:r')}_{N,\nu} \in C^\infty_{n,n-r'}(\mathbb{C}^n, (\mathbb{C}^{m_{r'}})^*), \quad \Omega^{(1:r'')}_{N,\nu} \in C^\infty_{n,n-(r''-1)}(\mathbb{C}^n, \mathrm{Hom}_{\mathbb{C}}(\mathbb{C}^{m_{r''}}, (\mathbb{C}^m)^*))$$

such that (7.266) leads, when ℓ tends to infinity, to the explicit full division formula

$$(7.267)\quad e^{-i\langle z,t\rangle}$$

$$= \sum_{|\nu|\leq N} (-iz)^\nu \Big(\sum_{r=1}^{\min(m,n)} \int_{\mathbb{C}^n} \widehat{\delta_{\nabla(H_{\mathrm{K}}*\rho_{\mathbb{R}})(\mathrm{Im}\,\zeta)}}(z)\, e^{i\langle \zeta, \nabla(H_{\mathrm{K}}*\rho_{\mathbb{R}})(\mathrm{Im}\,\zeta)-t\rangle} \Omega^{(0:r)}_{N,\nu}\, R^{\overline{\mathbf{F}^p_\bullet}}_{0,r}$$

$$+ 2i\pi \rfloor_{p(z)} \sum_{r=1}^{\min(m,n+1)} \int_{\mathbb{C}^n} \widehat{\delta_{\nabla(H_{\mathrm{K}}*\rho_{\mathbb{R}})(\mathrm{Im}\,\zeta)}}(z)\, e^{i\langle \zeta, \nabla(H_{\mathrm{K}}*\rho_{\mathbb{R}})(\mathrm{Im}\,\zeta)-t\rangle} \Omega^{(1:r)}_{N,\nu}\, P^{\overline{\mathbf{F}^p_\bullet}}_{0,r} \Big),$$

where the vector-valued distributions $T^{(0,r')}_{N,\nu}$ and $T^{(1:r'')}_{N,\nu}$, which are defined for each $r' = 1, \dots, \min(m,n)$, $r'' = 1, \dots, \min(m, n+1)$ and $\nu \in \mathbb{N}^n$ with $|\nu| \leq N$ by

$$(7.268)\quad \Omega^{(0:r')}_{N,\nu}\, R^{\overline{\mathbf{F}^p_\bullet}}_{0,r'} = T^{(0:r')}_{N,\nu}\, (dd^c\|\zeta\|^2)^n, \quad \Omega^{(1:r'')}_{N,\nu}\, P^{\overline{\mathbf{F}^p_\bullet}}_{0,r''} = T^{(1:r'')}_{N,\nu}\, (dd^c\|\zeta\|^2)^n,$$

are sums of distributions of the form $(\partial/\partial\zeta)^\alpha (\partial/\partial\overline{\zeta})^\beta (d\mu)$, where $\alpha, \beta \in \mathbb{N}^n$ and the positive measure $(1+\|\zeta\|)^{\deg p + |\alpha|+|\beta|}\, |d\mu(\zeta)|$ is a complex measure with finite total mass on $\mathbb{C}^n$. The fact that all derivatives up to a finite order of $H_{\mathrm{K}} * \rho_{\mathbb{R}}$ are bounded, together with the convexity inequality (7.263), ensures the absolute convergence of the integrals over $\mathbb{C}^n$ on the right-hand side of (7.267), once the action of the distributions in (7.268) is described in terms of integration by parts. It also ensures that for any $\nu \in \mathbb{N}^n$ with $|\nu| \leq N$, one has, since $(\nabla(H_{\mathrm{K}} * \rho_{\mathbb{R}}))(\mathbb{R}^n) \subset \mathrm{K}$, that

$$(7.269)\quad 2i\pi \rfloor_{p(z)} \sum_{r=1}^{\min(m,n+1)} \int_{\mathbb{C}^n} \widehat{\delta_{\nabla(H_{\mathrm{K}}*\rho_{\mathbb{R}})(\mathrm{Im}\,\zeta)}}(z)\, e^{i\langle \zeta, \nabla(H_{\mathrm{K}}*\rho_{\mathbb{R}})(\mathrm{Im}\,\zeta)-t\rangle} \Omega^{(1:r)}_{N,\nu}\, P^{\overline{\mathbf{F}^p_\bullet}}_{0,r}$$

$$= \sum_{j=1}^m p_j(z)\, \widehat{d\mu^{t,j}_{N,\nu}}(z) = \widehat{\sum_{j=1}^m d\mu^{t,j}_{N,\nu} * p_j(-i\partial/\partial t)(\delta_0)}(z),$$

where the $\mu^{t,j}_{N,\nu}$'s are measures supported by K. On the other hand, for each such $\nu \in \mathbb{N}^n$ with $|\nu| \leq N$, the entire function

$$z \longmapsto \sum_{r=1}^{\min(m,n)} \int_{\mathbb{C}^n} \widehat{\delta_{\nabla(H_{\mathrm{K}}*\rho_{\mathbb{R}})(\mathrm{Im}\,\zeta)}}(z)\, e^{i\langle \zeta, \nabla(H_{\mathrm{K}}*\rho_{\mathbb{R}})(\mathrm{Im}\,\zeta)-t\rangle} \Omega^{(0:r)}_{N,\nu}\, R^{\overline{\mathbf{F}^p_\bullet}}_{0,r}$$

is the analytic continuation to $\mathbb{C}^n$ of the Fourier transform of the averaged measure

$$(7.270)\qquad d\mu^t_{N,\nu} := \int_{\mathbb{C}^n} \delta_{\nabla(H_{\mathrm{K}}*\rho_{\mathbb{R}})(\mathrm{Im}\,\zeta)}\, e^{i\langle \zeta, \nabla(H_{\mathrm{K}}*\rho_{\mathbb{R}})(\mathrm{Im}\,\zeta)-t\rangle} \Omega^{(0:r)}_{N,\nu}\, R^{\overline{\mathbf{F}^p_\bullet}}_{0,r}$$

supported by K. Taking the antecedents through Fourier transform (acting on the space of compactly supported distributions on $\mathbb{R}^n$) of both sides in (7.267) leads

[53] One needs to repeat the argument used in the proof of Theorem 7.87 more precisely here: the proof of the full division formula (7.244) in Remark 7.88 with $\varepsilon = 0$ and $M = 0$.

then to
(7.271)

$$\delta_t = \sum_{|\nu|\le N'} \Big(d\mu^t_{N,\nu} * (-\partial/\partial t)^\nu(\delta_0) + \sum_{j=1}^m (-\partial/\partial t)^\nu (d\mu^{t,j}_{N,\nu}) * p_j(-i\partial/\partial t)(\delta_0)\Big).$$

It follows from (7.271) that

$$(7.272)\quad \int_{\mathbb{R}^n} y(\tau)d\delta_t(\tau) = y(t) = \sum_{|\nu|\le N} \int_{\mathbb{R}^n} \Big(\frac{\partial}{\partial\tau}\Big)^\nu [y](\tau)\, d\mu^t_{N,\nu}(\tau)$$
$$+ \sum_{|\nu|\le N}\sum_{j=1}^m \int_{\mathbb{R}^n} \Big(\frac{\partial}{\partial\tau}\Big)^\nu \Big[p_j\Big(i\frac{\partial}{\partial t}\Big)[y]\Big](\tau)\, d\mu^{t,j}_{N,\nu}(\tau).$$

Since

$$y \in \mathrm{Ker}_{C^\infty(\mathrm{K})}\Big(p_1\Big(i\frac{\partial}{\partial t}\Big),\dots,p_m\Big(i\frac{\partial}{\partial t}\Big)\Big),$$

one concludes from (7.272) that

$$y(t) = \sum_{|\nu|\le N} \int_{\mathbb{R}^n} \Big(\frac{\partial}{\partial\tau}\Big)^\nu [y](\tau)\, d\mu^t_{N,\nu}(\tau) = \sum_{r=1}^{\min(m,n)} \sum_{|\nu|\le N}$$
$$\int_{\mathbb{C}^n} e^{i\langle\zeta,\nabla(H_{\mathrm{K}}*\rho_{\mathbb{R}})(\mathrm{Im}\,\zeta)-t\rangle} \Big(\frac{\partial}{\partial\tau}\Big)^\nu [y]\big(\nabla(H_{\mathrm{K}}*\rho_{\mathbb{R}})(\mathrm{Im}\,\zeta)\big)\,\big(\Omega^{(0:r)}_{N,\nu}\, R^{\overline{\mathbf{F}^p_\bullet}}_{0,r}\big)(\zeta),$$

which is the integral representation formula (7.264). The structure of the distributions $T^{(0:r)}_{N,\nu}$ in (7.268), for every $r = 1,\dots,\min(m,n)$, shows that the integrals on the right-hand side of (7.264) can be differentiated at least $\deg p$ times by Lebesgue's differentiation theorem as functions of t in the integrand. It follows then from Theorem 3.38 that (7.264) reproduces the fact that $y \in \mathrm{Ker}_{C^\infty(\mathrm{K})}(p(i\partial/\partial t))$. □

Let us now come to the more classical (global) formulation of the Ehrenpreis–Palamodov fundamental principle. Theorem 5.5 admits the following transposition to the global setting. Let $p_1,\dots,p_m \in \mathbb{C}[X_1,\dots,X_n]$ and

$$(7.273)\quad \sum_{j=1}^m p_j\,\mathbb{C}[X_1,\dots,X_n] = \mathcal{Q}_1\cap\cdots\cap\mathcal{Q}_M$$

be a reduced primary decomposition of the ideal $\sum_{j=1}^m p_j\mathbb{C}[X_1,\dots,X_n]$ in the polynomial algebra $\mathbb{C}[X_1,\dots,X_n]$. It means that

$$(7.274)\quad \begin{aligned} &\iota'\neq\iota \Longrightarrow \sqrt{\mathcal{Q}_{\iota'}}\neq\sqrt{\mathcal{Q}_\iota}\\ &\mathcal{Q}_\iota \not\supset \bigcap_{\iota'\neq\iota}\mathcal{Q}_{\iota'} \quad \text{for any } \iota. \end{aligned}$$

Recall now that the list of associated primes $\mathcal{P}_\iota = \sqrt{\mathcal{Q}_\iota}$ is uniquely determined up to order. The associated primes, minimal for ordering by inclusion, are the isolated ones. The remaining elements of the list are the embedded ones. Nœther's preparation lemma, together with the existence of primitive elements for algebraic extensions (see [**Bj74**, Chapter 8]) implies that for any $\iota \in \{1,\dots,M\}$ there is a list

$$\Big\{\mathcal{N}_{\iota,\nu}\Big(\zeta,\frac{\partial}{\partial\zeta}\Big) : \mathcal{N}_{\iota,\nu}\in\mathbb{C}[X_1,\dots,X_n]\,\langle d/dX_1,\dots,d/dX_n\rangle\,,\nu=1,\dots,\nu_\iota\Big\}$$

of algebraic nœtherian operators such that the following statement holds for any $h \in \mathcal{O}(\Omega)$, where Ω is an open subset of $\mathbb{C}^n$:

$$
\begin{gathered}
h \in \sum_{j=1}^{m} p_j \mathcal{O}(\Omega) \Longleftrightarrow \\
\forall \iota = 1, \dots, M, \quad \mathscr{N}_{\iota,\nu}\Big(\zeta, \frac{\partial}{\partial \zeta}\Big)[h] = 0 \text{ on } V(\mathscr{P}_\iota) \cap \Omega \text{ for } \nu = 1, \dots, \nu_\iota.
\end{gathered}
\tag{7.275}
$$

The existence of such a collection of polynomial differential operators with polynomial coefficients goes back to Macaulay [**Mac**]. In addition to quoting [**Bj74**, Chapter 8], we refer the reader also to the pioneering monograph by V. Palamodov [**Pal**], as well as to [**ChHKL22**] for an updated bibliography and an algorithmic approach. To formulate the fundamental principle of Ehrenpreis and Palamodov [**Ehr, Pal**] in its classical (nonexplicit) global form, we need to attach to any convex domain $\mathrm{U} \subset \mathbb{R}^n$ the family $\mathscr{K}(\mathrm{U})$ of all nonvanishing positive continuous functions $\psi : \mathbb{C}^n \longrightarrow]0, +\infty[$ such that

$$
\forall\, u \in \mathcal{D}'(\mathbb{R}^n), \quad \operatorname{Supp} u \Subset \mathrm{U} \Longrightarrow \sup_{\mathbb{C}^n} \Big(\frac{|\widehat{u}|}{\psi}\Big) < +\infty. \tag{7.276}
$$

Here is the classical formulation of such a principle.[54]

THEOREM 7.93. *Let $p = (p_1, \dots, p_m)$ be a collection of polynomials in the polynomial algebra $\mathbb{C}[X_1, \dots, X_n]$ and let $\mathscr{N}^p = \big\{\mathscr{N}_{\iota,\nu}(\zeta, \partial/\partial\zeta),\ \iota = 1, \dots, M,\ \nu = 1, \dots, \nu_\iota\big\}$ be a list of algebraic nœtherian operators subordinate to a reduced primary decomposition (7.273) with length M of $\sum_1^m p_j \mathbb{C}[X_1, \dots, X_n]$ in $\mathbb{C}[X_1, \dots, X_n]$. Given any convex subset U and any*

$$
y \in \operatorname{Ker}_{C^\infty(\mathrm{U})} \Big(p_1\Big(i\frac{\partial}{\partial t}\Big), \dots, p_m\Big(i\frac{\partial}{\partial t}\Big)\Big)
$$

there exist $\psi \in \mathscr{K}(\mathrm{U})$ and complex measures $d\mu^y_{\iota,\nu}$ on $\mathbb{C}^n$, whose support satisfies

$$
\operatorname{Supp} d\mu^y_{\iota,\nu} \subset V(\mathscr{P}_\iota) \quad \forall\, \iota = 1, \dots, M,\ \forall\, \nu = 1, \dots, \nu_\iota,
$$

such that

$$
\begin{gathered}
\sum_{\iota=1}^{M} \sum_{\nu=1}^{\nu_\iota} \int_{V(\mathscr{P}_\iota)} \psi(\zeta)\, |d\mu^y_{\iota,\nu}(\zeta)| < \infty, \\
y(t) = \sum_{\iota=1}^{M} \sum_{\nu=1}^{\nu_\iota} \int_{\mathbb{C}^n} \mathscr{N}_{\iota,\nu}\Big(\zeta, \frac{\partial}{\partial \zeta}\Big)[e^{-i\langle \zeta\,,\,t\rangle}](\zeta)\, d\mu^y_{\iota,\nu}(\zeta) \quad \forall\, t \in \mathrm{U}.
\end{gathered}
\tag{7.277}
$$

We want to reformulate Theorem 7.93 in a form appealing to Bochner–Martinelli residue currents instead of nœtherian operators. For that purpose, let us assume that U is a convex open subset in $\mathbb{R}^n$ admitting the exhaustion

$$
\mathrm{U} = \bigcup_{\ell \geq 1} \mathrm{U}_\ell, \quad 0 \in \mathrm{U}_\ell \subset \mathrm{K}_\ell = \overline{\mathrm{U}_\ell} \subset \mathrm{U}_{\ell+1} \quad \text{for all } \ell \in \mathbb{N}^*,
$$

where the closure K_ℓ is bounded, strictly convex, and has a smooth boundary. Let

$$
y \in \operatorname{Ker}_{C^\infty(\mathrm{U})} \Big(p_1\Big(i\frac{\partial}{\partial t}\Big), \dots, p_m\Big(i\frac{\partial}{\partial t}\Big)\Big).
$$

[54] See [**Han81**, §3] for its extension through a functional analysis point of view.

Let us observe that, given any sequence of positive real numbers $(\gamma_\ell)_{\ell\geq 1}$ such that $\gamma_\ell \geq \sup_{\mathrm{K}_\ell} H_{\mathrm{K}}$ for any $\ell \in \mathbb{N}^*$, the convex function

$$\omega \in \mathbb{R}^n \longmapsto \sup_{\ell\geq 1}(H_{\mathrm{K}_\ell}(\omega) - \gamma_\ell)$$

is finite on $\mathbb{R}^n$. Actually, $H_{\mathrm{K}_\ell}(\omega) - \gamma_\ell \leq 0$ as soon as $\omega \in \mathrm{K}_\ell$, which occurs when $\ell \geq \ell(\omega)$ for each $\omega \in \mathrm{U}$ since the K_ℓ exhaust U. We can then state the following alternative form of the Ehrenpreis–Palamodov fundamental principle in the global setting.

THEOREM 7.94. *Let $\mathrm{U} \subset \mathbb{R}^n$ be open and convex. Let also the exhausting sequence $(\mathrm{K}_\ell)_{\ell\geq 1}$ and the p_j's be as above. There exists an integer $M = M(p)$ and for each $y \in \mathrm{Ker}_{C^\infty(\mathrm{U})}(p(i\partial/\partial t))$, there exists a sequence $(\gamma_{y,\ell})_{\ell\geq 1}$ of positive numbers satisfying $\gamma_{y,\ell} \geq \sup_{\mathrm{K}_\ell} H_{\mathrm{K}_\ell}$ for any $\ell \in \mathbb{N}^*$, so that y is reproduced in U as*

$$y(t) = \langle T_y(\zeta), e^{-i\langle\zeta,t\rangle}\rangle. \tag{7.278}$$

The distribution T_y is annihilated by the p_j's. Moreover, it is a finite sum of distributions of the form

$$\Big(\frac{\partial}{\partial\zeta}\Big)^\alpha \Big(\frac{\partial}{\partial\overline{\zeta}}\Big)^\beta (d\mu_y^{\alpha,\beta})$$

with $|\alpha| + |\beta| \leq M$ and complex measures $d\mu_y^{\alpha,\beta}$ depending on y such that

$$\int_{\mathbb{C}^n} (1 + \|\zeta\|)^{\deg p} e^{\limsup_{\ell\to+\infty}(H_{\mathrm{K}_\ell}(\mathrm{Im}\,\zeta)-\gamma_{y,\ell})}\, |d\mu_y^{\alpha,\beta}| < +\infty. \tag{7.279}$$

The representation (7.278) reproduces the fact that $y \in \mathrm{Ker}_{C^\infty(\mathrm{U})}(p(i\partial/\partial t))$.

PROOF. For each $\ell \in \mathbb{N}^*$, the following representation formulae hold for y in K_ℓ (see Theorem 7.92). For any $t \in \mathrm{K}_\ell$ and $\ell' \geq \ell$,

$$\begin{aligned} &y(t) = \\ &\sum_{|\nu|\leq N} \Big\langle e^{i\langle\zeta,\nabla(H_{\mathrm{K}_{\ell'}}*\rho_{\mathbb{R}})(\mathrm{Im}\,\zeta)\rangle} \Big(\frac{\partial}{\partial\tau}\Big)^\nu [y]\big(\nabla(H_{\mathrm{K}_{\ell'}} * \rho_{\mathbb{R}})(\mathrm{Im}\,\zeta)\big)\, T_{N,\nu}(\zeta)\,,\, e^{-i\langle\zeta,t\rangle}\Big\rangle,\end{aligned} \tag{7.280}$$

where $N = N(p) \in \mathbb{N}^*$ is conveniently chosen independently of y. Furthermore, the distributions $T_{N,\nu}$ involved in the representation (7.280) are independent of y and ℓ and such that

$$T_{N,\nu} = \sum_{\{\alpha,\beta\in\mathbb{N}^n\,:\,|\alpha|+|\beta|\leq M_N\}} \Big(\frac{\partial}{\partial\zeta}\Big)^\alpha \Big(\frac{\partial}{\partial\overline{\zeta}}\Big)^\beta d\mu_{N,\nu}^{\alpha,\beta},$$

where the complex measures $d\mu_{N,\nu}^{\alpha,\beta}$

$$\sum_{\{\alpha,\beta\in\mathbb{N}^n\,:\,|\alpha|+|\beta|\leq M_N\}} \int_{\mathbb{C}^n} (1 + \|\zeta\|)^{\deg p+|\alpha|+|\beta|} |d\mu_{N,\nu}^{\alpha,\beta}| < +\infty. \tag{7.281}$$

As a consequence, for each $\ell' \geq \ell$ and $\nu \in \mathbb{N}^n$ with $|\nu| \leq N$, the distribution

$$T_{N,\nu\,:\,y,\ell'} := e^{i\langle\zeta,\nabla(H_{\mathrm{K}_{\ell'}}*\rho_{\mathbb{R}})(\mathrm{Im}\,\zeta)\rangle} \Big(\frac{\partial}{\partial\tau}\Big)^\nu [y]\big(\nabla(H_{\mathrm{K}_{\ell'}} * \rho_{\mathbb{R}})(\mathrm{Im}\,\zeta)\big)\, T_{N,\nu}(\zeta),$$

acting on $\zeta \mapsto e^{-i\langle\zeta,t\rangle}$ in the bracket (7.280) and remaining annihilated by each polynomial map $\zeta \mapsto p_j(\zeta)$, splits as

$$T_{N,\nu\,:\,y,\ell'} = \sum_{\{\alpha,\beta\in\mathbb{N}^n\,:\,|\alpha|+|\beta|\le M_N\}} \Big(\frac{\partial}{\partial\zeta}\Big)^{\alpha}\Big(\frac{\partial}{\partial\overline{\zeta}}\Big)^{\beta} d\mu^{\alpha,\beta}_{N,\nu\,:\,y,\ell'}.$$

Here the complex measures $d\mu^{\alpha,\beta}_{N,\nu\,:\,y,\ell'}$ are such that

$$\sum_{|\nu|\le N}\sum_{|\alpha|+|\beta|\le M_N}\int_{\mathbb{C}^n}(1+\|\zeta\|)^{\deg p}e^{H_{\mathrm{K}_{\ell'}}(\mathrm{Im}\,\zeta)-\gamma_{y,\ell'}}\,|d\mu^{\alpha,\beta}_{N,\nu\,:\,y,\ell'}|\le 1, \tag{7.282}$$

where $\gamma_{y,\ell'}\ge\sup_{\mathrm{K}_{\ell'}}H_{\mathrm{K}_{\ell'}}$ is a constant depending on $\mathrm{K}_{\ell'}$, on the estimates for the derivatives of $\rho_{\mathbb{R}}$ with order less than M_N as well as on those of the derivatives of y with order less than $M_N+|\nu|$ on $\mathrm{K}_{\ell'}$, and on the positive constant (7.281). The convexity inequality (7.263) with $\mathrm{K}=\mathrm{K}_{\ell'}$ and $z=0$, plays here a crucial role. One has then, for any $t\in\mathrm{K}$ and $\ell'\ge\ell$,

$$y(t)=\sum_{|\nu|\le N}\sum_{|\alpha|+|\beta|\le M_N}\Big\langle\Big(\frac{\partial}{\partial\zeta}\Big)^{\alpha}\Big(\frac{\partial}{\partial\overline{\zeta}}\Big)^{\beta}(d\mu^{\alpha,\beta}_{N,\nu\,;\,y,\ell'})(\zeta)\,,\,e^{-i\langle\zeta,t\rangle}\Big\rangle \tag{7.283}$$

with estimates (7.282) for the positive measures $|d\mu^{\alpha,\beta}_{N,\nu\,;\,y,\ell'}|$ and

$$p_j(\zeta)\sum_{|\nu|\le N}\sum_{|\alpha|+|\beta|\le M_N}\Big(\frac{\partial}{\partial\zeta}\Big)^{\alpha}\Big(\frac{\partial}{\partial\overline{\zeta}}\Big)^{\beta}(d\mu^{\alpha,\beta}_{N,\nu\,;\,y,\ell'})=0 \tag{7.284}$$

for $j=1,\dots,m$. An argument of the Montel type shows that for any $\ell\in\mathbb{N}^*$, one can extract from the sequence (indexed by $\ell'=\ell,\ell+1,\ell+2,\dots$) of vectors (indexed by ν,α,β) of measures

$$\big((d\mu^{\alpha,\beta}_{N,\nu\,;\,y,\ell'})^{\alpha,\beta}_{\nu}\big)_{\ell'\ge\ell}$$

a subsequence, which converges weakly on $\mathbb{C}^n$ towards $(d\mu^{\alpha,\beta}_{N,\nu\,:\,y,\ge\ell})^{\alpha,\beta}_{\nu}$. Thus, for any $t\in\mathrm{K}_\ell$,

$$y(t)=\sum_{|\nu|\le N}\sum_{|\alpha|+|\beta|\le M_N}\Big\langle\Big(\frac{\partial}{\partial\zeta}\Big)^{\alpha}\Big(\frac{\partial}{\partial\overline{\zeta}}\Big)^{\beta}(d\mu^{\alpha,\beta}_{N,\nu\,:\,y,\ge\ell})(\zeta)\,,\,e^{-i\langle\zeta,t\rangle}\Big\rangle \tag{7.285}$$

with

$$\sum_{|\nu|\le N}\sum_{|\alpha|+|\beta|\le M_N}\int_{\mathbb{C}^n}(1+\|\zeta\|)^{\deg p}e^{\sup_{\ell'\ge\ell}(H_{\mathrm{K}_{\ell'}}(\mathrm{Im}\,\zeta)-\gamma_{y,\ell'})}\,|d\mu^{\alpha,\beta}_{N,\nu\,;\,y,\ge\ell}|\le 1 \tag{7.286}$$

and

$$p_j(\zeta)\sum_{|\nu|\le N}\sum_{|\alpha|+|\beta|\le M_N}\Big(\frac{\partial}{\partial\zeta}\Big)^{\alpha}\Big(\frac{\partial}{\partial\overline{\zeta}}\Big)^{\beta}(d\mu^{\alpha,\beta}_{N,\nu\,;\,y,\ge\ell})=0 \tag{7.287}$$

for $j=1,\dots,m$. We now repeat the Montel argument and extract from the sequence (indexed this time by $\ell=1,2,\dots$) of vectors (indexed by ν,α,β) of measures

$$\big((d\mu^{\alpha,\beta}_{N,\nu;\ge\ell,y})^{\alpha,\beta}_{\nu}\big)_{\ell\ge1}$$

a subsequence converging weakly on $\mathbb{C}^n$ towards $(d\mu^{\alpha,\beta}_{N,\nu\,:\,y})^{\alpha,\beta}_{\nu}$. One has then for any $t\in\mathrm{U}$ that

$$y(t)=\sum_{|\nu|\le N}\sum_{|\alpha|+|\beta|\le M_N}\Big\langle\Big(\frac{\partial}{\partial\zeta}\Big)^{\alpha}\Big(\frac{\partial}{\partial\overline{\zeta}}\Big)^{\beta}(d\mu^{\alpha,\beta}_{N,\nu\,:\,y})(\zeta)\,,\,e^{-i\langle\zeta,t\rangle}\Big\rangle$$

with

$$\sum_{|\nu|\leq N}\ \sum_{|\alpha|+|\beta|\leq M_N} \int_{\mathbb{C}^n} (1+\|\zeta\|)^{\deg p} e^{\limsup_{\ell\to\infty}(H_{K_\ell}(\operatorname{Im}\zeta)-\gamma_{y,\ell})}\, |d\mu_{N,\nu\,:\,y}^{\alpha,\beta}| \leq 1$$

and

$$p_j(\zeta) \sum_{|\nu|\leq N}\ \sum_{|\alpha|+|\beta|\leq M_N} \Big(\frac{\partial}{\partial\zeta}\Big)^{\alpha} \Big(\frac{\partial}{\partial\overline{\zeta}}\Big)^{\beta} (d\mu_{N,\nu\,:\,y}^{\alpha,\beta}) = 0$$

for $j=1,\dots,m$. The last assertion follows from the fact that the estimates (7.279) authorize the differentiation $\deg p$ times of the right-hand side of (7.278) as a function of t under the bracket, combined with the fact that the distribution T_y is annihilated by the polynomial maps $\zeta \longmapsto p_j(\zeta)$ for $j=1,\dots,m$. The proof of the theorem is now complete. □

The major difference between the representation formulae (7.277) and (7.278) (with constraints (7.279)) is that in the first case, the Ehrenpreis–Palamodov nœtherian operators are involved, while in the second case the Bochner–Martinelli residue current $R_0^{\overline{\mathbf{F}_\bullet^p}}$ is. The action of the residue current is realized through the action of the distributions $T_{N,\nu}^{(0:r)}$ introduced in (7.268), on which the materialization of T_y in (7.278) relies, as the proof of Theorem 7.94 shows. The Bochner–Martinelli current $T = R_0^{\overline{\mathbf{F}_\bullet^p}}$ involved there depends in general on the choice of metrics on the complex $\mathbf{F}_\bullet^p$ and fails in most cases to be a Coleff–Herrera current; see its extrinsic description in §3.4.2 or its intrinsic description in case the ideal generated by the p_j's is radical in §3.4.3. Nevertheless, there are two significant cases where such a current is Coleff–Herrera:

(1) The most important case is the case where $m = \operatorname{codim} V(p) \leq n$ and $\mathbf{F}_\bullet^p = \mathbf{K}_\bullet^p$,[55] thus $T = \overline{\partial}(1/p)\, \epsilon_1^* \wedge \cdots \wedge \epsilon_m^*$, if $p = p_1\,\epsilon_1 + \cdots + p_m\,\epsilon_m$.
(2) The other case is where $V(p)$ is purely dimensional and the quotient sheaf $\mathcal{O}_{\mathbb{P}^n_{\mathbb{C}}} / \sum_1^m P_j \mathcal{O}_{\mathbb{P}^n_{\mathbb{C}}}$, where the P_j's denote the homogenizations of the p_j's, is Cohen–Macaulay in $\mathbb{C}^n$. In this case, one can take the complex $\mathbf{F}_\bullet^p$ as (7.280) with length $\operatorname{codim} V(p)$ so that the corresponding Bochner–Martinelli current T is Coleff–Herrera, with pure bidegree $(0, \operatorname{codim} V(p))$, but of course matricial.[56]

Even in such particular cases, where the residue current is Coleff–Herrera, hence can be locally described by Proposition 2.37, the two global formulations (7.277) and (7.278) (with constraints (7.279)) of the Ehrenpreis–Palamodov fundamental principle in the global setting remain different. The main reason for that (see also §5.1.2) is the polar part $1/\delta$ appearing in the local presentations of Coleff–Herrera residue currents as

$$Q^*\Big(\Big[\frac{|\delta|^{2\lambda}}{\delta}\,[V(p)]\Big]_{\lambda=0}\Big)$$

(see Proposition 2.37), which is not the case in representation (7.277) involving nœtherian operators. Thus formulations (7.277) and (7.278) (with constraints (7.279)) remain intrinsically different. The main advantage of the current formulation of the fundamental principle is that semilocally speaking, it can be made explicit in strongly convex relatively compact domains with smooth boundaries

[55]This particular case was for example studied in [**BGVY**, Chapter 5, §3] or in [**BerP89**].
[56]Such is the case for example when $\operatorname{codim} V(p) = n$.

such as in Theorem 7.92. It is not the case for the representation (7.93), since its proof relies on its final step on a functional analysis argument. Moreover, it is not the case for the current formulation in Theorem 7.94 either, since its proof relies on a Montel-type argument too. As seen in Corollary 7.89, another advantage of the current approach is that sometimes it carries through in the transcendental context for the representation of mean-periodic solutions, where compactly supported distributions replace differential operators with constant coefficients.

7.7. Residue calculus outside the commutative setting

Although we did not explicitly mention that *commutativity* remained crucial to settle univariate or multidimensional residue theory, this is what we have seen until now throughout this monograph. Therefore, it would be a fascinating challenge to move towards the noncommutative world some of the concepts that the presented residue theory carries, and some of the tools it provides. As examples, being able to move towards the operator theory setting the strategy used to solve Bézout identity explicitly, hence Hilbert's nullstellensatz (as presented in §7.2.1) or Briançon–Skoda's problem in $\mathbb{C}[X_1, \dots, X_n]$ (see Theorem 7.47) would indeed be very exciting.[57] We only present in this last section a few modest attempts towards such a goal.

7.7.1. Univariate quaternionic setting and relevant regularity concepts. Consider the quaternionic algebra $\mathbb{H}$ as a model in this subsection. We begin by recalling some basic facts about various presentations, as well as various notions of regularity (Fueter, Fueter modified, slice regularity), since our goal is to analyze in §7.7.2 how the concepts of principal value or residue current could get a chance to be transposed in this noncommutative (but fundamental and well explored) setting. The most classical representation of the generic element in $\mathbb{H}$ is

$$Z = x_1 \mathbf{1} + y_1\, \boldsymbol{i} + x_2\, \boldsymbol{j} + y_2\, \boldsymbol{k} \tag{7.288}$$

where $(x_1, y_1, x_2, y_2) \in \mathbb{R}^4$ and the generators $\boldsymbol{i}, \boldsymbol{j}, \boldsymbol{k}$ follow the rules

$$\begin{aligned} \boldsymbol{i}^2 &= \boldsymbol{j}^2 = -\mathbf{1}, \\ \boldsymbol{i}\,\boldsymbol{j} &= -\boldsymbol{j}\,\boldsymbol{i} = \boldsymbol{k}, \end{aligned} \tag{7.289}$$

governing the univariate (noncommutative) quaternionic calculus. There exist several alternatives to such a presentation. A few of them follow.

(1) The quaternionic algebra $\mathbb{H}$ is the four-dimensional real Clifford algebra

$$\mathrm{Cl}_{0,2}(\mathbb{R}) = \mathrm{Cl}_2(\mathbb{R}) = \mathbb{R} + \mathbb{R}\,\epsilon_1 + \mathbb{R}\,\epsilon_2 + \mathbb{R}\,\epsilon_1 \cdot \epsilon_2,$$

where $\{\epsilon_1, \epsilon_2\}$ denotes the canonical basis of $\mathbb{R}^2$ and the rules

$$\epsilon_j \cdot \epsilon_{j'} + \epsilon_{j'} \cdot \epsilon_j = -2\, \delta_j^{j'} \tag{7.290}$$

govern the noncommutative calculus. The $\mathbb{R}$-isomorphism consisting of the identifications

$$\epsilon_1 \longleftrightarrow \boldsymbol{i},\ \epsilon_2 \longleftrightarrow \boldsymbol{j},\ \epsilon_1 \cdot \epsilon_2 \longleftrightarrow \boldsymbol{k}$$

realizes the equivalence with the classical presentation given initially by (7.288) and (7.289); see for example [**CSSS04**, Example 4.1.2 (2)].

[57] We should mention here that pertinent questions of such nature were frequently posed to us by Nikolai Nikolski.

(2) The quaternionic algebra $\mathbb{H}$ is the *even* subalgebra of the eight-dimensional real Clifford algebra

$$\mathrm{Cl}_{0,3}(\mathbb{R}) = \mathrm{Cl}_3(\mathbb{R}) = \mathbb{R}+\mathbb{R}\,\epsilon_1+\mathbb{R}\,\epsilon_2+\mathbb{R}\,\epsilon_3+\mathbb{R}\,\epsilon_1\cdot\epsilon_2+\mathbb{R}\,\epsilon_2\cdot\epsilon_3+\mathbb{R}\,\epsilon_3\cdot\epsilon_1+\mathbb{R}\,\epsilon_1\cdot\epsilon_2\cdot\epsilon_3,$$

where $\{\epsilon_1, \epsilon_2, \epsilon_3\}$ is the canonical basis of $\mathbb{R}^3$ and the rules (7.290) still govern noncommutative calculus. The $\mathbb{R}$-isomorphism consisting of the identifications

$$\epsilon_1\cdot\epsilon_2 \longleftrightarrow \boldsymbol{i},\ \epsilon_3\cdot\epsilon_1 \longleftrightarrow \boldsymbol{j},\ \epsilon_1\cdot\epsilon_2 = \boldsymbol{k}$$

realizes then the equivalence with the classical presentation given initially by (7.288) and (7.289).

(3) More interesting for us is the embedding

$$Z = x_1\,\mathbf{1}+y_1\,\boldsymbol{i}+x_2\,\boldsymbol{j}+y_2\,\boldsymbol{k} \hookrightarrow \begin{pmatrix} x_1 - iy_1 & -x_2+iy_2 \\ x_2+iy_2 & x_1+iy_1 \end{pmatrix} \in \mathrm{Hom}_{\mathbb{C}}(\mathbb{C}^2,\mathbb{C}^2); \tag{7.291}$$

see, for example, [**WaW17**] for an updated bibliography since it brings us to the complex $\mathbb{C}^2$ world. It is well known that multivariate residue theory in $\mathbb{C}^2$ profits extensively from the pairing between holomorphic coordinates (z_1, z_2) and their ghost companions $(\overline{z_1}, \overline{z_2})$. In view of such a presentation, we will use the notation

$$\boldsymbol{z}_1 = x_1\mathbf{1} + y_1\boldsymbol{i}, \quad \boldsymbol{z}_2 = x_2\mathbf{1} + y_2\,\boldsymbol{i}, \tag{7.292}$$

so that the embedding (7.291) is expressed as

$$Z = \boldsymbol{z}_1 + \boldsymbol{z}_2\,\boldsymbol{j} \hookrightarrow \begin{pmatrix} \overline{z_1} & -\overline{z_2} \\ z_2 & z_1 \end{pmatrix} \in \mathrm{Hom}_{\mathbb{C}}(\mathbb{C}^2,\mathbb{C}^2), \tag{7.293}$$

where naturally $z_1 := x_1 + iy_1$ and $z_2 := x_2 + iy_2$.

(4) Besides the three above presentations of $\mathbb{H}$, one should mention, inspired by the tropical considerations in §7.4.2 and §7.4.3, that $\mathbb{H}\setminus\mathbb{R}$ is diffeomorphic to $\mathbb{R}\times\mathbb{R}_{>0}\times\mathbb{S}^2 \simeq \mathbb{R}\times\mathbb{R}_{>0}\times\partial\,\overline{B_{\mathbb{R}^3}(0,1)}$ as a four-dimensional differentiable manifold. The corresponding diffeomorphism is

$$\begin{gathered} (x,r,\xi) \in \mathbb{R}\times\mathbb{R}_{>0}\times\partial\,\overline{\mathbb{B}_{\mathbb{R}^3}(0,1)} \longrightarrow x\mathbf{1} + r(\xi_1\,\boldsymbol{i} + \xi_2\,\boldsymbol{j} + \xi_3\,\boldsymbol{k}), \\ \Big(x_1, \sqrt{y_1^2+|z_2|^2}, \frac{(y_1,x_2,y_2)}{\sqrt{y_1^2+|z_2|^2}}\Big) \longleftarrow x_1\,\mathbf{1} + y_1\,\boldsymbol{i} + x_2\,\boldsymbol{j} + z_2\,\boldsymbol{k} \in \mathbb{H}\setminus\mathbb{R}. \end{gathered} \tag{7.294}$$

The involuting conjugation

$$Z = \boldsymbol{z}_1 + \boldsymbol{z}_2\,\boldsymbol{j} \longmapsto (x_1\,\mathbf{1} - y_1\,\boldsymbol{i}) - \boldsymbol{z}_2\,\boldsymbol{j} = x_1\,\mathbf{1} - (y_1\,\boldsymbol{i} + x_2\,\boldsymbol{j} + y_2\,\boldsymbol{k}), \tag{7.295}$$

where $\boldsymbol{z}_1 = x_1\mathbf{1} + y_1\,\boldsymbol{i}$ and $\boldsymbol{z}_2 = x_2\,\mathbf{1} + y_2\,\boldsymbol{j}$ are defined as in (7.292), together with the definition of the operator

$$\frac{\partial}{\partial\overline{z}} = \frac{1}{2}\Big(\frac{\partial}{\partial x} + i\,\frac{\partial}{\partial y}\Big), \tag{7.296}$$

which plays a central role in complex analysis and in multivariate residue theory as well, suggests the construction in the quaternionic setting of substitutes for the operator (7.296). It was done by Grigore Constantin Moisil [**Moi31**] and pursued by Rudolf Fueter [**Fu35, Fu36**]. It leads to the introduction of the so-called Cauchy–Riemann–Moisil–Fueter left (resp., right) differential operators acting on the left (resp.,right, depending on how φ is presented, see below) on smooth functions

$$\begin{aligned}\varphi : Z = x_1\mathbf{1} + y_1\,\boldsymbol{i} + x_2\,\boldsymbol{j} + y_2\,\boldsymbol{k} \in \mathbb{U} \\ \longmapsto \varphi(Z) &= X_1(x,y)\,\mathbf{1} + Y_1(x,y)\,\boldsymbol{i} + X_2(x,y)\,\boldsymbol{j} + Y_2(x,y)\,\boldsymbol{k} \\ &= \mathbf{1}\,X_1(x,y) + \boldsymbol{i}\,Y_1(x,y)\,\boldsymbol{i} + \boldsymbol{j}\,X_2(x,y) + \boldsymbol{k}\,Y_2(x,y) \in \mathbb{H}\end{aligned}$$

from an open subset $\mathbb{U}$ of the four-dimensional real manifold $\mathbb{H}$ to $\mathbb{H}$, with X_1, X_2, Y_1, Y_2 smooth real functions on $\mathbb{U}$.

DEFINITION 7.95. The Cauchy–Riemann-Moisil-Fueter left and right operators are respectively defined by

$$(7.297)\qquad \begin{aligned}\frac{\overrightarrow{\partial}}{\partial \overline{Z}} &= \mathbf{1}\,\frac{\partial}{\partial x_1} + \Big(\boldsymbol{i}\,\frac{\partial}{\partial y_1} + \boldsymbol{j}\,\frac{\partial}{\partial x_2} + \boldsymbol{k}\,\frac{\partial}{\partial y_2}\Big),\\ \frac{\overleftarrow{\partial}}{\partial \overline{Z}} &= \frac{\partial}{\partial x_1}\,\mathbf{1} + \Big(\frac{\partial}{\partial y_1}\,\boldsymbol{i} + \frac{\partial}{\partial x_2}\,\boldsymbol{j} + \frac{\partial}{\partial y_2}\,\boldsymbol{k}\Big).\end{aligned}$$

The subsequent notions of the left (resp., right) *regularity in the sense of Fueter* follow. We will consider only left regularity here since the concept of right regularity is analogous.

DEFINITION 7.96. A differentiable function $\varphi\ : \mathbb{U} \to \mathbb{H}$ is said to be (*left*) *regular in the sense of Fueter* if and only one has in $\mathbb{U}$ that

$$(7.298)\qquad \frac{\overrightarrow{\partial}}{\partial \overline{Z}}\,\varphi \equiv 0 \Longleftrightarrow \frac{\partial}{\partial z_2}\begin{pmatrix} Z_1(z) \\ \\ Z_2(z)\end{pmatrix} = \frac{\partial}{\partial z_1}\begin{pmatrix} -\overline{Z_2(z)} \\ \\ \overline{Z_1(z)}\end{pmatrix}$$

when $\varphi(z) := X_1(x,y)\,\mathbf{1} + Y_1(x,y)\,\boldsymbol{i} + X_2(x,y)\,\boldsymbol{j} + Y_2(x,y)\,\boldsymbol{k}$ and

$$z = (x_1 + iy_1, x_2 + iy_2),\ \ Z_1(z) = (X_1 + iY_1)(x,y),\ Z_2(z) = (X_2 + iY_2)(x,y)$$

according to the notations (7.292).

The role played by the Cauchy kernel defined $\mathscr{U} = U \times U$ by

$$(z,\zeta) \in \mathscr{U} \setminus \mathscr{D} \longmapsto \frac{1}{\zeta - z}$$

in univariate complex analysis, where U is an open subset of $\mathbb{C}$ and

$$\mathscr{D} \subset \mathscr{U} = \{(z,\zeta)\,:\,\zeta = z\},$$

if one keeps track of the notations introduced in §4.1, is now played by the archetypal example of a (both left and right) regular function in the sense of Fueter outside the diagonal $\mathscr{D}_{\mathbb{H}} = \{(\boldsymbol{z},\boldsymbol{\zeta}) \in \mathbb{H}\times\mathbb{H}\,:\,\boldsymbol{\zeta} = \boldsymbol{z}\}$. Namely, the Cauchy kernel becomes now the *Cauchy–Fueter* kernel, denoted following the notations from §4.2 and defined by

(7.299)

$$K_{\mathbb{H}}(Z,\boldsymbol{\zeta}) = \frac{1}{(Z-\boldsymbol{\zeta})(\overline{Z-\boldsymbol{\zeta}}))}\,(\boldsymbol{\zeta} - Z)^{-1} = \frac{\overline{\boldsymbol{\zeta} - Z}}{\|\zeta - z\|^4} = \frac{\overline{(\boldsymbol{\zeta}_1 - \boldsymbol{z}_1)} - (\boldsymbol{\zeta}_2 - \boldsymbol{z}_2)\,\boldsymbol{j}}{\|\zeta - z\|^4},$$

where $Z = \boldsymbol{z}_1 + \boldsymbol{z}_2\,\boldsymbol{j}$, $\boldsymbol{\zeta} = \boldsymbol{\zeta}_1 + \boldsymbol{\zeta}_2\,\boldsymbol{j}$, $z = (z_1, z_2) \in \mathbb{C}^2$, $\zeta = (\zeta_1,\zeta_2) \in \mathbb{C}^2$ according to (7.292). The analogy between the Cauchy–Fueter kernel in $(\mathbb{H}\times\mathbb{H}) \setminus \mathscr{D}_{\mathbb{H}}$ and the Cauchy kernel in $(\mathbb{C}\times\mathbb{C}) \setminus \mathscr{D}$ is reinforced by the fact that functions in $C^1(\overline{\mathbb{U}}, \mathbb{H})$, where $\mathbb{U}$ is a bounded open subset of $\mathbb{H}$ with piecewise smooth boundary, can be reproduced in $\mathbb{U}$ thanks to the *Cauchy–Fueter–Pompeiu formula*; see for example [**Su79**], [**CSSS04**, Theorem 3.1.7] and [**Pe07**, §2.5].[58]

THEOREM 7.97. *Let $\mathbb{U} \subset \mathbb{H}$ be as above. Any $\varphi \in C^1(\mathbb{U}, \mathbb{H})$ admits in $\mathbb{U}$ the following Cauchy–Pompeiu type integral representation formula*

$$(7.300)\quad \varphi(Z) =$$
$$\frac{1}{4}\Big(\int_{\partial\mathbb{U}} \mathbb{K}_{\mathbb{H}}(\boldsymbol{\zeta} - Z) \rfloor_{\frac{\vec{\partial}}{\partial\zeta}} (dd^c\|\zeta\|^2)^2 \, \varphi(\boldsymbol{\zeta}) - \int_{\mathbb{U}} \mathbb{K}_{\mathbb{H}}(\boldsymbol{\zeta} - Z)\, \frac{\vec{\partial}}{\partial\overline{\boldsymbol{\zeta}}}\, \varphi(\boldsymbol{\zeta})\, (dd^c\|\zeta\|^2)^2\Big).$$

As a consequence, if $\varphi \in C^0(\overline{\mathbb{U}}, \mathbb{H}) \cap C^1(\mathbb{U}, \mathbb{H})$ is (left) regular in $\mathbb{U}$ in the sense of Fueter, then the Cauchy–Fueter integral representation formula

$$(7.301)\qquad \varphi(Z) = \frac{1}{4}\int_{\partial\mathbb{U}} \mathbb{K}_{\mathbb{H}}(\boldsymbol{\zeta} - Z) \rfloor_{\frac{\vec{\partial}}{\partial\zeta}} (dd^c\|\zeta\|^2)^2 \, \varphi(\boldsymbol{\zeta})$$

holds for any $Z \in \mathbb{U}$.

Unfortunately, such an essential function as the identity $\mathrm{Id}_{\mathbb{H}} : Z \longmapsto Z$ fails to be (left) regular in the sense of Fueter in $\mathbb{H}$, since

$$\frac{\vec{\partial}}{\partial\overline{Z}}\, Z = 1 + \boldsymbol{i}^2 + \boldsymbol{j}^2 + \boldsymbol{k}^2 = -2.$$

Any monomial Z^k (or Laurent monomial Z^{-k} in $\mathbb{H} \setminus 0$) for $k \geq 1$ fails also to be left or right regular in the sense of Fueter. In the paper [**Fu39**] devoted to Hartogs's theorem, Fueter came up with the brilliant suggestion to modify the Moisil operators (7.297) in order that, given a holomorphic map $f = (f_1, f_2)$ from $U \subset \mathbb{C}^2$ to $\mathbb{C}^2$, the $\mathbb{H}$-valued map

$$(7.302)\qquad \mathbf{f} : Z \in \{\boldsymbol{z}_1 + \boldsymbol{z}_2\boldsymbol{j} : (z_1, z_2) \in U\} \longmapsto \boldsymbol{f}_1(\boldsymbol{z}) + \boldsymbol{f}_2(\boldsymbol{z})\,\boldsymbol{j} \in \mathbb{H}$$

inherits regularity with respect to left or right-modified operators.

DEFINITION 7.98. The modified Cauchy–Fueter left and right operators are respectively defined by

$$(7.303)\qquad \begin{aligned} \Big(\frac{\vec{\partial}}{\partial\overline{Z}}\Big)' &= \mathbf{1}\,\frac{\partial}{\partial x_1} + \Big(\boldsymbol{i}\,\frac{\partial}{\partial y_1} + \boldsymbol{j}\,\frac{\partial}{\partial x_2} - \boldsymbol{k}\,\frac{\partial}{\partial y_2}\Big), \\ \Big(\frac{\overleftarrow{\partial}}{\partial\overline{Z}}\Big)' &= \frac{\partial}{\partial x_1}\,\mathbf{1} + \Big(\frac{\partial}{\partial y_1}\,\boldsymbol{i} + \frac{\partial}{\partial x_2}\,\boldsymbol{j} - \frac{\partial}{\partial y_2}\,\boldsymbol{k}\Big). \end{aligned}$$

The subsequent notions of the left (resp., right) *regularity in the sense of Fueter* follow. We will consider again only left regularity here since the notion of right regularity is analogous. We recall now the terminology introduced in [**CLSSS07, Dol13**].

---

[58]Observe that $(dd^c\|\zeta\|^2)^2$ equals $2/\pi^2$ times the volume form $dx_1 \wedge dy_1 \wedge dx_2 \wedge dy_2$ on $\mathbb{H} \simeq \mathbb{R}^4 \simeq \mathbb{C}^2$ as a four-dimensional differential manifold.

DEFINITION 7.99. A differentiable function $\varphi : \mathbb{U} \to \mathbb{H}$ is said to be (*left*) *hyperholomorphic* if and only if one has in $\mathbb{U}$ that

$$\Big(\frac{\vec{\partial}}{\partial \overline{Z}}\Big)'\varphi \equiv 0 \Longleftrightarrow \frac{\partial}{\partial z_2}\begin{pmatrix} \overline{Z_1(z)} \\ \overline{Z_2(z)} \end{pmatrix} = \frac{\partial}{\partial \overline{z}_1}\begin{pmatrix} -Z_2(z) \\ Z_1(z) \end{pmatrix} \tag{7.304}$$

when $\varphi(z) := X_1(x,y)\,\mathbf{1} + Y_1(x,y)\,\boldsymbol{i} + X_2(x,y)\,\boldsymbol{j} + Y_2(x,y)\,\boldsymbol{k}$ and

$$z = (x_1 + iy_1, x_2 + iy_2),\;\; Z_1(z) = (X_1 + iY_1)(x,y),\; Z_2(z) = (X_2 + iY_2)(x,y)$$

according to (7.292).

EXAMPLE 7.100. Here is the major example motivating the modification of the Cauchy–Fueter–Moisil operators. As we already mentioned, given $U \subset \mathbb{C}^2$ and the map $f = (f_1, f_2)$ holomorphic from U to $\mathbb{C}^2$, then the map $\mathbb{f} \in C^\infty(\mathbb{U}, \mathbb{H})$ defined as in (7.302) is (left) hyperholomorphic in $\mathbb{U} = \{\boldsymbol{z}_1 + \boldsymbol{z}_2\,\boldsymbol{j} \,:\, (z_1, z_2) \in U\}$. Actually, it is enough to observe that the four entries to compare in the matrix equality on the right-hand side of (7.304) equal zero in view of the formal independence of holomorphic coordinates with respect to their antiholomorphic ghost companions; see (7.173).

The role devoted to the Cauchy–Fueter kernel $K_{\mathbb{H}}$ is played now by the archetypal example of left or right hyperholomorphic function. Namely, it is played by the *modified Cauchy–Fueter* kernel, denoted again by the notations compatible to those in §4.2 and defined by

$$K'_{\mathbb{H}}(Z, \boldsymbol{\zeta}) = \frac{\overline{(\boldsymbol{\zeta}_1 - \boldsymbol{z}_1)} - (\overline{\boldsymbol{\zeta}_2 - \boldsymbol{z}_2})\,\boldsymbol{j}}{\|\zeta - z\|^4}, \tag{7.305}$$

where $Z = \boldsymbol{z}_1 + \boldsymbol{z}_2\,\boldsymbol{j}$, $\boldsymbol{\zeta} = \boldsymbol{\zeta}_1 + \boldsymbol{\zeta}_2\,\boldsymbol{j}$, $z = (z_1, z_2) \in \mathbb{C}^2$, $\zeta = (\zeta_1, \zeta_2) \in \mathbb{C}^2$ according to (7.292). The analogy between the modified Cauchy–Fueter kernel in $(\mathbb{H} \times \mathbb{H}) \setminus \mathscr{D}_{\mathbb{H}}$ and that of the Cauchy kernel in $(\mathbb{C} \times \mathbb{C}) \setminus \mathscr{D}$ is reinforced once more by the fact that functions in $C^1(\overline{\mathbb{U}}, \mathbb{H})$, where $\mathbb{U}$ is a bounded open subset of $\mathbb{H}$ with piecewise smooth boundary, can be reproduced in $\mathbb{U}$ thanks to the *modified Cauchy–Fueter–Pompeiu formula*; see for example [**CSSS04**, Theorem 3.1.7].

THEOREM 7.101. *Let $\mathbb{U} \subset \mathbb{H}$ be as above. Any $\varphi \in C^1(\mathbb{U}, \mathbb{H})$ admits in $\mathbb{U}$ the following Cauchy–Pompeiu type integral representation formula*

$$\varphi(Z) = \tag{7.306}$$

$$\frac{1}{4}\Big(\int_{\partial\mathbb{U}} \mathbb{K}'_{\mathbb{H}}(\boldsymbol{\zeta} - Z)\rfloor_{\left(\frac{\vec{\partial}}{\partial \overline{\boldsymbol{\zeta}}}\right)'} (dd^c\|\zeta\|^2)^2 \varphi(\boldsymbol{\zeta}) - \int_{\mathbb{U}} \mathbb{K}'_{\mathbb{H}}(\boldsymbol{\zeta} - Z)\Big(\frac{\vec{\partial}}{\partial \overline{\boldsymbol{\zeta}}}\Big)'\varphi(\boldsymbol{\zeta})\,(dd^c\|\zeta\|^2)^2\Big).$$

As a consequence, if $\varphi \in C^0(\overline{\mathbb{U}}, \mathbb{H}) \cap C^1(\mathbb{U}, \mathbb{H})$ is hyperholomorphic in $\mathbb{U}$, then the modified Cauchy–Fueter integral representation formula

$$\varphi(Z) = \frac{1}{4}\int_{\partial\mathbb{U}} \mathbb{K}'_{\mathbb{H}}(\boldsymbol{\zeta} - Z)\rfloor_{\left(\frac{\vec{\partial}}{\partial \overline{\boldsymbol{\zeta}}}\right)'} (dd^c\|\zeta\|^2)^2\,\varphi(\boldsymbol{\zeta}) \tag{7.307}$$

holds for any $Z \in \mathbb{U}$.

REMARK 7.102. The clever idea of Rudolf Fueter in [**Fu39**], namely the idea to relate the theory of holomorphic functions in two (resp., more) complex variables to that of hyperholomorphic ones in one quaternionic variable Z (resp., in some more general hypercomplex variable) inspired Enzo Martinelli. He introduced the *Bochner–Martinelli kernel* $K_{\mathbb{C}^n}$ in [**Mar43**]. This kernel $K_{\mathbb{C}^n}$ (see §4.2.1) is a central object in the present monograph.

The identity $\mathrm{Id}_{\mathbb{H}} : \mathbb{H} \to \mathbb{H}$ is hyperholomorphic but $Z^{-1} : \mathbb{H}\setminus\{0\} \to \mathbb{H}\setminus\{0\}$ fails to be so. More generally, when $\mathbf{h} = \mathbf{h}_1 + \mathbf{h}_2 \boldsymbol{j}$, with $\mathbf{h}_1(Z) = X_1(x,y) + Y_1(x,y)\,\boldsymbol{i}$ and $\mathbf{h}_2(Z) = X_2(x,y) + Y_2(x,y)\,\boldsymbol{i}$, is hyperholomorphic in $\mathbb{U}$,

$$\mathbf{h}^{-1} : Z \longmapsto \frac{(X_1(x,y)\,\mathbf{1} - Y_1(x,y)\,\boldsymbol{i}) - (X_2(x,y)\,\mathbf{1} + Y_2(x,y)\,\boldsymbol{i})\,\boldsymbol{j}}{\overline{\mathbf{h}(Z)}\,\mathbf{h}(Z)}$$

fails in general to be hyperholomorphic in $\mathbb{U} \setminus \mathbf{h}^{-1}(\{0\})$. Similarly, $\boldsymbol{\zeta} \mapsto K'_{\mathbb{H}}(0,\boldsymbol{\zeta})$ is hyperholomorphic in $\mathbb{H} \setminus \{0\}$, while its inverse $\boldsymbol{\zeta} \mapsto \|\zeta\|^2(\boldsymbol{\zeta}_1 + \boldsymbol{j}\,\boldsymbol{\zeta}_2)$ fails to be hyperholomorphic in $\mathbb{H}$. It is a major stumbling block in view of univariate principal value or residue theory in a quaternionic context. Let us set the following definition from [**Dol13**, Definition 4.2].

DEFINITION 7.103. Let $\mathbb{U}$ be an open set of $\mathbb{H}$. A function $\mathbf{h}$ defined in $\mathbb{U}$ is called *hypermeromorphic* if it is a hyperholomorphic function defined almost everywhere in $\mathbb{U}$, whose inverse $\mathbf{h}^{-1}$ (see above) shares the same properties.

The following proposition is due to Pierre Dolbeault [**Dol13**, Proposition 3.1]. It emphasizes that hypermeromorphicity (when hyperholomorphicity of the inverse is satisfied) requires undeniably many constraints. For example, these constraints exclude in most cases $1/\mathbf{f}$, when $\mathbf{f} = \boldsymbol{f}_1 + \boldsymbol{f}_2\,\boldsymbol{j}$ with $f = (f_1, f_2) : U \to \mathbb{C}^2$ being holomorphic.

PROPOSITION 7.104. *Let $\mathbf{f} \in C^1(\mathbb{U}, \mathbb{H})$, where $\mathbb{U}$ is an open subset of $\mathbb{H}$ and $V(\mathbf{f})$ be its zero set in $\mathbb{U}$. Then $\mathbf{f}$ and $1/\mathbf{f}$ are both hyperholomorphic in $\mathbb{U}$ and $\mathbb{U} \setminus V(\mathbf{f})$, respectively, if and only if*

$$(7.308)\quad 2i\,\mathrm{Im}\, f_1(z)\,\frac{\partial}{\partial z_1}\begin{pmatrix}\overline{f_1(z)}\\ \\ \overline{f_2(z)}\end{pmatrix} = \overline{f_2(z)}\,\frac{\partial}{\partial z_1}\begin{pmatrix}-f_2(z)\\ \\ f_1(z)\end{pmatrix} - f_2(z)\,\frac{\partial}{\partial \overline{z}_2}\begin{pmatrix}\overline{f_1(z)}\\ \\ \overline{f_2(z)}\end{pmatrix},$$

where $f_1(z) = X_1(z) + iY_1(z)$, $f_2(z) = X_2(z) + iY_2(z)$, $z = (x_1 + iy_1, x_2 + iy_2)$, when

$$\mathbf{f}\,(x_1\,\mathbf{1} + y_1\,\boldsymbol{i} + (x_2\,\mathbf{1} + y_2\,\boldsymbol{i})\,\boldsymbol{j}) = X_1(z)\,\mathbf{1} + Y_1(z)\,\boldsymbol{i} + (X_2(z)\,\mathbf{1} + Y_2(z)\,\boldsymbol{i})\,\boldsymbol{j}.$$

EXAMPLE 7.105. If $\mathbf{f} = \boldsymbol{f}_1 + \boldsymbol{f}_2\,\boldsymbol{j}$, where $z = (z_1, z_2) \longmapsto f(z) = (f_1(z), f_2(z))$ is holomorphic from $U = \{z : \boldsymbol{z}_1 + \boldsymbol{z}_2\,\boldsymbol{j} \in \mathbb{U}\}$ to $\mathbb{C}^2$, conditions (7.308) reduce to

$$(7.309)\qquad \overline{f_2(z)}\,\frac{\partial}{\partial z_1}\begin{pmatrix}-f_2(z)\\ \\ f_1(z)\end{pmatrix} - f_2(z)\,\frac{\partial}{\partial \overline{z}_2}\begin{pmatrix}\overline{f_1(z)}\\ \\ \overline{f_2(z)}\end{pmatrix} = 0.$$

Suppose U is connected. The conditions (7.308) are only fulfilled when $f_2 \equiv 0$ or $f = \text{constant}$ in U.

EXAMPLE 7.106. If $\mathbf{f} = \boldsymbol{f}_1 + \boldsymbol{f}_2\,\boldsymbol{j}$ with f_1 and f_2 real, then the conditions (7.308) reduce to

$$(7.310)\qquad f_2(z)\,\Big(\frac{\partial}{\partial z_1}\begin{pmatrix}-f_2(z)\\ \\ f_1(z)\end{pmatrix} - \frac{\partial}{\partial \overline{z}_2}\begin{pmatrix}f_1(z)\\ \\ f_2(z)\end{pmatrix}\Big) = 0.$$

Note that such conditions are satisfied when $f_2 \equiv 0$ or $f_1(z) = x_1 + x_2 + \gamma_1$ and $f_2(z) = x_2 - x_1 + \gamma_2$ for real constants γ_1, γ_2; see [**Dol13**, Proposition 4].

As we have observed, important quaternionic functions, such as polynomial or Laurent polynomial functions

$$Z \in \mathbb{H} \longmapsto \sum_{\nu=0}^{d} Z^{d-\nu}\gamma_\nu, \quad Z \in \mathbb{H} \setminus 0 \longmapsto \sum_{\nu=0}^{d} Z^{\nu-d}\gamma_\nu, \tag{7.311}$$

where $\gamma_\nu \in \mathbb{H}$, fail to be either left regular in the sense of Fueter or (left) hyperholomorphic. For such reasons, Graziano Gentili and Daniele C. Struppa introduced in [**GeS06, GeS07**] (see also [**CSS16**] for an up-to-date bibliography) the concept of (left) *slice regularity* for quaternionic valued real-differentiable functions on some open subset $\mathbb{U} \subset \mathbb{H}$.

Definition 7.107. Let $\mathbf{f} : \mathbb{U} \longmapsto \mathbb{R}$ be a real differential function. The function $\mathbf{f}$ is said to be (left) *slice regular* in $\mathbb{U} \subset \mathbb{H}$ if and only if for any element $\boldsymbol{I}$ in the unit sphere of purely imaginary quaternions $y_1\,\boldsymbol{i} + x_2\,\boldsymbol{j} + y_2\,\boldsymbol{k}$ with $y_1^2 + |z_2|^2 = 1$, one has

$$\Big(\frac{\partial}{\partial u} + \boldsymbol{I}\,\frac{\partial}{\partial v}\Big) f(u + v\,\boldsymbol{I}) \equiv 0 \tag{7.312}$$

in $\Omega_{\boldsymbol{I}} := \{w = u + iv \in \mathbb{C} \,:\, u\,\mathbf{1} + v\,\boldsymbol{I} \in \mathbb{U}\}$.

Remark 7.108. Let us consider the action of the group $\mathrm{SO}(3, R)$ on the unit quaternionic sphere

$$\mathbb{S}_{\mathbb{H}} = \{y_1\,\boldsymbol{i} + x_2\,\boldsymbol{j} + y_2\,\boldsymbol{k} \,:\, (y_1, x_2, y_2) \in \partial\,\overline{\mathbb{B}_{\mathbb{R}^3}(0,1)}\}.$$

If $\mathbf{f}$ is a (left) slice regular function in $\mathbb{U}$, then for any $\sigma \in \mathrm{SO}(3,\mathbb{R})$ and any $u + iv \in \Omega_{\sigma(\boldsymbol{i})}$, the function

$$u + iv \in \Omega_{\sigma(\boldsymbol{i})} \longmapsto \mathbf{f}\,(u + \sigma(\boldsymbol{i})v)$$

decomposes as

$$\mathbf{f}\,(u + v\,\sigma(\boldsymbol{i})) = (U'_\sigma(u,v) + V'_\sigma(u,v)\,\sigma(\boldsymbol{i})) + (U''_\sigma(u,v) + V''_\sigma(u,v)\,\sigma(\boldsymbol{i}))\,\sigma(\boldsymbol{j}),$$

where the two complex-valued functions

$$u + iv \in \Omega_{\sigma(\boldsymbol{i})} \longmapsto U'_\sigma + iV'_\sigma, \quad u + iv \in \Omega_{\sigma(\boldsymbol{i})} \longmapsto U''_\sigma + iV''_\sigma$$

are holomorphic in $\Omega_{\sigma(\boldsymbol{i})}$; see [**CSS16**, Lemma 2.1].

Example 7.109. Polynomial functions or Laurent polynomial functions as in (7.311) are (left) slice regular. It is the case also for convergent power series

$$Z \in \mathbb{B}_{\mathbb{H}}(0,r) \longmapsto \sum_{\kappa=0}^{\infty} Z^\kappa \gamma_\kappa, \quad Z \in \mathbb{H} \setminus \mathbb{B}_{\mathbb{H}}(0,R) \longmapsto \sum_{\kappa=1}^{\infty} Z^{-\kappa}\gamma_\kappa.$$

Let us introduce the commutative multiplication $\circledast$ for quaternionic functions defined on a common domain as

$$\Big(\sum_\kappa Z^\kappa \gamma'_\kappa\Big) \circledast \Big(\sum_\kappa Z^\kappa \gamma''_\kappa\Big) := \sum_\kappa Z^\kappa \Big(\sum_{\{(\kappa',\kappa'')\,:\,\kappa'+\kappa''=\kappa\}} \gamma'_{\kappa'}\gamma''_{\kappa''}\Big). \tag{7.313}$$

The introduction of multiplication (7.313) allows us to formulate within the algebraic setting of a quaternionic polynomial with coefficients on the right as in (7.311) the following decomposition theorem; see [**LavR98, AlB19**] or [**CSS16**, Theorem 3.9].

THEOREM 7.110. *A quaternionic polynomial map* $Z \longmapsto p(Z) = \sum_{\nu=0}^{d} Z^{d-\nu}\gamma_\nu$ *with* $\gamma_\nu \in \mathbb{H}$ *and* $\gamma_0 \neq 0$ *admits the unique* $\circledast$ *decomposition*

$$\begin{aligned} p(Z) &= p_{\mathbb{S}}(Z) \circledast p_{\mathrm{O}}(Z) = p_{\mathrm{O}}(Z) \circledast p_{\mathbb{S}}(Z) \\ &= p_{\mathbb{S}}(Z)\, p_{\mathrm{O}}(Z) = p_{\mathrm{O}}(Z)\, p_{\mathbb{S}}(Z). \end{aligned} \tag{7.314}$$

The polynomials $p_{\mathbb{S}} \in \mathbb{R}[Z]$ *with degree* $2m \leq d$, $p_{\mathrm{O}}(Z) = \sum_{\nu=0}^{d-2m} Z^{d-2m-\nu}\gamma_{0,\nu}$ *with* $\gamma_{0,\nu} \in \mathbb{H}$ *and* $\gamma_{0,0} \neq 0$ *are such that* $V(p_{\mathrm{O}})$ *is a discrete or empty subset of* $\mathbb{H}$, $V(p_{\mathbb{S}})$ *is the finite union of at most* $m' \leq m$ *distinct spheres* $x_\iota + r_\iota \mathbb{S}_{\mathbb{H}}$ *with* $x_\iota \in \mathbb{R}$ *and* $r_\iota > 0$ *for* $\iota = 1, \dots, m'$. *Moreover, one has the Lelong–Poincaré type formula*

$$\boldsymbol{\Delta}_2\Big(- \frac{\big[\log |p_{\mathbb{S}}(Z)|^2\big]}{6}\Big) = \sum_{\iota=1}^{m'} \mu_\iota\, \delta_{x_\iota + r_\iota \mathbb{S}_{\mathbb{H}}} \tag{7.315}$$

in the sense of real-valued distributions on $\mathbb{H} \simeq \mathbb{R}^4$, *where* $\boldsymbol{\Delta}_2$ *is the bilaplacian*[59] *real differential operator with order* 4

$$\boldsymbol{\Delta}_2 = \Big(\frac{\partial^2}{\partial \overline{z}_1 \partial z_1} + \frac{\partial^2}{\partial \overline{z}_2 \partial z_2}\Big)^2 = \frac{1}{16}\Big(\Big(\frac{\partial}{\partial x_1}\Big)^2 + \Big(\frac{\partial}{\partial y_1}\Big)^2 + \Big(\frac{\partial}{\partial x_2}\Big)^2 + \Big(\frac{\partial}{\partial y_2}\Big)^2\Big)^2$$

and $\mu_\iota \in \mathbb{N}^*$ *with* $\sum_1^{m'} \mu_\iota = m$.

REMARK 7.111. Moreover, one can attach to each (isolated) zero of p_{O} in (7.314) a (nonlocal) multiplicity in terms of the $\circledast$ factorization. Namely, for each such zero

$$\boldsymbol{\alpha} \in x(\boldsymbol{\alpha}) + r(\boldsymbol{\alpha})\, \mathbb{S}_{\mathbb{H}} = \mathbb{S}_{\mathbb{H}}(\boldsymbol{\alpha})$$

according to the polar representation of quaternions (7.294), one can find the algorithmically unique sequence

$$[\boldsymbol{\alpha}_1 = \boldsymbol{\alpha}, \dots, \boldsymbol{\alpha}_{\mu_{\boldsymbol{\alpha}}(p_{\mathrm{O}})}]$$

with entries in $\mathbb{S}_{\mathbb{H}}(\boldsymbol{\alpha})$ and maximal length $\mu_{\boldsymbol{\alpha}}(p_{\mathrm{O}})$ such that

$$p_{\mathrm{O}}(Z) = (Z - \boldsymbol{\alpha}_1) \circledast \cdots \circledast (Z - \boldsymbol{\alpha}_{\mu_{\boldsymbol{\alpha}}(p_{\mathrm{O}})}) \circledast q_{\boldsymbol{\alpha}}(Z),$$

where $\boldsymbol{\alpha}_{j+1} \neq \overline{\boldsymbol{\alpha}_j}$ for $j = 1, \dots, \mu_{\boldsymbol{\alpha}}(p_{\mathrm{O}}) - 1$ and $q_{\boldsymbol{\alpha}}$ does not vanish on $\mathbb{S}_{\mathbb{H}}(\boldsymbol{\alpha})$; see [**CSS16**, Definition 3.4]. Furthermore, one has

$$\sum_{\boldsymbol{\alpha} \in V(p_{\mathrm{O}})} \mu_{\boldsymbol{\alpha}}(p_{\mathrm{O}}) = d - 2m = \deg p - \deg p_{\mathbb{S}}$$

in terms of decomposition (7.314).

7.7.2. Principal value and residue currents in the $\mathbb{H}$-setting. Let us first introduce the sheaf of $\mathbb{H}$-valued distributions on $\mathbb{H}$ through the representation of quaternions as $\boldsymbol{z}_1 + \boldsymbol{z}_2\, \boldsymbol{j}$ with $(z_1, z_2) = (x_1 + iy_1, x_2 + iy_2)$.

DEFINITION 7.112. Let $\mathbb{U}$ be an open subset of $\mathbb{H}$ and let

$$U = \{(z_1, z_2) \in \mathbb{C}^2 \,:\, \boldsymbol{z}_1\, \boldsymbol{1} + \boldsymbol{z}_2\, \boldsymbol{j} \in \mathbb{U}\}.$$

A distribution $\boldsymbol{T}$ on $\mathbb{U}$ corresponds to a pair of $\mathbb{C}$-valued distributions

$$T_1 = \mathrm{Re}\, T_1 + i\, \mathrm{Im}\, T_1\ ,\ T_2 = \mathrm{Re}\, T_2 + i\, \mathrm{Im}\, T_2$$

on U. The (left) distribution

$$\boldsymbol{T}_1 + \boldsymbol{j}\, \boldsymbol{T}_2 = (\mathrm{Re}\, T_1 + \mathrm{Im}\, T_1\, \boldsymbol{i}) + \boldsymbol{j}\, (\mathrm{Re}\, T_2 + \mathrm{Im}\, T_2\, \boldsymbol{i})$$

---

[59] Observe that the bilaplacian operator $16\, \boldsymbol{\Delta}_2$ plays here a central role, as it does also in [**Fu35**].

acts on smooth $\mathbb{H}$-valued functions $\boldsymbol{\varphi}+\boldsymbol{\psi}\,\boldsymbol{j}$ on $\boldsymbol{U}$, where $\boldsymbol{\varphi}=\operatorname{Re}\varphi+\operatorname{Im}\varphi\,\boldsymbol{i}$ and $\boldsymbol{\psi}=\operatorname{Re}\psi+\operatorname{Im}\psi\,\boldsymbol{i}$, φ and ψ being $\mathbb{C}$-valued test forms on U, as

$$\langle\!\langle \boldsymbol{T}_1+\boldsymbol{j}\,\boldsymbol{T}_2,\boldsymbol{\varphi}+\boldsymbol{\psi}\,\boldsymbol{j}\rangle\!\rangle=\big(\langle T_1,\varphi\rangle-\overline{\langle T_2,\psi\rangle}\big)_{i\to\boldsymbol{i}}+\big(\langle T_1,\psi\rangle+\overline{\langle T_2,\varphi\rangle}\big)_{i\to\boldsymbol{i}}\,\boldsymbol{j}.$$

REMARK 7.113. One can as well interpret $\boldsymbol{T}_1+\boldsymbol{j}\,\boldsymbol{T}_2$ as a $(0,0)$-current acting on the left on the 4-form $(\pi^2(dd^c\|z\|^2)/2)\,(\boldsymbol{\varphi}+\boldsymbol{\psi}\,\boldsymbol{j})$.

The sheaf of $\mathbb{H}$-valued $(0,0)$-currents acting on the left is denoted by $'\mathcal{D}^{0,0}_{\text{left}}(\cdot,\mathbb{H})$. Given a (left) hyperholomorphic function $\mathbf{f}:\mathbb{U}\longrightarrow\mathbb{H}$ of the form $\mathbf{f}=\boldsymbol{f}_1+\boldsymbol{f}_2\,\boldsymbol{j}$ with $z\mapsto(f_1(z),f_2(z))$ holomorphic in U (see Example 7.100) one associates to it a principal value current $P^{\mathbf{f}}$ as follows, according to what we did in one-dimensional complex setting; see Proposition 1.12.

DEFINITION 7.114. Let $\mathbb{U}$ and $\mathbf{f}=\boldsymbol{f}_1+\boldsymbol{f}_2\,\boldsymbol{j}$ be as above. Then

$$\lambda\longmapsto\|f\|^{2\lambda}\,\frac{\overline{\boldsymbol{f}_1}-\boldsymbol{j}\,\overline{\boldsymbol{f}_2}}{2i\pi\,\|f\|^2}\tag{7.316}$$

extends from $\{\lambda\in\mathbb{C}:\operatorname{Re}\lambda\gg 1\}$ to $\mathbb{C}$ as a $'\mathscr{D}^{0,0}_{\text{left}}(\mathbb{U},\mathbb{H})$-valued meromorphic map with poles in $-\mathbb{Q}^+$. Its value at $\lambda=0$ is called the principal value current $P^{\mathbf{f}}$.

For what concerns the approach to the notion of residue current in such a modified Fueter setting, with $\mathbf{f}:\mathbb{U}\longrightarrow\mathbb{H}$ as above, the noncommutativity generates a lot of complications. We will only write down a few computations below and then conclude with a suggestion. One has for $\operatorname{Re}\lambda\gg 1$ that

$$\begin{aligned}
&\Big(\frac{\vec{\partial}}{\partial\overline{Z}}\Big)'\Big(\|f\|^{2\lambda}\,\frac{\overline{\boldsymbol{f}_1}-\boldsymbol{j}\,\overline{\boldsymbol{f}_2}}{2i\pi\,\|f\|^2}\Big)\\
&=\Big(\Big(\frac{\vec{\partial}}{\partial\overline{Z}}\Big)'\|f\|^{2\lambda}\Big)\frac{\overline{\boldsymbol{f}_1}-\boldsymbol{j}\,\overline{\boldsymbol{f}_2}}{2i\pi\,\|f\|^2}+\|f\|^{2\lambda}\Big(\frac{\vec{\partial}}{\partial\overline{Z}}\Big)'\Big(\frac{\overline{\boldsymbol{f}_1}-\boldsymbol{j}\,\overline{\boldsymbol{f}_2}}{2i\pi\,\|f\|^2}\Big)\\
&=\lambda\|f\|^{2(\lambda-1)}(\boldsymbol{F}_1+\boldsymbol{j}\,\boldsymbol{F}_2)\,\frac{\overline{\boldsymbol{f}_1}-\boldsymbol{j}\,\overline{\boldsymbol{f}_2}}{i\pi\,\|f\|^2}+\|f\|^{2\lambda}\Big(\frac{\partial}{\partial\overline{z_1}}+\boldsymbol{j}\frac{\partial}{\partial\overline{z_2}}\Big)_{i\to\boldsymbol{i}}\Big(\frac{\overline{\boldsymbol{f}_1}-\boldsymbol{j}\,\overline{\boldsymbol{f}_2}}{2i\pi\,\|f\|^2}\Big)\\
&=\lambda\frac{\|f\|^{2(\lambda-2)}}{i\pi}\big(\overline{f_1}F_1+\overline{f_2}F_2+\boldsymbol{j}\,(\overline{f_1}F_2-\overline{f_2}F_1)\big)_{i\to\boldsymbol{i}}\\
&\quad+\frac{\|f\|^{2\lambda}}{i\pi}\Big(\frac{\partial}{\partial\overline{z_1}}\frac{\overline{f_1}}{\|f\|^2}+\frac{\partial}{\partial z_2}\frac{\overline{f_2}}{\|f\|^2}+\boldsymbol{j}\Big(\frac{\partial}{\partial\overline{z_2}}\,\frac{\overline{f_1}}{\|f\|^2}-\frac{\partial}{\partial z_1}\,\frac{\overline{f_2}}{\|f\|^2}\Big)\Big)_{i\to\boldsymbol{i}},
\end{aligned}\tag{7.317}$$

where

$$F_1=\frac{\partial}{\partial\overline{\zeta}_1}\|f\|^2\ ,\ F_2=\frac{\partial}{\partial\overline{\zeta}_2}\|f\|^2.$$

Since the $(2,2)$-complex valued pseudo-meromorphic currents

$$\Big[\frac{\|f\|^{2\lambda}}{i\pi}\Big(\frac{\partial}{\partial\overline{z_1}}\frac{\overline{f_1}}{\|f\|^2}+\frac{\partial}{\partial z_2}\frac{\overline{f_2}}{\|f\|^2}\Big)\,(dd^c\|z\|^2)^2\Big]_{\lambda=0},$$
$$\Big[\frac{\|f\|^{2\lambda}}{i\pi}\Big(\frac{\partial}{\partial\overline{z_2}}\,\frac{\overline{f_1}}{\|f\|^2}-\frac{\partial}{\partial z_1}\,\frac{\overline{f_2}}{\|f\|^2}\Big)\,(dd^c\|z\|^2)^2\Big]_{\lambda=0}$$

are clearly of standard extension on $U = \{z \in \mathbb{C}^2 : \boldsymbol{z}_1\, \mathbf{1} + \boldsymbol{z}_2\, \boldsymbol{j} \in \mathbb{U}\}$ with respect to their support U (see Definition 2.48) the $(2,2)$-current

$$
\begin{aligned}
(7.318)\quad R^{\mathbf{f}} &:= \mathbf{1}_{V(\mathbf{f})} \cdot \Big(\Big[\Big(\frac{\vec{\partial}}{\partial \overline{Z}} \Big)' P^{\mathbf{f},\lambda} \wedge dx_1 \wedge dy_1 \wedge dx_2 \wedge dy_2 \Big]_{\lambda=0} \Big) \\
&= \Big[\lambda \frac{\|f\|^{2(\lambda-2)}}{i\pi} \big(\overline{f_1} F_1 + \overline{f_2} \overline{F_2} + \boldsymbol{j}\, (\overline{f_1} F_2 - \overline{f_2} \overline{F_1}) \big)_{i \to \boldsymbol{i}}\, dx_1 \wedge dy_1 \wedge dx_2 \wedge dy_2 \Big]_{\lambda=0}
\end{aligned}
$$

could be a natural candidate for a so-called residue current attached to such hyperholomorphic function $\mathbf{f} : \mathbb{U} \longrightarrow \mathbb{H}$. Of course, such a candidate would be of a very particular type since f_1 and f_2 are both holomorphic in U. Observe in particular that

$$
\begin{aligned}
(7.319)\quad & \Big\langle\!\!\Big\langle R^{\mathbf{f}} , (\boldsymbol{f}_1 + \boldsymbol{f}_2\, \boldsymbol{j})\, \varphi(Z) \Big\rangle\!\!\Big\rangle = \\
& \qquad \Big[\Big\langle\!\!\Big\langle \lambda \|f\|^{2(\lambda-1}\, (\boldsymbol{F}_1 + \boldsymbol{j}\, \boldsymbol{F}_2)\, dx_1 \wedge dy_1 \wedge dx_2 \wedge dy_2 \,,\, \varphi(Z) \Big\rangle\!\!\Big\rangle \Big]_{\lambda=0} = 0
\end{aligned}
$$

for any $\varphi \in \mathcal{D}(\mathbb{H}, \mathbb{R})$.

To mimic constructions realized in the one-dimensional complex setting instead of the bivariate complex one in relation with hyperholomorphicity, let us finally consider the (basically univariate) concept of (left) slice regularity and profit from Remark 7.108. Suppose that $\mathbf{f}\ : \mathbb{U} \to \mathbb{H}$ is slice regular in $\mathbb{U}$ as described by Definition 7.107. Given any $\sigma \in \mathrm{SO}(3, \mathbb{R})$ (we assume that this group acts on the quaternionic unit sphere $\mathbb{S}_{\mathbb{H}}$), there is a holomorphic section of the trivial bundle $\mathbb{C}^2$ over

$$\Omega_{\sigma(\boldsymbol{i})} := \{u + iv \in \mathbb{C} : u\, \mathbf{1} + v\, \sigma(\boldsymbol{i}) \in \mathbb{U}\}$$

defined by

$$s_\sigma = u + iv \longmapsto (U'_\sigma + iV'_\sigma, U''_\sigma + iV''_\sigma)$$

such that

$$\mathbf{f}\, (u\, \mathbf{1} + v\, \sigma(\boldsymbol{i})) = (U'_\sigma + iV'_\sigma)_{i \to \sigma(\boldsymbol{i})} + (U'_\sigma + iV'_\sigma)_{i \to \sigma(\boldsymbol{i})}\, \sigma(\boldsymbol{j})$$

for any $u + iv \in \mathbb{C}$. The currents $P^{s_\sigma}_{\|\ \|}$ and $R^{s_\sigma}_{\|\ \|}$ constructed as in Proposition 1.27 are principal value and residue current attached to the restriction of $\mathbf{f}$ to the slice $\mathbb{U} \cap (\mathbb{R} + \mathbb{R}\, \sigma(\boldsymbol{i}))$; see also §1.2.1 for the particular case of trivial bundles (here $\mathbb{C}^2$) with the canonical metric. Any smooth $\mathbb{H}$-valued function in $\mathbb{U}$ splits on such slice as

$$\boldsymbol{\varphi}(u + v\, \sigma(\boldsymbol{i})) = \boldsymbol{\varphi}^{[1]}_\sigma(u + iv) + \boldsymbol{\varphi}^{[2]}_\sigma(u + iv)\, \sigma(\boldsymbol{j}),$$

where $\boldsymbol{\varphi}^{[1]}_\sigma(u + iv), \boldsymbol{\varphi}^{[2]}_\sigma(u + iv) \in \mathbb{R} + \mathbb{R}\, \sigma(\boldsymbol{i})$ for any $u + iv \in \Omega_{\sigma(\boldsymbol{i})}$. A natural principal value distribution $P^{\mathbf{f}}_{\text{slice}}$ attached to $\mathbf{f}$ in the slice regular context would be then

$$
\begin{aligned}
(7.320)\quad & \langle\!\langle P^{\mathbf{f}}_{\text{slice}} \,,\, \boldsymbol{\varphi} \rangle\!\rangle \\
&= \int_{\mathrm{SO}(3,\mathbb{R})} \Big\langle P^{s_\sigma}_{\|\ \|}(u+iv) \,,\, \big(\epsilon_1\, \varphi^{[1]}_\sigma(u+iv) + \epsilon_2\, \varphi^{[2]}_\sigma(u+iv)\, \sigma(\boldsymbol{j})\big)\, du \wedge dv \Big\rangle\, d\nu_{\text{haar}}(\sigma),
\end{aligned}
$$

where (ϵ_1, ϵ_2) is the canonical basis of the trivial bundle $\mathbb{C}^2$. Any smooth $\mathbb{H}$-valued 1-form splits on the slice $\mathbb{U} \cap (\mathbb{R} + \mathbb{R}\, \sigma(\boldsymbol{i}))$ as

$$\boldsymbol{\omega}(u + v\, \sigma(\boldsymbol{i})) = \boldsymbol{\omega}^{[1]}_\sigma(u + iv) + \boldsymbol{\omega}^{[2]}_\sigma(u + iv)\, \sigma(\boldsymbol{j}),$$

where this time $\boldsymbol{\omega}_\sigma^{[1]}(u+iv), \boldsymbol{\omega}_\sigma^{[2]}(u+iv)$ are $\mathbb{R} + \mathbb{R}\,\sigma(\boldsymbol{i})$ valued 1-forms. A natural residue current $R^{\mathbf{f}}_{\text{slice}}$ attached to $\mathbf{f}$ in such slice-regular context would be then

$$(7.321)\quad \langle\!\langle R^{\mathbf{f}}_{\text{slice}}\,,\, \boldsymbol{\varphi}\rangle\!\rangle$$
$$= \int_{\mathrm{SO}(3,\mathbb{R})} \Big\langle R^{s_\sigma}_{\|\ \|}(u+iv)\,,\, \epsilon_1\, \omega_\sigma^{[1]}(u+iv) + \epsilon_2\, \omega_\sigma^{[2]}(u+iv)\, \sigma(\boldsymbol{j})\Big\rangle\, d\nu_{\text{haar}}(\sigma).$$

APPENDIX A

Complex manifolds and analytic spaces

Throughout this monograph, we work on either a complex manifold and its structural sheaf[1] $(\mathcal{X}, \mathcal{O}_{\mathcal{X}})$ or on a complex reduced analytic space and its structural sheaf, denoted also by $(\mathcal{X}, \mathcal{O}_{\mathcal{X}})$. Furthermore, we assume that analytic subsets $V \subset \mathcal{X}$ are equipped with the structural sheaves $\mathcal{O}_V = \iota^* \mathcal{O}_{\mathcal{X}}$ inherited from $\mathcal{O}_{\mathcal{X}}$ through the embedding $\iota : V \hookrightarrow \mathcal{X}$. Thus Appendix A is devoted to a brief outline of these notions. For a detailed presentation of this material, we invite the reader to consult the extensive monographs by J. P. Demailly [**De**] and R. O. Wells [**W**].

A.1. Complex manifold, structural sheaf $\mathcal{O}_{\mathcal{X}}$, sheaves of $\mathcal{O}_{\mathcal{X}}$-modules

A.1.1. Definitions. A *complex manifold* $\mathcal{X}$ of dimension N consists of:

(1) a topological space $\mathcal{X}$, which is countable at infinity;[2]

(2) a collection of charts $\{(U_\iota, \zeta_\iota)_\iota\}$ on $\mathcal{X}$, where the family $(U_\iota)_\iota$ realizes an open covering of $\mathcal{X}$, namely $\mathcal{X} = \bigcup_\iota U_\iota$, and, for any ι, the mapping $\zeta_\iota : U_\iota \leftrightarrow \boldsymbol{U}_\iota \subset \mathbb{C}^N$ is a homeomorphism so that, for every pair of indices (ι_0, ι_1) satisfying $U_{\iota_0} \cap U_{\iota_1} \neq \emptyset$, the map $\zeta_{\iota_0,\iota_1} = \zeta_{\iota_0} \circ (\zeta_{\iota_1})^{-1}$ realizes a biholomorphism between the open subsets $\zeta_{\iota_1}(\boldsymbol{U}_{\iota_0} \cap \boldsymbol{U}_{\iota_1})$ and $\zeta_{\iota_0}(\boldsymbol{U}_{\iota_0} \cap \boldsymbol{U}_{\iota_1})$.

Such a collection of charts $\{(U_\iota, \zeta_\iota)_\iota\}$ is called an *atlas* for the complex manifold $\mathcal{X}$. Given an atlas for the complex manifold $\mathcal{X}$, the homeomorphism

$$\zeta_\iota : U_\iota \longleftrightarrow \boldsymbol{U}_\iota \subset \mathbb{C}^N$$

is called a *local system of coordinates* in the open subset U_ι. Moreover, if $0 \in \boldsymbol{U}_\iota$ and $z_\iota \in U_\iota$ is such that $\zeta_\iota(z_\iota) = 0$, then the local system of coordinates ζ_ι is said to be *centered* at $z_\iota \in U_\iota$.

Given a complex analytic manifold $\mathcal{X}$ equipped with an atlas $\{(U_\iota, \zeta_\iota)_\iota\}$, let us denote by $\zeta_\iota = \xi_\iota + i\eta_\iota$, where $\xi_\iota = \operatorname{Re} \zeta_\iota$ and $\eta_\iota = \operatorname{Im} \zeta_\iota$, are such that $\tau_\iota = (\zeta_\iota, \eta_\iota)$ realizes an homeomorphism between U_ι and $\boldsymbol{U}_{\iota,\mathbb{R}}$. Here $U_{\iota,\mathbb{R}}$ denotes the open subset of $\mathbb{R}^{2N}$ defined by

$$\boldsymbol{U}_{\iota,\mathbb{R}} =: \{(\xi, \eta) \in \mathbb{R}^{2N} : \xi + i\eta \in \boldsymbol{U}_\iota\}.$$

For every pair of indices (ι_0, ι_1), the map $\tau_{\iota_0,\iota_1} = \tau_{\iota_0} \circ \tau_{\iota_1}^{-1}$ is a C^∞ diffeomorphism (which is additionally real-analytic) between the open subsets $\tau_{\iota_1}(\boldsymbol{U}_{\iota_0,\mathbb{R}} \cap \boldsymbol{U}_{\iota_1,\mathbb{R}})$ and $\tau_{\iota_0}(\boldsymbol{U}_{\iota_0,\mathbb{R}} \cap \boldsymbol{U}_{\iota_1,\mathbb{R}})$. The collection $\{(U_\iota, \tau_\iota)_\iota\}$ defines a structure of a *differentiable*

[1] For those readers who are interested in knowing more about *sheaf theory*, they can consult the books by R. Godement [**God**] and by H. Grauert and R. Remmert [**GrtR**] (see also [**GuR**] and [**De**, II, §1]). Here, we present the elements from sheaf theory needed within this monograph.

[2] The existence of continuous (or of C^∞) partitions of unity on the underlying differentiable manifold $\mathcal{X}_{\mathbb{R}}$ is for us a major ingredient.

(in fact *real-analytic*) $2N$-dimensional *manifold* $\mathcal{X}_{\mathbb{R}}$, which is called the *underlying differentiable manifold* of $\mathcal{X}$.

Given an open subset $U \subset \mathcal{X}$, one denotes as $\mathcal{O}_{\mathcal{X}}(U)$ the commutative ring of *holomorphic functions* in U. Namely, $f \in \mathcal{O}_{\mathcal{X}}(U)$ if $f \ : U \to \mathbb{C}$ is such that for any atlas $\{(U_\iota, \zeta_\iota)_\iota\}$ and any index ι satisfying $U \cap U_\iota \neq 0$ the composed function $f \circ (\zeta_\iota^{-1})_{|\zeta_\iota(U \cap U_\iota)} : \ \zeta_\iota(U \cap U_\iota) \to \mathbb{C}$ is holomorphic. An element $f \in \mathcal{O}_{\mathcal{X}}(U)$ is called *regular* if it is not a zero-divisor in the ring $\mathcal{O}_{\mathcal{X}}(U)$. Regular elements in $\mathcal{O}_{\mathcal{X}}(U)$ constitute a multiplicative subset $\mathcal{S}_{\mathcal{X}}(U)$ and the commutative ring $(\mathcal{S}_{\mathcal{X}}(U))^{-1}\mathcal{O}_{\mathcal{X}}(U) = \mathcal{M}_{\mathcal{X}}(U)$ (which is canonically isomorphic as a commutative ring to $\mathcal{O}_{\mathcal{X}}(U)[(\mathcal{S}_{\mathcal{X}}(U))^{-1}]$) is called the commutative ring of *meromorphic functions* in U.

Let us recall briefly the definition of *sheaves on a topological space* $\mathcal{X}$; see for example [**De**, II, §1], or for more complete discussion [**God**, **GrtR**, **GuR**]. To define a *presheaf* $\mathcal{F}$ *of commutative rings* on a topological space $\mathcal{X}$, one needs:

(1) first to set a collection of commutative rings $\{\mathcal{F}(U)\}_{U \subset \mathcal{X}}$, where U is an arbitrary open subset of $\mathcal{X}$. The ring $\mathcal{F}(U)$ is called then *the ring of sections of the sheaf* $\mathcal{F}$ *over the open set* U;
(2) second to associate to every pair of open sets U, U' satisfying $U \subset U'$, a *restriction* map $\rho_{U,U'} : \mathcal{F}(U') \to \mathcal{F}(U)$, which is a homomorphism of rings, so that $\rho_{U,U'} \circ \rho_{U',U''} = \rho_{U,U''}$ whenever $U \subset U' \subset U''$ and $\rho_{U,U} = \mathrm{Id}_{\mathcal{F}(U)}$.

A *presheaf* $\mathcal{F}$ of commutative rings on the topological space $\mathcal{X}$ is said to be a *sheaf of commutative rings on* $\mathcal{X}$ if it satisfies in addition the following two *gluing* axioms. Given any open covering $\{U_\iota\}$ of an arbitrary open set U, i.e., $U = \bigcup_\iota U_\iota$ then:

(A) whenever $F, G \in \mathcal{F}(U)$ and $\rho_{U_\iota,U}(F) = \rho_{U_\iota,U}(G)$ for any index ι, then $F = G$;
(B) given any collection $(F_\iota)_\iota$ with $F_\iota \in \mathcal{F}(U_\iota)$ such that

$$\rho_{U_{\iota_0} \cap U_{\iota_1}, U_{\iota_0}}(F_{\iota_0}) = \rho_{U_{\iota_0} \cap U_{\iota_1}, U_{\iota_1}}(F_{\iota_1}) \quad \forall\, \iota_0, \iota_1,$$

then there is an element $F \in \mathcal{F}(U)$ such that $\rho_{U_\iota,U}(F) = F_\iota$ for any index ι.

For any $z \in \mathcal{X}$, one defines the *stalk* $\mathcal{F}_z$ of a sheaf of commutative rings at z as the direct limit

$$\mathcal{F}_z = \lim_{z \ni U} \mathcal{F}(U).$$

This means that an element of $\mathcal{F}_z$ can be represented as an element in $\mathcal{F}(U)$ for some neighborhood U of z, taking into account that two such sections in $\mathcal{F}(U')$ and $\mathcal{F}(U'')$ are identified if they coincide on a neighborhood U of z lying in $U' \cap U''$. Elements of the stalk $\mathcal{F}_z$ are called *germs at* z *of sections of the sheaf* $\mathcal{F}$, while elements of $\mathcal{F}(U)$ are called *sections of* $\mathcal{F}$ *on* $U \subset \mathcal{X}$, whenever U is an open subset of $\mathcal{X}$. The *support* of a sheaf of commutative rings on $\mathcal{X}$ is the set of points $z \in \mathcal{X}$ where $\mathcal{F}_z \neq 0$.

Given two sheaves of commutative rings $\mathcal{F}$ and $\mathcal{G}$ on a topological space $\mathcal{X}$ with restriction morphisms $\rho^{\mathcal{F}}_{U,U'}$ and $\rho^{\mathcal{G}}_{U,U'}$ respectively whenever $U \subset U'$, a *morphism* $\theta : \mathcal{F} \to \mathcal{G}$ *of sheaves of commutative rings* from $\mathcal{F}$ to $\mathcal{G}$ consists of a family $\{\theta_U\}_{U \subset \mathcal{X}}$ of morphisms $\theta_U : \mathcal{F}(U) \to \mathcal{G}(U)$ of commutative rings, defined for any open subset $U \subset \mathcal{X}$, satisfying $\theta_U \circ \rho^{\mathcal{F}}_{U,U'} = \rho^{\mathcal{G}}_{U,U'} \circ \theta_{U'}$ on $\mathcal{F}(U')$ for any $U \subset U'$.

Given a sheaf of commutative rings $\mathcal{F}$ on a topological space $\mathcal{X}$ and U an open subset of $\mathcal{X}$, there is a natural restriction operation $\mathcal{F} \to \mathcal{F}_{|U}$ as a sheaf on the topological space U equipped with the topology induced by that on $\mathcal{X}$.

DEFINITION A.1 (The structural sheaf $\mathcal{O}_{\mathcal{X}}$ and the sheaf $\mathcal{M}_{\mathcal{X}}$). Let $\mathcal{X}$ be a complex N-dimensional manifold. The collection $\{\mathcal{F}(U)\}_{U \subset \mathcal{X}}$, where $\mathcal{F}(U) = \mathcal{O}_{\mathcal{X}}(U)$ (respectively $\mathcal{M}_{\mathcal{X}}(U)$), indexed by open subsets of the underlying topological space $\mathcal{X}$, together with restriction maps $\rho_{U,U'} : f \in \mathcal{F}(U') \mapsto f_{|U} \in \mathcal{F}(U)$ whenever $U \subset U'$, defines a sheaf of commutative rings $\mathcal{O}_{\mathcal{X}}$ (respectively $\mathcal{M}_{\mathcal{X}}$) on the topological space $\mathcal{X}$. The sheaf of commutative rings $\mathcal{O}_{\mathcal{X}}$ is called the *structural sheaf* of the complex manifold $\mathcal{X}$, while the sheaf of commutative rings $\mathcal{M}_{\mathcal{X}}$ is called the *sheaf of meromorphic functions* on $\mathcal{X}$.

A sheaf $\mathcal{F} = \{\mathcal{F}(U)\}_{U \subset \mathcal{X}}$ with restriction maps $\rho_{U,U'}$ ($U \subset U'$) is said to be a *sheaf of* $\mathcal{O}_{\mathcal{X}}$*-modules* if, in addition to being a commutative ring, each $\mathcal{F}(U)$ inherits a structure of $\mathcal{O}_{\mathcal{X}}(U)$-module and each restriction map $\rho_{U,U'}$ realizes then a homomorphism of $\mathcal{O}_{\mathcal{X}}$-modules from $\mathcal{F}(U)$ to $\mathcal{F}(U')$, whenever $U \subset U' \subset \mathcal{X}$. Each stalk $\mathcal{F}_{X,z}$ inherits then a structure of $\mathcal{O}_{X,z}$-module. In the particular case where $\mathcal{F} = \mathcal{I}$ is a sheaf of $\mathcal{O}_{\mathcal{X}}$-modules on $\mathcal{X}$ such that $\mathcal{F}(U) = \mathcal{I}(U)$ is an ideal of the commutative ring $\mathcal{O}_{\mathcal{X}}(U)$ for any open subset $U \subset \mathcal{X}$, the sheaf of commutative rings is said to be an *ideal sheaf* on $\mathcal{X}$.

Given a morphism $\mathfrak{f}$ of sheaves of $\mathcal{O}_{\mathcal{X}}$-modules from $\mathcal{F}$ to $\mathcal{G}$, one defines naturally $\mathrm{Ker}(\mathfrak{f})$ as a subsheaf of $\mathcal{F}$, $\mathrm{Im}(\mathfrak{f})$ as a subsheaf of $\mathcal{G}$, and the quotient sheaf as $\mathrm{coker}(\mathfrak{f}) = \mathcal{G}/\mathrm{Im}(\mathfrak{f})$.

DEFINITION A.2 (Locally free sheaves of $\mathcal{O}_{\mathcal{X}}$-modules with constant rank). A sheaf $\mathcal{F}$ of $\mathcal{O}_{\mathcal{X}}$ modules over $\mathcal{X}$ is said to be locally free with (constant) rank r over some open subset $U \subset \mathcal{X}$ if for any $z \in \mathcal{X}$, there is a neighborhood U_z of z in $\mathcal{X}$ such that $\mathcal{F}_{|U} \simeq ((\mathcal{O}_{\mathcal{X}})^{\oplus^r})_{|U_z}$.

A.1.2. Examples ($\mathbb{C}^N$, $\mathbb{P}^N_{\mathbb{C}}$, $\mathcal{X}(\Sigma)$).

(1) The affine space $\mathbb{C}^N$ is the prototype of a complex manifold.

(2) Besides $\mathbb{C}^N$, the most common example of a complex manifold is the projective space $\mathbb{P}^N_{\mathbb{C}}$, which is realized as the geometric quotient of the open subset $\mathbb{C}^{N+1} \setminus \{(0,\dots,0)\} \subset \mathbb{C}^{N+1}$ by the equivalence relation of colinearity between vectors in $\mathbb{C}^{N+1} \setminus \{(0,\dots,0)\}$. The homogeneous coordinates $[z_0 : \cdots : z_N]$ represent points of the projective space. Open sets $U_j^{(N)} := \{[z_0 : \cdots : z_N] \,:\, z_j \neq 0\}$, $j = 0,\dots,N$, are the charts for this compact complex manifold with transition functions $\zeta_{i,j} = z_j/z_i$ on $U_i^{(N)} \cap U_j^{(N)}$. It is important to think here of the projective space $\mathbb{P}^N_{\mathbb{C}}$ as a hyperplane at *infinity*

$$U_0^{(N+1)} := \{[t : z_0 : \cdots : z_N] \,:\, t = 0\} \subset \mathbb{P}^{N+1}_{\mathbb{C}}$$

in the following sense: if $(t, z_0, \dots, z_N)$ is a point in $\mathbb{C}^{N+2} \setminus \{(0,\dots,0)\}$, the complex line $\{(\lambda t, \lambda\xi_0, \dots, \lambda\xi_N) \,:\, \lambda \in \mathbb{C}\}$ intersects the hyperplane at infinity $U_0^{(N+1)}$ of the projective space $\mathbb{P}^{N+1}_{\mathbb{C}}$ exactly at the point identified with the homogeneous coordinates $[z_0 : \cdots : z_N] \in \mathbb{P}^N_{\mathbb{C}}$. This point of view presides over the concept of perspective as developed by Italian Renaissance artists and appears to be a particular case of a more general fact; see Example 2.7 in the main body of the monograph.

(3) Let $\Sigma = \{\sigma_\iota\}_\iota$ be a *complete $\mathbb{Z}^N$-rational primitive fan in $\mathbb{R}^N$*. That is, Σ is a finite collection of strict[3] closed cones $\sigma_\iota \subset \mathbb{R}^N$ satisfying $\mathbb{R}^N = \bigcup_\iota \sigma_\iota$ and having the following properties:
- each cone σ_ι is generated by a set of nonzero vectors in $\mathbb{Z}^N$ with coprime coordinates (a set of *primitive* vectors), which can be completed as a basis of the lattice $\mathbb{Z}^N$;
- any face of any $\sigma_\iota \in \Sigma$ still belongs to Σ;
- any intersection $\sigma_{\iota_0} \cap \sigma_{\iota_1}$ of two elements of Σ is a common face to the cones σ_{ι_0} and σ_{ι_1}.

Let $\sigma_0, \dots, \sigma_{N'}$ be the distinct N-dimensional cones in Σ. The interiors of the cones σ_j, $j = 0, \dots, N'$, in $\mathbb{R}^N$ are pairwise disjoint and one has $\bigcup_{j=0}^{N'} \sigma_j = \bigcup_\iota \sigma_\iota = \mathbb{R}^N$. Let $\{\xi_j^1, \dots, \xi_j^N\}$, $j = 0, \dots, N'$, be a set of primitive vectors $\xi_j^\ell = (\xi_{j,1}^\ell, \dots, \xi_{j,N}^\ell) \in \mathbb{Z}^N$, $\ell = 1, \dots, N$, which on one hand generate σ_j as an N-dimensional cone and on the other hand form a basis of $\mathbb{Z}^N$ as a lattice. For each $j = 0, \dots, N'$, consider the invertible monomial map from $\mathbb{T}^N = (\mathbb{C}^*)^N$ into itself

$$\pi_j \ : \zeta \in \mathbb{T}^N \longmapsto \Big(\prod_{\ell=1}^N \zeta_\ell^{\xi_{j,1}^\ell}, \dots, \prod_{\ell=1}^N \zeta_\ell^{\xi_{j,N}^\ell} \Big) \in \mathbb{T}^N.$$

Let U_j, $j = 0, \dots, N'$, be the open dense subsets of $\mathbb{C}^N$ defined as

$$U_j = \big\{ z \in \mathbb{C}^N \ : \ z_\ell \neq 0, \ \text{whenever } \xi_{j,k}^\ell < 0 \text{ for some } k \in \{1, \dots, N\} \big\}.$$

Each monomial map π_j realizes a birational map between U_j and its image $\boldsymbol{U}_j = \pi_j(U_j)$ in $\mathbb{C}^N$. One can construct a compact complex manifold $\mathcal{X}(\Sigma)$ starting from the disjoint union $\bigsqcup_{j=0}^{N'} U_j$, then make the necessary identifications to glue the open subsets U_j (which were initially forced to be disjoint in $\bigsqcup_{j=0}^{N'} U_j$) along their open subsets $\pi_{j_0}^{-1}(\boldsymbol{U}_{j_0} \cap \boldsymbol{U}_{j_1})$ and $\pi_{j_1}^{-1}(\boldsymbol{U}_{j_0} \cap \boldsymbol{U}_{j_1})$ through the monomial maps $\pi_{j_0}^{-1} \circ \pi_{j_1}$, $0 \leq j_0 < j_1 \leq N'$ (see [**Ehl75, Da78, Ew**] for such construction). The projective space $\mathbb{P}_\mathbb{C}^N$ can be constructed in this way, starting with the $\mathbb{Z}^N$-rational primitive fan, whose one-dimensional cones are

$$\mathbb{R}^+\epsilon_1, \dots, \mathbb{R}^+\epsilon_N, -\mathbb{R}^+(\epsilon_1 + \cdots + \epsilon_N),$$

where $\{\epsilon_1, \dots, \epsilon_N\}$ is the canonical basis of $\mathbb{R}^N$. More generally, given a complete $\mathbb{Z}^N$-rational primitive fan Σ in $\mathbb{R}^N$, the complex manifold $X(\Sigma)$, realized as above, inherits a group action on itself by the complex torus $\mathbb{T}^N$. The complex torus $\mathbb{T}^N$ stands as a dense orbit for this action. The N-dimensional complex manifold $\mathcal{X}(\Sigma)$ is called the *toric manifold subordinate to the primitive fan* Σ. Since the transition functions are rational maps (even monomial maps), the complex structure of $\mathcal{X}(\Sigma)$ is algebraic and $\mathcal{X}(\Sigma)$ (such as $\mathbb{C}^N$ or $\mathbb{P}_\mathbb{C}^N$) is said to be an *algebraic manifold.* One can consider it either as being defined based on algebraically closed field $\mathbb{C}$ or based on the field $\mathbb{Q}$ or its algebraic closure $\overline{\mathbb{Q}}$.

[3]For more details about toric manifolds; see §3.1, which this example illustrates.

A.2. Coherence, Stein manifolds, free resolutions

A.2.1. Coherent sheaves of $\mathcal{O}_{\mathcal{X}}$-modules. Among the category of sheaves of $\mathcal{O}_{\mathcal{X}}$-modules on a complex manifold $\mathcal{X}$, coherent sheaves play an essential role in this monograph.

DEFINITION A.3 (Coherence of a sheaf). A sheaf $\mathcal{F}$ of $\mathcal{O}_{\mathcal{X}}$-modules on an N-dimensional complex manifold $\mathcal{X}$ is said to be *coherent* (or $\mathcal{O}_{\mathcal{X}}$-coherent) if and only if it fulfills the two following conditions.

(1) For any z in $\mathcal{X}$, there exists an open neighborhood U_z of z and q_z elements $s_1, \dots, s_{q_z}$ in $\mathcal{O}_{\mathcal{X}}(U_z)$ such that, for every $z' \in U_z$, the $\mathcal{O}_{\mathcal{X},z'}$-module $\mathcal{F}_{z'}$ is generated by the germs $s_{j,z'}$, $j = 1, \dots, q_z$, of the sections $s_j \in \mathcal{F}(U_z)$ at the point $z' \in U_z$ (one phrases this condition saying that $\mathcal{F}$ is *locally finitely generated*).

(2) For any open subset U' of $\mathcal{X}$, for any $s_1, \dots, s_q \in \mathcal{F}(U')$, the $(\mathcal{O}_{\mathcal{X}})_{|U'}$-subsheaf of $(\mathcal{O}_{\mathcal{X}}^{\oplus q})_{|U'}$ on U' called the *sheaf of relations* $\mathcal{R}_{U'}(s_1, \dots, s_q)$ on U', that is, the kernel of the sheaf homomorphism $\theta^{U'} : (\mathcal{O}_{\mathcal{X}}^{\oplus q})_{|U'} \to \mathcal{F}_{|U'}$, where, for any $U \subset U'$,

$$\theta_U^{U'} : (g_1, \dots, g_q) \in (\mathcal{O}_{\mathcal{X}}(U))^{\oplus q} \longmapsto \sum_{j=1}^{q} g_j \, s_{j|U} \in \mathcal{F}(U)$$

is also locally finitely generated.

The important fact that the structural sheaf $\mathcal{O}_{\mathcal{X}}$ is coherent is known as the *Oka coherence theorem.* Given a coherent sheaf of ideals $\mathcal{I}$ on $\mathcal{X}$, the sheaf $\mathcal{O}_{\mathcal{X}}/\mathcal{I}$ is also coherent on $\mathcal{X}$ since one has the following short exact sequence of sheaves

$$0 \longrightarrow \mathcal{I} \longrightarrow \mathcal{O}_{\mathcal{X}} \longrightarrow \mathcal{O}_{\mathcal{X}}/\mathcal{I} \longrightarrow 0.$$

Any finite locally free sheaf of $\mathcal{O}_{\mathcal{X}}$-modules on $\mathcal{X}$, in particular any locally free sheaf of $\mathcal{O}_{\mathcal{X}}$-modules with constant rank, inherits the coherence of the structural sheaf $\mathcal{O}_{\mathcal{X}}$ of the complex manifold $\mathcal{X}$.

A.2.2. Stein manifolds and Cartan Theorems A and B. We now turn to a natural class of complex manifolds known as *Stein manifolds.* Stein manifolds are natural in the sense that they carry enough nonconstant holomorphic functions. For example, this is not the case for the compact complex manifold $\mathbb{P}^N_{\mathbb{C}}$. More specifically, a complex manifold $\mathcal{X}$ of dimension N, with structural sheaf $\mathcal{O}_{\mathcal{X}}$, is a *Stein manifold* if satisfies the following two properties.

(1) It is holomorphically convex, that is, the *holomorphic convex hull*

$$\widehat{K} = \widehat{K}_{\mathcal{O}_{\mathcal{X}}} := \{z \in \mathcal{X}\,,\ |f(z)| \leq \sup_K |f|\,,\ \forall\, f \in \mathcal{O}_{\mathcal{X}}(\mathcal{X})\}$$

of any compact subset $K \Subset \mathcal{X}$ remains compact.

(2) It carries enough holomorphic functions to separate points. That is, for any points $z, z' \in \mathcal{X}$, $z \neq z'$, there exists an element $f \in \mathcal{O}_{\mathcal{X}}(\mathcal{X})$ such that $f(z) \neq f(z')$.

EXAMPLE A.4. The affine space $\mathbb{C}^N$ is the prototype of a Stein manifold. Any noncompact Riemann surface is a Stein manifold as well.

REMARK A.5. Any Stein manifold with dimension N admits an exhaustion $\{K_\nu\}_{\nu\geq 1}$ by holomorphically convex compact subsets $K_\nu \subset \mathcal{X}$ satisfying, in addition to properties $\mathcal{X} = \bigcup_{\nu\geq 1} K_\nu$ and $K_\nu \subset (K_{\nu+1})^0$, the property $\widehat{K}_\nu = K_\nu$. Here $(K_{\nu+1})^0$ denotes the interior of the set.

It is known that any Stein manifold $\mathcal{X}$ of dimension N can be holomorphically embedded into $\mathbb{C}^{2N+1}$. One has also two extremely powerful results, due to Henri Cartan.

THEOREM A.6 (Cartan Theorem A). *Let $\mathcal{X}$ be a Stein manifold of dimension N. Any coherent sheaf of $\mathcal{O}_{\mathcal{X}}$-modules on $\mathcal{X}$ is spanned by its global sections. That is, for any open set $U \subset \mathcal{X}$, $\rho_{U,\mathcal{X}}(\mathcal{F}(\mathcal{X}))$ generates $\mathcal{F}(U)$ as a $\mathcal{O}_{\mathcal{X}}(U)$-module. In particular, for each $z \in \mathcal{X}$, one can find global sections $f_{z,1}, \dots, f_{z,N}$ in $\mathcal{O}_{\mathcal{X}}(\mathcal{X})$ such that $(f_{z,1}, \dots, f_{z,N})$ defines a local system of coordinates in a neighborhood of z.*

THEOREM A.7 (Cartan Theorem B). *Let $\mathcal{X}$ be a Stein manifold of dimension N. If $\mathcal{F}$ is a coherent sheaf of $\mathcal{O}_{\mathcal{X}}$-modules on $\mathcal{X}$, then one has that $\check{H}^q(\mathcal{X}, \mathcal{F}) = 0$ for each $q > 0$, where $\check{H}^q(\mathcal{X}, \mathcal{F})$ denotes the Čech cohomology group*[4] *with values in the sheaf $\mathcal{F}$.*

One of the most remarkable consequences of Theorem A.6 for us will be the following: if $g, f_1, \dots, f_m$ are $m+1$ holomorphic functions on a Stein manifold $\mathcal{X}$, such that locally, at every point $z \in X$, $g_z \in (f_{1,z}, \dots, f_{m,z})\,\mathcal{O}_{\mathcal{X},z}$ at the level of germs, then there exist holomorphic functions $a_1, \dots, a_m$ on $\mathcal{X}$ such that one has $g \equiv \sum_{j=1}^m a_j f_j$ in $\mathcal{X}$. An important example occurs when $\mathcal{X} = \Omega \times \Omega \subset \mathbb{C}^N \times \mathbb{C}^N$, where Ω is a pseudo-convex domain of $\mathbb{C}^N$ (that is a Stein open subset of $\mathbb{C}^N$) and $f_j(z,\zeta) := \zeta_j - z_j$, $j = 1, \dots, N$. If g is a holomorphic function in Ω, then there exist N holomorphic functions $h_1, \dots, h_N$ on $\Omega \times \Omega$, such that

$$g(\zeta) - f(z) = \sum_{j=1}^{N} h_j(z,\zeta)\,(\zeta_j - z_j) \qquad \forall\,(z,w) \in \Omega \times \Omega. \tag{A.1}$$

This formula is called the *Hefer division formula.* The Hefer formula can easily be obtained when U is convex. Actually in this case one can use the Taylor formula with integral remainder, namely

(A.2)

$$f(\zeta) - f(z) = \int_0^1 \frac{d}{dt}[f(t\zeta + (1-t)z)]\,dt = \sum_{j=1}^{N} (\zeta_j - z_j) \int_0^1 \frac{\partial f}{\partial \zeta_j}(t\zeta + (1-t)z_j)\,dt.$$

Note that such a simple argument does not go through in the case of a pseudo-convex domain Ω.

[4]Let $\mathcal{F}$ be a presheaf of abelian groups on a topological space $\mathcal{X}$ and let $\mathcal{U} = \{U_\iota\}_\iota$ be a covering family of $\mathcal{X}$ by open subsets. The Čech groups $\check{H}^p(\mathcal{X}, \mathcal{U}, \mathcal{F}) = \operatorname{Ker}\delta_q / \operatorname{Im}\delta_{q-1}$, $q \in \mathbb{N}$, materialize the obstruction for the exactness of the complex

$$0 \longrightarrow \cdots \longrightarrow \check{\mathscr{C}}^q(\mathcal{X}, \mathcal{U}, \mathcal{F}) \xrightarrow{\delta_q} \check{\mathscr{C}}^{q+1}(\mathcal{X}, \mathcal{U}, \mathcal{F}) \longrightarrow \cdots,$$

where $\check{\mathscr{C}}^q(\mathcal{X}, \mathcal{U}, \mathcal{F})$ is the group of q-cochains of $\mathcal{U}$ with values in $\mathcal{F}$ and δ_q is the coboundary operator. δ_q-closed q-cocycles are called q-cocycles with values in $\mathcal{F}$. The Čech cohomology group $\check{H}^q(\mathcal{X}, \mathcal{F})$ is realized as the direct limit of the groups $\check{H}^q(\mathcal{X}, \mathcal{U}, \mathcal{F})$ along the refinements of covering families $\mathcal{U}$. See also §D.2.3 for more details, in particular for more details for the definition of the coboundary morphisms δ_0 and δ_1 in relation with the notion of Cartier divisor.

A.2.3. Free resolutions of coherent sheaves.

A.2.3.1. *Local aspects.* Let $\mathcal{X}$ be an N-dimensional complex manifold and let $\mathcal{F}$ be a coherent sheaf of $\mathcal{O}_{\mathcal{X}}$-modules. The following useful concept arises naturally from the two properties stated in Definition A.3.

DEFINITION A.8 (Free finite resolution for $\mathcal{F}_z$). A *finite free resolution* for $\mathcal{F}_z$ at a point $z \in \mathcal{X}$ is an exact[5] $\mathcal{O}_{\mathcal{X},z}$-complex

$$0 \longrightarrow \mathcal{O}_{\mathcal{X},z}^{\oplus^{r_{L_z,z}}} \xrightarrow{\mathfrak{f}_{L_z,z}} \mathcal{O}_{\mathcal{X},z}^{\oplus^{r_{L_z-1,z}}} \cdots \xrightarrow{\mathfrak{f}_{2,z}} \mathcal{O}_{\mathcal{X},z}^{\oplus^{r_{1,z}}} \xrightarrow{\mathfrak{f}_{1,z}} \mathcal{O}_{\mathcal{X},z}^{\oplus^{r_{0,z}}}, \tag{A.3}$$

with length $L_z \in \mathbb{N}^*$, such that

$$\operatorname{coker}(\mathfrak{f}_{1,z}) := \frac{\mathcal{O}_{\mathcal{X},z}^{\oplus^{r_{0,z}}}}{\operatorname{Im}(\mathfrak{f}_{1,z})} = \mathcal{F}_z. \tag{A.4}$$

Such a finite free resolution for $\mathcal{F}_z$ is said to be *minimal* if and only if

$$\operatorname{Im}(f_{\ell,z}) \subset \mathfrak{M}_{\mathcal{X},z}\, \mathcal{O}_{\mathcal{X},z}^{\oplus^{r_{\ell-1,z}}} \quad \text{for any } \ell \geq 1, \tag{A.5}$$

where $\mathfrak{M}_{\mathcal{X},z}$ denotes the maximal ideal in $\mathcal{O}_{\mathcal{X}}$.

Given a coherent sheaf of $\mathcal{O}_{\mathcal{X}}$-modules over a complex manifold $\mathcal{X}$ and $z \in \mathcal{X}$, there is a unique minimal finite free resolution for $\mathcal{F}_z$ up to an isomorphism of $\mathcal{O}_{\mathcal{X},z}$-complexes of $\mathcal{O}_{\mathcal{X},z}$-modules. More precisely, one has the following result [**Eis**, Theorem 20.2].

THEOREM A.9 (Uniqueness of minimal finite free resolutions for $\mathcal{F}_z$). *Let $\mathcal{F}$ be a coherent sheaf over $\mathcal{X}$ and $z \in \mathcal{X}$. Given a minimal finite free resolution $\mathfrak{R}_{z,\min}$ for $\mathcal{F}_z$, any finite free resolution $\mathfrak{R}_z$ for $\mathcal{F}_z$ is, as a complex of $\mathcal{O}_{\mathcal{X},z}$-modules, isomorphic to the direct sum of $\mathfrak{R}_{z,\min}$ with a trivial complex of $\mathcal{O}_{\mathcal{X},z}$-modules.*

As a consequence, one introduces the following definition.

DEFINITION A.10 (Depth of $\mathcal{F}$ at $z \in \mathcal{X}$). Let $\mathcal{F}$ be a coherent sheaf of $\mathcal{O}_{\mathcal{X}}$-modules over a complex N-dimensional manifold $\mathcal{X}$. The *depth* of $\mathcal{F}$ at $z \in \mathcal{X}$ is defined by

$$N - \operatorname{depth}_z(\mathcal{F}) = L_z, \tag{A.6}$$

where L_z denotes the length of the minimal finite free resolution of $\mathcal{F}_z$.

A.2.3.2. *Global aspects.* When $\mathcal{X}$ is additionally a Stein manifold, Theorem B implies that every coherent sheaf $\mathcal{F}$ of $\mathcal{O}_{\mathcal{X}}$-modules admits, in a convenient open neighborhood U_K of a given holomorphically convex compact subset $K \Subset \mathcal{X}$, a (global) *free resolution.* That is, there exists an exact complex of coherent sheaves of $\mathcal{O}_{\mathcal{X}}$-modules over U_K:

$$\cdots \xrightarrow{\mathfrak{f}_{\ell+1,K}} \mathscr{L}_{U_K}^{r_{\ell,K}} \xrightarrow{\mathfrak{f}_{\ell,K}} \mathscr{L}_{U_K}^{r_{\ell,K}} \cdots \xrightarrow{\mathfrak{f}_{2,K}} \mathscr{L}_{U_K}^{r_{1,K}} \xrightarrow{\mathfrak{f}_{1,K}} \mathscr{L}_{U_K}^{r_{0,K}}, \tag{A.7}$$

where $\operatorname{coker}(\mathfrak{f}_{1,K}) = \mathcal{F}_{|U_K}$ and

$$\mathscr{L}_{U_K}^{r_{\ell,K}} = (\mathcal{O}_{\mathcal{X}}^{\oplus^{r_{\ell,K}}})_{|U_K} \text{ for any } \ell \in \mathbb{N}^*.$$

Observe that the sequence above may be infinite to the left. The existence of the exact sequence (A.7) follows from the iterated use of [**Hor73**, Theorem 7.2.1] about the existence of generating global sections for $\mathcal{F}$ about a holomorphically convex

[5]Each $\mathfrak{f}_{\ell,z}$ is a homomorphism of $\mathcal{O}_{\mathcal{X},z}$-modules such that $\mathfrak{f}_{L_z,z}$ is injective and one has $\operatorname{Ker}\mathfrak{f}_{\ell,z} = \operatorname{Im}(\mathfrak{f}_{\ell+1,z})$ for any $\ell \geq 1$.

compact subset K. On the other hand, any holomorphically convex compact subset K can be covered with a finite number of open subsets U_{z_ι} above which there exists a locally free finite free resolution[6]

$$0 \longrightarrow (\mathcal{O}_{\mathcal{X}}^{\oplus r_{L_\iota,z_\iota}})_{|U_{z_\iota}} \xrightarrow{\mathfrak{f}_{L_\iota,z_\iota}} (\mathcal{O}_{\mathcal{X}}^{\oplus r_{L_\iota-1,z_\iota}})_{|U_{z_\iota}} \cdots \xrightarrow{\mathfrak{f}_{2,z_\iota}} (\mathcal{O}_{\mathcal{X}}^{\oplus r_{1,z_\iota}})_{|U_{z_\iota}} \xrightarrow{\mathfrak{f}_{1,z_\iota}} (\mathcal{O}_{\mathcal{X}}^{\oplus r_{0,z_\iota}})_{|U_{z_\iota}},$$

with $\operatorname{coker}(\mathfrak{f}_{1,z_\iota}) = \mathcal{F}_{|U_{z_\iota}}$ and $L_\iota = N - \operatorname{depth}_{z_\iota}\mathcal{F}$. Then, if one takes $L \gg \max_\iota L_\iota$ and invokes Theorem A.9, one can see that the coherent sheaf $\operatorname{Ker}(\mathfrak{f}_{L,K})$ in (A.7) is locally free with rank $r_{L+1,K}$ over U_K. Hence one can truncate the locally free resolution (A.7) to the left as

$$\text{(A.8)}\quad 0 \longrightarrow \operatorname{Ker}(\mathfrak{f}_{L,K}) = \mathscr{L}_{U_K}^{r_{L+1,K}} \xrightarrow{\mathfrak{f}_{L,K}} \mathscr{L}_{U_K}^{r_{L-1,K}} \xrightarrow{\mathfrak{f}_{L-1,K}} \cdots \xrightarrow{\mathfrak{f}_{2,K}} \mathscr{L}_{U_K}^{r_{1,K}} \xrightarrow{\mathfrak{f}_{1,K}} \mathscr{L}_{U_K}^{r_{0,K}},$$

where $\operatorname{coker}(\mathfrak{f}_{1,K}) = \mathcal{F}_{|U_K}$. Thus, one can formulate the following definition.

Definition A.11 (Finite locally free resolution for a coherent sheaf). A finite locally free resolution with length L for a coherent sheaf $\mathcal{F}$ over a complex manifold $\mathcal{X}$ in an open subset $U \subset \mathcal{X}$ is an exact complex of locally free coherent sheaves with constant rank

$$\text{(A.9)}\qquad 0 \longrightarrow \mathscr{L}_U^{r_L} \xrightarrow{\mathfrak{f}_L} \mathscr{L}_U^{r_{L-1}} \cdots \xrightarrow{\mathfrak{f}_2} \mathscr{L}_U^{r_1} \xrightarrow{\mathfrak{f}_1} \mathscr{L}_U^{r_0}$$

such that

$$\text{(A.10)}\qquad \operatorname{coker}(\mathfrak{f}_1) = \mathcal{F}_{|U}.$$

This definition leads to the following proposition holding in the context of Stein manifolds.

Proposition A.12. *Let $\mathcal{X}$ be an N-dimensional Stein manifold and let $\mathcal{F}$ be a coherent sheaf of $\mathcal{O}_{\mathcal{X}}$-modules. There exists an exhaustion of $\mathcal{X}$ by an increasing sequence of open subsets U_ν such that $\mathcal{F}$ admits in each U_ν a finite, locally free resolution with length L_ν as in* (A.9) *with*

$$\mathscr{L}_{U_\nu}^{r_{\nu,\ell}} = (\mathcal{O}_{\mathcal{X}}^{\oplus r_{\nu,\ell}})_{|U_\nu}$$

for all $\ell \geq 1$, except at $\ell = L_\nu$.

Example A.13. Let $\mathcal{X}$ be a Stein manifold and let $\mathcal{J}$ be a subsheaf of $\mathcal{O}_{\mathcal{X}}^{\oplus r}$. There exists an exhausting sequence of $\mathcal{X}$ by open subsets U_ν, such that for each ν, $\mathcal{F} = \mathcal{O}_{\mathcal{X}}^{\oplus r}/\mathcal{J}$ admits in U_ν a locally free resolution as in (A.9) with

$$\mathscr{L}_{U_\nu}^{r_0} = (\mathcal{O}_{\mathcal{X}}^{\oplus r})_{|U_\nu} \text{ and } \operatorname{Im}(\mathfrak{f}_1) = \mathcal{J}.$$

Such a finite, locally free resolution of $\mathcal{O}_{\mathcal{X}}^{\oplus r}/\mathcal{J}$ in U_ν is sometimes called by extension a *free resolution of* $\mathcal{J}$ in U_ν. The search for such a free resolution (locally in a neighborhood of a given point $z \in \mathcal{X}$ or, when $\mathcal{X}$ is Stein, globally in an open subset of $\mathcal{X}$) is known as the *determination of syzygies* problem for the quotient sheaf $\mathcal{O}_{\mathcal{X}}^{\oplus r}/\mathcal{J}$ or, by extension, for the subsheaf $\mathcal{J}$. In general, there is no algorithmic process that allows computing such a free resolution, even locally, within polynomial time. We refer the reader to [**Eis05**] for a complete overview of the important notion of syzygy going back to D. Hilbert. The fact that there exist in the local or global context-free resolutions an ideal sheaf $\mathcal{I}$ is known as *Hilbert's syzygy theorem* [**Hil1890**]. In particular, this theorem implies in the local setting of the regular

[6] One takes the minimal finite free resolution (A.3) at z_ι by taking U_{z_ι} sufficiently small around z_ι and then choose representatives in U_{z_ι} for the germs $\mathfrak{f}_{\ell,z_\iota}$.

local ring $\mathcal{O}_{\mathbb{C}^N,0}$ that, given any ideal $\mathcal{I}$ in $\mathcal{O}_{\mathbb{C}^N,0}$, $\mathcal{O}_{\mathbb{C}^N,0}/\mathcal{I}$ admits always a free resolution with length equal to N.

A.3. Closed analytic subsets of a complex manifold

DEFINITION A.14 (Closed analytic subset). Let $\mathcal{X}$ be a complex manifold of dimension N with $\mathcal{O}_{\mathcal{X}}$ as its structural sheaf. A subset $V \subset \mathcal{X}$ is called a *closed analytic subset* of $\mathcal{X}$ if the closed subset $V \subset \mathcal{X}$ is expressed locally in an open neighborhood U_z around any $z \in \mathcal{X}$ as the set of common zeros of a finite family of elements in $\mathcal{O}_{\mathcal{X}}(U_z)$. An analytic subset $V \subset \mathcal{X}$ is said to be *irreducible* if it cannot be decomposed into a union of two closed analytic subsets V' and V'' such that $V' \neq V$ and $V'' \neq V$.

EXAMPLE A.15. Given a coherent sheaf of $\mathcal{O}_{\mathcal{X}}$-modules $\mathcal{F}$ over a complex manifold $\mathcal{X}$, the *support* of $\mathcal{F}$ is defined as the subset of points $z \in \mathcal{X}$ such that $\mathcal{F}_z \neq 0$. It is a closed analytic subset of $\mathcal{X}$.

REMARK A.16. Suppose that a germ V_z of the closed analytic subset at a point $z \in \mathcal{X}$ is embedded in two ways through the embeddings

$$\iota' : V \cap U_z \hookrightarrow \mathrm{U}' \subset \mathbb{C}^{N_{\iota'}}, \quad \iota'' : V \cap U_z \hookrightarrow \mathrm{U}'' \subset \mathbb{C}^{N_{\iota''}}.$$

Here U' and U'' are open subsets of $\mathbb{C}^{N_{\iota'}}$ and $\mathbb{C}^{N_{\iota''}}$ correspondingly. Since the minimal embeddings $\iota'_{\min}, \iota''_{\min}$ are equivalent, and any embedding ι factorizes at the level of germs at z in a simple way over a minimal one $\iota_{\min}$, one can always take $\iota' = \iota_{\min}$ minimal (hence unique up to equivalence) and $\iota'' = \iota$, with $\mathrm{U}'' = \mathrm{U}' \times \mathbb{C}^{\nu}$, $\nu = N_\iota - N_{\iota_{\min}}$ and $\iota(z) = (\iota_{\min}(z), 0)$.

A.3.1. Local dimension. If $V \subset \mathcal{X}$ is a closed analytic subset of $\mathcal{X}$, one can associate with it a coherent sheaf of ideals $\mathcal{I}_V \subset \mathcal{O}_{\mathcal{X}}$, where

$$\mathcal{I}_V(U) := \{f \in \mathcal{O}_{\mathcal{X}}(U) : f(z) = 0, \quad \forall\, z \in V \cap U\} \tag{A.11}$$

for any open subset $U \subset \mathcal{X}$. For each $z \in \mathcal{X}$, the stalk $\mathcal{I}_{V,z}$ is a radical ideal of the stalk $\mathcal{O}_{\mathcal{X},z}$. Since $\mathcal{O}_{\mathcal{X},z}$ is a nœtherian regular local ring of dimension N, the stalk $\mathcal{I}_{V,z}$ admits a unique representation (up to the order of factors) as

$$\mathcal{I}_{V,z} = \bigcap_\iota \mathcal{P}_{z,\iota}\,. \tag{A.12}$$

The ideals $\mathcal{P}_{z,\iota}$ in (A.12) are prime ideals in $\mathcal{O}_{\mathcal{X},z}$, which are not contained pairwise in each other since $\mathcal{I}_{V,z}$ is radical. The finite set $\mathrm{Ass}(\mathcal{O}_{\mathcal{X},z}/\mathcal{I}_{V,z}) = \{\mathcal{P}_{z,\iota}\}_\iota$ is called the set of *associated primes* of the finitely generated $\mathcal{O}_{\mathcal{X},z}$-module $\mathcal{O}_{\mathcal{X},z}/\mathcal{I}_{V,z}$, which is also a reduced (in general nonregular) local ring. For each $\mathcal{P}_{z,\iota} \in \mathrm{Ass}(\mathcal{O}_{\mathcal{X},z}/\mathcal{I}_{V,z})$, the *Krull dimension* of the finitely generated $\mathcal{O}_{\mathcal{X},z}$-module $\mathcal{O}_{\mathcal{X},z}/\mathcal{P}_{z,\iota}$ is defined as the maximal length (possibly 0) of strictly increasing chains of distinct prime ideals

$$\mathcal{P}_{z,\iota} \subsetneq \mathcal{P}_1 \subsetneq \cdots \subsetneq \mathcal{P}_\kappa \subsetneq \mathcal{O}_{\mathcal{X},z}.$$

The *local dimension of V at the point z*, denoted as $\dim_z V$, is defined as the maximum of the Krull dimensions of the quotient rings $\mathcal{O}_{\mathcal{X},z}/\mathcal{P}_{z,\iota}$ for all associated primes $\mathcal{P}_{z,\iota}$ in $\mathrm{Ass}(\mathcal{O}_{\mathcal{X},z}/\mathcal{I}_{V,z})$.

To conclude the present section, let us mention the classical *box principle.* Assume that $I_1, \dots, I_r$ are r ideals in $\mathcal{O}_{\mathcal{X},z}$ and assume that $h_1, \dots, h_m$ are m elements in the same local ring such that, for any $k = 1, \dots, r$, there exists at least one h_j

so that $h_j \notin I_k$. Then, for generic complex coefficients $\lambda_1, \dots, \lambda_m$, $\sum_{j=1}^m \lambda_j h_j \notin I_k$ for any $k = 1, \dots, r$. The same result holds in the global setting (polynomial ideals in $\mathbb{K}[X_1, \dots, X_N]$ or homogeneous polynomial ideals in $\mathbb{K}[X_0, \dots, X_N]$) provided $\mathbb{K}$ is an infinite commutative field, for example, an algebraically closed field.

A.3.2. Splitting $V = V_{\text{sing}} \cup V_{\text{reg}}$, global dimension. Let V be a closed irreducible analytic subset of a complex manifold $\mathcal{X}$. Then the set V_{sing} of *singular points* of V, that is, the set of points z in V in a neighborhood U_z of which V cannot be described as a submanifold of U_z, is a proper closed analytic subset of V such that, for any $z \in V_{\text{sing}}$, $\dim(V_{\text{sing}})_z < \dim V_z$. It implies that $V \setminus V_{\text{sing}} = V_{\text{reg}}$ is dense in V. Nonsingular points of the set V are called *regular points* of V. Since V is irreducible, the function $z \mapsto \dim V_z$ is constant on V_{reg}. Its constant value on V_{reg} is defined as the *global dimension* of the irreducible closed analytic subset V. It takes values between 0 and $\dim \mathcal{X}$. The *global dimension* of a closed analytic subset $V \subset \mathcal{X}$ is the maximum of the dimensions of its irreducible components V_ι. A closed analytic subset $V \subset \mathcal{X}$ is said to be of *pure dimension* (or *purely dimensional*) if all its irreducible components have the same global dimension.

A.3.3. Free resolutions of sheaves and codimension of supports. Let $\mathcal{X}$ be an N-dimensional complex manifold and let $\mathcal{F}$ be a coherent sheaf over $\mathcal{X}$ with support the closed analytic subset $V_{\mathcal{F}}$. Let us recall in this section two important results related to the concept of finite locally free resolution introduced in §A.2.3. The first result is a local one. It concerns the notion of depth introduced in Definition A.10.

PROPOSITION A.17. *One has for any $z \in \mathcal{X}$ that*

$$\operatorname{depth}_z \mathcal{F} \le \dim V_{\mathcal{F}}. \tag{A.13}$$

In other words, the length of the minimal finite resolution of $\mathcal{F}_z$ equals at least the codimension of $V_{\mathcal{F}}$.

We formulate the second result in a global context since we need it within such a context in this monograph. We refer here to [**BuE74**] or [**Eis**, Theorem 20.9] for a formulation in the local context, from which our statement can be derived thanks to Theorem A.9.

THEOREM A.18. *Let $\mathfrak{R}$ be a finite locally free resolution of the coherent sheaf of $\mathcal{O}_{\mathcal{X}}$-modules $\mathcal{F}$ in $\mathcal{X}$ as in* (A.9) *with $U = \mathcal{X}$ and* (A.10) *fulfilled. Let for any $\ell \in \mathbb{N}^*$*

$$R_\ell = \sum_{\nu \ge 0} (-1)^\nu r_{\ell+\nu}, \quad \textit{where } r_{\ell'} = 0 \textit{ if } \ell' > L_{\mathfrak{R}} = \operatorname{length}(\mathfrak{R}).$$

Then, one has that

$$\operatorname{codim}\{z \in \mathcal{X} : \operatorname{rank}(\mathfrak{f}_\ell(z)) < R_\ell\} \ge \ell \quad \textit{for all } \ell \in \mathbb{N}^*. \tag{A.14}$$

Moreover, conditions (A.14) *are equivalent to the exactness of the complex $\mathfrak{R}$.*

REMARK A.19. The sequence of closed analytic subsets

$$V_\ell(\mathfrak{R}) := \{z \in \mathcal{X} : \operatorname{rank}(\mathfrak{f}_\ell(z)) < R_\ell\}, \quad \ell \in \mathbb{N}^*, \tag{A.15}$$

is a decreasing sequences of closed analytic subsets such that $V_1(\mathfrak{R}) = V_{\mathcal{F}}$ when $\mathfrak{f}_1$ is surjective. This follows from the following observation. One has

$$\begin{aligned}&\mathrm{rank}(\mathfrak{f}_{\ell+1}(z)) < R_{\ell+1} \Longrightarrow \\ &\qquad r_\ell - R_{\ell+1} = R_\ell > r_\ell - \mathrm{rank}(\mathfrak{f}_{\ell+1}(z)) = r_\ell - \dim(\mathrm{Ker}(\mathfrak{f}_\ell(z))) = \mathrm{rank}(\mathfrak{f}_\ell(z))\end{aligned}$$

and $V_1(\mathfrak{R})$ equals in general the closed analytic subset of points $z \in \mathcal{X}$ in a neighborhood in which $\mathcal{F}$ fails to be locally free (that is $V_{\mathcal{F}}$ as soon as $\mathfrak{f}_1$ is surjective).

As a consequence of Theorem A.9, the sequence of closed analytic subsets $(V_\ell(\mathfrak{R}))_{\ell\geq 1}$ does not depend on the finite locally free resolution $\mathfrak{R}$ of $\mathcal{F}$.

DEFINITION A.20. If the coherent sheaf of $\mathcal{O}_{\mathcal{X}}$-modules $\mathcal{F}$ admits a finite locally free resolution $\mathfrak{R}$ in $\mathcal{X}$, then the decreasing sequence $(V_\ell(\mathfrak{R}))_{\ell\geq 1} = (V_{\mathcal{F},\ell})_{\ell\geq 1}$ will be called because of [**BuE74**] the *Buchsbaum–Eisenbud–Fitting sequence* of the coherent sheaf $\mathcal{F}$.

Proposition A.17 suggests we introduce the following concept.

DEFINITION A.21 (Cohen–Macaulay sheaves and closed analytic subsets). A coherent sheaf of $\mathcal{O}_{\mathcal{X}}$-modules $\mathcal{F}$ over a complex manifold $\mathcal{X}$ is said to be *Cohen–Macaulay* if and only if for any z in the support of $\mathcal{F}$, the length L_z of the minimal free resolution of the sheaf $\mathcal{F}$ at z equals the codimension of the support of $\mathcal{F}$ at this point, which is the minimal value it can take. If $\mathcal{I}$ is a coherent sheaf of $\mathcal{O}_{\mathcal{X}}$ ideals such that $\mathcal{O}_{\mathcal{X}}/\mathcal{I}$ is Cohen–Macaulay, then we say that the coherent sheaf of ideals $\mathcal{I}$ is Cohen–Macaulay.

REMARK A.22. When the coherent sheaf of ideals $\mathcal{I}$ is Cohen–Macaulay at a point $z \in \mathrm{Supp}\,(\mathcal{O}_{\mathcal{X}}/\mathcal{I})$, all associated primes of $\mathcal{O}_{\mathcal{X},z}/\mathcal{I}_z$ are *isolated.*[7] That is, they are minimal elements in $\mathrm{Ass}(\mathcal{O}_{\mathcal{X},z}/\mathcal{I}_z)$ for the ordering induced by inclusion.

EXAMPLE A.23 (Zero-dimensional sheaves of ideals are Cohen–Macaulay). Let $\mathcal{X}$ be an N-dimensional complex manifold. Any coherent sheaf of ideals $\mathcal{I}$ such that $\mathcal{O}_{\mathcal{X}}/\mathcal{I}$ has zero-dimensional (that is discrete) support is Cohen–Macaulay.

A.3.4. Local presentation in the purely dimensional case. Let V be a purely dimensional closed analytic subset of N-dimensional complex manifold $\mathcal{X}$ such that $\dim V = N - M$, where $M \in \{1, \dots, N\}$. Let $z_0 \in V$, $\zeta\ : U_{z_0} \to \mathbb{C}^N$ be a local system of coordinates in a neighborhood of z_0, centered at z_0, and $\boldsymbol{U}_{z_0} = \zeta(U_{z_0})$. The following proposition ([**GrtR**, page 72]) plays an important role since it allows us to introduce geometrically the concept of local (Hilbert–Samuel) multiplicity for a purely dimensional closed analytic subset at a given point $z_0 \in V$.

PROPOSITION A.24 (Presentation of closed analytic sets). *There exist an open neighborhood of the origin $\mathfrak{U} \subset \boldsymbol{U}_{z_0}$ in $\mathbb{C}^N$ and holomorphic functions $f_j : \mathfrak{U} \to \mathbb{C}$ such that $\bigcap_{j=1}^M f_j^{-1}(\{0\})$ is a purely $(N-M)$-dimensional closed analytic subset of $\mathfrak{U}$, and the set $\zeta_{z_0}(V)$ can be described in $\mathfrak{U}$ as the finite union $\bigcup_\iota W_\iota$ of some irreducible components W_ι of the closed analytic subset $\bigcap_{j=1}^M f_j^{-1}(\{0\})$, with $df_1 \wedge \cdots \wedge df_M \not\equiv 0$ on any W_ι.*

Let us make this presentation of V in a neighborhood of z_0 more precise and thus introduce geometrically the concept of *local multiplicity*, known also as Hilbert–Samuel multiplicity. It follows from the normalization lemma of E. Nœther that

[7] In general, the associated primes that are not isolated are called *embedded.*

there is a linear change of coordinates $z \leftrightarrow (z'_1, \dots, z'_{N-M}, z''_1, \dots, z''_M)$ in a neighborhood of the origin in $\mathbb{C}^N$ (that one chooses here to be a generic point) such that if $\Delta \subset \mathfrak{U} \subset \mathbb{C}^{N-M}_{z'} \times \mathbb{C}^M_{z''}$ is a convenient open poly-cylinder $\Delta' \times \Delta''$ centered at $(0,0)$, the projection

$$\pi \,:(z', z'') \in \zeta_{z_0}(V) \cap (\Delta' \times \Delta'') \longmapsto z' \in \Delta$$

is a proper map. Since $df_1 \wedge \cdots \wedge df_M \not\equiv 0$ on any W_ι, the jacobian determinant $J(z', z'') = \partial(f_1, \dots, f_M)/\partial(z''_1, \dots, z''_M)$ is not identically zero on each $W_\iota \cap \Delta$ for any index ι. The discriminant locus $\{z' \in \Delta' \,:\, \sigma(z') = 0\}$ of the projection π equals $\pi\big(\zeta_{z_0}(V) \cap \{(z', z'') \in \Delta \,:\, J(z', z'') = 0\}\big)$. Above $\Delta' \setminus \sigma^{-1}(\{0\})$, $\zeta_{z_0}(V) \cap \Delta$ consists then in μ disjoint leaves ($\mu \in \mathbb{N}^*$), which are $(N-M)$-dimensional submanifolds parameterized by the coordinates $z' \mapsto w^{(\iota)}(z')$ for $\iota = 1, \dots, \mu$. One has $\mu \leq \tilde{\mu}$, where $\tilde{\mu}$ denotes the number of disjoint $(N-M)$-dimensional leaves $z' \mapsto w^{(\iota)}(z')$ of the proper projection $\tilde{\pi} \,: \bigcap_{j=1}^M f_j^{-1}(\{0\}) \cap ((\Delta' \setminus \tilde{\sigma}^{-1}(\{0\})) \times \Delta'')$, where

$$\{z' \in \Delta' \,:\, \tilde{\sigma}(z') = 0\} \supset \{z' \in \Delta' \,:\, \sigma(z') = 0\}$$

denotes its discriminant locus. The integers $\mu \leq \tilde{\mu}$ are independent of the choice of the generic change of coordinates as well as independent of the choice of the local centered coordinate system in a neighborhood of z_0. The integer μ defines the *local Hilbert–Samuel multiplicity* $\mathrm{mult}_{z_0}(V)$ of V at z_0. In the particular case where $M = 1$ and $\zeta_{z_0}(V) = f^{-1}(\{0\})$, $\mathrm{mult}_{z_0}(V)$ equals the degree of the homogeneous part of the lowest degree in the Taylor development in z in a neighborhood of the origin of f^{red}, where $f^{\mathrm{red}} = \prod_\iota f_\iota$ is the reduced (e.g., square-free) holomorphic function deduced from the splitting $f = \prod_\iota f_\iota^{\nu_\iota}$ of f in irreducible factors. Given a purely $(N-M)$-dimensional closed analytic subset $V \subset \mathcal{X}$ of a complex analytic manifold, the local Hilbert–Samuel multiplicity $\mathrm{mult}_{z_0}(V)$ at a point $z_0 \in V$ can also be interpreted *algebraically* instead of *geometrically* as follows. Let $\mathcal{I}_V$ be the coherent ideal sheaf attached to V and let $\mathfrak{M}_{V,z_0}$ be the maximal ideal of the local ring $\mathcal{O}_{\mathcal{X},z_0}/\mathcal{I}_{V,z_0}$. The *Hilbert function of the local ring* $\mathcal{O}_{\mathcal{X},z_0}/\mathcal{I}_{V,z_0}$ *with maximal ideal* $\mathfrak{M}_{V,z_0}$ is the function from $\mathbb{N}$ to $\mathbb{N}$ defined as

$$\nu \in \mathbb{N} \longmapsto H_{V,z_0}(\nu) = \dim_{\mathcal{O}_{\mathcal{X},z_0}/\mathcal{I}_{V,z_0}} \big((\mathcal{O}_{\mathcal{X},z_0}/\mathcal{I}_{V,z_0})/\mathfrak{M}^\nu_{V,z_0}\big).$$

It coincides for $\nu \gg 1$ with the *Hilbert polynomial function of the local ring* $\mathcal{O}_{\mathcal{X},z_0}/\mathcal{I}_{V,z_0}$ *with respect to its maximal ideal* $\mathfrak{M}_{V,z_0}$

$$\nu \in \mathbb{N} \longmapsto \boldsymbol{H}_{V,z_0}(\nu) = \mathrm{mult}_{z_0}(V)\, \frac{\nu^{N-M}}{(N-M)!} + \sum_{k=1}^{N-M} h_{V,z_0,k}\, \nu^{N-M-k} \qquad (h_{V,z_0,k} \in \mathbb{Q}).$$

We will later complete the definition of the Hilbert–Samuel local multiplicity of a closed analytic set at a point z_0 with that of the list of local Segre numbers; see Proposition C.25 and Example D.16.

A.3.5. Weakly holomorphic functions and Oka universal denominator. Let V be a closed analytic subset of an N-dimensional complex manifold $\mathcal{X}$. We introduce below three notions of holomorphicity for functions on V. It is not restrictive to assume that V is purely $(N-M)$-dimensional with $0 \leq M \leq N-1$, which we do.

DEFINITION A.25. Let $V \subset \mathcal{X}$ be a closed purely $(N-M)$-dimensional analytic subset in a N-dimensional complex manifold and h be a function from V to $\mathbb{C}$.

(1) The function h is said to be a *weakly holomorphic function* on V if and only if $h_{|V_{\rm reg}}$ is holomorphic (as a function defined on $V_{\rm reg}$ equipped with its structure of $(N-M)$-complex manifold $(V_{\rm reg}, (\mathcal{O}_{\mathcal{X}})_{|V_{\rm reg}})$) and h is locally bounded around any point $z \in V$.
(2) The function h is *c-holomorphic* ([**Wh**, **Ch**, **Lo**]) on V if and only if $h_{|V_{\rm reg}}$ is holomorphic (see above) and h is continuous on V.
(3) The function h is *strongly holomorphic* on V if and only if it is the restriction to V of a holomorphic function H in some open neighborhood of V in the ambient manifold $\mathcal{X}$.

One has the following useful characterizations for the two first concepts.

PROPOSITION A.26. *Let $\mathcal{X}$ be an N-dimensional complex manifold and let $V \subset \mathcal{X}$ be a closed analytic subset with pure codimension M.*

- *A holomorphic map $h : V_{\rm reg} \to \mathbb{C}$ extends as a weakly holomorphic function to the whole of V if and only if the closure in $\mathcal{X} \times \mathbb{C}$ of its graph*
$$\Gamma_{V_{\rm reg}}(h) = \{(z, h(z)) \,:\, z \in V_{\rm reg}\}$$
over $V_{\rm reg}$ is a closed analytic subset in $\mathcal{X} \times \mathbb{C}$.
- *A function $h : V \to \mathbb{C}$ is c-holomorphic on V if and only if it is continuous on V and its graph $\Gamma_V(h) = \{(z, h(z)) \,:\, z \in V\}$ is a closed analytic subset in $\mathcal{X} \times \mathbb{C}$.*

Weakly holomorphic functions, as well as c-holomorphic functions, on a closed analytic subset $V \subset \mathcal{X}$ are restrictions to V of meromorphic functions in a neighborhood of V in the ambient manifold $\mathcal{X}$. In this direction, we have a more precise and important result due to Kiyoshi Oka. This result is known as the *universal denominator theorem.*

THEOREM A.27. *Let $\mathcal{X}$ be a complex manifold, let $V \subset \mathcal{X}$ be a closed analytic subset with pure codimension M, and let h be a weakly holomorphic function on V. Then, for any $z_0 \in V$, h can be expressed in a neighborhood of z_0 as the restriction to V of a meromorphic function in a neighborhood of z_0 in the ambient manifold $\mathcal{X}$. For $z_0 \in V$ fixed, the denominator of this meromorphic extension can be chosen to be independent of h.*

PROOF. Since the proof of Oka's theorem depends on the local Nœther's presentation given in §A.3.4 and involves ideas appearing in many places in the present monograph, we outline its proof here. We take $\mathcal{X} = \Omega$ as a neighborhood of the origin in $\mathbb{C}^N$. We assume that V has pure codimension $M \in \{0, \dots, N-1\}$ and that the coordinates (z', z'') are taken such that the projection map
$$\pi : (z', z'') \in V \cap (\Delta' \times \Delta'') \to z'$$
is proper; see Proposition A.17 in §A.3.4. We respect such presentations and keep the notation from Proposition A.17 and the whole of §A.3.4 which follows it. There exist holomorphic functions $h_{jk} : \Delta' \times \Delta'' \times \Delta'' \to \mathbb{C}$ such that for any $z' = (z'_1, \dots, z'_{N-M}) \in \Delta'$, $u = (u_1, \dots, u_M), v = (v_1, \dots, v_M) \in \Delta''$,
$$f_j(z', u) - f_j(z', v) = \sum_{k=1}^{M} h_{jk}(z', u, v)\,(u_j - v_j), \; j = 1, \dots, M.$$

This comes from Hefer division formulae for polycylinders (see (A.1) in §A.2.2) since poly-cylinders are Stein manifolds. Let $\widetilde{V} = \bigcap_{j=1}^{M} f_j^{-1}(\{0\}) \supset V$. If h is a holomorphic function in the submanifold $V_{\rm reg} \cap (\Delta' \times \Delta'')$, then it can be extended holomorphically to a holomorphic function in the submanifold $\widetilde{V}_{\rm reg} \times (\Delta' \times \Delta'')$ by setting $h = 0$ on $(\widetilde{V}_{\rm reg} \setminus V_{\rm reg}) \cap (\Delta' \times \Delta'')$. For $z' \in \Delta'$,

$$\pi^{-1}(\{z'\}) \cap \widetilde{V} = \{w^{(1)}(z'), \dots, w^{(\tilde{\mu})}(z')\}$$

for some $\tilde{\mu} \in \mathbb{N}^*$ independent of z', namely the Hilbert–Samuel local multiplicity at z_0 (see §A.3.4) of the closed analytic $(N-M)$-dimensional subset

$$\bigcap_{j=1}^{M} f_j^{-1}(\{0\}) \supset V.$$

One can see then that the function

$$H \ : (z', z'') \in \Delta' \times \Delta'' \longmapsto \sum_{\iota=1}^{\tilde{\mu}} h(z, w^{(\iota)}(z')) \det[h_{jk}] \, (z', z'', w^{(\iota)}(z'))$$

is holomorphic in both variables $(z', z'') \in (\Delta' \backslash \sigma^{-1}(\{0\})) \times \Delta''$, where σ defines the discriminant locus of the proper projection π. The restriction of this holomorphic map to $V \cap ((\Delta' \setminus \sigma^{-1}(\{0\})) \times \Delta'')$ equals $hJ_{|V}$. If h is bounded in $V_{\rm reg}$, then Riemann's theorem shows that the function H has an analytic continuation to a holomorphic function in $\Delta' \times \Delta''$ and that

$$\forall \, (z', z'') \in V \cap (\Delta' \times \Delta'') \, , \; h(z', z'') = \frac{H(z', z'')}{J(z', z'')} \, .$$

The last equality proves that h coincides with the restriction to $V \cap (\Delta' \times \Delta'')$ of a meromorphic function in $\Delta' \times \Delta''$, with J (independent of h) as denominator. □

A.3.6. Example: closed analytic subsets of $\mathbb{P}^N_{\mathbb{C}}$, Chow's theorem. Since compact complex manifolds $\mathbb{P}^N_{\mathbb{C}}$, $N \in \mathbb{N}^*$, play a major role in the present monograph, we recall here the important theorem of Chow, which can be viewed as a consequence of the Remmert–Stein theorem [**RS53**]: if V is a purely dimensional closed analytic subset of the open affine subset $\mathbb{C}^{N+1} \setminus \{(0, \dots, 0)\}$ with $\dim V > 0$, then its closure $\overline{V}$ in $\mathbb{C}^{N+1}$ is a closed analytic subset of $\mathbb{C}^{N+1}$. Chow's theorem reflects the GAGA principle formulated by J. P. Serre [**Se56**].

THEOREM A.28. *Any closed analytic subset V of $\mathbb{P}^N_{\mathbb{C}}$ is the zero set of a homogeneous polynomial ideal in $\mathbb{C}[X_0, \dots, X_N]$. In particular, any n-dimensional submanifold $\mathcal{X}$ of $\mathbb{P}^N_{\mathbb{C}}$ is such that one can find homogeneous polynomials $P_1, \dots, P_m$ in $\mathbb{C}[X_0, \dots, X_N]$ so that*

$$\mathcal{X} = \{z \in \mathbb{P}^N_{\mathbb{C}}, ; \, P_1(z) = \cdots = P_m(z) = 0\},$$

$$\textit{with} \; \Big\{z \in \mathcal{X} \, : \, \bigwedge^{N-n} \Big(\sum_{j=1}^{m} \sum_{k=0}^{N} \frac{\partial P_j}{\partial z_k}(z) \, \epsilon_k \Big) = 0 \Big\} = \emptyset \, ,$$

where $\{\epsilon_0, \dots, \epsilon_N\}$ denotes the canonical basis of $\mathbb{C}^{N+1}$.

REMARK A.29. Since any projective algebraic manifold $\mathcal{X}$ admits an embedding $\iota : \mathcal{X} \hookrightarrow \mathbb{P}^{N'}_{\mathbb{C}}$ of complex manifolds in some projective space $\mathbb{P}^{N'}_{\mathbb{C}}$, the final assertion in Theorem A.28 holds for $\iota(\mathcal{X}) \subset \mathbb{P}^{N'}_{\mathbb{C}}$. If $\mathcal{X}$ is the reduced complex analytic space underlying a projective algebraic variety defined over $\mathbb{C}$, then there is an embedding $\iota : \mathcal{X} \hookrightarrow \mathbb{P}^{N'}_{\mathbb{C}}$ of $\mathcal{X}$ in some projective space $\mathbb{P}^{N'}_{\mathbb{C}}$ and the first assertion in Theorem A.28 applies to $\iota(\mathcal{X})$ as well.

A.4. Complex analytic spaces, normalization, and log resolutions

A.4.1. Complex analytic spaces. Let

$$V' \subset \Omega' \subset \mathbb{C}^{N'}, \quad V'' \subset \Omega'' \subset \mathbb{C}^{N''}$$

be two closed analytic subsets of the open sets Ω' and Ω'' of $\mathbb{C}^{N'}$ and $\mathbb{C}^{N''}$, respectively. A continuous function $f : V' \to V''$ is called a *morphism of analytic sets* from V' into V'' if and only if, for every $z \in V'$, there exists a neighborhood U'_z of z in Ω', a holomorphic function F_z from U'_z into $\mathbb{C}^{N''}$, such that $(F_z)_{|V' \cap U'_z} = f_{|V' \cap U'_z}$. If such is the case, then for every $z \in V'$, one can define the mapping

$$f_z^* : \mathcal{O}_{V'',f(z)} := \frac{\mathcal{O}_{\Omega'',f(z)}}{\mathcal{I}_{V'',f(z)}} \longrightarrow \mathcal{O}_{V',z} := \frac{\mathcal{O}_{\Omega',z}}{\mathcal{I}_{V',z}} \tag{A.16}$$

as $f_z^*(g_z) = (g \circ F_z)_z$. This map is called the *comorphism* of f at the point z. For a closed analytic subset $V' \subset \Omega' \subset \mathbb{C}^{N'}$, denote by $\mathcal{O}_{V'}$ the $\mathcal{O}_{\Omega'}$-coherent sheaf (see §A.2.1) defined as the quotient sheaf $\mathcal{O}_{V'} := \mathcal{O}_{\Omega'}/\mathcal{I}_{V'}$. Complex analytic spaces are realized on the model of complex manifolds (see §A.1.1), except that one needs to glue together closed analytic subsets $V_\iota \subset \Omega_\iota \subset \mathbb{C}^{N_\iota}$ instead of copies of open subsets of some fixed $\mathbb{C}^N$; see also [**De**, II, §5].

Definition A.30 (Complex analytic space). A *complex analytic space* (or, to be more precise, *reduced*),[8] $(\mathcal{X}, \mathcal{O}_{\mathcal{X}})$ consists of the following data:

(1) a separable and locally compact topological space $\mathcal{X}$, which is countable at infinity;
(2) a *structural sheaf* of rings of holomorphic functions $\mathcal{O}_{\mathcal{X}}$, together with a covering $(U_\iota, \tau_\iota)_\iota$ of $\mathcal{X}$, where τ_ι realizes a homeomorphism between U_ι and some closed analytic subset $V_\iota = \tau_\iota(U_\iota) \subset \Omega_\iota \subset \mathbb{C}^{N_\iota}$ (for some integer $N_\iota \in \mathbb{N}$), such that, for any index ι, the comorphism

$$\tau_\iota^* : g \in (\mathcal{O}_{V_\iota})_{\tau_\iota(z)} = \left(\frac{\mathcal{O}_{\Omega_\iota}}{\mathcal{I}_{V_\iota}}\right)_{\tau_\iota(z)} \longmapsto (g \circ \tau_\iota)_z \in (\mathcal{O}_{|U_\iota})_z, \quad z \in U_\iota,$$

is an isomorphism between sheaves of rings.

Remark A.31. Suppose that $\mathcal{X}$ is an N-dimensional complex manifold with $V \subset \mathcal{X}$ a purely $(N-M)$-dimensional closed analytic subset. Given $z \in V$, suppose that, besides the embedding $V_z \hookrightarrow \mathcal{X}_z$, V_z is embedded in two ways through the embeddings

$$\iota' : V \cap U_z \hookrightarrow \mathrm{U}' \subset \mathbb{C}^{N_{\iota'}}, \quad \iota'' : V \cap U_z \hookrightarrow \mathrm{U}'' \subset \mathbb{C}^{N_{\iota''}}$$

as in Remark A.16. If $\iota = \iota_{\min}$ is supposed to be minimal, hence $\iota = \iota''$ with $\mathrm{U}'' = \mathrm{U}' \times \mathbb{C}^\nu_w$ with $\nu = N_{\iota''} - N_{\iota_{\min}}$, and $\mathcal{I}_{\iota_{\min}(V),\mathrm{U}'}$ is a radical defining sheaf for $\iota_{\min}(V)$, then

$$\mathcal{I}_{\iota(V),\mathrm{U}''} = \mathcal{I}_{\iota_{\min}(V),\mathrm{U}'}\,\mathcal{O}_{\mathrm{U}''} + (w_1, \dots, w_\nu)\mathcal{O}_{\mathrm{U}''}$$

is a radical defining sheaf for $\iota''(V) = \iota(V)$ in U'' and the two isomorphic quotient sheaves

$$\mathcal{O}_{\mathrm{U}'}/\mathcal{I}_{\iota_{\min}(V),\mathrm{U}'} \simeq \mathcal{O}_{\mathrm{U}''}/\mathcal{I}_{\iota(V),\mathrm{U}''}$$

[8] In this monograph we avoid discussing the theory of schemes [**EGA**, **Ha**].

represent at the level of germs at z the structure sheaf $\mathcal{O}_{V,z}$ of the reduced analytic space with underlying space V around z.

If $\mathcal{X}$ is a complex analytic space, then the subset $\mathcal{X}_{\rm reg}$ of *regular points* of $\mathcal{X}$ (these are points in a neighborhood of which $(\mathcal{X}, \mathcal{O}_{\mathcal{X}})$ is a complex manifold with its structural sheaf of holomorphic functions) is dense in $\mathcal{X}$ (isolated points of $\mathcal{X}$ are considered as regular). One denotes by $\mathcal{X}_{\rm sing} = \mathcal{X} \setminus \mathcal{X}_{\rm reg}$ the set of *singular points* in $\mathcal{X}$. The closures of the connected components of $\mathcal{X}_{\rm reg}$ are called *irreducible components* of $\mathcal{X}$. The *dimension* of $\mathcal{X}$ at $z \in \mathcal{X}$, denoted as $\dim_z \mathcal{X}$, is by definition $\dim_{\tau_\iota(z)}(V_\iota)$, whenever (U_ι, τ_ι) is a local chart containing z . The function $z \mapsto \dim_z \mathcal{X}$ is constant on the irreducible components of $\mathcal{X}$ and thus it is enough to evaluate it at regular points in $\mathcal{X}$. The complex space $\mathcal{X}$ is *of pure dimension* N if all its irreducible components are of dimension N.

Given a complex analytic space $\mathcal{X}$, one defines on $\mathcal{X}$ the sheaf $\widetilde{\mathcal{O}}_{\mathcal{X}}$ of functions $h : \mathcal{X} \to \mathbb{C}$ such that $h_{|\mathcal{X}_{\rm reg}}$ is holomorphic on $\mathcal{X}_{\rm reg}$ and h is locally bounded on $\mathcal{X}$. Here, $\mathcal{X}_{\rm reg}$ is considered as a complex manifold equipped with the structural sheaf $(\mathcal{O}_{\mathcal{X}})_{|\mathcal{X}_{\rm reg}}$ and h being locally bounded on $\mathcal{X}$ means that, for any $z \in \mathcal{X}$, one can find an open neighborhood U_z of z such that $|h(\zeta)| \leq C(z)$ for any $\zeta \in U_z \cap \mathcal{X}_{\rm reg}$. Such functions are also called *weakly holomorphic functions* on the complex analytic space $\mathcal{X}$ referring to the notion of *weakly holomorphic function on a closed analytic subset* of a complex manifold introduced previously (Definition A.25 in §A.3.5). If $\mathcal{X}$ is irreducible (i.e., the set of regular points has only one connected component), then the local rings $\mathcal{O}_{\mathcal{X},z}$ are integral domains and one may consider their fraction fields $\mathcal{M}_{\mathcal{X},z}$, $z \in \mathcal{X}$. The corresponding sheaf $\mathcal{M}_{\mathcal{X}}$ is then called *the sheaf of meromorphic functions* (sometimes called *regular functions*) over $\mathcal{X}$. If $\mathcal{X}$ is not irreducible anymore, then the sheaf $\mathcal{M}_{\mathcal{X}}$ is a sheaf of rings. The stalk $\mathcal{M}_{\mathcal{X},z}$ is the ring of fractions of $\mathcal{O}_{\mathcal{X},z}$, that is, the quotient of $\mathcal{O}_{\mathcal{X},z}$ by the ideal of elements that are not zero-divisors. Oka's theorem (Theorem A.27 in §A.3.5) implies that $\widetilde{\mathcal{O}}_{\mathcal{X}} \subset \mathcal{M}_{\mathcal{X}}$. In fact, for $z \in \mathcal{X}$, $\widetilde{\mathcal{O}}_{\mathcal{X},z}$ is the *integral closure* of the subring $\mathcal{O}_{\mathcal{X},z}$ in the $\mathcal{O}_{\mathcal{X},z}$-algebra $\mathcal{M}_{\mathcal{X},z}$, that is, for every $z \in \mathcal{X}$, $\widetilde{\mathcal{O}}_{\mathcal{X},z}$ is the set of elements h_z in $\mathcal{M}_{\mathcal{X},z}$ satisfying a monic integral dependency relation:

$$h_z^\nu + a_1 h_z^{\nu-1} + \cdots + a_\nu = 0, \text{ with } a_1, \ldots, a_\nu \in \mathcal{O}_{\mathcal{X},z}.$$

A.4.2. Normalization and blowup of complex analytic spaces.

A.4.2.1. *Normal complex analytic spaces, normalization.* The notion of weakly holomorphic function on a complex analytic space $\mathcal{X}$ introduced in §A.4.1 appeals to *Riemann's analytic continuation theorem*: if $\mathcal{X}$ is a complex manifold of dimension N and V is a closed analytic subset of $\mathcal{X}$ with codimension 1, any function f from $\mathcal{X} \setminus V$ to $\mathbb{C}$, which is holomorphic in $\mathcal{X} \setminus V$ and locally bounded on $\mathcal{X}$, that is,

$$\forall z \in \mathcal{X}\,,\ \exists U_z \ni z,\ \sup_{\zeta \in U_z \setminus V} |f(\zeta)| < +\infty\,, \tag{A.17}$$

extends holomorphically over the whole manifold $\mathcal{X}$. In case $\operatorname{codim} V \geq 2$, the additional hypothesis (A.17) is redundant and the existence of the analytic continuation of f follows from Hartogs's theorem; see, e.g., [**GuR**].

When $\mathcal{X}$ is an irreducible complex analytic space of dimension N equipped with its structure sheaf $\mathcal{O}_{\mathcal{X}}$, Riemann analytic continuation theorem fails to be true in general. One may have holomorphic functions in the dense subset $\mathcal{X}_{\text{reg}}$ which are bounded in a neighborhood of a singular point $z \in \mathcal{X}_{\text{sing}}$, but which do not define elements of $\mathcal{O}_{\mathcal{X},z}$. Here is an example, namely that of the *cusp*: let $\varphi : t \in D(0,\varepsilon) \mapsto (t^3, t^2)$ be an injective (since 2 and 3 are coprime) holomorphic parametrization in a neighborhood of the origin in $\mathbb{C}^2$ of the analytic space $\mathcal{X}$ defined (as embedded in $\mathbb{C}^2$) by the equation $z_1^2 - z_2^3 = 0$. The function

$$h : z \mapsto \varphi^{-1}(z)$$

is holomorphic in $\mathcal{X}_{\text{reg}} = \mathcal{X} \setminus \{(0,0)\}$ and locally bounded in $\mathcal{X}$, but there is no holomorphic function $\widetilde{h}$ in a neighborhood of the origin $z = (0,0)$ in $\mathcal{X}$ such that $\widetilde{h}(t^3, t^2) = t$. This proves that φ^{-1} cannot be defined at the level of germs at $(0,0)$ as an element of $\mathcal{O}_{\mathcal{X},(0,0)}$. The fact that Riemann's analytic continuation theorem fails in general on a complex analytic space $(\mathcal{X}, \mathcal{O}_{\mathcal{X}})$ leads naturally to the introduction of the concept of *normality at a given point* $z \in \mathcal{X}_{\text{sing}}$ and of *normality* at any point $z \in \mathcal{X}_{\text{sing}}$. Namely, we have the following.

DEFINITION A.32. A complex analytic space $(\mathcal{X}, \mathcal{O}_{\mathcal{X}})$ is said to be *normal at a point* $z \in \mathcal{X}_{\text{sing}}$ if and only if $\widetilde{\mathcal{O}}_{\mathcal{X},z} = \mathcal{O}_{\mathcal{X},z}$. The space $\mathcal{X}$ is said to be *normal* if and only if it is normal at any point z of $\mathcal{X}_{\text{sing}}$.

Any complex analytic space $(\mathcal{X}, \mathcal{O}_{\mathcal{X}})$ can be *normalized* in the sense described by the following Oka's normalization theorem.

THEOREM A.33. *Let $\mathcal{X}$ be a complex analytic space. There exists a normal complex analytic space $(\widehat{\mathcal{X}}, \mathcal{O}_{\widehat{\mathcal{X}}})$, together with a proper projection $\pi : \widehat{\mathcal{X}} \to \mathcal{X}$, where $|(\pi^{-1}(\{z\})| < +\infty$ for any z, such that $\widehat{\mathcal{X}} \setminus \pi^{-1}(\mathcal{X}_{\text{sing}})$ is dense in $\widehat{\mathcal{X}}$ and π realizes an analytic isomorphism between $\widehat{\mathcal{X}} \setminus \pi^{-1}(\mathcal{X}_{\text{sing}})$ and $\mathcal{X}_{\text{reg}}$. Such a pair $((\widehat{\mathcal{X}}, \mathcal{O}_{\widehat{\mathcal{X}}}), \pi)$ (or, in short, $(\widehat{\mathcal{X}}, \pi)$), is said to be a normalization of the complex analytic space $\mathcal{X}$. It is unique in the following sense: if $(\widehat{\mathcal{X}_1}, \pi_1)$ and $(\widehat{\mathcal{X}_2}, \pi_2)$ are two normalizations of complex analytic space $\mathcal{X}$, then $(\widehat{\mathcal{X}_1}, \mathcal{O}_{\widehat{\mathcal{X}_1}})$ and $(\widehat{\mathcal{X}_2}, \mathcal{O}_{\widehat{\mathcal{X}_2}})$ are isomorphic as complex analytic spaces.*

A.4.2.2. *Normalized blowup of a complex manifold along a coherent sheaf of ideals.* The geometric operation consisting of *blowing up* plays an important role in analytic or algebraic geometry. In particular, given a $\mathcal{O}_{\mathcal{X}}$-coherent sheaf of ideals $\mathcal{I}$ over an N-dimensional complex manifold $(\mathcal{X}, \mathcal{O}_{\mathcal{X}})$, one will use extensively in this monograph the so-called *normalized blowup of $\mathcal{X}$ along the sheaf of ideals $\mathcal{I}$* (or also *with the center of the coherent ideal sheaf $\mathcal{I}$*). This is, up to an isomorphism between (reduced) complex analytic spaces, an intrinsic object (depending on $\mathcal{X}$ and $\mathcal{I}$); see Proposition A.34. Here are the two operations comprising the normalized blow-up procedure, once $\mathcal{X}$, its structural sheaf $\mathcal{O}_{\mathcal{X}}$ and the $\mathcal{O}_{\mathcal{X}}$-coherent sheaf of ideals $\mathcal{I}$ are given.

(1) The first one is the blowup of the complex manifold $\mathcal{X}$ along the coherent sheaf $\mathcal{I}$ (or with the center of the coherent sheaf $\mathcal{I}$). It leads, as described below, to an N-dimensional complex analytic space $(\mathcal{X}_{\mathcal{I}}, \mathcal{O}_{\mathcal{X}_{\mathcal{I}}})$, together with a holomorphic proper surjective projection $\pi : \mathcal{X}_{\mathcal{I}} \to \mathcal{X}$, such that the inverse image sheaf $\mathcal{I} \cdot \mathcal{O}_{\mathcal{X}_{\mathcal{I}}}$ of $\mathcal{I}$ by π is invertible. More precisely, the support of $\mathcal{O}_{\mathcal{X}_{\mathcal{I}}}/\mathcal{I} \cdot \mathcal{O}_{\mathcal{X}_{\mathcal{I}}}$ is a closed analytic hypersurface $H_{\mathcal{I}}$ of $\mathcal{X}_{\mathcal{I}}$, and π realizes a biholomorphism between $\mathcal{X}_{\mathcal{I}} \setminus H_{\mathcal{I}}$ and $\mathcal{X} \setminus \text{Supp}\,(\mathcal{O}_{\mathcal{X}}/\mathcal{I})$.

(2) The second one consists of the realization of a normalization

$$\widehat{\mathcal{X}}_{\mathcal{I}} \xrightarrow{\pi_{\text{norm}}} \mathcal{X}_{\mathcal{I}}$$

of the complex analytic space $\mathcal{X}_{\mathcal{I}}$ according to Theorem A.33.

The *normalized blowup* of $\mathcal{X}$ along $\mathcal{I}$ is then

$$\pi_{\mathcal{I},\text{norm}} : \widehat{\mathcal{X}}_{\mathcal{I}} \xrightarrow{\pi_{\text{norm}}} \mathcal{X}_{\mathcal{I}} \xrightarrow{\pi} \mathcal{X}. \tag{A.18}$$

The inverse image sheaf $\mathcal{I} \cdot \mathcal{O}_{\widehat{\mathcal{X}}_{\mathcal{I}}}$ *via* $\pi \circ \pi_{\text{norm}}$ is also invertible,[9] which means that its support

$$H = \text{Supp}\left(\frac{\mathcal{O}_{\widehat{\mathcal{X}}_{\mathcal{I}}}}{\mathcal{I} \cdot \mathcal{O}_{\widehat{\mathcal{X}}_{\mathcal{I}}}}\right) \subset \widehat{\mathcal{X}}_{\mathcal{I}} \tag{A.19}$$

is a closed hypersurface. Since $\widehat{\mathcal{X}}_{\mathcal{I}}$ is normal, its structural sheaf is *regular in codimension one*. This means that the generic point of any irreducible closed hypersurface is regular, or, which is equivalent, that $\operatorname{codim}(\widehat{\mathcal{X}}_{\mathcal{I}})_{\text{sing}} \geq 2$. To any irreducible component H_γ of such hypersurface H (defined by (A.19)), whose generic point is denoted by $\hat{z}_\gamma$, one can then attach a positive integer m_γ defined by

$$m_\gamma = \text{length}\Big((\mathcal{O}_{\widehat{\mathcal{X}}_{\mathcal{I}}})_{\hat{z}_\gamma} / (\mathcal{I} \cdot \mathcal{O}_{\widehat{\mathcal{X}}_{\mathcal{I}}})_{\hat{z}_\gamma}\Big). \tag{A.20}$$

The locally finite formal combination

$$E_{\mathcal{I}} = \sum_\gamma m_\gamma H_\gamma \tag{A.21}$$

is then a Weil divisor over $\mathcal{X}$ (see §D.3.1) called the Weil *exceptional divisor of the normalized blowup* $\pi_{\mathcal{I},\text{norm}}$. For a deeper understanding of these notions one can refer to [**Hir64**] or [**T73, T82**]. The reason why we denoted as $E_{\mathcal{I}}$ the exceptional divisor of the normalized blowup follows from the next important proposition, that ensures the uniqueness, up to isomorphism, of the normalized blowup.

PROPOSITION A.34. *Let $\mathcal{X}$ be a complex manifold and let $\mathcal{I}$ be a $\mathcal{O}_{\mathcal{X}}$-coherent sheaf of ideals on $\mathcal{X}$. The normalized blowup of $\mathcal{X}$ along $\mathcal{I}$ defined by* (A.18) *has the following universality property: if $\tau : \mathcal{X}' \longrightarrow \mathcal{X}$ is any morphism of complex analytic spaces such that $\mathcal{X}'$ is normal and the inverse image sheaf $\mathcal{I} \cdot \mathcal{O}_{\mathcal{X}'}$ is invertible, then τ factorizes in a unique manner as $\tau = \pi_{\mathcal{I},\text{norm}} \circ \tau_{\mathcal{I}}$, where $\tau_{\mathcal{I}}$ is a morphism of complex analytic spaces from $\mathcal{X}'$ into the normalized blowup of $\mathcal{X}$ along $\mathcal{I}$.*

It follows from the universality property of the normalized blowup that the following definition taken from ([**Ful, Ful84**]) describes intrinsically closed analytic subsets contained in $\text{Supp}(\mathcal{O}_{\mathcal{X}}/\mathcal{I})$.

DEFINITION A.35. Let $\mathcal{X}$ be a complex N-dimensional manifold and let $\mathcal{I}$ be a $\mathcal{O}_{\mathcal{X}}$ coherent sheaf of ideals. Let also $\{H_\gamma\}_\gamma$ be the collection of all irreducible hypersurfaces of the normalized blowup of $\mathcal{X}$ along $\mathcal{I}$, whose union is the hypersurface H defined by (A.19). Then the $Z_\gamma = \pi_{\mathcal{I},\text{norm}}(H_\gamma)$ are closed irreducible analytic subsets[10] of $\mathcal{X}$ lying in $\text{Supp}(\mathcal{O}_{\mathcal{X}}/\mathcal{I})$. For $\operatorname{codim}_{\mathcal{X}}(\mathcal{O}_{\mathcal{X}}/\mathcal{I}) \leq r \leq N = \dim \mathcal{X}$, the

[9] It is the sheaf of sections of a Cartier divisor on $\widehat{\mathcal{X}}$, which is called the *exceptional divisor* of the normalized blowup.

[10] The fact that Z_γ is a closed analytic subset of $\mathcal{X}$ follows from the properness of the map $\pi_{\mathcal{I},\text{norm}}$. The irreducibility of Z_γ is a consequence of the irreducibility of the hypersurface H_γ.

irreducible closed analytic subsets Z_γ^r, forming the (highly redundant in general) list

$$\{Z_\gamma \,:\, \mathrm{codim}_{\mathcal{X}} Z_\gamma = r\},$$

are called *distinguished irreducible varieties of codimension* r of $\mathrm{Supp}\,(\mathcal{O}_{\mathcal{X}}/\mathcal{I})$. The same definition holds if $\mathcal{X}$ is an N-dimensional complex analytic space and the notion of normalized blowup is extended to such a context too; see Remark A.38.

Let us make explicit the blowup of a complex analytic manifold $\mathcal{X}$ along a coherent sheaf of ideals $\mathcal{I}$ first in the case $\mathcal{I} = \mathcal{I}_{\mathcal{Y}}$, where $\mathcal{Y}$ is a closed submanifold in $\mathcal{X}$. Then we will make it explicit in the general case of a coherent $\mathcal{O}_{\mathcal{X}}$-sheaf $\mathcal{I}$ on $\mathcal{X}$ when the support of $\mathcal{O}_{\mathcal{X}}/\mathcal{I}$ is a submanifold; see, e.g., [**SmiKKT**] or [**De**, VII, §12]. To do so, it is convenient to invoke the theory of holomorphic bundles over a complex manifold; see Appendix B, in particular §B.1.3 to §B.1.5.

- Suppose first that $\mathcal{I} = \mathcal{I}_{\mathcal{Y}}$, where $\mathcal{Y}$ is a closed submanifold in $\mathcal{X}$. The blowup of $\mathcal{X}$ along $\mathcal{I}_{\mathcal{Y}}$ is an holomorphic vector bundle with rank equal to $\mathrm{codim}_{\mathcal{X}}\, \mathcal{Y}$ above $\mathcal{Y}$. Its fiber above $y \in \mathcal{Y}$ is the *projectivization* $\mathbb{P}(N_y)$ of the space N_y of all directions orthogonal to $\mathcal{Y}$ at the point y. That is, the $\mathbb{C}$-vector space obtained as the quotient of $N_y \setminus \{0\}$ by the colinearity relation. The holomorphic vector bundle as constructed above $\mathcal{Y}$ is the quotient bundle $(T_{\mathcal{X}})_{|\mathcal{Y}}/T_{\mathcal{Y}}$, where $T_{\mathcal{X}} \to \mathcal{X}$ is the holomorphic complex tangent bundle to $\mathcal{X}$ (see Example B.5 in §B.1.3) that one restricts above $\mathcal{Y}$ as a bundle above $\mathcal{Y}$, and $T_{\mathcal{Y}} \to \mathcal{Y}$ is the holomorphic tangent complex to $\mathcal{Y}$ considered here as a subbundle of $(T_{\mathcal{X}})_{|\mathcal{Y}} \to \mathcal{Y}$. Suppose the submanifold $\mathcal{Y}$ (here with dimension $N - M$) is defined in a local chart U around one of its points y as

  $$\mathcal{Y} := \{z \in \mathcal{X} \,:\, f_1(z) = \cdots = f_M(z) = 0\},$$

  where $f_1, \ldots, f_M$ are holomorphic functions in U, with $df_1 \wedge \cdots \wedge df_M \not\equiv 0$ over $\mathcal{Y} \cap U$. The blowup of U along $\mathcal{I}_{\mathcal{Y} \cap U}$ can be expressed, up to an isomorphism of complex manifolds, as the closure in $U \times \mathbb{P}_{\mathbb{C}}^{M-1}$ of the graph of the mapping

  $$z \in U \setminus \mathcal{Y} \mapsto [f_1(z) : \cdots : f_M(z)].$$

  This closure is denoted by $\Gamma_{\mathcal{Y},U}$ and is defined as the set of points

  $$(z, [w_1 : \cdots : w_M])$$

  in the product manifold $U \times \mathbb{P}_{\mathbb{C}}^{M-1}$ satisfying $w_j f_{j+1}(\zeta) - w_{j+1} f_j(\zeta) = 0$ for $j = 1, \ldots, M-1$. Then the proper projection $\pi_U \,: \Gamma_{\mathcal{Y},U} \to U$ is given by the projection

  $$(z, [w_1 : \cdots : w_M]) \mapsto z \in U.$$

  Note that the projection map π_U is proper, since $\mathbb{P}_{\mathbb{C}}^{M-1}$ is compact. Furthermore, the inverse image $H_{\mathcal{Y},U} := \pi_U^{-1}(\mathcal{Y}) =: (\mathcal{Y} \cap U) \times \mathbb{P}_{\mathbb{C}}^{M-1}$ is a smooth hypersurface in $\Gamma_{\mathcal{Y},U}$, and the projection π_U realizes a biholomorphism between the open subsets $\Gamma_{\mathcal{Y},U} \setminus H_{\mathcal{Y},U}$ and $U \setminus \mathcal{Y}$.
- Consider now the general case where $\mathcal{I}$ is an arbitrary $\mathcal{O}_{\mathcal{X}}$-coherent sheaf on $\mathcal{X}$. It is enough to blow up (along $\mathcal{I}_{|U}$) a local chart U around a point y in the support of $\mathcal{O}_{\mathcal{X}}/\mathcal{I}$. Let $(f_1, \ldots, f_M)$ be M elements in $\mathcal{O}(U)$, such

that $\mathcal{I}_z = ((f_1)_z, \ldots, (f_M)_z)$ for every $z \in U$. We introduce, as previously, the closure $\Gamma_{\mathcal{I},U}$ of the graph of the map

$$z \in U \setminus \mathrm{Supp}\,(\mathcal{O}_U/\mathcal{I}_{|U}) \mapsto [f_1(z) : \cdots : f_M(z)]$$

in $U \times \mathbb{P}^{M-1}(\mathbb{C})$. The projection $\pi : \Gamma_{\mathcal{I},U} \to U$ is described again by $(z, [w_1 : \cdots : w_M]) \mapsto z \in U$. The complex analytic space $\Gamma_{\mathcal{I},U}$ depends only (up to isomorphisms of complex analytic spaces) on the coherent sheaf $\mathcal{I}_{|U}$ and not on the system of generators $(f_1, \ldots, f_M)$ chosen in $\mathcal{O}_{\mathcal{X}}(U)$. Hartogs's theorem implies that the inverse image by π of the support of the quotient sheaf $\mathcal{O}_U/\mathcal{I}_U$ (i.e., the inverse image by π of the locus of the common zeroes of f_j, $j = 1, \ldots, M$, in U) cannot have components with a codimension larger than or equal to 2. Such an inverse image is therefore a hypersurface $H_{\mathcal{I},U}$ in $\Gamma_{\mathcal{I},U}$ and π realizes a biholomorphism between $\Gamma_{\mathcal{I},U} \setminus H_{\mathcal{I},U}$ and $U \setminus \bigcap_1^M f_j^{-1}(0) = U \setminus \mathrm{Supp}\,((\mathcal{O}_{\mathcal{X}})_{|U}/\mathcal{I}_U)$. This geometric description of the blowup $U_{\mathcal{I}_{|U}}$ of U along the coherent sheaf $\mathcal{I}_{|U}$ is sufficient for our needs in the core of the monograph.

A.4.2.3. *Integral closure and normalized blowup of a complex analytic space along a coherent sheaf of ideals.* The notion of *integral closure*[11] of an ideal in a commutative ring is closely related to the *normalized blowup*. We recall this algebraic definition in this subsection.

DEFINITION A.36. Let $\mathbb{A}$ be a commutative ring and let $\mathfrak{a}$ be an ideal in $\mathbb{A}$. We call *integral closure* of $\mathfrak{a}$ in $\mathbb{A}$ the ideal $\overline{\mathfrak{a}}$ in $\mathbb{A}$ consisting of elements h of $\mathbb{A}$ satisfying, for some $\nu \in \mathbb{N}^*$, a *homogeneous* integral dependence relation

$$h^\nu + u_1 h^{\nu-1} + \cdots + h_\nu = 0,$$

with $u_k \in \mathfrak{a}^k$, $k = 1, 2, \ldots, \nu$. An equivalent formulation is to say that hT, considered as an element of the graded algebra $\mathbb{A}[T]$, is *integral* over the Rees algebra

$$\widehat{\mathfrak{a}} := \mathbb{A} \oplus \mathfrak{a}T \oplus \mathfrak{a}^2T^2 \oplus \cdots = \bigoplus_{k=0}^{\infty} \mathfrak{a}^k T^k \subset \mathbb{A}[T],$$

i.e., it satisfies a monic integral dependence relation

$$(hT)^\nu + \sum_{j=1}^{\nu} (hT)^{\nu-j}\, \widehat{u}_j = 0\,,\ \widehat{u}_1, \ldots, \widehat{u}_\nu \in \widehat{\mathfrak{a}}.$$

If $\mathcal{X}$ is a complex analytic manifold and $\mathcal{I}$ is a $\mathcal{O}_{\mathcal{X}}$-coherent sheaf of ideals, then one can consider the coherent sheaf $\overline{\mathcal{I}}$, where, for every $z \in \mathcal{X}$, $\overline{\mathcal{I}}_z$ denotes the integral closure of $\mathcal{I}_z$ in $\mathcal{O}_{\mathcal{X},z}$. The coherence of the sheaf of ideals $\overline{\mathcal{I}}$ follows from the theorem of Grauert on the transport of coherence through direct image [**Grt60**] and Proposition A.37.

PROPOSITION A.37. *Let $\mathcal{X}$ be an N-dimensional complex manifold, let $\mathcal{I}$ be a coherent sheaf of ideals, let $U \subset \mathcal{X}$ be an open subset, and let $f \in \overline{\mathcal{I}}(U)$. If $\pi_{\mathcal{I},\mathrm{norm}} \longrightarrow \mathcal{X}$ denotes the normalized blowup of $\mathcal{X}$ along $\mathcal{I}$, then for any irreducible*

[11]One should not confuse the notion of the integral closure of a ring $\mathbb{A}$ in an $\mathbb{A}$-algebra (as introduced at the end of §A.4.1) with the subring $\mathcal{O}_{\mathcal{X},z}$ of the $\mathcal{O}_{\mathcal{X},z}$-algebra $\mathcal{M}_{\mathcal{X},z}$.

component H_γ of the hypersurface H defined by (A.19) *intersecting $\pi^{-1}_{\mathcal{I},\mathrm{norm}}(U)$ and for a generic point $\hat{z}_\gamma$ in $H \cap U$, one has that*

$$\text{(A.22)} \qquad \text{length}\, \frac{\big((\mathcal{O}_{\widehat{\mathcal{X}}_{\mathcal{I}}})_{|U}\big)_{\hat{z}_\gamma}}{\big(((\pi_{\mathcal{I},\mathrm{norm}})^* f)\, \mathcal{O}_{\widehat{\mathcal{X}}_{\mathcal{I}}})_{|U}\big)_{\hat{z}_\gamma}} \geq m_\gamma,$$

where m_γ is defined by (A.20).[12] *The above claim is also equivalent to the inclusion*

$$\text{(A.23)} \qquad (\pi^*_{\mathcal{I},\mathrm{norm}} f) \cdot (\mathcal{O}_{\overline{\mathcal{X}_{\mathcal{I}}}})_{|\pi^{-1}(U)} \subset \big(\mathcal{I} \cdot (\mathcal{O}_{\overline{\mathcal{X}_{\mathcal{I}}}})\big)_{|\pi^{-1}(U)}.$$

Conversely, if $f \in \mathcal{O}_{\mathcal{X}}(U)$ is such that either condition (A.22) *or condition* (A.23) *is fulfilled, then $f \in \overline{\mathcal{I}}(U)$.*

REMARK A.38. One can as well introduce the blowup of an N-dimensional complex analytic space $(\mathcal{X}, \mathcal{O}_{\mathcal{X}})$ (instead of an N-dimensional complex manifold) along a coherent $\mathcal{O}_{\mathcal{X}}$-sheaf of $\mathcal{O}_{\mathcal{X}}$ ideals $\mathcal{I}$, and then, normalize it to construct the normalized blowup of $(\mathcal{X}, \mathcal{O}_{\mathcal{X}})$ along $\mathcal{I}_{\mathcal{X}}$. Results analogous to Propositions A.34 and A.37 still hold. From the point of view of complex analytic scheme theory (see [**Ha**]) one should point out that the blowup $(\mathcal{X}_{\mathcal{I}}, \mathcal{O}_{\mathcal{X}_{\mathcal{I}}})$ of $(\mathcal{X}, \mathcal{O}_{\mathcal{X}})$ along the $\mathcal{O}_{\mathcal{X}}$-coherent sheaf $\mathcal{I}$ corresponds to the complex analytic scheme[13]

$$(X_{\mathcal{I}}, \mathcal{O}_{\mathcal{X}_{\mathcal{I}}}) = \mathrm{Proj}\,\Big(\bigoplus_{k=0}^{\infty} \mathcal{I}^k\Big).$$

The normalized blowup $(\overline{\mathcal{X}_{\mathcal{I}}}, \mathcal{O}_{\overline{\mathcal{X}_{\mathcal{I}}}})$ corresponds to the complex analytic scheme

$$(\widehat{\mathcal{X}}_{\mathcal{I}}, \mathcal{O}_{\widehat{\mathcal{X}}_{\mathcal{I}}}) = \mathrm{Proj}\,\overline{\Big(\bigoplus_{k=0}^{\infty} \mathcal{I}^k\Big)} = \mathrm{Proj}\,\Big(\bigoplus_{k=0}^{\infty} \overline{\mathcal{I}^k}\Big),$$

see for example to [**LejT08**].

A.4.3. Log resolution and the Hironaka theorem. Among closed singular hypersurfaces $\mathcal{H}$ in an N-dimensional complex manifold $\mathcal{X}$, *hypersurfaces with local normal crossings* play a major role, since their singularities can be explicitly described.

DEFINITION A.39. Let $\mathcal{X}$ be an N-dimensional manifold. A closed hypersurface $\mathcal{H} \subset \mathcal{X}$ is said to be *locally with normal crossings* if and only if, for any point $z \in \mathcal{H}$, there is a centered system of local coordinates in some open chart U around z such that

$$\mathcal{H} \cap U = \{\zeta \in U \,:\, \zeta_{i_{\nu_1}} \cdots \zeta_{i_{\nu_z}} = 0\,,\ \{i_{\nu_1}, \ldots, i_{\nu_x}\} \subset \{1, \ldots, N\}\},$$

i.e., $\mathcal{H}$ can be locally described in centered local coordinates ζ as the zero set of a monomial in ζ.

The important concept of *log resolution* is closely linked with that of closed hypersurface with normal crossings.

[12]In other words, if one refers to the notions of Cartier and Weil divisors on a complex analytic space (see respectively §D.2.1 and §D.3.1), one has $\mathrm{div}((\pi_{\mathcal{I},\mathrm{norm}})^* f) \geq E_{\mathcal{I}}$, where $\mathrm{div}((\pi_{\mathcal{I},\mathrm{norm}})^* f)$ denotes the Weil divisor which is attached to the principal Cartier divisor $(\pi^{-1}(U), f \circ \pi_{\mathcal{I},\mathrm{norm}})$.

[13]We are forced here to introduce the concept of scheme and just refer the reader to [**Ha**]. We quote this result for the sake of completeness, but we do not invoke it within our monograph, except for the fact that it is lying behind the statement of Theorem 7.8 which we quoted (without detailed proof) from [**Hi01**, Theorem 5.1].

DEFINITION A.40 (Log resolution). Let $(\mathcal{X}, \mathcal{O}_{\mathcal{X}})$ be an N-dimensional complex irreducible analytic space and let $\mathcal{Y}$ be a closed analytic subset in $\mathcal{X}$. A *log resolution* $\widetilde{\mathcal{X}} \overset{\pi}{\to} \mathcal{X}$ for $\mathcal{Y}$ consists of:

(1) a complex manifold $(\widetilde{\mathcal{X}}, \mathcal{O}_{\widetilde{\mathcal{X}}})$, where $\dim \widetilde{\mathcal{X}} = N$;
(2) a proper surjective holomorphic map $\pi : \widetilde{\mathcal{X}} \to \mathcal{X}$, such that the analytic subset $\pi^{-1}(\mathcal{Y}) \cup \mathrm{crit}(\pi)$ is a closed analytic hypersurface with normal crossings in $\widetilde{\mathcal{X}}$. Here $\mathrm{crit}(\pi)$ denotes the set of critical points of π, that is, the points of $\widetilde{\mathcal{X}}$ in a neighborhood of which π does not realize a local biholomorphism.

One of the main tools in developing multidimensional residue theory is the theorem on the resolution of singularities due to H. Hironaka [**Hir64**]. It justifies the existence of a log resolution for any analytic subset $\mathcal{Y}$ in an irreducible analytic complex space $(\mathcal{X}, \mathcal{O}_{\mathcal{X}})$.

THEOREM A.41 (Resolution of singularities). *Let $\mathcal{X}$ be an irreducible complex analytic space of dimension N and let $\mathcal{Y}$ be a complex analytic subset of $\mathcal{X}$ containing $\mathcal{X}_{\mathrm{sing}}$. There exists a complex manifold $\widetilde{\mathcal{X}}_{\mathcal{Y}}$ of dimension N (sometimes called a smooth modification of $\mathcal{X}$) and a proper holomorphic mapping $\pi_{\mathcal{Y}} : \widetilde{\mathcal{X}}_{\mathcal{Y}} \to \mathcal{X}$, such that, if $\widetilde{\mathcal{H}} = \pi_{\mathcal{Y}}^{-1}(\mathcal{Y})$:*

- *the mapping $\pi_{\mathcal{Y}}$ realizes a biholomorphism between $\widetilde{\mathcal{X}}_{\mathcal{Y}} \setminus \widetilde{\mathcal{H}}$ and $\mathcal{X} \setminus \mathcal{Y}$;*
- *the sheaf of ideals $\mathcal{I}_{\mathcal{Y}} \cdot \mathcal{O}_{\widetilde{\mathcal{X}}_{\mathcal{Y}}}$ (the inverse image by $\pi_{\mathcal{Y}}$ of the sheaf of ideals $\mathcal{I}_{\mathcal{Y}}$ on $\mathcal{X}$) is an invertible sheaf whose support is a hypersurface with locally normal crossings.*

REMARK A.42. The universality property of the normalized blowup (Proposition A.34 in §A.4.2) implies that if $\mathcal{I}$ is a coherent sheaf of ideals on a complex manifold $\mathcal{X}$ (or even a complex analytic space, see Remark A.38 in §A.4.2) and $\mathcal{Y}$ denotes the support of $\mathcal{O}_{\mathcal{X}}/\mathcal{I}$, then any resolution of singularities

$$\widetilde{\mathcal{X}}_{\mathcal{Y}} \overset{\widetilde{\pi}_{\mathcal{Y}}}{\to} \mathcal{X}$$

as in Theorem A.41 factorizes in a unique way as

$$\widetilde{\mathcal{X}}_{\mathcal{Y}} \overset{\theta}{\longrightarrow} \overline{\mathcal{X}_{\mathcal{I}}} \overset{\pi_{\mathcal{I},\mathrm{norm}}}{\longrightarrow} \mathcal{X},$$

where θ denotes a morphism of complex analytic spaces (depending on the log resolution).

APPENDIX B

Holomorphic bundles over complex analytic spaces

B.1. Analytic cocycles, holomorphic bundles, isomorphism classes

B.1.1. Cocycles on a complex analytic space, examples. All through this first subsection, $\mathcal{X}$ will be a (purely) N-dimensional complex analytic space with $\mathcal{O}_{\mathcal{X}}$ as its structural sheaf. Complex analytic 1-cocycles (or also, shortly, $\mathcal{O}_{\mathcal{X}}$-cocycles) with values in $\mathrm{GL}(\bullet, \mathbb{C})$ on $\mathcal{X}$ are defined as follows.[1]

DEFINITION B.1. Let $m \in \mathbb{N}^*$ and let $\mathcal{U} = \{U_\iota\}_\iota$ be an open covering of $\mathcal{X}$. A complex analytic 1-cocycle with values in $\mathrm{GL}(m, \mathbb{C})$ on $\mathcal{X}$ subordinate to the covering $\mathcal{U}$ is a collection of matrix-valued maps $\{g_{\iota_1,\iota_0}\}_{\iota_0,\iota_1}$, where g_{ι_1,ι_0} is a morphism of complex analytic spaces between $U_{\iota_0} \cap U_{\iota_1}$ (equipped with the restriction of the structure sheaf $\mathcal{O}_{\mathcal{X}}$) and $\mathrm{GL}(m, \mathbb{C})$ (equipped with its structure of complex submanifold of $\mathbb{C}^{m^2}$) such that the *cocycle conditions*

$$\forall\, z \in U_{\iota_0} \cap U_{\iota_1} \cap U_{\iota_2},\ g_{\iota_2,\iota_0}(z) = g_{\iota_2,\iota_1}(z) \circ g_{\iota_1,\iota_0}(z) \tag{B.1}$$

hold for any triplet of indices $\iota_0, \iota_1, \iota_2$.

EXAMPLE B.2. Let $\{U_\iota\}_\iota$ be a open covering of $\mathcal{X}$ and, for each index ι, let also $f_\iota \in \mathcal{M}_{\mathcal{X}}(U_\iota)$ be a nonidentically zero meromorphic function in U_ι. Suppose that any pair of indices ι_0, ι_1, f_{ι_1}/f_{ι_0} takes its values in $\mathbb{C}^*$. Then

$$\{g_{\iota_1,\iota_0} = f_{\iota_1}/f_{\iota_0}\}_{\iota_0,\iota_1}$$

is a complex analytic 1-cocycle with values in $\mathrm{GL}(1, \mathbb{C}) = \mathbb{C}^*$ on $\mathcal{X}$ which is called a *Cartier divisor*; see also §D.2.1. Let for example $\mathcal{X} = \mathbb{P}^N_{\mathbb{C}}$ and, for any $k = 0, \dots, n$, $U_k = \{[z_0 : \cdots : z_N] \in \mathcal{X} \,:\, z_j \neq 0\}$; for any $d \in \mathbb{Z}$, $\{[z_0 : \cdots : z_N] \in U_k \longmapsto z_0^d/z_k^d\}_k$ is a Cartier divisor on $\mathbb{P}^N_{\mathbb{C}}$. The representation of a Cartier divisor as $\{(U_\iota, f_\iota\}_\iota$ is not unique. Nevertheless, given two open coverings $\mathcal{U}$ and $\widetilde{\mathcal{U}}$, one identifies two Cartier divisors $\{(U_\iota, f_\iota)\}_\iota$ and $\{\widetilde{U}_{\tilde{\iota}}, \tilde{f}_{\tilde{\iota}}\}_{\tilde{\iota}}$ when the meromorphic section $f_\iota/\tilde{f}_{\tilde{\iota}}$ defines an invertible element in $\mathcal{O}_{\mathcal{X}}(U_\iota \cap \widetilde{U}_{\iota_\iota})$, whenever $U_\iota \cap \widetilde{U}_{\tilde{\iota}}$ is nonempty. Such identification allows us to call Cartier divisors $\{(\mathcal{X}, f)\}$ *principal Cartier divisors*, when $f \in \mathcal{M}_{\mathcal{X}}(\mathcal{X})$.

EXAMPLE B.3. Let $\mathcal{X}$ be an N-dimensional complex manifold with an atlas $\{(U_\iota, \zeta_\iota)\}_\iota$ as in §A.1.1. Then

$$\big\{g_{\iota_1,\iota_0} : z \in U_{\iota_0} \cap U_{\iota_1} \longmapsto \big(d(\zeta_{\iota_1} \circ \zeta_{\iota_0}^{-1})\big)(\zeta_{\iota_0}(z)) \in \mathrm{GL}(N, \mathbb{C})\big\}_{\iota_0,\iota_1}, \tag{B.2}$$

is a complex analytic 1-cocycle with values in $\mathrm{GL}(N, \mathbb{C})$ on $\mathcal{X}$. Such a complex analytic 1-cocycle allows us to preside over the construction of the holomorphic tangent bundle; see Example B.5.

---

[1] The terminology refers here to Čech cohomology (with respect to a covering $\mathcal{U}$) with values in the (multiplicative) sheaf of groups $\mathcal{F}_\bullet$ defined by $\mathcal{F}_\bullet(U) = \mathcal{O}(U, \mathrm{GL}(\bullet, \mathbb{C}))$; see the footnote in the statement of Theorem A.7 or also §D.2.3 for more details.

B.1.2. Linear algebra and construction of cocycles. Let E_1 and E_2 be two $\mathbb{C}$-vector spaces with dimension m_1, m_2, respectively. The $\mathbb{C}$-vector spaces E_1 and E_2 play naturally the role of elementary bricks to construct many other ones, such as E_1^* and E_2^*, $E_1 \oplus E_2$, $E_1 \otimes E_2$, $\mathrm{Hom}_{\mathbb{C}}(E_1, E_2) = E_2 \otimes E_1^*$, $\bigwedge^\ell E_1$ or $\bigwedge^\ell E_2$ for $\ell \in \mathbb{N}$, $\det E_1 = \bigwedge^{m_1} E_1$ or $\det E_2$, etc. Let us transpose such constructions in the world of cocycles.

- If $g = \{g_{\iota_1,\iota_0} : U_{\iota_0} \cap U_{\iota_1} \longrightarrow \mathrm{GL}(m, \mathbb{C})\}_{\iota_0,\iota_1}$ is a complex analytic 1-cocycle with values in $\mathrm{GL}(m, \mathbb{C})$ subordinate to the open covering $\mathfrak{U}$, the complex analytic 1-cocycle

$$\{g^*_{\iota_1,\iota_0} : z \in U_{\iota_0} \cap U_{\iota_1} \longmapsto ({}^t g_{\iota_1,\iota_0}(z))^{-1}\}, \tag{B.3}$$

  where ${}^t(\cdot)$ denotes the transposition, is also with values in $\mathrm{GL}(m, \mathbb{C})$ and is called the *dual cocycle* of g.
- When g^1 is a complex analytic 1-cocycle with values in $\mathrm{GL}(m_1, \mathbb{C})$ subordinate to an open covering $\mathcal{U}^1$ and g^2 is a complex analytic 1-cocycle with values in $\mathrm{GL}(m_2, \mathbb{C})$ subordinate to an open covering $\mathcal{U}^2$, then $g^1 \oplus g^2$ (respectively $g^1 \otimes g^2$) is the complex analytic 1-cocycle with values in $\mathrm{GL}(m_1 + m_2, \mathbb{C})$ (respectively $\mathrm{GL}(m_1 m_2, \mathbb{C})$) subordinate to the covering $\mathcal{U}^1 \cap \mathcal{U}^2 = \{U^1_{\iota^1} \cap U^2_{\iota^2}\}_{\iota^1,\iota^2} = \{U^{1,2}_\iota\}_\iota$ defined by

$$\begin{aligned} (g^1 \oplus g^2)_{\iota_0,\iota_1} &= \big(g^1_{\iota_0,\iota_1} \oplus g^2_{\iota_0,\iota_1}\big)_{|U^{1,2}_{\iota_0} \cap U^{1,2}_{\iota_1}} \\ \big(\text{resp., } (g^1 \otimes g^2)_{\iota_0,\iota_1} &= \big(g^1_{\iota_0,\iota_1} \otimes g^2_{\iota_0,\iota_1}\big)_{|U^{1,2}_{\iota_0} \cap U^{1,2}_{\iota_1}}\big), \end{aligned} \tag{B.4}$$

  where

$$G^1 \oplus G^2 = \begin{pmatrix} G^1 & 0 \\ 0 & G^2 \end{pmatrix},$$
$$(G^1 \otimes G^2)(\epsilon^1_{\nu_1} \otimes \epsilon^2_{\nu_2}) = G^1(\epsilon^1_{\nu_1}) \otimes G^2(\epsilon^2_{\nu_2}),$$

  when $G^j \in G(m_j, \mathbb{C})$ and $\epsilon^j = \{\epsilon^j_1, \dots, \epsilon^j_{m_j}\}$ denotes the canonical basis of $\mathbb{C}^{m_j}$. In particular, the complex analytic 1-cocycle $\mathrm{Hom}_{\mathbb{C}}(g^1, g^2)$ subordinate to the open covering $\mathcal{U}^{1,2}$ is defined as $g^2 \otimes (g^1)^*$ up to the canonical isomorphism $\mathrm{Hom}(\mathbb{C}^{m_1}, \mathbb{C}^{m_2}) \simeq \mathbb{C}^{m_2} \otimes (\mathbb{C}^{m_1})^*$.
- If g is a complex analytic 1-cocycle with values in $\mathrm{GL}(m, \mathbb{C})$ subordinate to the open covering $\mathcal{U}$ and $0 \le \ell \le m$, the complex analytic 1-cocycle $\bigwedge^\ell g$ with values in $\mathrm{GL}(\binom{m}{\ell}, \mathbb{C})$ subordinate to the same open covering is defined by

$$\Big(\bigwedge^{\ell} g_{\iota_1,\iota_0}\Big)(z)\,(\epsilon_{\nu_1} \wedge \cdots \wedge \epsilon_{\nu_\ell}) = \bigwedge_{j=1}^{\ell} \big(g_{\iota_1,\iota_0}(z)(\epsilon_{\nu_j})\big) \tag{B.5}$$

  for all $1 \le \nu_1 < \cdots < \nu_\ell \le m$, where $\epsilon = \{\epsilon_1, \dots, \epsilon_m\}$ is the canonical basis of $\mathbb{C}^m$, and $z \in U_{\iota_0} \cap U_{\iota_1}$. In particular, the complex analytic 1-cocycle $\bigwedge^m g$ with values in $\mathbb{C}^*$ subordinate to $\mathfrak{U}$ is denoted by $\det g$.

B.1.3. Holomorphic bundles over a complex analytic space. Let $\mathcal{X}$ be a (purely) N-dimensional complex analytic space, $m \in \mathbb{N}$, and let

$$\mathcal{E}_m = \{E_z : z \in \mathcal{X}, \quad \dim_{\mathbb{C}} E_z = m\} \tag{B.6}$$

be a collection of complex m-dimensional vector spaces indexed with $z \in \mathcal{X}$. The $\mathbb{C}$-vector space E_z labeled by z is called the *stalk* at the point z.

DEFINITION B.4. A holomorphic bundle $E \to \mathcal{X}$ with rank m over $\mathcal{X}$ is a structure (denoted by $(\mathbb{X}_{\mathcal{E}_m}, \mathcal{O}_{\mathbb{X}_{\mathcal{E}_m}})$) of $(N+m)$-purely dimensional complex analytic space on the disjoint union $\mathbb{X}_{\mathcal{E}_m} := \bigsqcup_{z\in\mathcal{X}}(\{z\}\times E_z)$ if $\mathcal{E}_m$ is as in (B.6), such that

- the projection map $\pi_{\mathcal{E}_m} : (z,\xi_z) \in \mathbb{X}_{\mathcal{E}_m} \to z \in \mathcal{X}$, where ξ_z denotes the current vector in E_z, is a morphism between complex analytic spaces;
- $\mathcal{X}$ admits an open covering $\{U_\iota\}_\iota$ such that for each index ι, there exists an analytic isomorphism of complex analytic spaces θ_ι between $\pi_{\mathcal{E}_m}^{-1}(U_\iota)$ and $U_\iota \times \mathbb{C}^m$ such that, for each $z \in U_\iota$,

$$\theta_\iota(z,\xi_z) = (z, \ell_{\iota,z}(\xi_z)) \in \{z\}\times\mathbb{C}^m \text{ with } \ell_{\iota,z} \in \mathrm{Isom}_{\mathbb{C}}(E_z,\mathbb{C}^m). \tag{B.7}$$

  Such an analytic isomorphism θ_ι between $\pi_{\mathcal{E}_m}^{-1}(U_\iota)$ and $U_\iota\times\mathbb{C}^m$ is called a *trivialization morphism* for the bundle $E\to\mathcal{X}$ over the open subset U_ι.

Holomorphic bundles with rank 1 over $\mathcal{X}$ are called *holomorphic line bundles.*

EXAMPLE B.5 (Holomorphic tangent and cotangent bundle over a complex manifold). Let $z \in \mathcal{X}$, where $\mathcal{X}$ is a complex N-dimensional manifold. Let $\mathbb{D}$ be the unit disc in $\mathbb{C}$. A centered complex disk γ at z on $\mathcal{X}$ is a holomorphic mapping γ from $\varepsilon\mathbb{D}$ (for some $\varepsilon > 0$ depending on γ) to $\mathcal{X}$ such that $\gamma(0) = z$. Two such discs $\gamma, \tilde\gamma$ are considered to be equivalent if $d(\zeta_z\circ\gamma)(0) = d(\zeta_z\circ\tilde\gamma)(0)$, where ζ_z is a local centered coordinate in a neighborhood of z (that is such that $\zeta_z(z)=0$), the choice of such a centered local coordinate in a neighborhood of z being irrelevant. The set of equivalence classes $\bar\gamma$ of centered complex discs at z modulo such an equivalence relation inherits a structure of N-dimensional $\mathbb{C}$-vector space and is called the *holomorphic tangent space*[2] at the point z. If $T^{(1,0)}_{\mathcal{X},z}$ denotes the *holomorphic* complex tangent space at $z \in \mathcal{X}$, then the family

$$\mathcal{E}_N = \{T^{1,0}_{\mathcal{X},z} \,:\, z\in\mathcal{X}\}$$

inherits a structure of the N-holomorphic bundle $T^{(1,0)}_{\mathcal{X}}$, which is called the *holomorphic tangent bundle* over $\mathcal{X}$. Moreover, for an atlas $\{(U_\iota,\zeta_\iota)\}$ of $\mathcal{X}$, the trivialization morphism θ_ι over U_ι is defined by

$$\forall\, z\in U_\iota,\ \forall \bar\gamma \in T^{1,0}_{\mathcal{X},z},\ \theta_\iota(z,\bar\gamma) = \big(z, (\zeta_\iota\circ\gamma)'(0)\big) \in U_\iota\times\mathbb{C}^N. \tag{B.8}$$

Given a holomorphic bundle $E\to\mathcal{X}$ with rank m, where $\mathcal{X}$ is a complex analytic space and $\{U_\iota\}_\iota$ is an open covering of $\mathcal{X}$ such that E admits a trivialization morphism θ_ι over U_ι, then the family $\{g_{\iota_1,\iota_0} : U_{\iota_0}\cap U_{\iota_1} \to \mathrm{GL}(m,\mathbb{C})\}_{\iota_0,\iota_1}$, defined by

$$\forall\, z\in U_{\iota_0}\cap U_{\iota_1},\ \big(\theta_{\iota_1}\circ\theta_{\iota_0}^{-1}\big)(z,\xi) = \big(z, (g_{\iota_1,\iota_0}(z))(\xi)\big) \tag{B.9}$$

is a complex analytic 1-cocycle on $\mathcal{X}$ with values in $\mathrm{GL}(m,\mathbb{C})$.

EXAMPLE B.6. When $\mathcal{X}$ is an N-dimensional complex manifold, the complex analytic 1-cocycle on $\mathcal{X}$ with values in $\mathrm{GL}(N,\mathbb{C})$ attached to the holomorphic tangent bundle (see Example B.5) has been described in Example B.3.

---

[2]The holomorphic tangent space at z admits two alternative presentations besides the geometric one we propose here: it is isomorphic in a canonical way to the $\mathbb{C}$-vector space of $\mathbb{C}$-derivations of the local ring $\mathcal{O}_{\mathcal{X},z}$, that is $\mathbb{C}$-linear maps $D:\mathcal{O}_{\mathcal{X},z}\to\mathbb{C}$ such that $D(fg) = f(z)\,Dg + g(z)\,Df$; it is also isomorphic in a canonical way to the $\mathbb{C}$-dual of the $\mathbb{C}$-vector space $\mathfrak{M}_{\mathcal{X},z}/\mathfrak{M}^2_{\mathcal{X},z}$, where $\mathfrak{M}_{\mathcal{X},z}$ is the maximal ideal in $\mathcal{O}_{\mathcal{X},z}$.

B.1.4. Isomorphism classes of holomorphic bundles.

DEFINITION B.7. Let $E_1 \to \mathcal{X}$ and $E_2 \to \mathcal{X}$ be two holomorphic bundles with respective ranks m_1 and m_2 over a complex analytic space $\mathcal{X}$. Assume that they are attached to $\mathcal{E}_{j,m_j} = \{E_{j,z} : z \in \mathcal{X}\}$ with $\dim E_{j,z} = m_j$ for $j = 1, 2$. A morphism of holomorphic bundles over $\mathcal{X}$ from $E_1 \to \mathcal{X}$ to $E_2 \to \mathcal{X}$ is an analytic morphism between complex analytic spaces $F : \mathbb{X}_{\mathcal{E}_1} \to \mathbb{X}_{\mathcal{E}_2}$ (with complex dimensions $\dim \mathcal{X} + m_1$ and $\dim \mathcal{X} + m_2$, respectively) satisfying

$$\text{(B.10)}\quad \forall z \in \mathcal{X},\ \forall \xi_{1,z} \in E_{1,z},$$
$$F(z, \xi_{1,z}) = (\pi_{\mathcal{E}_2}(F(z, 0)), L_z^{1,2}(\xi_{1,z})) \text{ with } L_z^{1,2} \in \mathrm{Hom}_{\mathbb{C}}\big(E_{1,z}, E_{2,\pi_{\mathcal{E}_2}(F(z,0))}\big).$$

Two holomorphic bundles $E_1 \to \mathcal{X}$ and $E_2 \to \mathcal{X}$ with the same rank are said to be isomorphic as holomorphic bundles if there are morphisms of holomorphic bundles $F^{1,2} : \mathbb{X}_{\mathcal{E}_1} \to \mathbb{X}_{\mathcal{E}_2}$ and $F^{2,1} : \mathbb{X}_{\mathcal{E}_2} \to \mathbb{X}_{\mathcal{E}_1}$ satisfying

$$F^{2,1} \circ F^{1,2} = \mathrm{Id}_{\mathbb{X}_{\mathcal{E}_1}}, \quad F^{1,2} \circ F^{2,1} = \mathrm{Id}_{\mathbb{X}_{\mathcal{E}_2}}.$$

As seen in §B.1.3, one can associate to any holomorphic bundle with rank m over an $\mathcal{X}$ a complex analytic 1-cocycle with values in $\mathrm{GL}(m, \mathbb{C})$. Conversely, given a complex analytic 1-cocycle $g = \{g_{\iota_1,\iota_0}\}_{\iota_0,\iota_1}$ with values in $\mathrm{GL}(m, \mathbb{C})$ subordinate to the open covering $\mathcal{U}$, one can realize a holomorphic bundle with rank m as follows. Consider the disjoint union

$$\mathbb{X}(\mathcal{U}) = \bigsqcup_{\iota} (U_\iota \times \mathbb{C}^m).$$

Then, whenever $U_{\iota_0} \cap U_{\iota_1} \neq \emptyset$, an element $(z_{\iota_0}, \xi_{\iota_0}) \in U_{\iota_0} \times \mathbb{C}^m$ is identified with the element $(z_{\iota_1}, \xi_{\iota_1}) \in U_{\iota_1} \times \mathbb{C}^m$ if and only if $z_{\iota_0} = z_{\iota_1} = z$ as elements in $\mathcal{X}$ and one has that $\xi_{\iota_1} = (g_{\iota_1,\iota_0}(z))\,(\xi_{\iota_0})$. The quotient of $\mathbb{X}(\mathcal{U})$ modulo such identifications provides a structure of holomorphic m-bundle $E_g \to \mathcal{X}$.

Before stating Proposition B.8, one needs to formulate an identification rule between complex analytic 1-cocycles with values in $\mathrm{GL}(m, \mathbb{C})$; see, e.g., [**Colon**, Definition 3.4.3]. Any such 1-cocycle g (with values in $\mathrm{GL}(m, \mathbb{C})$) consists in a family $\mathcal{U}(g)$ of open subsets $\{U_\iota\}_\iota$, together with a family of morphisms $\mathcal{F}(g) : \{g_{\iota_1,\iota_0} : U_{\iota_0} \cap U_{\iota_1} \to \mathrm{GL}(m, \mathbb{C})\}_{\iota_0,\iota_1}$. Two such cocycles g and $\tilde{g}$ are then identified[3] if and only if there exists a complex analytic 1-cocycle G (with valued in $\mathrm{GL}(m, \mathbb{C})$) such that $\mathcal{E}(g) \cup \mathcal{E}(\tilde{g}) \subset \mathcal{E}(G)$ and $\mathcal{F}(g) \cup \mathcal{F}(\tilde{g}) \subset \mathcal{F}(G)$.

PROPOSITION B.8. *There is a one-to-one correspondence between complex analytic* 1*-cocycles with values in* $\mathrm{GL}(m, \mathbb{C})$ *on* $\mathcal{X}$ *(with identifications as above) and equivalence classes of holomorphic m-bundles modulo isomorphisms between holomorphic bundles.*

EXAMPLE B.9. Let $\mathcal{X} = \mathbb{P}^N_{\mathbb{C}}$ and $d \in \mathbb{Z}$. The complex analytic 1-cocycle with values in $\mathbb{C}^*$ defined by $g([z_0 : \cdots : z_N]) = (z_0/z_k)^d$ for each $z = [z_0 : \cdots : z_N]$, such that $z_k \neq 0$, defines a class of isomorphic line bundles denoted by $\mathcal{O}(d)$. The class $\mathcal{O}(-1)$ admits as representing the tautological bundle corresponding to the family of $\mathbb{C}$-vector spaces $\mathcal{E} = \{(z_0, \ldots, z_N)\,\mathbb{C} : [z_0 : \cdots : z_N] \in \mathbb{P}^N_{\mathbb{C}}\}$ organized in such a way that it realizes a holomorphic line bundle over $\mathbb{P}^N_{\mathbb{C}}$.

[3]Observe that in this formulation, redundancy in the sets $\mathcal{E}(G)$, and therefore $\mathcal{F}(G)$, occurs in general in the definition of complex analytic 1-cocycles with values in $\mathrm{GL}(m, \mathbb{C})$.

Equivalence classes of line bundles $\bar{L}$ then form a multiplicative group through the tensorial multiplication of complex analytic 1-cocycles with values in $\mathbb{C}^*$.

DEFINITION B.10. The group of equivalence classes of holomorphic 1-bundles over a complex analytic space $\mathcal{X}$ is called the *Picard group of* $\mathcal{X}$ and is denoted by $\mathrm{Pic}(\mathcal{X})$. It is isomorphic to the quotient group of Cartier divisors modulo the subgroup of principal Cartier divisors, provided identifications between various representations of the same Cartier divisor are performed; see Example B.2.

EXAMPLE B.11. The Picard group of the projective space $\mathbb{P}^N_{\mathbb{C}}$ is isomorphic to $\mathbb{Z}$. Any isomorphism class of holomorphic 1-bundles equals $\mathcal{O}_{\mathbb{P}^n_{\mathbb{C}}}(d)$ for some unique $d \in \mathbb{Z}$ and $\bar{L} \in (\mathrm{Pic}(\mathbb{P}^N_{\mathbb{C}}), \otimes) \longmapsto d(\bar{L}) \in (\mathbb{Z}, +)$ realizes such an isomorphism.

Such construction suggests the *projectivization* of a holomorphic m-bundle.

DEFINITION B.12. Let $E \to \mathcal{X}$ be holomorphic with rank m over a complex analytic space $\mathcal{X}$ and let g be a complex analytic 1-cocycle with values in $\mathrm{GL}(m, \mathbb{C})$ attached to E and subordinate to an open covering $\mathcal{U}$ of $\mathcal{X}$. The *projectivization* $\mathrm{Proj}\, E$ of E is the $(\dim \mathcal{X} + m - 1)$-dimensional complex analytic space defined (up to isomorphism of complex analytic spaces) as follows. One considers the disjoint union $\bigsqcup_\iota (U_\iota \times \mathbb{P}^{m-1}_{\mathbb{C}})$ in which two points $(z_{\iota_0}, [\xi_{\iota_0}]) \in U_{\iota_0} \times \mathbb{P}^{m-1}_{\mathbb{C}}$ and $(z_{\iota_1}, [\xi_{\iota_1}]) \in U_{\iota_1} \times \mathbb{P}^{m-1}_{\mathbb{C}}$ are identified whenever $U_{\iota_0} \cap U_{\iota_1} \neq \emptyset$ and z_{ι_0}, z_{ι_1} represent the same point $z = z_{\iota_0} = z_{\iota_1}$ of $\mathcal{X}$ in the nonempty intersection $U_{\iota_0} \cap U_{\iota_1}$, and $\xi_{\iota_1} = g_{\iota_1, \iota_0}(z)\,(\xi_{\iota_0})$ in the open subset $\mathbb{C}^m \setminus \{(0, \dots, 0)\}$. Such a construction comes with a projection $\pi\ : \mathrm{Proj}\, E \to \mathcal{X}$.

All operations on complex analytic 1-cocycles described in §B.1.2 induce operations on holomorphic bundles (up to isomorphism) because of Proposition B.8. Given two holomorphic bundles $E_1 \to \mathcal{X}$ and $E_2 \to \mathcal{X}$, with respective ranks m_1 and m_2, one may then define the holomorphic dual bundle $E_1^* \to \mathcal{X}$ (with rank m_1), the holomorphic bundle $E_1 \oplus E_2$ with rank $m_1 + m_2$, the holomorphic bundles $E_1 \otimes E_2$ and $\mathrm{Hom}_{\mathbb{C}}(E_1, E_2) \simeq E_2 \otimes E_1^*$, both with rank $m_1 m_2$, as well as, for $\ell \in \mathbb{N}$, the holomorphic bundle $\bigwedge^\ell E_1$ with rank $\binom{m_1}{\ell}$.

EXAMPLE B.13 (Holomorphic cotangent bundle over a complex manifold). Let $\mathcal{X}$ be an N-dimensional complex manifold. The dual bundle of the holomorphic tangent bundle (see Example B.5) is a holomorphic bundle with rank N which is called the *holomorphic cotangent bundle* and denoted by $T_{\mathcal{X}}^{(1,0)*}$. The line bundle

$$\mathrm{can}(\mathcal{X}) = \det T_{\mathcal{X}}^{(1,0)*} = \bigwedge^N T_{\mathcal{X}}^{(1,0)*} \tag{B.11}$$

is called the *canonical bundle* of $\mathcal{X}$. In the particular case where $\mathcal{X} = \mathbb{P}^N_{\mathbb{C}}$, one has $\mathrm{can}(\mathbb{P}^N_{\mathbb{C}}) \simeq \mathcal{O}_{\mathbb{P}^N_{\mathbb{C}}}(-N-1)$.

B.1.5. Local holomorphic frames and sheaves of holomorphic sections.

DEFINITION B.14. Let $E \to \mathcal{X}$ be a holomorphic bundle with rank m over a complex analytic space and let U be an open subset of $\mathcal{X}$. A morphism s of complex analytic spaces from U to E such that $s(z) \in E_z$ for any $z \in U$ is called a *holomorphic section of E over U*.

Suppose in particular that $U = U_\iota$ is an open subset of $\mathcal{X}$ over which E admits a trivialization morphism $\theta_\iota = \theta : \pi^{-1}(U) \longleftrightarrow U \times \mathbb{C}^m$. Let $\{\epsilon_1, \dots, \epsilon_m\}$ be the canonical basis of $\mathbb{C}^m$.

DEFINITION B.15. The family of holomorphic sections

$$e_U = \{z \in U \mapsto \theta^{-1}(z, \epsilon_j) \, : \, j = 1, \dots, m\}$$

of E over U is said to be a *local holomorphic frame* for E in the open subset U over which E admits a trivialization morphism.

Given a complex analytic space $(\mathcal{X}, \mathcal{O}_\mathcal{X})$ and a holomorphic bundle $E \to \mathcal{X}$ with rank m, the $\mathbb{C}$-vector space $\mathcal{O}_\mathcal{X}(U, E)$ of holomorphic sections of E over U is a $\mathcal{O}_\mathcal{X}(U)$-module and $\{\mathcal{O}_\mathcal{X}(U, E) \, : \, U \subset \mathcal{X}\}$ inherits the structure of a locally free sheaf of $\mathcal{O}_\mathcal{X}$-modules with constant rank m on the complex analytic space $\mathcal{X}$.

B.1.6. The concepts of local versus global complete intersection. Given a holomorphic section s of a holomorphic bundle with rank m over a (purely) N-dimensional complex analytic space $\mathcal{X}$, the subset $s^{-1}(\{0\})$ is a closed analytic subset of $\mathcal{X}$. The concept of complete intersection (globally or locally) generalizes in a higher codimension than that of a (globally or locally defined) closed analytic hypersurface in the reduced complex analytic space $\mathcal{X}$.

DEFINITION B.16. Let $1 \le m \le N = \dim \mathcal{X}$. A closed analytic subset $V \subset \mathcal{X}$ with pure codimension m is said to be *globally defined as a complete intersection in* $\mathcal{X}$ if and only one can find a holomorphic bundle $E \to \mathcal{X}$ with rank m, together with a holomorphic global section s of E over $\mathcal{X}$, such that $V = s^{-1}(\{0\})$. It is said to be *locally defined as a complete intersection in* $\mathcal{X}$ if and only if, for any point $z \in V$, one can find $f_{z,1}, \dots, f_{z,m} \in \mathcal{O}_{\mathcal{X},z}$ such that the germ V_z of V at z can be realized as $V_z = \bigcap_{j=1}^m f_{j,z}^{-1}(\{0\})$.

A sequence $(a_1, \dots, a_m)$ in a commutative ring $\mathbb{A}$ is said to be *regular* in $\mathbb{A}$ if and only if $a_1 \neq 0$ and, for any $j = 1, \dots, m-1$, a_{j+1} is not a zero divisor in the quotient ring $\mathbb{A}/(a_1, \dots, a_j)$. If $1 \le m \le N$ and $f_{1,z}, \dots, f_{m,z}$ belong to the maximal ideal of the local ring $\mathcal{O}_{\mathcal{X},z}$, it is equivalent to saying that $(f_{1,z}, \dots, f_{m,z})$ is a regular sequence in $\mathcal{O}_{\mathcal{X},z}$ and that $\dim \bigcap_{j=1}^m f_{z,j}^{-1}(\{0\}) = N - m$. Therefore a closed analytic subset $V \subset \mathcal{X}$ with (pure) codimension m is locally defined in $\mathcal{X}$ as a complete intersection if and only if, for any $z \in V$, one can find a regular sequence of noninvertible elements $(f_{1,z}, \dots, f_{m,z})$ such that $V_z = \bigcap_{j=1}^m f_{j,z}^{-1}(\{0\})$. But one should be careful that the notions of complete intersection and regularity are not connected anymore in the global context; see Remark B.17.

REMARK B.17. The fact that a closed analytic subset V with codimension m in a complex analytic space $\mathcal{X}$ with dimension N is globally defined as a complete intersection, that is, as the zero set of a holomorphic section s of a m-holomorphic vector bundle $E \to \mathcal{X}$, does not imply that $(\sigma_{1,z}, \dots, \sigma_{m,z})$ (where $s = \sum_1^m \sigma_j \otimes \epsilon_j$ when expressed in a local holomorphic frame in a neighborhood of $z \in \mathcal{X}$), is a regular sequence in $\mathcal{O}_{\mathcal{X},z}$, unless $z \in s^{-1}(0)$. Take E for example as the trivial bundle over $\mathbb{C}^3$ and $V = s^{-1}(0) = \{(0,0,0)\}$, where

$$s(z) = \big(z_1(z_3+1), z_2(z_3+1), z_3\big).$$

The sequence $(s_{1,z}, s_{2,z}, s_{3,z})$ is not regular (in this order) in a neighborhood of any point in the complex plane $\{z_3 + 1 = 0\}$.

EXAMPLE B.18. Any closed analytic subset in $\mathbb{P}^N_{\mathbb{C}}$ is algebraic projective, which is defined as the zero set of a finite number of homogeneous polynomials in $z_0, \dots, z_N$; see Theorem A.28 in §A.3.6. If $P_1, \dots, P_m$ are m homogeneous polynomials in $z_0, \dots, z_N$ with respective total degrees $d_1, \dots, d_m$, then the section $(P_1, \dots, P_m)$ of the vector bundle $\mathcal{O}_{\mathbb{P}^N_{\mathbb{C}}}(d_1) \oplus \cdots \oplus \mathcal{O}_{\mathbb{P}^N_{\mathbb{C}}}(d_m)$ (with rank m) defines a globally complete intersection in $\mathbb{P}^N_{\mathbb{C}}$ if and only if $\{z \in \mathbb{P}^N_{\mathbb{C}} : P_1(z) = \cdots = P_m(z) = 0\}$ is purely $(N-m)$-dimensional.

We conclude this section with a heuristic remark. It is easier to describe the notion of a complete intersection in geometric terms than in algebraic ones, as the following equivalence shows. If $f_1, \dots, f_m$ are m holomorphic functions in an open pseudo-convex domain $\mathcal{X}$ of $\mathbb{C}^N$ (or more generally, m holomorphic functions over a *Stein manifold* $\mathcal{X}$, see §A.2.2), then $f_1, \dots, f_m$ either define an $(N-m)$-dimensional globally complete intersection in $\mathcal{X}$ or have no common zero in $\mathcal{X}$ if and only if, for every $\nu \in \mathbb{N}^*$, for every homogeneous relation

$$\sum_{\kappa_1 + \cdots + \kappa_m = \nu} a_\kappa \, f_1^{\kappa_1} f_2^{\kappa_2} \cdots f_m^{\kappa_m} \in \Big((f_1, \dots, f_m)\, \mathcal{O}_{\mathcal{X}}(\mathcal{X}) \Big)^{\nu+1},$$

where $a_\kappa \in \mathcal{O}_{\mathcal{X}}(\mathcal{X})$ are holomorphic functions in $\mathcal{X}$, one has that

$$\forall \kappa \in \mathbb{N}^m, \ a_\kappa \in (f_1, \dots, f_m)\, \mathcal{O}_{\mathcal{X}}(\mathcal{X}).$$

Thus one can see that the rather simple geometric concept of complete intersection (here global) corresponds, expressed in algebraic terms, to a more involved asymptotic characterization ("for every $\nu \in \mathbb{N}^*$, for every relation", etc.). Compare also the two approaches (geometric or algebraic) to the notion of a gap sheaf, where the same phenomenon occurs (see §D.1.1, in particular (D.4) and (D.5) where the cutting-out operation is introduced on complex analytic cycles): a simple geometric formulation on one side and an asymptotic more-involved algebraic formulation on the other side.

B.1.7. The bundles $T^{*(p,q)}_{\mathcal{X}}$ over a complex manifold $\mathcal{X}$.

B.1.7.1. *Complex bundles over $\mathcal{X}_{\mathbb{R}}$ and their isomorphism classes.* Let $\mathcal{X}$ be an N-dimensional complex manifold with underlying $2N$-dimensional real manifold $\mathcal{X}_{\mathbb{R}}$; see §A.1.1.

DEFINITION B.19. A $\mathbb{C}$-bundle $E \to \mathcal{X}$ with rank $m \in \mathbb{N}$ over $\mathcal{X}_{\mathbb{R}}$ is a structure (denoted by $(\mathbb{X}_{\mathcal{E}_m,\mathbb{R}}, C^\infty_{\mathbb{X}_{\mathcal{E}_m,\mathbb{R}}})$) of $2(N+m)$-dimensional differentiable manifold on the disjoint union $\mathbb{X}_{\mathcal{E}_m} := \bigsqcup_{z \in \mathcal{X}} \{z\} \times E_z$, where $\mathcal{E}_m$ is as in (B.6), such that

- the projection map $\boldsymbol{\pi}_{\mathcal{E}_m} : (z, \xi_z) \in \mathbb{X}_{\mathcal{E}_m} \to z \in \mathcal{X}_{\mathbb{R}}$ is a morphism between differentiable manifolds;
- $\mathcal{X}$ admits an open covering $\{U_\iota\}_\iota$ such that for each index ι, there exists an isomorphism of differential manifolds $\boldsymbol{\theta}_\iota$ between the differentiable manifolds $\boldsymbol{\pi}^{-1}((U_\iota)_{\mathbb{R}})$ and $(U_\iota \times \mathbb{C}^m)_{\mathbb{R}}$ such that, for each $z \in U_\iota$,

(B.12) $$\boldsymbol{\theta}_\iota(z, \xi_z) = (z, \boldsymbol{\ell}_{\iota,z}(\xi_z)) \in \{z\} \times \mathbb{C}^m \text{ with } \boldsymbol{\ell}_{\iota,z} \in \mathrm{Isom}_{\mathbb{C}}(E_z, \mathbb{C}^m).$$

  Such an isomorphism $\boldsymbol{\theta}_\iota$ of differentiable manifolds between $\boldsymbol{\pi}^{-1}_{\mathcal{E}_m}((U_\iota)_{\mathbb{R}})$ and $(U_\iota \times \mathbb{C}^m)_{\mathbb{R}}$ is called a *trivialization morphism* for the bundle $E \to \mathcal{X}$ over the open subset U_ι.

The notion of a *local frame* extends that of a *holomorphic local frame* already introduced in Definition B.15.

DEFINITION B.20. Assume that $U = U_\iota$ is an open subset of $\mathcal{X}$ over which E admits a trivialization morphism $\boldsymbol{\theta}_\iota = \theta \,:\, \pi^{-1}(U_\mathbb{R}) \longleftrightarrow U_\mathbb{R} \times \mathbb{C}^m$. Let $\{\epsilon_1, \dots, \epsilon_m\}$ be the canonical basis of $\mathbb{C}^m$. Then the family of sections

$$\{z \in U \mapsto e_j(z) = \boldsymbol{\theta}^{-1}(z, \epsilon_j) \,:\, j = 1, \dots, m\}$$

of E over U is said to be a *local frame* for E in the open subset U.

The concept of a complex bundle with rank m over the differentiable underlying manifold $\mathcal{X}_\mathbb{R}$ to a complex manifold $\mathcal{X}$ is closely linked with that of 1-complex differentiable cocycle with values in $\mathrm{GL}(m, \mathbb{C})$ over $\mathcal{X}_\mathbb{R}$.

DEFINITION B.21. Let $m \in \mathbb{N}^*$ and $\mathcal{U} = \{U_\iota\}_\iota$ be an open covering of $\mathcal{X}$. A complex differentiable 1-cocycle on $\mathcal{X}_\mathbb{R}$ with values in $\mathrm{GL}(m, \mathbb{C})$ subordinate to the covering $\mathcal{U}$ is a collection of matrix valued maps $\{g_{\iota_1,\iota_0}\}_{\iota_0,\iota_1}$, where g_{ι_1,ι_0} is a morphism of differential manifolds between $(U_{\iota_0} \cap U_{\iota_1})_\mathbb{R}$ and $(\mathrm{GL}(m, \mathbb{C}))_\mathbb{R}$ such that the *cocycle conditions*

$$\forall\, z \in U_{\iota_0} \cap U_{\iota_1} \cap U_{\iota_2},\ g_{\iota_2,\iota_0}(z) = g_{\iota_2,\iota_1}(z) \circ g_{\iota_1,\iota_0}(z),$$

hold for any triplet of indices $\iota_0, \iota_1, \iota_2$.

In a similar way to Definition B.7 (see §B.1.4) one may introduce a notion of *isomorphism between complex bundles* with the same rank over $\mathcal{X}_\mathbb{R}$, in place of that of an isomorphism between holomorphic bundles over a complex analytic space $\mathcal{X}$. In the same vein as in Proposition B.8 (see §B.1.4) there is a one-to-one correspondence between complex differentiable 1-cocycles on $\mathcal{X}_\mathbb{R}$ with values in $\mathrm{GL}(m, \mathbb{C})$ and isomorphism classes of complex bundles with rank m over $\mathcal{X}_\mathbb{R}$. Operations on complex differentiable 1-cocycles over $\mathcal{X}_\mathbb{R}$, such as described in §B.1.2, induce in this new context the realization the the correspondence of isomorphism classes of complex bundles over $\mathcal{X}_\mathbb{R}$. Observe that all such operations provide as output entries holomorphic bundles as soon as the input entries are holomorphic bundles; see §B.1.4.

B.1.7.2. *The $\mathbb{C}$-bundles $T_\mathcal{X}^{*(1,0)}$ and $T_\mathcal{X}^{*(0,1)}$ over $\mathcal{X}_\mathbb{R}$ and their exterior powers.* Let $\mathcal{X}$ be an N-dimensional complex manifold and let $\mathcal{X}_\mathbb{R}$ be its $2N$-dimensional underlying manifold. Let $\{(U_\iota, \zeta_\iota)\}_\iota$ be an atlas for $\mathcal{X}$ and $\tau_\iota \,:\, U_\iota \longleftrightarrow \boldsymbol{U}_{\iota,\mathbb{R}}$ as in §A.1.1. Then

$$\big\{\boldsymbol{g}_{\iota_1,\iota_0} \,:\, z \in U_{\iota_0} \cap U_{\iota_1} \longmapsto d(\tau_{\iota_1} \circ \tau_{\iota_0}^{-1})\big) \in \mathrm{GL}(2N, \mathbb{R})\big\}_{\iota_0,\iota_1}$$

is a C^∞ (even real analytic) 1-cocycle on the $2N$-dimensional underlying real manifold $\mathcal{X}_\mathbb{R}$, with values in $\mathrm{GL}(2N, \mathbb{R})$. For each $z \in \mathcal{X}$, let ζ_z be a local centered system of coordinates $\zeta_z : U \longleftrightarrow \boldsymbol{U}$ in a neighborhood of z and $\tau_z \,:\, U \longleftrightarrow \boldsymbol{U}_\mathbb{R}$ be the corresponding map from U to $\mathbb{R}^{2N}$ as in §A.1.1. The tangent real vector space $T_{\mathcal{X}_\mathbb{R},z}$ is defined (independent of the choice of ζ_z, hence of τ_z) as the set of equivalence classes $\dot{\gamma}_\mathbb{R}$ of germs of C^∞ maps $\gamma_\mathbb{R} :]-\varepsilon, \varepsilon[\to \mathcal{X}_\mathbb{R}$ such that $\gamma_\mathbb{R}(0) = z$ modulo the identification between $\gamma_\mathbb{R}$ and $\tilde{\gamma}_\mathbb{R}$ as soon as $d(\tau_z \circ \gamma_\mathbb{R})(0) = d(\tau_z \circ \tilde{\gamma}_\mathbb{R})(0)$. It inherits a structure of $2N$-dimensional $\mathbb{R}$-vector space. Let

$$\mathbb{C} \otimes_\mathbb{R} T_{\mathcal{X}_\mathbb{R},z} = T_{\mathcal{X}_\mathbb{R},z} \oplus i\, T_{\mathcal{X}_\mathbb{R},z} = T_{\mathcal{X},z} \oplus \overline{T_{\mathcal{X},z}} := T_{\mathcal{X},z}^{1,0} \oplus T_{\mathcal{X},z}^{0,1},$$

where $T_{\mathcal{X},z} = T_{\mathcal{X},z}^{1,0}$ is the holomorphic complex tangent space at z; see Example B.5. If $\partial/\partial\zeta_{z,1}, \dots, \partial/\partial\zeta_{z,N}$ is a holomorphic local frame for the holomorphic tangent complex bundle in a neighborhood of z, one has

$$T^{0,1}_{\mathcal{X},z} = \sum_{k=1}^{N} \mathbb{C}\, \frac{\partial}{\partial \bar{\zeta}_{k,z}}(z).$$

Both the holomorphic bundle (with rank N) $T^{1,0}_{\mathcal{X}}$ and its conjugate $T^{0,1}_{\mathcal{X}}$ can be considered as complex bundles over $\mathcal{X}_{\mathbb{R}}$, according to Definition B.19 in §B.1.7.1. Such are their duals, which we denote respectively as[4]

$$T^{*(1,0)}_{\mathcal{X}} = (T^{1,0}_{\mathcal{X}_{\mathbb{R}}})^* = (T^{1,0}_{\mathcal{X}})^*, \quad T^{*(0,1)}_{\mathcal{X}_{\mathbb{R}}} = (T^{0,1}_{\mathcal{X}_{\mathbb{R}}})^*.$$

DEFINITION B.22. Let E be a complex bundle with rank $m \in \mathbb{N}$ over $\mathcal{X}_{\mathbb{R}}$.[5] For any $0 \leq p, q \leq N$, the one-to-one correspondence between complex bundles and complex differentiable 1-cocycles on $\mathcal{X}_{\mathbb{R}}$ with values in $\mathrm{GL}(\bullet, \mathbb{C})$ allows us to define (up to isomorphisms between complex bundles over $\mathcal{X}_{\mathbb{R}}$ with the same rank, see §B.1.7.1), the complex bundle over $\mathcal{X}_{\mathbb{R}}$:

$$T^{*(p,q)}_{\mathcal{X}_{\mathbb{R}}} \otimes E = \bigwedge^{p} T^{*(1,0)}_{\mathcal{X}} \otimes \bigwedge^{q} T^{*(0,1)}_{\mathcal{X}_{\mathbb{R}}} \otimes E, \tag{B.13}$$

$$\text{with } \operatorname{rank}(T^{*(p,q)}_{\mathcal{X}_{\mathbb{R}}} \otimes E) = m \binom{N}{p}\binom{N}{q}.$$

B.2. Sheaves of bundle-valued differential forms or currents

B.2.1. Differential forms and currents on a complex manifold. In this subsection $\mathcal{X}$ is an N-dimensional complex manifold with underlying differentiable manifold $\mathcal{X}_{\mathbb{R}}$.

B.2.1.1. *Sheaves of complex bundle-valued forms on a complex manifold.*

DEFINITION B.23. Let E be a complex bundle with rank m over $\mathcal{X}_{\mathbb{R}}$ and let $0 \leq p, q \leq N$, $0 \leq r \leq 2N$. Smooth sections of the bundle $T^{*(p,q)}_{\mathcal{X}_{\mathbb{R}}} \otimes E$ are called E-valued differential (p,q)*-forms on* $\mathcal{X}_{\mathbb{R}}$, or *complex valued differential* (p,q)*-forms on* $\mathcal{X}_{\mathbb{R}}$ in case E is the trivial (holomorphic) bundle $\mathcal{X} \times \mathbb{C}$. For any open subset $U \subset \mathcal{X}$, one denotes as

$$(C^\infty_{p,q})_{\mathcal{X}}(U,E) := (C^\infty_{p,q})_{\mathcal{X}_{\mathbb{R}}}(U,E), \qquad \mathcal{D}^{(p,q)}_{\mathcal{X}}(U,E) := \mathcal{D}^{(p,q)}_{\mathcal{X}_{\mathbb{R}}}(U,E),$$

$$(C^\infty_{r})_{\mathcal{X}}(U,E) := \sum_{\substack{p+q=r \\ \max(p,q)\leq N}} (C^\infty_{p,q})_{\mathcal{X}}(U,E), \quad \mathcal{D}^{r}_{\mathcal{X}}(U,E) := \sum_{\substack{p+q=r \\ \max(p,q)\leq N}} \mathcal{D}^{(p,q)}_{\mathcal{X}}(U,E),$$

respectively, the $\mathbb{C}$-vector space of E-valued (p,q)-differential forms in U, its subspace of E-valued (p,q)-differential forms with compact support in U, the $\mathbb{C}$-vector space of E-valued r-differential forms in U, its subspace of E-valued r-differential forms with compact support in U. We will denote shortly the corresponding sheaves by $C^\infty_{p,q}(\cdot,E)$, $\mathcal{D}^{(p,q)}(\cdot,E)$, $C^\infty_r(\cdot,E)$, $\mathcal{D}^r(\cdot,E)$ and by $C^\infty_{p,q}$, $\mathcal{D}^{(p,q)}$, C^∞_r, $\mathcal{D}^r$ in the particular case where $E = \mathbb{C}$.

---

[4]Since the holomorphic cotangent complex bundle is holomorphic, we keep $\mathcal{X}$ instead of $\mathcal{X}_{\mathbb{R}}$ as an index in the notation.

[5]It could be in particular a holomorphic bundle with rank m over $\mathcal{X}$.

If $e = \{e_{z,1},\dots,e_{z,m}\}$ is a local frame[6] for E in a neighborhood of a trivialization open subset U, see Definition B.20, any $\varphi \in C^\infty_{p,q}(U,E)$ expresses in local coordinates $(\zeta_1,\dots,\zeta_N)$ in a neighborhood of $z \in U$ as

(B.14)
$$\varphi = \sum_{j=1}^{m}\Big(\sum_{1<i_1<\cdots<i_p\leq N}\ \sum_{1<j_1<\cdots<j_q\leq N} \varphi^j_{I,J}\Big(\bigwedge_{\kappa=1}^{p} d\zeta_{i_\kappa}\Big)\wedge\Big(\bigwedge_{\kappa=1}^{q} d\overline{\zeta}_{j_\kappa}\Big)\Big)\otimes e_{z,j}$$
$$= \sum_{j=1}^{m}\Big(\sum_{\#I=p,\ \#J=q} \varphi^j_{I,J}\, d\zeta_I \wedge d\overline{\zeta}_J\Big)\otimes e_{z,j} = \sum_{k=1}^{m}\varphi_j \otimes e_{z,j},$$

where the coordinate functions $\varphi^j_{I,J}$ are smooth functions and $\{d\zeta_1,\dots,d\zeta_N\}$ is the dual frame of the local holomorphic frame $\{(\partial/\partial\zeta_1)_z,\dots,(\partial/\partial\zeta_N)_z\}$ for the holomorphic tangent bundle $T^{1,0}_{\mathcal{X}}$ at z; see Example B.5. The $\mathbb{R}$-linear operators ∂, $\overline{\partial}$ (*Dolbeault* operator), $d = \partial + \overline{\partial}$ (*de Rham* operator), $dd^c = (i/2\pi)\,\partial\overline{\partial}$ (*Bott–Chern* operator[7]) operate along the well-known classical rules of differential geometric calculus in $\mathbb{C}^N \simeq \mathbb{R}^{2N}$, which we assume the reader is acquainted with, hence we will not repeat it here. We have

(B.15)
$$\begin{aligned}
\partial : C^\infty_{p,\bullet} &\longrightarrow C^\infty_{p+1,\bullet} \quad (0 \text{ if } p = N),\\
\overline{\partial} : C^\infty_{\bullet,q} &\longrightarrow C^\infty_{\bullet,q+1} \quad (0 \text{ if } q = N),\\
d : C^\infty_{r} &\longrightarrow C^\infty_{r+1} \quad (r = 0,\dots,2N,\ 0 \text{ if } r = N),\\
dd^c : C^\infty_{p,q} &\longrightarrow C^\infty_{p+1,q+1} \quad (0 \text{ if } p \text{ or } q = N).
\end{aligned}$$

One has, with respect to such actions, $d\circ d = 0$, hence $\partial\circ\partial = 0$, $\overline{\partial}\circ\overline{\partial} = 0$ and $\partial\circ\overline{\partial} = -\overline{\partial}\circ\partial$.

B.2.1.2. *Connection on a complex bundle over $\mathcal{X}_{\mathbb{R}}$ and its curvature tensor.* It follows from the existence of smooth partitions of unity on the differentiable manifold $\mathcal{X}_{\mathbb{R}}$ that the $\mathbb{C}$-vector space $\boldsymbol{C}^\infty(\mathcal{X},E) := \bigoplus_{r\in\mathbb{N}} C^\infty_r(\mathcal{X},E)$ is nontrivial. The notion of *connection on a complex bundle* E over the differentiable manifold $\mathcal{X}_{\mathbb{R}}$ allows us to perform differential calculus on $\mathbb{C}$-spaces of E-valued differential forms instead of $\mathbb{C}$-valued differential forms (as the de Rham operator with its splitting $d = \partial + \overline{\partial}$ allows one to do in a such particular case). Since the de Rham operator will remain the main ingredient, respecting Leibniz's rule inherently leads to introducing the notion of *connection* on a complex bundle E over $\mathcal{X}_{\mathbb{R}}$.

DEFINITION B.24. A (smooth) connection D on E is a $\mathbb{R}$- linear operator[8] from $\boldsymbol{C}^\infty(\mathcal{X},E)$ into itself which obeys to Leibniz's rule

(B.16)
$$D(\theta\wedge\varphi) = d\theta\wedge\boldsymbol{\varphi} + (-1)^r\theta\wedge D\boldsymbol{\varphi}$$

for any $\theta \in C^\infty_r$ and any $\boldsymbol{\varphi} \in \boldsymbol{C}^\infty_{\mathcal{X}}(\mathcal{X},E)$.

[6]In the case where the complex bundle $E \to \mathcal{X}$ is a holomorphic bundle with rank m (when considered over $\mathcal{X}$ instead of $\mathcal{X}_{\mathbb{R}}$), it is convenient to chose e_z as a holomorphic local frame; see Definition B.15, §B.1.5.

[7]It may happen (in particular in the Russian mathematical community) that the operator dd^c is defined as $(i/\pi)\,\partial\overline{\partial}$. In the present monograph, dd^c will always be $(i/2\pi)\,\partial\overline{\partial}$.

[8]The $\mathbb{R}$-linearity, instead of $\mathbb{C}$-linearity, even though $\mathbf{C}^\infty(\mathcal{X},E)$ is here a $\mathbb{C}$-vector space, is in this definition a major point.

REMARK B.25. It follows from the $\mathbb{R}$-linearity of such a connection D on the complex bundle E that D splits as $D = D' + D''$, where

$$\begin{aligned} D' &: \boldsymbol{C}^\infty_{p,\bullet}(\mathcal{X}, E) \longrightarrow \boldsymbol{C}^\infty_{p+1,\bullet}(\mathcal{X}, E), \\ D'' &: \boldsymbol{C}^\infty_{\bullet,q}(\mathcal{X}, E) \longrightarrow \boldsymbol{C}^\infty_{\bullet,q+1}(\mathcal{X}, E) \end{aligned} \tag{B.17}$$

for any $0 \leq p, q \leq N$.

EXAMPLE B.26. When E is a *holomorphic bundle over* $\mathcal{X}$ with rank m (instead of just being a complex bundle with rank m over $\mathcal{X}_{\mathbb{R}}$), the Dolbeault operator $\overline{\partial}$ acts as a $\mathbb{R}$-linear operator from $C^\infty_{\bullet,q}(\cdot, E)$ to $C^\infty_{\bullet,q+1}(\cdot, E)$ for any $q = 0, \dots, N$. It realizes a connection $\overline{\partial}_E : \boldsymbol{C}^\infty(\mathcal{X}, E) \longrightarrow \boldsymbol{C}^\infty(\mathcal{X}, E)$. This follows immediately from the fact that if $\{g_{\iota_1,\iota_0}\}_{\iota_0,\iota_1}$ is a 1-cocycle (with values in $\mathrm{GL}(m, \mathbb{C})$) for E subordinate to a covering of $\mathcal{X}$ with open subsets U_ι over which E admits a trivialization morphism (see Definition B.4 together with B.9), then $\overline{\partial} g_{\iota_1,\iota_0} = 0$ in $U_{\iota_0} \cap U_{\iota_1}$ for any ι_0, ι_1. *Such a connection* $\overline{\partial}_E$ *is in fact intrinsically attached to the complex structure of the holomorphic bundle* E. It is called the *canonical connection* of the holomorphic bundle E.

Let E be a complex bundle with rank m over $\mathcal{X}_{\mathbb{R}}$. It is convenient to describe the action of the connection D locally, in an open trivialization open subset $U_{\mathbb{R}} = U$ around z. Let e_z be a local frame for E in U. Let $\varphi \in C^\infty_r(\mathcal{X}, E)$ which local expression in U is $\varphi = \sum_{j=1}^m \varphi_j \otimes e_{z,j}$ with $\varphi_j \in C^\infty_r(U)$. If one writes

$$De_{z,k} = \sum_{j=1}^m \omega_{k,j} \otimes e_{z,j}$$

with $\omega_{k,j} \in C^\infty_1(U, \mathbb{C})$, it follows from Leibniz's rule (B.16) that

$$\omega_{k,j} = d\varphi_k + \sum_{j=1}^m \omega_{k,j} \wedge \varphi_j, \tag{B.18}$$

that is

$$D\varphi \overset{e_z}{=} D \begin{pmatrix} \varphi_1 \\ \vdots \\ \varphi_m \end{pmatrix} = \left(d \circ \mathrm{Id}_{\mathbb{C}^m} + \begin{pmatrix} \omega_{1,1} & \cdots & \omega_{1,m} \\ \vdots & \vdots & \vdots \\ \omega_{m,1} & \cdots & \omega_{m,m} \end{pmatrix} \right) \wedge \begin{pmatrix} \varphi_1 \\ \vdots \\ \varphi_m \end{pmatrix}. \tag{B.19}$$

The matrix $\mathcal{M}(D; e_z; e_z) = [\omega_{j,k}]_{1 \leq j,k \leq m}$ (which materializes the corrective terms for the de Rham connection on $E_{|U_{\mathbb{R}}}$ over $U_{\mathbb{R}}$) is called the *matrix of the connection* D *expressed in the local frame* e_z *in the trivialization open subset* U *around* z. Changing the local frame for E above the trivialization open subset U around z induces the following *gauge transformation formula*.

PROPOSITION B.27. *Let* e_z *and* e'_z *be two local frames for the complex bundle* $E \to \mathcal{X}_{\mathbb{R}}$ *in the trivialization open subset* U *around* z, *such that* $e'_z = P^{e_z}_{e'_z} e_z$, *where* $P^{e_z}_{e'_z}$ *is an invertible* (m, m) *matrix with entries in* $C^\infty(U)$, *one has the gauge transformation law*

$$\mathcal{M}(D; e_z; e_z) = (P^{e_z}_{e'_z})^{-1} \mathcal{M}(D; e'_z; e'_z) + (P^{e_z}_{e'_z})^{-1} dP^{e_z}_{e'_z}. \tag{B.20}$$

DEFINITION B.28. Let E be a complex bundle with rank m over the differential underlying manifold $\mathcal{X}_{\mathbb{R}}$ to a complex manifold $\mathcal{X}$, and D be a connection on E. The

$\mathbb{R}$-linear operator $D^2 = D \circ D : \boldsymbol{C}^\infty(\mathcal{X}, E) \longrightarrow \boldsymbol{C}^\infty(\mathcal{X}, E)$ is called the *curvature* of the connection D, while the differential bundle-valued form

$$\Theta(D) = \Theta_E(D) \in C_2^\infty(\mathcal{X}, \mathrm{Hom}_{\mathbb{C}}(E, E)),$$

such that $D^2\boldsymbol{\varphi} = \Theta(D) \wedge \boldsymbol{\varphi}$ for any $\boldsymbol{\varphi} \in \boldsymbol{C}_{\mathcal{X}}^\infty(\mathcal{X}, E)$, is called its *curvature tensor.*

If U is a trivialization open subset for the complex bundle E around z and e_z is a local frame for E over U, one has in U, for any $\boldsymbol{\varphi} \in \boldsymbol{C}^\infty(\mathcal{X}, E)$ expressed in U as $\boldsymbol{\varphi} = \sum_{j=1}^m \boldsymbol{\varphi}_j \otimes e_{z,j}$, that

$$\text{(B.21)} \quad D^2\boldsymbol{\varphi} = \sum_{j=1}^m (D^2\boldsymbol{\varphi})_j \otimes e_{z,j}, \text{ where}$$

$$\begin{pmatrix} (D^2\boldsymbol{\varphi})_1 \\ \vdots \\ (D^2\boldsymbol{\varphi})_m \end{pmatrix} = \Big(d\,\mathcal{M}(D; e_z; e_z) + \mathcal{M}(D; e_z; e_z) \wedge \mathcal{M}(D; e_z; e_z)\Big) \begin{pmatrix} \boldsymbol{\varphi}_1 \\ \vdots \\ \boldsymbol{\varphi}_m \end{pmatrix},$$

where $\mathcal{M}(D; e_z; e_z)$ denotes the matrix of the connection D expressed in the local frame e_z. The matrix with entries in $C_2^\infty(U, \mathbb{C})$ defined by

$$\text{(B.22)} \qquad \Theta(D; e_z; e_z) := d\,\mathcal{M}(D; e_z; e_z) + \mathcal{M}(D; e_z; e_z) \wedge \mathcal{M}(D; e_z; e_z)$$

is called the *curvature matrix* of the connection D expressed in the local frame e_z; it is also the matrix of the curvature tensor $\Theta(D)$ expressed in the local frame e_z.

B.2.1.3. *Operations on complex bundles on $\mathcal{X}_{\mathbb{R}}$ and connections.* We explain in this subsection how operations on complex bundles over $\mathcal{X}_{\mathbb{R}}$ (where $\mathcal{X}$ is an N-dimensional complex manifold, see §B.1.7.1) such as $(E_1, E_2) \mapsto E_1 \bullet E_2$ induce realizations of connections once a connection on each E_j is given. Let E_1 and E_2 be paired with connections D_1 and D_2.

- The $\mathbb{R}$-linear map

  $$D_1 \oplus D_2 \ : \boldsymbol{\varphi}_1 \oplus \boldsymbol{\varphi}_2 \in \boldsymbol{C}^\infty(\mathcal{X}, E_1) \oplus \boldsymbol{C}^\infty(\mathcal{X}, E_2) \longmapsto \boldsymbol{\varphi}_1 \oplus \boldsymbol{\varphi}_2$$

  defines a connection on $E_1 \oplus E_2$ with the curvature tensor

$$\text{(B.23)} \qquad \Theta_{E_1 \oplus E_2}(D_1 \oplus D_2) = \Theta_{E_1}(D_1) \oplus \Theta_{E_2}(D_2).$$

- There is a unique connection $D_1 \otimes D_2$ on $E_1 \otimes E_2$ such that for any $r \in \mathbb{N}$, for any $\varphi_1 \in C_r^\infty(\mathcal{X}, E_1)$ and any $\boldsymbol{\varphi}_2 \in \boldsymbol{C}^\infty(\mathcal{X}, E_2)$,

  $$(D_1 \otimes D_2)(\varphi_1 \otimes \boldsymbol{\varphi}_2) = D_1\varphi_1 \wedge \boldsymbol{\varphi}_2 + (-1)^r \varphi_1 \otimes D_2\boldsymbol{\varphi}_2.$$

  Moreover, one has the identity

$$\text{(B.24)} \qquad \Theta_{E_1 \otimes E_2}(D_1 \otimes D_2) = \Theta_{E_1}(D_1) \otimes \mathrm{Id}_{E_2} + \mathrm{Id}_{E_1} \otimes \Theta_{E_2}(D_2).$$

- There is a unique connection $D_{E_1 \to E_2}$ on $\mathrm{Hom}_{\mathbb{C}}(E_1, E_2) \simeq E_2 \otimes E_1^*$ such that for any $r \in \mathbb{N}$, for any $\varphi \in C_r^\infty(\mathcal{X}, \mathrm{Hom}_{\mathbb{C}}(E_1, E_2))$ and any form $\boldsymbol{\theta}_1 \in \boldsymbol{C}^\infty(\mathcal{X}, E_1)$,

  $$D_{E_1 \to E_2}\,\varphi \ : \boldsymbol{\theta}_1 \longmapsto D_2\left(\boldsymbol{\varphi}(\boldsymbol{\theta}_1)\right) - (-1)^r \varphi\,(D_1\boldsymbol{\theta}_1).$$

  Moreover, one has the identity

$$\text{(B.25)} \qquad \Theta_{\mathrm{Hom}_{\mathbb{C}}(E_1, E_2)}(D_{E_1 \to E_2}) = \Theta_{E_2}(D_2) \otimes \mathrm{Id}_{E_1^*} - \mathrm{Id}_{E_2} \otimes {}^t(\Theta_{E_1}(D_1)).$$

The important *Bianchi identity* for a complex bundle E over $\mathcal{X}_{\mathbb{R}}$ equipped with a connection D can be formulated within such context as

(B.26) $$D_{E\to E}(\Theta_E(D)) = D\big(\Theta_E(D)(\cdot)\big) - \Theta_E(D)\,\big(D(\cdot)\big) = D^3(\cdot) - D^3(\cdot) = 0.$$

- The *dual connection* of a connection D on a complex bundle E over $\mathcal{X}_{\mathbb{R}}$ is as a consequence of the definition of $D_{E_1\to E_2}$ the connection on the $\mathbb{C}$-dual bundle $E^* = \mathrm{Hom}(E,\mathbb{C})$ defined as follows. It is the unique connection D^* on E^* such that for any $r\in\mathbb{N}$, for any $\varphi\in C^\infty_r(\mathcal{X},E^*)$, and for any $\boldsymbol{\theta}\in \boldsymbol{C}^\infty(\mathcal{X},E)$,

(B.27) $$D^*_{E\to\mathbb{C}}\,\varphi = D^*\,\varphi = d(\varphi(\boldsymbol{\theta})) - (-1)^r\varphi(D\,\boldsymbol{\theta}).$$

Moreover, one has the identity

(B.28) $$\Theta_{E^*}(D^*) = -{}^t(\Theta_E(D)).$$

With respect to wedge powers, one has the following. Let E be a complex bundle over $\mathcal{X}_{\mathbb{R}}$ equipped with a connection D and $\ell\in\mathbb{N}^*$. One defines inductively a unique connection $D_{\wedge^r}$ on $\bigwedge^\ell E$ such that for any $\varphi_\kappa\in C^\infty_{r_\kappa}(\mathcal{X},E)$ with $r_\kappa\in\mathbb{N}$ for $\kappa=1,\dots,\ell$,

$$D_{\wedge^\ell}\Big(\bigwedge_{\kappa=1}^{\ell}\varphi_\kappa\Big) = \sum_{\kappa=1}^{\ell}(-1)^{\sum_{\kappa'<\kappa} r_{\kappa'}}\Big(\bigwedge_{\kappa'<\kappa}\varphi_{\kappa'}\Big)\wedge D\,\varphi_\kappa\wedge\Big(\bigwedge_{\kappa'>\kappa}\varphi_{\kappa'}\Big).$$

In the particular case where $\ell = m = \operatorname{rank} E$, the curvature tensor $\Theta_{\det E}(D_{\wedge^m})$ can be identified with the element of $C^\infty(\mathcal{X})$ defined as the trace of $\Theta(D)$ as an element of $C^\infty(\mathcal{X},\mathrm{Hom}_{\mathbb{C}}(E,E))$.

B.2.1.4. *Sheaves of bundle-valued currents on a complex manifold.*

DEFINITION B.29. Let E be a complex bundle with rank m over $\mathcal{X}_{\mathbb{R}}$, where $\mathcal{X}$ is an N-dimensional complex manifold. For any open subset $U\subset\mathcal{X}$, the dual space of the $\mathbb{C}$-vector space $\mathcal{D}^{(N-p,N-q)}(U,E)$, which is denoted by $'\mathcal{D}^{(p,q)}(U,E^*)$, is called the $\mathbb{C}$-space of E^*-valued *currents of bidegree* (p,q) (or with bi-dimension $(N-p,N-q)$) over U. The sheaf $'\mathcal{D}^{(p,q)}_{\mathcal{X}}(\cdot,E^*)$ thus defined is called the *sheaf of* E^**-valued currents with bidegree* (p,q) *over* $\mathcal{X}_{\mathbb{R}}$. When E is the trivial (holomorphic) bundle $\mathbb{C}$, one denotes this sheaf by $'\mathcal{D}^{(p,q)}$.

If e_z is a local frame for E over a trivialization open subset U around z, with dual frame e^*_z, and U is a local chart for $\mathcal{X}$ with local holomorphic coordinates $(\zeta_1,\dots,\zeta_N)$, it is convenient to represent in U any $T\in {}'\mathcal{D}^{(p,q)}(U,E^*)$ by

(B.29)
$$\begin{aligned} T &= \sum_{j=1}^{m}\Big(\sum_{1<i_1<\cdots<i_p\le N}\ \sum_{1<j_1<\cdots<j_q\le N} T_j^{I,J}\,\Big(\bigwedge_{\kappa=1}^{p} d\zeta_{i_\kappa}\Big)\wedge\Big(\bigwedge_{\kappa=1}^{q} d\overline{\zeta}_{j_\kappa}\Big)\Big)\otimes e^*_{z,j}\\ &= \sum_{j=1}^{m}\Big(\sum_{\#I=p,\ \#J=q} T_j^{I,J}\,d\zeta_I\wedge d\overline{\zeta}_J\Big)\otimes\, e^*_{z,j} = \sum_{j=1}^{m} T_j\,\otimes e^*_{z,j}, \end{aligned}$$

where the $T_j^{I,J}$ are $\mathbb{C}$-valued distributions in U, that is, they belong to the $\mathbb{C}$-dual of $\mathcal{D}(U)$. If $\varphi\in\mathcal{D}^{(N-p,N-q)}(U,E)$ is an E-valued test form in U, represented as in

(B.14) where the coordinate functions $\varphi^j_{I,J}$ belong to $\mathcal{D}^{(N-p,N-q)}(U)$, the duality bracket $\langle T, \varphi\rangle$ is expressed as

$$\text{(B.30)} \qquad \langle T\,,\,\varphi\rangle = (2i)^n \sum_{j=1}^{m} \sum_{1\leq i_1<\cdots<i_p\leq N} \sum_{1\leq j_1<\cdots<j_q\leq N} \epsilon^{I,J}\,\langle \tau_* T^{I,J}_j\,,\,\tau^*\,\varphi^j_{I^c,J^c}\rangle,$$

where I^c and J^c denotes the multi-indices with respective supports $\{1,\dots,N\}\setminus I$, $\{1,\dots,N\}\setminus J$ ordered in strictly increasing order, $\epsilon^{I,J} = \pm 1$ is defined by

$$d\zeta_I \wedge d\overline{\zeta}_J \wedge d\zeta_{I^c} \wedge d\overline{\zeta}_{J^c} = \epsilon_{I,J} \bigwedge_{k=1}^{N} d\overline{\zeta}_k \wedge d\zeta_k = (2i)^N\,\epsilon_{I,J} \bigwedge_{k=1}^{N} (dx_k \wedge dy_k),$$

and $\tau : z' \in U \mapsto (\mathrm{Re}\,\zeta(z'), \mathrm{Im}\,\zeta(z')) \in U_{\mathbb{R}}$; see §A.1.1.

B.2.2. Differential forms and currents on a complex analytic space. We consider now in this subsection a complex analytic space $\mathcal{X}$ with pure dimension N instead of an N-dimensional complex manifold.

B.2.2.1. *Sheaves of bundle-valued complex differential forms on a complex analytic space.* Let $(\mathcal{X}, \mathcal{O}_\mathcal{X})$ be an N-purely dimensional complex analytic space[9] and let p, q be two positive integers.[10] Given any local embedding

$$\iota : U \hookrightarrow \Omega \subset \mathbb{C}^{N'},$$

where U is an open subset of $\mathcal{X}$ and Ω an open subset of the ambient manifold $\mathbb{C}^{N'}$, the sheaf $(C^\infty_{p,q})_U$ of (p,q) differential forms in U is defined by

$$\text{(B.31)} \qquad (C^\infty_{p,q})_U = \frac{(C^\infty_{p,q})_{|\Omega}}{\{\phi \in (C^\infty_{p,q})_{|\Omega} \,:\, \iota^*_{|U\setminus \mathcal{X}_{\rm sing}}\phi = 0\}},$$

where $(C^\infty_{p,q})_\Omega$ denotes the sheaf of complex-valued differential (p,q)-forms on the open subset $\Omega \subset \mathbb{C}^{N'}$. Since all embeddings are essentially equivalent, such a definition does not depend, once U is fixed, on the embedding $\iota : U \hookrightarrow \Omega \subset \mathbb{C}^{N'}$. It follows from such observation that there is a unique sheaf $(C^\infty_{p,q})_\mathcal{X}$ such that, for any such $U \subset \mathcal{X}$ with an embedding $U \hookrightarrow \Omega \subset \mathbb{C}^{N'}$, one has

$$((C^\infty_{p,q})_\mathcal{X})_{|U} = (C^\infty_{p,q})_U.$$

The sheaf $(C^\infty_{p,q})_\mathcal{X}$ is the *sheaf of complex-valued differential (p,q)-forms on $\mathcal{X}$*. We set $C^\infty = (\mathbb{C}^\infty_{0,0})_\mathcal{X}$, $C^\infty_{p,q} = (\mathbb{C}^\infty_{p,q})_\mathcal{X}$. The sheaf $C^\infty_{p,q}$ is a sheaf of C^∞ modules on $\mathcal{X}$. For any $r = 0, \dots, 2N$, let also

$$C^\infty_r := \bigoplus_{p+q=r,\max(p,q)\leq N} C^\infty_{p,q}.$$

Let now $E \to \mathcal{X}$ be a holomorphic[11] complex bundle with rank m over the complex analytic space $\mathcal{X}$; see Definition B.4, §B.1.3. Let U be a trivialization open subset around $z \in \mathcal{X}$ such that one has also a local embedding $\iota : U \hookrightarrow \Omega \subset \mathbb{C}^{N'}$ as

[9]We keep always considering the *reduced* case, even if we avoid repeating it.

[10]Only the case where $p \leq N$ and $q \leq N$ leads to nontrivial $\mathbb{C}$ vector spaces.

[11]Within the context of complex analytic spaces, we restrict ourselves to the context of holomorphic complex bundles over $\mathcal{X}$ instead of that of complex bundles over $\mathcal{X}_\mathbb{R}$, as we did in the case where $\mathcal{X}$ was a complex manifold; compare with §B.2.1.1.

before. The sheaf $(C^\infty_{p,q})_U(\cdot, E)$ of E-valued (p,q) differential forms in U is defined as

$$(C^\infty_{p,q})_U(\cdot, E) = \sum_{j=1}^{m} (C^\infty_{p,q})_U \otimes e_{z,j},$$

where e_z is an arbitrary holomorphic local frame for E over the trivialization open subset U around z. Once U is fixed, the definition of such a sheaf is independent both of the local embedding $\iota : U \hookrightarrow \Omega \subset \mathbb{C}^{N'}$ and of the local holomorphic frame e_z for E over U. It follows from such observation that there is a unique sheaf $C^\infty_{p,q}(\cdot, E)$ such that, for any such $U \subset \mathcal{X}$ with an embedding $U \hookrightarrow \Omega \subset \mathbb{C}^{N'}$ and such that E is trivialized over U, one has

$$C^\infty_{p,q}(\cdot, E)_{|U} = (C^\infty_{p,q})_U(\cdot, E).$$

The sheaf $C^\infty_{p,q}(\cdot, E)$ on $\mathcal{X}$ uniquely defined in this way is the *sheaf of* (p,q) *E-valued differential forms on* $\mathcal{X}$. It is a sheaf of C^∞-modules on $\mathcal{X}$. The $\mathbb{R}$-linear operators ∂, $\overline{\partial}$, $d = \partial + \overline{\partial}$, $dd^c = (i/2\pi)\,\partial\overline{\partial}$ introduced previously in the context of complex manifolds (see (B.15)) still act as follows

$$\begin{aligned}
\partial : C^\infty_{p,\bullet} &\longrightarrow C^\infty_{p+1,\bullet} \quad (0 \text{ if } p \geq N),\\
\overline{\partial} : C^\infty_{\bullet,q} &\longrightarrow C^\infty_{\bullet,q+1} \quad (0 \text{ if } q \geq N),\\
d : \boldsymbol{C}^\infty_{r} &\longrightarrow \boldsymbol{C}^\infty_{r+1} \quad (r = 0, \dots, 2N,\ 0 \text{ if } r = N),\\
dd^c : C^\infty_{p,q} &\longrightarrow C^\infty_{p+1,q+1} \quad (0 \text{ if } p \text{ or } q \geq N),
\end{aligned}$$

since ∂ and $\overline{\partial}$ commute with local embeddings $\iota : U \hookrightarrow \Omega \subset \mathbb{C}^N$. Moreover, given any holomorphic bundle E, the Dolbeault operator acts $\mathbb{R}$-linearly from $C^\infty_{\bullet,q}(\cdot, E)$ to $C^\infty_{\bullet,q+1}(\cdot, E)$ for the same reasons as those given in Example B.26. The operator $\overline{\partial}_E$ is still called the *canonical connection* of the holomorphic bundle E over the complex analytic space $\mathcal{X}$.

B.2.2.2. *Sheaves of complex-valued currents on a complex analytic space.* As usual, the sheaf ${}'\mathcal{D}^{(p,q)} = {}'\mathcal{D}^{(p,q)}_{\mathcal{X}}$ of complex-valued currents of bidegree (p,q) on $\mathcal{X}$ is realized as

$$(B.32) \qquad {}'\mathcal{D}^{(p,q)} = \big(\mathcal{D}^{(N-p,N-q)}\big)',$$

where $\mathcal{D}^{(N-p,N-q)}(U)$ denotes when U is an open subset of $\mathcal{X}$ the $\mathbb{C}$ space of sections in U of $C^\infty_{N-p,N-q}$ with compact support in U, that is by duality. It is convenient to proceed to the following identifications: given $\iota : U \hookrightarrow \Omega \subset \mathbb{C}^{N'}$ a local embedding of $\mathcal{X}$ as in §B.2.2.1, one may identify

$$(B.33) \quad T \in {}'\mathcal{D}^{(p,q)}(U) \longleftrightarrow \iota_* T \in \big\{\mathrm{T} \in {}'\mathcal{D}^{(N'-(N-p),N'-(N-q))}(\Omega) \text{ such that } \forall \phi \in C^\infty_{N-p,N-q}(\Omega),\ \iota^*_{|U\setminus \mathcal{X}_{\mathrm{sing}}}\phi = 0 \Longrightarrow \langle \mathrm{T}, \phi\rangle = 0\big\}.$$

The sheaf ${}'\mathcal{D}^{(p,q)}$ is the *sheaf of* (p,q)*-currents on* $\mathcal{X}$. It is a sheaf of C^∞-modules on $\mathcal{X}$. One can as well define by duality the sheaves ${}'\mathcal{D}^{(p,q)}(\cdot, E^*)$ of (p,q) E^*-valued currents on $\mathcal{X}$, where the $\mathbb{C}$-vector spaces ${}'\mathcal{D}^{(p,q)}(\cdot, E^*)$ are dual to the $\mathbb{C}$-vector spaces $\mathcal{D}^{(N-p,N-q)}(\cdot, E)$. If U is an open subset of $\mathcal{X}$ containing z over which E admits a holomorphic frame e_z, it is convenient to represent in U a section $T \in {}'\mathcal{D}^{(p,q)}(U, E^*)$ as

$$T = \sum_{j=1}^{m} e^*_{z,j} \otimes T_j,$$

where $T_j \in {}'\mathcal{D}^{p,q}(U)$ for $j = 1,\dots,m$. If $\varphi = \sum_{j=1}^m \varphi_j\, e_{z,j} \in \mathcal{D}^{(N-p,N-q)}(U,E)$, the duality bracket is then given in U by

$$\langle T,\varphi\rangle = \sum_{j=1}^m \langle T_j\,,\,\varphi_j\rangle.$$

The canonical connection $\overline{\partial}_{E^*}$ transforms a current $T \in {}'\mathcal{D}^{(\bullet,q)}(U,E^*)$ into the current $\overline{\partial}_{E^*}T \in {}'\mathcal{D}^{(\bullet,q+1)}(U,E^*)$, where $\overline{\partial}_{E^*}T$ is represented in the open subset U as

$$\overline{\partial}_{E^*}T = \sum_{j=1}^m e^*_{z,j} \otimes \overline{\partial} T_{z,j}.$$

For a more detailed introduction to the sheaves of currents on a complex analytic space instead of on a complex manifold, we refer the reader to [**HL71**, §4.2].

B.2.3. Dolbeault complex, Serre duality on a complex manifold.

DEFINITION B.30. Let $(\mathcal{X},\mathcal{O}_{\mathcal{X}})$ be a complex analytic manifold with pure dimension N and let E be a holomorphic m-bundle. Let $0 \le p \le N$. The complex of sheaves

$$\text{(B.34)}\quad 0 \longrightarrow \Omega^p(\cdot,E) \stackrel{\iota}{\longrightarrow} C^\infty_{p,0}(\cdot,E) \stackrel{\overline{\partial}_E}{\longrightarrow} C^\infty_{p,1}(\cdot,E) \stackrel{\overline{\partial}_E}{\longrightarrow} \cdots$$
$$\cdots \stackrel{\overline{\partial}_E}{\longrightarrow} C^\infty_{p,N-1}(\cdot,E) \stackrel{\overline{\partial}_E}{\longrightarrow} C^\infty_{p,N}(\cdot,E) \stackrel{\overline{\partial}_E}{\longrightarrow} 0,$$

where $\Omega^p(\cdot,E)$ denotes the sheaf of holomorphic sections of $T^{*(p,0)} \otimes E$, is locally exact. It is called the $(p,\bullet)$ *Dolbeault complex of sheaves* attached to the holomorphic bundle E over $(\mathcal{X},\mathcal{O}_{\mathcal{X}})$. Such a complex provides a *fine resolution* of the sheaf $\Omega^p(\cdot,E)$ since all sheaves $C^\infty_{p,q}(\cdot,E)$, $q \in \mathbb{N}$, are *fine sheaves*, that is sheaves of C^∞-modules, which makes possible the use of smooth partitions of unity.

Given U an open subset of $\mathcal{X}$, the cohomology groups

$$H^{p,q}(U,E) = \frac{\operatorname{Ker}\overline{\partial}_E : C^\infty_{p,q}(U,E) \longrightarrow C^\infty_{p,q+1}(U,E)}{\operatorname{Im}\overline{\partial}_E : C^\infty_{p,q-1}(U,E) \longrightarrow C^\infty_{p,q}(U,E)},\quad p\in\mathbb{N},\ q\in\mathbb{N}^*,$$

are the *Dolbeault cohomology groups* $H^{p,q}(U,E)$. *Dolbeault isomorphism theorem*, asserts, as a consequence of abstract de Rham theorem, that for any $p \in \mathbb{N}$ and $q \in \mathbb{N}^*$,

$$\text{(B.35)}\qquad H^{p,q}(\mathcal{X},E) \simeq \check{H}^q(\mathcal{X},\Omega^p(\cdot,E)),$$

where $\check{H}^q(\mathcal{X},\Omega^p(\cdot,E))$ denotes the Čech cohomology group (see §A.2.2) of the sheaf $\Omega^p(\cdot,E) = \Omega^p \otimes E$ of $(p,0)$ E-valued holomorphic forms on $\mathcal{X}$ (in particular $\Omega^0_{\mathcal{X}}(\cdot,E) = \mathcal{O}(\cdot,E)$); see, e.g., [**De**, §V.11]. Let us also introduce the following (dual each other) complexes of sheaves

(B.36)

$$0 \to \mathcal{D}^{(p,0)}(\cdot,E) \stackrel{\overline{\partial}_E}{\to} \mathcal{D}^{(p,1)}(\cdot,E) \cdots \stackrel{\overline{\partial}_E}{\to} \mathcal{D}^{(p,N-1)}(\cdot,E) \stackrel{\overline{\partial}_E}{\longrightarrow} \mathcal{D}^{(p,N)}_{\mathcal{X}}(\cdot,E) \stackrel{\overline{\partial}_E}{\to} 0,$$

$$0 \to {}'\mathcal{D}^{(p,0)}(\cdot,E^*) \stackrel{\overline{\partial}_{E^*}}{\to} {}'\mathcal{D}^{(p,1)}(\cdot,E^*) \cdots \stackrel{\overline{\partial}_{E^*}}{\to} {}'\mathcal{D}^{(p,N-1)}(\cdot,E^*) \stackrel{\overline{\partial}_{E^*}}{\to} {}'\mathcal{D}^{(p,N)}(\cdot,E^*) \stackrel{\overline{\partial}_{E^*}}{\to} 0.$$

If

$$H^{N,N-q}_c(\mathcal{X},E) = \frac{\operatorname{Ker}\overline{\partial}_E : \mathcal{D}^{(N,N-q)}(\mathcal{X},E) \longrightarrow \mathcal{D}^{(N,N-q+1)}(\mathcal{X},E)}{\operatorname{Im}\overline{\partial}_E : \mathcal{D}^{(N,N-q-1)}(\mathcal{X},E) \longrightarrow \mathcal{D}^{(N,N-q)}(\mathcal{X},E)},$$

then

$$(\bar{\varphi}, \bar{\psi}) \in H^{0,q}(\mathcal{X}, E^*) \times H_c^{N,N-q}(\mathcal{X}, E) \longmapsto \int_{\mathcal{X}} \varphi \wedge \psi \tag{B.37}$$

provides a nondegenerate pairing under the condition that both cohomology groups $H^{0,q}(\mathcal{X}, E^*)$ and $H^{0,q+1}(\mathcal{X}, E^*)$ are Hausdorff when considered with their structure of topological spaces; see, e.g., [**RSW17**, §1]. This is an analytic formulation of *Serre's duality theorem* [**Se55**], which can also be interpreted thanks to the Dolbeault isomorphism theorem as follows: the cup product realizes a nondegenerate pairing

$$\check{H}^q(\mathcal{X}, \mathcal{O}_{\mathcal{X}}(\cdot, E^*)) \times \check{H}_c^{N-q}(\mathcal{X}, \Omega^N(\cdot, E)) \to \mathbb{C}, \tag{B.38}$$

once both $\check{H}^q(\mathcal{X}, \mathcal{O}(\cdot, E^*))$ and $\check{H}^{q+1}(\mathcal{X}, \mathcal{O}(\cdot, E^*))$ are Hausdorff.

EXAMPLE B.31.

(1) If $\mathcal{X}$ is a Stein manifold, one has $H^{p,q}(\mathcal{X}, \mathbf{C}) = 0$ for any $p \geq 0$ and $q \geq 1$.
(2) When $\mathcal{X} = \mathbb{P}^N_{\mathbb{C}}$, one has the following:

$$\begin{aligned} 1 \leq q \leq N-1 \text{ and } d \in \mathbb{Z} &\Longrightarrow H^{0,q}(\mathbb{P}^N_{\mathbb{C}}, \mathcal{O}(d)) = 0, \\ q = N \text{ and } d + N \in \mathbb{N} &\Longrightarrow H^{0,q}(\mathbb{P}^N_{\mathbb{C}}, \mathcal{O}(d)) = 0. \end{aligned} \tag{B.39}$$

The first implication is a particular case of the situation, where $q \neq 0, N, p$ (since $p = 0$); see [**De**, VII-Theorem 10.7 (a)]. On the other hand, the second implication is a particular case of the situation where $q = N$, $d \geq N + p$ and $(d, p) \neq (0, N)$ since $p = 0 < N$; see [**De**, VII-Theorem 10.7(c)]. The other situations where $H^{p,q}(\mathbb{P}^N_{\mathbb{C}}, \mathcal{O}(d)) = 0$ are those where $q = 0$, $d \leq p$ and $(d, p) \neq (0, 0)$ (which is equivalent to the second assertion in (B.39) thanks to Serre duality) or $q = p \neq 0, N$ and $d \in \mathbb{Z}^*$ [**De**, VII-Theorem 10.7 (d)]. Such vanishing results quoted in [**De**, VII] are due to Joseph Le Potier; see [**LeP75**] and [**LeP77**, Theorem 2] for more general results when $\mathbb{P}^N_{\mathbb{C}}$ is replaced by a grassmannian.

B.2.4. Strongly holomorphic $(p, 0)$ forms on $V \hookrightarrow \mathcal{X}$. Let V be a closed analytic subset of an N-dimensional complex manifold with pure codimension M. For any $p = 0, \ldots, N - M$, let $\Omega^p_{\mathcal{X}}$ be the sheaf of holomorphic (or *abelian*) $(p, 0)$-forms, that is, the sheaf of $\overline{\partial}$-closed $(p, 0)$-forms on $\mathcal{X}$. Let also $\mathcal{J}^{[p]}_{V,\mathcal{X}}$ be the subsheaf of $\Omega^p_{\mathcal{X}}$ whose elements ω are such that $\iota^*_{|V_{\mathrm{reg}}} \omega$ vanishes on V_{reg}, where $\iota : V \hookrightarrow \mathcal{X}$ denotes the (local) embedding. Strongly holomorphic $(p, 0)$ forms on $V \hookrightarrow \mathcal{X}$ are defined as follows; see [**Fe70**, Definition 1.2], [**HeP99**, §2], [**Sam21**, §3].

DEFINITION B.32. The sheaf of $(0, 0)$-strongly holomorphic forms on $V \hookrightarrow \mathcal{X}$ is the structural sheaf $\mathcal{O}_{\mathcal{X}}/\mathcal{I}_{V,\mathcal{X}}$ of the $(N - M)$-dimensional reduced analytic space $(V, \mathcal{O}_{\mathcal{X}}/\mathcal{I}_{V,\mathcal{X}})$. For any $1 \leq p \leq N - M$, the sheaf of $(p, 0)$-strongly holomorphic forms on $V \hookrightarrow \mathcal{X}$ is defined as the restriction to V of the sheaf $\Omega^p_{\mathcal{X}}/\mathcal{J}^{[p]}_{V,\mathcal{X}}$.

REMARK B.33. For any $p = 1, \ldots, \dim V$, the sheaf of Kähler–Grothendieck differentials on $V \hookrightarrow \mathcal{X}$ is defined as

$$(\mathcal{J}^p_{\text{Käh}})_{V \hookrightarrow \mathcal{X}} = \frac{\Omega^p_{\mathcal{X}}}{\mathcal{I}_{V,\mathcal{X}}\, \Omega^p_{\mathcal{X}} + d\mathcal{I}_{V,\mathcal{X}} \wedge \Omega^{p-1}_{\mathcal{X}}}, \tag{B.40}$$

where $\mathcal{I}_{V,\mathcal{X}}$ denotes as usual the $\mathcal{O}_{\mathcal{X}}$ defining radical sheaf of $V \hookrightarrow \mathcal{X}$, such that $\mathcal{O}_V = \mathcal{O}_{\mathcal{X}}/\mathcal{I}_{V,\mathcal{X}}$. The sheaf Ω^p_V is obtained then from the sheaf $(\mathcal{J}^p_{\text{Käh}})_{V\hookrightarrow\mathcal{X}}$ of Kähler-differentials by quotienting out the torsion; see [**Sam21**, §3]. The comparison of the local embeddings

$$\iota' : V \hookrightarrow \mathrm{U}' \subset \mathbb{C}^{N_{\iota'}}, \quad \iota'' : V \hookrightarrow \mathrm{U}'' \subset \mathbb{C}^{N_{\iota''}}$$

as in Remark A.16, with $\iota' = \iota_{\min}$ minimal, $\iota'' = \iota$, $\mathrm{U}'' = \mathrm{U}' \times \mathbb{C}^\nu$ and with $\nu = N_\iota - N_{\iota_{\min}}$, so that $\iota : z \mapsto (\iota_{\min}(z), 0)$, leads to the following observation: for any $p = 1, \ldots, \dim V$, the sheaves $\Omega^p_{V\hookrightarrow\mathcal{X}}$ of strongly holomorphic $(p,0)$-forms on $V \hookrightarrow \mathcal{X}$ are (as in the case where $p = 0$) intrinsic with respect to V, that is they do not depend on the embedding $\iota : V \hookrightarrow \mathcal{X}$. This follows from the isomorphism between sheaves of Kähler differentials

$$\text{(B.41)} \qquad \frac{\Omega^p_{\mathrm{U}'}}{\mathcal{I}_{\iota'(V),\mathrm{U}'}\,\Omega^p_{\mathrm{U}'} + d\mathcal{I}_{\iota'(V),\mathrm{U}'} \wedge \Omega^{p-1}_{\mathrm{U}'}} \simeq \frac{\Omega^p_{\mathrm{U}''}}{\mathcal{I}_{\iota''(V),\mathrm{U}''}\,\Omega^p_{\mathrm{U}''} + d\mathcal{I}_{\iota''(V),\mathrm{U}''} \wedge \Omega^{p-1}_{\mathrm{U}''}}$$

when $\iota' : V \hookrightarrow \mathrm{U}' \subset \mathbb{C}^{N_{\iota'}}$ is minimal and thus $\iota'' = \iota : V \to \mathrm{U}'' \subset \mathbb{C}^{N_{\iota''}}$ can be supposed such that $\mathrm{U}'' = \mathrm{U}' \times \mathbb{C}^\nu$ and $\iota(z) = (\iota'(z), 0)$; see Remarks A.16 and A.31.

B.2.5. The Barlet sheaves on $V \hookrightarrow \mathcal{X}$. Given a complex N-dimensional manifold $\mathcal{X}$, one defines the sheaf $\mathcal{M}^{p,0}_{\mathcal{X}}$ of meromorphic $(p,0)$-forms for any $p = 0, \ldots, N$. Its stalk at z^0 equals $((\mathcal{O}_{\mathcal{X},z^0})^*)^{-1}(\Omega^p_{z^0})_{\mathcal{X}}$. Here $\Omega^p_{\mathcal{X}}$ denotes the sheaf of holomorphic (or *abelian*) $(p,0)$-forms as in §B.2.4. Given a closed, purely $(N-M)$-dimensional analytic subset $V \subset \mathcal{X}$, we introduce the following definition.

DEFINITION B.34. Let $\mathcal{X}$ be a complex N-dimensional manifold $\mathcal{X}$ and let V be a closed analytic set embedded in $\mathcal{X}$, namely, $V \hookrightarrow \mathcal{X}$.[12] Given $0 \le p \le N-M$, a holomorphic differential $(p,0)$-form ω on the complex $(N-M)$-dimensional submanifold $V_{\text{reg}} \subset \mathcal{X}$ is said to define a holomorphic differential $(p,0)$-form on the closed analytic subset V *in the sense of Barlet*[13] if the $(p+M,M)$-current T_ω is defined on the open subset $\mathcal{X} \setminus V_{\text{sing}}$ by

$$T_\omega : \varphi \in \mathcal{D}^{(N-M-p,N-M)}_{\mathcal{X}}(\mathcal{X} \setminus V_{\text{sing}}) \longmapsto \int_{V_{\text{reg}}} \omega \wedge \varphi$$

and extends as a $\bar\partial$-closed $(p+M,M)$-current to the whole ambient manifold $\mathcal{X}$. One defines in such a way a sheaf $\omega^p_{V\hookrightarrow\mathcal{X}}$ of holomorphic $(p,0)$-forms on the complex analytic space $(V, \mathcal{O}_{\mathcal{X}}/\mathcal{I}_{\mathcal{X},V})$ in the sense of Barlet, that is considered as embedded in the ambient complex manifold $(\mathcal{X}, \mathcal{O}_{\mathcal{X}})$. Such a sheaf is called the p-Barlet sheaf on $(V, \mathcal{O}_{\mathcal{X}}/\mathcal{I}_{\mathcal{X},V})$.

REMARK B.35. In the particular case where $p = N-M = \dim V$, the sheaf $\omega^{N-M}_{V\hookrightarrow\mathcal{X}} = \omega^{N-M}_V$ is independent of the embedding $V \hookrightarrow \mathcal{X}$ and is called the *dualizing sheaf* of $(V, \mathcal{O}_V) \simeq (V, \mathcal{O}_{\mathcal{X}}/\mathcal{I}_{\mathcal{X},V})$. It is intrinsically defined (see, e.g., [**PeR94**, §5.3]) in terms of the Ext functor[14] by

$$\text{(B.42)} \qquad \omega^{N-M}_V = \mathscr{E}xt^{N-M}_{\mathcal{O}_{\mathcal{X}}}(\mathcal{O}_{\mathcal{X}}/\mathcal{I}_{V,\mathcal{X}}, \Omega^N_{\mathcal{X}}).$$

[12] Unless $p = N-M$, such a definition is not intrinsic with respect to V considered with its structure of purely $(N-M)$-dimensional (reduced) complex analytic space $(V, \mathcal{O}_{\mathcal{X}}/\mathcal{I}_V)$. It depends on the embedding $V \hookrightarrow \mathcal{X}$.

[13] The terminology refers here to the pioneering work of Daniel Barlet [**Bar78**]; see also [**HeP99**, Definition 5.1], where such a definition is proposed for the notion of holomorphicity on the closed analytic subset V considered as embedded in $\mathcal{X}$.

[14] We refer the reader for example to [**PeR94**, §5.2] for the definitions of Ext groups and $\mathscr{E}xt$-sheaves.

Its sections are called *regular differential forms* in the algebraic context; refer to [**KuW88**]. When $0 \le p < \dim V = N - M$, the stalk $\omega^p_{V\hookrightarrow \mathcal{X},z}$ of $\omega^p_{V\hookrightarrow\mathcal{X}}$ at $z \in V$ consists of all germs at z on V of meromorphic $(p,0)$-forms ω about z in the ambient manifold $\mathcal{X}$ such that

$$\omega_z \wedge \Omega^{N-M-p}_{V,z} \in \omega^{N-M}_{V,z}, \tag{B.43}$$

where Ω^{N-M-p}_V denotes the sheaf of strongly $N-M-p$ holomorphic forms on V; see §B.2.4. That is, in terms of the $\mathscr{H}om$-functor,

$$\omega^p_{V\hookrightarrow\mathcal{X}} = \omega^p_V = \mathscr{H}om_{\mathcal{O}_{\mathcal{X}}/\mathcal{I}_{\mathcal{X},V}}(\Omega^{N-M-p}_V, \omega^{N-M}_V); \tag{B.44}$$

see [**Bar78**, Proposition 3], [**TsY04**, Definition 5.2], [**AlT08**, Definition 2.1] or also [**Sam21**, §4]. As proved in [**Bar78**], the sheaf ω^p_V, $0 \le p \le \dim V$, where $\dim V = N - M$, can also be interpreted as the sheaf of $\bar\partial$-closed $(p,0)$-currents on the reduced complex analytic space $(V, \mathcal{O}_{\mathcal{X}}/\mathcal{I}_{\mathcal{X},V})$ satisfying the standard extension property (Definition 2.48, transposed to the setting of complex analytic spaces according to Remark 2.29). That is, the sheaf of $\bar\partial$-closed, $(p,0)$-pseudomeromorphic currents on the reduced complex analytic space $(V, \mathcal{O}_{\mathcal{X}}/\mathcal{I}_{\mathcal{X},V})$. We also refer the reader to [**AndL19, AndLLS20**] for the extension of such results to the case where $(V, \mathcal{O}_{\mathcal{X}}/\mathcal{I}_{\mathcal{X},V} = \mathcal{O}_V)$ is a nonreduced complex analytic space, which means that $(\mathcal{I}_{\mathcal{X},V})_z$ fails to be radical for some $z \in V$.

EXAMPLE B.36. If $V \subset \mathrm{U} \subset \mathbb{C}^N$ contains 0 as a singular point and is defined in U as $f^{-1}(\{0\})$, where $f = (f_1, \dots, f_M)$ is a vector of holomorphic function in U such that $(df_1 \wedge \cdots \wedge df_M)_0 \not\equiv 0$ on any irreducible component $V_{0,\iota}$ of V_0, then the stalk at 0 of $\omega^{N-M}_{V\hookrightarrow \mathrm{U}}$ is

$$\omega^{N-M}_{V\hookrightarrow\mathrm{U},0} = \Big(\frac{\mathcal{O}_{\mathrm{U}}}{\sum_j f_j \mathcal{O}_{\mathrm{U}}}\Big)_0 \Big(\frac{dz_1\wedge\cdots\wedge dz_N}{df_1\wedge\cdots\wedge df_M}\Big),$$

where the meromorphic form $\bigwedge_1^N dz_i / \bigwedge_1^M df_j$ is defined as follows. In the open subset U_I $(1 \le i_1 < \cdots < i_M \le N)$ of U where the minor $\det[\partial f_j/\partial z_{i_\kappa}]_{1\le\kappa,j\le M}$ does not vanish (such open subsets U_I realize a covering of V_{reg} because of the hypothesis), one can use $f_1, \dots, f_M, z_j$ for $j \in \{1,\dots,N\}\setminus I$, as coordinates on $\mathcal{X}$ and thus rewrite $\bigwedge_1^N dz_i = \bigwedge_1^M df_j \wedge \Omega_I$, where $\Omega_I \in \Omega^{N-M}(\mathrm{U}_I)$. If $\iota_{V_{\mathrm{reg}}} : V_{\mathrm{reg}} \hookrightarrow \mathcal{X}$, the forms $\iota^*\Omega_I$ define the meromorphic form on V denoted by $\bigwedge_1^N dz_i / \bigwedge_1^M df_j$.

PROPOSITION B.37. *Any $\omega \in \omega^p_{V\hookrightarrow\mathcal{X}}(U)$, where U is an open subset of $V \subset \mathcal{X}$, is such that for any $z \in U \cap V$, there is a neighborhood U_z of z in $\mathcal{X}$, a germ of meromorphic form*

$$\boldsymbol{\omega}_z = \frac{h_z}{\delta_z} \in \mathcal{M}^{(p,0)}_{\mathcal{X}}(\mathrm{U}_z) \text{ with } h_z \in \Omega^p_{\mathcal{X}}(\mathrm{U}_z),\ \delta_z \in \mathcal{O}_{\mathcal{X}}(\mathrm{U}_z)\ s.t.\ (V_{\mathrm{sing}})_z \subset \delta_z^{-1}(\{0\}), \tag{B.45}$$

and moreover one has in U_z that in the sense of $(p+M, M)$-currents,

$$\bar\partial\Big(\Big[|\delta|^{2\lambda}\, h \wedge [V]\Big]_{\lambda=0}\Big)_z = 0, \tag{B.46}$$

where the current $[\ ]_{\lambda=0}$ is defined via analytic continuation (and expressed in terms of integration by parts) through the Bernstein–Sato identity (C.27)*; see Remark* C.34 *following Proposition* C.33 *in* §C.2.5. *Elements in any stalk $\omega^p_{V\subset\mathcal{X},z}$ are in fact characterized in such a way, letting $\delta_z = 1$ if $z \in V_{\mathrm{reg}}$.*

EXAMPLE B.38. If $\mathcal{X} = \mathbb{C}^2$ and V is the cusp $z^2 - w^3 = 0$ which is parameterized by $t \mapsto (t^3, t^2)$, then $V_{\rm sing} = (0,0)$, one can check easily that (B.46) holds with $p = 0$, $g = z$ and $\omega = w/z$ in a neighborhood of $(0,0)$. One defines in this way in a neighborhood of $(0,0)$ a section of the Barlet sheaf $\omega^0_{V\hookrightarrow\mathbb{C}^2}$ which is not weakly holomorphic on V (see §A.3.5) since it is unbounded on V in a neighborhood of the origin.

When V is locally defined in $\mathcal{X}$ as a reduced complete intersection, that is, for each $z \in V$

$$V_z = \bigcap_{j=1}^{M} f_{j,z}^{-1}(\{0\}) = \bigcup_{\iota} V_{z,\iota}$$

with $f_{j,z} \in \mathcal{O}_{\mathcal{X},z}$ such that $df_{1,z} \wedge \cdots \wedge df_{M,z} \not\equiv 0$ on any irreducible component $V_{z,\iota}$, then[15] for any $p = 0, \dots, N-M$, there is a one-to-one correspondence between elements in $\omega^p_{V\hookrightarrow U_z}$ and iterated Leray residues[16]

$$\mathrm{Res}_{V(f_{1,z})\cap(\mathrm{U}_z\setminus V_{\rm sing}),\dots,V(f_{M,z})\cap(\mathrm{U}_z\setminus V_{\rm sing})}\left[\frac{\omega_z}{J_z}\right],$$

where $\omega \in \mathcal{M}^{(p+M,0)}_{\mathcal{X}}(\mathrm{U}_z)$ has poles along $\bigcup_{j=1}^M V(f_{j,z})$ and J_z is a local Oka universal denominator for the reduced complete intersection $V \cap \mathrm{U}_z$ as defined in Theorem A.27.

B.2.6. Dolbeault complex, Serre duality on a complex analytic space. Let us suppose that $\mathcal{X}$ is a complex analytic space with pure dimension N. An ad hoc intrinsic resolution of $\mathcal{O} = \mathcal{O}_{\mathcal{X}}$:

$$0 \to \mathcal{O} \xrightarrow{i} \mathcal{A}_{0,0} \xrightarrow{\bar\partial} \mathcal{A}_{0,1} \xrightarrow{\bar\partial} \cdots \xrightarrow{\bar\partial} \mathcal{A}_{0,N} \to 0, \tag{B.47}$$

with fine sheaves $\mathcal{A}_{0,q}$, whose restriction to $\mathcal{X}_{\rm reg}$ coincides with that of $C^\infty_{0,q}$ and such that $\bigoplus_q \mathcal{A}_{0,q}$ is a sheaf of $C^\infty_{0,\bullet}$-modules, arises from the combination of multivariate residue theory with Koppelman's formulae for the resolution of $\bar\partial$ on complex analytic spaces [**AndS11**]. Such an ad hoc fine resolution was introduced in [**AndS12**, Theorem 1.2]. As described in [**AndS11**, §7], the stalk at $z_0 \in \mathcal{X}$ of the sheaf $\mathcal{A}_{0,q}$ is realized when taking the germ at z_0 after finitely many successive operations

$$(\cdot) \longrightarrow \Big(z \mapsto \varphi(z) \wedge \int k(z,\zeta) \wedge (\cdot)(\zeta)\Big), \tag{B.48}$$

$$\varphi \in \mathcal{D}^{(0,\bullet)}(U_{z_0}),\ \mathrm{Supp}(\varphi) \subset \mathrm{Def}\big(z \mapsto \int k(z,\cdot)(\cdot)\big),$$

where $k(\cdot,\cdot)$ is an integral kernel that can be expressed in terms of the structure form ω for $\mathcal{X} \hookrightarrow \Omega \subset \mathbb{C}^{N'}$ around z_0, such as introduced in §3.4.3. Proposition 4.29, where $\mathbf{F}_\bullet$ is the generally exact complex of hermitian bundles attached to a free resolution of $\mathcal{O}_\Omega/\mathcal{I}_{\Omega,\mathcal{X}\hookrightarrow\Omega}$ around z_0, is the cornerstone towards the construction of such integral kernels $k(\cdot,\cdot)$ involved in the iteration of the operation (B.39); see [**AndS12**, §5 and §6] or also [**AndS11**] for details. Such construction of an explicit fine resolution for $\mathcal{O}$ should be compared to the concrete fine resolutions introduced (under some restrictions on $\mathcal{X}$, among then that $\mathcal{X}$ normal and $\mathcal{X}_{\rm sing}$ smooth, as

[15] See [**AlT01**, Theorem 2.4], [**AlT08**, Theorem 4.1] or also [**TsY04**, Theorem 5.4].

[16] This was introduced in (2.26), then extended in Proposition 2.16.

well as the existence of a log resolution $\pi : \tilde{\mathcal{X}} \to \mathcal{X}$ with extra requirements) by Vincenzo Ancona and Bernard Gaveau.[17] Given a holomorphic bundle E over $\mathcal{X}$, one can as well introduce the exact fine resolution

$$0 \to \mathcal{O}(\cdot, E) \xrightarrow{\overline{\partial}_E} \mathcal{A}_{0,0}(\cdot, E) \xrightarrow{\overline{\partial}_E} \mathcal{A}_{0,1}(\cdot, E) \cdots \xrightarrow{\overline{\partial}_E} \mathcal{A}_{0,N}(\cdot, E) \to 0, \tag{B.49}$$

where the fine sheaves $\mathcal{A}_{0,q}$ are those which realize the fine resolution (B.47) of the structural sheaf $\mathcal{O}$. As in (B.35), one has, for $q = 1, \ldots, N = \dim \mathcal{X}$,

$$\check{H}^q(\mathcal{X}, \mathcal{O}(\cdot, E)) = \frac{\operatorname{Ker} \overline{\partial}_E : \mathcal{A}_{0,q}(\mathcal{X}, E) \longrightarrow \mathcal{A}_{0,q+1}(\mathcal{X}, E)}{\operatorname{Im} \overline{\partial}_E : \mathcal{A}_{0,q-1}(\mathcal{X}, E) \longrightarrow \mathcal{A}_{0,q}(\mathcal{X}, E)}; \tag{B.50}$$

see [**AndS12**, Corollary 1.3]. Serre's duality theorem (see (B.37) or (B.38) for the context of N-dimensional complex manifolds) can also be formulated within the context of complex analytic spaces (see [**RSW17**, Theorem 1.3]) once ad hoc coherent sheaves $'\mathcal{B}^{N,\bullet}$ of $(N, \bullet)$-currents have been defined (as the $\mathcal{A}_{0,\bullet}$ previously were, to provide the fine resolution (B.47) of $\mathcal{O}$). The particular case when $\mathcal{X}$ is locally Cohen–Macaulay at any of its points is of most interest; see [**RSW17**, Theorem 1.2]. Such ad hoc coherent sheaves provide in this particular case an exact resolution

$$0 \to \omega_{\mathcal{X}}^N \xrightarrow{i} {}'\mathcal{B}^{(N,0)} \xrightarrow{\overline{\partial}} {}'\mathcal{B}^{(N,1)} \xrightarrow{\overline{\partial}} \cdots \xrightarrow{\overline{\partial}} {}'\mathcal{B}^{(N,N)} \to 0, \tag{B.51}$$

where the dualizing sheaf $\omega_{\mathcal{X}}^N$ is introduced in Remark B.2.5 and the action of $\overline{\partial}$ is here understood in the sense of currents (instead of forms).

B.3. Hermitian bundles

B.3.1. Hermitian complex bundles over a complex manifold. Let E be a complex bundle with rank m over $\mathcal{X}_{\mathbb{R}}$, where $\mathcal{X}$ is a complex N-dimensional complex manifold.

DEFINITION B.39. A (smooth) hermitian metric on E is a positive global section $|\ |^2$ of $(C^\infty)_E(E)$ (where E is considered as a $2(N+m)$ differentiable manifold) such that its restriction $\xi \mapsto |\xi|_z^2$ to each fiber $E_z = \pi^{-1}(\{z\})$ defines a hermitian metric $\langle \xi, \eta \rangle_z$ on the m-dimensional $\mathbb{C}$-vector space E_z. The metric is said to be continuous if the positive function $|\ |$ is continuous on E.

Let $r', r'' \in \mathbb{N}$. The hermitian metric $|\ |$ on E induces a sesquilinear mapping

$$\langle\ ,\ \rangle^{r',r''} : C^\infty_{r'}(\mathcal{X}, E) \times C^\infty_{r''}(\mathcal{X}, E) \longrightarrow C^\infty_{r'+r''}(\mathcal{X}). \tag{B.52}$$

Because of the existence of smooth partitions of unity, it is convenient to express it above a trivialization open subset U around z, which is also a local chart for $\mathcal{X}$ around z with ζ as a coordinate system. Let e_z be a local frame for E above U. Given two differential forms

$$\varphi = \sum_{j=1}^m \varphi_j \otimes e_{z,j} \in C^\infty_{r'}(U, E),\ \psi = \sum_{j=1}^m \psi_j \otimes e_{z,j} \in C^\infty_{r''}(U, E),$$

[17] See [**AnG92A**, **AnG92B**, **AnG94**, **AnG10**] as well as [**AnG06**] for a more complete bibliography as well as relations with Leray residue calculus; sections of the fine sheaves $(\mathcal{A}_{0,\bullet})_{\mathcal{X},\pi}$ constructed there are differential forms on the log resolution with specified jet conditions along the exceptional divisor.

one defines $\langle \varphi , \psi \rangle^{(r',r'')} \in C^\infty_{r'+r''}(U)$ by

$$\langle \varphi , \psi \rangle^{r',r''} : z \in U \longmapsto \sum_{j=1}^{m} \sum_{j'=1}^{m} \varphi_j(w) \wedge \overline{\psi_{j'}(z)} \, \langle e_{z,j} , e_{z,j'} \rangle_z , \tag{B.53}$$

where $\overline{d\zeta_j} = d\bar{\zeta}_j$ and $\overline{d\bar{\zeta}_j} = d\zeta_j$ for $j = 1, \dots, N$. Given a connection D on E (see Definition B.24 in §B.2.1.2 above) one can define the notion of *compatibility of* D *with the hermitian metric* $|\ |$.

DEFINITION B.40. A connection D on the complex bundle E is compatible with the smooth hermitian metric $|\ |$ if and only if, for any $r', r'' \in \mathbb{N}$, for any $\varphi \in C^\infty_{r'}(\mathcal{X}, E)$, and any $\psi \in C^\infty_{r''}(\mathcal{X}, E)$, one has

$$d \, \langle \varphi, \psi \rangle^{r',r''} = \langle D \, \varphi, \psi \rangle^{r'+1,r''} + (-1)^{r'} \, \langle \varphi , D \, \psi \rangle^{r',r''+1}. \tag{B.54}$$

Let $E \to \mathcal{X}$ be a complex bundle over $\mathcal{X}_{\mathbb{R}}$ equipped with a (smooth) hermitian metric $|\ |$ and let D be a connection on E which is compatible with such metric $|\ |$. Given $z \in E$, let $e_z^\perp$ be a local frame around z which is orthonormal with respect to the metric $|\ |$, that is $\langle e^\perp_{z,j}(z) , e^\perp_{z,k}(z) \rangle_z \equiv \delta_j^k$ in a neighborhood of z for any $1 \leq j, k \leq m$. The compatibility condition (B.55) implies that

$$\mathcal{M}(D; e_z^\perp; e_z^\perp) = -{}^t\big(\overline{\mathcal{M}(D; e_z^\perp; e_z^\perp)}\big), \tag{B.55}$$

that is

$$i \, \mathcal{M}(D; e_z^\perp; e_z^\perp) \in \mathrm{Herm}(C^\infty_1(U)), \tag{B.56}$$

where $\mathrm{Herm}(C^\infty_1(\cdot))$ is the sheaf of hermitian (m, m) matrices with entries in C^∞_1. It follows from the fact that $d \circ d = 0$ that

$$i \, \Theta_E(D) \in C^\infty_2(\mathcal{X} , \mathrm{Herm}(E, E)) \subset C^\infty_2(\mathcal{X} , \mathrm{Hom}_{\mathbb{C}}(E, E)). \tag{B.57}$$

REMARK B.41. When $E = L$ is a complex bundle with rank 1 but is not necessarily holomorphic, this means that

$$i \, \Theta_L(D) \in C^\infty_2(\mathcal{X}, \mathbb{R}) \subset C^\infty_2(\mathcal{X}, \mathbb{C}) = C^\infty_2(\mathcal{X}, \mathrm{Hom}_{\mathbb{C}}(L, L)). \tag{B.58}$$

B.3.2. Hermitian holomorphic bundles over a complex analytic space. Let $\mathcal{X}$ be an N-dimensional complex analytic space and let $E \to \mathcal{X}$ be a holomorphic bundle with rank m over $\mathcal{X}$; see Definition B.4 in §B.1.3.

DEFINITION B.42. The holomorphic bundle $E \to \mathcal{X}$ is equipped with a smooth hermitian metric $|\ |$ if each stalk $E_z = \pi^{-1}(\{z\})$ is equipped with a hermitian metric $|\ |_z^2$ such that for any $z_0 \in \mathcal{X}$, $z \longmapsto |s(z)|_z^2$ is a local section of C^∞ around z_0 as soon as s is a local holomorphic section of E in a neighborhood of this point.

REMARK B.43. If the map $z \mapsto |s(z)|^2_{z_0}$ is just continuous in a neighborhood of z_0 for any $z_0 \in \mathcal{X}$ and any holomorphic section of E in a neighborhood of z_0, $(E, |\ |)$ is said to be equipped with a continuous metric.

B.3.2.1. *Conjugation.* Let $(E \to \mathcal{X}, |\ |)$ be a holomorphic vector bundle that is equipped with a hermitian metric. This metric induces a linear isometry (called *complex conjugation*) between $E \to \mathcal{X}$ and $E^* \to \mathcal{X}$: if s is a holomorphic section of $E \to \mathcal{X}$, the conjugate section $s^* : \mathcal{X} \to E^*$ of $E^* \to \mathcal{X}$ is defined by

$$s^*(z)(\xi) = \langle \xi, s(z) \rangle_z , \ \forall \, z \in \mathcal{X} , \ \forall \xi \in E_z. \tag{B.59}$$

This isometry extends naturally to an isometry between spaces of differential forms with values in E and corresponding spaces of differential forms with values in E^*.

B.3.2.2. *Chern connection of a holomorphic bundle $(E, |\ |)$ over a complex manifold.*

DEFINITION B.44. Let E be a holomorphic bundle with rank m over an N-dimensional complex manifold $\mathcal{X}$. Suppose that E is equipped with a hermitian metric $|\ |$. The *compatibility of the connection D with the hermitian metric* (see Definition B.40) together with the additional constraint[18] that $D'' = \bar{\partial}_E$ (see Remark B.25 and §B.2.2.1) impose the unicity of D. Such connection $D = D_{E,|\ |}$ exists and is called the *Chern connection* of the hermitian holomorphic bundle $(E, |\ |)$ over $\mathcal{X}$.

PROPOSITION B.45. *Let $(E, |\ |)$ be a holomorphic m bundle over a complex manifold $\mathcal{X}$, $z \in \mathcal{X}$, and let e_z be a holomorphic frame for E around z. Let $G_{|\ |, e_z}$ be the Gram matrix $[\langle e_{z,j}, e_{z,k}\rangle]_{1\leq j,k\leq m}$. Then the matrix of the Chern connection $D_{E,|\ |}$ expressed in the holomorphic frame e_z is*

$$\mathcal{M}(D_{E,|\ |}; e_z; e_z) = G^{-1}_{|\ |,e_z}\, \partial G_{|\ |,e_z} \tag{B.60}$$

while its curvature matrix expressed in the same holomorphic local frame is

$$\Theta(D_{E,|\ |}; e_z; e_z) = \bar{\partial}\big(G^{-1}_{|\ |,e_z}\, \partial G_{|\ |,e_z}\big). \tag{B.61}$$

It also follows from (B.57) that if $e_z^{\perp}$ is a local orthonormal frame[19] for E in a neighborhood of $z \in \mathcal{X}$, then one can express $i\,\Theta(D_{E,|\ |})$ in a neighborhood of z as

$$i\,\Theta(D_{E,|\ |}) = \sum_{j=1}^{m}\sum_{j'=1}^{m}\Big(\sum_{k=1}^{N}\sum_{k'=1}^{N}\varphi_{j,k}^{j',k'}\, d\zeta_{k'} \wedge d\bar{\zeta}_{k'}\Big) \otimes \big(e^{\perp}_{z,j} \otimes (e^{\perp}_{z,j'})^*\big), \tag{B.62}$$

with

$$\varphi_{j',k'}^{j,k} = \overline{\varphi_{j,k}^{j',k'}} \tag{B.63}$$

for all $1 \leq j, j' \leq m$ and $1 \leq k, k' \leq N$.

B.3.3. Chern and Segre forms and classes. Let L be a (holomorphic) line bundle over an N-dimensional complex analytic space $\mathcal{X}$. Suppose that L is equipped with a hermitian metric $|\ |$ which for the time being is assumed to be smooth. Given a point $z \in \mathcal{X}$, the $(1,1)$-smooth form $dd^c \log |s|^2$ does not depend on the choice of the holomorphic section s of L in a neighborhood of z. Then, one defines globally a $(1,1)$-form $c_1(L, |\ |)$ over $\mathcal{X}$ by setting $c_1(L, |\ |)(z) = -dd^c \log |e_z|^2$, where e_z is a local holomorphic frame in a neighborhood of z.

DEFINITION B.46. Given a holomorphic line bundle L equipped with a smooth hermitian metric $|\ |$ over the complex analytic space $\mathcal{X}$, its (total) *Chern form* is the closed form $c(L, |\ |) = 1 + c_1(|\ |, L)$, where $c_1(|\ |, L)(z) = -dd^c \log |e_z|^2$, e_z being any holomorphic local frame around the current point[20] $z \in \mathcal{X}$. The $(1,1)$-form $c_1(L, |\ |)$ is called the 1-*Chern form of* $(L, |\ |)$. The de Rham cohomology class of

[18] Such constraint is called *compatibility with the complex structure of E*.

[19] Observe that, unless the hermitian metric is trivial, this is impossible to reconcile with the fact that e_z could be a local holomorphic frame around z.

[20] In case the metric is continuous instead of being smooth, the notion of Chern form needs to be replaced by that of *Chern current* $c(L, |\ |) = [1] + c_1(L, |\ |)$, where $c_1(L, |\ |) = -dd^c([\log |e_z|^2])$, where the action of dd^c is understood in the sense of currents.

the Chern form $c(L, |\ |)$ is independent on the metric $|\ |$ and is called the *Chern class* of the holomorphic line bundle L, while the de Rham cohomology class of $c_1(L, |\ |)$ is called the 1-*characteristic class* of L.

When L is a holomorphic line bundle equipped with a smooth hermitian metric $|\ |$, one can define as well its *Segre form* $s(L, |\ |)$ as the closed form

$$s(L, |\ |) = \frac{1}{c(L, |\ |)} := 1 + \sum_{\ell=1}^{N} (-1)^\ell \big(c_1(L, |\ |)\big)^\ell. \tag{B.64}$$

For each $\ell = 1, \ldots, N$, $s_\ell(L, |\ |) := (-1)^\ell (c_1(L, |\ |))^\ell$ is called the ℓ-*Segre form* of $(L, |\ |)$. When $(E, |\ |)$ is a holomorphic m bundle with $m \in \mathbb{N}^*$ over $\mathcal{X}$, one can attach to it the $(N + m - 1)$-dimensional complex analytic space $\operatorname{Proj} E$ with its projection $\pi : \operatorname{Proj} E \longmapsto \mathcal{X}$; see Definition B.12. The tautological line bundle $\boldsymbol{L} = \mathcal{O}_{\operatorname{Proj} E}(-1)$, whose definition is analogous to that of the tautological bundle $\mathcal{O}_{\mathbb{P}^N_{\mathbb{C}}}(-1)$ (see Example B.9)[21] inherits a metric $|\ |_{\mathrm{proj}}$ from the metric $|\ |$ on E as follows. One can view $\boldsymbol{L}$ as a subbundle of the m bundle over the complex analytic space $\operatorname{Proj} E$ which is realized by pulling back E through the morphism π. The metric on $\boldsymbol{L}$ is then defined by

$$|\ |_{\mathrm{proj}} = \big(\pi^*(|\ |)\big)_{|\boldsymbol{L}}. \tag{B.65}$$

Given the first (formal) quality in (B.64), it is natural to profit from the fact that π is a submersion to define the total Segre form of the metrized holomorphic bundle $(E, |\ |)$ as follows, hence the total Chern class of $(E, |\ |)$.

DEFINITION B.47. Given a holomorphic m bundle E equipped with a smooth hermitian metric $|\ |$ over the N-dimensional complex analytic space $\mathcal{X}$, its *total Segre form* $s(E, |\ |)$ is the smooth closed form on $\mathcal{X}$ defined by

$$\begin{aligned} s(E, |\ |) &= \pi_*\Big(\frac{1}{c(\boldsymbol{L}, |\ |_{\mathrm{proj}})}\Big) = 1 + \sum_{\ell=1}^{N} s_\ell(E, |\ |) \\ &:= \sum_{\ell=0}^{N} (-1)^{\ell+m-1} \big(\pi_* \big(c_1(\boldsymbol{L}, |\ |_{\mathrm{proj}})\big)\big)^{\ell+m-1}, \end{aligned} \tag{B.66}$$

where $\boldsymbol{L} = \mathcal{O}_{\operatorname{Proj}(E)}(-1)$ and the metric $|\ |_{\mathrm{proj}}$ on this holomorphic line bundle $\boldsymbol{L}$ over $\operatorname{Proj}(E)$ is the restricted metric defined by (B.65). Its (total) *Chern form* is then defined as the smooth closed form

$$c(E, |\ |) = \frac{1}{1 + \sum_{\ell=1}^{N} s_\ell(E, |\ |)} = 1 - \sum_{\ell=1}^{N} s_\ell(E, |\ |) + \cdots = 1 + \sum_{\ell=1}^{N} c_\ell(|\ |, E). \tag{B.67}$$

For each $\ell = 1, \ldots, \dim \mathcal{X}$, $s_\ell(E, |\ |)$ (respectively $c_\ell(E, |\ |)$) is called the ℓ-*Segre form* (respectively, ℓ-Chern form)[22] of $(E, |\ |)$. The de Rham cohomology classes of $s(E, |\ |)$ and $c(E, |\ |)$ are independent on the metric $|\ |$ and are called respectively the (total) *Segre class* and *Chern class* of the m-bundle E. The de Rham cohomology classes $c_\ell(E, |\ |)$, $\ell \in \mathbb{N}^*$, are called the *characteristic classes* of the holomorphic bundle E.

---

[21] In particular $\operatorname{Proj} E = \mathcal{X}$ and $\boldsymbol{L} = E$ when E is a line bundle over $\mathcal{X}$.

[22] One has $c_\ell(E, |\ |) = 0$ as soon as $\ell > m$; see Proposition B.48 and Remark B.49. Therefore the characteristic classes $c_\ell(E, |\ |)$ equal 0 as soon as $\ell > \min(m, N)$.

We now suppose that $\mathcal{X}$ is a complex N-dimensional manifold. Given an m holomorphic bundle $(E, |\ |)$ equipped with a smooth hermitian metric, one can introduce the Chern connection $D_{E,|\ |}$ of the metrized bundle $(E, |\ |)$; see Definition B.44 in §B.3.2.2. Consider the Chern curvature tensor $\Theta_{E,|\ |}$ of $(E, |\ |)$. Given any holomorphic local frame e_z in a neighborhood of $z \in \mathcal{X}$, one can express $\Theta_{E,|\ |}$ around z in such frame as

$$\Theta_{E,|\ |} = \sum_{1 \leq j,j' \leq m} \omega_{j,j'}\, e_{z,j} \otimes e^*_{z,j'};$$

see more precisely Proposition B.45, (B.61), for the explicit expression of the $(1,1)$-forms $\omega_{j,j'}$ in terms of the Gram matrix of the $e_{z,j}$, $j = 1, \ldots, m$. The following proposition provides a re-interpretation of the Chern form within the context of differentiable complex geometry; see [**Mou04**, Proposition 6].

PROPOSITION B.48. *Let $(E, |\ |)$ be a hermitian holomorphic bundle over a complex manifold $\mathcal{X}$. The differential form*

$$\det\Big(\frac{i}{2\pi}\,\Theta_{E,|\ |} + \mathrm{Id}_E\Big) \overset{\text{locally}}{=} \det\Big(\sum_{1 \leq j,j' \leq m} \Big(\frac{i}{2\pi}\,\omega_{j,j'} + \delta_j^{j'}\Big)\, e_{z,j} \otimes e^*_{z,j'}\Big)$$

is globally defined as a characteristic polynomial and equals the Chern form $c(E, |\ |)$.

REMARK B.49. It follows from Proposition B.48 that $c_\ell(E, |\ |) = 0$ as soon as $\ell > \min(m, N)$.

APPENDIX C

Positivity on complex analytic spaces

C.1. Positive forms, positive currents

C.1.1. Positivity of differential forms on a $\mathbb{C}$-vector space. Let W be a complex vector space of dimension N and let $(\xi_1,\dots,\xi_N)$ be complex coordinates on W once a basis $\{e_1,\dots,e_N\}$ has been chosen. Denote as $\overline{W}$ its conjugate vector space. Then the vector space W admits a canonical orientation, the rule being that the form

$$\frac{1}{(2i)^N}\bigwedge_{j=1}^{N} d\overline{\xi}_j \wedge d\xi_j = \bigwedge_{k=1}^{N} (du_k \wedge dv_k)$$

($\xi_j = u_j + iv_j$, $j = 1,\dots,N$) is a *volume form* on $\mathbb{R}^{2N}$. Any holomorphic change of coordinates preserves the canonical orientation.

EXAMPLE C.1. Take as $W = E_z$ the holomorphic complex tangent space $T^{1,0}_{\mathcal{X},z}$ at the point z of a complex manifold $\mathcal{X}$ of dimension N. The previous observation induces the existence of an orientation on $\mathcal{X}$. That is, it induces the existence of a nonvanishing section of $(C^{\infty}_{N,N})_{\mathcal{X}}$ considered as a volume form on $\mathcal{X}_{\mathbb{R}}$, hence as a *positive* (N,N)-form. It is convenient to materialize such a notion of positivity on a N-dimensional complex manifold through the interplay between holomorphic and antiholomorphic coordinate systems. This point of view will be of great importance to us. Note, for example, that real differential geometry, or geometry within the nonarchimedean setting,[1] etc., lacks this useful tool.

DEFINITION C.2. Let W be a complex vector space of dimension N equipped with its canonical orientation and let

$$\omega \in \bigwedge^{p,p} W^* := \bigwedge^{p} W^* \otimes \bigwedge^{p} \overline{W^*}$$

be a differential (p,p)-form on W, where $0 \le p \le N$. The form ω is said to be *positive* or *nonnegative* if and only if one has

$$\omega \wedge \bigwedge_{j=1}^{N-p} (i\,\alpha_j^* \wedge \overline{\alpha_j^*}) \ge 0 \tag{C.1}$$

for any $\alpha_1^*,\dots,\alpha_{N-p}^* \in W^*$. It is called *strongly positive* if it expresses as a finite sum

$$\omega = \sum_{\iota} \lambda_\iota \bigwedge_{\kappa=1}^{p} (i\beta_{\iota_\kappa}^* \wedge \overline{\beta_{\iota_\kappa}^*}), \quad \lambda_\iota \ge 0,\ \beta_{\iota_\kappa}^* \in W^* \text{ for } \kappa = 1,\dots,N-p. \tag{C.2}$$

[1] That is geometry on analytic Berkovich spaces [**ChD12, GubK17**], which stands as an obligatory companion to complex analytic geometry when dealing with intersection theory in diophantine geometry; see §6.6.2.

The (p,p)-form ω is said to be *negative* (respectively *strongly negative*) if and only if $-\omega$ is positive (respectively *strongly positive*).

REMARK C.3. The notions of positivity and strong positivity for (p,p)-forms on E indeed differ when $2 \le p \le N-2$. A positive (p,p)-form $\omega = i^{p^2}\beta^* \wedge \overline{\beta^*}$ is strongly positive if and only if β splits as the wedge product of p elements of W^*. It is not always the case as soon as $2 \le p \le N-2$.

REMARK C.4. Any strongly positive (p,p)-form ω on W or also, by duality, any positive (p,p)-form ω on W^* is real, that is, it satisfies $\overline{\omega} = \omega$. Moreover, it is equivalent to saying that a $(1,1)$-form expressed in a system of complex coordinates as $\omega = i\sum_{j,k} h_{j,k}\, d\xi_j \wedge d\overline{\xi}_k$ is positive and that the sesquilinear form

$$\sum_{1\le j,k\le N} h_{j,k}\, d\xi_j \otimes d\overline{\xi}_k : (w_1,\dots,w_N)\in \mathbb{C}^N \longmapsto \sum_{1\le j,k\le N} h_{j,k}\, w_j\, \overline{w_k} \tag{C.3}$$

defines a hermitian semipositive metric on $\mathbb{C}^N$.

Any wedge product of strongly positive(p_ι,p_ι)-forms on W is strongly positive. If all wedge products, but one, which is assumed to be positive, are strongly positive, then the wedge product remains positive.

C.1.2. Positive currents on a complex manifold.

DEFINITION C.5. Let $\mathcal{X}$ be an N-dimensional complex manifold and let T be a current in ${}'\mathcal{D}^{(p,p)}(\mathcal{X})$ for $0 \le p \le N$. The current T is said to be *positive* (respectively *strongly positive*) if and only if $\langle T, \varphi\rangle \ge 0$ for any test form $\varphi \in \mathcal{D}^{(N-p,N-p)}(\mathcal{X})$ is a strongly positive (respectively positive) element[2] in $\bigwedge^{p,p}(T^{1,0})^*_z$ for any z in its support.

Let T be a positive (p,p)-current on the complex manifold $\mathcal{X}$ and let $z\in\mathcal{X}$. If

$$\begin{aligned} T &= \sum_{1\le i_1<\cdots<i_p\le N}\ \sum_{1\le j_1<\cdots<j_p\le N} T^{I,J}\, d\zeta_I \wedge d\overline{\zeta}_J \\ &= i^{p^2} \sum_{|I|=|J|=p} \mathrm{T}^{I,J}\, d\zeta_I \wedge d\overline{\zeta}_J \end{aligned} \tag{C.4}$$

is the local representation of T in a local coordinate system $\zeta = \zeta_z$ in U about z (see (B.29)), then any distribution $\mathrm{T}^{K,K} = i^{-p^2} T^{K,K} \in \mathcal{D}'(U)$ is a positive measure for any ordered multi-index K with length p. Moreover, for any pair (I,J) of such ordered multi-indices with respective cardinalities $|I|,|J| \subset \{1,\dots,N\}$, $T^{I,J}$ is a complex measure and one has

$$\mathrm{T}^{I,J} = \overline{\mathrm{T}^{J,I}}. \tag{C.5}$$

Furthermore, for any positive real parameters $\tau_1,\dots,\tau_N$,

$$\begin{aligned} \tau_I\tau_J|\mathrm{T}^{I,J}| = \tau_I\tau_J|\mathrm{T}^{I,J}| &\le 2^{N-p} \sum_{\substack{1\le k_1<\cdots<k_p\le N\\ |I|\cap|J|\subset|K|\subset|I|\cup|J|}} \tau_K^2\, \mathrm{T}^{K,K} \\ &\le 2^{N-p} \sum_{\substack{1\le k_1<\cdots<k_p\le N\\ |I|\cap|J|\subset|K|\subset|I|\cup|J|}} \tau_K^2\, |\mathrm{T}^{K,K}|, \end{aligned} \tag{C.6}$$

---

[2]See Definition C.2 with $E = (T^{1,0}_{\mathcal{X}})_z$ and $\omega = \varphi(z)$.

where $\tau_{K'} = \prod_{\ell \in K'} \tau_\ell$ for any subset L with cardinal $|K'| = p \in \{1, \ldots, N\}$; see, e.g., [**De**, III, Proposition 1.14]. Every (p,p)-current of order zero[3] (such is the case, for example, for positive currents) is said to be normal as soon as dT remains a current of order 0. In particular, any closed positive current is normal (since T has order 0 and $dT = 0$ also). Proposition C.6 will be quite useful;[4] see [**De**, III, Proposition 2.1].

PROPOSITION C.6 (Dimension principle). *Any normal current $T \in {}'\mathcal{D}^{(p,p)}(\mathcal{X})$, whose support is contained in a closed analytic subset $V \subset \mathcal{X}$ with* $\operatorname{codim}_{\mathcal{X}} V > p$, *is necessarily the null current on $\mathcal{X}$.*

Closed, positive $(1,1)$-currents on a complex manifold $\mathcal{X}$ are represented locally in terms of locally integrable *plurisubharmonic functions.*

DEFINITION C.7. A function $\phi : \Omega \subset \mathbb{C}^N \to [-\infty, +\infty[$ is said to be *plurisubharmonic* (psh) in Ω if and only if it is upper semicontinuous in Ω and its restriction to the intersection of Ω with any complex line of $\mathbb{C}^N$ is subharmonic.[5] A function ϕ is psh in an open subset U of an N-dimensional complex manifold $\mathcal{X}$ if, for any given $z \in U$, there is a local system of coordinates ζ_z in U_z with $z \in U_z \subset U$ such that $\phi \circ \zeta_z^{-1}$ is psh in $\zeta_z(U_z) \subset \mathbb{C}^N$.

PROPOSITION C.8. *Any locally integrable psh function ϕ in a complex manifold $\mathcal{X}$ defines a closed positive $(1,1)$-current T as $T = dd^c[\phi]$. Conversely, any closed, positive $(1,1)$-current on $\mathcal{X}$ can be locally expressed about each $z \in \mathcal{X}$ as $dd^c[\phi_z]$, where ϕ_z is psh and locally integrable about z.*

Given closed positive $(1,1)$-currents $T_1, \ldots, T_m \in {}'\mathcal{D}^{(1,1)}(\mathcal{X})$ and a closed positive (p,p)-current T with $m + p \leq N$, one can define a closed positive current $(T_m \wedge \cdots \wedge T_1) \wedge T$ under the additional condition that each of the T_j's can be locally defined as $dd^c u_{j,z}$, where $u_{j,z}$ is a psh function that defines an element of $L^\infty_{\text{loc}}(\mathcal{X}_{\mathbb{R}})$ about z. In this case, the current $T_m \wedge \cdots \wedge T_1$ is defined about $z \in \mathcal{X}$ by iteration

$$\text{(C.7)}\quad (T_{j+1} \wedge \cdots \wedge T_1) \wedge T = dd^c\big(u_{z,j+1}\,\big((T_j \wedge \cdots \wedge T_1) \wedge T\big)\big), \quad j = 0, \ldots, m-1.$$

Iteration (C.7) is valid since locally bounded Borel functions operate by multiplication on complex measures. This procedure goes back to the work of Eric Bedford and Alan B. Taylor [**BedT76, BedT82**]; see also [**De**, III, Proposition 3.2]. If the functions $u_{z,j}$ cannot be chosen locally to be bounded about z, then we introduce for each of them its unbounded locus $\operatorname{sing}_{U_z}(u_{z,j})$. In other words, we consider the closed subset of U_z consisting of points in U_z about which $u_{z,j}$ fails to be locally bounded. The inductive procedure in (C.7) can still be carried through under the assumption

$$\text{(C.8)}\quad \forall\, J \subset \{1, \ldots, m\}, \quad \mathscr{H}_{2(N-p-|J|)+1}\Big(U_z \cap \operatorname{Supp} T \cap \bigcap_{j \in J} \operatorname{sing}_{U_z}(u_{z,j})\Big) = 0,$$

where $\mathscr{H}_{2(N-p-|J|)+1}$ means the $2(N-p-|J|)+1$-dimensional Hausdorff measure. One needs to invoke as a justification for this *Chern–Nirenberg inequalities*; see [**De**, III, Theorem 4.5].

[3]This means that current is expressed with measure coefficients in local coordinates about any point in $\mathcal{X}$.

[4]It will be specified in Proposition C.26 that will come next.

[5]For example, if ρ is a convex (hence continuous) function in an open subset $u \subset \mathbb{R}^N$, the function $\phi : z \mapsto \rho(\operatorname{Re} z)$ is plurisubharmonic (and continuous) in the tubular domain $u + i\mathbb{R}^N$.

EXAMPLE C.9. Let f be a holomorphic section of the trivial bundle $\mathbb{C}^m$ equipped with its canonical hermitian metric $\| \ \|$. The function $\log \|f\|^2$ is the archetypal example of a psh function not locally bounded about any point in the closed analytic subset $V(f)$, though it is locally integrable.[6] The iterative procedure (C.7) carries through to define the positive closed currents $(dd^c \log \|f\|^2)^r$ as far as $r \leq \operatorname{codim}_{\mathcal{X}} V(f)$. But condition (C.8) cannot be invoked anymore in order to define (at least in such a way) higher wedge powers of the positive closed $(1,1)$-current $dd^c \log \|f\|^2$.

C.1.3. Trivial extension of positive closed (p,p)-currents. The following notion (arising from pluricomplex potential theory) is directly related to that of plurisubharmonicity and the definition of $\operatorname{sing}_{\mathcal{X}}(u)$ for a psh function as given in §C.1.2.

DEFINITION C.10. A subset of complex analytic manifold $\mathcal{X}$ is called *complete pluripolar* if it can be locally defined by $u^{-1}(\{-\infty\})$, where u is a locally integrable psh function in $\mathcal{X}$. It is called pluripolar if it is subset of a complete pluripolar subset.

EXAMPLE C.11. Every closed analytic subset V of a complex analytic manifold $\mathcal{X}$, is pluripolar since the function $\log \|f\|^2$, where f is a local section of the trivial bundle $\mathbb{C}^m$ (equipped with its canonical metric) about $z \in \mathcal{X}$ is locally integrable and psh; see Example C.9.

The extension theorem by Hassine El Mir [**EM84**] asserts that pluripolar sets do not constitute an obstacle for the (trivial) extension of positive closed currents.

THEOREM C.12. *Let $\mathcal{X}$ be a complex analytic manifold and let T be a positive closed (p,p)-current in the complement of a closed complete pluripolar subset V°. If any point a in V° admits a compact neighborhood $\overline{U}_a$ with*[7]

$$\sum_{|K|=p} |T_a^{K,K}|\,(\overline{U}_a) < +\infty,$$

then the positive current obtained locally in U_a through trivial extension by 0 at the point a of the measure coordinate coefficients $T_a^{I,J}$ for any $a \in V^\circ$ is a positive closed (p,p)-current on $\mathcal{X}$.

EXAMPLE C.13. If T is a positive closed (p,p)-current $\mathcal{X}$ and if $V^\circ \subset \mathcal{X}$ is a pluripolar closed subset, then Theorem C.12 implies that the current

$$T_{|\mathcal{X}\setminus V^\circ} \in {}'\mathcal{D}^{(p,p)}(\mathcal{X} \setminus V^\circ)$$

admits a trivial extension by 0 on W to the whole complex manifold $\mathcal{X}$. This extended current is denoted by $\mathbf{1}_{\mathcal{X}\setminus V^\circ} \cdot T \in {}'\mathcal{D}^{(p,p)}(\mathcal{X})$. Then, one has the decomposition of T in ${}'\mathcal{D}^{(p,p)}(\mathcal{X})$ given by

(C.9)
$$T = \mathbf{1}_{\mathcal{X}\setminus V^\circ} \cdot T + \big(T - \mathbf{1}_{\mathcal{X}\setminus V^\circ} \cdot T\big) := \mathbf{1}_{\mathcal{X}\setminus V^\circ} \cdot T + \mathbf{1}_{V^\circ} \cdot T, \text{ with } \operatorname{Supp}(\mathbf{1}_{V^\circ} \cdot T) \subset V^\circ.$$

[6] Since $\log \|f\|^2 \geq \log |f_1|^2$, it is enough to check local integrability of $\log \|f\|$ when $m = 1$. As a consequence of Weierstrass's preparation theorem, one may assume that in local coordinates, $f = f_1$ expresses as $f(\zeta) = \zeta_1^\mu + \sum_1^\mu a_\ell(\zeta_2, \dots, \zeta_N)\, \zeta_1^{\mu-\ell} = \prod_1^\mu(\zeta_1 - \xi_\ell(\zeta_2, \dots, \zeta_N))$, where μ is the local Hilbert–Samuel multiplicity; see §A.3.4. In such a well settled situation, local integrability of $\log |f|^2 = \sum_{\ell=1}^\mu \log |\zeta_1 - \xi_\ell(\zeta_2, \dots, \zeta_N)|^2$ follows from Fubini's theorem.

[7] See the local expression (C.4) with complex measure coordinates $T_a^{I,J}$ of T.

C.2. Lelong numbers and integration currents

C.2.1. Lelong numbers of positive closed currents on Stein manifolds.

DEFINITION C.14. Let $\mathcal{X}$ be an N-dimensional Stein manifold and let T be a current on $\mathcal{X}$. A psh function $\phi : \mathcal{X} \to [-\infty, +\infty[$ such that e^{ϕ} is continuous on $\mathcal{X}$ is said to be *semi-exhaustive with respect to* T if and only if $\operatorname{Supp} T \cap \phi^{-1}([-\infty, R[)$ is relatively compact in $\mathcal{X}$ for some $R \in \mathbb{R}$.

The following concept was introduced by Jean-Pierre Demailly [**De87**]; see also [**De**, III, Definition 5.4].

DEFINITION C.15. Let $\mathcal{X}$ be as above, let T be a closed positive (p,p)-current on $\mathcal{X}$, and let ϕ be a psh semi-exhaustive function with respect to T. For R as in Definition C.14 define the function

$$r \in]-\infty, R[\longmapsto \int_{\phi^{-1}([-\infty,r[)} (dd^c\phi)^{N-p} \wedge T, \tag{C.10}$$

where the wedge product of positive currents under the integral sign is defined inductively by (C.7). The function (C.10) is a positive, increasing function, and its limit at $r = -\infty$ equals

$$\nu_{\mathcal{X}}(T, \phi) = \int_{\phi^{-1}(\{-\infty\})} (dd^c\phi)^{N-p} \wedge T. \tag{C.11}$$

The limit (C.11) is called the *generalized Lelong number of* T *for the semi-exhaustive psh function* ϕ.

EXAMPLE C.16. Let $\mathcal{X}$ be an N-dimensional complex manifold, let T be a closed positive (p,p)-current in a relatively compact local chart U_a about $a \in \operatorname{Supp} T$ (with centered local coordinate $\zeta = U_a \leftrightarrow \Omega_a \subset \mathbb{C}^N$), and let $\phi_a : U_a \to [-\infty, +\infty[$ be the psh function $\log \|\zeta(z) - \zeta(a)\|^2$, which is semi-exhaustive with respect to T in U_a. Then

(C.12)

$$\begin{aligned}
\nu_a(T) &= \lim_{r\to 0_+} \frac{1}{r^{2(N-p)}} \int_{\zeta^{-1}(\zeta(U_a)\cap B_{\mathbb{C}^N}(0,r))} T \wedge (dd^c\phi_a)^{N-p} \\
&= \lim_{r\to 0_+} \Big(\frac{\pi^{N-p}}{(N-p)!}\, r^{2(N-p)}\Big)^{-1} \frac{\pi^{N-p}}{(N-p)!} \int_{\zeta^{-1}(\zeta(U_a)\cap \mathbb{B}_{\mathbb{C}^N}(0,r))} T \wedge (dd^c\phi_a)^{N-p} \\
&= \lim_{r\to 0_+} \frac{1}{\operatorname{vol}_{2(N-p)}(B_{N-p}(0,r))} \frac{\pi^{N-p}}{(N-p)!} \int_{\zeta^{-1}(\zeta(U_a)\cap \mathbb{B}_{\mathbb{C}^N}(0,r))} T \wedge (dd^c\phi_a)^{N-p}
\end{aligned}$$

is called the *ordinary Lelong number of the closed positive* (p,p)*-current* T *at the point* a *of its support* [**Lel**].

The fact that the inductive process (C.7) with $m = N - p$ carries under the hypothesis (C.8) implies that the ordinary Lelong number $\nu_a(T)$ of a positive closed (p,p)-current on a complex manifold at a point a of its support also interprets through the current identity

$$\mathbf{1}_{\{a\}} \cdot \big(T \wedge (dd^c\phi_a)^{N-p}\big) = \nu_a(T)\, [\{a\}] \tag{C.13}$$

(see (C.9) for the meaning of the current on the left-hand side), where

$$\langle [\{a\}], \varphi\rangle = \varphi(a) \tag{C.14}$$

for any test function $\varphi \in \mathcal{D}(\mathcal{X}) = \mathcal{D}^{(0,0)}(\mathcal{X})$. Since one has in $'\mathcal{D}^{(N,N)}(U_a)$ that

$$\begin{aligned} T \wedge (dd^c\phi_a)^{N-p} &= dd^c\big(\phi_a\,\big((dd^c\phi_a)^{N-p-1} \wedge T\big)\big) \\ &= dd^c\Big(\lim_{\lambda\to 0^+}\Big(\frac{e^{\lambda\phi_a}-1}{\lambda}\Big)\,\big((dd^c\phi_a)^{N-p-1}\wedge T\big)\Big) \\ &= \lim_{\lambda\to 0^+} dd^c\Big(\Big(\frac{e^{\lambda\phi_a}-1}{\lambda}\Big)\,\big((dd^c\phi_a)^{N-p-1}\wedge T\big)\Big) \\ &= \lim_{\lambda\to 0^+}\Big(e^{\lambda\phi_a}\,(dd^c\phi_a)^{N-p}\wedge T \\ &\qquad + \overline{\partial}\, e^{\lambda\phi_a} \wedge \frac{\partial\,\phi_a}{2i\pi} \wedge (dd^c\phi_a)^{N-p-1}\wedge T\Big), \end{aligned}$$

one also has, in the sense of currents on U_a,

$$\begin{aligned} \nu_a(T)\,[\{a\}] &= T\wedge(dd^c\phi_a)^{N-p} - \lim_{\lambda\to 0^+}|\zeta-\zeta(a)|^{2\lambda}\,(dd^c\phi_a)^{\wedge^{N-p}}\wedge T \\ &= \lim_{\lambda\to 0^+}\Big(\bar\partial\,(e^{\lambda\phi_a})\wedge\frac{\partial\,\phi_a}{2i\pi}\wedge(dd^c\phi_a)^{N-p-1}\wedge T\Big) \qquad \text{(C.15)} \\ &= \lim_{\lambda\to 0^+}\Big(\overline{\partial}\,|\zeta-\zeta(a)|^{2\lambda}\wedge\frac{\partial\,\phi_a}{2i\pi}\wedge(dd^c\phi_a)^{N-p-1}\wedge T\Big); \end{aligned}$$

see [**AndSWY17B**, Lemma 2.1].

C.2.2. Lelong numbers of positive closed currents on analytic spaces.

DEFINITION C.17. Let $\mathcal{X}$ be a purely N-dimensional complex analytic space, and let $0 \le p \le N$ and $T \in {}'\mathcal{D}^{(p,p)}(\mathcal{X})$. The current T is said to be *positive* (respectively *closed positive*) on $\mathcal{X}$ if and only if for any $a \in \operatorname{Supp} T$, there is a local embedding $\iota\ : U_a \hookrightarrow \Omega \subset \mathbb{C}^{N'}$ about a such that the current $\iota_* T \in {}'\mathcal{D}_\Omega^{(N'-(N-p),N'-(N-p))}(\Omega)$ identified with $T_{|U_a}$ though the identification (B.33) is positive (respectively closed positive).

Given a closed positive (p,p)-current on $\mathcal{X}$, one can then define its Lelong number at $a \in \operatorname{Supp} T$ as follows; see, e.g., [**AndSWY17B**, §2.2].

DEFINITION C.18. Let T be a closed positive (p,p)-current on $\mathcal{X}$ and let $z \in \mathcal{X}$. If $\iota\ : U_a \subset \mathcal{X} \hookrightarrow \Omega \subset \mathbb{C}^{N'}$ is a local embedding of $\mathcal{X}$ about a, the Lelong number $\nu_{\iota(a)}(\iota_* T)$[8] does not depend on the local embedding ι and is called the *Lelong number* $\nu_a(T)$ *of* T *at* a.

Siu's analyticity theorem, proved within the context of complex manifolds, admits an extension to the setting of complex analytic spaces. It is the following key result [**Siu74**].

THEOREM C.19. *Let T be a closed positive (p,p)-current on an N-dimensional complex analytic space $\mathcal{X}$. For any $c > 0$, the level subset*

$$\mathrm{LV}_c(T) = \{z \in \mathcal{X}\ :\ \nu_z(T) \ge c\} \qquad \text{(C.16)}$$

is a closed analytic subset of $\mathcal{X}$, whose codimension is greater than or equal to p.

For complex Stein manifolds, Siu's analyticity theorem has been extended by J. P. Demailly [**De87**] to Lelong numbers $\nu_{\mathcal{X}}(T,\phi)$, where ϕ is a psh continuous function.

[8] See Definition C.15 or also (C.15) with $\iota_* T$ instead of T and $N' - (N-p)$ instead of p.

THEOREM C.20. *Let T be a closed positive (p,p)-current on an N-dimensional complex Stein manifold $\mathcal{X}$, let $\mathcal{T}$ be a complex manifold, acting as a parameter space, and let $\phi : \mathcal{X} \times \mathcal{T} \to [-\infty, +\infty[$ be a psh function such that e^{ϕ} is continuous and ϕ is exhaustive with respect to T.*[9] *For any $c > 0$, the level subset*

$$\mathrm{LV}_c(T, \phi) = \{t \in \mathcal{T} : \nu(T, \phi(\cdot, t)) \geq c\} \tag{C.17}$$

is a closed analytic subset of the complex manifold $\mathcal{T}$.

EXAMPLE C.21. Taking $U_{z_0} \subset \mathcal{X}$ a Stein open chart about $z^0 \in \operatorname{Supp} T$, $\mathcal{T} = U_{z^0}$ and $\phi_{z^0}(z,t) = \log \|\zeta_{z^0}(z) - \zeta_{z^0}(t)\|^2 : U_{z^0} \times U_{z^0} \to [-\infty, +\infty[$, where ζ_{z^0} is a local holomorphic system of coordinates in U_{z^0}, one recovers locally Siu's analyticity theorem on a complex manifold (namely Theorem C.19) as a particular case of Theorem C.20.

C.2.3. Analytic subvarieties and corresponding integration currents. Let $\mathcal{X}$ be a purely N-dimensional complex analytic space and let $V \subset \mathcal{X}$ be a purely $(N-M)$-dimensional closed analytic subset of $\mathcal{X}$, that we shall call also an *analytic subvariety* of $\mathcal{X}$ of pure codimension M. For any $z \in V$, let $\iota : U_z \subset \mathcal{X} \hookrightarrow \Omega \subset \mathbb{C}^{N'}$ be a local embedding of $\mathcal{X}$ about z, so that $\iota(V) \subset \iota(U_z)$ is a purely $(N-M)$-dimensional closed analytic subset of Ω lying in the purely N-dimensional complex analytic space $\iota(U_z) \subset \Omega$. Theorem C.12 implies that the mapping

$$\varphi \in {}'\mathcal{D}_{\Omega}^{(N-M,N-M)}(\Omega \setminus (\iota(V))_{\mathrm{sing}}) \longmapsto \int_{(\iota(V))_{\mathrm{reg}}} \varphi := \int_{\iota(V)} \varphi$$

has trivial extension by 0 across $(\iota(V))_{\mathrm{sing}}$. Such extension defines a current of bidegree $(N' - (N-M), N' - (N-M))$ on Ω denoted by $[\iota(V)]_{\Omega}$ and is called the *integration current on the closed purely $(N-M)$-dimensional analytic subset $\iota(V)$*. Then it follows from the identification (B.33) that such a current $[\iota(V)]_{\Omega}$ can be identified with an element in ${}'\mathcal{D}_{U_z}^{(M,M)}(U_z)$ denoted as $[V \cap U_z]_{U_z}$. The existence of smooth partitions of unity on $\mathcal{X}$ implies that such local constructions globalize to obtain an element $[V] \in {}'\mathcal{D}^{(M,M)}(\mathcal{X}) = {}'\mathcal{D}_{\mathcal{X}}^{(M,M)}(\mathcal{X})$ characterized by

$$\forall \varphi \in \mathcal{D}^{(N-M,N-M}(\mathcal{X}),\ \langle [V], \varphi \rangle = \int_{V_{\mathrm{reg}}} \varphi. \tag{C.18}$$

DEFINITION C.22. Given a purely $(N-M)$-dimensional subvariety of the purely N-dimensional complex space, the (M,M)-current $[V]$ on $\mathcal{X}$ defined by (C.18) is called the *integration current on V*.

REMARK C.23. One has $\partial [V] = \bar{\partial} [V] = 0$, hence $[V]$ is a closed positive (M,M)-current on $\mathcal{X}$, in view of Definition C.17. Therefore, one can attach to it a Lelong number $\nu_z(T)$ at any $z \in \mathcal{X}$; see Definition C.18.

The following important proposition relates Lelong numbers of integration currents on subvarieties of a complex analytic space with Hilbert–Samuel local multiplicities of subvarieties of a complex manifold; see [**Ch**, Proposition 3.15.1.2].

PROPOSITION C.24. *Let V be a purely $(N-M)$-dimensional subvariety of a purely N-dimensional complex analytic space $\mathcal{X}$. Then $\nu_z([V]) = \mathrm{mult}_{\iota(z)} \iota(V)$, for $z \in V$ and any local embedding $\iota : U_z \hookrightarrow \Omega \subset \mathbb{C}^{N'}$, where the local Hilbert–Samuel*

[9] That is, given any compact subset $K \Subset \mathcal{T}$, there is $R_K \in \mathbb{R}$ such that $\phi^{-1}([-\infty, R_K])$ is relatively compact in $\mathcal{X} \times \mathcal{T}$.

multiplicity of a closed analytic subset in a complex manifold was introduced in §A.3.4. *If one takes in particular* $M = 0$ *and* $[V] = [1]$, *then one gets*

$$\nu_z([1]) = \mathrm{mult}_{\iota(z)}\iota(\mathcal{X}) := \mathrm{mult}_z\mathcal{X}.$$

It follows from (C.18) that no multiplicity accounts in the definition of the integration current $[V]$ attached to a purely dimensional analytic subvariety of a complex analytic space. Note that the analytic subvariety $V_{\mathcal{I}}$ fails, in general, to be purely dimensional. We consider now, in place of a complex analytic purely dimensional subvariety V of $\mathcal{X}$, a coherent $\mathcal{O}_{\mathcal{X}}$-sheaf of ideals $\mathcal{I}_{\mathcal{X}}$, whose support $\mathrm{Supp}\,\mathcal{O}_{\mathcal{X}}/\mathcal{I}$ is an analytic subvariety $V_{\mathcal{I}}$. We need then to take into account multiplicities, which is done as follows.

PROPOSITION C.25. *Let* $z \in \mathrm{Supp}\,(\mathcal{O}_{\mathcal{X}}/\mathcal{I})$ *and let* f *be a finite system of elements in* $\mathcal{O}_{\mathcal{X}}(U_z)$, *where* U_z *is a neighborhood of* z, *whose germs at* z *generate* $\mathcal{I}_z$. *For any nonnegative integer* r, *the closed positive current*

$$\mathbf{1}_{V_{\mathcal{I}}} \cdot \lim_{\varepsilon \to 0_+} \left(dd^c \log(\|f\|^2 + \varepsilon)\right)^r = [M^{f,\lambda}_{|\ |,r}]_{\lambda=0} \tag{C.19}$$

$$:= \begin{cases} \left[\overline{\partial}\,\|f\|^{2\lambda} \wedge \dfrac{\partial \log\|f\|^2}{2i\pi} \wedge \left(dd^c \log(\|f\|^2)\right)^{r-1}\right]_{\lambda=0} & \text{if } 0 < r \le N \\ \left[1 - \|f\|^{2\lambda}\right]_{\lambda=0} \ \text{if } r = 0 \end{cases}$$

admits as Lelong number at z *a nonnegative integer* $e_r(\mathcal{I}, \mathbf{1}_{\mathcal{X}}, z)$ *called the* r*-Segre number of the sheaf of* $\mathcal{O}_{\mathcal{X}}$ *ideals* $\mathcal{I}$. *The* r*-Segre number of the sheaf of* $\mathcal{O}_{\mathcal{X}}$ *ideals* $\mathcal{I}$ *is independent of the choice of the generating system* f. *Moreover, one has that* $e_r(\mathcal{I}, \mathbf{1}_{\mathcal{X}}, z) = e_r(\overline{\mathcal{I}}, \mathbf{1}_{\mathcal{X}}, z)$, *where* $\overline{\mathcal{I}}$ *denotes the integral closure of the sheaf of ideals* $\mathcal{I}$, *see* §A.4.2.3, *in particular Proposition* A.37. *One has also that* $e_r(\mathcal{I}, \mathbf{1}_{\mathcal{X}}, z) = 0$, *whenever* $r < \mathrm{codim}_z(\mathcal{O}_{\mathcal{X},z}/\mathcal{I}_z)$.

Integration currents play a fundamental role in the key *support result* for normal currents; see, e.g., [**AndSWY17B**, Lemma 2.2].

PROPOSITION C.26 (Dimension principle). *Let* T *be a normal* (p,p)*-current on a complex* N*-dimensional analytic space* $\mathcal{X}$. *If* $\mathrm{Supp}\,T$ *lies in a closed analytic subset of* $\mathcal{X}$, *whose codimension is strictly larger than* p, *then one has that* $T = 0$. *If in addition the normal* (p,p)*-current* T *is* d*-closed and* $\mathrm{Supp}\,T$ *lies in a closed analytic subset* W *of pure codimension equal to* p, *then there are complex numbers* $\nu_{T,\iota}$ *such that* $T = \sum_\iota \nu_{T,\iota}\,[W_\iota]$, *where* W_ι *denote the irreducible components of* W.

EXAMPLE C.27. If T is a closed positive (p,p)-current and V is an irreducible closed analytic subset of codimension p, then one has that $\mathbf{1}_V \cdot T = \inf_{a \in V} \nu_a(T)\,[V]$. If T is a closed positive (p,p)-current, whose support is the closed analytic subset $W = \bigcup_\iota W_\iota$ of pure codimension p and the analytic sets W_ι are irreducible, then $T = \sum_\iota \min_{a \in W_\iota} \nu_a(T)\,[W_\iota]$.

Combining Proposition C.26 with Siu's analyticity theorem (namely Theorem C.19) leads to the so-called *Siu's stratification theorem* for closed positive (p,p)-currents on a complex analytic space.

THEOREM C.28. *Let* T *be a closed positive* (p,p)*-current on an* N*-dimensional complex analytic space* $\mathcal{X}$. *The decomposition of* T *as the weak limit (in the sense of currents) holds*

$$T = N_T + \lim_{\ell \to +\infty} \sum_{\ell' \le \ell} \nu_{T,\ell}[V_{T,\ell}], \tag{C.20}$$

where $(\{V_{T,\ell}\})_{\ell\in\mathbb{N}}$ *is an (at most) countable family of irreducible closed subsets of* $\mathcal{X}$ *of codimension* p*, the coefficients* $\nu_{T,\ell}$ *are strictly positive real numbers, and* N_T *is a closed positive* (p,p)*-current. The current* N_T *is negligible in the sense that for any* $c>0$*, one has* $\operatorname{codim}_{\mathcal{X}}(\mathrm{LV}_c(N_T))>p$*. Moreover, the stratification* (C.20) *is unique up to the reordering of the (at most) countable family* $\{V_{T,\ell}\}_{\ell\in\mathbb{N}}$*. The* (p,p)*-current*

$$T-N_T=\lim_{\ell\to+\infty}\sum_{\ell'\leq\ell}\nu_{T,\ell}[V_{T,\ell}]\tag{C.21}$$

is called the singular part of the closed positive (p,p)*-current* T*.*

C.2.4. Integration currents attached to sections of holomorphic bundles. In this subsection let $\mathcal{X}$ be an N-dimensional irreducible complex analytic space and let E be a holomorphic $m\geq 1$-bundle over $\mathcal{X}$.

DEFINITION C.29. Let $s:\mathcal{X}\to E$ be a global nonzero holomorphic section of E and let $V(s)=\bigcup_\iota V_\iota(s)$ be the locally finite decomposition of the closed analytic subset $V(s)$ (assumed to be nonempty) in irreducible closed analytic subsets $V_\iota(s)$. The vanishing order $\mu_\iota(s)$ of s along V_ι at a generic point $a_\iota\in(V_\iota(s))_{\mathrm{reg}}$ is well defined independent of the choice of such a generic point, and the positive current $\sum_\iota\mu_\iota(s)\,[V_\iota]$ is called the *integration current attached to the holomorphic section s.* Given $p=\operatorname{codim}_{\mathcal{X}}V(s)>0$, one denotes as $(\sum_\iota\mu_\iota(s)\,[V_\iota])_p$ the $(N-p,N-p)$-dimensional component of $\sum_\iota\mu_\iota(s)\,[V_\iota(s)]$.

When $\mathcal{X}$ is an N-dimensional manifold, there are different ways to realize the integration current $(\sum_\iota\mu_\iota(s)\,[V_\iota])_p$ through the analytic continuation process. All of the above rely on the choice of a smooth hermitian metric $|\;|$ on E. Let $E_{|\mathcal{X}\setminus V(s)}$ be the restriction of E to the open subset $\mathcal{X}\setminus V(s)$ over which s defines a line bundle $L(s)$. Namely, assume that $L_z(s)=\mathbb{C}\,s(z)\subset E_z$, when $z\in\mathcal{X}\setminus V(s)$. The $(m-1)$-holomorphic quotient bundle $E_{|\mathcal{X}\setminus V(s)}/L(s)$ over the open subset $\mathcal{X}\setminus V(s)$ inherits a hermitian metric $|\;|_s$ induced by the original metric $|\;|$ prescribed on E. The (total) Chern form $c(E_{|\mathcal{X}\setminus V(s)}/L(s),|\;|_s)$ (see §B.3.3) is a smooth form on the complex manifold $\mathcal{X}\setminus V(s)$ and extends as a locally integrable form over the whole complex manifold $\mathcal{X}$; see [**And07**, Proposition 4.2 and 4.3]. Moreover, one has the following result; see [**And07**, Theorem 1.1].

PROPOSITION C.30. *Let* $\mathcal{X},E\to\mathcal{X},|\;|,|\;|_s$ *and let* $s:\mathcal{X}\to E$ *be as above. If* $p=\operatorname{codim}_{\mathcal{X}}V(s)$*, then*

$$\Big(\sum_\iota\mu_\iota(s)\,[V_\iota(s)]\Big)_p=\Big[\bar\partial|s|^{2\lambda}\wedge\frac{\partial\log|s|^2}{2i\pi}\wedge c_{p-1}\big(E_{|\mathcal{X}\setminus V(s)}/L(s),|\;|_s\big)\Big]_{\lambda=0},\tag{C.22}$$

where

$$\lambda\longmapsto\bar\partial|s|^{2\lambda}\wedge\frac{\partial\log|s|^2}{2i\pi}\wedge c_{p-1}\big(E_{|\mathcal{X}\setminus V(s)}/L(s),|\;|_s\big)$$

is a ${}'\mathcal{D}^{(p,p)}(\mathcal{X})$*-valued meromorphic function in the complex parameter* λ *having strictly negative rational poles.*

REMARK C.31. When E is the trivial bundle $\mathbb{C}^m$, $|\ | = \|\ \|$ is the canonical hermitian metric $\|\ \|$ and $p = \mathrm{codim}_{\mathcal{X}} V(s)$, then one has also the important formula

$$\begin{aligned}(\sum_\iota \mu_\iota(s)\,[V_\iota(s)])_p &= \Big[\bar\partial \|s\|^{2\lambda} \wedge \frac{\partial \log \|s\|^2}{2i\pi} \wedge (dd^c \log \|s\|^2)^{p-1}\Big]_{\lambda=0} \\ &= \mathbf{1}_{V(s)} \cdot \lim_{\varepsilon\to 0_+} (dd^c \log(\|s\|^2+\varepsilon))^p,\end{aligned} \tag{C.23}$$

due to J. R. King [**Ki70**].

Let s^* be the conjugate section of s with respect to the smooth hermitian metric $|\ |$ on E (see §B.3.2.1) and let $D_{|\ |}$ be the Chern connection on $(E, |\ |)$; see Definition B.44. Besides (C.22), an alternative way to realize the integration current $(\sum_\iota \mu_\iota(s)\,[V_\iota])_p$, where $p = \mathrm{codim}_{\mathcal{X}} V(s)$, is given as follows; see [**And05**, Theorem 1.1] and also [**Me01**, **Me06**].

PROPOSITION C.32. *Let $\mathcal{X}, E \to \mathcal{X}, |\ |$, $s : \mathcal{X} \to E$ and p be as in Proposition C.30. Then one has that*[10]

(C.24)

$$\begin{aligned}\Big(\sum_\iota \mu_\iota(s)\,[V_\iota(s)]\Big)_p &= \Big[\Big(\bar\partial |s|^{2\lambda} \wedge \frac{s^*}{2i\pi\,|s|^2} \wedge \Big(\bar\partial\Big(\frac{s^*}{2i\pi|s|^2}\Big)\Big)^{p-1}\Big) \rfloor \frac{(D_{|\ |}s)^{\wedge^p}}{p!}\Big]_{\lambda=0} \\ &= \Big[R^{s,\lambda}_{|\ |,p} \rfloor \frac{(D_{|\ |}s)^{\wedge^p}}{p!}\Big]_{\lambda=0} = R^s_{|\ |,p} \rfloor \frac{(D_{|\ |}s)^{\wedge^p}}{p!},\end{aligned}$$

where $\rfloor$ means here the interior action of $\bigwedge^p E^*$ on $\bigwedge^p E$ and

$$\lambda \longmapsto \Big(\bar\partial |s|^{2\lambda} \wedge \frac{s^*}{2i\pi\,|s|^2} \wedge \Big(\bar\partial\Big(\frac{s^*}{2i\pi|s|^2}\Big)\Big)^{p-1}\Big) \rfloor \frac{(D_{|\ |}s)^{\wedge^p}}{p!}$$

is a ${}'\mathcal{D}^{(p,p)}_{\mathcal{X}}(\mathcal{X})$-valued meromorphic function in the complex parameter λ having strictly negative rational poles.

C.2.5. Holonomy of integration currents and consequences. Given a purely dimensional closed analytic subset V in an N-dimensional complex manifold $\mathcal{X}$ and a local chart U_a about $a \in V$, the measure coordinate coefficients $([V])^{I,J}_a$, where $1 \le i_1 < \cdots < i_p \le N$ and $1 \le j_1 < \cdots < j_p \le N$ (see (C.4)) are *regular holonomic.*[11] Since we do not want to enter into the theory of $\mathcal{O}_{\mathcal{X}}\langle \partial/\partial\zeta\rangle$-modules within this monograph, let us only quote here some operational aspects inherited from regular holonomicity that will be of crucial importance to us; see [**Bj93**, VI.1, VII.6].

PROPOSITION C.33. *Let $V \subset \Omega \subset \mathbb{C}^N$ be a closed irreducible analytic subset of Ω with $\mathrm{codim}_\Omega V = p$. For $m \in \mathbb{N}^*$, let h be a section of the trivial bundle $\mathbb{C}^m$ over Ω assumed to be equipped with its canonical metric $\|\ \|$. Let μ be any coordinate measure coefficient of $[V] \in {}'\mathcal{D}^{(p,p)}(\Omega)$ along the frame $d\zeta_I \wedge d\bar\zeta_J$, $|I| = |J| = p$. For each $a \in V$, there is an open neighborhood U_a of a in Ω and a functional equation*

$$b_{\mu,\boldsymbol{h},a}(\lambda)\,[\,\|h\|^{2\lambda} \otimes \mu\,] = \mathcal{Q}_{\mu,h,a}(\lambda, \zeta, \bar\zeta, \partial/\partial\zeta, \partial/\partial\bar\zeta)\,([\,\|h\|^{2(\lambda+1)} \otimes \mu\,]), \tag{C.25}$$

[10] We invite the reader to refer here to Proposition 3.15 for what concerns the notations used for Bochner–Martinelli residue currents (or their approximations through a complex parameter λ) attached to the section s of E when E is metrized with the smooth metric $|\ |$.

[11] The concept was introduced by Jan-Erik Björk [**Bj93**, **Bj96**, **Bj04**].

understood as an identity in $\mathcal{D}'(U_a)$*, when* $\operatorname{Re}\lambda > 0$*. In* (C.25)*,* $\mathcal{Q}_{\mu,\boldsymbol{h},a}$ *is a differential operator in* $\partial/\partial\zeta$*,* $\partial/\partial\overline{\zeta}$ *with analytic coefficients in* $\zeta, \overline{\zeta}, \lambda$ *(where* $\zeta \in U_a$ *and* $\lambda \in \mathbb{C}$*) depending polynomially on* λ*, and* $b_{\mu,h,a} \in \mathbb{Q}[\lambda]$ *is a product of factors* $\lambda + \gamma$*,* $\gamma \in \mathbb{Q}^{+*}$*. The polynomial* $b_{\mu,h,a}$ *can be chosen of minimal degree so that the identity* (C.25) *holds. In this case* $b_{\mu,h,a}$ *is called the Bernstein–Sato polynomial attached to the regular holonomic measure* μ *and to the holomorphic section* h *at the point* $a \in \operatorname{Supp}\mu$*, while* (C.25) *is called a Bernstein–Sato equation for* $\|h\|^{2\lambda} \otimes \mu$ *about* a*.*

REMARK C.34. If $m = 1$, one has a formal holomorphic equation

$$b_{\mu,h,a}(\lambda)\big(h^{\lambda} \otimes \mu\big) = \mathcal{Q}_{\mu,h,a}(\lambda, \zeta, \partial/\partial\zeta)\,\big(h^{\lambda+1} \otimes \mu\big), \tag{C.26}$$

where $b_{\mu,h,a}$ is a product of factors $\lambda + \gamma$, $\gamma \in \mathbb{Q}^{+*}$ and $\mathcal{Q}_{\mu,h,a}$ is a differential operator with analytic coefficients in $\zeta \in U_a$ and polynomial coefficients in λ. The formal identity (C.26) induces the identity

$$b_{\mu,h,a}(\lambda)\Big(\frac{|h|^{2\lambda}}{h} \otimes \mu\Big) = \overline{\mathcal{Q}_{\mu,h,a}}(\lambda, \overline{\zeta}, \partial/\partial\overline{\zeta})\,\Big(|h|^{2\lambda}\frac{\overline{h}}{h} \otimes \mu\Big) \tag{C.27}$$

in the sense of distributions in U_a, when $\operatorname{Re}\lambda > 1/2$.

C.3. Positivity of holomorphic hermitian bundles

We summarize and compare in this section various concepts of *positivity* for holomorphic hermitian vector bundles $(E \to \mathcal{X}, |\ |)$ over an N-dimensional complex manifold $\mathcal{X}$. Let $(E \to \mathcal{X}, |\ |)$ be a holomorphic vector bundle with rank m, equipped with a smooth hermitian metric $|\ |$. Keeping the notation introduced previously, it follows from (B.62) and (B.63), together with Proposition B.48 that the first Chern form $c_1(E, |\ |)$ induces a hermitian form $\theta_{E,|\ |}$ on the $N \times m$ bundle $T_{\mathcal{X}} \otimes E$. The form $\theta_{E,|\ |}$ is expressed locally about z in local coordinates $\zeta_1, \dots, \zeta_N$ on $\mathcal{X}$ (given a local orthonormal frame $e_z^{\perp}$ for E about such point) as

$$\begin{aligned} \theta_{E,|\ |} &= \sum_{1\le j,j'\le m}\ \sum_{1\le k,k'\le N} \varphi_{j,k}^{j',k'}\,(d\zeta_k \otimes e_{z,j}^{\perp}) \otimes (d\overline{\zeta}_{k'} \otimes (e_{z,j'}^{\perp})^*) \\ &= \sum_{1\le j,j'\le m}\ \sum_{1\le k,k'\le N} \varphi_{j,k}^{j',k'}\,(d\zeta_k \otimes e_{z,j}^{\perp})\,\overline{d\zeta_{k'} \otimes e_{z,k'}^{\perp}}. \end{aligned} \tag{C.28}$$

Namely, for any $1 \le j, j' \le m$ and any sections ξ, η of $T_{\mathcal{X}}$ about z, one has

$$\theta_{E,|\ |}(\xi \otimes e_{z,j}^{\perp}, \eta \otimes e_{z,j}^{\perp}) = \Big(\sum_{1\le k,k'\le N} \varphi_{j,k}^{j',k'}\, d\zeta_k \wedge \overline{d\zeta_{k'}}\Big)(\xi, \overline{\eta}). \tag{C.29}$$

In other words

$$\begin{aligned} \theta_E\Big(\sum_{k=1}^{N}\sum_{j=1}^{m} \xi_{j,k}\frac{\partial}{\partial\zeta_k} \otimes e_{z,j}^{\perp},\ \sum_{k=1}^{N}\sum_{j=1}^{m} \eta_{j,k}\frac{\partial}{\partial\zeta_k} \otimes e_{z,j}^{\perp}\Big) \\ = \sum_{1\le j,j'\le m}\ \sum_{1\le k,k'\le N} \varphi_{j',k'}^{j,k}\, \xi_{j,k}\overline{\eta_{j',k'}}\,. \end{aligned} \tag{C.30}$$

REMARK C.35. When $E = L$ is a holomorphic line bundle, the fact that $\theta_{L,|\ |}$ defines an hermitian semipositive form on $T_{\mathcal{X}}$ is equivalent (see Remark C.4) to the fact that $c_1(L, |\ |)$ is a positive form. Similarly, it is equivalent to the fact that $c_1(L^*, |\ |_*)$ (with the metric $|\ |_*$ induced from $|\ |$ by duality) is a negative form.

We may now introduce the notions of Griffiths and Nakano positivity for holomorphic hermitian bundles over an N-dimensional complex manifold $\mathcal{X}$; see, e.g., [**Gri69**, **Na55**] and [**De**, VII.6, Definitions 6.3 and 6.4].

DEFINITION C.36. The holomorphic hermitian bundle $(E, |\ |)$ over the complex manifold $\mathcal{X}$ is said to be *Griffiths positive* and denoted by $(E, |\ |) >_{\mathrm{G}} 0$, if and only if

$$\forall z \in \mathcal{X},\ \forall (\xi, e_z) \in (T_{\mathcal{X},z} \times E_z) \setminus \{(0,0)\},\quad \theta_{E,|\ |}(\xi \otimes e_z, \xi \otimes e_z) > 0. \tag{C.31}$$

It is said to be *Griffiths semipositive*, which is denoted by $(E, |\ |) \geq_{\mathrm{G}} 0$, if and only if

$$\forall z \in \mathcal{X},\ \forall (\xi, e_z) \in T_{\mathcal{X},z} \times E_z,\quad \theta_{E,|\ |}(\xi \otimes e_z, \xi \otimes e_z) \geq 0. \tag{C.32}$$

When $>$ and $\geq$ are replaced respectively by $<$ and $\leq$ in (C.31) or (C.32), $(E, |\ |)$ is said to be *Griffiths negative*, (which is denoted by $(E, |\ |) <_{\mathrm{G}} 0$), respectively *Griffiths seminegative* (which is denoted by $(E, |\ |) \leq_{\mathrm{G}} 0$).

DEFINITION C.37. The holomorphic hermitian bundle $(E, |\ |)$ over the complex manifold $\mathcal{X}$ is said to be *Nakano positive*, which is denoted by $(E, |\ |) >_{\mathrm{N}} 0$, if and only if the hermitian form $\theta_{E,|\ |}$ defines a scalar product on each fiber $(T_{\mathcal{X}} \otimes E)_z$, $z \in \mathcal{X}$. It is said to be *Nakano semipositive* and denoted by $(E, |\ |) \geq_{\mathrm{N}} 0$, if and only if $\theta_{E,|\ |}$ defines a positive form on each fiber $(T_{\mathcal{X}} \otimes E)_z$, $z \in \mathcal{X}$. The hermitian holomorphic bundle $(E, |\ |)$ is said to be *Nakano negative* (respectively *Nakano seminegative*) and denoted by $(E, |\ |) <_{\mathrm{N}} 0$ (respectively $(E, |\ |) \leq_{\mathrm{N}} 0$), if and only if $-\theta_{E,|\ |}$ induces a scalar product (respectively a positive hermitian form) on each fiber $(T_{\mathcal{X}} \otimes E)_z$, $z \in \mathcal{X}$.

REMARK C.38. The notions of Griffiths positivity (respectively semipositivity) and Nakano positivity (respectively Nakano semipositivity) coincide for hermitian (holomorphic) line bundles. In general, Nakano positivity (respectively Nakano seminegativity) implies Griffiths positivity (respectively Griffiths seminegativity).

EXAMPLE C.39. Consider the projective space $\mathbb{P}^N_{\mathbb{C}}$. The line bundle $\mathcal{O}_{\mathbb{P}^N_{\mathbb{C}}}(1)$ equipped with the Fubini–Study metric $|\ |_{\mathrm{FS}}$, whose first Chern form is the so-called positive Kähler form $dd^c \log \|z\|^2$, where $z = [z_0 : \cdots : z_N]$ (see §3.1.3) is positive. On the other hand, the tautological line bundle $(\mathcal{O}_{\mathbb{P}^N_{\mathbb{C}}}(-1), |\ |_{\mathrm{FS}})$ is negative, since its first Chern form equals $-dd^c \log \|z\|^2$.

Remark C.35 suggests introducing an alternative notion of semipositivity for hermitian holomorphic bundles $(E, |\ |)$; see [**And07**, §7].

DEFINITION C.40. A hermitian holomorphic bundle $(E, |\ |)$ over a complex manifold $\mathcal{X}$ is said to be *Bott–Chern semipositive*, which one formulates this time by $(E, |\ |) \geq_{\mathrm{BC}} 0$, if and only if there are finitely many elements $\Phi_\iota \in C^\infty_{1,0}(\mathcal{X}, E)$ such that

$$c_1(E, |\ |) = i \sum_\iota \Phi_\iota \wedge \Phi_\iota^*. \tag{C.33}$$

The conjugation isometry of sections of $T^{*(1,0)}_{\mathcal{X}} \otimes E$ into sections of $T^{*(0,1)}_{X} \otimes E^*$ is induced by the conjugation isometry of sections of E into sections of E^* as introduced in §B.3.2.1.

REMARK C.41. Let $|\ |^*$ be the hermitian metric on E^* induced by the metric $|\ |$ on E. Since $\theta_{E^*,|\ |^*} = -\theta_{E,|\ |}$ (see (B.28)) the following equivalence holds: $(E, |\ |) \geq_{\mathrm{BC}} 0 \Longleftrightarrow (E^*, |\ |^*) \leq_{\mathrm{N}} 0$; [**And07**, Lemma 7.2].

The reason why the concept of Bott–Chern semipositivity seems to be quite useful is that whenever

$$0 \longrightarrow (E', |\ |') \longrightarrow (E, |\ |) \longrightarrow (E'', |\ |'') \longrightarrow 0$$

is a short exact sequence of hermitian holomorphic bundles, where the induced metrics $|\ |'$ and $|\ |''$ depend on the hermitian metric $|\ |$ on E as described for example in [**BotC65**, §4] or also [**De**, V.14], the positivity implication

$$E \geq_{\mathrm{BC}} 0 \Longrightarrow E'' \geq_{\mathrm{BC}} 0$$

holds; [**And07**, Proposition 7.1]. Another important fact is that $(E, |\ |) \geq_{\mathrm{BC}} 0$ implies that all Chern forms $c_\ell(E, |\ |)$ are positive for $\ell \geq 1$, as it follows from Proposition B.48.

We conclude §C.3 with Kodaira's theorem [**Kod54**], which provides a characterization of projective algebraic varieties (see Remark A.29) based on the concept of positivity for line bundles.

THEOREM C.42 (Kodaira theorem). *A compact analytic manifold $\mathcal{X}$ is projective algebraic if and only if $\mathcal{X}$ carries a positive hermitian holomorphic line bundle $(E \to \mathcal{X}, |\ |)$.*

In order to conclude this section about positivity, let us point out that we assumed here that all considered hermitian metrics on bundles are smooth metrics. Resorting to semipositivity for hermitian line bundles, it may be meaningful to extend it to the case where the hermitian metric is just continuous instead of being smooth. Let us then enlarge the concept of semipositivity of hermitian line bundles with the following definition, that proved to be quite fruitful in diophantine geometry and more specifically in logarithmic height theory; see [**BGPS14**, §1.4], [**Y21**, Definition 2.1 and §4] or §6.6.2 in the present monograph.

DEFINITION C.43. Let L be a holomorphic line bundle over a complex analytic space $\mathcal{X}$. The metrized bundle $\overline{L} = (L, |\ |)$ defined as L equipped with a continuous metric $|\ |$ is called *semipositive* (SP) if and only if there exists a sequence $(|\ |^{(\nu)})_{\nu \in \mathbb{N}^*}$ of smooth metrics on L such that the $(1,1)$-current $[c_1(L, |\ |^{(\nu)})]$ induced by the smooth form $c_1(L, |\ |^{(\nu)})$ is positive and the sequence of smooth metrics $|\ |^{(\nu)}$ converges uniformly to $|\ |$ on any compact subset of $\mathcal{X}$. Such a metrized bundle $\overline{L}$ is said to be *difference semipositive* (DSP) if and only if there are two semipositive metrized bundles $\bar{L}' = (L', |\ |')$ and $\bar{L}'' = (L'', |\ |'')$ such that $\overline{L}$ is the line bundle $L' \otimes (L'')^{-1}$ equipped with the continuous metric $|\ |' \otimes (|\ |'')^*$, where $(|\ |'')^*$ denotes the dual metric of $|\ |''$.

EXAMPLE C.44. The bundle $\mathcal{O}_{\mathbb{P}^N_{\mathbb{C}}}(1)$ equipped with the canonical metric (that is $\max_{0 \leq j \leq N} |z_j|$ replaces $\|z\|_{\ell^2}$ if one compares it with the Fubini–Study smooth metric) is the archetypal example of a semipositive (SP) metrized line bundle (with a continuous but not smooth metric) over the projective space $\mathbb{P}^N_{\mathbb{C}}$.

REMARK C.45. The point here is that first Chern forms of DSP metrized bundles can be multiplied (in terms of the exterior multiplication) thanks to the

Bedford–Taylor iterative procedure (C.7) introduced previously. This is feasible since the mentioned Bedford–Taylor iterative procedure applies in the framework of complex analytic spaces as well as in the setting of complex analytic manifolds.

APPENDIX D

Various concepts in algebraic or analytic geometry

In Appendix D, we assume that $\mathcal{X}$ is a purely N-dimensional reduced complex analytic space. We may also consider algebraic varieties $\mathbb{X}$ from a scheme point of view, but we insist more on geometric aspects (points) than on functional ones (schemes).

D.1. Cycles on a complex analytic space

D.1.1. Cycles, effectivity, splitting along a dimension and cutting out.

DEFINITION D.1. Let $\mathcal{X}$ be a complex analytic space. The additive group $\mathcal{Z}(\mathcal{X}) = \mathcal{Z}(\mathcal{X}, \mathbb{Z})$ of complex analytic cycles on $\mathcal{X}$ consists of locally finite formal sums $Z = \sum_\iota \nu_\iota \{V_\iota\}$, where the weights ν_ι are nonzero integers and the V_ι's are irreducible closed analytic subsets of $\mathcal{X}$. The realization of the cycle $\sum_\iota \nu_\iota \{V_\iota\}$ as a current is provided by the corresponding integration current $\sum_\iota \nu_\iota [V_\iota]$; see §C.2.3. The support $|Z|$ of the cycle $Z = \sum_\iota \nu_\iota \{V_\iota\}$ is defined by $|Z| = \bigcup_\iota V_\iota$. The cycle Z is said to be *effective* if all of its weights ν_ι belong to $\mathbb{N}^*$.

Given an effective cycle $Z = \sum_\iota \nu_\iota \{V_\iota\} \in \mathcal{Z}(\mathcal{X})$, one can attach to it a coherent sheaf of $\mathcal{O}_\mathcal{X}$ ideals $\mathcal{I}(Z)$ with support $\mathrm{Supp}(\mathcal{O}_\mathcal{X}/\mathcal{I}) = |Z|$ by

$$\mathcal{I}(Z) = \bigcap_\iota \mathcal{I}_{V_\iota}^{\nu_\iota}. \tag{D.1}$$

Direct images of cycles through proper morphisms between complex analytic spaces, corresponding to direct images of related currents through such proper morphisms, play an important role in our monograph. The direct images are realized in terms of currents as follows. Let $\pi \ : \widetilde{\mathcal{X}} \longmapsto \mathcal{X}$ be a proper morphism between complex analytic spaces. If $\widetilde{Z} = \sum_\iota \nu_\iota \{\widetilde{V}_\iota\} \in Z(\widetilde{\mathcal{X}})$, then the direct image of the associate current $[\widetilde{Z}]$, corresponding to $\widetilde{Z}$, equals the locally finite $\mathbb{Z}$-combination of integration currents

(D.2)

$$\pi_*([\widetilde{Z}]) = \pi_*\Big(\sum_\iota \nu_\iota [\widetilde{V}_\iota]\Big) = \sum_\iota \nu_\iota \delta(\pi_{|\widetilde{V}_\iota}) [\pi(\widetilde{V}_\iota)], \quad \delta(\pi_{|\widetilde{V}_\iota}) := \deg \big(\widetilde{V}_\iota \xrightarrow{\pi_{\widetilde{V}_\iota}} \pi(\widetilde{V}_\iota)\big)$$

on $\mathcal{X}$. Such a current corresponds then to a cycle $\pi_* Z$ on $\mathcal{X}$. The cycle $\pi_* Z$ is called the *direct image* of Z through the proper morphism π.

Let $0 \leq r \leq N$. A cycle $Z \in \mathcal{Z}(\mathcal{X})$ is said to be purely $(N-r)$-dimensional if and only if all the irreducible, closed analytic sets V_ι involved in its decomposition $Z = \sum_\iota \nu_\iota \{V_\iota\}$ share the same dimension $N - r$. This is denoted by $Z \in \mathcal{Z}_{N-r}(\mathcal{X})$.

Any element in $\mathcal{Z}(\mathcal{X})$ admits then a unique decomposition as

$$Z = \sum_{r=0}^{N} Z_{N-r}, \quad Z_{N-r} \in \mathcal{Z}_{N-r}(\mathcal{X}) \text{ for } r = 0, \dots, N. \tag{D.3}$$

The decomposition (D.3) is called the *splitting of Z along dimension.* There is another splitting of cycles that will prove to play an important role in our monograph. When V is a fixed closed irreducible analytic subset of $\mathcal{X}$, any $Z = \sum_\iota \nu_\iota \{V_\iota\}$ in $\mathcal{Z}(X)$ admits a unique splitting

$$Z = Z^V + Z^{\mathcal{X}\setminus V} := \sum_{\{\iota\,:\,V_\iota \subset V\}} \nu_\iota \{V_\iota\} + Z^{\mathcal{X}\setminus V}. \tag{D.4}$$

One can re-interpret the splitting (D.4) in terms of currents as

$$[Z] = \mathbf{1}_V \cdot [Z] + ([Z] - \mathbf{1}_V \cdot [Z]); \tag{D.5}$$

see Example C.13. The splitting (D.4) or (D.5) is called the *cutting out* with respect to the closed irreducible analytic subset V. Observe that one has

$$\mathcal{I}(Z^{\mathcal{X}\setminus V}) = \bigcup_{\ell=1}^{\infty} [\mathcal{I}(Z) : \mathcal{I}_V^\ell], \tag{D.6}$$

where, for any $z \in |Z|$,

$$[\mathcal{I}(Z) : \mathcal{I}_V^\ell]_z = [(\mathcal{I}(Z))_z : \mathcal{I}_{V,z}^\ell] = \{h \in \mathcal{O}_{\mathcal{X},z} \,:\, h\mathcal{I}_{V,z}^\ell \subset (\mathcal{I}(Z))_z\} \tag{D.7}$$

is the *carrier* of the ideal $\mathcal{I}_{V,z}^\ell$ inside $(\mathcal{I}(Z))_z$ for any $\ell \in \mathbb{N}^*$. Such a cutting-out operation appeals to the important notion of *gap sheaf* in analytic geometry; see [**Mas03**, Chapter 1].

D.1.2. Lelong–Poincaré equation and classes of generalized cycles. Unfortunately, analytic cycles $Z \in Z(\mathcal{X})$, when interpreted as integration currents $[Z]$ in $'\mathcal{D}_{\mathcal{X}}^{(\bullet,\bullet)}(\mathcal{X})$, prove to be too rigid objects. This prevents them, for example, from playing the role of markers for both local and global intersection theory (in some unified way) in familiar algebraic environments as $\mathbb{P}^N_{\mathbb{C}}$. The Lelong–Poincaré equation stands as the cornerstone suggesting a natural enlarging of the group $Z(\mathcal{X})$, which would fit with the flexibility inherent within complex analysis, where complex conjugation plays a major role. Let us formulate it here as in [**AndESWY21A**, Proposition 2.1].

THEOREM D.2. *Let s be a meromorphic section of a hermitian line bundle $(L, |\ |)$ over $\mathcal{X}$ and let $\pi : \widetilde{\mathcal{X}} \to \mathcal{X}$ be any smooth modification of $\mathcal{X}$; see Theorem* A.41. *Then one has in the sense of currents on $\mathcal{X}$ that*

$$dd^c(-\log|s|^2) + [\mathrm{div}\,(s)] := dd^c(-\log|s|^2) + [\pi_*(\mathrm{div}\,(\pi^* s))] = c_1(L, |\ |), \tag{D.8}$$

where

$$\mathrm{div}\,(\pi^* s) = \sum_{\{W \subset \widetilde{X}\,:\, W \text{ irr. closed analytic, } \dim W = N-1\}} \mathrm{ord}_W(\pi^* f)\,\{W\} \tag{D.9}$$

and the cycle $\mathrm{div}(s) := \pi_*(\mathrm{div}\,(\pi^* s)) \in \mathcal{Z}_{N-1}(\mathcal{X})$ *is independent of the smooth modification* $\pi : \widetilde{\mathcal{X}} \to \mathcal{X}$.

The following definition appeared, as suggested by the Lelong–Poincaré equation, in [**AndESWY21A**, §3]. It is deeply connected with pluripotential complex theory through the realization of Green currents in the complex analytic setting and thus to intersection theory at archimedean places; see §6.6.2 or also [**Y21**, §4].

DEFINITION D.3. A current $T \in {}'\mathcal{D}_{\mathcal{X}}^{(\bullet,\bullet)}(\mathcal{X})$ is said to be a *generalized cycle* on $\mathcal{X}$ if and only if it is a locally finite combination of currents of the form $(\pi_\iota)_*(C_\iota)$ with integral coefficients. Here π_ι is a proper morphism of complex analytic spaces from a connected complex manifold $\mathcal{Y}_\iota$ to $\mathcal{X}$, and C_ι is either $\mathbf{1}_{\mathcal{Y}_\iota}$ or a finite wedge product of first Chern forms of hermitian line bundles with smooth metrics over $\mathcal{Y}_\iota$. Generalized cycles on $\mathcal{X}$ form a $\mathbb{Z}$ -module denoted by $\mathcal{GZ}(\mathcal{X})$.

The above definition suggests the corresponding concept of *irreducibility.*

DEFINITION D.4. A generalized cycle T is said to be irreducible if the Zariski closure $|T|$ of the set $\operatorname{Supp} T$ is irreducible as a closed analytic subset of $\mathcal{X}$ and, for any proper closed analytic subset W of $|T|$ (locally defined in a neighborhood of $z \in |T|$ as $W = w_z^{-1}(\{0\})$), one has

$$\text{(D.10)}\quad \mathbf{1}_W \cdot T \overset{\text{locally in a neighborhood of } z\in|T|}{=} \big[(1-\|w_z\|^{2\lambda})\,T\big]_{\lambda=0} = 0 \text{ in } {}'\mathcal{D}_{\mathcal{X}}^{(\bullet,\bullet)}(\mathcal{X}).$$

Any generalized cycle $T = \sum_\iota (\pi_\iota)_*(C_\iota)$ admits then a unique decomposition (up to the ordering of the summands) as a sum of irreducible generalized cycles with distinct supports. Namely, if $V_{\iota'}$ are the *distinct* irreducible components appearing in the decomposition of at least one $|(\pi_\iota)_*(C_\iota)|$, then this decomposition is given by

$$T = \sum_{\iota'} \Big(\sum_{\{\iota \,:\, \pi_\iota(Y_\iota) = V_{\iota'}\}} (\pi_\iota)_*(C_\iota) \Big) = \sum_{\iota'} T_{\iota'};$$

see [**AndESWY21A**, Proposition 3.9]. A fundamental *dimension principle* also holds for generalized cycles, in the same vein as it does for cycles in Proposition C.26; see [**AndESWY21A**, Proposition 3.4].

PROPOSITION D.5. *If T is a generalized cycle of pure codimension r, that is, T is an (r,r)-current, then* $\operatorname{codim}(|T|) > r \Longrightarrow T = 0$, *and* $T = [Z]$ *with* $Z \in \mathcal{Z}_{N-r}(\mathcal{X})$ *as soon as* $\operatorname{codim}(|T|) = r$.

EXAMPLE D.6. Given any holomorphic section s of a hermitian holomorphic bundle $\bar{E} = (E, |\ |)$ over $\mathcal{X}$, the *Segre current*

(D.11)

$$\widehat{S}(\bar{E}\,;\, s\,;\, \mathbf{1}_{\mathcal{X}}) := \Big[(1-|s|^{2\lambda}) + \bar\partial\,|s|^{2\lambda} \wedge \frac{d\log|s|^2}{2i\pi} \wedge \Big(\sum_{r=1}^{N} (dd^c \log|s|^2)^{r-1}\Big)\Big]_{\lambda=0}$$

provides an archetypal example of a generalized cycle; see [**AndESWY21A**, Theorem 5.2]. Given $m+1$ holomorphic sections $s_0, \dots, s_m$ of a hermitian holomorphic line bundle $\overline{L} = (L, |\ |)$ over $\mathcal{X}$, the Stückrad–Vogel[1] current

(D.12)

$$\widehat{V}(L\,;\, s\,;\, \mathbf{1}_{\mathcal{X}}) := \Big[(1-|s|^{2\lambda}) + \overline{\partial}\,|s|^{2\lambda} \wedge \frac{d\log|s|^2}{2i\pi} \wedge \Big(\sum_{r=1}^{n} (dd^c \log\|\sigma\|^2)^{r-1}\Big)\Big]_{\lambda=0},$$

[1] The terminology used here arises from the algorithmic approach towards intersection theory introduced in [**StV82**], as we shall justify it in §D.5.4; see Definition D.55 and Theorem D.56 in §D.5.4.

where $|s| = \sqrt{|s_0|^2 + \cdots + |s_m|^2}$ and $s = (\sigma_0, \dots, \sigma_m)\, e_z$ locally (e_z being a local holomorphic frame for L), is a generalized cycle on $\mathcal{X}$. The generalized cycle (D.12) does not depend on the choice of the hermitian metric $|\ |$ on L; see [**AndESWY21A**, Theorem 9.7]. This is a second archetypal example of a generalized cycle over $\mathcal{X}$ given the content of this monograph.

Let $\mathcal{Y}$ be a smooth, connected complex manifold. Any smooth differential form $b = c(E, |\ |) - c(E', |\ |') \wedge c(\bar{E}'', |\ |'')$, where

$$0 \longrightarrow (E', |\ |') \longrightarrow (E, |\ |) \longrightarrow (E'', |\ |'') \longrightarrow 0$$

is a short exact sequence of metrized hermitian bundles, with the hermitian metrics on E' and E'' induced by that on E (see [**BotC65**, §4] or [**De**, V.14]), is called a *Bott–Chern form* b (in view of [**BotC65**]). Its (ℓ, ℓ) components β_ℓ, $\ell \in \mathbb{N}^*$, such that $b = \sum_{\ell \geq 1} \beta_\ell$, are dd^c exact; see [**BotC65**, Proposition 4.2].

DEFINITION D.7. A generalized cycle T on the complex analytic space $\mathcal{X}$ is said to be equivalent to 0, which we denote by $T \sim 0$, if and only if T is a linear combination of currents of the form $(\pi_\iota)_*(\beta_{\ell_\iota} \wedge C_\iota)$ with integer coefficients, where $\pi_\iota : \mathcal{Y}_\iota \to \mathcal{X}$ and C_ι are as in Definition D.3 and β_{ℓ_ι} is a component of a Bott–Chern form on the connected complex manifold $\mathcal{Y}_\iota$.

One can state now the following proposition [**AndESWY21A**, Proposition 4.4].

PROPOSITION D.8. *The morphism of $\mathbb{Z}$-modules*

$$Z \in \mathcal{Z}(\mathcal{X}) \longmapsto [Z] \in \mathcal{GZ}(\mathcal{X}) \longmapsto \big([Z] \text{ modulo } \sim\big) \in \mathcal{GZ}(\mathcal{X})/\sim\, := \mathcal{B}(\mathcal{X})$$

is injective. Thus the $\mathbb{Z}$-module $\mathcal{B}(\mathcal{X}) = \mathrm{GZ}(\mathcal{X})/\sim$, where $\mathcal{Z}(\mathcal{X}) \hookrightarrow \mathcal{B}(\mathcal{X})$, can be interpreted as an extension of $\mathcal{Z}(\mathcal{X})$. Its elements μ are called classes of generalized cycles on $\mathcal{X}$.

REMARK D.9. When U is an open subset of $\mathcal{X}$, the natural restriction

$$T \in \mathcal{GZ}(\mathcal{X}) \longmapsto T_{|U} \in \mathcal{GZ}(U)$$

induces a restriction map $\mu \in \mathcal{B}(\mathcal{X}) \longmapsto \mu_{|U} \in \mathcal{B}(U)$; see [**AndESWY21A**, Proposition 4.5].

EXAMPLE D.10 (Segre class of coherent ideal sheaves). Let $\mu \in \mathcal{B}(\mathcal{X})$ with representant $\widehat{\mu} \in \mathcal{GZ}(\mathcal{X})$ (an archetypal example being $\mu = [1_{\mathcal{X}}]$). Let also $\overline{E} = (E, |\ |)$ be a hermitian holomorphic bundle over $\mathcal{X}$ with $s\ : \mathcal{X} \to E$ being a holomorphic section. Then the generalized cycle
(D.13)

$$\widehat{S}(\overline{E}\,;\, s\,;\, \widehat{\mu}) := \Big[\Big((1 - |s|^{2\lambda}) + \bar{\partial}\, |s|^{2\lambda} \wedge \frac{d \log |s|^2}{2i\pi} \wedge \Big(\sum_{r=1}^{N} (dd^c \log |s|^2)^{r-1}\Big)\Big) \wedge \widehat{\mu}\Big]_{\lambda=0}$$

is well defined through the analytic continuation process. Its Segre class $S(\overline{\mathcal{I}}_s; \mu)$ in $\mathcal{B}(\mathcal{X})$ depends only on μ and the coherent sheaf $\overline{\mathcal{I}}_s$; see §A.4.2.3. Recall that $\mathcal{I}_s$ is the coherent sheaf of ideals attached to the section s; [**AndESWY21A**, Theorem 5.2].

EXAMPLE D.11 (Stückrad–Vogel classes). Let $\mu \in \mathcal{B}(\mathcal{X})$ with representant $\widehat{\mu}$ in $\mathcal{GZ}(\mathcal{X})$ as in Example D.10. For $m \in \mathbb{N}$, let $s = (s_0, \dots, s_m)$ be a global holomorphic section of $L \oplus \cdots \oplus L$ ($m + 1$ times), where L is a given line bundle over $\mathcal{X}$. For any metric $|\ |$ on L, inducing the metric $\|\ \| = \sqrt{|\cdot|^2 + \cdots + |\cdot|^2}$ on $L \oplus \cdots \oplus L$, the generalized cycle

(D.14)

$$\widehat{V}(L\,;\,s\,;\,\widehat{\mu}) := \Big[\Big((1 - \|s\|^{2\lambda}) + \overline{\partial}\,\|s\|^{2\lambda} \wedge \frac{d\log\|s\|^2}{2i\pi} \wedge \Big(\sum_{r=1}^{N}(dd^c \log\|\sigma\|^2)^{r-1}\Big)\Big) \wedge \widehat{\mu}\Big]_{\lambda=0},$$

where $s = (\sigma_0, \ldots, \sigma_m)\, e_z$ locally and e_z is an arbitrary local frame for L, is well defined through the analytic continuation process (independently of the choice of the metric $|\ |$ on L). Its *Stückrad–Vogel class* $V(L, \overline{\mathcal{I}}_s; \mu)$ in $\mathcal{B}(\mathcal{X})$ depends on L, μ and on the coherent sheaf $\overline{\mathcal{I}}_s$; see again [**AndESWY21A**, Theorem 9.7].

Any element $\mu \in \mathcal{B}(\mathcal{X})$ splits in a unique way as $\mu = \sum_\iota \mu_\iota$, where, for each ι, $\mu_\iota \in \mathcal{B}(\mathcal{X})$ is irreducible. That is, μ_ι admits a representant $\widehat{\mu}_\iota \in \mathcal{GZ}(\mathcal{X})$, which is irreducible as a generalized cycle according to Definition D.4. The support $|\mu|$ of μ is defined as $|\mu| = \bigcup_\iota |\widehat{\mu}_\iota|$, where $\widehat{\mu}_\iota$ is an irreducible representant of μ_ι in $\mathcal{GZ}(\mathcal{X})$.[2] Let δ_ι be the dimension of the (irreducible) support of $\widehat{\mu}_\iota$. According to the dimension principle for normal currents (see Proposition C.6 and Proposition D.5) any such irreducible component μ_ι splits as $\mu_\iota = [Z_\iota] + \sum_{\delta<\delta_\iota} \mu_{\iota,\delta}$, where Z_ι belongs to $\mathcal{Z}_{\delta_\iota}(\mathcal{X})$. Furthermore, for any $\delta < \delta_\iota$, $\mu_{\iota,\delta} \in \mathcal{B}_\delta(\mathcal{X})$ admits as representant a (δ,δ)-dimensional generalized cycle. Then, one has the decomposition of $\mu \in \mathcal{B}(\mathcal{X})$ as

$$\mu = \sum_\iota \mu_\iota = \sum_\iota \Big([Z_\iota] + \sum_{\delta<\delta_\iota} \mu_{\iota,\delta}\Big) = \sum_\iota [Z_\iota] + \sum_\iota \sum_{\delta<\delta_\iota} \mu_{\iota,\delta}. \tag{D.15}$$

The *fixed component* $\mu_{\text{fix}} := [Z] = \sum_\iota [Z_\iota]$ corresponds to a true cycle, while the *moving component* is defined by $\mu_{\text{mov}} = \sum_\iota \sum_{\delta<\delta_\iota} \mu_{\iota,\delta}$.

EXAMPLE D.12. Given a coherent ideal sheaf $\mathcal{I}$, there is a unique element $S(\overline{\mathcal{I}}; \mathbf{1}_\mathcal{X}) \in \mathcal{B}(\mathcal{X})$ such that for any $z \in \mathcal{X}$, for any open neighborhood U_z of z satisfying $\mathcal{I}_{|U_z} = \mathcal{I}_{s_z}$ as coherent sheaves of $\mathcal{O}_{U_z}$-ideals over U_z, where s_z is a section of a trivial bundle over U_z, one has that

$$(S(\overline{\mathcal{I}}\,;\,\mathbf{1}_\mathcal{X}))_{|U_z} = S(\overline{\mathcal{I}_{s_z}}\,;\,\mathbf{1}_{U_z});$$

see Example D.10. The fixed part of $S(\overline{\mathcal{I}}\,;\,\mathbf{1}_\mathcal{X})$ is an effective cycle of the form

$$\big(S(\overline{\mathcal{I}}\,;\,\mathbf{1}_\mathcal{X})\big)_{\text{fix}} = \sum_{r=\operatorname{codim}(\operatorname{Supp}(\mathcal{O}_\mathcal{X}/\mathcal{I})}^{N} \Big(\sum_\gamma \nu_{r,\gamma}\{Z_\gamma^r\}\Big), \tag{D.16}$$

where $\{Z_\gamma^r = \pi_{\mathcal{I},\text{norm}}(H_\gamma^r)\}$ is the collection of distinguished irreducible varieties with codimension r of $\operatorname{Supp}(\mathcal{O}_\mathcal{X}/\mathcal{I})$ (see Definition A.35) and the weights $\nu_{r,\gamma}$ are positive integers ([**AndESWY21A**, Theorem 1.1]).

REMARK D.13. Given a coherent sheaf of ideals $\mathcal{I}$ on an N-dimensional complex manifold $\mathcal{X}$, one can attach to $\mathcal{I}$ a cycle

$$Z = Z(\mathcal{O}_\mathcal{X}/\mathcal{I}) \quad \text{such that } \mathcal{I}\big(Z(\mathcal{O}_\mathcal{X}/\mathcal{I}) \subset \mathcal{I};$$

[2] If V is a closed analytic subset of $\mathcal{X}$, then one can consider the embedding $\iota_V : V \hookrightarrow \mathcal{X}$ and thus the complex analytic space $(V, \mathcal{O}_V) = (V, \iota_V^* \mathcal{O}_\mathcal{X})$. Elements in $\mathcal{B}(V)$ can be identified with elements $\mu \in \mathcal{B}(\mathcal{X})$ such that $|\mu| \subset V$; see [**AndESWY21A**, Proposition 4.10].

see [**Ko99**, Definition 2.9]. We do it as follows. For each $\ell = 0, \dots, N-1$, let $\mathcal{F}_\ell$ be the sheaf of sections σ of $\mathcal{O}_{\mathcal{X}}/\mathcal{I}$ such that $\sigma^{-1}(0)$ is a closed analytic subset of $\mathcal{X}$ with dimension at least equal to ℓ. Let, for each $\ell = 0, \dots, N-1$, $\{V_{\ell,\iota}\}_\iota$ be the locally finite family of irreducible components of $\operatorname{Supp} \mathcal{F}_\ell/\mathcal{F}_{\ell+1}$ and, for each ι, $z_{\ell,\iota}$ be a generic point in $(V_{\ell,\iota})_{\text{reg}}$. Take then

$$Z(\mathcal{O}_{\mathcal{X}}/\mathcal{I}) = \sum_{\ell=0}^{N-1} \sum_{\iota} \operatorname{length}\big((\mathcal{F}_\ell)_{z_{\ell,\iota}}\big)\, \{V_{\ell,\iota}\}, \tag{D.17}$$

where the length is the usual finite length[3] of a (necessarily artinian and nœtherian) finitely generated module $\mathscr{M}$ over the nœtherian ring $\mathscr{R} = \mathcal{O}_{\mathcal{X}, z_{\ell,\iota}}$. This is indeed a truly effective cycle supported by $\operatorname{Supp} \mathcal{O}_{\mathcal{X}}/\mathcal{I}$. The correspondence between coherent sheaves of ideals $\mathcal{I}$ on $\mathcal{X}$ and effective cycles $Z \in \mathcal{Z}(\mathcal{X})$ given by

$$Z \longrightarrow \mathcal{I}(Z), \qquad \mathcal{I} \longrightarrow Z(\mathcal{O}_{\mathcal{X}}/\mathcal{I}) \tag{D.18}$$

is an imperfect one, since we have only the inclusion $\mathcal{I}\big(Z(\mathcal{O}_{\mathcal{X}}/\mathcal{I})\big) \subset \mathcal{I}$. When $\mathcal{X}$ is an N-dimensional complex manifold, Examples D.10 and D.12 suggest another correspondence that reconciles better the algebraic and geometric points of view. This alternative correspondence is given by

$$Z = \sum_{\iota} \nu_\iota \{V_\iota\} \longrightarrow (I_{V_\iota}^{\text{Chow}})^{\nu_\iota} \qquad \mathcal{I} \longrightarrow \big(S(\overline{\mathcal{I}}; \mathbf{1}_{\mathcal{X}})\big)_{\text{fix}}, \tag{D.19}$$

where the ideal sheaf $\mathcal{J} = \mathcal{I}_V^{\text{Chow}}$, for an irreducible closed analytic subset V of $\mathcal{X}$, is defined as follows[4] (see [**Ko99**, §4]):

- when $z \in \mathcal{X} \setminus V$, take $\mathcal{J}_z = \mathcal{O}_{\mathcal{X},z}$;
- when $z \in V$ and U_z is a local chart around z with a centered system of local coordinates ζ_z, take $\mathcal{J}_z$ as the ideal of $\mathcal{O}_{\mathcal{X},z}$ generated by all germs at z of functions $z' \mapsto \sigma_\pi(\pi(z'))$. Here $\pi : U_z \simeq \zeta_z(U_z) \to \mathbb{C}^{\dim V+1}$ is a linear projection, whose restriction to $V \cap U_z$ is proper, and σ_π is a reduced equation for the germ at 0 of the hypersurface $\pi(U_z \cap V)$.

Such a sheaf $\mathcal{I}_V^{\text{Chow}}$ is called the coherent *sheaf of Chow ideals of* V.

D.1.3. Local multiplicities of classes of generalized cycles.

DEFINITION D.14 (local multiplicity of a class of generalized cycle). Let μ be a class of generalized cycle with pure codimension $r \leq N-1$ on the N-dimensional complex analytic space $\mathcal{X}$ and let $\widehat{\mu} \in \mathcal{GZ}(\mathcal{X})$ be an arbitrary representant of μ with pure bidegree (r,r). In particular, it means that $\widehat{\mu} = [Z]$ if $\mu = Z$ is a cycle. For any $z \in \mathcal{X}$, let $\mathfrak{M}_{\mathcal{X},z} = \sum_{j=0}^{m_z} u_{z,j} \mathcal{O}_{\mathcal{X},z}$ be the maximal ideal of the local ring $\mathcal{O}_{\mathcal{X},z}$, U_z be a neighborhood in which the germ u_z admits a representant, and $|\ |$ be an arbitrary hermitian metric on the trivial bundle $\mathbb{C}^{m_z+1}$ over U_z. Then, the generalized cycle

$$\Big[\overline{\partial}\, |u_z|^{2\lambda} \wedge \frac{d\log|u_z|^2}{2i\pi} \wedge (dd^c \log|u_z|^2)^{N-r-1} \wedge \widehat{\mu}_{|U_z}\Big]_{\lambda=0} \in \mathcal{GZ}(U_z) \tag{D.20}$$

with support $\{z\}$ is of the form $\operatorname{mult}_z(\mu)\, [\{z\}]$. The coefficient $\operatorname{mult}_z(\mu)$ is independent of the representant $\widehat{\mu}$, also independent of the generating set u_z for $\mathfrak{M}_z$ and

[3] That is, the largest length $L_{\max}$ for increasing chains $\mathscr{M}_0 \subset \mathscr{M}_1 \subset \cdots \subset \mathscr{M}_{L_{\max}} = \mathscr{M}$ of distinct $\mathscr{R}$-submodules.

[4] Such a geometric concept is reminiscent of that of *contour apparent* introduced by Gaspard Mone [**Mon**].

finally, independent of both the open set U_z and of the metric $|\ |$ on $\mathbb{C}^{m_z+1}$. The integer $\mathrm{mult}_z(\mu) \in \mathbb{Z}$ is called the *local multiplicity of μ at z.*[5]

EXAMPLE D.15 (The case of cycles). When

$$\mu = Z = \sum_\iota \nu_\iota \{V_\iota\} \quad \text{with } \dim V_\iota = N - r,$$

one has $\mathrm{mult}_z(Z_r) = \sum_\iota \nu_\iota\, \nu_z([V_\iota])$, where $\nu_z([V_\iota])$ is the ordinary Lelong number of the integration current $[V_\iota]$ at point z. Recall Definition C.14, when $\mathcal{X}$ is a complex manifold and Definition C.17 in the general case (for the definition of Lelong numbers), together with Definition C.22 (for that of integration currents).

EXAMPLE D.16 (Segre class and local Segre numbers of a coherent sheaf of ideals). Given a coherent sheaf of $\mathcal{O}_\mathcal{X}$ ideals $\mathcal{I}$ and a point $z \in \mathcal{X}$, the nonnegative integer $\mathrm{mult}_z(S_r(\overline{\mathcal{I}}\,;\, \mathbf{1}_\mathcal{X}))$, where $S_r(\overline{\mathcal{I}}\,;\, \mathbf{1}_\mathcal{X})$ denotes the $(N-r)$-dimensional component of the Segre class $S(\overline{\mathcal{I}}\,;\, \mathbf{1}_\mathcal{X})$, is the r-Segre number $e_r(\mathcal{I}, \mathbf{1}_\mathcal{X}, z)$ of the coherent sheaf of ideals $\mathcal{I}$, as introduced in Proposition C.25; see [**AndESWY21A**, Theorem 1.1]. Moreover, one has $e_r(\mathcal{I}, \mathbf{1}_\mathcal{X}, z) = 0$ for any $z \in \mathcal{X}$ and any $r < \mathrm{codim}_\mathcal{X}(\mathcal{O}_\mathcal{X}/\mathcal{I})$, since $S_r(\overline{\mathcal{I}}\,;\, \mathbf{1}_\mathcal{X}) = 0$ for such r.

D.2. Cartier divisors and Čech cohomology

D.2.1. Cartier divisors and Picard group on a complex analytic space. Cartier divisors $\{(U_\iota, f_\iota)\}_\iota$ on a (reduced) complex analytic space have been previously introduced in Example B.2. Given two open coverings $\mathcal{U}$ and $\widetilde{\mathcal{U}}$, one identifies two Cartier divisors $\{(U_\iota, f_\iota)\}_\iota$ and $\{\widetilde{U}_{\tilde\iota}, \widetilde{f}_{\tilde\iota}\}_{\tilde\iota}$ when the meromorphic section $f_\iota/\widetilde{f}_{\tilde\iota}$ defines an invertible element in $\mathcal{O}_\mathcal{X}(U_\iota \cap \widetilde{U}_{\iota_\iota})$ whenever $U_\iota \cap \widetilde{U}_{\tilde\iota}$ is nonempty. The addition of Cartier divisors on $\mathcal{X}$ is defined as follows: if D and $\widetilde{D}$ are represented by the coverings $\{U_\iota, f_\iota\}_\iota$ and $\{\widetilde{U}_{\tilde\iota}, \widetilde{f}_{\tilde\iota}\}_{\tilde\iota}$, respectively, $D + \widetilde{D}$ is represented by $\{U_\iota \cap \widetilde{U}_{\tilde\iota}, f_\iota \widetilde{f}_{\tilde\iota}\}_{\iota,\tilde\iota}$.

PROPOSITION D.17. *The addition above equips the set* $\mathrm{Div}_{\mathrm{Cart}}(\mathcal{X})$ *with an additive group structure. The group of Cartier divisors* $(\mathrm{Div}_{\mathrm{Cart}}(\mathcal{X}), +)$ *admits as subgroup the subgroup* $\mathrm{Pr}_{\mathrm{Cart}}(\mathcal{X})$*, whose elements, called principal Cartier divisors, are represented by a covering reduced to the singleton* $\{(\mathcal{X}, f)\}$*, where f is a global meromorphic section of the trivial line bundle* $\mathbb{C}$*. That is, f is a global section of the sheaf $\mathcal{M}_\mathcal{X}$ of meromorphic functions on $\mathcal{X}$. Moreover, since $\mathcal{X}$ is assumed to be reduced (see, e.g.,* [**Laz**, Example 1.1.5]*) the quotient group* $\mathrm{Div}_{\mathrm{Cart}}(\mathcal{X})/\mathrm{Pr}_{\mathrm{Cart}}(\mathcal{X})$ *is isomorphic to the Picard group* $\mathrm{Pic}(\mathcal{X})$*. The elements of* $\mathrm{Pic}(\mathcal{X})$ *are holomorphic line bundles L over $\mathcal{X}$, uniquely determined up to isomorphisms of holomorphic line bundles (see Definition* B.10 *in* §B.1.4).

For any $D \in \mathrm{Div}_{\mathrm{Cart}}(\mathcal{X})$, admitting the nonunique (see Example B.2) representation $\{(U_\iota, f_\iota)\}_\iota$, let $\mathcal{O}(D)$ (denoted also by $[-D]$, see below) be the isomorphism class of holomorphic line bundles, which is associated through the one-to-one correspondence in Proposition B.8 to the cocycle $\{g_{\iota_1,\iota_0} = f_{\iota_1}/f_{\iota_0} : U_{\iota_0} \cap U_{\iota_1} \to \mathbb{C}^*\}_{\iota_0,\iota_1}$ (up to identifications of cocycle as described in §B.1.4). The morphism $D \in \mathrm{Pic}_{\mathrm{Cart}}(\mathcal{X}) \mapsto \mathcal{O}(D)$ is surjective and its kernel equals $\mathrm{Pr}_{\mathrm{Cart}}(\mathcal{X})$. Talking

[5]In the case $\mathrm{codim}\,\mu = N$, $\mu = \sum_\iota \nu_\iota \{z_\iota\}$ and the multiplicity at z equals ν_ι if $z = z_\iota$ for some index ι, 0 elsewhere.

about locally free sheaves on sections instead of holomorphic line bundles, the formulation of the correspondence between $D \in \mathrm{Pic}_{\mathrm{Cart}}(\mathcal{X})$ (modulo $\mathrm{Pr}_{\mathrm{Cart}}(\mathcal{X})$) and the isomorphism class of invertible sheaves $\mathcal{L}(D) = \mathcal{L}(\mathcal{O}(D))$ goes as follows. The invertible sheaf of a line bundle L representing $\mathcal{O}(D)$ is identified, up to isomorphism between invertible sheaves, with the locally free sheaf $\mathcal{L}$ with rank 1 such that $\mathcal{L}(U_\iota) = \mathcal{O}_{\mathcal{X}}(U_\iota)/f_\iota$ for any index ι. This means that the trivialization of $\mathcal{L}$ over the open subset U_ι is

$$\sigma \in \mathcal{L}_{|U_\iota} \longrightarrow f_\iota\, \sigma \in (\mathcal{O}_{\mathcal{X}})_{|U_\iota}. \tag{D.21}$$

REMARK D.18 (Cartier algebraic divisors). A Cartier divisor on a complex analytic space $\mathcal{X}$, whose structural sheaf is the sheaf of holomorphic functions $\mathcal{O}_{\mathcal{X}}$, can also be interpreted as a global section of $\mathcal{M}^*_{\mathcal{X}}/\mathcal{O}^*_{\mathcal{X}}$, where $\mathcal{M}^*_{\mathcal{X}}$ and $\mathcal{O}^*_{\mathcal{X}}$ are the sheaves of multiplicative abelian groups of invertible meromorphic functions and invertible holomorphic functions, respectively. The concept of a Cartier divisor can be transposed to the setting of algebraic varieties interpreted in terms of schemes. When $\mathbb{X}$ is an algebraic variety defined over the algebraically closed field $\mathbb{K}$, a Cartier divisor $\mathbb{D}$ (in the algebraic sense) on $\mathbb{X}$ is a global section on $\mathbb{X}$ of the sheaf $\mathcal{M}^*_{\mathbb{X}}/\mathcal{O}^*_{\mathbb{X}}$, where $\mathcal{O}^*_{\mathbb{X}}$ is the sheaf of invertible regular functions, while $\mathcal{M}^*_{\mathbb{X}}$ is the sheaf of invertible rational functions.

D.2.2. Toric Cartier divisors on a toric variety. Let $\boldsymbol{N}$ be an N-dimensional lattice and let Σ be a rational fan on $N_{\mathbb{R}} = \boldsymbol{N} \otimes_{\mathbb{Z}} \mathbb{R}$; see for example §3.1 where, for the sake of simplicity, one takes $\boldsymbol{N} = \mathbb{Z}^N$. Denote as $|\Sigma| \subset \boldsymbol{N}_{\mathbb{R}} = \mathbb{R}^N$ the union of all cones in the fan Σ. Let $\mathbb{X} = \mathbb{X}_\Sigma$ be the N-dimensional algebraic toric variety (defined over $\mathbb{Q}$) attached to the fan Σ and let $\mu : \mathbb{T}^N \times \mathbb{X}_\Sigma \longrightarrow \mathbb{X}_\Sigma$ be the torus action on it; see [**Ful93**, §1.4], or also, for a brief recap of the construction, Remark 3.9 in the core of the present monograph.

DEFINITION D.19. A function $\Psi : |\Sigma| \longmapsto \mathbb{R}$ is said to be a *virtual support function for* Σ if and only if for any $\sigma \in \Sigma$, there is m_σ in the dual lattice $(\mathbb{Z}^N)^\star$ such that $\Psi = \langle m_\sigma, \cdot\rangle$ on σ for any $\sigma \in \Sigma$. The collection $\{m_\sigma\}_{\sigma\in\Sigma}$ is then called a *set of defining vectors* for Ψ.

Among Cartier divisors $\mathbb{D}$ on $\mathbb{X}_\Sigma$ (in the algebraic sense, see Remark D.18), the *toric* ones play a very important role since they can be described in a combinatorial way.

DEFINITION D.20. A Cartier divisor $\mathbb{D}$ (in the algebraic sense) on $\mathbb{X}_\Sigma$ is said to be *toric* if and only if it is invariant under the action μ of the torus $\mathbb{T}^N$ on $\mathbb{X}_\Sigma$, that is, $\mu^*\mathbb{D} = \pi^*\mathbb{D}$ as Cartier divisors on $\mathbb{T}^N \times \mathbb{X}_\Sigma$, where $\pi : \mathbb{T}^N \times \mathbb{X}_\Sigma \to \mathbb{X}_\Sigma$ denotes the projection on the second factor $\mathbb{X}_\Sigma$.

Let Ψ be a virtual support function with respect to Σ. Since

$$(\langle m_\sigma, \cdot\rangle)_{|\sigma\cap\sigma'} = (\langle m_{\sigma'}, \cdot\rangle)_{|\sigma\cap\sigma'} = \Psi_{|\sigma\cap\sigma'}$$

for all pairs of elements $\sigma, \sigma' \in \Sigma$, the characters χ^{-m_σ} on the affine toric variety X_σ and $\chi^{-m_{\sigma'}}$ on the affine toric variety $X_{\sigma'}$ are such that $\chi^{-m_\sigma}/\chi^{-m_{\sigma'}}$ is a regular invertible function on $X_\sigma \cap X_{\sigma'}$. This implies that $\{m_\sigma\}_\sigma$ induces the realization of a toric Cartier divisor $\mathbb{D}_\Psi$ on $\mathbb{X}_\Sigma$. In fact, all toric divisors $\mathbb{D}$ on $\mathbb{X}_\Sigma$ are realized in such a way in [**KKMS73**, §I.2, Theorem 9]; see also [**BGPS14**, §3.3].

PROPOSITION D.21. *The above construction provides a one-to-one correspondence between virtual support functions Ψ with respect to the fan Σ and toric Cartier divisors on the toric algebraic variety $\mathbb{X}_\Sigma$. Two such toric divisors $\mathbb{D}_\Psi$ and $\mathbb{D}_{\widetilde{\Psi}}$ are congruent modulo principal toric divisors if and only if $\Psi - \widetilde{\Psi}$ is a linear function.*

D.2.3. Picard group in terms of Čech cohomology with values in $\mathcal{O}^*_{\mathcal{X}}$. Čech cohomology over a topological space $\mathcal{X}$ is defined relatively to a sheaf of abelian groups over $\mathcal{X}$. In our particular setting, where $\mathcal{X}$ is a complex analytic space, the sheaves we are interested in will be, for example, the constant sheaves $\mathbb{Z}, \mathbb{Q}, \mathbb{R}, \mathbb{C}, \mathbb{C}^*$ (which means that for any open subset $U \subset \mathcal{X}$, $\mathcal{F}(U)$ is one of the groups $(\mathbb{Z}, +)$, $(\mathbb{Q}, +)$, $(\mathbb{R}, +)$, $(\mathbb{C}, +)$, or $(\mathbb{C}^*, \times)$). They could be as well the sheaves of rings $\mathcal{O}_{\mathcal{X}}$ when $\mathcal{X}$ is a complex analytic space, or $C^\infty_{\mathcal{X}_\mathbb{R}}$ in the particular case when $\mathcal{X}$ is a complex manifold. In the last two examples, both sheaves are considered as sheaves of additive groups. Alternatively, they could be the sheaves $\mathcal{O}^*_{\mathcal{X}}$ when $\mathcal{X}$ is a complex analytic space, or $(C^\infty_{\mathcal{X}_\mathbb{R}})^*$ in the particular case when $\mathcal{X}$ is a complex manifold; both are then considered as sheaves of multiplicative groups. Given such a sheaf $\mathcal{F}$ of abelian groups, together with a covering $\mathcal{U} = \{U_\iota\}_\iota$ of $\mathcal{X}$ and $\ell \in \mathbb{N}$, a *Čech cochain* $g \in \check{\mathscr{C}}^\ell(\mathcal{X}, \mathcal{U}, \mathcal{F})$ is a mapping associating to any nonvoid intersection $U_{\iota_0} \cap \cdots \cap U_{\iota_\ell}$ an element $g_{\iota_\ell,\dots,\iota_0} \in \mathcal{F}(U_{\iota_0} \cap \cdots \cap U_{\iota_\ell})$. Since $\mathcal{F}$ is a sheaf of abelian groups (where the operation is denoted by $\bullet$), any $\check{\mathscr{C}}^\ell(\mathcal{X}, \mathcal{U}, \mathcal{F})$, $\ell \in \mathbb{N}$, inherits a structure of an abelian group $(\check{\mathscr{C}}^\ell(\mathcal{X}, \mathcal{U}, \mathcal{F}), \bullet)$. For any $\ell \in \mathbb{N}$, one defines a *coboundary* morphism of abelian groups $\delta^{\mathcal{F},\mathcal{U}}_\ell$ from $(\check{\mathscr{C}}^\ell(\mathcal{X}, \mathcal{U}, \mathcal{F}), \bullet)$ to $(\check{\mathscr{C}}^{\ell+1}(\mathcal{X}, \mathcal{U}, \mathcal{F}), \bullet)$. For example, if we denote the operation additively, for the sake of simplicity, then

$$
\begin{aligned}
(\delta^{\mathcal{F},\mathcal{U}}_0 g)_{\iota_1,\iota_0} &= (g_{\iota_1} - g_{\iota_0})_{|U_{\iota_0} \cap U_{\iota_1}} \quad \forall g \in \check{\mathscr{C}}^0(\mathcal{X}, \mathcal{U}, \mathcal{F}), \\
(\delta^{\mathcal{F},\mathcal{U}}_1 g)_{\iota_2,\iota_1,\iota_0} &= \big(g_{\iota_2,\iota_1} - g_{\iota_2,\iota_0} + g_{\iota_1,\iota_0}\big)_{|U_{\iota_0} \cap U_{\iota_1} \cap U_{\iota_2}} \quad \forall g \in \check{\mathscr{C}}^1(\mathcal{X}, \mathcal{U}, \mathcal{F}),
\end{aligned}
\tag{D.22}
$$

and so on. For each $\ell \in \mathbb{N}^*$, the inductive limit of quotient groups

$$
\check{\mathrm{H}}^\ell(\mathcal{X}, \mathcal{F}) = \varinjlim_{\mathcal{U}} \frac{\operatorname{Ker} \delta^{\mathcal{F},\mathcal{U}}_\ell}{\operatorname{Im} \delta^{\mathcal{F},\mathcal{U}}_{\ell-1}}
\tag{D.23}
$$

along the set of all coverings $\mathcal{U}$ of the topological space $\mathcal{X}$ ordered by refinement, is called the *ℓ-Čech cohomology group of $\mathcal{X}$ with values in $\mathcal{F}$*. For Cartier divisors on a complex analytic space, one has the following representation of the Picard group $\mathrm{Pic}(\mathcal{X}) \simeq \mathrm{Div}_{\mathrm{Cart}}(\mathcal{X})/\mathrm{Pr}_{\mathrm{Cart}}(\mathcal{X})$.

PROPOSITION D.22. *Given a complex analytic space $\mathcal{X}$, one has the isomorphisms of groups*

$$
\frac{\mathrm{Div}_{\mathrm{Cart}}(\mathcal{X})}{\mathrm{Pr}_{\mathrm{Cart}}(\mathcal{X})} \simeq \mathrm{Pic}(\mathcal{X}) \simeq \check{\mathrm{H}}^1(\mathcal{X}, \mathcal{O}^*_{\mathcal{X}}),
\tag{D.24}
$$

where the second isomorphism is realized by taking the inductive limit of Čech cohomology classes of $c_{\mathcal{U}}(L)$. The line bundle L represents $\bar{L} \in \mathrm{Pic}(\mathcal{X})$ and $c_{\mathcal{U}}(L)$ is the complex analytic 1-cocycle (with values in $\mathbb{C}^*$) attached to L and a sufficiently refined covering $\mathcal{U}$ of $\mathcal{X}$ as in §B.1.3; *see* (B.9).

REMARK D.23 (The algebraic setting). The notion of the Picard group exists both in the category of algebraic varieties $\mathbb{X}$ defined over $\mathbb{C}$ and the category of complex (reduced) analytic spaces $\mathcal{X}$, since both of them are categories of locally ringed spaces, as recalled for example in [**HaL05**, §1]. For a complete complex algebraic variety, which means that the corresponding complex analytic space is compact, the algebraic and analytic concepts of Picard group do coincide as a consequence of Serre's GAGA principle; see [**Se56**] and [**Gro71**, Chapter XII, Theorem 4.4]. This fails to be true in the case where the complex algebraic variety $\mathbb{X}$ is not complete; see, e.g., [**HaL05**, Corollary 1.3]. The concept of Picard group exists more generally in the category of integral schemes (hence of algebraic varieties defined on a field $\mathbb{K}$). It remains true in this category that such a concept is related to Cartier divisors (as defined in §D.18 in the algebraic context) through the isomorphism $\mathrm{Pic}(\mathbb{X}) \simeq \mathrm{Div}_{\mathrm{Cart}}(\mathbb{X})/\mathrm{Pr}_{\mathrm{Cart}}(\mathbb{X})$; see [**Ha**, II, Proposition 6.15].

D.2.4. The first Chern characteristic class of a complex line bundle. Suppose now that $\mathcal{X}$ is a complex manifold. In the same vein as in the result stated as Proposition D.22 for holomorphic line bundles on a complex analytic space, the (multiplicative) group of isomorphism classes of complex line bundles over $\mathcal{X}_{\mathbb{R}}$ (see §B.1.7.1) happens to be isomorphic this time to the Čech cohomology group $\check{\mathrm{H}}^1(\mathcal{X}_{\mathbb{R}}, (C^\infty_{\mathcal{X}_{\mathbb{R}}})^*)$. On the other hand, one has the short exact sequence of sheaves over $\mathcal{X}_{\mathbb{R}}$

$$0 \longrightarrow \mathbb{Z} \longrightarrow C^\infty_{\mathcal{X}_{\mathbb{R}}} \overset{\exp(2i\pi(\cdot))}{\longrightarrow} (C^\infty_{\mathcal{X}_{\mathbb{R}}})^* \longrightarrow 1. \tag{D.25}$$

Note that $\check{\mathrm{H}}^1(\mathcal{X}_{\mathbb{R}}, C^\infty_{\mathcal{X}_{\mathbb{R}}}) = \check{\mathrm{H}}^2(\mathcal{X}_{\mathbb{R}}, C^\infty_{\mathcal{X}_{\mathbb{R}}}) = 0$ as a consequence of the flatness of the sheaf $C^\infty_{\mathcal{X}_{\mathbb{R}}}$. If one forms from (D.25) the long Čech cohomology sequence, then one concludes that the Čech cohomology groups $\check{\mathrm{H}}^1(\mathcal{X}_{\mathbb{R}}, C^\infty_{\mathcal{X}_{\mathbb{R}}})$ and $\check{\mathrm{H}}^2(\mathcal{X}_{\mathbb{R}}, \mathbb{Z})$ are isomorphic.

DEFINITION D.24. Let $\mathcal{X}$ be a complex manifold. Given a complex line bundle L over $\mathcal{X}_{\mathbb{R}}$, its first Chern characteristic class is the image in $\check{\mathrm{H}}^2(\mathcal{X}_{\mathbb{R}}, \mathbb{Z})$ of the element in $\check{\mathrm{H}}^1(\mathcal{X}_{\mathbb{R}}, (C^\infty_{\mathcal{X}_{\mathbb{R}}})^*)$ corresponding to the isomorphism class of L.

REMARK D.25. Given a holomorphic line bundle L over the complex manifold $\mathcal{X}$, one already introduced in §B.3.3 the first Chern class $c_1(L) \in H^2_{\mathrm{DR}}(\mathcal{X}_{\mathbb{R}}, \mathbb{C})$ as the de Rham cohomology class of the first Chern form $c_1(L, |\ |)$, where $|\ |$ denotes an arbitrary hermitian metric on L. In fact, the two groups $\check{\mathrm{H}}^2(\mathcal{X}_{\mathbb{R}}, \mathbb{C})$ and $H^2_{\mathrm{DR}}(\mathcal{X}_{\mathbb{R}}, \mathbb{C})$ are isomorphic and the image of $c_1(L)$ in $\check{\mathrm{H}}^2(\mathcal{X}_{\mathbb{R}}, \mathbb{C})$ through such isomorphism belongs to $\check{\mathrm{H}}^2(\mathcal{X}_{\mathbb{R}}, \mathbb{Z})$ and corresponds to the first characteristic class of L as introduced in the Definition D.24.

D.3. Weil divisors and Chow groups

D.3.1. Weil divisors on complex analytic spaces or algebraic varieties.

DEFINITION D.26. Let $\mathcal{X}$ be an N-dimensional complex analytic space. Any $(N-1)$-dimensional cycle $\sum_\iota \nu_\iota \{V_\iota\}$, where $\nu_\iota \in \mathbb{Z}^*$, V_ι is an irreducible hypersurface and the sum is locally finite, is called a *Weil divisor* on $\mathcal{X}$. The additive group of Weil divisors on $\mathcal{X}$ is denoted by $\mathrm{Div}_{\mathrm{Weil}}(\mathcal{X})$. A Weil divisor $\sum_\iota \nu_\iota \{V_\iota\}$ is said to be effective if and only if $\nu_\iota \in \mathbb{N}^*$ for any ι.

REMARK D.27 (Algebraic Weil divisors). Given an algebraic variety $\mathbb{X}$ of pure dimension N defined over an algebraically closed field $\mathbb{K}$, a Weil divisor on $\mathbb{X}$ is a locally finite sum $\sum_\iota \nu_\iota \{\mathbb{V}_\iota\}$, where $\mathbb{V}_\iota$ is an irreducible subvariety of the algebraic variety $\mathbb{X}$ having dimension $N-1$.

Suppose that $\mathcal{X}$ is a normal complex analytic space. This fact implies that $\mathcal{X}$ is regular in codimension 1. Then the homomorphism

(D.26)
$$\chi\ : D \in \mathrm{Div}_{\mathrm{Cart}}(\mathcal{X}) \longmapsto \sum_{\{V \text{ irreducible}\,:\, \dim V = N-1\}} \mathrm{order}_V(D)\,\{V\} \in \mathrm{Div}_{\mathrm{Weil}}(\mathcal{X}),$$

where order_V denotes the valuation in the discrete valuation ring $\mathcal{O}_{\mathcal{X},V}$, is injective, but fails to be surjective in general;[6] see [**Ful**, §2.1]. Then

$$\chi(\mathrm{Pr}_{\mathrm{Cart}}(\mathcal{X})) = \mathrm{Pr}_{\mathrm{Weil}}(\mathcal{X})$$

is a subgroup of $\mathrm{Div}_{\mathrm{Weil}}(\mathcal{X})$, which is isomorphic to $\mathrm{Pr}_{\mathrm{Cart}}(\mathcal{X})$ through χ. For this reason, it is called the *subgroup of principal Weil divisors.* If s is a holomorphic section of the line bundle $\mathcal{O}(D)$, then the principal Weil divisor

$$\mathrm{div}(s) := \sum_{\{V \text{ irreducible}\,:\, \dim V = N-1\}} \mathrm{order}_V(s)\,\{V\}$$

is such that $\mathrm{div}(s) + \chi(D)$ is effective.[7]

When the complex analytic space $\mathcal{X}$ fails to be normal, one can still define a homomorphism $\tilde{\chi}$ from $\mathrm{Div}_{\mathrm{Cart}}(\mathcal{X})$ to $\mathrm{Div}_{\mathrm{Weil}}(\mathcal{X})$, but it fails, in general, to be injective. Such a homomorphism is defined as follows (see [**Ful75**, §1.1]): one can attach to any Cartier divisor D on $\mathcal{X}$ a coherent sheaf of $\mathcal{O}_{\mathcal{X}}$-ideals $\mathcal{I}(D)$ such that $\mathrm{Supp}(\mathcal{O}_{\mathcal{X}}/\mathcal{I}(D)) = \bigcup_\iota V_\iota$, where the V_ι are irreducible closed analytic hypersurfaces of $\mathcal{X}$ (with generic point z_ι), and the sum is locally finite. Then, according to (D.17), one defines $\tilde{\chi}$ by[8]

(D.27)
$$\tilde{\chi} : D \in \mathrm{Pic}_{\mathrm{Cart}}(\mathcal{X}) \longmapsto \tilde{\chi}(D) = \mathcal{Z}(\mathcal{O}_{\mathcal{X}}/\mathcal{I}(D)) = \sum_\iota \mathrm{length}\,(\mathcal{O}_{\mathcal{X}}/\mathcal{I}(D))_{z_\iota})\,\{V_\iota\} \in \mathrm{Pic}_{\mathrm{Weil}}(\mathcal{X}).$$

D.3.2. Chow groups on complex analytic spaces or algebraic varieties. Following [**Ful75**, Proposition 1.8 (2)], we introduce the following equivalence relation on the group $\mathcal{Z}(\mathcal{X})$ of cycles on a complex analytic space $\mathcal{X}$.

DEFINITION D.28. A cycle $Z \in \mathcal{Z}(\mathcal{X})$ is said to be rationally equivalent to zero in $\mathcal{Z}(\mathcal{X})$ if and only if there are irreducible (reduced) complex analytic spaces $\mathcal{Y}_\iota$, together with proper analytic morphisms $\pi_\iota : \mathcal{Y}_\iota \to \mathcal{X}$ and principal Cartier divisors $D_\iota \in \mathrm{Pr}_{\mathrm{Cart}}(\mathcal{Y}_\iota)$, such that

(D.28)
$$[Z] = \sum_\iota (\pi_\iota)_*([\tilde{\chi}(D_\iota)]) \quad \text{in} \quad {}'\mathcal{D}^{(\bullet,\bullet)}(\mathcal{X}).$$

For any $k = 0, \ldots, N = \dim \mathcal{X}$, the quotient group of $\mathcal{Z}_k(\mathcal{X})$ modulo rational equivalence is called the *k-Chow group* $A_k(\mathcal{X})$ of the complex analytic space $\mathcal{X}$.

---

[6]Nevertheless, the homomorphism (D.26) is an isomorphism, when $\mathcal{X}$ is nonsingular. That is, the groups $\mathrm{Div}_{\mathrm{Cart}}(\mathcal{X})$ and $\mathrm{Div}_{\mathrm{Weil}}(\mathcal{X})$ can be identified through such isomorphism when $\mathcal{X}$ is a complex manifold.

[7]This can be reformulated as $\mathrm{div}(s) \geq -D$ and explains why the notation $[-D]$ could be used instead of $\mathcal{O}(D)$.

[8]Such a definition agrees with that of χ in case $\mathcal{X}$ is normal.

One can define the Chow groups $A_k(\mathbb{X})$, $k = 0, \ldots, \dim \mathbb{X}$, of an algebraic variety defined over the algebraically closed field $\mathbb{K}$, that is, a reduced irreducible scheme of finite type, along similar terms. Elements in $\mathcal{Z}_k(\mathbb{X})$, $k = 0, \ldots, \dim \mathbb{X}$ are now formal combinations $\sum_\iota \nu_\iota \{\mathbb{V}_\iota\}$, where $\nu_\iota \in \mathbb{Z}$ and $\mathbb{V}_\iota$ is an irreducible algebraic subvariety of $\mathbb{X}$ of dimension k. An element $\mathbf{Z}$ in $\mathcal{Z}_k(\mathbb{X})$ is rationally equivalent to 0 if and only if there are algebraic varieties $\mathbb{Y}_\iota$ defined over $\mathbb{K}$, together with proper morphisms (in the algebraic sense) $\Pi_\iota : \mathbb{Y}_\iota \to \mathbb{X}$ and global invertible rational maps $r_\iota \in \mathcal{M}^*_{\mathbb{Y}_\iota}(\mathbb{Y}_\iota)$ such that $\mathbf{Z} = \sum_\iota (\Pi_\iota)_*(\widetilde{\chi}(\mathbb{D}_{r_\iota}))$, where $\mathbb{D}_{r_\iota}$ is the principal Cartier divisor attached to r_ι. See [**Ful75**, §1.1 to 1.3, Proposition §1.8 (2)] for the formulation in terms of schemes, the definition both of $\widetilde{\chi}$ and of the pushforward $(\Pi_\iota)_*$ of cycles by proper morphisms. The Chow group $A_k(\mathbb{X})$ is defined in terms of $\mathcal{Z}_k(\mathbb{X})$ modulo rational equivalence as introduced above.

EXAMPLE D.29 (Complete toric varieties). Let Σ be a complete simplicial $\mathbb{Z}^N$-rational fan in $\mathbb{R}^N = \mathbb{Z}^N \otimes_{\mathbb{Z}} \mathbb{R}$ and let $\mathbb{X}_\Sigma$ be the corresponding simplicial toric variety (see [**Ful93**, §1.4]) defined over $\mathbb{Q}$, but which will be considered here as a complex complete algebraic variety, with corresponding complex analytic space $\mathcal{X}_\Sigma$. Such a complex algebraic variety inherits the structure of a complex *orbifold*, which becomes the structure of a complex manifold when the fan is assumed to be simple) from the fact that transition maps are monomials. Let $\mathbb{H}_1, \ldots, \mathbb{H}_{N+r}$ be the irreducible toric (e.g., globally invariant under the torus action)[9] hypersurfaces defined as the closed orbits (under the torus action on $\mathbb{X}_\Sigma$, which correspond to $d = N + r$ elements in $\Sigma(1)$); see also Remark 3.9. That is, the irreducible toric hypersurfaces correspond to the $N + r$ rays $\xi_1\mathbb{R}, \ldots, \xi_d\mathbb{R}$ of the fan Σ, where each ξ_j is a primitive vector for the ray $\xi_j\mathbb{R}$ in $\mathbb{Z}^N \setminus \{(0, \ldots, 0)\}$. The classes of $\mathbb{H}_1, \ldots, \mathbb{H}_{N+r}$ modulo rational equivalence generate $A_{N-1}(\mathbb{X}_\Sigma) \otimes_{\mathbb{Z}} \mathbb{Q}$ over $\mathbb{Q}$. Although $\operatorname{rank}(A_{N-1}(\mathbb{X}_\Sigma)) = r$, the Chow group $A_{N-1}(\mathbb{X}_\Sigma)$ splits as the direct sum of a free group L with rank r and a finite abelian group $(A_{N-1}(\mathbb{X}_\Sigma))_{\text{tor}}$ corresponding to its torsion part. If Σ is a primitive fan[10] or $\Sigma(N)$ contains at least one primitive cone, the Chow group $A_{N-1}(\mathbb{X}_\Sigma)$ of the complete toric algebraic variety $\mathbb{X}_\Sigma$ is free of rank $r = |\Sigma(1)| - N$. For the combinatorial description of the Chow groups $A_k(\mathbb{X}_\Sigma)$ with $k < N - 1$, see [**FulS97**, Proposition 2.1].

D.4. Ampleness versus positivity for holomorphic line bundles

D.4.1. Globally generated holomorphic bundles. Let $\mathcal{X}$ be a complex analytic space and let $\mathcal{F}$ be a sheaf of $\mathcal{O}_{\mathcal{X}}$-modules on $\mathcal{X}$. The sheaf $\mathcal{F}$ is said to be *globally generated* if there exists a set of global sections $\{s_\iota\}_\iota$, $s_\iota \in \mathcal{F}(\mathcal{X}) = \Gamma(\mathcal{X}, \mathcal{F})$ such that the map

$$\bigoplus_\iota h_\iota \in \bigoplus_\iota \mathcal{O}_{\mathcal{X}} \longmapsto \sum_\iota h_\iota s_\iota \in \mathcal{F} \tag{D.29}$$

is surjective. This is equivalent to saying that given any open subset U of $\mathcal{X}$, any element ϕ in $\mathcal{F}(U) = \Gamma(U, \mathcal{F})$ can be expressed as a finite sum $\phi = \sum_\iota h_\iota\, s_\iota$, where $h_\iota \in \mathcal{O}_{\mathcal{X}}(U)$. A holomorphic bundle E over $\mathcal{X}$ is said to be *globally generated* if and only if its locally free sheaf of sections is globally generated.

---

[9]A Weil divisor on $\mathbb{X}_\Sigma$ is said to be toric if the irreducible closed hypersurfaces involved in its decomposition are toric.

[10]Observe that any $\mathbb{Z}^N$-rational fan can be algorithmically refined to become simplicial (see [**KKMS73**] or [**Ful93**, §2.6]), then to become primitive (introducing additional cones through additional rays in the fan).

EXAMPLE D.30 (The case of line bundles). A holomorphic line bundle L over $\mathcal{X}$ is globally generated if and only if the intersection of the zero sets of all its holomorphic global sections is empty. Such a holomorphic line bundle L is also said to be *basepoint free*, which means that, given any $z \in \mathcal{X}$, there is at least one global section $s \in \mathcal{O}_{\mathcal{X}}(\mathcal{X}, L)$ such that $s(z) \neq 0$.

When $\mathcal{X}$ is a complex N-dimensional manifold and E is a globally generated holomorphic m-bundle over $\mathcal{X}$, it follows from Baire's theorem (see [**De**, VII, Proposition 11.2]) that there is a finite-dimensional subspace of $\mathcal{O}_{\mathcal{X}}(\mathcal{X}, E)$ with dimension at most $m + N$ which generates all fibers E_z for $z \in \mathcal{X}$. As a consequence, one has the following result (see [**De**, VII, Proposition 6.10 and Corollary 11.5]) for the concepts of Griffiths or Nakano semipositivity for holomorphic bundles on a complex manifold, as introduced in §C.3.

PROPOSITION D.31. *Any globally generated holomorphic bundle on a complex manifold $\mathcal{X}$ inherits a smooth hermitian metric on E such that $(E^*, |\ |) \leq_{\mathrm{N}} 0$, hence $(E, |\ |) \geq_{\mathrm{G}} 0$.*

D.4.2. Semi-ampleness of holomorphic bundles. Given a holomorphic m-bundle $E \to \mathcal{X}$ over the complex analytic space, one introduces for any $p \in \mathbb{N}^*$ the p symmetric power[11] $S^{\otimes^p} E$ of the m-holomorphic bundle E.

DEFINITION D.32. A holomorphic bundle E over a complex analytic space $\mathcal{X}$ is said to be *semi-ample* if and only if $S^{\otimes^p} E$ is finitely generated for $p \in \mathbb{N}^*$ sufficiently large.

REMARK D.33. When L is a line bundle, one has $S^{\otimes^p} L = L^{\otimes^p}$. It follows then from Proposition D.31 that any semi-ample holomorphic line bundle over a complex manifold inherits a hermitian metric such that $(L, |\ |) \geq 0$.

D.4.3. Very ampleness/ampleness of line bundles (compact setting). Let us assume in addition that the complex analytic space $\mathcal{X}$ is topologically compact.[12] Let L be a basepoint free line bundle over $\mathcal{X}$. Since $\mathcal{X}$ is compact, $\mathcal{O}_{\mathcal{X}}(\mathcal{X}, L)$ is a finite dimensional $\mathbb{C}$-vector space with dimension $N_L \in \mathbb{N}^*$. Then there exists a morphism ι_L of complex compact analytic spaces from $\mathcal{X}$ to the projective $(N_L - 1)$-dimensional space $\mathbb{P}(\mathcal{O}_{\mathcal{X}}(\mathcal{X}, L))$ of $\mathbb{C}$-hyperplanes of the N_L-dimensional vector space $\mathcal{O}_{\mathcal{X}}(\mathcal{X}, L)$.

DEFINITION D.34. A holomorphic line bundle L over a complex compact analytic space $\mathcal{X}$ is said to be *very ample* if and only if it is basepoint free and the morphism ι_L between complex analytic spaces defined as above realizes a closed immersion[13] from the complex analytic space $(\mathcal{X}, \mathcal{O}_{\mathcal{X}})$ to the $(N_L - 1)$-dimensional

[11] If W is an m-dimensional $\mathbb{C}$-vector space, its p symmetric power $S^{\otimes^p} W$ is the $\mathbb{C}$-vector space defined as the quotient of $W^{\otimes^p}$ under the natural action of the symmetric group $\mathscr{S}_p$:

$$\mathfrak{s} \bullet (e_0 \otimes \cdots \otimes e_{p-1}) = e_{\mathfrak{s}(0)} \otimes \cdots \otimes e_{\mathfrak{s}(p-1)}$$

for $\mathfrak{s} \in \mathscr{S}_p$ and $e_j \in W$, $j = 0, \ldots, p-1$. Its dimension equals $N(m,p) = \binom{m+p-1}{p}$. The construction of the bundle $S^{\otimes^p} E$ (up to isomorphism between $N(m,p)$-holomorphic vector bundles) is then realized via cocycles as in §B.1.4.

[12] In the algebraic setting, one could consider more generally a complete scheme, for example, a complete algebraic variety.

[13] A morphism ι between complex analytic spaces $\mathcal{X}$ and $\mathcal{X}'$ is a closed immersion (or embedding) if $\iota(\mathcal{X})$ is identified with a closed analytic subset $V' = \iota(\mathcal{X})$ of $\mathcal{X}'$ such that any section

complex manifold $(\mathbb{P}(\mathcal{O}_{\mathcal{X}}(\mathcal{X},L)),\mathcal{O}_{\mathbb{P}(\mathcal{O}_{\mathcal{X}}(\mathcal{X},L))})$. The holomorphic line bundle L is said to be *ample* if and only if $L^{\otimes^p}$ is very ample for $p \in \mathbb{N}^*$ sufficiently large.

REMARK D.35. An alternative way to formulate the definition of a very ample line bundle over a compact complex analytic space is the following; see [**Ha**, Definition, p. 120]. A line bundle L over a compact complex analytic space $\mathcal{X}$ is very ample if and only if there is a closed immersion $\iota : \mathcal{X} \to \mathbb{P}^{N'}_{\mathbb{C}}$ such that one has $L = \iota^*(\mathcal{O}_{\mathbb{P}^{N'}_{\mathbb{C}}}(1))$. As a consequence, any very ample line bundle over a compact complex manifold $\mathcal{X}$ inherits (through pullback) the smooth Fubini–Study metric, making $(\mathcal{O}_{\mathbb{P}^{N'}_{\mathbb{C}}}(1), |\ |_{\mathrm{FS}})$ a positive bundle; see Definition C.36, Remark C.38 and Example C.39. Extracting pth roots, one sees that any ample holomorphic line bundle on a compact complex manifold admits a smooth metric with definite positive curvature, hence such that $(L,|\ |) > 0$.

One can state the following important result; see e.g [**De**, VII, Corollary 13.3].

PROPOSITION D.36. *A holomorphic bundle L over a compact complex manifold is ample if and only if L can be equipped with a smooth hermitian metric such that* $(L,|\ |) > 0$.

D.4.4. Very ampleness and ampleness of higher rank bundles. When $\mathcal{X}$ is a compact complex analytic space,[14] a m-bundle over $\mathcal{X}$ is classically said to be *ample* if and only if the so-called *Serre line bundle* $\mathcal{O}_{\mathrm{Proj}\,E}(1) = (\mathcal{O}_{\mathrm{Proj}\,E}(-1))^*$, see Definition B.12 and §B.3.3, is ample[15] according to Definition D.34. We will not enter into such considerations here, but recall instead, for bundles with rank $m > 1$ over a compact[16] complex manifold $\mathcal{X}$, the concepts of *very ampleness* and then of *ampleness*, as presented in [**De**, VII, §11.B]. To do so, one needs to introduce a bundle E over the compact complex manifold $\mathcal{X}$ the bundle J^1E of its 1-jets, whose sheaf of sections is defined as follows: $\phi \in \mathcal{O}_{\mathcal{X}}(U, J^1E)$ if and only if $\phi = j^1\Phi$ is the Taylor development at order one of $\Phi \in \mathcal{O}_{\mathcal{X}}(U,E)$, U being an arbitrary open subset of $\mathcal{X}$. Then, one can state the following definition [**De**, VII, Definition 11.7].

DEFINITION D.37. A holomorphic m-bundle (with $m > 1$) over a compact complex manifold $\mathcal{X}$ is said to be *very ample* if and only if both maps

$$\begin{aligned} &s \in \mathcal{O}_{\mathcal{X}}(\mathcal{X},E) \longmapsto (j^1s)(z) \in (J^1E)_z, \\ &s \in \mathcal{O}_{\mathcal{X}}(\mathcal{X},E) \longmapsto (s(z'),s(z'')) \in E_{z'} \times E_{z''} \end{aligned} \tag{D.30}$$

are surjective for any z,z',z'' with $z' \neq z''$ in $\mathcal{X}$. It is said to be *ample* if $S^{\otimes^p}E$ is very ample for $p \in \mathbb{N}^*$ sufficiently large.

One can state within this context the following result, in the same vein as Proposition D.31; see [**De**, VII, Corollary 11.13].

---

of $\mathcal{O}_{\mathcal{X}}$ in a neighborhood of z can be identified with the restriction of a section of $\mathcal{O}_{\mathcal{X}'}$ in a neighborhood of $\iota(z)$, that is $\iota^\sharp : \mathcal{O}_{\mathcal{X}'} \to \iota_*\mathcal{O}_{\mathcal{X}}$ is surjective.

[14]In the algebraic setting, one could consider more generally a projective scheme over $\mathbb{C}$ (for example a projective algebraic variety over $\mathbb{C}$).

[15]For an introduction to the concept of ampleness for holomorphic bundles with a higher rank on a projective scheme, we refer the reader to [**Laz**, II, §6.1].

[16]The compactness hypothesis is not, in fact, necessary anymore, but we keep it so that the definition proposed here is coherent with the previous comments, as well as with Definition D.34 for line bundles over a compact complex analytic space, according to [**De**, VII, Proposition 11.10].

PROPOSITION D.38. *If E is a very ample bundle over a compact complex manifold $\mathcal{X}$ according to Definition* D.37, *then E^* inherits an hermitian metric such that $(E^*, |\ |^*) <_{\mathrm{N}} 0$, hence $(E, |\ |) >_{\mathrm{G}} 0$.*

D.4.5. Examples.

EXAMPLE D.39 (Ampleness on $\mathbb{P}^N_{\mathbb{C}}$). The line bundle $\mathcal{O}_{\mathbb{P}^N_{\mathbb{C}}}(1) \to \mathbb{P}^N_{\mathbb{C}}$ is the archetypical example of a very ample line bundle over $\mathbb{P}^N_{\mathbb{C}}$. Observe that the two notions of ampleness and very ampleness for line bundles do coincide here since $\mathbb{P}^N_{\mathbb{C}}$ is a smooth complete toric variety; see [**Dem70**]. If $\mathbb{X}$ is a projective complex algebraic manifold (i.e., is embedded via ι in some $\mathbb{P}^{N'}_{\mathbb{C}}$ for N' large enough), then the line bundle $\big(\mathcal{O}_{\mathbb{P}^{N'}_{\mathbb{C}}}(1)\big)_{|\mathbb{X}} \to \mathbb{X}$ is a very ample holomorphic bundle over $\mathbb{X}$.

EXAMPLE D.40 (Ampleness of toric Cartier divisors on a complete toric variety). Let $\mathbb{D}$ be a toric Cartier divisor on the toric algebraic variety $\mathbb{X}_\Sigma$ attached to a complete $\boldsymbol{N}$-rational fan Σ in $\boldsymbol{N}_{\mathbb{R}} = \boldsymbol{N} \otimes_{\mathbb{Z}} \mathbb{R}$, when $\boldsymbol{N}$ denotes a N-lattice (for example $\boldsymbol{N} = \mathbb{Z}^N$) as in §D.2.2. The Cartier divisor $\mathbb{D}$ corresponds to an isomorphism class $\mathcal{O}(\mathbb{D})$ of basepoint free line bundles if and only if the corresponding support function $\Psi_{\mathbb{D}}$, through the one-to-one correspondence stated in Proposition D.21, is concave. If this is the case, global holomorphic sections of $\mathcal{O}(\mathbb{D})$ are expressed in the dense torus $\mathbb{T}^N$ as $\zeta \longmapsto F(\zeta)$, where F is a Laurent polynomial in N variables with support in $\mathrm{Stab}(\Psi_{\mathbb{D}}) := \{\xi^\star \in (\boldsymbol{N}^\star)_{\mathbb{R}} : \langle \xi^\star, \cdot\rangle \geq \Psi_{\mathbb{D}} \text{ in } \boldsymbol{N}_{\mathbb{R}}\}$. The isomorphism class $\mathcal{O}(\mathbb{D})$ is a class of ample line bundles if and only if $\Psi_{\mathbb{D}}$ is strictly concave; see [**Ful93**, §3.4] or [**BGPS14**, §3.4]. Let us give another interpretation of these equivalences in the particularly important case where $\boldsymbol{N} = \mathbb{Z}^N$ and the fan Σ is primitive. The assumptions on the fan imply that the toric variety $\mathbb{X}_\Sigma$ is nonsingular, in which case the notions ampleness and very ampleness for (algebraic) line bundles on $\mathbb{X}_\Sigma$, in particular toric ones, coincide; see [**Dem70**]. For $j = 1, \dots, d = N + r$, let $\mathbb{D}_j$ be the Cartier divisor such that $\chi(\mathbb{D}_j) = \{\mathbb{H}_j\}$ (χ as in (D.26)), where $\mathbb{H}_j$ denotes the closed orbit corresponding to the ray $\xi_j\mathbb{R} \in \Sigma(1)$ (with primitive vector $\xi_j \in \mathbb{Z}^N \setminus \{(0, \dots, 0)\}$). Given the Cartier toric divisor $\mathbb{D} = \sum_1^d a_j \mathbb{D}_j$, $a_j \in \mathbb{Z}$, one can associate to it the convex polyhedron

$$P_{\mathbb{D}} := \{\xi^\star \in (\mathbb{R}^N)^\star : \langle \xi^\star, \xi_j\rangle + a_j \geq 0,\ j = 1, \dots, d\}. \tag{D.31}$$

Let $z_1, \dots, z_d$ be the homogeneous coordinates attached to the rays $\xi_1\mathbb{R}, \dots, \xi_d\mathbb{R}$ in $\Sigma(1)$ respectively. If $F \in \mathbb{C}[X_1^{\pm 1}, \dots, X_N^{\pm 1}]$ is a generic Laurent polynomial with support equal to $P_{\mathbb{D}}$, then

$$F\Big(\prod_{j=1}^d z_j^{\xi_{j,1}}, \dots, \prod_{j=1}^d z_j^{\xi_{j,N}}\Big) = \frac{\mathrm{F}(z_1, \dots, z_d)}{\prod_{j=1}^d z_j^{a_j}}, \tag{D.32}$$

where the $\mathbb{X}_\Sigma$-homogenization $\mathrm{F} \in \mathbb{C}[z_1, \dots, z_d]$ (see, e.g., [**CaD97**, §2]) is a polynomial, whose support equals a convex polyhedron $\Delta_{\mathbb{D}} \subset \mathbb{R}^d_{\geq 0}$. The line bundle $\mathcal{O}(\mathbb{D})$ is semi-ample if and only if for any $\sigma \in \Sigma(N)$ with $\xi_{j_1}, \dots, \xi_{j_N}$ as primitive generators, the projection $\Delta^\sigma_{\mathbb{D}}$ of $\Delta_{\mathbb{D}}$ on the space of coordinates $t_{j_1}, \dots, t_{j_N}$ admits the origin as a vertex. This amounts to saying if $\{j_1, \dots, j_N\} = \{1, \dots, N\}$ in order to fix ideas, that the Weil divisor

$$\sum_{\rho=1}^r \Big(a_{N+\rho} - \Big\langle \sum_{j=1}^N \xi_j\,,\, \xi_{N+\rho}\Big\rangle\Big)\{H_{N+\rho}\} \tag{D.33}$$

is effective. The line bundle $\mathcal{O}(\mathbb{D})$ is ample if and only if all Weil divisors (D.33), for any $\sigma \in \Sigma(N)$ with primitive generators $\xi_{j_1}, \dots, \xi_{j_N}$, are strictly effective. In the case where $\mathcal{O}(\mathbb{D})$ is ample, the convex polyhedron $P_{\mathbb{D}}$ defined by (D.31) is combinatorially dual to the fan Σ.

D.5. Local versus global intersection theory

D.5.1. Local extended multi-index of intersection, Crofton formula. Let $\mathcal{X}$ be an N-dimensional complex manifold. For every $m \in \mathbb{N}$ consider the N^{m+1}-dimensional complex product manifold $\mathscr{X}_{m+1} = \mathcal{X}^{m+1}$ with structural sheaf $\mathscr{O}_{m+1}$, and N-dimensional diagonal submanifold

$$\mathscr{D}_{m+1} := \{(z^0, \dots, z^m) \in \mathcal{X}^{m+1} \; : \; z^0 = \cdots = z^m\}. \tag{D.34}$$

Let $\mathscr{I}_{m+1}$ be the coherent sheaf of $\mathscr{O}_{m+1}$ radical ideals attached to $\mathscr{D}_{m+1} \subset \mathscr{X}_{m+1}$, with $\overline{\mathscr{I}_{m+1}}$ denoting its integral closure; see Proposition A.37. Given $m+1$ classes of generalized cycles $\mu_0, \dots, \mu_m \in \mathscr{B}(\mathcal{X})$ with pure dimensions $\delta_0, \dots, \delta_m$ so that one lets $|\delta| = \delta_0 + \cdots + \delta_m$ (see Definition D.7) one defines as follows local (extended) multi-indexes of intersection $(\mu_0, \dots, \mu_m)_z$ at any $z \in |\mu_0| \cap \cdots \cap |\mu_m|$. Let $\boldsymbol{\mu} \in \mathscr{B}_{|\delta|}(\mathscr{X}_{m+1})$ be the generalized cycle with representative the $(\delta_0 + \cdots + \delta_m, \delta_0 + \cdots + \delta_m)$-dimensional current defined as

$$\widehat{\boldsymbol{\mu}}(z^0, \dots, z^m) = \widehat{\mu}_0(z^0) \wedge \cdots \wedge \widehat{\mu}_m(z^m),$$

where $\widehat{\mu}_j \in \mathcal{GZ}_{\delta_j}(\mathcal{X})$ represents $\mu_j \in \mathcal{B}(\mathcal{X})$. Let

$$S(\overline{\mathscr{I}_{m+1}}\,;\boldsymbol{\mu}) = \sum_{\max(0,|\delta|-\dim(|\boldsymbol{\mu}|\,\cap\,\mathscr{D}_{m+1}))}^{\min(|\delta|,\dim(|\boldsymbol{\mu}|\,\cap\,\mathscr{D}_{m+1})} S_r(\overline{\mathscr{I}_{m+1}}\,;\boldsymbol{\mu}) \in \mathcal{B}(\mathscr{X}_{m+1}) \tag{D.35}$$

be represented on $\mathscr{X}_{m+1}$ by the generalized cycle realized in a neighborhood of $\boldsymbol{z} \in \mathscr{D}_{m+1}$ as

$$\Big[(1 - \|\boldsymbol{s}\|^{2\lambda})\,\widehat{\boldsymbol{\mu}}\Big]_{\lambda=0} + \sum_{r=\max(1,|\delta|-\dim(|\boldsymbol{\mu}|\,\cap\,\mathscr{D}_{m+1}))}^{\min(|\delta|,\dim(|\boldsymbol{\mu}|\,\cap\,\mathscr{D}_{m+1})} \Big[\overline{\partial}\,|\boldsymbol{s}|^{2\lambda} \wedge \frac{d\log|\boldsymbol{s}|^2}{2i\pi} \wedge (dd^c \log|\boldsymbol{s}|^2)^{r-1} \wedge \widehat{\boldsymbol{\mu}}\Big]_{\lambda=0},$$

where $\mathcal{I}_{\boldsymbol{s}} = \mathscr{I}_{m+1}$ in a neighborhood of $\boldsymbol{z}$; see Example D.10. It is supported by $\mathscr{D}_{m+1}$. Definition D.14 leads to the following one.

Definition D.41. Given $\mu_0, \dots, \mu_m$ as above and $z \in |\mu_0| \cap \cdots \cap |\mu_m|$, the *(extended) multi-index of intersection* of the classes of generalized cycles μ_j at z is the ordered sequence of nonnegative integers $(\mu_0 \cap \cdots \cap \mu_m)_z(\kappa)$, where

$$\begin{aligned}(\mu_0 \cap \cdots \cap \mu_m)_z(\kappa) &= \operatorname{mult}_{\boldsymbol{z}} S_{|\delta|-\kappa}(\overline{\mathscr{I}_{m+1}}\,;\boldsymbol{\mu}) \\ &= \begin{cases} \operatorname{mult}_{\boldsymbol{z}} \Big[\overline{\partial}\,|\boldsymbol{s}|^{2\lambda} \wedge \frac{d\log|\boldsymbol{s}|^2}{2i\pi} \wedge (dd^c \log|\boldsymbol{s}|^2)^{|\delta|-\kappa-1} \wedge \widehat{\boldsymbol{\mu}}\Big]_{\lambda=0} & \text{if } \kappa < |\delta| \\ \operatorname{mult}_{\boldsymbol{z}} \Big[(1 - \|\boldsymbol{s}\|^{2\lambda})\,\widehat{\boldsymbol{\mu}}\Big]_{\lambda=0} \quad \text{if } \kappa = |\delta| \end{cases},\end{aligned} \tag{D.36}$$

where $\kappa = 0, \dots, \min(|\delta|, \dim(|\boldsymbol{\mu}| \cap \mathscr{D}_{m+1}))$ and $\boldsymbol{z} = (z, \dots, z)$.

REMARK D.42. The (extended) multi-index of intersection $(\mu_0, \dots, \mu_m)_z$ was introduced by P. Tworzewski [**Tw95**, Definition 4.2]; see also [**AndSWY17B**, §1 and §10]. This ordered list of integers can also be interpreted algebraically as the list of Hilbert–Samuel generalized multiplicities introduced, for example, in [**AcM93**, §2] and [**AcM97**]; see also [**AcR01**, §5].

A very important averaging formula in integral geometry lies behind the realization of the extended multi-index of intersection as the list of Segre numbers (D.36). It goes back to Morgan Crofton (1826–1915) and its simplest formulation in complex analytic geometry is the following; see [**De**, III.7] or [**AndSWY17B**, Lemma 6.3].

PROPOSITION D.43. *Let $\mathcal{X}$ be a purely N-dimensional complex analytic space and let $s = (s_0, \dots, s_m)$ be a nonvanishing section of the trivial bundle $\mathbb{C}^{m+1}$, equipped with its canonical hermitian metric $\| \ \|$. Then the smooth $(1,1)$-current $dd^c \log \|s\|^2$ can be realized independently of the choice of a smooth metric $| \ |$ on $\mathcal{O}_{\mathbb{P}^m_{\mathbb{C}}}(1)$ as*

(D.37)
$$\begin{aligned} dd^c \log \|s\|^2 &= \pi_* \Big(\Big[\bar\partial \, |\langle u, s(z)\rangle|^{2\lambda} \wedge \frac{\partial \log |\langle u\,,\, s(z)\rangle|^2}{2i\pi} \wedge \big(dd^c \log \|u\|^2\big)^m \Big]_{\lambda=0} \Big) \\ &= \pi_* \Big([\mathrm{div}\, \langle u\,,\, s(z)\rangle] \wedge \big(dd^c \log \|u\|^2\big)^m \Big), \end{aligned}$$

where the proper projection $\pi_{\mathcal{X}} \ : (u,z) \in \mathbb{P}^m_{\mathbb{C}} \times \mathcal{X} \to z \in \mathcal{X}$ transforms by direct image currents of bidimension $(N-1, N-1)$ on $\mathbb{P}^m_{\mathbb{C}} \times \mathcal{X}$ into $(1,1)$-currents on the complex analytic space $\mathcal{X}$.

REMARK D.44. The invariance of the Monge–Ampère measure $(dd^c \log \|u\|^2)^{\wedge^m}$ under the action of the unitary group $U(m+1, \mathbb{C})$ acting on $\mathbb{P}^m_{\mathbb{C}}$ justifies Proposition D.43. It also implies the identities

(D.38)
$$\begin{aligned} (\pi_{\mathcal{X}})_* \Big(\mathrm{PV} \Big[\frac{u_j}{\langle u, s(z)\rangle} \Big] \wedge \big(dd^c \log \|\xi\|^2\big)^m \Big) &= \frac{\overline{u_j}}{\|s\|^2}, \\ (\pi_{\mathcal{X}})_* \Big(u_j \, \bar\partial_z \Big(\frac{1}{\langle u_j\,,\, s(z)\rangle} \Big) \wedge \big(dd^c \log \|u\|^2\big)^m \Big) &= \bar\partial \Big(\frac{\overline{u_j}}{\|u\|^2} \Big), \end{aligned}$$

for $j = 0, \dots, m$ [**VidY22**, §2]. These identities will prove to be useful when averaging multivariate residue calculus to build a bridge from multiplicative Coleff–Herrera residue current calculus, as in §2.3.2, to Bochner–Martinelli residue current calculus developed in §3.2.1.

Let us now suppose that $s = (s_0, \dots, s_m)$ is a holomorphic section of $L^{\oplus^{m+1}}$, where L is a holomorphic line bundle over $\mathcal{X}$ equipped with an arbitrary smooth metric $| \ |$ that induces, as in the Example D.11, the metric $\| \ \| = \sqrt{| \ |^2 + \cdots + | \ |^2}$ on $L^{\oplus^{m+1}}$. If one considers the complex analytic space $\mathbb{P}^m_{\mathbb{C}} \times \mathcal{X}$ with the two projections $\pi_u : (u,z) \mapsto u \in \mathbb{P}^m_{\mathbb{C}}$ and $\pi_{\mathcal{X}} : (u,z) \mapsto z \in \mathcal{X}$, then the line bundle $\pi_{\mathcal{X}}^* L \otimes \pi_u^* \big(\mathcal{O}_{\mathbb{P}^m_{\mathbb{C}}}(1)\big)$ inherits a metric also denoted by $| \ |$ by picking up again an arbitrary metric on $\mathcal{O}_{\mathbb{P}^m_{\mathbb{C}}}(1)$. A natural example of such a choice is the Fubini–Study metric. Assume also $\widehat{\mu}$ is a generalized cycle with pure codimension $0 \le M \le \dim \mathcal{X}$, representing the class $\mu \in \mathcal{B}_{N-M}(\mathcal{X})$. Let r be an integer such that

$$\max(0, N - M - \dim(|\mu| \cap V(s)) \le r \le \dim \mu.$$

Then, for any such r, the component $\widehat{V}_r(L\,;\,s\,;\,\mu)$ of bidegree $(M+r, M+r)$ of the Stückrad–Vogel generalized cycle (D.14) introduced in Example D.11 can be realized independently of the choice of the metrics on L and $\mathcal{O}_{\mathbb{P}^m_{\mathbb{C}}}(1)$ as the following direct image; see [**AndESWY21A**, Proposition 9.3] and [**AndSWY17B**, Theorem 1.2].

PROPOSITION D.45. *One has the averaged formula*

$$\widehat{V}_r(L\,;\,s\,;\,\widehat{\mu}) = (\pi^r_{\mathcal{X}})_* \Big(\Big[(1-|s(z)|^{2\lambda}) \Big(\bigwedge_{\rho=1}^{r} \Gamma^{\lambda^{\rho+1}}_{L,s,|\ |}(u^\rho, z) \Big) \wedge \widehat{\mu}(z) \Big]_{\lambda=0} \Big), \tag{D.39}$$

where $\pi^r_{\mathcal{X}}$ denotes the proper projection from the product space $(\mathbb{P}^m_{\mathbb{C}})^r \times \mathcal{X}$ (with current point $(u^1, \dots, u^r, z)$) onto $\mathcal{X}$ and

$$\Gamma^{\lambda}_{L,s,|\ |}(u,z) = \overline{\partial}\,\big|\langle u\,,\,s(z)\rangle\big|^{2\lambda} \wedge \partial\Big(\frac{\log\big|\langle u\,,\,s(z)\rangle\big|^2}{2i\pi}\Big) \wedge (dd^c \log\|u\|^2)^m \tag{D.40}$$

is an $(m+1, m+1)$ form on $\mathbb{P}^m_{\mathbb{C}} \times \mathcal{X}$, which is regular up to any order ℓ provided the complex parameter λ is such that $\mathrm{Re}\,\lambda \gg_\ell 1$.

REMARK D.46. When $\mu = Z$ is a cycle on $\mathcal{X}$ of pure dimension and $\widehat{\mu} = [Z]$, the current $\widehat{V}_r(L\,;\,s\,;\,[Z])$ is the direct image by the proper projection $\pi^r_{\mathcal{X}}$ as in (D.39) of

$$\big[\mathcal{Z}_{s,r}\big] \wedge (dd^c \log\|u^1\|^2 \wedge \cdots \wedge dd^c \log\|u^r\|^2),$$

where $\mathcal{Z}_{s,r}$ is a cycle with pure codimension r on $\mathcal{X} \times (\mathbb{P}^m_{\mathbb{C}})^r$. Such a cycle is obtained from $Z \times (\mathbb{P}^m_{\mathbb{C}})^r$ after r iterations of the cutting-out transient process $Z \mapsto Z^V$ described in (D.4) and (D.5),[17] followed by a cutting-out final operation which consists in keeping, with their weights ν_ι, the closed irreducible subsets which lie entirely in some irreducible component of $s^{-1}(\{0\}) \times (\mathbb{P}^m_{\mathbb{C}})^r$.

D.5.2. Crofton's averaging and metrized toric divisors. The intimate connection between Crofton's formula and the Fubini–Study metric on the line bundle $\mathcal{O}(1) \to \mathbb{P}^N_{\mathbb{C}}$, discussed in §D.5.1, plays a central role in many places within our monograph. Let us briefly present here this connection from a real *tropical*[18] point of view. Let Σ be a complete $\mathbb{Z}^N$-rational fan in $\mathbb{R}^N = \mathbb{Z}^N \otimes_{\mathbb{Z}} \mathbb{R}$, $\mathbb{X}_\Sigma$ be the corresponding complete algebraic toric variety considered here as defined over $\mathbb{C}$. That is, we consider $\mathbb{X}_\Sigma$ as a complex algebraic variety, even though it is defined over $\mathbb{Q}$, with $(t,z) \in \mathbb{T}^N \times \mathbb{X}_\Sigma \to t \bullet z \in \mathbb{X}_\Sigma$ being the action on it of the complex dense torus $\mathbb{T}^N = (\mathbb{C}^*)^N$; see [**Ful93**]. Let $\mathbb{D}$ be a toric Cartier divisor on $\mathbb{X}_\Sigma$, with concave virtual support function $\Psi_{\mathbb{D}}$ (see Definition D.19 and Proposition D.21) which means that $\mathcal{O}(\mathbb{D})$ is basepoint free (Example D.40). Then, there is a one-to-one correspondence between semipositive (SP) continuous hermitian metrics on $L = \mathcal{O}(\mathbb{D})$ in the sense of Definition C.43 and concave functions $\psi : \mathbb{R}^N \to \mathbb{R}$ such that $\psi - \Psi_{\mathbb{D}}$ is a bounded real function on $\mathbb{R}^N$. This one-to-one correspondence goes as follows; see [**BGPS14**, Theorem 4.8 (1)]: if $s_{\mathbb{D}} = s_{\Psi_{\mathbb{D}}}$ denotes the

[17] One needs to extend this process by $\mathbb{Z}$-linearity to the case where $V = \sum_\iota \nu_\iota \{V_\iota\}$ is a cycle instead of just a closed irreducible subset such as $V = V_\iota$.

[18] We refer here to *tropical geometry*, in relation with the $(\max, +)$ or $(\min, +)$ calculus in $\mathbb{R} \cup \{-\infty\}$ or $\mathbb{R} \cup \{+\infty\}$; see, e.g., [**RST05**] for an elementary introduction. Such operations are denoted within this monograph by $\boxplus$ for max and $\boxtimes$ for $+$, where the tropical real line (here with respect to $(\max, +)$ calculus), is denoted by $\mathrm{Trop} = \mathbb{R} \cup \{-\infty\}$; see also §7.4.1 in the present monograph.

distinguished toric meromorphic global section of $\mathcal{O}(\mathbb{D})$,[19] then the semipositive metric $|\ |$ corresponds to the unique concave function $\psi_{|\ |}$ (which differs from $\Psi_{\mathbb{D}}$ by a bounded real function) such that $-\log|s_{\mathbb{D}}| = \psi_{|\ |} \circ (-\mathrm{Log})$, where $-\mathrm{Log}$ is the so-called *tropicalization map* defined on the dense orbit $(\mathbb{C}^*)^N = \mathbb{T}^N$ by $-\mathrm{Log}\,\zeta = \mathrm{trop}(\zeta) = (-\log|\zeta_1|, \ldots, -\log|\zeta_N|)$.

EXAMPLE D.47. In the archetypal situation where $\mathbb{X}_\Sigma = \mathbb{P}^N_{\mathbb{C}}$ (see Example A.1.2(3) and $L = \mathcal{O}(1)$) the Fubini–Study (toric) smooth positive metric $|\ |_{\mathrm{FS}}$ on L corresponds to the concave function

$$\psi_{|\ |_{\mathrm{FS}}} : x \in \mathbb{R}^N \longmapsto \frac{1}{2}\log\Big(1 + \sum_{k=1}^N e^{-2x_k}\Big).$$

DEFINITION D.48 (Crofton averaging of (SP) toric metrics). Let $\mathbb{D}$ be a toric Cartier divisor on $\mathbb{X}_\Sigma$ with concave virtual support function and σ be a global holomorphic (e.g., regular) section of $L = \mathcal{O}(\mathbb{D})$. Given a semipositive (SP) continuous metric $|\ |$ on L (see Definition C.43) its Crofton's averaging with respect to σ is the unique SP continuous metric $|\ |_{\mathrm{Croft},\sigma}$ on L defined by the convergent integral

$$\log|s_{\mathbb{D}}|_{\mathrm{Croft},\sigma} = \frac{1}{\deg_L \mathbb{X}_\Sigma}\int_{\mathbb{T}^N}(-\log)\Big(\frac{|\sigma(t\bullet z)|}{|s_{\mathbb{D}}(t\bullet z)|}\Big)\,(c_1(L,|\ |))^N, \tag{D.41}$$

where the wedge powers of the 1-Chern current $c_1(L,|\ |)$ are computed iteratively via the Bedford–Taylor procedure (C.7); see §C.1.2. Such an SP continuous metric $|\ |_{\mathrm{Croft},\sigma}$ is also called the *croftonization of the metric* $|\ |$ *with respect to the global regular section* σ.

Any concave real function ψ on $\mathbb{R}^N$ defines a real positive Borel measure on $\mathbb{R}^N_x$, namely its Monge–Ampère measure $d\mathrm{MA}(\psi,\ldots,\psi)$; see [**RaT77**] or also [**BGPS14**, §2.7].[20] We have from a direct computation based on Fubini's theorem the following proposition.

PROPOSITION D.49. *Let* $|\ |$ *be any* SP *continuous metric on* $\mathcal{O}(\mathbb{D})$ *and let* $\sigma = F s_{\mathbb{D}}$, *where* $F \in \mathbb{C}[X_1^{\pm 1}, \ldots, X_N^{\pm 1}]$ *is a global section with support in*

$$\{\xi \in \mathbb{R}^N : \langle \xi, \cdot\rangle \geq \Psi_{\mathbb{D}} \text{ on } \mathbb{R}^N\} = \mathrm{Stab}\,(\Psi_{\mathbb{D}}).$$

Then one has that

$$\psi_{|\ |_{\mathrm{Croft},\sigma}}(x) = N!\int_{\mathbb{R}^N}\rho_F(x+y)\,(d\mathrm{MA}(\psi,\ldots,\psi))(y), \tag{D.42}$$

where the convex function

$$R_F : x \mapsto -\rho_F(-x) = \frac{1}{(2\pi)^N}\int_{(\mathbb{R}/2\pi\mathbb{Z})^N}\log|F(e^{x_1+i\theta_1}, \ldots, e^{x_N+i\theta_N})|\,d\theta_1\wedge\cdots\wedge d\theta_N \tag{D.43}$$

is known as the Ronkin function[21] *of the Laurent polynomial* F [**Ro00**].

---

[19]See, e.g., [**BGPS14**, Remark 3.3.6 and Theorem 3.3.7]. This section is invariant, as $\mathbb{D}$ is, under the torus action $(t,z) \to t\bullet z$.

[20]When ψ is smooth $d\mathrm{MA}(\psi,\ldots,\psi) = (-1)^N H_\psi(x)\,dx_1\wedge\cdots\wedge dx_N$, where H_ψ denotes the hessian determinant of ψ. When ψ is just continuous and A is a Borel subset of $\mathbb{R}^N$, $\int_A (d\mathrm{MA}(\psi,\ldots,\psi))(y)$ is the N-dimensional Lebesgue measure of the image of A by the gradient map $\nabla\psi$.

[21]The Ronkin function plays an important role in the explicit expression of the global logarithmic height of the Zariski closure of $\{\zeta \in \mathbb{T}^N : F(\zeta) = 0\}$ in $\mathbb{X}_\Sigma$ when the coefficients of F belong

REMARK D.50. In the particular case where $\mathbb{X}_\Sigma = \mathbb{P}^N_{\mathbb{C}}$ and $L = \mathcal{O}(1)$, taking into account Proposition D.43, Crofton's formula (D.42) admits within such a setting the alternative realization

$$\text{(D.44)} \qquad \psi_{|\ |_{\rm FS}}(x) = N! \int_{\mathbb{R}^N} \rho_{1+X_1+\cdots+X_N}(x+y)\,(d\mathrm{MA}(\psi_{|\ |_{\rm FS}},\dots,\psi_{|\ |_{\rm FS}}))(y),$$

since this amounts to saying that the croftonization of the global section $(1+\sum_1^N \zeta_k)\, z_0$ (where $\mathbb{D} = \mathcal{O}(1)$, hence $s_{\mathbb{D}} = z_0$, and $\zeta_k = z_k/z_0$ for $k = 1,\dots,N$) leaves unchanged the Fubini–Study metric on $\mathcal{O}(1)$.

D.5.3. The join product in the projective setting. Concerning global intersection questions, let us focus on the archetypal example of the projective complex algebraic variety $\mathbb{X} = \mathbb{P}^N_{\mathbb{C}}$. Given $m \in \mathbb{N}^*$, one needs first to define within the frame of classes of algebraic generalized cycles on $\mathbb{P}^N_{\mathbb{C}}$ the important notion of $(m+1)$-*join product of* $m+1$ *cycles or classes of generalized cycles.* What makes the construction essential is that the diagonal subvariety $\mathscr{D}_{m+1} \subset (\mathbb{P}^N_{\mathbb{C}})^{m+1}$ fails to be globally defined as the proper intersection of supports of regular sections of algebraic line bundles. Although the construction is classical, let us recall it for the sake of completeness. Consider in

$$\text{(D.45)} \quad \begin{aligned} &\mathbb{P}^{(N+1)(m+1)-1}_{\mathbb{C}} :\\ &= \mathrm{Spec}(\mathrm{Proj}(\mathbb{C}[z_0^{[0]} : \cdots : z_N^{[0]} \ :\ z_0^{[1]} : \cdots : z_N^{[1]} \ : \cdots :\ z_0^{[m]} : \cdots : z_N^{[m]}]))\\ &\qquad\qquad = \mathrm{Spec}(\mathrm{Proj}(\mathbb{C}[z^{[0]} : \cdots : z^{[m]}]))\end{aligned}$$

the $((N+1)m-1)$-linear subspaces $\Pi^{[0]},\dots,\Pi^{[m]}$, whose intersection is empty, defined as $\Pi^{[j]} = (z^{[j]})^{-1}(\{0\})$ for $j = 0,\dots,m$. Let γ_m, π_m, ι_m be the following algebraic morphisms:[22]

$$\text{(D.46)} \quad \begin{aligned} \gamma_m &: [z^{[0]} : \cdots : z^{[m]}] \in \mathbb{P}^{(N+1)(m+1)-1}_{\mathbb{C}} \setminus \bigcup_{j=0}^m \Pi^{[j]}\\ &\qquad\qquad \longmapsto ([z^{[0]}],\dots,[z^{[m]}]) \in (\mathbb{P}^N_{\mathbb{C}})^{m+1},\\ \pi_m &: \widehat{\mathbb{P}^{(N+1)(m+1)-1}_{\mathbb{C}}} \xrightarrow{\text{blow up along } \bigcup_0^m \Pi^{[j]}} \mathbb{P}^{(N+1)(m+1)-1}_{\mathbb{C}},\\ \iota_m &: z = [z_0 : \cdots : z_N] \in \mathbb{P}^N_{\mathbb{C}} \longmapsto [z,\dots,z] \in \mathbb{P}^{(N+1)(m+1)-1}_{\mathbb{C}}.\end{aligned}$$

Let $\mu_0,\dots,\mu_m \in \mathcal{B}(\mathbb{P}^N_{\mathbb{C}})$ be $(m+1)$-classes of generalized cycles over $\mathbb{P}^N_{\mathbb{C}}$ with respective dimensions $\delta_0,\dots,\delta_m$, with $|\delta| = \delta_0+\cdots+\delta_m$. Let also $\widehat{\boldsymbol{\mu}}$ be a generalized cycle of bidegree $((m+1)N - |\delta|, (m+1)N - |\delta|)$ representing the product class

$$\boldsymbol{\mu} = \mu_0 \times \cdots \times \mu_m \in \mathcal{B}_{|\delta|}((\mathbb{P}^N_{\mathbb{C}})^{m+1}).$$

---

to $\mathbb{Q}$ or more generally to a number field; see, e.g., **[Gua18, GuaS23]**. When $F \in \mathbb{Z}[X_1,\dots,X_N]$, $R_F(0)$ is the logarithm of the *Mahler measure* of $F \in \mathbb{Z}[X]$. The logarithm of such a Mahler measure provides the global logarithmic height of the Zariski closure of $\{\zeta \in \mathbb{C}^N : F(\zeta) = 0\}$ in $\mathbb{P}^N_{\mathbb{C}}$ with respect to the line bundle $\mathcal{O}(1)$ equipped (at any place, archimedean or arithmetic) with its canonical metric.

[22] See §A.4.2.2 for the notion of blowup with center at the ideal sheaf attached to the union of the linear subspace $\Pi^{[j]}$ for $j = 0,\dots,m$.

As explained in [**AndESWY21B**, §6], the pullback of the generalized cycle $\widehat{\boldsymbol{\mu}}$ via the submersion $\gamma_m \circ \pi_m$ (outside $\bigcup_0^m \Pi^{[j]}$) is well defined as a global section of

$$'\mathcal{D}\Big(\widehat{\mathbb{P}_{\mathbb{C}}^{(N+1)(m+1)-1}} \setminus \pi_m^{-1}(\textstyle\bigcup_0^m \Pi^{[j]}), \mathbb{C}\Big)$$

and extends as a generalized cycle

$$(\gamma_m \circ \pi_m)^*(\widehat{\boldsymbol{\mu}}) \in \mathcal{G}Z\left(\widehat{\mathbb{P}_{\mathbb{C}}^{(N+1)(m+1)-1}}\right).$$

Since the restriction of $\gamma_m \circ \pi_m$ to $\pi_m^{-1}(\bigcup_0^m \Pi^{[j]})$ operates from a complex manifold with dimension $((m+1)(N+1)-1)$ to a complex manifold with dimension $N(m+1)$, one has

$$(\gamma_m \circ \pi_m)^*(\widehat{\boldsymbol{\mu}}) \in \mathcal{G}Z_{|\delta|+m}\left(\widehat{\mathbb{P}_{\mathbb{C}}^{(N+1)(m+1)-1}}\right).$$

Moreover, the class of such a generalized cycle does not depend on the choice of the representative $\widehat{\boldsymbol{\mu}}$ for $\boldsymbol{\mu}$. Such is the case for the direct image

$$(\pi_m)_* \, (\gamma_m \circ \pi_m)^*(\widehat{\boldsymbol{\mu}}) \in \mathcal{G}Z_{|\delta|+m}\big(\mathbb{P}_{\mathbb{C}}^{(N+1)(m+1)-1}\big).$$

This leads us to the following definition.

DEFINITION D.51. The class of generalized cycle defined as

$$\text{(D.47)} \quad \mu_0 \times_J \cdots \times_J \mu_m := \mathrm{class}\big((\pi_m)_* \, (\gamma_m \circ \pi_m)^*(\widehat{\boldsymbol{\mu}})\big) \in \mathcal{B}_{|\delta|+m}\big(\mathbb{P}_{\mathbb{C}}^{(N+1)(m+1)-1}\big)$$

is called the $(m+1)$-*join product of the classes of generalized cycles* $\mu_0, \dots, \mu_m$ with respective dimensions $\delta_0, \dots, \delta_m$ such that $|\delta| = \delta_0 + \cdots + \delta_m$. It is an algebraic cycle defined over $\mathbb{K} \subset \mathbb{C}$, whenever all the μ_j's are true purely dimensional cycles on $\mathbb{P}_{\mathbb{C}}^N$ also defined over $\mathbb{K}$.

REMARK D.52. When m is fixed,

$$(\mu_0, \dots, \mu_N) \longmapsto \mu_0 \times_J \cdots \times_J \mu_m$$

extends by $\mathbb{Z}$-linearity to a $\mathbb{Z}$-linear map from $(\mathcal{B}(\mathbb{P}_{\mathbb{C}}^N))^{m+1}$ to $\mathcal{B}(\mathbb{P}_{\mathbb{C}}^{(N+1)(m+1)-1})$. Moreover, such map sends $(\mathcal{Z}(\mathbb{P}_{\mathbb{C}}^N))^{m+1}$ to $\mathcal{Z}(\mathbb{P}_{\mathbb{C}}^{(N+1)(m+1)-1})$ and respects the subfield $\mathbb{K}$ of $\mathbb{C}$ over which cycles are defined.

D.5.4. Global intersection in $\mathbb{P}_{\mathbb{C}}^N$ and classes of generalized cycles. For fixed $m \in \mathbb{N}^*$, let us consider the elements $\mu_0, \dots, \mu_m$ belonging respectively to $\mathcal{B}_{\delta_j}(\mathbb{P}_{\mathbb{C}}^N)$ for $j = 0, \dots, m$. Let us recall first the concept of *proper intersection* for $m+1$ purely dimensional cycles Z_j in $\mathbb{P}_{\mathbb{C}}^N$.

DEFINITION D.53. Let $m \in \mathbb{N}^*$ and let $Z_j \in \mathcal{Z}_{\delta_j}(\mathbb{P}_{\mathbb{C}}^N)$ with $0 \leq \delta_j \leq N$ for $j = 0, \dots, m$. The m cycles $Z_0, \dots, Z_m$ are said to *intersect properly* in the manifold $\mathbb{P}_{\mathbb{C}}^N$ if and only if[23]

$$\text{(D.48)} \qquad |\delta| = \sum_{j=0}^m \delta_j \geq Nm \quad \text{and} \quad \mathrm{codim}\,(|Z_0| \cap \cdots \cap |Z_m|) = m+1.$$

[23]The first condition in (D.48) can be reformulated in terms this time of codimensions as $\sum_{j=0}^m (N - \delta_j) \leq N$.

A marker for the geometric complexity of a purely dimensional cycle Z in $\mathcal{Z}_{N-(m+1)}(\mathbb{P}^N_{\mathbb{C}})$ is its geometric degree for the line bundle $\mathcal{O}_{\mathbb{P}^N_{\mathbb{C}}}(1)$. Such a degree is defined by

$$\deg Z = \deg_{\mathcal{O}_{\mathbb{P}^N_{\mathbb{C}}}(1)}(Z) = \int_{\mathbb{P}^N_{\mathbb{C}}} [Z] \wedge (dd^c \log \|\zeta\|^2)^{N-(m+1)}, \tag{D.49}$$

since the natural choice of a smooth metric on the bundle $\mathcal{O}_{\mathbb{P}^N_{\mathbb{C}}}(1)$ is the Fubini–Study metric, which makes it a positive line bundle. The degree of an element $\mu \in \mathcal{B}_{N-(m+1)}(\mathbb{P}^N_{\mathbb{C}})$ with respect to the holomorphic line bundle $\mathcal{O}_{\mathbb{P}^n_{\mathbb{C}}}(1)$ is defined similarly, as

$$\deg \mu = \deg_{\mathcal{O}_{\mathbb{P}^N_{\mathbb{C}}}(1)} \mu = \int_{\mathbb{P}^N_{\mathbb{C}}} \widehat{\mu} \wedge (dd^c \log \|\zeta\|^2)^{N-(m+1)}, \tag{D.50}$$

where $\widehat{\mu}$ is any $(N-(m+1))$-dimensional generalized cycle, that is, a $(m+1, m+1)$-current, representing μ.

THEOREM D.54 (Bézout geometric intersection theorem in $\mathbb{P}^N_{\mathbb{C}}$, proper case).[24] *When $Z_0, \dots, Z_m$ are $m+1$ purely dimensional cycles on a $\mathbb{P}^N_{\mathbb{C}}$ intersecting properly, the intersection cycle $\bigcap_0^m Z_j$ is well defined through the cutting-out operation and its degree equals the product of the degrees of the Z_j.*

Consider now the case where $Z_0, \dots, Z_m$ are still $m+1$ purely dimensional cycles in $\mathbb{P}^N_{\mathbb{C}}$ satisfying the so-called defect condition (D.48)

$$|\delta| = \sum_{j=0}^{m} \dim \mu_j \geq mN,$$

but which are such that $\operatorname{codim}(|Z_0| \cap \cdots \cap |Z_m|) < m+1$. What happens in such a situation is that the class of cycles on $\mathbb{P}^N_{\mathbb{C}}$ consists in too *rigid* objects, hence does not offer enough *flexibility* to reconcile local intersection theory (whose local markers are the extended multi-indexes of intersection, see Definition D.41 with $\mu_j = Z_j$ for $j = 0, \dots, m$) with global intersection theory (ruled this time by some analogue of Bézout's geometric intersection theorem, Theorem D.54, $\deg Z_j$ being the global geometric marker of each Z_j). Fortunately, the class of generalized cycles is sufficiently large to authorize such a reconciliation. To explain how to proceed once $m \in \mathbb{N}^*$ is fixed, let us introduce the line bundle

$$\boldsymbol{L}_m = \mathcal{O}_{\mathbb{P}_{\mathbb{C}}^{(N+1)(m+1)-1}}(1),$$

where homogeneous coordinates on the projective $((N+1)(m+1)-1)$-dimensional space are as in (D.45). Let also

$$\boldsymbol{s}_m = \left(z_0^{[1]} - z_0^{[0]}, \dots, z_0^{[m]} - z_0^{[m-1]}, \cdots, z_N^{[1]} - z_N^{[0]}, \dots, z_N^{[m]} - z_N^{[m-1]}\right) \tag{D.51}$$

be the global section of $\boldsymbol{L}_m^{\otimes (N+1)m}$, which defines in $\mathbb{P}_{\mathbb{C}}^{(N+1)(m+1)-1}$ the N-dimensional submanifold $\iota_m(\mathbb{P}^N_{\mathbb{C}})$ as a complete intersection. One may now formulate, following Example D.11 and Remark D.52, the next definition [**AndESWY21B**, Definition 6.6].

---

[24] We refer the reader to the excellent didactically presentation involving the correspondence between cycles Z and integration currents $[Z]$ in [**De92**], see also [**De93**].

DEFINITION D.55. Given $(m+1)$-classes $\mu_0, \dots, \mu_m$ of generalized cycles in $\mathcal{B}(\mathbb{P}^N_{\mathbb{C}})$, there is a unique element $\mu_0 \cap \cdots \cap \mu_m \in \mathcal{B}(\mathbb{P}^N_{\mathbb{C}})$ satisfying

$$(\iota_m)_* (\mu_0 \cap \cdots \cap \mu_m) = V(\boldsymbol{L}_m \,;\, \overline{\mathcal{I}_{\boldsymbol{s}_m}} \,;\, \mu_0 \times_J \cdots \times_J \mu_m); \tag{D.52}$$

see (D.46) above for the definition of the embedding ι_m. The map

$$(\mu_0, \dots, \mu_N) \in (\mathcal{B}(\mathbb{P}^N_{\mathbb{C}}))^{m+1} \longmapsto \mu_0 \cap \cdots \cap \mu_m \in \mathcal{B}(\mathbb{P}^N_{\mathbb{C}})$$

defined this way[25] is $\mathbb{Z}$-linear in any one of $(m+1)$ intersection components. Furthermore, the general intersection cycle $\mu_0 \cap \cdots \cap \mu_m$ does not depend on the ordering of the entries μ_j provided m is fixed. In addition,

$$|\mu_0 \cap \cdots \cap \mu_m| \subset |\mu_0| \cap \cdots \cap |\mu_N|,$$

which implies that $\mu_0 \cap \cdots \cap \mu_m$ can be considered in $\mathcal{B}(|\mu_0| \cap \cdots \cap |\mu_m|)$.

The following generalization [**AndESWY21B**, Theorem 1.1, Assertions (1.5) and (1.7)] of Theorem D.54 emphasizes why currents, because of their inherent flexibility, are some of the ad hoc tools when realizing in a concrete and explicit way the intersection of cycles.

THEOREM D.56 (Bézout geometric intersection theorem in $\mathbb{P}^N_{\mathbb{C}}$, general case). *Let m in $\mathbb{N}^*$. For any μ_j in $\mathcal{B}(\mathbb{P}^N_{\mathbb{C}})$, $j = 0, \dots, m$, the class $\mu_0 \cap \cdots \cap \mu_m \in \mathcal{B}(\mathbb{P}^N_{\mathbb{C}})$ uniquely defined by* (D.52) *fits with the local intersection rules. That is, if*

$$\bigcap_{j=0}^m \mu_j = \sum_{\kappa=0}^{\dim(|\mu_0| \cap \cdots \cap |\mu_m|)} \Big(\bigcap_{j=0}^m \mu_j \Big)_\kappa$$

is the splitting of the class of a generalized cycle $\mu_0 \cap \cdots \cap \mu_m$ along dimension, then one has for any point $z \in |\mu_0| \cap \cdots \cap |\mu_m|$ and any $\kappa \in \mathbb{N}$,

$$\operatorname{mult}_z \Big(\bigcap_{j=0}^m \mu_j \Big)_\kappa = (\mu_0 \cap \cdots \cap \mu_m)_z(\kappa), \tag{D.53}$$

where the local extended multi-index of intersection $(\mu_0 \cap \cdots \cap \mu_m)_z$ has beed defined in Definition D.41*. If each μ_j is purely δ_j dimensional and $|\delta| := \sum_{j=0}^m \delta_j \geq mN$, then the class $\mu_0 \cap \cdots \cap \mu_m$ fits also with the Bézout's global intersection rule:*

$$\deg \Big(\bigcap_{j=0}^m \mu_j \Big) = \deg_{\mathcal{O}_{\mathbb{P}^N_{\mathbb{C}}}(1)} \Big(\bigcap_{j=0}^m \mu_j \Big) = \prod_{j=0}^m \deg_{\mathcal{O}_{\mathbb{P}^N_{\mathbb{C}}}(1)} (\mu_j) = \prod_{j=0}^m \deg \mu_j. \tag{D.54}$$

REMARK D.57. When $\mu_0, \dots, \mu_m$ have respective pure dimensions δ_j with in addition $|\delta| \geq mN$, the three reasons why $\mu_0 \cap \cdots \cap \mu_m$ satisfies (D.54) are the following:

- $\deg((\iota_m)_* T) = \deg T$ for all $T \in \mathcal{GZ}(\mathbb{P}^N_{\mathbb{C}})$ with pure bidegree;

[25] This map realizes in an effective way the cup intersection product of $m+1$ generalized cycles, when m is fixed. Observe nevertheless that the cup intersection product operates classically on cycles, but modulo rational equivalence. That is, it operates on Chow classes; see [**Ful, Ful84**]. Our so-called *cup intersection product* operates here, which is the price to pay for going from $(\mathcal{B}(\mathbb{P}^N_{\mathbb{C}}))^{m+1}$ to $\mathcal{B}(\mathbb{P}^N(\mathbb{C}))$ instead of from $(A_\bullet(\mathbb{P}^N_{\mathbb{C}}))^{m+1}$ to $A_\bullet(\mathbb{P}^N_{\mathbb{C}})$. It is of course nonassociative since the choice of $m \in \mathbb{N}^*$ precedes its definition.

- one has
$$\prod_{j=0}^{m} \deg \mu_j = \deg_{\mathbb{P}_{\mathbb{C}}^{(N+1)(m+1)-1}} (\mu_0 \times_J \cdots \times_J \mu_m)$$
(see [**AndESWY21B**, Proposition 6.5]);
- the defect condition $|\delta| \geq mN$ implies that
$$(N+1)m = \operatorname{rank} \boldsymbol{s}_m \leq \dim(\mu_0 \times_J \cdots \times_J \mu_m),$$
which in turn implies that
$$\deg_{\mathbb{P}_{\mathbb{C}}^{(N+1)(m+1)-1}} (\mu_0 \times_J \cdots \times_J \mu_m) \\ = \deg_{\mathbb{P}_{\mathbb{C}}^{(N+1)(m+1)-1}} V(\boldsymbol{L}_m; \boldsymbol{s}_m; \mu_0 \times_J \cdots \times_J \mu_m)$$
(see [**AndESWY21B**, Proposition 4.4]).

REMARK D.58. Even when $\mu_j = Z_j \in \mathcal{Z}(\mathbb{P}_j^N)$ for any $j = 0, \ldots, m$, the intersection $\mu_0 \cap \cdots \cap \mu_m$ is not, in general, a cycle, except in the case of proper intersection. In that case, the intersection coincides with the intersection cycle in Theorem D.54. If all μ_j are purely dimensional with $\dim \mu_j = \delta_j$ for $j = 0, \ldots, m$ and $|\delta| \geq mN$, then, for any κ between 0 and $\dim(|\mu_0| \cap \cdots \cap |\mu_m|)$, the class of generalized cycle $(\iota_m)_*((\mu_0 \cap \cdots \cap \mu_m)_\kappa)$ admits as representative the generalized cycle
$$\widehat{V}_{d-\kappa}(\boldsymbol{L}_m\,;\, \boldsymbol{s}_N\,;\, \mu_0 \overbrace{\times_J \cdots \times_J}\, \mu_m)$$
as defined (and realized as a direct image) in (D.39), where
$$d = \dim(\mu_0 \times_J \cdots \times_J \mu_m).$$

A class of generalized cycle μ is said to be *effective* if it can be represented by a generalized cycle which at the same time is a positive current. The following result [**AndESWY21B**, Theorem 1.1, (1.8)] is important, when dealing with the intersection of effective cycles in $\mathbb{P}_{\mathbb{C}}^N$, with the intersection being restricted, for example, to the affine space $\mathbb{C}^N \subset \mathbb{P}_{\mathbb{C}}^N$.

PROPOSITION D.59. *Let $m \in \mathbb{N}^*$ and let $\mu_0, \ldots, \mu_m$ be $m+1$ effective classes of generalized cycles on $\mathbb{P}_{\mathbb{C}}^N$. Then $\mu_0 \cap \cdots \cap \mu_m$ remains effective, with degree at most $\prod_0^m \deg \mu_j$, hence $\deg \left(\mathbf{1}_{\mathbb{C}^N} \cdot (\mu_0 \cap \cdots \cap \mu_m)\right) \leq \prod_0^m \deg \mu_j$.*

Bibliography

[AAC99] Kenzō Adachi, Mats Andersson, and Hong Rae Cho, *L^p and H^p extensions of holomorphic functions from subvarieties of analytic polyhedra*, Pacific J. Math. **189** (1999), no. 2, 201–210, DOI 10.2140/pjm.1999.189.201. MR1696119

[AcM93] Rüdiger Achilles and Mirella Manaresi, *Multiplicity for ideals of maximal analytic spread and intersection theory*, J. Math. Kyoto Univ. **33** (1993), no. 4, 1029–1046, DOI 10.1215/kjm/1250519127. MR1251213

[AcM97] R. Achilles and M. Manaresi, *Multiplicities of a bigraded ring and intersection theory*, Math. Ann. **309** (1997), no. 4, 573–591, DOI 10.1007/s002080050128. MR1483824

[AcR01] Rüdiger Achilles and Sławomir Rams, *Intersection numbers, Segre numbers and generalized Samuel multiplicities*, Arch. Math. (Basel) **77** (2001), no. 5, 391–398, DOI 10.1007/PL00000509. MR1858883

[AkN] Yasuo Akizuki and Shigeo Nakano, *Note on Kodaira-Spencer's proof of Lefschetz theorems*, Proc. Japan Acad. **30** (1954), 266–272. MR66694

[AKS13] Carlos D'Andrea, Teresa Krick, and Martín Sombra, *Heights of varieties in multiprojective spaces and arithmetic Nullstellensätze* (English, with English and French summaries), Ann. Sci. Éc. Norm. Supér. (4) **46** (2013), no. 4, 549–627 (2013), DOI 10.24033/asens.2196. MR3098424

[Alf] Liliane Alfonsi, *Étienne Bézout (1730–1783): Mathématicien des Lumières; [Mathematician of the Enlightenment]* (French), Histoire des Sciences (Paris). [History of Sciences (Paris)], L'Harmattan, Paris, 2011. MR3286429

[AlB19] Amedeo Altavilla and Cinzia Bisi, *Log-biharmonicity and a Jensen formula in the space of quaternions*, Ann. Acad. Sci. Fenn. Math. **44** (2019), no. 2, 805–839, DOI 10.5186/aasfm.2019.4447. MR3973543

[AlPV20] Daniel Alpay, Ariel Pinhas, and Victor Vinnikov, *de Branges spaces on compact Riemann surfaces and a Beurling-Lax type theorem*, Adv. Math. **373** (2020), 107315, 66, DOI 10.1016/j.aim.2020.107315. MR4130461

[AlT01] Aleksandr G. Aleksandrov and Avgust K. Tsikh, *Théorie des résidus de Leray et formes de Barlet sur une intersection complète singulière* (French, with English and French summaries), C. R. Acad. Sci. Paris Sér. I Math. **333** (2001), no. 11, 973–978, DOI 10.1016/S0764-4442(01)02166-8. MR1872457

[AlT08] A. Aleksandrov and A. Tsikh, *Multilogarithmic differential forms on complete intersections*, Journal of Siberian Federal University. Mathematics & Physics **2** (2008), 105–124.

[AlV02] Daniel Alpay and Victor Vinnikov, *Finite dimensional de Branges spaces on Riemann surfaces*, J. Funct. Anal. **189** (2002), no. 2, 283–324, DOI 10.1006/jfan.2000.3623. MR1891851

[AlY18] Daniel Alpay and Alain Yger, *About a non-standard interpolation problem*, Comput. Methods Funct. Theory **19** (2019), no. 1, 97–115, DOI 10.1007/s40315-018-0255-1. MR3922295

[AlY21] Daniel Alpay and Alain Yger, *Cauchy-Weil formula, Schur-Agler type classes, new Hardy spaces of the polydisk and interpolation problems*, J. Math. Anal. Appl. **504** (2021), no. 2, Paper No. 125437, 29, DOI 10.1016/j.jmaa.2021.125437. MR4277104

[Am96] F. Amoroso, *On a conjecture of C. Berenstein and A. Yger*, Algorithms in algebraic geometry and applications (Santander, 1994), Progr. Math., vol. 143, Birkhäuser, Basel, 1996, pp. 17–28. MR1414443

[And92] Mats Andersson, *L_p estimates for the $\overline{\partial}$-equation in analytic polyhedra in Stein manifolds*, Several complex variables (Stockholm, 1987/1988), Math. Notes, vol. 38, Princeton Univ. Press, Princeton, NJ, 1993, pp. 34–47. MR1207854

[And03] Mats Andersson, *Integral representation with weights. I*, Math. Ann. **326** (2003), no. 1, 1–18, DOI 10.1007/s00208-003-0382-4. MR1981609

[And04] Mats Andersson, *Residue currents and ideals of holomorphic functions*, Bull. Sci. Math. **128** (2004), no. 6, 481–512, DOI 10.1016/j.bulsci.2004.03.003. MR2074610

[And05] Mats Andersson, *Residues of holomorphic sections and Lelong currents*, Ark. Mat. **43** (2005), no. 2, 201–219, DOI 10.1007/BF02384777. MR2172988

[And06A] Mats Andersson, *Integral representation with weights. II. Division and interpolation*, Math. Z. **254** (2006), no. 2, 315–332, DOI 10.1007/s00209-006-0949-3. MR2262705

[And06B] Mats Andersson, *The membership problem for polynomial ideals in terms of residue currents* (English, with English and French summaries), Ann. Inst. Fourier (Grenoble) **56** (2006), no. 1, 101–119. MR2228682

[And06C] Mats Andersson, *Explicit versions of the Briançon-Skoda theorem with variations*, Michigan Math. J. **54** (2006), no. 2, 361–373, DOI 10.1307/mmj/1156345600. MR2252765

[And06D] Mats Andersson, *Residue currents of holomorphic morphisms*, J. Reine Angew. Math. **596** (2006), 215–234, DOI 10.1515/CRELLE.2006.060. MR2254813

[And07] Mats Andersson, *A generalized Poincaré-Lelong formula*, Math. Scand. **101** (2007), no. 2, 195–218, DOI 10.7146/math.scand.a-15040. MR2379285

[And09] Mats Andersson, *Uniqueness and factorization of Coleff-Herrera currents* (English, with English and French summaries), Ann. Fac. Sci. Toulouse Math. (6) **18** (2009), no. 4, 651–661. MR2590383

[And10] Mats Andersson, *A residue criterion for strong holomorphicity*, Ark. Mat. **48** (2010), no. 1, 1–15, DOI 10.1007/s11512-009-0100-x. MR2594583

[And11] Mats Andersson, *Coleff-Herrera currents, duality, and Noetherian operators* (English, with English and French summaries), Bull. Soc. Math. France **139** (2011), no. 4, 535–554, DOI 10.24033/bsmf.2618. MR2869304

[AndB82] B. Berndtsson and M. Andersson, *Henkin-Ramirez formulas with weight factors* (English, with French summary), Ann. Inst. Fourier (Grenoble) **32** (1982), no. 3, v–vi, 91–110. MR688022

[AndESWY21A] Mats Andersson, Dennis Eriksson, Håkan Samuelsson Kalm, Elizabeth Wulcan, and Alain Yger, *Global representation of Segre numbers by Monge-Ampère products*, Math. Ann. **380** (2021), no. 1-2, 349–391, DOI 10.1007/s00208-020-01973-y. MR4263687

[AndESWY21B] Mats Andersson, Dennis Eriksson, Håkan Samuelsson Kalm, Elizabeth Wulcan, and Alain Yger, *Nonproper intersection products and generalized cycles*, Eur. J. Math. **7** (2021), no. 4, 1337–1381, DOI 10.1007/s40879-021-00473-w. MR4340940

[AndG11] Mats Andersson and Elin Götmark, *Explicit representation of membership in polynomial ideals*, Math. Ann. **349** (2011), no. 2, 345–365, DOI 10.1007/s00208-010-0524-4. MR2753825

[AndL19] Mats Andersson and Richard Lärkäng, *The $\overline{\partial}$-equation on a non-reduced analytic space*, Math. Ann. **374** (2019), no. 1-2, 553–599, DOI 10.1007/s00208-018-1678-8. MR3961321

[AndLLS20] M. Andersson, R. Lärkäng, M. Lennartsson, H. Samuelsson Kalm, *The $\overline{\partial}$-equation for (p, q) forms on a non-reduced analytic space*, `arXiv:2002.01797`, 2020.

[AndP88] Mats Andersson and Mikael Passare, *A shortcut to weighted representation formulas for holomorphic functions*, Ark. Mat. **26** (1988), no. 1, 1–12, DOI 10.1007/BF02386104. MR948276

[AndS11] Mats Andersson and Håkan Samuelsson, *Weighted Koppelman formulas and the $\overline{\partial}$-equation on an analytic space*, J. Funct. Anal. **261** (2011), no. 3, 777–802, DOI 10.1016/j.jfa.2011.02.018. MR2799580

[AndS12] Mats Andersson and Håkan Samuelsson, *A Dolbeault-Grothendieck lemma on complex spaces via Koppelman formulas*, Invent. Math. **190** (2012), no. 2, 261–297, DOI 10.1007/s00222-012-0380-9. MR2981816

[AndSS10] Mats Andersson, Håkan Samuelsson, and Jacob Sznajdman, *On the Briançon-Skoda theorem on a singular variety* (English, with English and French summaries), Ann. Inst. Fourier (Grenoble) **60** (2010), no. 2, 417–432. MR2667781

[AndSWY17A] Mats Andersson, Håkan Samuelsson Kalm, Elizabeth Wulcan, and Alain Yger, *One parameter regularizations of products of residue currents*, Analysis meets geometry, Trends Math., Birkhäuser/Springer, Cham, 2017, pp. 81–90, DOI 10.1007/978-3-319-52471-9_5. MR3773609

[AndSWY17B] Mats Andersson, Håkan Samuelsson Kalm, Elizabeth Wulcan, and Alain Yger, *Segre numbers, a generalized King formula, and local intersections*, J. Reine Angew. Math. **728** (2017), 105–136, DOI 10.1515/crelle-2014-0109. MR3668992

[AndW07] Mats Andersson and Elizabeth Wulcan, *Residue currents with prescribed annihilator ideals* (English, with English and French summaries), Ann. Sci. École Norm. Sup. (4) **40** (2007), no. 6, 985–1007, DOI 10.1016/j.ansens.2007.11.001. MR2419855

[AndW10] Mats Andersson and Elizabeth Wulcan, *Decomposition of residue currents*, J. Reine Angew. Math. **638** (2010), 103–118, DOI 10.1515/CRELLE.2010.004. MR2595337

[AndW15] Mats Andersson and Elizabeth Wulcan, *Global effective versions of the Briançon-Skoda-Huneke theorem*, Invent. Math. **200** (2015), no. 2, 607–651, DOI 10.1007/s00222-014-0544-x. MR3338011

[AndW18] Mats Andersson and Elizabeth Wulcan, *Direct images of semi-meromorphic currents* (English, with English and French summaries), Ann. Inst. Fourier (Grenoble) **68** (2018), no. 2, 875–900. MR3803121

[AndW21] Mats Andersson and Elizabeth Wulcan, *Regularity of pseudomeromorphic currents*, Bull. Sci. Math. **173** (2021), Paper No. 103055, 29, DOI 10.1016/j.bulsci.2021.103055. MR4320970

[AnG92A] Vincenzo Ancona and Bernard Gaveau, *Differential forms and resolutions on certain analytic spaces. I. Irreducible exceptional divisor*, Bull. Sci. Math. **116** (1992), no. 3, 307–324. MR1177284

[AnG92B] Vincenzo Ancona and Bernard Gaveau, *Differential forms and resolutions on certain analytic spaces. II. Flat resolutions*, Canad. J. Math. **44** (1992), no. 4, 728–749, DOI 10.4153/CJM-1992-043-7. MR1178565

[AnG94] Vincenzo Ancona and Bernard Gaveau, *Differential forms and resolution on certain analytic spaces. III. Spectral resolution*, Ann. Mat. Pura Appl. (4) **166** (1994), 175–202, DOI 10.1007/BF01765634. MR1313804

[AnG06] Vincenzo Ancona and Bernard Gaveau, *Differential forms on singular varieties: De Rham and Hodge theory simplified*, Pure and Applied Mathematics (Boca Raton), vol. 273, Chapman & Hall/CRC, Boca Raton, FL, 2006. MR2168647

[AnG10] Vincenzo Ancona and Bernard Gaveau, *Bochner-Martinelli formulas on singular complex spaces*, Internat. J. Math. **21** (2010), no. 2, 225–253, DOI 10.1142/S0129167X10005994. MR2650368

[AN64] Aldo Andreotti and François Norguet, *Problème de Levi pour les classes de cohomologie* (French), C. R. Acad. Sci. Paris **258** (1964), 778–781. MR159960

[Ar74] S. Ju. Arakelov, *An intersection theory for divisors on an arithmetic surface* (Russian), Izv. Akad. Nauk SSSR Ser. Mat. **38** (1974), 1179–1192. MR0472815

[As04] Matthias Aschenbrenner, *Ideal membership in polynomial rings over the integers*, J. Amer. Math. Soc. **17** (2004), no. 2, 407–441, DOI 10.1090/S0894-0347-04-00451-5. MR2051617

[At70] M. F. Atiyah, *Resolution of singularities and division of distributions*, Comm. Pure Appl. Math. **23** (1970), 145–150, DOI 10.1002/cpa.3160230202. MR256156

[AVGZ] V. I. Arnol′d, S. M. Guseĭn-Zade, and A. N. Varchenko, *Singularities of differentiable maps. Vol. I: The classification of critical points, caustics and wave fronts*, Monographs in Mathematics, vol. 82, Birkhäuser Boston, Inc., Boston, MA, 1985. Translated from the Russian by Ian Porteous and Mark Reynolds, DOI 10.1007/978-1-4612-5154-5. MR777682

[Ax71] James Ax, *On Schanuel's conjectures*, Ann. of Math. (2) **93** (1971), 252–268, DOI 10.2307/1970774. MR277482

[AY] I. A. Aĭzenberg and A. P. Yuzhakov, *Integral representations and residues in multidimensional complex analysis*, Translations of Mathematical Monographs, vol. 58, American Mathematical Society, Providence, RI, 1983. Translated from the Russian by H. H. McFaden; Translation edited by Lev J. Leifman, DOI 10.1090/mmono/058. MR735793

[Bab14] F. Babaee, *Complex tropical currents, extremality, and approximations*, `arXiv:1403.7456v1`, 2014.

[BabH17] Farhad Babaee and June Huh, *A tropical approach to a generalized Hodge conjecture for positive currents*, Duke Math. J. **166** (2017), no. 14, 2749–2813, DOI 10.1215/00127094-2017-0017. MR3707289

[Bar78] Daniel Barlet, *Le faisceau $\omega^{\cdot}_X$ sur un espace analytique X de dimension pure* (French), Fonctions de plusieurs variables complexes, III (Sém. François Norguet, 1975), Lecture Notes in Math., vol. 670, Springer, Berlin, 1978, pp. 187–204. MR521919

[Bar82] D. Barlet, *Développement asymptotique des fonctions obtenues par intégration sur les fibres* (French), Invent. Math. **68** (1982), no. 1, 129–174, DOI 10.1007/BF01394271. MR666639

[BarM85] D. Barlet and H.-M. Maire, *Développements asymptotiques, transformation de Mellin complexe et intégration sur les fibres* (French), Séminaire d'Analyse P. Lelong–P. Dolbeault–H. Skoda, Années 1985/1986, Lecture Notes in Math., vol. 1295, Springer, Berlin, 1987, pp. 11–23, DOI 10.1007/BFb0081975. MR1047719

[Bat93] Victor V. Batyrev, *Quantum cohomology rings of toric manifolds: Journées de Géométrie Algébrique d'Orsay (Orsay, 1992)*, Astérisque **218** (1993), 9–34. MR1265307

[Be72] I. N. Bernšteĭn, *Analytic continuation of generalized functions with respect to a parameter*, Funkcional. Anal. i Priložen. **6** (1972), no. 4, 26–40. MR0320735

[Be75] D. N. Bernstein, *The number of roots of a system of equations* (Russian), Funkcional. Anal. i Priložen. **9** (1975), no. 3, 1–4. MR0435072

[BedT76] Eric Bedford and B. A. Taylor, *The Dirichlet problem for a complex Monge-Ampère equation*, Invent. Math. **37** (1976), no. 1, 1–44, DOI 10.1007/BF01418826. MR445006

[BedT82] Eric Bedford and B. A. Taylor, *A new capacity for plurisubharmonic functions*, Acta Math. **149** (1982), no. 1-2, 1–40, DOI 10.1007/BF02392348. MR674165

[Ber88] Bo Berndtsson, *Integral formulas on projective space and the Radon transform of Gindikin-Henkin-Polyakov*, Publ. Mat. **32** (1988), no. 1, 7–41, DOI 10.5565/PUBLMAT_32188_01. MR939766

[BerP89] Bo Berndtsson and Mikael Passare, *Integral formulas and an explicit version of the fundamental principle*, J. Funct. Anal. **84** (1989), no. 2, 358–372, DOI 10.1016/0022-1236(89)90103-1. MR1001466

[BG] Carlos A. Berenstein and Roger Gay, *Complex variables: An introduction*, Graduate Texts in Mathematics, vol. 125, Springer-Verlag, New York, 1991, DOI 10.1007/978-1-4612-3024-3. MR1107514

[BGHM01] B. Bank, M. Giusti, J. Heintz, and G. M. Mbakop, *Polar varieties and efficient real elimination*, Math. Z. **238** (2001), no. 1, 115–144, DOI 10.1007/PL00004896. MR1860738

[BGPS14] José Ignacio Burgos Gil, Patrice Philippon, and Martín Sombra, *Arithmetic geometry of toric varieties. Metrics, measures and heights* (English, with English and French summaries), Astérisque **360** (2014), vi+222. MR3222615

[BGS94] J.-B. Bost, H. Gillet, and C. Soulé, *Heights of projective varieties and positive Green forms*, J. Amer. Math. Soc. **7** (1994), no. 4, 903–1027, DOI 10.2307/2152736. MR1260106

[BGVY] Carlos A. Berenstein, Roger Gay, Alekos Vidras, and Alain Yger, *Residue currents and Bezout identities*, Progress in Mathematics, vol. 114, Birkhäuser Verlag, Basel, 1993, DOI 10.1007/978-3-0348-8560-7. MR1249478

[BGY89] Carlos A. Berenstein, Roger Gay, and Alain Yger, *Analytic continuation of currents and division problems* (English, with French summary), Forum Math. **1** (1989), no. 1, 15–51, DOI 10.1515/form.1989.1.15. MR978974

[BiM98] Hélène Biosca, Joël Briancon, Philippe Maisonobe, and Hélène Maynadier, *Espaces conormaux relatifs. II. Modules différentiels* (French), Publ. Res. Inst. Math. Sci. **34** (1998), no. 2, 123–134, DOI 10.2977/prims/1195144757. MR1617064

[Bj74] J.-E. Björk, *Rings of differential operators*, North-Holland Mathematical Library, vol. 21, North-Holland Publishing Co., Amsterdam-New York, 1979. MR549189

[Bj93] Jan-Erik Björk, *Analytic $\mathcal{D}$-modules and applications*, Mathematics and its Applications, vol. 247, Kluwer Academic Publishers Group, Dordrecht, 1993, DOI 10.1007/978-94-017-0717-6. MR1232191

[Bj96] J. E. Björk, *Residue calculus and $\mathcal{D}$-modules on complex manifolds*, preprint Stockholm University, 1996.

[Bj04] Jan-Erik Björk, *Residues and $\mathcal{D}$-modules*, The legacy of Niels Henrik Abel, Springer, Berlin, 2004, pp. 605–651. MR2077588

[BjS10] Jan-Erik Björk and Håkan Samuelsson, *Regularizations of residue currents*, J. Reine Angew. Math. **649** (2010), 33–54, DOI 10.1515/CRELLE.2010.087. MR2746465

[Bl04] Lenore Blum, *Computing over the reals: where Turing meets Newton*, Notices Amer. Math. Soc. **51** (2004), no. 9, 1024–1034. MR2089092

[BlCSS96] Lenore Blum, Felipe Cucker, Michael Shub, and Steve Smale, *Complexity and real computation*, Springer-Verlag, New York, 1998. With a foreword by Richard M. Karp, DOI 10.1007/978-1-4612-0701-6. MR1479636

[BlSS89] Lenore Blum, Mike Shub, and Steve Smale, *On a theory of computation and complexity over the real numbers: NP-completeness, recursive functions and universal machines*, Bull. Amer. Math. Soc. (N.S.) **21** (1989), no. 1, 1–46, DOI 10.1090/S0273-0979-1989-15750-9. MR974426

[Boc75] J. Bochnak and J. J. Risler, *Sur les exposants de Lojasiewicz*, Comment. Math. Helv. **50** (1975), no. 4, 493–507, DOI 10.1007/BF02565764. MR404674

[Boo97] Jörgen Boo, *The H^p corona theorem in analytic polyhedra*, Ark. Mat. **35** (1997), no. 2, 225–251, DOI 10.1007/BF02559968. MR1478779

[Bor87] A. Borel, P.-P. Grivel, B. Kaup, A. Haefliger, B. Malgrange, and F. Ehlers, *Algebraic D-modules*, Perspectives in Mathematics, vol. 2, Academic Press, Inc., Boston, MA, 1987. MR882000

[Bos92] J. B. Bost, *Introduction to compact Riemann surfaces, Jacobian and algebraic varieties* in From Number Theory to Physics, (M. Waldschmidt, P. Moussa, J.M. Luck, C. Itzykson eds.), pp. 64–211, Springer-Verlag, 1992.

[BotC65] Raoul Bott and S. S. Chern, *Hermitian vector bundles and the equidistribution of the zeroes of their holomorphic sections*, Acta Math. **114** (1965), 71–112, DOI 10.1007/BF02391818. MR185607

[BoyH97] Jean-Yves Boyer and Michel Hickel, *Une généralisation de la loi de transformation pour les résidus* (French, with English and French summaries), Bull. Soc. Math. France **125** (1997), no. 3, 315–335. MR1605465

[BoyH99] J.-Y. Boyer and M. Hickel, *Extension dans un cadre algébrique d'une formule de Weil* (French, with English summary), Manuscripta Math. **98** (1999), no. 2, 195–223, DOI 10.1007/s002290050135. MR1667604

[Bri97] J. Briançon, *Espaces conormaux relatifs. I. Conditions de transversalité* (French, with English and French summaries), Ann. Sci. École Norm. Sup. (4) **30** (1997), no. 5, 675–692, DOI 10.1016/S0012-9593(97)89935-1. MR1474808

[BriS74] Henri Skoda and Joël Briançon, *Sur la clôture intégrale d'un idéal de germes de fonctions holomorphes en un point de* $\mathbf{C}^n$ (French), C. R. Acad. Sci. Paris Sér. A **278** (1974), 949–951. MR340642

[Bro87A] W. Dale Brownawell, *Bounds for the degrees in the Nullstellensatz*, Ann. of Math. (2) **126** (1987), no. 3, 577–591, DOI 10.2307/1971361. MR916719

[Bro87B] W. Dale Brownawell, *Borne effective pour l'exposant dans le théorème des zéros* (French, with English summary), C. R. Acad. Sci. Paris Sér. I Math. **305** (1987), no. 7, 287–290. MR910361

[Bro88] W. Dale Brownawell, *Local Diophantine Nullstellen inequalities*, J. Amer. Math. Soc. **1** (1988), no. 2, 311–322, DOI 10.2307/1990919. MR928261

[BT79] Carlos A. Berenstein and B. A. Taylor, *A new look at interpolation theory for entire functions of one variable*, Adv. in Math. **33** (1979), no. 2, 109–143, DOI 10.1016/S0001-8708(79)80002-X. MR544846

[BT80] C. A. Berenstein and B. A. Taylor, *Interpolation problems in* $\mathbf{C}^n$ *with applications to harmonic analysis*, J. Analyse Math. **38** (1980), 188–254, DOI 10.1007/BF03033881. MR600786

[BuE74] David A. Buchsbaum and David Eisenbud, *Some structure theorems for finite free resolutions*, Advances in Math. **12** (1974), 84–139, DOI 10.1016/S0001-8708(74)80019-8. MR340240

[BVY05] Carlos A. Berenstein, Alekos Vidras, and Alain Yger, *Analytic residues along algebraic cycles*, J. Complexity **21** (2005), no. 1, 5–42, DOI 10.1016/j.jco.2004.03.006. MR2112741

[BY86] Carlos A. Berenstein and Alain Yger, *Ideals generated by exponential-polynomials*, Adv. in Math. **60** (1986), no. 1, 1–80, DOI 10.1016/0001-8708(86)90002-2. MR839482

[BY91] Carlos A. Berenstein and Alain Yger, *Effective Bézout identities in* $\mathbf{Q}[z_1, \cdots, z_n]$, Acta Math. **166** (1991), no. 1-2, 69–120, DOI 10.1007/BF02398884. MR1088983

[BY95] C. A. Berenstein and A. Yger, *Exponential polynomials and* $\mathcal{D}$*-modules*, Compositio Math. **95** (1995), no. 2, 131–181. MR1313872

[BY98] Carlos A. Berenstein and Alain Yger, *Green currents and analytic continuation*, J. Anal. Math. **75** (1998), 1–50, DOI 10.1007/BF02788690. MR1655822

[BY99] Carlos A. Berenstein and Alain Yger, *Residue calculus and effective Nullstellensatz*, Amer. J. Math. **121** (1999), no. 4, 723–796. MR1704477

[BY00] Carlos A. Berenstein and Alain Yger, *Analytic residue theory in the non-complete intersection case*, J. Reine Angew. Math. **527** (2000), 203–235, DOI 10.1515/crll.2000.083. MR1794023

[CaCD97] Eduardo Cattani, David Cox, and Alicia Dickenstein, *Residues in toric varieties*, Compositio Math. **108** (1997), no. 1, 35–76, DOI 10.1023/A:1000180417349. MR1458757

[CaD97] Eduardo Cattani and Alicia Dickenstein, *A global view of residues in the torus: Algorithms for algebra (Eindhoven, 1996)*, J. Pure Appl. Algebra **117/118** (1997), 119–144, DOI 10.1016/S0022-4049(97)00008-X. MR1457836

[CaDS96] E. Cattani, A. Dickenstein, and B. Sturmfels, *Computing multidimensional residues*, Algorithms in algebraic geometry and applications (Santander, 1994), Progr. Math., vol. 143, Birkhäuser, Basel, 1996, pp. 135–164. MR1414449

[CaM00] Julien Cassaigne and Vincent Maillot, *Hauteur des hypersurfaces et fonctions zêta d'Igusa* (French, with French summary), J. Number Theory **83** (2000), no. 2, 226–255, DOI 10.1006/jnth.1999.2490. MR1772614

[CGHMP03] D. Castro, M. Giusti, J. Heintz, G. Matera, and L. M. Pardo, *The hardness of polynomial equation solving*, Found. Comput. Math. **3** (2003), no. 4, 347–420, DOI 10.1007/s10208-002-0065-7. MR2009683

[Ch] E. M. Chirka, *Complex analytic sets*, Mathematics and its Applications (Soviet Series), vol. 46, Kluwer Academic Publishers Group, Dordrecht, 1989. Translated from the Russian by R. A. M. Hoksbergen, DOI 10.1007/978-94-009-2366-9. MR1111477

[ChD12] A. Chambert-Loir and A. Ducros, *Formes différentielles réelles et courants sur les espaces de Berkovich*, `arXiv:1204.6277`, 2012.

[ChHKL22] Justin Chen, Marc Härkönen, Robert Krone, and Anton Leykin, *Noetherian operators and primary decomposition*, J. Symbolic Comput. **110** (2022), 1–23, DOI 10.1016/j.jsc.2021.09.002. MR4322323

[ChM20] Huayi Chen and Atsushi Moriwaki, *Arakelov geometry over adelic curves*, Lecture Notes in Mathematics, vol. 2258, Springer, Singapore, [2020] ©2020, DOI 10.1007/978-981-15-1728-0. MR4292529

[CLO] David Cox, John Little, and Donal O'Shea, *Using algebraic geometry*, Graduate Texts in Mathematics, vol. 185, Springer-Verlag, New York, 1998, DOI 10.1007/978-1-4757-6911-1. MR1639811

[CLSSS07] Fabrizio Colombo, M. Elena Luna-Elizarrarás, Irene Sabadini, Michael Shapiro, and Daniele C. Struppa, *A new characterization of a class of pseudoconvex domains in* $\mathbb{C}^2$ (English, with English and French summaries), C. R. Math. Acad. Sci. Paris **344** (2007), no. 11, 677–680, DOI 10.1016/j.crma.2007.04.014. MR2334674

[CoH78] Nicolas R. Coleff and Miguel E. Herrera, *Les courants résiduels associés à une forme méromorphe* (French), Lecture Notes in Mathematics, vol. 633, Springer, Berlin, 1978. MR492769

[Col80] Robert F. Coleman, *On a stronger version of the Schanuel-Ax theorem*, Amer. J. Math. **102** (1980), no. 4, 595–624, DOI 10.2307/2374090. MR584463

[Colon] Lawrence Conlon, *Differentiable manifolds*, 2nd ed., Modern Birkhäuser Classics, Birkhäuser Boston, Inc., Boston, MA, 2008. MR2413709

[CSS16] Fabrizio Colombo, Irene Sabadini, and Daniele C. Struppa, *Entire slice regular functions*, SpringerBriefs in Mathematics, Springer, Cham, 2016, DOI 10.1007/978-3-319-49265-0. MR3585395

[CSSS04] Fabrizio Colombo, Irene Sabadini, Franciscus Sommen, and Daniele C. Struppa, *Analysis of Dirac systems and computational algebra*, Progress in Mathematical Physics, vol. 39, Birkhäuser Boston, Inc., Boston, MA, 2004, DOI 10.1007/978-0-8176-8166-1. MR2089988

[Cx95] David A. Cox, *The homogeneous coordinate ring of a toric variety*, J. Algebraic Geom. **4** (1995), no. 1, 17–50. MR1299003

[Cx96] David A. Cox, *Toric residues*, Ark. Mat. **34** (1996), no. 1, 73–96, DOI 10.1007/BF02559508. MR1396624

[Cx97] David A. Cox, *Recent developments in toric geometry*, Algebraic geometry—Santa Cruz 1995, Proc. Sympos. Pure Math., vol. 62, Amer. Math. Soc., Providence, RI, 1997, pp. 389–436, DOI 10.1090/pspum/062.2/1492541. MR1492541

[CxLS11] David A. Cox, John B. Little, and Henry K. Schenck, *Toric varieties*, Graduate Studies in Mathematics, vol. 124, American Mathematical Society, Providence, RI, 2011, DOI 10.1090/gsm/124. MR2810322

[Cy98] Ewa Cygan, *Intersection theory and separation exponent in complex analytic geometry*, Ann. Polon. Math. **69** (1998), no. 3, 287–299, DOI 10.4064/ap-69-3-287-299. MR1665012

[Cy05] Ewa Cygan, *A note on separation of algebraic sets and the Łojasiewicz exponent for polynomial mappings* (English, with English and French summaries), Bull. Sci. Math. **129** (2005), no. 2, 139–147, DOI 10.1016/j.bulsci.2004.10.001. MR2123264

[CyKT99] Ewa Cygan, Tadeusz Krasiński, and Piotr Tworzewski, *Separation of algebraic sets and the Łojasiewicz exponent of polynomial mappings*, Invent. Math. **136** (1999), no. 1, 75–87, DOI 10.1007/s002220050304. MR1681109

[Da78] V. I. Danilov, *The geometry of toric varieties* (Russian), Uspekhi Mat. Nauk **33** (1978), no. 2(200), 85–134, 247. MR495499

[De] J. P. Demailly, *Complex analytic and differential geometry*, `http://www-fourier.ujf-grenoble.fr/~demailly/manuscripts/agbook.pdf`.

[De87] Jean-Pierre Demailly, *Nombres de Lelong généralisés, théorèmes d'intégralité et d'analyticité* (French), Acta Math. **159** (1987), no. 3-4, 153–169, DOI 10.1007/BF02392558. MR908144

[De92] Jean-Pierre Demailly, *Courants positifs et théorie de l'intersection* (French), Gaz. Math. **53** (1992), 131–159. MR1175540

[De93] Jean-Pierre Demailly, *Monge-Ampère operators, Lelong numbers and intersection theory*, Complex analysis and geometry, Univ. Ser. Math., Plenum, New York, 1993, pp. 115–193. MR1211880

[Dem70] Michel Demazure, *Sous-groupes algébriques de rang maximum du groupe de Cremona* (French), Ann. Sci. École Norm. Sup. (4) **3** (1970), 507–588. MR284446

[Den92] Jan Denef, *Report on Igusa's local zeta function*, Astérisque **201-203** (1991), Exp. No. 741, 359–386 (1992). Séminaire Bourbaki, Vol. 1990/91. MR1157848

[Dol93] Pierre Dolbeault, *On the structure of residual currents*, Several complex variables (Stockholm, 1987/1988), Math. Notes, vol. 38, Princeton Univ. Press, Princeton, NJ, 1993, pp. 258–273. MR1207865

[Dol13] Pierre Dolbeault, *On quaternionic functions* (English, with English and Polish summaries), Bull. Soc. Sci. Lett. Łódź Sér. Rech. Déform. **63** (2013), no. 1, 9–22. MR3220305

[DP99] Sinnou David and Patrice Philippon, *Minorations des hauteurs normalisées des sous-variétés des tores* (French, with English summary), Ann. Scuola Norm. Sup. Pisa Cl. Sci. (4) **28** (1999), no. 3, 489–543. MR1736526

[DR32] G. de Rham, *Sur la notion d'homologie et les résidus des intégrales multiples*, Verhandlungen des Internationalen Mathematike-Kongresse, Zürich (1932), vol. II, p. 195.

[DR36] G. de Rham, *Relations entre la topologie et la théorie des intégrales muliples*, Ens. Math. **35** (1936), 213–228.

[DR55] Georges de Rham, *Variétés différentiables. Formes, courants, formes harmoniques* (French), Publ. Inst. Math. Univ. Nancago, III, Hermann & Cie, Paris, 1955. MR0068889

[DS85] A. Dickenstein and C. Sessa, *Canonical representatives in moderate cohomology*, Invent. Math. **80** (1985), no. 3, 417–434, DOI 10.1007/BF01388723. MR791667

[DS96] Alicia Dickenstein and Carmen Sessa, *Résidus de formes méromorphes et cohomologie modérée* (French, with English summary), Géométrie complexe (Paris, 1992), Actualités Sci. Indust., vol. 1438, Hermann, Paris, 1996, pp. 35–59, DOI 10.1016/j.brainres.2011.12.039. MR1487899

[EaN62] J. A. Eagon and D. G. Northcott, *Ideals defined by matrices and a certain complex associated with them*, Proc. Roy. Soc. London Ser. A **269** (1962), 188–204, DOI 10.1098/rspa.1962.0170. MR142592

[EGA] A. Grothendieck and J. Dieudonné, *Élements de géométrie algébrique. I, II, III, IV*, Publ. Math. I.H.E.S, **4,8,11,17,20,24,28,32**, 1960–1967.

[Ehl75] Fritz Ehlers, *Eine Klasse komplexer Mannigfaltigkeiten und die Auflösung einiger isolierter Singularitäten* (German), Math. Ann. **218** (1975), no. 2, 127–156, DOI 10.1007/BF01370816. MR492378

[Ehr] Leon Ehrenpreis, *Fourier analysis in several complex variables*, Pure and Applied Mathematics, Vol. XVII, Wiley-Interscience [A division of John Wiley & Sons, Inc.], New York-London-Sydney, 1970. MR0285849

[EinL99] Lawrence Ein and Robert Lazarsfeld, *A geometric effective Nullstellensatz*, Invent. Math. **137** (1999), no. 2, 427–448, DOI 10.1007/s002220050332. MR1705839

[Eis] David Eisenbud, *Commutative algebra: With a view toward algebraic geometry*, Graduate Texts in Mathematics, vol. 150, Springer-Verlag, New York, 1995, DOI 10.1007/978-1-4612-5350-1. MR1322960

[Eis05] David Eisenbud, *The geometry of syzygies: A second course in commutative algebra and algebraic geometry*, Graduate Texts in Mathematics, vol. 229, Springer-Verlag, New York, 2005. MR2103875

[El93] M. Elkadi, *Bornes pour les degrés et les hauteurs dans le problème de division* (French, with English summary), Michigan Math. J. **40** (1993), no. 3, 609–618, DOI 10.1307/mmj/1029004841. MR1236182

[El94] M. Elkadi, *Une version effective du théorème de Briançon-Skoda dans le cas algébrique discret* (French), Acta Arith. **66** (1994), no. 3, 201–220, DOI 10.4064/aa-66-3-201-220. MR1276988

[ElM06] Mohamed Elkadi and Bernard Mourrain, *Introduction à la résolution des systèmes polynomiaux* (French), Mathématiques & Applications (Berlin) [Mathematics & Applications], vol. 59, Springer, Berlin, 2007, DOI 10.1007/978-3-540-71647-1. MR2322065

[ElY07] Mohamed Elkadi and Alain Yger, *Residue calculus and applications*, Publ. Res. Inst. Math. Sci. **43** (2007), no. 1, 55–73. MR2317112

[EM84] Hassine El Mir, *Sur le prolongement des courants positifs fermés* (French), Acta Math. **153** (1984), no. 1-2, 1–45, DOI 10.1007/BF02392374. MR744998

[Ew] Günter Ewald, *Combinatorial convexity and algebraic geometry*, Graduate Texts in Mathematics, vol. 168, Springer-Verlag, New York, 1996, DOI 10.1007/978-1-4612-4044-0. MR1418400

[Fa73] John D. Fay, *Theta functions on Riemann surfaces*, Lecture Notes in Mathematics, Vol. 352, Springer-Verlag, Berlin-New York, 1973. MR0335789

[Fe70] Aldo Ferrari, *Cohomology and holomorphic differential forms on complex analytic spaces*, Ann. Scuola Norm. Sup. Pisa Cl. Sci. (3) **24** (1970), 65–77. MR274810

[Fi36] H. Fitting, *Die Determinantenideale eines Moduls*, Jahresbericht der Deutschen Mathematiker-Vereinigung **46** (1936), 195–228.

[FK] H. M. Farkas and I. Kra, *Riemann surfaces*, 2nd ed., Graduate Texts in Mathematics, vol. 71, Springer-Verlag, New York, 1992, DOI 10.1007/978-1-4612-2034-3. MR1139765

[Fu35] Run Fueter, *Die Funktionentheorie der Differentialgleichungen* $\Theta u = 0$ *und* $\Theta\Theta u = 0$ *mit vier reellen Variablen* (German), Comment. Math. Helv. **7** (1934), no. 1, 307–330, DOI 10.1007/BF01292723. MR1509515

[Fu36] Rud Fueter, *Über die analytische Darstellung der regulären Funktionen einer Quaternionenvariablen* (German), Comment. Math. Helv. **8** (1935), no. 1, 371–378, DOI 10.1007/BF01199562. MR1509533

[Fu39] Rud Fueter, *Über einen Hartogs'schen Satz* (German), Comment. Math. Helv. **12** (1939), 75–80, DOI 10.1007/BF01620640. MR693

[Ful] William Fulton, *Intersection theory*, 2nd ed., Ergebnisse der Mathematik und ihrer Grenzgebiete. 3. Folge. A Series of Modern Surveys in Mathematics [Results in Mathematics and Related Areas. 3rd Series. A Series of Modern Surveys in Mathematics], vol. 2, Springer-Verlag, Berlin, 1998, DOI 10.1007/978-1-4612-1700-8. MR1644323

[Ful75] William Fulton, *Rational equivalence on singular varieties*, Inst. Hautes Études Sci. Publ. Math. **45** (1975), 147–167. MR404257

[Ful84] William Fulton, *Introduction to intersection theory in algebraic geometry*, CBMS Regional Conference Series in Mathematics, vol. 54, Published for the Conference Board of the Mathematical Sciences, Washington, DC; by the American Mathematical Society, Providence, RI, 1984, DOI 10.1090/cbms/054. MR735435

[Ful93] William Fulton, *Introduction to toric varieties*, Annals of Mathematics Studies, vol. 131, Princeton University Press, Princeton, NJ, 1993. The William H. Roever Lectures in Geometry, DOI 10.1515/9781400882526. MR1234037

[FulS97] William Fulton and Bernd Sturmfels, *Intersection theory on toric varieties*, Topology **36** (1997), no. 2, 335–353, DOI 10.1016/0040-9383(96)00016-X. MR1415592

[GeS06] Graziano Gentili and Daniele C. Struppa, *A new approach to Cullen-regular functions of a quaternionic variable* (English, with English and French summaries), C. R. Math. Acad. Sci. Paris **342** (2006), no. 10, 741–744, DOI 10.1016/j.crma.2006.03.015. MR2227751

[GeS07] Graziano Gentili and Daniele C. Struppa, *A new theory of regular functions of a quaternionic variable*, Adv. Math. **216** (2007), no. 1, 279–301, DOI 10.1016/j.aim.2007.05.010. MR2353257

[GHMMP98] M. Giusti, J. Heintz, J. E. Morais, J. Morgenstern, and L. M. Pardo, *Straight-line programs in geometric elimination theory*, J. Pure Appl. Algebra **124** (1998), no. 1-3, 101–146, DOI 10.1016/S0022-4049(96)00099-0. MR1600277

[GHMMPM97] M. Giusti, J. Heintz, K. Hägele, J. E. Morais, L. M. Pardo, and J. L. Montaña, *Lower bounds for Diophantine approximations: Algorithms for algebra (Eindhoven, 1996)*, J. Pure Appl. Algebra **117/118** (1997), 277–317, DOI 10.1016/S0022-4049(97)00015-7. MR1457843

[GKZ] I. M. Gel′fand, M. M. Kapranov, and A. V. Zelevinsky, *Discriminants, resultants, and multidimensional determinants*, Mathematics: Theory & Applications, Birkhäuser Boston, Inc., Boston, MA, 1994, DOI 10.1007/978-0-8176-4771-1. MR1264417

[Gl63] Andrew M. Gleason, *The Cauchy-Weil theorem*, J. Math. Mech. **12** (1963), 429–444. MR0148938

[Gl64] Andrew M. Gleason, *Finitely generated ideals in Banach algebras*, J. Math. Mech. **13** (1964), 125–132. MR0159241

[God] R. Godement, *Théorie des faisceaux*, Hermann, Paris, 1958.

[Got08] Elin Götmark, *Weighted integral formulas on manifolds*, Ark. Mat. **46** (2008), no. 1, 43–68, DOI 10.1007/s11512-007-0056-7. MR2379683

[GrH] Phillip Griffiths and Joseph Harris, *Principles of algebraic geometry*, Pure and Applied Mathematics, Wiley-Interscience [John Wiley & Sons], New York, 1978. MR507725

[Gri69] Phillip A. Griffiths, *Hermitian differential geometry, Chern classes, and positive vector bundles*, Global Analysis (Papers in Honor of K. Kodaira), Univ. Tokyo Press, Tokyo, 1969, pp. 185–251. MR0258070

[Gri76] Phillip A. Griffiths, *Variations on a theorem of Abel*, Invent. Math. **35** (1976), 321–390, DOI 10.1007/BF01390145. MR435074

[Gro71] Alexander Grothendieck, *Revêtements étales et groupe fondamental. Fasc. I: Exposés 1 à 5*, Institut des Hautes Études Scientifiques, Paris, 1963. Troisième édition, corrigée; Séminaire de Géométrie Algébrique, 1960/61. MR0217087

[Grt60] Hans Grauert, *Ein Theorem der analytischen Garbentheorie und die Modulräume komplexer Strukturen* (German), Inst. Hautes Études Sci. Publ. Math. **5** (1960), 64. MR121814

[GrtR] Hans Grauert and Reinhold Remmert, *Coherent analytic sheaves*, Grundlehren der mathematischen Wissenschaften [Fundamental Principles of Mathematical Sciences], vol. 265, Springer-Verlag, Berlin, 1984, DOI 10.1007/978-3-642-69582-7. MR755331

[GS90] Henri Gillet and Christophe Soulé, *Arithmetic intersection theory*, Inst. Hautes Études Sci. Publ. Math. **72** (1990), 93–174 (1991). MR1087394

[Gua18] Roberto Gualdi, *Heights of hypersurfaces in toric varieties*, Algebra Number Theory **12** (2018), no. 10, 2403–2443, DOI 10.2140/ant.2018.12.2403. MR3911135

[GuaS23] R. Gualdi and M. Sombra, *Limit heights and special values of the Riemann zeta function*, `arXiv:2304.019666v1`, 2023.

[Gub02] W. Gubler, *Basic properties of heigths of subvarieties*, Hab. Thesis, Regensburg, 2002, `http://www.mathematik.uni-regensburg.de/gubler/Publikationen/habi+.pdf`.

[GubK17] Walter Gubler and Klaus Künnemann, *A tropical approach to nonarchimedean Arakelov geometry*, Algebra Number Theory **11** (2017), no. 1, 77–180, DOI 10.2140/ant.2017.11.77. MR3602767

[GuR] Robert C. Gunning and Hugo Rossi, *Analytic functions of several complex variables*, Prentice-Hall, Inc., Englewood Cliffs, N.J., 1965. MR0180696

[Gy93] Akihiko Gyoja, *Bernstein-Sato's polynomial for several analytic functions*, J. Math. Kyoto Univ. **33** (1993), no. 2, 399–411, DOI 10.1215/kjm/1250519266. MR1231750

[Ha] Robin Hartshorne, *Residues and duality*, Lecture Notes in Mathematics, No. 20, Springer-Verlag, Berlin-New York, 1966. Lecture notes of a seminar on the work of A. Grothendieck, given at Harvard 1963/64; With an appendix by P. Deligne. MR0222093

[Ha66] Robin Hartshorne, *Algebraic geometry*, Graduate Texts in Mathematics, No. 52, Springer-Verlag, New York-Heidelberg, 1977. MR0463157

[HaL05] Helmut A. Hamm and Dũng Tráng Lê, *On the Picard group for non-complete algebraic varieties* (English, with English and French summaries), Singularités Franco-Japonaises, Sémin. Congr., vol. 10, Soc. Math. France, Paris, 2005, pp. 71–86. MR2145949

[Han81] Sönke Hansen, *Localizable analytically uniform spaces and the fundamental principle*, Trans. Amer. Math. Soc. **264** (1981), no. 1, 235–250, DOI 10.2307/1998422. MR597879

[He1926] Grete Hermann, *Die Frage der endlich vielen Schritte in der Theorie der Polynomideale* (German), Math. Ann. **95** (1926), no. 1, 736–788, DOI 10.1007/BF01206635. MR1512302

[Hef50] Hans Hefer, *Zur Funktionentheorie mehrerer Veränderlichen. Über eine Zerlegung analytischer Funktionen und die Weilsche Integraldarstellung* (German), Math. Ann. **122** (1950), 276–278, DOI 10.1007/BF01342970. MR39082

[HeP99] Gennadi Henkin and Mikael Passare, *Abelian differentials on singular varieties and variations on a theorem of Lie-Griffiths*, Invent. Math. **135** (1999), no. 2, 297–328, DOI 10.1007/s002220050287. MR1666771

[Hi01] Michel Hickel, *Solution d'une conjecture de C. Berenstein–A. Yger et invariants de contact à l'infini* (French, with English and French summaries), Ann. Inst. Fourier (Grenoble) **51** (2001), no. 3, 707–744. MR1838463

[Hi08] M. Hickel, *Une note à propos du jacobien de n fonctions holomorphes à l'origine de* $\mathbb{C}^n$ (French, with English summary), Ann. Polon. Math. **94** (2008), no. 3, 245–264, DOI 10.4064/ap94-3-4. MR2453516

[Hil1893] David Hilbert, *Ueber die vollen Invariantensysteme* (German), Math. Ann. **42** (1893), no. 3, 313–373, DOI 10.1007/BF01444162. MR1510781

[Hil1890] David Hilbert, *Ueber die Theorie der algebraischen Formen* (German), Math. Ann. **36** (1890), no. 4, 473–534, DOI 10.1007/BF01208503. MR1510634

[Hir64] Heisuke Hironaka, *Resolution of singularities of an algebraic variety over a field of characteristic zero. I, II*, Ann. of Math. (2) **79** (1964), 109–203; ibid. (2) **79** (1964), 205–326, DOI 10.2307/1970547. MR0199184

[HK90] Reinhold Hübl and Ernst Kunz, *Integration of differential forms on schemes*, J. Reine Angew. Math. **410** (1990), 53–83, DOI 10.1515/crll.1990.410.53. MR1068799

[HL71] M. Herrera and D. Lieberman, *Residues and principal values on complex spaces*, Math. Ann. **194** (1971), 259–294, DOI 10.1007/BF01350129. MR296352

[HM93] Joos Heintz and Jacques Morgenstern, *On the intrinsic complexity of elimination theory*, J. Complexity **9** (1993), no. 4, 471–498, DOI 10.1006/jcom.1993.1031. MR1250549

[HoH90] Melvin Hochster and Craig Huneke, *Tight closure, invariant theory, and the Briançon-Skoda theorem*, J. Amer. Math. Soc. **3** (1990), no. 1, 31–116, DOI 10.2307/1990984. MR1017784

[Hor] Lars Hörmander, *The analysis of linear partial differential operators. I: Distribution theory and Fourier analysis*, Classics in Mathematics, Springer-Verlag, Berlin, 2003. Reprint of the second (1990) edition [Springer, Berlin; MR1065993 (91m:35001a)], DOI 10.1007/978-3-642-61497-2. MR1996773

[Hor73] Lars Hörmander, *An introduction to complex analysis in several variables*, Second revised edition, North-Holland Mathematical Library, Vol. 7, North-Holland Publishing Co., Amsterdam-London; American Elsevier Publishing Co., Inc., New York, 1973. MR0344507

[Hu92] Craig Huneke, *Uniform bounds in Noetherian rings*, Invent. Math. **107** (1992), no. 1, 203–223, DOI 10.1007/BF01231887. MR1135470

[Igu] Jun-ichi Igusa, *An introduction to the theory of local zeta functions*, AMS/IP Studies in Advanced Mathematics, vol. 14, American Mathematical Society, Providence, RI; International Press, Cambridge, MA, 2000, DOI 10.1090/amsip/014. MR1743467

[J1835] C. G. J. Jacobi, *Theoremata nova algebraica circa systema duarum aequationum, inter duas variabiles propositarum* (Latin), J. Reine Angew. Math. **14** (1835), 281–288, DOI 10.1515/crll.1835.14.281. MR1578080

[Jea74] Pierre Jeanquartier, *Développement asymptotique de la distribution de Dirac attachée à une fonction analytique* (French), C. R. Acad. Sci. Paris Sér. A-B **201** (1970), A1159–A1161. MR420695

[Jel05] Zbigniew Jelonek, *On the effective Nullstellensatz*, Invent. Math. **162** (2005), no. 1, 1–17, DOI 10.1007/s00222-004-0434-8. MR2198324

[JKS92] Shanyu Ji, János Kollár, and Bernard Shiffman, *A global Łojasiewicz inequality for algebraic varieties*, Trans. Amer. Math. Soc. **329** (1992), no. 2, 813–818, DOI 10.2307/2153965. MR1046016

[Jo83] Jean-Pierre Jouanolou, *Théorèmes de Bertini et applications* (French), Progress in Mathematics, vol. 42, Birkhäuser Boston, Inc., Boston, MA, 1983. MR725671

[Ka76] Masaki Kashiwara, *B-functions and holonomic systems. Rationality of roots of B-functions*, Invent. Math. **38** (1976/77), no. 1, 33–53, DOI 10.1007/BF01390168. MR430304

[Kh77] A. G. Hovanskiĭ, *Newton polyhedra, and toroidal varieties* (Russian), Funkcional. Anal. i Priložen. **11** (1977), no. 4, 56–64, 96. MR0476733

[Kh78] A. G. Hovanskiĭ, *Newton polyhedra and the Euler-Jacobi formula* (Russian), Uspekhi Mat. Nauk **33** (1978), no. 6(204), 237–238. MR526036

[KhG96] O. A. Gel′fond and A. G. Khovanskiĭ, *Newton polyhedra and Grothendieck residues* (Russian), Dokl. Akad. Nauk **350** (1996), no. 3, 298–300. MR1444043

[KhG02] O. A. Gelfond and A. G. Khovanskii, *Toric geometry and Grothendieck residues* (English, with English and Russian summaries), Mosc. Math. J. **2** (2002), no. 1, 99–112, 199, DOI 10.17323/1609-4514-2002-2-1-99-112. MR1900586

[Ki70] James R. King, *A residue formula for complex subvarieties*, Proc. Carolina Conf. on Holomorphic Mappings and Minimal Surfaces (Chapel Hill, N.C., 1970), University of North Carolina, Department of Mathematics, Chapel Hill, N.C., 1970, pp. 43–56. MR0273061

[KK87] Martin Kreuzer and Ernst Kunz, *Traces in strict Frobenius algebras and strict complete intersections*, J. Reine Angew. Math. **381** (1987), 181–204. MR918848

[KKMS73] G. Kempf, Finn Faye Knudsen, D. Mumford, and B. Saint-Donat, *Toroidal embeddings. I*, Lecture Notes in Mathematics, Vol. 339, Springer-Verlag, Berlin-New York, 1973. MR0335518

[Ko88] János Kollár, *Sharp effective Nullstellensatz*, J. Amer. Math. Soc. **1** (1988), no. 4, 963–975, DOI 10.2307/1990996. MR944576

[Ko99] János Kollár, *Effective Nullstellensatz for arbitrary ideals*, J. Eur. Math. Soc. (JEMS) **1** (1999), no. 3, 313–337, DOI 10.1007/s100970050009. MR1714736

[Kod54] K. Kodaira, *On Kähler varieties of restricted type*, Proc. Nat. Acad. Sci. U.S.A. **40** (1954), 313–316, DOI 10.1073/pnas.40.5.313. MR68870

[Kop67] Walter Koppelman, *The Cauchy integral for functions of several complex variables*, Bull. Amer. Math. Soc. **73** (1967), 373–377, DOI 10.1090/S0002-9904-1967-11757-9. MR209519

[Kou75] A. G. Kouchnirenko, *Polyèdres de Newton et nombres de Milnor* (French), Invent. Math. **32** (1976), no. 1, 1–31, DOI 10.1007/BF01389769. MR419433

[KP94] T. Krick and L. M. Pardo, *A computational method for Diophantine approximation*, Algorithms in algebraic geometry and applications (Santander, 1994), Progr. Math., vol. 143, Birkhäuser, Basel, 1996, pp. 193–253. MR1414452

[KPS01] Teresa Krick, Luis Miguel Pardo, and Martín Sombra, *Sharp estimates for the arithmetic Nullstellensatz*, Duke Math. J. **109** (2001), no. 3, 521–598, DOI 10.1215/S0012-7094-01-10934-4. MR1853355

[Ku92] E. Kunz, *Über den n-dimensionalen Residuensatz* (German), Jahresber. Deutsch. Math.-Verein. **94** (1992), no. 4, 170–188. MR1190210

[Ku08] Ernst Kunz, *Residues and duality for projective algebraic varieties*, University Lecture Series, vol. 47, American Mathematical Society, Providence, RI, 2008. With the assistance of and contributions by David A. Cox and Alicia Dickenstein, DOI 10.1090/ulect/047. MR2464546

[KuW88] Ernst Kunz and Rolf Waldi, *Regular differential forms*, Contemporary Mathematics, vol. 79, American Mathematical Society, Providence, RI, 1988, DOI 10.1090/conm/079. MR971502

[Ky89] A. M. Kytmanov, *A formula for the transformation of the Grothendieck residue and some of its applications* (Russian), Sibirsk. Mat. Zh. **29** (1988), no. 3, 198–202, 223, DOI 10.1007/BF00969664; English transl., Siberian Math. J. **29** (1988), no. 3, 495–499 (1989). MR953040

[Ky05] A. A. Kytmanov, *On an analogue of the Bochner-Martinelli representation in d-circled polyhedra in the space* $\mathbb{C}^d$ (Russian), Izv. Vyssh. Uchebn. Zaved. Mat. **3**

(2005), 52–58; English transl., Russian Math. (Iz. VUZ) **49** (2005), no. 3, 49–55. MR2179843

[KyS10] A. A. Kytmanov and A. Y. Semusheva, *Averaging of the Cauchy kernels and integral realization of the local residue*, Math. Z. **264** (2010), no. 1, 87–98, DOI 10.1007/s00209-008-0453-z. MR2564933

[L] Serge Lang, *Algebra*, 3rd ed., Graduate Texts in Mathematics, vol. 211, Springer-Verlag, New York, 2002, DOI 10.1007/978-1-4613-0041-0. MR1878556

[L88] Serge Lang, *Introduction to Arakelov theory*, Springer-Verlag, New York, 1988, DOI 10.1007/978-1-4612-1031-3. MR969124

[Lag12] Aron Lagerberg, *Super currents and tropical geometry*, Math. Z. **270** (2012), no. 3-4, 1011–1050, DOI 10.1007/s00209-010-0837-8. MR2892935

[Lar12] Richard Lärkäng, *Residue currents associated with weakly holomorphic functions*, Ark. Mat. **50** (2012), no. 1, 135–164, DOI 10.1007/s11512-010-0141-1. MR2890348

[Lar19] Richard Lärkäng, *A comparison formula for residue currents*, Math. Scand. **125** (2019), no. 1, 39–66, DOI 10.7146/math.scand.a-113032. MR4009324

[LarW18] Richard Lärkäng and Elizabeth Wulcan, *Residue currents and fundamental cycles*, Indiana Univ. Math. J. **67** (2018), no. 3, 1085–1114, DOI 10.1512/iumj.2018.67.7285. MR3820242

[Las1905] E. Lasker, *Zur Theorie der moduln und Ideale* (German), Math. Ann. **60** (1905), no. 1, 20–116, DOI 10.1007/BF01447495. MR1511288

[LavR98] Guy Laville and Ivan Ramadanoff, *Holomorphic Cliffordian functions*, Adv. Appl. Clifford Algebras **8** (1998), no. 2, 323–340, DOI 10.1007/BF03043103. MR1697976

[Laz] Robert Lazarsfeld, *Positivity in algebraic geometry.: I. Classical setting: line bundles and linear series; II. Positivity for vector bundles, and multiplier ideals*, Ergebnisse der Mathematik und ihrer Grenzgebiete. 3. Folge. A Series of Modern Surveys in Mathematics [Results in Mathematics and Related Areas. 3rd Series. A Series of Modern Surveys in Mathematics], vol. 49, Springer-Verlag, Berlin, 2004, DOI 10.1007/978-3-642-18808-4. MR2095472

[Le59] Jean Leray, *Le calcul différentiel et intégral sur une variété analytique complexe. (Problème de Cauchy. III)* (French), Bull. Soc. Math. France **87** (1959), 81–180. MR125984

[Lef53] Solomon Lefschetz, *Algebraic geometry*, Princeton University Press, Princeton, N. J., 1953. MR0056950

[Lej82] M. Lejeune-Jalabert, *Liaison et résidu* (French), Algebraic geometry (La Rábida, 1981), Lecture Notes in Math., vol. 961, Springer, Berlin, 1982, pp. 233–240, DOI 10.1007/BFb0071285. MR708336

[LejT08] Monique Lejeune-Jalabert and Bernard Teissier, *Clôture intégrale des idéaux et équisingularité* (French, with English summary), Ann. Fac. Sci. Toulouse Math. (6) **17** (2008), no. 4, 781–859. With an appendix by Jean-Jacques Risler. MR2499856

[Lel] P. Lelong, *Plurisubharmonic functions and positive differential forms*, Gordon and Breach, New York, 1968.

[LeP75] J. Le Potier, *Annulation de la cohomolgie à valeurs dans un fibré vectoriel holomorphe positif de rang quelconque* (French), Math. Ann. **218** (1975), no. 1, 35–53, DOI 10.1007/BF01350066. MR385179

[LeP77] J. le Potier, *Cohomologie de la grassmannienne à valeurs dans les puissances extérieures et symétriques du fibré universel* (French), Math. Ann. **226** (1977), no. 3, 257–270, DOI 10.1007/BF01362429. MR460347

[Lic88] Ben Lichtin, *Generalized Dirichlet series and b-functions*, Compositio Math. **65** (1988), no. 1, 81–120. MR930148

[Lip87] Joseph Lipman, *Residues and traces of differential forms via Hochschild homology*, Contemporary Mathematics, vol. 61, American Mathematical Society, Providence, RI, 1987, DOI 10.1090/conm/061/05. MR868864

[LipS81] Joseph Lipman and Avinash Sathaye, *Jacobian ideals and a theorem of Briançon-Skoda*, Michigan Math. J. **28** (1981), no. 2, 199–222. MR616270

[LipT81] Joseph Lipman and Bernard Teissier, *Pseudorational local rings and a theorem of Briançon-Skoda about integral closures of ideals*, Michigan Math. J. **28** (1981), no. 1, 97–116. MR600418

[Lo] Stanisław Łojasiewicz, *Introduction to complex analytic geometry*, Birkhäuser Verlag, Basel, 1991. Translated from the Polish by Maciej Klimek, DOI 10.1007/978-3-0348-7617-9. MR1131081

[Lo59] S. Łojasiewicz, *Sur le problème de la division* (French), Studia Math. **18** (1959), 87–136, DOI 10.4064/sm-18-1-87-136. MR107168

[Lund] Johannes Lundqvist, *A local Grothendieck duality theorem for Cohen-Macaulay ideals*, Math. Scand. **111** (2012), no. 1, 42–52, DOI 10.7146/math.scand.a-15212. MR3001357

[Mac] F. S. Macaulay, *The algebraic theory of modular systems*, Cambridge Mathematical Library, Cambridge University Press, Cambridge, 1994. Revised reprint of the 1916 original; With an introduction by Paul Roberts. MR1281612

[Mac1902] F. S. MacAulay, *Some Formulae in Elimination*, Proc. Lond. Math. Soc. **35** (1903), 3–27, DOI 10.1112/plms/s1-35.1.3. MR1577000

[Mai00] Vincent Maillot, *Géométrie d'Arakelov des variétés toriques et fibrés en droites intégrables* (French, with English and French summaries), Mém. Soc. Math. Fr. (N.S.) **80** (2000), vi+129, DOI 10.24033/msmf.393. MR1775582

[MaM82] Ernst W. Mayr and Albert R. Meyer, *The complexity of the word problems for commutative semigroups and polynomial ideals*, Adv. in Math. **46** (1982), no. 3, 305–329, DOI 10.1016/0001-8708(82)90048-2. MR683204

[Mar43] Enzo Martinelli, *Sopra una dimostrazione di R. Fueter per un teorema di Hartogs* (Italian), Comment. Math. Helv. **15** (1943), 340–349, DOI 10.1007/BF02565649. MR10729

[Mas03] David B. Massey, *Numerical control over complex analytic singularities*, Mem. Amer. Math. Soc. **163** (2003), no. 778, xii+268, DOI 10.1090/memo/0778. MR1962934

[Mat18] Saiei-Jaeyeong Matsubara-Heo, *Residue current approach to Ehrenpreis-Malgrange type theorem for linear differential equations with constant coefficients and commensurate time lags*, Adv. Math. **331** (2018), 170–208, DOI 10.1016/j.aim.2018.04.004. MR3804675

[Me01] Michel Meo, *Résidus dans le cas non nécessairement intersection complète* (French, with English and French summaries), C. R. Acad. Sci. Paris Sér. I Math. **333** (2001), no. 1, 33–38, DOI 10.1016/S0764-4442(01)01948-6. MR1844937

[Me06] Michel Méo, *Courants résidus et formule de King* (French), Ark. Mat. **44** (2006), no. 1, 149–165, DOI 10.1007/s11512-005-0003-4. MR2237218

[Mi04] Grigory Mikhalkin, *Decomposition into pairs-of-pants for complex algebraic hypersurfaces*, Topology **43** (2004), no. 5, 1035–1065, DOI 10.1016/j.top.2003.11.006. MR2079993

[MiR18] G. Mikhalkin and J. Rau, *Tropical geometry*, `https://www.math.uni-tuebingen.de/user/jora/downloads/main.pdf`.

[Moi31] Gr.C. Moisil, *Sur les quaternions monogènes*, Bull. Sci. Math. Ser. **55** (1931), 168–194.

[Mon] Gaspard Monge, *Géométrie descriptive* (French), Éditions Jacques Gabay, Sceaux, 1989. Reprint of the 1799 original. MR1190271

[Mou04] Christophe Mourougane, *Computations of Bott-Chern classes on $\mathbb{P}(E)$*, Duke Math. J. **124** (2004), no. 2, 389–420, DOI 10.1215/S0012-7094-04-12425-X. MR2079253

[Mu84] David Mumford, *Tata lectures on theta. II: Jacobian theta functions and differential equations*, Progress in Mathematics, vol. 43, Birkhäuser Boston, Inc., Boston, MA, 1984. With the collaboration of C. Musili, M. Nori, E. Previato, M. Stillman and H. Umemura, DOI 10.1007/978-0-8176-4578-6. MR742776

[MW83] D. W. Masser and G. Wüstholz, *Fields of large transcendence degree generated by values of elliptic functions*, Invent. Math. **72** (1983), no. 3, 407–464, DOI 10.1007/BF01398396. MR704399

[Na55] Shigeo Nakano, *On complex analytic vector bundles*, J. Math. Soc. Japan **7** (1955), 1–12, DOI 10.2969/jmsj/00710001. MR73263

[Ne] E. Netto, *Vorlesungen über Algebra, II*, Teubner, Leipzig, 1900.

[Ne1910] E. Netto, *Die determinanten*, Teubner, Leipzig, Berlin, 1910.

[No1873] M. Nöther, *Ueber einen Satz aus der Theorie der algebraischen Functionen* (German), Math. Ann. **6** (1873), no. 3, 351–359, DOI 10.1007/BF01442793. MR1509826

[Nor74] François Norguet, *Introduction aux fonctions de plusieurs variables complexes: représentations intégrales* (French), Fonctions de plusieurs variables complexes (Sém. François Norguet, 1970–1973; à la mémoire d'André Martineau), Lecture Notes in Math., Vol. 409, Springer, Berlin, 1974, pp. 1–97. MR0369729

[Nor96] François Norguet, *Résidus: de Poincaré à Leray un siècle de suspense* (French, with English and French summaries), Géométrie complexe (Paris, 1992), Actualités Sci. Indust., vol. 1438, Hermann, Paris, 1996, pp. 295–319. MR1487913

[NR54] D. G. Northcott and D. Rees, *Reductions of ideals in local rings*, Proc. Cambridge Philos. Soc. **50** (1954), 145–158, DOI 10.1017/s0305004100029194. MR59889

[Oda] Tadao Oda, *Convex bodies and algebraic geometry: An introduction to the theory of toric varieties*, Ergebnisse der Mathematik und ihrer Grenzgebiete (3) [Results in Mathematics and Related Areas (3)], vol. 15, Springer-Verlag, Berlin, 1988. Translated from the Japanese. MR922894

[Os19] Winnie Ossete Ingoba, *A new insight on Ronkin functions or currents*, Complex Anal. Oper. Theory **13** (2019), no. 2, 525–562, DOI 10.1007/s11785-018-0845-y. MR3935210

[P1885] H. Poincaré, *Sur une généralisation du théorème d'Abel*, Comptes rendus de l'Académie des Sciences Paris, t. **100** (1885), pp. 40-42.

[P1887] H. Poincaré, *Sur les résidus des intégrales doubles* (French), Acta Math. **9** (1887), no. 1, 321–380, DOI 10.1007/BF02406742. MR1554721

[P1906] H. Poincaré, *Sur les périodes des intégrales doubles*, Journal de Mathématiques, 6-ème série, t. **2** (1906), pp. 135-189.

[Pal] V. P. Palamodov, *Linear differential operators with constant coefficients*, Die Grundlehren der mathematischen Wissenschaften, Band 168, Springer-Verlag, New York-Berlin, 1970. Translated from the Russian by A. A. Brown. MR0264197

[Pas88A] Mikael Passare, *Residues, currents, and their relation to ideals of holomorphic functions*, Math. Scand. **62** (1988), no. 1, 75–152, DOI 10.7146/math.scand.a-12211. MR961584

[Pas88B] Mikael Passare, *A calculus for meromorphic currents*, J. Reine Angew. Math. **392** (1988), 37–56, DOI 10.1515/crll.1988.392.37. MR965056

[PasR04] Mikael Passare and Hans Rullgård, *Amoebas, Monge-Ampère measures, and triangulations of the Newton polytope*, Duke Math. J. **121** (2004), no. 3, 481–507, DOI 10.1215/S0012-7094-04-12134-7. MR2040284

[PasTY00] Mikael Passare, August Tsikh, and Alain Yger, *Residue currents of the Bochner-Martinelli type*, Publ. Mat. **44** (2000), no. 1, 85–117, DOI 10.5565/PUBLMAT_44100_02. MR1775747

[Pe07] A. Perotti, *Quaternionic regularity and the $\overline{\partial}$-Neumann problem in $\mathbb{C}^2$*, Complex Var. Elliptic Equ. **52** (2007), no. 5, 439–453, DOI 10.1080/17476930601178392. MR2321840

[Per] Oskar Perron, *Algebra. I. Die Grundlagen* (German), Walter de Gruyter & Co., Berlin, 1951. 3d ed. MR0038319

[PeR94] Th. Peternell and R. Remmert, *Differential calculus, holomorphic maps and linear structures on complex spaces*, Several complex variables, VII, Encyclopaedia Math. Sci., vol. 74, Springer, Berlin, 1994, pp. 97–144, DOI 10.1007/978-3-662-09873-8_3. MR1326619

[Ph91] Patrice Philippon, *Sur des hauteurs alternatives. I* (French), Math. Ann. **289** (1991), no. 2, 255–283, DOI 10.1007/BF01446571. MR1092175

[PhS08] Patrice Philippon and Martín Sombra, *Hauteur normalisée des variétés toriques projectives* (French, with English and French summaries), J. Inst. Math. Jussieu **7** (2008), no. 2, 327–373, DOI 10.1017/S1474748007000138. MR2400725

[Pl90] Arkadiusz Płoski, *On the Noether exponent* (English, with Polish summary), Bull. Soc. Sci. Lett. Łódź **40** (1990), no. 1-10, 23–29 (1991). MR1162169

[Pl05] Arkadiusz Płoski, *Algebraic dependence of polynomials after O. Perron and some applications*, Computational commutative and non-commutative algebraic geometry, NATO Sci. Ser. III Comput. Syst. Sci., vol. 196, IOS, Amsterdam, 2005, pp. 167–173. MR2179198

[Pol72] Jean-Baptiste Poly, *Sur un théorème de J. Leray en théorie des résidus* (French), C. R. Acad. Sci. Paris Sér. A-B **274** (1972), A171–A174. MR289807

[Pol74] Jean-Baptiste Poly, *Formule des résidus et intersection des chaînes sous-analytiques* (French), Université de Poitiers, Poitiers, 1974. Thèse présentée à l'Université de Poitiers pour obtenir le grade de Docteur ès Sciences Mathématiques; Faculté des Sciences de Poitiers, No. 181. MR0481078

[Pom1909] D. Pompeiu, *Sur la représentation des fonctions analytiques par des intégrales définies*, C. R. Acad. Sc. Paris **149** (1909), 1355–1357.

[Ra30] J. L. Rabinowitsch, *Zum Hilbertschen Nullstellensatz* (German), Math. Ann. **102** (1930), no. 1, 520, DOI 10.1007/BF01782361. MR1512592

[RaT77] Jeffrey Rauch and B. A. Taylor, *The Dirichlet problem for the multidimensional Monge-Ampère equation*, Rocky Mountain J. Math. **7** (1977), no. 2, 345–364, DOI 10.1216/RMJ-1977-7-2-345. MR454331

[Ro00] L. I. Ronkin, *On zeros of almost periodic functions generated by functions holomorphic in a multicircular domain* (Russian), Complex analysis in modern mathematics (Russian), FAZIS, Moscow, 2001, pp. 239–251. MR1833516

[RS53] Reinhold Remmert and Karl Stein, *Über dei wesentlichen Singularitäten analytischer Mengen* (German), Math. Ann. **126** (1953), 263–306, DOI 10.1007/BF01343164. MR60033

[RST05] Jürgen Richter-Gebert, Bernd Sturmfels, and Thorsten Theobald, *First steps in tropical geometry*, Idempotent mathematics and mathematical physics, Contemp. Math., vol. 377, Amer. Math. Soc., Providence, RI, 2005, pp. 289–317, DOI 10.1090/conm/377/06998. MR2149011

[RSW17] Jean Ruppenthal, Håkan Samuelsson Kalm, and Elizabeth Wulcan, *Explicit Serre duality on complex spaces*, Adv. Math. **305** (2017), 1320–1355, DOI 10.1016/j.aim.2016.10.013. MR3570160

[Rud] Walter Rudin, *Function theory in polydiscs*, W. A. Benjamin, Inc., New York-Amsterdam, 1969. MR0255841

[Sa42] Arthur Sard, *The measure of the critical values of differentiable maps*, Bull. Amer. Math. Soc. **48** (1942), 883–890, DOI 10.1090/S0002-9904-1942-07811-6. MR7523

[Sab87] C. Sabbah, *Proximité évanescente. II. Équations fonctionnelles pour plusieurs fonctions analytiques* (French, with English summary), Compositio Math. **64** (1987), no. 2, 213–241. MR916482

[Sam09] Håkan Samuelsson, *Analytic continuation of residue currents*, Ark. Mat. **47** (2009), no. 1, 127–141, DOI 10.1007/s11512-008-0086-9. MR2480918

[Sam21] Håkan Samuelsson Kalm, *The $\overline{\partial}$-equation, duality, and holomorphic forms on a reduced complex space*, J. Geom. Anal. **31** (2021), no. 2, 1786–1820, DOI 10.1007/s12220-019-00325-w. MR4215277

[SamL13] Richard Lärkäng and Håkan Samuelsson Kalm, *Various approaches to products of residue currents*, J. Funct. Anal. **264** (2013), no. 1, 118–138, DOI 10.1016/j.jfa.2012.10.004. MR2995702

[Sc55] Laurent Schwartz, *Division par une fonction holomorphe sur une variété analytique complexe* (French), Summa Brasil. Math. **3** (1955), 181–209 (1955). MR139937

[ScS75] Günter Scheja and Uwe Storch, *Über Spurfunktionen bei vollständigen Durchschnitten* (German), J. Reine Angew. Math. **278(279)** (1975), 174–190. MR393056

[Se55] Jean-Pierre Serre, *Un théorème de dualité* (French), Comment. Math. Helv. **29** (1955), 9–26, DOI 10.1007/BF02564268. MR67489

[Se56] Jean-Pierre Serre, *Géométrie algébrique et géométrie analytique* (French), Ann. Inst. Fourier (Grenoble) **6** (1955/56), 1–42. MR82175

[Shc07] A. Shchuplev, *Toric varieties and residues*, Doctoral dissertation, Department of Mathematics, Stockholm University 2007.

[ShcTY06] A. Shchuplev, A. Tsikh, and A. Yger, *Residual kernels with singularities on coordinate planes*, Tr. Mat. Inst. Steklova **253** (2006), no. Kompleks. Anal. i Prilozh., 277–295, DOI 10.1134/s0081543806020210; English transl., Proc. Steklov Inst. Math. **2(253)** (2006), 256–274. MR2338703

[Shu93] Michael Shub, *Some remarks on Bezout's theorem and complexity theory*, From Topology to Computation: Proceedings of the Smalefest (Berkeley, CA, 1990), Springer, New York, 1993, pp. 443–455. MR1246139

[ShuS96] Michael Shub and Steve Smale, *On the intractability of Hilbert's Nullstellensatz and an algebraic version of "NP ≠ P?"*, Duke Math. J. **81** (1995), no. 1, 47–54 (1996), DOI 10.1215/S0012-7094-95-08105-8. A celebration of John F. Nash, Jr. MR1381969

[Siu74] Yum Tong Siu, *Analyticity of sets associated to Lelong numbers and the extension of closed positive currents*, Invent. Math. **27** (1974), 53–156, DOI 10.1007/BF01389965. MR352516

[Sm98] Steve Smale, *Mathematical problems for the next century*, Math. Intelligencer **20** (1998), no. 2, 7–15, DOI 10.1007/BF03025291. MR1631413

[Smi01] Karen E. Smith, *Tight closure and vanishing theorems*, School on Vanishing Theorems and Effective Results in Algebraic Geometry (Trieste, 2000), ICTP Lect. Notes, vol. 6, Abdus Salam Int. Cent. Theoret. Phys., Trieste, 2001, pp. 149–213. MR1919458

[SmiKKT] Karen E. Smith, Lauri Kahanpää, Pekka Kekäläinen, and William Traves, *An invitation to algebraic geometry*, Universitext, Springer-Verlag, New York, 2000, DOI 10.1007/978-1-4757-4497-2. MR1788561

[Sol77] Jorge E. Solomin, *Le résidu logarithmique dans les intersections non complètes* (French, with English summary), C. R. Acad. Sci. Paris Sér. A-B **284** (1977), no. 17, A1061–A1064. MR442277

[Som99] Martín Sombra, *A sparse effective Nullstellensatz*, Adv. in Appl. Math. **22** (1999), no. 2, 271–295, DOI 10.1006/aama.1998.0633. MR1659402

[SomY21] Martín Sombra and Alain Yger, *Bounds for multivariate residues and for the polynomials in the elimination theorem*, Mosc. Math. J. **21** (2021), no. 1, 129–173, DOI 10.17323/1609-4514-2021-21-1-129-173. MR4219038

[SoZ16] Mads Søgaard and Yang Zhang, *Scattering equations and global duality of residues*, Phys. Rev. D **93** (2016), no. 10, 105009, 11, DOI 10.1103/PhysRevD.93.105009. MR3597956

[Sp89] Stanisław Spodzieja, *On some property of the Jacobian of a homogeneous polynomial mapping* (English, with Polish summary), Bull. Soc. Sci. Lett. Łódź **39** (1989), no. 5, 6. MR1050809

[StV82] Jürgen Stückrad and Wolfgang Vogel, *An algebraic approach to the intersection theory*, The curves seminar at Queens, Vol. II (Kingston, Ont., 1981/1982), Queen's Papers in Pure and Appl. Math., vol. 61, Queen's Univ., Kingston, ON, 1982, pp. Exp. No. A, 32. MR783085

[Su79] A. Sudbery, *Quaternionic analysis*, Math. Proc. Cambridge Philos. Soc. **85** (1979), no. 2, 199–224, DOI 10.1017/S0305004100055638. MR516081

[SwH06] Craig Huneke and Irena Swanson, *Integral closure of ideals, rings, and modules*, London Mathematical Society Lecture Note Series, vol. 336, Cambridge University Press, Cambridge, 2006. MR2266432

[Sz14] Jacob Sznajdman, *The Briançon-Skoda number of analytic irreducible planar curves* (English, with English and French summaries), Ann. Inst. Fourier (Grenoble) **64** (2014), no. 1, 177–187, DOI 10.5802/aif.2843. MR3330545

[Sz18] Jacob Sznajdman, *A Briançon-Skoda-type result for a non-reduced analytic space*, J. Reine Angew. Math. **742** (2018), 1–16, DOI 10.1515/crelle-2015-0099. MR3849620

[T73] Bernard Teissier, *Cycles évanescents, sections planes et conditions de Whitney* (French), Singularités à Cargèse (Rencontre Singularités Géom. Anal., Inst. Études Sci., Cargèse, 1972), Astérisque, Nos. 7 et 8, Soc. Math. France, Paris, 1973, pp. 285–362. MR0374482

[T82] Bernard Teissier, *Variétés polaires. II. Multiplicités polaires, sections planes, et conditions de Whitney* (French), Algebraic geometry (La Rábida, 1981), Lecture Notes in Math., vol. 961, Springer, Berlin, 1982, pp. 314–491, DOI 10.1007/BFb0071291. MR708342

[T90] Bernard Teissier, *Résultats récents d'algèbre commutative effective* (French), Astérisque **189-190** (1990), Exp. No. 718, 107–131. Séminaire Bourbaki, Vol. 1989/90. MR1099873

[Ts] A. K. Tsikh, *Multidimensional residues and their applications*, Translations of Mathematical Monographs, vol. 103, American Mathematical Society, Providence, RI, 1992. Translated from the 1988 Russian original by E. J. F. Primrose, DOI 10.1090/mmono/103. MR1181199

[TsY04] August Tsikh and Alain Yger, *Residue currents: Complex analysis*, J. Math. Sci. (N.Y.) **120** (2004), no. 6, 1916–1971, DOI 10.1023/B:JOTH.0000020710.57247.b7. MR2085502

[Tw95] Piotr Tworzewski, *Intersection theory in complex analytic geometry*, Ann. Polon. Math. **62** (1995), no. 2, 177–191, DOI 10.4064/ap-62-2-177-191. MR1356791

[Va92] Wolmer V. Vasconcelos, *The top of a system of equations*, Bol. Soc. Mat. Mexicana (2) **37** (1992), no. 1-2, 549–556. Papers in honor of José Adem (Spanish). MR1317602

[VdW] B. L. van der Waerden, *Algebra* (German), Die Grundlehren der mathematischen Wissenschaften in Einzeldarstellungen mit besonderer Berücksichtigung der Anwendungsgebiete, Band XXXIV, Springer-Verlag, Berlin-Göttingen-Heidelberg, 1955. 3te Aufl. MR0072829

[Vid19] Alekos Vidras, *Locally explicit fundamental principle for homogeneous convolution equations* (English, with English and Russian summaries), Zh. Sib. Fed. Univ. Mat. Fiz. **12** (2019), no. 4, 466–474, DOI 10.17516/1997-1397-2019-12-4-466-474. MR3998069

[VidY98] Alekos Vidras and Alain Yger, *On asymptotic approximations of the residual currents*, Trans. Amer. Math. Soc. **350** (1998), no. 10, 4105–4125, DOI 10.1090/S0002-9947-98-02332-0. MR1608506

[VidY01] Alekos Vidras and Alain Yger, *On some generalizations of Jacobi's residue formula* (English, with English and French summaries), Ann. Sci. École Norm. Sup. (4) **34** (2001), no. 1, 131–157, DOI 10.1016/S0012-9593(00)01056-9. MR1833092

[VidY12] Alekos Vidras and Alain Yger, *Coleff-Herrera currents revisited*, The mathematical legacy of Leon Ehrenpreis, Springer Proc. Math., vol. 16, Springer, Milan, 2012, pp. 327–351, DOI 10.1007/978-88-470-1947-8_21. MR3289693

[VidY16] Alekos Vidras and Alain Yger, *Briançon-Skoda theorem for a quotient ring*, Complex analysis and dynamical systems VI. Part 2, Contemp. Math., vol. 667, Amer. Math. Soc., Providence, RI, 2016, pp. 253–278, DOI 10.1090/conm/667/13545. MR3511266

[VidY22] Alekos Vidras and Alain Yger, *Bergman-Weil expansion for holomorphic functions*, Math. Ann. **382** (2022), no. 1-2, 383–419, DOI 10.1007/s00208-020-02137-8. MR4377308

[Vin98] Victor Vinnikov, *Commuting operators and function theory on a Riemann surface*, Holomorphic spaces (Berkeley, CA, 1995), Math. Sci. Res. Inst. Publ., vol. 33, Cambridge Univ. Press, Cambridge, 1998, pp. 445–476. MR1634421

[W] Raymond O. Wells Jr., *Differential analysis on complex manifolds*, 3rd ed., Graduate Texts in Mathematics, vol. 65, Springer, New York, 2008. With a new appendix by Oscar Garcia-Prada, DOI 10.1007/978-0-387-73892-5. MR2359489

[WaW17] Dongrui Wan and Wei Wang, *On the quaternionic Monge-Ampère operator, closed positive currents and Lelong-Jensen type formula on the quaternionic space*, Bull. Sci. Math. **141** (2017), no. 4, 267–311, DOI 10.1016/j.bulsci.2015.03.001. MR3657957

[We35] André Weil, *L'intégrale de Cauchy et les fonctions de plusieurs variables* (French), Math. Ann. **111** (1935), no. 1, 178–182, DOI 10.1007/BF01472212. MR1512987

[Wei08] Martin Weimann, *An interpolation theorem in toric varieties* (English, with English and French summaries), Ann. Inst. Fourier (Grenoble) **58** (2008), no. 4, 1371–1381. MR2427963

[Wh] Hassler Whitney, *Complex analytic varieties*, Addison-Wesley Publishing Co., Reading, Mass.-London-Don Mills, Ont., 1972. MR0387634

[Wi69] Hartmut Wiebe, *Über homologische Invarianten lokaler Ringe* (German), Math. Ann. **179** (1969), 257–274, DOI 10.1007/BF01350771. MR255531

[Wu09] Elizabeth Wulcan, *Residue currents constructed from resolutions of monomial ideals*, Math. Z. **262** (2009), no. 2, 235–253, DOI 10.1007/s00209-008-0371-0. MR2504876

[Y87] Alain Yger, *Formules de division et prolongement méromorphe* (French), Séminaire d'Analyse P. Lelong–P. Dolbeault–H. Skoda, Années 1985/1986, Lecture Notes in Math., vol. 1295, Springer, Berlin, 1987, pp. 226–283, DOI 10.1007/BFb0081984. MR1047728

[Y10] Alain Yger, *The concept of "residue" after Poincaré: cutting across all of mathematics*, The scientific legacy of Poincaré, Hist. Math., vol. 36, Amer. Math. Soc., Providence, RI, 2010, pp. 225–241, DOI 10.1090/hmath/036/12. MR2647627

[Y12] Alain Yger, *Averaging residue currents and the Stückrad-Vogel algorithm*, The mathematical legacy of Leon Ehrenpreis, Springer Proc. Math., vol. 16, Springer, Milan, 2012, pp. 367–388, DOI 10.1007/978-88-470-1947-8_23. MR3289695

[Y13] Alain Yger, *A conjecture by Leon Ehrenpreis about zeroes of exponential polynomials*, From Fourier analysis and number theory to Radon transforms and geometry, Dev. Math., vol. 28, Springer, New York, 2013, pp. 517–535, DOI 10.1007/978-1-4614-4075-8_26. MR2986975

[Y21] Alain Yger, *Flexibility of analysis serving computational polynomial algebra or arithmetics*, Complex Anal. Oper. Theory **15** (2021), no. 3, Paper No. 61, 40, DOI 10.1007/s11785-021-01080-z. MR4244838

[Z05] Hai Zhang, *Formules de Jacobi et méthodes analytiques* (French, with French summary), Colloq. Math. **102** (2005), no. 2, 229–243, DOI 10.4064/cm102-2-5. MR2152663

[Zo89] Georg Zöll, *Regular n-forms in Clifford analysis, their behavior under change of variables and their residues*, Complex Variables Theory Appl. **11** (1989), no. 1, 25–38, DOI 10.1080/17476938908814321. MR998239

Index

absolute value, on a field, 273
admissible path, 100
algebraic complexity, 362
ample
  higher rank holomorphic bundle, 496
  holomorphic line bundle, 495
analytic spread, 326
arithmetic
  complexity, 308
  elimination theorem, 308, 340
  nullstellensatz, 309, 339
associated prime of $\mathcal{O}_{\mathcal{X},z}/\mathcal{I}_{V,z}$, 429
atlas, for a complex manifold, 421

balanced, weighted polyhedral complex, 374
balancing conditions, 374
Barlet, sheaf, 86, 460
basepoint free, line bundle, 495
Bedford–Taylor, recursive process, 303, 471
Bergman–Weil expansion
  in the multivariate setting, 249, 251, 313
  in the univariate setting, 27, 267
  local, of the Bochner–Martinelli type, 239
Berkovich, analytification, 305
Bernstein–Kouchnirenko theorem, 299
Bernstein–Sato
  for $P(X)$ or $P(e^X)$, 391
  formal equation, 2, 68, 479
Bézout
  arithmetic intersection theorem, 307
  geometric intersection theorem, 303, 307
  identity, 41, 323, 327
  identity *via* G. Hermann's algorithm, 324
  identity, parametric, 350
  toric geometric intersection theorem, 299
Bézout, Étienne, 41
bézoutian, 40
Bianchi identity, 455
blowup
  along a coherent sheaf of $\mathcal{O}_{\mathcal{X}}$ ideals, 439
  along the ideal sheaf of a manifold, 439
  normalized, 150, 438
  universal property of the normalized, 438
Bochner–Martinelli
  current, King type splitting of, 151
  current, of a bundle section, 125, 150
  current, of a complex of bundles, 136
  kernel, in $\mathbb{C}^N$, 67, 172
  kernel, in $\mathbb{P}^N_{\mathbb{C}}$, 66, 174
  residue current, extrinsic, 153
  residue current, intrinsic, 158
  residue, principal value, of a bundle section, 128
Bott–Chern
  form, 486
  operator, 452
  semipositivity in the sense of, 480
box principle, 111, 429
Briançon–Skoda
  effective, global in $\mathbb{C}[X_1, \ldots, X_n]$, 370
  exponent, for a germ of curve, 229
  in a nonregular local ring, 221
  in a regular local ring, 215, 218, 240
  in the nonreduced setting, 230
Briançon–Skoda–Huneke, effective, on $\mathbb{X} \subset \mathbb{C}^N$, 365
Buchsbaum–Eisenbud–Fitting
  conditions, 147, 430
  sequence, 154, 431

canonical
  $\mathbb{Z}$ basis, of $H^1(\mathcal{X}, \mathbb{Z})$, 44
  bundle, 447
  metric, on $\mathcal{O}_{\mathbb{P}^N_{\mathbb{C}}}(1)$, 56, 304
Cartan Theorem A, 426
Cartan Theorem B, 426
Cartier
  divisor, 443, 489
  exceptional divisor, of blowup, 438
  toric divisor, 490
Castelnuovo–Mumford regularity, 362
Cauchy formula
  in the univariate affine setting, 40
  in the univariate projective setting, 43
Cauchy–Fueter
  formula modified, 413, 414
  kernel modified, 412, 414
  operators modified, 412, 413

Cauchy–Fueter–Pompeiu formula, 413, 414
Cauchy–Pompeiu formula, 23
Cauchy–Weil
algebraic formula, 265
integral representation formula, 249, 251
operator, 253
Cech
cochains, cocycles and coboundaries, 491
cohomology, 426, 443, 491
long cohomology sequence, 492
characteristics, theta function with, 45
Chern
1-characteristic class, 466
1-form or current, 465
characteristic classes, class, 466
class, of a holomorphic line bundle, 466
connection, 465
form, of a metrized holomorphic bundle, 466
Chow
arithmetic product, 51
coherent ideal sheaf of a cycle, 488
form, 328
group, 493
theorem, 434
closed analytic subset, 429
coboundary, Leray morphism, 70
cocycle
1, with values in $\mathrm{GL}(m,\mathbb{C})$, on $\mathcal{X}$, $\mathcal{X}_{\mathbb{R}}$, 443, 450
Cech, 491
conditions, 443, 450
Cohen–Macaulay, ideal or sheaf, 161, 431
cohomological
dualizing local residue, 58
residue formula, 71
Coleff–Herrera
current, of a complete intersection, 115
currents, with poles, 90
local structure of currents, 87
principal value-residue currents, 109
residue current, 96, 97
sheaf, 84
comorphism, 435
compactification with corners, 382
complete intersection, 115, 128
completion, of a commutative ring, 254
complex
bundle, over $\mathcal{X}_{\mathbb{R}}$, 449
generically exact, 136
holomorphic bundle, 445
manifold, 421
reduced analytic space, 435
complexity
algebraic, 362
arithmetic, 308
geometric, 303
cone, rational with respect to $\mathbb{Z}^N$, 117
conjugation, isometry, 464
connection
canonical, of a holomorphic bundle, 453
dual, 455
on a complex bundle over a complex manifold, 452
convolutor, 400
corner locus, 376
Crofton
formula, 303, 499
formula and croftonization, 501
formula and Stückrad–Vogel class, 500
croftonization, 501
cup product, of intersection, in $\mathbb{P}^N_{\mathbb{C}}$, 505
current
on a complex analytic space, 457
on a complex manifold, 455
curvature of a connection, matrix, tensor, 454
cusp, 437, 462
cutting out, 484
cycle
class of generalized, 486
direct image of a, 483
fundamental, of a primary ideal, 165
generalized, 485
on a complex analytic space, 483

d-bar ($\overline{\partial}$) connection, 453
de Rham
division lemma, 63
operator, 452
decision problem, 340
degree
of a complex algebraic projective variety, 302
of a cycle or class of cycle in $\mathbb{P}^N_{\mathbb{C}}$, 504
depth, of a coherent sheaf, 427
determinantal, sheaf of ideals, 139
deviation, of an ideal, 385
differential form
on a complex analytic space, 456
on a complex manifold, 451
semimeromorphic, 63
dimension
Krull, 429
of a complex analytic space, 436
of a rational generalized polyhedron, 373
of an oriented entire ray of a lattice, 373
dimension principle
for generalized cycles or classes of generalized cycles, 485
for normal (p,p)-currents, 471, 476
for pseudo-meromorphic currents, 81
direct image, of a cycle, 483
distinguished, variety, 151, 438
divisor
Cartier, 443, 489

effective, 13
Weil, 13, 492
Weil, toric, on a toric variety, 494
Dolbeault
cohomology groups, 458
complex, over a complex manifold, 458
isomorphism theorem, 458
operator, 452
dual
$\mathbb{Z}^N$-rational cone, 117
fan, to a $\mathbb{Z}^N$-rational convex polyhedron, 298
of an integral polyhedron, 119
duality
theorem for complete intersections, 116
theorems, geometric versions, 145
dualizing sheaf, 85, 460

Eagon–Northcott
complex, 144, 147, 216
residue, 144, 220
effective
class of generalized cycle, 506
cycle, 483
divisor, 13
Ehrenpreis–Palamodov
fundamental principle, global, 406
fundamental principle, local, 87, 214
El Mir, extension theorem, 472
elimination theorem, arithmetic, 308, 340
embedded, associated prime ideal, 431
euclidean division, univariate setting, 41, 301
Euler form, 66
Euler–Jacobi formula
in the affine algebraic setting, 277, 288
in the toric algebraic setting, 299
in the univariate setting, 39
extension, of a current, 472

face
of a $\mathbb{Z}^N$-rational cone, 117
of a generalized polyhedron, 373
factorization of integration currents
in $U \subset \mathbb{C}$, 18
in the complete intersection case, 97
on a complex manifold, 477, 478
on a Riemann surface, 36
fan
convex, 118
dual to a $\mathbb{Z}^N$-rational polyhedron, 298
primitive, 118, 424
rational, 117, 424
simplicial, 118
fine, sheaf, 458
finite free resolution, 427
Fitting ideal of a coherent sheaf, 154
Fourier analysis, on $\mathcal{D}'((\mathbb{C}^*)^N)$ or ${}'\mathcal{D}^{(p,q)}((\mathbb{C}^*)^N)$, 378
Fourier–Borel transform, 390
frame, changing of, 453
free resolution
finite, 427
for a coherent sheaf, 428
minimal, 427
minimal, of $\mathbb{C}[X_0, \dots, X_N]/I$, 206
of a coherent sheaf, 427
Fubini–Study
algebraic Hefer form, in $\mathbb{P}^N_{\mathbb{C}}$, 201
algebraic Hefer morphism, in $\mathbb{P}^N_{\mathbb{C}}$, 203
metric, 51, 480
weights, for $\mathbb{C}^N$ and $\mathbb{P}^N_{\mathbb{C}}$, 180
fundamental principle
global, 406
local, 87, 214

gauge transformation law, 453
generalized
cycle or class of cycle, 371, 485, 486
transformation law, 134
generically exact, complex, 136
Gentili–Struppa, slice regularity in the sense of, 416
geometric
complexity, 303
nullstellensatz, 327, 334
geometric Bezout theorem in $\mathbb{P}^N_{\mathbb{C}}$
general case, 505
proper case, 303, 504
globally generated, sheaf, 494
Gorenstein algebra, 58
Green
current, with respect to (L, s), 50
equation, 50

half-order differential, 46
Hardy space, of a strongly nondegenerate Weil polyhedron, 253
Hefer
(or Gleason) problem, for a complex of bundles, 186
differential form, affine context, 331
division formula, 331, 426
Fubini–Study form, in $\mathbb{P}^N_{\mathbb{C}}$, 201
Fubini–Study morphism, in $\mathbb{P}^N_{\mathbb{C}}$, 203
morphism, of a complex, 186
morphism, on a Stein manifold, 196
height
canonical, logarithmic global, 306
projective, logarithmic global, 55, 306
Hilbert
nullstellensatz, 323
polynomial function, 432
syzygy theorem, 206, 429
Hironaka theorem, 442

holomorphic
  c, weakly, function, 433
  bundle, 445
  function, in $U \subset \mathcal{X}$, 422
  strongly, $(p, 0)$ form, 459
  strongly, function, 433
  tangent, cotangent bundle, 445, 447
holonomic regular, 478
homogeneous coordinate Cox ring, 119

ideal sheaf, 423
improper
  division *versus* intersection, 371
  intersection, in $\mathbb{P}^N_{\mathbb{C}}$, 505
incidence, 53
integral closure
  of a coherent sheaf of ideals, 440
  of a subring in a ring, 436
  of an ideal in a ring, 211, 440
integration current, 13
interpolation, 22, 345
intersection
  improper, in $\mathbb{P}^N_{\mathbb{C}}$, 505
  proper, in $\mathbb{P}^N_{\mathbb{C}}$, 503
  proper, on a Riemann surface, 51
  transversal, 70
irreducibility
  of a complex analytic space, 436
  of a cycle or class of cycle, 485
irrelevant ideal, of a simplicial fan, 119
isolated, associated prime ideal, 431

Jacobi–Lagrange–Kronecker type, formula, and method, 345, 346, 360, 400
jacobian
  determinant, 245, 258, 259, 388, 389
  ideal, 386
  matrix, 245
join product, in $\mathbb{P}^N_{\mathbb{C}}$, 502
Jordan lemma in a half-plane, 6

Kahler
  cone, of a complete toric manifold, 122
  form, in $\mathbb{P}^N_{\mathbb{C}}$, 480
  form, on a complete projective toric variety, 124
King theorem, 153, 478
Kodaira theorem, 481
Koszul complex
  multivariate setting, 137, 144, 146, 255
  univariate setting, 32
Kronecker, 345
Krull, dimension, 429

Lagrange, 22, 345
Laurent
  expansion as series of, 1
  polynomial, 58
  polynomial, tropical, 375
Legendre–Fenchel transform, 376
Leibniz, transition laws, 263
Lelong number, ordinary, 473
Lelong–Poincaré
  factorization formula, 18, 97, 165, 477, 478
  Green equation, 50, 304, 484
  Green equation, in Berkovich spaces, 305
Leray, coboundary morphism, 70
Lipman–Teissier theorem, 326
local
  coordinate, on a complex manifold, 421
  dimension, of an analytic subset, 429
  frame, 450
  Hilbert–Samuel multiplicity, 432
  holomorphic frame, 448
log resolution, 442
logarithmic, meromorphic form, 67
Lojasiewicz
  exponent, along the hyperplane at infinity, 326, 366
  multi-exponent for an algebraic variety, 289
  multipolyhedron, 298
Lojasiewicz, Stanislas, 289

Macaulay
  exponent, 353
  resultant, 352, 358
Mahler measure, 4, 57, 307, 502
markers
  arithmetic, in arithmetic intersection theory, 308
  geometric, in geometric intersection theory, 303
max,+ or min,+ calculus, 500
mean periodicity, 400
Mellin
  inversion formula, 8
  transform, on $\mathbb{R}^+$, 7
meromorphic function
  in $U \subset \mathbb{C}$, 1
  in $U \subset \mathcal{X}$, 422
  on an analytic space, 436
metric, canonical, on $\mathcal{O}_{\mathbb{P}^N_{\mathbb{C}}}(1)$, 304
Milnor number, 230
minimal
  embedding, 429
  inverse, 140
modification, smooth, 442
Monge–Ampère measure, real, 501
morphism, of analytic sets, 435
multilogarithmic, meromorphic form, 72
multiplicity, local, of a cycle or its class, 488

normal
  (p, p)-current, 471
  complex analytic space, 436

crossings, 441
fan, 118
locally at a point, 437
normalization, 229, 437
normalized
Green current, with respect to $(\overline{L}, s)$, 50
projective logarithmic height, 56
normalized blowup
and exceptional Weil divisor, 438
versus geometric nullstellensatz, 325
NP, class, completeness, 341
nullstellen inequalities, 348
nullstellensatz
arithmetic, 309, 339
classical formulation in $\mathbb{K}[X]$, 323
geometric, 334
geometric, versus normalized blowup, 327
Jacobi–Lagrange–Kronecker type formula for the, 350
Nœther
exponent, 323
Max, AF + BG division theorem, 308, 336, 340
nœtherian operator, 86, 215

Oka
coherence theorem, 425
normalization theorem, 437
universal denominator, 88, 433
oriented ray, 373

Paley–Wiener algebra, 390
Perron theorem
geometric, 270
parametric version, 308, 339
perspective, in Italian Renaissance, 66, 423
Picard group
and Čech cohomology, 491
of a complex analytic space, 447, 489
pluripolar, subset, 472
Poincaré–Leray residue, 64, 65
polar part, of a meromorphic section, 48
polar set, 63
pole, simple, 63
polyhedral rational complex, 374
polyhedron
generalized, with respect to a lattice, 373
integral with respect to $\mathbb{Z}^N$, reflexive, 119
polynomial morphism, of graded free modules, 205
polynomial or Laurent polynomial tropical function, 375
positive
current, 470
differential form, 469
semi (SP) or difference semi (DSP), metric on a line bundle, 481
positivity of a hermitian holomorphic bundle
in the sense of Griffiths, 480
in the sense of Nakano, 480
semi, in the sense of Bott–Chern, 480
presheaf, of commutative rings, 422
prime form, 47
primitive, vector in $\mathbb{Z}^N$, 118, 424
principal value in the univariate case
current, distribution, 2
current, of a bundle section, 33
principal value-residue, mixed current, 98, 109
projective
height, logarithmic global, 55, 306
space, 423
projectivization
of an m-holomorphic bundle, 447
of the normal bundle for a submanifold, 439
proper intersection
in $\mathbb{P}^N_{\mathbb{C}}$, 503
on a compact Riemann surface, 51
properness
of a polynomial morphism, 271
rational multi-exponent of, 273
pseudo-meromorphic
currents, sheaf $\mathrm{PM}_{\mathcal{X}}$ and sections, 78
germ of a current, 77

quasi-regularity, in a commutative ring, 254

Rabinowitsch trick, 323
reduced, complex analytic space, 435
Rees algebra, 440
reflexive, integral polyhedron, 119
regular
element in $\mathcal{O}_{\mathcal{X}}(U)$, 422
function on an analytic space, 436
holonomy, 79, 478
in codimension 1, 438
point in a complex space, 436
point in an analytic subset, 430
sequence, 109, 110, 255, 448
Renaissance, perspective in Italian, 66, 423
residue current
attached to a complex, 142
of a bundle section, 128
of a section over a Riemann surface, 33
residue formula on a Riemann surface, 38
residue symbols
algebraic, as traces, 258
in the $\mathbb{K}[X]$ setting, 264
in the $\mathbb{K}[X^{\pm 1}]$ setting, 269
residue, local, of $h(\zeta)\, d\zeta$, 4
resolution
finite, for a stalk of coherent sheaf, 427
locally free, for a coherent sheaf, 428

Riemann analytic continuation theorem, 436
Ronkin function, 382, 501

section
  germ of, given a sheaf of commutative rings, 422
  of a m-holomorphic bundle, 447
  of a sheaf, 422
Segre
  class of a coherent ideal sheaf, 486
  class, of a holomorphic bundle, 466
  current, of a holomorphic section, 485
  form, of a hermitian holomorphic bundle, 466
  generalized cycle or class, 485
  number, of a coherent sheaf of ideals at a point, 476, 489
semi-ample, holomorphic bundle, 495
semi-analytic chain, 99
semimeromorphic
  function, 17
  section of an holomorphic bundle, 37
  with polar set on a codimension 1 submanifold, 63
Serre, duality theorem, 459
sheaf
  dualizing, 460
  fine, 458
  of $\mathcal{O}_\mathcal{X}$-modules, 423
  of commutative rings, 422
  of meromorphic functions on a complex manifold, 423
  structural, 423, 435
Shilov boundary, of a Weil polyhedron, 245
singular
  part, of a closed positive (p,p)-current, 477
  point in a complex space, 436
  point in an analytic subset, 430
singularities, resolution of, 442
Siu
  analyticity theorem, 474
  stratification theorem, 153, 476
skeleton, of a Weil polyhedron, 245
slice regularity, 416
slowly decreasing family, 400
smooth, differential form, 451, 456
socle, and jacobian determinant, 389
splitting
  along dimensions, 484
  along irreducible components, 436
stalk, of a sheaf at a point, 422
standard extension property (SEP), 2, 37, 69, 84, 95
Stein, complex manifold, 425
straight line program with parameters, 343
strongly holomorphic
  $(p,0)$ form, 459
  function, 433
strongly positive
  current, 470
  differential form, 469
structural sheaf, 423
structure, form or current, 161
Stuckrad–Vogel
  class and Crofton formula, 500
  class of generalized cycle, 486
supercurrent, on $\mathbb{R}^N$ or a compactification with corners, 379, 382
support
  of a cycle, 483
  of a sheaf, 422
support function
  of a compact convex subset in $\mathbb{R}^N$, 182
  of a toric Cartier divisor, 491, 497
  virtual, with respect to a fan, 383, 490
syzygies, determination of, 428

tautological bundle, over $\mathbb{P}^N_\mathbb{C}$, 446
Taylor series of an analytic trace function, 248, 259, 310
tempered distribution, with respect to a weighted algebra A_ϕ, 391
theta function, 45
top radical, of an ideal, 385
toric
  action group, 119, 120
  Cartier divisor, 383, 490
  manifold of a primitive fan, 424
  simplicial complete variety, 119
  variety or manifold, 120, 424
toric residue
  global, in the multivariate setting, 269
  global, in the univariate setting, 60
trace
  arithmetic complexity of, 310
  formula, 238, 259
  formula, in the univariate setting, 22
trace function
  algebraic, in the $\mathbb{K}[X]$ setting, 271
  analytic, 248
  Taylor series of an analytic, 248
transformation law
  algebraic version, 260
  gauge, 453
  geometric version, 133
  geometric, generalized, 134
  in the Cohen–Macaulay setting, 162
  in the univariate setting, 35
transversal intersection, 70
tropical
  current, 378
  geometry, 500
  Laurent polynomial, of a tropical hypersurface, 375

operations, 375, 500
variety, 375
tropicalization map, 501

underlying real manifold, 422

valuation, general, 273
valuative, criterion, 212
very ampleness
for a holomorphic line bundle, 495
in the higher rank case, 496
volume form, 469

weakly holomorphic, function, 433
weighted
algebra A_ϕ of entire functions, 390
polyhedral rational complex, 374
weights
attached to a Hefer morphism, 187
attached to psh positive functions, 182
construction by assembling, 181, 184
global, in Bochner–Martinelli type formulae, 180
involving support functions and the exponential, 182
local, attached to a cut-off function, 183, 184
local, in Bochner–Martinelli formulae, 183
plurisubharmonic, and corresponding algebras, 390
Weil
analytic polyhedron, 243
divisor, 13, 492
exceptional divisor, of normalized blowup, 438
integral formula, 238
strongly nondegenerate analytic polyhedron, 244
toric divisor, on a complete toric variety, 494
Whitney type, division argument of the, 111
Wiebe theorem, 260
Wirtinger formula, 302

SELECTED PUBLISHED TITLES IN THIS SERIES

275 **Alekos Vidras and Alain Yger,** Multidimensional Residue Theory and Applications, 2023

274 **Willi Freeden and M. Zuhair Nashed,** Recovery Methodologies: Regularization and Sampling, 2023

273 **Gennadiy Feldman,** Characterization of Probability Distributions on Locally Compact Abelian Groups, 2023

272 **Tadashi Ochiai,** Iwasawa Theory and Its Perspective, Volume 1, 2023

271 **Hideto Asashiba,** Categories and Representation Theory, 2022

270 **Leslie Hogben, Jephian C.-H. Lin, and Bryan L. Shader,** Inverse Problems and Zero Forcing for Graphs, 2022

269 **Ralph S. Freese, Ralph N. McKenzie, George F. McNulty, and Walter F. Taylor,** Algebras, Lattices, Varieties, 2022

268 **Ralph S. Freese, Ralph N. McKenzie, George F. McNulty, and Walter F. Taylor,** Algebras, Lattices, Varieties, 2022

267 **Dragana S. Cvetković Ilić,** Completion Problems on Operator Matrices, 2022

266 **Kate Juschenko,** Amenability of Discrete Groups by Examples, 2022

265 **Nabil H. Mustafa,** Sampling in Combinatorial and Geometric Set Systems, 2022

264 **J. Scott Carter and Seiichi Kamada,** Diagrammatic Algebra, 2021

263 **Vugar E. Ismailov,** Ridge Functions and Applications in Neural Networks, 2021

262 **Ragnar-Olaf Buchweitz,** Maximal Cohen–Macaulay Modules and Tate Cohomology, 2021

261 **Shiri Artstein-Avidan, Apostolos Giannopoulos, and Vitali D. Milman,** Asymptotic Geometric Analysis, Part II, 2021

260 **Lindsay N. Childs, Cornelius Greither, Kevin P. Keating, Alan Koch, Timothy Kohl, Paul J. Truman, and Robert G. Underwood,** Hopf Algebras and Galois Module Theory, 2021

259 **William Heinzer, Christel Rotthaus, and Sylvia Wiegand,** Integral Domains Inside Noetherian Power Series Rings, 2021

258 **Pramod N. Achar,** Perverse Sheaves and Applications to Representation Theory, 2021

257 **Juha Kinnunen, Juha Lehrbäck, and Antti Vähäkangas,** Maximal Function Methods for Sobolev Spaces, 2021

256 **Michio Jimbo, Tetsuji Miwa, and Fedor Smirnov,** Local Operators in Integrable Models I, 2021

255 **Alexandre Boritchev and Sergei Kuksin,** One-Dimensional Turbulence and the Stochastic Burgers Equation, 2021

254 **Karim Belabas and Henri Cohen,** Numerical Algorithms for Number Theory, 2021

253 **Robert R. Bruner and John Rognes,** The Adams Spectral Sequence for Topological Modular Forms, 2021

252 **Julie Déserti,** The Cremona Group and Its Subgroups, 2021

251 **David Hoff,** Linear and Quasilinear Parabolic Systems, 2020

250 **Bachir Bekka and Pierre de la Harpe,** Unitary Representations of Groups, Duals, and Characters, 2020

249 **Nikolai M. Adrianov, Fedor Pakovich, and Alexander K. Zvonkin,** Davenport–Zannier Polynomials and Dessins d'Enfants, 2020

248 **Paul B. Larson and Jindrich Zapletal,** Geometric Set Theory, 2020

247 **István Heckenberger and Hans-Jürgen Schneider,** Hopf Algebras and Root Systems, 2020

For a complete list of titles in this series, visit the
AMS Bookstore at **www.ams.org/bookstore/survseries/**.